GENERAL RELATIVITY FOR THE GIFTED AMATEUR

General Relativity for the Gifted Amateur

Tom Lancaster
Durham University

Stephen J. Blundell
University of Oxford

OXFORD
UNIVERSITY PRESS

OXFORD
UNIVERSITY PRESS

Great Clarendon Street, Oxford, OX2 6DP,
United Kingdom

Oxford University Press is a department of the University of Oxford.
It furthers the University's objective of excellence in research, scholarship,
and education by publishing worldwide. Oxford is a registered trade mark of
Oxford University Press in the UK and in certain other countries

Published in the United States of America by Oxford University Press
198 Madison Avenue, New York, NY 10016, United States of America

British Library Cataloguing in Publication Data
Data available

Library of Congress Control Number: 2024946925

ISBN 9780192867407

ISBN 9780192867414 (pbk.)

DOI: 10.1093/oso/9780192867407.001.0001

Printed and bound by CPI Group (UK) Ltd, Croydon, CR0 4YY

Cover image: The authors

Links to third party websites are provided by Oxford in good faith and
for information only. Oxford disclaims any responsibility for the materials
contained in any third party website referenced in this work.

The manufacturer's authorised representative in the EU for product safety is Oxford University
Press España S.A. of El Parque Empresarial San Fernando de Henares, Avenida de Castilla, 2 –
28830 Madrid (www.oup.es/en or product.safety@oup.com). OUP España S.A. also acts as
importer into Spain of products made by the manufacturer.

Preface

I saw Eternity the other night
Like a great Ring of pure and endless light,
  All calm as it was bright,
And round beneath it, Time in hours, days, years
    Driv'n by the spheres
Like a vast shadow mov'd. In which the world
    And all her train were hurl'd.
Henry Vaughan (1621–1695) *The World*

You sometimes speak of gravity as essential and inherent to
matter. Pray do not ascribe that notion to me, for the cause
of gravity is what I do not pretend to know and therefore
would take more time to consider of it.
Sir Isaac Newton (1642–1726) *Letter to Richard Bentley*

Albert Einstein's crowning theoretical achievement was his formulation of his general theory of relativity, a theory of gravity that superseded Isaac Newton's approach and transformed our view of the Universe. It took a century for one of its key predictions, the existence of gravitational waves, to be verified, but it is a measure of how persuasive its underlying principles have been that no-one seriously doubted that gravitational waves would eventually be detected. General relativity engages profoundly with the nature of space and time (even more so than Henry Vaughan did in his magnificent poem quoted above) and provides the key ideas missing from Newton's theory (the deficiency of which Newton was keenly aware, as indicated in the second quotation). Our point of view in writing this book is that everyone should have the opportunity to engage with this beautiful theory which, conceptually, is based on simple ideas from the physics of fields. The mastery of the machinery of general relativity does however require some facility with mathematics that is likely to be unfamiliar to many students of physics, but the payoff is so great and the material so stimulating that we hope the reader will join us in exploring one of the greatest achievements in physics.

The text follows the same approach as our earlier *Quantum Field Theory for the Gifted Amateur* (QFTGA) (OUP, 2014) and this is perhaps a good moment to restate what we mean by the slightly tongue-in-cheek term 'gifted amateur'. We are not writing for the mathematically uninitiated and do assume that our reader has a background in physics. However, we are not writing for experts either and aim to provide an entry point to a profound topic that we hope readers will find both entertaining and useful. The use of the term 'gifted amateur' encourages the potential reader to have a go for themselves, conveying the feeling that a difficult subject is open to those who considered themselves non-experts. We adopt the same approach used in writing QFTGA, dividing the material into short and easily digestible chapters, spelling out mathematical steps in worked examples and illustrating the arguments with hand-drawn figures. In response to reader requests from QFTGA, we also include a large number of problems with worked solutions. We have both had a long-standing fascination with the subject and a conviction that it belongs more centrally in the physics curriculum. However, we have not lost the memory of finding some of this material difficult and so hope that our book will give a curious reader a more patient and illuminating guide to this subject that they would find in some of the more weighty and established tomes.

The first two parts of the book introduce the main concepts that lead to the formulation of the Einstein field equations, and this material concludes with an outline description of the most important implications of the theory. These implications are worked out in much more detail in the middle section of the book, the third part covering cosmology and the fourth part detailing the consequences for orbits and black holes. General relativity is a theory about the geometry of spacetime and a more mathematical treatment of geometry is given in the fifth part of the book for those with an appetite to explore these aspects in more detail. The final part of the book returns to field theory, framing general relativity as a classical field theory and looking forward to how it might be formulated as a quantum field theory in a future theory of quantum gravity.

In writing this volume we are particularly grateful to the following individuals who have helped us: Rodrigo Alonso, Nathan Bentley, Katherine Blundell, Theo Breeze, Harvey Brown, Andrei Constantin, Felix Flicker, Martin Galpin, Matjaž Gomilšek, Thomas Hicken, Ben Huddart, Ifan Hughes, Baojiu Li, Guillaume Mahler, and Trevor Wishart. These individuals have been generous with their time and have helped improve the book, but any errors that are found post-publication will be posted on the book's website:

`http://tomlancaster.webspace.durham.ac.uk/grgabook`

We are also grateful to our copy editor Aravind Kannankara. Finally, we thank Cally, Eden, and Katherine for their patience, love, and support.

TL & SJB
Durham & Oxford
November 11, 2024

Contents

Overture

0

Our Theory of Gravitation is as good as perfect: Lagrange, it is well known, has proved that the Planetary System, on this scheme, will endure forever; Laplace, still more cunningly, even guesses that it could not have been made on any other scheme.
Thomas Carlyle (1795–1881) *Sartor Restartus*

General relativity is one of the most profound statements in science. It is a theory of gravity that allows us to model the large-scale structure of the Universe; to understand and explain the workings of black holes; to reveal how gravity interacts with light waves and even how the Universe hosts its own, gravitational, waves. It is central to our notions of where the Universe comes from and what its eventual fate might be. The theory's conception was largely the work of one remarkable scientist.[1] General relativity is often viewed as a fearsomely difficult theory whose mastery is a rite of passage into the world of advanced physics. However, as we will show, the theory is based on simple principles which are straightforward to grasp. This initial chapter will outline the path we will take through the book and will introduce some important bits of jargon. We start with the word *relativity*.

[1] Albert Einstein (1879–1955).

0.1 What is relativity?

Newton's[2] first law states that a body with no force acting on it will move in a straight line with a uniform velocity. This statement would be true if viewed in any **inertial reference frame** ('inertial' here means that the reference frame, which defines the coordinates used, is not accelerating). There are lots of inertial reference frames to choose from (all moving at different speeds and in different directions with respect to each other), but in all of them, Newton's first law holds. Even before Einstein came on the scene it was possible to formulate a principle of relativity:

[2] Sir Isaac Newton (1642–1726).

> **The principle of relativity:**
> Physical laws are the same in all inertial reference frames.

This implies that there is no absolute rest frame in Newtonian physics.[3] Any inertial reference frame will do, and we then have to describe motion *relative* to the inertial reference frame we have chosen.

[3] A rest frame of a particle is that frame of reference in which a particle is measured to be at rest.

(a)

(b)

Fig. 1 Juggling is best performed in (a) an inertial reference frame, or one which is accelerating constantly, rather than (b) one which has a time-varying acceleration $a(t)$.

Example 0.1

Physical processes follow simple laws in inertial frames, because we can then apply Newton's laws in their simplest form.

- A juggler will prefer to carry out their juggling when they are standing on a fixed floor [Fig. 1(a)]. They are then in an inertial rest frame and the juggler can effectively calculate the parabolic Newtonian trajectories of all the balls in his or her head, just assuming the effect of gravity.

- However, you can juggle a set of balls equally well on a moving train or in a moving plane, as long as you are travelling at a *constant velocity* (i.e. that you are in an inertial frame). The same Newtonian laws apply as before. Einstein's special theory of relativity is concerned with these inertial frames of reference.

- In fact, juggling will also be possible if the train or plane is in a state of *constant* acceleration. In that case, the juggler would not be in an inertial frame but the uniform acceleration would be indistinguishable from an additional gravitational field, and the juggler would be able to correct for this effect without difficulty, again using Newtonian laws. This idea is at the root of the equivalence principle that underlies Einstein's general theory of relativity.

- Juggling is very difficult though if the acceleration is rapidly time-varying (i.e. if the train suddenly jolts forward or shakes backwards and forwards) because additional time-varying forces would then act on the balls [Fig. 1(b)].

The principle of relativity is, in effect, a symmetry principle. It tells us that physics works in the same way, however we choose our coordinates, as long as our coordinates are described relative to an inertial reference frame. We can transform from one inertial set of coordinates to another by rotating, translating, or even what we will call 'boosting'. A **boost** is a transformation to another coordinate system moving with uniform velocity with respect to the initial one.

Example 0.2

An example of the independence of physics to boosts that is familiar to many is the sensation one experiences when seated on a train in a station and observing a neighbouring train moving forward. For a moment, you might think that your train is moving backward, and you need to check some fixed object on the station platform before you are sure which situation has occurred, and in effect whether your train is still in the station reference frame or in a new boosted reference frame.

The principle of relativity has been understood for a long time; Newton and Galileo accepted it. As we will explore in more detail in Chapter 1, Einstein's first revolutionary step, made in 1905, was to add an additional postulate:

The principle of invariant light speed:
As measured in any inertial reference frame, light propagates in empty space with a definite speed c, that is independent of the state of motion of the emitting body.

This principle has all manner of strange consequences that form the basis of Einstein's **special theory of relativity**. Why is it *special*? Because it is a theory that focuses on inertial reference frames and ignores gravity. Thus, it is restricted to some special (but important) cases. A good physical theory is said to be **covariant** if it transforms sensibly[4] under coordinate transformations. Special relativity is a theory which is covariant with respect to translations, rotations, reflections, and boosts. The boosts have to be carried out consistently with respect to the principle of invariant light speed and we will see in Chapter 1 that this must be carried out using a **Lorentz transformation**. Thus special relativity is said to be a theory which possesses **Lorentz covariance**.

Special relativity tells us that nothing can go faster than light. Thus, on a **spacetime diagram**, that is, a graph with time running up the page with spatial coordinates perpendicular, an observer's future and past can be represented as being inside a forward and backward light cone (see Fig. 2). Anything the observer can do now (throw a stone, shine a torch) can only influence the region of spacetime inside, or on, the forward light cone; anything that influences the observer now (the appearance of the night sky, an assassin's bullet) can only originate from inside, or on, the backward light cone. Moreover, if we populate spacetime with lots and lots of observers at different points, each will have their own light cone and all these light cones will be oriented in the same way [see Fig. 3(a)]. We shall see that this is a description of what is known as **flat spacetime** and is the situation that we assume to hold in special relativity.

[4]Clearly the notion of a 'sensible' transformation requires some explanation. For now, it can be thought of as the requirement that equations take the same form after transformation. This implies that no new terms should appear in an equation upon transformation to a different system of coordinates.

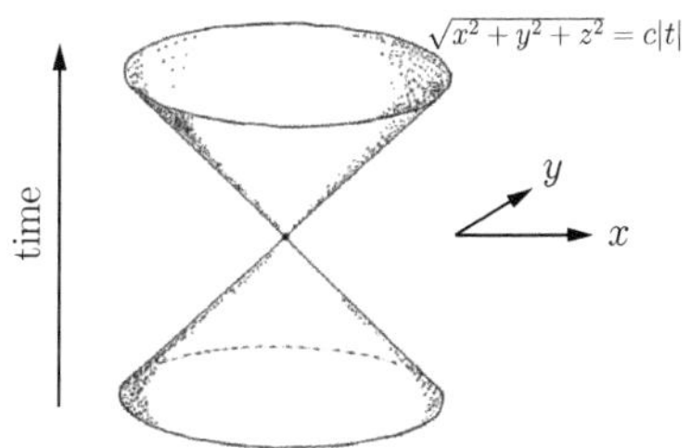

Fig. 2 The light cone in a spacetime diagram. Time is plotted vertically and the horizontal plane represents two of the three orthogonal spatial directions. An observer at the origin has the potential to influence any event inside her forward light cone and be influenced by any event inside her backward light cone.

(a) (b)

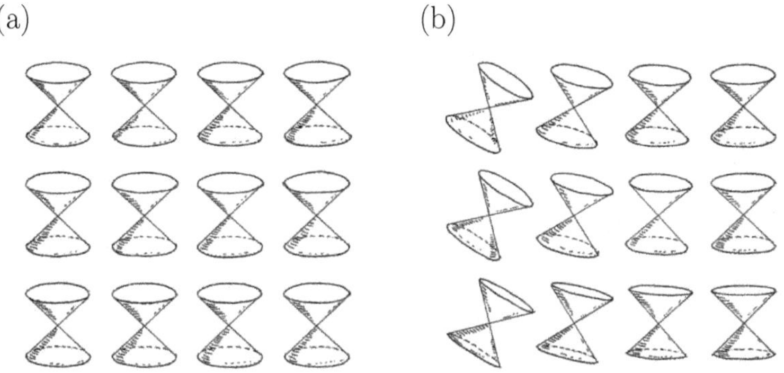

Fig. 3 (a) The light cones in flat spacetime all line up at different points, like soldiers on parade. (b) The light cones in curved spacetime look much more disorderly, as if some of the soldiers on parade now have too much alcohol in their bloodstream.

0.2 What is general relativity?

This book is about Einstein's *general* theory of relativity in which gravity is described. To understand the significance of what Einstein did, it is helpful to first take a step back. Newton constructed a theory of gravity,

[5] G is the gravitational constant, $6.6741 \times 10^{-11}\,\mathrm{N\,kg^{-2}\,m^2}$, measured first by Henry Cavendish (1731–1810) in 1798.

[6] Hence, the attitude of Thomas Carlyle in the quotation (written in 1836) that opened this chapter.

[7] Newton knew this, as can be seen in the quotation heading the Preface to this book on page v. Newton has described the force produced by a distant mass, but a real force was felt to require a cause, and Newton couldn't come up with one. In the 1717 preface to his book on '*Opticks*', he stated that he would not be taking gravity as an essential property of matter because he didn't know its cause 'because I am not yet satisfied about it for want of experiments'.

[8] John Archibald Wheeler (1911–2008)

[9] General relativity is a classical field theory. By *classical* we mean that the theory is not compatible with quantum mechanics. The search for a quantum theory of gravitation is still ongoing, a matter we will return to in Chapter 49.

[10] By *matter fields* we mean those fields describing massive particles or massive fluids, and also those describing phenomena such as electromagnetism, which is represented by a field with energetic, but massless, excitations.

published in his *Principia* in 1686, which meant that for the first time it was possible to appreciate that the same force that caused the Moon to orbit the Earth also caused the famous (and probably apocryphal) apple to fall from the tree. Newtonian gravity could be described by an equation, $F = GMm/r^2$, relating[5] the magnitude of the force F between masses M and m separated by distance r. This inverse-square relationship beautifully explains the elliptical motions of the planets and led to many people thinking that gravity was a done deal.[6] However, there was a fly in the ointment. Newton had shown how gravity behaves, but *he had not explained what it was*.[7] Mechanical explanations were popular in the seventeenth century (it was, after all, the golden age of clockwork mechanisms) and in Newton's theory of gravity it was not possible to see where the gear wheels were located in this theory; there was no mechanism, no machinery, just an influence teleporting itself through empty space; *it made no sense*. What was transmitting the gravity through space? And what even was gravity anyway? It took Einstein's genius to realize that gravity isn't something that just gets transmitted through space. Space, or more accurately *spacetime*, is a structural property of the gravitational field, with the *curvature* in the very fabric of spacetime itself [see Fig. 3(b)] being directly determined by the matter within it. In the beautiful phrase coined by Wheeler[8]:

> Spacetime tells matter how to move;
> matter tells spacetime how to curve.

General relativity is a field theory that describes gravity. A **field** is a machine that takes a position in spacetime and outputs an object representing the amplitude of something at that point in spacetime. The amplitude could be a scalar, a vector, a tensor etc.[9] Field theories describe matter, such that we speak of the electromagnetic field as describing light and charges, of particle fields as describing elementary particles and of the fluid field as describing the dynamics of continuous fluids. General relativity tells us that the effects we call gravitational reflect the energy content of all of the matter fields[10] in the Universe. What makes general relativity unique as a field theory is that the energy of these matter fields, and hence gravitation itself, is inextricably linked to another, very special, field: the **metric field** that describes the **geometry** of space and time.

The clearest expression of how general relativity describes gravitation is the **Einstein equation**. This may be written conceptually as

$$\begin{pmatrix} \text{Curvature of} \\ \text{spacetime} \end{pmatrix} = \begin{pmatrix} \text{Energy density} \\ \text{of matter fields} \end{pmatrix}. \tag{1}$$

The left-hand side of the Einstein equation is geometrical. The *curvature* is a geometrical property of space and time that follows from the metric field. The right-hand side of the Einstein equation is physical and reflects fields that describe the content of the Universe.

In formulating general relativity, Einstein began from this intuition, but initially struggled with the details of how curvature can be described mathematically using geometrical techniques that were unfamiliar to him. In the century since Einstein's monumental work, there has been a great deal of progress in both the techniques and presentation of geometry, not least following the work of Élie Cartan,[11] but the reputation for difficulty that general relativity enjoys can still be traced back to the mathematical barrier this material presents to new students of gravitation. In fact, Einstein was helped by a friend, the mathematician Marcel Grossman,[12] to master geometry, but despite this, Einstein worked tirelessly for a further decade before the theory was complete. In this spirit of friendly help, the opening sections of this book are designed to help the gifted amateur understand the mathematical language of the left-hand (geometrical) side of the Einstein equation, but in due course we will fill in the details of both sides.

[11] Élie Joseph Cartan (1869–1951).

[12] Marcel Grossmann (1878–1936).

0.3   What is a metric?

General relativity concerns the metric field, but what is that? The metric field can be thought of as a set of rules that allow us to work out the distances and angles between points in space and time. The geometrical description links space and time so inseparably that we refer to them as a single entity[13] **spacetime**. The metric field then describes geometry by providing the distances and angles between points in spacetime, known as **events**. The metric itself can be expressed by writing down the metric **line element** which is an equation for the interval between two closely spaced events. This allows us to carry out thought experiments where we imagine that spacetime has various particular curvatures and then investigate the consequences.

[13] This notion of a single entity requires another conceptual leap: the coordinates in the metric are of no intrinsic significance. The symbol t, to which we have grown accustomed for representing time, becomes less important.

Ancient Alexandria's great mathematician Euclid[14] was never able to prove his parallel postulate: the intuition that two lines that start parallel will continue to be parallel out to infinity. It was realized by geometers in the eighteenth and nineteenth centuries that this is only true for a flat plane and that consistent geometries where parallel lines converge or diverge are possible in curved spaces. The deviation of parallel lines from parallelism gives us a test for, and measure of, **curvature**. In other words, any relative motion of two small, uncharged test particles, set off at the same speeds on parallel paths, must be the consequence of a gravitational field. The information about curvature is encoded in the metric. Einstein's equation is a differential equation that, when solved for a distribution of matter, gives us access to a metric field.

[14] Euclid (who lived around 300 BC).

The metric is a field because, in general, it varies throughout spacetime. That is to say we insert a position in spacetime into the metric field and we are returned with a metric that allows us to compute the distance between events in that part of spacetime. The left-hand side of the Einstein equation can be thought of as a differential equation describing the variation of the metric field in spacetime and hence we obtain our notion of the curvature of spacetime [see Fig. 3(b)].

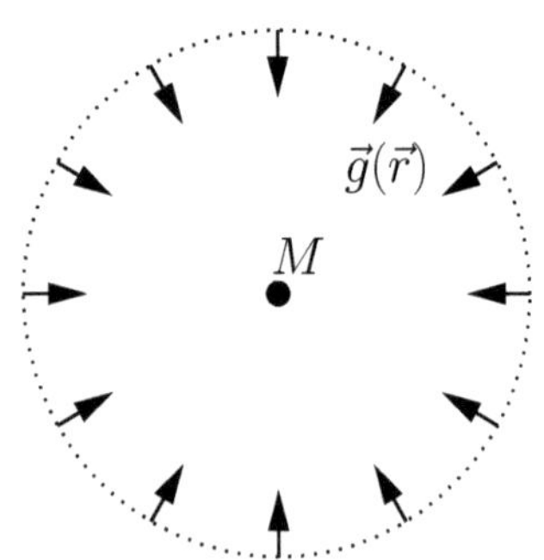

Fig. 4 The gravitational field $\vec{g}(\vec{r})$ around a particle of mass M.

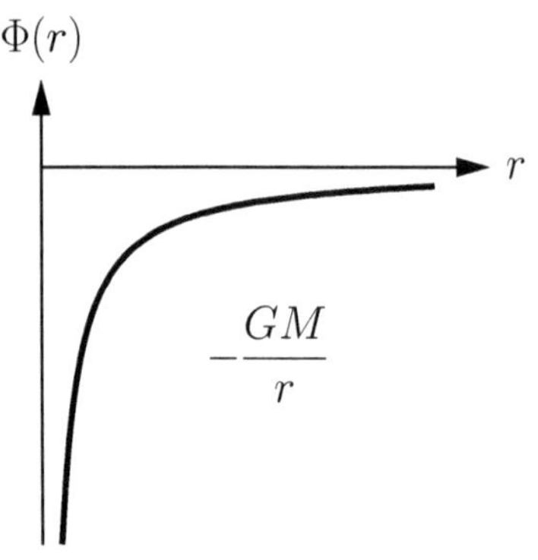

Fig. 5 The gravitational potential $\Phi(r)$ (a scalar field) at a distance r from a particle of mass M at the origin.

[16]Henry Cavendish used a torsion balance to measure the tiny gravitational attraction between lead spheres.

0.4 What are we building on?

General relativity supersedes Newton's theory of gravity, but the two theories should agree if the gravitational fields are weak.[15] Therefore, it is worth restating the older Newtonian theory: Newton asserted that the force $\vec{F}$ on a point mass m at position $\vec{r}$ due to a point mass M at the origin is given by the vector equation

$$\vec{F} = -\frac{GMm}{r^2}\hat{r},\tag{2}$$

in which arrows denote three-dimensional vectors, and the minus sign expresses the fact that gravity is an *attractive* force. Since the force scales with the mass m, we can define a gravitational field vector $\vec{g}$ as the force per unit mass, i.e.

$$\vec{F} = m\vec{g},\tag{3}$$

and this is in fact a vector field $\vec{g}(\vec{r})$ that depends on position $\vec{r}$. For a point mass, we then have (see Fig. 4)

$$\vec{g}(\vec{r}) = -\frac{GM\hat{r}}{r^2}.\tag{4}$$

The gravitational field is a conservative field of force (meaning the net work done in moving a point mass around a closed loop is zero, basically that the work done in rolling a ball up a hill is equivalent to the energy liberated when it rolls back down again), and hence we can write it as the gradient of a scalar potential. Conventionally, we include a minus sign and so write

$$\vec{g} = -\vec{\nabla}\Phi,\tag{5}$$

where $\Phi(\vec{r})$ is a scalar field known as the gravitational potential. For the case of a point mass M at the origin, $\Phi(\vec{r}) = -GM/r$ (see Fig. 5). Gauss' theorem (to be discussed below) shows that if the mass at the origin is not point-like, but is spherically symmetric, then outside the radius of the mass distribution the same results still hold.

Example 0.3

For a test mass on the surface of Earth, the gravitational force $F = mg$, where $g = 9.81\,\mathrm{m\,s^{-2}}$. Following the Cavendish experiment[16] of 1798 (and later improvements on it), the gravitational constant G was measured to be $6.6741 \times 10^{-11}\,\mathrm{N\,kg^{-2}\,m^2}$. Cavendish described this experiment as 'weighing the world' because we can then use eqn 4 to deduce that

$$M_\oplus = \frac{gR_\oplus^2}{G},\tag{6}$$

where $R_\oplus = 6.378 \times 10^6$ m is the radius of the Earth. This then gives the mass of the Earth as $M_\oplus = 5.97 \times 10^{24}$ kg. Here we are using the commonly used symbol $\oplus$ to denote the Earth. With these two numbers, we can also work out the mean density of the Earth by dividing mass $M_\oplus$ by volume $\frac{4}{3}\pi R_\oplus^3$, which yields a value of $5.5 \times 10^3\,\mathrm{kg\,m^{-3}}$, just over a factor of 5 greater than water.

We can play a similar game with our nearest star, the Sun. The Earth's orbit around the Sun is elliptical, but it's not far from circular, so for an estimate we can equate the gravitational force on the Earth due to the Sun $GM_\odot M_\oplus/R_{\rm ES}^2$ to the centripetal force $M_\oplus v^2/R_{\rm ES}$, where v is the speed of the Earth, $M_\odot$ is the mass of the Sun ($\odot$ being the symbol we use for denoting the Sun) and $R_{\rm ES}$, the separation of the Sun and Earth is called the **astronomical unit** (abbreviated A.U.). The value of $R_{\rm ES}$ was first estimated by the Greeks by measuring the angle between a half-moon and the Sun (see Fig. 6), although subsequently improved in the seventeenth century and later by measuring the solar parallax using the transit of Venus. A modern value is 1.496×10^{11} m. The period τ of the circular orbit is related to v and $R_{\rm ES}$ by $\tau = 2\pi R_{\rm ES}/v$, where our equating of centripetal and gravitational forces yields $v^2 = GM_\odot/R_{\rm ES}$. We can hence deduce from $\tau = 1$ year that $M_\odot = 1.99 \times 10^{30}$ kg. The density of the Sun, using $R_\odot = 6.96 \times 10^8$ m, then works out to be around 1.4×10^3 kg m^{-3}, just a bit larger than that of water.

One conclusion from all of this is that, from a nineteenth-century perspective, the idea of a black hole (an object with such intense surface gravity that even light could not escape) seems highly unlikely. The escape velocity $v_{\rm esc}$ from a spherical object of radius R, mass $M = \frac{4}{3}\pi\rho R^3$ and density ρ is simply worked out by equating the kinetic energy $\frac{1}{2}mv_{\rm esc}^2$ of a launching test mass to the depth $m|\Phi| = GMm/R$, of the potential energy well in which it starts its journey. This yields $v_{\rm esc} = \sqrt{2GM/R} = R\sqrt{8\pi G\rho}$. This result scales linearly with R and would reach $v_{\rm esc} = c$ only when

$$R = \frac{c}{\sqrt{8\pi G\rho}}. \tag{7}$$

Since the best-studied objects in the Universe were those in our own Solar System, and these have mean densities that don't exceed that of water by more than a factor of about five, and since the rest of the Universe seems to be filled with stars that look somewhat similar to the Sun, then eqn 7 would only be likely to be satisfied by an object with radius of more than an astronomical unit, the distance from the Earth to the Sun. No normal stars were thought to be this big. Thus, it didn't seem as if eqn 7 would hold.[17]

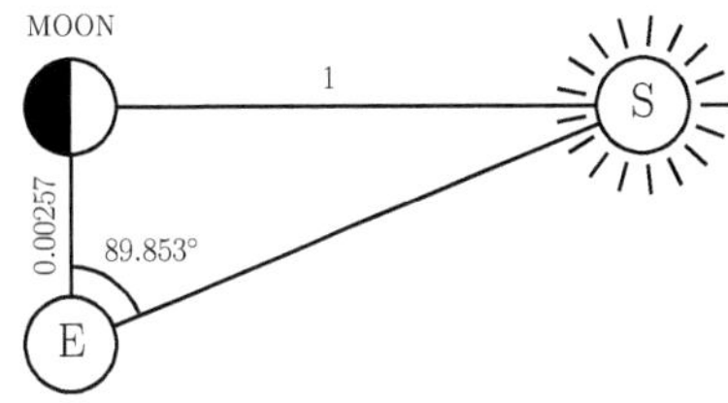

Fig. 6 Diagram (not to scale) showing how the distance to the Sun can be estimated by measuring the angle between the Sun and the half-Moon. The distances are given in A.U. The calculation relies on an estimate of the distance to the Moon which can be estimated from measurements of lunar parallax.

The Earth $\oplus$
Mass: $M_\oplus = 5.97 \times 10^{24}$ kg
Radius: $R_\oplus = 6.378 \times 10^6$ m

The Sun $\odot$
Mass: $M_\odot = 1.99 \times 10^{30}$ kg
Radius: $R_\odot = 6.96 \times 10^8$ m

$$\frac{M_\odot}{M_\oplus} = 3.33 \times 10^5$$

$$\frac{R_\odot}{R_\oplus} = 1.09 \times 10^2$$

[17]As we shall see later, it is possible to have compact objects such as neutron stars which have enormous densities. This only became possible to understand after the development of quantum mechanics.

Because $\vec{g}$ points inwards to any point mass M at the origin, we deduce that the integral of $\vec{g}$ over any spherical surface S of radius R surrounding the origin is

$$\int_S \vec{g} \cdot {\rm d}\vec{S} = -\frac{GM}{R^2} \cdot 4\pi R^2 = -4\pi GM. \tag{8}$$

The divergence theorem is a result from vector calculus and says that the left-hand side of this equation, a surface integral of the flux of the vector $\vec{g}$ out of the surface, can be rewritten as an integral over the volume of the divergence of $\vec{g}$, written as $\vec{\nabla} \cdot \vec{g}$. Hence, we have

$$\int_S \vec{g} \cdot {\rm d}\vec{S} = \int_V \vec{\nabla} \cdot \vec{g}\,{\rm d}V, \tag{9}$$

where here the volume element ${\rm d}V = {\rm d}^3 r$, i.e. the gravitational flux out of a surface is equal to the integral of the divergence of the gravitational field inside the volume enclosed by the surface. From this, we can deduce that[18]

$$\vec{\nabla} \cdot \vec{g} = -4\pi GM\delta(\vec{r}). \tag{12}$$

The Newtonian results for a point mass can be generalized for the field due to a distribution of mass since Newtonian theory is linear. Thus, for example,

$$\Phi(\vec{r}) = \int -\frac{G\rho(\vec{r}')}{|\vec{r} - \vec{r}'|}\,{\rm d}^3 r' \tag{13}$$

[18]The **Dirac delta function** $\delta(x)$ is a function localized at the origin and which has integral unity. It is the perfect model of a localized particle, and is used here to fix the point mass M at the origin. We have written a three-dimensional delta function $\delta(\vec{r}) \equiv \delta(x)\delta(y)\delta(z)$, often denoted $\delta^{(3)}(\vec{x})$. The integral of a d-dimensional Dirac delta function $\delta^{(d)}(\vec{x})$ is given by

$$\int {\rm d}^d x\, \delta^{(d)}(\vec{x}) = 1. \tag{10}$$

It is defined by

$$\int {\rm d}^d x\, f(\vec{x})\delta^{(d)}(\vec{x}) = f(0). \tag{11}$$

[19]An arbitrary distribution of mass can be written as an integral of point masses using

$$\rho(\vec{r}) \equiv \int \rho(\vec{r}')\,\delta(\vec{r} - \vec{r}')\,\mathrm{d}^3 r'.$$

The integral form of eqn 12 is

$$\int_S \vec{g} \cdot \mathrm{d}\vec{S} = -4\pi G M,$$

where $M = \int \rho(\vec{r}')\,\mathrm{d}^3 r'$ is the total mass enclosed inside the surface S. This result, which generalizes eqn 8 to an arbitrary distribution of mass, is often known as **Gauss' theorem** for gravitational fields.

[20]In electrostatics, the force on charge q is $\vec{F} = q\vec{E}$ where $\vec{E} = -\vec{\nabla}\phi$ is the electric field and ϕ is the electrostatic potential. Gauss' theorem for electrostatics is (in SI units)

$$\int_S \vec{E} \cdot \mathrm{d}\vec{S} = Q/\epsilon_0,$$

where Q is the charge enclosed by the surface S, and $\vec{\nabla} \cdot \vec{E} = \rho/\epsilon_0$ where ρ here is the charge density, and

$$\nabla^2 V = -\rho/\epsilon_0$$

is Poisson's equation.

[21]Carlyle may have said that 'Our Theory of Gravitation is as good as perfect...' in the quote that opened the chapter, but this discrepancy turned out to be rather significant!

is the gravitational potential at position $\vec{r}$ from a distribution of masses with density $\rho(\vec{r}')$. The divergence of the gravitational field can then be written (generalizing eqn 12) as[19]

$$\vec{\nabla} \cdot \vec{g}(\vec{r}) = -4\pi G \rho(\vec{r}). \tag{14}$$

Equivalently, this can be written using the gravitational potential Φ, via $\vec{g} = -\vec{\nabla}\Phi$, to yield

$$\nabla^2 \Phi = 4\pi G \rho, \tag{15}$$

which is analogous to Poisson's equation in electrostatics.[20]

The dimensions of the gravitational potential Φ are $(\text{velocity})^2$ so one might wonder what happens when $|\Phi|$ becomes of the same order as c^2, where c is the speed of light. This would be equivalent to the size of the gravitational potential energy $m|\Phi|$ of a mass m becoming of the same order as the rest mass energy mc^2. This gives a rough criterion for when Newton's law of gravitation is likely to break down and the effects of general relativity to become extremely important. However, as we shall see in this book, the effects of general relativity can become significantly important, even before this point is reached.

Example 0.4

- Effects such as gravitational time dilation are certainly measurable, if not dramatic, on the surface of planet Earth (where $|\Phi| \ll c^2$) and are important in accurate operation of the global positioning system (GPS).

- The orbit around the Sun of Mercury, the innermost planet in the Solar System, gives Mercury an orbital speed larger than that of any other planet (though at $47\,\mathrm{km\,s^{-1}}$ it's less than $0.0002c$ and so you wouldn't have thought relativistic effects would be that important). Its orbit axes slightly precesses around, by about 575 arcseconds per century, and most of this (about 532 arcseconds per century) is due to the gravitational effects of other bodies in the Solar system, perfectly calculable by Newtonian gravity. However, despite careful calculations, a discrepancy[21] of about 43 arcseconds per century stubbornly resisted explanation, until Einstein's general relativity came to the rescue.

0.5 Who is this book for?

As with our earlier book on quantum field theory, our imagined reader is an amateur. We have written this book for someone wanting to learn general relativity without (at least initially) joining the ranks of professional relativists; but (s)he is gifted, possessing a curious and adaptable mind and willing to embark on a significant intellectual challenge; (s)he has abundant curiosity about the physical world, a basic grounding in undergraduate physics, and a desire to be told an entertaining and intellectually stimulating story, but will not feel patronized if a few mathematical niceties are spelled out in detail. We appreciate that some readers will want to get to the physical predictions of the theory as soon as possible, as their primary concern is with understanding what the Universe is actually like. Others will have more interest in the mathematical structure of the theory; such readers will want to know more about how some more advanced geometric formalism can yield additional insights. We have tried to cater for both types of readers and have designed the book so that it is possible to dip in and out of sections that may be more or less to a reader's taste, though we recommend all beginners to persevere with at least the first thirteen chapters.

The book is structured as follows. We begin in Part I with an introduction to the geometry of flat spacetime, reviewing special relativity and setting up the mathematics of the metric. Part II introduces the mathematics of curvature and sets up the physics of general relativity and finishes with the Einstein field equation. Part III applies these ideas to the Universe and studies various models used in cosmology. Part IV turns to smaller structures inside the Universe: stars, black holes and their orbits. Part V contains a more formal treatment of geometry which may be of more interest to those with more mathematical inclinations. Part VI considers general relativity as a type of field theory and examines how one might link the ideas in our best theory of gravitation to our most successful theories of quantum fields. Before we get going, we will say a few words about units.

> ⤳ **It is certainly not necessary to read the book in order. In fact, we would recommend skipping several sections on a first reading. Boxes like this one are intended to allow you to navigate a path through the text.**

0.6   Units in this book

Most readers will be familiar with SI units and we will begin the book using them. However, once we get going, we will switch over to what are known as **geometrized units** in which we set $G = c = 1$.[22] This has the great advantage of simplifying equations into more memorable forms since they will no longer be encumbered with unnecessary factors of c and G whose presence, to the experts, is 'obvious'. It of course has the great disadvantage of creating some confusion whenever a numerical result it needed, but after a bit of practice this does become second nature. Because of the potential confusion for newcomers to the field, we will frequently translate back to SI units (which we will tend to call 'real-world' units) when we need to. Here is an explanation of how to translate between the two systems.

[22] The reader will be let in gently to geometrized units. We will not begin to set $c = 1$ until Chapter 2, and will not set $G = 1$ as well until Part III. In addition, Appendix B contains a summary of the units we use to discuss electromagnetism, along with a summary of useful notation.

Example 0.5

Conversion factors to convert from quantities expressed in real-world units into geometrized units can be computed by noting that the dimension[23] of c is L/T in the real world, while the dimension of G/c^2 is L/M. To convert a quantity with real-world dimension time into geometrized units, multiply by c. To convert a quantity with real-world dimension mass multiply by G/c^2. Both of these quantities then have units of length in the geometrized system.

[23] We denote dimension of length by L, time by T and mass by M.

The generalized version of the above argument says that if a quantity has units $\mathsf{L}^n\mathsf{T}^m\mathsf{M}^p$ in the real world, then it has units L^{n+m+p} in the geometrized system and the conversion factor is $c^m(G/c^2)^p$. The table gives some examples.

Quantity	real world	geometrized	conversion
Length	L	L	1
Time	T	L	c
Mass	M	L	G/c^2
Velocity	$\mathsf{L}\mathsf{T}^{-1}$	1	c^{-1}
Energy	$\mathsf{L}^2\mathsf{T}^{-2}\mathsf{M}$	L	G/c^4
Energy density	$\mathsf{L}^{-1}\mathsf{T}^{-2}\mathsf{M}$	L^{-2}	G/c^4
Mass density	$\mathsf{L}^{-3}\mathsf{M}$	L^{-2}	G/c^2
Pressure	$\mathsf{L}^{-1}\mathsf{T}^{-2}\mathsf{M}$	L^{-2}	G/c^4

If you want to convert an equation expressed in geometrized units into real-world units, multiply the quantities in the table by their respective conversion factors.

Example 0.6

In geometrized units, the Einstein equation is $\boldsymbol{G} = 8\pi\boldsymbol{T}$, where $\boldsymbol{G}$ and $\boldsymbol{T}$ are tensors that we will define later in the book (and should not be confused with the gravitational constant G and temperature T). The left-hand side of the Einstein equation has units L^{-2}, and the right-hand side has units of energy density ($\mathsf{L}^{-1}\mathsf{T}^{-2}\mathsf{M}$). The left-hand side is multiplied by unity; the right by G/c^4 and we obtain

$$\boldsymbol{G} = \tfrac{8\pi G}{c^4}\boldsymbol{T} \quad \text{(real world)}. \tag{16}$$

Exercises

(0.1) Show using Newtonian theory that the escape velocity from the surface of a star of mass M and radius r is $v_{\text{esc}} = \sqrt{2GM/r} = \sqrt{2|\Phi|}$. Show that the condition $v_{\text{esc}} = c$ will occur if $r = 2GM/c^2$, which is known as the Schwarzschild radius

(0.2) Estimate the surface gravity g and the escape velocity v_{esc} for (i) the surface of the Earth ($R_\oplus = 6.378 \times 10^6$ m, $M_\oplus = 5.97 \times 10^{24}$ kg), (ii) the surface of the Sun ($R_\odot = 6.96 \times 10^8$ m, $M_\odot =$ 1.99 $\times 10^{30}$ kg), and (iii) the surface of a $1.4M_\odot$ neutron star with radius 10 km.

(0.3) Evaluate the tidal force (the difference in gravitational forces from one end [head] to the other [feet]) on a 1.8 m tall human being (i) standing on the Earth, (ii) at the Schwarzschild radius of a $3M_\odot$ black hole with her body aligned in a radial direction, and (iii) the same as (ii) but for a $10^6 M_\odot$ black hole.

Part I

Geometry and mechanics in flat spacetime

In this introductory part of the book, we trace the development of the picture of the Universe which underpins relativity.

- Based upon the principle that light travels at c in all inertial frames, we describe the geometry of spacetime in Chapter 1 and show that the consequences that stem from this are surprisingly far-reaching.

- In Chapter 2, we show how vectors are treated in special relativity and how the dynamics of particles in flat spacetime can be obtained from the principle of least action.

- Chapter 3 is concerned with coordinates. Sometimes we choose a geometric, coordinate-free approach, but often we have to choose a particular coordinate system. We consider Cartesian and non-Cartesian bases and how to transform from one to the other.

- We introduce tensors in Chapter 4, describing them as 'linear slot machines' into which you insert a number of vectors and their dual objects which are called 1-forms; the slot machine then spits out a number. Vectors and 1-forms are themselves both tensors, as is the energy-momentum tensor which we also introduce.

- In Chapter 5, we consider a very special tensor: the metric tensor. The metric tensor encodes information about the spacetime, how distance is measured and also whether the spacetime is curved.

1

Special relativity

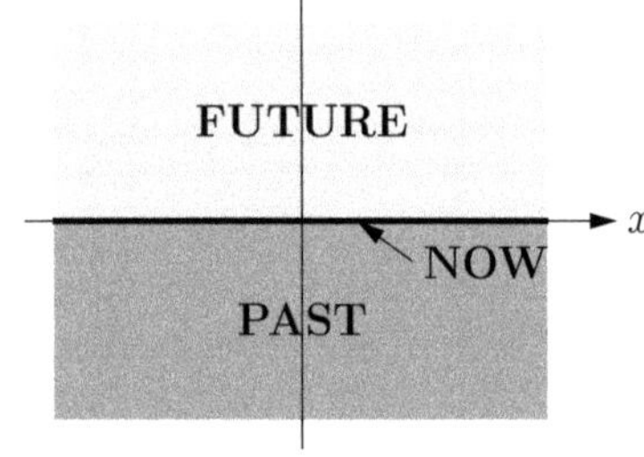

Fig. 1.1 Spacetime diagram for our naive conception of the past, present and future. In particular, the present 'now' is represented by a horizontal line.

[1] There might need to be a few calculations made to correct for light-travel-time-effects (estimating the time delay in getting signals from you and your aunt to the space station), but after doing this it will make perfect sense for everyone to talk about those two sandwich-biting events occurring at precisely the same instant.

[2] Although for the latter case we will need to be sent a signal of when the train left Paris, and will have to make a correction for the time taken for the signal to get to us.

Nowadays most people die of a sort of creeping common sense, and discover when it is too late that the only things one never regrets are one's mistakes.
Oscar Wilde (1854–1900) *The picture of Dorian Gray*

1.1 A common sense start

Einstein revolutionized our thinking about reality. To appreciate why, let's start with confirming some basic, obvious notions that would be self-evident to anyone who hadn't been exposed to Einstein's ideas. These are so straightforward that they might seem unnecessary to state, but we will do so because they turn out, in fact, to be wrong.

(1) *The notion of now:* For a start, we all understand how time rolls on inexorably for all of us. We all live in 'now', we leave the 'past' behind, and march into the 'future'. This is something we all experience, and as we look out of the window we see what others in the world are doing *right now*. Of course, if we train our telescopes on a distant galaxy, we might be observing it as it was, several billion years ago. But that's just a light-travel-time-effect. We can sensibly talk about what the inhabitants of the Andromeda galaxy might be doing *right now*, even if we can't see them. We could draw a spacetime diagram of this picture of reality and it would look like the one in Fig. 1.1.

(2) *The notion of simultaneity:* Because time is a quantity that we all experience identically (we all march to the same beat of the drum) you can make statements about *simultaneity*, such as 'at the exact same moment that I took my first bite of the sandwich in London, my aunt in Melbourne took the first bite of her sandwich'. We expect this statement to be universally true, agreed upon by all observers, so that if it is true for you and your aunt, it will be true for an observer of whatever standpoint (even if they are on the international space station).[1]

(3) *Time intervals:* Next, if we measure the time that something lasts, like a particular journey from Paris to Strasbourg, then we will get the same answer whether we are on the train or standing at Strasbourg station.[2] Moreover, the rate at which time elapses surely doesn't depend on your altitude above sea level. It would be ridiculous for time to go at a different rate on the top floor of a building

than at the basement. Time intervals are therefore something that everyone can agree on, irrespective of their frame of reference.

(4) *Spatial intervals:* Moreover, intervals in space are similarly universal. If you measure the length of a moving train carriage as a passenger you should get the same answer as an observer standing on the station platform.[3] Again, completely self-evident.

These concepts are all intuitively obvious. It was Einstein's particular genius to understand that, amazingly, our 'common sense' intuition is at fault and that these supposedly self-evident concepts must be abandoned.

1.2 The speed of light

By the start of the twentieth century, physicists were faced with a series of rather profound questions about how light propagates that put many accepted notions of physics at risk.[4] Einstein was motivated by wanting to save Maxwell's equations of electromagnetism which showed that the speed of light, c, could be related to electric and magnetic constants via the famous equation linking c to free space's permittivity ϵ_0 and permeability μ_0

$$c = \frac{1}{\sqrt{\epsilon_0 \mu_0}}. \tag{1.1}$$

But speed is a relative quantity. A car travels at 50 miles per hour *with respect to the road*. What does light travel with respect to? If you are travelling at speed $c/2$ with respect to a laser which emits a beam of light travelling in opposite direction to you, does the light travel with respect to you at a speed $\frac{c}{2} - (-c) = \frac{3c}{2}$? If you then measured the speed of light to be $\frac{3c}{2}$, how could you reconcile that with Maxwell's equations? Einstein concluded that eqn 1.1 was a universal concept and that *the speed of light was the same for all observers* in all inertial reference frames.[5] The consequences of this bold assumption on spacetime geometry are far-reaching. Before we get to these, let's start with a review of some notions of ordinary geometry.

Example 1.1

The two-dimensional xy plane is shown in Fig. 1.2(a). The point (x, y) is a distance $d = \sqrt{x^2 + y^2}$ from the origin. If we rotate the coordinates [Fig. 1.2(b)] so that $x \to x'$ and $y \to y'$, we want this distance to be unchanged, so that $x^2 + y^2 = x'^2 + y'^2$. A linear transform that accomplishes this is given by

$$\begin{pmatrix} x' \\ y' \end{pmatrix} = \begin{pmatrix} \cos\theta & \sin\theta \\ -\sin\theta & \cos\theta \end{pmatrix} \begin{pmatrix} x \\ y \end{pmatrix}, \tag{1.2}$$

which works because $\cos^2\theta + \sin^2\theta = 1$. The matrix in this equation is known as a **rotation matrix**. If you ask what are the set of points which are equidistant from the origin then, obviously, you will end up with concentric circles centred on the origin. The shortest distance between the origin and a point (x, y) is, of course, a straight line and that straight line will intersect with all of those circles at right angles.

---

[3]It is easier to make the measurement as a passenger on the train (just run a very long tape measure from one end to the other). On the platform, you would need to measure where the front of the moving train and the back of the train are at some simultaneous instant. Harder to do in practice, but perfectly possible in principle. You would naively expect to get the same answer in both cases.

[4]Specifically, the assumption that light was a mechanical wave propagating in an ether raised several troubling issues. See the book by Cheng (Appendix A) for the history.

[5]Reminder: An inertial reference frame, or **inertial frame**, is a reference frame that is not accelerating.

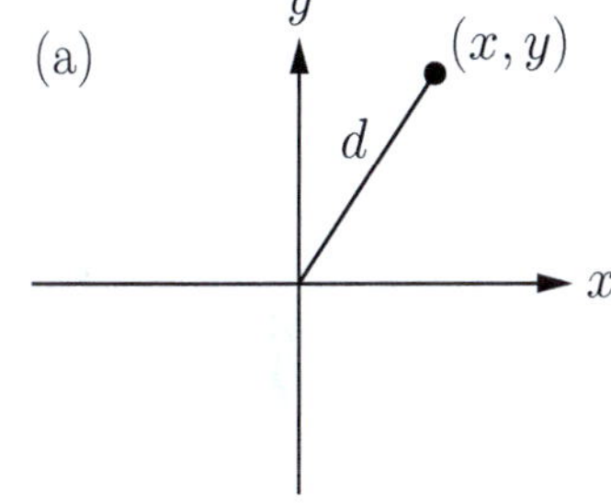

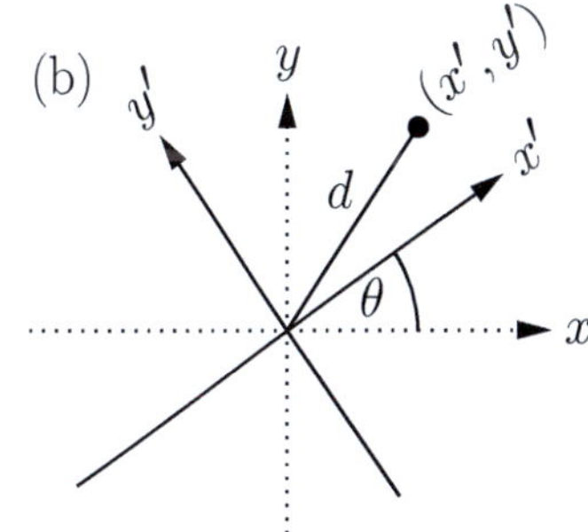

Fig. 1.2 The xy plane. The distance between a point (x, y) and the origin is d and is (of course) independent of whether the coordinates used are (a) x and y or (b) the rotated x' and y'.

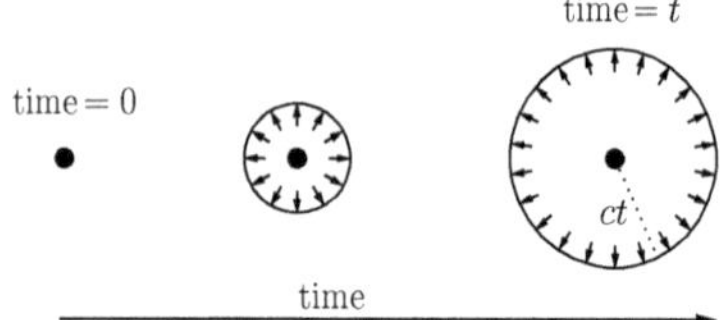

Fig. 1.3 A light source flashes at the origin at $t = 0$ and a spherical wave front, with radius ct, expands outward.

[6]We refer to points in spacetime as **events**. An event is something which happens at a particular place and particular time: a photon is emitted, a photon is absorbed, a gun is fired, a balloon bursts. Each event is characterized by a single point in spacetime.

[7]By dx^2 we really mean $(dx)^2$, but we write this so often the convention is to leave out the brackets to save on notational clutter.

[8]We will work with ds^2, rather than taking the square root, in order to avoid dealing with square roots of negative numbers. For brevity, people sometimes refer to the square of the interval ds^2 as simply 'the interval' (even though strictly the term refers only to ds). Note also that in quantum field theory it is conventional to define ds^2 with the opposite sign to what we have done here, i.e. to write

$$ds^2 \equiv c^2 dt^2 - dx^2 - dy^2 - dz^2,$$

and indeed we have done so in our own *Quantum Field Theory for the Gifted Amateur*. In this book, we adopt the convention used by most textbooks on general relativity. One might wish there was a common convention between the two fields, but this is the price you pay for exploring a number of topics in physics. It's the same with international motoring: you have to get used to driving on both the left and the right.

[9]This follows from the fact, discussed above, that if $ds = 0$ in one inertial frame, then $ds' = 0$ in any other system.

[10]That is, $v_{12} = |\vec{v}_1 - \vec{v}_2|$.

Let's now consider not just space but spacetime. As in the previous example, we want to have some notion of length which is unchanged under a rotation in spacetime (whatever that might mean). How do we define a length? Einstein's postulate gives us a clue, because if a light source flashes at $x = y = z = 0$ and $t = 0$ it will send out a beam of light travelling at speed c in all directions. There will therefore be a spherical wave front (Fig. 1.3) defined by

$$x^2 + y^2 + z^2 = c^2 t^2. \tag{1.3}$$

Let's now consider two points in spacetime[6] which are separated only by infinitesimal distances but connected by a light pulse, so that[7]

$$dx^2 + dy^2 + dz^2 = c^2 dt^2. \tag{1.4}$$

Another way of writing this equation is to put the $c^2 dt^2$ on the left-hand side so that the quantity that we will call ds^2, the square of the **spacetime interval** or **invariant line element**[8] is

$$ds^2 \equiv -c^2 dt^2 + dx^2 + dy^2 + dz^2 = 0. \tag{1.5}$$

This has been written using the coordinates of some inertial frame we can call S. In another inertial frame S' our coordinates will change, but Einstein insists that light travels at the same speed in all inertial frames and so the interval between the same two events is given by

$$ds'^2 \equiv -c^2 dt'^2 + dx'^2 + dy'^2 + dz'^2 = 0, \tag{1.6}$$

or $ds^2 = ds'^2$. Remarkably, we can now show in the following example that the spacetime interval is the same in all inertial frames, even if the two points are not connected by a light pulse.

Example 1.2

For intervals separated by infinitesimal distances in space and time, ds^2 and ds'^2 can be related using some function[9] $a(v)$ by $ds^2 = a(v)ds'^2$. Note that a can't be a function of position or time without violating the principle of the homogeneity of spacetime (every point in spacetime is like any other point). The function a can depend on the velocity $\vec{v}$ between frames S and S', but can't depend on the direction of $\vec{v}$, only on its magnitude $v = |\vec{v}|$, otherwise it would violate the principle of the isotropy of space (no special directions). Now consider three frames: S, S_1, which moves at a speed v_1 relative to S, and S_2 which moves at a speed v_2 relative to S. We have $ds^2 = a(v_1)ds_1^2$ and $ds^2 = a(v_2)ds_2^2$, but we must also have $ds_1^2 = a(v_{12})ds_2^2$, where v_{12} is the relative speed[10] of S_1 and S_2. Comparing, we must have

$$a(v_{12}) = \frac{a(v_2)}{a(v_1)}. \tag{1.7}$$

However, $v_{12} = \sqrt{v_1^2 + v_2^2 - 2v_1 v_2 \cos\theta}$ where θ is the angle between $\vec{v}_1$ and $\vec{v}_2$. Since θ appears on the left-hand side of eqn 1.7, but not on the right-hand side, $a(v)$ cannot depend on v and must be a constant (call it a). However, eqn 1.7 now becomes $a = a/a$ which is only true if $a = 1$. Thus we conclude that, in general,

$$ds^2 = ds'^2. \tag{1.8}$$

1.3 Light cones and the Lorentz transformation

This book is about general relativity in which spacetime can be curved, but for this chapter and the next three we will be considering the simplest case[11] of **flat spacetime**, in which the geometry considered above extends over all space. Thus, we can consider not just infinitesimal intervals $\mathrm{d}s$ but also the interval Δs between more distant points in spacetime, i.e. we can write

$$\Delta s^2 \equiv -c^2\Delta t^2 + \Delta x^2 + \Delta y^2 + \Delta z^2. \tag{1.9}$$

A sketch of a spacetime diagram near the origin for this flat spacetime is shown in Fig. 1.4. The set of points that satisfy $\Delta s^2 = 0$ are said to be on the **light cone** defined by eqn 1.5 since they can be connected to the origin by light rays. Points inside the light cone have $\Delta s^2 < 0$ and can be connected to the origin by particles travelling less than the speed of light. The light cone actually contains two sections, a past light cone and a future light cone. Physical processes at the origin can be affected by anything on or within the past light cone and processes at the origin can affect anything on or within the future light cone. On the other hand, the set of points outside the light cone (which have $\Delta s^2 > 0$) cannot be causally connected to the origin. We introduce some jargon for these three classes of interval:

$$\begin{array}{lll}\text{Spacelike separation} & \Delta s^2 > 0, & \\ \text{Null separation} & \Delta s^2 = 0, & \text{(1.10)} \\ \text{Timelike separation} & \Delta s^2 < 0. & \end{array}$$

[11]This is all that was envisaged in Einstein's 1905 paper on special relativity.

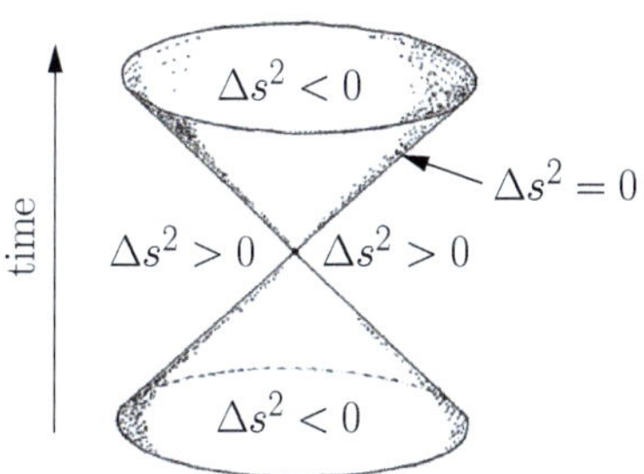

Fig. 1.4 A spacetime diagram near the origin, showing points which are spacelike ($\Delta s^2 > 0$) and timelike ($\Delta s^2 < 0$) separated from the origin. The set of points with $\Delta s^2 = 0$ are on the light cone.

Example 1.3

Let's see how to transform between inertial frames. We shall deal with the xt plane as shown in Fig. 1.5. The point (x, t) is now at an interval $\sqrt{-c^2 t^2 + x^2}$ from the origin. The interval is somewhat like distance in Example 1.1, but the minus sign in the definition will change things. The analogue of rotating the coordinates, mapping $x \to x'$ and $t \to t'$, which preserves the squared interval $(-c^2 t^2 + x^2 = -c^2 t'^2 + x'^2)$, is the linear transform given by

$$\begin{pmatrix} x' \\ ct' \end{pmatrix} = \begin{pmatrix} \cosh\theta & \sinh\theta \\ \sinh\theta & \cosh\theta \end{pmatrix} \begin{pmatrix} x \\ ct \end{pmatrix}, \tag{1.11}$$

which works because $\cosh^2\theta - \sinh^2\theta = 1$. This is known as a **Lorentz transformation**. If frame S' moves[12] at speed $v \equiv \beta c$ with respect to frame S, a particle located at a point in space which is stationary in S, is moving in S' at speed $-v$. If we then set $x = 0$ we have that $x' = ct\sinh\theta$ and $t' = t\cosh\theta$, but $x'/t' = -v$, so we deduce that $v = -c\tanh\theta$, or equivalently $\beta = -\tanh\theta$. This means that with the definition

$$\gamma = (1 - \beta^2)^{-1/2} \tag{1.12}$$

we have that $\gamma = (1 - \beta^2)^{-1/2} = (1 - \tanh^2\theta)^{-1/2} = \cosh\theta$ and $\beta\gamma = -\tanh\theta\cosh\theta = -\sinh\theta$. This puts the Lorentz transformation into the more familiar form[13]

$$\begin{pmatrix} x' \\ ct' \end{pmatrix} = \begin{pmatrix} \gamma & -\beta\gamma \\ -\beta\gamma & \gamma \end{pmatrix} \begin{pmatrix} x \\ ct \end{pmatrix}. \tag{1.13}$$

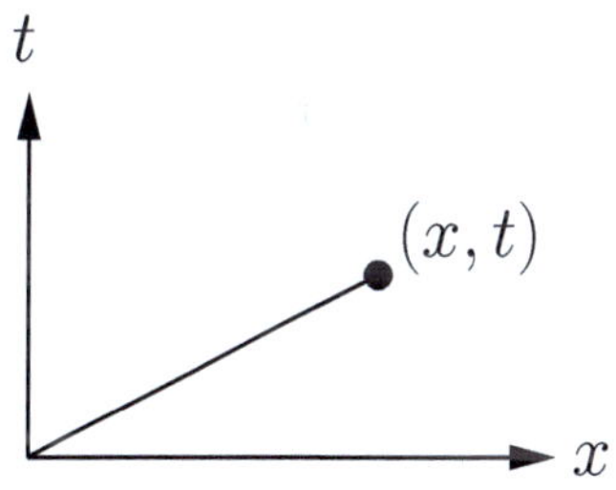

Fig. 1.5 The xt plane.

[12]We define the quantity β using

$$\beta = \frac{v}{c}.$$

[13]Writing this out in components as

$$x' = \gamma(x - vt),$$

$$t' = \gamma\left(t - \frac{vx}{c^2}\right),$$

might be even more familiar.

[14]In general it preserves the square of the interval $-c^2t^2 + x^2 + y^2 + z^2$ in all frames, but here we are just considering one spatial dimension.

[15]The proof is straightforward. Setting $\Delta t' = 0$ in eqn 1.13 gives $\beta = -c\Delta t/\Delta x$ which only has a sensible solution ($|\beta| < 1$) for spacelike intervals.

[16]We are therefore forced to drop our common-sense points 1 and 2 from the start of the chapter.

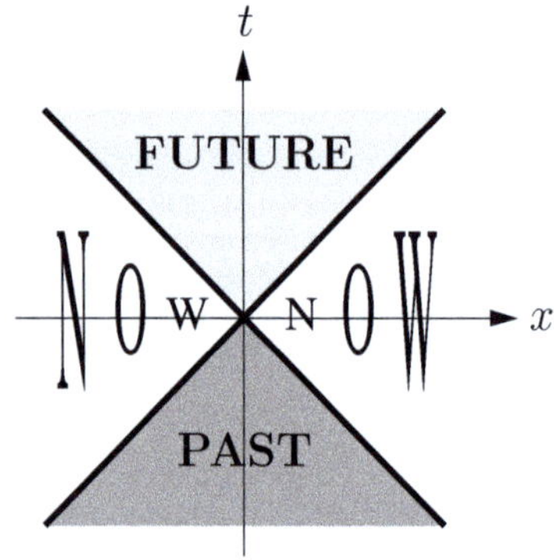

Fig. 1.6 A spacetime diagram in special relativity leads to a notion in which the region of spacetime outside the light cone is an *extended present*.

[17]This also kills off common-sense point 4.

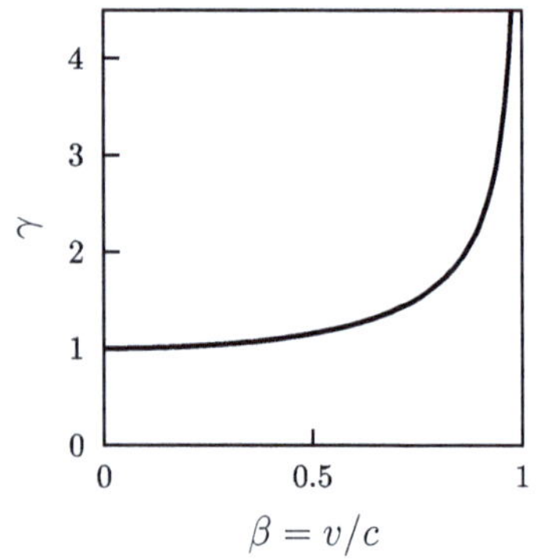

Fig. 1.7 The quantity $\gamma = (1-\beta^2)^{-1/2}$ as a function of $\beta = v/c$.

The Lorentz transformation preserves the square of the interval[14] $-c^2t^2 + x^2$ in all frames. Let's now state a couple of important consequences of this transformation for different types of intervals.

(1) For any two points separated by any **spacelike interval**, one can find a reference frame[15] for which their separation $\Delta t' = 0$, i.e. the two events separated by that interval occur *simultaneously*. Therefore, one can think of the set of points outside the light cone as an **extended present**, a region of spacetime which is not causally connected to the origin but is potentially simultaneous to it (in some reference frame). We now realize that our notion of 'now' is not a horizontal plane in spacetime as in Fig. 1.1 but forms everything outside the light cone (see Fig. 1.6). Strangely we have access to our past and our future, but it is the extended present, the 'now', which we have no access to! Our notions of simultaneity have been dramatically altered.[16]

Example 1.4

Spacelike intervals can be measured using rulers. A **ruler** is a device for measuring a spacelike length Δx. (Length being the difference in two spatial coordinates evaluated at the same value of the time coordinate.) If the ruler is stationary in frame S' and has length L then it doesn't matter *when* you measure the location of its two ends. If the ruler is moving then it can still be used to measure distances but it is then critical you measure its two ends *at the same time*. Thus eqn 1.13 yields

$$\begin{pmatrix} L \\ c\Delta t' \end{pmatrix} = \begin{pmatrix} \gamma & -\beta\gamma \\ -\beta\gamma & \gamma \end{pmatrix} \begin{pmatrix} \Delta x \\ 0 \end{pmatrix}, \tag{1.14}$$

and hence $\Delta x = L/\gamma$ and the moving ruler is shorter than it is in its rest frame (remember, $\gamma \geq 1$; see Fig. 1.7). This effect is known as **Lorentz contraction**. The ruler's length when it is stationary, L, is called the rest length or **proper length**.[17]

(2) For two points separated by any **timelike interval** (which has negative Δs^2), the straight-line path between those two points represents the longest 'distance' (i.e. interval) between them, so that small deviations from this path result in a shorter interval. This surprising result is related to the famous **twin paradox** and we will explore this in Example 1.6. Before that, we will explain how time is measured in special relativity.

Example 1.5

For a timelike interval $\Delta s^2 < 0$ it is helpful to define a real quantity $\Delta\tau$ (with units of time) by

$$\Delta\tau^2 = -\frac{\Delta s^2}{c^2}. \tag{1.15}$$

We call τ the **proper time** because it yields the time in the rest frame of a particular particle; it is measured using a **clock** in that reference frame. In a general frame, we define the interval by eqn 1.9 ($\Delta s^2 \equiv -c^2\Delta t^2 + \Delta x^2 + \Delta y^2 + \Delta z^2$), but by the invariance of the interval then

$$\Delta s^2 \equiv -c^2\Delta t^2 + \Delta x^2 + \Delta y^2 + \Delta z^2 = -c^2\Delta\tau^2, \tag{1.16}$$

where τ measures the time elapsed in the rest frame. Moving back to infinitesimal changes we can use eqn 1.16 to show that

$$d\tau = \left[(dt)^2 - \frac{(dx)^2 + (dy)^2 + (dx)^2}{c^2} \right]^{1/2}$$

$$= dt \left\{ 1 - \frac{1}{c^2} \left[\left(\frac{dx}{dt} \right)^2 + \left(\frac{dy}{dt} \right)^2 + \left(\frac{dz}{dt} \right)^2 \right] \right\}^{\frac{1}{2}}$$

$$= \frac{dt}{\gamma}. \tag{1.17}$$

This demonstrates an effect known as **time dilation**,[18] showing that the time elapsed between two events is longest in the rest frame of a clock. This effect is sometimes remembered using the slogan 'moving clocks run slow'. This phrase sometimes causes confusion. Clocks run in their rest frames at a particular rate; it's just when viewed from reference frames in which the clocks are moving is it deduced that the clocks are slowed down.[19]

For any deviation from the straight-line path the elapsed time will be shorter because additional segments of spatial-like motion will reduce the value of the elapsed time. We can treat this in general using eqn 1.17 by writing the time elapsed τ along a path in spacetime (between two points α and β) as

$$\tau = \int_{\tau_\alpha}^{\tau_\beta} d\tau = \int_{\tau_\alpha}^{\tau_\beta} \left[(dt)^2 - \frac{(dx)^2 + (dy)^2 + (dx)^2}{c^2} \right]^{1/2}$$

$$= \int_{t_\alpha}^{t_\beta} dt \left\{ 1 - \frac{1}{c^2} \left[\left(\frac{dx}{dt} \right)^2 + \left(\frac{dy}{dt} \right)^2 + \left(\frac{dz}{dt} \right)^2 \right] \right\}^{1/2}$$

$$= \int_{t_\alpha}^{t_\beta} \frac{dt}{\gamma(t)}, \tag{1.18}$$

where[20] $\gamma(t) = \left[1 - v^2(t)/c^2 \right]^{-1/2}$.

Example 1.6

The ideas from the last example can be used to resolve the famous **twin paradox**.[21] Consider two twins A and B whose clocks are synchronized. Twin A remains on Earth, while twin B is briefly accelerated to speed v and travels to Proxima Centauri at a distance x^* from Earth (journey time x^*/v in A's frame). B then is briefly deaccelerated and made to return home with velocity $-v$ (arriving home after a total journey time of $2x^*/v$ in A's frame). Both twins age at the same rate, according to their own individual clocks. However, when they meet at the end of the B's journey they find that twin A has aged more than twin B. From A's perspective, B's clock runs slow (time dilation), so that one hour experienced by B is $\gamma > 1$ hours for A. But, couldn't B argue that from their perspective it was B that remained stationary and A did all the travelling? The resolution is A and B do not have identical experiences; while A has remained at rest in a single inertial frame, B has not, as the accelerometer in B's spacecraft will have recorded. Thus, there is no paradox because the situations are not symmetric. Because the interval is frame-independent, it suffices to work it out in A's frame (see Fig. 1.8). The straight-line path of A yields $\Delta s_A^2 = -c^2(2x^*/v)^2$ corresponding to a total time of $2x^*/v$. The more circuitous path taken by B has two segments, each of which has $\Delta s_B^2 = x^{*2} - c^2(x^*/v)^2 = -c^2(x^*/v\gamma)^2$, leading to a total time interval of $2x^*/v\gamma$, which is indeed a factor of γ down from A's time interval (as we deduced from appreciating that B's clocks run slow in A's frame). The fact that B's world line (the path through spacetime) *appears* longer than A's world line in Fig. 1.8, and yet takes less time to travel, is all due to the minus sign in the expression for the interval.

[18]The same result as in the previous example can also be obtained directly from the Lorentz transformation. Any timelike interval (involving Δx, Δt) can be turned into one involving zero spatial distance using an appropriate Lorentz transformation into a frame with $\beta = -\Delta x/c\Delta t$ (so that $\Delta x' = 0$, using eqn 1.13) leading to $\Delta\tau \equiv \Delta t' = (\Delta t)/\gamma$ (in agreement with eqn 1.17). The squared interval for this is then $\Delta s^2 = \Delta x^2 - c^2\Delta t^2 = -c^2\Delta\tau^2 = -\frac{c^2\Delta t^2}{\gamma^2}$.

[19]A famous example is the cosmic ray muon which is generated in the upper atmosphere and makes it down to ground level. Muons have a lifetime of 2.2 μs in their rest frames which serves as their clock. Even if they travelled at the speed of light, they should only make it down about 2.2 μs$\times c \approx 0.7$ km down from the atmosphere. The fact that many arrive on the ground is due to the time dilation effect; their clocks seem to be running slowly due to their high speed (large γ). In the muon's reference frame, the effect is due to the Lorentz contraction of the atmosphere which is rushing towards it!

[20]Common-sense point 3 must now bite the dust too!

[21]The twin paradox, as we shall see, is only an apparent paradox.

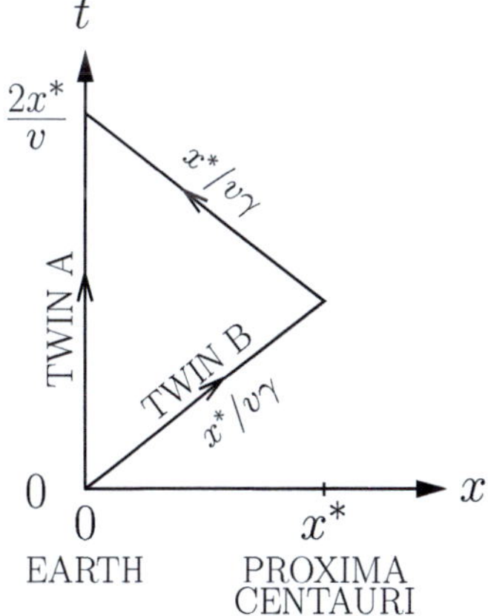

Fig. 1.8 A spacetime diagram for the twin paradox.

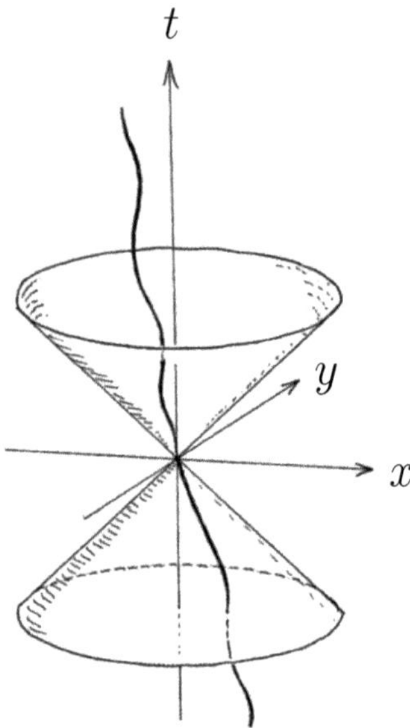

Fig. 1.9 A path through spacetime.

[22]This clock may be a wristwatch, a radioactive source that measurably decays in activity as time increases, or may simply be the fact that the observer is slowly ageing.

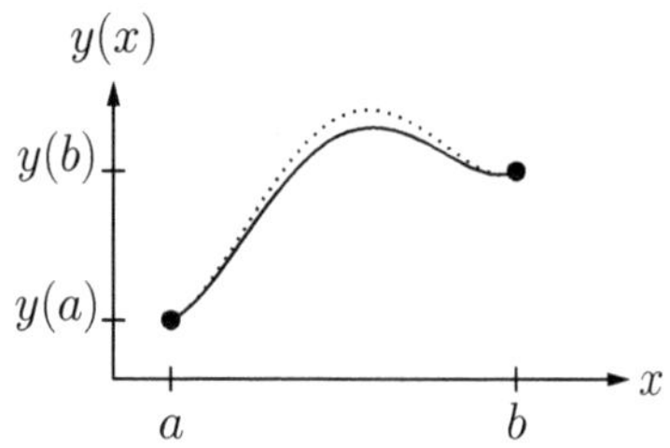

Fig. 1.10 The function $y(x)$ minimizes I. We consider small deviations from $y(x)$ given by $y(x)+\epsilon\eta(x)$ (dashed line) where $\eta(x)$ vanishes at $x = a$ and $x = b$ and ϵ is a small parameter.

[23]Though note that this condition will give us the extremal value of the integral, which could be a minimum *or* a maximum (*or*, and this becomes important in higher dimensions, a saddle point).

1.4　Paths through spacetime

We have seen that *events* are points in spacetime, and thus are instantaneous in time and localized in space. Events are witnessed and recorded by observers, who are each equipped with some kind of clock[22] which tracks the time in the observer's reference frame (i.e. measures the observer's *proper time*). The path the observer takes through spacetime (Fig. 1.9) is a chain of events connected by infinitesimal timelike intervals (the observer's speed through spacetime has to be less than c) and this path is known as the observer's **world line**.

We can now ask a simple question about paths through spacetime: what is the shortest distance between two points? This can be worked out using a technique in mathematics known as the **calculus of variations** and we review this in the following example for the simple case of usual flat (or Euclidean) space.

Example 1.7

In the calculus of variations, one deals with an integral of the form $I = \int_a^b F(y(x), y'(x), x)\,dx$, where $y' = dy/dx$. We want to find the form of $y(x)$ that minimizes I, while ensuring that $y(a)$ and $y(b)$ are fixed (see Fig. 1.10). The method assumes that you can make small variations to $y(x)$ by adding a tiny bit of another function to it, so that

$$y(x) \to y(x) + \epsilon\eta(x), \tag{1.19}$$

where ϵ is a small number and $\eta(x)$ must vanish at $x = a$ and $x = b$. Then we look for the condition[23]

$$\left.\frac{dI}{d\epsilon}\right|_{\epsilon=0} = 0 \quad \text{for all } \eta(x). \tag{1.20}$$

We can then write

$$I = \int_a^b F(y + \epsilon\eta, y' + \epsilon\eta', x)\,dx$$

$$= \int_a^b F(y, y', x)\,dx + \epsilon\int_a^b \left(\frac{\partial F}{\partial y}\eta + \frac{\partial F}{\partial y'}\eta'\right)dx + O(\epsilon^2)$$

$$= \int_a^b F(y, y', x)\,dx + \epsilon\delta I + O(\epsilon^2), \tag{1.21}$$

and our condition will be satisfied if $\epsilon\delta I = 0$ for all $\eta(x)$. One of the integrals can be done by parts

$$\int_a^b \frac{\partial F}{\partial y'}\eta'\,dx = \left[\frac{\partial F}{\partial y'}\eta\right]_a^b - \int_a^b \frac{d}{dx}\left(\frac{\partial F}{\partial y'}\right)\eta\,dx, \tag{1.22}$$

and the term in square brackets vanishes because $\eta(a) = \eta(b) = 0$. Thus,

$$\delta I = \int_a^b \left[\frac{\partial F}{\partial y} - \frac{d}{dx}\left(\frac{\partial F}{\partial y'}\right)\right]\eta(x)\,dx, \tag{1.23}$$

and this will be zero for all $\eta(x)$ if we satisfy

$$\frac{\partial F}{\partial y} - \frac{d}{dx}\left(\frac{\partial F}{\partial y'}\right) = 0, \tag{1.24}$$

which is known as the **Euler–Lagrange equation**.

We can now apply this to the case of the shortest distance between two points in Euclidean space. The length ℓ of a path between two points is given by

$$\ell = \int ds = \int \sqrt{(dx)^2 + (dy)^2} = \int \sqrt{1 + y'^2}\,dx = \int F(y', x)\,dx. \tag{1.25}$$

The integrand F is a function of y', not y, and so $\partial F/\partial y = 0$ and we can work out that $\partial F/\partial y' = -y'/\sqrt{1+y'^2}$. The Euler–Lagrange equation then gives $\mathrm{d}/\mathrm{d}x(y'/\sqrt{1+y'^2}) = 0$ which is solved by $y' =$ constant (let's call it m). The solution is then $y = mx + c$, where c is another constant, and so is evidently a straight line.

We can now use this technique for working out the shortest interval between two points in spacetime (in the special case of a single spatial dimension). The proper time elapsed along a path between two points is

$$\tau = \int \mathrm{d}\tau = \int \sqrt{(\mathrm{d}t)^2 - \frac{1}{c^2}(\mathrm{d}x)^2} = \int \mathrm{d}t \sqrt{1 - \frac{1}{c^2}\left(\frac{\mathrm{d}x}{\mathrm{d}t}\right)^2}, \quad (1.26)$$

and so is a bit different from the Euclidean case. However, application of the Euler–Lagrange equation also gives a straight line solution. Here we have to remember that the Euler–Lagrange equation identifies a stationary solution and in this case the solution is a maximum, not a minimum. We can prove that very simply: consider a timelike interval between the origin $(0,0)$ and the point (x,t). One can move to a frame[24] in which this interval is purely along the time axis, whereupon it becomes $(0, t/\gamma)$. The straight-line path thus corresponds to an elapsed time of t/γ. Any deviation from this straight line path will result in a *shorter* elapsed time because excursions along the x-axis carry a *reduced* elapsed proper time because $\mathrm{d}\tau = \sqrt{(\mathrm{d}t)^2 - \frac{1}{c^2}(\mathrm{d}x)^2}$. This is, of course, the twin paradox all over again.

We will often **parametrize** paths using the proper time τ as a way of recording how far along a path an observer has travelled. Of course, you can set the zero of proper time any way you wish, and you can measure time in units of seconds, hours, or months as you please. For this reason, any **affine**[25] scaling of τ will do. An affine transformation of τ can be written as

$$\lambda = a\tau + b, \quad (1.27)$$

where a and b are real numbers and our new **affine parameter** λ is just the proper time in different units with a different zero of time. We will have cause to use affine parameters later on when we tackle general relativity.

[24] A frame moving with velocity $-x/t$, so that γ is given by $[1 - (x/ct)^2]^{-1/2}$.

[25] The word *affine* comes from the Latin *affinis* meaning 'related to' or 'connected with'.

1.5 Experiments

In this chapter, we have outlined the consequences of Einstein's bold vision of 1905 that led to the formulation of special relativity. Why should we believe any of this? The answer is that this theory agrees spectacularly well with experiment, although the experiments were mostly all done after 1905. In this section, we briefly summarize some of these.

- *The speed of light is absolute and constant*: The Michelson–Morley experiment (1887) demonstrated that the total time for light to traverse, in free space, a distance ℓ and to return back again is independent of its direction. This was accomplished by allowing light to travel back and forth along two perpendicular arms of equal length in a Michelson interferometer.[26] The Kennedy–Thorndike experiment (1932) was a modification in which the arms of the interferometer are of unequal length. This experiment shows the time for light to traverse a closed path is independent of not only the orientation of the apparatus but also its velocity. Modern versions[27] of this experiment frequently use two lasers, one locked to a well-known transition (such as a molecular absorption line, with frequency ν_{ref}) and the other locked to a very stable Fabry–Pérot reference cavity (with frequency $\nu_{\text{cav}} = nc(v)/(2\ell)$, where n is the mode number and ℓ is the length of the cavity, the speed of light $c(v)$ being allowed the possibility of depending on v). The difference between these two frequencies is measured precisely and monitored over time (as the laboratory velocity changes as the Earth rotates around the Sun).

- *Time dilation does occur*: The Ives–Stillwell experiment (1938) used the Doppler shift in light from a moving source (accelerated ions) to infer time dilation. Time dilation is also used to interpret the flux of cosmic muons, as discussed earlier, though modern experiments use muon beams in accelerators (and from the muons' perspective, where the accelerator beamline is Lorentz contracted, this demonstrates Lorentz contraction). Modern Ives–Stillwell-type experiments have used heavy ion storage rings and laser spectroscopy to improve precision. A particularly elegant version[28] uses very slowly moving ions together with extremely accurate spectroscopy. Two optical clocks based on laser-cooled Al^+ ions are operated but in one of them the Al^+ ion is given a velocity by an applied static electric field. The frequency emitted by the two clocks can be measured (to an accuracy of 10^{-17}) and accurately compared, providing agreement with Einstein's theory even though the velocity of one of the ions is only a rather sluggish $\approx 10\,\text{m}\,\text{s}^{-1}$.

- *Lorentz invariance holds*: Numerous experiments have been performed to test Lorentz invariance to a high level of precision. No significant departure from Lorentz invariance has yet been found.[29]

- *Relativity has been used for more than a century*: This is not a good argument, as Newtonian physics had survived unscathed for more than two centuries but was eventually superseded. However, we still find that Newtonian physics still has a very wide domain of applicability (and we now understand the limits of that domain). Relativity may one day meet its match (and we expect an as-yet unformulated theory of quantum gravity will take its place), but it has so far proved reliable in the design and operation of particle accelerators, the understanding of phenomena in astrophysics, telecommunications, the space programme and condensed matter

[26]Although justly lauded as a landmark experiment and known by Einstein, it is not clear that the Michelson–Morley experiment was a major influence on his thinking. The book by Cheng discusses the Einstein's motivations.

[27]A good example can be found in C. Braxmaier *et al.*, Phys. Rev. Lett. **88**, 010401 (2002).

[28]C. W. Chou, D. B. Hume, T. Rosenband and D. J. Wineland, Science **329**, 1630 (2010).

[29]A review of recent results can be found in S. Liberati, Class. Quantum Grav. **30**, 133001 (2013).

physics. The experiments we've described above have stringently tested many aspects of relativity, and we have now accumulated ample evidence that it works across many branches of physics.

Chapter summary

- The speed of light is the same in all inertial frames. This has profound consequences for the nature of reality, including time dilation, length contraction, and a revolution in our notion of simultaneity and the meaning of the present.
- In relativity we deal with events. The history of a particle, given in terms of events, forms its world line.
- The square of the invariant interval $\mathrm{d}s^2$ between two events will be identical, no matter which coordinate system is used to evaluate it.
- The straight-line world line between two timelike separated points maximizes the interval. Deviations from this result in a smaller interval and hence elapsed time (which helps explain the twin paradox).
- A light cone is defined by $\mathrm{d}s^2 = 0$.
- The predictions of special relativity have been tested in detail and the theory is strongly supported by substantial experimental evidence.

Exercises

(1.1) Review the theory of special relativity and the derivations for the breakdown of simultaneity, the extended present, time dilation, the Lorentz contraction and the twin paradox. Give a critique of the 'common sense' statements in Section 1.1.

(1.2) The proper mean lifetime of a muon is 2.2 μs. Muons are formed in the upper atmosphere due to the collision of cosmic rays with molecules in the atmosphere. If such muons travel down to the Earth's surface with a speed of $0.995c$, calculate their mean distance travelled before decaying (a) ignoring the effect of time dilation and (b) including the effect of time dilation.

(1.3) We would like to measure the interval Δs between events p (on our world line) and q (not on our world line), using only a clock and a light pulse. To do this we emit a light pulse at event r which strikes event q and is reflected back, meeting our world line at event u. We measure the proper time interval between r and p, which we call τ_2, and the proper time interval between p and u, which we call τ_1. Show that $\Delta s^2 = c^2 \tau_1 \tau_2$.

(1.4) The quantity $(\gamma - 1)$ provides a measure of the difference between special-relativistic and Newtonian mechanics. What values of β are needed to obtain a value of $(\gamma - 1)$ equal to (a) 0.01, (b) 0.1, (c) 1, (d) 10, (e) 100?

2

Vectors in flat spacetime

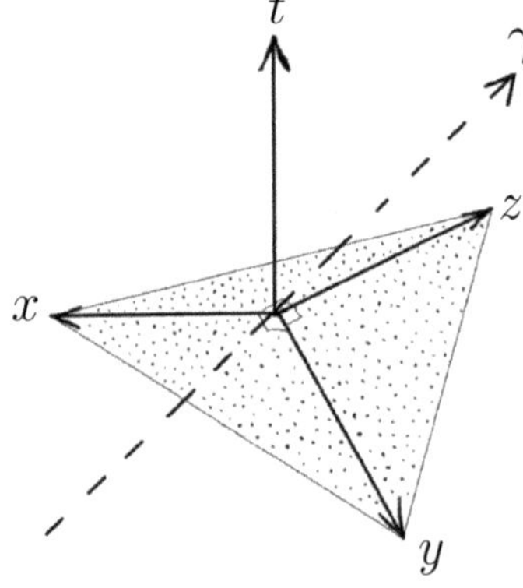

Fig. 2.1 We can't draw spacetime very accurately since it has $3+1$ dimensions, but here is an attempt. In this diagram, the three spatial dimensions have been flattened, unceremoniously, into a plane (shaded). The path of a photon (γ) is also shown.

[1]The convention used is to write the index as a superscript, i.e. it goes in the upstairs position. Keep an eye out for whether an index goes upstairs or downstairs because this will have a significance that we will explain later in this chapter.

[2]In other words, they depend on the reference frame used.

Whether 'tis nobler in the mind to suffer
The slings and arrows of outrageous fortune,
Or to take Arms against a Sea of troubles . . .
William Shakespeare (1564–1616) *Hamlet* (Act III, Scene I)

In special relativity, we are dealing with **flat spacetime** because gravity is ignored. Let's consider what kind of physical quantities might exist in such a spacetime (see Fig. 2.1). The first type we might think of is a **scalar**. A scalar is simply a number, and takes the same value in every inertial frame. It is thus said to be **Lorentz invariant**. Examples of scalars include the electric charge and rest mass of a particle.

The second type of quantity is the subject of this chapter: a **vector**. This quantity can be thought of *geometrically* as an arrow in spacetime. However, we might also wish to choose a particular reference frame and describe the *components* of this vector with respect to a particular basis. To do this we will need to specify a *coordinate system* in which to work. Because we are dealing (for now) with flat spacetime, a choice of coordinates made in one part of spacetime will work throughout the whole of spacetime. As we shall see later, this rather convenient property will not work in a curved spacetime, and there our coordinates will generally only apply *locally*. (In the same way, a local map of New York, printed on a two-dimensional sheet of paper, cannot be extended to the whole Earth because the planet is spherical.)

Example 2.1

We have seen that the basic currency of relativity is the **event**. Examples of events include the emission of a photon, receiving a photon, hearing a loud noise or being shot by an arrow. Events are witnessed and recorded by **observers**. The simplest class of events occurs directly at the point in space occupied by the observer carrying a clock. The observer assigns a time, as measured on their clock, to the event. Once we have coordinate frames at our disposal, we can record events that occur at different points in spacetime as well as the intervals that separate them. We record the events in terms of the position on the coordinate grid and the time on the clock at that position. Events can therefore be expressed in the coordinates[1] of some frame

$$x^\mu = (x^0, x^1, x^2, x^3) = (ct, x, y, z), \tag{2.1}$$

which will sometimes be written as $x^\mu = (ct, \vec{x})$, using the notion of the 3-vector $\vec{x}$ with spatial coordinates $x^i = (x, y, z)$, taken from the end of the alphabet. The location of an event in spacetime can be described by a 4-vector $\boldsymbol{x}$ considered as an arrow in spacetime (stretching, say, from the origin to the event). The particular coordinates x^μ relating to $\boldsymbol{x}$ depend on the basis chosen.[2]

Note that in this chapter, and from now on unless otherwise indicated, we choose units such that $c = 1$.

A vector isn't just any old collection of components. It is an object that has to transform appropriately under coordinate transformations.[3] In flat spacetime, 4-vectors are made from a timelike part and a spacelike part and are displayed in bold italics, so a position in spacetime is written as $\boldsymbol{x}$ where $\boldsymbol{x}$ has components $x^\mu = (t, \vec{x})$. Components for 4-vectors are given a Greek index, so for example x^μ where $\mu = 0, 1, 2, 3$. In the jargon, x^0 is the timelike component, x^1, x^2 and x^3 are the spacelike components. The spacelike components themselves form a 3-vector, whose components are given a Roman index such as x^i, where $i = 1, 2, 3$.

2.1   Vectors

In this chapter, we are going to consider the role of vectors in special relativity. We can think of a **vector** in special relativity as an arrow in spacetime. If we have two events at points $\mathcal{A}$ and $\mathcal{B}$ in flat spacetime, then we can define a vector[4] that points from $\mathcal{A}$ to $\mathcal{B}$ by

$$\boldsymbol{X} = \mathcal{B} - \mathcal{A}. \tag{2.2}$$

Defined in this way, a vector lives independently of any coordinate system. The vector points from the event at point $\mathcal{A}$ to an event at $\mathcal{B}$, no matter what time and space coordinates we assign to the events (see Fig. 2.2). In order to express the vector in terms of coordinates, we need to define a set of **basis vectors** which we shall denote[5] by $\boldsymbol{e}_\mu$.

Example 2.2

Old-fashioned 3-vectors in Euclidean three-dimensional space are written as

$$\vec{a} = a^x \vec{e}_x + a^y \vec{e}_y + a^z \vec{e}_z. \tag{2.3}$$

Note that components are given upstairs indices, while basis vectors are given downstairs indices. As a result, the **scalar product** (or **dot product**) is written as

$$\vec{a} \cdot \vec{b} = a^x b^x + a^y b^y + a^z b^z. \tag{2.4}$$

In a Cartesian coordinate system, the basis vectors are orthonormal, expressed as[6]

$$\vec{e}_i \cdot \vec{e}_j = \delta_{ij}. \tag{2.5}$$

A useful trick to note is that components of a vector can be projected out, that is, they can be extracted using

$$\vec{a} \cdot \vec{e}_x = a^x. \tag{2.6}$$

By analogy with Example 2.2, a 4-vector in spacetime is written as

$$\boldsymbol{X} = X^0 \boldsymbol{e}_0 + X^1 \boldsymbol{e}_1 + X^2 \boldsymbol{e}_2 + X^3 \boldsymbol{e}_3 = X^\mu \boldsymbol{e}_\mu, \tag{2.7}$$

where in the last equality we have used the **Einstein summation convention**, by which index variables (like μ) repeated in both the upstairs and downstairs positions are assumed to be summed.

[3]A good counterexample is the two–component 'shopping vector' that contains the price of fish and the price of bread in each component. If you approach the supermarket checkout with the trolley at 45° to the vertical, you will soon discover that the prices of your shopping will not transform appropriately. To use the jargon introduced earlier, vectors have to transform *covariantly* (see the discussion on page 3), and our 'shopping vector' fails. It isn't a vector at all, just a couple of numbers surrounded by brackets.

[4]This makes it look like vectors and intervals are very similar, and so they are, at this stage. We'll see, however, that they lose this similarity when we start to look at curved spacetimes.

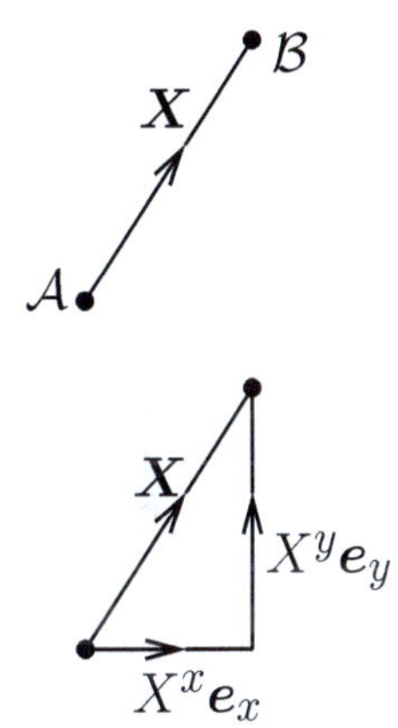

Fig. 2.2 A vector $\boldsymbol{X}$ lives free of any coordinate system. We can, however, impose a coordinate system and express a vector in terms of basis vectors $\boldsymbol{e}_\mu$ and its components X^μ.

[5]The μ in $\boldsymbol{e}_\mu$ tells us which basis vector we're dealing with, rather than telling us which component of a vector we're talking about.

[6]We define the symbol δ_{ij} such that $\delta_{ij} = 1$ when $i = j$ and $\delta_{ij} = 0$ otherwise. It is known as the **Kronecker delta**.

2.2 Coordinate transformations

Since vectors $\boldsymbol{X}$ exist independently of bases and coordinates, they can be expressed in different coordinate systems (see Fig. 2.3) via a different set of basis vectors[7]

$$\boldsymbol{X} = X^{\mu}\boldsymbol{e}_{\mu} = X^{\sigma'}\boldsymbol{e}_{\sigma'}. \tag{2.8}$$

Special relativity is based on the observation that the components of 4-vectors transform between inertial frames according to the Lorentz transformations

$$X^{\mu'} = \Lambda^{\mu'}{}_{\nu}X^{\nu}, \tag{2.9}$$

where we represent the Lorentz transformations in component form using $\Lambda^{\mu'}{}_{\nu}$, which are functions of the relative velocity $v(=\beta)$ of the frames.

[7]Although the prime is written on the subscript, the components $X^{\sigma'}$ and basis vectors $\boldsymbol{e}_{\sigma'}$ refer to a different coordinate system (the primed coordinate system) from that of X^{μ} and $\boldsymbol{e}_{\mu}$. Putting the prime on the indices, rather than on the variables themselves, might seem like an odd choice, but it will turn out to be very useful when we start dealing with more complicated equations.

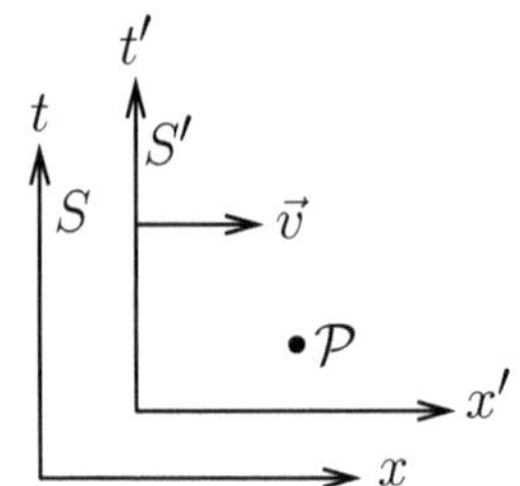

Fig. 2.3 The unprimed and primed coordinate system, showing just one spatial direction.

Example 2.3

The Lorentz transformation for the coordinates of an event in a frame S and a frame S' (moving relative to frame S at speed β along the x-axis) can be rewritten in matrix form as

$$\begin{pmatrix} X^{0'} \\ X^{1'} \\ X^{2'} \\ X^{3'} \end{pmatrix} = \begin{pmatrix} \gamma & -\beta\gamma & 0 & 0 \\ -\beta\gamma & \gamma & 0 & 0 \\ 0 & 0 & 1 & 0 \\ 0 & 0 & 0 & 1 \end{pmatrix} \begin{pmatrix} X^{0} \\ X^{1} \\ X^{2} \\ X^{3} \end{pmatrix}, \tag{2.10}$$

or for short as

$$X^{\mu'} = \Lambda^{\mu'}{}_{\nu}X^{\nu}, \tag{2.11}$$

where $\Lambda^{\mu'}{}_{\nu}$ is the **Lorentz transformation matrix**. Here we have again used the Einstein summation convention, and the twice-repeated index which is assumed to be summed is ν.[8]

[8]This equation can be written as

$$X^{\mu'} = \sum_{\nu} \Lambda^{\mu'}{}_{\nu}X^{\nu}.$$

Note that all the coordinate transformations we are considering are ones that preserve the origin, so that an event at $\vec{x} = 0$ and $t = 0$ in frame S is mapped into $\vec{x}' = 0$ and $t' = 0$ in frame S'.

An example of a vector is the infinitesimal translation $\mathrm{d}\boldsymbol{x}$ which has components $\mathrm{d}x^{\nu}$ in frame S. In frame S', the components then change to $\mathrm{d}x^{\mu'} = \Lambda^{\mu'}{}_{\nu}\mathrm{d}x^{\nu}$. Noting how each component resembles a **differential** of a function $x^{\mu'}$, we recall that the ordinary rules of calculus also give us a rule for manipulating differentials that reads[9]

$$\mathrm{d}x^{\mu'} = \frac{\partial x^{\mu'}}{\partial x^{\nu}}\mathrm{d}x^{\nu}. \tag{2.12}$$

Thus, we conclude that

$$\Lambda^{\mu'}{}_{\nu} = \frac{\partial x^{\mu'}}{\partial x^{\nu}}. \tag{2.13}$$

[9]This is an example of the famous relation for differentials

$$\mathrm{d}f = \frac{\partial f}{\partial x}\mathrm{d}x + \frac{\partial f}{\partial y}\mathrm{d}y + \frac{\partial f}{\partial z}\mathrm{d}z,$$

for a function $f(x,y,z)$.

[10]In this chapter, the only transformation we shall consider is the Lorentz transformation. It will turn out that this rule applies more generally to components of vectors although, as discussed in the next chapter, only to the position vector in special cases.

Thus, transforming components from an unprimed to a primed frame uses this partial derivative which varies a coordinate in the primed frame with respect to a coordinate in the unprimed frame, keeping other coordinates in the unprimed frame fixed. We say that the components of vectors transform like differentials.[10]

A key property of the Lorentz transformation is that it preserves the length of a vector, which is a quantity obtained by taking the scalar product of a vector with itself. The scalar product is a rule for combining vectors that we write

$$\begin{aligned}
\boldsymbol{X} \cdot \boldsymbol{Y} &= (X^\mu \boldsymbol{e}_\mu) \cdot (Y^\nu \boldsymbol{e}_\nu) \\
&= (\boldsymbol{e}_\mu \cdot \boldsymbol{e}_\nu) X^\mu Y^\nu .
\end{aligned} \tag{2.14}$$

The object $(\boldsymbol{e}_\mu \cdot \boldsymbol{e}_\nu)$ is a matrix giving a rule for combining vectors. In flat space, this matrix is defined to be $\eta_{\mu\nu} \equiv \boldsymbol{e}_\mu \cdot \boldsymbol{e}_\nu$ and written out in full as

$$\eta_{\mu\nu} = \begin{pmatrix} -1 & 0 & 0 & 0 \\ 0 & 1 & 0 & 0 \\ 0 & 0 & 1 & 0 \\ 0 & 0 & 0 & 1 \end{pmatrix} . \tag{2.15}$$

This matrix is called the **Minkowski metric tensor** or Minkowski metric, for short.[11] The rule for combining two 4-vectors is, therefore,

$$\begin{aligned}
\boldsymbol{X} \cdot \boldsymbol{Y} &= \eta_{\mu\nu} X^\mu Y^\nu \\
&= -X^0 Y^0 + X^1 Y^1 + X^2 Y^2 + X^3 Y^3 .
\end{aligned} \tag{2.16}$$

The minus sign in η_{00} is chosen to fit with our definition of $\mathrm{d}s^2$, so that we have $\mathrm{d}s^2 = \mathrm{d}\boldsymbol{x} \cdot \mathrm{d}\boldsymbol{x}$ which, when written in terms of components, becomes $\mathrm{d}s^2 = \eta_{\mu\nu} \mathrm{d}x^\mu \mathrm{d}x^\nu$.

We can summarize the key expressions involving the Minkowski tensor as follows:

$$\begin{aligned}
\eta_{\mu\nu} &= \boldsymbol{e}_\mu \cdot \boldsymbol{e}_\nu , & (2.17) \\
\boldsymbol{X} \cdot \boldsymbol{Y} &= \eta_{\mu\nu} X^\mu Y^\nu , & (2.18) \\
\mathrm{d}s^2 &= \eta_{\mu\nu} \mathrm{d}x^\mu \mathrm{d}x^\nu . & (2.19)
\end{aligned}$$

Just as an interval $\mathrm{d}s$ can be timelike, spacelike or null, we can classify a vector $\boldsymbol{X}$ in terms of its square $\boldsymbol{X}^2 = \boldsymbol{X} \cdot \boldsymbol{X}$ by saying

$$\begin{aligned}
\text{Spacelike vector} \quad & \boldsymbol{X}^2 > 0, \\
\text{Null vector} \quad & \boldsymbol{X}^2 = 0, \\
\text{Timelike vector} \quad & \boldsymbol{X}^2 < 0.
\end{aligned} \tag{2.20}$$

Consider a light cone (see Fig. 2.4) based at a point $\mathcal{P}$. Timelike vectors starting from $\mathcal{P}$ can only exist within the forward or backward light cones. Spacelike vectors exist outside of the light cones while null vectors lie on the light cones. Light cones are sometimes called **absolute surfaces** as they always allow us to separate intervals and vectors in this way.

[11]Some jargon: The **signature** of the metric tensor is the number of positive, negative or zero eigenvalues. Here, our metric is diagonal and so the eigenvalues can be read off very simply. The Minkowski metric tensor in eqn 2.15 has eigenvalues -1, 1, 1 and 1 and so its signature can be written as $(3, 1, 0)$ (3 plusses, 1 minus, no zeros) or, more commonly, as $(-, +, +, +)$ (enumerating the signs of the eigenvalues). Positive definite metrics $(+, +, +, +)$ are called **Riemannian**. A **Lorentzian metric** has a signature of one minus and the rest plusses, such as $(-, +, +, +)$ as in eqn 2.15, or one plus and the rest minuses, as in $(+, -, -, -)$. These latter metrics are known as **pseudo Riemannian**.

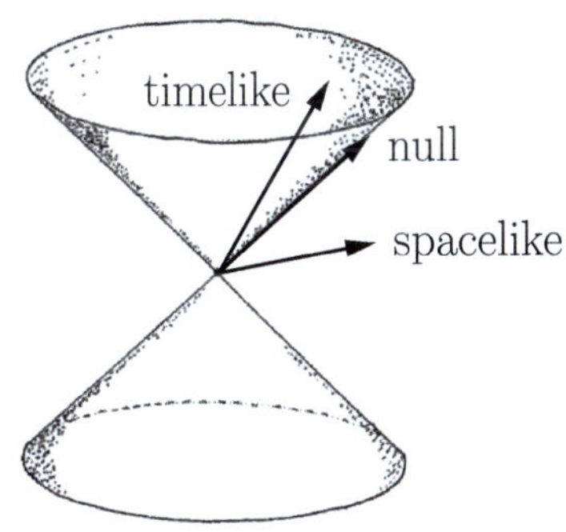

Fig. 2.4 The anatomy of a light cone, showing timelike, spacelike and null vectors.

Example 2.4

The Lorentz transformation preserves the length of a vector, which is therefore a Lorentz invariant. This means that we can write

$$\boldsymbol{X} \cdot \boldsymbol{X} = \eta_{\mu\nu} X^\mu X^\nu = \eta_{\mu'\nu'} X^{\mu'} X^{\nu'} . \tag{2.21}$$

However, this equation implies that

$$\eta_{\mu'\nu'} X^{\mu'} X^{\nu'} = \eta_{\mu'\nu'} (\Lambda^{\mu'}{}_{\mu} X^{\mu})(\Lambda^{\nu'}{}_{\nu} X^{\nu}) = \eta_{\mu\nu} X^{\mu} X^{\nu}, \tag{2.22}$$

which is true for any choice of $\boldsymbol{X}$ and hence

$$\eta_{\mu'\nu'} \Lambda^{\mu'}{}_{\mu} \Lambda^{\nu'}{}_{\nu} = \eta_{\mu\nu}. \tag{2.23}$$

This identity can be used to then show that for two different vectors that

$$\boldsymbol{X} \cdot \boldsymbol{Y} = \eta_{\mu\nu} X^{\mu} Y^{\nu} = \eta_{\mu'\nu'} X^{\mu'} Y^{\nu'}. \tag{2.24}$$

and hence the scalar product $\boldsymbol{X} \cdot \boldsymbol{Y}$ is Lorentz invariant. This will be very useful.

Example 2.5

Vectors exist independently of any coordinate system. Therefore, the object $\boldsymbol{X} = X^{\mu} \boldsymbol{e}_{\mu}$ doesn't change as it is transformed. This allows us to work out how the basis vectors $\boldsymbol{e}_{\mu}$ themselves transform. We write the transformation

$$\boldsymbol{e}_{\alpha'} = \Lambda^{\mu}{}_{\alpha'} \boldsymbol{e}_{\mu}, \tag{2.25}$$

by analogy with eqn 2.9. In order for the vector $\boldsymbol{X}$ itself to be coordinate independent, the product of the transformations of the components X^{μ} and basis vectors $\boldsymbol{e}_{\mu}$ must yield the identity. That is to say that $\boldsymbol{X}$ can be written equivalently as

$$\boldsymbol{X} = X^{\mu} \boldsymbol{e}_{\mu} = X^{\alpha'} \boldsymbol{e}_{\alpha'}, \tag{2.26}$$

and application of eqns 2.9 and 2.25 implies that[12]

$$\boldsymbol{X} = X^{\alpha'} \boldsymbol{e}_{\alpha'} = (\Lambda^{\alpha'}{}_{\mu} X^{\mu})(\Lambda^{\nu}{}_{\alpha'} \boldsymbol{e}_{\nu}) = (\Lambda^{\alpha'}{}_{\mu} \Lambda^{\nu}{}_{\alpha'}) X^{\mu} \boldsymbol{e}_{\nu} = X^{\mu} \boldsymbol{e}_{\mu}, \tag{2.27}$$

where the last equality relies on $\Lambda^{\alpha'}{}_{\mu} \Lambda^{\nu}{}_{\alpha'} = \delta^{\nu}{}_{\mu}$, the identity operation.[13] Thus, we identify the **inverse** of the Lorentz transformation $\Lambda^{\alpha'}{}_{\mu}$ which we write

$$(\Lambda^{\alpha'}{}_{\mu})^{-1} = \Lambda^{\mu}{}_{\alpha'}. \tag{2.28}$$

To summarize: the inverse of the matrix $\Lambda^{\alpha'}{}_{\mu}$ is the matrix[14] $\Lambda^{\mu}{}_{\alpha'}$. We deduce that the basis vectors transform using the inverse of the Lorentz transformation used for the vector components. The key equations for identifying inverses are

$$\Lambda^{\alpha'}{}_{\beta} \Lambda^{\beta}{}_{\gamma'} = \delta^{\alpha'}{}_{\gamma'} \quad \text{and} \quad \Lambda^{\beta}{}_{\alpha'} \Lambda^{\alpha'}{}_{\gamma} = \delta^{\beta}{}_{\gamma}. \tag{2.29}$$

Example 2.6

We saw earlier that the components of vectors, carrying an up index, transform like differentials. The rule for transforming objects with down indices, such as the basis vectors, is that they transform like derivatives

$$\frac{\partial}{\partial y^{\alpha'}} = \left(\frac{\partial x^{\mu}}{\partial y^{\alpha'}} \right) \frac{\partial}{\partial x^{\mu}}, \tag{2.30}$$

where the derivative involves varying a coordinate in the unprimed frame with respect to a coordinate in the primed frame, keeping other coordinates in the primed frame fixed. Thus, another down-indexed object, the **gradient vector** $\partial_{\mu} \phi \equiv \partial \phi / \partial x^{\mu}$, transforms as

$$\frac{\partial \phi}{\partial x^{\mu'}} = \left(\frac{\partial x^{\nu}}{\partial x^{\mu'}} \right) \frac{\partial \phi}{\partial x^{\nu}}. \tag{2.31}$$

[12]Note that, in these expressions full of components, we are free to reorder the terms for convenience.

[13]The symbol $\delta^{\nu}{}_{\mu}$ is another version of the Kronecker delta which is defined by

$$\delta^{\nu}{}_{\mu} = \begin{cases} 1 & \mu = \nu \\ 0 & \mu \neq \nu. \end{cases}$$

This particular form of the Kronecker delta, with one index up and the other down, is needed for reasons that will be explored after we have considered tensors in Chapter 4 (see in particular Exercise 4.2).

[14]You can check this is true using the matrix in eqn 2.10, noting that the sign of the velocity $\beta = v$ is flipped in the inverse operation which transforms back from the primed frame to the unprimed frame.

The jargon is that a^μ transforms like a **contravariant vector** and $\partial\phi/\partial x^\mu \equiv \partial_\mu\phi$ transforms like a **covariant vector**,[15] though we avoid these terms and just note that the component a^μ has its indices 'upstairs' and $\partial_\mu\phi$ has them 'downstairs' and they then transform accordingly. We can also construct the 4-vector analogue of $\vec{\nabla}^2$ using[16] $\partial^2 = \eta^{\mu\nu}\partial_\mu\partial_\nu$, and this will turn out to be very useful in the theory of gravitational waves.

In summary, we will insist that an object with indices in a downstairs position like a_μ transforms as

$$a_{\mu'} = \Lambda^\nu{}_{\mu'}\, a_\nu, \tag{2.32}$$

where $\Lambda^\nu{}_{\mu'} \equiv (\partial x^\nu/\partial x^{\mu'})$ is the inverse of the Lorentz transformation matrix $\Lambda^{\mu'}{}_\nu$.

It is a good moment to summarize some of our key results so far:

- A vector $\boldsymbol{X}$ is an arrow in space. It can be written in components as $\boldsymbol{X} = X^\mu \boldsymbol{e}_\mu$. The components transform according to $X^{\mu'} = \Lambda^{\mu'}{}_\nu X^\nu$, in the same way as a differential $\mathrm{d}x^\mu$.

- The basis vectors have downstairs components and transform according to $\boldsymbol{e}_{\mu'} = \Lambda^\nu{}_{\mu'}\boldsymbol{e}_\nu$ (i.e. the inverse transformation), in the same way as a gradient $\partial_\mu = \partial/\partial x^\mu$.

- The scalar product $\boldsymbol{X}\cdot\boldsymbol{Y} = \eta_{\mu\nu}X^\mu Y^\nu$ is Lorentz invariant.

2.3   Examples of vectors

In ordinary Euclidean space we have vectors such as $\vec{r}$, the position vector, which is obviously an arrow in space, but also current density $\vec{J}$ and acceleration $\vec{a}$ which are also vectors but somehow live in different spaces. For spacetime vectors we have an analogous situation and some commonly used vectors are listed in Table 2.1, all of which transform appropriately under Lorentz transformations.

Physical quantity	4-vector $\boldsymbol{v}$	Invariant $\boldsymbol{v}\cdot\boldsymbol{v}$
Interval	$\mathrm{d}\boldsymbol{s}$	$-\mathrm{d}\tau^2$
Position*	$\boldsymbol{x}$	$-\tau^2$
Velocity	$\boldsymbol{u} = \mathrm{d}\boldsymbol{x}/\mathrm{d}\tau$	-1
Momentum	$\boldsymbol{p} = m\boldsymbol{u}$	$-m^2$
Force	$\boldsymbol{f} = \mathrm{d}\boldsymbol{p}/\mathrm{d}\tau$	$m^2|\vec{a}|^2$
Acceleration	$\boldsymbol{a} = \mathrm{d}\boldsymbol{u}/\mathrm{d}\tau$	$|\vec{a}|^2$
Current	$\boldsymbol{J} = n_0\boldsymbol{u}$	$-n_0^2$

Table 2.1 Commonly used 4-vectors. *The position vector transforms correctly under Lorentz transformations, but does not transform correctly under general coordinate transformations (see Chapter 3).

(a) Interval and position: The first one in our list is our old friend the interval $\mathrm{d}\boldsymbol{s}$. We can also define a spacetime position 4-vector $\boldsymbol{x}$ which

---

[15]These unfortunate terms are due to the English mathematician J. J. Sylvester (1814–1897). Both types of vectors transform covariantly, in the sense of 'properly', and we wish to retain this sense of the word 'covariant' rather than using it to simply label one type of object that transforms properly. Thus, we usually specify whether the indices on a particular object are 'upstairs' (like a^μ) or 'downstairs' (like $\partial_\mu\phi$) and their transformation properties can then be deduced accordingly.

[16]We define ∂^2 as the scalar product $\eta^{\mu\nu}\partial_\mu\partial_\nu$ so that

$$\partial^2 = -\frac{\partial^2}{\partial t^2} + \frac{\partial^2}{\partial x^2} + \frac{\partial^2}{\partial y^2} + \frac{\partial^2}{\partial z^2}$$

$$= -\frac{\partial^2}{\partial t^2} + \vec{\nabla}^2.$$

We will show in Chapter 4 (page 45) that $\eta^{\mu\nu}$ behaves as the same matrix as $\eta_{\mu\nu}$.

The position vector x^μ works fine in special relativity, but is the one vector in our list in Table 2.1 which will not upgrade nicely to general relativity. We will still be able to work with infinitesimal displacements, such as $\mathrm{d}x^\mu$. The other vectors in our list will still be extremely useful in general relativity.

[17]That is, it does transform properly under Lorentz transformations.

Velocity component trick: from the velocity vector $\boldsymbol{u}$ with components $(u^0, u^i) = (\mathrm{d}t/\mathrm{d}\tau, \mathrm{d}x^i \mathrm{d}\tau)$ we can extract a spatial part by saying

$$u^i = \frac{\mathrm{d}x^i}{\mathrm{d}\tau} = \frac{\mathrm{d}x^i}{\mathrm{d}t}\frac{\mathrm{d}t}{\mathrm{d}\tau} = v^i u^0.$$

[18]For a slow-moving object (i.e. slow compared to light speed) we have $\gamma \approx 1$ and this reduces to

$$u^\mu \approx (1, \vec{v}). \qquad (2.35)$$

[19]Remember that massive particles, which we assume we're describing here, have timelike velocity vectors with negative $|\boldsymbol{u}|^2$, as we find.

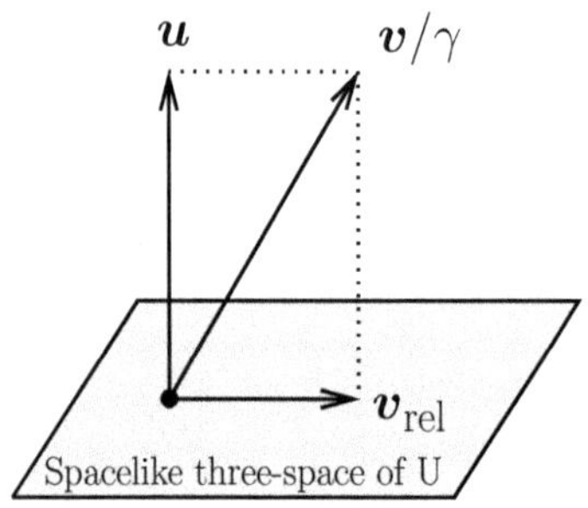

Fig. 2.5 The velocities $\boldsymbol{u}$, $\boldsymbol{v}/\gamma$ and $\boldsymbol{v}_{\mathrm{rel}}$ in the reference frame of observer U.

is the non-infinitesimal version of the same thing. In coordinates, we could write $x^\mu = (t, \vec{x})$. As we have seen before, its invariant (the scalar product of itself with itself) is $\boldsymbol{x} \cdot \boldsymbol{x} = -t^2 + x^2 + y^2 + z^2 = -\tau^2$, where τ is the proper time.

(b) Velocity: Next, let's try and find the velocity. Its tempting to write this as $\mathrm{d}\boldsymbol{x}/\mathrm{d}t$ with components $\mathrm{d}x^\mu/\mathrm{d}t = (1, \mathrm{d}\vec{x}/\mathrm{d}t)$ but this is not a 4-vector because it does not transform properly. You can see this easily by taking the scalar product of it with itself

$$\frac{\mathrm{d}\boldsymbol{x}}{\mathrm{d}t} \cdot \frac{\mathrm{d}\boldsymbol{x}}{\mathrm{d}t} = -1 + v^2 = -\frac{1}{\gamma^2}, \qquad (2.33)$$

where $v = |\mathrm{d}\vec{x}/\mathrm{d}t|$ is the magnitude of the 3-velocity. This clearly depends on which frame you are in and is not an invariant. The solution is to differentiate $\boldsymbol{x}$ not with respect to time t but with respect to *proper time* τ. This gives us a velocity that is Lorentz covariant[17] defined by

$$\boldsymbol{u} = \frac{\mathrm{d}\boldsymbol{x}}{\mathrm{d}\tau}. \qquad (2.34)$$

The velocity vector can be thought of as being the **tangent** to the world line of a particle. Using the equation $\mathrm{d}t/\mathrm{d}\tau = \gamma$ from the last chapter, we can deduce that the velocity $\boldsymbol{u}$ has components[18]

$$u^\mu = \left(\frac{\mathrm{d}t}{\mathrm{d}\tau}, \frac{\mathrm{d}x^i}{\mathrm{d}\tau}\right) = \left(\gamma, \gamma\frac{\mathrm{d}\vec{x}}{\mathrm{d}t}\right) = (\gamma, \gamma\vec{v}). \qquad (2.36)$$

The useful invariant is[19]

$$\boldsymbol{u} \cdot \boldsymbol{u} = (\gamma, \gamma\vec{v}) \cdot (\gamma, \gamma\vec{v}) = \gamma^2(-1 + v^2) = -1. \qquad (2.37)$$

This latter expression, confirmed in the next example, is used in computations throughout the book.

Example 2.7

Three examples to illustrate 4-vector velocity:

(i) Define the 4-vector velocity of an observer by $\boldsymbol{u}$. In the observer's rest frame, by definition, the 3-velocity is zero. The observer's time is then the proper time $x^0 = \tau$. As a result, the components of the 4-velocity in the observer's rest frame are $u^\mu = (1, 0, 0, 0)$. Since $\eta_{00} = -1$, we also have $\boldsymbol{u}^2 = \eta_{00}u^0 u^0 = -1$, as required.

(ii) Take any 4-vector $\boldsymbol{X}$. A useful result is that the timelike component of $\boldsymbol{X}$ in the observer's frame is then given by

$$X^0_{\mathrm{obs}} = -\boldsymbol{X} \cdot \boldsymbol{u}, \qquad (2.38)$$

where $\boldsymbol{u}$ describes the observer's 4-velocity. Why is this? The great thing about 4-vector dot products is if you work them out in one, easy frame, the result holds for all frames. So let's choose the observer's frame in which $u^\mu = (1, 0, 0, 0)$ and $X^\mu = (X^0_{\mathrm{obs}}, X^1_{\mathrm{obs}}, X^2_{\mathrm{obs}}, X^3_{\mathrm{obs}})$ and $\boldsymbol{X} \cdot \boldsymbol{u} = -X^0_{\mathrm{obs}}$ as required.

(iii) Two observers U and V have velocity 4-vectors $\boldsymbol{u}$ and $\boldsymbol{v}$. Let's move into U's frame of reference in which the velocities have components $u^\mu = (1, 0)$ and $v^\mu = (\gamma, \gamma\vec{v}_{\mathrm{rel}})$, where $\gamma = (1 - v^2_{\mathrm{rel}})^{-1/2}$ is appropriate for the relative 3-velocity $\vec{v}_{\mathrm{rel}}$ between the two observers. There is an elegant geometrical construction we can make by looking at $v^\mu/\gamma = (1, \vec{v}_{\mathrm{rel}})$ which can be written as

$$\frac{\boldsymbol{v}}{\gamma} = \boldsymbol{u} + \boldsymbol{v}_{\mathrm{rel}}, \qquad (2.39)$$

where $\boldsymbol{v}_{\text{rel}}$, which has components $(0, \vec{v}_{\text{rel}})$, lies in the spacelike 3-space[20] of observer U (see Fig. 2.5). This means that $\boldsymbol{u} \cdot \boldsymbol{v}_{\text{rel}} = 0$ [since $u^\mu = (1, 0)$ and $v^\mu_{\text{rel}} = (0, \vec{v}_{\text{rel}})$]. Taking the scalar product of each side of eqn 2.39 with itself we have $-1/\gamma^2 = -1 + v^2_{\text{rel}}$, or

$$\gamma = (1 - v^2_{\text{rel}})^{-\frac{1}{2}}, \tag{2.40}$$

as might be expected. The point is that we can take the scalar product of eqn 2.39 with $\boldsymbol{u}$ to obtain

$$\boldsymbol{u} \cdot \boldsymbol{v} = -\gamma, \tag{2.41}$$

which is a useful result.

[20] Here *spacelike 3-space* simply means the parts of the space where vectors are written in terms of spacelike components v^i with $i = 1, 2, 3$. Equation 2.39 can be checked by substituting in the components given in the example.

(c) Momentum: From the definition of velocity $\boldsymbol{u}$, it's a short step to the momentum $\boldsymbol{p} = m\boldsymbol{u}$ which then has invariant $\boldsymbol{p} \cdot \boldsymbol{p} = -m^2$ and components $p^\mu = (\gamma m, \gamma m \vec{v}) = (E, \vec{p})$ using $E = \gamma m$ and $\vec{p} = \gamma m \vec{v}$.

Example 2.8

It's useful to remember that the 3-momentum $\vec{p} = \gamma m \vec{v}$ and energy $E = \gamma m$ are related via

$$\vec{p} = E\vec{v}. \tag{2.42}$$

Hence, the 4-momentum $\boldsymbol{p} = m\boldsymbol{u}$ can also be written as

$$p^\mu = (E, Ev^x, Ev^y, Ev^z). \tag{2.43}$$

This is helpful as this expression also applies to massless particles such as the photon (whose velocity is a null vector). We therefore take this latter equation to be true for light, giving us an expression for the photon momentum.

(d) Force: Newton's second law may be written in terms of our new language as $\boldsymbol{f} = m\frac{\mathrm{d}\boldsymbol{u}}{\mathrm{d}\tau}$, or

$$\boldsymbol{f} = \frac{\mathrm{d}\boldsymbol{p}}{\mathrm{d}\tau}. \tag{2.44}$$

Note that since $\boldsymbol{u} \cdot \boldsymbol{u} = -1$, the result of differentiation with respect to τ is

$$\frac{\mathrm{d}\boldsymbol{u}}{\mathrm{d}\tau} \cdot \boldsymbol{u} + \boldsymbol{u} \cdot \frac{\mathrm{d}\boldsymbol{u}}{\mathrm{d}\tau} = 2\boldsymbol{u} \cdot \frac{\mathrm{d}\boldsymbol{u}}{\mathrm{d}\tau} = 0, \tag{2.45}$$

or equivalently $\boldsymbol{f} \cdot \boldsymbol{u} = 0$. That is, the 4-force is perpendicular to the 4-velocity.

Example 2.9

The condition $\boldsymbol{f} \cdot \boldsymbol{u} = 0$ provides a useful relation if we recall that $u^\mu = (\gamma, \gamma\vec{v})$. Writing $f^\mu = (f^0, \vec{f})$, the dot product $\boldsymbol{f} \cdot \boldsymbol{u}$ yields

$$(f^0, \vec{f}) \cdot (\gamma, \gamma\vec{v}) = \gamma f^0 - \gamma \vec{f} \cdot \vec{v} = 0,$$

which implies

$$f^0 = \vec{f} \cdot \vec{v}. \tag{2.46}$$

Writing out $\boldsymbol{f} = \frac{\mathrm{d}\boldsymbol{p}}{\mathrm{d}\tau}$ in components

$$f^\mu = \left(\frac{\mathrm{d}t}{\mathrm{d}\tau} \frac{\mathrm{d}p^0}{\mathrm{d}t}, \frac{\mathrm{d}t}{\mathrm{d}\tau} \frac{\mathrm{d}p^i}{\mathrm{d}t} \right), \tag{2.47}$$

[21]It is often thought that special relativity cannot treat acceleration because it only deals with inertial frames. That is not the case. At each moment of time, an accelerating object can be thought of as in an **instantaneous rest frame** moving at speed v, but that speed varies along the trajectory.

[22]In words: the acceleration as measured in the rest frame, $\mathrm{d}\vec{v}/\mathrm{d}t$, sometimes known as the **proper acceleration**, is found by evaluating the invariant $\boldsymbol{a}^2$, which gives precisely the square of this proper acceleration.

[23]The third scalar product in eqn 2.55 gives us $-a^0 a^0 + a^1 a^1 = g^2$ while the second one gives us $a^1 = (u^0/u^1)a^0$. Putting these together gives $(a^0)^2[-1+(u^0/u^1)^2] = g^2$. Rearranging gives

$$a^0 = \frac{gu^1}{\sqrt{-u^1 u^1 + u^0 u^0}} = gu^1,$$

(using $-u^0 u^0 + u^1 u^1 = -1$ in the final step). The equation $a^1 = gu^0$ is produced similarly. The final hyperbolic solution comes from differentiating eqn 2.56 with respect to τ giving $\mathrm{d}^2 u^0/\mathrm{d}\tau^2 = g^2 u^0$ and $\mathrm{d}^2 u^1/\mathrm{d}\tau^2 = g^2 u^1$, and then choosing solutions so that at proper time $\tau = 0$ we have $t = 0$ and x is non-zero.

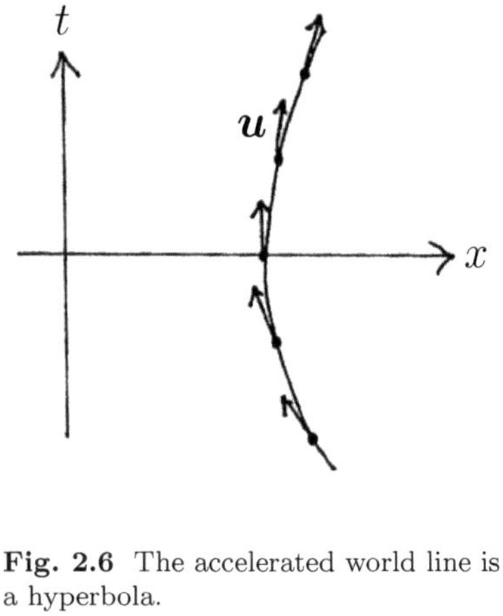

Fig. 2.6 The accelerated world line is a hyperbola.

and since $\mathrm{d}t/\mathrm{d}\tau = \gamma$, this simplifies to

$$f^\mu = \gamma \left(\frac{\mathrm{d}p^0}{\mathrm{d}t}, \frac{\mathrm{d}p^i}{\mathrm{d}t} \right). \tag{2.48}$$

The 3-force $\vec{F}$ is $\vec{F} = \frac{\mathrm{d}\vec{p}}{\mathrm{d}t}$, and hence we deduce that

$$f^\mu = (f^0, \gamma \vec{F}), \tag{2.49}$$

implying that $\vec{f} = \gamma \vec{F}$. Using eqn 2.46 we can write this as

$$f^\mu = (\gamma \vec{F} \cdot \vec{v}, \gamma \vec{F}). \tag{2.50}$$

This result is consistent with the power dissipated being given by $\mathrm{d}E/\mathrm{d}t = \mathrm{d}p^0/\mathrm{d}t = \vec{F} \cdot \vec{v}$, familiar from classical mechanics.

Using $\boldsymbol{f} = m\frac{\mathrm{d}\boldsymbol{u}}{\mathrm{d}\tau}$, we can express Newton's first law in terms of the velocity 4-vector and the proper time as

$$\frac{\mathrm{d}\boldsymbol{u}}{\mathrm{d}\tau} = 0, \quad \text{or in component form} \quad \frac{\mathrm{d}u^\mu}{\mathrm{d}\tau} = 0. \tag{2.51}$$

(e) The **acceleration**[21] is given by $\boldsymbol{a} = \frac{\mathrm{d}\boldsymbol{u}}{\mathrm{d}\tau}$, and has components $a^\mu = (a^0, \vec{a})$. From eqn 2.45, we have

$$\boldsymbol{a} \cdot \boldsymbol{u} = 0, \tag{2.52}$$

which implies $a^0 = 0$ in the rest frame of the observer [where $u^\mu = (1,0,0,0)$]. This means that, in the observer's instantaneous rest frame,[22]

$$\boldsymbol{a} \cdot \boldsymbol{a} = \left(\frac{\mathrm{d}\vec{v}}{\mathrm{d}t} \right)^2 = |\vec{a}|^2. \tag{2.53}$$

Example 2.10

A body is subjected to uniform acceleration g in its instantaneous rest frame and g is applied along x^1. In the instantaneous rest frame we have equations of motion

$$\frac{\mathrm{d}t}{\mathrm{d}\tau} = u^0, \quad \frac{\mathrm{d}x^1}{\mathrm{d}\tau} = u^1, \quad \frac{\mathrm{d}u^0}{\mathrm{d}\tau} = a^0, \quad \frac{\mathrm{d}u^1}{\mathrm{d}\tau} = a^1, \tag{2.54}$$

where $t = x^0$. Taking scalar products tells us three things

$$\boldsymbol{u} \cdot \boldsymbol{u} = -1, \quad \boldsymbol{u} \cdot \boldsymbol{a} = 0, \quad \boldsymbol{a} \cdot \boldsymbol{a} = g^2. \tag{2.55}$$

Solving,[23] we obtain

$$a^0 = \frac{\mathrm{d}u^0}{\mathrm{d}\tau} = gu^1, \quad a^1 = \frac{\mathrm{d}u^1}{\mathrm{d}\tau} = gu^0. \tag{2.56}$$

The solutions to these equations are hyperbolic sines and cosines

$$t = \tfrac{1}{g} \sinh g\tau, \quad x = \tfrac{1}{g} \cosh g\tau. \tag{2.57}$$

We conclude that the accelerated world line is the **hyperbola** $x^2 - t^2 = g^{-2}$ (see Fig. 2.6). The velocity along the world line is

$$u^0 = \frac{\mathrm{d}t}{\mathrm{d}\tau} = \cosh g\tau, \quad u^1 = \frac{\mathrm{d}x^1}{\mathrm{d}\tau} = \sinh g\tau, \tag{2.58}$$

which satisfies $\boldsymbol{u} \cdot \boldsymbol{u} = -u^0 u^0 + u^1 u^1 = -1$. The particle's 3-velocity is

$$v = \frac{\mathrm{d}x^1}{\mathrm{d}t} = \frac{\mathrm{d}x^1/\mathrm{d}\tau}{\mathrm{d}t/\mathrm{d}\tau} = \tanh g\tau. \tag{2.59}$$

This never exceeds $v = 1$, but approaches it for $\tau = \pm\infty$. The 4-acceleration is

$$a^0 = g \sinh g\tau, \quad a^1 = g \cosh g\tau, \tag{2.60}$$

and the magnitude is $|\boldsymbol{a}| = \sqrt{-(a^0)^2 + (a^1)^2} = g$. The 4-force required for this acceleration is given by $f^\mu = ma^\mu$.

(f) Particle current: This example refers to a cloud of particles. The **particle current** $\boldsymbol{J} = n_0\boldsymbol{u}$, where n_0 is the number of density of particles in their rest frame and $\boldsymbol{u}$ is their velocity. In the rest-frame of the particles, we can write $J^\mu = n_0(1,0,0,0)$; in a general frame $J^\mu = \gamma n_0(1,\vec{u})$. The timelike component gives the number density n [and in a general frame the density increases according to $n = \gamma n_0$ because of Lorentz contraction (Fig. 2.7)]. The spacelike components give the flux of particles along that direction: e.g. $J^x = \gamma n_0 u^x = n u^x$ is the number of particles crossing the yz plane in unit time.

2.4   Principle of least action

In the last section of this chapter, we shall show how some deep ideas in classical mechanics can be adapted for use in relativity. In contrast to using Newton's laws to work out now a system behaves, an alternative procedure was developed by the mathematician Joseph-Louis Lagrange to derive equations of motion from a variational principle. Thus, rather than starting with Newton's laws, we start with Hamilton's principle of least action, which we state below. The idea is to suppose that every mechanical system is characterized by a function called the **Lagrangian**, written[24] $L(q_1, q_2, ..., q_n, \dot{q}_1, \dot{q}_2, ..., \dot{q}_n, t)$, which is a function of the positions q_i of each of the n particles in the system, their velocities $v = \mathrm{d}q_i/\mathrm{d}t = \dot{q}_i$ and the time t. For simplicity we'll consider a single particle moving in one dimension, whose Lagrangian is then written $L(q, \dot{q}, t)$. Consider the trajectory of this particle as it travels from point $\mathcal{A}$ with coordinate $q(t_1)$ at time t_1 to a point $\mathcal{B}$ with coordinate $q(t_2)$ at time t_2. The **action** for this trajectory is defined to be

$$S = \int_{t_1}^{t_2} \mathrm{d}t\, L(q, \dot{q}, t). \tag{2.61}$$

That is, we evaluate the Lagrangian at each point along the trajectory and add these up in the integral. What *is* the Lagrangian? In classical mechanics, Lagrange showed that it takes the form

$$L = (\text{Kinetic energy}) - (\text{Potential energy}), \tag{2.62}$$

and we shall work this out in some particular cases in Example 2.11.

Hamilton's principle of least action says that the action, when describing the motion that actually takes place subject to the laws of physics, takes an extremal value (i.e. a maximum, stationary or minimum value). That is, if we find the path $q(t)$ that extremizes the action, we have found the path the particle takes in travelling from $\mathcal{A}$ to $\mathcal{B}$. There are many possible paths, and some of these are drawn in Fig. 2.8. Finding the path that extremizes the action is a simple problem in the calculus of variations (see Section 1.4) and from that we can immediately conclude that the **equations of motion** governing the motion of any particle in the Universe, is given by plugging the Lagrangian into the **Euler–Lagrange equation**.

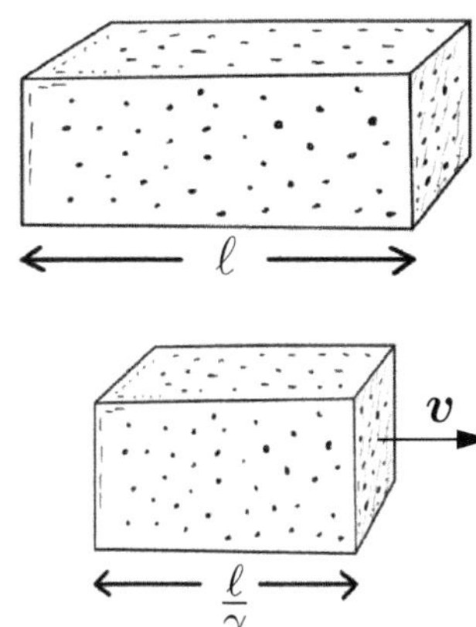

Fig. 2.7 Length contraction increases the density of particles in a box owing to the shortening of the box length along the direction of travel.

Joseph-Louis Lagrange (1736–1813)

[24]We have assumed one-dimensional motion for each of the particles in this expression. In three dimensions, we write $L(\vec{q}_1, \vec{q}_2, ..., \vec{q}_n, \dot{\vec{q}}_1, \dot{\vec{q}}_2, ..., \dot{\vec{q}}_n, t)$.

William Rowan Hamilton (1805–1865)

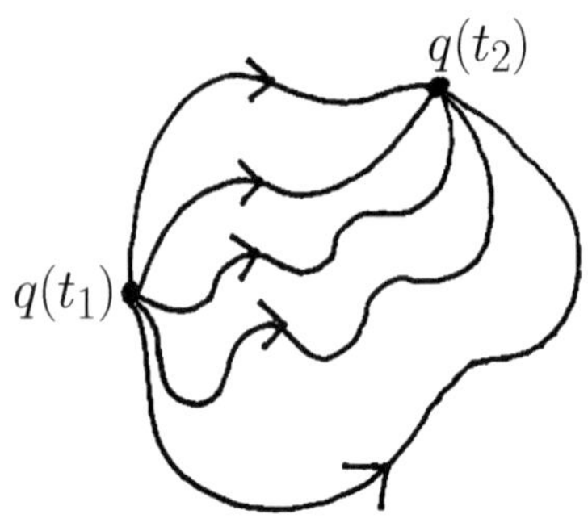

Fig. 2.8 Different trajectories with $\delta q(t_1) = \delta q(t_2) = 0$.

$$\frac{\mathrm{d}}{\mathrm{d}t}\frac{\partial L}{\partial \dot{q}} = \frac{\partial L}{\partial q}. \tag{2.63}$$

If the motion is in several dimensions, and consequently described by several coordinates $x^i = (x^1, x^2, x^3 ...)$, we have an Euler–Lagrange equation for each coordinate, giving a set of equations of motion which are known as the **Euler–Lagrange equations**

$$\frac{\mathrm{d}}{\mathrm{d}t}\frac{\partial L}{\partial \dot{x}^i} = \frac{\partial L}{\partial x^i}. \tag{2.64}$$

A solution can then be found by solving the entire set of equations. For our purposes, the fact that the Euler–Lagrange equations pick out the extremal values of the action S also makes them useful in a geometrical context. We now turn to some very simple examples and applications of this concept.

Example 2.11

- A free, non-relativistic particle has kinetic energy $\frac{1}{2}m\dot{x}^2$ and so has Lagrangian $L = \frac{1}{2}m\dot{x}^2$. The particle obeys the Euler–Lagrange equations and here this reduces to

$$0 = \frac{\partial L}{\partial x} - \frac{\mathrm{d}}{\mathrm{d}t}\frac{\partial L}{\partial \dot{x}} = 0 - \frac{\mathrm{d}}{\mathrm{d}t}(m\dot{x}), \tag{2.65}$$

  and we obtain $\ddot{x} = 0$, implying that $\dot{x}$ is a constant of the motion.

- Repeating this in three dimensions means the Lagrangian is $L = \frac{1}{2}m(\dot{x}^2 + \dot{y}^2 + \dot{z}^2)$ and we find that $\ddot{x} = \ddot{y} = \ddot{z} = 0$, implying that $\dot{x}$, $\dot{y}$ and $\dot{z}$ are each individually constant, and that $\vec{v} = (\dot{x}, \dot{y}, \dot{z})$ is a constant vector. We conclude that, in an inertial frame, free motion takes place with a velocity that is constant in magnitude and direction. This is known as the **law of inertia**.

- For a particle in a potential $V(x)$ we have $L = \frac{1}{2}m\dot{x}^2 - V(x)$, and we obtain an equation of motion

$$m\ddot{x} = -\frac{\partial V}{\partial x}, \tag{2.66}$$

  which expresses Newton's second law.

- For a simple harmonic oscillator, $V(x) = \frac{1}{2}kx^2$, and so $L = \frac{1}{2}mv^2 - \frac{1}{2}kx^2$, giving an equation of motion $m\ddot{x} = -kx$.

- For Newtonian gravitation

$$L = \frac{1}{2}m\left(\dot{x} + \dot{y}^2 + \dot{z}^2\right) + \frac{GMm}{|\vec{r}|}, \tag{2.67}$$

  where $r^2 = x^2 + y^2 + z^2$. We derive equations of motion

$$\ddot{x} = -GM\tfrac{x}{r^3}, \quad \ddot{y} = -GM\tfrac{y}{r^3}, \quad \ddot{z} = -GM\tfrac{z}{r^3}. \tag{2.68}$$

Hamilton's principle of least action is closely related to **Fermat's principle of least time**, the idea that light chooses a route that minimizes the travel time. This gives us a way of thinking about the Lagrangian for a relativistic particle in flat spacetime. We write the action as

$$S = -m \int \mathrm{d}\tau. \tag{2.69}$$

This is mass m (which is an energy mc^2 if you put the factors of c back in) multiplied by the path length in time. There is also a minus sign which expresses that when we minimize S we maximize $\int d\tau$ (and we have already argued from the twin paradox that the straight-line path is the longest, not shortest path). As we shall see, these choices give us the correct dynamics. First note that since $d\tau = dt/\gamma$, eqn 2.69 gives us a very simple Lagrangian[25]

$$L = -\frac{m}{\gamma}. \tag{2.70}$$

[25] Hence, $S = \int L\,dt = -m \int d\tau$ as required.

Example 2.12

More usefully for later discussion, we can rewrite the action in eqn 2.69 in terms of the components of the metric, so that

$$S = -m \int \sqrt{-ds^2} = -m \int (-\eta_{\mu\nu} dx^\mu dx^\nu)^{\frac{1}{2}}. \tag{2.71}$$

We can show that the Lagrangian we have identified in eqn 2.70 makes sense by taking a non-relativistic limit for small velocities

$$L = -\frac{m}{\gamma} = -m\left[1 - \frac{1}{2}v^2 + ...\right] = -m + \frac{1}{2}mv^2 - ... \tag{2.72}$$

Ignoring the constant $-m$, which vanishes as soon as we differentiate the Lagrangian, we have the Lagrangian $L = \frac{1}{2}mv^2$ for a free particle in Newtonian mechanics. Having passed this test, we realize that eqn 2.70 is a fully working Lagrangian for free particles in special relativity: $L = -m/\gamma = -m(1-v^2)^{\frac{1}{2}}$. To analyse the mechanics of a system we need to deal with the forces that act on particles owing to the potential V that they feel. In Newtonian physics, we would simple write $L = \frac{1}{2}mv^2 - V$. However, the role of the potential energy in relativity turns out to be more subtle and depends on which potential we are considering.

[26] That is, we write

$$ds^2 = -\left[(1+\Phi) - \frac{v^2}{2}\right]^2 dt^2$$

$$= -\left[(1+\Phi)^2 - (1+\Phi)v^2 + \frac{v^4}{4}\right] dt^2$$

$$= -\big[(1 + 2\Phi + \Phi^2)$$

$$-v^2 - v^2\Phi + \frac{v^4}{4}\big] dt^2, \tag{2.74}$$

and drop higher order terms proportional to Φ^2, $v^2\Phi$ and v^4.

Example 2.13

Inserting a potential Φ into the Lagrangian in eqn 2.72 would give $L = -m + \frac{1}{2}mv^2 - m\Phi + \cdots$, so that the action $S = \int L\,dt = -m \int \sqrt{-ds^2}$, and this would imply

$$\sqrt{-ds^2} = (1 - \frac{1}{2}v^2 + \Phi)\,dt. \tag{2.73}$$

Hence, using $v^2 = (dx/dt)^2 + (dy/dt)^2 + (dz/dt)^2$ we have (to leading order)[26]

$$ds^2 = -(1 + 2\Phi)dt^2 + dx^2 + dy^2 + dz^2. \tag{2.75}$$

This is our first hint that a gravitational potential can change the metric of spacetime. However, this expression has been obtained in a non-relativistic limit so one should exercise some caution.[27]

[27] The $v \ll c$ limit used in eqn 2.72 to derive eqn 2.75 means that this expression is only expected to work for timelike intervals close to the time axis, and so one might suspect that the spatial coordinates in eqn 2.75 might need tweaking due to the effects of the potential. That turns out to be the case (see eqn 5.22).

This brings us on to a very important idea. In order to describe gravitation, Einstein did not simply put together a potential to contribute to a Lagrangian, but instead showed how gravitation changes the very properties of spacetime itself. This is done by having matter directly affect the metric and, by extension, the basic rules governing the intervals in spacetime. In general relativity, the action $S = -m \int \left(-\eta_{\mu\nu}\mathrm{d}x^\mu \mathrm{d}x^\nu\right)^{\frac{1}{2}}$ becomes

$$S = -m \int \left(-g_{\mu\nu}\mathrm{d}x^\mu \mathrm{d}x^\nu\right)^{\frac{1}{2}}, \tag{2.76}$$

where the gravitating masses determine the form of the metric $g_{\mu\nu}$ (which, as we shall see, can then be different from the flat spacetime metric $\eta_{\mu\nu}$).

Chapter summary

- Vectors in special relativity are four-dimensional. Their components can be transformed by the Lorentz transformations, but scalar products are invariant.
- The Minkowski metric tensor, with components $\eta_{\mu\nu}$ allows us to make scalar products in flat spacetime.
- The relativistic action is given in terms of the interval $\mathrm{d}s$.
- The velocity vector is tangent to the world line of a particle.
- An observer with 4-velocity $\boldsymbol{u}$ measures the timelike component of 4-vector $\boldsymbol{X}$ to be $-\boldsymbol{X}\cdot\boldsymbol{u}$.

Exercises

(2.1) Show that an observer travelling with velocity 4-vector $\boldsymbol{u}$ will deduce that, in their frame, the energy of a particle with 4-vector $\boldsymbol{p}$ is $E = -\boldsymbol{p}\cdot\boldsymbol{u}$.

(2.2) Show that eqn 2.41 $(\boldsymbol{u}\cdot\boldsymbol{v} = -\gamma)$ implies that

$$\gamma = \gamma(\vec{u})\gamma(\vec{v})(1 - \vec{u}\cdot\vec{v}), \tag{2.77}$$

and that in the non-relativistic limit this reduces to $\vec{v}_{\mathrm{rel}} = \vec{u} - \vec{v}$ or $\vec{v}_{\mathrm{rel}} = \vec{v} - \vec{u}$. Interpret these results physically.

(2.3) The momentum vector can be written in components as

$$p^\mu = (E, \vec{p}). \tag{2.78}$$

We can define the same object with a downstairs index by $p_\mu = \eta_{\mu\nu}p^\nu$. Show that

$$p_\mu = (-E, \vec{p}). \tag{2.79}$$

Show also that

$$\boldsymbol{p}\cdot\boldsymbol{p} = \eta_{\mu\nu}p^\mu p^\nu = p_\nu p^\nu = -E^2 + p^2 = -m^2. \tag{2.80}$$

The gradient operator has components

$$\partial_\mu = \left(\frac{\partial}{\partial t}, \vec{\nabla}\right). \tag{2.81}$$

Prove the following relations:
(i) $\partial^\mu = (-\partial/\partial t, \vec{\nabla})$;
(ii) $\partial^2 = \partial_\mu \partial^\mu = -\partial^2/\partial t^2 + \vec{\nabla}^2$;

(iii) $\partial_\mu J^\mu = -\partial\rho/\partial t + \vec{\nabla} \cdot \vec{J}$;

(iv) $a_\mu u^\mu = 0$.

(2.4) The Galaxy is about 10^5 light years across and the most energetic cosmic rays known have energies of the order of 10^{19} eV. How long would it take a proton (rest mass ≈ 1 GeV) with this energy to cross the Galaxy as measured in the rest frame of (i) the Galaxy and (ii) the proton?

(2.5) In its rest frame, a particle of mass m will have 4-vector $p^\mu = (m, 0, 0, 0)$. Using the Lorentz transformation on this 4-vector, find the energy and momentum of a particle in a frame moving so that the particle has speed v. Check that the original and transformed 4-vector components give the same invariant.

(2.6) Use 4-vectors to show that an electron in free space cannot absorb a single photon.

(2.7) Using the principle of least action, calculate the shape traced out by a hanging string.

(2.8) The generalized momentum in Lagrangian mechanics is $\vec{p} = \partial L/\partial\vec{v}$. Show that with $L = -m/\gamma$, this yields $\vec{p} = m\gamma\vec{v}$.

(2.9) The Hamiltonian H in Lagrangian mechanics is given by $H = \vec{p} \cdot \vec{v} - L$ where $\vec{p}$ is the momentum and $\vec{v}$ is the velocity. Show that this agrees with $E = \gamma mc^2$.

3

Coordinates

I'll put a girdle round about the earth in forty minutes
William Shakespeare *A Midsummer Night's Dream* (Act II, Scene I)

Do we actually need coordinates? In many cases, we are better off not using them. Take a relationship like the one that expresses Newton's second law

$$ \boldsymbol{f} = \frac{\mathrm{d}\boldsymbol{p}}{\mathrm{d}\tau}. \tag{3.1} $$

This is a relationship between two vectors (arrows in spacetime) and holds irrespective of the coordinates chosen. Yes, you could choose a frame in which you can write $f^\mu = \mathrm{d}p^\mu/\mathrm{d}\tau$, but if you transform into a second frame which is rotating with respect to the first, then eqn 3.1 will emerge with extra terms (centrifugal and Coriolis forces) and will look a lot more complicated, even though the same physics is being described. It's much better, whenever possible, to stay above the fray and focus on a coordinate-free approach in which you only deal with statements expressed in purely geometrical terms.

However, particular physical problems have a very nasty habit of requiring us to dive back down into the murky world of coordinates, rather than staying aloof in our heavenly geometrical realm. Sometimes, like Puck in *A Midsummer Night's Dream*, we need to put a coordinate girdle around the Earth. There are often occasions when we need coordinates to express the value quantities take in particular frames, or to allow us to exploit a particular symmetry. Therefore, in this chapter, we develop a few ideas about coordinates and, for a start, we will need to define some terms: *Euclidean space, Cartesian coordinates, coordinate and non-coordinate bases, non-Euclidean space.*

3.1 Coordinates in Euclidean space

Euclidean space is a set of points in n-dimensions in which the scalar product between two vectors $\boldsymbol{X}$ and $\boldsymbol{Y}$ is[1] $\boldsymbol{X} \cdot \boldsymbol{Y} = \sum_\mu X^\mu Y^\mu$, so that the length of a vector $\boldsymbol{X}$ is $|\boldsymbol{X}| = \sqrt{\boldsymbol{X} \cdot \boldsymbol{X}} = \left(\sum_\mu X^\mu X^\mu\right)^{\frac{1}{2}}$ and the angle between $\boldsymbol{X}$ and $\boldsymbol{Y}$ is $\cos^{-1}(\boldsymbol{X} \cdot \boldsymbol{Y}/|\boldsymbol{X}||\boldsymbol{Y}|)$. In other words, it is the flat space you have been using all your life, equipped with geometric axioms that date back to Euclid's *Elements* around 300 BC Euclid focussed on the geometry of the plane ($n = 2$), showing that the

---

[1] That is to say that, employing the Einstein summation convention, the metric tensor $\delta_{\mu\nu}$ in the expression

$$ \boldsymbol{X} \cdot \boldsymbol{Y} = \delta_{\mu\nu} X^\mu Y^\nu $$

is the identity matrix. In three dimensions, we have therefore that

$$ \delta_{\mu\nu} = \begin{pmatrix} 1 & 0 & 0 \\ 0 & 1 & 0 \\ 0 & 0 & 1 \end{pmatrix}. $$

sum of the internal angles in a triangle adds up to 180° and so forth, and so we will also choose $n = 2$ for now.

Euclidean space is most often described using **Cartesian coordinates**, an innovation of René Descartes in 1637 and so, following his example, we describe any point in the plane using two numbers x and y that encode where on the Cartesian plane in Fig. 3.1(a) a particular point happens to lie. Of course, that's not the only way of describing the Euclidean plane. Polar coordinates[2] are another option where the same point can be described by a distance r from the origin and an angle θ, as shown in Fig. 3.1(b).

These two sets of coordinates are related by the familiar equations

$$x = r\cos\theta, \quad y = r\sin\theta, \quad r = (x^2 + y^2)^{\frac{1}{2}}, \quad \tan\theta = y/x. \tag{3.2}$$

To express the components of a vector $\boldsymbol{X} = X^\mu \boldsymbol{e}_\mu$ in terms of the new coordinates we can use the formula for coordinate transformations (eqn 2.11, reserved until now for Lorentz transformations)

$$X^{\mu'} = \Lambda^{\mu'}{}_\nu X^\nu = \left(\frac{\partial x^{\mu'}}{\partial x^\nu}\right) X^\nu, \tag{3.3}$$

as demonstrated in the following example.

Example 3.1

We will consider the transformation between two-dimensional Cartesian coordinates and polar coordinates on the components of the infinitesimal-displacement vector $d\boldsymbol{x} = dx^\mu \boldsymbol{e}_\mu$ with components dx^μ. For the unprimed coordinates we write $x^\mu = (x, y)$. For the primed coordinates we write $x^{\alpha'} = (r, \theta)$. The transformations are given by a matrix formed from the partial derivatives computed below

$$\left(\frac{\partial x}{\partial r}\right)_\theta = \cos\theta, \qquad \left(\frac{\partial y}{\partial r}\right)_\theta = \sin\theta,$$
$$\left(\frac{\partial x}{\partial \theta}\right)_r = -r\sin\theta, \qquad \left(\frac{\partial y}{\partial \theta}\right)_r = r\cos\theta, \tag{3.4}$$

and

$$\left(\frac{\partial r}{\partial x}\right)_y = \frac{x}{\sqrt{x^2+y^2}}, \qquad \left(\frac{\partial r}{\partial y}\right)_x = \frac{y}{\sqrt{x^2+y^2}},$$
$$\left(\frac{\partial \theta}{\partial x}\right)_y = -\frac{y}{x^2+y^2}, \qquad \left(\frac{\partial \theta}{\partial y}\right)_x = \frac{x}{x^2+y^2}. \tag{3.5}$$

We can write the derivatives in matrix form $dx^{\alpha'} = \left(\partial x^{\alpha'}/\partial x^\mu\right) dx^\mu$, which gives a transformation matrix

$$\begin{pmatrix} dr \\ d\theta \end{pmatrix} = \begin{pmatrix} \left(\frac{\partial r}{\partial x}\right)_y & \left(\frac{\partial r}{\partial y}\right)_x \\ \left(\frac{\partial \theta}{\partial x}\right)_y & \left(\frac{\partial \theta}{\partial y}\right)_x \end{pmatrix} \begin{pmatrix} dx \\ dy \end{pmatrix}$$
$$= \begin{pmatrix} \cos\theta & \sin\theta \\ -\frac{1}{r}\sin\theta & \frac{1}{r}\cos\theta \end{pmatrix} \begin{pmatrix} dx \\ dy \end{pmatrix}, \tag{3.6}$$

where we have reexpressed the terms in the matrix using the cylindrical polar coordinates.

Basis vectors transform the opposite way, as we found in eqn 2.25 which stated that

$$\boldsymbol{e}_{\mu'} = \Lambda^\nu{}_{\mu'} \boldsymbol{e}_\nu = \left(\frac{\partial x^\nu}{\partial x^{\mu'}}\right) \boldsymbol{e}_\nu, \tag{3.7}$$

and we illustrate the use of this in the following example.

[2]Polar coordinates are our first example of **curvilinear coordinates**, which are sets of coordinates where Pythagoras' theorem doesn't hold simply. That is, $s^2 \neq r^2 + \theta^2$.

(a)

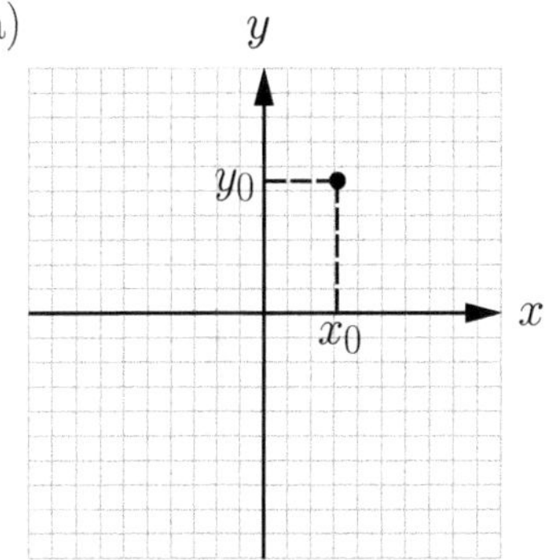

(b)

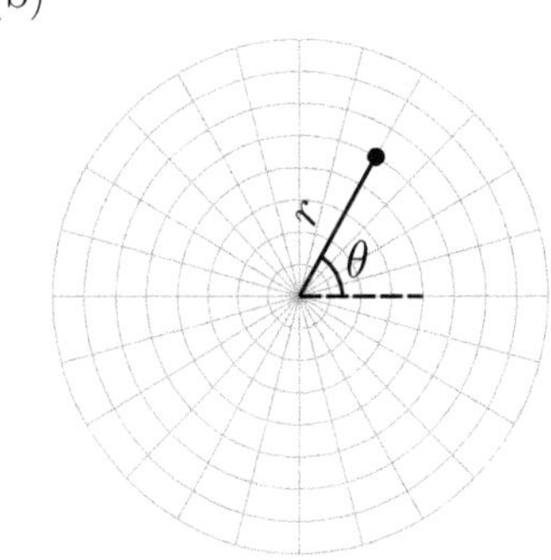

Fig. 3.1 (a) The point (x_0, y_0) in the Euclidean plane. (b) In polar coordinates, the same point is at (r, θ).

(a)

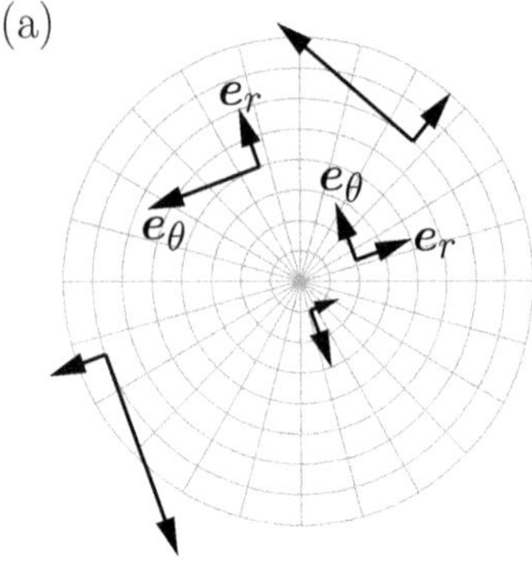

(b)

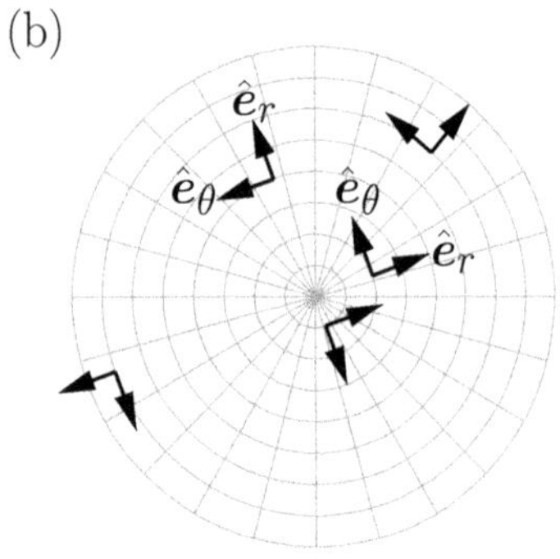

Fig. 3.2 (a) The coordinate basis e_r and e_θ has the feature that the basis vectors do not stay a uniform size. In particular, the length of e_θ increases with r, the distance from the origin. (b) The non-coordinate basis $\hat{e}_r$ and $\hat{e}_\theta$ remains normalized.

[3] The word holonomy comes from *holo* (entire) + *nomy* (law).

(a)   (b)

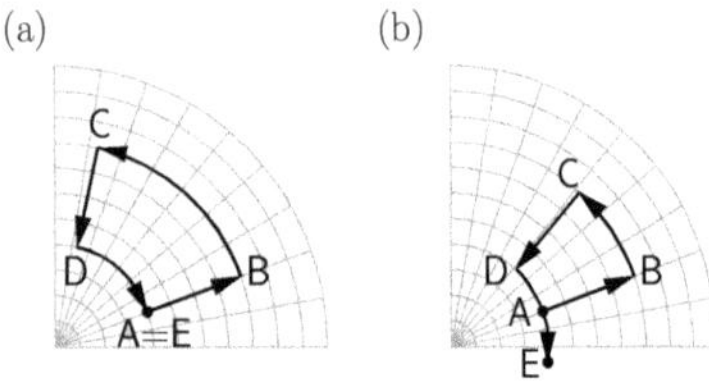

Fig. 3.3 (a) With the coordinate basis, a closed loop ABCDE comes back to its starting point A=E. We now see the reason why e_θ has to grow in length as r increases (so that the path BC is longer than DA). (b) For the non-coordinate basis, you pay the price for having your basis vectors normalized and the close path starting at A terminates at a different point E.

Example 3.2

Plugging our expressions for polar coordinates into eqn 3.7 gives

$$e_r = \Lambda^x{}_r e_x + \Lambda^y{}_r e_y = \left(\frac{\partial x}{\partial r}\right)_\theta e_x + \left(\frac{\partial y}{\partial r}\right)_\theta e_y = \cos\theta\, e_x + \sin\theta\, e_y, \qquad (3.8)$$

$$e_\theta = \Lambda^x{}_\theta e_x + \Lambda^y{}_\theta e_y = \left(\frac{\partial x}{\partial \theta}\right)_r e_x + \left(\frac{\partial y}{\partial \theta}\right)_r e_y = -r\sin\theta\, e_x + r\cos\theta\, e_y.$$

We can also express these results as a matrix equation

$$\begin{pmatrix} e_r & e_\theta \end{pmatrix} = \begin{pmatrix} e_x & e_y \end{pmatrix} \begin{pmatrix} \left(\frac{\partial x}{\partial r}\right)_\theta & \left(\frac{\partial x}{\partial \theta}\right)_r \\ \left(\frac{\partial y}{\partial r}\right)_\theta & \left(\frac{\partial y}{\partial \theta}\right)_r \end{pmatrix}. \qquad (3.9)$$

The method we have used hasn't produced the usual basis vectors that we might have expected. Elementary treatments of polar coordinates usually give *normalized* basis vectors $\hat{e}_r$ and $\hat{e}_\theta$ given by

$$\hat{e}_r = \cos\theta\, e_x + \sin\theta\, e_y,$$
$$\hat{e}_\theta = -\sin\theta\, e_x + \cos\theta\, e_y. \qquad (3.10)$$

The ones we have found (without the hats) have the disquieting feature that they are not all normalized. In fact

$$e_r \cdot e_r = 1 \quad \text{and} \quad e_\theta \cdot e_\theta = r^2. \qquad (3.11)$$

So e_r looks fine, but e_θ grows the further out you go (see Fig. 3.2). We will show that this seemingly odd property is not a bug but a feature! It's actually exactly what you need. It's helpful for two reasons:

(1) e_r and e_θ were easy to derive. We just had to plug straight into eqn 2.25 ($\Lambda^\mu{}_{\alpha'} e_\mu$) and out they popped.

(2) More importantly, e_r and e_θ form a **coordinate basis** (also known as a **holonomic basis**[3]) whereas $\hat{e}_r$ and $\hat{e}_\theta$ form a **non-coordinate basis** (also known as an **anholonomic basis**). What does that mean? We will return to the notion of coordinate and non-coordinate bases later, but (loosely) the idea is to take a walk around a closed loop in your space and to see if you return to the starting point in the same geometric state as you started. In a coordinate basis, your basis vectors are truly independent and don't depend on each other. This means that they commute (in technical language, the Lie bracket $[e_r, e_\theta] = e_r e_\theta - e_\theta e_r = 0$) which means that you can make a closed path by travelling one unit along e_r, one unit along e_θ, then *minus* one unit along e_r and *minus* one unit along e_θ and you will get back to your starting point [see Fig. 3.3(a)]. This doesn't work if you use the normalized vectors (where $[\hat{e}_r, \hat{e}_\theta] \neq 0$) and you don't get back to your starting point [see Fig. 3.3(b)].

Example 3.3

Consider a function $f(x^\mu)$ which assigns a number to any spacetime point x^μ. Now consider a path through spacetime $x^\mu(\lambda)$ where λ is a number between 0 and 1. Then $f(x^\mu(\lambda))$ represents a function of that path parameter, giving the value that f takes for every spacetime point along the path. How does f change with λ? That is given by the derivative of f along the path, written as

$$\frac{\mathrm{d}f}{\mathrm{d}\lambda} = \frac{\mathrm{d}x^\mu}{\mathrm{d}\lambda}\frac{\partial f}{\partial x^\mu}. \tag{3.12}$$

This is true for any function f, and so we could write in general that

$$\frac{\mathrm{d}}{\mathrm{d}\lambda} = \frac{\mathrm{d}x^\mu}{\mathrm{d}\lambda}\frac{\partial}{\partial x^\mu}. \tag{3.13}$$

This expression looks a little like that of a vector $\boldsymbol{X} = X^\mu \boldsymbol{e}_\mu$, with $\mathrm{d}x^\mu/\mathrm{d}\lambda$ playing the role of the components of the vector and $\partial/\partial x^\mu$ playing the role of the basis vectors. Consequently, we shall identify $\partial/\partial x^\mu$ with $\boldsymbol{e}_\mu$, and so in our example of polar coordinates we would write

$$\boldsymbol{e}_r = \frac{\partial}{\partial r} \quad \text{and} \quad \boldsymbol{e}_\theta = \frac{\partial}{\partial \theta}. \tag{3.14}$$

Using this trick, it is clear that these basis vectors commute ($[\boldsymbol{e}_r, \boldsymbol{e}_\theta] = 0$) and serve as a coordinate basis. If we had used the non-coordinate basis

$$\hat{\boldsymbol{e}}_r = \frac{\partial}{\partial r} \quad \text{and} \quad \hat{\boldsymbol{e}}_\theta = \frac{1}{r}\frac{\partial}{\partial \theta}, \tag{3.15}$$

then we would have found that they do not commute, since

$$[\hat{\boldsymbol{e}}_r, \hat{\boldsymbol{e}}_\theta]f = \frac{\partial}{\partial r}\left(\frac{1}{r}\frac{\partial f}{\partial \theta}\right) - \frac{1}{r}\frac{\partial}{\partial \theta}\frac{\partial f}{\partial r} = -\frac{1}{r^2}\frac{\partial f}{\partial \theta} = -\frac{\hat{\boldsymbol{e}}_\theta}{r}f, \tag{3.16}$$

and so $[\hat{\boldsymbol{e}}_r, \hat{\boldsymbol{e}}_\theta] = -\frac{1}{r^2}\frac{\partial}{\partial \theta} = -\frac{\hat{\boldsymbol{e}}_\theta}{r} \neq 0$.

> ↷ **This relationship between vectors and derivatives is explored in detail in Chapter 31. We return to non-coordinate bases in Chapter 10.**

3.2 Farewell to the position vector

When we first encounter vectors as students, the simplest vector that we usually start with is the position (or displacement) vector $\boldsymbol{x} = x^\mu \boldsymbol{e}_\mu$. In Chapter 2, we found that the position vector does transform appropriately under Lorentz transformations, making it possible to use it in special relativity. However, from this point onwards we will not be using it in general relativity. The reason is that it does not, in general, transform according to our rule for coordinate transformations[4]

$$X^{\mu'} = \frac{\partial x^{\mu'}}{\partial x^\nu} X^\nu. \tag{3.17}$$

The problem is that coordinates are related by a transformation of the form

$$x^{\mu'} = A^{\mu'}{}_\nu x^\nu, \tag{3.18}$$

and in curved spacetime the coefficients $A^{\mu'}{}_\nu$ will depend on the coordinates x^ν. It is only in the special case[5] that $A^{\mu'}{}_\nu$ is *independent* of x^ν that we can write that $\partial x^{\mu'}/\partial x^\nu = A^{\mu'}{}_\nu$ and eqn 3.17 will then

[4]We can see this in Example 3.1, where the matrix in eqn 3.6 does not allow us to transform between components $x^{\mu'} = (r, \theta)$ and $x^\alpha = (x, y)$.

[5]In flat spacetime, as assumed in special relativity, this condition holds and the displacement vector then presents no problem.

Fig. 3.4 A circle of radius r on the surface of a sphere of radius R (in a galaxy far, far away).

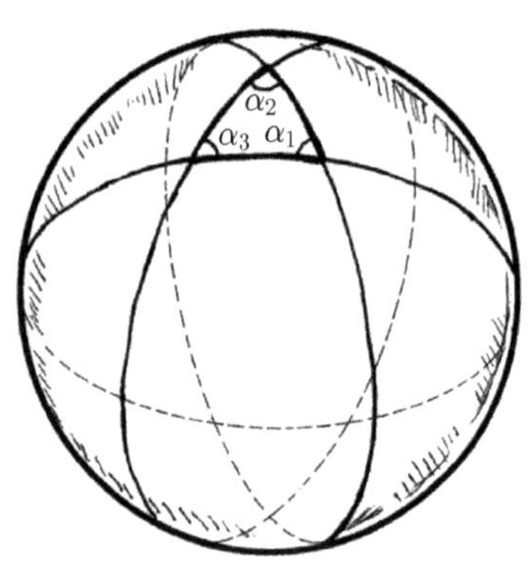

Fig. 3.5 A spherical triangle is constructed by three great circles. The sum of the internal angles, $\alpha_1 + \alpha_2 + \alpha_3$, is *greater* than π.

[7] Girard's theorem was originally written down by Thomas Harriot (1560–1621), who was also the first person to make a drawing of the moon through a telescope, several months before Galileo, and worked out Snell's law of refraction nearly two decades before Snell, though six centuries after Ibn Sahl. Credit for first discovery is not always apportioned fairly!

hold for $X^\mu = x^\mu$. This will also work in the case of linear, homogeneous transformations, such as the Lorentz transformations or spatial rotations. Another way of seeing the same thing is to notice that in a curved spacetime a displacement vector is not very well defined; it may not even live in that space. For example, if we consider only the space describing the Earth's surface, then a displacement vector from New York to Tokyo will be an arrow that ploughs through the interior of the Earth. What is well defined though is a path from New York to Tokyo made up of lots of infinitesimal displacements which all can lie on the Earth's surface.

As a result, we shall now drop the position vector $\boldsymbol{x}$ from our list of well-behaved vectors that transform appropriately, since the transformation we want to use is the one described by eqn 3.17. However, we will make still use of the coordinates x^μ describing particular events in spacetime.[6] We might now be concerned that not having a displacement vector prevents us from defining a velocity vector, which was previously the derivative of $\boldsymbol{x}$ with respect to the proper time τ. As we'll see in Chapter 7, this concern is unfounded as the velocity vector can be constructed geometrically from the tangent to the world line of a particle. In any case, as we have been explaining, the vector corresponding to an *infinitesimal* displacement in spacetime does transform correctly.

3.3 Non-Euclidean space

Non-Euclidean space is any space which is not Euclidean, i.e. not equipped with the Euclidean metric with components $\delta_{\mu\nu}$ (so that $ds^2 = \delta_{\mu\nu}dx^\mu dx^\nu$). An example of a non-Euclidean space is the Minkowski spacetime of special relativity in which

$$ds^2 = \eta_{\mu\nu}dx^\mu dx^\nu = -dt^2 + dx^2 + dy^2 + dz^2. \tag{3.19}$$

Minkowski space is known as a (3+1)-dimensional space (meaning events are described by three spatial coordinates and one time coordinate).

One of the consequences of a non-Euclidean space is that some of Euclid's famous results don't always hold.

Example 3.4

In two-dimensional Euclidean space, the circumference C of a circle of radius r is given by $C = 2\pi r$ and the internal angles in a triangle add up to $180°$ (or π radians). However, these results don't work on the surface of a sphere. The circumference of a circle of radius r on the surface of a sphere of radius R (see Fig. 3.4) is given by

$$C = 2\pi r \operatorname{sinc}\frac{r}{R}, \tag{3.20}$$

where $\operatorname{sinc} x = (\sin x)/x$, so $C \to 2\pi r$ when $r \ll R$. Moreover, from Girard's theorem,[7] the sum of the internal angles $\sum \alpha_1$ of a spherical triangle on the surface of a sphere (see Fig. 3.5) is given by

$$\sum \alpha_i = \pi + \frac{A}{R^2}, \tag{3.21}$$

where A is the area of the triangle. Thus, the sum of the angles is greater than π, although if $A \ll R^2$ Euclid's result is good enough. These two results are proved in Exercises 3.3 and 3.4.

Chapter summary

- Euclidean space uses a metric $\delta_{\mu\nu}$ and gives us the familiar results from Euclidean geometry. It can be described using a Cartesian coordinate system $(x, y, z,$ etc.$)$, but also by other coordinate systems (e.g. plane polar coordinates in two dimensions).

- The basis vectors for another coordinate basis can be derived using a transformation from those from another coordinate basis (such as from Cartesian coordinates). These basis vectors are independent from one another so that they commute (their Lie bracket is zero). A non-coordinate basis does not have this property.

- A non-Euclidean space has a non-Euclidean metric, but it can still be flat (i.e. not curved), and an example is the Minkowski space with metric $\eta_{\mu\nu}$.

Exercises

(3.1) Show that for polar coordinates in two dimensions

$$\boldsymbol{e}_x = \frac{x}{r}\boldsymbol{e}_r - \frac{y}{r^2}\boldsymbol{e}_\theta, \qquad (3.22)$$

and

$$\boldsymbol{e}_y = \frac{y}{r}\boldsymbol{e}_r + \frac{x}{r^2}\boldsymbol{e}_\theta. \qquad (3.23)$$

(3.2) For plane polar coordinates, show that

$$\begin{array}{ll} \frac{\partial \boldsymbol{e}_r}{\partial \theta} = \frac{\boldsymbol{e}_\theta}{r}, & \frac{\partial \boldsymbol{e}_\theta}{\partial \theta} = -r\boldsymbol{e}_r, \\ \frac{\partial \boldsymbol{e}_r}{\partial r} = 0, & \frac{\partial \boldsymbol{e}_\theta}{\partial r} = \frac{\boldsymbol{e}_\theta}{r}. \end{array} \qquad (3.24)$$

(3.3) Using simple geometry, prove eqn 3.20. By defining the curvature $K = 1/R^2$ for the sphere, eqn 3.20 becomes $C = 2\pi r \operatorname{sinc}(r\sqrt{K})$. Hence, show that

$$K = \lim_{r \to 0} \frac{3}{\pi r^3}(2\pi r - C), \qquad (3.25)$$

and hence the curvature of a sphere can be calculated by comparing the circumference to 2π times the radius for circles of ever-decreasing size.

(3.4) To prove Girard's theorem (i.e. to prove eqn 3.21), Fig. 3.6 may be helpful. Three great circles produce a spherical triangle of area A but they also produce another circular triangle on the other side of the sphere. Without loss of generality, you can take the radius of the sphere to be unity, so the total surface area is then 4π. With two spherical triangles, the remaining area is then $4\pi - 2A$. That remaining area is made up of strips like the two shown shaded in Fig. 3.6. You should be able to argue that each of those strips has area $2\alpha_1 - A$. Putting that together, you should then be able to deduce that $\alpha_1 + \alpha_2 + \alpha_3 = \pi + A$ and hence prove the theorem.

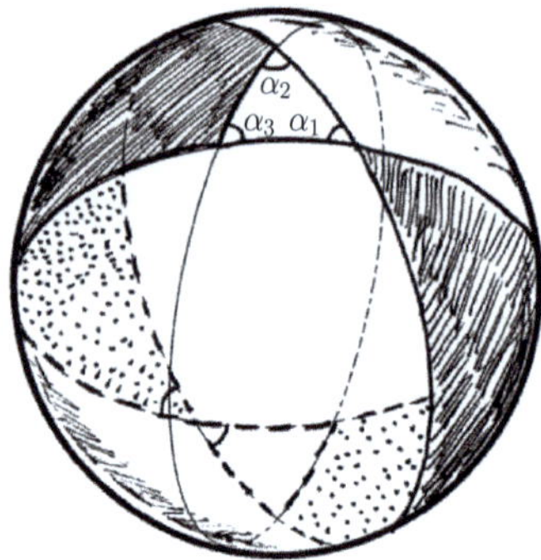

Fig. 3.6 Construction for the proof of Girard's theorem.

(3.5) A transformation to a flat, uniformly rotating frame can be achieved via the transformation

$$
\begin{aligned}
t &= t' \\
x &= x' \cos \Omega t' - y' \sin \Omega t', \\
y &= x' \sin \Omega t' + y' \cos \Omega t', \\
z &= z',
\end{aligned}
\tag{3.26}
$$

where Ω is the angular speed of the rotation. What form does the Minkowski metric line element $ds^2 = d\boldsymbol{x} \cdot d\boldsymbol{x}$ take in this rotating frame?

Linear slot machines

4

Thou, silent form, dost tease us out of thought
As doth eternity: Cold Pastoral!
John Keats (1795–1821) *Ode on a Grecian Urn* (1820)

A vector can be thought of as an arrow in spacetime, but when spacetime is curved some odd things start to happen. If you travel due North from one city to another over a curved surface (see Fig. 4.1) then you might think you are following a vector in the space of that curved surface. However, following the vector takes you out of the curved surface and leaves you hovering in mid-air, suspended over your final destination! This simple example demonstrates the fact that the vectors defined at a point in a curved space don't necessarily live in that space. In fact, the vectors defined at a point in a particular space live in what is called the **tangent space**. For the example of the Earth's surface, the space is the sphere (S^2, in the language used by mathematicians) and the tangent space is the two-dimensional (flat) plane ($\mathbb{R}^2$, in the language used by mathematicians). Thus, when travelling between two cities on a curved space, the journey is best thought of as a *path* through the space, not a *vector* between the end points. Vectors are really things that tell you about the local behaviour at a point (because they exist only in the tangent space [see Fig. 4.2]). For now, it is enough to remember that vectors like $\boldsymbol{X}$ are independent of coordinates, but can be described in a particular coordinate system using basis vectors $\boldsymbol{e}_\mu$ and components X^μ in an expression $\boldsymbol{X} = X^\mu \boldsymbol{e}_\mu$. A vector has a direction and a magnitude, or length.[1]

Vectors are only one of the sorts of objects that we require to produce a geometrical description of Nature. In this chapter, we introduce another object that, in many ways, complements the notion of a vector. It has a rather odd name which comes about because the subject of differential geometry [pioneered by the French mathematician Élie Cartan (1869–1951)] contains the notion of what are called 'differential forms'. These can be of increasing 'degree' p and are then called p-forms. Here we only want to consider the simplest such object ($p = 1$) which is called a **1-form**. Like a vector, a 1-form $\tilde{\sigma}$ exists independently of coordinates. It can be expressed in a particular coordinate system via its components and a set of **basis 1-forms** ω^μ, in an expression $\tilde{\sigma} = \sigma_\mu \omega^\mu$. Notice how the positions of the indices in the components and basis are reversed compared to vectors. Notice also our notation: $\boldsymbol{X}$ is a vector, $\tilde{\sigma}$ is a 1-form. The tilde (the wiggly line above the symbol) signifies the 1-form.

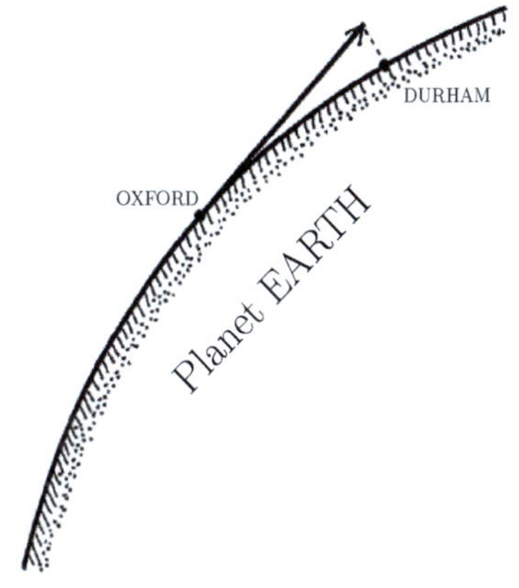

Fig. 4.1 Oxford and Durham are two cities in the UK, with Durham 337 km (only about 200 miles) due North of Oxford. Travelling due North from Oxford on a straight line leaves you in mid-air, suspended about 9 km above Durham, due to the curvature of the Earth. (Travelling due South from Durham would have the same effect when arriving at Oxford, so the sense of superiority felt by the inhabitants of each city would be the same!) The diagram exaggerates the curvature of the Earth for clarity.

[1]The length of a vector is something about which all observers agree and gives rise to the notion of an invariant equal to $\boldsymbol{X}^2$.

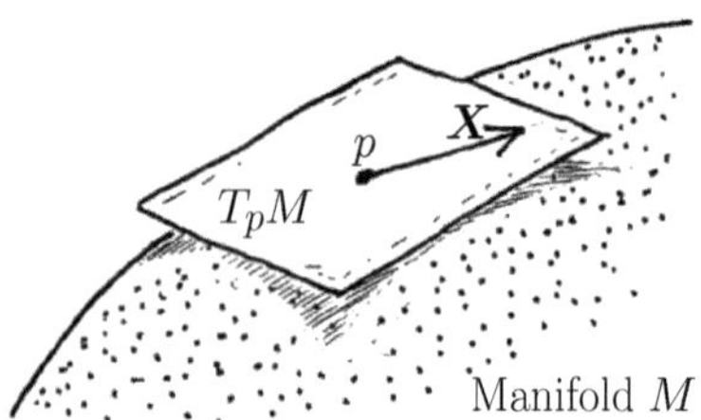

Fig. 4.2 A vector $\boldsymbol{X}$ lives in a special space called the tangent space. Points in spacetime can be described by what is called a **manifold** M. In general, this will be curved and therefore a vector cannot live in it, but only in a space which is tangent to it. For some point p in the manifold, there will be a tangent space which (in the notation of differential geometry which we generally avoid in this book) is denoted by T_pM (which you can read as 'the tangent space at point p of the manifold M').

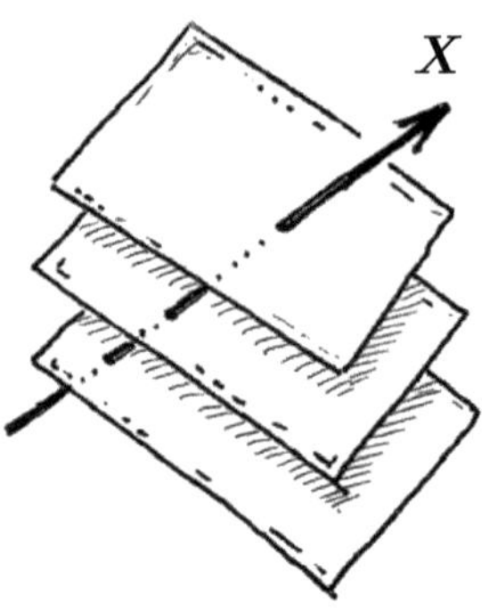

Fig. 4.3 A 1-form can be described as a set of planes. The inner product between a vector $\boldsymbol{X}$ and the 1-form can then be thought of as the number of planes skewered by the vector.

[2]We saw these in the definition of the 1-form $\tilde{\boldsymbol{Y}} = Y_\mu \boldsymbol{\omega}^\mu$ in the last section. The link to 1-forms will be made shortly.

Why do we need this additional object? The reason is that when we combine vectors and 1-forms which, as we discuss in this chapter, involves forming an **inner product**, we have to produce a number (i.e. a scalar), and numbers are invariant with respect to coordinate transformations. Thus, if the components of vectors transform in one particular way due to a change of coordinates then we need the components of the object that they combine with to transform in the opposite way so that the result of their combination is independent of the coordinate transformation. This idea might be already familiar as it appears in quantum mechanics; a vector can be represented by a ket $|\psi\rangle$ and it combines with a bra $\langle\phi|$ to make a number $\langle\phi|\psi\rangle$. The kets and the bras live in different spaces. Mathematicians think of vectors living in a vector space and the 1-forms live in the **dual space** to that vector space. Thus, the 1-forms can be thought of as objects that map vectors onto real numbers. (In quantum mechanics, bras live in a dual space to the ket space and can be thought of as objects that map kets onto complex numbers.)

If a vector can be thought of as an arrow, what geometric object does a 1-form resemble? One answer is a set of equally spaced plane surfaces, as shown in Fig. 4.3. The magnitude of a 1-form corresponds to the spatial frequency of the planes (that is, the reciprocal of the distance between planes). The direction of the 1-form tells us how the planes are arranged. This is most easily seen via the basis 1-forms, which are planes arranged perpendicular to the axes of the coordinate system.

In order to understand objects like vectors and 1-forms, we shall examine the various ways that they can be combined to make numbers. What unites the methods of combining these objects is that they can be represented as machines that generate scalars. We call the machines **tensors** or, more colourfully, **linear slot machines**, since the notation we employ features slots in which to insert vectors and 1-forms (and, as we shall see, the operations are linear ones, see Section 4.3). We start by returning to a familiar way of combining two vectors to make a number: the **dot product**.

4.1   Dot products and down vectors

In Chapter 2, we wrote the dot product as the component equation

$$\boldsymbol{X} \cdot \boldsymbol{Y} = \eta_{\mu\nu} X^\mu Y^\nu, \tag{4.1}$$

where $\eta_{\mu\nu}$ are the components of the Minkowski metric and we sum over repeated indices. The result of evaluating a dot product using eqn 4.1 is a scalar, which is to say that the result is the same, no matter which coordinate system we consider. Let's consider some new ways of writing the dot product. We can simplify eqn 4.1 by absorbing the $\eta_{\mu\nu}Y^\nu$ part into a new object which has components with indices in the down position,[2] and we shall call these components Y_μ, so that

$$Y_\mu = \eta_{\mu\nu} Y^\nu. \tag{4.2}$$

The dot product is now an expression in which we have to sum over one up-index and one down-index. That is to say

$$\boldsymbol{X} \cdot \boldsymbol{Y} = X^{\mu} Y_{\mu}. \tag{4.3}$$

Another way of looking at eqn 4.2 is that the components of the metric take the index in the up position and replace it with the index in the down position. We say that the metric $\eta_{\mu\nu}$ **lowers the index**.

Example 4.1

Let's take a dot product of a basis vector $\boldsymbol{e}_{\lambda}$ with a vector $\boldsymbol{X} = X^{\mu}\boldsymbol{e}_{\mu}$. We have

$$
\begin{aligned}
\boldsymbol{X} \cdot \boldsymbol{e}_{\lambda} = \ & X^{\mu}\boldsymbol{e}_{\mu} \cdot \boldsymbol{e}_{\lambda} \\
= \ & X^{\mu}\eta_{\mu\lambda} \quad \text{(using eqn 2.15: } \boldsymbol{e}_{\mu} \cdot \boldsymbol{e}_{\lambda} = \eta_{\mu\lambda}) \\
= \ & X_{\lambda} \quad \text{(lowering an index).}
\end{aligned} \tag{4.4}
$$

If we assume that we're working in space with $(3+1)$ dimensions and consider the $\lambda = 2$ component, we see that

$$X_2 = X^{\mu}\eta_{\mu 2} = X^0\eta_{02} + X^1\eta_{12} + X^2\eta_{22} + X^3\eta_{32}. \tag{4.5}$$

In the usual Minkowski space, where $\eta_{\mu\nu}$ is diagonal,[3] we have

$$X_2 = X^2\eta_{22} = X^2. \tag{4.6}$$

This is true for all of the spatial components (i.e. we have $X_1 = X^1$ and $X_3 = X^3$), but we also have that $X_0 = -X^0$, since $\eta_{00} = -1$.

We define the inverse of $\eta_{\mu\nu}$ as $\eta^{\mu\nu}$, which is to say[4]

$$\eta_{\mu\nu}\eta^{\nu\lambda} = \delta^{\lambda}{}_{\mu}. \tag{4.7}$$

By the same logic that led to eqn 4.2, we then also have the action of **raising an index**

$$X^{\nu} = \eta^{\mu\nu}X_{\mu}. \tag{4.8}$$

This allows us to write the dot product in terms of down-vector components as

$$\boldsymbol{X} \cdot \boldsymbol{Y} = \eta^{\mu\nu}X_{\mu}Y_{\nu}. \tag{4.9}$$

Example 4.2

A simple way to understand the geometry of up and down components is to consider Fig. 4.4, showing a vector $\boldsymbol{X}$ expressed in a coordinate system in which the basis vectors are not orthogonal. As usual, the vector is written as

$$\boldsymbol{X} = X^1\boldsymbol{e}_1 + X^2\boldsymbol{e}_2. \tag{4.10}$$

From the figure, we see that vectors $X^1\boldsymbol{e}_1$ and $X^2\boldsymbol{e}_2$ form the usual parallelogram describing the addition of two vectors, with sides of length X^1 and X^2. We saw from the previous example that X_1 is simply the projection of $\boldsymbol{X}$ along $\boldsymbol{e}_1$, achieved using the dot product $\boldsymbol{X} \cdot \boldsymbol{e}_1$. So we have

$$X_j = \boldsymbol{X} \cdot \boldsymbol{e}_j, \tag{4.11}$$

with $j = 1, 2$ (in the lowered position), as shown in the figure.

The metric has components $\eta_{\mu\nu} = \boldsymbol{e}_{\mu} \cdot \boldsymbol{e}_{\nu}$, which in this coordinate system is not diagonal. We now use the metric to raise an index, and we obtain

$$X^i = \eta^{ij}(\boldsymbol{X} \cdot \boldsymbol{e}_j). \tag{4.12}$$

So for X^1 we have

$$X^1 = \eta^{11}X_1 + \eta^{12}X_2. \tag{4.13}$$

---

[3]Recall eqn 2.15 for the components of the Minkowski metric

$$\eta_{\mu\nu} = \begin{pmatrix} -1 & 0 & 0 & 0 \\ 0 & 1 & 0 & 0 \\ 0 & 0 & 1 & 0 \\ 0 & 0 & 0 & 1 \end{pmatrix},$$

which means that $\eta_{00} = -1$, $\eta_{11} = 1$, $\eta_{22} = 1$, $\eta_{33} = 1$, and all other elements are zero.

[4]This means that

$$\eta^{\mu\nu} = \begin{pmatrix} -1 & 0 & 0 & 0 \\ 0 & 1 & 0 & 0 \\ 0 & 0 & 1 & 0 \\ 0 & 0 & 0 & 1 \end{pmatrix},$$

and hence $\eta^{00} = -1$, $\eta^{11} = 1$, $\eta^{22} = 1$, $\eta^{33} = 1$, and all other elements are zero. Thus, $\eta^{\mu\nu}$ and $\eta_{\mu\nu}$ act like the same matrix. This property will *not* hold for most second-rank tensors (i.e. for objects with two indices, to be defined later in this chapter).

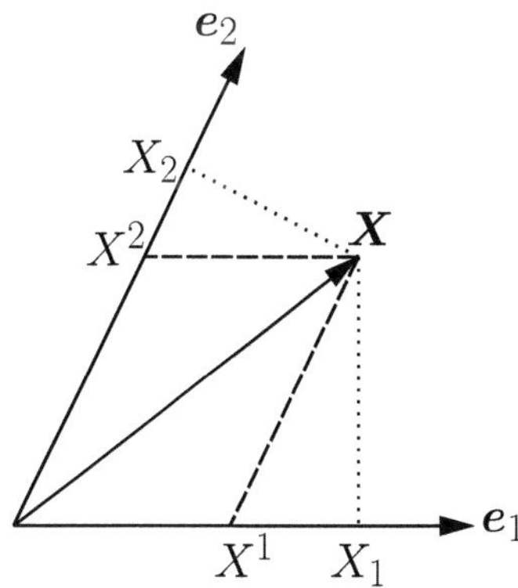

Fig. 4.4 The geometry of the up and down components.

[5]Despite the route we have taken, it is not the case that 1-forms owe their existence to the metric or to vectors. In fact, they can exist more generally in a system where a metric is not defined. Although we shall not abandon the metric until later in the book, we turn to the more general properties of 1-forms in the next section.

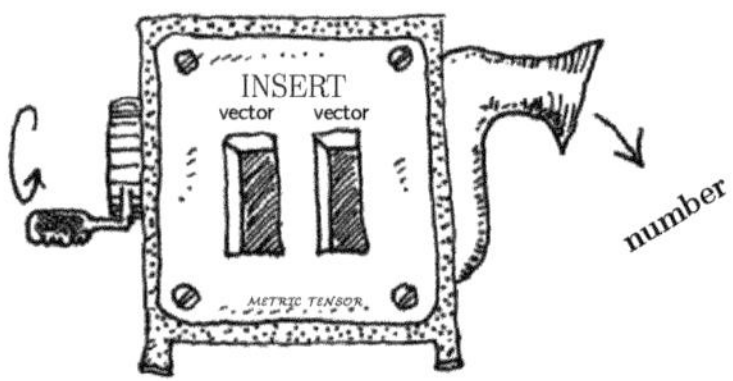

Fig. 4.5 The metric tensor can be thought of as a kind of 'slot machine', written as $\eta(\ ,\ )$ in mathematical symbols, but here is a mental picture of this object. The machine has two slots into which you have to insert vectors. Once you have inserted them, then turn the handle (meaning evaluate eqn 4.17), and out pops a number which is the output of the machine.

[6]This corresponds to the procedure of summing over all up and down components in expressions like $\eta_{\mu\nu}X^\mu u^\nu$.

The inner product is a linear object, which is to say that, if a and b are constants, we have

$$\langle a\tilde{\boldsymbol{\sigma}}, b\boldsymbol{X}\rangle = ab\langle \tilde{\boldsymbol{\sigma}}, \boldsymbol{X}\rangle,$$

and also, if $\boldsymbol{Y}$ is another vector and $\tilde{\boldsymbol{\zeta}}$ another 1-form, that

$$\langle \tilde{\boldsymbol{\sigma}}, (\boldsymbol{X}+\boldsymbol{Y})\rangle = \langle \tilde{\boldsymbol{\sigma}}, \boldsymbol{X}\rangle + \langle \tilde{\boldsymbol{\sigma}}, \boldsymbol{Y}\rangle,$$
$$\langle (\tilde{\boldsymbol{\sigma}}+\tilde{\boldsymbol{\zeta}}), \boldsymbol{X}\rangle = \langle \tilde{\boldsymbol{\sigma}}, \boldsymbol{X}\rangle + \langle \tilde{\boldsymbol{\zeta}}, \boldsymbol{X}\rangle.$$

[7]Note that we are writing that $\boldsymbol{X}(\tilde{\boldsymbol{\sigma}}) = \tilde{\boldsymbol{\sigma}}(\boldsymbol{X})$, namely that a vector operating on a 1-form gives the same result as a 1-form operating on a vector. We needn't do that (the mathematics doesn't insist upon it), and for example in quantum mechanics the analogue doesn't hold: $\langle \sigma|X\rangle$ is the *complex conjugate* of $\langle X|\sigma\rangle$, and the two are only equal if $\langle \sigma|X\rangle$ is real. In general relativity, the quantities we use are real and we will always be able to assume that these give the same result.

The existence of down components implies that, just as we have vectors built from up components and basis vectors, there exist objects whose components are the down components. These objects are the 1-forms and are written as

$$\tilde{\boldsymbol{X}} = X_\mu \boldsymbol{\omega}^\mu, \tag{4.14}$$

where the basis is made up[5] from basis 1-forms $\boldsymbol{\omega}^\mu$.

4.2 Vectors and 1-forms

Previously, the dot product of two vectors was built using the metric components via

$$\boldsymbol{X} \cdot \boldsymbol{Y} = \eta_{\mu\nu}X^\mu Y^\nu = X^\mu Y_\mu. \tag{4.15}$$

We can think of the metric in a different way. We take the metric to be the slot machine $\eta(\ ,\ )$. This machine has two slots into which we can input vectors (see Fig. 4.5). The machine outputs a scalar, which is the dot product of the two vectors we have inserted. So take the metric $\eta(\ ,\ )$ and fill in the slots with vectors $\boldsymbol{X}$ and $\boldsymbol{Y}$ to obtain $\eta(\boldsymbol{X},\boldsymbol{Y})$. This can be written in components as

$$\begin{aligned}
\eta(\boldsymbol{X},\boldsymbol{Y}) &= \eta_{\mu\nu}X^\mu Y^\nu \\
&= \eta_{00}X^0 Y^0 + \eta_{11}X^1 Y^1 + \eta_{22}X^2 Y^2 + \eta_{33}X^3 Y^3,
\end{aligned} \tag{4.16}$$

just as we had before. The slot machine is **linear**, which is to say that, if a and b are scalars, then the following rules hold:

$$\begin{aligned}
\eta(a\boldsymbol{X}, b\boldsymbol{Y}) &= ab\eta(\boldsymbol{X},\boldsymbol{Y}), \\
\eta(\boldsymbol{X}+\boldsymbol{Y}, \boldsymbol{Z}) &= \eta(\boldsymbol{X},\boldsymbol{Z}) + \eta(\boldsymbol{Y},\boldsymbol{Z}).
\end{aligned} \tag{4.17}$$

We call the metric slot machine $\eta(\ ,\ )$ a (0,2) **tensor**. The notation (m, n) gives the **valence** of a tensor: how many indices the components have in the up (m) and down (n) positions. Since the components of the metric tensor $\eta(\ ,\ )$ are $\eta_{\mu\nu}$, we have two down indices and so $m = 0, n = 2$.

Next we identify vectors as valence $(1,0)$ tensors and 1-forms, such as $\tilde{\boldsymbol{\sigma}} = \sigma_\mu \boldsymbol{\omega}^\mu$, as $(0,1)$ tensors. In filling the slots to make a scalar, the sum of valences of all objects involved must make m and n equal.[6] So inputting two $(1,0)$ vectors into a $(0,2)$ tensor gives $(1,0)+(1,0)+(0,2) = (2,2)$, so that $m = n$, and this then yields a scalar.

What does the slot machine interpretation imply for vectors and 1-forms? A vector, taken as a $(1,0)$ tensor, has a slot that can be filled with a $(0,1)$ tensor to make a number. We rewrite the vector to show its slot as $\boldsymbol{X}(\ )$. Insert a 1-form into the slot of a vector $\boldsymbol{X}(\tilde{\boldsymbol{\sigma}})$. This is equivalent to inserting a vector into the slot of a 1-form $\tilde{\boldsymbol{\sigma}}(\boldsymbol{X})$. To put things on an equal footing we write this as a linear operation known as an **inner product** (known sometimes as a contraction) using angle brackets as follows:[7]

$$\langle \tilde{\boldsymbol{\sigma}}, \boldsymbol{X}\rangle = \boldsymbol{X}(\tilde{\boldsymbol{\sigma}}) = \tilde{\boldsymbol{\sigma}}(\boldsymbol{X}). \tag{4.18}$$

To calculate the inner product, we expand the components and use linearity to find

$$\langle \tilde{\boldsymbol{\sigma}}, \boldsymbol{X} \rangle = \langle \sigma_\nu \boldsymbol{\omega}^\nu, X^\mu \boldsymbol{e}_\mu \rangle = \sigma_\nu X^\mu \langle \boldsymbol{\omega}^\nu, \boldsymbol{e}_\mu \rangle. \tag{4.19}$$

To compute this, we need a rule for the inner product of the basis 1-forms and basis vectors $\langle \boldsymbol{\omega}^\nu, \boldsymbol{e}_\mu \rangle$. This is perhaps the most important rule for manipulating tensors and is given by

$$\langle \boldsymbol{\omega}^\nu, \boldsymbol{e}_\mu \rangle = \delta^\nu{}_\mu. \tag{4.20}$$

Using this rule, we have

$$\langle \tilde{\boldsymbol{\sigma}}, \boldsymbol{X} \rangle = \sigma_\nu X^\mu \langle \boldsymbol{\omega}^\nu, \boldsymbol{e}_\mu \rangle = \sigma_\nu X^\mu \delta^\nu{}_\mu = \sigma_\mu X^\mu. \tag{4.21}$$

As promised at the start of this chapter, we see that the components of vectors and 1-forms are combined to make a scalar.

Example 4.3

Having basis vectors and 1-forms available allows us a simple method to extract components. An up component of a vector can be extracted by feeding a basis 1-form $\boldsymbol{\omega}^\mu$ into the vector's slot

$$\boldsymbol{X}(\boldsymbol{\omega}^\mu) = \langle \boldsymbol{\omega}^\mu, \boldsymbol{X} \rangle = X^\nu \langle \boldsymbol{\omega}^\mu, \boldsymbol{e}_\nu \rangle = X^\nu \delta^\mu{}_\nu = X^\mu. \tag{4.22}$$

Similarly for the 1-form, we extract its components by inserting a basis vector into its slot

$$\tilde{\boldsymbol{\sigma}}(\boldsymbol{e}_\mu) = \langle \tilde{\boldsymbol{\sigma}}, \boldsymbol{e}_\mu \rangle = \sigma_\nu \langle \boldsymbol{\omega}^\nu, \boldsymbol{e}_\mu \rangle = \sigma_\nu \delta^\nu{}_\mu = \sigma_\mu. \tag{4.23}$$

We introduced the 1-form geometrically as a set of planes and the vector as an arrow. The inner product also has a geometrical interpretation: we think of the vector arrow piercing the 1-form planes, as described in the next example.

Example 4.4

In 1924, Louis Victor Pierre Raymond, 7th duc de Broglie, proposed that all particles have wave-like properties. A particle's momentum $\boldsymbol{p}$ is related to its wavevector $\boldsymbol{k}$ via $\boldsymbol{p} = \hbar \boldsymbol{k}$. Here, the magnitude of the wavevector is related to the particle's wavelength λ via $|\boldsymbol{k}| = 2\pi/\lambda$. The amplitude ψ of a wave is written as a complex exponential with a phase ϕ

Louis de Broglie (1892–1987)

$$\psi(x) = A e^{i\phi} = A e^{i\boldsymbol{k} \cdot \boldsymbol{x}} = A e^{i\boldsymbol{p} \cdot \boldsymbol{x}/\hbar}. \tag{4.24}$$

We can describe the quantum wave/particle by its momentum vector. If we want to know the phase difference $\Delta\phi$ between the wave at two positions $\boldsymbol{x}_1$ and $\boldsymbol{x}_2$, separated by a vector $\boldsymbol{x} = \boldsymbol{x}_2 - \boldsymbol{x}_1$ we can evaluate $\Delta\phi = \boldsymbol{k} \cdot \boldsymbol{x}$, that is, the dot product of $\boldsymbol{k}$ and the vector linking the two points $\boldsymbol{x}$. This works elegantly in Minkowski space. The 4-vector $\boldsymbol{k}$ is related to $\boldsymbol{p}$ by $\boldsymbol{p} = \hbar \boldsymbol{k}$, where $\boldsymbol{p} = (E, \vec{p})$ and $\boldsymbol{k} = (\omega, \vec{k})$, and now the phase $\Delta\phi = \boldsymbol{k} \cdot \boldsymbol{x} = \vec{k} \cdot \vec{x} - \omega t$.

There is another way. Instead of a momentum vector, we can imagine equally spaced, parallel surfaces separated by a distance proportional to the wavelength of the wave. We'll call this set of surfaces the wave's 1-form $\tilde{k}$. They are in fact the surfaces of constant phase in the wave. Now if we want to know the phase difference between two points we simply evaluate the inner product $\langle \tilde{k}, x \rangle$ which we can think of as a machine that counts the number of the surfaces of $\tilde{k}$ that the vector x pierces (see Fig. 4.3). We have

$$\Delta\phi = \langle \tilde{k}, x \rangle = (\text{number of surfaces pierced}). \tag{4.25}$$

See from Fig. 4.3 how a vector appears to pierce some number of the 1-form's planes: this number is equal to the inner product. We input a 1-form into the inner product's first slot, and a vector into the second. The inner product slot machine outputs a number telling us how many 1-form planes are pierced by the vector or

$$\langle \tilde{\sigma}, X \rangle = \left(\begin{array}{c} \text{Number of planes of the 1-form } \tilde{\sigma} \\ \text{pierced by the vector } X \end{array} \right). \tag{4.26}$$

With 1-forms as part of our machinery, a natural question is what physical quantities they represent. We examine this in the next example.

Example 4.5

The momentum of a particle[8] can be represented by a 1-form, whose components are given from the Lagrangian by $p_\mu = \partial L/\partial \dot{x}^\mu$. So a particle has momentum 1-form $\tilde{p}(\ )$. Insert the velocity 4-vector u into its slot and we output a number. This quantity is[9] minus the energy $-E$ of the particle, as measured by an observer O_u with velocity vector u tangent to their world line. That is

$$\tilde{p}(u) = -E = - \left(\begin{array}{c} \text{Energy of particle} \\ \text{measured by } O_u \end{array} \right). \tag{4.27}$$

We can test this equation in the particle's rest frame,[10] in which O_u has $u = e_0$, which means $E = -p_\mu \langle \omega^\mu, e_0 \rangle = -p_0$. Using the Minkowski metric, we have $-p_0 = p^0$, which is indeed the particle's energy.

An observer with velocity u passes through a cloud of dust carrying a small, permeable box of a known spatial volume (known as a **3-volume**). The observer makes measurements by counting the number of particles in the box. We define the **particle current** 1-form $\tilde{J}(\ )$. Insert[11] the velocity vector of the observer u and output (minus) the number density of particles $-n$ measured by the observer with velocity u. That is

$$\tilde{J}(u) = -n = - \left(\begin{array}{c} \text{number density of particles} \\ \text{measured by } O_u \end{array} \right). \tag{4.28}$$

At the start of the chapter, we described vectors as living in a tangent space. 1-forms live in a different space, known as a dual space. Equation 4.20 ($\langle \omega^\nu, e_\mu \rangle = \delta^\nu{}_\mu$) gives the relationship between these spaces. An interesting question is whether it is possible to map objects between these two spaces, such that one could take a vector and then find an equivalent 1-form. Such a mapping is carried out using the metric tensor.[12] Notice that the inner product $\langle \tilde{\sigma}, X \rangle = \sigma_\mu X^\nu$ is just as if we

---

[8] Recap: We previously defined a momentum vector p with components $p^\mu = (E, p^x, p^y, p^z)$ and a velocity vector u with components $u^\mu = \gamma(1, v^1, v^2, v^3)$. The Minkowski tensor can be used to produce down versions $p_\mu = (-E, p^x, p^y, p^z)$ and $u_\mu = \gamma(-1, v^1, v^2, v^3)$.

[9] See eqn 2.38 [$X_0^{\text{obs}} = -X \cdot u$] or, using this chapter's ideas, $X_0^{\text{obs}} = -\langle \tilde{X}, u \rangle \equiv -\tilde{X}(u)$.

[10] Recall that in the particle's rest frame, we have components $u^\mu = (1, 0, 0, 0)$, so we can write $u = u^\mu e_\mu = e_0$.

[11] Once again, we can use eqn 2.38 but this time with J as a 4-vector.

Note that in these examples, we could turn things around. For example, we could treat $\tilde{u}$ as a 1-form and have J as a vector and then write the final answer as $\tilde{u}(J) = -n$. In component form, eqns 4.27 and 4.28 can be written as $p_\mu u^\mu = -E$ and $J_\mu u^\mu = -n$.

[12] This should be unsurprising since we have already done this using components in eqn 4.2, using the metric tensor to convert an object with up-indices into one with down-indices. Here though, we are doing it entirely *geometrically*, without worrying about the components.

had taken the dot product of two *vectors* $\boldsymbol{X} = X^\mu \boldsymbol{e}_\mu$ and $\boldsymbol{\sigma} = \sigma^\nu \boldsymbol{e}_\nu$ (i.e. the components of the 1-form with the index raised). In fact, we have

$$\boldsymbol{\sigma} \cdot \boldsymbol{X} = \boldsymbol{\eta}(\boldsymbol{\sigma}, \boldsymbol{X}) \equiv \langle \tilde{\boldsymbol{\sigma}}, \boldsymbol{X} \rangle. \tag{4.29}$$

This allows us to read off what happens if we fill in just one slot in the metric $\boldsymbol{\eta}(\boldsymbol{\sigma}, \ )$. Since, upon doing this, we still have one slot left to input a vector, the output must be a $(0,1)$ tensor, also known as a 1-form. We conclude that

$$\boldsymbol{\eta}(\boldsymbol{\sigma}, \ ) = \tilde{\boldsymbol{\sigma}}. \tag{4.30}$$

That is, the metric slot machine maps vectors onto 1-forms.

4.3 Transformations

We have stressed the role of transformations between sets of coordinates. The inner product $X_{\alpha'} Y^{\alpha'}$ is a scalar and should, therefore, be coordinate invariant. Recall that the up components of a vector transform as

$$X^{\alpha'} = \Lambda^{\alpha'}{}_\mu X^\mu = \frac{\partial x^{\alpha'}}{\partial x^\mu} X^\mu, \tag{4.31}$$

where $x^{\alpha'}$ and x^μ are two sets of coordinates.[13] Since we have that $\Lambda^\nu{}_{\alpha'} \Lambda^{\alpha'}{}_\mu = \delta^\nu{}_\mu$ it must be the case that the down components should transform as

$$X_{\beta'} = \Lambda^\nu{}_{\beta'} X_\nu = \frac{\partial x^\nu}{\partial x^{\beta'}} X_\nu \tag{4.32}$$

with the result that

$$X_{\alpha'} Y^{\alpha'} = X_\mu \Lambda^\mu{}_{\alpha'} \Lambda^{\alpha'}{}_\nu Y^\nu = X_\mu \delta^\mu{}_\nu Y^\nu = X_\mu Y^\mu. \tag{4.33}$$

Compared with the results from the previous chapter, we see that the down components transform in the same way as the basis vectors $\boldsymbol{e}_\mu$.[14] It should then come as no surprise that the basis 1-forms transform in the same way as the vector components, as demonstrated in the next example.

Example 4.6

We can see how to transform basis 1-forms by considering the contraction $\langle \boldsymbol{\omega}^\beta, \boldsymbol{e}_\alpha \rangle = \delta^\beta{}_\alpha$. First, multiply through by a coordinate transformation $\Lambda^\alpha{}_{\gamma'}$ to find

$$\langle \boldsymbol{\omega}^\beta, \Lambda^\alpha{}_{\gamma'} \boldsymbol{e}_\alpha \rangle = \Lambda^\alpha{}_{\gamma'} \delta^\beta{}_\alpha = \Lambda^\beta{}_{\gamma'} \tag{4.34}$$

Now multiply through by $\Lambda^{\sigma'}{}_\beta$ and write

$$\langle \Lambda^{\sigma'}{}_\beta \boldsymbol{\omega}^\beta, \Lambda^\alpha{}_{\gamma'} \boldsymbol{e}_\alpha \rangle = \Lambda^{\sigma'}{}_\beta \Lambda^\beta{}_{\gamma'} = \delta^{\sigma'}{}_{\gamma'} \tag{4.35}$$

Comparing against $\langle \boldsymbol{\omega}^{\sigma'}, \boldsymbol{e}_{\gamma'} \rangle = \delta^{\sigma'}{}_{\gamma'}$, we conclude

$$\boldsymbol{\omega}^{\sigma'} = \Lambda^{\sigma'}{}_\beta \boldsymbol{\omega}^\beta. \tag{4.36}$$

Once more, with feeling: the basis 1-forms transform in the same way as the vector components.

Written out in components, eqn 4.30 is

$$\eta_{\mu\nu} X^\mu = X_\nu,$$

where X_ν are the components of the 1-form $\tilde{\boldsymbol{X}} = X_\alpha \boldsymbol{\omega}^\alpha$. This means that in Example 4.4, the 1-form $\tilde{\boldsymbol{k}}$ has components $k_\mu = \eta_{\mu\nu} k^\nu = (-\omega, \vec{k})$ and so $\langle \tilde{\boldsymbol{k}}, \boldsymbol{x} \rangle$ yields the required phase.

[13]As usual we denote one coordinate system with primed indices and one without primes.

[14]The reason is the same: we want both (i) the vector $X^\mu \boldsymbol{e}_\mu$ and (ii) the scalar $X_\mu Y^\mu$ to be independent of coordinates. This also explains our notation, with 1-form components and basis vectors both carrying an index in the down position: this tells us that they transform the same way.

We can now summarize how to transform components and basis vectors:

$$X^{\beta'} = \Lambda^{\beta'}{}_{\alpha} X^{\alpha}, \quad \sigma_{\beta'} = \Lambda^{\alpha}{}_{\beta'} \sigma_{\alpha},$$
$$e_{\beta'} = \Lambda^{\alpha}{}_{\beta'} e_{\alpha}, \quad \omega^{\beta'} = \Lambda^{\beta'}{}_{\alpha} \omega^{\alpha}. \tag{4.37}$$

[15] One thing that the bold-symbol notation for a tensor $T(\ ,\ ,\ )$ lacks is clear guidance on the valence of the tensor, which must be given separately in the form (m, n). One solution to this is to use **abstract index notation**, invented by Roger Penrose (1931–). The idea here is to specify the slots using indices, so a $(2, 1)$ tensor would be written as $T^{ab}{}_{c}$. The indices here are not the components; to express those we need to ensure that we specify components and slots with different letters. One common convention is to use Roman letters for slots and Greek letters for components, so that the components of T would be $T^{\mu\nu}{}_{\rho}$. Clearly, there is potential for confusion here, so it's necessary to know the convention being adopted.

To extract a number from a tensor, we insert 1-forms $\tilde{Z}_a$ and $\tilde{Z}_b$ and a vector A^c, balancing Roman indices to obtain

$$T^{ab}{}_{c} \tilde{Z}_a \tilde{Z}_b A^c, \tag{4.38}$$

where we note that the letters denote the relevant slot, rather than an instruction to sum on an index. We don't use abstract index notation here, although some of the more advanced textbooks in the subject (e.g. Wald) do use it.

[16] A mixed object is a tensor with a valence (m, n) where $m, n \neq 0$, that is, a tensor whose components carry both up and down indices.

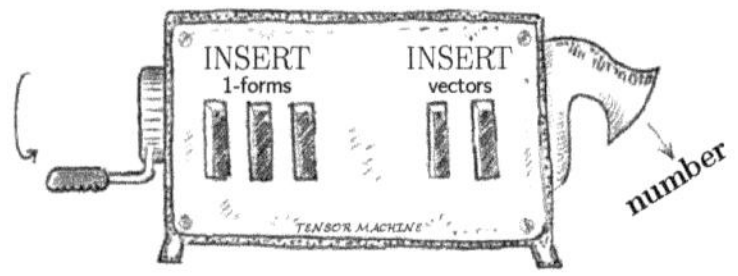

Fig. 4.6 The tensor as a slot machine. It has m slots for 1-forms and n slots for vectors. If you insert those and turn the handle (metaphorically) then it spits out a number.

4.4   Tensors

Let's now look at the general concept of a **tensor**. Generally speaking, the tensor T is a linear slot machine with m slots for inputting 1-forms and n slots to input vectors (see Fig. 4.6). We have to specify how many of each by specifying the valence (m, n) of the tensor.[15]

For vectors we can write an expression relating the vector to its components and basis vectors $X = X^{\mu} e_{\mu}$, and an analogous expression for 1-forms $\tilde{\sigma} = \sigma_{\mu} \omega^{\mu}$. To write a similar expression for tensors we need to use the **outer product** between basis vectors, denoted by $\otimes$. This symbol is simply a means of denoting the slot machine character of the tensor. Its key property is that it maintains the ordering of the slots. This idea is best understood by considering an example.

Example 4.7

Consider a tensor $e_1 \otimes e_2$: in words, the outer product of the basis vector in the 1 direction and the basis vector in the 2 direction. This is an object with two slots that takes two 1-forms. The outer product symbol $\otimes$ simply tells us that the first slot refers to e_1 and the second to e_2. Inserting 1-forms $\tilde{\alpha} = \alpha_{\mu} \omega^{\mu}$ and $\tilde{\beta} = \beta_{\nu} \omega^{\nu}$, we have

$$
\begin{aligned}
e_1 \otimes e_2(\tilde{\alpha}, \tilde{\beta}) &= e_1(\tilde{\alpha}) e_2(\tilde{\beta}) \\
&= \alpha_{\lambda} \langle \omega^{\lambda}, e_1 \rangle \beta_{\rho} \langle \omega^{\rho}, e_2 \rangle \\
&= \alpha_{\lambda} \delta^{\lambda}{}_1 \beta_{\rho} \delta^{\rho}{}_2 \\
&= \alpha_1 \beta_2.
\end{aligned} \tag{4.39}
$$

For a mixed object[16] like $\omega^2 \otimes e_3$, that is, a tensor formed from the outer product of the basis 1-form for the 2 direction and the basis vector for the 3 direction, let's enter a vector $v = v^{\mu} e_{\mu}$ in the first slot and a 1-form $\tilde{\alpha}$ in the second to find

$$
\begin{aligned}
\omega^2 \otimes e_3(v, \tilde{\alpha}) &= \omega^2(v) e_3(\tilde{\alpha}) \\
&= v^{\mu} \langle \omega^2, e_{\mu} \rangle \alpha_{\lambda} \langle \omega^{\lambda}, e_3 \rangle \\
&= v^2 \alpha_3.
\end{aligned} \tag{4.40}
$$

Example 4.8

We could write a (3,1) tensor

$$S(\ ,\ ,\ ). \tag{4.41}$$

To find the components of the (3,1) tensor, we simply fill its slots with three basis 1-forms and one basis vector

$$S^{\mu\nu\lambda}{}_{\sigma} = S(\omega^{\mu}, \omega^{\nu}, \omega^{\lambda}, e_{\sigma}). \tag{4.42}$$

An expression for $\boldsymbol{S}$ in terms of its components can be written as

$$\boldsymbol{S}(\ ,\ ,\ ,\ ) = S^{\mu\nu\lambda}{}_{\sigma}\, \boldsymbol{e}_\mu(\ ) \otimes \boldsymbol{e}_\nu(\ ) \otimes \boldsymbol{e}_\lambda(\ ) \otimes \boldsymbol{\omega}^\sigma(\ ). \tag{4.43}$$

This is an object into which we can insert three 1-forms and a vector. Inserting 1-forms $\tilde{\boldsymbol{\zeta}}$, $\tilde{\boldsymbol{\eta}}$ and $\tilde{\boldsymbol{\chi}}$ and the vector $\boldsymbol{u}$, into our example tensor $\boldsymbol{S}$, we find

$$\begin{aligned}
\boldsymbol{S}(\tilde{\boldsymbol{\zeta}},\tilde{\boldsymbol{\eta}},\tilde{\boldsymbol{\chi}},\boldsymbol{u}) &= S^{\mu\nu\lambda}{}_{\sigma}\, \zeta_\alpha \underbrace{\langle \boldsymbol{\omega}^\alpha, \boldsymbol{e}_\mu \rangle}_{\delta^\alpha{}_\mu}\, \eta_\beta \underbrace{\langle \boldsymbol{\omega}^\beta, \boldsymbol{e}_\nu \rangle}_{\delta^\beta{}_\nu}\, \chi_\gamma \underbrace{\langle \boldsymbol{\omega}^\gamma, \boldsymbol{e}_\lambda \rangle}_{\delta^\gamma{}_\lambda}\, u^\delta \underbrace{\langle \boldsymbol{\omega}^\sigma, \boldsymbol{e}_\delta \rangle}_{\delta^\sigma{}_\delta} \\
&= S^{\mu\nu\lambda}{}_{\sigma}\, \zeta_\mu \eta_\nu \chi_\lambda u^\sigma.
\end{aligned} \tag{4.44}$$

Tensors are independent of coordinate system, but their components depend on the details of the coordinates. How do tensor components transform? We use the **tensor transformation law** that says that the transformation is carried out by a multiplication of transformation matrices, one for each index. Specifically, every up index μ is transformed by a matrix $\Lambda^{\alpha'}{}_\mu = \partial x^{\alpha'}/\partial x^\mu$ and every down index σ is transformed by a matrix $\partial x^\sigma/\partial x^{\beta'}$. Our example tensor therefore transforms as

$$S^{\mu'\nu'\lambda'}{}_{\sigma'} = \frac{\partial x^{\mu'}}{\partial x^\mu}\frac{\partial x^{\nu'}}{\partial x^\nu}\frac{\partial x^{\lambda'}}{\partial x^\lambda}\frac{\partial x^\sigma}{\partial x^{\sigma'}} S^{\mu\nu\lambda}{}_{\sigma}. \tag{4.45}$$

In coordinate-based treatments, the tensor transformation law is used to *define* tensors, but we prefer to use the slot-machine definition which is much cleaner.[17]

Example 4.9

The $(0,2)$ metric tensor is written as

$$\boldsymbol{\eta}(\ ,\ ) = \eta_{\mu\nu}\boldsymbol{\omega}^\mu(\ ) \otimes \boldsymbol{\omega}^\mu(\ ). \tag{4.46}$$

This tensor has components that can be extracted: $\boldsymbol{\eta}(\boldsymbol{e}_\alpha, \boldsymbol{e}_\beta) = \eta_{\alpha\beta}$. We can now see explicitly what happens if we insert a vector $\boldsymbol{v}$ into one of the slots

$$\begin{aligned}
\boldsymbol{\eta}(\boldsymbol{v},\ ) &= \eta_{\mu\nu}\boldsymbol{\omega}^\mu(\boldsymbol{v})\boldsymbol{\omega}^\nu(\ ) \\
&= \eta_{\mu\nu}v^\sigma\langle \boldsymbol{e}_\sigma, \boldsymbol{\omega}^\mu \rangle\boldsymbol{\omega}^\nu(\ ) \\
&= \eta_{\mu\nu}v^\mu\boldsymbol{\omega}^\nu(\ ) \\
&= v_\nu\boldsymbol{\omega}^\nu(\ ),
\end{aligned} \tag{4.47}$$

where in the penultimate line, we've used the components of the metric tensor to lower an index. We see that the output is, as we predicted, a 1-form with components v_ν.

The tensor above has valence $(0,2)$, but we can also define a $(2,0)$ version

$$\boldsymbol{\eta}(\ ,\ ) = \eta^{\mu\nu}\boldsymbol{e}_\mu \otimes \boldsymbol{e}_\nu, \tag{4.48}$$

where $\eta_{\mu\nu}\eta^{\nu\sigma} = \delta^\sigma{}_\mu$ (implying $\eta_{\mu\nu} = \eta^{\mu\nu}$). The $(2,0)$ version of the tensor inputs two 1-forms and can map a 1-form to a vector.

Finally, note that we can use the metric tensor on the components of a tensor to raise or lower them, one at a time. So we have, for example, that

$$S^{\mu\nu}\eta_{\nu\beta} = S^\mu{}_\beta, \quad T_{\mu\nu}\eta^{\mu\alpha} = T^\alpha{}_\nu, \tag{4.49}$$

[17]There is an unfortunate tendency for general relativity to become 'death by indices'. Our use of coordinate-free objects, such as $\boldsymbol{S}(\ ,\ ,\ ,\ )$, is intended to avoid this and this way of writing equations is sometimes called **index-free notation**. Imagine if you had learnt electromagnetism just in terms of coordinates, but never having seen vector notation. Efficient notation can declutter equations and (hopefully) make the physics more transparent.

and so on.

This section has contained a lot of formalism, but let's finish it with a couple of very simple corollaries that shouldn't be forgotten amidst all the mathematical manipulations.

- If two tensors $\boldsymbol{A}$ and $\boldsymbol{B}$ are equal to each other in one frame, they will be equal to each other in all frames. This is obvious if you think of the tensors in a coordinate-free way. Alternatively, construct the tensor $\boldsymbol{C} = \boldsymbol{A} - \boldsymbol{B}$, which is identically zero, and so all its components are zero. Its components will clearly all be zero if multiplied by any transformation matrix.

- A scalar [which is a $(0,0)$ tensor] takes the same numerical value in all frames. (An example is the Ricci scalar, to be introduced in Chapter 11.) Therefore, if you evaluate a scalar in the most convenient frame, you have got it for all frames.

4.5   Energy-momentum tensor

As a payoff for all of this formalism, we introduce one of the most important tensors in all of physics: the energy-momentum tensor. This tensor gives the (physical) right-hand side of the Einstein equation of general relativity.[18]

Let's start off by considering a set of dust particles in spacetime,[19] each of mass m, and imagine that in some frame S these particles are all distributed in space but are at rest. Their energy will just be m per particle (remember that, if we reinstate the factors of c, this would be mc^2 per particle), and if there are n_0 particles per unit volume the energy density will be $n_0 m$. In another inertial frame S', the energy becomes γm per particle [acquiring a factor of γ because the particles are now moving with speed v, and $\gamma = (1 - v^2)^{-1/2}$] and the energy density becomes $\gamma^2 n_0 m$ (acquiring a second factor of γ because the region containing the particles in S will have become Lorentz contracted in S' by a factor of γ, increasing the density). Energy density therefore transforms with two factors of γ and this indicates that it is part of a **second-rank tensor**.

To understand what this second-rank tensor could be, let's take a step back and think about particle current. Recall that the particle current can be expressed as a 4-vector $\boldsymbol{J} = n_0 \boldsymbol{u}$, where here n_0 is the density of dust particles in their rest frame and $\boldsymbol{u}$ is the 4-velocity of the assembly of dust particles. The time-component of this current tells us about the number density $n = \gamma n_0$ of the particles [remember from Chapter 2 that $\boldsymbol{u} = \gamma(1, \vec{v})$ so $\boldsymbol{J} = \gamma n_0(1, \vec{v})$] and each spatial component of this current tells us the flux of particles along that direction (e.g. in Cartesian coordinates, J^x tells us about the number of particles crossing the yz plane, per unit area, per unit time).

This is all useful for thinking about the flux of particles, but what if we want to understand the flux of 4-momentum? That's a really interesting question because we would like to know how energy and momentum are transported across spacetime. The problem is that, unlike the number

---

[18]Spoiler alert: the Einstein equation (which we will get to properly at the end of Part II) has the form

$$\begin{pmatrix} \text{Curvature} \\ \text{of} \\ \text{spacetime} \end{pmatrix} = \begin{pmatrix} \text{Mass-energy} \\ \text{density at} \\ \text{this point} \end{pmatrix}. \tag{4.50}$$

The right-hand side of this equation will be related to the energy-momentum tensor.

[19]Astrophysicists like talking about dust, as there's a lot of it about in the Universe. The term refers to solid particles that can be anything from a few molecules up to macroscopic size, and for our purposes we are going to assume that they are just bits of mass, distributed in space, at a low enough density that they don't interact with each other.

of particles, which is a scalar, the 4-momentum is a 4-vector, and so its flux has to be a more complicated object than a 4-vector. This confirms that the object needed to describe the flux of momentum will need to be a[20] second-rank tensor since it depends on the 4-momentum *and* the 4-current. We call this object the **energy-momentum tensor**[21] $\boldsymbol{T}(\ ,\ )$, and it has two slots (or in components, it will be a second-rank tensor). We will define it (for now) as the symmetric tensor

$$\boldsymbol{T}(\ ,\ ) = \tilde{\boldsymbol{J}}(\ ) \otimes \tilde{\boldsymbol{p}}(\ ), \tag{4.51}$$

and from this it's readily apparent that $\boldsymbol{T}(\ ,\ )$ is a (0,2) object that inputs two vectors and has components $T_{\mu\nu} = \boldsymbol{T}(\boldsymbol{e}_\mu, \boldsymbol{e}_\nu) = J_\mu \rho_\nu$. We can, of course, rewrite this tensor in other ways and we could define it as a symmetric (2,0) object, with upstairs indices on its components $T^{\mu\nu}$.[22]

Example 4.10

To see what $\boldsymbol{T}$ looks like in practice, let's stick with components to begin with and evaluate everything in a frame in which the number density $J^0 \equiv n = \gamma n_0$. The (2,0) version of the energy-momentum tensor for the cloud of particles can then be written as[23] $T^{\mu\nu} = J^\mu p^\nu = (n_0 u^\mu)(m u^\nu)$, where $\boldsymbol{u}$ is the velocity of the cloud with components u^μ.

- The time-time element T^{00} is then just the energy $p^0 = \gamma m$ multiplied by $J^0 = \gamma n_0 = n$, and hence $T^{00} = \gamma nm$ is equal to the energy density.

- The space-time and time-space elements T^{i0} and T^{0i} are $n\gamma mv^i$ and hence correspond to the density of the ith component of the momentum.

- The space-space elements T^{ij} are $n\gamma mv^i v^j$ and are momentum fluxes which, as discussed below, correspond to stresses.

Another way of looking at the components is to say that the energy-momentum tensor $T^{\mu\nu}$ tells us the flux of the 4-momentum p^μ that crosses a surface of constant x^ν. In particular, this means that

- T^{00} is the energy density, since it is the flux of p^0 (energy) crossing a surface of constant time (i.e. filling space).

- $T^{0i} = T^{i0}$ is the mass flux across a surface of constant x^i, which is equivalent to the density of the ith component of linear momentum.

- T^{ij} is the ij component of the usual stress tensor, meaning that the off diagonal terms are shear stresses and the diagonal terms (T^{ii}) correspond to pressures.

Two very simple examples of this tensor are as follows:
(1) A set of dust particles at rest. These only have energy density, and are not moving and so have no linear momentum. Hence, in their rest frame, we have

$$T^{\mu\nu} = T_{\mu\nu} = \begin{pmatrix} \rho & 0 & 0 & 0 \\ 0 & 0 & 0 & 0 \\ 0 & 0 & 0 & 0 \\ 0 & 0 & 0 & 0 \end{pmatrix}. \tag{4.53}$$

(2) An isotropic fluid in equilibrium. The particles in the fluid exert[24] a pressure p, but have no preferred direction (meaning that $T^{i0} = 0$ and T^{ij} has no off-diagonal components). Hence, in the rest frame of the fluid, we have

$$T^{\mu\nu} = T_{\mu\nu} = \begin{pmatrix} \rho & 0 & 0 & 0 \\ 0 & p & 0 & 0 \\ 0 & 0 & p & 0 \\ 0 & 0 & 0 & p \end{pmatrix}. \tag{4.54}$$

We will return to this problem in much more detail in Chapter 12.

---

[20] In general, a second-rank tensor has $(m,n) = (2,0)$ or $(m,n) = (0,2)$.

[21] This is also known as the **stress-energy tensor**.

[22] That is, we transform

$$T^{\mu\nu} = \eta^{\alpha\mu}\eta^{\beta\nu}T_{\alpha\beta}. \tag{4.52}$$

Since $\eta_{\mu\nu} = \eta^{\mu\nu}$ is a diagonal tensor with components $\mathrm{diag}(-1,1,1,1)$, we see that raising or lowering a timelike component earns us a minus sign. This allows us to see immediately that $T^{00} = T_{00}$, $T^{ii} = T_{ii}$, $T^{0i} = -T_{0i}$ and $T^{i0} = -T_{i0}$. This is a good example of a tensor for which the components *can* change when you move the indices from upstairs to downstairs (in contrast to $\eta_{\mu\nu}$, for which they do not, as explained on page 45).

[23] Reminder: We use $u^\mu = (\gamma, \gamma v^i)$, $J^\mu = n_0 u^\mu = (n, nv^i)$ and $p^\mu = mu^\mu = (\gamma m, \gamma mv^i)$.

[24] Do not confuse pressure p with momentum. The context should always make it clear, but it's a shame the two quantities have the same symbol.

Finally, let's use the elegant formalism of our slot machines to consider the energy-momentum tensor as a machine with two slots in it. This will allow us to read off the properties of our set of dust particles in the frame of an observer travelling with velocity vector $\boldsymbol{u}$. Our dust particles are described with a momentum 1-form $\tilde{p}(\ )$ and so, following the results proved in Example 4.5, if we insert a velocity vector $\boldsymbol{u}$ into $\tilde{p}(\ )$ then we will output (with a minus sign) the energy of the particle E, as measured by an observer (with velocity vector $\boldsymbol{u}$)

$$E = -\tilde{p}(\boldsymbol{u}). \tag{4.55}$$

The particle current 1-form is $\tilde{J}(\ )$. Insert a velocity $\boldsymbol{u}$ and, with a minus sign, we output the number density of particles measured by the observer with velocity $\boldsymbol{u}$

$$n = -\tilde{J}(\boldsymbol{u}). \tag{4.56}$$

Example 4.11

A swarm of massive particles has a particle current $\boldsymbol{J} = n_0\boldsymbol{v}$, where n_0 is the number density of particles in the rest frame of the swarm, each particle has rest mass m, velocity $\boldsymbol{v}$ and momentum $\boldsymbol{p} = m\boldsymbol{v}$. The density of particles in the swarm's rest frame is $\rho_0 = mn_0$. The (0,2) energy-momentum tensor in this case is given by

$$\boldsymbol{T}(\ ,\ ) = \tilde{J}(\ ) \otimes \tilde{p}(\ ) = \rho_0 \tilde{v}(\ ) \otimes \tilde{v}(\ ). \tag{4.57}$$

Here $\tilde{v} = v_\mu \boldsymbol{\omega}^\mu$ are velocity 1-forms for the fluid. This energy-momentum tensor has components $T_{\mu\nu} = \rho_0 v_\mu v_\nu$. We can insert some vectors into the slots in order to understand the physical meaning of the components of $\boldsymbol{T}$.[25]
(i) We start by inserting the observer's velocity vector $\boldsymbol{u}$ in both slots

$$\begin{aligned} \boldsymbol{T}(\boldsymbol{u},\boldsymbol{u}) &= \tilde{J}(\boldsymbol{u}) \otimes \tilde{p}(\boldsymbol{u}) \\ &= nE, \end{aligned} \tag{4.58}$$

the output[26] is the energy density in the observer's rest frame.
(ii) Now enter a dimensionless vector $\boldsymbol{a}$ and the velocity to find

$$\begin{aligned} \boldsymbol{T}(\boldsymbol{u},\boldsymbol{a}) &= \tilde{J}(\boldsymbol{u}) \otimes \tilde{p}(\boldsymbol{a}) \\ &= -n\tilde{p}(\boldsymbol{a}), \end{aligned} \tag{4.59}$$

which is (minus) the momentum density pointing along vector $\boldsymbol{a}$, as measured by the observer.
(iii) Putting the vectors in the other way round, we find

$$\begin{aligned} \boldsymbol{T}(\boldsymbol{a},\boldsymbol{u}) &= \tilde{J}(\boldsymbol{a}) \otimes \tilde{p}(\boldsymbol{u}) \\ &= -E\tilde{J}(\boldsymbol{a}), \end{aligned} \tag{4.60}$$

which is (minus) the energy transported along the direction $\boldsymbol{a}$, according to the observer. This expression is equal to the particle momentum density transported along $\boldsymbol{a}$ (in eqn 4.59).[27]
(iv) Now try entering a single velocity vector into one slot

$$\boldsymbol{T}(\boldsymbol{u},\ ) = -n\tilde{p}(\ ), \tag{4.61}$$

which[28] gives the 4-momentum density 1-form in the rest frame of the observer.

---

[25] We need our previous results that for an observer with velocity $\boldsymbol{u}$, $\tilde{p}(\boldsymbol{u}) = -E$ and $\tilde{J}(\boldsymbol{u}) = -n$.

[26] In components, eqn 4.58 is $T_{\mu\nu}u^\mu u^\nu = nE$.

[27] This is guaranteed because the tensor is symmetric, i.e. $\boldsymbol{T}(\boldsymbol{u},\boldsymbol{a}) = \boldsymbol{T}(\boldsymbol{a},\boldsymbol{u})$. In components, one could write eqns 4.59 and 4.60 as $T_{\mu\nu}u^\mu a^\nu = -np_\mu a^\mu$ and $T_{\mu\nu}a^\mu u^\nu = -EJ_\mu a^\mu$. These quantities are equal because, recalling that the observer is travelling with velocity $\boldsymbol{u}$, we have $J = n_0 v = np/\gamma(u)m$ and $E = \gamma(u)m$, where u is the speed of the observer relative to the measurement frame.

[28] In components, $T_{\mu\nu}u^\mu = -np_\nu$.

Chapter summary

- 1-forms can be viewed as a set of equally spaced planes. They combine with vectors to form numbers

- Tensors are linear slot machines that input vectors and 1-forms and output numbers.

- The energy-momentum tensor $\boldsymbol{T}$ is a (0,2) [or (2,0)] symmetric tensor with components $T_{\mu\nu}$ [or $T^{\mu\nu}$]. It tells us about the flux of the 4-momentum, and the time-time component T^{00} gives us access to the energy density.

Exercises

(4.1) A tensor $\boldsymbol{W}$ has components $W^{\alpha\beta} = u^\alpha v^\beta$ where $\boldsymbol{u}$ and $\boldsymbol{v}$ are 4-vectors. Show that $\boldsymbol{W}$ transforms properly as a tensor.

(4.2) Show that $\delta^\alpha{}_\beta$ transforms properly as a tensor. What about $\delta_{\alpha\beta}$ and $\delta^{\alpha\beta}$?

(4.3) Show that if you take a tensor $\boldsymbol{S}(\ ,\ )$ with components $S^\mu{}_\nu$ you can construct a Lorentz invariant scalar by evaluating $S^\mu{}_\mu$. (Remember that using the summation convention, this involves evaluating $\sum_{\mu=0}^4 S^\mu{}_\mu$.)

(4.4) Consider flat space in spherical polar coordinates.
(a) Compute the basis 1-forms $\boldsymbol{\omega}^r$, $\boldsymbol{\omega}^\theta$ and $\boldsymbol{\omega}^\phi$ in terms of the Cartesian basis 1-forms $\boldsymbol{\omega}^x$, $\boldsymbol{\omega}^y$ and $\boldsymbol{\omega}^z$.
(b) Write basis vectors $\boldsymbol{e}_r$, $\boldsymbol{e}_\theta$ and $\boldsymbol{e}_\phi$ in terms of the usual Cartesian basis $\boldsymbol{e}_x$, $\boldsymbol{e}_y$ and $\boldsymbol{e}_z$.
(c) Using these results, show that $\langle \boldsymbol{\omega}^\mu, \boldsymbol{e}_\nu \rangle = 0$ for $\mu \neq \nu$, where the indices are r, θ and ϕ.

(4.5) Consider a coordinate system (u, v, w) related to Cartesian coordinates via

$$
\begin{aligned}
x &= u - v, \\
y &= u + v, \\
z &= -2uv + w.
\end{aligned}
\tag{4.62}
$$

Compute (a) basis vectors and (b) basis 1-forms, in terms of the Cartesian $\boldsymbol{e}_\mu$ and $\boldsymbol{\omega}^\mu$.

(4.6) Consider the invariant

$$
M = \boldsymbol{a} \cdot \boldsymbol{a} = \eta_{\mu\nu} a^\mu a^\nu,
\tag{4.63}
$$

where $\eta_{\mu\nu}$ are the components of the metric and a^μ are the components of the vector $\boldsymbol{a}$.
(a) Explain why

$$
\frac{\partial M}{\partial a^\mu} = 2\eta_{\mu\nu} a^\nu.
\tag{4.64}
$$

(b) Show that

$$
a^\mu \frac{\partial M}{\partial a^\mu} = 2M.
\tag{4.65}
$$

(4.7) Prove the useful relation that, for a vector $\boldsymbol{A}$,

$$
\frac{\mathrm{d}}{\mathrm{d}\tau} \boldsymbol{A} \cdot \boldsymbol{A} = 2\eta_{\mu\nu} A^\mu \frac{\mathrm{d}A^\nu}{\mathrm{d}\tau}.
\tag{4.66}
$$

(4.8) *The Doppler effect in special relativity.* By considering the Lorentz-transformation properties of the wavevector 4-vector $\boldsymbol{k}$ with components $k^\mu = (\omega, \vec{k})$ derive the expression for the Doppler effect on the frequency of the wave, as predicted by special relativity.

5

The metric

...with the measure you use, it will be measured to you
(Matthew 7[2])

In the previous chapter, we learned that a tensor is a machine for turning vectors and 1-forms into scalars. We have also seen that the metric tensor $\boldsymbol{\eta}(\ ,\ )$ is a (0,2) tensor (meaning that you feed it two vectors and it spits out a number) that tells you how to obtain the scalar product of two vectors ($\boldsymbol{X} \cdot \boldsymbol{Y} = \eta_{\mu\nu}X^\mu Y^\nu$, from eqn 4.1). However, everything so far has been for flat Minkowski spacetime (i.e. for special, not general, relativity). It's time now to tackle gravity and when we include that, spacetime becomes curved. We are working up to Einstein's theory of gravitation that can be succinctly stated as

$$\left(\begin{array}{c} \text{Curvature of} \\ \text{spacetime} \end{array}\right) = \left(\begin{array}{c} \text{Energy density of} \\ \text{matter in spacetime} \end{array}\right). \tag{5.1}$$

With curvature included, we need a more general metric and we shall use the symbol $\boldsymbol{g}(\ ,\ )$ for our general (0,2) metric tensor [retaining $\boldsymbol{\eta}(\ ,\ )$ for the special case of flat Minkowski spacetime]. The metric tensor $\boldsymbol{g}(\ ,\ )$ will feed into the left-hand side of eqn 5.1. This chapter will focus on the form of $\boldsymbol{g}(\ ,\ )$ for several different types of spacetime.

In flat spacetime, $\boldsymbol{g}(\ ,\ ) = \boldsymbol{\eta}(\ ,\ )$ *everywhere* throughout spacetime. In curved spacetime, $\boldsymbol{g}(\ ,\ )$ varies from point to point. This means that the metric is a **field**. A field is a quantity where we input a position in spacetime and output a tensor valid for that position in spacetime. For the metric field, we input a position in spacetime and output the appropriate metric tensor that allows us to take dot products at that point in space.[1] Of course the tensors at closely spaced points in spacetime will be related and this relationship gives rise to a **field theory**: a theory that allows us to describe and predict changes in the fields as a function of space and time and also to examine the consequences the field has on the physics. General relativity is the field theory of gravity. The left-hand side of its governing equation is based on the metric field: the field describing the geometry of spacetime.[2] We begin by restating some simple general facts about metric tensors and their components.

[1] There's scope for confusion here as a metric is an object into which we input two vectors. The order here is that we input a position in spacetime into the metric field and output a metric for that point in spacetime. We can then insert two vectors into that metric tensor and find their scalar product at that point in spacetime.

[2] It will turn out that the metric also features on the right-hand side of this equation too, with the result that the equation is difficult to solve.

5.1 Metrics in general

For a general space, the metric at a point in spacetime $\boldsymbol{g}(\ ,\ )$ is a (0,2) slot machine that takes two vectors as input and outputs their scalar product.[3] Inserting vectors $\boldsymbol{X}$ and $\boldsymbol{Y}$ into the metric, we obtain

$$g(\boldsymbol{X}, \boldsymbol{Y}) = \boldsymbol{X} \cdot \boldsymbol{Y}. \tag{5.2}$$

If we prefer to work in components, then these can be extracted by inputting the basis vectors into the metric

$$\boldsymbol{g}(\boldsymbol{e}_\mu, \boldsymbol{e}_\nu) = \boldsymbol{e}_\mu \cdot \boldsymbol{e}_\nu \equiv g_{\mu\nu}. \tag{5.3}$$

The metric encodes distances in spacetime via an expression in terms of infinitesimal intervals between coordinates. Previously, we wrote the invariant line element in terms of the Minkowski metric as $\mathrm{d}s^2 = \eta_{\mu\nu}\mathrm{d}x^\mu\mathrm{d}x^\nu$. This hints at a simple way to write down the invariant infinitesimal length of a line element in any space, so we adopt it and write[4]

$$\begin{aligned} \mathrm{d}s^2 &= \mathrm{d}\boldsymbol{x} \cdot \mathrm{d}\boldsymbol{x} = \boldsymbol{g}(\mathrm{d}\boldsymbol{x}, \mathrm{d}\boldsymbol{x}) = \mathrm{d}x^\mu\mathrm{d}x^\nu\boldsymbol{g}(\boldsymbol{e}_\mu, \boldsymbol{e}_\nu) \\ &= g_{\mu\nu}\mathrm{d}x^\mu\mathrm{d}x^\nu. \end{aligned} \tag{5.4}$$

This equation is a simple statement of how long an infinitesimal interval is in a particular spacetime geometry, specified by the components of the metric. It's often useful to write down this line-element equation and, since it contains all of the components of the metric $g_{\mu\nu}$, we often say that the line element *is* the metric. We can integrate $\mathrm{d}s$ to work out the total interval s between two events. For example, the magnitude of the interval along a curve between events at points $\mathcal{A}$ and $\mathcal{B}$ can be worked out using the metric via the prescription

$$s = \int_\mathcal{A}^\mathcal{B} |g_{\mu\nu}\mathrm{d}x^\mu\mathrm{d}x^\nu|^{\frac{1}{2}}. \tag{5.5}$$

Example 5.1

Suppose we have two different coordinate systems. We can use the invariance of the line element to relate the components of the metric together. The line element is written as

$$\mathrm{d}s^2 = g_{\mu\nu}\mathrm{d}x^\mu\mathrm{d}x^\nu = g'_{\alpha\beta}\mathrm{d}x^{\alpha'}\mathrm{d}x^{\beta'}. \tag{5.6}$$

Using the standard transformation law for differentials[5]

$$g'_{\alpha\beta}\mathrm{d}x^{\alpha'}\mathrm{d}x^{\beta'} = g_{\mu\nu}\frac{\partial x^\mu}{\partial x^{\alpha'}}\frac{\partial x^\nu}{\partial x^{\beta'}}\mathrm{d}x^{\alpha'}\mathrm{d}x^{\beta'}. \tag{5.7}$$

We conclude that we have the transformation law

$$g'_{\alpha\beta} = g_{\mu\nu}\frac{\partial x^\mu}{\partial x^{\alpha'}}\frac{\partial x^\nu}{\partial x^{\beta'}}. \tag{5.8}$$

That is, the components of the metric transform as we expect the components of a $(0,2)$ tensor to transform.

---

[3]We have also seen that it is possible to define the metric as a $(2,0)$ tensor, which is a slot machine that takes two 1-forms as input and outputs a scalar. In components, this gives the 'up' form of the metric $g^{\mu\nu}$. As explained in the previous chapter, the 'up' form of the metric is the inverse of the 'down' form, or

$$g^{\mu\nu}g_{\nu\lambda} = \delta^\mu{}_\lambda.$$

[4]This is a useful example of the metric tensor acting as a slot machine. If we want to find the squared infinitesimal interval between positions we insert $\mathrm{d}\boldsymbol{x}$ into both slots of the metric tensor and the output is exactly the invariant interval $\mathrm{d}s^2$ that we seek.

[5]Note that the line element in this form features a multiplication of the factors $\mathrm{d}x^\mu$. In this expression, it does *not* represent an infinitesimal area, but rather the square of a length, and so the transformation simply requires a multiplication of the individual transformations. Later in the chapter we shall see how a transformation of a product that *does* represent an area or volume requires us to consider a Jacobian. This feature is discussed in more detail in Part V of the book (see, in particular, Chapter 38).

5.2   Meet some metrics

With some general rules recapped, let's now meet some metric line elements.[6]

- We begin with the simplest and most familiar line element. This is the metric for Euclidean space in three dimensions, whose line element is simply

$$ds^2 = dx^2 + dy^2 + dz^2. \tag{5.9}$$

- Special relativity is founded on the Minkowski metric in (3+1) dimensions (i.e. a combination of three spatial dimensions and one time dimension). In Cartesian coordinates, the line element for this metric is written as

$$ds^2 = -dt^2 + dx^2 + dy^2 + dz^2. \tag{5.10}$$

We can also take the coordinate components to form a vector with components $dx^\mu = (dt, dx, dy, dz)$ and write the line element as $ds^2 = g_{\mu\nu}dx^\mu dx^\nu$ where

$$ g_{\mu\nu} = \begin{pmatrix} -1 & 0 & 0 & 0 \\ 0 & 1 & 0 & 0 \\ 0 & 0 & 1 & 0 \\ 0 & 0 & 0 & 1 \end{pmatrix}. \tag{5.11}$$

- There's nothing stopping us working in cylindrical polar coordinates and describing the same flat Minkowski space using these.[7] The Minkowski line element in cylindrical polars is therefore

$$ds^2 = -dt^2 + dr^2 + r^2 d\theta^2 + dz^2. \tag{5.15}$$

We can also take the coordinate components to form a column vector with components $dx^\mu = (dt, dr, d\theta, dz)$ and then the line element is $ds^2 = g_{\mu\nu}dx^\mu dx^\nu$, where

$$ g_{\mu\nu} = \begin{pmatrix} -1 & 0 & 0 & 0 \\ 0 & 1 & 0 & 0 \\ 0 & 0 & r^2 & 0 \\ 0 & 0 & 0 & 1 \end{pmatrix}. \tag{5.16}$$

- In the same way, we can write the matrix representing the metric components for flat space in spherical polar coordinates.[8]

$$ g_{\mu\nu} = \begin{pmatrix} -1 & 0 & 0 & 0 \\ 0 & 1 & 0 & 0 \\ 0 & 0 & r^2 & 0 \\ 0 & 0 & 0 & r^2 \sin^2\theta \end{pmatrix}, \tag{5.19}$$

or

$$ds^2 = -dt^2 + dr^2 + r^2 d\theta^2 + r^2 \sin^2\theta \, d\phi^2. \tag{5.20}$$

---

[6] Although we're used to the very useful practice that r always represents a radius, t a time and θ an angle, in the world of metrics, general relativity and curved spacetime, we shall need to abandon this unsafe assumption, which turns out not to be true. The slogan that embodies this is that coordinates don't have any intrinsic **metric significance**, which just means t isn't necessarily time etc. Einstein's opinion was that this was one of the most difficult thing to give up in getting to grips with the consequences of his theory. We shall return to this idea.

[7] Recall that cylindrical coordinates are related to Cartesian ones via

$$ x = r\cos\theta, \qquad y = r\cos\theta, $$
$$ r = (x^2 + y^2)^{\frac{1}{2}}, \quad \tan\theta = y/x, $$

with z the same in both systems. Using the usual relationship between Cartesians and cylindrical polars that $x = r\cos\theta$, we have the rule for differentials

$$ \begin{aligned} dx &= \left(\frac{\partial x}{\partial r}\right)_\theta dr + \left(\frac{\partial x}{\partial \theta}\right)_r d\theta \\ &= \cos\theta \, dr - r\sin\theta \, d\theta, \end{aligned} \tag{5.12} $$

and also

$$ dy = \sin\theta \, dr + r\cos\theta \, d\theta. \tag{5.13} $$

We conclude that

$$ dx^2 + dy^2 = dr^2 + r^2 \, d\theta. \tag{5.14} $$

[8] The transformation in this case is

$$ \begin{aligned} x &= r\sin\theta\cos\phi, \\ y &= r\sin\theta\sin\phi, \\ z &= r\cos\theta, \end{aligned} $$

(which reduces down to the cylindrical coordinates if we set $\theta = \pi/2$). In dealing with spherical polars, we shall often write

$$ r^2(d\theta^2 + \sin^2\theta \, d\phi^2) = r^2 d\Omega^2, \tag{5.17} $$

so that the flat-space line element is

$$ ds^2 = -dt^2 + dr^2 + r^2 d\Omega^2. \tag{5.18} $$

Example 5.2

Although the space that's described here is still flat, if we fix r we have our first curved space: the two-dimensional surface of a sphere. Consider the surface of a sphere of circumference $2\pi a$. From the spherical polar example, we need only fix $r = a$ and the metric line element for this two-dimensional surface becomes

$$\mathrm{d}s^2 = -\mathrm{d}t^2 + a^2\left(\mathrm{d}\theta^2 + \sin^2\theta\mathrm{d}\phi^2\right). \tag{5.21}$$

- A final example of a metric is the Newtonian limit of the metric that emerges from general relativity as a limiting case of the Einstein equation. This Newtonian-metric line element is written as

$$\mathrm{d}s^2 = -(1+2\Phi)\mathrm{d}t^2 + (1-2\Phi)(\mathrm{d}x^2 + \mathrm{d}y^2 + \mathrm{d}z^2), \tag{5.22}$$

  where Φ is the gravitational potential. We have already motivated the $(1+2\Phi)$ term multiplying $\mathrm{d}t^2$ in eqn 2.75 (Example 2.13) where by working in the limit $v \ll c$ we were only able to obtain the correction to the time-dependent part of the metric. Equation 5.22 has included a very similar correction to the spatial coordinates as well. Note that if the potential Φ is set to zero, this metric reverts to the Minkowski metric of flat space (eqn 5.10), as expected. Equation 5.22 can also be expressed by writing down the non-zero components of the metric tensor $g_{00} = -(1 + 2\Phi)$, $g_{11} = g_{22} = g_{33} = 1 - 2\Phi$, or by writing down the components of the tensor as

$$g_{\mu\nu} = \begin{pmatrix} -(1+2\Phi) & 0 & 0 & 0 \\ 0 & 1-2\Phi & 0 & 0 \\ 0 & 0 & 1-2\Phi & 0 \\ 0 & 0 & 0 & 1-2\Phi \end{pmatrix}. \tag{5.23}$$

  Note that the potential Φ is assumed to be a function of position, but this limit works for static solutions in which the potential is not time-varying.

- The Newtonian-limit line element can also be written in spherical polars

$$\mathrm{d}s^2 = -(1+2\Phi)\mathrm{d}t^2 + (1-2\Phi)(\mathrm{d}r^2 + r^2\mathrm{d}\theta^2 + r^2\sin^2\theta\mathrm{d}\phi^2). \tag{5.24}$$

Notice how in the previous examples of metrics (other than the simple Minkowski metric line element in Cartesian coordinates) the components of the metric vary in space. That is, the metric is a function of the underlying coordinates. We should therefore write the metric components as a function $g_{\mu\nu}(x)$. The metric is now manifestly an example of a field. The **metric field** takes a spacetime coordinate x^μ as an input and outputs the metric tensor at that point.[9]

↳ See Chapter 14 for a derivation of eqn **5.22**.

---

[9]Here's a simple example of the use of the metric using the slot-machine picture. Working in flat space with cylindrical coordinates $X^\mu = (t, r, \theta, z)$, the distance between points separated by an interval $\mathrm{d}\boldsymbol{X}$ with coordinates $\mathrm{d}X^\mu = (0, 0, \mathrm{d}\theta, 0)$ is found by evaluating

$$\begin{aligned} \mathrm{d}s^2 &= \boldsymbol{g}(\mathrm{d}\boldsymbol{X}, \mathrm{d}\boldsymbol{X}) \\ &= g_{\mu\nu}\mathrm{d}X^\mu\mathrm{d}X^\nu \\ &= g_{\theta\theta}\mathrm{d}\theta^2 = r^2\mathrm{d}\theta^2. \end{aligned} \tag{5.25}$$

Notice how the interval $\mathrm{d}s^2$ changes its size depending on the value of r at which we evaluate the interval. This follows from $\boldsymbol{g}$ being a field: we input a position in spacetime and output the interval appropriate for that position. To compute an interval between points θ_1 and θ_2, separated by a larger interval (i.e. the separation is not infinitesimal), we then evaluate

$$\Delta s = \int \mathrm{d}s = \int_{\theta_1}^{\theta_2} \sqrt{g_{\theta\theta}}\mathrm{d}\theta = \int_{\theta_1}^{\theta_2} r\mathrm{d}\theta. \tag{5.26}$$

We'll use these ideas in the following sections.

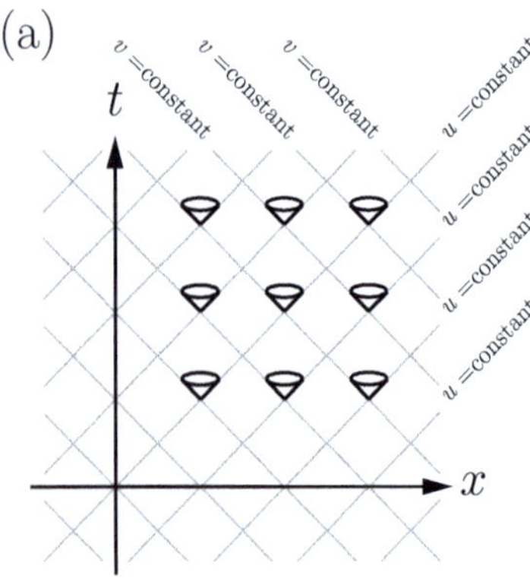

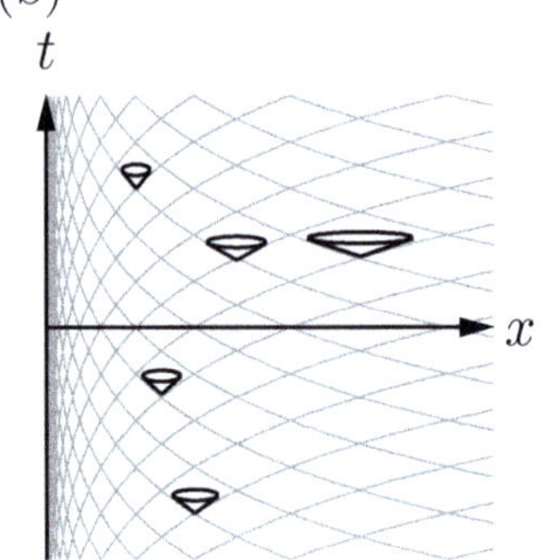

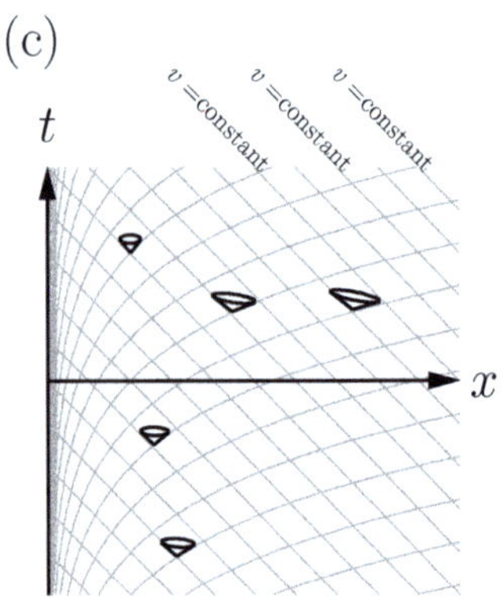

Fig. 5.1   Light cones in (a) flat
spacetime, (b) Rindler spacetime and
(c) baby-Eddington–Finkelstein coordinates.

5.3   Light and light cones

One way of visualizing a metric is to draw the light cones in the spacetime that it describes. Light cones are absolute surfaces that separate
timelike and spacelike intervals. We use *local* light cones to visualize
spaces because, in curved spacetime, the light cones change their orientation as a function of position in spacetime. The pattern of local light
cones represents almost all of the structure of spacetime.[10] Recall that
the infinitesimal interval between two events on the photon's world line
satisfies

$$ds^2 = 0. \tag{5.27}$$

Using this condition, we can work out the orientation of the light cones in
whichever coordinates we've chosen to use. The orientation is important
as it reveals the signals that observers can send and receive at each point.
We investigate some examples below.

Example 5.3

(a) Light in **flat Minkowski spacetime** obeys

$$ds^2 = -dt^2 + dx^2 + dy^2 + dz^2 = 0, \tag{5.28}$$

from which we find

$$\frac{dt}{d|\vec{r}|} = \pm 1, \tag{5.29}$$

and so $dt = \pm d|\vec{r}|$, which is integrated to give an equation for the light cones emerging
from a point (t_0, r_0), with the result that light cones obey the coordinate equation

$$t - t_0 = \pm(|\vec{r}| - |\vec{r}_0|). \tag{5.30}$$

The light cones look the same everywhere [see Fig. 5.1(a), where we just draw the
forward light cones for simplicity]. This uniformity of the light cones throughout
spacetime is exactly as we stated earlier [see Fig. 3(a)]. A useful innovation at this
point are the so-called **light-cone coordinates**

$$u = t - |\vec{r}|, \quad v = t + |\vec{r}|, \tag{5.31}$$

giving

$$ds^2 = -du\,dv, \tag{5.32}$$

from which we can immediately read off that the light cones are coincident with the
lines of constant u and v [see Fig. 5.1(a)].

  Let's try other spacetimes. (For this particular exercise, we simply pluck these
from thin air, without derivation.)

(b) The first one is known as **Rindler spacetime** (it comes from a coordinate choice
appropriate for accelerated observers) and has a metric line element

$$ds^2 = -x^2 dt^2 + dx^2. \tag{5.33}$$

Setting $ds^2 = 0$ for light, we have the equation of the light cone

$$\frac{dt}{dx} = \pm \frac{1}{x}. \tag{5.34}$$

Integrating, we find

$$t - t_0 = \pm \ln|x|. \tag{5.35}$$

We see that the cones change their shape as a function of position as shown in
Fig. 5.1(b).

(c) Let's try another spacetime. This one has a metric line element we'll call the
baby-Eddington–Finkelstein metric

$$ds^2 = -x\,dv^2 + 2dv\,dx. \tag{5.36}$$

Light cones must cause $\mathrm{d}s^2$ to vanish and so we spot that they have $v = \text{constant}$ and also $\mathrm{d}v/\mathrm{d}x = 2/x$, so we find

$$v - v_0 = 2\ln|x|. \tag{5.37}$$

The light cones are shown in Fig. 5.1(c).

(d) Finally, we consider the slightly more complicated **Eddington–Finkelstein** metric[11]

$$\mathrm{d}s^2 = -\left(1 - \frac{2GM}{r}\right)\mathrm{d}v^2 + 2\mathrm{d}v\mathrm{d}r. \tag{5.38}$$

Light cones have $v = \text{const}$ again and also

$$\frac{\mathrm{d}v}{\mathrm{d}r} = 2\left(1 - \frac{2GM}{r}\right)^{-1}. \tag{5.39}$$

[11]We shall see this again when we examine the geometry of stars. It arises in the theory of spherically symmetric black holes.

Being able to visualize spacetimes in this way allows us to understand some exotic spacetimes, such as that of the next example.

[12]Miguel Alcubierre (1964–). The warp drive proposal originated in M. Alcubierre, Class. Quantum Grav. **11**, L73 (1994).

Example 5.4

The Alcubierre metric[12] is an attempt to build the spacetime that would result from the action of a **warp drive**. A warp drive, much discussed in science fiction, is a device that appears to allow faster-than-light travel (judged from the point of view of an observer in flat spacetime). Of course, the propagation of signals (and travellers) faster than c is not allowed. Instead, the warp drive works by making a bubble of curved spacetime where the light cones are oriented differently to those in flat space.

The mathematical construction of the bubble starts with a curve $x = x_{\mathrm{s}}(t), y = z = 0$, which has a tangent $v_{\mathrm{s}} = \mathrm{d}x/\mathrm{d}t$. Now we construct a smooth bubble function $f(r_{\mathrm{s}})$, where $r_{\mathrm{s}} = \sqrt{(x - x_{\mathrm{s}})^2 + y^2 + z^2}$. By construction, this function has the property $f(0) = 1$ and decreases as we move from the origin, vanishing for $r_{\mathrm{s}} > R$, where R is some distance that sets the edge of the bubble. The Alcubierre metric is then written as

$$\mathrm{d}s^2 = -\left[1 + v_{\mathrm{s}}(t)^2 f(r_{\mathrm{s}})^2\right]\mathrm{d}t^2 + 2v_{\mathrm{s}}(t)f(r_{\mathrm{s}})\mathrm{d}x\mathrm{d}t + \mathrm{d}x^2 + \mathrm{d}y^2 + \mathrm{d}z^2, \tag{5.40}$$

where $v_{\mathrm{s}} = \frac{\mathrm{d}x_{\mathrm{s}}}{\mathrm{d}t}$. This can be simplified to read

$$\mathrm{d}s^2 = -\mathrm{d}t^2 + \left[\mathrm{d}x - v_{\mathrm{s}}(t)f(r_{\mathrm{s}})\mathrm{d}t\right]^2 + \mathrm{d}y^2 + \mathrm{d}z^2. \tag{5.41}$$

This all looks rather complicated, but is best understood using Fig. 5.2. Figure 5.2(a) shows the world line of the traveller between two distant points in spacetime with different values of the coordinate x. The warp drive creates a bubble in spacetime shown by the dotted region. We can examine the light-cone structure inside the bubble by setting $\mathrm{d}s^2 = 0$. We find that the light cones are given by

$$\frac{\mathrm{d}x}{\mathrm{d}t} = \pm 1 + v_{\mathrm{s}}(t)f(r_{\mathrm{s}}). \tag{5.42}$$

In regions outside the bubble, where $f = 0$, the light cones are the normal ones of flat spacetime. Inside the bubble, the function f causes the light cones to tip over, as shown in Fig. 5.2(b). The traveller must always move inside her forward light cone, but we see that inside the bubble, the tipping of the light cones means that the tangent of the world line appears to give rise to a velocity greater than c, judged by the light cones outside of the warp bubble. As a result of the warp in spacetime, the traveller is able to travel vast distances that would require superluminal velocities in flat spacetime.[13]

(a)

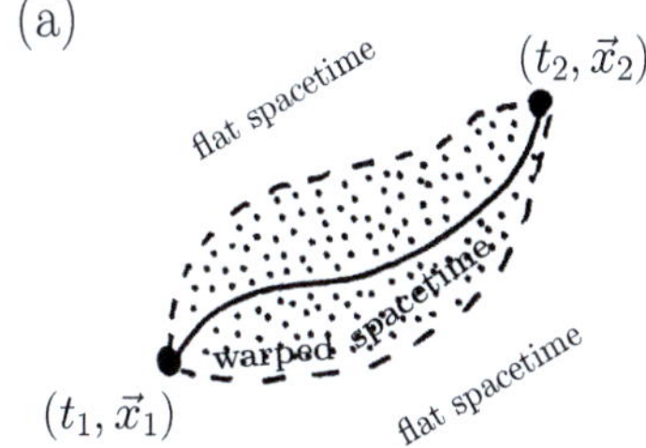

(b)

Fig. 5.2 (a) The world line of a trip in spacetime from $(t_1, \vec{x}_1)$ to $(t_2, \vec{x}_2)$, surrounded by a warped bubble of spacetime. (b) The light-cone structure in the warped region of spacetime. The light cones tip over in the warped region.

[13]It's worth noting that warping spacetime in this way would require a source of negative energy, so is not something that is readily achievable!

5.4   Lengths, areas, volumes

Once we have a metric in the form of a line element, we can use it to calculate not only lengths (or intervals), but also areas and volumes.

Example 5.5

Consider the metric for flat space, given in terms of spherical polars, with line element $ds^2 = -dt^2 + dr^2 + r^2 d\theta^2 + r^2 \sin^2\theta d\phi^2$.[14] We can use this to find the radius of a sphere. We fix $t = \theta = \phi = \text{constant}$ so that $dt = d\theta = d\phi = 0$, with the result that $ds^2 = dr^2$. We then integrate the line element ds from $r = 0$ to $r = R$

$$\int_0^R ds = \int_0^R dr = R. \tag{5.43}$$

This is no surprise, but shows the general principle of how to manipulate the metric to extract a length. In the same way, we can find the radius and circumference of a circle (Fig. 5.3) which has a fixed value of $\theta = \theta_0$ on the sphere. First, let's work out r_0 which is the distance from the North pole to the circle. We start by fixing t, r and ϕ (so that $ds^2 = R^2 d\theta^2$) and then integrating ds to find

$$\int_{\theta=0}^{\theta_0} ds = \int_{\theta=0}^{\theta_0} \sqrt{g_{\theta\theta}(r=R)}\, d\theta = \int_{\theta=0}^{\theta_0} R d\theta = \theta_0 R. \tag{5.44}$$

Again, an unsurprising result (elementary geometry tells us that $\theta_0 R = r_0$). The circumference of this circle is calculated using $ds^2 = g_{\phi\phi}(r = R, \theta = \theta_0) d\phi^2 = R^2 \sin^2\theta_0 d\phi^2$ and so

$$\int_{\phi=0}^{2\pi} ds = \int_{\phi=0}^{2\pi} \sqrt{g_{\phi\phi}(R, \theta_0)}\, d\phi = 2\pi R \sin\theta_0 = 2\pi r_0 \operatorname{sinc}\frac{r_0}{R}, \tag{5.45}$$

agreeing with eqn 3.20.

We notice from these examples that the (proper) length of infinitesimal segments of coordinate x^1 are given by $dl^1 = \sqrt{g_{11}}\, dx^1$. We can use this fact to work out how to calculate areas and volumes.

Example 5.6

Consider the special case of a diagonal[15] metric, with line element

$$ds^2 = g_{00}(dx^0)^2 + g_{11}(dx^1)^2 + g_{22}(dx^2)^2 + g_{33}(dx^3)^2. \tag{5.46}$$

An element of area (sometimes called a 2-volume) can be written as[16]

$$\begin{aligned} dA &= dl^2 dl^3 \\ &= \sqrt{g_{22}g_{33}}\, dx^2 dx^3. \end{aligned} \tag{5.47}$$

Similarly, an element of 3-volume is written as

$$\begin{aligned} dV &= dl^1 dl^2 dl^3 \\ &= \sqrt{g_{11}g_{22}g_{33}}\, dx^1 dx^2 dx^3. \end{aligned} \tag{5.48}$$

As an example, let's consider the metric in spherical polar coordinates. The area element is, using eqn 5.47, given by

$$dA = r^2 \sin\theta d\theta d\phi, \tag{5.49}$$

and the volume, using eqn 5.48, is

$$dV = r^2 \sin\theta dr d\theta d\phi. \tag{5.50}$$

[14]Recall that we can write this $ds^2 = g_{tt}dt^2 + g_{rr}dr^2 + g_{\theta\theta}d\theta^2 + g_{\phi\phi}d\phi^2$. In this example, we use the fact that if $dt = d\theta = d\phi = 0$, then $ds = \sqrt{g_{rr}}dr$, and so on.

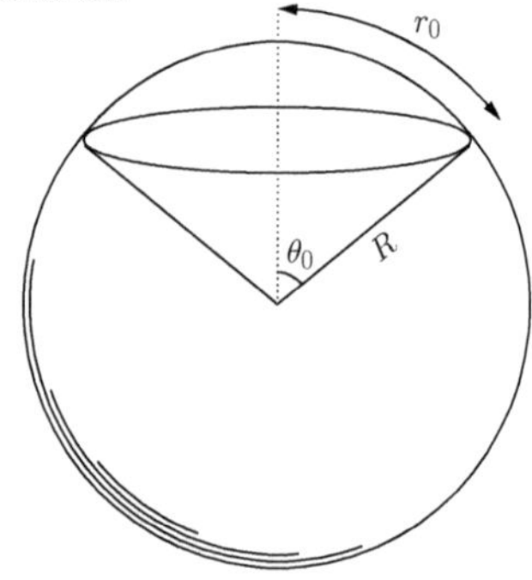

Fig. 5.3 A sphere of radius R. A circle on the surface of the sphere has a radius r_0 and is a line of fixed $\theta = \theta_0$.

[15]In this context, *diagonal* refers to the matrix of metric components $g_{\mu\nu}$. A diagonal metric has a line element that features the squares of intervals such as $(dx^\mu)^2$, but not mixed components such as $dx^\mu dx^\nu$ with $\mu \neq \nu$.

[16]This is one example of an area. We can also form areas from other pairs of coordinates too. For our example of a sphere, this choice turns out to be a sensible one since it yields an element of surface area.

For an element of 4-volume [i.e. a volume of (3+1)-dimensional spacetime], we need to take account of the fact that the timelike component of all Lorentz metrics comes with a minus sign. We therefore write

$$
\begin{aligned}
\mathrm{d}\mathcal{V} &= \mathrm{d}l^0 \mathrm{d}l^1 \mathrm{d}l^2 \mathrm{d}l^3 \\
&= \sqrt{-g_{00}g_{11}g_{22}g_{33}}\, \mathrm{d}x^0 \mathrm{d}x^1 \mathrm{d}x^2 \mathrm{d}x^3,
\end{aligned}
\tag{5.51}
$$

and since g_{00} is negative, we take the square root of a positive quantity. Note that the product $g_{00}g_{11}g_{22}g_{33}$ is the determinant of the diagonal metric matrix.[17]

We now show that this result generalizes to cases where we don't have a diagonal metric, which is to say that an element of (3+1)-dimensional volume, known as a 4-volume, is given by

$$
\mathrm{d}\mathcal{V} = \sqrt{-\det \boldsymbol{g}}\, \mathrm{d}^4 x,
\tag{5.53}
$$

where we note that the determinant $\det \boldsymbol{g}$ is often simply written g, so we would write $\mathrm{d}\mathcal{V} = \sqrt{-g}\, \mathrm{d}^4 x$, with $\mathrm{d}^4 x = \mathrm{d}x^0 \mathrm{d}x^1 \mathrm{d}x^2 \mathrm{d}x^3$. As we shall see below, the 4-volume $\mathrm{d}\mathcal{V}$ is an invariant.

To prove that the 4-volume is an invariant, let's note that volumes transform using an object that is called the **Jacobian**.[18] For our case of (3+1)-dimensional spacetime, the Jacobian is given by the determinant of the transformation matrix and so

$$
J = \frac{\partial(x^{0'}, x^{1'}, x^{2'}, x^{3'})}{\partial(x^0, x^1, x^2, x^3)} = \det(\Lambda^{\alpha'}{}_{\beta}).
\tag{5.55}
$$

Volume elements in different coordinate systems are then transformed using the Jacobian by writing

$$
\mathrm{d}x^{0'} \mathrm{d}x^{1'} \mathrm{d}x^{2'} \mathrm{d}x^{3'} = \frac{\partial(x^{0'}, x^{1'}, x^{2'}, x^{3'})}{\partial(x^0, x^1, x^2, x^3)} \mathrm{d}x^0 \mathrm{d}x^1 \mathrm{d}x^2 \mathrm{d}x^3.
\tag{5.56}
$$

We can now prove the general rule for finding the volume of an element of 4-space. We assume that the volume on the left is the usual volume of an infinitesimal element in a Cartesian flat space. We need to show that the Jacobian (i.e. the determinant of the transformation matrix) is equal to $\sqrt{-g}$. The argument proceeds from a principle called **local flatness**.[19] An observer will perceive spacetime to be flat at the point at which they reside, in much the same way that (unless you are living on a hillside) we tend to perceive the Earth as locally flat and only abandon our locally Euclidean street-maps when we look further afield. The following example fills in the details.

Example 5.7

Proof: Arguing from local flatness, we transform the Minkowski metric tensor $\boldsymbol{\eta}$ (for an observer's locally flat spacetime) into a general tensor $\boldsymbol{g}$ (for curved spacetime) at the particular point in spacetime of the observer's location. This can be done using[20]

$$
\boldsymbol{g} = \boldsymbol{\Lambda}\boldsymbol{\eta}\boldsymbol{\Lambda}^{\mathrm{T}}.
\tag{5.58}
$$

[17] The determinant of an $n \times n$ matrix $\boldsymbol{A}$ is computed using the rule

$$
\det \boldsymbol{A} =
$$

$$
\sum_{i_1, i_2, \dots, i_n = 1}^{n} \varepsilon_{i_1 \dots i_n} A_{1, i_1} \dots A_{n, i_n},
\tag{5.52}
$$

where $\varepsilon_{i_1 \dots i_n}$ is the **Levi-Civita symbol**, which is defined as $\varepsilon_{i_1 \dots i_n} = 1$ for an even permutation of the indices and $= -1$ for an odd permutation. If the matrix is diagonal, the determinant is simply the product of the n non-zero elements.

[18] The Jacobian is named after the German mathematician C. G. J. Jacobi (1804–1851). A general coordinate transformation may be written $x^{\mu'} = x^{\mu'}(x^1, x^2, x^3, \dots, x^n)$ where $(\mu = 1, 2\dots, n)$. If we now arrange the $n \times n$ partial derivatives $\partial x^{\mu'}/\partial x^{\nu}$ into the transformation matrix

$$
\begin{aligned}
\Lambda^{\mu'}{}_{\nu} &= \frac{\partial x^{\mu'}}{\partial x^{\nu}} \\
&= \begin{pmatrix} \frac{\partial x^{1'}}{\partial x^1} & \cdots & \frac{\partial x^{1'}}{\partial x^n} \\ \vdots & \cdots & \vdots \\ \frac{\partial x^{n'}}{\partial x^1} & \cdots & \frac{\partial x^{n'}}{\partial x^n} \end{pmatrix},
\end{aligned}
\tag{5.54}
$$

then the Jacobian J of the transformation matrix is then defined as the determinant of this matrix. In symbols, we write

$$
J = \frac{\partial(x^{1'} \dots x^{n'})}{\partial(x^1 \dots x^n)} = \det\left(\frac{\partial x^{\mu'}}{\partial x^{\nu}}\right),
$$

where we've introduced the commonly used notation for J. The Jacobian tells us how volume elements transform (and is discussed in more detail in Chapter 38).

[19] The concept of local flatness is explained in more detail in Chapter 6 and will be used regularly from here onwards.

[20] We write a matrix equation here where $\boldsymbol{\eta}$ is the Minkowski matrix (i.e. the matrix with components $\eta_{\mu\nu}$) and $\boldsymbol{\Lambda}$ is a matrix with components $\Lambda^{\mu}{}_{\nu}$. The equation therefore is simply a rewriting of eqn 5.8:

$$
g_{\alpha\beta} = \Lambda^{\mu}{}_{\alpha} \Lambda^{\nu}{}_{\beta} \eta_{\mu\nu}.
\tag{5.57}
$$

Taking determinants we find

$$\det \underline{\boldsymbol{g}} = \det \underline{\boldsymbol{\Lambda}} \det \underline{\boldsymbol{\eta}} \det \underline{\boldsymbol{\Lambda}}^{\mathrm{T}}. \tag{5.59}$$

However, $\det \underline{\boldsymbol{A}} = \det \underline{\boldsymbol{A}}^{\mathrm{T}}$ and also $\det \underline{\boldsymbol{\eta}} = -1$. We conclude therefore that

$$\det \underline{\boldsymbol{g}} = -(\det \underline{\boldsymbol{\Lambda}})^2, \tag{5.60}$$

or, writing $\det \underline{\boldsymbol{g}} = g$,

$$\det \underline{\boldsymbol{\Lambda}} = (-g)^{\frac{1}{2}}. \tag{5.61}$$

The result is that the elementary volume in flat Cartesian spacetime (the primed coordinates) is given by

$$
\begin{aligned}
\mathrm{d}x^{0'}\mathrm{d}x^{1'}\mathrm{d}x^{2'}\mathrm{d}x^{3'} &= \frac{\partial(x^{0'},x^{1'},x^{2'},x^{3'})}{\partial(x^0,x^1,x^2,x^3)}\mathrm{d}x^0\mathrm{d}x^1\mathrm{d}x^2\mathrm{d}x^3 \\
&= (-g)^{\frac{1}{2}}\mathrm{d}x^0\mathrm{d}x^1\mathrm{d}x^2\mathrm{d}x^3.
\end{aligned} \tag{5.62}
$$

The right-hand side of this equation is just the invariant 4-volume $\mathrm{d}\mathcal{V}$ that we met in eqn 5.53. Thus, the volume of an infinitesimal element in locally flat Cartesian space is equal to the invariant 4-volume, nicely illustrating the principle of local flatness.

Chapter summary

- The metric is a field that encodes the geometry of spacetime. It allows us to compute intervals between events. The $(0,2)$ metric tensor takes two vectors and outputs a number

$$g(\boldsymbol{X},\boldsymbol{Y}) = g_{\mu\nu}X^\mu Y^\nu. \tag{5.63}$$

- The components of a metric are often given via a line element

$$\mathrm{d}s^2 = g_{\mu\nu}\mathrm{d}x^\mu\mathrm{d}x^\nu. \tag{5.64}$$

- A metric can be visualized by working out its light cone structure. Light cones are defined by $\mathrm{d}s^2 = 0$.
- The invariant volume element is given by

$$\mathrm{d}\mathcal{V} = \sqrt{-g}\,\mathrm{d}^4x, \tag{5.65}$$

where g is the determinant of the metric tensor $\boldsymbol{g}$.

- The principle of local flatness says that an observer will perceive spacetime to be flat Minkowski spacetime at the point at which they reside.

Exercises

(5.1) By finding the determinant $g = \det \boldsymbol{g}$ of the relevant metric, compute the invariant volume element for the following systems:
(a) A flat two-dimensional surface in cylindrical coordinates.
(b) A two-dimensional spherical surface in spherical coordinates.
(c) The two-dimensional surface of a torus.
Hint: A torus has a line element

$$\mathrm{d}s^2 = (c + a\cos v)^2 \mathrm{d}u^2 + a^2 \mathrm{d}v^2. \tag{5.66}$$

(5.2) Show that the Rindler metric

$$\mathrm{d}s^2 = -x^2 \mathrm{d}t^2 + \mathrm{d}x^2, \tag{5.67}$$

represents flat space in coordinates (T, X) by investigating the coordinate transformation

$$X = x\cosh t, \quad T = x\sinh t. \tag{5.68}$$

(5.3) Consider transforming into a reference frame moving at a constant speed v along the x-axis.
(a) Using the transformation $x = x' - vt$, $y = y'$, $z = z'$, $t = t'$, show that the Minkowski metric line element becomes, in a moving-coordinate reference frame,

$$\mathrm{d}s^2 = -\mathrm{d}t'^2 \left(1 - v^2\right) + \mathrm{d}x'^2 + \mathrm{d}y'^2 + \mathrm{d}z'^2 - 2v\mathrm{d}x'\mathrm{d}t'. \tag{5.69}$$

(b) What is the light cone structure of this metric?
(c) What is the proper time interval measured along an observer's world line?
(d) Writing the line element as a matrix with components $g_{\mu\nu}$, compute the inverse matrix with components $g^{\mu\nu}$.

(5.4) Consider a metric line element

$$\mathrm{d}s^2 = -\mathrm{d}t^2 + a(t)^2 \left[\frac{\mathrm{d}r^2}{1 - kr^2} + r^2 \left(\mathrm{d}\theta^2 + \sin^2\theta\mathrm{d}\phi^2\right) \right], \tag{5.70}$$

where k is a constant.
(a) What is the proper length between events that occur at (r, θ, ϕ) and $(r + \mathrm{d}r, \theta, \phi)$?
(b) A light pulse is sent from $(t_{\mathrm{em}}, \chi, 0, 0)$ to $(t_{\mathrm{ob}}, 0, 0, 0)$. Show that the photon's path can be described by

$$\int_{t_{\mathrm{em}}}^{t_{\mathrm{ob}}} \frac{\mathrm{d}t}{a(t)} = \int_\chi^0 \frac{\mathrm{d}r}{(1 - kr^2)^{\frac{1}{2}}}. \tag{5.71}$$

(5.5) Consider again the metric line element for Rindler space

$$\mathrm{d}s^2 = -x^2 \mathrm{d}t^2 + \mathrm{d}x^2 \tag{5.72}$$

(a) What is the proper length interval between events at $x = x_a$ and $x_a + \mathrm{d}x$?
(b) What is the proper time interval between events at t_b and $t_b + \mathrm{d}t_b$?

Part II

Curvature and general relativity

In this part of the book, we introduce the tools needed to understand the curvature of spacetime and its relationship to matter, culminating in the Einstein field equation.

- In Chapter 6, we introduce some of the key ideas behind general relativity, and in particular the *equivalence principle.*

- In Chapter 7, we describe the notion of what it means for vectors to be parallel in curved spaces. We introduce *connection coefficients* which allow us to take derivatives of vectors in curved space.

- In Chapters 8 and 9, we investigate geodesics: the paths that particles fall along in spacetime.

- Although we study curved space, we make physical observations locally, in the flat space of our experience. The method to translate between different frames of reference is described in Chapter 10.

- In Chapter 11, we explain how curvature of spaces are described using the *Riemann tensor* and the *Ricci tensor.* These will supply the left-hand side of the Einstein equation.

- The right-hand side of Einstein's equation is supplied by the energy-momentum tensor, discussed in Chapter 12.

- In Chapter 13, we write down the *Einstein field equation*, which is the foundation of general relativity.

- In Chapter 14, we review some of the successes of general relativity that follow from the formalism described in this part of the book. Many of the topics described in this chapter will then be unpacked in more detail in the rest of the book.

6 Finding a theory of gravitation

A little reflection will show that the law of the equality of the inertial and gravitational mass is equivalent to the assertion that the acceleration imparted to a body by a gravitational field is independent of the nature of the body... It is only when there is numerical equality between the inertial and gravitational mass that the acceleration is independent of the nature of the body.

Albert Einstein

Newtonian gravitation acts instantaneously across the Universe and therefore picks out a unique time for all observers when the gravitational interaction occurs. This is inconsistent with relativity and its treatment of simultaneity. We must therefore look beyond Newton's theory for a complete description of gravitation. We begin our search with two claims. (1) A person falling under gravity can't feel their own weight. This is a statement of the principle of equivalence. (2) The metric alone describes the role of spacetime in the laws of physics. This is an expression of general covariance. Taken together, these two ideas, which turn out to be closely linked, will guide us towards a theory of gravitation.

6.1 Free fall and the equivalence principle

Central to general relativity is the notion that the inertial mass m_i of a particle and its gravitational mass m_g are identical. If we write an equation of motion for a particle in a gravitational field as

$$m_i \ddot{\vec{x}} = m_g \vec{g}(\vec{x}), \tag{6.1}$$

then, taking $m_i = m_g$, the equation of motion tells us that the gravitational field $\vec{g}(\vec{x})$ acting on a particle is equal to the acceleration $\ddot{\vec{x}}$ of the coordinates of the particle. Einstein's great insight was to grasp that gravity and an accelerating coordinate system are actually *the same thing*. A freely falling[1] astronaut, sees, by definition, no change in her coordinates in her local rest frame and so concludes that there is no gravitational field. To put it another way, a freely falling observer cannot feel their own weight.

The **weak principle of equivalence**[2] is a statement that gravitational and inertial mass are identical. This implies that it is possible to

---

[1] **A freely falling body** is one that experiences only the effects of gravity.

[2] This is sometimes called the principle of weak equivalence, but we will take the view that the weakness is an attribute of the principle. The corresponding strong principle of equivalence will be introduced on the following page. Why is this principle *weak*? Because it only applies to mechanical forces.

choose a coordinate system in which the laws of motion of a freely falling particle take the same form as in unaccelerated Cartesian coordinates in the absence of gravity. An observer O on Earth and her freely falling astronaut friend O' detect no difference in the laws of *mechanics*, except that O observes the effect of, and herself feels, a gravitational field, while freely falling O' does not.[3]

Example 6.1

Consider a cloud of N test particles,[4] each with mass m, subject to a uniform gravitational field $\vec{g}$. Let's fix our attention on one particular particle, which will have an equation of motion of the form

$$m\frac{\mathrm{d}^2\vec{x}}{\mathrm{d}t^2} = m\vec{g} + \sum_{p=1}^{N-1} \vec{F}(\vec{x} - \vec{x}_p), \tag{6.2}$$

which is to say that it feels the force of gravity and the non-gravitational interaction forces from the $N-1$ other particles. This equation of motion describes the dynamics within the coordinate system of a particular observer O. Next we make the coordinate transformation to another frame of reference that is uniformly accelerating with acceleration $\vec{g}$ in the $-x$-direction. The coordinate transformation is

$$\vec{x}' = \vec{x} - \tfrac{1}{2}\vec{g}t^2, \quad t = t'. \tag{6.3}$$

These coordinates describe the viewpoint of the freely falling observer O'. The equation of motion becomes

$$m\frac{\mathrm{d}^2\vec{x}'}{\mathrm{d}t'^2} = \sum_{p=1}^{N-1} \vec{F}(\vec{x}' - \vec{x}'_p), \tag{6.4}$$

which looks like the equation of motion for the particle in the absence of the gravitational field $\vec{g}$.

Example 6.2

The principle of equivalence can also be illustrated by considering the acceleration measuring device in Fig. 6.1. It consists of a mass in a box suspended by springs. If the box is carried by an observer then it provides a measure of acceleration: an observer accelerating in the positive x-direction, will see the mass displaced in the negative x-direction. The mass will be also affected by the presence of a gravitational field, which also causes a displacement, but there is no way to distinguish this from the effect of an acceleration.

We can strengthen the equivalence principle to generalize it beyond the realm of mechanics. One form of the **strong principle of equivalence** says that it is possible to find a frame of reference where *all* non-gravitational laws of physics take on their special relativistic forms. In more detail:

The strong principle of equivalence: at every spacetime point in an arbitrary gravitational field it is possible to choose a local coordinate system such that, within a sufficiently small region of the point in question, *all* of the laws of nature take the same form as in unaccelerated Cartesian coordinate systems in the absence of gravitation.

[3]There are some caveats. One is that no difference is detected over a small region of space and time. The size of the region is described below.

[4]In this chapter, test particles have a low mass compared to the source of the gravitational field. As a result, we neglect any gravitational interaction between them.

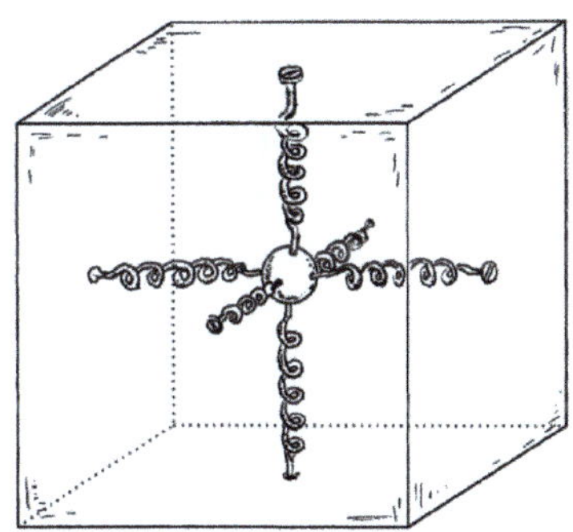

Fig. 6.1 A machine to measure acceleration consisting of a mass suspended in a box by light springs.

[5]Tidal forces arise due to the difference in gravitational field strength across a body. The gravitational field due to the Sun and Moon varies across the Earth and this results in a tidal force on our planet and its oceans. The tides (the rise and fall of sea level due to the motion of the Moon and the Sun) originate due to the greater responsiveness to tidal forces of fluid water (oceans) than solid rock (the Earth). Even the solid Earth responds a little bit to tidal forces (the so-called 'Earth tide') and this is responsible for the Large Electron-Positron Collider at CERN stretching a few millimetres from its circumference of about 27 km as the Earth stretches, something the CERN scientists have to correct for.

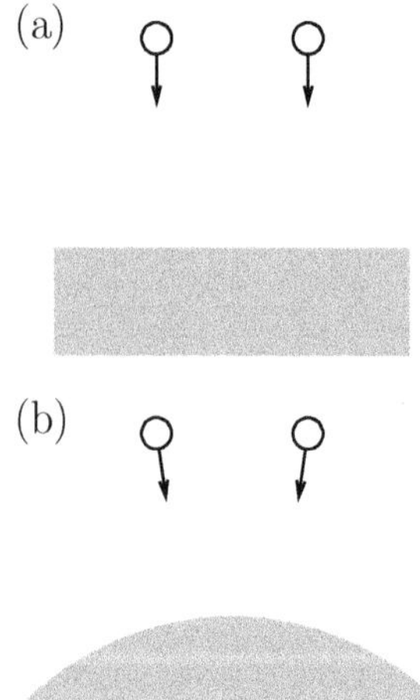

Fig. 6.2 (a) In a uniform field, two test particles released along parallel paths at different positions will both accelerate in the same direction. A coordinate transformation can be used to remove this acceleration. (b) A real gravitational field, such as that generated by a planet, is never uniform and so test particles initially released along parallel paths will approach each other. The apparent force causing acceleration between them is called a tidal force.

Example 6.3

Special relativity tells us that a particle at rest in an inertial frame moves along the time axis. The strong principle of equivalence tells us that the same must be true in general relativity, so that free-falling particles follow a curve whose tangent vector is always timelike (known as a timelike curve). Such timelike curves are called **geodesics**. The equivalence principle has ended up being surprisingly powerful because it has allowed us to take a result from special relativity (in which gravity is completely absent) and deduce from it an important result in general relativity (which includes gravity): *free-falling particles follow timelike curves!*

Let's emphasize an important idea: at the particular point where the observer is localized, there is no way to distinguish between a gravitational field and acceleration. Therefore, we can transform away the apparent effect of acceleration to give the physics expected from special relativity *in the absence of gravity*. As a result of the strong principle of equivalence, no experiment can distinguish between a **homogeneous** gravitational field and an accelerating reference frame. That is, if all points experience a uniform gravitational field (as they did in Example 6.1), then we can find a coordinate system where we transform away the equivalent effects of the gravitational field and acceleration for all points, so that it appears that gravity is not acting. In less technical language: *a uniform gravitational field is equivalent to there being no gravitational field*. This would seem to make hunting for the effects of gravitation a hopeless endeavour, since we could never be sure that an apparent gravitational effect wasn't simply the effect of an accelerating coordinate system. All is not lost, however, because real gravitational fields are never homogeneous! Over sufficiently large distance, a real gravitational field *can* be distinguished from an accelerating reference frame. Distinguishing them can be done by noticing the presence of **tidal forces**,[5] as shown in the following example.

Example 6.4

An observer awakes in a spaceship feeling the apparent effect of gravity holding them in bed. Is this due to the acceleration of the ship or to the gravitational field of a nearby planet? A planet's field, assumed spherically symmetric, is not homogeneous, so can, in principle, be detected. The observer releases two test particles, initially moving parallel to each other (Fig. 6.2). If the particles start to move towards each other (or away from each other) we attribute this to a tidal force. The presence of a tidal force tells the observer that (s)he is not simply in an accelerating frame and therefore gravity must be acting.

Our investigation of the equivalence principle allows us to learn a number of lessons that will be very important in the rest of this book as we describe general relativity.

- **Lesson 1**: A freely falling observer does not feel any force and so can't tell if a gravitational field is present. They fall along a timelike curve called a geodesic.[6]

- **Lesson 2**: Locally, the freely falling observer can set up a laboratory covered by the unaccelerated coordinate system familiar from special relativity. This is an example of a **local inertial frame** (LIF).[7] All laws of physics described in a LIF are those from special relativity.

- **Lesson 3**: If measurements are made over a sufficiently small time frame and length scale, the observer can never detect the effect of gravitation. However, a real gravitational field will be inhomogeneous, so an observer can detect the effects of gravitation by making measurements at different points in spacetime that show the effects of tidal forces.[8]

Let's now get a quick physics payoff from the equivalence principle and show that light is bent by a gravitational field.

[6]We can contrast the difference in philosophy between the Newtonian and Einsteinian views of gravitation. In Newtonian gravitation, the Sun exerts a force on the Earth; in Einsteinian gravitation, the Earth feels no force and simply falls freely along a geodesic which is a path that orbits the Sun.

[7]If orthonormal basis vectors are used to describe a LIF, as in the case of the usual conventions of special relativity, the frame is sometimes called a local Lorentz frame.

[8]This notion of 'sufficiently small' provides the caveat for the earlier sidenote.

Example 6.5

Consider the experiment shown in Fig. 6.3(a) in which a rocket accelerates upwards with an acceleration g. A particle passes through the windows of the rocket and travels through the interior of the rocket, but by the time it has passed through the width of the rocket, the rocket has moved upwards and so it ends up landing on the rocket wall at a position which is lower than the point at which it entered. Figure 6.3(b) illustrates that from the point of view of an astronaut inside the rocket the photon follows a curved trajectory.[9] The parabolic path would follow the equations $x = x_0 + vt$, where v is the velocity of the particle, and $y = y_0 - \frac{1}{2}gt^2$, where x and y are the coordinates [horizontal and vertical respectively, in Fig. 6.3(b)] measured in the rocket frame and (x_0, y_0) are the coordinates of the point where the particle enters the rocket at $t = 0$. Thus, $y = y_0 - gx^2/(2v^2)$.

However, so far, this analysis has shown nothing remarkable. But now we can deploy the equivalence principle, which implies that the astronaut cannot tell whether her rocket is accelerating upwards, with an acceleration of g, or simply that she is experiencing a gravitational field equal to g. So it is possible that her rocket is parked on a planet with a gravitational field of g and the bending of the particle trajectory would still occur. We can make the experiment particularly vivid by imagining that the particle is a photon, so that it would suggest that light is bent by gravitational fields! Our analysis shows that, for our rocket problem, the light beam changes from purely horizontal to travelling at an angle of $\approx |dy/dx| = gx/c^2$ to the horizontal.

This argument is only part of the story however. It is quantitatively correct for a particle beam travelling at a non-relativistic velocity, but it turns out that the bending effect for photons in the field of a spherical mass is actually *three times larger* than suggested by this analysis. The problem for the photon case (or any particle travelling at relativistic speeds) is in the incorrect assumption that our experiment is carried out over a sufficiently small distance to allow our straightforward application of the equivalence principle. However, the intuition that light should bend is correct, even if the size of the bending is not, and our argument does account for one third of the total deflection effect.

[9]This is nothing particularly to do with special relativity (the rocket is only starting to accelerate and so its speed is much less than c) and the same effect would be seen with an accelerating car in vertical rain (with the diagram rotated).

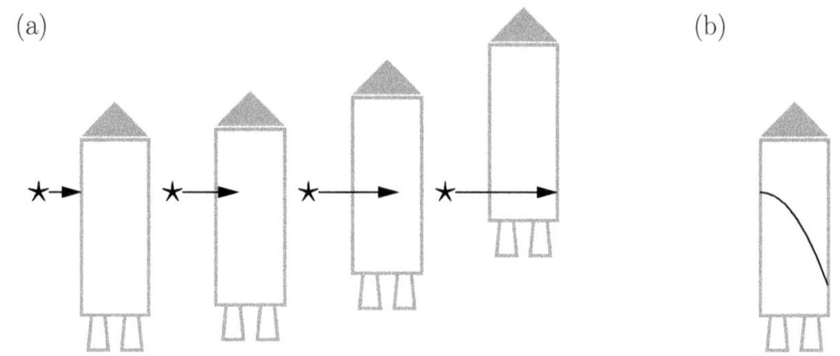

(a)
(b)

Fig. 6.3 (a) Time snapshots in an inertial frame in which a rocket accelerates (upwards in this diagram). A particle passes through the windows of the rocket in a direction perpendicular to the direction of acceleration of the rocket. (b) In the rocket's frame, the path of the particle follows a parabolic trajectory.

We'll see later how gravitation follows from the curvature of spacetime encoded in the components of the metric tensor. This curvature can affect intervals in space and also in time; the argument presented above only takes the time part into account. However, the measurement of the effect for (highly relativistic) photons in a gravitational field from a spherically symmetric mass requires us to consider the space part too, since the deflection from this enters at the same order as that of the time part, owing to the photon exploring a spatial distance large enough to experience the spatial contribution to the curvature. The point is that this measurement is made over a distance where the curvature of space can be discerned, and is therefore not local enough to allow the straightforward application of the equivalence principle. If you treat the spatial part of the curvature as well, you recover the factor of three missing from the present analysis.

In spite of these difficulties, let's now use our result to have a first, hand-waving attempt at the famous problem of the bending of starlight around the Sun, visible during a total solar eclipse.[10] We will treat the problem properly later, but we will here make a crude estimate where we discard numerical factors with wild abandon. Starlight passing close to the surface of the Sun will experience a gravitational field equal to $g = GM_\odot/R_\odot^2$. This will be when the most bending will occur, but there will also be bending when the light is further away. Crudely, we can say that the starlight will be bent over a distance which must scale as $R_\odot$ and the gravitational field that it will experience will be of order $g = GM_\odot/R_\odot^2$. Therefore, the angle of deflection θ (using our previous result that light is bent by an angle given very roughly by $|\mathrm{d}y/\mathrm{d}x| = gx/c^2$) will be

$$\theta \approx \frac{gR_\odot}{c^2} = \frac{GM_\odot}{R_\odot c^2}. \tag{6.5}$$

Remarkably, this turns out to be the correct answer apart from a factor of 4.

[10]This was the effect that was measured very early in the history of relativity and gave an initial vindication of Einstein's ideas.

↳ **Section 24.1 of Chapter 24 presents the calculation done properly.**

We can now attempt to find gravitational fields by transforming away all effects of uniform acceleration from our coordinate systems and conclude that whatever is left over at the end must be gravity. However, we still lack a clear guide to help us formulate a theory of relativistic gravitation. Before moving on, we present a brief historical interlude. Einstein's own motivation in invoking the principle of equivalence is often explained in terms of the influence of **Mach's principle** on his thinking, as discussed in the next example.

↳ **Readers eager to get on to the features of General Relativity can skip to the next section at this point.**

Example 6.6

Newton's laws appear to have restricted applicability: they apply in inertial frames. In non-inertial frames, new **inertial forces** appear.[11] These present a philosophical difficulty if we accept the principle of relativity, since they imply that mysterious new forces appear in non-inertial frames, but it is not immediately clear if these are exerted by space itself or by other bodies.

How do we define an inertial or non-inertial frame if there is no absolute space against which to judge whether the frame is accelerating or not? Ernst Mach[12] came up with a solution that was a great influence on Einstein. Mach says we can judge a frame to be inertial if it is unaccelerated relative to some suitable average of all of the motion of all of the matter in the Universe. We say an inertial frame is unaccelerated with respect to *fixed stars*.[13] This is helpful in that it allows us to use relativity to propose that inertial forces arise because of the acceleration of a mass relative to this fixed frame or, equivalently, the acceleration of the fixed frame with respect to the mass. This saves Newton's laws: they apply in *all* frames of reference with the extra non-inertial forces being real, physical forces that arise from the motion of the stars. We can then attempt to formulate a theory of inertial forces as arising from the interaction of inertial masses. By analogy with electromagnetism this would have a static contribution like Coulomb's law that varies according to $\alpha m_i m_j / r^2$, with α a constant. It might also have more complicated parts that contribute. For example, when masses are accelerating relative to each other[14] we would expect a force of the form $\beta m_1 m_2 / r$, with β a constant.

Now, if we accept the principle of equivalence, then (the static part of) the inertial interaction can be directly identified with the gravitational force. The equivalence principle is then a principle of the equivalence of the gravitational force and the static part of the inertial force. We would then conclude that the relative acceleration of masses gives rise to a $1/r$ contribution to the force. This does not exhaust the possible contributions to the inertial force, however, and so we should expect further contributions to the inertial part of the force law.

So why not continue along this line and describe gravitation by considering the interactions between particles in this way? The reason is that we are faced with a very complicated nonlinear problem. This is because mass and energy are interchangeable, and mass-energy is the source of gravitation. For example, when two particles act as a source of gravitation then to correctly evaluate the total size of the source of the gravitational effect, we must compute not only their individual gravitational potential energies but also the interaction potential energy between them. Each of these makes a contribution to the gravitational force. Owing to this complication of nonlinearity, we must abandon the discussion in terms of the interaction of individual masses. Fields then become important as they simplify our task. They allow us to avoid the question of how particles directly influence each other. Instead, a particle contributes to, and is acted on by, a field that is defined locally. In the field viewpoint, we therefore restrict our view to a local point in spacetime, and compute the effect of gravitation at the point of interest.[15]

[11] An example is the centrifugal force experienced in a rotating frame. See Chapter 8 for a discussion of inertial forces.

[12] Ernst W. J. W. Mach (1838–1916)

[13] '*When the subway jerks, it's the fixed stars that throw you down.*' Ernst Mach, attributed by Philipp Frank (1884–1966).

[14] An accelerating charge in electromagnetism exerts a force that varies as $1/r$. This is closely related to the interaction that gives rise to the emission of electromagnetic radiation from an accelerating charge.

[15] Even using fields, nonlinearity continues to make the problem a complicated one. This can be made clearer by considering the contrasting case of electromagnetism. In electromagnetism, charges are the sources of electromagnetic fields and these fields add linearly. The charges determine the fields and the field determines the motion of the charges. An electromagnetic wave passes though another electromagnetic wave without scattering precisely because the theory is linear: the (electrically neutral) waves are not sources of electric field. This is not the case for gravitation. For example, gravitational waves interact with other gravitational waves since a gravitational wave is also a source of gravitation.

6.2 Why *general* relativity?

Einstein's theory of gravitation is called *general* relativity because it is founded on the principle of general covariance.[16]

The principle of **general covariance**: the laws of physics must be invariant under all coordinate transformations, so the laws must hold in different coordinate systems.

This principle is, in a way, a restatement of the equivalence principle, but framing it this way is very helpful because it puts constraints on the

[16] As pointed out by several authors, 'general relativity' is something of a misnomer: it doesn't describe a form of relativity that is more general than special relativity.

[17]The law

$$F = \frac{\mathrm{d}p}{\mathrm{d}t},$$

is a relationship between two geometrical quantities (in this case vectors) and no preferred basis is singled out. On the other hand, if the equation were something like

$$F = p\frac{\mathrm{d}p_y}{\mathrm{d}t},$$

we would smell a rat, not only because it's not dimensionally correct but because it somehow makes the y-direction special and so it wouldn't transform sensibly.

[18]Indeed, we could attempt to regard this equation as a component of a more general tensor equation. As we will discover later, the left-hand side of the correct equation will turn out to be something to do with spacetime derivatives (but we will need a more sophisticated notion of curvature to get this right) and the right-hand source term is indeed something to do with the mass density, but in fact we need the energy-momentum tensor introduced in Section 4.5.

[19]A valid tensor equation can often be identified as having a component version whose indices match on both sides of the equation. Such an equation is sometimes called **manifestly covariant**.

[20]By $\mathrm{diag}(-1,1,1,1)$ we mean a 4×4 matrix with these diagonal elements and all other elements equal to zero.

sorts of equations we can write down that will be physically valid. For a start, we can't have any preferred bases in our theory (any more than in Newtonian mechanics[17] could claim that the y-axis was particularly special). This means that we can't write down a law which singles out some specific components of a vector or a tensor, for the simple reason that this would look very different in different frames of reference.

Example 6.7

Failed theory of gravitation: Based on what we have learned in the book so far, here is a misguided attempt at a theory of gravitation and then an explanation of why it won't work. We know Newton's approach to gravitation ended up with $\vec{\nabla}^2\Phi = 4\pi G\rho$ (eqn 15 of Chapter 0), so let's try and fix it up by using the 4-vector generalization of $\vec{\nabla}^2$ which is $\partial^2 = \partial_\mu\partial^\mu = -\partial_t^2 + \vec{\nabla}^2$. So our patched up theory of gravitation could be written down as

$$\partial^2\Phi = -\frac{1}{c^2}\frac{\partial^2\Phi}{\partial t^2} + \vec{\nabla}^2\Phi = 4\pi G\rho. \tag{6.6}$$

This equation looks like a worthy guess, but it fails at the first hurdle — general covariance! It doesn't work in different frames of reference. The source term ρ is not a scalar and will transform when you go into different inertial frames. So this equation will not do, but it is heading in the right direction.[18]

The equivalence principle has told us that any physical law that can be expressed in special relativity without gravity will hold in a LIF, even when gravity is present. General covariance then tells us that the law holds in the same form in different coordinate system. We choose to express our laws using tensors and so a valid tensor equation[19] compatible with special relativity also holds in a LIF even when gravity is present, and this same equation should take the same form in any coordinate system in the presence of gravity. In short: *a valid tensor equation in the absence of gravity is a valid tensor equation in the presence of gravity.* But something must change in the mathematics to herald the presence of gravitation! That something is *geometry*. The geometry of spacetime is altered by the presence of gravity, causing it to become curved. This is expressed in terms of the metric tensor. In the LIF of special relativity, the metric tensor, which can be thought of as a box of clocks and rulers that tells us how to measure vectors, is given by the Minkowski metric tensor $\boldsymbol{\eta}$, whose components are[20] $\mathrm{diag}(-1,1,1,1)$. However, in a general frame of reference, the metric is written as a tensor $\boldsymbol{g}$, whose components (i) are different to those of $\boldsymbol{\eta}$ and (ii) will vary in space and time. This provides our next lesson.

- **Lesson 4**: The effect of gravitation on our tensor equations describing physical laws is to change the Minkowski metric $\boldsymbol{\eta}$ to a new metric $\boldsymbol{g}$.

Our strategy to find physical theories in the presence of gravitation will be to take a valid tensor equation that works in special relativity and upgrade it, simply by changing the Minkowski metric $\boldsymbol{\eta}$ to the general metric tensor g. The next step in our search for a theory of gravity is therefore to seek a theory that determines this metric g in the presence of gravitation and also obeys the principles of equivalence and of general covariance.

6.3 A differential equation to describe gravity

The metric is a rulebook that allows us access to the distances and angles between points in spacetime. For examples, the interval in spacetime along a curve $x^\mu(\lambda)$ from $\lambda = a$ to $\lambda = b$ may be worked out using the metric via the prescription

$$s = \int_a^b \mathrm{d}\lambda \left| g_{\mu\nu} \frac{\mathrm{d}x^\mu}{\mathrm{d}\lambda} \frac{\mathrm{d}x^\nu}{\mathrm{d}\lambda} \right|^{\frac{1}{2}}. \tag{6.7}$$

In the absence of a gravitational field, we might evaluate this path length in any number of coordinate systems, but always obtain the same answer to questions such as the ratio of a circle's circumference to its diameter being π, or the angles in a triangle adding up to π. The only way in which this is possible is if g at one point is related to g at another point. This implies that the tensor g, a function of position x,[21] should satisfy a differential equation. It is this differential equation, the equation that tells us how the metric $g(x)$ varies in spacetime, that allows us to separate the true effects of gravity, from those effects that result from a particular choice of coordinates. From this point of view, the metric is a field. We can think of a field[22] as a machine into which we input a position in spacetime $x^\mu = y^\mu$. The field outputs the value of the tensor $g(y)$ appropriate for the point y. The field g must obey a differential equation of motion that we'll call a **field equation**.

The equation in question arranges the components of g and their derivatives into a new tensor that describes the *curvature* of spacetime[23] called the **Riemann tensor R**. It is curvature that is the true effect of gravitating mass and which can never simply be the results of having chosen a perverse set of coordinates. It will transpire that non-zero components of R tell us about curvature, and curvature means gravitation.

Despite these pointers, we still have relatively little guidance on how to put together the field theory of gravitation. However, there is one other clue: the theory must be **compatible** with (i) Newtonian gravitation and (ii) with special relativity. That is to say that, in the limit of weak gravitational fields and low velocities, the predictions of the field theory must recreate those of Newton's universal theory of gravitation. In the limit of vanishing gravitational field, the theory must agree with special relativity. Conceptually, therefore, the field theory of gravity (that is, general relativity) fits into the scheme shown in Fig. 6.4. We can summarize this section as follows:

[21]We will usually write tensor fields in the form $T(x)$, showing that they are functions of position x. This position could be a general point in spacetime $\mathcal{P}$ or the point specified by its coordinates $x^\mu(\mathcal{P}) = (t(\mathcal{P}), x(\mathcal{P}), y(\mathcal{P}), z(\mathcal{P}))$.

[22]As a result of the geometric nature of gravitation, embodied in the exchange $\eta \to g$, it is unhelpful to talk about a gravitational field in the sense of a force field permeating all spacetime. However, gravitation is a field theory: the field in question is the metric itself.

[23]We will define this properly later in Chapter 11.

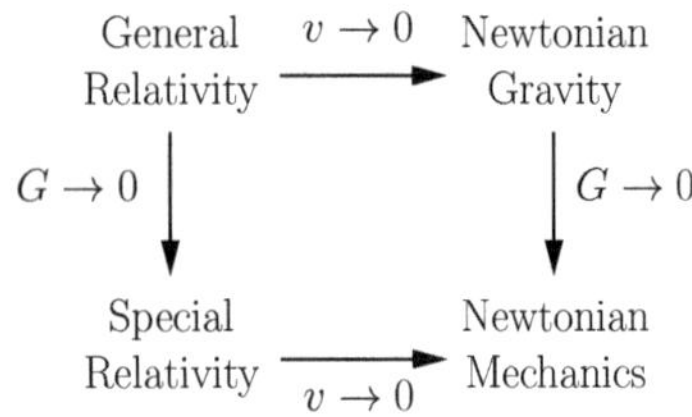

Fig. 6.4 The relationship between general relativity, special relativity, Newtonian gravity and Newtonian mechanics as a function of velocity v and gravitational constant G.

[24]The observer will need to set up their local frame to have the Lorentz signature [i.e. the $(-1, 1, 1, 1)$ pattern of signs on the diagonal of the Minkowski metric.]

[25]There is a distinction between a frame of reference and a set of coordinates. A frame of reference is defined by some basis *vectors* and so has an existence independent of a coordinate system.
• A general LIF will be covered in coordinates for which $g_{\mu\nu} (x = x^\alpha(\mathcal{P})) = \eta_{\mu\nu}$ and $\partial g_{\mu\nu}/\partial x^\alpha|_{x=x^\alpha(\mathcal{P})} = 0$ in an infinitesimal region around the point $(t(\mathcal{P}), x(\mathcal{P}), y(\mathcal{P}), z(\mathcal{P}))$. This can be achieved using **Riemann normal coordinates**, described in Chapter 35.
• The freely falling frame has its time direction e_t tangent to a geodesic. It remains freely falling as a function of the local (proper) time so continues to be a LIF for the whole time it is falling. In terms of coordinates, it is only flat in a very small spatial region around the origin of the frame, but for long intervals of proper time.

• **Lesson 5**: General relativity provides a field theory of the metric field $g(x)$ which encodes gravitation through its effect in providing a curvature to spacetime.

6.4   Local flatness

The Minkowski metric can be used by any observer who is not subject to a gravitational field. Such a region of spacetime has no curvature and so Minkowski spacetime is **flat**. By the principle of equivalence, a freely falling observer does not feel a gravitational field providing they make measurements over a small enough region of spacetime. This implies that locally, all observers can treat spacetime as flat if they use a LIF. This is the content of the **local flatness theorem**.

It is always possible to reduce a metric field $g(x)$, evaluated at a single point $x = \mathcal{P}$, to the Minkowski metric η. That is, we can introduce coordinates $x^\alpha(\mathcal{P})$ such that the components of the tensors obey the equation

$$g_{\mu\nu} (x = x^\alpha(\mathcal{P})) = \eta_{\mu\nu}. \tag{6.8}$$

This is possible since g is represented by a symmetric 4×4 matrix which can always be diagonalized. This implies that there are potentially lots of local frames (i.e. not just the special freely falling LIF we have described so far) that appear flat at a single point and, in these frames, the observer uses the Minkowski metric of flat spacetime to manipulate vectors.[24] The point of the local flatness theorem is that it is also possible to find coordinates such that the derivatives of the metric vanish at this point

$$\left. \frac{\partial g_{\mu\nu}}{\partial x^\alpha} \right|_{x=x^\alpha(\mathcal{P})} = 0. \tag{6.9}$$

This means that spacetime will be described by the Minkowski metric in an infinitesimal region around $x = x^\alpha(\mathcal{P})$, making the notion of local flatness mathematically respectable. A local inertial frame (LIF) is a frame where this requirement is satisfied at some point $x = \mathcal{P}$. Note that in the presence of gravity it is not possible to find a vanishing *second* derivative of g, since second derivatives will turn out to be related to spacetime curvature.[25]

The freely falling frame of reference (used by the falling observer) that we have described in this chapter is an important example of a LIF. LIFs are very useful in that (i) as we've said, the laws of physics are identical in LIFs and in general frames, subject to the change $\eta \to g$; and (ii) analysing physics in LIFs is invariably far easier than doing so in curved spacetime.

We shall also use the idea that we can straightforwardly identify frames in which $g = \eta$ at the origin. By design, an observer erects an orthogonal set of basis vectors where they are situated, and normalizes this basis. The reason we're interested in these **local orthonormal frames** is that observations and measurements can be thought of as being made in them. This provides the final lesson in this chapter:

- **Lesson 6**: An observation is made and interpreted by an observer in a local orthonormal frame, who uses the Minkowski tensor at the point they inhabit in spacetime.

The picture to have in mind is of the observer in a laboratory using their set of orthonormal axes, constructed from short, rigid rods, as an instrument to interpret the components of vectors locally.

6.5 Time dilation in a gravitational field

Our discussion so far has been very general and, consequently, a little abstract. We have yet to see how the curvature of spacetime that encodes gravity via the metric g has any effect beyond the possibility of Newtonian-style gravitational attraction. To give an idea of how g affects measurements we shall conclude this chapter by illustrating the influence of the metric on measurements made by two distant observers in an inhomogeneous gravitational field. Here gravitation leads to **time dilation** and a **gravitational shift** of the frequencies of light signals.

Example 6.8

Consider a clock at rest in some coordinate system. The clock ticks (which we will take to be infinitesimally separated by a coordinate interval dt) will then be separated by the proper time interval $d\tau$ where

$$-d\tau^2 = g_{\alpha\beta}\, dx^\alpha dx^\beta = g_{00}\, dt^2, \tag{6.10}$$

where we have upgraded the flat metric $\eta_{\alpha\beta}$ to the curved-space metric $g_{\alpha\beta}$. If the clock were sitting 'at infinity', well away from any sources of gravitational field, it would indeed be in flat space, so $g_{00} = \eta_{00} = -1$ and $dt = d\tau$. However, if the clock was a distance r from a star of mass M then we could use the metric of eqn 5.22 (assuming the Newtonian limit holds) and hence $g_{00} = -(1 - 2GM/r)$. In this case,[26]

$$d\tau = (-g_{00})^{1/2} dt = \left(1 - \frac{2GM}{r}\right)^{1/2} dt. \tag{6.11}$$

That is say that the interval $d\tau$ between ticks of a clock measured in a particular frame depends on the details of the metric and hence, on the gravitational field. Since $(1 - 2GM/r)^{1/2} < 1$, we have $d\tau < dt$ and so this is gravitational time dilation. Note that the factor $2GM/r$ that enters this expression is just the square of the classical escape velocity v_{esc} at distance r from the star, which appears when you equate the kinetic energy $\frac{1}{2}mv_{\text{esc}}^2$ to the gravitational potential energy GMm/r for a test mass m. Thus we could write eqn 6.11 as $d\tau = dt\sqrt{1 - (v_{\text{esc}}/c)^2}$. If instead we wish to write this expression in terms of the **Schwarzschild radius**[27] $r_{\text{S}} = 2GM/c^2$, then we could write it as

$$d\tau = dt\sqrt{1 - \frac{r_{\text{S}}}{r}}. \tag{6.12}$$

[26] Remember that we are using units for which $c = 1$. The gravitational time-dilation factor is $[1 - 2GM/(c^2 r)]^{1/2}$ if you put the factors of c back in.

[27] This quantity will be discussed in detail in Part IV of the book.

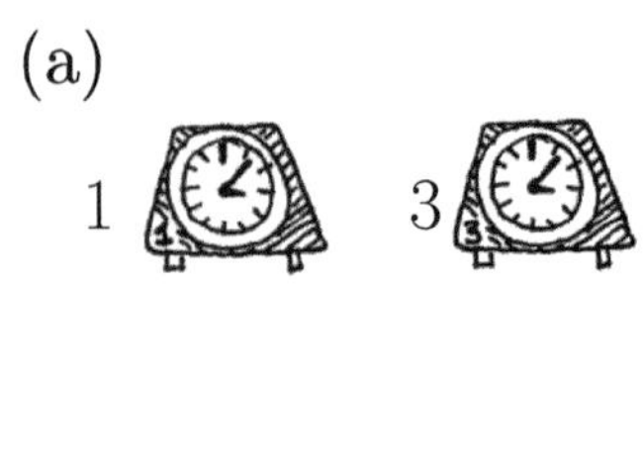

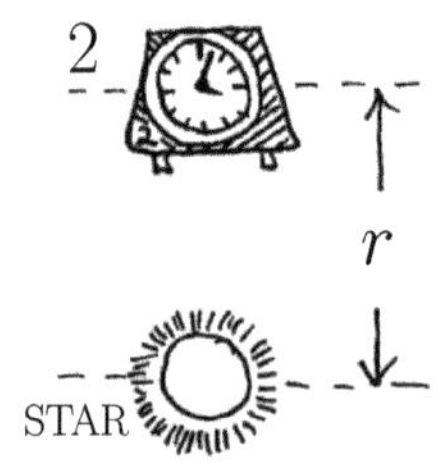

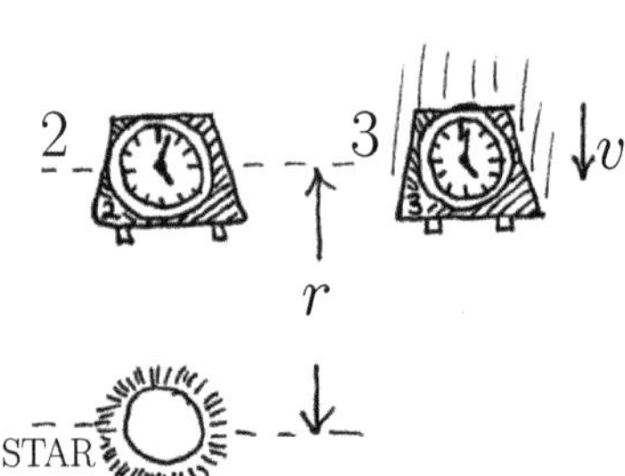

Fig. 6.5 Identical clocks 1 and 2 are held fixed, a long way from the star and at distance r respectively. (a) A third clock is released when it is next to clock 1. (b) Clock 3 is travelling at speed v by the time it ends up next to clock 2. The diagram is schematic, so all three clocks and the star should be in a straight line (and of course the star will be much bigger!).

Let's derive eqn 6.11 a different way. Consider two identical clocks, the first held well away from the star and a second at a distance r from it (see Fig. 6.5). Take a third identical clock, to measure the first two, and release it at the position of clock 1 [see Fig. 6.5(a)]. Clock 3 is in free-fall in the gravitational field of the star, and so the interval between its clicks can be taken to be $\mathrm{d}\tau$ in its LIF. Immediately after releasing clock 3, we find that clock 1 and clock 3 agree (their ticks are in sync) because clock 3 is initially barely moving and so no relativistic corrections are needed (just as we found above: $\mathrm{d}t = \mathrm{d}\tau$). However, clock 3 starts to accelerate towards the star as it drawn inexorably towards it and by the time it reaches clock 2 it will be moving much faster, let's say at a speed v [see Fig. 6.5(b)]. We are in the Newtonian limit so its kinetic energy $\frac{1}{2}mv^2$ has been obtained from releasing gravitational potential energy GMm/r, implying that $v^2 = 2GM/r$. Because clock 3 is instantaneously in free fall, gravity is absent in its inertial reference frame and so it has the same time interval between clicks $\mathrm{d}\tau$ as it had previously. However, the interval between ticks for clock 2 will be $\mathrm{d}t = \gamma\,\mathrm{d}\tau$ where $\gamma = (1 - v^2)^{-1/2}$, so that once again

$$\mathrm{d}\tau = \left(1 - \frac{2GM}{r}\right)^{1/2} \mathrm{d}t. \tag{6.13}$$

We have centred our discussion around clocks, but we could have framed the argument around atoms emitting light of a well-defined frequency due to some atomic transition. In this case, the frequency of the detected light from an atom in a gravitational field is found to be lower than that from the process occurring at infinity. In terms of wavelength, the light has been shifted towards the red end of the spectrum and, as a result, we call the effect **gravitational redshift**.

Bear in mind though that the calculation in this example was carried out in the Newtonian limit, meaning that $GM/Rc^2 \ll 1$, and so one needs to check that this limit holds before using eqn 6.11 to perform a calculation. However, if this limit does hold we can simplify eqn 6.11 and write the time dilation factor as $1 - GM/R$ using the binomial theorem to expand the square root.

In Exercise 6.1, you can put numbers into these formulae, but suffice to say that the effect of gravitational redshift between an observer on the Earth's surface compared to one in deep space is negligible. For an observer on the surface of a neutron star (whose radius might only be 10 km, but the mass could be something like $1.4M_\odot$) the effect is extremely significant. However, even though the effect on Earth is extremely tiny, it is needed to take into account for the proper working of the satellite navigation methods based on the Global Positioning System (GPS). This relies on accurate timing of signals coming from a network of satellites and received by an observer who wants to know where on the Earth's surface she is. Relativistic effects need to be taken into account for this to be accurate, first because the satellites are in motion with respect to the ground-based observer (special relativity correction) and second because the satellites experience a lower gravitational field than the ground-based observer (general relativity correction due to the gravitational redshift).

This is the second form of redshift we have encountered. The first (seen in Exercise 4.8) was due to the Doppler effect in flat spacetime. We will encounter a third form of redshift when we discuss cosmology; that form results from the expansion of spacetime itself over very large distances. Gravitational and cosmological redshift are effects due to the change in metric in spacetime, which is different to the special relativistic Doppler effect which is due to the velocity of sources and observers.

In the next chapter, we will continue to explore the properties of curved spacetime and find out how to define a derivative of a vector in a generally covariant way. As we have learnt in this chapter, it is only generally covariant quantities that will be admissible in any theory that aspires to describe the physical universe.

Chapter summary

- The principle of equivalence tells us that in every local inertial frame *all* non-gravitational laws of physics must take on their *special relativistic* forms.
- The principle of general covariance tells us that laws must be preserved in different coordinate systems.
- We have used these principles to describe time dilation in a gravitational field.

Exercises

(6.1) Estimate the gravitational redshift (the factor by which a clock in a gravitational field runs slow compared to one subject to zero gravitational field) for the following cases: (a) a clock on the surface of the Earth; (b) a clock on the surface of the Sun; (c) a clock on the surface of a solar mass white dwarf with radius 10^3 km.

(6.2) A recent experiment uses clouds of ^{87}Sr atoms at around 100 nK, loaded into an optical lattice and operated as a sophisticated atomic clock [T. Bothwell *et al.*, Nature **602**, 420 (2022)]. It is possible to measure the gravitational redshift across the millimetre scale of this system, and the laboratory experiment gives a value of the frequency gradient of around $-1.0(2) \times 10^{-19}$ mm^{-1}. Is this consistent with what you would expect from general relativity?

(6.3) A satellite is in a circular orbit of radius r around a planet of radius R and mass m. Show that a clock on the satellite runs faster than a clock on the surface of the planet, located at one of the poles, by a factor of approximately $1 + GM/c^2[1/R - 3/(2r)]$. Hence show that there is one possible orbit radius for which the two clocks run at the same rate.
Hint: You not only need the gravitational time dilation but also the effect due to the satellite moving (i.e. the special relativity time dilation), which is (at least instantaneously) in a straight line.
Estimate the factor for a geostationary satellite orbiting around the Earth.

(6.4) Consider the Schwarzschild metric line element which describes the spacetime around spherically symmetric stars (here we have taken $G = c = 1$)

$$ds^2 = -\left(1 - \frac{2M}{r}\right)dt^2 + \left(1 - \frac{2M}{r}\right)^{-1}dr^2$$
$$+ r^2\left(d\theta^2 + \sin^2\theta d\phi^2\right). \tag{6.14}$$

(a) What is the proper time interval, measured by an observer at rest, between events at coordinate time t and $t+dt$ that both occur at a point (r, θ, ϕ)?
(b) Now consider two observers at rest in this spacetime. An atom undergoes an atomic transition at position (r_2, θ, ϕ). What is the time interval between two successive wavefronts measured at point (r_2, θ, ϕ)?
(c) What is the interval between wavefronts measured at r_2 from the experiment that takes place at r_1?

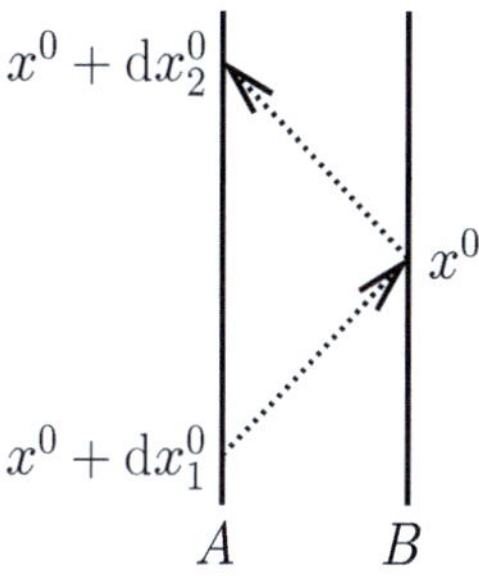

Fig. 6.6 Light signal send from A to B and back to A (Exercises 6.5 and 6.6).

(6.5) Consider a measurement of length involving a light signal being sent from point A to B and then back to A, as shown in Fig. 6.6. Multiplying c by the time that the observer at A measures for this process gives twice the distance between points.

(a) By considering the interval of coordinate time that elapses for a signal sent between A and B, show that we obtain

$$\mathrm{d}t = \frac{1}{g_{00}} \left\{ -g_{0i}\mathrm{d}x^i \pm \left[(g_{0i}g_{0j} - g_{ij}g_{00})\,\mathrm{d}x^i\mathrm{d}x^j \right]^{\frac{1}{2}} \right\}. \tag{6.15}$$

(b) What do the two roots correspond to?

(c) Show that the corresponding proper time interval measured by the observer at A for the signal to be sent and received back is

$$\mathrm{d}\tau = -\frac{2}{g_{00}} \left[(g_{0i}g_{0j} - g_{ij}g_{00})\,\mathrm{d}x^i\mathrm{d}x^j \right]^{\frac{1}{2}}. \tag{6.16}$$

(d) Show that this leads to a measured length interval of

$$\mathrm{d}l^2 = \left(g_{ij} - \frac{g_{0i}g_{0j}}{g_{00}} \right) \mathrm{d}x^i\mathrm{d}x^j. \tag{6.17}$$

If the g_{ij} depend on x^0 so that the spatial components are time dependent, it would not make sense to integrate this expression to obtain a general expression for proper time, since the integral would depend on the world line between the two points in space.

(6.6) Consider again the set up in Exercise 6.5 with light signals sent between A and B and call the time on B's world line when the light signal is received x^0. We define the time on A's world line that is simultaneous to this to be half way between emission and reception of the light signals.

(a) Show that this time is given by

$$x^0 - \int \frac{g_{0i}}{g_{00}}\,\mathrm{d}x^i. \tag{6.18}$$

Attempting to use this formula to synchronize clocks on a closed path, such as a rotating disc, will fail, since the integral will not vanish.

(b) Using the metric for the rotating reference frame from Exercise 3.5, show that the discrepancy over one circuit is

$$\Delta t = \oint \frac{\Omega r^2}{1 - \Omega^2 r^2} \approx 2\pi \Omega r^2, \tag{6.19}$$

when $\Omega r \ll 1$. To the same level of approximation, the discrepancy in proper time is $\Delta\tau = \sqrt{-g_{tt}}\Delta t \approx \Delta t$.

(c) By comparing the optical path lengths of two counter-propagating beams along a rotating circular fibre, show that the rotation causes a shift in their interference pattern of

$$\Delta N = \frac{4\pi\Omega r^2}{\lambda}, \tag{6.20}$$

where λ is the wavelength of the light. This shift is known as the **Sagnac effect** after George Sagnac (1869–1928).

Parallel lines and the covariant derivative

7

We never remark any passion or principle in others, of which, in some degree or other, we may not find a parallel in ourselves.
David Hume (1711–1776) *A Treatise of Human Nature*

Comparisons are odorous
William Shakespeare (1564–1616)
Much Ado About Nothing III:5

In order to describe physical quantities in general relativity, we shall need to evaluate mathematical objects (functions, vector, and tensor fields and so forth) at particular points in spacetime. We shall also identify differential equations for these objects to understand how they change with position in spacetime. This requires the notion of a derivative. When spacetime is curved, some of our basic assumptions about vectors and their derivatives break down. This chapter is concerned with finding a method to take derivatives of vectors with respect to position in spacetime in cases where the spacetime is curved.

Example 7.1

Probably the most familiar example of a curved space is the one in which we live: the Earth's surface. In navigating around our home town, we might use a street map, the coordinates for which are based on a two-dimensional rectangular grid based on north-south and east-west axes. But this street map only works locally and can't be extended to the whole planet because of the curvature of the Earth. Nevertheless, we can imagine smoothly transitioning between lots of small rectangular maps to cover the whole surface of the globe.

In much the same way, (3+1)-dimensional spacetime can be covered smoothly using lots of locally flat maps based on four coordinates. A spacetime that can be covered by a smoothly changing set of coordinates is known in mathematics as a **manifold**.[1] The study of smoothly changing spaces is known as **differential geometry** and is the subject of Part V of this book. Owing to its smoothness, a manifold describing spacetime is necessarily **flat** over a sufficiently small region, across which it looks identical to the Minkowski spacetime encountered in the first part of this book. Over larger distances however, the spacetime might be curved, and it is this curvature that we describe over the next few chapters, with the derivative formulated in this chapter an essential first step.

[1] We can think of a two-dimensional manifold as being made of rubber. It can be a sphere or a torus, but the crucial thing about it is that it shouldn't be punctured in any way: it should be smooth and continuous. Smoothness is a property we'll return to regularly.

[2] For example, if we think of the surface of the Earth then a straight-line path from New York to Paris would plough through the Earth, burrowing hundreds of kilometres underground. This directed straight-line path is then outside the space we are trying to describe.

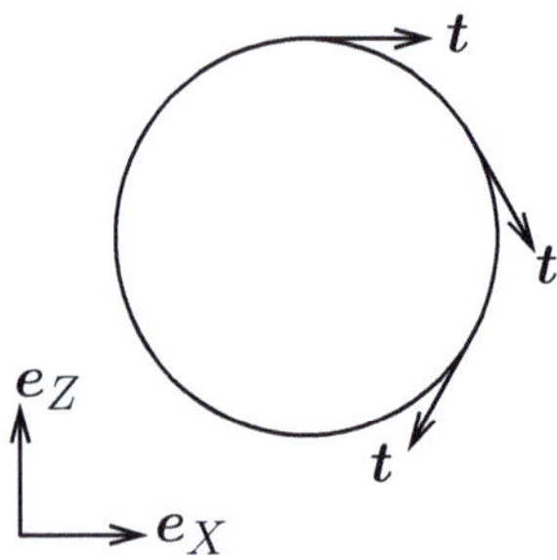

Fig. 7.1 A tangent vector *t* is parallel transported around a surface. Its components are always the same in the local coordinate system, but *t* changes when viewed in the (X, Z) coordinate system set up by observers able to embed the space in higher dimensions.

[3] Imagine a tourist using a street map in London. In absolute terms, their North-direction is rather different to the North-direction of an analogous tourist in Tokyo, even though it is analogously defined in both city street maps as a particular vector tangent to the Earth. The North-direction could be thought of as being parallel transported between the two cities.

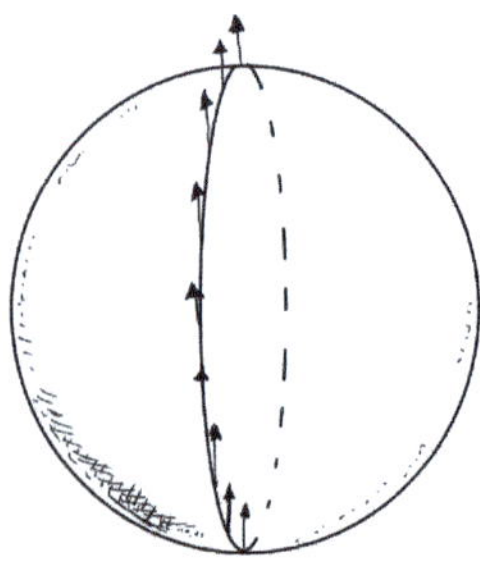

Fig. 7.2 Failure of parallelism for sphere embedded in $\mathbb{R}^3$. The vectors drawn on the surface of the sphere all point in the same absolute direction, according to the embedding in $\mathbb{R}^3$, but they only lie in the tangent plane at one point; more often, they are pointing out of the tangent plane.

7.1 Parallelism

Consider a curved spacetime where observers are confined. The picture to have in mind is of ants confined to a two-dimensional surface such as a football, or of humans confined to the surface of the earth. All measurements are to be made in the surface, so the observers are not allowed to float above it to take advantage of its being embedded in three-dimensional space. The idea of a vector as a directed straight line joining two points is fine for flat space, but ceases to be of much use in a curved space.[2] The notion of a vector as a **tangent** to a path is, however, of much more use. Picture how the **tangent vector** to a path on a two-dimensional surface embedded in three-dimensional space will change its direction as we move it around the curved space, in order for it to still lie in the tangent plane of the surface. This behaviour of the tangent vector will provide a measure of vectors being **parallel**.

Next, we imagine that the path in the surface is one that doesn't change direction according to the observers (e.g. a great circle on a sphere, as shown in Fig. 7.1). The tangent vector of this path should, according to the trapped ants, be 'the same', or parallel, at all points on the path. A tangent vector to the path at some point may then be transported to a different point on the path and, if it is identical to the tangent vector determined at its new position, then we say that the vector has been **parallel transported** (Fig. 7.1). From the point of view of the observers confined to the surface, two vectors can then be compared at two different points in spacetime.

To generalize beyond tangent vectors: our trapped ants set up a coordinate system with which to make measurements. Measurements are always made locally, so they parallel transport their set of axes to the point where they want to measure the orientation of a vector. (The local coordinate system might, for example, use the tangent vector described above, and another axis in the surface orthogonal to this direction.) From this point of view, any vector that has the same components in each of the local coordinate systems is judged to be parallel at the different points. However, these vectors *appear* to change directions when we view the surface as embedded in a higher dimensional space, just as the coordinate system used by the ants on the surfaces appears to change with position (Fig. 7.1).[3] This concept of parallelism will be included when we describe derivatives in curved spacetime, since the derivative of a field of parallel vectors should come out to be zero.

Example 7.2

The vectors in Fig. 7.2 are all parallel in three-dimensional space $\mathbb{R}^3$. However, for observers living on the surface of the sphere, the vectors are not parallel: they all make different angles to the tangent plane of the sphere's surface. The result of correctly parallel transporting a vector along several paths on a spherical surface is shown in Fig. 7.3. The components of the vector are identical in each of the local coordinate systems that the confined observers set up.

7.2 Derivatives and connections

We now turn to a method to evaluate the change in a vector with position. In order to evaluate, via a derivative, the change in a vector as it is transported around, we need to disentangle two effects. The first is the intrinsic change in the vector with position, which is what we want the derivative to output. The second is the change in the vector reflecting the fact that the coordinates (or, equivalently, the basis vectors) change in different parts of space. To extract the intrinsic change we define a new kind of derivative of the vector. This is the **covariant derivative** which evaluates the intrinsic change in the vector $\boldsymbol{v}$. We can make sense of the covariant derivative conceptually as

$$\left(\begin{array}{c} \text{Covariant} \\ \text{derivative of } \boldsymbol{v} \end{array} \right)_{\boldsymbol{u}} = \left(\begin{array}{c} \text{Change in} \\ \text{vector } \boldsymbol{v} \end{array} \right) - \left(\begin{array}{c} \text{Change due to} \\ \text{coordinate system} \end{array} \right).$$

(7.1)

This derivative is **directional**: it tells us the change in $\boldsymbol{v}$ as we move along the vector $\boldsymbol{u}$ (hence the subscript in the previous equation).

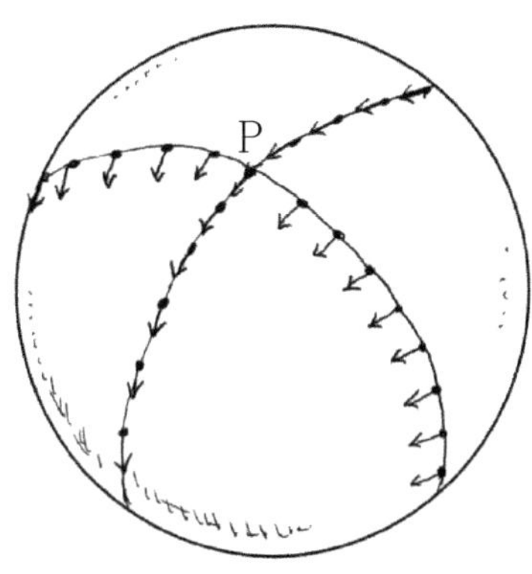

Fig. 7.3 Parallel transport of a vector on a spherical surface.

Example 7.3

The covariant derivative in a curved space generalizes the notion of a directional derivative in ordinary calculus. The gradient of a surface of constant f in Euclidean 3-space is given by

$$\vec{\nabla} f(x, y, z) = \frac{\partial f}{\partial x} \vec{e}_x + \frac{\partial f}{\partial y} \vec{e}_y + \frac{\partial f}{\partial z} \vec{e}_z.$$

(7.2)

This is interpreted as a vector normal to the tangent plane of the surface of constant f. If we want to know the change of $f(x, y, z)$ along a particular vector $\vec{u}$ we use the directional derivative, defined as

$$\vec{u} \cdot \vec{\nabla} f = u^x \frac{\partial f}{\partial x} + u^y \frac{\partial f}{\partial y} + u^z \frac{\partial f}{\partial z}.$$

(7.3)

This can be thought of as

$$\vec{u} \cdot \vec{\nabla} f = \left(\begin{array}{c} \text{Value of } f \\ \text{at tip of } \vec{u} \end{array} \right) - \left(\begin{array}{c} \text{Value of } f \\ \text{at base of } \vec{u} \end{array} \right).$$

(7.4)

Now to take the derivative. Consider the vector $\boldsymbol{v} = v^\mu \boldsymbol{e}_\mu$. We take a derivative with respect to the coordinates, allowing both the components v^μ and the basis vectors $\boldsymbol{e}_\mu$ to change in spacetime. The derivative we shall take is $\partial/\partial x^\alpha$, which should be thought of as the directional derivative along the direction $\boldsymbol{e}_\alpha$. Employing the Leibniz product rule, we have

$$\frac{\partial \boldsymbol{v}}{\partial x^\alpha} = \frac{\partial v^\mu}{\partial x^\alpha} \boldsymbol{e}_\mu + v^\mu \frac{\partial \boldsymbol{e}_\mu}{\partial x^\alpha}.$$

(7.5)

The tricky second term on the right is due to the change in basis vectors with position. The derivative of the basis vector $\boldsymbol{e}_\mu$ can have components along any of the basis vectors, so to express this we define **connection coefficients**, also known as **Christoffel symbols**,[4] denoted $\Gamma^\mu{}_{\alpha\beta}$, and

[4]Elwin Bruno Christoffel (1829–1900). The mathematics described here were invented by Christoffel in 1869 and further explored by Gregorio Ricci-Curbastro (1853–1925) who, in the years leading to 1900, developed the mathematical machinery employed by Einstein in developing general relativity. We shall follow the modern convention of calling the symbols 'connection coefficients' in this book.

then write the change in basis vectors as

$$\frac{\partial e_\mu}{\partial x^\alpha} = \Gamma^\lambda{}_{\alpha\mu} e_\lambda. \tag{7.6}$$

The connection coefficients encode all of the information describing how the coordinates change as we move around spacetime. Another way of thinking about this is that, since there are different local coordinate systems at different points in space, the connection coefficients tell us how the coordinate systems are *connected*, that is, how to translate between coordinate system as we move around.[5]

Example 7.4

In the following two chapters, we shall find a simple and efficient means of extracting connection coefficients. Before we get to that, here is a simple, 'brute force and ignorance' example, based on eqn 7.6. We saw in Chapter 3 that for a plane-polar coordinate system, the basis vectors have derivatives[6]

$$\begin{aligned} \frac{\partial e_r}{\partial r} = 0, \qquad & \frac{\partial e_r}{\partial \theta} = \frac{e_\theta}{r}, \\ \frac{\partial e_\theta}{\partial r} = \frac{e_\theta}{r}, \qquad & \frac{\partial e_\theta}{\partial \theta} = -r e_r. \end{aligned} \tag{7.7}$$

This allows us to write down the connection coefficients. Using eqn 7.6 we can read off

$$\begin{aligned} \Gamma^r{}_{rr} = 0, \quad \Gamma^\theta{}_{rr} = 0, \quad \Gamma^r{}_{\theta r} = 0, \quad \Gamma^\theta{}_{\theta r} = \frac{1}{r}, \\ \Gamma^r{}_{r\theta} = 0, \quad \Gamma^\theta{}_{r\theta} = \frac{1}{r}, \quad \Gamma^r{}_{\theta\theta} = -r, \quad \Gamma^\theta{}_{\theta\theta} = 0. \end{aligned} \tag{7.8}$$

Although Euclidean space described by the plane polar coordinates is flat, we still have non-zero connection coefficients. The presence of connection coefficients therefore does not alone tell us whether a space is curved.[7]

In all of the coordinate frames that we examine in this book, the connection coefficients have the property

$$\Gamma^\mu{}_{\alpha\beta} = \Gamma^\mu{}_{\beta\alpha}. \tag{7.9}$$

A connection with this property is often called symmetric or **torsion free**.

Finally, we ask why we call the Γs connection *coefficients* or Christoffel *symbols*? The answer is because, unlike most of the objects we deal with in relativity, they are not the components of a tensor.[8]

Example 7.5

Usually, we expect a tensor's components to transform as

$$T^{\alpha'}{}_{\gamma'\beta'} = \frac{\partial x^{\alpha'}}{\partial x^\alpha} \frac{\partial x^\beta}{\partial x^{\beta'}} \frac{\partial x^\gamma}{\partial x^{\gamma'}} T^\alpha{}_{\gamma\beta}. \tag{7.10}$$

We'll see in Part V that the components of the connection actually transform as

$$\Gamma^{\alpha'}{}_{\gamma'\beta'} = \frac{\partial x^{\alpha'}}{\partial x^\alpha} \left(\frac{\partial^2 x^\alpha}{\partial x^{\beta'} \partial x^{\gamma'}} + \frac{\partial x^\beta}{\partial x^{\beta'}} \frac{\partial x^\gamma}{\partial x^{\gamma'}} \Gamma^\alpha{}_{\gamma\beta} \right). \tag{7.11}$$

This shows us that the first term in the braces spoils the tensor transformation law. Therefore, the Γs are not tensors in general.

---

[5] Over the next few chapters, we shall see how the connection coefficients, whose values are functions of position of spacetime, can be used to tell us about the curvature of spacetime itself. They can be derived from the metric, which is ultimately the source of curvature.

[6] Recall that

$$e_r = \left(\frac{\partial x}{\partial r}\right)_\theta e_x + \left(\frac{\partial y}{\partial r}\right)_\theta e_y,$$

which yields, from $x = r\cos\theta$ and $y = r\sin\theta$, the expression

$$e_r = \cos\theta\, e_x + \sin\theta\, e_y.$$

Similarly

$$e_\theta = \left(\frac{\partial x}{\partial \theta}\right)_r e_x + \left(\frac{\partial y}{\partial \theta}\right)_r e_y,$$

or

$$e_\theta = -r\sin\theta\, e_x + r\cos\theta\, e_y.$$

[7] We shall see in Chapter 11 that a combination of the coefficients and their derivatives is needed to tell us whether the space is curved or not.

[8] In summary: $\Gamma^\lambda{}_{\alpha\mu}$ is the rate at which the vector e_μ rotates towards e_λ as we travel along a curve whose tangent vector is e_α. Notice that we choose the order of the downstairs indices in $\Gamma^\lambda{}_{\alpha\mu}$ to start with the component with respect to which we take the derivative in eqn 7.6. This convention is not universal: Hawking and Ellis use it; Misner, Thorne, and Wheeler do not. Since $\Gamma^\lambda{}_{\alpha\mu} = \Gamma^\lambda{}_{\mu\alpha}$ in all of the coordinate frames we shall deal with, this is not a crucial point.

7.3 The covariant derivative

We now have the tools to finally take a covariant derivative. Substituting eqn 7.6 into eqn 7.5 we have a derivative

$$\frac{\partial \boldsymbol{v}}{\partial x^\alpha} = \frac{\partial v^\mu}{\partial x^\alpha} \boldsymbol{e}_\mu + v^\mu \Gamma^\lambda{}_{\alpha\mu} \boldsymbol{e}_\lambda, \tag{7.12}$$

and by relabelling the dummy indices

$$\frac{\partial \boldsymbol{v}}{\partial x^\alpha} = \left(\frac{\partial v^\mu}{\partial x^\alpha} + v^\lambda \Gamma^\mu{}_{\alpha\lambda} \right) \boldsymbol{e}_\mu. \tag{7.13}$$

From here on we shall write[9] this quantity as $\boldsymbol{\nabla}_\alpha \boldsymbol{v}$, which we call the covariant derivative of $\boldsymbol{v}$ along the direction $\boldsymbol{e}_\alpha$. This new notation allows us to replace the left-hand side of eqn 7.13 with $\boldsymbol{\nabla}_\alpha \boldsymbol{v}$, rather than as $\partial \boldsymbol{v}/\partial x^\alpha$ which turns out to be a much more convenient.[10] While we are in the process of introducing notation, a commonly used labour-saving shorthand for writing derivatives using commas and semicolons is given in the shaded box in the margin. Putting everything together, we have

$$\boldsymbol{\nabla}_\alpha \boldsymbol{v} = \left(\frac{\partial v^\mu}{\partial x^\alpha} + v^\lambda \Gamma^\mu{}_{\alpha\lambda} \right) \boldsymbol{e}_\mu. \tag{7.18}$$

In terms of our classification of tensors, notice that $\boldsymbol{\nabla}_\alpha \boldsymbol{v}$ is a $(1,0)$ object, just like a vector.

Example 7.6

We can immediately note that in Minkowski spacetime the Cartesian basis vectors do not change with position and so all of the Γ coefficients are zero, giving the result that

$$\boldsymbol{\nabla}_\alpha \boldsymbol{v} = \frac{\partial v^\mu}{\partial x^\alpha} \boldsymbol{e}_\mu \quad \text{(flat spacetime)}. \tag{7.19}$$

We have worked out the covariant derivative along the direction of the basis vector $\boldsymbol{e}_\alpha$. What about the covariant derivative along an arbitrary vector? For this purpose we define the **connection operator** $\boldsymbol{\nabla}$ by writing $\boldsymbol{e}_\alpha \cdot \boldsymbol{\nabla} = \boldsymbol{\nabla}_\alpha$. We then define the action of our covariant derivative $\boldsymbol{\nabla}_\alpha$ on a scalar function simply as the derivative with respect to x^α, or

$$\boldsymbol{e}_\alpha \cdot \boldsymbol{\nabla} f = \boldsymbol{\nabla}_\alpha f = \frac{\partial f}{\partial x^\alpha}. \tag{7.20}$$

Generalizing to different directions by replacing $\boldsymbol{e}_\alpha$ with an arbitrary vector $\boldsymbol{u}$, we have the directional derivative

$$\boldsymbol{\nabla}_{\boldsymbol{u}} f = \boldsymbol{u} \cdot \boldsymbol{\nabla} f = u^\alpha \boldsymbol{e}_\alpha \cdot \boldsymbol{\nabla} f = u^\alpha \frac{\partial f}{\partial x^\alpha}. \tag{7.21}$$

If we now interpret the action of the connection operator on vectors in the same way, $\boldsymbol{\nabla}_{\boldsymbol{u}} \boldsymbol{v}$ is the covariant derivative of the vector $\boldsymbol{v}$ along the direction $\boldsymbol{u}$. We write this as

$$\boldsymbol{\nabla}_{\boldsymbol{u}} \boldsymbol{v} = \boldsymbol{u} \cdot \boldsymbol{\nabla} \boldsymbol{v} = u^\alpha \boldsymbol{e}_\alpha \cdot \boldsymbol{\nabla} \boldsymbol{v} = u^\alpha \boldsymbol{\nabla}_\alpha \boldsymbol{v}, \tag{7.22}$$

---

[9] We take $\boldsymbol{\nabla}_\alpha$ to be short for the more cumbersome

$$\boldsymbol{\nabla}_\alpha \equiv \boldsymbol{\nabla}_{\boldsymbol{e}_\alpha}. \tag{7.14}$$

In words, this is the directional derivative along the direction $\boldsymbol{e}_\alpha$.

[10] Relating the old notation to the new notation, we have

$$\frac{\partial \boldsymbol{v}}{\partial x^\alpha} \equiv (\boldsymbol{\nabla}_\alpha \boldsymbol{v})^\mu \boldsymbol{e}_\mu. \tag{7.15}$$

Important here is that the components of the derivative are not necessarily equal to the derivatives of the components. This is due to the non-zero connection coefficients. In fact, we can rewrite our definition of the derivative of components and basis vectors in the new notation

$$\boldsymbol{\nabla}_\alpha (v^\mu) = \frac{\partial v^\mu}{\partial x^\alpha}, \tag{7.16}$$

and

$$\boldsymbol{\nabla}_\alpha (\boldsymbol{e}_\mu) = \Gamma^\lambda{}_{\alpha\mu} \boldsymbol{e}_\lambda. \tag{7.17}$$

Commas and semicolons: Ordinary derivatives of functions can be written in **comma notation** as

$$\frac{\partial f}{\partial x^\alpha} \equiv f_{,\alpha}.$$

The covariant derivative is written in **semicolon notation** so that the μ component of eqn 7.18 becomes

$$(\boldsymbol{\nabla}_\alpha \boldsymbol{v})^\mu \equiv v^\mu{}_{;\alpha} = v^\mu{}_{,\alpha} + v^\lambda \Gamma^\mu{}_{\alpha\lambda},$$

and so $v^\mu{}_{;\alpha}$ are the components of the covariant derivative. This notation has the virtue of allowing some equations to be written more compactly, though this comes at the expense of leaving expressions littered with punctuation marks.

[11] In semicolon notation

$$\nabla_{\!u}v = u^\alpha v^\mu{}_{;\alpha}e_\mu,$$

where as before we have

$$v^\mu{}_{;\alpha} = v^\mu{}_{,\alpha} + v^\lambda\Gamma^\mu{}_{\alpha\lambda}.$$

and conclude that the covariant derivative along a vector $\boldsymbol{u}$ is given in terms of components by[11]

$$\nabla_{\!u}v = u^\alpha\left(\frac{\partial v^\mu}{\partial x^\alpha} + v^\lambda\Gamma^\mu{}_{\alpha\lambda}\right)e_\mu. \tag{7.23}$$

The covariant derivative selects out the change in a vector with position owing to its genuine change, eliminating the contribution due to the changing of coordinates with position. If a vector is parallel transported along a path then the only change should be due to the coordinates changing. So parallel transport of a vector $\boldsymbol{v}$ along the direction $\boldsymbol{u}$ implies

$$\nabla_{\!u}v = 0 \quad \text{(parallel transport)}. \tag{7.24}$$

Example 7.7

This latter expression means that the components obey

$$u^\alpha\frac{\partial v^\mu}{\partial x^\alpha} + \Gamma^\mu{}_{\alpha\beta}v^\beta u^\alpha = 0, \tag{7.25}$$

or

$$\frac{\partial v^\mu}{\partial x^\alpha} = -\Gamma^\mu{}_{\alpha\beta}v^\beta. \tag{7.26}$$

In words, the change in the components $\partial v^\mu/\partial x^\nu$ is, in this case, entirely due[12] to the change in coordinates $-\Gamma^\mu{}_{\alpha\beta}v^\beta$. That is to say that when a vector is parallel transported we have

$$\begin{pmatrix} \text{Change in a} \\ \text{vector's components} \end{pmatrix} = \begin{pmatrix} \text{Change due to} \\ \text{coordinate system} \end{pmatrix}. \tag{7.27}$$

[12] We therefore identify the derivative $\partial v^\mu/\partial x^\alpha$ as having contributions from both the intrinsic change in $\boldsymbol{v}$ and the change in coordinates. To fix this in the covariant derivative, we must subtract off the change from the coordinates $-\Gamma^\mu{}_{\alpha\beta}v^\beta$.

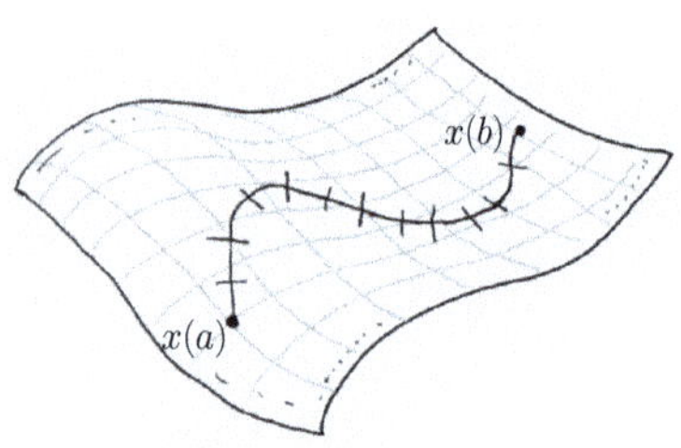

Fig. 7.4 The path $x(\lambda)$ is a curve through spacetime as a function of the parameter λ, stretching from the points associated with $\lambda = a$ to $\lambda = b$. The path marks off regular intervals, so we know how far along the curve we are.

7.4 Parametrized paths

We now have the covariant derivative at our disposal in the form of a directional derivative of some vector $\boldsymbol{v}$ taken along a vector $\boldsymbol{u}$. We shall also need the derivative in a form more suitable to apply to curves in spacetime such as the world lines of particles.

We saw in Chapter 1 that the most general way to describe a curve is to parametrize it by introducing a quantity that varies monotonically along its length. That is, a curve stretching from point $\lambda = a$ to $\lambda = b$ is written as $x(\lambda)$, where λ **parametrizes** the curve. It marks off regular intervals, so we know how far along the curve we are, as shown in Fig. 7.4.

Example 7.8

For the two-dimensional space $\mathbb{R}^2$ a curve is given by $(x(\lambda), y(\lambda))$, i.e. with x and y both functions of λ.
- A straight line $y = mx + c$ can be parametrized with $x(\lambda) = \lambda$ and $y(\lambda) = m\lambda + c$.
- A parabola $y = x^2$ can be parametrized with $x(\lambda) = \lambda$ and $y(\lambda) = \lambda^2$.
- A circle $x^2 + y^2 = a^2$ can be parametrized with $x(\lambda) = a\cos\lambda$ and $y(\lambda) = a\sin\lambda$.

The precise choice of parametrization isn't crucial. If an allowable parametrization is given by regular intervals of λ, we could equally well choose a different parametrization η such that $\lambda = \alpha\eta + \beta$, where α and β are constants. Such a parametrization is called an **affine parametrization**.

With this in mind, we can return to the covariant derivative itself. In many cases, we are interested in the rate of change of a **vector field**, $\boldsymbol{v}(x)$, an object where we input a position $x = \mathcal{P}$ and output a vector $\boldsymbol{v}$ appropriate for that point $\mathcal{P}$. We then ask how rapidly the vector field $\boldsymbol{v}(x)$ changes along a curve $x^\mu(\lambda)$. This involves parametrizing the curve, and then we seek

$$\left(\begin{array}{c} \text{Rate of change of} \\ \boldsymbol{v} \text{ with respect to } \lambda \end{array} \right) \equiv \frac{\mathrm{D}\boldsymbol{v}}{\mathrm{d}\lambda}. \tag{7.28}$$

Here we've introduced some new notation: the covariant derivative with respect to an affine parameter is denoted $\mathrm{D}/\mathrm{d}\lambda$.[13]

In order to use the covariant derivative as we've defined it so far, we seek a vector telling us the direction along which to take the derivative. This is provided by the tangent vector to the curve $x^\mu(\lambda)$, given by[14]

$$\boldsymbol{u} = \left(\frac{\mathrm{d}x^\mu(\lambda)}{\mathrm{d}\lambda} \right) \boldsymbol{e}_\mu. \tag{7.29}$$

That is, at every point λ along the curve we have a tangent vector $\boldsymbol{u}$ (Fig. 7.5). This tangent vector is one of the most useful tools in this book. We then have

$$\left(\begin{array}{c} \text{Rate of change of} \\ \boldsymbol{v} \text{ with respect to } \lambda \end{array} \right) \equiv \frac{\mathrm{D}\boldsymbol{v}}{\mathrm{d}\lambda} \equiv \boldsymbol{\nabla}_{\boldsymbol{u}}\boldsymbol{v} \equiv \left(\begin{array}{c} \text{Covariant derivative of} \\ \boldsymbol{v} \text{ along } \boldsymbol{u} \end{array} \right), \tag{7.30}$$

where $\boldsymbol{u}$ us the tangent vector field to the curve. For massive particles, which follow timelike curves, we shall usually choose λ to be the proper time τ, which allows us to interpret the tangent as the particle's velocity, and provides the useful constraint $\boldsymbol{u} \cdot \boldsymbol{u} = -1$.

Example 7.9

This way of thinking about the covariant derivative makes it similar to an ordinary derivative defined in terms of evaluating a function at two points, $f(x)$ and $f(x+\delta x)$, and taking the difference in the limit of small δx. However, key to the definition here is the notion of parallelism, which allows us to remove the change caused by the changing coordinate system. In order to do this, the instructions of how to take the covariant derivative, in these terms, are as follows:
(i) Take the vector $\boldsymbol{v}$ at $\lambda = \lambda_0 + \varepsilon$.
(ii) Parallel transport it back to λ_0.
(iii) Evaluate $\delta\boldsymbol{v}$, which measured how different it is from $\boldsymbol{v}$ at λ_0.
(iv) Divide by ε and take the limit.
In equations, we have

$$\frac{\mathrm{D}\boldsymbol{v}}{\mathrm{d}\lambda} = \boldsymbol{\nabla}_{\boldsymbol{u}}\boldsymbol{v} = \lim_{\varepsilon \to 0} \left(\frac{\boldsymbol{v}(\lambda_0 + \varepsilon)_{\text{(parallel transport to } \lambda_0)} - \boldsymbol{v}(\lambda_0)}{\varepsilon} \right). \tag{7.31}$$

This is illustrated in Fig. 7.6.

[13]The notation reminds us that owing to the changes in the coordinate systems, the components of the covariant derivative $(\mathrm{D}\boldsymbol{v}/\mathrm{d}\lambda)^\mu$, will not generally be equivalent to the derivatives of components $\mathrm{d}v^\mu/\mathrm{d}\lambda$. However, for a scalar field f we do have $\mathrm{d}f/\mathrm{d}\lambda = \mathrm{D}f/\mathrm{d}\lambda$.

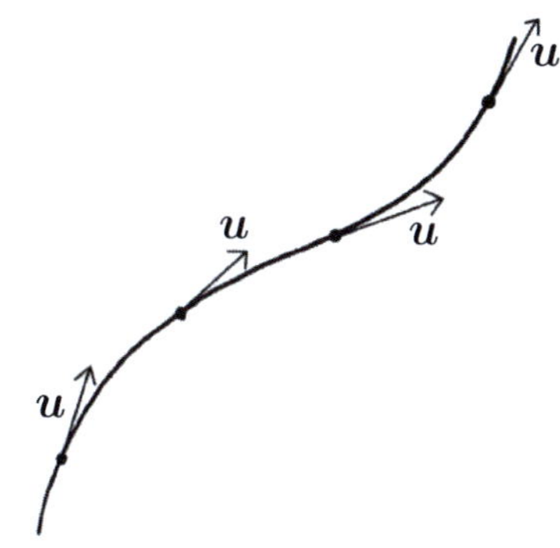

Fig. 7.5 The tangent vectors $\boldsymbol{u} = (\mathrm{d}x^\mu(\lambda)/\mathrm{d}\lambda)\,\boldsymbol{e}_\mu$ along the curve parametrized by λ. For $\lambda = \tau$ this provide a velocity vector (which itself varies along the path).

[14]We do not write $\mathrm{d}\boldsymbol{x}/\mathrm{d}\lambda$ as we sometime do in special relativity. As discussed in Chapter 3, the displacement vector $\boldsymbol{x} = x^\mu\boldsymbol{e}_\mu$, thought of as pointing a distance $|\boldsymbol{x}|$ from the origin to coordinate point x^μ, does not transform appropriately, and so we won't use it in this form (e.g. by taking its derivative). Note also that the tangent vector is given by $\boldsymbol{u} = (\mathrm{D}x^\mu(\lambda)/\mathrm{d}\lambda)\boldsymbol{e}_\mu = (\mathrm{d}x^\mu(\lambda)/\mathrm{d}\lambda)\boldsymbol{e}_\mu$, since $x^\mu(\lambda)$ is a set of scalar functions.

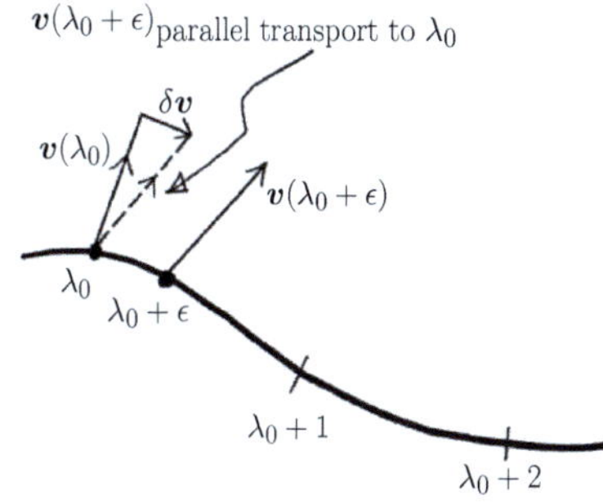

Fig. 7.6 Taking the covariant derivative using eqn 7.31.

Example 7.10

We can check our new version of the covariant derivative of a vector $\boldsymbol{A}$ in the case of flat spacetime, where the connection coefficients expressed in Cartesian coordinates vanish. Writing out all the components, we have

$$\left(\frac{\mathrm{D}\boldsymbol{A}}{\mathrm{d}\lambda}\right)^{\alpha} = (\boldsymbol{\nabla}_{\boldsymbol{u}}\boldsymbol{A})^{\alpha} = u^{\mu}\left(\frac{\partial A^{\alpha}}{\partial x^{\mu}} + \Gamma^{\alpha}{}_{\mu\nu}A^{\nu}\right)$$

$$= u^{\mu}\frac{\partial A^{\alpha}}{\partial x^{\mu}} \quad (\text{the connection } \Gamma^{\alpha}{}_{\mu\nu} = 0)$$

$$= \frac{\mathrm{d}x^{\mu}}{\mathrm{d}\lambda}\frac{\partial A^{\alpha}}{\partial x^{\mu}} = \frac{\mathrm{d}A^{\alpha}}{\mathrm{d}\lambda}, \tag{7.32}$$

where we've used the chain rule in the final step. We conclude that for flat spacetime the components of the derivative are simply the derivatives of the components with respect to the parameter λ.

The covariant derivative notation $\mathrm{D}/\mathrm{d}\lambda$ proves very useful, not least because of its resemblance to the ordinary derivative.

7.5 Enter the metric

After formulating a covariant derivative, we might ask if this is the only way we could have constructed it. It turns out that our freedom to formulate it was restricted by the metric field $\boldsymbol{g}$, which is the foundation of our physical description of spacetime, and it is exactly this metric field that forces this version of the covariant derivative upon us. This idea is encapsulated in the notion of what is called the **compatibility of the connection** which requires that the covariant derivative obeys

$$\boldsymbol{\nabla}_{\alpha}\boldsymbol{g} = 0 \quad \text{or, in components,} \quad g_{\mu\nu;\alpha} = 0, \tag{7.33}$$

for all α. This equation inseparably joins the metric and the covariant derivative.[15] The importance of this condition is that, if it did not hold, then the lengths of vectors would change as we parallel transport them.[16] This would be highly undesirable for a description of the physics of the real world.

[15] We do not yet have an explicit expression for how to compute the covariant derivative of a $(0,2)$ tensor like $\boldsymbol{g}$. We delay deriving the explicit expression until Part V. At this stage we state that it is given in components by

$$g_{\mu\nu;\alpha} = g_{\mu\nu,\alpha} - \Gamma^{\beta}{}_{\alpha\mu}g_{\beta\nu} - \Gamma^{\beta}{}_{\alpha\nu}g_{\mu\beta}. \tag{7.34}$$

It is also helpful at this stage to note that we can also write the derivative for a $(2,0)$ tensor like $\boldsymbol{T}$ in components as

$$T^{\mu\nu}{}_{;\alpha} = T^{\mu\nu}{}_{,\alpha} + \Gamma^{\mu}{}_{\alpha\beta}T^{\beta\nu} + \Gamma^{\nu}{}_{\alpha\beta}T^{\mu\beta}. \tag{7.35}$$

[16] Another consequence of this equation is that it provides the long-awaited explanation of what affine parametrizations actually are. They are those smooth parametrizations of a curve that have the property that the length of a vector doesn't change as we parallel transport the vector along the curve.

[17] The Leibniz product rule does indeed hold, as discussed in Part V.

Example 7.11

We shall prove the compatibility condition. The length of a vector is given by (the square root of) $\boldsymbol{A}\cdot\boldsymbol{A} = \boldsymbol{g}(\boldsymbol{A},\boldsymbol{A})$, or $g_{\mu\nu}A^{\mu}A^{\nu}$. Take $\boldsymbol{A}$ to be a covariant constant (i.e. parallel) such that $\boldsymbol{\nabla}_{\alpha}\boldsymbol{A} = 0$ (or $A^{\mu}{}_{;\alpha} = 0$). If the length of the vector is constant we expect the covariant derivative of $\boldsymbol{g}(\boldsymbol{A},\boldsymbol{A})$ to vanish. Assuming the Leibniz product rule[17] we have

$$g_{\mu\nu;\alpha}A^{\mu}A^{\nu} + g_{\mu\nu}A^{\mu}{}_{;\alpha}A^{\nu} + g_{\mu\nu}A^{\mu}A^{\nu}{}_{;\alpha} = 0. \tag{7.36}$$

Since $A^{\mu}{}_{;\alpha} = 0$, then we must have $g_{\mu\nu;\alpha} = 0$, as claimed.

Finally, we note that the compatibility condition is the basis of other links between the metric and the covariant derivative. The connection coefficients $\Gamma^{\mu}{}_{\alpha\beta}$ may be derived directly from the components of the metric.[18] We saw how the connection coefficients arose due to the change in the basis vectors with position in spacetime and could be calculated via derivatives like $\partial e_{\mu}/\partial x^{\alpha} = \Gamma^{\lambda}{}_{\mu\alpha}e_{\lambda}$. Recall also that the components of the metric reflect the basis vectors via $g_{\mu\nu} = g(e_{\mu}, e_{\nu})$ or, more simply $g_{\mu\nu} = e_{\mu} \cdot e_{\nu}$. It therefore comes as little surprise that the connection coefficients are formed from a combination of first derivatives of the metric components as enshrined in the conceptual expression

$$\boxed{\partial g} \to \boxed{\Gamma}. \tag{7.38}$$

We shall expand on this point in the coming chapters.

In the next two chapters, we turn to the use of the covariant derivative in understanding how a particle moves under the influence of gravitation.

[18]We shall see in Chapter 9 that the expression we will need is

$$g_{\rho\lambda}\Gamma^{\rho}{}_{\mu\sigma} = \frac{1}{2}\left(\frac{\partial g_{\lambda\mu}}{\partial x^{\sigma}} + \frac{\partial g_{\lambda\sigma}}{\partial x^{\mu}} - \frac{\partial g_{\mu\sigma}}{\partial x^{\lambda}}\right). \tag{7.37}$$

Chapter summary

- Parallel transport provides a method of comparing vectors at different points in curved space. A vector that is moved such that it has the same components in different local coordinate systems has been parallel transported.

- The covariant derivative can be used to measure how vector fields change with position in spacetime. It is a directional derivative given by

$$\nabla_{u}v = u^{\alpha}\left(\frac{\partial v^{\mu}}{\partial x^{\alpha}} + v^{\lambda}\Gamma^{\mu}{}_{\alpha\lambda}\right)e_{\mu}. \tag{7.39}$$

- The tangent vector to the world line of a massive particle parametrized by the proper time is the timelike velocity vector u, with the property $u \cdot u = -1$.

Exercises

(7.1) (a) Translate these expressions from component notation into index-free (i.e. tensor) notation
(i) $v^{\alpha}{}_{;\beta}u^{\beta} - u^{\alpha}{}_{;\beta}v^{\beta}$.
(ii) $u^{\alpha}v^{\beta}w_{\alpha;\beta}$.
If T is a symmetric $(0,2)$ tensor while u, v and w are vectors, translate the following into component notation:
(iii) $v \cdot (\nabla_{u}T) \cdot w$.

(7.2) Consider a unit circle in the Euclidean plane. Parametrize the curve using an affine parameter s by writing $(x, y) = (\cos s, \sin s)$. Here s simple measures the distance on the curve, implying it must indeed be affine.
(a) Compute the components of the tangent vector to the curve u, and its magnitude.
(b) Compute the components of $Du/ds \equiv \nabla_{u}u$

and verify $\boldsymbol{u} \cdot D\boldsymbol{u}/ds$ vanishes.

Now consider the reparametrization $t = \sin s$. This is not in the form $t = as + b$, so is not an affine parametrization.

(c) Recompute the components of the tangent vector $\boldsymbol{u}$, the derivative $D\boldsymbol{u}/dt$, and $\boldsymbol{u} \cdot D\boldsymbol{u}/dt$ using this new parametrization.

(7.3) Consider the vector field $\boldsymbol{v}$ in two-dimensional flat space with Cartesian components $(v^x, v^y) = (0, Cx)$, with C a constant.

(a) Compute the vector $\boldsymbol{\nabla}_\mu \boldsymbol{v}$ for $\mu = x$ and y.

(b) Convert the components of the vector into cylindrical polar coordinates using the transformations from Chapter 3.

(c) Using the connection coefficients given in the chapter, compute the vectors $\boldsymbol{\nabla}_\mu \boldsymbol{v}$ for $\mu = r$ and θ.

(d) Treat the quantity $(\boldsymbol{\nabla}_\mu \boldsymbol{v})^\nu$ as components of a $(1,1)$ tensor. Using the tensor transformation law, show that the results from part (c) are consistent with those of part (a).

(7.4) The covariant derivative of a $(0,2)$ tensor is written as

$$(\boldsymbol{\nabla}_\alpha \boldsymbol{\xi})_{\mu\nu} = \frac{\partial \xi_{\mu\nu}}{\partial x^\alpha} - \Gamma^\lambda{}_{\alpha\mu} \xi_{\lambda\nu} - \Gamma^\lambda{}_{\alpha\nu} \xi_{\mu\lambda}. \quad (7.40)$$

Apply this to the components of the metric tensor and show, using eqn 7.37, that $(\boldsymbol{\nabla}_\alpha \boldsymbol{g})_{\mu\nu} = 0$.

Free fall and geodesics

8

The bigger they come, the harder they fall
Barbados Joe Walcott (1873–1935) and Bob Fitzsimmons (1863–1917)

A **geodesic** can be thought of geometrically as the straightest possible path in a curved spacetime. Geodesics are the paths that extremize the interval between spacetime events. Physically, a particle in free fall has a world line that follows a geodesic.[1] The equation of motion for such a particle, known as the **geodesic equation**, therefore tells us about the motion of a particle that is not subject to external forces. In this chapter, we investigate geodesics and derive the geodesic equation that tells us how curvature causes particles to move. Our task here is to introduce some key ideas in geodesic motion. In the next chapter, we look at the details of how to extract the connection coefficients required to compute geodesics.

Example 8.1

In pre-relativity physics, a particle subject to no forces does not accelerate and has an equation of motion given by $\ddot{\vec{x}} = 0$. In relativity, a particle follows a path $x^\mu(\tau) = (t(\tau), x(\tau), y(\tau), z(\tau))$ parametrized by some affine parameter, such as its proper time τ. If a particle is in a flat, Minkowski spacetime, we have an acceleration

$$\frac{\mathrm{d}^2 x^\mu}{\mathrm{d}\tau^2} = 0 \quad \text{(flat spacetime).} \tag{8.1}$$

We can immediately write down the set of possible geodesics as

$$x^\mu(\tau) = b^\mu \tau + c^\mu, \tag{8.2}$$

where b^μ and c^μ are a set of constant components. We see that in Minkowski spacetime, the particles fall along straight lines.

8.1 Extremal intervals

Hamilton's principle tells us that the action for the trajectory that is realized by a system, is the one that takes a *stationary* value when the action is varied. The action often takes a minimum value (e.g. for straight line motion in a two-dimensional plane), but examples of why and when it takes a saddle point or maximum value are less well known. However, they can be important in relativity.

[1] One way to justify this is to recall from mechanics that the action for free, massive particles is given by $S = -m \int \left(-\mathrm{d}s^2\right)^{\frac{1}{2}}$. The result of extremizing the action is the equations of motion for the particle. Since m in a scalar, extremizing the action amounts to extremizing the timelike interval $\Delta\tau = \int \left(-\mathrm{d}s^2\right)^{\frac{1}{2}}$ between events with timelike separation. This means that a solution of the equations of motion gives us a geodesic (which is timelike for a massive particle). It is worth stressing that it is only free particles that travel along geodesics. Particles subject to other interactions have additional terms in their Lagrangian and so give rise to equations of motion whose solutions are not the geodesics of spacetime. Although this argument applies to massive particles and their timelike geodesics, we can also discuss null and spacelike geodesics. To compute spacelike geodesics we extremize the proper length $\Delta l = \int \left(+\mathrm{d}s^2\right)^{\frac{1}{2}}$ between events with spacelike separation. Although nothing travels on spacelike geodesics (except hypothetical tachyon particles, which travel faster than light), spacelike geodesics are often interesting and useful. Photons travel along null geodesics and are discussed in Section 8.4.

[2]More generally, a conjugate point is a point where the matrix N_{ij}^{-1} is singular, and this occurs at q_1^* since a non-zero value of δp^j gives no variation in δq^i. The rule is that the trajectory is a minimum in the action if the trajectory does not pass though a conjugate point. This was the case for path (i). The trajectory is not a minimum in the action if the trajectory passes through a conjugate point, which is the case for path (ii). In general, if two geodesics are sent out from a point $\mathcal{P}$ and later cross at a point $\mathcal{Q}$, then $\mathcal{Q}$ is a conjugate point to $\mathcal{P}$. This argument has a very important use in singularity theorems, such as the one discussed in Chapter 50.

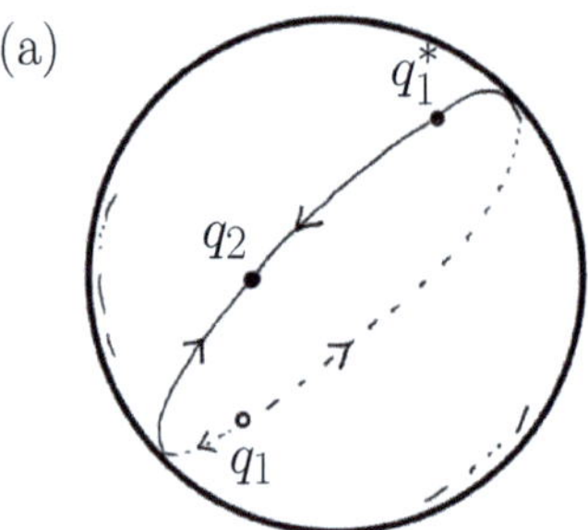

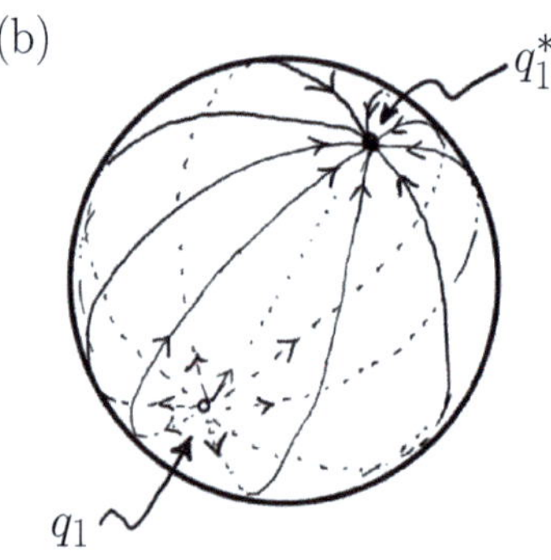

Fig. 8.1 (a) Points q_1 and q_2 on the sphere. The conjugate point q_1^* is at the antipode of q_1. Paths (i) and (ii), which lie on a great circle, are shown. (b) Setting off a swarm of particles at q_1 results in their trajectories realizing a focus at q_1^*.

Example 8.2

Consider non-relativistic motion between two points on the surface of a sphere, starting at q_1 and ending at q_2, as shown in Fig. 8.1(a). We assume that these points are not antipodal (i.e. not opposite points on the sphere). There are two geodesics: (i) a trajectory that represents the shortest distance between the points; and (ii) a trajectory that lies on the same great circle, but which heads off from q_1 in the other direction, through its antipodal point and then to q_2. For particles set off by an observer along these two paths with the same momentum, path (i) takes least time and gives the minimum action; path (ii) takes longer and gives a saddle point action. There is a way to use this example to find work out if the trajectory is a minimum or not. We set in motion a swarm of free particles from q_1, almost along path (i), but with slightly different directions of their momentum, as shown in Fig. 8.1(b). The swarm spreads out initially, with each particle following its own geodesic. In general, the variation in initial momenta δp^j (where j labels the particle in question) will lead to variations in position δq^i at some final time, given by a matrix equation $\delta q^i = N_{ij}\delta p^j$. However, on the sphere, the result of this thought experiment is that all of the trajectories eventually collapse down and focus at the antipode of q_1. This focal point at q_1^* is known as a **conjugate point**[2] to q_1. The trajectory that passes through q_1^* is the saddle point; the trajectory that does not is the minimum.

Turning now to geometry, the line element is given in terms of metric components by $\mathrm{d}s^2 = g_{\mu\nu}\mathrm{d}x^\mu\mathrm{d}x^\nu$. We shall be interested in the extremal value of the total interval $s = \int \sqrt{|\mathrm{d}s^2|}$ for the path between two points. Whether this extremal interval represents a maximum or minimum interval now depends on the signature of the metric.

Example 8.3

With a **Riemannian metric** (that is, one with signature $++++$) it is always possible to find arbitrarily long paths between two points, but the path length is bounded from below at a minimum value representing the *shortest* path between the two points. This path is a geodesic and represents the straightest possible curve in the space represented by the metric. (However, owing to the discussion in the last example, we cannot say that some given geodesic is necessarily the shortest distance between the two points.)

If we have a Lorentz metric (with signature $-+++$) then the interval between two points is positive if the interval is spacelike, negative if the interval is timelike and zero if the interval is null. It is always possible to find timelike curves with arbitrarily small intervals of proper time linking two points. If a curve of *maximum proper time* exists, it will be a timelike geodesic and represent the straightest path between the two points. This might seem the wrong way round, but follows from the minus sign in front of the timelike component in the metric. As a sanity check, we can confirm that a timelike geodesic does not minimize the length of a curve using a graphical method. Consider Fig. 8.2, showing a timelike curve approximated by a series of null paths. The timelike interval $\Delta\tau = \int \mathrm{d}\tau = \int(-\mathrm{d}s^2)^{\frac{1}{2}}$ is positive, but is infinitesimally close to a path formed from a series of null paths (Fig. 8.2) which, by definition, each have $\mathrm{d}s = 0$, giving a vanishing total interval. It is therefore possible to make the timelike interval arbitrarily small.

We shall parametrize paths in spacetime using an affine parameter λ. To find the geodesic curve $x^\mu(\lambda)$ that extremizes the interval s between two points, we split the interval into elements of length $\mathrm{d}s$ and write

$$s = \int \sqrt{|\mathrm{d}s^2|} = \int \mathrm{d}\lambda \left| \left(\frac{\mathrm{d}s}{\mathrm{d}\lambda} \right)^2 \right|^{\frac{1}{2}} = \int \mathrm{d}\lambda\, L(x^\mu, \dot{x}^\mu), \qquad (8.3)$$

which gives us a method for identifying the function $L(x^\mu, \dot{x}^\mu)$ in this problem as $L = |(\mathrm{d}s/\mathrm{d}\lambda)^2|^{\frac{1}{2}}$. This function must obey the Euler–Lagrange (EL) equation from Chapter 2, whose solution will allow us to identify the geodesic. Let's discuss a simple example.

Example 8.4

In Cartesian coordinates, in two dimensions we write the distance between points
$$\mathrm{d}s^2 = \mathrm{d}x^2 + \mathrm{d}y^2, \qquad (8.4)$$
and so
$$L = \frac{\mathrm{d}s}{\mathrm{d}\lambda} = \left[\left(\frac{\mathrm{d}x}{\mathrm{d}\lambda} \right)^2 + \left(\frac{\mathrm{d}y}{\mathrm{d}\lambda} \right)^2 \right]^{\frac{1}{2}}. \qquad (8.5)$$
The task is to find the shortest path between points in the plane (a spacelike geodesic). We know the answer: the path is, of course, a straight line. Applying the E-L equations for the variable x, we find
$$\frac{\partial L}{\partial \left(\frac{\mathrm{d}x}{\mathrm{d}\lambda} \right)} = \frac{1}{L} \frac{\mathrm{d}x}{\mathrm{d}\lambda}, \qquad \frac{\partial L}{\partial x} = 0. \qquad (8.6)$$
Similar expressions are found for the variable y.

Before powering ahead to solve these equations, it's useful to take a closer look at the idea of parametrizing a path. Notice that the particular recipe for choosing λ along the curve hasn't been specified. Considering the interval from the previous example, we see that the choice of λ is arbitrary: simply scaling $\lambda \to a\lambda + b$ has no effect on the action.[3]

We shall exploit the freedom to choose λ to make life easy for us. We vary the Lagrangian the first time, to find $\partial L/\partial x$ and $\partial L/\partial \dot{x}$, with an as-yet-unspecified parametrization λ. This tells us how the action changes with x and $\dot{x} = \mathrm{d}x/\mathrm{d}\lambda$. After this stage, the dependence of the interval on the parameter λ has been determined, but the precise choice of λ is still unspecified.[4] Next, we make a choice of λ. The labour-saving parametrization to choose is called **length parametrization**. This choice is simply that $\mathrm{d}\lambda = \mathrm{d}s$, which implies that the parameter λ simply measures the interval along the length of curve. The physical interpretation of this choice is that the parameter λ represents the proper time for timelike paths or proper length for spacelike paths. So for timelike paths, λ is the (proper) time measured by the observer in their locally flat spacetime as they fall along the geodesic. Length parametrization is therefore not just convenient, it is necessary to allow us to interpret s as the interval between events in spacetime.

In the case of Cartesian coordinates discussed above, we would write $\mathrm{d}\lambda = \mathrm{d}s = \left(\mathrm{d}x^2 + \mathrm{d}y^2 \right)^{\frac{1}{2}}$, which, inserted into the Lagrangian yields

$$L = \left[\left(\frac{\mathrm{d}x}{\mathrm{d}\lambda} \right)^2 + \left(\frac{\mathrm{d}y}{\mathrm{d}\lambda} \right)^2 \right]^{\frac{1}{2}} = 1. \qquad (8.10)$$

Fig. 8.2 A smooth timelike curve can be represented as being approximated by a series of null paths. As the number of null paths is increased we get closer to the timelike curve. A timelike curve is in this sense infinitesimally close to a series of curves of zero length.

[3]If we have another parameter η, such that $\lambda = \lambda(\eta)$, we write
$$\frac{\mathrm{d}x}{\mathrm{d}\lambda} = \frac{\mathrm{d}x}{\mathrm{d}\eta} \frac{\mathrm{d}\eta}{\mathrm{d}\lambda}, \qquad (8.7)$$
and also the differential
$$\mathrm{d}\lambda = \frac{\mathrm{d}\lambda}{\mathrm{d}\eta} \mathrm{d}\eta. \qquad (8.8)$$
We see that the interval becomes
$$s = \int \mathrm{d}\lambda \left[\left(\frac{\mathrm{d}x}{\mathrm{d}\lambda} \right)^2 + \left(\frac{\mathrm{d}y}{\mathrm{d}\lambda} \right)^2 \right]^{\frac{1}{2}}$$
$$= \int \frac{\mathrm{d}\lambda}{\mathrm{d}\eta} \mathrm{d}\eta \left[\left(\frac{\mathrm{d}x}{\mathrm{d}\eta} \frac{\mathrm{d}\eta}{\mathrm{d}\lambda} \right)^2 \right.$$
$$\left. + \left(\frac{\mathrm{d}y}{\mathrm{d}\eta} \frac{\mathrm{d}\eta}{\mathrm{d}\lambda} \right)^2 \right]^{\frac{1}{2}}$$
$$= \int \mathrm{d}\eta \left[\left(\frac{\mathrm{d}x}{\mathrm{d}\eta} \right)^2 + \left(\frac{\mathrm{d}y}{\mathrm{d}\eta} \right)^2 \right]^{\frac{1}{2}}.$$
This implies that we can just as well use η as λ and expect no change in the form of the action integral.

[4]One can think of λ as cancelling in the left-hand side of the E-L equation
$$\frac{\mathrm{d}}{\mathrm{d}\lambda} \frac{\partial L}{\partial \left(\frac{\mathrm{d}x}{\mathrm{d}\lambda} \right)} = \frac{\partial L}{\partial x}. \qquad (8.9)$$

This does not imply that we are varying unity (with the inevitable result that the terms in the equation vanish): we have already taken a first set of derivatives and so the dependence on λ is now fixed. A consequence of the choice of parametrization is that, in addition to the equations of motion derived from the Euler–Lagrange equations, we also have an addition constraint equation given by, for our choice, by $L = 1$. There is some redundancy here: the Euler–Lagrange equations give us enough information to find the extremal path. However, the extra equation frequently simplifies the algebra and is therefore a valuable help.

After this detour, let's now return to our simple example.

Example 8.5

Use our freedom to choose λ such that $L = 1$ for the remainder of the calculation. We then have

$$\frac{\mathrm{d}}{\mathrm{d}\lambda}\left(\frac{\partial L}{\partial\left(\frac{\mathrm{d}x}{\mathrm{d}\lambda}\right)}\right) = \frac{\mathrm{d}}{\mathrm{d}\lambda}\left(\frac{1}{L}\frac{\mathrm{d}x}{\mathrm{d}\lambda}\right) = \frac{\mathrm{d}^2 x}{\mathrm{d}\lambda^2} = 0. \tag{8.11}$$

and also, of course, $\mathrm{d}^2 y/\mathrm{d}\lambda^2 = 0$.

Since the double derivative of each of the coordinates is zero, the path must represent a straight line. We can check that these three equations are solved with the equation for a straight line

$$y(\lambda) = mx(\lambda) + c, \tag{8.12}$$

with m constant, by using the parametrization $x = \lambda$ and $y = m\lambda + c$. We find

$$\tfrac{\mathrm{d}x}{\mathrm{d}\lambda} = 1, \quad \tfrac{\mathrm{d}y}{\mathrm{d}\lambda} = m, \tag{8.13}$$

and so both of these equations will, of course, give zero if differentiated again with respect to λ.

Now for a more interesting example. We can use the same techniques to find the equations describing the geodesics on the surface of a unit sphere.

Example 8.6

The line element on the surface of a sphere with fixed radius $r = 1$ is given by

$$\mathrm{d}s^2 = \mathrm{d}\theta^2 + \sin^2\theta\,\mathrm{d}\phi^2. \tag{8.14}$$

The parametrized interval is then given by

$$s = \int \mathrm{d}\lambda \left[\left(\frac{\mathrm{d}\theta}{\mathrm{d}\lambda}\right)^2 + \sin^2\theta\left(\frac{\mathrm{d}\phi}{\mathrm{d}\lambda}\right)^2\right]^{\frac{1}{2}}. \tag{8.15}$$

Feeding the integrand into the E-L equations we obtain, at a first stage[5]

$$\frac{\partial L}{\partial\dot\theta} = \frac{\dot\theta}{L}, \qquad \frac{\partial L}{\partial\dot\phi} = \sin^2\theta\frac{\dot\phi}{L},$$
$$\frac{\partial L}{\partial\theta} = \sin\theta\cos\theta\frac{\dot\phi^2}{L}, \qquad \frac{\partial L}{\partial\phi} = 0. \tag{8.16}$$

Invoking length parametrization, we set $L = 1$ and we have that

$$\tfrac{\mathrm{d}}{\mathrm{d}\lambda}\tfrac{\partial L}{\partial\dot\theta} = \ddot\theta, \quad \tfrac{\mathrm{d}}{\mathrm{d}\lambda}\tfrac{\partial L}{\partial\dot\phi} = 2\sin\theta\,\dot\theta\dot\phi + \sin^2\theta\,\ddot\phi. \tag{8.17}$$

[5] The dot notation is used here to denote a derivative with respect to the parameter λ.

The equations of motion can then be arranged[6]

$$\frac{\mathrm{d}^2\theta}{\mathrm{d}\lambda^2} - \sin\theta\cos\theta\left(\frac{\mathrm{d}\phi}{\mathrm{d}\lambda}\right)^2 = 0, \tag{8.19}$$

$$\frac{\mathrm{d}}{\mathrm{d}\lambda}\left(\sin^2\theta\frac{\mathrm{d}\phi}{\mathrm{d}\lambda}\right) = 0. \tag{8.20}$$

These equations are solved if ϕ is constant and θ increases linearly with λ (path A in Fig. 8.3). There is a similar solution where $\theta = \pi/2$ and ϕ increases linearly with λ (path B in Fig. 8.3). These solutions are familiar as the shortest distances between points on a sphere since they are arcs of great circles. Generally, $\theta = \mathrm{const.}$ is not a solution and so path C in Fig. 8.3, which has $\theta = \pi/4$, is not a solution.[7]

With some geodesics under our belt, we now turn to the more general problem of finding the geodesic representing path of a particle in free fall.

8.2 A geodesic equation

We define the covariant acceleration vector for a massive particle as $\boldsymbol{a} = \mathrm{D}\boldsymbol{u}/\mathrm{d}\tau$, where $\boldsymbol{u}$ is the particle's velocity and the proper time τ parametrizes the world line. A particle in free fall follows a geodesic. It feels no force (by definition of free fall), so has no covariant acceleration, giving $\mathrm{D}\boldsymbol{u}/\mathrm{d}\tau = 0$. Geometrically, the particle's velocity is tangent to its world line, so an equivalent expression is $\boldsymbol{a} = \boldsymbol{\nabla}_{\boldsymbol{u}}\boldsymbol{u} = 0$. Recall that parallel transport of a vector $\boldsymbol{v}$ along a path with tangent $\boldsymbol{u}$ implies that the covariant derivative $\boldsymbol{\nabla}_{\boldsymbol{u}}\boldsymbol{v}$ vanishes. So we now arrive at a geometrical definition that *the geodesic is a path in spacetime that parallel transports its own tangent vector.* Although we've discussed the dynamics of a particle, this geometric definition applies to any geodesic (e.g. a spacelike one) parametrized by an arbitrary affine parameter λ. Therefore, we can test if a curve with tangent vector $\boldsymbol{u}$ is a geodesic using

$$\frac{\mathrm{D}\boldsymbol{u}}{\mathrm{d}\lambda} = \boldsymbol{\nabla}_{\boldsymbol{u}}\boldsymbol{u} = 0. \tag{8.21}$$

Using the expression from the previous chapter, we have an equation for components

$$(\boldsymbol{\nabla}_{\boldsymbol{u}}\boldsymbol{u})^\mu = u^\alpha\left(\frac{\partial u^\mu}{\partial x^\alpha} + u^\beta\Gamma^\mu{}_{\alpha\beta}\right) = 0. \tag{8.22}$$

We then have, using $u^\alpha = \mathrm{d}x^\alpha/\mathrm{d}\lambda$, that

$$\frac{\mathrm{d}x^\alpha}{\mathrm{d}\lambda}\frac{\partial u^\mu}{\partial x^\alpha} + \frac{\mathrm{d}x^\alpha}{\mathrm{d}\lambda}\frac{\mathrm{d}x^\beta}{\mathrm{d}\lambda}\Gamma^\mu{}_{\alpha\beta} = 0. \tag{8.23}$$

We notice that the first term can be written as $\mathrm{d}u^\mu/\mathrm{d}\lambda$, which is simply the acceleration in the ordinary, flat, Cartesian system: the double derivative of x^μ with respect to λ. This allows us to write a differential equation for the path of a geodesic, known as the **geodesic equation**

$$\frac{\mathrm{d}^2 x^\mu}{\mathrm{d}\lambda^2} + \frac{\mathrm{d}x^\alpha}{\mathrm{d}\lambda}\frac{\mathrm{d}x^\beta}{\mathrm{d}\lambda}\Gamma^\mu{}_{\alpha\beta} = 0. \tag{8.24}$$

[6]The latter equation can be expanded to read

$$\frac{\mathrm{d}^2\phi}{\mathrm{d}\lambda^2} + 2\frac{\cos\theta}{\sin\theta}\frac{\mathrm{d}\theta}{\mathrm{d}\lambda}\frac{\mathrm{d}\phi}{\mathrm{d}\lambda} = 0. \tag{8.18}$$

[7]For example, set $\theta(\lambda) = \lambda$ and $\phi(\lambda) = 0$ and the equations are solved. Similarly $\theta = \pi/2$ and $\phi = \lambda$ solve the equations. Setting $\theta = \pi/4$ and $\phi = \lambda$ does not solve the equations.

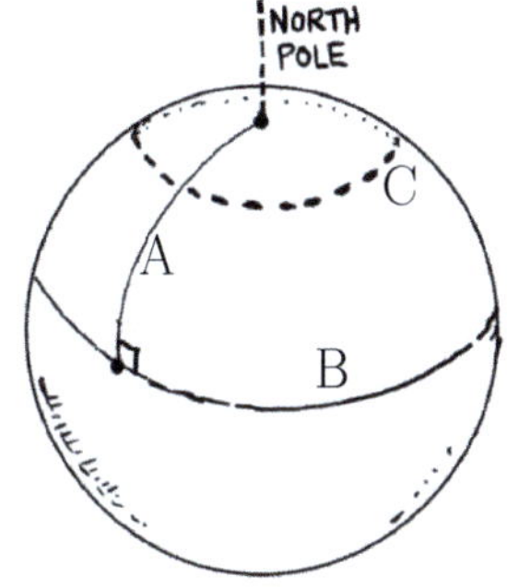

Fig. 8.3 Paths on the surface of a sphere. A (which runs from the North Pole to the equator) and B (which runs round the equator) are geodesics; C (dashed curve) is not.

[8] Keep in mind that instead of τ, we are free to use any affine parameter λ related to τ via $\lambda = a\tau + b$, where a and b are constants. In fact, this provides us with another definition of an affine parameter: they are those parameters for which the description of the world line has the form of the geodesic equation. Note that intervals for photons cannot be assigned a proper time (or length) since they travel along null geodesics. In terms of an affine parameter σ, the infinitesimal interval between two closely spaced events on a photon's world line is

$$g_{\mu\nu}\frac{\mathrm{d}x^\mu}{\mathrm{d}\sigma}\frac{\mathrm{d}x^\nu}{\mathrm{d}\sigma} = 0, \tag{8.25}$$

where $\mathrm{d}x^\sigma$ is the coordinate interval between the events. We discuss photons further at the end of this chapter.

[9] For photons, we take the momentum to be tangent to their null world line, and so $\boldsymbol{\nabla_p p} = 0$ also encodes momentum conservation in this case.

Taking $\lambda = \tau$, this is the equation of motion for a massive particle in curved spacetime in the absence of an external force. We can interpret the geodesic equation as telling us that particles that aren't subject to an external force freely fall along a geodesic, following what a local observer would interpret to be straight lines.[8]

Example 8.7

The expression for velocity $\boldsymbol{u} \cdot \boldsymbol{u}$, involving the tangent vector of a geodesic $\boldsymbol{u}$, determines whether it is timelike ($\boldsymbol{u} \cdot \boldsymbol{u} = -1$) null ($= 0$), or spacelike (where, if we take the affine parameter λ to be the proper length, we will have $\boldsymbol{u} \cdot \boldsymbol{u} = 1$). We can see how this quantity changes along a geodesic by evaluating

$$\boldsymbol{\nabla_u}(\boldsymbol{u} \cdot \boldsymbol{u}) = 2\boldsymbol{u} \cdot (\boldsymbol{\nabla_u u}) = 0, \tag{8.26}$$

where the zero on the right-hand side of the last expression follows because $\boldsymbol{\nabla_u u} = 0$ for a geodesic. We conclude that a timelike tangent vector is always timelike along a timelike geodesic, and similarly for null and spacelike vectors.

The geodesic equation in its geometric version, $\boldsymbol{\nabla_u u} = 0$, can also be used to motivate a geometric form of **momentum conservation**, which can be expressed as

$$\boldsymbol{\nabla_p p} = 0, \tag{8.27}$$

where $\boldsymbol{p}$ is the momentum.

Example 8.8

Proof: For massive particles the components of $\boldsymbol{\nabla_p p}$ may be written as

$$(\boldsymbol{\nabla_p p})^\beta = mu^\alpha(p^\beta{}_{,\alpha} + \Gamma^\beta{}_{\alpha\sigma}p^\sigma)$$

$$= m\left(\frac{\mathrm{d}p^\beta}{\mathrm{d}\tau} + u^\alpha p^\sigma \Gamma^\beta{}_{\alpha\sigma}\right). \tag{8.28}$$

We can always find a locally flat local inertial frame (LIF) coordinate system, where all of the Γ's vanish at a point, and therefore flat-space momentum conservation (or $\frac{\mathrm{d}\boldsymbol{p}}{\mathrm{d}\tau} = 0$) can be written as $\boldsymbol{\nabla_p p} = 0$. As this is a valid tensor equation in a flat frame, by the principle of general covariance, the same equation must also be true in any frame.[9]

8.3 Inertial forces

The geodesic equation tells us that if spacetime gives non-zero connection coefficients $\Gamma^\mu{}_{\alpha\beta}$, we observe an acceleration, even in the absence of an external force. This acceleration could be due to spacetime being curved, or simply to our choice of coordinates. We are used to accelerations resulting from forces and so we interpret the acceleration $\ddot{x}^\mu$ that results from the connections as corresponding to an **inertial force**. The inertial force on particle of mass m is given by the geodesic equation as

$$f^\mu_{\text{inertial}} = m\ddot{x}^\mu = -m\Gamma^\mu{}_{\alpha\lambda}\dot{x}^\alpha\dot{x}^\lambda, \tag{8.29}$$

where dot notation means $\dot{x}^\mu = \mathrm{d}x^\mu/\mathrm{d}\tau$.

Example 8.9

In classical mechanics, if you move in an accelerating frame of reference such as the rotating earth, you feel an inertial force. Although these are sometimes known as fictional forces, there is little fictional about them to the person experiencing them.

Consider the game swingball[10] where a tennis ball is connected by a string to a post (as shown in Fig. 8.4) and executes circular motion in the horizontal plane. In the frame of the players, the ball, with mass m accelerates as it swings around the pole. Referring to the figure, the tension T in the string supplies a vertical component $T\cos\theta = mg$ that balances the gravitational force. The horizontal component $T\sin\theta$ is not cancelled, but instead supplies the centripetal force that maintains the circular motion, equal to mv^2/r, where v is the velocity and r the radius of the circular motion. From the local frame of the tennis ball the picture is rather different. The ball is not accelerating horizontally in this frame, so the inward-directed force $T\sin\theta$ must be balanced by a new force $\vec{F}$. This is the **centrifugal force**: an outward-directed force felt by the ball as a real force, but not evident in the players' frame.

The principle of equivalence teaches us that the acceleration due to gravitation, for a single particle, is indistinguishable from the acceleration due to a particular choice of coordinates. Since free particles fall along geodesics, the geodesic equation says that acceleration a^μ is given by $\ddot{x}^\mu = -\dot{x}^\alpha \dot{x}^\lambda \Gamma^\mu{}_{\alpha\lambda}$, which is to say that gravitation gives rise to an inertial force.

Example 8.10

Consider the metric that describes a weak gravitational field

$$\mathrm{d}s^2 = -[1 + 2\Phi(x,y,z)]\,\mathrm{d}t^2 + (\mathrm{d}x^2 + \mathrm{d}y^2 + \mathrm{d}z^2), \tag{8.30}$$

where $\Phi(x,y,z) = -GM/(x^2+y^2+z^2)^{\frac{1}{2}}$ is the gravitational potential. This can be used to gives an equation of motion[11]

$$\frac{\mathrm{d}^2 x^i}{\mathrm{d}\tau^2} + \frac{\partial\Phi}{\partial x^i}\left(\frac{\mathrm{d}t}{\mathrm{d}\tau}\right)^2 = 0, \tag{8.31}$$

where $x^i = (x,y,z)$. Comparing this to the geodesic equation, we read off the (only non-zero) connection coefficients are

$$\Gamma^i{}_{tt} = \frac{\partial\Phi}{\partial x^i} = GM\frac{x^i}{r^3}, \tag{8.32}$$

where $r^2 = x^2 + y^2 + z^2$. In a non-relativistic limit, we have $\tau \approx t$. The connection therefore supplies the components of the gravitational force $-\Gamma^i_{tt}$ in the geodesic equation.

If a particle is subject to an externally applied 4-force $\boldsymbol{f}$ with components f^μ, then this supplies a non-zero right-hand side of the geodesic equation and we have, in components,

$$m\left(\frac{\mathrm{d}^2 x^\mu}{\mathrm{d}\tau^2} + \frac{\mathrm{d}x^\alpha}{\mathrm{d}\tau}\frac{\mathrm{d}x^\lambda}{\mathrm{d}\tau}\Gamma^\mu{}_{\alpha\lambda}\right) = f^\mu. \tag{8.33}$$

[10]Swingball is more usually known as tetherball outside Britain.

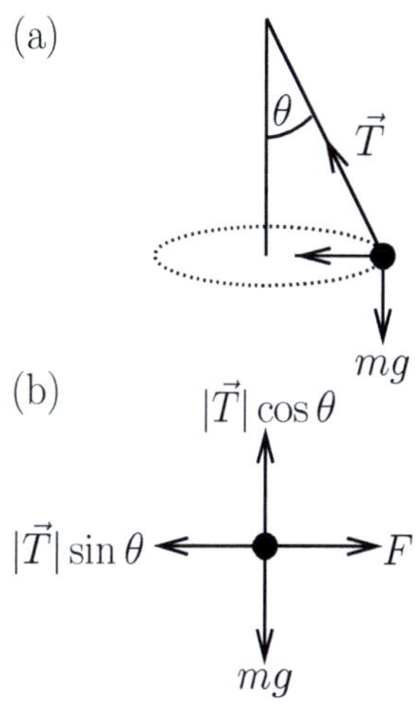

Fig. 8.4 Swingball in (a) the frame of the players and (b) the frame of the ball.

[11]See the next chapter for details of the method.

[12]Recall that in two-dimensional cylindrical coordinates we have non-zero connection coefficients $\Gamma^\theta{}_{r\theta} = \frac{1}{r}$ and $\Gamma^r{}_{\theta\theta} = -r$.

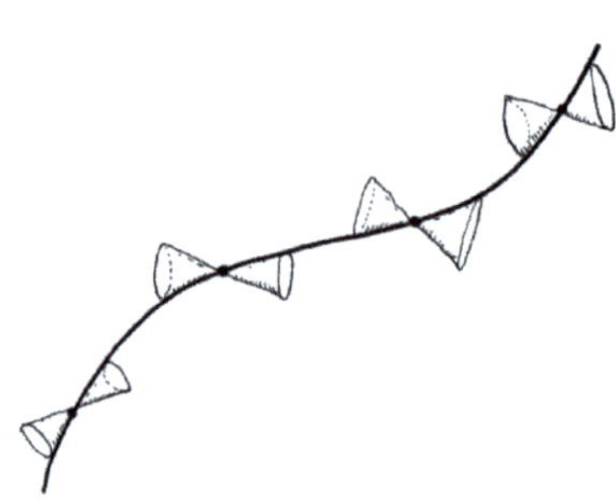

Fig. 8.5 A null world line. The light cones are always tangent to the line.

[13]We summarize the key distinction between the paths of massive particles and massless photons:
- Free massive particles follow *timelike* geodesics
- Photons follow *null* geodesics

The behaviour of free massive particles is sometimes called the **geodesic principle**.

[14]See Section 14.1 for a derivation of eqn 8.37, though a more detailed derivation will be given in Chapter 45.

Example 8.11

We can use the connection coefficients from the last chapter for motion in cylindrical coordinates to derive equations of motion using the geodesic equation. In the non-relativistic limit, we take $\tau \approx t$ and write[12]

$$\frac{\mathrm{d}^2 x^\mu}{\mathrm{d}t^2} + \Gamma^\mu{}_{\alpha\beta}\frac{\mathrm{d}x^\alpha}{\mathrm{d}t}\frac{\mathrm{d}x^\beta}{\mathrm{d}t} = \frac{f^\mu}{m}, \qquad (8.34)$$

which gives us two equations. The radial part is

$$f^r/m = \frac{\mathrm{d}^2 r}{\mathrm{d}t^2} + \Gamma^r{}_{\alpha\beta}u^\alpha u^\beta$$
$$= a^r + \Gamma^r{}_{\theta\theta}u^\theta u^\theta = a^r - r\left(u^\theta\right)^2, \qquad (8.35)$$

where the angular velocity $u_\theta = \dot\theta$. The angular part is

$$f^\theta/m = \frac{\mathrm{d}^2\theta}{\mathrm{d}t^2} + \Gamma^\theta{}_{\alpha\beta}u^\alpha u^\beta$$
$$= a^\theta + 2\Gamma^\theta{}_{r\theta}u^r u^\theta = a^\theta + \frac{2u^r u^\theta}{r}. \qquad (8.36)$$

The trajectory of a particle following uniform circular motion in flat spacetime is not a geodesic: the geodesics are straight lines. From eqn 8.35 we see that if a particle is undergoing uniform circular motion in this coordinate system we have $a^r = 0$ and so an inward-directed external force f^r (the centripetal force) must be applied in order to balance the inertial force $mr(u^\theta)^2$. If we want the circular motion to be uniform with $a^\theta = 0$, then having $u^r = 0$ guarantees no component of force is needed in the θ direction.

8.4   Geodesics for photons

One of the most striking predictions of general relativity is that curved spacetime affects the motion of light. Photons, the particles of light, are massless and in the absence of interactions, fall along **null geodesics**.[13] At each point along null geodesic a light cone will be tangent to the curve, as shown in Fig. 8.5. A simply way to analyse the paths of light is, therefore, to consider directly the constraint $\mathrm{d}s^2 = 0$ for photons, as demonstrated in the following example.

Example 8.12

Consider a light ray travelling in a weak, Newtonian gravitational field. Earlier we used the line element in eqn 8.30 to describe the geometry. Recall that the more accurate expression, which includes the correction to the spacelike parts, is given by[14]

$$\mathrm{d}s^2 = -\left[1 + 2\Phi(x,y,z)\right]\mathrm{d}t^2 + \left[1 - 2\Phi(x,y,z)\right]\left(\mathrm{d}x^2 + \mathrm{d}y^2 + \mathrm{d}z^2\right), \qquad (8.37)$$

where $\Phi(x,y,z) = -GM/r$. Setting $\mathrm{d}s^2 = 0$ for photons, we find an expression for the null geodesics in terms of the coordinates r and t which is

$$\frac{\mathrm{d}r}{\mathrm{d}t} = \left(\frac{r + GM}{r - GM}\right)^{\frac{1}{2}}, \qquad (8.38)$$

which, as we saw in Chapter 5, can be used to investigate the light cone structure of this geometry.

This expression can also be used to demonstrate a famous relativistic effect known as **Shapiro time delay**. A light pulse is sent from a distant planet to the Earth, passing close to the Sun, with a distance of closest approach of b, as shown in Fig. 8.6. If we imagine light travelling between two coordinate points $(-d_\mathrm{p}, y_0, z_0)$ and (d_e, y_0, z_0) along the x-direction, then $\mathrm{d}r = \mathrm{d}x$. Expanding the equation describing the geodesic in the limit of small[15] GM/r we find the coordinate time elapsed between the events is

$$t = \int \mathrm{d}x \left(1 + \frac{2GM}{r}\right) = \int \mathrm{d}x \left[1 + \frac{2GM}{(x^2 + y^2 + z^2)^{\frac{1}{2}}}\right]. \qquad (8.39)$$

For a path from $x = 0$ to $x = x_0$, we can integrate[16] to find, for small b/x_0,

$$t \approx x_0 + 2GM \ln \frac{2|x_0|}{b}, \qquad (8.40)$$

to leading order. The time taken by the light pulse is the usual time $t = x_0/c$ plus a relativistic delay, caused by the gravitational field of the Sun. For the geometry in the figure, in which the delay for each leg of the journey adds, we find a total relativistic time delay of

$$\Delta t = 2GM \ln \frac{4d_\mathrm{p}d_\mathrm{e}}{b^2}. \qquad (8.41)$$

Irwin Shapiro (1929–) suggested this as a fourth test of general relativity. The other three so-called **classical solar-system tests of general relativity** are: (i) the perihelion precession of Mercury; (ii) the deflection of light by the Sun; (both described in Chapter 23) (iii) the Gravitational redshift of light (Chapter 13).

[15] Restoring factors of c this is the limit of small $GM/c^2 r$.

[16] Use the result

$$\int \frac{\mathrm{d}x}{\sqrt{x^2 + a^2}} = \ln\left(x + \sqrt{x^2 + a^2}\right).$$

Chapter summary

- Geodesics are the paths that free particles fall along in general relativity. They can be found by extremizing the path between two events using the calculus of variations, which is much simplified if length parametrization is used. *Free massive particles follow timelike geodesics.*

- The equation of motion

$$m\left(\frac{\mathrm{d}^2 x^\mu}{\mathrm{d}\tau^2} + \Gamma^\mu{}_{\alpha\beta}\frac{\mathrm{d}x^\alpha}{\mathrm{d}\tau}\frac{\mathrm{d}x^\beta}{\mathrm{d}\tau}\right) = f^\mu, \qquad (8.42)$$

relates the acceleration to the geometry and any external forces. The term that includes the connection coefficients gives rise to inertial forces. When $f^\mu = 0$ the equation describes a geodesic.

- *Light travels along null geodesics* and this is easiest to analyse using the null condition $\mathrm{d}s^2 = 0$.

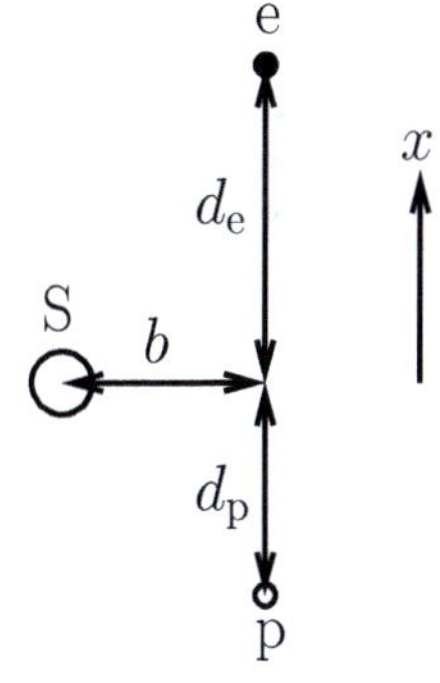

Fig. 8.6 The geometry for the Shapiro time delay, showing the planet (p), and the Earth (e), with the Sun (S) at the origin.

Exercises

(8.1) We will find the shortest distance between two points in flat space, expressed in cylindrical polar coordinates, where interval is written as

$$s = \int \left(\mathrm{d}r^2 + r^2\mathrm{d}\theta^2\right)^{\frac{1}{2}}. \qquad (8.43)$$

(a) Show that the equations of motion are

$$\frac{\mathrm{d}^2 r}{\mathrm{d}\lambda^2} = r\left(\frac{\mathrm{d}\theta}{\mathrm{d}\lambda}\right)^2, \quad \frac{\mathrm{d}}{\mathrm{d}\lambda}\left(r^2 \frac{\mathrm{d}\theta}{\mathrm{d}\lambda}\right) = 0. \tag{8.44}$$

(b) Using the equations of motion, show that the equation for the geodesic obeys

$$r^2 = \lambda^2 + a^2, \tag{8.45}$$

and

$$\frac{\mathrm{d}\theta}{\mathrm{d}\lambda} = \frac{a}{\lambda^2 + a^2}, \tag{8.46}$$

leading to a solution

$$a \tan(\theta - \theta_0) = \lambda. \tag{8.47}$$

(8.2) Despite its complicated appearance, the solution in the previous problem does represent a straight line expressed in cylindrical coordinates. We can show this by referring to Fig. 8.7, which shows how a and θ_0 should be interpreted.

(a) Eliminate λ to show

$$r = \frac{a}{\cos(\theta - \theta_0)}. \tag{8.48}$$

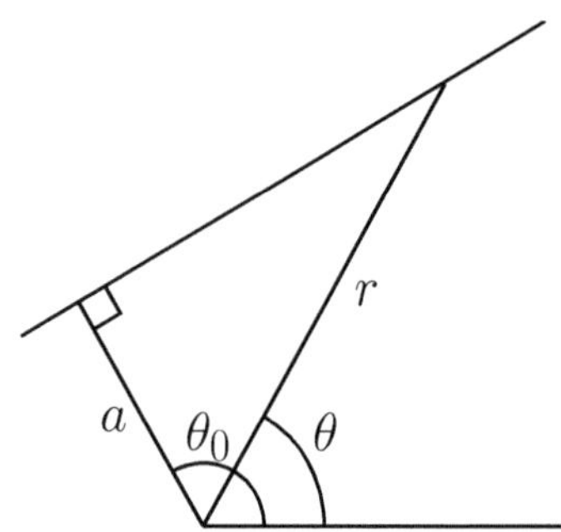

Fig. 8.7 The geometry of a straight line in polar coordinates.

(b) In Cartesian coordinates, a straight line can be written as $\alpha x + \beta y = \gamma$. Using the substitutions

$$a = \frac{\gamma}{(\alpha^2 + \beta^2)^{\frac{1}{2}}}, \quad \cos\theta_0 = \frac{\alpha}{(\alpha^2 + \beta^2)^{\frac{1}{2}}},$$
$$\sin\theta_0 = \frac{\beta}{(\alpha^2 + \beta^2)^{\frac{1}{2}}},$$

show that eqn 8.48 is, indeed, a description of a straight line.

(8.3) *Suppose we did not know about spacetime curvature nor the details of geometry. We would still need to use the geodesic equation, as we shall demonstrate.* Consider a particle which is accelerating. In a frame that moves along with the particle, its position is ξ^μ, the particle can't feel its own weight, which is to say that no forces act on it and it undergoes no acceleration. As a result, in this frame, $d^2\xi^\mu/d\tau^2 = 0$. Now consider how this particle's trajectory appears in some other frame with coordinates x^ν. By using the chain rule, show

$$\frac{\mathrm{d}^2 x^\nu}{\mathrm{d}\tau^2} + \frac{\partial x^\nu}{\partial \xi^\mu}\frac{\partial^2 \xi^\mu}{\partial x^\lambda \partial x^\sigma}\frac{\mathrm{d}x^\lambda}{\mathrm{d}\tau}\frac{\mathrm{d}x^\sigma}{\mathrm{d}\tau} = 0, \tag{8.49}$$

and interpret this equation.

(8.4) The covariant derivative of a 1-form $\tilde{\sigma}$ will be discussed in Part V. For now we can simply note that it can be written as

$$\nabla_\mu \tilde{\sigma} = \sigma_{\nu;\mu} \boldsymbol{\omega}^\nu, \tag{8.50}$$

with

$$\sigma_{\nu;\mu} = \sigma_{\nu,\mu} - \Gamma^\lambda{}_{\mu\nu}\sigma_\lambda. \tag{8.51}$$

(a) Use this to compute a geodesic equation for a velocity 1-form and hence for acceleration $\ddot{x}_\mu$.

(b) Use the result of part (a) to prove that the equation of motion in eqn 8.33 for massive particles has the property that $\boldsymbol{f} \cdot \boldsymbol{u} = 0$, for a force $\boldsymbol{f}$ and instantaneous velocity $\boldsymbol{u}$.

Geodesic equations and connection coefficients

9

I know now that if I break my neck by falling off a cliff, my death is not to be blamed on the force of gravity (what does not exist is necessarily guiltless), but on the fact that I did not maintain the first curvature of my world-line, exchanging its security for a dangerous geodesic.
John Lighton Synge (1897–1995)

Synge's words, quoted above, remind us that our new geometric perspective on gravity motivates us to think about the curvature of our world line.[1] In the last two chapters, we saw that how basis vectors change as we move through spacetime is reflected in the connection coefficients $\Gamma^\mu{}_{\alpha\beta}$ which feature in the geodesic equation, which is the equation of motion for a particle in free fall. The connection coefficients are important, not only for the role they play in the geodesic equation, but because they tell us about the **curvature** of spacetime itself. In this chapter, we describe a method to extract the connection coefficients. As shown in Fig. 9.1, the idea is to input the metric and outputs the connection. The most important point of this chapter is the following.

> The metric field of spacetime, via the line element $\mathrm{d}s^2 = g_{\mu\nu}\mathrm{d}x^\mu\mathrm{d}x^\nu$, generates the geodesics that freely falling particles follow and, therefore, the connection coefficients.

9.1 Finding connection coefficients

Let's formulate our method. The interval between spacetime points a and b can we written as the integral[2]

$$s = \int_a^b \mathrm{d}\lambda \left| g_{\mu\nu} \frac{\mathrm{d}x^\mu}{\mathrm{d}\lambda} \frac{\mathrm{d}x^\nu}{\mathrm{d}\lambda} \right|^{\frac{1}{2}}. \tag{9.3}$$

We use the Euler–Lagrange equations on the integrand to find the equations of motion. We saw in the last chapter that our expressions can be simplified by choosing length parametrization after the first set of derivatives have been taken and this is necessary to interpret s as the interval between spacetime points.[3] The equations of motion can be written in the form of the geodesic equation

$$\frac{\mathrm{d}^2 x^\mu}{\mathrm{d}\lambda^2} + \frac{\mathrm{d}x^\alpha}{\mathrm{d}\lambda} \frac{\mathrm{d}x^\beta}{\mathrm{d}\lambda} \Gamma^\mu{}_{\alpha\beta} = 0. \tag{9.4}$$

[1]A world line which, we very much hope, will avoid any cliff falls.

$$\boxed{\; \boldsymbol{g} \implies \mathrm{d}s^2 \implies \Gamma \;}$$

Fig. 9.1 The metric generates the connection coefficients.

[2]For spacelike curves we have $\mathrm{d}s^2 > 0$, and we can write the interval as

$$\Delta l = \int_a^b \mathrm{d}\lambda \left(g_{\mu\nu} \frac{\mathrm{d}x^\mu}{\mathrm{d}\lambda} \frac{\mathrm{d}x^\nu}{\mathrm{d}\lambda} \right)^{\frac{1}{2}}. \tag{9.1}$$

This is the proper length along the curve. Massive particles traverse timelike curves which have $\mathrm{d}s^2 < 0$. We can write the timelike interval

$$\Delta\tau = \int_a^b \mathrm{d}\lambda \left(-g_{\mu\nu} \frac{\mathrm{d}x^\mu}{\mathrm{d}\lambda} \frac{\mathrm{d}x^\nu}{\mathrm{d}\lambda} \right)^{\frac{1}{2}}. \tag{9.2}$$

This equation gives us the proper time that elapses for an observer travelling along the world line.

[3]We described the need for this in the previous chapter. For massive particles, which follow timelike geodesics, length parametrization, involving setting $\lambda = \tau$, is also needed to ensure that the velocity vector $\boldsymbol{u}$, which is the tangent of the particle's world line, is constrained according to $\boldsymbol{u} \cdot \boldsymbol{u} = -1$.

By comparing the equations of motion we can simply read off the connection coefficients.

The method formalizes the method used in the examples in the previous chapter. It can be summarized as follows:

> **Step I**: From the metric line element $\mathrm{d}s^2$, write a parametrized expression for the spacetime interval $s = \int L\,\mathrm{d}\lambda$, where λ is the parameter.
> **Step II**: Calculate $\frac{\partial L}{\partial\left(\frac{\mathrm{d}x^\mu}{\mathrm{d}\lambda}\right)}$ and $\frac{\partial L}{\partial x^\mu}$.
> **Step III**: Choose length parametrization such that $L = 1$.
> **Step IV**: Calculate $\frac{\mathrm{d}}{\mathrm{d}\lambda}\frac{\partial L}{\partial\left(\frac{\mathrm{d}x^\mu}{\mathrm{d}\lambda}\right)}$ and insert the values into the E-L equations.
> **Step V**: Read off the connection coefficients, remembering that $\Gamma^\mu_{\ \alpha\beta} = \Gamma^\mu_{\ \beta\alpha}$.

We shall work through a number of examples demonstrating how to extract connection coefficients. In each case, we start with a metric and end with connection coefficients.[4]

> ↷ **The rest of this chapter goes through in detail how to extract connection coefficients and calculate geodesics. A reader impatient to get on to the heart of general relativity can skip the rest of this chapter on first reading.**

[4]These geometries reappear in several of the exercises and examples later in the book.

Tips for these calculations:
■ To save on writing, in step I it's sometimes useful to write $\frac{\mathrm{d}x^\mu}{\mathrm{d}\lambda}$ as $\dot{x}^\mu$.
■ In step IV, we often employ the chain rule, noting that $\frac{\mathrm{d}}{\mathrm{d}\lambda} f(x^\mu) = \frac{\partial f(x^\mu)}{\partial x^\mu}\frac{\mathrm{d}x^\mu}{\mathrm{d}\lambda}$.
■ In step V, equations of motion of the form $\ddot{x}^\mu + 2F\dot{x}^\alpha\dot{x}^\beta = 0$ for $\alpha \neq \beta$ yield $\Gamma^\mu_{\ \alpha\beta} = F$, owing to the summation convention in the geodesic equation.
■ The length parametrization condition $L = 1$ is often a useful, additional constraint when solving the equations of motion.

Example 9.1

Let's try the two-dimensional space on the surface of a unit sphere. The interval (step I) is

$$s = \int \left(\mathrm{d}\theta^2 + \sin^2\theta\,\mathrm{d}\phi^2\right)^{\frac{1}{2}}$$

$$= \int \mathrm{d}\lambda \left[\left(\frac{\mathrm{d}\theta}{\mathrm{d}\lambda}\right)^2 + \sin^2\theta\left(\frac{\mathrm{d}\phi}{\mathrm{d}\lambda}\right)^2\right]^{\frac{1}{2}}. \tag{9.5}$$

Therefore, the metric has components $g_{\theta\theta} = 1$ and $g_{\phi\phi} = \sin^2\theta$. We saw [in Exercise 8.1 from the last chapter] that (following steps II–IV) the equations of motion are

$$\frac{\mathrm{d}^2\theta}{\mathrm{d}\lambda^2} - \sin\theta\cos\theta\left(\frac{\mathrm{d}\phi}{\mathrm{d}\lambda}\right)^2 = 0,$$

$$\frac{\mathrm{d}^2\phi}{\mathrm{d}\lambda^2} + 2\cot\theta\frac{\mathrm{d}\phi}{\mathrm{d}\lambda}\frac{\mathrm{d}\theta}{\mathrm{d}\lambda} = 0, \tag{9.6}$$

and we read off the connection coefficients (step V)

$$\Gamma^\theta_{\ \phi\phi} = -\sin\theta\cos\theta, \quad \Gamma^\phi_{\ \theta\phi} = \cot\theta. \tag{9.7}$$

We can examine different curved surfaces, such as the parabolic space of the next example.

Example 9.2

A parabolic surface with line element $\mathrm{d}s^2 = (1+a^2r^2)\mathrm{d}r^2 + r^2\mathrm{d}\theta^2$ has interval (step I)

$$s = \int \mathrm{d}\lambda \left[(1+a^2r^2)\left(\frac{\mathrm{d}r}{\mathrm{d}\lambda}\right)^2 + r^2\left(\frac{\mathrm{d}\theta}{\mathrm{d}\lambda}\right)^2\right]^{\frac{1}{2}}. \tag{9.8}$$

Step II of the method gives us

$$\frac{\partial L}{\partial(\frac{\mathrm{d}r}{\mathrm{d}\lambda})} = \frac{1}{L}(1+a^2r^2)\frac{\mathrm{d}r}{\mathrm{d}\lambda}, \qquad \frac{\partial L}{\partial(\frac{\mathrm{d}\theta}{\mathrm{d}\lambda})} = \frac{1}{L}r^2\frac{\mathrm{d}\theta}{\mathrm{d}\lambda},$$
$$\frac{\partial L}{\partial r} = \frac{1}{L}\left[a^2r\frac{\mathrm{d}r}{\mathrm{d}\lambda} + r\left(\frac{\mathrm{d}\theta}{\mathrm{d}\lambda}\right)^2\right], \qquad \frac{\partial L}{\partial \theta} = 0. \tag{9.9}$$

Now use length parametrization (step III) and find the equations of motion (step IV). Here's the first

$$\frac{\mathrm{d}^2r}{\mathrm{d}\lambda^2} + \frac{a^2r}{1+a^2r^2}\left(\frac{\mathrm{d}r}{\mathrm{d}\lambda}\right)^2 - \frac{r}{1+a^2r^2}\left(\frac{\mathrm{d}\theta}{\mathrm{d}\lambda}\right)^2 = 0. \tag{9.10}$$

And the second

$$\frac{\mathrm{d}^2\theta}{\mathrm{d}\lambda^2} + \frac{2}{r}\frac{\mathrm{d}r}{\mathrm{d}\lambda}\frac{\mathrm{d}\theta}{\mathrm{d}\lambda} = 0. \tag{9.11}$$

We read off the connection coefficients (step V)

$$\Gamma^{\theta}{}_{r\theta} = \frac{1}{r}, \quad \Gamma^{r}{}_{rr} = \frac{a^2r}{1+a^2r^2}, \quad \Gamma^{r}{}_{\theta\theta} = -\frac{r}{1+a^2r^2}. \tag{9.12}$$

As expected, these reduce down to the flat-plane connection coefficients in the case that $a = 0$.

We can examine more exotic spaces still, such as the interesting Poincaré half plane.[5]

[5]Henri Poincaré (1854–1912). The Poincaré half plane provides a model of hyperbolic geometry. See Exercise 9.8 for an introduction to the Poincaré half plane and Chapters 16 and 19 for more discussion of hyperbolic spaces.

Example 9.3

The Poincaré half plane has a metric

$$\mathrm{d}s^2 = \frac{1}{r^2}\left(\mathrm{d}r^2 + \mathrm{d}x^2\right), \tag{9.13}$$

which is defined for $r > 0$. The interval (step I) is

$$s = \int \mathrm{d}\lambda \left[\frac{1}{r^2}\left(\frac{\mathrm{d}r}{\mathrm{d}\lambda}\right)^2 + \frac{1}{r^2}\left(\frac{\mathrm{d}x}{\mathrm{d}\lambda}\right)^2\right]^{\frac{1}{2}}. \tag{9.14}$$

Put this through the standard machine (step II)

$$\frac{\partial L}{\partial(\frac{\mathrm{d}r}{\mathrm{d}\lambda})} = \frac{1}{L}\frac{1}{r^2}\frac{\mathrm{d}r}{\mathrm{d}\lambda}, \qquad \frac{\partial L}{\partial \frac{\mathrm{d}x}{\mathrm{d}\lambda}} = \frac{1}{L}\frac{1}{r^2}\frac{\mathrm{d}x}{\mathrm{d}\lambda},$$
$$\frac{\partial L}{\partial r} = -\frac{1}{L}\frac{1}{r^3}\left[\left(\frac{\mathrm{d}r}{\mathrm{d}\lambda}\right)^2 + \left(\frac{\mathrm{d}x}{\mathrm{d}\lambda}\right)^2\right], \qquad \frac{\partial L}{\partial x} = 0.$$

Now use length parametrization (step III) so that we make $L = 1$. We then find

$$\frac{\mathrm{d}}{\mathrm{d}\lambda}\left(\frac{1}{r^2}\frac{\mathrm{d}r}{\mathrm{d}\lambda}\right) = \frac{1}{r^2}\frac{\mathrm{d}^2r}{\mathrm{d}\lambda^2} - \frac{2}{r^3}\frac{\mathrm{d}r}{\mathrm{d}\lambda}\frac{\mathrm{d}r}{\mathrm{d}\lambda},$$
$$\frac{\mathrm{d}}{\mathrm{d}\lambda}\left(\frac{1}{r^2}\frac{\mathrm{d}x}{\mathrm{d}\lambda}\right) = \frac{1}{r^2}\frac{\mathrm{d}^2x}{\mathrm{d}\lambda^2} - \frac{2}{r^3}\frac{\mathrm{d}r}{\mathrm{d}\lambda}\frac{\mathrm{d}x}{\mathrm{d}\lambda}. \tag{9.15}$$

We obtain equations of motion (step IV)

$$\frac{\mathrm{d}^2r}{\mathrm{d}\lambda^2} - \frac{2}{r}\frac{\mathrm{d}r}{\mathrm{d}\lambda}\frac{\mathrm{d}r}{\mathrm{d}\lambda} = -\frac{1}{r}\left[\left(\frac{\mathrm{d}r}{\mathrm{d}\lambda}\right)^2 + \left(\frac{\mathrm{d}x}{\mathrm{d}\lambda}\right)^2\right],$$
$$\frac{\mathrm{d}^2x}{\mathrm{d}\lambda^2} - \frac{2}{r}\frac{\mathrm{d}r}{\mathrm{d}\lambda}\frac{\mathrm{d}x}{\mathrm{d}\lambda} = 0. \tag{9.16}$$

From the first of these we obtain

$$\frac{\mathrm{d}^2r}{\mathrm{d}\lambda^2} - \frac{1}{r}\left(\frac{\mathrm{d}r}{\mathrm{d}\lambda}\right)^2 + \frac{1}{r}\left(\frac{\mathrm{d}x}{\mathrm{d}\lambda}\right)^2 = 0. \tag{9.17}$$

We read off connection coefficients (step V)

$$\Gamma^{x}{}_{xr} = \Gamma^{x}{}_{rx} = -\frac{1}{r}, \quad \Gamma^{r}{}_{rr} = -\frac{1}{r}, \quad \Gamma^{r}{}_{xx} = \frac{1}{r}. \tag{9.18}$$

The geodesics themselves turn out to be circular arcs and are examined in Exercise 9.8.

9.2 The geodesic equation from the action

[6]Lots more examples can be found in the exercises.

So far we have looked at a selection of special cases, evaluating spacelike intervals for space-only metrics with a $(+++)$ signature.[6] However, using the Euler–Lagrange equations, we should be able to derive the equation of motion for a massive particle in a general spacetime, once and for all, from the action

$$S = -m \int d\tau \left(-g_{\mu\nu} \frac{dx^\mu}{d\tau} \frac{dx^\nu}{d\tau} \right)^{\frac{1}{2}}. \tag{9.19}$$

This action is proportional to the proper time interval along a world line parametrized by the proper time τ. We can therefore extremize the action using the same procedure as before. We already know what the answer must be: the geodesic equation from the previous chapter. However, this procedure will also provide a useful and simple formula for extracting the connection coefficients directly from the metric.[7]

[7]Although the resulting expression is useful, it is often quicker in practice to use the five-point method to extract the coefficients directly from the action.

Example 9.4

We identify (step I)

$$L = -m \left(-g_{\mu\nu} \frac{dx^\mu}{d\tau} \frac{dx^\nu}{d\tau} \right)^{\frac{1}{2}}. \tag{9.20}$$

Step II yields

$$\frac{\partial L}{\partial \dot{x}^\mu} = m g_{\mu\nu} \frac{dx^\nu}{d\tau} \frac{1}{L}, \tag{9.21}$$

along with a force term

$$\frac{\partial L}{\partial x^\nu} = \frac{m}{2} \frac{\partial g_{\mu\sigma}}{\partial x^\nu} \frac{dx^\mu}{d\tau} \frac{dx^\sigma}{d\tau} \frac{1}{L}. \tag{9.22}$$

Choose the parametrization such that $L = 1$ (Step III) and we find, at step IV, that

$$\frac{d}{d\tau} \frac{\partial L}{\partial \dot{x}^\mu} = m \frac{d}{d\tau} \left(g_{\mu\nu} \frac{dx^\mu}{d\tau} \right). \tag{9.23}$$

The Euler–Lagrange equation then reads

$$\frac{d}{d\tau} \left(g_{\mu\nu} \frac{dx^\mu}{d\tau} \right) = \frac{1}{2} \frac{\partial g_{\mu\sigma}}{\partial x^\nu} \frac{dx^\mu}{d\tau} \frac{dx^\sigma}{d\tau}. \tag{9.24}$$

We can evaluate the left-hand side to find

$$\frac{d}{d\tau} \left(g_{\mu\nu} \frac{dx^\mu}{d\tau} \right) = g_{\mu\nu} \frac{d^2 x^\mu}{d\tau^2} + \frac{\partial g_{\mu\nu}}{\partial x^\sigma} \frac{dx^\sigma}{d\tau} \frac{dx^\mu}{d\tau}. \tag{9.25}$$

The Euler–Lagrange equation becomes

$$g_{\mu\nu} \frac{d^2 x^\mu}{d\tau^2} + \left(\frac{\partial g_{\mu\nu}}{\partial x^\sigma} - \frac{1}{2} \frac{\partial g_{\mu\sigma}}{\partial x^\nu} \right) \frac{dx^\mu}{d\tau} \frac{dx^\sigma}{d\tau} = 0. \tag{9.26}$$

We now tidy up by noting that, since the metric components are symmetric,

$$\dot{x}^\mu \dot{x}^\sigma g_{\mu\nu,\sigma} = \frac{\dot{x}^\mu \dot{x}^\sigma}{2} \left(g_{\mu\nu,\sigma} + g_{\sigma\nu,\mu} \right), \tag{9.27}$$

which allows us to conclude (step V):

$$g_{\mu\nu} \frac{d^2 x^\mu}{d\tau^2} + \frac{1}{2} \left(\frac{\partial g_{\mu\nu}}{\partial x^\sigma} + \frac{\partial g_{\sigma\nu}}{\partial x^\nu} - \frac{\partial g_{\mu\sigma}}{\partial x^\nu} \right) \frac{dx^\mu}{d\tau} \frac{dx^\sigma}{d\tau} = 0. \tag{9.28}$$

We define the all-down-index connection coefficients $\Gamma_{\lambda\mu\sigma} = g_{\rho\lambda}\Gamma^{\rho}{}_{\mu\sigma}$, and then we have the following.[8]

$$\Gamma_{\lambda\mu\sigma} = \frac{1}{2}\left(\frac{\partial g_{\lambda\mu}}{\partial x^{\sigma}} + \frac{\partial g_{\lambda\sigma}}{\partial x^{\mu}} - \frac{\partial g_{\mu\sigma}}{\partial x^{\lambda}}\right). \tag{9.29}$$

The conclusion of this lengthy exercise[9] is that, given only the metric, we can work out connection coefficients and have access to the equation of motion of the freely falling particle. Note, however, that since the geodesic equation applies beyond the timelike geodesics followed by massive particles, eqn 9.29 is a general, geometrical expression linking the metric with the connection coefficients.

Example 9.5

Consider the metric line element

$$ds^2 = \frac{1}{t^2}\left(dx^2 - dt^2\right). \tag{9.33}$$

This corresponds to a metric with components $g_{tt} = -1/t^2$ and $g_{xx} = 1/t^2$. The connection coefficients can be calculated using eqn 9.28, to yield

$$\Gamma_{ttt} = \tfrac{1}{t^3}, \quad \Gamma_{xxt} = \Gamma_{xtx} = -\Gamma_{txx} = -\tfrac{1}{t^3}. \tag{9.34}$$

Using $g^{tt} = -t^2$ and $g^{xx} = t^2$, we have $\Gamma^{t}{}_{tt} = \Gamma^{t}{}_{xx} = \Gamma^{x}{}_{xt} = -1/t$.

---

[8]In comma notation:

$$\Gamma_{\lambda\mu\sigma} = \frac{1}{2}\left(g_{\lambda\mu,\sigma} + g_{\lambda\sigma,\mu} - g_{\mu\sigma,\lambda}\right). \tag{9.29a}$$

[9]There are some other useful expressions that can be employed in calculations. One useful identity is

$$g_{\alpha\beta,\gamma} = \Gamma_{\alpha\beta\gamma} + \Gamma_{\beta\alpha\gamma}. \tag{9.30}$$

Another is

$$\Gamma^{\alpha}{}_{\beta\alpha} = \frac{\partial}{\partial x^{\beta}}\ln\sqrt{-g}, \tag{9.31}$$

where g is the determinant of the metric tensor. For a diagonal metric the following hold

$$\begin{aligned}
\Gamma^{\mu}{}_{\nu\lambda} &= 0,\\
\Gamma^{\mu}{}_{\lambda\lambda} &= -\frac{1}{2g_{\mu\mu}}\frac{\partial g_{\lambda\lambda}}{\partial x^{\mu}},\\
\Gamma^{\mu}{}_{\mu\lambda} &= \frac{\partial}{\partial x^{\lambda}}\left(\ln|g_{\mu\mu}|^{\frac{1}{2}}\right),\\
\Gamma^{\mu}{}_{\mu\mu} &= \frac{\partial}{\partial x^{\mu}}\left(\ln|g_{\mu\mu}|^{\frac{1}{2}}\right).
\end{aligned} \tag{9.32}$$

Here $\mu \neq \nu \neq \lambda$ and we don't sum over repeated indices.

Chapter summary

- The connection coefficients may be extracted using a simple routine based on extremizing the action.
- The metric leads directly to the connection coefficients.

Exercises

(9.1) Using the methods described in the chapter, extract the connection coefficients for two-dimensional plane polar coordinates.

(9.2) The torus has a line element

$$ds^2 = (c + a\cos v)^2 du^2 + a^2 dv^2. \tag{9.35}$$

Show that we obtain the connection coefficients

$$\Gamma^{v}{}_{uu} = \frac{\sin v}{u}(c + a\cos v), \quad \Gamma^{u}{}_{uv} = -\frac{a\sin v}{(c + a\cos v)}. \tag{9.36}$$

(9.3) Consider the non-diagonal metric

$$ds^2 = du^2 + dv^2 + 2dudv\cos\theta(u, v). \tag{9.37}$$

Show that the non-zero connection coefficients are given by

$$\begin{aligned}
\Gamma^{u}{}_{uu} &= \frac{\cos\theta}{\sin\theta}\theta_{,u}, & \Gamma^{v}{}_{uu} &= -\frac{1}{\sin\theta}\theta_{,u},\\
\Gamma^{v}{}_{vv} &= \frac{\cos\theta}{\sin\theta}\theta_{,v}, & \Gamma^{u}{}_{vv} &= -\frac{1}{\sin\theta}\theta_{,v}
\end{aligned} \tag{9.38}$$

(9.4) Consider Rindler spacetime with metric line element

$$ds^2 = -x^2 dt^2 + dx^2. \qquad (9.39)$$

(a) Show that the equations of motion are

$$2\dot{x}\dot{t} + x\ddot{t} = 0, \quad \ddot{x} + x\dot{t}^2 = 0, \qquad (9.40)$$

and extract the connection coefficients.
(b) We can use the physical interpretation of the length parametrization to allow some insight into this spacetime. Define the velocity in the x-direction as $v = dx/dt$ and show that

$$\ddot{x} = -\frac{x}{x^2 - v^2}. \qquad (9.41)$$

Now suppose that the particle does not have a velocity in the x-direction, so that $v = 0$. The equation of motion says that in order to have $v = 0$ in this spacetime we must have an observer undergoing a uniform acceleration $\ddot{x} = -1/x$, which diverges as $x \to 0$. We shall see this property again when we examine the spherically symmetric Schwarzschild geometry.

(9.5) Consider the rotating-frame line element

$$ds^2 = - \left[1 - \Omega^2 (x^2 + y^2)\right] dt^2 + dx^2 + dy^2$$
$$+ dz^2 - 2\Omega y \, dx \, dt + 2\Omega x \, dy \, dt. \qquad (9.42)$$

(a) Find the matrix $g^{\mu\nu}$.
(b) Compute the connection coefficients for the space described by this line element.

(9.6) (a) Express the line element from the previous question in cylindrical polars.
(b) Compute the connection coefficients in cylindrical polars.

(9.7) The Schwarzschild metric gives an interval

$$s = \int d\lambda \left[e^{2\Phi} \left(\frac{dt}{d\lambda}\right)^2 - e^{2\Lambda} \left(\frac{dr}{d\lambda}\right)^2 \right.$$
$$\left. -r^2 \left(\frac{d\theta}{d\lambda}\right)^2 - r^2 \sin^2\theta \left(\frac{d\phi}{d\lambda}\right)^2 \right]^{\frac{1}{2}}, \qquad (9.43)$$

where Φ and Λ are functions of r. Find the connection coefficients.

(9.8) *We can start to understand the space represented by the Poincaré half plane in Example 9.3 by computing its geodesics. The geometry is defined by its metric in the upper half plane, $r > 0$, only.*
(a) Consider the equation of motion $\ddot{x} - 2\dot{r}\dot{x}/r = 0$, where the dot indicates a derivative with respect to

the affine parameter λ. Show that this equation is solved by

$$\dot{x} = \frac{r^2}{a}, \qquad (9.44)$$

where a is a constant.
(b) Consider the length parametrization condition $L = 1$ and show that this yields

$$d\lambda = \frac{dt}{\sin t}, \qquad (9.45)$$

where $r = a \sin t$.
(c) Use these results to show that x is given by

$$x = -a \cos t + x_0, \qquad (9.46)$$

where x_0 is a constant offset.
(d) Argue that this shows that the geodesics are circular arcs, centred on $(x, r) = (x_0, 0)$ with radius a, as shown in Fig. 9.2.
(e) Compute the length $\int d\lambda$ of a geodesic starting at $t = a$ and finishing at $t = b$. Use this to show that the length of a geodesic starting at $t = 0$ and ending at $t = \pi$ is infinite.
You can see why this is the case by considering a ruler of interval length Δs parallel to the x axis. Since r is constant we have $\Delta s = \Delta x/r$. As a result, rulers of an equivalent interval length Δs must have larger coordinate length Δx if they are at a larger height r, as shown in Fig. 9.2.

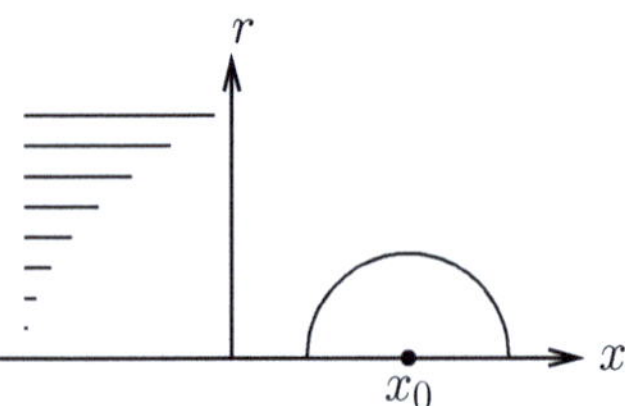

Fig. 9.2 The Poincaré half plane from Exercise 9.8 and Example 9.3. An example geodesic is shown on the right. On the left several lines of equivalent interval length Δs are shown.

(9.9) We end up with the same geodesics if we extremize $L = \sqrt{-g_{\mu\nu}\dot{x}^\mu \dot{x}^\nu}$, and if we extremize $L = \frac{1}{2}g_{\mu\nu}\dot{x}^\mu \dot{x}^\nu$. Show this by finding the Euler–Lagrange equation for a function

$$L = F\left(\sqrt{g_{\mu\nu}\dot{x}^\mu \dot{x}^\nu}\right), \qquad (9.47)$$

where $\dot{x}^\mu = \frac{dx^\mu}{d\lambda}$, λ is the proper length and F is any monotonic function.
This means we can equally well use a Lagrangian $L = \frac{1}{2}g_{\mu\nu}\dot{x}^\mu \dot{x}^\nu$ which resembles the kinetic energy of a non-relativistic particle.

(9.10) Consider the moving-coordinate metric

$$ds^2 = -\left(1 - v^2\right)dt^2 + dx^2 + dy^2 + dz^2 - 2v\,dx\,dt. \tag{9.48}$$

By extremizing the world line, show that the geodesics are straight lines.

(9.11) Consider a two-dimensional space with metric line element

$$ds^2 = (1 + r)dr^2 + r^2 d\phi^2. \tag{9.49}$$

By computing the acceleration, determine whether the curve $r(\lambda) = (3\lambda/2)^{2/3} - 1$, $\phi(\lambda) = 0$, with λ an affine parameter, is a geodesic.

10

Making measurements in relativity

Why were another seven years required for the construction of the general theory of relativity? The main reason lies in the fact that it is not easy to free oneself from the idea that coordinates must have an immediate metrical meaning.
Einstein quoted in P. A. Schilpp (ed.) *Albert Einstein – Philosopher Scientist* (1969).

The stage on which the drama of general relativity is played out is curved spacetime. However, measurements are made *locally* by observers in laboratories. Over the small distances involved in a typical experiment, an observer will experience spacetime as if it were **flat spacetime** with the Minkowski metric. We therefore need to know how to relate the observations made by observers in their local spacetime to the objects we manipulate in the curved spacetime of general relativity. The key is that measurements are made in local orthonormal frames: frames of reference set up by observers where the basis vectors are usually orthogonal and normalized. In this book, component labels in such frames will be given with hats, so that the basis vectors, for example, will be written as $e_{\hat{\alpha}}$. By definition, the metric of the flat, local frame is simply the Minkowski metric, with components $\eta_{\hat{\mu}\hat{\nu}} = e_{\hat{\mu}} \cdot e_{\hat{\nu}} = \mathrm{diag}(-1, 1, 1, 1)$.

A further point to note in these discussions stems from Einstein's observation in the quotation above. When presented with a t coordinate or an r coordinate, it is tempting to assume that t must represent time and r the radius. This is not correct. Coordinates are intrinsically meaningless labels which are only given meaning by relating them to measurements and intervals determined by observers in their local, inertial frames of reference. This can be summed up using the slogan that *coordinates have no immediate metrical significance.*[1]

[1] Or, perhaps more memorably, in the words of the Time Traveller, '*There is no difference between Time and any of the three dimensions of Space except that our consciousness moves along it. But some foolish people have got hold of the wrong side of that idea.*' H. G. Wells (1866–1946) *The Time Machine.*

10.1 Observers and their observations

A particle passes through a laboratory as shown in Fig. 10.1. The particle has a momentum p which, expressed as a vector, is a quantity independent of any set of coordinates. An observer in the laboratory makes measurements by carrying around their own orthonormal axes $e_{\hat{0}}, e_{\hat{1}}, e_{\hat{2}}$ and $e_{\hat{3}}$. Referred to these axes, the vector can be expressed in coordinates as $p = p^{\hat{\mu}} e_{\hat{\mu}}$. To make a measurement of a particular

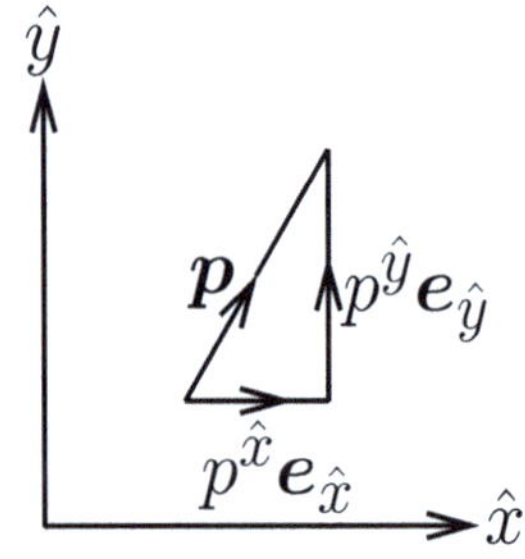

Fig. 10.1 A measurement of momentum p. In this figure, we use the familiar $\hat{x}$ and $\hat{y}$ axes of two-dimensional space (with basis vectors $e_{\hat{x}}$ and $e_{\hat{y}}$), but the idea carries over to the four-dimensional spacetime axes of an orthonormal frame with basis vectors $e_{\hat{0}}, e_{\hat{1}}, e_{\hat{2}}$ and $e_{\hat{3}}$.

component of a vector the observer projects out the component using their local axes. For example, measuring the momentum along the $\hat{\alpha}$ direction means that the observer makes the projection via a dot product $\boldsymbol{p} \cdot \boldsymbol{e}_{\hat{\alpha}}$.

Example 10.1

The observer makes the projection

$$\boldsymbol{p} \cdot \boldsymbol{e}_{\hat{\alpha}} = p^{\hat{\mu}} \boldsymbol{e}_{\hat{\mu}} \cdot \boldsymbol{e}_{\hat{\alpha}}$$
$$= p^{\hat{\mu}} \eta_{\hat{\mu}\hat{\alpha}} = p_{\hat{\alpha}}, \tag{10.1}$$

showing that the observer has access to the component $p_{\hat{\alpha}}$. The observer can use $\eta^{\hat{\alpha}\hat{\beta}}$ to raise the index, if they want the up-index form of the component via $p^{\hat{\beta}} = \eta^{\hat{\alpha}\hat{\beta}} p_{\hat{\alpha}}$.

If we spot an observer, how do we know what their local orthonormal axes will look like? That is, how will they orient their orthonormal coordinates? Start by noting that the observer's world line is characterized by their velocity vector $\boldsymbol{u}_{\text{obs}}$, which is tangent to the world line (Fig. 10.2). The key is that the timelike vector of the local basis $\boldsymbol{e}_{\hat{0}}$ will also be tangent to the observer's world line, since this is the direction that a clock at rest in the observer's frame moves in spacetime. We therefore have

$$\boldsymbol{e}_{\hat{0}} = \boldsymbol{u}_{\text{obs}}. \tag{10.2}$$

Therefore, expressed in some coordinate frame, the observer's timelike axis $\boldsymbol{e}_{\hat{0}}$ has the components we would ascribe to their tangent vector $\boldsymbol{u}_{\text{obs}}$. We write these components of the observer's timelike basis vector

$$(\boldsymbol{e}_{\hat{0}})^{\mu} = u^{\mu}_{\text{obs}}. \tag{10.3}$$

The other components of the observer's orthonormal system can then be picked out, subject to being orthogonal to $\boldsymbol{e}_{\hat{0}}$ and to each other.

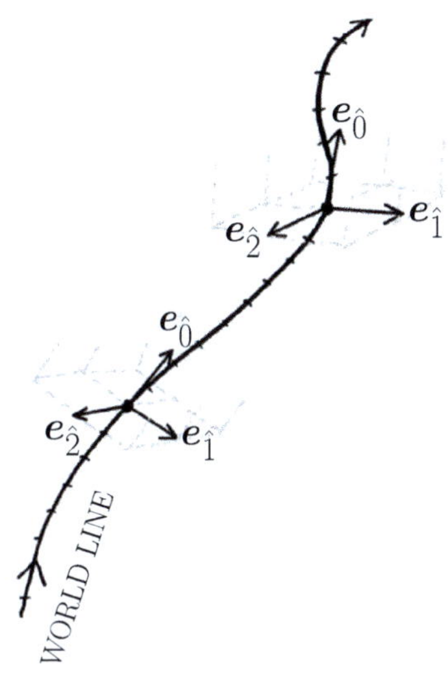

Fig. 10.2 Local orthonormal frames picked out along a world line by setting $\boldsymbol{e}_{\hat{0}} = \boldsymbol{u}$.

Example 10.2

Consider the constantly accelerated observer in Minkowski space from Chapter 2. They have a timelike basis vector with components

$$(\boldsymbol{e}_{\hat{0}})^{\mu} = u^{\mu}_{\text{obs}}(\tau) = (\cosh(g\tau), \sinh(g\tau), 0, 0). \tag{10.4}$$

Pick $\boldsymbol{e}_{\hat{2}}$ and $\boldsymbol{e}_{\hat{3}}$ to point along the y- and z-directions. The remaining 4-vector $\boldsymbol{e}_{\hat{1}}$ has the form $(f(\tau), g(\tau), 0, 0)$. We require orthogonality of the observer's basis vectors, which is to say that

$$\eta_{\mu\nu}(\boldsymbol{e}_{\hat{0}})^{\mu}(\boldsymbol{e}_{\hat{1}})^{\nu} = -\cosh(g\tau)f(\tau) + \sinh(g\tau)g(\tau) = 0. \tag{10.5}$$

We also require that the vectors are normalized in the observer's frame, so that

$$-f^{2}(\tau) + g^{2}(\tau) = 1. \tag{10.6}$$

This allows us to solve for $(\boldsymbol{e}_{\hat{1}})^{\mu}$ and we end up with

$$(\boldsymbol{e}_{\hat{0}})^{\mu} = (\cosh(g\tau), \sinh(g\tau), 0, 0),$$
$$(\boldsymbol{e}_{\hat{1}})^{\mu} = (\sinh(g\tau), \cosh(g\tau), 0, 0),$$
$$(\boldsymbol{e}_{\hat{2}})^{\mu} = (0, 0, 1, 0),$$
$$(\boldsymbol{e}_{\hat{3}})^{\mu} = (0, 0, 0, 1). \tag{10.7}$$

One particularly helpful tool is that the energy of a particle measured by an observer with velocity $\boldsymbol{u}_{\text{obs}}$ is given by $E = -\boldsymbol{p} \cdot \boldsymbol{u}_{\text{obs}}$. This is easily confirmed by noting that

$$-\boldsymbol{p} \cdot \boldsymbol{u}_{\text{obs}} = -p^{\hat{\mu}} \boldsymbol{e}_{\hat{\mu}} \cdot \boldsymbol{e}_{\hat{0}}$$
$$= -p^{\hat{\mu}} \eta_{\hat{\mu}0} = -p_{\hat{0}} = E, \tag{10.8}$$

where, in the final step we remember that $E = p^{\hat{0}} = -p_{\hat{0}}$ because $\eta_{\hat{0}\hat{0}} = -1$ in the orthonormal frame.

Example 10.3

Consider Minkowski space. In a frame where a particle is at rest, the particle has $p^{\mu} = (m, 0, 0, 0)$. Relative to this frame the observer travels with constant speed v along the x-axis, so the 4-velocity of the observer has components $u^{\mu}_{\text{obs}} = (\gamma, \gamma v, 0, 0)$, which are therefore also the components of the local basis vector $(\boldsymbol{e}_{\hat{0}})^{\mu}$. The energy of the particle measured by the observer, when the world lines of particle and observer intersect, is[2]

$$E = -\boldsymbol{p} \cdot \boldsymbol{u}_{\text{obs}} = m\gamma. \tag{10.9}$$

So the particle has energy $E = \gamma mc^2$ (restoring factors of c) as we expect.

Now consider the accelerated observer from the previous example, measuring light from a star in Minkowski space. The star gives out light at a frequency ω. The wave 4-vector of a photon reaching the observer has components[3] $k^{\mu} = (\omega, \omega, 0, 0)$ in the star's rest frame. The method for finding the frequency the observer measures is the same as for the energy, to which frequency is proportional. We evaluate $\omega = -\boldsymbol{k} \cdot \boldsymbol{u}_{\text{obs}}$. Computing, we find

$$\omega(\tau) = -\boldsymbol{k} \cdot \boldsymbol{u}_{\text{obs}}$$
$$= k^0 u^0 - k^1 u^1$$
$$= \omega \left[\cosh(g\tau) - \sinh(g\tau) \right]$$
$$= \omega \exp(-g\tau), \tag{10.10}$$

which demonstrates that the shift in observed frequency varies exponentially.[4]

The procedure above allows us to understand what an observer will measure. However, this is complicated by the fact that the natural, orthonormal coordinate system that observers employ has an unpleasant mathematical property: it is a **non-coordinate basis**, as we now describe.

10.2 Coordinate and non-coordinate bases

Recall from Chapter 3, that plane polar coordinates had the property that $|\boldsymbol{e}_r| = 1$ but that $|\boldsymbol{e}_{\theta}| = r$. That is, the length of the $\boldsymbol{e}_{\theta}$ basis vector is proportional to the distance away from the origin. This coordinate system was derived from the Cartesian one by expressing components as partial derivatives with respect to the Cartesian coordinates. We call such a coordinate system a **coordinate basis**.

We could choose to **normalize** $\boldsymbol{e}_{\theta}$ so that we have $\boldsymbol{e}_{\hat{\theta}} = \boldsymbol{e}_{\theta}/r$, which yields an orthonormal basis set $\boldsymbol{e}_{\hat{r}}(= \boldsymbol{e}_r)$ and $\boldsymbol{e}_{\hat{\theta}}$. Although this is the

[2] We can evaluate the dot product using the Minkowski tensor in the rest frame of the particle. The result is a scalar, so is true in any frame.

[3] In our units, the vector $\boldsymbol{k}$ has components $k^{\mu} = (\omega, k^x, k^y, k^z)$ and $\omega = |\vec{k}|$ for light. We also assume the quantum mechanical relationship $E = \hbar\omega$, but set $\hbar = 1$.

[4] This result, discussed in the book by Hartle, will be seen again in the discussion of black holes in Chapters 26 and 27.

↷ **The rest of this chapter goes through further detail on coordinates and their bases, and how to transform between them. A reader impatient to get on to the topic of curvature may skip to the start of the next chapter on a first reading.**

coordinate set we would most probably want to choose when plotting the position of events in the laboratory, it does not have the property that it is derivable directly from the Cartesian coordinates. That is to say that the basis vectors $\boldsymbol{e}_r$ and $\boldsymbol{e}_\theta$ cannot be written as an expansion in Cartesian basis vectors with prefactors given in terms of derivatives of to the Cartesian components with respect to r and θ (see eqn 3.7). We call such a basis a[5] **non-coordinate basis**.

[5]As described in Chapter 3, a non-coordinate basis can be identified because its basis vectors don't commute, in contrast to the basis vectors of a co-ordinate basis which do commute.

Example 10.4

Non-coordinate bases are useful for many problems. Consider, for example, the Kepler problem, which is conventionally discussed using an orthonormal basis and cylindrical coordinates. In such a frame, the velocity $\boldsymbol{v} = v^{\hat{r}}\boldsymbol{e}_{\hat{r}} + v^{\hat{\theta}}\boldsymbol{e}_{\hat{\theta}}$ is given by

$$\boldsymbol{v} = \frac{\mathrm{d}r}{\mathrm{d}t}\boldsymbol{e}_{\hat{r}} + r\frac{\mathrm{d}\boldsymbol{e}_{\hat{r}}}{\mathrm{d}t}, \tag{10.11}$$

that is, we take a time derivative of $\boldsymbol{r} = r\boldsymbol{e}_{\hat{r}}$. Acceleration is then given by

$$\boldsymbol{a} = \frac{\mathrm{d}\boldsymbol{v}}{\mathrm{d}t} = \frac{\mathrm{d}v^{\hat{r}}}{\mathrm{d}t}\boldsymbol{e}_{\hat{r}} + v^{\hat{r}}\frac{\mathrm{d}\boldsymbol{e}_{\hat{r}}}{\mathrm{d}t} + \frac{\mathrm{d}v^{\hat{\theta}}}{\mathrm{d}t}\boldsymbol{e}_{\hat{\theta}} + v^{\hat{\theta}}\frac{\mathrm{d}\boldsymbol{e}_{\hat{\theta}}}{\mathrm{d}t}. \tag{10.12}$$

The basis vectors, as shown in Fig. 10.3, obey

$$\frac{\mathrm{d}\boldsymbol{e}_{\hat{r}}}{\mathrm{d}t} = \frac{\mathrm{d}\theta}{\mathrm{d}t}\boldsymbol{e}_{\hat{\theta}} = \omega\boldsymbol{e}_{\hat{\theta}}, \tag{10.13}$$

$$\frac{\mathrm{d}\boldsymbol{e}_{\hat{\theta}}}{\mathrm{d}t} = -\frac{\mathrm{d}\theta}{\mathrm{d}t}\boldsymbol{e}_{\hat{r}} = -\omega\boldsymbol{e}_{\hat{r}}, \tag{10.14}$$

where $\omega = \mathrm{d}\theta/\mathrm{d}t$. The components of the acceleration are then written as

$$a^{\hat{r}} = \frac{\mathrm{d}v^{\hat{r}}}{\mathrm{d}t} - v^{\hat{\theta}}\frac{\mathrm{d}\theta}{\mathrm{d}t} = \frac{\mathrm{d}^2 r}{\mathrm{d}t^2} - r\left(\frac{\mathrm{d}\theta}{\mathrm{d}t}\right)^2 = \ddot{r} - r\dot{\theta}^2, \tag{10.15}$$

and

$$a^{\hat{\theta}} = \frac{\mathrm{d}v^{\hat{\theta}}}{\mathrm{d}t} + v^{\hat{r}}\frac{\mathrm{d}\theta}{\mathrm{d}t} = \frac{\mathrm{d}}{\mathrm{d}t}\left(r\frac{\mathrm{d}\theta}{\mathrm{d}t}\right) + \frac{\mathrm{d}r}{\mathrm{d}t}\frac{\mathrm{d}\theta}{\mathrm{d}t} = r\ddot{\theta} + 2\dot{r}\dot{\theta}. \tag{10.16}$$

We shall usually work in coordinate bases owing to their neat geometrical properties. However, observers work in local frames, which is where their measurements are made. These are chosen to be orthonormal and hence usually have non-coordinate bases. We therefore need to be able to transform between coordinate bases (where we do our calculations) and non-coordinate bases (where measurements are made).

The basis vectors in a coordinate basis are related to the metric via the definition

$$\boldsymbol{e}_\mu \cdot \boldsymbol{e}_\nu = g_{\mu\nu}. \tag{10.17}$$

Measurements are made in the local orthonormal frame, with Minkowski metric whose components are related to the dot product of the basis vectors via

$$\boldsymbol{e}_{\hat{\alpha}} \cdot \boldsymbol{e}_{\hat{\beta}} = \eta_{\hat{\alpha}\hat{\beta}}. \tag{10.18}$$

Any vector can be decomposed in either system:

$$\boldsymbol{a} = a^\alpha \boldsymbol{e}_\alpha = a^{\hat{\beta}}\boldsymbol{e}_{\hat{\beta}}. \tag{10.19}$$

See Chapter 20 for a discussion of the Kepler problem.

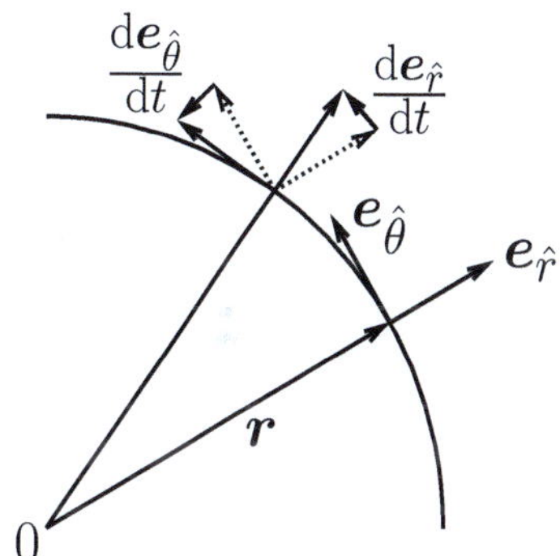

Fig. 10.3 The change of the vector $\boldsymbol{e}_{\hat{r}}$ is in the direction $\boldsymbol{e}_{\hat{\theta}}$, which the change in $\boldsymbol{e}_{\hat{\theta}}$ is in the direction $-\boldsymbol{e}_{\hat{r}}$.

In order to transform between these two descriptions, we write the components of the orthonormal basis vectors in the coordinate frame as $(e_{\hat{\beta}})^{\alpha}$, giving us an expression

$$a^{\alpha} = a^{\hat{\beta}}(e_{\hat{\beta}})^{\alpha}. \tag{10.20}$$

Objects such as $(e_{\hat{\beta}})^{\alpha}$ are a set of matrices known as the components of a **vielbein**.[6] These turn out to be very useful.[7] We can also write the components of the coordinate basis vectors in the orthonormal frame, which leads to the expression

$$a^{\hat{\beta}} = a^{\alpha}(e_{\alpha})^{\hat{\beta}}. \tag{10.21}$$

[6]Which translates from German into English as many-leg. Since the vielbein with which we're concerned describes (3+1)-dimensional spacetime, it is sometimes called a vierbein ($\equiv$ 4-leg). The bracket in the notation $(e_{\mu})^{\hat{\alpha}}$ is really just there for aesthetic reasons to remind us that a vielbein combines information about two different sorts of coordinate systems. We could write the components $e_{\mu}^{\hat{\alpha}}$ if we prefer.

[7]One way to think of the action of a vielbein is that the coordinate frame possesses a set of global coordinates [e.g. (t, r, θ, ϕ)] and that we make them local using the vielbein.

Example 10.5

Consider cylindrical-polar coordinates. The coordinate basis consists of vectors e_r and e_{θ} and we write components (r, θ). The metric is given by the line element

$$ds^2 = dr^2 + r^2 d\phi^2. \tag{10.22}$$

The off-diagonal elements of the metric are zero; while the diagonal components are $g_{rr} = 1$ and $g_{\theta\theta} = r^2$. From $g_{\mu\nu} = e_{\mu} \cdot e_{\nu}$, the coordinate basis vectors obey

$$|e_r| = 1, \quad |e_{\theta}| = r, \tag{10.23}$$

and we argued that a good set of orthonormal basis vectors are given by

$$e_{\hat{r}} = e_r, \quad e_{\hat{\theta}} = \tfrac{1}{r} e_{\theta}. \tag{10.24}$$

Let's first look at how vielbein notation works. Quite trivially, we can say that the components of the coordinate basis vectors in the coordinate basis are, by definition

$$(e_r)^{\mu} = (1, 0), \quad (e_{\theta})^{\mu} = (0, 1). \tag{10.25}$$

Similarly, in the orthornormal frame we write the trivial equation

$$(e_{\hat{r}})^{\hat{\mu}} = (1, 0), \quad (e_{\hat{\theta}})^{\hat{\mu}} = (0, 1). \tag{10.26}$$

More interestingly, the components of the orthonormal basis vectors in the coordinate basis can be written down and are[8]

$$(e_{\hat{r}})^{\mu} = (1, 0), \quad (e_{\hat{\theta}})^{\mu} = (0, 1/r). \tag{10.27}$$

[8]Our use of this notation here follows Hartle. We will record vielbein components in margin notes throughout the book. The vielbein components in this case are

$$(e_{\hat{r}})^r = 1, \quad (e_{\hat{\theta}})^{\theta} = \tfrac{1}{r},$$
$$(e_r)^{\hat{r}} = 1, \quad (e_{\theta})^{\hat{\theta}} = r$$

How do we know that the orthonormal vectors we have selected are correct? The key is that they must obey the defining relationship $\eta_{\hat{\mu}\hat{\nu}} = e_{\hat{\mu}} \cdot e_{\hat{\nu}} = g_{\alpha\beta}(e_{\hat{\mu}})^{\alpha}(e_{\hat{\nu}})^{\beta}$ and this is quickly checked using $g_{rr} = 1$ and $g_{\theta\theta} = r^2$. For example

$$e_{\hat{r}} \cdot e_{\hat{r}} = g_{rr}(e_{\hat{r}})^r(e_{\hat{r}})^r + g_{\theta\theta}(e_{\hat{r}})^{\theta}(e_{\hat{r}})^{\theta} = 1 + 0 = 1,$$
$$e_{\hat{\theta}} \cdot e_{\hat{\theta}} = g_{rr}(e_{\hat{\theta}})^r(e_{\hat{\theta}})^r + g_{\theta\theta}(e_{\hat{\theta}})^{\theta}(e_{\hat{r}})^{\theta} = 0 + r^2 \frac{1}{r}\frac{1}{r} = 1. \tag{10.28}$$

This shows that our choice is correct, since $\eta_{\hat{r}\hat{r}} = \eta_{\hat{\theta}\hat{\theta}} = 1$. Similarly, it follows that the coordinate basis vectors in the orthonormal basis are

$$(e_r)^{\hat{\mu}} = (1, 0), \quad (e_{\theta})^{\hat{\mu}} = (0, r). \tag{10.29}$$

These must obey the defining relationship $e_{\mu}(x) \cdot e_{\nu}(x) = g_{\mu\nu}(x)$, which they do.

A vielbein can be presented as a matrix as we now demonstrate.

Example 10.6

Consider the metric with line element $\mathrm{d}s^2 = \mathrm{d}\theta^2 + \sin^2\theta\,\mathrm{d}\phi^2$. Using the method above, we can compute the matrix representing the components of the coordinate basis vectors in the orthonormal basis[9]

$$(e_\mu)^{\hat\alpha} = \begin{pmatrix} (e_1)^{\hat 1} & (e_1)^{\hat 2} \\ (e_2)^{\hat 1} & (e_2)^{\hat 2} \end{pmatrix} = \begin{pmatrix} 1 & 0 \\ 0 & \sin\theta \end{pmatrix}. \tag{10.30}$$

As the vielbein is simply a square matrix, the usual rules for inverses apply and so we have

$$(e_\mu)^{\hat\alpha}(e_{\hat\alpha})^\nu = \delta^\nu{}_\mu \quad (e_\mu)^{\hat\alpha}(e_{\hat\beta})^\mu = \delta^{\hat\alpha}{}_{\hat\beta}. \tag{10.31}$$

The inverse matrix follows representing the components of the orthonormal basis vectors in the coordinate basis

$$(e_{\hat\alpha})^\mu = \begin{pmatrix} (e_{\hat 1})^1 & (e_{\hat 1})^2 \\ (e_{\hat 2})^1 & (e_{\hat 2})^2 \end{pmatrix} = \begin{pmatrix} 1 & 0 \\ 0 & \frac{1}{\sin\theta} \end{pmatrix}. \tag{10.32}$$

Once we have the vielbein we need the general rule[10] that the vielbein components $(e_{\hat\mu})^\beta$ remove a down coordinate component β and replaces it with an orthonormal component $\hat\mu$. It also replaces the up component $p^{\hat\mu}$ with component p^β. That is to say

$$(e_{\hat\mu})^\beta p_\beta = p_{\hat\mu}, \quad (e_{\hat\mu})^\beta p^{\hat\mu} = p^\beta. \tag{10.38}$$

We also have

$$(e_\mu)^{\hat\beta} p_{\hat\beta} = p_\mu, \quad (e_\mu)^{\hat\beta} p^\mu = p^{\hat\beta}. \tag{10.39}$$

These equations also apply to tensor components, so we might have, for example

$$T_{\alpha\beta} = (e_\alpha)^{\hat\mu}(e_\beta)^{\hat\nu} T_{\hat\mu\hat\nu}. \tag{10.40}$$

To summarize, a vielbein allows us to set up orthonormal frames across all spacetime according to the defining rules

$$g_{\mu\nu} = \eta_{\hat\alpha\hat\beta}(e_\mu)^{\hat\alpha}(e_\nu)^{\hat\beta}, \tag{10.41}$$

$$\eta_{\hat\alpha\hat\beta} = g_{\mu\nu}(e_{\hat\alpha})^\mu(e_{\hat\beta})^\nu. \tag{10.42}$$

Example 10.7

The use of vielbein components generalizes the rule that energy measured by an observer with velocity $\boldsymbol{u}_{\mathrm{obs}}$ is given by $E = -\boldsymbol{p}\cdot\boldsymbol{u}_{\mathrm{obs}}$, where $\boldsymbol{p}$ is the momentum vector. This is because we always choose $\boldsymbol{e}_{\hat 0} = \boldsymbol{u}_{\mathrm{obs}}$. To prove this we write

$$E = -g_{\mu\nu}p^\mu(e_{\hat 0})^\nu = -p_\nu(e_{\hat 0})^\nu. \tag{10.43}$$

We know from the definitions of how a vielbein works that $-p_\nu(e_{\hat 0})^\nu = -p_{\hat 0}$. Finally, since in the orthonormal frame, indices are manipulated with the Minkowski tensor $\boldsymbol{\eta}$, we have that $E = -p_{\hat 0} = -\eta_{\hat 0\hat 0}p^{\hat 0} = p^{\hat 0}$, as we require.

Vielbein components will be very useful to us. Next, we need to identify some examples of frames in which observers make their measurements.

---

[9] The vielbein components in this case can be written as

$$(e_\theta)^{\hat\theta} = 1, \quad (e_\phi)^{\hat\phi} = \sin\theta.$$

[10] Life is made easier in understanding the action of a vielbein if we also employ our knowledge of 1-forms. Formally we define the action of the vielbein on basis vectors and basis 1-forms via

$$\begin{aligned} \boldsymbol{e}_{\hat\mu} &= (e_{\hat\mu})^\alpha \boldsymbol{e}_\alpha, \quad \boldsymbol{\omega}^{\hat\mu} = (e_\alpha)^{\hat\mu}\boldsymbol{\omega}^\alpha, \\ \boldsymbol{e}_\alpha &= (e_\alpha)^{\hat\mu}\boldsymbol{e}_{\hat\mu}, \quad \boldsymbol{\omega}^\alpha = (e_{\hat\mu})^\alpha \boldsymbol{\omega}^{\hat\mu}. \end{aligned} \tag{10.33}$$

From which the rules in the text can be confirmed. Equivalently, we can return to the definition

$$\boldsymbol{v} = v^\mu \boldsymbol{e}_\mu = v^{\hat\nu} \boldsymbol{e}_{\hat\nu}, \tag{10.34}$$

and use the inner product $\langle\boldsymbol{\omega}^\alpha, \boldsymbol{e}_\mu\rangle = \delta^\alpha{}_\mu$. To remove a vector like $\boldsymbol{e}_\mu$, from the second term in eqn 10.34, we note that, on taking an inner product with basis 1-form $\boldsymbol{\omega}^\mu$, we have

$$v^\mu = v^{\hat\nu}\langle\boldsymbol{\omega}^\mu, \boldsymbol{e}_{\hat\nu}\rangle = v^{\hat\nu}(e_{\hat\nu})^\mu, \tag{10.35}$$

where we've fixed the vielbein components by

$$\langle\boldsymbol{\omega}^\mu, \boldsymbol{e}_{\hat\nu}\rangle = \langle\boldsymbol{\omega}^\mu, \boldsymbol{e}_\alpha\rangle(e_{\hat\nu})^\alpha = (e_{\hat\nu})^\mu. \tag{10.36}$$

Note here that $(e_{\hat\nu})^\mu = \langle\boldsymbol{\omega}^\mu, \boldsymbol{e}_{\hat\nu}\rangle \neq \delta^\mu{}_{\hat\nu}$, since here we're working with the basis vectors of two different coordinate systems. The same method also yields

$$v^{\hat\mu} = v^\nu\langle\boldsymbol{\omega}^{\hat\mu}, \boldsymbol{e}_\nu\rangle = v^\nu(e_\nu)^{\hat\mu}. \tag{10.37}$$

The other relationships can be confirmed using the idea that a 1-form can be written as $\tilde{\boldsymbol{u}} = u_\mu\boldsymbol{\omega}^\mu = u_{\hat\mu}\boldsymbol{\omega}^{\hat\mu}$.

10.3   The orthonormal frame

In many of the cases we'll consider later in the book, we shall find that a very convenient orthonormal frame in which to carry out computations, particularly of curvature, is one where the observer is at rest relative to the coordinate frame. That is, the velocity expressed in the coordinate frame is $u^\mu = \mathrm{d}x^\mu/\mathrm{d}\tau = (u^0, 0, 0, 0)$, with u^0 fixed such that $g_{\mu\nu}u^\mu u^\nu = g_{00}(u^0)^2 = -1$. This observer then chooses $e_{\hat{t}} = u$. We will call this rather natural choice[11] **the stationary orthonormal frame**. In such a frame, we have a metric that looks locally like the Minkowski metric, but there is no reason to believe that the connection coefficients should vanish.[12]

To find the orthonormal frame we effectively diagonalize the metric and normalize the components. That is, we are trying to solve

$$g(e_{\hat{\alpha}}, e_{\hat{\beta}}) = \eta_{\hat{\alpha}\hat{\beta}}, \tag{10.44}$$

which is equivalent to the component equation

$$g_{\mu\nu}(e_{\hat{\alpha}})^\mu (e_{\hat{\beta}})^\nu = \eta_{\hat{\alpha}\hat{\beta}}. \tag{10.45}$$

Example 10.8

The matrix form for (1+1)-dimensional spacetime is

$$\begin{pmatrix} (e_{\hat{0}})^0 & (e_{\hat{0}})^1 \\ (e_{\hat{1}})^0 & (e_{\hat{1}})^1 \end{pmatrix} \begin{pmatrix} g_{00} & g_{01} \\ g_{10} & g_{11} \end{pmatrix} \begin{pmatrix} (e_{\hat{0}})^0 & (e_{\hat{1}})^0 \\ (e_{\hat{0}})^1 & (e_{\hat{1}})^1 \end{pmatrix} = \begin{pmatrix} -1 & 0 \\ 0 & 1 \end{pmatrix}. \tag{10.46}$$

In the most commonly encountered case of a metric that is already diagonal, we can simply normalize the components as discussed in the next example.

Example 10.9

For an observer at rest u has a single non-zero component u^0. Its value is given via the normalization of the velocity, by

$$g_{00}(u^0)^2 = -1, \tag{10.47}$$

so we must have $u^0 = (-g_{00})^{-1/2}$. Our rule $e_{\hat{0}} = u$ then mandates $e_{\hat{0}} = e_0/\sqrt{-g_{00}}$. For the diagonal metric we can then pick out an orthonormal basis

$$e_{\hat{0}} = \frac{1}{\sqrt{-g_{00}}} e_0, \quad e_{\hat{i}} = \frac{1}{\sqrt{g_{ii}}} e_i, \tag{10.48}$$

with components in the coordinate frame of

$$(e_{\hat{0}})^0 = \frac{1}{\sqrt{-g_{00}}}, \quad (e_{\hat{i}})^i = \frac{1}{\sqrt{g_{ii}}}. \tag{10.49}$$

We can check this works by considering the rule $g_{\mu\nu}(e_{\hat{\alpha}})^\mu (e_{\hat{\beta}})^\nu = \eta_{\hat{\alpha}\hat{\beta}}$. If the metric g is diagonal then, by inspection, we have

$$g_{\mu\mu}(e_{\hat{0}})^\mu (e_{\hat{0}})^\mu = -1,$$
$$g_{\mu\mu}(e_{\hat{i}})^\mu (e_{\hat{i}})^\mu = 1. \tag{10.50}$$

Note that for a diagonal metric we also have[13] $(e_\mu)^{\hat{\mu}} = 1/(e_{\hat{\mu}})^\mu$.

---

[11] Later in the book, if we don't specify that the orthonormal frame is that of a specific observer, one can assume we're referring to the stationary orthonormal frame of an observer who is stationary in the coordinate frame.

[12] It is a local inertial frame (LIF) where the connection coefficients vanish at the point in question. This more restrictive condition is dealt with in the next section.

[13] In expressions like these, involving an index μ both with and without a hat, no summation is implied.

The normalization procedure in the last example makes identifying the vielbein components for the orthonormal frame trivial for the diagonal metric. We simply normalize by writing

$$(e_{\hat{\mu}})^{\mu} = 1/\sqrt{|g_{\mu\mu}|}, \quad (e_{\mu})^{\hat{\mu}} = \sqrt{|g_{\mu\mu}|}. \tag{10.51}$$

In this way, we can think of the vielbein components as the square roots of the metric components.

Example 10.10

As we shall see later, a spherically symmetric gravitating object of mass M gives rise to the **Schwarzschild metric** with line element given by[14]

$$ds^2 = -\left(1 - \frac{2M}{r}\right)dt^2 + \left(1 - \frac{2M}{r}\right)^{-1} dr^2 + r^2\left(d\theta^2 + \sin^2\theta\, d\phi^2\right). \tag{10.52}$$

We can identify an orthonormal frame in this so-called Schwarzschild geometry, which has coordinates ordered (t, r, θ, ϕ). An observer at rest in this geometry has a velocity vector $\boldsymbol{u}$ with components

$$u^{\mu} = \left[\left(1 - \frac{2M}{r}\right)^{-\frac{1}{2}}, 0, 0, 0\right], \tag{10.53}$$

so that $\boldsymbol{u} \cdot \boldsymbol{u} = g_{tt}\left(1 - \frac{2M}{r}\right)^{-1} = -1$, as it must. We set $e_{\hat{t}} = \boldsymbol{u}$. We then choose

$$e_{\hat{t}} = \left(1 - \tfrac{2M}{r}\right)^{-\frac{1}{2}} e_t, \quad e_{\hat{r}} = \left(1 - \tfrac{2M}{r}\right)^{\frac{1}{2}} e_r, \quad e_{\hat{\theta}} = \tfrac{1}{r} e_\theta, \quad e_{\hat{\phi}} = \tfrac{1}{r\sin\theta} e_\phi. \tag{10.54}$$

It's not hard to see that these must obey the defining rules above. Alternatively, writing non-zero components of the vielbein explicitly[15]

$$(e_{\hat{t}})^t = \left(1 - \tfrac{2M}{r}\right)^{-\frac{1}{2}}, \quad (e_{\hat{r}})^r = \left(1 - \tfrac{2M}{r}\right)^{\frac{1}{2}}, \quad (e_{\hat{\theta}})^\theta = \tfrac{1}{r}, \quad (e_{\hat{\phi}})^\phi = \tfrac{1}{r\sin\theta}. \tag{10.55}$$

Of course the point of identifying vielbein components is to use them, in ways such as that shown in the next example.

Example 10.11

Spacetime is described by a metric line element[16]

$$ds^2 = -dt^2 + d\chi^2 + \chi^2(d\theta^2 + \sin^2\theta\, d\phi^2). \tag{10.56}$$

In the coordinate frame, the energy-momentum tensor for the fluid has components

$$T_{tt} = \rho, \quad T_{\chi\chi} = p, \quad T_{\theta\theta} = p\chi^2, \quad T_{\phi\phi} = p\chi^2\sin^2\theta, \tag{10.57}$$

where ρ is an energy density and p is a pressure. We can express these in the orthonormal frame. So, for example, $T_{\theta\theta}$ becomes

$$T_{\hat{\theta}\hat{\theta}} = (e_{\hat{\theta}})^\theta (e_{\hat{\theta}})^\theta T_{\theta\theta} = p. \tag{10.58}$$

We find that in the orthonormal frame, we have

$$T_{\hat{t}\hat{t}} = \rho, \quad T_{\hat{\chi}\hat{\chi}} = p, \quad T_{\hat{\theta}\hat{\theta}} = p, \quad T_{\hat{\phi}\hat{\phi}} = p, \tag{10.59}$$

which is a useful simplification.

> ⤳ **The Schwarzschild metric will be introduced in Chapter 21.**

[14] We are here using units in which $G = 1$. The metric is diagonal and has components

$$g_{tt} = -\left(1 - \frac{2M}{r}\right),$$

$$g_{rr} = \left(1 - \frac{2M}{r}\right)^{-1},$$

$$g_{\theta\theta} = r^2,$$

$$g_{\phi\phi} = r^2\sin\theta.$$

[15] Since the metric is diagonal, we have

$$(e_t)^{\hat{t}} = \left(1 - \frac{2M}{r}\right)^{\frac{1}{2}},$$

$$(e_r)^{\hat{r}} = \left(1 - \frac{2M}{r}\right)^{-\frac{1}{2}},$$

$$(e_\theta)^{\hat{\theta}} = r,$$

$$(e_\phi)^{\hat{\phi}} = r\sin\theta.$$

[16] This metric implies that the vielbein is diagonal, with components given by the components

$$(e_t)^{\hat{t}} = 1, \qquad (e_\chi)^{\hat{\chi}} = 1,$$
$$(e_\theta)^{\hat{\theta}} = \chi, \qquad (e_\phi)^{\hat{\phi}} = \chi\sin\theta.$$

and

$$(e_{\hat{t}})^t = 1, \qquad (e_{\hat{\chi}})^\chi = 1,$$
$$(e_{\hat{\theta}})^\theta = \tfrac{1}{\chi}, \qquad (e_{\hat{\phi}})^\phi = \tfrac{1}{\chi\sin\theta}.$$

10.4 Freely falling frames

There are several possible orthonormal frames that can be identified. In Chapter 6, we discussed the possibility of finding locally inertial frames (LIFs) in which, in addition to the metric being identical to the Minkowski metric, the frame also has the property that the first derivatives of the components $\partial g_{\mu\nu}/\partial x^\alpha$ vanish at the point considered. This implies that the connection coefficients $\Gamma^\mu{}_{\alpha\beta}$ also vanish at that point.

There are a few methods for identifying LIFs, but one of the most useful is the **freely falling frame**. Recall that a body that is freely falling follows a (timelike) geodesic curve in spacetime. The equivalence principle tells us that a sufficiently small laboratory in free fall should not be able to detect any gravitation. As a result of this, we might expect the laboratory's coordinate system has vanishing connection coefficients. This is indeed the case.

Freely falling frames are therefore defined to possess a system of coordinates in which the connection coefficients vanish along the geodesic that describes their free fall. The frame is described via a set of orthonormal basis vectors $\boldsymbol{e}_{\hat{a}}(\tau)$ that we should determine in order to be able to understand the results of measurements.

Consider the geodesic of the falling observer with proper time τ, which is the curve $x^\mu(\tau)$. The observer's four velocity is $\boldsymbol{u}(\tau) = \frac{\mathrm{d}x^\mu}{\mathrm{d}\tau}\boldsymbol{e}_\mu$. This vector is identified with the zeroth basis vector $\boldsymbol{e}_{\hat{0}}(\tau) = \boldsymbol{u}(\tau)$. We can find the spatial basis vectors at some point along the geodesic by identifying a set of orthonormal vectors that are perpendicular to $\boldsymbol{u}$. The basis vectors at other points could be found by parallel transporting the basis vectors along the geodesic. Since the connection coefficients vanish, the freely falling frame is then defined by[17]

$$\boldsymbol{\nabla}_{\boldsymbol{u}}\boldsymbol{e}_{\hat{\alpha}} = 0, \tag{10.60}$$

for all α. Notice that this is automatically satisfied for $\boldsymbol{e}_{\hat{0}} = \boldsymbol{u}$ by definition for a geodesic.

[17]We can also identify orthonormal basis 1-forms $\boldsymbol{\omega}^\alpha$ with the property $\boldsymbol{\nabla}_{\boldsymbol{u}}\boldsymbol{\omega}^{\hat{\alpha}} = 0$.

Example 10.12

We shall see in Chapter 22 that falling radially inwards from rest at infinity in the Schwarzschild geometry, a particle has velocity $\boldsymbol{u}$ with components in the coordinate frame of

$$u^\alpha = \left[\left(1 - \frac{2M}{r}\right)^{-1}, -\left(\frac{2M}{r}\right)^{\frac{1}{2}}, 0, 0\right]. \tag{10.61}$$

We write down that $\boldsymbol{e}_{\hat{t}} = \boldsymbol{u}(\tau)$. As far as it's possible to identify them, diagonal vielbein components are most simple to use. So, as in the previous examples, we write $(\boldsymbol{e}_{\hat{\theta}})^\theta = 1/r$ and $(\boldsymbol{e}_{\hat{\phi}})^\phi = 1/(r\sin\theta)$. For $(\boldsymbol{e}_{\hat{r}})^\alpha$ we have $g_{\mu\nu}(\boldsymbol{e}_{\hat{r}})^\mu(\boldsymbol{e}_{\hat{r}})^\nu = \eta_{\hat{r}\hat{r}}$ or

$$-\left(1 - \frac{2M}{r}\right)(\boldsymbol{e}_{\hat{r}})^t(\boldsymbol{e}_{\hat{r}})^t + \left(1 - \frac{2M}{r}\right)^{-1}(\boldsymbol{e}_{\hat{r}})^r(\boldsymbol{e}_{\hat{r}})^r = 1. \tag{10.62}$$

Choose $(e_{\hat{r}})^r = 1$, and conclude that the result is

$$
\begin{aligned}
(e_{\hat{t}})^\alpha = (e_{\hat{0}})^\alpha &= \left((1 - 2M/r)^{-1}, -(2M/r)^{\frac{1}{2}}, 0, 0\right), \\
(e_{\hat{r}})^\alpha = (e_{\hat{1}})^\alpha &= \left(-(2M/r)^{\frac{1}{2}}(1 - 2M/r)^{-1}, 1, 0, 0\right), \\
(e_{\hat{\theta}})^\alpha = (e_{\hat{2}})^\alpha &= (0, 0, 1/r, 0), \\
(e_{\hat{\phi}})^\alpha = (e_{\hat{3}})^\alpha &= \left(0, 0, 0, (r \sin\theta)^{-1}\right).
\end{aligned}
\tag{10.63}
$$

We saw in Chapter 7, Example 7.10, that for a system in which the connection coefficients vanish, we should have $D\chi/d\tau = d\chi/d\tau$. We should check that χ in a freely falling frame (defined by $\nabla_u e_{\hat{\alpha}} = 0$) has this property.

Example 10.13

Note that in the freely falling frame we have the defining fact that the basis vector and basis 1-forms are parallel transported.

$$
\nabla_u e_{\hat{\alpha}} = 0, \quad \nabla_u \omega^{\hat{\alpha}} = 0.
\tag{10.64}
$$

Our first task is to express the covariant derivative vector $\nabla_u \chi$ in the freely falling frame. To do this we use a vielbein $(e_\alpha)^{\hat{\mu}}$ to bring its components into the orthonormal freely falling system. The components are $(\nabla_u \chi)^{\hat{\mu}} = (e_\alpha)^{\hat{\mu}} (\nabla_u \chi)^\alpha$, which can be rewritten as[18]

$$
(e_\alpha)^{\hat{\mu}} (\nabla_u \chi)^\alpha = \nabla_u \left((e_\alpha)^{\hat{\mu}} \chi^\alpha\right).
\tag{10.68}
$$

But this is just the covariant derivative of the $\hat{\mu}$ component of χ, or $\nabla_u(\chi^{\hat{\mu}})$. The covariant derivative of a component of a vector is the same as the covariant derivative of some scalar function, which (as we saw in the earlier sidenote) is just the directional derivative, which is to say

$$
\begin{aligned}
\nabla_u \left(\chi^{\hat{\mu}}\right) &= u^\beta \frac{\partial}{\partial x^\beta} \chi^{\hat{\mu}} \\
&= \frac{dx^\beta}{d\tau} \frac{\partial}{\partial x^\beta} \chi^{\hat{\mu}} = \frac{d\chi^{\hat{\mu}}}{d\tau},
\end{aligned}
\tag{10.69}
\tag{10.70}
$$

where τ is the proper time. Putting things together, we conclude

$$
\left(\frac{D\chi}{d\tau}\right)^{\hat{\mu}} \equiv (\nabla_u \chi)^{\hat{\mu}} = \frac{d\chi^{\hat{\mu}}}{d\tau}.
\tag{10.71}
$$

[18] A useful step in seeing this is to write the vielbein components as $(e_\alpha)^{\hat{\mu}} = \langle \omega^{\hat{\mu}}, e_\alpha \rangle$, so we have

$$
\langle \omega^{\hat{\mu}}, e_\alpha \rangle (\nabla_u \chi)^\alpha = \langle \omega^{\hat{\mu}}, \nabla_u \chi \rangle.
\tag{10.65}
$$

Next, we note that

$$
\nabla_u \langle \omega^{\hat{\mu}}, \chi \rangle = \langle \nabla_u \omega^{\hat{\mu}}, \chi \rangle + \langle \omega^{\hat{\mu}}, \nabla_u \chi \rangle,
\tag{10.66}
$$

but $\nabla_u \omega^{\hat{\mu}} = 0$ by definition of the freely falling frame. So the right-hand side of eqn 10.65 becomes

$$
\nabla_u \langle \omega^{\hat{\mu}}, \chi \rangle = \nabla_u \chi^{\hat{\mu}}.
\tag{10.67}
$$

At various points in the remainder of the book, we will deploy vielbein components in order to efficiently perform certain calculations. In the next chapter, we turn to the long-awaited method of determining the curvature of spacetime.

Chapter summary

- Measurements in general relativity are carried out by observers who carry around an orthonormal coordinate system with basis vectors $e_{\hat{\alpha}}$. An observer with velocity u has $e_{\hat{0}} = u$.

- A vielbein, with components $(e_\mu)^{\hat{\alpha}}$, is a matrix that allows the transformation between orthonormal frames and coordinate frames.

- A particularly convenient frame is an orthonormal one where the observer is at rest relative to the coordinate frame. An alternative is the freely falling frame (defined by $\nabla_u e_{\hat{\alpha}} = 0$ for all α) in which the connection coefficients also vanish.

- In the (stationary) orthonormal frame, the observer's 3-velocity vanishes and so, using $u \cdot u = g_{00}(u^0) = -1$, we have $u^0 = (-g_{00})^{-\frac{1}{2}}$. The observer orients their orthonormal frame with $e_{\hat{0}} = u$, so $e_{\hat{0}} = (-g_{00})^{-\frac{1}{2}} e_0$, or $(e_{\hat{0}})^0 = (-g_{00})^{-\frac{1}{2}}$.

- In the orthonormal frame for a diagonal metric, we have $(e_0)^{\hat{0}} = (-g_{00})^{\frac{1}{2}}$ and $(e_i)^{\hat{i}} = g_{ii}^{\frac{1}{2}}$.

Exercises

(10.1) Consider spacetime with a line element

$$ds^2 = -\,dt^2 + a(t)^2 \left(d\chi^2 + \sinh^2\chi\,d\theta^2 \right.$$
$$\left. + \sinh^2\chi \sin^2\phi\,d\phi^2\right). \qquad (10.72)$$

(a) Using a coordinate system (t, χ, θ, ϕ), a vector has components in the coordinate frame if $V^\mu = (V^t, V^\chi, V^\theta, V^\phi)$. What are the vector's components in the orthonormal frame?

(b) A $(1, 2)$ tensor has a non-zero component $G^\theta{}_{\chi\phi}$. What does this become in the orthonormal frame?

(10.2) (a) Working in the orthonormal frame, find the connection coefficients for flat space represented in cylindrical polar coordinates.

Hint: Remember that the connection coefficients do not transform like tensors, so you cannot simply use the vielbein. You can compute the coefficients directly from the definitions of the basis vectors, or transform using eqn 7.11.

(b) Show that the connection coefficients you have derived obey the rule $\Gamma^{\hat{\mu}}{}_{\hat{\alpha}\hat{\beta}} - \Gamma^{\hat{\mu}}{}_{\hat{\beta}\hat{\alpha}} = \left\langle \omega^{\hat{\mu}}, \left[e_{\hat{\alpha}}, e_{\hat{\beta}}\right] \right\rangle$.

(10.3) *This problem combines several ideas from the last few chapters and is a useful warm up for some of the physics in Part IV of the book.*

A particle travels radially in a static, spherically symmetric gravitational field described by diagonal metric components $g_{\mu\nu}$ and a velocity vector u.

(a) If the timelike component of the particle's velocity 1-form is given in the static frame of the potential by a constant $u_t = a$, give the other components in terms of a and the components of the metric.

(b) Compute the coordinate velocity dr/dt.

(c) What is the coordinate velocity, as measured by a local observer?

(10.4) Consider flat spacetime expressed in an orthogonal coordinate system (x^1, x^2, x^3) with a diagonal metric.

(a) Show that the gradient operator acting on a function f becomes

$$\frac{\partial f}{\partial x^{\hat{\mu}}} = \left(\frac{1}{(g_{11})^{\frac{1}{2}}} \frac{\partial f}{\partial x^1}, \frac{1}{(g_{22})^{\frac{1}{2}}} \frac{\partial f}{\partial x^2}, \frac{1}{(g_{33})^{\frac{1}{2}}} \frac{\partial f}{\partial x^3}\right).$$
$$(10.73)$$

(b) Prove that, in the coordinate frame with this diagonal metric, the divergence can be rewritten as

$$(\nabla_\mu v)^\mu = \frac{\partial v^\mu}{\partial x^\mu} + \frac{1}{\sqrt{g}} \frac{\partial \sqrt{g}}{\partial x^\mu} v^\mu. \tag{10.74}$$

(c) Use the result from (b) to show that, expressed in terms of the components in the orthonormal frame, the divergence can be written as

$$\nabla \cdot v = \frac{1}{\sqrt{g}} \left[\frac{\partial}{\partial x^1} \left(\sqrt{g_{22}g_{33}} v^{\hat{1}} \right) \right.$$
$$+ \frac{\partial}{\partial x^2} \left(\sqrt{g_{33}g_{11}} v^{\hat{2}} \right)$$
$$\left. + \frac{\partial}{\partial x^3} \left(\sqrt{g_{11}g_{22}} v^{\hat{3}} \right) \right]. \tag{10.75}$$

What does this formula yield for (d) orthonormal cylindrical polar coordinates and (e) orthonormal spherical polar coordinates?

(10.5) A light signal is emitted by a source on the rim of a centrifuge and detected by a detector at another point on the rim, separated by an angle α. Use the metric for the rotating frame

$$ds^2 = - \left(1 - \Omega^2 r^2\right) dt^2 + dr^2 + r^2 d\theta^2$$
$$+ dz^2 + 2\Omega r^2 d\theta dt, \tag{10.76}$$

to show that there is no shift in the frequency of the signal.

(10.6) An alternative to orthonormal local basis vectors was suggested by Newman and Penrose. They considered a pair of real null vectors l and n and a pair of complex-conjugate null vectors m and $\bar{m}$ obeying

$$l \cdot m = l \cdot \bar{m} = n \cdot m = n \cdot \bar{m} = 0. \tag{10.77}$$

The vectors are normalized according to $l \cdot n = 1$ and $m \cdot \bar{m} = -1$.
(a) If we take the local basis to be

$$e_{\hat{1}} = l, \quad e_{\hat{2}} = n,$$
$$e_{\hat{3}} = m, \quad e_{\hat{4}} = \bar{m}, \tag{10.78}$$

find the components of the local metric $\eta_{\hat{\mu}\hat{\nu}}$.
(b) Find the local basis 1-forms $\omega^{\hat{\mu}}$, assuming the usual relationship $\langle \omega^{\hat{\mu}}, e_{\hat{\nu}} \rangle = \delta^{\hat{\mu}}_{\hat{\nu}}$.

(10.7) Suggest vielbein components for a $(1+1)$-dimensional metric with line element $ds^2 = -dudv$.

11

Riemann curvature and the Ricci tensor

[1]Our problem in finding the curvature of spacetime is more acute, of course, since we can't use any method analogous to noting the change in elevation of distant stars to work out the nature of the curvature.

[2]G. F. Bernhard Riemann (1826–1866)

[3]The motion of particles in curved spacetime is often described in terms of a **rubber sheet model**. This involves picturing space (in two dimensions) as being akin to an elastic sheet. Massive objects make indentations in the sheet and so all objects have their motion affected by having to negotiate these indentations as they move around. Although useful as a picture, it should also be kept in mind that this model describes the curvature of *space* and not *spacetime*. In some cases (such as the precession of the orbit of a planet around a star), the metric component describing the timelike variable has a larger effect on the corrections to Newtonian motion than those relating to the spatial coordinates. The combined effect is therefore certainly a *spacetime* one. This is another manifestation of the feature that space and time are inextricably linked in relativity.

From the point of view of general relativity, humans are condemned to be trapped in spacetime. We cannot, for example, lift ourselves out of spacetime and look at its structure from the outside, by embedding it in some higher dimensional space. If gravitation is manifested in terms of the curvature of spacetime then how are we to measure, understand and explain this curvature? Of course humans have experienced a similar problem before: the surface of the Earth is a two-dimensional space and there are several means of working out that the Earth is, to a good approximation, a spherical ball.[1]

Carl Friedrich Gauss made some of the most significant progress in finding that there was a way to convert measurements of distances, made by an observer trapped on a two-dimensional surface, into an objective description of the curvature of that space. However, Bernhard Riemann[2] was engaged in an even more ambitious programme to evaluate the curvature of higher dimensional spaces. He solved the problem in 1854 and introduced a mathematical description that would be built up by Christoffel, Ricci, Levi-Civita and others. It is this description that was incorporated into physics by Einstein. Riemann's description of curvature is encoded into a tensor $\boldsymbol{R}$. We shall see that a non-zero $\boldsymbol{R}$ is the sure-fire way of telling mathematically that a spacetime is truly curved.[3]

11.1 What is curvature?

Euclid's fifth postulate, known as the parallel postulate, may be paraphrased as saying that parallel lines always remain parallel. Euclid did not prove this, which is why it remained a postulate. It turns out only to be the case in flat space.

The surface of a cylinder appears curved, but this curvature is **extrinsic**: we can recreate (or wrap) the curved surface of the cylinder using a flat piece of paper with two of its opposite edges identified. We can therefore unravel the cylinder, turning it into a flat plane. Parallel lines on a cylindrical surface never meet, therefore, as they are equivalent to parallel lines on the flat plane. The sphere is different, its curvature is **intrinsic**: it is impossible to wrap the sphere in a flat piece of paper in the way that we did the cylinder. Parallel lines on the surface of a sphere don't remain parallel; they eventually meet.

Curved surfaces can be classified in terms of the paths of such parallel lines. On a flat surface parallel lines remain parallel. On a surface with positive curvature initially parallel lines converge; on a surface with negative curvature they diverge (Fig. 11.1). An example of a surface with positive curvature is the surface of a sphere; surfaces with negative curvature can be thought of as being saddle-like.

There are two simple ways of telling whether curvature is intrinsic or extrinsic. The first, motivated by the discussion so far, is to set two free, spatially separated particles moving parallel. If the space is intrinsically curved, the particles' paths will deviate from parallelism. This effect, shown in Fig. 11.1 is known as **geodesic deviation**.

The second method is to parallel transport a vector in a closed loop. If the space has intrinsic curvature, then the vector will rotate. If the space has extrinsic curvature, there will be no change in direction. The rotation is most easily seen by examining parallel transport on the surface of a sphere as shown in Fig. 11.2.

In summary, to measure the amount of curvature present we choose one of the two methods: (i) identify a fiducial geodesic and compare it to another geodesic whose tangent vector was originally parallel. The extent of the deviation tells us the amount by which the space is curved. (ii) Take a vector and parallel transport it around a loop in spacetime. The amount by which the vector changes its direction provides a measure of curvature. As we shall show, these two methods both provide the same measure of curvature: the components of the **Riemann curvature tensor *R*.**

The ultimate source of curvature is the metric field. We saw in the previous chapters how the connection coefficients were derived from the first derivatives of the components of the metric (with the rule of thumb expression that $\boxed{\partial g} \to \boxed{\Gamma}$). The Riemann curvature tensor is built from the first and second derivatives of the metric and so we have the rule of thumb

$$\boxed{\partial g} + \boxed{\partial^2 g} \to \boxed{R}. \tag{11.1}$$

This leads us to expect that the components of this tensor can also be written in terms of the connection coefficients and their derivatives (i.e. $\boxed{\Gamma} + \boxed{\partial \Gamma} \to \boxed{R}$), which, as we shall see, turns out to be the case.

11.2 Tidal forces

The simplest gravitating system might be expected to have a static metric, completely independent of time. This might represent the geometry around a star. Such a geometry would allow us to identify a static frame of reference (i.e. the rest frame of the star). We could then perform experiments in which we placed a single particle at rest in this frame and watched to see if it started accelerating. This would allow us to identify a gravitational field. However, this is not a good description of our Universe, where the principles of relativity teach us that it is not generally possible to identify a static frame against which to identify accelera-

Euclid's postulates are:

I: A straight line segment can be drawn joining any two points.
II: Any straight line segment can be extended indefinitely in a straight line.
III: Given any straight line segment, a circle can be drawn having the segment as radius and one endpoint as centre.
IV: All right angles are congruent.
V: If two lines are drawn which intersect a third in such a way that the sum of the inner angles on one side is less than two right angles, then the two lines inevitably must intersect each other on that side if extended far enough. This postulate is equivalent to the parallel postulate.

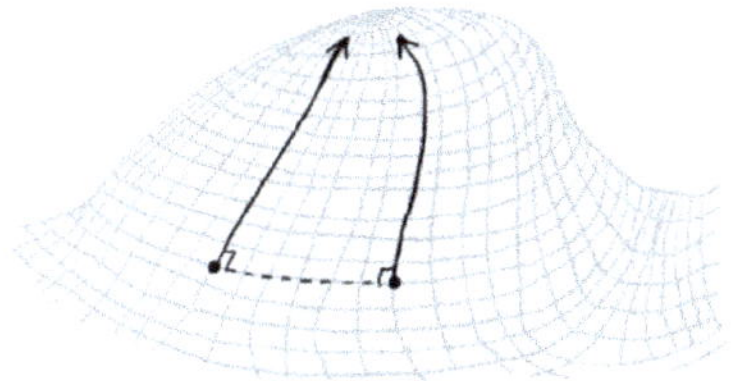

Fig. 11.1 Geodesic deviation. In a curved spacetime, particles fall along geodesics that are parallel at some point in spacetime. Extending the paths into the future and past, the world lines of the particles meet if the space is positively curved.

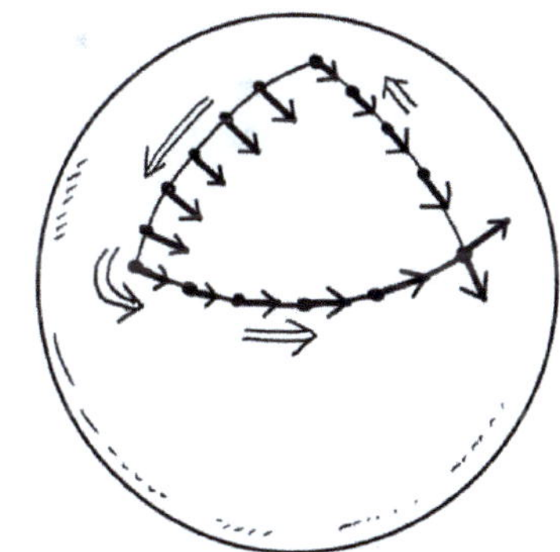

Fig. 11.2 Transporting a vector around a closed loop on the surface of a sphere. The vector rotates as a result of the intrinsic curvature of the spherical surface.

tion. We saw that taking relativity into account leads to the equivalence principle, telling us that from the motion of a single particle we cannot tell the difference between acceleration due to a choice of coordinates and acceleration due to spacetime curvature. The best we can do, is to examine the relative motion of two particles.

Let's briefly return to the world of Newtonian gravity and discuss geodesic deviation in this limit. Consider a spherical cluster of test particles falling towards the earth. Each particle moves on a straight line through the Earth's centre, but those that are closer fall faster than those further away, owing to the greater force on the closer particles. As a result of the motion, the sphere won't remain spherical, but will be distorted into an ellipse of the same volume.[4] In the rest frame of an observer falling with the particles, the forces that cause the change to the shape of bodies in a gravitational field are known as **tidal forces**. They are illustrated in Fig. 11.3.

[4]We prove this statement about the volume staying constant at the end of this section.

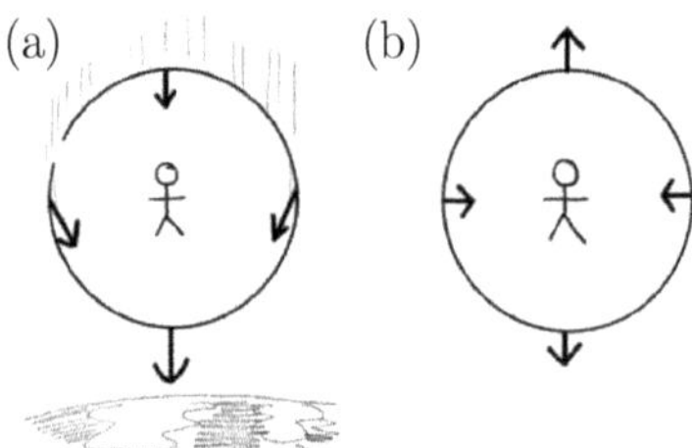

Fig. 11.3 Tidal forces. (a) The forces on a cloud of particles surrounding an observer that are falling towards the Earth, shown in the Earth's rest frame. (b) The same forces in the rest frame of the observer.

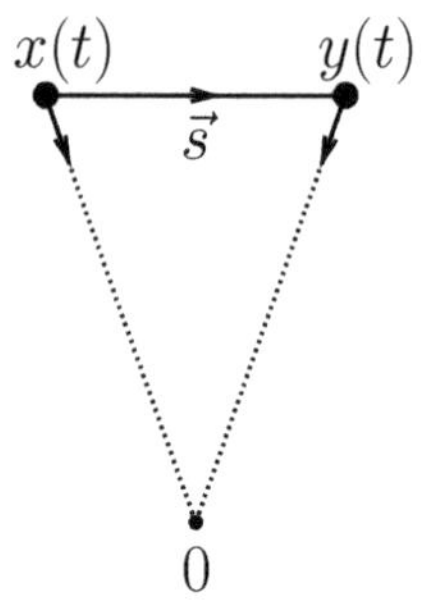

Fig. 11.4 Particles falling in a Newtonian potential, with separation vector $\vec{s}$.

Example 11.1

Consider a cloud of dust particles that are initially spherical, that are released in the vicinity of a gravitating mass. Newton's second law gives us the equation of motion for each particle in a gravitational potential $\Phi(\vec{x}) = -GM/|\vec{x}|$ (temporarily relaxing our rules on balanced up and down tensor components)

$$m\frac{\mathrm{d}^2 x^i}{\mathrm{d}t^2} = -m\frac{\partial\Phi(\vec{x})}{\partial x^i}. \tag{11.2}$$

We imagine the trajectories of two particles in the cloud, shown in Fig. 11.4, with a vector $\vec{s}(t) = \vec{y}(t) - \vec{x}(t)$ stretching between them. (The connecting vector is assumed small compared to the distance from the gravitating mass in what follows.) We can take the difference in their equations of motion

$$\begin{aligned}\left(\frac{\mathrm{d}^2 y^i}{\mathrm{d}t^2} - \frac{\mathrm{d}^2 x^i}{\mathrm{d}t^2}\right) &= -\frac{\partial\Phi(\vec{y})}{\partial x^i} + \frac{\partial\Phi(\vec{x})}{\partial x^i} \\ &= -\frac{\partial\Phi(\vec{x}+\vec{s})}{\partial x^i} + \frac{\partial\Phi(\vec{x})}{\partial x^i} \\ &\approx -\frac{\partial^2\Phi(\vec{x})}{\partial x^i \partial x^j}s^j. \end{aligned} \tag{11.3}$$

Define a tensor with components $R^i{}_j = R_{ij} = \frac{\partial^2\Phi(\vec{x})}{\partial x^i \partial x^j}$, leading to the equation of motion of the components of the deviation vector

$$\frac{\mathrm{d}^2 s^i}{\mathrm{d}t^2} = -R^i{}_j s^j. \tag{11.4}$$

Let's examine the case where the particle starts with $\vec{x} = (0,0,z)$, and falls towards a planet, whose centre is the origin. The tensor $R^i{}_j$ may be evaluated, giving

$$R^i{}_j = \frac{GM}{z^3}\begin{pmatrix} 1 & 0 & 0 \\ 0 & 1 & 0 \\ 0 & 0 & -2 \end{pmatrix}. \tag{11.5}$$

The cloud becomes elongated along z, because $\frac{\mathrm{d}^2 s^z}{\mathrm{d}t^2} = 2\frac{GM}{r^3}z$. The cloud is compressed along x and y, this is because $\frac{\mathrm{d}^2 s^i}{\mathrm{d}t^2} = -\frac{GM}{r^3}s^i$, for $i = x$ and y.

After this discussion of Newtonian mechanics, we turn to relativistic gravitation. We use the same strategy, but this time using the **Riemann tensor** $\boldsymbol{R}$. The technique is, once again, to assess the geodesic deviation. We identify some freely falling particle on a fiducial geodesic and note its 4-velocity $\boldsymbol{u}$. We compare the geodesic of a different freely falling particle on a different geodesic, whose velocity was originally parallel to the fiducial one. The separation of the two geodesics is $\boldsymbol{\xi}$. If there's curvature present then $\boldsymbol{\xi}$ will accelerate. The equation of motion for the separation vector is[5]

$$\frac{\mathrm{D}^2\boldsymbol{\xi}}{\mathrm{d}\tau^2} + \boldsymbol{R}(\ ,\boldsymbol{u},\boldsymbol{\xi},\boldsymbol{u}) = 0. \tag{11.6}$$

The $(1,3)$ tensor $\boldsymbol{R}(\ ,\ ,\ ,\ )$ is the key object for assessing curvature. Filling the latter three slots with vectors as shown in eqn 11.6 gives rise to a $(1,0)$ object (i.e. a vector) that measures the relative acceleration of the geodesics. If $\boldsymbol{R} = 0$ (that is, its components vanish) then space is **flat**. If the components aren't zero then we are able to say that the space is **curved**.[6]

We shall return to the use of geodesic deviation at the end of the chapter. As a measure of curvature it is conceptually clear, but computations rely on some mathematical tricks we haven't yet encountered. In the next section we make use of the other approach, parallel transport around a loop, in order to obtain an explicit form of the tensor $\boldsymbol{R}$.

[5]This equation is discussed in detail in Chapter 35 where it is justified more thoroughly and compared against the parallel transport method discussed in the chapter.

[6]The various methods of measuring curvature are discussed in detail in Chapter 30. As a rough measure, we say that the amount of curvature induced by the non-zero Riemann tensor is, in the orthonormal frame, of order $\left(1/R^{\hat{\mu}}{}_{\hat{\nu}\hat{\alpha}\hat{\beta}}\right)^{\frac{1}{2}}$.

Example 11.2

Proof that the volume of the cluster of particles is constant as it deforms. Recall that we can write Coulomb's law for the electrostatic force between charges and justify Gauss' law for the divergence of the $\vec{E}$-field from charge density ρ_{c}, which reads $\vec{\nabla} \cdot \vec{E} = \rho_{\mathrm{c}}/\varepsilon_0$. In the same way, we can write a differential equation relating the gravitational field[7] $\vec{g}$ to a density of matter ρ, which reads $\vec{\nabla} \cdot \vec{g} = -4\pi G\rho$. In integral form, this is written as

$$\int \vec{g} \cdot \mathrm{d}\vec{S} = -4\pi GM, \tag{11.7}$$

where M is the total mass in a volume bounded by the closed surface with area vector $\vec{S}$. In words, this expression says that the strength of the gravitational field is given by the mass inside the surface $\vec{S}$. If, instead of $\vec{g}$ we use the potential Φ, via $\vec{g} = -\vec{\nabla}\Phi$, then we obtain Poisson's equation[8] for gravitational fields,

$$\vec{\nabla}^2\Phi = 4\pi G\rho. \tag{11.8}$$

We can use this expression to justify the claim that a sphere of test particles falling towards the earth does not change its volume. Note first that $\vec{\nabla}^2\Phi$ is the sum of eigenvalues of the matrix $\frac{\partial^2\Phi}{\partial x^i \partial x^j}$ that we considered above. We can then use the intervals s^i from the previous example to build a small sphere of particles. We then have, as a result of our expression $m\ddot{s}^i = -\frac{\partial^2\Phi}{\partial x^i \partial x^j}s^j$, that the second time derivative of the volume of the sphere of particles is determined by the density ρ of matter inside the sphere. Therefore, a sphere of free particles falling towards the earth forms an ellipsoid of the same volume as the original sphere, as there is (always) no gravitating mass inside the sphere. (If, on the other hand, we built a large sphere around the earth, we would expect the sphere to reduce its volume as the particles fell towards the earth.)

[7]Defined by $\vec{F} = m\vec{g}$, where $\vec{F}$ is the gravitation force on a particle with mass m.

[8]See Section 0.4.

11.3 Riemann curvature

Although geodesic deviation provides a physical and intuitive means of working out if space is curved, there is a mathematically simpler[9] method of determining the form of the Riemann tensor, involving the parallel transport of a vector. Recall that parallel transport is a method to work out how a vector changes due to the change in the underlying coordinate system. This isn't curvature necessarily, we may just be using a hopelessly complicated coordinate system. In order to work out if a spacetime is truly curved, we recall the fate of a vector parallel transported on a spherical surface from Fig. 11.2. We can see from this figure the effect of parallel transportation around a closed loop: the vector *rotates*. This is a smoking gun, telling us that the surface is truly curved. If it were not, the vector would point in the same direction after parallel transport. In order to assess the curvature of a spacetime we shall follow the same procedure and move a 4-vector in an infinitesimal loop to see if its direction changes. If it does, we will be able to identify the intrinsic curvature of the spacetime.

So let's be specific and take a vector $\boldsymbol{V}$ and move it round a closed, infinitesimal parallelogram formed by a vectors $\delta a\, \boldsymbol{e}_1$ and $\delta b\, \boldsymbol{e}_2$, where δa and δb are (small) constants (see Fig. 11.5). The idea is that we track the change in a component of $\boldsymbol{V}$ as follows:

$$\delta V^\alpha = \left(\begin{array}{c} \text{Change in } V^\alpha \text{ transported along } \delta a\, \boldsymbol{e}_1, \\ \text{then } \delta b\, \boldsymbol{e}_2, \text{ then } - \delta a\, \boldsymbol{e}_1, \text{ then } - \delta b\, \boldsymbol{e}_2 \end{array} \right)$$

$$= \left(\begin{array}{c} \text{Measure of the curvature} \\ \text{of spacetime} \end{array} \right)_\mu V^\mu\, \delta a\, \delta b. \tag{11.9}$$

This measure of curvature is, once again, supplied by the components of the Riemann tensor $\boldsymbol{R}(\ ,\ ,\ ,\ )$. How does this work? In the preceding equation, we have made the deceptively simple, but mathematically significant, claim that the right-hand side of the equation is linear in the sides of the parallelogram δa and δb. It is this that enables us to identify the part in brackets as the components of a tensor (i.e. a linear slot-machine object). To show that this is the case, we will dive ahead and use this method to derive an expression for the curvature tensor.

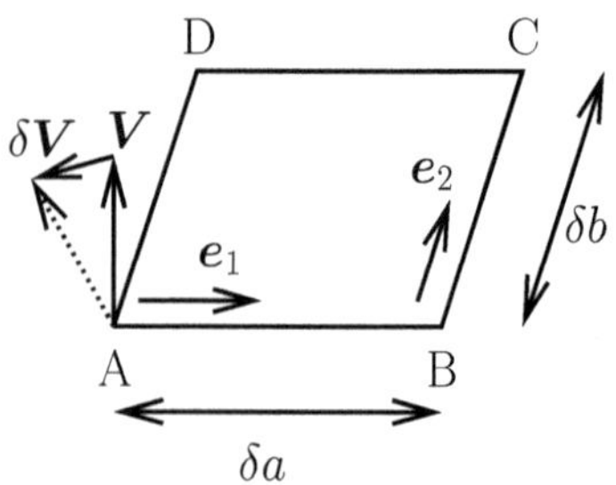

Fig. 11.5 Transporting a vector around an infinitesimal closed loop.

[9] It is slightly simpler in terms of computational sophistication, but significantly more tedious.

Example 11.3

We start with the vector at position A, with coordinates (a, b) and parallel transport it along $\boldsymbol{e}_1$ to position B. Parallel transport along $\boldsymbol{e}_1$ implies $\boldsymbol{\nabla}_{\boldsymbol{e}_1} \boldsymbol{V} = 0$, so we must have

$$\frac{\partial V^\alpha}{\partial x^1} = -\Gamma^\alpha{}_{1\beta} V^\beta. \tag{11.10}$$

At position B, the vector has components

$$V^\alpha(B) = V^\alpha(A) + \int_A^B dx^1\, \frac{\partial V^\alpha}{\partial x^1}$$

$$= V^\alpha(A) - \int_{x^2 = b} dx^1\, \Gamma^\alpha{}_{1\beta} V^\beta. \tag{11.11}$$

We repeat, transporting to points C and D and back to A[10]

$$V^\alpha(C) = V^\alpha(B) - \int_{x^1=a+\delta a} \mathrm{d}x^2\, \Gamma^\alpha{}_{2\beta} V^\beta,$$

$$V^\alpha(D) = V^\alpha(C) + \int_{x^2=b+\delta b} \mathrm{d}x^1\, \Gamma^\alpha{}_{1\beta} V^\beta,$$

$$V^\alpha(A_{\text{final}}) = V^\alpha(D) + \int_{x^1=a} \mathrm{d}x^2\, \Gamma^\alpha{}_{2\beta} V^\beta. \tag{11.12}$$

The net change δV^α going round the loop is, therefore,

$$\delta V^\alpha = V^\alpha(A_{\text{final}}) - V^\alpha(A_{\text{initial}})$$

$$= \int_{x^1=a} \mathrm{d}x^2\, \Gamma^\alpha{}_{2\beta} V^\beta - \int_{x^1=a+\delta a} \mathrm{d}x^2\, \Gamma^\alpha{}_{2\beta} V^\beta$$

$$+ \int_{x^2=b+\delta b} \mathrm{d}x^1\, \Gamma^\alpha{}_{1\beta} V^\beta - \int_{x^2=b} \mathrm{d}x^1\, \Gamma^\alpha{}_{1\beta} V^\beta. \tag{11.13}$$

Collecting terms and expanding to lowest order, we obtain

$$\delta V^\alpha \approx -\int_b^{b+\delta b} \mathrm{d}x^2 \delta a \frac{\partial}{\partial x^1}\left(\Gamma^\alpha{}_{2\beta} V^\beta\right) + \int_a^{a+\delta a} \mathrm{d}x^1 \delta b \frac{\partial}{\partial x^2}\left(\Gamma^\alpha{}_{1\beta} V^\beta\right)$$

$$\approx \delta a \delta b \left[-\frac{\partial}{\partial x^1}\left(\Gamma^\alpha{}_{2\beta} V^\beta\right) + \frac{\partial}{\partial x^2}\left(\Gamma^\alpha{}_{1\beta} V^\beta\right)\right]. \tag{11.14}$$

Finally, replacing the derivatives of V^α using eqn 11.10, we obtain

$$\delta V^\alpha \approx \delta a \delta b \left[\frac{\partial \Gamma^\alpha{}_{1\beta}}{\partial x^2} - \frac{\partial \Gamma^\alpha{}_{2\beta}}{\partial x^1} + \Gamma^\alpha{}_{2\sigma}\Gamma^\sigma{}_{1\beta} - \Gamma^\alpha{}_{1\sigma}\Gamma^\sigma{}_{2\beta} \right] V^\beta. \tag{11.15}$$

The change in the vector $\boldsymbol{V}$ can be summarized as

$$\delta V^\alpha = \left(\begin{array}{c}\text{Change in } V^\alpha \text{ transported along } \delta a\, \boldsymbol{e}_\nu, \\ \text{then } \delta b\, \boldsymbol{e}_\mu, \text{ then } -\delta a\, \boldsymbol{e}_\nu, \text{ then } -\delta b\, \boldsymbol{e}_\mu \end{array}\right) \tag{11.16}$$

$$= \left[\frac{\partial \Gamma^\alpha{}_{\nu\beta}}{\partial x^\mu} - \frac{\partial \Gamma^\alpha{}_{\mu\beta}}{\partial x^\nu} + \Gamma^\alpha{}_{\mu\sigma}\Gamma^\sigma{}_{\nu\beta} - \Gamma^\alpha{}_{\nu\sigma}\Gamma^\sigma{}_{\mu\beta} \right] V^\beta\, \delta a\, \delta b.$$

As claimed, the part in square brackets depends linearly on the sides of the parallelogram δa and δb. This enables us to define the **Riemann curvature tensor** as the $(1,3)$ tensor $\boldsymbol{R}(\ ,\ ,\ ,\ )$ with components[11]

$$R^\alpha{}_{\beta\mu\nu} = \left[\frac{\partial \Gamma^\alpha{}_{\nu\beta}}{\partial x^\mu} - \frac{\partial \Gamma^\alpha{}_{\mu\beta}}{\partial x^\nu} + \Gamma^\alpha{}_{\mu\sigma}\Gamma^\sigma{}_{\nu\beta} - \Gamma^\alpha{}_{\nu\sigma}\Gamma^\sigma{}_{\mu\beta} \right]. \tag{11.19}$$

Example 11.4

Recall for the weak gravitational field metric $\mathrm{d}s^2 = -[1+2\Phi(x^i)]\mathrm{d}t^2 + (\mathrm{d}x^2 + \mathrm{d}y^2 + \mathrm{d}z^2)$ we had non-zero connections

$$\Gamma^i{}_{tt} = \frac{\partial \Phi}{\partial x^i}. \tag{11.20}$$

Plugging in, we find that the only non-zero components of $\boldsymbol{R}$ are given by[12]

$$R^i{}_{tjt} = \frac{\partial \Gamma^i{}_{tt}}{\partial x^j} = \frac{\partial^2 \Phi}{\partial x^j \partial x^i}. \tag{11.21}$$

These are identical to the components that we identified earlier in the chapter in eqn 11.4.

[10]The notation here records the constant coordinate, rather than the limits of the integral.

[11]Equation 11.19 has the dubious distinction of being one of the more lengthy equations in this subject that it is useful to remember. A helpful aide-mémoire that is often employed is to use the **matrix form** of the equations to say

$$R^\bullet{}_{\bullet\mu\nu} = \partial_\mu \Gamma^\bullet{}_{\nu\bullet} - \partial_\nu \Gamma^\bullet{}_{\mu\bullet} + [\Gamma^\bullet{}_{\mu\bullet}, \Gamma^\bullet{}_{\nu\bullet}], \tag{11.17}$$

where square brackets denote a commutator (i.e. $[A,B] = AB - BA$). If the connection coefficients are written out as matrices, eqn 11.19 has the additional advantage of simplifying the sums over indices. For example, in two dimensions we have

$$\Gamma^\bullet{}_{\mu\bullet} = \begin{pmatrix} \Gamma^1{}_{\mu 1} & \Gamma^1{}_{\mu 2} \\ \Gamma^2{}_{\mu 1} & \Gamma^2{}_{\mu 2} \end{pmatrix}. \tag{11.18}$$

The matrix notation then allows the products of the connection coefficients to be computed using matrix algebra.

[12]Strictly we should write the final term on the right as $\frac{\partial^2 \Phi}{\partial x^j \partial x^k} g^{ki}$ if we were being fussy about tensor components.

We stress finally that the curvature tensor is built from connection coefficients and the connection coefficients are build from the metric. The metric field is the source of curvature.

11.4 Symmetries of the Riemann tensor

The Riemann tensor is vitally important object in general relativity. The tensor has a number of important **symmetries**.[13] Let's now find these. We'll consider the all-down index version of $\boldsymbol{R}$, achieved by saying $R_{\alpha\beta\gamma\delta} = g_{\alpha\mu} R^{\mu}{}_{\beta\gamma\delta}$.

[13] By 'symmetries' here we mean cases where components are, within a sign change, identical to other components. These mean we have to search for fewer distinct components of the tensor to characterize the curvature of spacetime. This is especially important for a $(1,3)$ tensor like $\boldsymbol{R}$, since is potentially has $4^4 = 256$ different components in $(3+1)$-dimensional spacetime in the absence of symmetry.

Example 11.5

The important local flatness theorem tells us it is always possible to find a local inertial frame (LIF) coordinate system such that the metric is identical to the Minkowski metric and the first derivative of the metric field vanishes at a point $\mathcal{P}$. We can therefore work at this point $\mathcal{P}$ where, as a consequence, the connection vanishes such that $\Gamma^{\mu}{}_{\alpha\beta}(x = \mathcal{P}) = 0$. However, the derivatives of the connection do not vanish (since local flatness certainly does *not* imply the vanishing of the second derivatives of the metric field). Therefore, the Riemann tensor has components given by the derivatives of the connections only. These can, in turn, be expressed as derivatives of the metric components, with the result that, in the LIF,[14]

$$R_{\hat{\mu}\hat{\nu}\hat{\alpha}\hat{\beta}} = \frac{1}{2}\left(\frac{\partial^2 g_{\hat{\alpha}\hat{\nu}}}{\partial x^{\hat{\beta}}\partial x^{\hat{\mu}}} - \frac{\partial^2 g_{\hat{\alpha}\hat{\mu}}}{\partial x^{\hat{\beta}}\partial x^{\hat{\nu}}} + \frac{\partial^2 g_{\hat{\beta}\hat{\mu}}}{\partial x^{\hat{\alpha}}\partial x^{\hat{\nu}}} - \frac{\partial^2 g_{\hat{\beta}\hat{\nu}}}{\partial x^{\hat{\alpha}}\partial x^{\hat{\mu}}} \right). \tag{11.23}$$

This equation, built from the symmetric metric tensor $\boldsymbol{g}$ makes the symmetries of the curvature tensor $\boldsymbol{R}$ manifest.

[14] In a general coordinate frame, $\boldsymbol{R}$ can be computed more directly as follows:

$$R_{\alpha\beta\mu\nu} =$$
$$\frac{1}{2}\left(g_{\alpha\nu,\beta\mu} - g_{\alpha\mu,\beta\nu} \right.$$
$$+ g_{\beta\mu,\alpha\nu} - g_{\beta\nu,\alpha\mu} \right)$$
$$+ g_{\sigma\rho}\left(\Gamma^{\sigma}{}_{\beta\mu}\Gamma^{\rho}{}_{\alpha\nu} - \Gamma^{\sigma}{}_{\beta\nu}\Gamma^{\rho}{}_{\alpha\mu} \right).$$
$$\tag{11.22}$$

We'll have little need to apply this expression in anger, but it's comforting to know that it exists. Note that the symmetries apply to both frames.

By considering eqn 11.23 we can identify the symmetries of $R_{\mu\nu\alpha\beta}$. Swapping indices within the first pair $(\mu\nu)$ and second pair $(\alpha\beta)$ results in picking up a minus sign, so we have[15]

$$R_{\alpha\beta\gamma\delta} = - R_{\beta\alpha\gamma\delta},$$
$$R_{\alpha\beta\gamma\delta} = - R_{\alpha\beta\delta\gamma},$$
$$R_{\alpha\beta\gamma\delta} = + R_{\gamma\delta\alpha\beta}. \tag{11.24}$$

[15] The symmetries mean that $R_{\mu\mu\mu\mu} = R_{\mu\mu\nu\sigma} = R_{\mu\nu\sigma\sigma} = 0$.

In words, these say that swaps within the first or second pair of indices earn a minus sign; swapping the pairs together does not. We also can check that we have a cyclic identity

$$R_{\alpha\beta\gamma\delta} + R_{\alpha\delta\beta\gamma} + R_{\alpha\gamma\delta\beta} = 0. \tag{11.25}$$

One consequence of these symmetries is on the makeup of the Riemann tensor. At first glance, it looks like the $\boldsymbol{R}$ tensor in N dimensions has N^4 components. However, these symmetries of the tensor strongly restrict the number of degrees of freedom, so that (as we will prove in Chapter 35) there are only $N^2(N^2 - 1)/12$ independent components. We have for $R_{\mu\nu\alpha\beta}$ that the number of independent components are as follows.

Dimensions	1	2	3	4
Independent components	0	1	6	20

Some immediate consequences are considered in the following example.

Example 11.6

In **one dimension**, there is always a coordinate transformation that reduces the metric on a line to the Euclidean form. There is never any curvature.[16]
In **two dimensions** we have a single non-zero component of $\mathbf{R}$. According to the symmetries, this must be R_{1212}.

11.5   The Ricci tensor and Ricci scalar

Once we have the Riemann tensor $\mathbf{R}$ and its symmetries then this also unlocks two simpler tensors that turn out to be essential to building the Einstein equation, linking the curvature and the matter fields of the Universe. It is perhaps to be expected that a $(1,3)$ tensor does not feature as raw material in a law of nature. Physics is full of $(2,0)$ and $(0,2)$ valent tensors, but very few $(3,1)$ valent tensors. To build a $(0,2)$ tensor from a $(3,1)$ tensor the simplest operation would be to **contract** the up index with a down index.[17] The symmetries constrain the possible choices, with the result that we define the **Ricci tensor**[18] as the $(0,2)$ valent tensor with components

$$R_{\nu\beta} = R^{\mu}{}_{\nu\mu\beta}$$
$$= R^{0}{}_{\nu0\beta} + R^{1}{}_{\nu1\beta} + R^{2}{}_{\nu2\beta} + R^{3}{}_{\nu3\beta}. \tag{11.26}$$

The symmetries of $\mathbf{R}$ imply that the Ricci tensor is symmetric, such that $R_{\mu\nu} = R_{\nu\mu}$. We can now make a further contraction, and turn the Ricci tensor into another tensor which is exceptionally simple: a $(0,0)$ tensor, otherwise known of course as a scalar. We will discuss the resulting object in more detail in the following chapter, but let's define it now. It's called the **Ricci scalar** R and is obtained by tracing over the indices of the Ricci tensor. Hence,[19] the Ricci scalar R is given by

$$R = g^{\mu\nu} R_{\mu\nu}. \tag{11.27}$$

The significance of the Ricci tensor will become clear as we try to relate this measure of curvature to the mass-energy of the fields filling the Universe. For now, you can think of it as a kind of average over components of the Riemann tensor.[20]

[16]The one-dimensional example is an interesting, but trivial, case. It has no intrinsic curvature. You can of course embed a one-dimensional space in a higher dimension, and make it *appear* curved by tying it into a bow or winding it around a cylinder, but that curvature is always extrinsic and never intrinsic. The one-dimensional space is always isometric to a perfectly straight line. Things are more complicated in two dimensions; an ant confined to the surface of a large ball only knows it is living in a curved space if it goes for a walk *around* the surface and measures how a vector is parallel transported around its walk and compares the result on returning to its initial position. Such closed-loop walks are not possible in one dimension.

[17]A contraction is an operation on a tensor that involves setting an upstairs index equal to a downstairs one and summing. The result is to reduce the number of indices by two.

[18]Gregorio Ricci-Curbastro (1853–1925). The Ricci tensor is one of the few important tensors to which we do not assign a bold symbol. The reason is that it does not appear in the field theory of gravity alone, but in a so-called trace-reversed form $R_{\mu\nu} - \frac{1}{2}g_{\mu\nu}R$, where R is the Ricci scalar (the trace over $R_{\mu\nu}$, see below). These are the components of the Einstein tensor $\mathbf{G}$, which is the quantity given a bold symbol. Some texts do not follow this convention and assign the Ricci tensor a bold symbol such as $\mathbf{Ri}$.

[19]A trace is often written as $A^{\mu}{}_{\mu} = A^{0}{}_{0} + A^{1}{}_{1} + A^{2}{}_{2} + A^{3}{}_{3}$. In general relativity, it is best written $g_{\mu\nu}A^{\mu\nu}$ as an aide-memoir that the metric must be used to construct the trace in curved spacetime.

[20]It will turn out that it is this average that captures the ability of gravitational curvature to cause volumes to shrink.

11.6　Example computations

We shall now carry out two brute force (and ignorance) calculations of the components of the Riemann tensor, using the equation

$$R^{\mu}{}_{\nu\alpha\beta} = \frac{\partial \Gamma^{\mu}{}_{\beta\nu}}{\partial x^{\alpha}} - \frac{\partial \Gamma^{\mu}{}_{\alpha\nu}}{\partial x^{\beta}} + \Gamma^{\mu}{}_{\alpha\rho}\Gamma^{\rho}{}_{\beta\nu} - \Gamma^{\mu}{}_{\beta\rho}\Gamma^{\rho}{}_{\alpha\nu}. \tag{11.28}$$

Example 11.7

First consider flat space with line element $ds^2 = dr^2 + r^2 d\theta^2$. We expect that all of the components of the Riemann tensor should vanish. We saw that this metric has connections $\Gamma^{r}{}_{\theta\theta} = -r$ and $\Gamma^{\theta}{}_{r\theta} = \frac{1}{r}$. We know from the symmetries that $R^{r}{}_{rrr} = R^{\theta}{}_{\theta\theta\theta} = 0$. Let's try some of the mixed-index components

$$R^{\theta}{}_{r\theta r} = \frac{\partial \Gamma^{\theta}{}_{rr}}{\partial x^{\theta}} - \frac{\partial \Gamma^{\theta}{}_{\theta r}}{\partial x^{r}} + \Gamma^{\theta}{}_{\theta\rho}\Gamma^{\rho}{}_{rr} - \Gamma^{\theta}{}_{r\rho}\Gamma^{\rho}{}_{\theta r}$$

$$= 0 + \frac{1}{r^2} + 0 - \frac{1}{r}\frac{1}{r} = 0. \tag{11.29}$$

Another one is

$$R^{r}{}_{\theta r\theta} = \frac{\partial \Gamma^{r}{}_{\theta\theta}}{\partial x^{r}} - \frac{\partial \Gamma^{r}{}_{r\theta}}{\partial x^{\theta}} + \Gamma^{r}{}_{r\rho}\Gamma^{\rho}{}_{\theta\theta} - \Gamma^{r}{}_{\theta\rho}\Gamma^{\rho}{}_{r\theta}$$

$$= -1 + 0 + 0 + 1 = 0. \tag{11.30}$$

As expected, the components all vanish in flat space.[21]

[21]Consequently, the Ricci tensor and Ricci scalar both vanish. This example may not have seemed that exciting, but it's good to check that our formalism works. Flat space is indeed not curved.

Now the (considerably less dull) example of a two-dimensional spherical space on the surface of a sphere of radius a.

Example 11.8

The spherical space has a line element

$$ds^2 = a^2 d\theta^2 + a^2 \sin^2\theta\, d\phi^2. \tag{11.31}$$

This gives connection coefficients

$$\Gamma^{\theta}{}_{\phi\phi} = -\sin\theta\cos\theta, \quad \Gamma^{\phi}{}_{\theta\phi} = \cot\theta. \tag{11.32}$$

The components may be calculated. Here's an example

$$R^{\theta}{}_{\phi\theta\phi} = \frac{\partial \Gamma^{\theta}{}_{\phi\phi}}{\partial x^{\theta}} - \frac{\partial \Gamma^{\theta}{}_{\theta\phi}}{\partial x^{\phi}} + \Gamma^{\theta}{}_{\theta\rho}\Gamma^{\rho}{}_{\phi\phi} - \Gamma^{\theta}{}_{\phi\rho}\Gamma^{\rho}{}_{\theta\phi}$$

$$= \sin^2\theta - \cos^2\theta - 0 + 0 + \cos^2\theta = \sin^2\theta. \tag{11.33}$$

[22]In fact, from the symmetries we know that $R_{1212} = R_{2121}$.

Another one is[22]

$$R^{\phi}{}_{\theta\phi\theta} = \frac{\partial \Gamma^{\phi}{}_{\theta\theta}}{\partial x^{\phi}} - \frac{\partial \Gamma^{\phi}{}_{\phi\theta}}{\partial x^{\theta}} + \Gamma^{\phi}{}_{\phi\rho}\Gamma^{\rho}{}_{\theta\theta} - \Gamma^{\phi}{}_{\theta\rho}\Gamma^{\rho}{}_{\phi\theta}$$

$$= 0 + \frac{1}{\sin^2\theta} + 0 - \frac{\cos^2\theta}{\sin^2\theta} = 1. \tag{11.34}$$

We have for the components of the Ricci tensor that

$$R_{\phi\phi} = \sin^2\theta, \quad R_{\theta\theta} = 1, \tag{11.35}$$

and the Ricci scalar is, therefore,

$$R = g^{\theta\theta}R_{\theta\theta} + g^{\phi\phi}R_{\phi\phi}$$

$$= \frac{1}{a^2} + \frac{\sin^2\theta}{a^2\sin^2\theta} = \frac{2}{a^2}. \tag{11.36}$$

These two examples show how the components of Riemann tensor $\boldsymbol{R}$ can be evaluated via direct (admittedly, rather tedious) computation. Fortunately, there is a more efficient method to achieve this, although it does rely on a little more geometrical technology, as we'll see in Part V.

11.7 Geodesic deviation revisited

We now have access to the tensor $\boldsymbol{R}$ that allows us to evaluate the curvature of spacetime. However, as we saw in Chapter 10, our measurements are made in orthonormal frames. By employing a specially-chosen orthonormal frame we can link the components of the Riemann tensor more directly to the notion of geodesic deviation we discussed earlier. The local frame we will choose is the freely falling local inertial frame.

We imagine setting a swarm of particles free from rest and allowing them to follow their geodesics. We assign one the status of being the fiducial geodesic and note its tangent $\boldsymbol{u}$. If the particles accelerate relative to each other then we can compute $\boldsymbol{R}$. We therefore need to work out the acceleration of a vector $\boldsymbol{\xi}$ linking a neighbouring geodesic to the fiducial one, which is related to the Riemann tensor via[23]

$$\frac{D^2\boldsymbol{\xi}}{d\tau^2} = \boldsymbol{R}(\ ,\boldsymbol{u},\boldsymbol{\xi},\boldsymbol{u}), \tag{11.38}$$

or, in components

$$\left(\frac{D^2\boldsymbol{\xi}}{d\tau^2}\right)^\alpha = -R^\alpha{}_{\beta\gamma\delta}u^\beta\xi^\gamma u^\delta. \tag{11.39}$$

In the coordinates of a freely falling frame we have[24] $u^{\hat{\beta}} = (1,0,0,0)$, the geodesic deviation equation simplifies to[25]

$$\frac{d^2\xi^{\hat{\alpha}}}{d\tau^2} = -R^{\hat{\alpha}}{}_{\hat{t}\hat{\beta}\hat{t}}\xi^{\hat{\beta}}. \tag{11.40}$$

Using the vielbein, we relate the coordinates of $\boldsymbol{R}$ in the local freely falling frame to a coordinate frame via $R^{\hat{\alpha}}{}_{\hat{\beta}\hat{\gamma}\hat{\delta}} = R^\mu{}_{\nu\lambda\sigma}(e_\mu)^{\hat{\alpha}}(e_{\hat{\beta}})^\nu(e_{\hat{\gamma}})^\lambda(e_{\hat{\delta}})^\sigma$. We can use the simplified form of the deviation equation in eqn 11.40 to investigate a simple example.

Example 11.9

An important example is the Schwarzschild geometry of a spherically symmetric gravitating object.[26] As we shall see, the metric for this spacetime gives rise to a Riemann curvature tensor $\boldsymbol{R}$ with components in the freely falling frame of

$$R_{\hat{t}\hat{r}\hat{t}\hat{r}} = -\frac{2M}{r^3}, \qquad R_{\hat{\theta}\hat{\phi}\hat{\theta}\hat{\phi}} = +\frac{2M}{r^3},$$
$$R_{\hat{t}\hat{\theta}\hat{t}\hat{\theta}} = R_{\hat{t}\hat{\phi}\hat{t}\hat{\phi}} = +\frac{M}{r^3}, \quad R_{\hat{r}\hat{\theta}\hat{r}\hat{\theta}} = R_{\hat{r}\hat{\phi}\hat{r}\hat{\phi}} = -\frac{M}{r^3}. \tag{11.41}$$

Since we're in an orthonormal frame, we use the Minkowski tensor to raise indices and the geodesic deviation equations become

$$\frac{d^2\xi^{\hat{r}}}{d\tau^2} = \frac{2M}{r^3}\xi^{\hat{r}}, \quad \frac{d^2\xi^{\hat{\theta}}}{d\tau^2} = -\frac{M}{r^3}\xi^{\hat{\theta}}, \quad \frac{d^2\xi^{\hat{\phi}}}{d\tau^2} = -\frac{M}{r^3}\xi^{\hat{\phi}}. \tag{11.42}$$

Notice the resemblance of these equations of motion to those of the Newtonian problem discussed in Example 11.1.

---

[23] This expression is sometimes called the Jacobi equation. The double derivative can be written as

$$\frac{D^2\boldsymbol{\xi}}{d\tau^2} = \boldsymbol{\nabla}_{\boldsymbol{u}}\boldsymbol{\nabla}_{\boldsymbol{u}}\boldsymbol{\xi}. \tag{11.37}$$

It is also useful to recall that in an orthonormal frame we take $\boldsymbol{u} = \boldsymbol{e}_{\hat{t}}$.

[24] That is, the observer is at rest and so the only non-zero component of $\boldsymbol{u}$ is $u^{\hat{t}}$. Since in the local coordinate system the observer uses the Minkowski tensor $\boldsymbol{\eta}$ to form products, the normalization $\boldsymbol{u}\cdot\boldsymbol{u} = -1$ becomes $\eta_{\hat{t}\hat{t}}(u^{\hat{t}})^2 = -1$ and so, since $\eta_{\hat{t}\hat{t}} = -1$, we have $u^{\hat{t}} = 1$.

[25] The use of the freely falling frame allows the swap $(D^2\boldsymbol{\xi}/d\tau^2)^\alpha$ to $d^2\xi^{\hat{\alpha}}/d\tau^2$. Notice how the resulting expression is analogous to a simple harmonic oscillator: $\ddot{x} = -\omega_0^2 x$ (and it is identical to this equation of motion for $\hat{\alpha} = \hat{\beta}$). The analogue of the natural frequency of the oscillator is provided by the components of the Riemann tensor. In this way, the curvature of spacetime can be thought of as analogous to a spring constant.

[26] Discussed in Part IV.

Chapter summary

- Geodesic deviation allows us to identify tidal forces that result from the curvature of spacetime due to gravity.
- The Riemann tensor $\boldsymbol{R}$ is non-zero if there is intrinsic curvature in spacetime.
- The tensor $\boldsymbol{R}$ can be derived by parallel transporting a vector around a loop. The components of $\boldsymbol{R}$ are given by

$$R^{\alpha}{}_{\beta\mu\nu} = \left[\frac{\partial \Gamma^{\alpha}{}_{\nu\beta}}{\partial x^{\mu}} - \frac{\partial \Gamma^{\alpha}{}_{\mu\beta}}{\partial x^{\nu}} + \Gamma^{\alpha}{}_{\mu\sigma}\Gamma^{\sigma}{}_{\nu\beta} - \Gamma^{\alpha}{}_{\nu\sigma}\Gamma^{\sigma}{}_{\mu\beta}\right]. \quad (11.43)$$

Exercises

(11.1) Using the connection coefficients computed for the Schwarzschild geometry in Exercise 9.7, show the following:

$$
\begin{aligned}
R^{t}{}_{rtr} &= -\Phi'' + \Phi'\Lambda' - (\Phi')^2, \\
R^{t}{}_{\theta t\theta} &= -r e^{-2\Lambda}\Phi', \\
R^{t}{}_{\phi t\phi} &= -r e^{-2\Lambda}\sin^2\theta\,\Phi', \\
R^{r}{}_{\theta r\theta} &= r e^{-2\Lambda}\Lambda', \\
R^{r}{}_{\phi r\phi} &= r e^{-2\Lambda}\Lambda'\sin^2\theta, \\
R^{\theta}{}_{\phi\theta\phi} &= (1 - e^{-2\Lambda})\sin^2\theta.
\end{aligned}
\quad (11.44)
$$

(11.2) We shall see in Chapter 21 that the Schwarzschild metric line element outside a star of mass M can be written as

$$
\begin{aligned}
\mathrm{d}s^2 = &-\left(1 - \frac{2M}{r}\right)\mathrm{d}t^2 + \left(1 - \frac{2M}{r}\right)^{-1}\mathrm{d}r^2 \\
&+ r^2\left(\mathrm{d}\theta^2 + \sin^2\theta\,\mathrm{d}\phi^2\right).
\end{aligned}
\quad (11.45)
$$

In terms of these variables, the components of $\boldsymbol{R}$ from the last problem can be written as

$$
\begin{aligned}
R^{t}{}_{rtr} &= -\frac{2M}{r^2(2M - r)}, \\
R^{t}{}_{\theta t\theta} &= -\frac{M}{r}, \\
R^{t}{}_{\phi t\phi} &= -\frac{M\sin^2\theta}{r}, \\
R^{r}{}_{\theta r\theta} &= -\frac{M}{r}, \\
R^{r}{}_{\phi r\phi} &= -\frac{M\sin^2\theta}{r}, \\
R^{\theta}{}_{\phi\theta\phi} &= \frac{2M\sin^2\theta}{r}.
\end{aligned}
\quad (11.46)
$$

(a) Write expressions for (i) $R^{t}{}_{rrt}$ and (ii) $R^{r}{}_{trt}$.

(b) Compute the components of $\boldsymbol{R}$ in the orthonormal frame.

(11.3) Using the components in Exercise 11.2, show that the components of the Ricci tensor vanish for the Schwarzschild metric outside a star. Why must this be the case?

(11.4) Demonstrate the spacetime described by the rotating frame metric in Exercise 3.5 is flat.

The energy-momentum tensor

12

The curvature that we discussed in the last chapter provides the left-hand, geometrical side of Einstein's field equation. The right-hand, physical side is supplied by the mass-energy content of the matter fields that fill time and space.[1] The mass-energy content of spacetime is expressed using the energy-momentum tensor field $\boldsymbol{T}(x)$. We met an example of the energy-momentum tensor briefly in Part I of the book. In this chapter, we meet the tensor again in a little more detail, along with its all-important constraint $\boldsymbol{\nabla} \cdot \boldsymbol{T} = 0$. These tools will enable us, in the next chapter, to put everything together and write down the Einstein field equation for general relativity.

12.1 Another look at the energy-momentum tensor

The energy-momentum tensor field $\boldsymbol{T}(x)$ gives us access to a tensor $\boldsymbol{T}(\ ,\ )$ at each point in spacetime that can be used to evaluate the energy and momentum content of matter fields at that point. If in doubt, you can think of $\boldsymbol{T}(x)$ as describing the current of momentum for a distribution of mass-energy. We can equally well consider the tensor in (2,0) or (0,2) form. We start by considering the (0,2) version of the tensor

$$\boldsymbol{T}(x) = T_{\mu\nu}(x)\, \boldsymbol{\omega}^{\mu} \otimes \boldsymbol{\omega}^{\nu}, \tag{12.2}$$

where, since the tensor is symmetric, $T_{\mu\nu}(x) = T_{\nu\mu}(x)$.[2]

Take the 4-velocity of an observer to be $\boldsymbol{u}$. There is a recipe for extracting physical information by filling in slots of the tensor $\boldsymbol{T}(\ ,\ )$. Start by inserting the velocity into one slot (it doesn't matter which, since the tensor is symmetric), to obtain a 1-form

$$\boldsymbol{T}(\ ,\boldsymbol{u}) = -\left(\begin{array}{c}\text{4-momentum density in}\\ \text{the observer's rest frame}\end{array}\right). \tag{12.4}$$

Now insert a unit displacement vector $\boldsymbol{a}$ into the second slot. Since the slots are full, the output is a number. In this case, we obtain

$$\boldsymbol{T}(\boldsymbol{a},\boldsymbol{u}) = -\left(\begin{array}{c}\text{4-momentum density component directed}\\ \text{along } \boldsymbol{a} \text{ in the observer's rest frame}\end{array}\right). \tag{12.5}$$

[1] We expect the Einstein field equation to read (in words)

$$\begin{array}{c}\text{Curvature of}\\ \text{spacetime}\end{array} = \begin{array}{c}\text{Energy density}\\ \text{of matter fields}\end{array}. \tag{12.1}$$

We will (finally!) justify this in the next chapter, using the tools we have developed.

[2] We saw in Chapter 4, how the (0,2) tensor $\boldsymbol{T}$ could be understood by describing it as an outer product of 1-forms

$$\boldsymbol{T}(\ ,\ ) = \tilde{\boldsymbol{p}}(\ ) \otimes \tilde{\boldsymbol{J}}(\ ), \tag{12.3}$$

where $\tilde{\boldsymbol{p}}$ is a momentum 1-form and $\tilde{\boldsymbol{J}}$ is a number current 1-form for particles. We previously took $\tilde{\boldsymbol{J}}$ to describe a swarm of non-interacting particles. Although the tensor $\boldsymbol{T}$ describes energy-momentum more generally than for the case of swarms of particles, this description is useful in understanding $\boldsymbol{T}$. For example, if an observer has velocity $\boldsymbol{u}$, then we can interpret $\boldsymbol{T}(\ ,\boldsymbol{u}) = \tilde{\boldsymbol{p}} \otimes \tilde{\boldsymbol{J}}(\boldsymbol{u}) = -\tilde{\boldsymbol{p}}n$ as (minus) the momentum density of particles in the observer's frame. We would interpret $\boldsymbol{T}(\boldsymbol{u},\ ) = \tilde{\boldsymbol{p}}(\boldsymbol{u}) \otimes \tilde{\boldsymbol{J}} = -E\tilde{\boldsymbol{J}}$ as (minus) the energy flux. Since the tensor is symmetric, these are the same thing.

[3]Since we use the local basis vectors in the observer's orthonormal frame, denoted with hats on the indices, the interpretations here will always apply to measurements made in this observer's frame. We can turn these into up components using the Minkowski tensor $\eta^{\hat{\mu}\hat{\nu}}$, so we pick up a factor of -1 for each of the two timelike components we raise. This gives us

$$
\begin{aligned}
T^{\hat{0}\hat{0}} &= T_{\hat{0}\hat{0}}, \\
T^{\hat{0}\hat{i}} &= -T_{\hat{0}\hat{i}}, \\
T^{\hat{i}\hat{j}} &= T_{\hat{i}\hat{j}}.
\end{aligned}
\tag{12.7}
$$

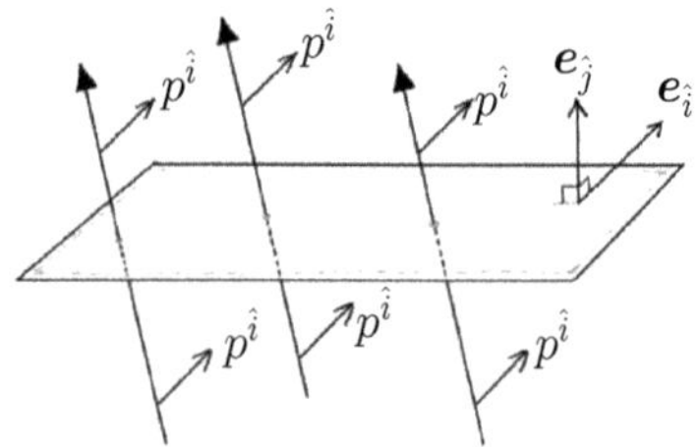

Fig. 12.1 Flux of the $\hat{i}$th component of momentum across a surface parallel to $\boldsymbol{\omega}^{\hat{j}}$ (that is, a surface with normal $\boldsymbol{e}_{\hat{j}}$).

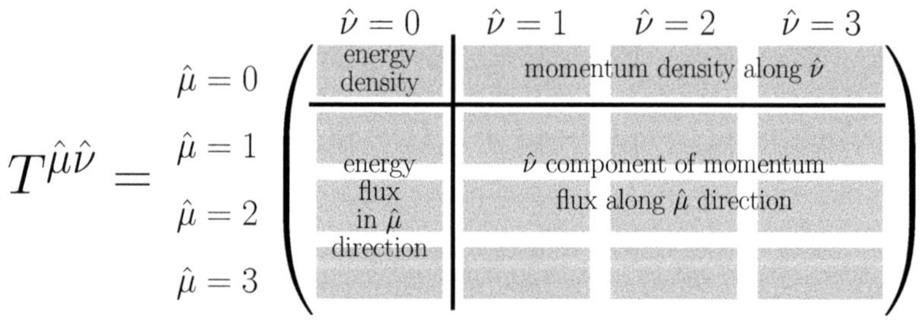

[4]A more systematic version from the viewpoint of fields is discussed in Chapter 40.

Filling, instead, both slots with $\boldsymbol{u}$, we obtain

$$
\boldsymbol{T}(\boldsymbol{u}, \boldsymbol{u}) = \left(\begin{array}{c} \text{Energy density} \\ \text{in the observer's rest frame} \end{array} \right).
\tag{12.6}
$$

Example 12.1

We shall also discuss components $T_{\mu\nu} = \boldsymbol{T}(\boldsymbol{e}_\mu, \boldsymbol{e}_\nu)$. Timelike components can be extracted by noting that in the observer's orthonormal frame $\boldsymbol{u} = \boldsymbol{e}_{\hat{0}}$ and so, and perhaps most importantly,[3]

$$
\boldsymbol{T}(\boldsymbol{e}_{\hat{0}}, \boldsymbol{e}_{\hat{0}}) = T_{\hat{0}\hat{0}} = T^{\hat{0}\hat{0}} = (\text{ energy density }).
\tag{12.8}
$$

We also have

$$
\boldsymbol{T}(\boldsymbol{e}_{\hat{i}}, \boldsymbol{e}_{\hat{0}}) = T_{\hat{i}\hat{0}} = -T^{\hat{i}\hat{0}} = -\left(\begin{array}{c} i\text{th component of} \\ 4\text{-momentum density} \end{array} \right).
\tag{12.9}
$$

which is also equivalent to an energy flux in the ith direction. We can also interpret the purely spatial components.

$$
\boldsymbol{T}(\boldsymbol{e}_{\hat{i}}, \boldsymbol{e}_{\hat{j}}) = T_{\hat{i}\hat{j}} = T^{\hat{i}\hat{j}} = \left(\begin{array}{c} i\text{th component of 4-momentum flux} \\ \text{crossing a surface parallel to } \boldsymbol{\omega}^{\hat{j}} \end{array} \right).
\tag{12.10}
$$

The idea of the latter is shown in Fig. 12.1. From the version of the energy-momentum tensor in the sidenote, we have that $T_{\hat{i}\hat{j}} = \tilde{\boldsymbol{p}}(\boldsymbol{e}_{\hat{i}}) \tilde{\boldsymbol{J}}(\boldsymbol{e}_{\hat{j}}) = p_{\hat{i}} J_{\hat{j}}$, which is consistent with this interpretation. However, another interpretation is possible. Since rate of momentum is equivalent to force, we can also write

$$
\boldsymbol{T}(\boldsymbol{e}_{\hat{i}}, \boldsymbol{e}_{\hat{j}}) = T_{\hat{i}\hat{j}} = T^{\hat{i}\hat{j}} = \left(\begin{array}{c} i\text{th component of force} \\ \text{across a surface parallel to } \boldsymbol{\omega}^{\hat{j}} \end{array} \right).
\tag{12.11}
$$

A cartoon of the components of the energy-momentum tensor is shown in Fig. 12.2.

$$
T^{\hat{\mu}\hat{\nu}} =
\begin{array}{c|c|ccc}
 & \hat{\nu}=0 & \hat{\nu}=1 & \hat{\nu}=2 & \hat{\nu}=3 \\
\hline
\hat{\mu}=0 & \text{energy density} & \multicolumn{3}{c}{\text{momentum density along } \hat{\nu}} \\
\hline
\hat{\mu}=1 & \text{energy flux in } \hat{\mu} \text{ direction} & \multicolumn{3}{c}{\nu \text{ component of momentum flux along } \hat{\mu} \text{ direction}} \\
\hat{\mu}=2 & & & & \\
\hat{\mu}=3 & & & &
\end{array}
$$

Fig. 12.2 The components of the energy-momentum tensor in the orthonormal frame of an observer.

12.2 Example energy-momentum tensors

The general discussion of the energy-momentum tensor is all very well, but of limited use without some idea of what the various components of $\boldsymbol{T}$ actually are. We now examine some examples of how to build[4] $\boldsymbol{T}$.

Since gravity works over long length scales where the granularity of matter is often not too important, we will mostly be concerned with continuous fields of matter and so we need the energy-momentum tensor for this purpose. The continuous version of the particle matter we have considered so far is a **fluid**. We will consider a so-called **perfect fluid**, which is continuous matter with energy density ρ and an isotropic **pressure**[5] p in its rest frame, but no other interactions.[6]

Example 12.2

First, consider the case of the fluid at rest in a local inertial reference frame, which has metric $\boldsymbol{\eta}$. We have an energy density $T^{\hat{0}\hat{0}} = T_{\hat{0}\hat{0}} = \rho$, where ρ is the total energy-density of the fluid, as measured in the fluid's rest frame. The force in the $\hat{i}$th direction, perpendicular to the surface whose normal is in the $\hat{i}$ direction, is simply the definition of the pressure p. We therefore have $T^{\hat{i}\hat{i}} = T_{\hat{i}\hat{i}} = p$ and can write

$$T_{\hat{\mu}\hat{\nu}} = \begin{pmatrix} \rho & 0 & 0 & 0 \\ 0 & p & 0 & 0 \\ 0 & 0 & p & 0 \\ 0 & 0 & 0 & p \end{pmatrix}. \tag{12.12}$$

The velocity 4-vector $\boldsymbol{u}$ of the fluid fixes the timelike basis vector according to $\boldsymbol{e}_{\hat{0}} = \boldsymbol{u}$. In the fluid's rest frame, the velocity 1-form $\tilde{\boldsymbol{u}} = u_{\hat{\mu}}\boldsymbol{\omega}^{\hat{\mu}}$ has components $u_{\hat{\mu}} = (-1,0,0,0)$. The (0,2) tensor product $\tilde{\boldsymbol{u}} \otimes \tilde{\boldsymbol{u}}$ has components

$$(\tilde{\boldsymbol{u}} \otimes \tilde{\boldsymbol{u}})_{\hat{\mu}\hat{\nu}} = \tilde{\boldsymbol{u}}(\boldsymbol{e}_{\hat{\mu}})\tilde{\boldsymbol{u}}(\boldsymbol{e}_{\hat{\nu}}) = u_{\hat{\mu}}u_{\hat{\nu}}, \tag{12.13}$$

and so we have

$$(\tilde{\boldsymbol{u}} \otimes \tilde{\boldsymbol{u}})_{\hat{\mu}\hat{\nu}} = \begin{pmatrix} 1 & 0 & 0 & 0 \\ 0 & 0 & 0 & 0 \\ 0 & 0 & 0 & 0 \\ 0 & 0 & 0 & 0 \end{pmatrix}. \tag{12.14}$$

We therefore find we can recreate the energy-momentum matrix via the decomposition

$$T_{\hat{\mu}\hat{\nu}} = \begin{pmatrix} \rho & 0 & 0 & 0 \\ 0 & 0 & 0 & 0 \\ 0 & 0 & 0 & 0 \\ 0 & 0 & 0 & 0 \end{pmatrix} + \begin{pmatrix} 0 & 0 & 0 & 0 \\ 0 & p & 0 & 0 \\ 0 & 0 & p & 0 \\ 0 & 0 & 0 & p \end{pmatrix}$$

$$= \rho(\tilde{\boldsymbol{u}} \otimes \tilde{\boldsymbol{u}})_{\hat{\mu}\hat{\nu}} + p\left[\eta_{\hat{\mu}\hat{\nu}} + (\tilde{\boldsymbol{u}} \otimes \tilde{\boldsymbol{u}})_{\hat{\mu}\hat{\nu}}\right]. \tag{12.15}$$

The components of this equation balance[7] and so it represents a valid tensor equation.

We conclude that for a perfect fluid in a local inertial frame (or in flat spacetime) we have a equation for the (0,2) tensor $\boldsymbol{T}$, which we can write as

$$\boldsymbol{T}(\ ,\ ) = (\rho + p)\tilde{\boldsymbol{u}}(\ ) \otimes \tilde{\boldsymbol{u}}(\ ) + p\boldsymbol{\eta}(\ ,\ ), \tag{12.16}$$

where we take $\tilde{\boldsymbol{u}}$ to be the velocity 1-form of the fluid itself. The tensor $\boldsymbol{T}$ inputs two vectors to generate a number. It has components

$$T_{\hat{\mu}\hat{\nu}} = (\rho + p)u_{\hat{\mu}}u_{\hat{\nu}} + p\eta_{\hat{\mu}\hat{\nu}}. \tag{12.17}$$

Example 12.3

Taking the trace T of the tensor gives us a number

$$T = T^{\hat{\nu}}{}_{\hat{\nu}} = T_{\hat{\mu}\hat{\nu}}\eta^{\hat{\mu}\hat{\nu}} = -(\rho + p) + 4p = -\rho + 3p. \tag{12.18}$$

Thermodynamics says that a photon gas has the equation of state $\rho = 3p$, so we conclude that the energy-momentum tensor for the photon gas is traceless ($T = 0$).[8]

---

[5] Do not confuse pressure p with momentum, which regrettably has the same symbol.

[6] The total energy density ρ will generally include a contribution from the mass density ρ_0 of the fluid, along with a contribution from the internal energy density. The latter results from internal motion and interactions between the constituents of the fluid. If there is no internal energy then pressure $p = 0$ and we call the matter **dust**, which has $\rho = \rho_0$. In contrast, a **perfect fluid** has an isotropic pressure p in its rest frame, but no viscosity or shear stresses (and now $\rho \neq \rho_0$). The relationship between ρ and p for the fluid is known as its **equation of state**.

[7] That is, we have $\hat{\mu}\hat{\nu}$ in the down position on both sides. Recall that we say such equations are *manifestly covariant* and are therefore 'valid tensor equations'.

[8] A field with an energy-momentum tensor whose trace vanishes is expected to have massless excitations. This is true for the electromagnetic field, whose excitations, photons, are certainly massless. The vanishing trace can be linked to scale invariance, as discussed in the book by Zee.

The equivalence principle tells us that a valid tensor equation in a flat space is a valid tensor equation in a curved space as long as we replace the Minkowski metric tensor $\boldsymbol{\eta}$ with the generalized metric tensor $\boldsymbol{g}$. We can then upgrade eqn 12.16 to the generalized version in curved space, where it becomes

$$\begin{aligned}
\boldsymbol{T}(\ ,\ ) &= (\rho + p)\tilde{\boldsymbol{u}}(\ ) \otimes \tilde{\boldsymbol{u}}(\ ) + p\boldsymbol{g}(\ ,\ ), \\
T_{\mu\nu} &= (\rho + p)u_\mu u_\nu + p g_{\mu\nu}.
\end{aligned} \tag{12.19}$$

Since tensor equations in physics are generally valid with respect to transforming all indices from up to down, and vice versa, we shall also use the (2,0) version of $\boldsymbol{T}$ whose components in a general coordinate frame are $T^{\mu\nu} = (\rho + p)u^\mu u^\nu + p g^{\mu\nu}$. In comparing quantities with up and down indices, remember that, as usual, indices are raised and lowered using the components of the metric tensor.

12.3   Classical particles

Einstein's equation is expressed in terms of classical fields. We shall find that the matter field corresponding to a perfect fluid is the most useful for our applications. However, in addition to fluids, we can also treat the motion of point-like particles. The most simple example is introduced below.

Example 12.4

Consider a swarm of n (non-interacting) dust particles per unit volume in flat space, each of rest mass m, moving with the same 4-velocity with components $(\gamma, \gamma v^x, \gamma v^y, 0)$. Temporarily restoring factors of c, we have total energy density $T^{00} = nm\gamma c^2$. The momentum density in the i-direction is $T^{0i} = T^{i0} = nm\gamma c v^i$. The momentum flux in, say, the x-direction, crossing the surface parallel to $\boldsymbol{\omega}^y$ is $T^{xy} = nm\gamma v^x v^y$. This is enough for us to construct the tensor, whose components are

$$T^{\mu\nu} = nm\gamma \begin{pmatrix} c^2 & v^x c & v^y c & 0 \\ v^x c & (v^x)^2 & v^x v^y & 0 \\ v^y c & v^x v^y & (v^y)^2 & 0 \\ 0 & 0 & 0 & 0 \end{pmatrix}. \tag{12.20}$$

In order to treat particles more generally using the techniques of field theory, we can use the fact that particles appear on a world line $z^\mu(\tau)$, where τ is the proper time parametrizing the world line.[9] In order to select that part of spacetime that the particle intersects, we use a delta function to pick out the curve $z^\mu(\tau)$. The particles are assumed non-interacting here, so the only contribution to their energy is their mass. To capture the mass density field $\rho_0(x)$ of a particle of mass m in flat spacetime we write

$$\rho_0(x) = m \int d\tau\, \delta^{(4)}[x - z(\tau)], \tag{12.21}$$

[9] Remember that in flat space the particle has velocity $\boldsymbol{u} = (dz^\mu/d\tau)\boldsymbol{e}_\mu$, which is tangent to its world line.

where, in Cartesian coordinates, we have defined a four-dimensional delta-function

$$\delta^{(4)}[x - z(\tau)] = \delta[x^0 - z^0(\tau)]\delta[x^1 - z^1(\tau)]\delta[x^2 - z^2(\tau)]\delta[x^3 - z^3(\tau)]. \tag{12.22}$$

Example 12.5

To see how the definition works, we change the integration variable from τ to $z^0(\tau)$ using $u^0 = \mathrm{d}z^0/\mathrm{d}\tau$ to write

$$
\begin{aligned}
\rho_0(x) =& m \int \mathrm{d}z^0\, \frac{\delta^{(4)}[x - z(z^0(\tau))]}{u^0} \\
=& m \int \mathrm{d}z^0\, \frac{\delta^{(3)}[x - z(z^0(\tau))]}{u^0} \delta[t - z^0(\tau)] \\
=& \frac{m}{u^0} \delta^{(3)}[x - z(\tau)],
\end{aligned}
\tag{12.23}
$$

where τ in the final line solves the equation $z^0(\tau) = t$. This is now simply a function that finds the particle in three-dimensional space. Notice how the factor u^0, which in flat space is simply γ, deals with the length contraction of the three-dimensional volume that determines the value of the mass density (see Fig. 2.7).

We can also write the mass current, which is most generally written as a field $\boldsymbol{J}_m(x) = \rho_0(x)\boldsymbol{u}(x)$, where $\boldsymbol{u}(x)$ is the velocity of the matter at x. In terms of particles, we have the component expression[10]

$$J_m^\alpha(x) = m \int \mathrm{d}\tau\, u^\alpha \delta^{(4)}[x - z(\tau)]. \tag{12.25}$$

In the same way, we write a flat-space energy-momentum tensor with components

$$T^{\alpha\beta}(x) = m \int \mathrm{d}\tau\, u^\alpha u^\beta \delta^{(4)}[x - z(\tau)], \tag{12.26}$$

which is simply a version of eqn 12.19 for particles with $p = 0$.

[10]By the same token we can also define a charge current with components in flat space of

$$J_q^\alpha(x) = q \int \mathrm{d}\tau\, u^\alpha \delta^{(4)}[x - z(\tau)], \tag{12.24}$$

where q is the scalar charge.

Example 12.6

In curved spacetime, we must recall that $\sqrt{-g}\,\mathrm{d}^4x$ is the invariant volume element, so $\delta^{(4)}(x - y)/\sqrt{-g}$ is a scalar. The expressions above become, in curved spacetime,

$$
\begin{aligned}
\rho_0(x) =& m \int \mathrm{d}\tau\, \frac{\delta^{(4)}[x - z(\tau)]}{\sqrt{-g}}, \\
J_m^\alpha(x) =& m \int \mathrm{d}\tau\, u^\alpha \frac{\delta^{(4)}[x - z(\tau)]}{\sqrt{-g}}, \\
T^{\alpha\beta}(x) =& m \int \mathrm{d}\tau\, u^\alpha u^\beta \frac{\delta^{(4)}[x - z(\tau)]}{\sqrt{-g}}.
\end{aligned}
\tag{12.27}
$$

[11] Restoring factors of c we have the 4-vector $J^\mu = \left(\rho c, \vec{J}\right)$. The space-like part $\vec{J}$ therefore has units of charge density times velocity.

[12] Integrating eqn 12.30 over a flat-space 3-volume and invoking the divergence theorem gives the conservation law in three-dimensional Cartesian space as

$$-\frac{\partial Q}{\partial t} = \int_S \vec{J} \cdot d\vec{S}, \qquad (12.29)$$

where S is the boundary of the volume V, whose area element is written as $d\vec{S}$. In words, the rate of change of charge in the volume is equal to the flux of particles through its surface.

[13] The 3-surface could be oriented, so that we have a sense of direction defined from one side of the surface to the other, just as one has for the 2-surface in three-dimensions.

[14] This can be interpreted as saying that all the charge that enters the volume in the past part of its boundary (where J and $d\Sigma$ might point in opposite directions) has to exit out of the future part of its boundary (where J and $d\Sigma$ might point in the same direction).

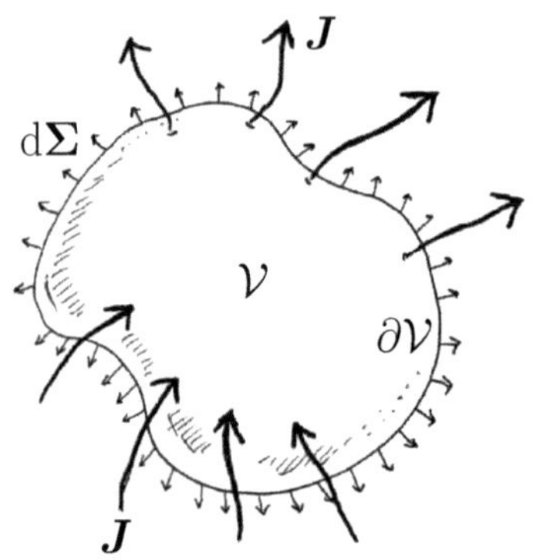

Fig. 12.3 A volume $\mathcal{V}$ in flat spacetime is bounded by a 3-surface $\partial\mathcal{V}$ which can be described by lots of infinitesimal surface vectors $d\Sigma$. The current density 4-vector J flows into and out of the surface, subject to a global conservation law (eqn 12.32).

12.4   Conservation laws

Example 12.7

In general, a current vector J has components $J^\mu = (\rho, \vec{J})$. The timelike component ρ is called a **charge density**. In flat space, the integral of this quantity over the 3-volume V is the conserved charge

$$Q = \int_V d^3x \, \rho. \qquad (12.28)$$

This charge might be electrical charge, or might be the number density of particles. The spacelike components represents the flux of the charge, that is, the number of charges that cross unit area per unit time.[11] Again working in flat spacetime, local conservation implies that J obeys the equation[12]

$$\frac{\partial\rho}{\partial t} + \vec{\nabla} \cdot \vec{J} = 0, \qquad (12.30)$$

or,

$$\partial_\mu J^\mu = 0, \quad \text{or equivalently,} \quad J^\mu{}_{,\mu} = 0. \qquad (12.31)$$

We shall often write this in vector notation as $\boldsymbol{\nabla} \cdot \boldsymbol{J} = 0$ (i.e. the divergence of $\boldsymbol{J}$ is zero). This expression guarantees that the rate of change of charge is equal to the amount of the charge flowing into, or out of, an element of volume. We call this **local conservation** of charge. We can't therefore have charge disappear on one point and then appear at some other arbitrary point in the Universe, which is what we mean by 'local' in our definition. (Conversely, if we wanted to guarantee conservation of charge, but allow charges to arbitrarily transport across the universe, we would demand only **global conservation** of a charge.)

We can think about this problem another way. Imagine a flat 3-surface which lies in the xyt plane, perpendicular to the z-axis. Then J^z represents the total charge that flows across this surface per unit area, per unit time. If this 3-surface had dimensions $\Sigma_z = a \times b \times \tau$ (i.e. $a \times b$ in the xy plane and τ in the time dimension) then $J^z \Sigma_z$ would represent the total charge flowing through this 3-surface (we've multiplied through by Σ_z so it is no longer 'per unit area, per unit time'). We could generalize this idea (which has privileged a particular slice in space time) and think of $J^\alpha \Sigma_\alpha$ as the total charge that flows across any flat 3-surface described by a 4-vector $\boldsymbol{\Sigma}$ normal to the 3-surface.[13] We could make this even more general by allowing the 3-surface to be curved, and then the total charge that flows across it could be written as $\int J^\mu \, d\Sigma_\mu$; here, both $\boldsymbol{J}$ and $\boldsymbol{\Sigma}$ are now functions of spacetime.

Now consider some 4-volume of spacetime $\mathcal{V}$, bounded by a closed 3-surface which we can write as $\partial\mathcal{V}$ (see Fig. 12.3). Integrating over the closed surface $\partial\mathcal{V}$, we can then express the global law of charge conservation as[14]

$$\int_{\partial\mathcal{V}} J^\mu \, d\Sigma_\mu = 0. \qquad (12.32)$$

This expresses global charge conservation in flat spacetime, since our 4-volume $\mathcal{V}$ can be of any size.

In setting this problem up, we have been thinking about the conservation of *electric charge*, but this approach would work with any conserved scalar quantity. For example, it could be a number-flux of baryons (which one could write as $\boldsymbol{N} = (N^0, \vec{N})$, with N_0 the baryon number density and $\vec{N}$ the baryon number 3-flux) or the flux of rest mass $[\rho_0(1, \vec{u})$, with ρ_0 the rest mass density and $\vec{u}$ the 3-velocity]. For both these cases, an equation analogous to eqn 12.32 would follow.

Energy and momentum are conserved, so we might expect a local conservation law for the energy-momentum tensor (which describes the flow of 4-momentum), just as we found for $\boldsymbol{J}$ (which describes the flow of charge). By analogy with the previous example, the local conservation law in flat spacetime can be written in three (equivalent) ways

$$\text{(i)}\ \partial_\mu T^{\mu\nu} = 0, \quad \text{(ii)}\ T^{\mu\nu}{}_{,\mu} = 0, \quad \text{(iii)}\ \boldsymbol{\nabla} \cdot \boldsymbol{T} = 0, \tag{12.33}$$

as justified in the following example.

Example 12.8

- Putting $\nu = 0$ into $T^{\mu\nu}{}_{,\mu} = 0$ gives $T^{00}{}_{,0} + T^{0i}{}_{,i} = 0$, which expands into $\partial\rho/\partial t + \vec{\nabla} \cdot (\rho\vec{v}) = 0$, which is the conservation law given in eqn 12.30. This part therefore expresses *local conservation of energy* in flat spacetime.
- Putting $\nu = i$ into $T^{\mu\nu}{}_{,\mu} = 0$ gives $T^{i0}{}_{,0} + T^{ij}{}_{,j} = 0$, which expands into $\partial(\rho v^i)/\partial t + \vec{\nabla} \cdot (\rho v^i \vec{v}) = 0$, which is a very similar conservation law, but now the ith component of the momentum density ρv^i has replaced the energy. This part therefore expresses *local conservation of momentum* in flat spacetime.

All this has been for *flat* spacetime. The next natural step is check if a version of $T^{\mu\nu}{}_{,\mu} = 0$ is valid for *curved* spacetime. If an observer makes measurements in a local inertial frame (LIF), they will conclude that

$$T^{\hat\mu\hat\nu}{}_{,\hat\mu} = 0. \tag{12.34}$$

Recall that, in a LIF, the connection coefficients $\Gamma^\mu{}_{\alpha\beta}$ vanish at a point,[15] making ordinary and covariant derivatives identical. As a result, the observer will conclude that

$$T^{\hat\mu\hat\nu}{}_{;\hat\mu} = 0. \tag{12.35}$$

This is a valid tensor equation in flat space so, by the principle of general covariance, should also work in curved space, if properly upgraded. The upgrade simply involves removing the hats

$$T^{\mu\nu}{}_{;\mu} = 0, \tag{12.36}$$

and we now have our curved space version. This argument, based on the principle of general covariance, is often called **comma goes to semicolon** and is equivalent to the swap $\eta \to g$. Probably the most memorable way to write this important result, eqn 12.36, is[16]

$$\boldsymbol{\nabla} \cdot \boldsymbol{T} = 0. \tag{12.38}$$

Example 12.9

We can take the divergence of the perfect fluid tensor part by part by noting that the covariant derivative obeys the Leibniz ($\equiv$ chain) rule. Start with components $T^{\mu\nu} = (\rho + p)u^\mu u^\nu + pg^{\mu\nu}$ and using the semicolon notation and the Leibniz rule, we find[17]

$$T^{\mu\nu}{}_{;\nu} = (\rho+p)_{;\nu}u^\mu u^\nu + (\rho+p)u^\mu{}_{;\nu}u^\nu + (\rho+p)u^\mu u^\nu{}_{;\nu} + p_{;\nu}g^{\mu\nu} + pg^{\mu\nu}{}_{;\nu}. \tag{12.39}$$

---

[15] This argument is the same one we rehearsed in Example 8.8.

[16] We have not yet worried about how to take the covariant derivative of anything more complicated than a vector. We return to this in Part V, but for now note that, written explicitly, the component equation is

$$T^{\mu\nu}{}_{;\mu} = \frac{\partial T^{\mu\nu}}{\partial x^\mu} + \Gamma^\mu{}_{\mu\alpha}T^{\alpha\nu} + \Gamma^\nu{}_{\mu\alpha}T^{\mu\alpha}. \tag{12.37}$$

[17] Remember that the tensor $\boldsymbol{T}$ is symmetric, so $T^{\mu\nu} = T^{\nu\mu}$ and there is no difference if we take the derivative with respect to μ or ν. Note also that (i) for a scalar function f we have

$$f_{;\mu} = f_{,\mu},$$

and (ii) we have the important equation

$$g^{\mu\nu}{}_{;\mu} = 0.$$

This quantity must, of course, vanish since $\boldsymbol{\nabla} \cdot \boldsymbol{T} = 0$. Noting the comments in the sidenote, we find

$$0 = T^{\mu\nu}{}_{;\nu} = \left[(\rho+p)_{,\nu}u^{\nu} + (\rho+p)u^{\nu}{}_{;\nu}\right]u^{\mu} + (\rho+p)u^{\nu}u^{\mu}{}_{;\nu} + p_{,\nu}g^{\mu\nu}. \tag{12.40}$$

We can write this in tensor notation as

$$0 = \boldsymbol{\nabla} \cdot \boldsymbol{T} = \left[\boldsymbol{\nabla}_{\boldsymbol{u}}\rho + \boldsymbol{\nabla}_{\boldsymbol{u}}p + (\rho+p)(\boldsymbol{\nabla} \cdot \boldsymbol{u})\right]\boldsymbol{u} + (\rho+p)\boldsymbol{\nabla}_{\boldsymbol{u}}\boldsymbol{u} + \boldsymbol{\nabla}p. \tag{12.41}$$

We shall see in Chapter 39 that this complicated expression gives us an equation of motion for the fluid.[18]

We therefore have that $\boldsymbol{\nabla} \cdot \boldsymbol{T} = 0$ is a valid constraint on the energy-momentum tensor $\boldsymbol{T}$ in curved space. It's tempting to interpret $\boldsymbol{\nabla}\cdot\boldsymbol{T} = 0$ as an equation of energy conservation for curved spacetime, but we need to be careful. The conservation of energy over anything other than small distances in curved spacetime is a knotty problem.

To see this, let's first discuss what we mean by the energy E in general relativity. We define the energy of an object with momentum 1-form $\tilde{p}$ via $E = -\tilde{p}(\boldsymbol{u})$, where $\boldsymbol{u}$ is the velocity[19] of an observer who is located at the site of the object whose energy we are measuring. In flat space, we then define the energy measured by an observer who isn't present, as being equal to the energy measured by the local observer if their velocity is *parallel* to that of the distant observer. Geometrically, parallel observers have no relative velocity in flat spacetime, so their world lines are parallel straight lines (Fig. 12.4). We can define a velocity field $\boldsymbol{u}(x)$ for all observers, which when we input a position in spacetime, will output a vector tangent to the world line of whichever observer is at that spacetime point. Mathematically, the condition that these flat-spacetime observers have parallel velocity may be written as

$$u^{\nu}{}_{,\mu} = 0, \tag{12.42}$$

which tells us that the tangent vectors to the world lines don't vary in space, and so are parallel. The mathematical need for parallel observers defined in this way is justified in the next example.

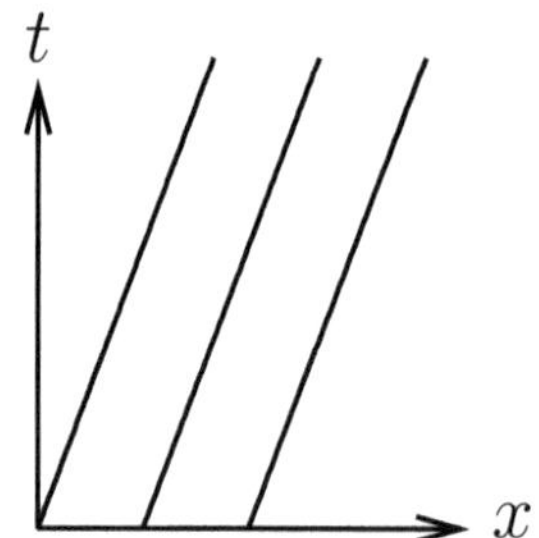

Fig. 12.4 The world lines of three parallel observers in flat spacetime. With respect to the coordinate frame of the figure, they all have identical velocity.

[18]This can be reduced to the geodesic equation in the case of constant pressure and conserved mass density. See the exercises at the end of this chapter and also Chapter 39.

[19]As always, the geometrical interpretation of velocity is the vector tangent to the world line of a observer.

Example 12.10

In flat spacetime, the vanishing divergence is written as $T^{\mu\nu}{}_{,\mu} = 0$. Recall that for dust we had $\boldsymbol{T} = \tilde{\boldsymbol{J}} \otimes \tilde{p}$ and so the energy density can be written as

$$-\boldsymbol{T}(\boldsymbol{u}, \boldsymbol{u}) = -T_{\mu\nu}u^{\mu}u^{\nu} = E\tilde{\boldsymbol{J}}(\boldsymbol{u}). \tag{12.43}$$

The quantity $\tilde{\boldsymbol{M}}(\ ) = E\tilde{\boldsymbol{J}}(\ )$ can be thought of as the current 1-form of energy carried by the particles in the dust cloud. If this is to be locally conserved, we must have $M^{\mu}{}_{,\mu} = 0$, where M^{μ} are the components of vector $\boldsymbol{M}$ which, from eqn 12.43 are

$$M^{\mu} = -T^{\mu\nu}u_{\nu}. \tag{12.44}$$

The condition that $M^{\mu}{}_{,\mu} = 0$ therefore requires that the velocity field obeys

$$0 = (-T^{\mu\nu}u_{\nu})_{,\mu} = -(T^{\mu\nu}{}_{,\mu})u_{\nu} - T^{\mu\nu}(u_{\nu,\mu}) = -T^{\mu\nu}u_{\nu,\mu}, \tag{12.45}$$

where we have used the Leibniz rule and $T^{\mu\nu}{}_{,\mu} = 0$. To guarantee that this vanishes, we require $u^{\nu}{}_{,\mu} = 0$. Physically, this requires us to have access to a family of inertial, parallel observers if we are to have a notion of conservation of energy.

The argument above is appropriate for flat spacetime. However, in curved spacetime we lose the ability to globally define a family of parallel observers, and this will frustrate our attempts to interpret $\boldsymbol{\nabla} \cdot \boldsymbol{T} = 0$ as statement of energy conservation. This is because 'being parallel' is a path-dependent property in curved spacetime, as we see in the next example.[20]

Example 12.11

Replaying the last example for curved spacetime, we apply a condition $\boldsymbol{\nabla} \cdot \boldsymbol{T} = 0$ using the covariant derivative, and so we need

$$(T^{\mu\nu}u_\nu)_{;\mu} = 0. \tag{12.46}$$

Follow the same steps as before, we conclude that this is so if $u_{\nu;\mu} = 0$, which, owing to the symmetry of $\boldsymbol{T}$, can be rewritten as

$$u_{\nu;\mu} + u_{\mu;\nu} = 0. \tag{12.47}$$

This condition is known as **Killing's equation** (Chapter 33) and is not generally true of velocity vector fields in curved spacetime.[21]

The previous example implies that in curved spacetime, although $\boldsymbol{\nabla}\cdot\boldsymbol{T} = 0$ holds, it can't be strictly interpreted in terms of energy conservation. Physically, this is to be expected since general relativity shows that matter and gravity are coupled: the gravitational field (described by the curvature of spacetime) can do work on matter, and matter can do work on the gravitational field. However, over small distances they don't do *much* work on each other and so, approximately at least, we have energy conservation[22] since $u_{\nu;\mu} \approx 0$.

We now have the object $\boldsymbol{T}$ that, in some form, lives on the right-hand side of Einstein's equation, along with its key property[23] $\boldsymbol{\nabla} \cdot \boldsymbol{T} = 0$ (which is true, in spite of its rather subtle interpretation). Our next task is to marry the geometric left-hand side of Einstein's equation with the physical right-hand side.

Chapter summary

- The right-hand side of Einstein's equation encodes the energy-momentum of matter fields using the tensor $\boldsymbol{T}$.

- In flat spacetime, the local conservation of energy-momentum can be expressed via the important constraint $\boldsymbol{\nabla} \cdot \boldsymbol{T} = 0$. The same expression holds in curved spacetime.

[20]Another way of seeing this point is that, to globally conserve momentum, we need to demonstrate

$$\int_{\partial \mathcal{V}} T^{\mu\nu}\, \mathrm{d}\Sigma_\nu = 0,$$

where $\partial \mathcal{V}$ is the boundary of some closed region of spacetime $\mathcal{V}$. But to work this out you have to combine lots of individual 4-vectors $T^{\mu\nu}\,\mathrm{d}\Sigma_\nu$ from different parts of the surface and bring them to a common location to add them all up. No problem in flat spacetime, but in curved spacetime they all live in different tangent spaces and this process cannot be done in a well-defined manner.

[21]Special cases that do obey Killing's equation are very interesting and we examine them in detail from Part IV onwards.

[22]We shall therefore call $\boldsymbol{\nabla} \cdot \boldsymbol{T} = 0$ a conservation equation in this book, in a slight abuse of language. The law $\boldsymbol{\nabla} \cdot \boldsymbol{T} = 0$ can be proven geometrically from the invariance of the manifold with respect to a very general type of translation known as a diffeomorphism. (See Appendix C for more details.) So although it does not represent conservation of energy in the strict sense, $\boldsymbol{\nabla}\cdot\boldsymbol{T} = 0$ is still a significant constraint on our equations.

[23]We also remark that $\boldsymbol{\nabla} \cdot \boldsymbol{T} = 0$ is connected with the **geodesic principle**, namely that *free massive point particles follow timelike geodesics*. The conservation equation captures the idea that the massive particle is *free*, that is, it is not exchanging energy-momentum with the fields and matter in its environment. Together with an energy condition (see page 145 for more details) to capture the idea that energy propagates within the body in a timelike or null manner, the geodesic principle can be proved (see e.g. J. Ehlers and R. Geroch, Ann. Phys. **309**, 232 (2004) for a proof, and J. O. Weatherall, Studies in the History and Philosophy of Modern Physics **42**, 276 (2011) for a reflection on the debate that has ensued about the status of the principle).

Exercises

(12.1) Consider the transformations between Cartesian coordinates (t, x, y, z) and spherical polar coordinates (t, r, θ, ϕ). In the Cartesian system, the (0,2) tensor $\boldsymbol{T}$ is diagonal with non-zero components

$$T_{tt} = \rho, \quad T_{ii} = p, \tag{12.48}$$

where $i = x, y, z$. Transform the components of this tensor to spherical polars.

(12.2) Consider dust with energy density $\boldsymbol{T}$ with components $T^{\mu\nu} = \rho_0 u^\mu u^\nu$.
(a) Explain why $\boldsymbol{\nabla} \cdot (\rho_0 \boldsymbol{u}) = 0$.
(b) Show that

$$0 = \boldsymbol{\nabla} \cdot \boldsymbol{T} = u^\mu (\rho_0 u^\nu)_{;\nu} + \rho_0 u^\nu u^\mu{}_{;\nu}. \tag{12.49}$$

(c) Using the result of part (a), show that the conservation of mass-energy guarantees that the dust particles obey the geodesic equation and hence that dust particles follow geodesic world lines.

(12.3) We write an energy-momentum tensor for a particle as

$$T^{\mu\nu}(x) = \int d\tau \, \frac{m}{\sqrt{-g}} \frac{dz^\mu}{d\tau} \frac{dz^\nu}{d\tau} \delta^{(4)}[x - z(\tau)], \tag{12.50}$$

where $z(\tau)$ is the world line of the particle parametrized by proper time τ. A useful expression when using this tensor, is that for functions $f(x)$ and $g(x)$ we have

$$\int dx \, \delta[f(x)] g(x) = \sum_a \frac{g(x_a)}{|f'(x_a)|}, \tag{12.51}$$

where the sum over a is over all values of x_a that have the property that $f(x_a) = 0$.
Using the tools above, justify the following expressions for the components of the energy-momentum tensor for a swarm of particles in flat space.
(a) The momentum density

$$T^{\alpha 0}(x) = \sum_n p_n^\alpha(\tau_n) \delta^{(3)}[x - z_n(\tau_n)]; \tag{12.52}$$

(b) The momentum current

$$T^{\alpha i}(x) = \sum_n p_n^\alpha(\tau_n) \frac{dz_n^i(\tau_n)}{dt} \delta^{(3)}[x - z_n(\tau_n)]; \tag{12.53}$$

(c) The energy-momentum tensor

$$T^{\alpha\beta}(x) = \sum_n \frac{p_n^\alpha(\tau_n) p_n^\beta(\tau_n)}{E_n(\tau_n)} \delta^{(3)}[x - z_n(\tau_n)]; \tag{12.54}$$

where the index n labels a specific particle and τ_n solves the equation $z^0(\tau_n) = t$ and t is the coordinate time [i.e. particle n's proper time at the coordinate time t that is inputted into the tensor $T(x)$ as part of the argument x with components $x^\mu = (t, \vec{x})$].

(12.4) By applying $\boldsymbol{\nabla} \cdot \boldsymbol{T} = 0$ to the energy-momentum tensor for a single particle in the previous question, show that we obtain the geodesic equation.

(12.5) In flat space, define a (1,1) energy-momentum tensor for dust particles as

$$\boldsymbol{T}(\ ,\ ) = \boldsymbol{p} \otimes \tilde{\boldsymbol{J}}. \tag{12.55}$$

This (1,1) object has two slots: one for a 1-form and one for a vector. It has components $T^\mu{}_\nu$.
(a) Find the eigenvalue of this tensor when the velocity eigenvector $\boldsymbol{v}$ is inserted. That is, determine the constants α for the equation

$$\boldsymbol{T}(\ ,\boldsymbol{v}) = \alpha \boldsymbol{v}. \tag{12.56}$$

Note that the other three eigenvectors are orthogonal to $\boldsymbol{v}$ and therefore in the 3-space we can call $\boldsymbol{v}^\perp$. By isotropy these three eigenvectors are degenerate and have some eigenvalue p, such that

$$\boldsymbol{T}(\ ,\boldsymbol{e}_i) = p\boldsymbol{e}_i. \tag{12.57}$$

(b) Show that we can write

$$\boldsymbol{T}(\tilde{\boldsymbol{Y}}, \boldsymbol{X}) = -X^{\hat{0}} Y_{\hat{0}}(\rho + p) + X^{\hat{\mu}} Y_{\hat{\mu}} p. \tag{12.58}$$

(c) Use the result of part (b) result to justify the curved space expression

$$T_{\mu\nu} = (\rho + p) v_\mu v_\nu + p g_{\mu\nu}. \tag{12.59}$$

See the book by Ludvigsen for a discussion of this approach.

The gravitational field equations

13

The key idea of general relativity is that the curvature of spacetime, which gives rise to gravitation, is determined by the energy density of the matter fields of the Universe, such that we can write the Einstein equation

$$\left(\begin{array}{c} \text{Curvature of} \\ \text{spacetime} \end{array} \right) = \left(\begin{array}{c} \text{Energy density} \\ \text{of matter fields} \end{array} \right). \tag{13.1}$$

At last, in this chapter, we are able to formulate the details of this equation! Our task is, on the (geometrical) left-hand side, to identify a suitable expression for the curvature of spacetime and, on the (physical) right, a suitable expression for the density of energy in matter fields.

In order to get to Einstein's equation, we take the route shown in Fig. 13.1. Starting with the physical principles of fields and of eqn 13.1, we shall review the key elements of geometry that supply the left-hand side of the equation. We then examine the physics of the right-hand side, in particular the energy-momentum tensor field $\boldsymbol{T}(x)$ that encodes the energy density of the matter fields that fill the Universe. Finally we link these to form the **Einstein field equation**, the tensor field version of eqn 13.1.

13.1 Geometry: a recap of the key ingredients

The metric tensor field $\boldsymbol{g}(x)$ is the foundation of the curvature of spacetime. As we've written it, it is a classical field: a function of position in spacetime x that gives us a tensor valid at the point x. We often express the components of the metric field $g_{\mu\nu}(x)$ via a line element

$$ds^2 = g_{\mu\nu}dx^\mu dx^\nu. \tag{13.2}$$

There is no way to tell whether a single particle accelerates due to the effects of a homogeneous gravitational field or because of the choice of coordinates. If a gravitational field is present, and all real-life gravitational fields are necessarily inhomogeneous, spacetime is curved and this gives rise to the acceleration of a particle in the absence of forces external to spacetime. However, a particle might accelerate merely due to the perversity of the coordinate system that we've chosen, rather than due to curvature. The method to unambiguously detect curvature

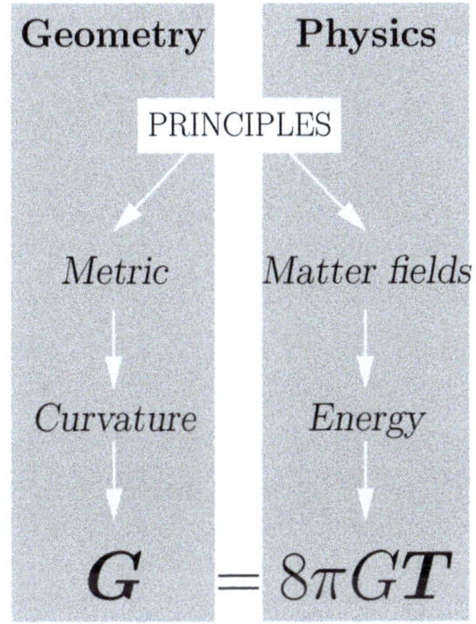

Fig. 13.1 A conceptual route to Einstein's equation. The left-hand side leads to the Einstein tensor field $\boldsymbol{G}$, related to the Riemann tensor field as described later in this chapter. The right-hand side leads to a constant $8\pi G$ (where G here is the gravitational constant) multiplied by the energy-momentum tensor field $\boldsymbol{T}$.

involves transforming away the effects due to the coordinate system to arrive at what's left behind: the curvature due to gravitation. This is the motivation behind the Riemann curvature tensor $\boldsymbol{R}(x)$, which evaluates the presence of this stuff that's left behind. We expect that this tensor should play a key role in the left-hand side of the equation, which indeed it does. It depends on the (first and second) derivatives of the metric field. The first derivatives of $\boldsymbol{g}(x)$ provide the connection coefficients $\Gamma^{\mu}{}_{\alpha\beta}(x)$,[1] and reflect how the metric causes the coordinates to change as we move through spacetime. They are easily manipulated, so we often describe $\boldsymbol{R}(x)$ in terms of them, rather directly through $\boldsymbol{g}(x)$.

To obtain $\boldsymbol{R}(x)$ and thus determine the curvature of spacetime, we evaluate the geodesic deviation of two freely falling particles whose motion was initially parallel and extract $\boldsymbol{R}(x)$ from the equation of motion of their separation.[2] The resulting Riemann tensor field has components

$$R^{\alpha}{}_{\beta\gamma\delta}(x) = \frac{\partial \Gamma^{\alpha}{}_{\delta\beta}(x)}{\partial x^{\gamma}} - \frac{\partial \Gamma^{\alpha}{}_{\gamma\beta}(x)}{\partial x^{\delta}} + \Gamma^{\alpha}{}_{\gamma\mu}(x)\Gamma^{\mu}{}_{\delta\beta}(x) - \Gamma^{\alpha}{}_{\delta\mu}(x)\Gamma^{\mu}{}_{\gamma\beta}(x).$$
(13.5)

This gives us access to the curvature at each point in spacetime. Two quantities derived from the Riemann tensor field will be important in setting up the Einstein equation. The first is the Ricci tensor field, which at a spacetime point x has components[3]

$$R_{\mu\nu}(x) = R^{\alpha}{}_{\mu\alpha\nu}(x).$$
(13.6)

The second is the Ricci scalar field[4]

$$R(x) = g_{\mu\nu}R^{\mu\nu}(x).$$
(13.7)

Using these two questions, we will be able to gain access to the curvature of spacetime. This will be achieved by taking a double derivative of the metric field $\boldsymbol{g}(x)$, which is the basis of the left-hand side of the Einstein equation.[5] After having reviewed the geometrical left-hand side of the equation, let's now turn to the right-hand side and the physical content of the theory.

13.2   Physics: the key ingredients

We start with a principle: *a physical relativistic/geometrical field theory of gravitation must be compatible with Newton's theory in the non-relativistic limit.* It is therefore useful to revisit Newton's law of gravitation from the point of view of fields and see what it teaches us about the shape that our new theory must take.[6] Newton's law says that the force $\vec{F}$ experienced by a mass m from a gravitating object with mass M separated by a displacement vector $\vec{r}$, is given by

$$\vec{F}(r) = -\frac{GMm}{r^{3}}\vec{r},$$
(13.8)

where $G = 6.672(4) \times 10^{-11}$ m^{3}kg^{-1}s^{-2} is the gravitational constant. We also find it useful to deal with a scalar potential energy function

---

[1] The connection coefficients may be related to the metric tensor field via the component equation

$$\Gamma^{\mu}{}_{\alpha\beta} = \frac{1}{2}g^{\mu\lambda}\left(\frac{\partial g_{\lambda\alpha}}{\partial x^{\beta}} + \frac{\partial g_{\lambda\beta}}{\partial x^{\alpha}} - \frac{\partial g_{\alpha\beta}}{\partial x^{\lambda}}\right).$$
(13.3)

[2] We use

$$\frac{D^{2}\boldsymbol{\xi}}{d\tau^{2}} = \boldsymbol{R}(\ ,\boldsymbol{u},\boldsymbol{\xi},\boldsymbol{u}),$$
(13.4)

where $\boldsymbol{\xi}$ is the particles' separation and $\boldsymbol{u}$ is the velocity of the particle following the fiducial geodesic.

[3] Note for later in the chapter that this is symmetric: $R_{\mu\nu}(x) = R_{\nu\mu}(x)$.

[4] Recall that a scalar is the same, independent of the coordinate system in which it is evaluated.

[5] One point of view is that the (1,3) Riemann curvature tensor field $\boldsymbol{R}(x)$ *is* the gravitational field. So although we have the equivalence principle telling us that a measurement at a point cannot tell us whether it's gravitation or acceleration causing an effect, we can identify the physical field that causes gravitation. We cannot simply slot this 'gravitational field' $\boldsymbol{R}$ into the Einstein equation (curvature) = (energy), owing to the (1,3) valence of $\boldsymbol{R}$ compare to the (2,0) valence of the energy-momentum tensor $\boldsymbol{T}$. However, if we are looking for *the* physical manifestation of gravitation, this is as good a candidate as any.

[6] See also Section 0.4.

$U(r)$ experienced by mass m due to mass M and given by

$$U(r) = -\frac{GMm}{r},\qquad(13.9)$$

where $\vec{F} = -\vec{\nabla}U$. The function is $U(r)$ is proportional to m, so it is better to think about the potential $\Phi(r) = U(r)/m$, the potential energy per unit mass. Newton's law of gravitation can therefore be turned into a field equation straightforwardly. The field version of the law describes the force per unit mass $\vec{g}(\vec{x})$ experienced at a position $\vec{x}$ due to the presence of a distribution of mass with density $\rho(\vec{x}')$. The generalization of the force law is

$$\vec{g}(\vec{x}) = -G \int \mathrm{d}^3x'\, \rho(\vec{x}')\frac{(\vec{x} - \vec{x}')}{|\vec{x} - \vec{x}'|^3}.\qquad(13.10)$$

We also define a gravitational potential

$$\Phi(\vec{x}) = -G \int \mathrm{d}^3x'\, \frac{\rho(\vec{x}')}{|\vec{x} - \vec{x}'|}.\qquad(13.11)$$

We can then show that (i) the function Φ obeys the important rule $\vec{g}(\vec{x}) = -\vec{\nabla}\Phi(\vec{x})$, and, most importantly, (ii) Φ can be computed from a given distribution of mass. This is the purpose of the next example.

Example 13.1

(i) Using the potential $\Phi(\vec{x})$ we can write[7]

$$\vec{g}(\vec{x}) = \vec{\nabla}_{\vec{x}} \int \mathrm{d}^3x'\, \frac{G\rho(\vec{x}')}{|\vec{x} - \vec{x}'|} = -\vec{\nabla}\Phi(\vec{x}).\qquad(13.13)$$

(ii) Next, we use the result[8]

$$\vec{\nabla}_{\vec{x}} \cdot \left(\frac{\vec{x} - \vec{x}'}{|\vec{x} - \vec{x}'|^3}\right) = 4\pi\delta^{(3)}(\vec{x} - \vec{x}'),\qquad(13.15)$$

which allows us to take the divergence of $\vec{g}$ in eqn 13.10 to find

$$\vec{\nabla} \cdot g(\vec{x}) = -\,4\pi G \int \mathrm{d}^3x'\, \rho(\vec{x}')\delta^{(3)}(\vec{x} - \vec{x}')$$

$$= -\,4\pi G\rho(\vec{x}).\qquad(13.16)$$

Since we also have $\vec{\nabla} \cdot g(\vec{x}) = -\vec{\nabla}^2\Phi$, we conclude that

$$\vec{\nabla}^2\Phi(\vec{x}) = 4\pi G\rho(\vec{x}).\qquad(13.17)$$

This is the gravitational version of Poisson's equation.[9]

To recap the main results of the last example, we have that, expressed as a field equation, Newton's force law is written as[10]

$$-\vec{\nabla} \cdot \vec{g}(\vec{x}) = 4\pi G\rho(\vec{x}),\qquad(13.18)$$

where $\vec{g}$ is the gravitational field. The law can be expressed in terms of the potential Φ (via $\vec{g} = -\vec{\nabla}\Phi$). The potential field is the solution of Poisson's equation

[7] *Proof:* Check by a direct computation that

$$\vec{\nabla}_{\vec{x}} \left(\frac{1}{|\vec{x} - \vec{x}'|}\right) = -\frac{\vec{x} - \vec{x}'}{|\vec{x} - \vec{x}'|^3}.\qquad(13.12)$$

[8] See exercises for the proof that

$$\vec{\nabla}_{\vec{x}}^2 \left(\frac{1}{|\vec{x} - \vec{x}'|}\right) = -4\pi\delta^{(3)}(\vec{x} - \vec{x}').\qquad(13.14)$$

[9] Recall that Laplace's equation for a field in the absence of sources says $\vec{\nabla}^2\Phi = 0$, but Poisson's equation for the field in the presence of sources says $\vec{\nabla}^2\Phi = 4\pi G\rho$, where $4\pi G\rho$ is the density of the source.

[10] Note that eqn 13.18 is the same as presented in eqn 0.12, and eqn 13.19 is the same as presented in eqn 0.15.

$$\vec{\nabla}^2\Phi(\vec{x}) = 4\pi G\rho(\vec{x}). \tag{13.19}$$

Note the form of this equation that says that two derivatives of the potential field are proportional to the mass density ρ, which is the source of the field. We therefore have a field theory of Newtonian gravitation that deals in a scalar field $\Phi(\vec{x}, t)$. We input a position in spacetime and output a function Φ. This can be used to compute the gravitational field and hence the force on any mass.

[11]The idea of the Green's function is that for a differential equation $\hat{L}A(x) = f(x)$, with $\hat{L}$ representing a differential operator, we define the Green's function as $\hat{L}G(x,y) = \delta(x-y)$. This is useful, as the Green's function can then be used to build a solution using the prescription

$$A(x) = \int \mathrm{d}y\, G(x,y)f(y). \tag{13.20}$$

[12]We know that both the energy density and metric field are tensor fields, so we might expect Newton's law to be equivalent to one component of the Einstein equation.

Example 13.2

Green's functions are a very useful tool in solving equations like the Poisson equation.[11] For our problem, we seek a solution to the equation

$$\vec{\nabla}^2 G(\vec{x}, \vec{x}') = 4\pi\delta^{(3)}(\vec{x} - \vec{x}'). \tag{13.21}$$

We can then use the resulting Green's function $G(\vec{x}, \vec{x}')$ to build a solution to Poisson's equation using the prescription

$$\Phi(\vec{x}) = G \int \mathrm{d}^3x'\, G(\vec{x}, \vec{x}')\rho(\vec{x}'). \tag{13.22}$$

Comparing to the steps above, we can identify the Green's function we need to solve Poisson's equation: it is simply

$$G(\vec{x}, \vec{x}') = -\frac{1}{|\vec{x} - \vec{x}'|}. \tag{13.23}$$

We shall try to fashion our theory of gravitation following the pattern set by Newton's field theory. In place of the Newtonian potential, we shall substitute the metric. In place of the mass density, we substitute the energy density. If we postulate that the Einstein equation has the same form as Poisson's equation for the Newtonian potential field, then Einstein's equation must look like

$$\boxed{\partial^2 g} = \boxed{\kappa T}, \tag{13.24}$$

where κ is a constant.[12] That is to say that two derivatives of the metric field g are proportional to the energy density T.

Although these steps might make us optimistic, there must also be differences between Einstein and Newton's pictures of gravitation. First, Newton's law is instantaneous, running counter to the rules of special relativity. In addition, Fermat's principle says that light propagates in straight lines in flat spacetime but experiments show the bending of light from stars by gravitation, implying that gravity must warp spacetime. This means that, for Einstein's gravity, the geometry of curved space is caused by the energy density of matter which acts as a source (just like ρ is a source of Φ). However, the arrangement of energy density is itself determined by the geometry encoded in the metric field, leading to the situation shown in Fig. 13.2. We conclude that, in contrast to Newton, Einstein's gravitation is necessarily nonlinear: gravitational energy is acted on by gravity itself.[13]

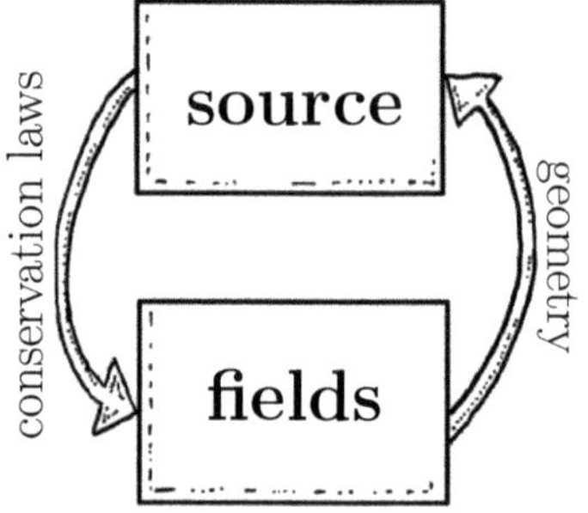

Fig. 13.2 The relationship between the source of energy-momentum T and the metric field g in the Einstein equation. The source is constrained by conservation laws (via $\nabla \cdot T = 0$) affecting its coupling to the metric field that, in turn, determines the geometry. The metric field also determines the form of T itself.

[13]We can be a little more specific here. In electromagnetism (another linear theory), we often specify the currents and charges and then solve Maxwell's equations to find the electromagnetic fields. The same can be done in Newtonian gravity: specify the mass distribution and solve for Φ. The same cannot be done in general relativity since the source of gravitation is the energy-momentum tensor T, which is described in terms of the metric g, precisely the field we are trying to find. This forces us to attempt to find consistent values of g and T simultaneously, which is very difficult.

We have now reached the point where we must fit the geometrical and physical fields together. The key physical principles determining the content on the right-hand side of a *relativistic* equation of motion are as follows. (i) **Local causality**: A signal can be sent between two points if, and only if, their separation is not spacelike. Colloquially: nothing travels faster than light. Our fields must therefore be consistent with special relativity. (ii) **Local conservation of energy and momentum**: All matter fields in the Universe have energy-momentum. This energy-momentum is locally conserved. That is to say, it must obey a continuity equation.

We also require energy to obey another condition. (iii) The **dominant energy condition** says that to any observer the local energy density appears non-negative and the local energy flow vector is not spacelike (i.e. it is timelike or null). We therefore demand that positive energy, which cannot move around a system at a speed faster than that of light, is locally conserved.[14]

All of the physical principles relating to the behaviour of mass energy can be embodied in an **energy-momentum tensor field** $\boldsymbol{T}(x)$, as described in the previous chapter. This tensor must describe positive energy and encodes local conservation of energy-momentum (at least approximately) via the constraint $\boldsymbol{\nabla} \cdot \boldsymbol{T} = 0$. The energy-momentum tensor vanishes only if all of the matter fields vanish. This means that all fields cost energy, and this is the source of gravitation.

In the next section, we shall, finally, attempt to marry the geometry of curvature embodied in $\boldsymbol{R}$ and the mass energy of the fields of the Universe, expressed in $\boldsymbol{T}$ in a tensor equation.

[14]There is a less stringent **weak energy condition**, which says that the energy density measured by any observer must be non-negative.

13.3 An incorrect guess

The apparently obvious field equation turns out to be incorrect.

The energy-momentum tensor has two slots or, equivalently, its components have two indices. We are therefore tempted to guess that the Einstein equation, that fixes the relationship of curvature (on the left) to the energy density of the matter fields (on the right) has a (2,0) object on the left and $T^{\mu\nu}$ on the right. The obvious, but incorrect, natural candidate would be that the curvature-defining object is the Ricci tensor, with components $R^{\mu\nu}$. This would give us the field equation

$$R^{\mu\nu} = \kappa T^{\mu\nu} \quad \text{(incorrect guess)}, \tag{13.25}$$

with κ a constant.[15] But why is this wrong? The key is to remember that we must have the conservation equation $\boldsymbol{\nabla} \cdot \boldsymbol{T} = 0$. We examine the consequence of this on our candidate field equation by taking the divergence

$$R^{\mu\nu}{}_{;\mu} = \kappa T^{\mu\nu}{}_{;\mu} \quad \text{(incorrect guess)}. \tag{13.26}$$

Since $\boldsymbol{\nabla} \cdot \boldsymbol{T} = 0$ (or, equivalently $T^{\mu\nu}{}_{;\mu} = 0$), the right-hand side of eqn 13.26 must vanish. What happens to the geometrical left-hand side?

[15] This seems a good idea since the Ricci tensor is a sort of average of $\boldsymbol{R}$ that retains those parts of curvature that, in a gravitational equation, can cause volumes to shrink. This was Einstein's original suggestion in October 1915.

To answer this, we need a mathematical result to constrain our equations.

Example 13.3

As we'll discuss in Chapter 43, the components of the Riemann tensor satisfy a geometrical constraint known as the **Bianchi identity**, whose component form is written as

$$R^{\alpha}{}_{\beta\gamma\delta;\lambda} + R^{\alpha}{}_{\beta\lambda\gamma;\delta} + R^{\alpha}{}_{\beta\delta\lambda;\gamma} = 0. \tag{13.27}$$

For now, it is enough to say that this identity comes from a general principle that 'the boundary of a boundary is zero', an idea that leads to the well-known vector identity that the divergence of the curl of a vector field vanishes. We can contract indices in this identity, setting $\alpha = \lambda$, to find[16]

$$R^{\alpha}{}_{\beta\gamma\delta;\alpha} + R_{\beta\gamma;\delta} - R_{\beta\delta;\gamma} = 0. \tag{13.28}$$

Now raise the β index and contract with the δ index to get

$$R^{\alpha}{}_{\gamma;\alpha} + R^{\beta}{}_{\gamma;\beta} - R_{;\gamma} = 0$$
$$2R^{\alpha}{}_{\gamma;\alpha} = R_{;\gamma}. \tag{13.29}$$

This equation is known as the **contracted Bianchi identity** and we will use it below.

---

[16]We make use of the symmetries of $R_{\alpha\beta\gamma\delta}$, where swaps within the first and second pairs of indices result in a minus sign.

To examine the consequence of the identity discussed in the last example, we now define the **trace-reversed Ricci tensor**, also known as the **Einstein tensor field**, as the (2,0) tensor $\boldsymbol{G}(x)$ with components[17]

$$G^{\mu\nu}(x) = R^{\mu\nu}(x) - \frac{1}{2}g^{\mu\nu}(x)R(x). \tag{13.30}$$

Taking the divergence of the Einstein tensor $\boldsymbol{\nabla} \cdot \boldsymbol{G}$, we find that[18]

$$G^{\mu\nu}{}_{;\mu} = R^{\mu\nu}{}_{;\mu} - \frac{1}{2}g^{\mu\nu}R_{;\mu}. \tag{13.31}$$

[17]Do not confuse the Einstein tensor field $\boldsymbol{G}$ with the gravitational constant G. It's unfortunate that they have the same symbol, but context should make clear which is which. In later chapters, we'll use units in which $G = 1$.

[18]We use the compatibility condition $g_{\mu\nu;\alpha} = 0$ here.

[19]If you don't see this immediately, the steps are (i) raise the μ index in the second term to find $R^{\mu\nu}{}_{;\mu} - R^{\alpha\nu}{}_{;\alpha}$; (ii) rename the α index μ and the expression vanishes.

Using the contracted Bianchi identity $(R_{;\mu} = 2R^{\alpha}{}_{\mu;\alpha})$ on the second term on the right, we find[19]

$$G^{\mu\nu}{}_{;\mu} = R^{\mu\nu}{}_{;\mu} - g^{\mu\nu}R^{\alpha}{}_{\mu;\alpha} = 0. \tag{13.32}$$

We conclude that the Einstein tensor has the property

$$\boldsymbol{\nabla} \cdot \boldsymbol{G} = \left(R^{\mu\nu} - \frac{1}{2}g^{\mu\nu}R\right)_{;\mu} = 0. \tag{13.33}$$

Although this doesn't look particularly scandalous, it causes a major problem for the divergence of our candidate equation: $R^{\mu\nu}{}_{;\mu} = \kappa T^{\mu\nu}{}_{;\mu} = 0$. We combine this expression with eqn 13.33 to conclude

$$-\frac{1}{2}g^{\mu\nu}R_{;\mu} = -\frac{1}{2}R^{;\nu} = 0. \tag{13.34}$$

[20]Remember that the effect of the covariant derivative on a scalar field is simply

$$\nabla f = f_{;\mu} = f_{,\mu} = \frac{\partial f}{\partial x^{\mu}}.$$

The guts of this latter equation[20] $(R^{;\nu} = 0)$ forces the Ricci scalar field $R(x)$ to be a constant throughout the Universe. This on its own is not a problem. However, the trouble comes when we examine the candiate equation (eqn 13.25) again, since we find that

$$R(x) = g^{\mu\nu}(x)R_{\mu\nu}(x) = \kappa g^{\mu\nu}(x)T_{\mu\nu}(x) = \text{const.} \qquad (13.35)$$

That is, the trace $g^{\mu\nu}T_{\mu\nu}$ of $T_{\mu\nu}$ must be constant throughout the Universe. *This* is unreasonable! It implies a uniform distribution of matter and energy throughout the Universe which is at odds with our experience, where matter is certainly non-uniform. We are forced to drop the candidate equation.

13.4 Einstein's field equation

The problem encountered in the last section leads to the solution.

On our way to proving that our first attempt at a field equation was inadequate, we identified a new (2,0) [or (0,2)] tensor $\boldsymbol{G}$, which had components $G_{\mu\nu} = R_{\mu\nu} - \frac{1}{2}g_{\mu\nu}R$. It is this object that provides the solution to our problem. Therefore, having discarded our first guess, we now consider instead a new candidate field equation[21]

$$G_{\mu\nu} = \kappa T_{\mu\nu}. \qquad (13.36)$$

We know from the last section that the divergence of this equation is zero on both sides. This, it turns out, is exactly the field equation for which we've been searching. However, we have yet to work out the constant of proportionality κ. This can be evaluated using the condition that this equation is compatible with the results of Newtonian gravitation in the limit of weak fields (i.e. with the equation $\vec{\nabla}^2\Phi = 4\pi G\rho$).

[21] In case this mathematics is a distraction, here's a physics recap. The Ricci tensor represents an average of the Riemann curvature tensor that will encode those parts of curvature that cause volumes to shrink owing to the gravitational interaction. However, equating the Ricci tensor to the energy-momentum tensor gives a theory that doesn't conserve energy (i.e. it is incompatible with $\boldsymbol{\nabla}\cdot\boldsymbol{T} = 0$). The Bianchi identity (geometrically) encodes the desired energy conservation. Using a trace-reversed Ricci tensor, known as the Einstein tensor, builds energy conservation back into the theory since, via the Bianchi identity, it allows the resulting equations to conserve energy.

In this example, we will temporarily reinsert the factors of c so it's a bit more clear how large various factors are.

Example 13.4

In the weak-field limit, spacetime is only very slightly curved and therefore the metric is very close to the flat Minkowski metric, so $g_{00} \approx g^{00} \approx -1$ and $g_{ii} \approx g^{ii} \approx 1$. As a result, we have for any tensor $\boldsymbol{A}$ in this geometry, that $A^{00} \approx A_{00}$, along with $A^{ii} = A_{ii}$. Moreover, the energy-momentum tensor is dominated by the energy density part (because $\rho \gg p/c^2$) so the only significant element of $\boldsymbol{T}$ is $T^{00} = \rho c^2$, with all other elements vanishingly small. This means that, for all of the spatial (ij) components of the Einstein tensor, we have

$$G^{ij} = R^{ij} - \frac{1}{2}g^{ij}R = 0, \qquad (13.37)$$

from which we find $R^{11} = R^{22} = R^{33} = R/2$ and so

$$R^{11} + R^{22} + R^{33} = 3R/2. \qquad (13.38)$$

As a result, $R = g_{\mu\nu}R^{\mu\nu} = -R^{00} + \frac{3}{2}R$ and therefore $R = 2R^{00}$. This allows us to calculate $G^{00} = R^{00} - \frac{1}{2}g^{00}R$ which gives

$$G^{00} = 2R^{00} \approx 2R_{00}. \qquad (13.39)$$

Recall that, with the weak-field metric, the only appreciable connection coefficients are (replacing factors of c) $\Gamma^i_{00} = \partial(\Phi/c^2)/\partial x^i$ and so $R^i{}_{0j0} = \partial^2(\Phi/c^2)/\partial x^i \partial x^j$. We can then read off[22] that $R_{00} = (1/c^2)\vec{\nabla}^2\Phi$ and so

$$G^{00} = \frac{2}{c^2}\vec{\nabla}^2\Phi. \qquad (13.40)$$

[22] Explicitly we have $R_{00} = \sum_{i=1}^{3} R^i{}_{0i0}.$

Putting this together with $T^{00} = \rho c^2$, the Einstein equation $\boldsymbol{G} = \kappa \boldsymbol{T}$ predicts a universal field equation of

$$\vec{\nabla}^2 \Phi = \frac{\kappa c^4}{2} \rho. \tag{13.41}$$

This can be compared with Poisson's equation for gravitation (the field equation that corresponds to Newton's universal law of gravitation), which reads $\vec{\nabla}^2 \Phi = 4\pi G \rho$. Hence, we deduce that

$$\kappa = \frac{8\pi G}{c^4}. \tag{13.42}$$

In SI units, the factor is $\kappa = 8\pi G/c^4 \approx 2.08 \times 10^{-43}$ N^{-1}.

[23]This is the equation Einstein presented in November 1915, after having rejected his first attempt.

The result of the previous example, setting $c = 1$ again, is that $\kappa = 8\pi G$. We are now in a position to write a field equation for gravity, known as the Einstein field equation.[23]

The Einstein field equation, version 1

In coordinates,

$$\left(R_{\mu\nu} - \frac{1}{2} g_{\mu\nu} R \right) = 8\pi G T_{\mu\nu}. \tag{13.43}$$

In coordinate-free notation, this becomes

$$\boldsymbol{G}(x) = 8\pi G \boldsymbol{T}(x). \tag{13.44}$$

Example 13.5

As a further manipulation, let's take the trace of eqn 13.43. This is achieved by multiplying by $g^{\mu\nu}$, so that we get

$$g^{\mu\nu} R_{\mu\nu} - \frac{1}{2} g^{\mu\nu} g_{\mu\nu} R = 8\pi G g^{\mu\nu} T_{\mu\nu}. \tag{13.45}$$

Let's now consider the trace T of the energy-momentum tensor $T = g^{\mu\nu} T_{\mu\nu}$, and hence eqn 13.45 gives (using $g^{\mu\nu} g_{\mu\nu} = 4$ and $R = g_{\mu\nu} R^{\mu\nu}$) $R - 2R = -R = 8\pi G T$. This implies we can also write

$$R_{\mu\nu} + \frac{1}{2} g_{\mu\nu} (8\pi G T) = 8\pi G T_{\mu\nu}. \tag{13.46}$$

The previous example therefore leads to a second form for the Einstein field equation:

The Einstein field equation, version 2

$$R_{\mu\nu} = 8\pi G \left(T_{\mu\nu} - \frac{1}{2} g_{\mu\nu} T \right). \tag{13.47}$$

Our method of arriving at the Einstein equation has been rather haphazard. We might ask if we can add other terms to the right-hand side of the Einstein equation. In fact, it is permissible to add any scalar multiple of the metric field tensor, so we can take our (version 1) equation $\boldsymbol{G}(x) = 8\pi G \boldsymbol{T}(x)$ and put an extra $-\Lambda \boldsymbol{g}(x)$ on the right-hand side and if Λ is just a constant then our new equation will still work. In fact, the property that $\boldsymbol{\nabla}_\alpha \boldsymbol{g} = g_{\mu\nu;\alpha} = 0$ means that the divergence arguments we have been using won't be troubled by this extra term. We therefore write our third version of the Einstein equation as follows:

The Einstein field equation, version 3

$$\left(R_{\mu\nu} - \frac{1}{2}g_{\mu\nu}R\right) = 8\pi G T_{\mu\nu} - \Lambda g_{\mu\nu}, \tag{13.48}$$

$$\boldsymbol{G}(x) = 8\pi G \boldsymbol{T}(x) - \Lambda \boldsymbol{g}(x). \tag{13.49}$$

The constant of proportionality Λ is known as the **cosmological constant**. Physically, the cosmological constant represents an extra source of energy in the Universe, in addition to the matter fields. Owing to the difference in sign compared to positive $T_{\mu\nu}$ (as demanded by the dominant energy condition), a positive value of Λ will give a repulsive gravitational interaction.

In order to retain the simplicity of the equation $\boldsymbol{G} = 8\pi G \boldsymbol{T}$, we could treat the cosmological constant term as if it were a source of energy-momentum and write

$$\boldsymbol{G} = 8\pi G \left(\boldsymbol{T}^{\mathrm{mat}} + \boldsymbol{T}^{\mathrm{vac}}\right), \tag{13.50}$$

where $\boldsymbol{T}^{\mathrm{mat}}$ is the energy-momentum tensor for matter that we have described in this chapter and

$$\boldsymbol{T}^{\mathrm{vac}} = -\frac{\Lambda}{8\pi G}\boldsymbol{g} \tag{13.51}$$

is a contribution to the energy-momentum tensor from sources other than the matter fields. The introduction and rejection of Λ by Einstein has become a famous story in physics folklore. We shall discuss its consequences in Part III of the book.[24]

We have arrived at an equation of motion for the metric field that determines gravitation. Our task is now to explore its consequences on our Universe. In the coming chapters, it is useful to keep in mind a few helpful heuristics we have seen already in dealing with general relativity.

[24] A brief history of solutions involving Λ is given in Chapter 18.

- A metric is visualized via its light cones. These are found using $\mathrm{d}s^2 = 0$ along null geodesics.

- We use two types of reference frame: (i) a coordinate frame, where our geometric arguments are often simplest; (ii) an orthonormal frame, where observers make measurements.

- Coordinates have no intrinsic metric significance but, in cases of spherical symmetry, useful coordinates to have in mind are spherical polars (t, r, θ, ϕ).

- Indices are raised/lowered in the coordinate frame using the metric. Indices in orthonormal frames are manipulated with the Minkowski tensor. The connection coefficients don't vanish in the orthonormal frame. If we want this we need to shift to a local inertial frame.

■ General covariance says that a valid tensor equation in flat space is a valid tensor equation in curved space. To upgrade tensor equations, exchange $\boldsymbol{\eta} \to \boldsymbol{g}$. To upgrade derivatives, use the *comma goes to semicolon* rule.

■ When manipulating vectors, it's often helpful to use (i) $g_{\mu\nu} = \boldsymbol{e}_\mu \cdot \boldsymbol{e}_\nu$; (ii) $g_{\mu\nu;\alpha} = 0$. For timelike velocities $\boldsymbol{u} \cdot \boldsymbol{u} = -1$.

■ The perfect fluid, in the orthonormal frame for spherical polars, has components

$$T_{\hat{t}\hat{t}} = \rho, \quad T_{\hat{r}\hat{r}} = p, \quad T_{\hat{\theta}\hat{\theta}} = p, \quad T_{\hat{\phi}\hat{\phi}} = p.$$

Chapter summary

• The left-hand side of the Einstein field equation encodes geometry using the Ricci tensor and scalar to form the Einstein tensor $\boldsymbol{G}$ with components

$$G_{\mu\nu} = R_{\mu\nu} - \frac{1}{2}g_{\mu\nu}R. \tag{13.52}$$

• The right-hand side of Einstein's equation encodes the energy-momentum of matter fields using the tensor $\boldsymbol{T}$.

• These are tied together in Einstein's equation, subject to the key constraint that energy-momentum is locally conserved. The Einstein equation is $\boldsymbol{G}(x) = 8\pi G \boldsymbol{T}(x) - \Lambda \boldsymbol{g}(x)$ and has components

$$\left(R_{\mu\nu} - \frac{1}{2}g_{\mu\nu}R\right) = 8\pi G T_{\mu\nu} - \Lambda g_{\mu\nu}. \tag{13.53}$$

Exercises

(13.1) Show that Newton's second law and his law of gravitation are invariant with respect to Galilean transformations.

(13.2) Consider the equation

$$\vec{\nabla}^2 \left(\frac{1}{r}\right) = -4\pi\delta(\vec{x}), \tag{13.54}$$

where $r = |\vec{x}|$. We shall prove this. (a) Evaluate $\frac{\partial}{\partial x^i}r$ and $\frac{\partial}{\partial x^i}r^{-1}$. (b) Show that

$$\frac{\partial^2}{\partial x^i \partial x^j}\left(\frac{1}{r}\right) = \frac{(3n^i n^j - \delta^{ij})}{r^3}, \tag{13.55}$$

where $n^i = x^i/r$. (c) Use the previous result to show that $\vec{\nabla}^2(r^{-1}) = 0$ for $|\vec{x}| \neq 0$. (d) Defining $\vec{j} = \vec{\nabla}(r^{-1})$, use Gauss' theorem to show that

$$\int_V \mathrm{d}^3x\, \vec{\nabla} \cdot \vec{j} = -4\pi, \tag{13.56}$$

where V is a volume that encloses $\vec{x} = 0$. (e) Use the previous result to complete the proof.

(13.3) Verify that taking the trace $\mathrm{Tr}(G^{\mu\nu})$ of the Einstein tensor with components $G^{\mu\nu}$ gives $\mathrm{Tr}(G^{\mu\nu}) = -R$, where R is the trace of the Ricci tensor.

The triumphs of general relativity

14

The ascent to greatness, however steep and dangerous, may entertain an active spirit with the consciousness and exercise of its own power: but the possession of a throne could never yet afford a lasting satisfaction to an ambitious mind.
Edward Gibbon (1737–1794)

In our journey through general relativity's intellectual landscape, we have now finished a long, arduous, uphill trek and reached the mountain summit, on the top of which is engraved the Einstein field equation, which we write here in SI units as

$$G = \frac{8\pi G}{c^4} T. \tag{14.1}$$

After all the hard effort expended in the steep climb of the previous chapters, discovering the constituents of the equation, their meaning and how they fit together, it's time to enjoy a well-earned breather to take in the view! Therefore, in this chapter we will review some of the triumphs of the theory, that is, areas where general relativity makes accurate predictions and provides convincing explanations of physical behaviour. All of the ideas outlined here will be discussed in far more detail throughout the remainder of the book.

14.1 Weak fields and the Newtonian limit

In our search for a relativistic theory of gravity, one thing we required in Chapter 6 was that the Einstein equation reproduces the results of Newton's theory in the non-relativistic limit.[1] Our first task is to show that this is indeed the case. By 'non-relativistic' we usually mean the limit of velocities $|\vec{u}|$ that are small compared to c, and so it makes sense to restore the factors of c in this section. More precisely, we find relativistic corrections are not required when $|\vec{u}|^2/c^2 \ll 1$ which means that $\gamma = (1 - |\vec{u}|^2/c^2)^{-\frac{1}{2}} \approx 1$. Since in a flat frame the relativistic velocity has components $u^\mu = (\gamma c, \gamma u^i)$, we also expect that $u^0 \gg u^i$ in this limit.

In a weak gravitational field, the acceleration due to gravity is not expected to lead to particles attaining a high velocity compared to c. Another way to say this using dimensional analysis is that small $|\vec{u}|$ is equivalent to small GM/r, where G is the gravitational constant for a

[1] The consequences of taking the weak-field limit are discussed in more detail in Chapter 45.

source of mass M at distance r from a test particle. The non-relativistic limit, therefore, is the limit that $(GM/rc)^2 \ll 1$, and we will use this condition to define what we mean by a weak-field limit.

Example 14.1

Consider replacing the factors of c in the components of the energy-momentum tensor for a perfect fluid. In a frame in which the elements of the fluid move with a velocity with components u^μ, we have[2] $T^{\mu\nu} = (\rho + p/c^2)u^\mu u^\nu + pg^{\mu\nu}$. Since[3] the pressure $p \approx \rho v^2$, where ρ is the mass density, then $\rho \gg p/c^2$ in the weak-field limit, and so we have $T^{\mu\nu} \approx \rho u^\mu u^\nu$ and, since $u^0 \gg u^i$, we also have $T^{00} \gg T^{ii}$.

---

[2] Factors of c can be restored here either using straightforward dimensional analysis, or by consulting the table in Chapter 0.

[3] Because pressure is due to particles with kinetic energy $\frac{1}{2}mv^2$ bouncing off the container walls, the energy density $\approx \rho v^2$ and this, apart from a numerical factor, gives the pressure.

In the weak-field limit, the gravitational metric field can be modelled as making a small perturbation to the Minkowski metric for flat space. We then have

$$g_{\mu\nu} = \eta_{\mu\nu} + h_{\mu\nu}, \tag{14.2}$$

with $|h_{\mu\nu}| \ll 1$. Note that, as we shall see later, $h_{\mu\nu}$ here are not actually the components of a tensor, but we raise and lower all components with $\eta_{\mu\nu}$. Equation 14.2 can be inserted into the Einstein equation to deduce an equation for $h_{\mu\nu}$ in terms of the energy-momentum tensor $T_{\mu\nu}$. It turns out that the equation we need comes out most cleanly in terms of $\bar{h}_{\mu\nu}$, an object related[4] to $h_{\mu\nu}$ defined by

[4] This is the trace reversal of $h_{\mu\nu}$ (see the previous chapter), with the property $\bar{h} = \eta^{\mu\nu}\bar{h}_{\mu\nu} = -h$.

$$\bar{h}_{\mu\nu} = h_{\mu\nu} - \frac{1}{2}\eta_{\mu\nu}h, \tag{14.3}$$

and h is the trace $h = \eta^{\mu\nu}h_{\mu\nu}$. We then find that

$$\partial^2 \bar{h}_{\mu\nu} = -\frac{16\pi G}{c^4}T_{\mu\nu}. \tag{14.4}$$

Example 14.2

Let's evaluate eqn 14.4 in the weak-field limit for $T_{\mu\nu}$ which means that, setting $c = 1$,[5] the only appreciable term is $T_{00} \approx T^{00} \approx \rho$. We deduce that $-\partial^2 \bar{h}_{00} = 16\pi GT_{00} = 16\pi G\rho$, but we note that $\partial^2 = -\partial^2/\partial t^2 + \nabla^2$ and so the time dependence can be neglected (because all speeds are much less than c). Hence, we have

[5] Restoring factors of c, the relevant equations here become

$$T^{00} = \rho c^2,$$
$$\nabla^2 \bar{h}_{00} = -\frac{16\pi G}{c^2}\rho,$$
$$\bar{h}_{00} = -4\Phi/c^2,$$
$$g_{00} = -\left(1 + 2\Phi/c^2\right). \tag{14.5}$$

$$\nabla^2 \bar{h}_{00} = -16\pi G\rho, \tag{14.6}$$

with the other components of $\bar{h}_{\mu\nu}$ obeying a Laplace equation (i.e. with no source), so will be zero (assuming that these terms should fall off to zero at infinity anyway). Comparing with the Newtonian result $\nabla^2\Phi = 4\pi G\rho$ we deduce that $\bar{h}_{00} = -4\Phi$ with all other components of $\bar{h}_{\mu\nu}$ equal to zero. Hence, the trace of $\bar{h}_{\mu\nu}$ is $\bar{h} = 4\Phi$ and we can therefore deduce the perturbation $h_{\mu\nu}$ to the metric using

$$h_{\mu\nu} = \bar{h}_{\mu\nu} - \frac{1}{2}\eta_{\mu\nu}\bar{h}, \tag{14.7}$$

yielding

$$h_{\mu\nu} = \begin{pmatrix} -4\Phi & 0 & 0 & 0 \\ 0 & 0 & 0 & 0 \\ 0 & 0 & 0 & 0 \\ 0 & 0 & 0 & 0 \end{pmatrix} - \frac{1}{2}(4\Phi)\begin{pmatrix} -1 & 0 & 0 & 0 \\ 0 & 1 & 0 & 0 \\ 0 & 0 & 1 & 0 \\ 0 & 0 & 0 & 1 \end{pmatrix}$$

$$= \begin{pmatrix} -2\Phi & 0 & 0 & 0 \\ 0 & -2\Phi & 0 & 0 \\ 0 & 0 & -2\Phi & 0 \\ 0 & 0 & 0 & -2\Phi \end{pmatrix}. \tag{14.8}$$

The full metric (Minkowski plus the small perturbation) then becomes

$$g_{\mu\nu} = \eta_{\mu\nu} + h_{\mu\nu} = \begin{pmatrix} -1 & 0 & 0 & 0 \\ 0 & 1 & 0 & 0 \\ 0 & 0 & 1 & 0 \\ 0 & 0 & 0 & 1 \end{pmatrix} + \begin{pmatrix} -2\Phi & 0 & 0 & 0 \\ 0 & -2\Phi & 0 & 0 \\ 0 & 0 & -2\Phi & 0 \\ 0 & 0 & 0 & -2\Phi \end{pmatrix}. \quad (14.9)$$

Writing the metric in terms of the line element gives

$$ds^2 = g_{\mu\nu}\,dx^\mu\,dx^\nu = -\left(1+2\Phi\right)dt^2 + \left(1-2\Phi\right)\left(dx^2 + dy^2 + dz^2\right), \quad (14.10)$$

finally proving the result given in eqn 5.22.

> Chapter 45 contains a more detailed derivation of the equations in Section 14.1, and further discussion of the weak-field limit.

14.2 Gravitational waves

The metric field, just like the electromagnetic field, can host waves in space and time. These **gravitational waves** cause periodic distortions in the metric field that can tell masses how to move. Their effects are, however, very small, and escaped detection until 2015 when a burst of waves lasting 150 ms were detected by the LIGO gravitational wave detectors.[6] This burst was deduced to have resulted from the collision of two black holes: an event with such an immense output of power that the gravitational waves could finally pass the threshold for detection.

[6]LIGO stands for Laser Interferometer Gravitational-Wave Observatory.

Example 14.3

We saw in Chapter 11 that the effect of the tidal forces caused by a massive planet is to distort a spherical distribution of mass into an ellipsoid of the same volume. The distribution is stretched in the direction of the planet and contracted in the other two orthogonal directions. The reason behind this conservation of volume can actually be traced back to the Ricci tensor, whose components are $R_{\mu\nu}$. The Einstein equation given in terms of these tensor components reads

$$R_{\mu\nu}(x) - \frac{1}{2}g_{\mu\nu}R(x) = \frac{8\pi G}{c^4}T_{\mu\nu}(x). \quad (14.11)$$

This is a local equation (hence, the dependence on the position in spacetime x), and so, if $T(x)$ vanishes at a point, then so does the right-hand side of the equation. This will be the case in vacuum, where there are no sources of energy-momentum. This, in turn, implies that $R_{\mu\nu}(x)$ *vanishes in the vacuum*, and provides an explanation for the volume-preserving property of the tidal forces from the planet: in order for $R_{\mu\nu}(x)$ to vanish,[7] the tidal forces must have positive parts (that cause contraction of the ellipse) exactly cancelling the negative parts (that stretch out the ellipse in the orthogonal direction).[8]

Since, in free space, gravitational waves are also governed by $R_{\mu\nu}(x) = 0$, there is a similar balance between stretching (negative curvature) and contraction (positive curvature) owing to the existence of the wave. In this case, there is a stretching effect and a compressive effect in *one* orthogonal direction (not two directions, as we had for the planet example above). To achieve the cancellation, these must be equal and opposite.

[7]Recall that $R_{\mu\nu}(x) = R^{\alpha}{}_{\mu\alpha\nu}(x)$, so can be thought of as a sort of average of the Riemann curvature at a point, retaining the parts of the curvature that cause volumes to shrink.

[8]This makes sense if we refer back to Chapter 11, where we saw that the compressive forces were half as large as the stretching ones, but since the compression occurs in two directions, we achieve the cancellation we have claimed. The properties of the Ricci tensor are discussed in more detail in Chapter 35.

From the last example, we can draw a picture of some typical gravitational waves. These are shown in terms of their effect on a circle of

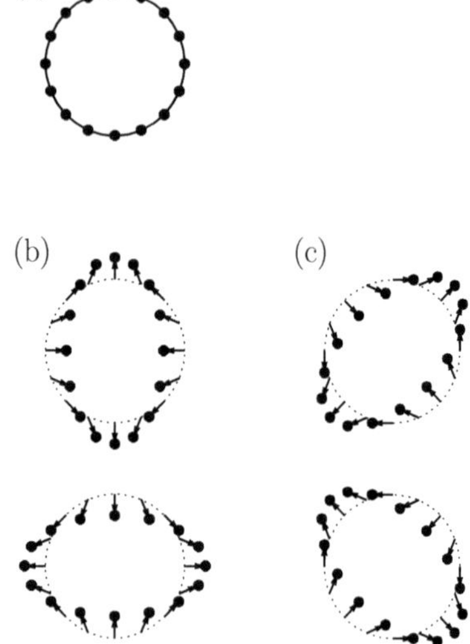

Fig. 14.1 (a) A circle of test masses in flat space. (b) The effect of a gravitational wave on the circle of test masses at two different times, corresponding to half a period of the gravitational wave. (c) The same as (b) but with a different polarization.

[9]The polarization of an electromagnetic wave can be linear or circular, or a superposition expressed in a linear or circular basis.

[10]See Chapter 42.

[11]For example, the fact that $h_{\mu\nu}$ is symmetrical immediately reduces the potential number of independent components from sixteen to ten.

↷ **Chapter 46 goes into more detail on gravitational waves, taking advantage of some of the geometrical ideas presented in Parts V and VI.**

masses [Fig. 14.1(a)] in Fig. 14.1(b) and (c) for two different polarizations. In (b), for example, we see the effect at some point in spacetime of expansion and contraction of the circle into an ellipse. Half a wavelength further along the direction of propagation of the wave, the stretching and compressive forces have swapped over, causing the circle of masses to be distorted in the opposite sense.

In order to understand where the waves come from, we must look at the behaviour of the metric field, and in particular we can look at the perturbation to the Minkowski metric $\bar{h}_{\mu\nu}$ introduced earlier in the chapter. In the absence of any matter, the components of the energy-momentum tensor vanish, and eqn 14.4 reduces to

$$-\partial^2 \bar{h}_{\mu\nu} = 0. \tag{14.12}$$

This is a **wave equation** for the components $\bar{h}_{\mu\nu}$.

Example 14.4

A wave equation for a scalar field $\phi(t, \vec{x})$ looks like $\partial^2 \phi = 0$, or, in component form

$$-\frac{1}{c^2}\frac{\partial^2 \phi}{\partial t^2} + \frac{\partial^2 \phi}{\partial x^2} + \frac{\partial^2 \phi}{\partial y^2} + \frac{\partial^2 \phi}{\partial z^2} = 0. \tag{14.13}$$

The solutions to this equation of motion are plane waves such as $\phi(t, \vec{x}) = Ce^{i\boldsymbol{k}\cdot\boldsymbol{x}} = Ce^{-i(\omega t - \vec{k}\cdot\vec{x})}$, with $\omega = c|\vec{k}|$ and C a constant. The waves are **excitations** of the scalar field, requiring a non-zero, but finite energy to come into existence. Similarly, a wave equation for a 1-form field with components A_μ can be written as $\partial^2 A_\mu = 0$ with solutions

$$A_\mu = \mathrm{Re}\left[C_\mu e^{-i(\omega t - \vec{k}\cdot\vec{x})}\right]. \tag{14.14}$$

The relationship between the different components give us the idea of the polarization of the wave.[9] In the absence of sources, we can write such a wave equation for the electromagnetic field $\tilde{\boldsymbol{A}}(x)$, from which the more familiar electric and magnetic fields can be extracted using[10]

$$\vec{E} = -\vec{\nabla}A^0 - \frac{\partial \vec{A}}{\partial t}, \quad \vec{B} = \vec{\nabla} \times \vec{A}. \tag{14.15}$$

A wave equation for the components $\bar{h}_{\mu\nu}$, describing the weak gravitational field, means that this field can also support wave-like excitations, just like those in the last example. These gravitational waves have the potential to move masses in periodic motions. The solutions to the wave equation look like

$$\bar{h}_{\mu\nu} = \mathrm{Re}\left[A_{\mu\nu}e^{i\boldsymbol{k}\cdot\boldsymbol{x}}\right]. \tag{14.16}$$

A cause of complication here is that the field has two indices, implying sixteen components in (3+1)-dimensional spacetime. Although these components are not all independent (so several are identical),[11] this feature leads to the orthogonal stretching and compressive forces along with the polarizations shown in Fig. 14.1.

14.3 Stars, trajectories, and orbits

Not too many solutions of the Einstein equation are known, but one of the most useful exact solutions is the Schwarzschild metric, which describes the geometry of spacetime near a static, spherically symmetric distribution of mass (with total mass M). The line element of his metric is written in spherical coordinates as

$$\mathrm{d}s^2 = -\left(1 - \frac{2GM}{rc^2}\right)c^2\mathrm{d}t^2 + \left(1 - \frac{2GM}{rc^2}\right)^{-1}\mathrm{d}r^2 + r^2\left(\mathrm{d}\theta^2 + \sin^2\theta\mathrm{d}\phi^2\right).$$

(14.17)

In the free space outside of the mass distribution, we have no local sources of energy momentum, and so $\boldsymbol{T} = 0$. The line element then solves the Einstein equation $\boldsymbol{G} = 0$. The motion of masses in this geometry allows us to see the affects of general relativity on the planets in our solar system. There are several famous solar-system-based tests of general relativity, the three known as the 'classical' ones are listed in the margin. Here's we'll concentrate on the first of these.

One triumph of Newton's theory was the prediction of elliptical orbits of planets around the Sun. General relativity provides a richer array of orbits and motions of massive particles in a gravitational field. We can describe the motion as that of a particle in an effective potential $V_{\mathrm{eff}}(r)$. We write

$$\mathcal{E} = \frac{1}{2}\left(\frac{\mathrm{d}r}{\mathrm{d}\tau}\right)^2 + V_{\mathrm{eff}}(r),$$

(14.18)

where the effective potential for a particle is given by

$$V_{\mathrm{eff}}(r) = -\frac{GM}{rc^2} + \frac{\tilde{L}^2}{2r^2c^2} - \frac{GM\tilde{L}^2}{r^3c^4},$$

(14.19)

where $\tilde{L}$ refers to the angular momentum per unit mass of the orbiting particle. The first two terms are also found in Newtonian gravitation; the final term is the result of general relativity, providing a correction to the motion.

Some quite general allowed trajectories in this potential are shown in Fig. 14.2. They depend on the value of the effective potential, which is dependent on the (conserved) value of angular momentum of the particle. The most striking of these is the precession of the near-elliptical orbits shown in Fig. 14.3, by which we mean the apparent rotation of the elliptical orbit of a planet about the sun. The accurate prediction of the perihelion shift (i.e. the rate of this rotation) of the planet mercury due to this precession effect was historically one of the most significant of the early triumphs of Einstein's theory. The calculated value of precession is approximately 43.0 arc seconds per century, a small but observable quantity, which agrees very well with experiment.

In contrast to the Newtonian theory, light is also predicted to be bent by gravitational fields in general relativity. A similar approach to that outlined above for massive particles, yields an effective potential for light rays. This leads to the prediction of gravitational-lensing effects.

The three classical solar system tests of relativity are:
- Precession of the perihelion of mercury;
- Bending of light by the sun;
- Gravitational redshift.

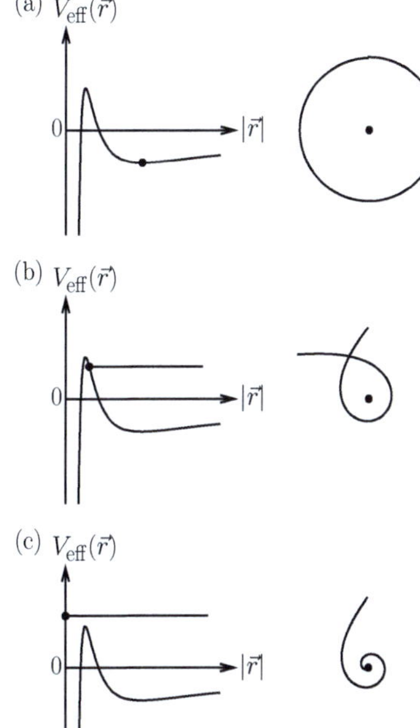

Fig. 14.2 Some allowed trajectories in the relativistic potential. (a) Circular; (b) an encounter with the star leading to scattering; (c) a spiral into the star.

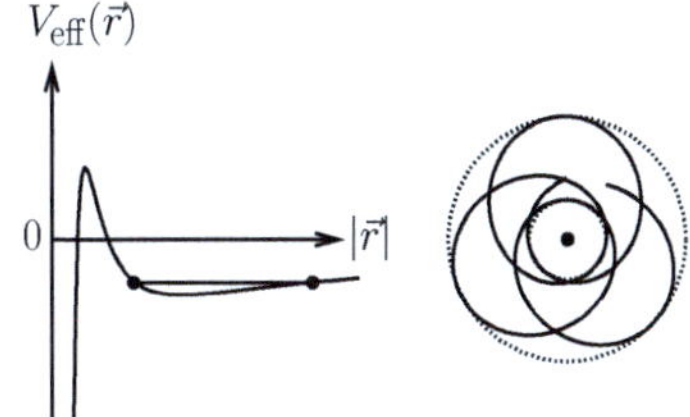

Fig. 14.3 The precessing orbit in the relativistic potential. The motion is bounded at the points shown on the left, represented by the dotted lines on the right.

↰ **Part IV of the book discusses orbits, stars and blackholes, further developing the ideas outlined in Section 14.3.**

The most extreme gravitational effects occur in this geometry close to a source of the gravitational field which is very dense. For particularly dense mass distributions, we obtain the prediction of a black hole, which is a region of spacetime where the gravitational field is so strong that not even light can escape its clutches.

14.4   Cosmology

In addition to treating the physics of the trajectories of individual bodies, general relativity can describe the spacetime structure of the entire Universe. This is the field of cosmology, which gives us an insight into the origin and fate of our Universe.

Many model universes have a geometry described by the Robertson–Walker metric, with line element

$$ds^2 = -c^2 dt^2 + a(t)^2 \left[\frac{dr^2}{1 - kr^2} + r^2 \left(d\theta^2 + \sin^2 \theta d\phi^2 \right) \right], \qquad (14.20)$$

where the parameter $k = 1, 0$ or -1, depending on the curvature of spacetime. This line element solves the Einstein equation and describes universes with geometries that, when filled with different sorts of matter, have distinct origins and fates. Perhaps the most significant feature of these model universes is that the spatial parts of the metric vary with the time coordinate t, giving us a notion of cosmic evolution. In fact, the mechanics of spacetime can be determined via equations of motion for $a(t)$, which is a quantity that functions as a sort of effective radius of the Universe. Notably, many of these model universes start from a Big Bang: an initial state where the whole universe is shrunk to a point. (The universe then undergoes a period of expansion.) The Big Bang represents an example of a singularity, which is a point in spacetime where our classical field theories, such as general relativity, break down.

↰ **Part III of the book, coming up next, is concerned with cosmology, giving much more detail on the subject outlined in Section 14.4.**

Clearly, there is a lot to unpack here and it is to cosmology that we turn in the next part of the book.

Chapter summary

- The Einstein equation matches the Newtonian theory in the weak-field limit.
- Gravitational waves are excitations of the gravitational field.
- Newtonian theory predicted the existence of elliptical orbits in central potentials, but general relativity leads to precession of elliptical-like orbits.
- Cosmology, the application of the Einstein equation to the entire Universe, results in descriptions of the origin and fate of the Universe.

Part III

Cosmology

Having formulated a theory of gravity, we are now going to use it on the largest conceivable problem: the origin and fate of our Universe.

- In Chapter 15, we introduce the ideas behind the field of cosmology and investigate two model universes.
- In Chapter 16, we investigate the possible geometries describing homogeneous, isotopic spacetimes with constant curvature and in Chapter 17 we fill these with matter.
- We apply all of these ideas in Chapter 18 to a number of different model universes.
- In the final chapter of Part III (Chapter 19), we study some very general ideas that allow us to think about infinity in the context of the geometry of spacetime.

15

An introduction to cosmology

[1] In addition to gravity, these are electromagnetism, the weak nuclear force and the strong nuclear force.

To see a World in a Grain of Sand
And a Heaven in a Wild Flower
Hold Infinity in the palm of your hand
And Eternity in an hour
William Blake (1757–1827)

I don't pretend to understand the Universe — it's a great deal bigger than I am ... People ought to be modester.
Thomas Carlyle (1795–1881)

After several chapters spent building a field theory of gravitation, we are now ready to use it. Our first task will be the, admittedly ambitious, application of Einstein's equation to the entire Universe! Gravitation is the weakest of the four so-called fundamental interactions.[1] However, unlike the strong and weak nuclear interactions, gravity is a long-range force. Moreover, unlike the (similarly long-range) electromagnetic interaction, gravitation is always attractive and so cannot cancel out in the same way that, owing to a balance in the number of positive and negative electric charges, electromagnetism frequently does. Gravitation is therefore always attractive, additive, and acts over large distances, causing it to be the dominant interaction in determining the large-scale behaviour of the Universe.

In this chapter, we begin an investigation of **cosmology** by applying Einstein's equation to describe models of the Universe in terms of the large-scale structure of spacetime. First, some good news: there are several solvable models of the Universe. In general, these have the property that the spatial parts of the metric field depend on the time coordinate, leading to the idea of cosmic evolution, a universe changing with time. Although this disturbed early relativists who expected a steady-state universe, we are now more comfortable with this idea. Second, the bad news: many of the universes appear to start, and some end, in a **singularity**. This is a point in spacetime that cannot be dealt with using classical field theory, which is designed to deal with smoothly evolving spacetime. A singularity is a point in spacetime where the known laws of physics break down, and that doesn't sound like a good thing. However, in later chapters we shall learn how to accommodate them. In any case, let's begin by setting out the nature of the problem and some first attempts at solutions.

15.1 The cosmological principle

The central idea of cosmology is to solve the Einstein field equation[2]

$$R_{\mu\nu} - \frac{1}{2}g_{\mu\nu}R = 8\pi T_{\mu\nu} - \Lambda g_{\mu\nu}, \tag{15.1}$$

for all of the matter in the Universe.[3] A solution involves a complete description of spacetime, achieved by specifying the metric field at each point in space and time.[4] Cosmology therefore promises to provide us with a complete picture of all space in the past, present, and future.

To make progress in cosmology, we need a simplifying idea to get us started. We start with the **Copernican principle**[5] which states that the view of the Universe that we have from our vantage point on Earth is not special; the Universe would look the same from any other viewpoint, at this epoch in cosmological history. This idea is closely related to the cosmological principle, which is stated as follows:

The **cosmological principle** says that the Universe is spatially homogeneous and isotropic.

- The property of **spatial homogeneity** means that the properties of the Universe are not a function of position in the Universe; wherever you are in the Universe, everything looks the same.

- The property of **spatial isotropy** means that there are no special directions; in whichever direction you look from your vantage point in the Universe, everything looks the same.

Although it is a great simplifying assumption, the cosmological principle is clearly wrong. If you look up, you see the sky; if you look down you see the ground; the Universe doesn't look isotropic. What's more, it doesn't seem homogeneous, as we can change our environment from mountains to valleys simply by changing our spatial position, and they are clearly different. Even looking at larger length scales doesn't seem to help as, for example, the Solar System (which is filled with planets, asteroids, comets and with a large star in the centre) seems very different from the intergalactic medium (which is largely empty). The problem here is that we are still looking at too small a length scale and we need to broaden our horizons.

Example 15.1

In Example 0.3, we introduced the Astronomical Unit, the distance from the Earth to the Sun, which is 1.496×10^{11} m. However, this is a relatively small distance compared to the light-year which is the distance light travels in a year, roughly 9.46×10^{15} m or more than 60,000 Astronomical Units. Astrophysicists tend to use, more often, the **parsec** (abbreviated to pc), which is 3.26 light-years:[6]

$$1\,\mathrm{pc} = 3.086 \times 10^{16}\,\mathrm{m}. \tag{15.2}$$

[2]Reminder: In this part of the book, we will take $G = c = 1$, unless otherwise indicated.

[3]It might be worth keeping in mind that our basic strategy in this part of the book will be to solve the Einstein equation $\boldsymbol{G} = \kappa\boldsymbol{T} - \Lambda\boldsymbol{g}$, by matching the components of $\boldsymbol{G}$ with those of the energy-momentum contributions from $\boldsymbol{T}$ and $\Lambda\boldsymbol{g}$.

[4]This is a feature of the spacetime view of cosmology: the behaviour of spacetime is completely predetermined, and is therefore, in a sense, unchanging. This statement about spacetime as a whole does not contradict our assertion above that the spatial parts of the metric depend on the time coordinate.

[5]Nicolaus Copernicus' (1473–1543) book *De Revolutionibus* described a heliocentric cosmology, removing the Earth from the centre of the Universe. Arthur Koestler (1905–1983) described the book as an 'all-time worst-seller', claiming it had not been widely read after its publication. The claim was disproved by Owen Gingerich (1930–), as described in his *The Book Nobody Read*, after examination of almost every surviving copy of Copernicus' book. The Copernican principle, in its modern cosmological form, was named by Hermann Bondi (1919–2005).

[6]A parsec is defined as the distance at which an astronomical unit subtends an angle of one arc second. Its name comes from *parallax*, the technique by which distances of nearby stars can be measured, and is essentially an abbreviation of 'parallax of one second'.

The nearest star, Proxima Centauri, is 1.3 parsecs from us, and the brightest star in the night sky, Sirius, is 2.7 parsecs away. Thus, a parsec gives a rough estimate for the distance between stars in our Galaxy, the Milky Way. There are about 10^{11} stars in the Milky Way, shaped like a disc with a radius of something like 25 kpc which is around 0.3 kpc thick (kpc means kiloparsec). The Andromeda Galaxy, the nearest large galaxy to the Milky Way, is about 770 kpc away from us. Thus, even on the scale of a megaparsec, the Universe looks quite lumpy and not at all homogeneous and isotropic. However, if you zoom out further, say to a scale of 100 Mpc (which is 3×10^{24} m, or more than thirteen orders of magnitude above the astronomical unit), the graininess and lumpiness of the Universe starts to disappear and the cosmological principle starts to hold.

Our theories of cosmology will therefore be constructed to understand the Universe on length scales longer than ≈ 100 Mpc, and they will clearly not work on length scales shorter than this. Therefore, when we do cosmology we will be applying the Einstein field equations to a model Universe in which all the matter has been smeared out (the technical term is **coarse-grained**) so that it becomes a fluid of uniform density.[7] The averaged-out mass content of the Universe is, in the absence of internal interactions, equivalent to the dust we discussed in Chapter 12. However, we shall often allow internal interactions too. In either case, the matter constitutes a perfect **cosmological fluid** (Chapter 12) that fills the Universe. We can allow the cosmic fluid to move, expand, and contract.

15.2　The Hubble flow

An important observational fact about the Universe was discovered by Edwin Hubble. From measurements of redshifts of emission lines in astrophysical objects we now know that there is a relationship between the velocity $\vec{v}$ and position[8] $\vec{r}$ of an object in the Universe which is given by **Hubble's law**, which can be written as

$$\vec{v} = H_0 \vec{r}, \tag{15.3}$$

where $H_0 = 70 \ \mathrm{km\,s^{-1}\,Mpc^{-1}}$ is **Hubble's constant**.[9] The interpretation of this law is that the Universe is expanding as a function of time. Two important points follow:

- Hubble's law makes it look like our own position in the Universe is special, because everything in the Universe is receding from us. Does this violate the Copernican principle? Not at all, because Hubble's law holds viewed from everywhere in the Universe. One way of thinking about this is that for two objects, currently (time t_0) at positions $\vec{r}_i$ and $\vec{r}_j$, have a separation given by the vector $\vec{r}_{ij} = \vec{r}_i - \vec{r}_j$ (and this description subtracts out the choice of coordinate origin that you have when you write down a particular $\vec{r}_i$ or $\vec{r}_j$). We now suppose that the Universe expands in such a way that, at some later time t, we have

$$\vec{r}(t) = a(t)\vec{r}(t_0), \tag{15.4}$$

[7]This is like looking at a uniform grey rectangle on a computer screen; only by close examination with a magnifying glass will you notice that it is made up of tiny dots. Similarly, only by close examination of the Universe (on scales much less than 100 Mpc) do you notice that it is actually made up of discrete galaxies, themselves composed of individual stars.

[8]You can think of this as the distance to a galaxy from the Earth, where the latter sits at the origin of the coordinate system.

[9]Edwin Hubble (1889–1953). Hubble's original measurement was nearly an order of magnitude larger than this, but the value has been refined over the intervening period and we now believe we have it to within about 10% or so. Is it really a *constant*, taking the same value for all time? As we will see later, various cosmological models will show that it isn't, and so we will later refer to it as the **Hubble parameter** $H(t)$.

where $a(t)$ is the **scale factor** of the Universe. The function $a(t)$ follows the expansion of the Universe and we can define it so that $a(t_0) = 1$ (i.e. the Universe is not scaled up or down at the current time t_0). Then the separation of the two objects will be given by

$$\vec{r}_{ij}(t) = a(t)\vec{r}_{ij}(t_0), \tag{15.5}$$

and so the separation of any two objects increases according to the scale factor $a(t)$, irrespective of any choice of origin. Moreover, this equation can be used to work out the relative velocity between these two objects as

$$\vec{v}_{ij}(t) = \dot{\vec{r}}_{ij}(t) = \dot{a}(t)\vec{r}_{ij}(t_0) = \frac{\dot{a}(t)}{a(t)}\vec{r}_{ij}(t), \tag{15.6}$$

and so we deduce that the Hubble parameter is

$$H(t) = \frac{\dot{a}(t)}{a(t)}. \tag{15.7}$$

We will call $\vec{r}(t)$ $\{= [x^i(t), y^i(t), z^i(t)]\}$ the real, physical, coordinates and $\vec{r}(t_0)$ the **comoving coordinates**.[10] Galaxies generally remain fixed in place when measured in the comoving coordinates, but becomes increasingly separated from one another in the physical coordinates, as $a(t)$ increases.

- We use the term **Hubble flow** to denote the outward motion of objects due to the expansion of the Universe, as expressed by the scale factor $a(t)$. There can be other so-called **proper motions** superimposed on top of the Hubble flow. For example, some galaxy may be gravitationally bound to some others in a cluster. Its velocity, resulting from forces within that system of galaxies, will be superimposed upon the velocity resulting from the Hubble flow. Another example is that the cosmic microwave background (CMB) radiation, observable from the Earth, is remarkably uniform across the sky but has a small dipole anisotropy superimposed on it. This results from the motion of the Earth with respect to the reference frame of the CMB,[11] meaning that the CMB viewed from Earth is very slightly redshifted or blueshifted depending on which direction you look.

Note that the expansion of the Universe, governed by $a(t)$, doesn't affect the Solar System.[12] The physics of things on smaller length scales, such as the orbits of gravitationally bound objects in the Solar System, can be treated by general relativity (see Part IV of this book), but are dominated by the effects of the local interactions between the objects; the weak curvature of the Universe plays no role. Thus, in this part of the book (Part III) we are keeping our eyes on the big picture and ignoring the tiny details that only affect processes on minuscule length scales.[13]

[10] A simple working definition of comoving coordinates is that a particle at rest in a comoving system continues to be at rest as a function of coordinate time t. In this case, we have $\dot{\vec{r}}_{ij}(t_0) = 0$.

[11] Sometimes described as the echo of the Big Bang, the CMB is electromagnetic radiation resulting from recombination processes that occurred when the first atoms were formed, during a relatively early epoch of the Universe. Although special relativity teaches us that there's no privileged frame of reference, the CMB frame is an interesting case of a rather physically unique frame! Of course, the laws of physics are the same in that frame as in any other inertial frame. The small anisotropy in the CMB that we measure results from the motion of the Earth around the Sun (a time-dependent term with a period of one year), and the motion of the Solar System with respect to the CMB, which includes the motion of the Local Group of galaxies, including our own Milky Way, towards the centre of the constellation Hydra.

[12] Or, for that matter, the expansion of your waistline. That cannot be blamed on Edwin Hubble.

[13] By minuscule, we here mean less than 100 Mpc!

15.3 Cosmic time

[14]This idea assumes that the Universe is evolving monotonically towards some final state and that there are characteristic motions that can be used as a standard of time. This give us access to some scalar S that decreases monotonically as the Universe evolves forward in time. An example might be the black-body radiation temperature T of the Universe: since the Universe is assumed to be cooling monotonically, T should steadily decrease as a function of the age of the Universe. The time t assigned to an event is then defined to be a function $t(T)$ where T is evaluated when and where the event takes place. Since the scalar S used to characterize cosmic standard time t is a function of t only, this implies that for an event that is assigned time t in cosmic standard, and t' in the other coordinate system, we have $S(t) = S'(t')$. However, since S is a scalar, we have $S' = S$ and so $S(t) = S(t')$. We see then that $t = t'$, and that all equivalent coordinate systems must use cosmic standard time.

We shall now set up a coordinate system with which to investigate cosmology. We start with the notion of a **cosmic time** coordinate t that could reasonably be used by sensible observational astronomers.[14] We can then define a **cosmic standard coordinate system** in terms of a stack of **spacelike hypersurfaces** at various fixed values of t. A hypersurface is a slice of spacetime formed by our keeping one of the four spacetime coordinates fixed so that the remaining coordinates describe a three-dimensional space. A spacelike hypersurface for a particular time t is then described by three spatial coordinates (x, y, z), so that one spacelike hypersurface passes through each and every point in 3-space. The spacelike hypersurfaces are said to **foliate** all of spacetime, meaning that spacetime is carved up into a stack of non-intersecting spacelike hypersurfaces, as shown in Fig. 15.1. The cosmological principle (homogeneity and isotropy) holds in each of these spacelike hypersurfaces, and so at each moment in time (the coordinate which labels which slice you are considering) the Universe has the same density, pressure and curvature everywhere. This means that in each hypersurface the Universe looks the same everywhere (satisfying our assumption of homogeneity) and in every direction (satisfying our assumption of isotropy, see Fig. 15.2), at least on length scales greater than about 100 Mpc (as discussed before). Of course, if you zoom along in a spacecraft travelling at some large speed with respect to this cosmic standard coordinate system (in other words, if you travel with some proper motion with respect to the Hubble flow) then the Universe will not look isotropic and homogeneous. Thus, when doing cosmology we will restrict our attention to the cosmic standard coordinate system where the cosmological principle works.

We met the idea of comoving coordinates in the last section. We shall see that a good choice of cosmic coordinate system is a comoving one. An observer who is at rest with respect to comoving coordinates (and who therefore 'goes with the flow' of the cosmic evolution, which you could describe colloquially as 'going with the Hubble flow') is known as a **comoving observer**. The comoving observers have world lines that coincide with the elements of the cosmological fluid, that pierce the spacelike hypersurfaces, as shown in Fig. 15.3.

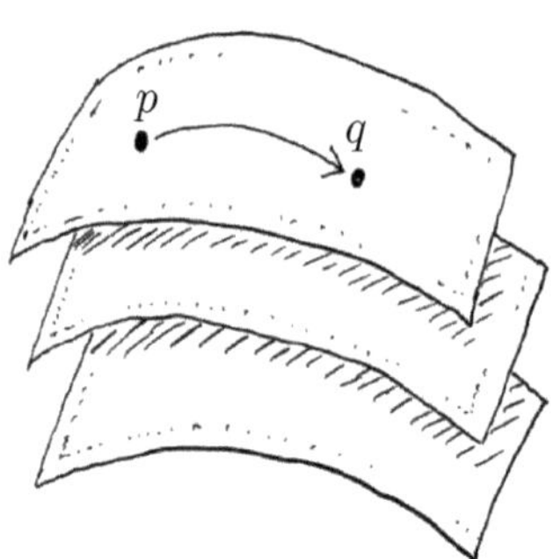

Fig. 15.1 Spacelike hypersurfaces foliate the whole of spacetime. Homogeneity is defined in terms of spacelike hypersurfaces, so each spatial coordinate p on a given hypersurface is equivalent to every other one on that surface, e.g. point q. Points p and q on the same hypersurface are related by an **isometry** of spacetime which keeps time t fixed. An isometry is a distance-preserving transformation between parts of a spacetime.

Example 15.2

Putting the notions of homogeneity and isotropy together, it seems reasonable that spacelike hypersurfaces must be orthogonal to the world lines of the cosmological fluid. If they weren't we would be able to identify a preferred direction, contradicting the assumption of isotropy. This construction allows us geometrically to assert that the time coordinate in the comoving coordinates has mathematical properties:

$$e_0 \cdot e_i = g_{0i} = 0, \qquad (15.8)$$

(i.e. the timelike vector is perpendicular to the spacelike ones);

$$e_0 = u, \qquad (15.9)$$

(i.e. the comoving observer has a velocity that coincides with the timelike vector)

$$\boldsymbol{e}_0 \cdot \boldsymbol{e}_0 = g_{00} = -1, \tag{15.10}$$

(i.e. the timelike vector is normalized).

If a metric yields $\Gamma^i{}_{00} = 0$ then, by the geodesic equation ($\ddot{x}^\mu + \Gamma^\mu{}_{\alpha\beta}\dot{x}^\alpha\dot{x}^\beta = 0$), a particle at rest in this coordinate system stays at rest. This is then a comoving coordinate system. This connection coefficient will vanish if the derivatives $g_{00,i}$ and $g_{0i,0}$ vanish, giving us a quick way of identifying comoving coordinates. So in the present case, we can conclude that the conditions for the coordinates to be comoving ($g_{0i,0} = g_{00,i} = 0$) are met.

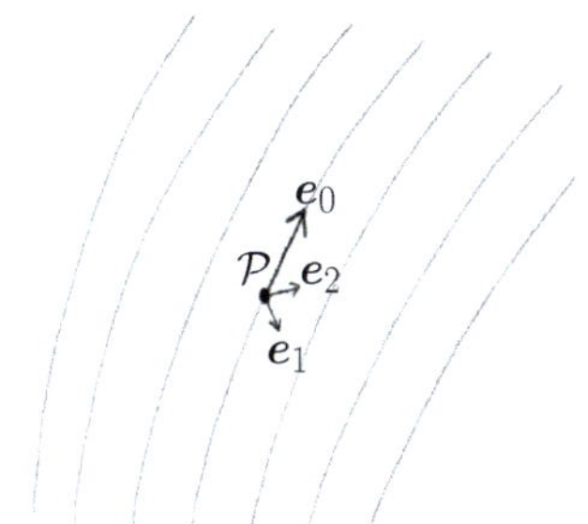

Comoving observers have a special place in our cosmological theories as they form an important part of the underlying geometry, as enshrined in **Weyl's postulate**, which states that:[15]

> In cosmic spacetime, there exists a set of privileged fundamental observers whose world lines form a smooth bundle of timelike geodesics. These geodesics never meet (apart from at an initial, past singularity or a final, future singularity).

This state of affairs is shown in Fig. 15.3. In comoving coordinates, each galaxy is at rest (ignoring proper motions) and each hypersurface is assigned a universal cosmic time coordinate. This time coincides with the proper time of the privileged observer at that point.

Finally, we consider the consequences of the principle of cosmology on the form of the fundamental field in general relativity: the metric. This will be expressed in our comoving coordinates $x^\mu = (t, x^i)$. To retain homogeneity and isotropy, the metric line element $\mathrm{d}\sigma$ of each of the three-dimensional, spacelike hypersurfaces must take the form $\mathrm{d}\sigma^2 = \gamma_{ij}\mathrm{d}x^i\mathrm{d}x^j$, where γ_{ij} is a function of spatial coordinates x^1, x^2 and x^3 and encodes information about the curvature of the space.[16] Also including the time, the metric line element of spacetime in the general comoving system is given by

$$\mathrm{d}s^2 = -\mathrm{d}t^2 + a(t)^2\gamma_{ij}\mathrm{d}x^i\mathrm{d}x^j. \tag{15.11}$$

Here, we have included $a(t)$, the time-dependent scaling factor, that describes the expansion of space as a function of time.[17] This kinematic approach to the metric was pioneered by Robertson and Walker in the 1930s[18] and we shall return to its details in the next chapter. To obtain the form of $a(t)$ we must make assumptions about the content of the Universe. This is the subject to which we now turn. It's time to make some model universes! Remember that our goal is to obtain a solution to the Einstein field equation, which could be that for a given geometry or for a given distribution of mass and energy.

15.4 Universe 0: an empty universe

The first and simplest model describes an empty universe, containing no matter and no energy, so that $T^{\mu\nu} = 0$. We therefore have an Einstein equation

$$G_{\mu\nu} = R_{\mu\nu} - \frac{1}{2}g_{\mu\nu}R = 0. \tag{15.12}$$

Fig. 15.2 World lines run perpendicular to the spacelike hypersurfaces and isotropy means that, at some point $\mathcal{P}$, the Universe looks the same along any two spacelike vectors $\boldsymbol{e}_1$ and $\boldsymbol{e}_2$. We've drawn them to be roughly perpendicular to each other, but they needn't be. However, both have to be spacelike and so both must be perpendicular to $\boldsymbol{e}_0$, the timelike vector that is tangent to the world line. There exists an isometry in spacetime which leaves $\mathcal{P}$ fixed but which rotates $\boldsymbol{e}_1$ into $\boldsymbol{e}_2$.

[15] Hermann Weyl (1885–1955). 'Smoothness' in this statement guarantees that our classical field theory can be used to describe the physics. The word 'bundle' here has a technical meaning examined in Appendix C, but you can think of it in the usual sense of the word, as 'a collection of things fastened together for convenient handling'.

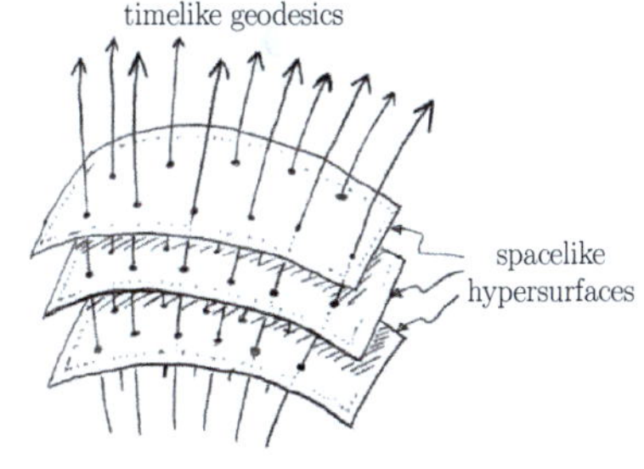

Fig. 15.3 Timelike geodesics of cosmic observers. These are orthogonal to all of the spacelike hypersurfaces.

[16] For flat space we have $\gamma_{11} = \gamma_{22} = \gamma_{33} = 1$ and other components zero.

[17] Check: $g_{0i,0} = g_{00,i} = 0$, so the system is confirmed as comoving.

[18] Howard P. 'Bob' Robertson (1903–1961); Arthur G. Walker (1909–2001).

[19]However, the condition $R_{\mu\nu} = 0$ *on its own* does not mean that the components of the Riemann tensor must also vanish (giving a flat universe, as they do in this model); rather that the positive curvature parts cancel out the negative curvature parts when the Ricci tensor is formed. Physically, the Ricci tensor describes the tendency of gravitational fields to make volumes of mass contract. This is described in more detail in Chapter 35. The photograph of Einstein writing the equation '$R_{ik} = 0$?' on a blackboard has become iconic and might, after $E = m$ (or $E = mc^2$ in units that Einstein didn't generally use), have a claim to being his second most-famous equation.

[20]Expressed in spherical polars, this metric is

$$ds^2 = -dt^2 + dr^2 + r^2 \left(d\theta^2 + \sin^2\theta d\phi^2\right). \tag{15.13}$$

This can be rewritten using light-cone coordinates $u = t - r$ and $v = t + r$ as

$$ds^2 = -dudv + \frac{1}{4}(v - u)^2 \left(d\theta^2 + \sin^2\theta d\phi^2\right). \tag{15.14}$$

We shall make use of each of these forms in the coming chapters.

[21]Willem de Sitter (1872–1934). We discuss a neat geometrical description of the this model's spacetime in the exercises of Chapter 18.

[22]Fixing the time at $t = t_0$ results in a spacelike hypersurface with $d\sigma^2 = a(t_0)\left(dx^2 + dy^2 + dz^2\right)$, which is the homogeneous, isotropic metric we had in Universe 0, scaled by a factor $a(t_0)$. Notice how the expansion factor $a(t)$ affects the behaviour of light cones, which obey $dt/dr = \pm a(t)$. For a monotonically increasing $a(t)$ this causes light cones to become narrower with time. This effect is examined in Chapter 25.

Since $R = g_{\mu\nu}R^{\mu\nu}$, we see that $R_{\mu\nu} = 0$ guarantees a solution to the Einstein equation.[19]

A simple solution to the Einstein equation that is consistent with an empty, isotropic universe is flat, Minkowski spacetime with metric[20]

$$ds^2 = -dt^2 + dx^2 + dy^2 + dz^2. \tag{15.15}$$

This flat spacetime, which is written in comoving coordinates (i.e. $g_{00,i} = g_{0i,0} = 0$), is consistent with eqn 15.11, and has $R_{\mu\nu\alpha\beta} = 0$ and therefore $R_{\mu\nu} = 0$. We conclude that flat space is a solution to this problem. Hypersurfaces of this spacetime are spatially flat with $d\sigma^2 = dx^2 + dy^2 + dz^2$. Geodesics in this universe take the form of straight lines

$$x^\mu(\tau) = b^\mu\tau + c^\mu, \tag{15.16}$$

with b^μ and c^μ constants and τ the proper time parametrizing the geodesic.

So far so good, but this energy/matter-free model is rather unrealistic. We can go much further by investigating a more interesting scenario.

15.5   Universe 1: flat and expanding

We next discuss a universe where space is flat, but expands as a function of time with expansion factor $a(t)$. We'll call this the **de Sitter model**.[21] The model's metric line element can be written as

$$\begin{aligned} ds^2 &= -dt^2 + a(t)^2\left(dx^2 + dy^2 + dz^2\right) \\ &= -dt^2 + a(t)^2\left[dr^2 + r^2(d\theta^2 + \sin^2\theta d\phi^2)\right], \end{aligned} \tag{15.17}$$

where, in the second line we've employed spherical polar coordinates. This is another isotropic and homogeneous model of spacetime.[22] The dimensionless factor $a(t)$, if positive, causes space itself to expand, such that all of the coordinate points (or all of the galaxies) move away from each other. Although the spacelike hypersurfaces of this model are flat, spacetime is not (i.e. there will be some non-zero components of $\boldsymbol{R}$). Therefore, our first task it to work out how the scaling factor $a(t)$ affects the curvature of spacetime.

Example 15.3

We can use the metric field of the de Sitter model to compute the curvature of spacetime. The metric components are

$$\begin{aligned} g_{00} &= -1, \quad g_{ii} = a^2, \\ g^{00} &= -1, \quad g^{ii} = \tfrac{1}{a^2}. \end{aligned} \tag{15.18}$$

The derivation of the connection coefficients and the components of the Riemann tensor will be covered in the Exercises. Here we'll get some practice using the orthonormal frame and the vielbein. The metric field yields up the components of the Riemann tensor in the orthonormal frame of

$$R^{\hat{0}}{}_{\hat{i}\hat{0}\hat{i}} = \frac{\ddot{a}}{a}, \quad R^{\hat{i}}{}_{\hat{j}\hat{i}\hat{j}} = \left(\frac{\dot{a}}{a}\right)^2, \tag{15.19}$$

where $\dot{a} = da(t)/dt$ and we (temporarily) do not sum over repeated indices.

We can identify the vielbein[23] allowing us to move between the orthonormal and coordinate frames. The Ricci tensor's 00th component in the orthonormal frame is

$$R_{\hat{0}\hat{0}} = \sum_{\hat{i}=1}^{3} R^{\hat{i}}{}_{\hat{0}\hat{i}\hat{0}}, \tag{15.21}$$

where we write the sum explicitly. Using the idea that components are raised and lowered in the orthonormal frame using the Minkowski tensor $\boldsymbol{\eta}$, we have from eqn 15.19 that

$$\begin{aligned}\frac{\ddot{a}}{a} &= -R_{\hat{0}\hat{i}\hat{0}\hat{i}} && (\text{using } \eta_{00} = -1)\\ &= -R_{\hat{i}\hat{0}\hat{i}\hat{0}} && (\text{two } - \text{signs from swaps within pairs})\\ &= -R^{\hat{i}}{}_{\hat{0}\hat{i}\hat{0}} && (\text{using } \eta^{ii} = 1).\end{aligned} \tag{15.22}$$

Doing the sum in eqn 15.21, we find

$$R_{\hat{0}\hat{0}} = -3\frac{\ddot{a}}{a}. \tag{15.23}$$

It will be useful later to have the components in the coordinate frame, and so we transform out of the orthonormal frame using the vielbein

$$R_{00} = (\boldsymbol{e}_0)^{\hat{0}}(\boldsymbol{e}_0)^{\hat{0}} R_{\hat{0}\hat{0}} = -3\frac{\ddot{a}}{a}. \tag{15.24}$$

Note that owing to the symmetries of $\boldsymbol{R}$ we have, as always, that $R^{\mu}{}_{\mu\mu\mu} = 0$. As a result, to avoid a contribution from the case when $i = j$ we only have two contributions from the spatial sum in the expression

$$\begin{aligned}R_{\hat{i}\hat{i}} &= R^{\hat{0}}{}_{\hat{i}\hat{0}\hat{i}} + \sum_{j=1}^{2} R^{\hat{j}}{}_{\hat{i}\hat{j}\hat{i}}\\ &= \frac{\ddot{a}}{a} + 2\left(\frac{\dot{a}}{a}\right)^2,\end{aligned} \tag{15.25}$$

and so, transforming into the coordinate frame,

$$\begin{aligned}R_{ii} &= (\boldsymbol{e}_i)^{\hat{i}}(\boldsymbol{e}_i)^{\hat{i}} R_{\hat{i}\hat{i}} = a^2 R_{\hat{i}\hat{i}}\\ &= \ddot{a}a + 2\dot{a}^2.\end{aligned} \tag{15.26}$$

Working in any frame, we must obtain the same Ricci scalar. In the coordinate frame, we find

$$\begin{aligned}R = g^{\mu\nu}R_{\mu\nu} &= 3\frac{\ddot{a}}{a} + \frac{3}{a^2}\left(2\dot{a}^2 + a\ddot{a}\right)\\ &= 6\left(\frac{\ddot{a}}{a} + \frac{\dot{a}^2}{a^2}\right).\end{aligned} \tag{15.27}$$

What matter distribution could be responsible for this spacetime metric and provide the right-hand side of the Einstein equation? It turns out that **pure vacuum energy** is consistent with this metric as we shall now see. If we allow the presence of vacuum energy but not matter, then our task is to solve an Einstein equation

$$R_{\mu\nu} - \frac{1}{2}g_{\mu\nu}R = -\Lambda g_{\mu\nu}, \tag{15.28}$$

with the Ricci tensor specified above.

[23]Recall that the orthonormal frame we choose is the one in which the observer's velocity vector $\boldsymbol{u}$ only has a timelike component $u^{\mu} = (u^0, 0, 0, 0)$. Since we need $\boldsymbol{u} \cdot \boldsymbol{u} = g_{\mu\nu}u^{\mu}u^{\nu} = -1$, we must have $u^0 = 1$. We then set $\boldsymbol{e}_{\hat{0}} = \boldsymbol{u}$ and $\boldsymbol{e}_{\hat{i}} = \frac{1}{a}\boldsymbol{e}_i$ to give us orthonormal basis vectors. The vielbein components are

$$\begin{aligned}(\boldsymbol{e}_0)^{\hat{0}} &= 1, & (\boldsymbol{e}_i)^{\hat{i}} &= a,\\ (\boldsymbol{e}_{\hat{0}})^0 &= 1, & (\boldsymbol{e}_{\hat{i}})^i &= \frac{1}{a}.\end{aligned} \tag{15.20}$$

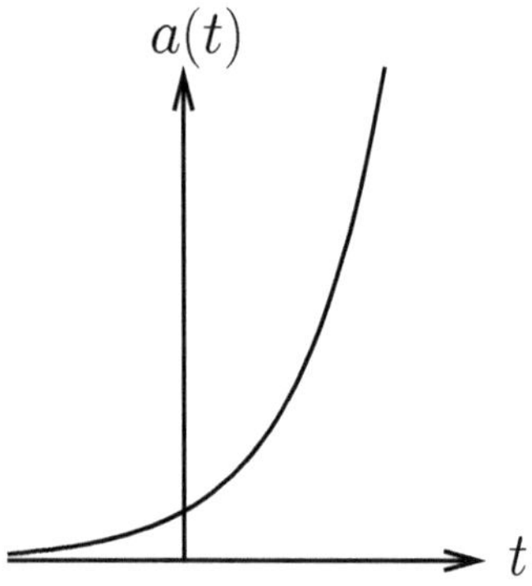

Fig. 15.4 The function $a(t)$ for **Universe 1**: the de Sitter model.

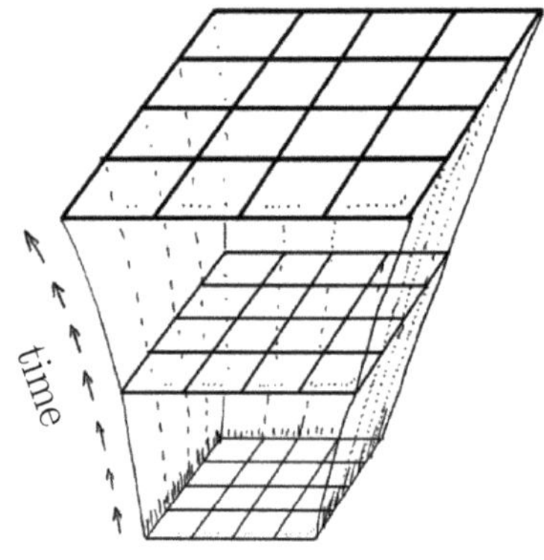

Fig. 15.5 Universe 1 is spatially flat with a scaling factor $a(t)$ that increases exponentially as a function of time.

Example 15.4

Consider the 00th component $R_{00} - \frac{1}{2}g_{00}R = -\Lambda g_{00}$, which becomes, on substitution,

$$-\frac{3\ddot{a}}{a} - \frac{g_{00}}{2} \times 6\left(\frac{\ddot{a}}{a} + \frac{\dot{a}^2}{a^2}\right) = -\Lambda g_{00}. \tag{15.29}$$

This simplifies to

$$3\frac{\dot{a}^2}{a^2} = \Lambda. \tag{15.30}$$

We can solve this straightforward first-order differential equation to find an equation for the scaling factor. The solution is

$$a(t) = \mathrm{e}^{Ht}, \tag{15.31}$$

with $H^2 = \Lambda/3$. As can be checked, this also solves the other components of the Einstein equation. The behaviour of $a(t)$ is shown in Fig. 15.4. The increasing value reflects an accelerating expansion of this universe.

This is our first kinematic model of a universe. Granted, there's no matter in it yet (just energy), but we have found that adding uniform cosmological-constant energy density causes the universe to expand at an accelerating rate. The expansion constant H is an example of a Hubble parameter. The metric-field line element describing the geometry of this universe is, on substituting for $a(t)$ and using polar coordinates,

$$\mathrm{d}s^2 = -\mathrm{d}t + \mathrm{e}^{2Ht}\left[\mathrm{d}r^2 + r^2(\mathrm{d}\theta^2 + \sin^2\theta\mathrm{d}\phi^2)\right], \tag{15.32}$$

with $H = (\Lambda/3)^{\frac{1}{2}}$. The evolution of Universe 1 is shown in Fig. 15.5, which shows the spatially flat universe expanding as a function of time.

Looked at another way, the (positive) vacuum-energy density described by Λ has the curious property of causing an extra expansion to the universe. We can treat the vacuum energy as part of an energy-momentum tensor, just as we did in Chapter 13 where we found that $T_{\mu\nu}^{\mathrm{vac}} = -\Lambda g_{\mu\nu}/8\pi$. Comparing this to the components $T^\mu = (\rho, p, p, p)$ describing a perfect fluid in flat spacetime, we find an effective pressure $p = -\Lambda/8\pi$. We conclude that vacuum energy with positive Λ gives rise to a negative pressure and so, if squeezed, must expand.

We shall return to this model in Chapter 18. In the next two chapters, we shall put in place the groundwork that enables us to treat a far wider class of model universes.

Chapter summary

- The cosmological principle says that the Universe is spatially homogeneous and isotropic.

- An empty universe (Universe 0) has flat spacetime described by a Minkowski metric field.

- The de Sitter model (Universe 1) is spatially flat, but has a spacetime with a constant curvature. It expands exponentially as a result of the vacuum energy content of the universe.

Exercises

(15.1) Taking a typical interstellar distance in the Milky Way as 1 pc, and the approximate dimensions of the Milky Way as given in Example 15.1, *estimate* the number of stars in the Milky Way. (This is astrophysics, so you only need to do this to the nearest power of ten.)

(15.2) Consider the de Sitter model.
(a) Show that the connection coefficients are given by

$$\Gamma^0{}_{ij} = a\dot{a}\delta_{ij}, \quad \Gamma^i{}_{0j} = \Gamma^i{}_{j0} = \frac{\dot{a}}{a}\delta^i{}_j, \tag{15.33}$$

where $\dot{a} = da/dt$.
(b) Compute the components of the curvature tensor $\boldsymbol{R}$.

(15.3) Another way to think about the de Sitter model is that it is defined to be a model with a **constant curvature** of spacetime, where the Ricci scalar R is sufficient to determine the Riemann tensor. The condition for such a constant curvature in $(3+1)$-dimensional spacetime is that

$$R_{\mu\nu} = \frac{1}{4}Rg_{\mu\nu}. \tag{15.34}$$

(a) Compute the components of the Einstein tensor in terms of R and $g_{\mu\nu}$.
(b) Using this result, show that a solution to the Einstein equation requires $\Lambda = R/4$.
(c) Check the consistency of this last result using the expression $R = 6\left(\frac{\ddot{a}}{a} + \frac{\dot{a}^2}{a^2}\right)$, that we determined previously.

(15.4) A two-dimensional spacetime of constant curvature can be found by considering the metric line element we examined in Exercise 9.3

$$ds^2 = du^2 + dv^2 + 2dudv\cos\theta(u,v). \tag{15.35}$$

It is possible to write the line element of any two-dimensional spacetime in this form.
(a) Using the connection coefficients computed in Exercise 9.3, show that we have (using comma notation for derivatives)

$$R^u{}_{vuv} = -\frac{\theta_{,uv}}{\sin\theta}, \quad R^u{}_{uuv} = -\frac{\cos\theta}{\sin\theta}\theta_{,uv}, \tag{15.36}$$

and that, consequently,

$$R_{uu} = -\frac{\theta_{,uv}}{\sin\theta}, \quad R_{uv} = -\frac{\cos\theta}{\sin\theta}\theta_{,uv}. \tag{15.37}$$

(b) Finally, show that

$$R = -2\frac{\theta_{,uv}}{\sin\theta}. \tag{15.38}$$

(c) If the spacetime is of constant curvature K, then it has the property that $R_{\mu\nu} = Kg_{\mu\nu}$. Show that this implies that we have

$$\theta_{,uv} = \alpha\sin\theta, \tag{15.39}$$

where α is a constant to be determined.
This is the equation of motion for the **sine-Gordon field theory**, *which is widely used in quantum field theory.*

(15.5) In quantum field theory, the vacuum has an energy density owing to the zero-point energy of the matter fields. This energy of often removed (or *renormalized*) from the theory upon quantization. Let's interpret the vacuum energy realistically, and see what the effect is on the cosmology of the Universe. The strong nuclear interaction operates at energies of order 1 GeV over characteristic lengths of 10^{-15} m.
(a) Compute the characteristic energy density of the nuclear matter field.
(b) Make an order of magnitude estimate of the radius of curvature of a universe with this energy density and comment on the result.

(15.6) Consider a particle propagating radially in the de Sitter universe with Lagrangian

$$L = \left[\left(\frac{dt}{d\tau}\right)^2 - a(t)^2\left(\frac{dr}{d\tau}\right)^2\right]^{\frac{1}{2}}, \tag{15.40}$$

where τ is the proper time.
(a) By eliminating the dependence on $dr/d\tau$ in the equations of motion, show that the t coordinate obeys

$$\frac{d^2t}{d\tau^2} + \frac{\dot{a}}{a}\left[\left(\frac{dt}{d\tau}\right)^2 - 1\right] = 0. \tag{15.41}$$

(b) Verify that this latter equation is solved by

$$a(t)^2\left[\left(\frac{dt}{d\tau}\right)^2 - 1\right] = \text{const.} \tag{15.42}$$

(c) Use the result from (b) and the relation $g_{\mu\nu}u^\mu u^\nu = -1$ to show that the 3-velocity of the particle varies as $|\vec{u}|^2 \propto a(t)^{-2}$.

This demonstrates that the magnitude of the momentum of a particle falls off due to the expansion of the Universe.

(15.7) Consider sending a light signal from the origin of the de Sitter universe at a time t_s, which is received at a time t_r at radial position ℓ.

(a) Show that

$$\ell = \frac{1}{H}\left(e^{-Ht_\mathrm{s}} - e^{-Ht_\mathrm{r}}\right). \qquad (15.43)$$

(b) Explain why this implies the existence of position ℓ' beyond which the light signal cannot propagate owing to the expansion of the universe.

As a result of this, the de Sitter universe is said to have an event horizon. *See Chapter 19 for the full story.*

(15.8) Consider de Sitter spacetime, with metric line element

$$ds^2 = -dt^2 + e^{2Ht}\left(dx^2 + dy^2 + dz^2\right). \qquad (15.44)$$

Show that the geodesics for massive particles are straight lines.

Robertson–Walker spaces

16

The universe ought to be presumed too vast to have any character
C. S. Peirce (1839–1914)

In the last chapter, we looked at two spatially flat universes. We now relax the geometrical constraint of spatial flatness. This will allow us to identify a large class of geometries that can serve as the basis for modelling universes. These will not yet be fully fledged models, since we are only treating the left-hand (geometrical) half of Einstein's equation. It will be the job of the next chapter to fill the universes with matter to complete the modelling process.

The geometries in this chapter have constant spatial curvature.[1] We shall find that these geometries can be divided into three different types, according to whether the curvature is positive, zero, or negative.[2] The three types of geometry share many features, perhaps the most dramatic is a consequence of time dependence: the three geometries can shrink to a point at some value of the time coordinate t. At this point the curvature of spacetime becomes infinite and we have a singularity in the geometry. Singularities have a number of worrying features for the description of Nature in terms of fields that will occupy us in several contexts in the remainder of this book. However, perhaps we should not be too jumpy: it is likely that our Universe emerged from an initial singularity, the **Big Bang**, and in identifying it in these geometries, we might well have taken a step closer to uncovering the origin of spacetime itself.

[1] That is, the curvature is the same at all points in space and so can be characterized using a single number. This feature is examined further below and also in Exercises 16.5 and 15.3.

[2] Reminder: A simple example of a two-dimensional surface with positive curvature is a ball. The circumference of a circle at fixed θ is $2\pi r \sin\theta$, which is less than $2\pi r$. As a result, flattening the surface onto a flat plane causes it to tear, with gaps left in its projection. In contrast, a surface with negative curvature has circles with circumference larger than $2\pi r$, so that a projection onto a plane causes folds in the surface owing to the overlaps in the projection.

16.1 Spaces with constant curvature

In order to allow the spatial part of the metric to take different functional forms, we write the line element in the form, identified in the last chapter, demanded by homogeneity and isotropy:

$$\mathrm{d}s^2 = -\mathrm{d}t^2 + a(t)^2 \gamma_{ij}\mathrm{d}x^i\mathrm{d}x^j. \tag{16.1}$$

It's worth keeping in mind that the constraints on the geometry demanded by homogeneity and isotropy, rather than anything about general relativity, give us this metric. The flat spatial universes (i.e. Universes 0 and 1) turn out to be special cases of the more general spaces described by this metric.

[3] This can only be built out of γ_{ij} and the only other 3-tensor available that maintains homogeneity and isotropy, the Levi-Civita tensor, which has components ε_{ijk}. All other tensors at our disposal lead to a preferred direction and therefore violate isotropy. In fact, we find that we only need γ_{ij}.

[4] See Chapter 11.

> ↷ **We discuss Gaussian curvature in detail in Chapter 30. For now, we can simply regard K as a constant whose form ultimately describes the possible classes of spacetime.**

[5] This relies on the computation

$$\gamma^{ik}(\gamma_{ik}\gamma_{jl} - \gamma_{il}\gamma_{jk}) = 3\gamma_{jl} - \gamma_{jl} = 2\gamma_{jl}. \quad (16.3)$$

[6] Be careful to note these are three-dimensional expressions for the spatial part of this spacetime. They are not the spacetime expressions that feed into the Einstein equation.

[7] Note here that we're discussing space only, and not spacetime. In the last chapter (Exercise 15.3), we said that the de Sitter universe (Universe 1) was a space*time* with constant curvature. There we described a spacetime of constant curvature as having a Riemann curvature tensor with components determined by the Ricci scalar, and obeying the expression $R_{\mu\nu} = \frac{1}{4}Rg_{\mu\nu}$. An equivalent description is that a spacetime with curvature tensor has components

$$R_{\mu\nu\alpha\beta} = C(g_{\mu\alpha}g_{\nu\beta} - g_{\mu\beta}g_{\nu\alpha}), \quad (16.6)$$

with C a constant. (This is usually called K, but we distinguish it here from the value of the constant in the three-dimensional case to avoid any confusion.) The link between the two descriptions is made in the exercises for this chapter.

[8] See Exercise 16.4.

Consider the part of the geometry defined by the line element of three-dimensional space $\mathrm{d}\sigma^2 = \gamma_{ij}\mathrm{d}x^i\mathrm{d}x^j$. Recall that this is the geometry of the three-dimensional, spacelike hypersurfaces we have constructed at each fixed value of the time coordinate t. We shall construct the three-dimensional Riemann curvature tensor, denoted $^{(3)}\boldsymbol{R}$, that describes the curvature of these slices of spacetime.[3] The form of the components of $^{(3)}\boldsymbol{R}$ is fixed, not only by homogeneity and isotropy, but also by the usual symmetries of the Riemann tensor.[4] There is only one arrangement that obeys all of these symmetries and it is written as

$$^{(3)}R_{ijkl} = K(\gamma_{ik}\gamma_{jl} - \gamma_{il}\gamma_{jk}), \quad (16.2)$$

where K is the **Gaussian curvature**: a constant with dimensions $1/(\text{length})^2$.

If we contract the components $^{(3)}R_{ijkl}$ using the 3-metric γ^{ik} we obtain the components of the three-dimensional Ricci tensor, $^{(3)}R_{jl}$, which are[5]

$$^{(3)}R_{jl} = 2K\gamma_{jl}. \quad (16.4)$$

It follows that the three-dimensional Ricci scalar is $^{(3)}R = 6K$.[6]

Having argued what the three-curvature must be from homogeneity, isotropy and the properties of the Riemann tensor, we next need to find the metric that gives this three-curvature. Robertson and Walker showed that there is only one metric that is consistent with the constraints. It is the one where the function γ_{ij} takes the form

$$\gamma_{ij} = \left[1 + \frac{K}{4}(x^2 + y^2 + z^2)\right]^{-2}\delta_{ij}. \quad (16.5)$$

Spaces with this metric are called **spaces of constant curvature**.[7]

Example 16.1

The metric line element using the above form of γ_{ij} is written as

$$\mathrm{d}s^2 = -\mathrm{d}t^2 + a(t)^2 \frac{(\mathrm{d}x^2 + \mathrm{d}y^2 + \mathrm{d}z^2)}{\left[1 + \frac{K}{4}(x^2 + y^2 + z^2)\right]^2}. \quad (16.7)$$

Written in polar coordinates, this becomes

$$\mathrm{d}s^2 = -\mathrm{d}t^2 + a(t)^2 \frac{(\mathrm{d}R^2 + R^2\mathrm{d}\Omega^2)}{\left(1 + \frac{KR^2}{4}\right)^2}, \quad (16.8)$$

where $\mathrm{d}\Omega^2 = \mathrm{d}\theta^2 + \sin^2\theta\mathrm{d}\phi^2$ and the radial coordinate here is written as $R(= \sqrt{x^2 + y^2 + z^2})$. This expression can be simplified if we exploit the freedom to choose a new radial coordinate r. This choice is constrained by our wanting the proper circumference that looks like $a(t) \times (2\pi r)$, which we can achieve if the θ-dependent term in the metric has the form $a^2r^2\mathrm{d}\theta^2$. We are therefore motivated to choose

$$r = \frac{R}{1 + \frac{KR^2}{4}}. \quad (16.9)$$

Substituting for this coordinate, we find, after some algebra,[8] that

$$\mathrm{d}s^2 = -\mathrm{d}t^2 + a(t)^2 \left(\frac{\mathrm{d}r^2}{1 - Kr^2} + r^2\mathrm{d}\Omega^2\right). \quad (16.10)$$

The resulting **Robertson–Walker metric** is written in terms of a line element as

$$ds^2 = -dt^2 + a(t)^2 \left[\frac{dr^2}{1 - Kr^2} + r^2 \left(d\theta^2 + \sin^2\theta d\phi^2 \right) \right]. \qquad (16.11)$$

Note that this metric has the useful property that if the Gaussian curvature vanishes, i.e. if $K = 0$, it reduces back to the flat spacetime of the de Sitter model (or Universe 1) of the previous chapter.

The Gaussian curvature can be zero, but can also take positive or negative values. If it is positive, the geometry is spherical (the Gaussian curvature of a sphere of radius r is $K = +1/r^2$); if it is negative, the geometry is hyperbolic. We will deal with these geometrical aspects in more detail below (and also in Part V), but for now we would like to simplify the metric a bit further by rescaling variables so that rescaled curvature k can only take three values: $k = 0$ (flat), $k = +1$ (spherical) and $k = -1$ (hyperbolic). We do this by setting $k = K/|K|$ if $K \neq 0$, and setting $k = 0$ if $K = 0$. Next we introduce a rescaled, dimensionless radial coordinate $\bar{r} = |K|^{\frac{1}{2}} r$, so we have

$$ds^2 = -dt^2 + \frac{a(t)^2}{|K|} \left[\frac{d\bar{r}^2}{1 - k\bar{r}^2} + \bar{r}^2 \left(d\theta^2 + \sin^2\theta d\phi^2 \right) \right]. \qquad (16.12)$$

The part in the square brackets is now dimensionless and so the prefactor has dimensions (length)^2. Finally, we redefine the scale factor $a(t)$ so that it becomes

$$a(t) := \begin{cases} a(t)/\sqrt{|K|} & K \neq 0 \\ a(t) & K = 0. \end{cases} \qquad (16.13)$$

This implies that $a(t)$, which was previously a dimensionless scaling factor, has taken on the dimensions of length. The curvature of the universe is now given by $K = k/a(t)^2$. Motivated by the result that Gaussian curvature of a sphere is $K = 1/r^2$, the factor $a(t)$ is sometimes known as the radius of the universe in this context.

Dropping the bar from $\bar{r}$ in the last example (purely to save on writing), we have the final result for the Robertson–Walker metric

$$ds^2 = -dt^2 + a(t)^2 \left[\frac{dr^2}{1 - kr^2} + r^2 \left(d\theta^2 + \sin^2\theta d\phi^2 \right) \right], \qquad (16.14)$$

where $k = 1, 0$ or -1.

In order to visualize the three possibilities presented by the values of the curvature parameter k (which may be ± 1 or 0), we shall, in Section 16.2, consider the result of embedding the 3-spaces described by line element $d\sigma^2$ into four-dimensional flat space.

Example 16.2

Before we get there, we can make some progress by noting the similarity between the Robertson–Walker spaces and the 2-spaces discussed in Appendix D. There we present an example analogous to the $k = 1$ Robertson–Walker space, which is a sphere. We also meet an example analogous to the $k = -1$ Robertson–Walker space case, which can't, as a whole, be embedded in Euclidean space, but can be represented as a 2-sheet hyperboloid when embedded in Minkowski space. As might be expected, we shall see below how we can also characterize the $k = 1$ Robertson–Walker case as a sort of spherical space, and the $k = -1$ case as a hyperbolic space. The $k = 0$ case is a flat space. Also similar to what we present in Appendix D, the factor $(1 - kr^2)$ can be removed from eqn 16.14 if we replace the (now dimensionless) variable r with a convenient alternative variable χ,[9] which leads to the substitution $d\chi^2 = dr^2/(1 - kr^2)$.

Robertson–Walker coordinates are comoving, so a galaxy has fixed (r, θ, ϕ). We have metric components

$$g_{tt} = -1, \qquad g_{rr} = \frac{a^2}{1 - kr^2},$$
$$g_{\theta\theta} = a^2 r^2, \qquad g_{\phi\phi} = a^2 r^2 \sin^2\theta.$$

Vielbein components are given by

$$(e_t)^{\hat{t}} = 1, \qquad (e_r)^{\hat{r}} = \frac{a}{(1 - kr^2)^{\frac{1}{2}}},$$
$$(e_\theta)^{\hat{\theta}} = ar, \qquad (e_\phi)^{\hat{\phi}} = ar\sin\theta.$$

The connection coefficients are

$$\Gamma^t{}_{rr} = \frac{a\dot{a}}{1 - kr^2},$$
$$\Gamma^t{}_{\theta\theta} = a\dot{a}r^2,$$
$$\Gamma^t{}_{\phi\phi} = a\dot{a}r^2\sin^2\theta,$$
$$\Gamma^r{}_{rr} = \frac{kr}{1 - kr^2},$$
$$\Gamma^i{}_{tj} = \frac{\dot{a}}{a}\delta^i{}_j,$$
$$\Gamma^r{}_{\theta\theta} = -r(1 - kr^2),$$
$$\Gamma^r{}_{\phi\phi} = -r(1 - kr^2)\sin^2\theta,$$
$$\Gamma^\theta{}_{r\theta} = \Gamma^\phi{}_{r\phi} = \frac{1}{r},$$
$$\Gamma^\theta{}_{\phi\phi} = -\sin\theta\cos\theta,$$
$$\Gamma^\phi{}_{\phi\theta} = \cot\theta.$$

The Ricci tensor has non-zero components in the orthonormal frame of

$$R_{\hat{t}\hat{t}} = -3\frac{\ddot{a}}{a},$$
$$R_{\hat{i}\hat{i}} = \frac{\ddot{a}}{a} + 2\left(\frac{\dot{a}}{a}\right)^2 + 2\frac{k}{a},$$

The Ricci scalar is

$$R = 6\left(\frac{\ddot{a}}{a} + \frac{\dot{a}^2}{a^2} + \frac{k}{a^2}\right).$$

[9]In Appendix D, it is shown that for the $k = 1$ case, $r = \sin\chi$ simplifies the line element. The corresponding choice for $k = -1$ is $r = \sinh\chi$, while we simply take $r = \chi$ for $k = 0$.

The metric has non-zero components

$$g_{tt} = -1,$$
$$g_{\chi\chi} = a(t)^2,$$
$$g_{\theta\theta} = a(t)^2 \Sigma(\chi)^2,$$
$$g_{\phi\phi} = a(t)^2 \Sigma(\chi)^2 \sin^2\theta.$$

We have vielbein components

$$(e_t)^{\hat{t}} = 1,$$
$$(e_\chi)^{\hat{\chi}} = a(t),$$
$$(e_\theta)^{\hat{\theta}} = a(t)\Sigma(\chi),$$
$$(e_\phi)^{\hat{\phi}} = a(t)\Sigma(\chi)\sin\theta.$$

[10]Recall that comoving coordinates must have $g_{00,i} = g_{0i,0} = 0$, which these certainly do.

In terms of the new dimensionless variable χ from the last example, we are able to summarize the Robertson–Walker spaces in a form that's very useful for applications, where the coordinates resemble ordinary spherical coordinates. These are given in the box below.

In spherical coordinates, the three Robertson–Walker spacetimes are written as

$$ds^2 = -dt^2 + a(t)^2 \left[d\chi^2 + \Sigma(\chi)^2 (d\theta^2 + \sin^2\theta d\phi^2) \right], \qquad (16.15)$$

where

$$\begin{aligned}
\Sigma(\chi) &= \sin\chi \quad &(k = +1), \\
\Sigma(\chi) &= \chi \quad &(k = 0), \\
\Sigma(\chi) &= \sinh\chi \quad &(k = -1).
\end{aligned} \qquad (16.16)$$

The spatial hypersurfaces are swept out with coordinates

$$\begin{aligned}
0 \leq \chi \leq \pi, \quad 0 \leq \theta \leq \pi, \quad -\pi \leq \phi \leq \pi, \quad &(k = 1), \\
0 \leq \chi < \infty, \quad 0 \leq \theta \leq \pi, \quad -\pi \leq \phi \leq \pi, \quad &(k = 0, -1).
\end{aligned} \qquad (16.17)$$

Notice the difference in the range of χ for $k = 1$ compared to the other cases. This tells us that the properties of $k = 1$ space are rather different to the others. Of course, to understand the effect of a metric field we insert it into the Einstein equation. For this purpose we extract the components of the Riemann tensor $\boldsymbol{R}$ and use these to compute the Ricci tensor and scalar. These are given in the margin on the previous page.

The Robertson–Walker metric is written in terms of comoving coordinates: the time t is the cosmic time variable, and the spatial variables are carried along with galaxies.[10] In the next example, we deal with the proper length as a measure of the spacelike interval between two points on a spacelike hypersurface. This interval varies with time (so is different on different hypersurfaces) owing to the presence of the radius factor $a(t)$. It also differs for different choices of the constant k.

Example 16.3

Consider two points along a radial line: (r, θ, ϕ) and $(r + dr, \theta, \phi)$. The element of proper distance along the line between the points is

$$dl = a(t)\frac{dr}{(1 - kr^2)^{\frac{1}{2}}}. \qquad (16.18)$$

If the two points are separated by fixed radial comoving coordinate ξ, they have a total spacelike interval between them of

$$l(t) = \int_0^\xi a(t)\frac{dr}{(1 - kr^2)^{\frac{1}{2}}}, \qquad (16.19)$$

which, on integration with the different choices of k, gives intervals

$$l(t) = \begin{cases} a(t)\sin^{-1}\xi & (k = 1), \\ a(t)\xi & (k = 0), \\ a(t)\sinh^{-1}\xi & (k = -1). \end{cases} \qquad (16.20)$$

These proper distances are shown in the Fig. 16.1. Notice that for $k = 1$ the proper distance reaches a finite limit for $\xi \to 1$ where $\sin^{-1}\xi = \pi/2$.

We can also define a proper radial velocity, measuring the rate of change of the interval, via the time derivative of $l(t)$, from which we have

$$\frac{\mathrm{d}l}{\mathrm{d}t} = \begin{cases} \frac{\mathrm{d}a}{\mathrm{d}t}\sin^{-1}\xi & (k=1), \\ \frac{\mathrm{d}a}{\mathrm{d}t}\xi & (k=0), \\ \frac{\mathrm{d}a}{\mathrm{d}t}\sinh^{-1}\xi & (k=-1). \end{cases} \tag{16.21}$$

Notice how the part that depends on ξ can be replaced by $l(t)/a(t)$ in each case. Following from this, we have the result for the proper length that

$$\frac{\mathrm{d}l(t)}{\mathrm{d}t} = \frac{1}{a(t)}\frac{\mathrm{d}a(t)}{\mathrm{d}t}l(t). \tag{16.22}$$

The equation describes the change in the proper spacelike interval as a function of time and therefore describes the expansion of space itself. The equation is usually rewritten in terms of a proper velocity $v_\mathrm{p} = \mathrm{d}l(t)/\mathrm{d}t$, and proper distance $d_\mathrm{p} = l(t)$ as Hubble's law

$$v_\mathrm{p} = H(t)d_\mathrm{p}, \tag{16.23}$$

where the function $H(t)$ is the Hubble parameter $H(t) = \frac{1}{a}\frac{\mathrm{d}a}{\mathrm{d}t}$, as introduced in Chapter 15.

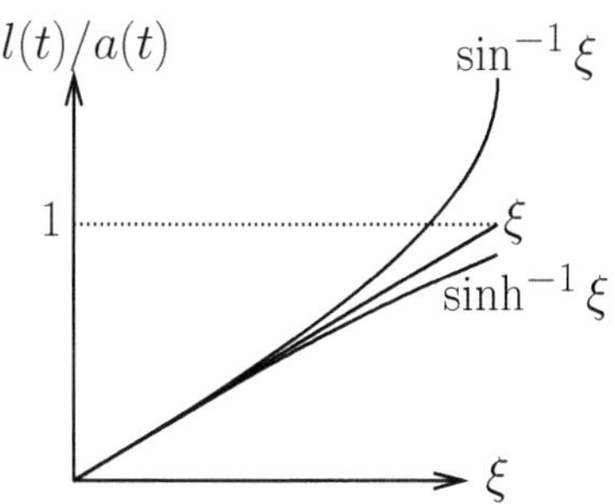

Fig. 16.1 The ratio of the proper distance $l(t)$ to the expansion factor $a(t)$ as a function of radial coordinate ξ.

These notions of proper distance are useful in giving a more complete geometrical description of the Robertson–Walker spaces, which is our next topic.

16.2   Three Robertson–Walker spaces

The Robertson–Walker metric provides us with three classes of expanding spacetime with line element

$$\mathrm{d}s^2 = -\mathrm{d}t^2 + a(t)^2\left[\frac{\mathrm{d}r^2}{1-kr^2} + r^2(\mathrm{d}\theta^2 + \sin^2\theta\,\mathrm{d}\phi^2)\right]. \tag{16.24}$$

The three spaces are characterized by the choice of curvature parameters $k = 0$ or $k = \pm 1$. Next, we examine each of the spaces in turn in terms of their geometrical properties. To do this we compare the proper circumference of a circle to its proper radius. This yields 2π if the space is flat, but differs from this value if it is not. The idea is exactly that of Sidenote 2, that on a surface with positive curvature the ratio of circumference to radius is less than 2π (so the surface tears when projected onto a flat plane), whereas for a negative curvature surface the ratio is more than 2π (and the surface folds up as it's projected onto a flat plane). Recall that a galaxy at radial coordinate ξ is separated from a galaxy at the origin by a spacelike interval $l(t) = a(t)\xi$. Taking $\theta = \pi/2$, the integral of the line element over a circle of coordinate radius ξ gives a circumference $2\pi a(t)\xi$ for each of the spaces.

Classical curvature of spherical, flat and hyperbolic spaces is described in Chapter 30. Embedding is discussed in Appendix D.

The case $k = 0$
We describe $k = 0$ universes as **flat**. The line element is given by

$$\mathrm{d}s^2 = -\mathrm{d}t^2 + a(t)^2\left[\mathrm{d}r^2 + r^2(\mathrm{d}\theta^2 + \sin^2\theta\,\mathrm{d}\phi^2)\right]. \tag{16.25}$$

This is identical to Universe 1 discussed in the last chapter. If $k = 0$ and $a(t)$ is a constant, then we have a Riemann tensor $\boldsymbol{R} = 0$ for all time

[11]The embedding required is

$$
\begin{aligned}
w &= a\cos\chi,\\
x &= a\sin\chi\sin\theta\cos\phi,\\
y &= a\sin\chi\sin\theta\sin\phi,\\
z &= a\sin\chi\cos\theta.
\end{aligned}
\tag{16.28}
$$

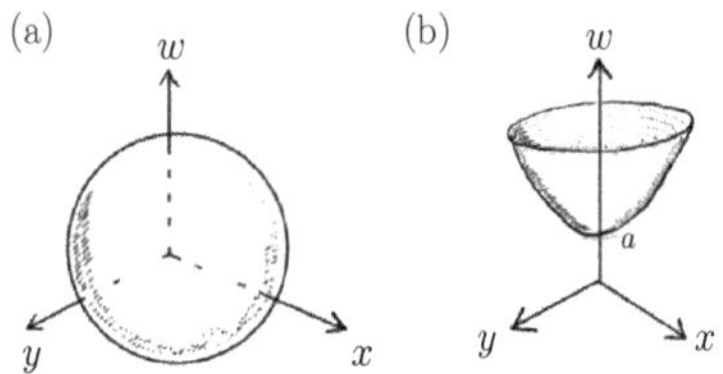

Fig. 16.2 (a) The three-dimensional $k = 1$ (closed) universe in four-dimensional Euclidean space, with one rotational degree of freedom suppressed. The embedding is $w^2 + x^2 + y^2 + z^2 = a^2$ (the equation of a four-dimensional sphere). (b) The three-dimensional $k = -1$ (hyperbolic) universe in four-dimensional Minkowski space with one rotational degree of freedom suppressed. The embedding is $w^2 - x^2 - y^2 - z^2 = a^2$ (the equation of a four-dimensional hyperboloid).

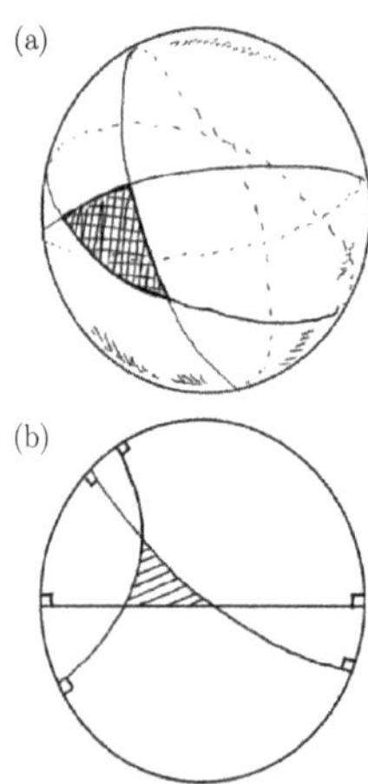

Fig. 16.3 (a) A triangle on a 2-sphere, whose angles sum to $> \pi$. The geodesics of this space are great circles on the sphere. (b) A triangle on the Poincaré disc representation of hyperbolic space, whose angles add to $< \pi$. The geodesics meet the boundary of the disc at right angles.

and the result is that the spacetime (as well as space) is flat, which is to say it has zero curvature. On the other hand, if $k = 0$ and $a(t) \neq$ const., then three-dimensional *space* is flat, but this flat space expands as a function of the time coordinate. In this case, some components of $\boldsymbol{R}$ are non-zero and so spacetime is curved.

Example 16.4

The geometry to have in mind for the $k = 0$ case is the flat, Euclidean plane, with all of its familiar features such as (i) parallel lines remaining parallel forever, and (ii) triangles with internal angles summing to π.

Since, for our three-dimensional space, we have $l(t) = a(t)\xi$ then the ratio

$$
\left(\frac{\text{Proper circumference of circle}}{\text{proper radius}}\right) = 2\pi,
\tag{16.26}
$$

just as we have in flat two-dimensional space.

The case $k = 1$

If $k = 1$ then the line element is given by

$$
\mathrm{d}s^2 = -\mathrm{d}t^2 + a(t)^2\left[\frac{\mathrm{d}r^2}{1 - r^2} + r^2(\mathrm{d}\theta^2 + \sin^2\theta\,\mathrm{d}\phi^2)\right].
\tag{16.27}
$$

The curvature of $k = 1$ space is positive. In Appendix D, we show how the spatial part of this metric can be embedded in four-dimensional Euclidean space as a sphere.[11] It is shown with a dimension suppressed in Fig. 16.2(a).

Example 16.5

The $k = 1$ geometry is sometime called **elliptic** or **spherical**. The example to have in mind is the two-dimensional surface of a sphere. Parallel lines in this geometry will eventually meet. A triangle in a two-dimensional space with elliptic geometry has internal angles that add up to greater than π. In fact, we saw in Chapter 3 that if we draw a triangle on a 2-sphere of radius R [Fig. 16.3(a)] its area obeys Girard's theorem

$$
(\text{Area}) = R^2\left(\alpha_1 + \alpha_2 + \alpha_3 - \pi\right),
\tag{16.29}
$$

where α_i are the internal angles of the triangle.

Returning to three spatial dimensions and using the result that the proper circumference of a circle in the plane $\theta = \pi/2$ is $2\pi a(t)\xi$, we have, on substituting for $\ell(t)$ from eqn 16.20,

$$
\left(\frac{\text{Proper circumference of circle}}{\text{proper radius}}\right) = \frac{2\pi a(t)\xi}{a(t)\sin^{-1}\xi} \leq 2\pi.
\tag{16.30}
$$

This is also the case if we take this ratio for a circle drawn on a 2-sphere, just as we did in Chapter 3, justifying the intuition of our embedding argument.

The volume of the $k = 1$ universe is given by

$$
V = \int_{\chi=0}^{\pi} 4\pi a(t)^3 \sin^2\chi\,\mathrm{d}\chi = 2\pi^2 a(t)^3.
\tag{16.31}
$$

Since the space has this finite proper volume (owing to the limits on the variable χ in eqn 16.17, which ranges from $0 \leq \chi \leq \pi$), it is said to be **closed**. However, what it does not have is a well-defined boundary, edge, or centre. This is similar to the situation for beings confined to the surface of a 2-sphere. It's only when we embed the 2-sphere in three-dimensional space that we're able to characterize the ball as having a centre or boundary.

Example 16.6

The oddest feature of the closed space is that the dimensionless radial coordinate has become an angle-like variable. As a result, purely radial light rays from distant points in a closed universe can be thought of as traversing the surface of a unit circle to reach the observer, as shown in Fig. 16.4(a). This creates the intriguing possibility that if you wait long enough, light bouncing off the back of your head could eventually reach your eyes after traversing the whole of the universe [Fig. 16.4(b)], with the result that you can see the back of your own head.[12] Keep in mind also that a scaling factor $a(t)$ that increases with t implies that the geometrical distance between comoving points on the unit sphere is increasing with time.

The case $k = -1$

If $k = -1$ then the line element is given by

$$\mathrm{d}s^2 = -\mathrm{d}t^2 + a(t)^2 \left[\frac{\mathrm{d}r^2}{1+r^2} + r^2(\mathrm{d}\theta^2 + \sin^2\theta\,\mathrm{d}\phi^2) \right]. \tag{16.32}$$

The spatial part of this metric can be embedded in Minkowski space where it becomes a 2-sheet hyperboloid as shown in Fig. 16.2.[13] This is a space with a negative curvature.

Example 16.7

The geometry for the $k = -1$ case is sometimes called **hyperbolic**.[14] Parallel lines in this geometry will eventually diverge. A triangle in a two-dimensional space with hyperbolic geometry has angles that add up to less than π. In this case, the area of a triangle obeys Lambert's[15] formula:

$$(\text{Area}) = -R^2\left(\alpha_1 + \alpha_2 + \alpha_3 - \pi\right), \tag{16.34}$$

with R a constant. Comparing with eqn 16.29 we realize that the rule can be interpreted as giving the area of a triangle on a 2-sphere of radius $\sqrt{-R^2}$, implying a negative curvature. The triangle is visualized in Fig. 16.3(b) in a representation of hyperbolic geometry known as the **Poincaré disc**.[16] This is a representation of a hyperbolic space where the geodesics meet the boundary at right angles. The Poincaré disc can be thought of as a projection from a two-sheet hyperboloid, as shown in Fig. 16.5.

For our circle in three-dimensional $k = -1$ space we now have

$$\left(\frac{\text{Proper circumference of circle}}{\text{proper radius}} \right) = \frac{2\pi a(t)\xi}{a(t)\sinh^{-1}\xi} \geq 2\pi. \tag{16.35}$$

This is also the state of affairs for two-dimensional surfaces with negative curvature embedded in Euclidean 3-space.

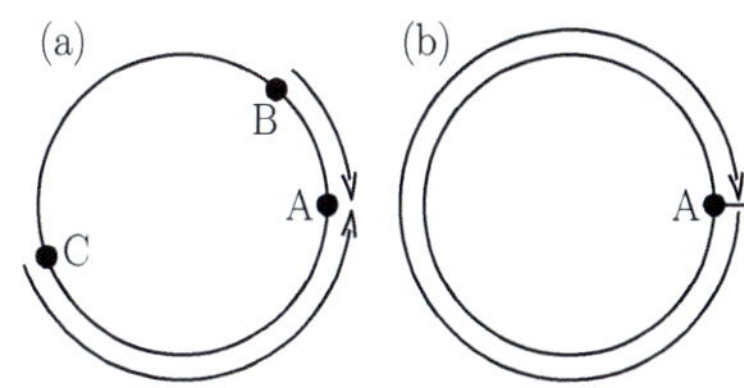

Fig. 16.4 (a) Radial light rays in a closed universe reach an observer at A, who sees a star at B in front of them and one at C behind them. (b) If we wait long enough, radial light can potentially traverse the whole universe.

[12] We show in Chapter 19 that in a physically realistic model there is not enough time for this to actually occur!

[13] The embedding required is

$$\begin{aligned} w &= a\cosh\chi, \\ x &= a\sinh\chi\sin\theta\cos\phi, \\ y &= a\sinh\chi\sin\theta\sin\phi, \\ z &= a\sinh\chi\cos\theta. \end{aligned} \tag{16.33}$$

The hyperboloid is defined by $w^2 - x^2 - y^2 - z^2 = a^2$. Remember that it is possible to embed parts of the space in Euclidean space, but not all of it.

[14] Hyperbolic geometry is described in Penrose (2004) and Needham (1997). It was independent discovered by several people, but is sometimes called Lobachevskian geometry after Nikolai I. Lobachevsky (1792–1856). Lobachevsky is the subject of a song by satirist and mathematician Tom Lehrer, which was described as 'dazzlingly inventive in its shameless promotion of plagiarism', although Lehrer stressed the song was not meant as a slur, with Lobachevsky's name chosen 'solely for prosodic reasons'.

[15] Johann Heinrich Lambert (1728–1777)

[16] This is an example of a *conformal* representation. Conformal geometry is discussed in Chapter 19. Despite its name, the Poincaré disc was originally formulated by Eugenio Beltrami (1835–1900) in 1868, but was rediscovered by Henri Poincaré fourteen years later. See Penrose (2004) for a short account of the history of hyperbolic space.

[17] We shall generally refer to hyperbolic ($k = -1$), flat ($k = 0$) and spherical ($k = 1$) spaces below. They are sometimes respectively called open, flat and closed in the literature, although this is slightly ambiguous since flat space with $k = 0$ is also open.

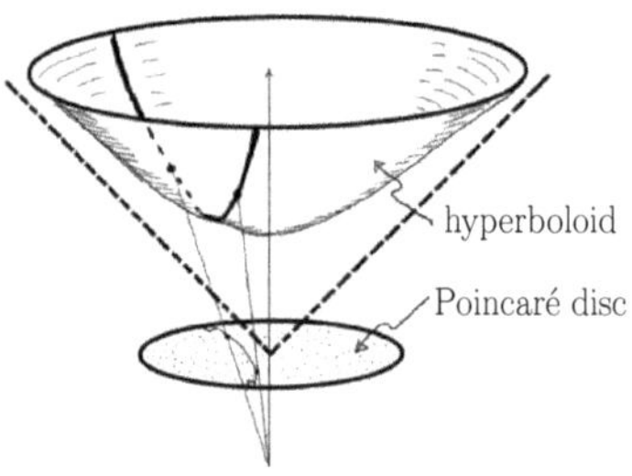

Fig. 16.5 Curves on a hyperboloid can be projected down on to the Poincaré disc where the geodesics meet the boundary at right angles.

> ⟲ **The second cause of redshift, due to light climbing out of a relativistic potential, is discussed further in Chapter 22 in the next part of the book.**

[18] It's useful to note that a time-independent measure of the **comoving distance** η to the galaxy from which the photon was emitted is

$$\eta = \int_{t_{\rm em}}^{t_{\rm ob}} \frac{{\rm d}t}{a(t)} = \int_0^{\chi} \frac{{\rm d}r}{(1 - kr^2)^{\frac{1}{2}}}.$$
(16.37)

This is a constant, unlike the geometrical distance $a(t)\eta$, which changes with time.

It is notable that $k = -1$ space does not have a finite proper volume: the integral for the volume diverges. This is because χ ranges from $0 \le \chi \le \infty$ in eqn 16.17 for $k = -1$. As a result of its infinite volume, $k = -1$ space is said to be **open**. (The $k = 0$ space is also open.)[17]

Although useful for understanding intervals in the spacelike hypersurfaces, proper distances are not directly measurable quantities. In the next section, we turn to the key observable in cosmology: the redshift.

16.3   Redshift and cosmic expansion

In the last section, we saw how the evolution of a model universe is determined by the expansion factor $a(t)$. The expansion of a universe is not something we can simply measure with a ruler. The problem, of course, is that a measuring device of a suitable size to measure the scales over which space expands will also be subject to the same expansion of space itself. In this section, we investigate the key observable in cosmology: a shift in frequency of light signals propagating over large distances that occurs due to changes in the geometry of spacetime. The shift is to lower frequencies for an expanding universe, and so is called a **redshift**. We have seen two reasons for redshift previously: the Doppler effect in special relativity, and the gravitational effect described in Chapter 6. Here we meet a third cause of redshift, the expansion of the universe itself. Such a redshift allows us to make measurements of the cosmological radius $a(t)$.

An observer sits at the origin of our comoving coordinates at coordinate $x^\mu = (t_{\rm ob}, 0, 0, 0)$, observing light signals sent from a distant galaxy at a (fixed) radial coordinate $r = \chi$, such that emission of photons occurs at coordinates $(t_{\rm em}, \chi, 0, 0)$. Light travels on null geodesics so that ${\rm d}s^2 = 0$ and, consequently, the crests of the light wave have an equation of motion given by

$$\mathrm{d}t^2 - a(t)^2 \frac{\mathrm{d}r^2}{1 - kr^2} = 0.$$
(16.36)

We examine the consequences of this in the following example.

Example 16.8

Rearranging and integrating along the null geodesic taken by the light ray, we have for a photon that[18]

$$\int_{t_{\rm em}}^{t_{\rm ob}} \frac{{\rm d}t}{a(t)} = \int_0^{\chi} \frac{{\rm d}r}{(1 - kr^2)^{\frac{1}{2}}}.$$
(16.38)

Now we imagine a second photon emitted at a time $t_{\rm em} + \delta t_{\rm em}$ and detected at $t_{\rm ob} + \delta t_{\rm ob}$. As above, we have an equation that describes the null geodesic

$$\int_{t_{\rm em}+\delta t_{\rm em}}^{t_{\rm ob}+\delta t_{\rm ob}} \frac{{\rm d}t}{a(t)} = \int_0^{\chi} \frac{{\rm d}r}{(1 - kr^2)^{\frac{1}{2}}}.$$
(16.39)

The integral over the radial coordinate on the right-hand side of these two equations is identical (it should be: we're just describing two identical photon emission and detection events). Equating the left-hand side of the equations we have

$$\int_{t_{\rm em}}^{t_{\rm ob}} \frac{{\rm d}t}{a(t)} = \int_{t_{\rm em}+\delta t_{\rm em}}^{t_{\rm ob}+\delta t_{\rm ob}} \frac{{\rm d}t}{a(t)}.$$
(16.40)

Next, we use a trick: splitting the integrals up and summing. The idea is illustrated in Fig. 16.6. We write the left-hand side of eqn 16.40 as

$$\int_{t_{\rm em}}^{t_{\rm ob}} \frac{{\rm d}t}{a(t)} = \int_{t_{\rm em}}^{t_{\rm em}+\delta t_{\rm em}} \frac{{\rm d}t}{a(t)} + \int_{t_{\rm em}+\delta t_{\rm em}}^{t_{\rm ob}} \frac{{\rm d}t}{a(t)}. \tag{16.41}$$

Likewise, the integral on the right-hand side of eqn 16.40 can be written as

$$\int_{t_{\rm em}+\delta t_{\rm em}}^{t_{\rm ob}+\delta t_{\rm ob}} \frac{{\rm d}t}{a(t)} = \int_{t_{\rm em}+\delta t_{\rm em}}^{t_{\rm ob}} \frac{{\rm d}t}{a(t)} + \int_{t_{\rm ob}}^{t_{\rm ob}+\delta t_{\rm ob}} \frac{{\rm d}t}{a(t)}. \tag{16.42}$$

This sleight of hand allows us to cancel out the identical parts [i.e. the limits represent the long period between the emission of the second photon and the observation of the first one, which is common to both sides of the equation, as shown in Fig. 16.6]. This leaves

$$\int_{t_{\rm em}}^{t_{\rm em}+\delta t_{\rm em}} \frac{{\rm d}t}{a(t)} = \int_{t_{\rm ob}}^{t_{\rm ob}+\delta t_{\rm ob}} \frac{{\rm d}t}{a(t)}, \tag{16.43}$$

which, in words, tells us that the comoving distance covered by the first photon before the emission of the second, is the same as the comoving distance covered by the second photon after the detection of the first. The time between emission (or detection) of the two light photons is assumed very small on cosmological scales, and so each of the integrands is very nearly constant over the limits. We can therefore replace the integrals with the approximate expression

$$\frac{\delta t_{\rm em}}{a(t_{\rm em})} = \frac{\delta t_{\rm ob}}{a(t_{\rm ob})}. \tag{16.44}$$

This provides us with a relationship between intervals at the observer and emitter of the signal and the expansion factor evaluated at their respective coordinate times.

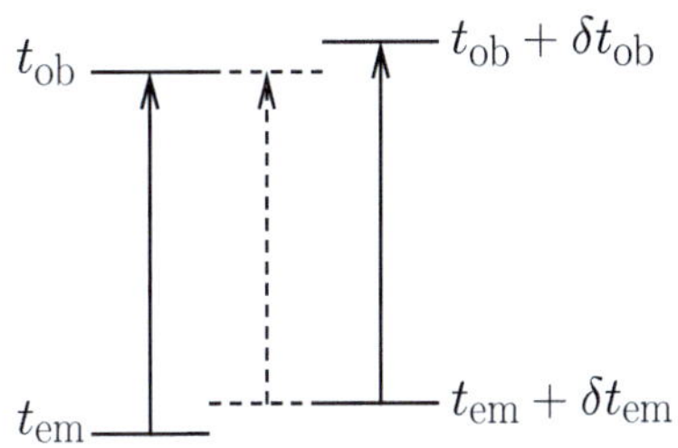

Fig. 16.6 The derivation of the cosmological redshift relies on equating the comoving distance $\int {\rm d}t/a(t)$ for a ray emitted at $t_{\rm em}$ and observed at $t_{\rm ob}$ to one emitted at $t_{\rm em} + \delta t_{\rm em}$ and observed at $t_{\rm ob} + \delta t_{\rm em}$. The part between $t_{\rm em}+\delta t_{\rm em}$ and $t_{\rm ob}$ (dashed) is comment to both sides of the equation, and therefore cancels.

Rearranging the previous expression we have

$$\frac{\delta t_{\rm em}}{\delta t_{\rm ob}} = \frac{a(t_{\rm em})}{a(t_{\rm ob})}. \tag{16.45}$$

If δt represents a proper period for oscillations of the light (i.e. the crests of a light wave, separated in emission time by $\delta t_{\rm em}$ and in observation time by $\delta t_{\rm ob}$), we can convert this ratio to a ratio of frequencies and write

$$\frac{\omega_{\rm ob}}{\omega_{\rm em}} = \frac{a(t_{\rm em})}{a(t_{\rm ob})}. \tag{16.46}$$

Notice that the shift depends only on the radius function $a(t)$ evaluated at the time of emission and of observation.

Example 16.9

For an expanding universe, we have $a(t_{\rm ob}) > a(t_{\rm em})$ and so $\omega_{\rm ob} < \omega_{\rm em}$. Each point is receding away from every other point and light signals are redshifted. Conversely, for a contracting universe, light signals will be blueshifted. Referring back to our closed universe in Fig. 16.4(a), if $a(t)$ increases with time, the light received via the route shown between the stars at B and C and the observer at A is redshifted. If $a(t)$ decreases with time, these light rays will be blueshifted.

Very often, we will take the scale factor at the time of observation to be unity, so this simplifies to

$$1 + z = \frac{1}{a(t_{\rm em})}.$$

We have found that the frequency of the photon is inversely proportional to the scale factor, meaning that the wavelength of the photon is proportional to the scale factor. Another way of thinking of the cosmological redshift is that it is as if the wavelength of a photon is frozen into the Universe and stretches with it.[19] Conventionally, redshifts are written in terms of a **redshift parameter** z defined in terms of the wavelength at emission and observation as

$$z = \frac{\lambda_{\rm ob} - \lambda_{\rm em}}{\lambda_{\rm em}}, \tag{16.47}$$

from which we find

$$1 + z = \frac{\lambda_{\rm ob}}{\lambda_{\rm em}} = \frac{a(t_{\rm ob})}{a(t_{\rm em})}. \tag{16.48}$$

We conclude that a measurement of z gives us access to information about the expansion factor $a(t)$. Note that when discussing expanding model universes, we often define the zero of time such that light emitted then has infinite redshift z [i.e. $a(t = 0) = 0$].

16.4   The initial singularity

Robertson–Walker spacetimes can be used to describe expanding universes that can be spherical, flat or hyperbolic. In the next chapter, we look at the consequences of filling these spacetimes with homogeneous, isotropic distributions of matter. This results in universes whose densities decrease as time runs forward. The natural question arises about what happens if we run time backwards, with the density of the universe increasing. We shall find that for most of our models, the factor $a(t)$ is zero at some time t_0, a finite length of time ago. The density of the universe increases without bound as t_0 is approached, and this results in a **singularity** in the physical fields at time t_0, which can reasonably be assigned to be the time of the beginning of the universe. This is, of course, the Big Bang. The singularity is even worse than it would appear in a Newtonian model of this physics, where the density of the universe becomes infinite. Since the metric field is the key field in general relativity, a singularity in the density of matter implies a singularity in the very fabric of space and time, which amounts to an infinite curvature. This is something with which our classical field theory of gravitation cannot cope, and so we are forced to cut this point out of our description of spacetime. No known physical laws can apply at this **initial singularity**.

The initial singularity in spacetime is one of the most interesting things about the Robertson–Walker spacetimes. We shall not dwell on it in the next chapters, where we solve the Einstein equation for several distributions of matter filling these spaces. We return to the singularities in Chapter 19, where we shall find a way for treating the singularity as an effective edge of spacetime.

Chapter summary

- There are three Robertson–Walker geometries described by the metric

$$ds^2 = -dt^2 + a(t)^2 \left[\frac{dr^2}{1 - kr^2} + r^2 \left(d\theta^2 + \sin^2\theta d\phi^2 \right) \right]. \quad (16.49)$$

  They are all isotropic and homogeneous and are classified as spherical ($k = 1$), flat ($k = 0$) and hyperbolic ($k = -1$). They evolve as a function of time.

- The evolution of the universes, given in terms of the effective radius $a(t)$, which is related to the measured redshift z via

$$1 + z = \frac{a(t_{\mathrm{ob}})}{a(t_{\mathrm{em}})}. \quad (16.50)$$

- The Robertson–Walker universes have, at their start, the potential for an initial Big-Bang singularity.

Exercises

(16.1) Verify that $^{(3)}R_{ijkl} = K(\gamma_{ik}\gamma_{jl} - \gamma_{il}\gamma_{jk})$ has the expected symmetries of the Riemann tensor.

(16.2) Verify eqn 16.10.

(16.3) Compute the connection coefficients for the Robertson–Walker spacetimes and verify they are those given in this chapter.

(16.4) Starting with $d\sigma^2 = \gamma_{ij}dx^i dx^j$ and transforming to spherical coordinates results in the general three-dimensional metric

$$d\sigma^2 = e^{2\Lambda(r)}dr^2 + r^2(d\theta^2 + \sin^2\theta d\phi^2), \quad (16.51)$$

where $e^{2\Lambda(r)}$ is some function of r that we shall determine.

(a) Identify vielbein that transform from the coordinate frame to the orthonormal frame.

(b) Comparing with the results from Exercise 11.1 in Chapter 11, show that the components of the Riemann tensor in the orthonormal frame can be written as

$$R^{\hat\theta\hat\phi}{}_{\hat\theta\hat\phi} = F,$$
$$R^{\hat\phi\hat r}{}_{\hat r\hat\phi} = -\bar{F}, \quad (16.52)$$
$$R^{\hat r\hat\theta}{}_{\hat r\hat\theta} = \bar{F},$$

where

$$F = \frac{1}{r^2}\left(1 - e^{-2\Lambda}\right),$$
$$\bar{F} = \frac{1}{r}e^{-2\Lambda}\Lambda'. \quad (16.53)$$

(c) What are the components of the Ricci tensor in the orthonormal frame? Show that in the coordinate frame they are

$$R_{rr} = \frac{\Lambda'}{r}, \quad R_{\theta\theta} = \frac{R_{\phi\phi}}{\sin^2\theta} = (1 - e^{-2\Lambda}) + r\Lambda' e^{-2\Lambda}. \quad (16.54)$$

(d) Show that for a space of constant curvature, where $R_{ij} = 2Kg_{ij}$, we have

$$\Lambda' = 2Kre^{2\Lambda}, \quad (1 - e^{-2\Lambda}) + r\Lambda' e^{-\Lambda} = 2Kr^2. \quad (16.55)$$

Hint: It's easiest to work in the orthonormal frame.

(e) By solving the equations given in (d), show that we must have a line element

$$d\sigma^2 = \frac{dr^2}{1 - Kr^2} + r^2\left(d\theta^2 + \sin^2\theta d\phi^2\right). \quad (16.56)$$

(16.5) Using the arguments from this chapter, we can say that if the curvature of a spacetime is isotropic, the components of the Riemann tensor can always be written in the form

$$R_{\mu\nu\alpha\beta} = C(g_{\mu\alpha}g_{\nu\beta} - g_{\mu\beta}g_{\nu\alpha}). \qquad (16.57)$$

Use the Bianchi identity (from Chapter 13), along with the compatibility condition that $g_{\mu\nu;\alpha} = 0$ (from Chapter 7), to show that C must be a constant.

(16.6) In the last chapter, we described $(3+1)$-dimensional spaces of constant curvature as having $R_{\mu\nu} = \frac{1}{4}Rg_{\mu\nu}$. Show this is the case by contracting the expression in the previous problem. Generalize the result for d-dimensional spacetime.

(16.7) What is the three-dimensional Ricci scalar $^{(3)}R$ for the spatial part of the Robertson–Walker spacetime, written in terms of $a(t)$? How can this be used to write $^{(3)}R_{ij}$ in terms of γ_{ij}?

(16.8) A galaxy with redshift z and is observed to be at a proper distance $\ell_{\rm ob}$ away from us. What was the proper distance $\ell_{\rm em}$ to the galaxy at the time when its light was emitted?

(16.9) Show that if we differentiate $1 + z = a(t_{\rm ob})/a(t_{\rm em})$ with respect to the time at the observer $t_{\rm ob}$, we find that the rate of expansion of the Universe at the time of light emission is given by

$$H(t_{\rm em}) = H(t_{\rm ob})(1 + z) - \frac{\partial z}{\partial t_{\rm ob}}. \qquad (16.58)$$

The Friedmann equations

17

Now hast thou but one bare hour to live,
And then thou must be damn'd perpetually!
Stand still, you ever-moving spheres of heaven,
That time may cease, and midnight never come.
Christopher Marlowe (1564–1593) *Doctor Faustus*

Margaret Fuller: *I accept the Universe*
Thomas Carlyle: *Gad! she'd better!*

Our mission remains to determine the origin and fate of model universes using the Einstein equation. With the geometry of the Robertson–Walker spaces supplied by the metric fields of the last chapter, we have access to the left-hand side of Einstein's equation. Now we need to fill in the right-hand side of the equation by specifying the matter content of the universes. This is our task in this chapter. Filling model universes with different idealized distributions of matter will lead us to two (equivalent) equations of motion for the expansion factor $a(t)$, known as the **Friedmann equations**. These allow us access to a wealth of different models with which to amuse ourselves in the next chapter.

Example 17.1

Before ploughing into the general-relativistic version, here is a simple argument that gives some physical insight, albeit working things out in the Newtonian limit. The total energy $U = T + V$ of a particle at radius r from the centre of a expanding homogeneous medium of density ρ is the sum of kinetic energy $T = \frac{1}{2}m\dot{r}^2$ and potential energy $U = -GMm/r$. Writing $\vec{r} = a(t)\vec{x}$, where $\vec{x}$ is a comoving coordinate, and $M = \frac{4}{3}\pi r^3 \rho$, we have

$$U = \frac{1}{2}m\dot{a}^2 x^2 - \frac{4\pi G}{3}\rho a^2 x^2 m, \tag{17.1}$$

which rearranges to

$$\left(\frac{\dot{a}}{a}\right)^2 + \frac{k}{a^2} = \frac{8\pi G\rho}{3}, \tag{17.2}$$

where $k = -\frac{2U}{mx^2}$ is a time-independent constant. Remarkably, eqn 17.2 is essentially the first Friedmann equation (that we shall shortly derive), and so one can already appreciate that it arises from energy considerations. But after this warm-up exercise, it's time to derive the Friedmann equations using the general-relativistic techniques that we've built up.

17.1   Enter energy-momentum

[1]Alexander Friedmann (1888–1925). Friedmann died tragically young from typhoid. Legend has it that this was brought on by eating an unwashed pear that he had bought at a station on the way back from his honeymoon. The conventional spelling of Friedmann's surname is apparently due to a mistake of Einstein's. In a note Einstein published in 1923, Einstein referred to him as 'Friedmann', although he had signed himself 'Friedman' in his own paper of 1922. According to Ari Belenkiy: *'It seemed Friedman followed the "advice" and submitted his 1924 paper with two n's.'* (See A. Belenkiy, arXiv:1302.1498.)

[2]We saw the same components for the energy-momentum tensor in the orthonormal frame in Chapter 10 for the $k = 0$ case. See the Exercise 17.1 for verification that they're appropriate for $k = \pm 1$.

[3]The Robertson–Walker line element is

$$ds^2 = -\,dt^2 + a(t)^2 \left[\frac{dr^2}{1-kr^2} \right.$$
$$\left. +r^2 \left(d\theta^2 + \sin^2 \theta d\phi^2 \right) \right].$$

We have metric components

$$g_{tt} = -1, \qquad g_{rr} = \frac{a^2}{1-kr^2},$$
$$g_{\theta\theta} = a^2 r^2, \qquad g_{\phi\phi} = a^2 r^2 \sin^2 \theta.$$

The vielbein components are

$$(e_t)^{\hat{t}} = 1, \qquad (e_r)^{\hat{r}} = \frac{a}{\left(1-kr^2\right)^{\frac{1}{2}}},$$
$$(e_\theta)^{\hat{\theta}} = ar, \qquad (e_\phi)^{\hat{\phi}} = ar \sin \theta.$$

The Ricci tensor has non-zero components in the orthonormal frame of

$$R_{\hat{t}\hat{t}} = -\,3\frac{\ddot{a}}{a},$$

$$R_{\hat{i}\hat{i}} = \frac{\ddot{a}}{a} + 2\left(\frac{\dot{a}}{a}\right)^2 + 2\frac{k}{a}.$$

The Ricci scalar is

$$R = 6\left(\frac{\ddot{a}}{a} + \frac{\dot{a}^2}{a^2} + \frac{k}{a^2}\right).$$

The vielbein suggests that the spacelike components of the energy-momentum tensor can be thought of as comprising time-dependent effective pressures $p(t)$ proportional to $1/a(t)^2$. This makes sense as the expansion of space caused by $a(t)$ should lead to a reduction in pressure (i.e. the force per unit area) in the fluid.

Following the pioneering work of Alexander Friedmann,[1] we shall fill the Robertson–Walker spaces with both vacuum energy (with energy density Λ) and a perfect fluid (i.e. a fluid without any dissipation). Written in spherical polars in the orthonormal frame, the energy-momentum tensor for a perfect fluid is[2]

$$T_{\hat{t}\hat{t}} = \rho, \quad T_{\hat{r}\hat{r}} = T_{\hat{\theta}\hat{\theta}} = T_{\hat{\phi}\hat{\phi}} = p, \tag{17.3}$$

where ρ is the mass-energy density and p is the pressure. Armed with this, and the Ricci tensor for the Robertson–Walker spaces from the previous chapter, we can write down the components of the Einstein equation that governs the kinematics of these spaces. This results in two famous equations of motion known as the **Friedmann equations**.

Example 17.2

We work in the orthonormal frame with basis vectors $e_{\hat{t}}, e_{\hat{r}}, e_{\hat{\theta}}, e_{\hat{\phi}}$. Equating the $\hat{0}\hat{0}$th components of the Einstein equation[3] we have

$$R_{\hat{0}\hat{0}} - \frac{g_{\hat{0}\hat{0}}}{2} R = 8\pi T_{\hat{0}\hat{0}} - \Lambda g_{\hat{0}\hat{0}}$$

$$-3\frac{\ddot{a}}{a} + 3\left[\frac{\ddot{a}}{a} + \left(\frac{\dot{a}}{a}\right)^2 + \frac{k}{a^2}\right] = 8\pi\rho + \Lambda$$

$$\left(\frac{\dot{a}}{a}\right)^2 + \frac{k}{a^2} = \frac{8\pi\rho}{3} + \frac{\Lambda}{3}. \tag{17.4}$$

Next, we equate the $\hat{i}\hat{i}$th components and find

$$\left[\frac{\ddot{a}}{a} + 2\left(\frac{\dot{a}}{a}\right)^2 + 2\frac{k}{a}\right] - 3\left[\frac{\ddot{a}}{a} + \left(\frac{\dot{a}}{a}\right)^2 + \frac{k}{a^2}\right] = 8\pi p - \Lambda$$

$$2\frac{\ddot{a}}{a} + \left(\frac{\dot{a}}{a}\right)^2 + \frac{k}{a^2} = -\,8\pi p + \Lambda. \tag{17.5}$$

We are left with two simple equations for the evolution of the Robertson–Walker spaces.

The two equations of motion from the last example for $a(t)$ in terms of energy density, pressure, and constant curvature are repeated below, with their usual names.

Friedmann equation 1: the initial value equation

$$\left(\frac{\dot{a}}{a}\right)^2 + \frac{k}{a^2} = \frac{8\pi}{3}\rho + \frac{\Lambda}{3}. \tag{17.6}$$

Friedmann equation 2: the dynamic equation

$$2\frac{\ddot{a}}{a} + \left(\frac{\dot{a}}{a}\right)^2 + \frac{k}{a^2} = -8\pi p + \Lambda. \tag{17.7}$$

So far the relationship between ρ and p has not been given. These quantities are linked by an **equation of state** which is determined by the thermodynamics of the matter comprising the perfect fluid that fills the model universe. Since the scale of space itself depends on the radius-like expansion factor $a(t)$, this will also come into play in determining the volume in which our matter is confined (and therefore also its density). This is our next topic.

17.2 Enter thermodynamics

In order to make further progress, we need to consider how the effective radius of the universe a is linked to the thermodynamic variables ρ and p. General relativity is not the only place where geometry and mass-energy are linked: they are also tied together in the thermodynamic laws of physics. In the next example, we consider the first law of thermodynamics.

> ↶ **Thermodynamics is explored in more detail in Chapter 39.**

Example 17.3

The first law of thermodynamics says

$$dU = T dS - p dV, \tag{17.8}$$

linking the internal energy U, temperature T and entropy S in a volume V. We shall often assume an adiabatic evolution of the universe,[4] so we can set $dS = 0$. Taking the volume $V = ba(t)^3$, where b is a dimensionless constant, we then have $U = \rho V = b\rho a(t)^3$. Taking a time derivative of the first law of thermodynamics, we find (cancelling factors of b)

$$\frac{dU}{dt} = -p\frac{dV}{dt}$$

$$\frac{d}{dt}\left(\rho a^3\right) = -p\frac{d}{dt}a^3$$

$$\frac{d\rho}{dt}a^3 + 3\rho a^2\frac{da}{dt} = -3pa^2\frac{da}{dt}, \tag{17.9}$$

or, on rearranging,

$$\frac{1}{\rho + p}\frac{d\rho}{dt} = -\frac{3}{a}\frac{da}{dt}. \tag{17.10}$$

[4]This amounts to an assumption that the elements of the cosmic fluid remain in thermal equilibrium and there are no phase transitions or shockwaves. In the case of a shockwave, for example, the kinetic energy of the fluid elements can be converted to heat, leading to an increase in entropy.

Using the result of the previous example, we have an alternative statement of the first law of thermodynamics, in a form directly applicable to the Friedmann cosmologies.

The first law of thermodynamics can be written as

$$\frac{d\rho}{\rho + p} = -3\frac{da}{a}. \tag{17.11}$$

Armed with this, we can show that the two Friedmann equations actually represent the same information.

Example 17.4

Start with a slight rearrangement of the first Friedmann equation

$$\dot{a}^2 + k = \left(\frac{\Lambda}{3} + \frac{8\pi\rho}{3}\right)a^2, \tag{17.12}$$

and differentiate to find

$$2\dot{a}\ddot{a} = 2a\dot{a}\left(\frac{\Lambda}{3} + \frac{8\pi\rho}{3}\right) + a^2\frac{8\pi\dot{\rho}}{3}. \tag{17.13}$$

Combine this with the first law of thermodynamics in the form

$$\dot{\rho} = -3(\rho + p)\frac{\dot{a}}{a}, \tag{17.14}$$

to give

$$2\frac{\ddot{a}}{a} = \frac{2\Lambda}{3} - \frac{8\pi\rho}{3} - 8\pi p. \tag{17.15}$$

Finally, substitute for $8\pi\rho/3$ using the first Friedmann equation and we find the second Friedmann equation

$$2\frac{\ddot{a}}{a} + \left(\frac{\dot{a}}{a}\right)^2 + \frac{k}{a^2} = -8\pi p + \Lambda. \tag{17.16}$$

Conclusion: The two Friedmann equations are actually telling us the same thing.

The first law of thermodynamics links ρ, p and a. In order to specify the equation of state, we must decide what sort of matter we wish to model. Two types of matter are discussed in the next section.

17.3 Dust and radiation

All matter contributes to the energy-momentum tensor and hence affects the geometry of the universe. On the length scales over which we are working, the matter is continuous and so is well described as a classical fluid giving rise to an energy-momentum field. We now divide the matter in the fluid that fills the universe into two sorts:

(1) **matter with mass**: This is also called **dust**. It is simply massive, continuous matter with $p = 0$. The mass-energy density of dust is written as ρ_d.

(2) **massless matter**: This is also called **radiation**, and is assumed to be a gas of photons with uniform energy density ρ_r and non-zero pressure.

The total energy density of the universe is a sum $\rho = \rho_\mathrm{d} + \rho_\mathrm{r}$.

Considering dust first, setting $p = 0$, the first law of thermodynamics (eqn 17.11) implies the equation of state for dust is

$$\rho_\mathrm{d}(t) \propto a(t)^{-3}, \tag{17.17}$$

or, if we make measurements relative to some state at a time $t = t_0$,[5] we have

$$\frac{\rho_\mathrm{d}(t)}{\rho_\mathrm{d}^{(0)}} = \left[\frac{a(t_0)}{a(t)}\right]^3, \tag{17.18}$$

[5] This is often taken to be the time right now, when we make our measurements. The time t at which we evaluate our expressions can be greater or less than t_0. The Hubble parameter evaluated at t_0 is often called the Hubble constant $H_0 = H(t = t_0)$.

where we use the notation $\rho_{\mathrm{d}}^{(0)} = \rho_{\mathrm{d}}(t_0)$. We deduce[6] $\rho_{\mathrm{d}} \propto a(t)^{-3}$.

Now for radiation.[7] The equation of state for radiation is a fundamental result of thermodynamics and is given by

$$p = \frac{\rho}{3}. \tag{17.19}$$

Eqn 17.11 then gives us

$$\frac{\mathrm{d}\rho}{\rho} = -4\frac{\mathrm{d}a}{a}, \tag{17.20}$$

or, again making measurements relative to conditions at t_0,

$$\frac{\rho_{\mathrm{r}}(t)}{\rho_{\mathrm{r}}^{(0)}} = \left[\frac{a(t_0)}{a(t)}\right]^4, \tag{17.21}$$

where $\rho_{\mathrm{r}}^{(0)} = \rho_{\mathrm{r}}(t_0)$. We deduce[8] $\rho_{\mathrm{r}} \propto a(t)^{-4}$. We conclude that in both cases, dust and radiation, an increasing $a(t)$ leads to a power-law drop in the energy density of the universe. In other words, the universe becomes emptier as a function of time, but the energy density due to radiation drops off fastest.

Using the thermodynamic information, we can then rewrite the Friedmann equations for a universe filled with a mixture of dust, radiation, and vacuum energy. They become

$$\left[\frac{\dot{a}(t)}{a(t)}\right]^2 + \frac{k}{a(t)^2} = \frac{8\pi}{3}\left\{\rho_{\mathrm{d}}^{(0)}\left[\frac{a(t_0)}{a(t)}\right]^3 + \rho_{\mathrm{r}}^{(0)}\left[\frac{a(t_0)}{a(t)}\right]^4\right\} + \frac{\Lambda}{3}, \tag{17.22}$$

$$\frac{\ddot{a}}{a} = -\frac{4\pi}{3}\left\{\rho_{\mathrm{d}}^{(0)}\left[\frac{a(t_0)}{a(t)}\right]^3 + \rho_{\mathrm{r}}^{(0)}\left[\frac{a(t_0)}{a(t)}\right]^4\right\} + \frac{\Lambda}{3}. \tag{17.23}$$

These equations tell us that if we have values for the initial densities $\rho^{(0)}$ at some time t_0 when we know the value of the radius $a = a(t_0)$, we can find the behaviour of the effective radius function $a(t)$ at all times.

With the Friedmann equations in hand we can continue our survey of possible universes. The Friedmann equations now allow us to fill each of the Robertson–Walker spaces with different sorts of matter (that is, radiation, dust, and vacuum energy) and examine the consequences.

Example 17.5

It is sometimes useful to express the energy density as a fraction of a **critical density** ρ_{c}, defined by[9]

$$\rho_{\mathrm{c}} = \frac{3}{8\pi}\left(\frac{\dot{a}}{a}\right)^2. \tag{17.25}$$

Substituting for $\dot{a}/a$ in the first Friedmann equation, we find

$$\frac{8\pi}{3}\rho_{\mathrm{c}} + \frac{k}{a^2} = \frac{8\pi}{3}(\rho_{\mathrm{d}} + \rho_{\mathrm{r}}) + \frac{\Lambda}{3}. \tag{17.26}$$

Next, we define a density of vacuum energy $\rho_\Lambda = \Lambda/8\pi$ and obtain

$$1 + \frac{k}{a^2 H^2} = \frac{\rho_d}{\rho_{\mathrm{c}}} + \frac{\rho_r}{\rho_{\mathrm{c}}} + \frac{\rho_\Lambda}{\rho_{\mathrm{c}}}. \tag{17.27}$$

[6]We can rationalize this result as follows: in thermal equilibrium, particle number is conserved, and so the density of matter must be inversely proportional to the volume which is proportional to $a(t)^3$. The energy of the matter is just its time-independent mass. Hence, the energy density $\rho_{\mathrm{d}} \propto a(t)^{-3}$.

[7]See also Exercise 17.2.

[8]We can rationalize this result as follows: in thermal equilibrium, particle number is conserved, and so the density of photons must be inversely proportional to the volume which is proportional to $a(t)^3$. However, the energy of each photon is inversely proportional to $a(t)$ because of the effect of redshift. Hence, the energy density $\rho_{\mathrm{r}} \propto a(t)^{-4}$.

[9]The name is justified below. Note that, equivalently, we could write

$$\rho_{\mathrm{c}} = \frac{3}{8\pi}H(t)^2. \tag{17.24}$$

One way forward is to substitute for a^2 from eqn 17.25, giving

$$1 + \frac{k}{\dot{a}^2} = \frac{\rho_d}{\rho_c} + \frac{\rho_r}{\rho_c} + \frac{\rho_\Lambda}{\rho_c}. \tag{17.28}$$

We conclude that the first Friedmann equation can be rewritten to read

$$\begin{aligned}
\frac{k}{\dot{a}^2} &= \frac{\rho_d}{\rho_c} + \frac{\rho_r}{\rho_c} + \frac{\rho_\Lambda}{\rho_c} - 1 \\
&= \Omega_d + \Omega_r + \Omega_\Lambda - 1,
\end{aligned} \tag{17.29}$$

where in the last step we have written the ratio $\Omega_j = \rho_j/\rho_c$. The usefulness of this equation is that it provides a straightforward means of telling us which of the three possible geometries ($k = 0, \pm 1$) must result from a universe containing a particular amount of mass-energy. That is, for a total density ratio $\Omega = \Omega_d + \Omega_r + \Omega_\Lambda$ we have $k/\dot{a}^2 = \Omega - 1$, from which (assuming $\dot{a} > 0$) we can read off the following:
- If $\Omega < 1$ then $k = -1$ and the space is hyperbolic.
- If $\Omega = 1$ then $k = 0$ and the space is flat.
- If $\Omega > 1$ then $k = 1$ and the space is spherical.

We can now see why ρ_c is called a critical density: it's only if the total density $\rho = \rho_c$ that we have the result that $k = 0$.

Another way of expressing things is to relate everything to measurements taken at $t = t_0$, as we did in eqn 17.22. First we write $\rho_c^{(0)} = 8\pi H_0/3$ and $\Omega_j^{(0)} = \rho_j^{(0)}/\rho_c^{(0)}$. Then define $\Omega_k^{(0)} = -k/a(t_0)^2 H_0^2$, so that we have, from eqn 17.27, that

$$\Omega_d^{(0)} + \Omega_r^{(0)} + \Omega_\Lambda^{(0)} + \Omega_k^{(0)} = 1. \tag{17.30}$$

Returning to eqn 17.22 we then write

$$\left[\frac{\dot{a}(t)}{a(t)}\right]^2 + \frac{k}{a(t)^2} = \left\{ \Omega_d^{(0)} \left[\frac{a(t_0)}{a(t)}\right]^3 + \Omega_r^{(0)} \left[\frac{a(t_0)}{a(t)}\right]^4 + \Omega_\Lambda^{(0)} \right\}. \tag{17.31}$$

Simplifying by writing $x = a(t)/a(t_0)$ we have

$$\frac{1}{x^2} \left(\frac{dx}{dt}\right)^2 = H_0^2 \left(\frac{\Omega_k^{(0)}}{x^2} + \frac{\Omega_d^{(0)}}{x^3} + \frac{\Omega_r^{(0)}}{x^4} + \Omega_\Lambda^{(0)} \right). \tag{17.32}$$

The usefulness of this expression is that it can be related to light reaching us with some value of redshift z. If light was emitted at a time t and observed at the current time, t_0, then we have, by eqn 16.48, that $x = a(t)/a(t_0) = (1+z)^{-1}$. This means that if we define the zero of time as corresponding to infinite redshift, then the time at which light was emitted that reaches us with redshift z is given by

$$t(z) = \frac{1}{H_0} \int_0^{(z+1)^{-1}} \frac{dx}{x \left(\Omega_k^{(0)} x^{-2} + \Omega_d^{(0)} x^{-3} + \Omega_r^{(0)} x^{-4} + \Omega_\Lambda^{(0)} \right)^{\frac{1}{2}}}. \tag{17.33}$$

Notably, if we set $z = 0$, this equation outputs the present age of the Universe.

We now have equations that link the matter content of the universe and its geometry. This is a central goal of cosmology. We are, therefore, now ready for the payoff in the following chapter: a description of the possible Robertson–Walker universes, their evolution, origin, and fate.

Chapter summary

- Filling the Robertson–Walker universes with a perfect fluid gives us the Friedmann equations for the evolution of $a(t)$ with time. These can be given in terms of the dust, radiation and vacuum-energy content of the universe.

- The Friedmann equations are given by

$$\left(\frac{\dot{a}}{a}\right)^2 + \frac{k}{a^2} = \frac{8\pi}{3}\rho + \frac{\Lambda}{3}, \tag{17.34}$$

$$2\frac{\ddot{a}}{a} + \left(\frac{\dot{a}}{a}\right)^2 + \frac{k}{a^2} = -8\pi p + \Lambda. \tag{17.35}$$

- The energy density of matter $\rho_{\mathrm{d}} \propto a(t)^{-3}$ and the energy density of radiation $\rho_{\mathrm{r}} \propto a(t)^{-4}$.

Exercises

(17.1) We had in Chapter 10 that $T_{\hat{\chi}\hat{\chi}} = p$.

(a) Using the transformation from the previous chapter that said $\mathrm{d}\chi^2 = \mathrm{d}r^2/(1 - kr^2)$, compute $\Lambda^{\chi}{}_r = \partial\chi/\partial r$.

(b) Verify that

$$(e_\chi)^{\hat{\chi}}(e_{\hat{r}})^r = (1 - kr^2)^{\frac{1}{2}}.$$

(c) Use the previous results to show that $T_{\hat{r}\hat{r}} = T_{\hat{\chi}\hat{\chi}} = p$.

(17.2) We can check the thermodynamic results for photons using statistical mechanics. The single particle partition function for a gas of ultra-relativistic indistinguishable particles (i.e. photons) in a 3-volume V is

$$Z = \frac{V}{\pi^2}\left(\frac{k_{\mathrm{B}}T}{\hbar}\right)^3, \tag{17.36}$$

where T is the temperature.

(a) Compute the internal energy of the gas.

(b) Compute its pressure.

(c) Verify that $p = \rho/3$ as claimed in the chapter.

(17.3) Consider a flat universe filled only with matter with an equation of state $\rho \propto a(t)^{-2}$. How does $a(t)$ evolve?

(17.4) For a flat, dust-dominated model universe, calculate the age of the universe when an observed photon was emitted from a galaxy with redshift z.

18

Universes of the past and future

We will now discuss in a little more detail the struggle for existence

Charles Darwin (1809–1882)

After determining the geometry of the Robertson–Walker spaces and combining them with matter to derive the Friedmann equations, we now continue our survey of possible universes. Each of the models we discuss in this chapter makes use of the metric field described by the Robertson–Walker line element

$$\mathrm{d}s^2 = -\mathrm{d}t^2 + a(t)^2 \left[\frac{\mathrm{d}r^2}{1 - kr^2} + r^2 \left(\mathrm{d}\theta^2 + \sin^2 \theta \mathrm{d}\phi^2 \right) \right]. \tag{18.1}$$

Friedmann equation 1:

$$\left(\frac{\dot{a}}{a} \right)^2 + \frac{k}{a^2} = \frac{8\pi}{3} \rho + \frac{\Lambda}{3}.$$

Friedmann equation 2:

$$2\frac{\ddot{a}}{a} + \left(\frac{\dot{a}}{a} \right)^2 + \frac{k}{a^2} = -8\pi p + \Lambda.$$

Recall that we also define a Hubble parameter

$$H(t) = \frac{\dot{a}(t)}{a(t)}.$$

This is a useful quantity as it tells us the rate of change of proper distance $l(t)$ via $\dot{l}(t) = H(t)l(t)$.

We choose to fill the universes with perfect fluid and vacuum energy, so can use the Friedmann equations to compute the behaviour of the expansion factor $a(t)$, which can be thought of as the effective radius of the universe. The universes can be broadly classified by the value of the curvature constant k that they take. [Recall that the two universes (numbered 0 and 1) that we have already examined had $k = 0$ and so were spatially flat.] We start this chapter by exploring some more spatially flat universes before going on to look at some $k = 1$ and $k = -1$ cases. Throughout we assume that we know the value of the expansion factor $a(t = t_0)$ at some time t_0.

18.1 Spatially flat universes

We start by filling spatially flat ($k = 0$) universes with different sorts of matter. Recall that **Universe 0** is empty and flat. As a result, it is identical to static, Minkowski spacetime. **Universe 1** (the de Sitter model) is flat by construction, and contains only vacuum energy Λ. The result is that this latter universe expands with $a(t) = a(t_0)e^{H(t-t_0)}$, where the Hubble parameter $H \equiv \dot{a}/a = \sqrt{\Lambda/3}$ is a constant. This universe is not badly behaved at $t = 0$, where we expect a non-zero radius $a(t = 0)$ and finite mass-energy density $\rho(t = 0)$. Universe 1 does not, therefore, have any singularity at $t = 0$ and so has no Big Bang.

Now let's consider models without any vacuum energy, but where we allow cosmological fluid with a time-dependent energy density $\rho(t)$ to fill a flat universe. We then have (using the first Friedmann equation, with $k = \Lambda = 0$) that

$$\left[\frac{\dot{a}(t)}{a(t)}\right]^2 = \frac{8\pi\rho(t)}{3} = H(t)^2, \tag{18.2}$$

where now the Hubble parameter is time dependent and is given by $H(t) = [8\pi\rho(t)/3]^{\frac{1}{2}}$. We saw in the last chapter that it is thermodynamic considerations that cause the energy density ρ to be time dependent and that $\rho(t)$ is itself a function of $a(t)$. This is to say that, as the universe evolves, the energy density ρ changes as the expansion factor $a(t)$ changes. Let's use these features to create two new, spatially flat universes. We shall fill them, not with vacuum energy density, but with either radiation (**Universe 2**) or dust (**Universe 3**).

Universe 2 is a **flat, radiation-filled** model. That is, we fill the universe with radiation, which has an energy density

$$\rho_{\rm r}(t) = \left[\frac{a(t_0)}{a(t)}\right]^4 \rho_{\rm r}^{(0)}, \tag{18.3}$$

where measurements are made relative to the state $\rho_{\rm r}^{(0)}$ that occurs at time t_0. The Friedmann equation becomes

$$\left[\frac{\dot{a}(t)}{a(t)}\right]^2 = \frac{8\pi\rho_{\rm r}^{(0)}}{3}\left[\frac{a(t_0)}{a(t)}\right]^4. \tag{18.4}$$

For ease of writing, let's also write $a(t_0) = a_0$, and then we have

$$\dot{a}(t) = \frac{H_0 a_0^2}{a(t)}, \tag{18.5}$$

where $H_0 = \left[8\pi\rho_{\rm r}^{(0)}/3\right]^{\frac{1}{2}}$. This equation is solved by an expansion factor

$$a(t) = a_0\sqrt{2H_0 t}. \tag{18.6}$$

This spatially flat, expanding universe grows without bound as illustrated in Fig. 18.1. Notice also that at $t = 0$ we have $a(t = 0) = 0$. That is, the universe begins at zero spatial size, implying an initially infinite density of the radiation field. This is the Big Bang: an initial singularity in the model of the universe. We confirm that the density is indeed infinite in the next example.

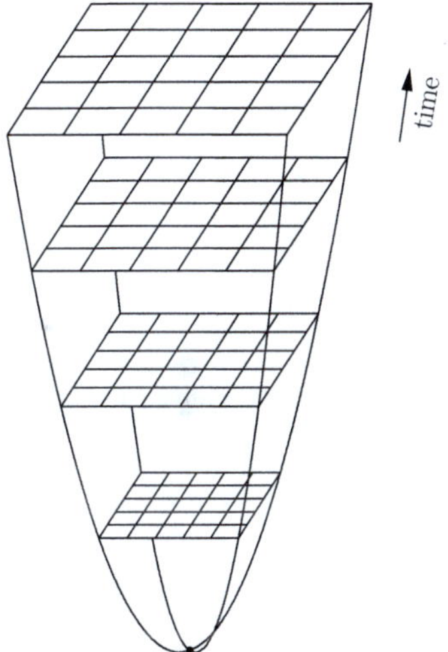

Fig. 18.1 The evolution of **Universes 2** and **3**, which have $k = 0$ and $\Lambda = 0$. These spatially flat universes expand from an initial Big-Bang singularity without bound. The flat planes represent flat three-dimensional space.

Example 18.1

With the solution in eqn 18.5, we can evaluate the Hubble parameter $H(t) = \frac{\dot{a}(t)}{a(t)}$. Plugging in, we find

$$H(t) = \frac{\frac{1}{2}a_0(2H_0/t)^{1/2}}{a_0(2H_0 t)^{1/2}} = \frac{1}{2t}. \tag{18.7}$$

So we have that $H_0 = 1/2t_0$, and that the Hubble parameter (i) is badly behaved at $t = 0$, and (ii) decreases as the universe expands. This is also the case for $\rho(t)$ [since $H(t) \propto \rho(t)^{\frac{1}{2}}$], and so we deduce that, for Universe 2, the energy density starts infinite and decreases with time to zero.

Universe 3 is also known as the **Einstein–de Sitter model**. This spatially flat ($k = 0$) universe is filled only with dust, with energy density

$$\rho_{\mathrm{d}}(t) = \left[\frac{a_0}{a(t)}\right]^3 \rho_{\mathrm{d}}^{(0)}. \tag{18.8}$$

With this equation of state, we have a Friedmann equation

$$\dot{a} = \frac{H_0 a_0^{\frac{3}{2}}}{a^{\frac{1}{2}}}, \tag{18.9}$$

implying that

$$a(t) = a_0 \left[\frac{3}{2} H_0 t\right]^{\frac{2}{3}}. \tag{18.10}$$

From this we find that $H(t) \equiv \dot{a}/a = 2/(3t)$. Again, the universe starts from a Big-Bang singularity and grows without bound, evolving as shown in Fig. 18.1.

We have now examined three spatially flat models filled with different manifestations of energy. In each case, the universe is found to expand without limits. We summarize our findings in the table below:

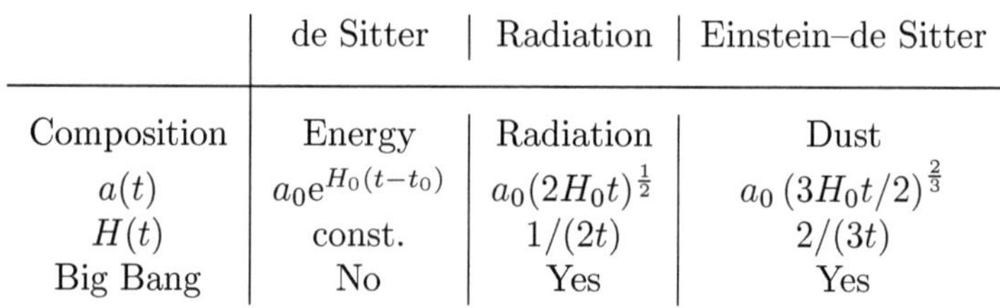

	de Sitter	Radiation	Einstein–de Sitter
Composition	Energy	Radiation	Dust
$a(t)$	$a_0 e^{H_0(t-t_0)}$	$a_0(2H_0 t)^{\frac{1}{2}}$	$a_0 \left(3H_0 t/2\right)^{\frac{2}{3}}$
$H(t)$	const.	$1/(2t)$	$2/(3t)$
Big Bang	No	Yes	Yes

The evolution of $a(t)$ for each model is compared in Fig. 18.2.

Let's return to Universe 3 [the ($k = 0$) Einstein–de Sitter model], which contains only dust, and add in some vacuum energy Λ. The first Friedmann equation for this flat, dust-and-vacuum-energy-universe (**Universe 4**) now takes the form[1]

$$\dot{a}^2 = \frac{C}{a} + \frac{\Lambda}{3}a^2, \tag{18.12}$$

where $C = \frac{8\pi}{3} a_0^3 \rho_{\mathrm{d}}^{(0)}$, and we assume that the vacuum energy $\Lambda > 0$. In the initial stages of the expansion, where the radius $a(t)$ of the universe is small, we expect the dust term to be large compared to the vacuum energy term and so the universe has the behaviour described above for Universe 3, with $a \propto t^{2/3}$ (implying a Big Bang). However, with $\Lambda > 0$, at some point when $a(t)$ is large enough, the vacuum energy must dominate and we have the behaviour of the de Sitter universe (Universe 1) that $a \propto e^{Ht}$. Actually, the full behaviour of Universe 4 can be calculated exactly, with the result that

$$a(t) = \left(\frac{3C}{2\Lambda}\right)^{\frac{1}{3}} \left[\cosh\sqrt{3\Lambda}t - 1\right]^{\frac{1}{3}}. \tag{18.13}$$

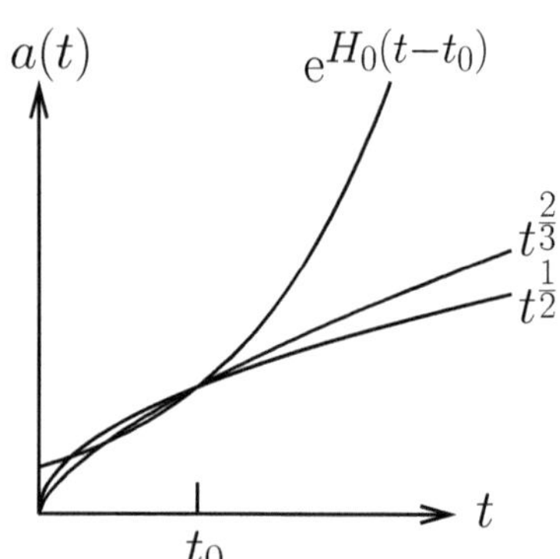

Fig. 18.2 The radius function $a(t)$ for the de Sitter (Universe 1), Radiation (Universe 2) and Einstein–de Sitter (Universe 3) models.

[1] For the case of vacuum energy and dust, the more general case of this equation for $k = \pm 1$ and 0 is

$$\dot{a}^2 + k = \frac{C}{a} + \frac{\Lambda}{3}a^2, \tag{18.11}$$

which turns out to be a useful expression.

The general behaviour of this universe is therefore that it starts from a singularity and expands without bound.

Finally, we can consider the intriguing possibility (**Universe 5**) that the universe is filled with vacuum energy with a negative cosmological constant (i.e. $\Lambda < 0$). Recall that positive Λ vacuum energy exerts a negative pressure, so having $\Lambda < 0$ should be expected to provide something closer to positive-pressure mass energy, which causes gravitational attraction, and hence arrests the expansion of the universe if it is large enough. This does indeed happen: the right-hand side of the Friedmann equation vanishes when $a(t)^3 = 3C/(-\Lambda)$ and the expansion must stop. The full behaviour of the model is given by the exact solution

$$a(t) = \left[\frac{3C}{2(-\Lambda)} \right]^{\frac{1}{3}} \left[1 - \cos \sqrt{3(-\Lambda)}t \right]^{\frac{1}{3}}. \tag{18.14}$$

This solution tells us that the universe initially expands from a Big-Bang singularity, reaches a maximum radius, and then contracts, ending in the possibility of a final **Big-Crunch singularity** as shown in Fig. 18.3.

18.2   Curved universes with $\Lambda = 0$

So far, all of the universes we have considered are spatially flat, with $k = 0$. We now turn to some spatially curved cases, and we'll start with examining the behaviour of dust-filled universes with $\Lambda = 0$. Equation 18.11, then gives us a Friedmann equation

$$\dot{a}^2 + k = \frac{C}{a}. \tag{18.15}$$

We already have the Einstein–de Sitter model of Universe 3 for the flat ($k = 0$) case, where the Universe was found to expand without bound from a singularity as shown in Fig. 18.1. What about the cases $k = \pm 1$? First, consider the case that the space is spherical ($k = 1$). This is **Universe 6** ($k = 1$, $\Lambda = 0$), examined in the next example.

Example 18.2

Start with $\dot{a}^2 = C/a - 1$ and make a change of variable $u^2 = a/C$ and the equation of motion becomes

$$\dot{u}^2 = \frac{1}{4C^2 u^2} \left(\frac{1}{u^2} - 1 \right). \tag{18.16}$$

We shall integrate this equation, employing so-called **Big-Bang initial conditions** that $a = 0$ when $t = 0$, and hence $u(0) = 0$, and we find

$$2 \int_0^{u(t)} \frac{u^2 \mathrm{d}u}{(1 - u^2)^{\frac{1}{2}}} = \frac{t}{C}. \tag{18.17}$$

The left-hand side can be integrated with the result

$$2 \int_0^{u(t)} \frac{u^2 \mathrm{d}u}{(1 - u^2)^{\frac{1}{2}}} = \sin^{-1} u(t) - u(t) \left[1 - u^2(t) \right]^{\frac{1}{2}}. \tag{18.18}$$

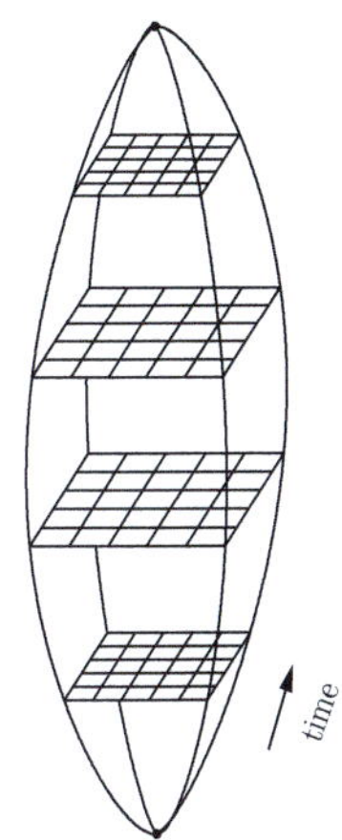

Fig. 18.3 The evolution of **Universe 5** which has $k = 0$ and negative vacuum energy.

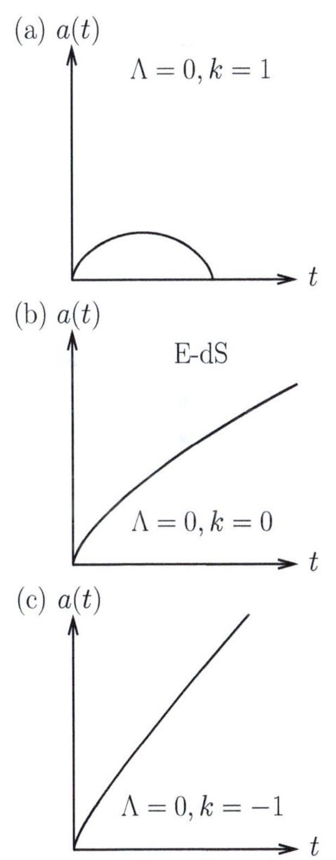

Fig. 18.4 The radius function $a(t)$ for the $\Lambda = 0$, dust-filled universes. (a) **Universe 6** with $k = 1$; (b) **Universe 3** (the Einstein–de Sitter model) with $k = 0$; (c) **Universe 7** with $k = -1$.

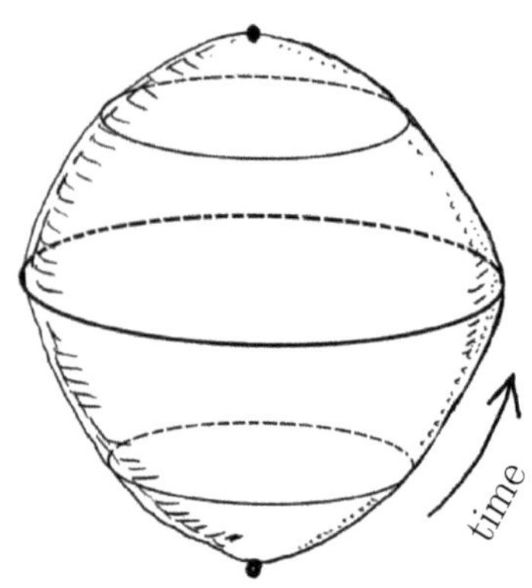

Fig. 18.5 The evolution of **Universe 6**, which has $k = 1$. The circles represent 3-spheres.

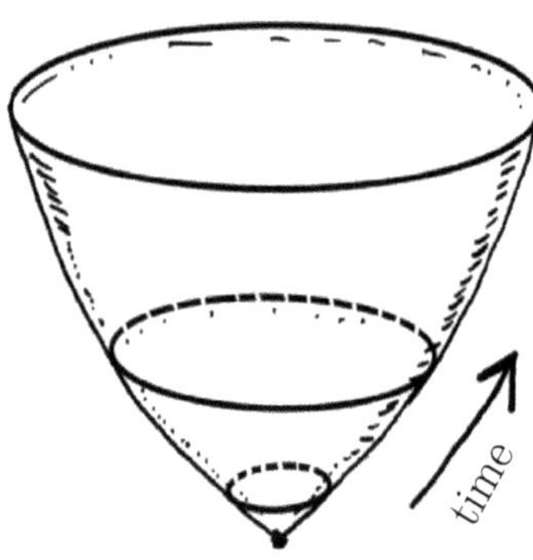

Fig. 18.6 The evolution of **Universe 7**, which has $k = -1$ and $\Lambda = 0$. The surfaces represent three-dimensional hyperbolic space.

We conclude that

$$t = C \left[\sin^{-1} \left(\frac{a}{C} \right)^{\frac{1}{2}} - \left(\frac{a}{C} \right)^{\frac{1}{2}} \left(1 - \frac{a}{C} \right)^{\frac{1}{2}} \right]. \tag{18.19}$$

The behaviour of $a(t)$ is shown in Fig. 18.4(a), where we see it expand from zero, reach a maximum and finally collapse to a zero.

The result of the last example is that Universe 6 starts from a Big Bang, expands and then collapses into a big crunch as illustrated in Fig. 18.5, where each of the spatial hypersurfaces represents a 3-sphere.

For **Universe 7**, we carry out an analogous computation for a spatially hyperbolic case ($k = -1$). We find that

$$t = C \left[\left(\frac{a}{C} \right)^{\frac{1}{2}} \left(1 + \frac{a}{C} \right)^{\frac{1}{2}} - \sinh^{-1} \left(\frac{a}{C} \right)^{\frac{1}{2}} \right]. \tag{18.20}$$

This equation, which is graphed in Fig 18.4(c), tells us that Universe 7 grows without bound. [It can be compared with the slower expansion of the Einstein–de Sitter universe shown in Fig. 18.4(b).] Universe 7 is also represented in Fig. 18.6, where each of the spatial hypersurfaces represents a hyperbolic space.

To summarize, the $\Lambda = 0$, dust-filled universes we have examined all have an initial expansion from a Big Bang. The flat ($k = 0$) case (Universe 3) and the hyperbolic $k = -1$ (Universe 7) case both grow without bound. The spherical ($k = 1$) case (Universe 6) turns over and re-collapses into a Big-Crunch singularity. The expansion factors for these universes are summarized in Fig. 18.4.

18.3 Einstein, Lemaître and Eddington

The universes so far have all had the potentially disturbing feature of being time dependent. The static **Universe 8** is constructed such that it does not have this apparent problem. This model is known as the **Einstein Universe**. It contains dust and positive vacuum energy Λ_E. We seek a *static* solution and see what happens.

Example 18.3

We start with the first Friedmann equation, written in the form for dust $\rho_d(t)$ and an amount of positive energy density Λ_E

$$\dot{a}^2 + k = \left(\frac{8\pi \rho_d}{3} + \frac{\Lambda_E}{3} \right) a^2. \tag{18.21}$$

Static implies that $\dot{a}(t) = 0$, $a(t) = a_0$ and $\rho(t) = \rho_d^{(0)}$ and so the first Friedmann equation becomes

$$k = \left(\frac{8\pi}{3} \rho_d^{(0)} + \frac{\Lambda_E}{3} \right) a_0^2. \tag{18.22}$$

It's convenient to also make use of the second Friedmann equation here with $\ddot{a} = \dot{a} = 0$, which then reads

$$\frac{k}{a_0^2} = \Lambda_E. \tag{18.23}$$

From the positivity of Λ_E we conclude that for a static universe we need to select a spatially spherical ($k = 1$) universe since this is the only way to have a positive k. The second Friedmann equation then tells us that $a_0 = \Lambda_E^{-\frac{1}{2}}$ and substituting this into eqn 18.22 we obtain $\rho_d^{(0)} = \Lambda_E/4\pi$. In words: a universe with dust density $\rho_d^{(0)}$ is kept static by the negative pressure of vacuum energy, provided that $\rho_d^{(0)} = \Lambda_E/4\pi$. It follows that the radius of the universe is given in terms of the amount of dust matter by $a_0 = (4\pi\rho_d^{(0)})^{-\frac{1}{2}}$.

The static Einstein Universe (Universe 8) is shown in Fig. 18.7 where, because $k = 1$, each spatial hypersurface represents a 3-sphere. The Einstein Universe is static, but this doesn't mean it is stable. In fact, the next example shows that it cannot withstand a small perturbation in the density of the dust that fills it.

Example 18.4

We examine the stability of this universe[2] by introducing a perturbation in both $a(t)$ and ρ_d. Subtracting the first Friedmann equation from the second we obtain

$$2\frac{\ddot{a}}{a} = -\frac{8\pi}{3}\rho_d + \frac{2}{3}\Lambda_E. \tag{18.24}$$

We perturb around the equilibrium values by setting $a = \Lambda_E^{-\frac{1}{2}} + \delta a$ and $\rho_d = \frac{\Lambda_E}{4\pi} + \delta\rho_d$, giving, to order δa, the equation of motion

$$2\Lambda_E^{\frac{1}{2}}(\ddot{\delta a}) = -\frac{8\pi}{3}\delta\rho_d. \tag{18.25}$$

If the total amount of dust is constant, then we also have the thermodynamic result that

$$\rho_d a^3 = \text{const.}, \tag{18.26}$$

and so we can find an equation for $\delta\rho_d$ by differentiating:

$$\frac{\delta\rho_d}{\rho_d} = -3\frac{\delta a}{a}. \tag{18.27}$$

Substituting into eqn 18.25, we conclude that

$$(\ddot{\delta a}) = \Lambda_E(\delta a), \tag{18.28}$$

whose solution is $\delta a \propto e^{\sqrt{\Lambda_E}t}$. The result, therefore, is that the perturbation grows exponentially in time, causing the universe to evolve. The static universe is not in a stable equilibrium.

We conclude that the Einstein Universe is only just static, since it contains a critical amount of energy density $\Lambda_E = 4\pi\rho_d$. The vacuum energy tries to make the universe expand; the matter tends to cause the universe to contract and they end up in unstable equilibrium.

In **Universe 9**, known as the **Lemaître solution**,[3] we have $k = 1$ and a universe full of dust and vacuum energy, but this time with $\Lambda > \Lambda_E$. Since the vacuum energy tends to make universes expand, we should expect that this universe expands without bound. However, there is a surprise: the model has a **coasting period** where it stays at roughly the same size. This coasting period can last for arbitrarily long time if the energy density is set appropriately. This is represented in Fig. 18.8, where each spatial hypersurface represents a 3-sphere.

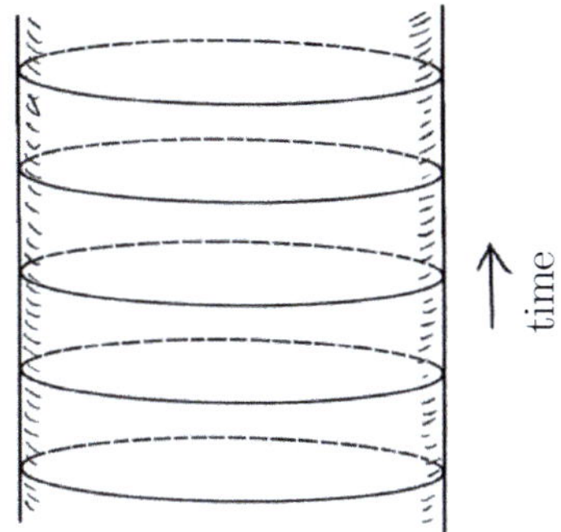

Fig. 18.7 The evolution of **Universe 8**, the Einstein Universe, which has $k = 1$ and vacuum energy $\Lambda = \Lambda_E$. Each circle represents a 3-sphere.

[2]Here we follow the approach described in the problem book by Lightman *et al.*

[3]George Lemaître (1894–1966)

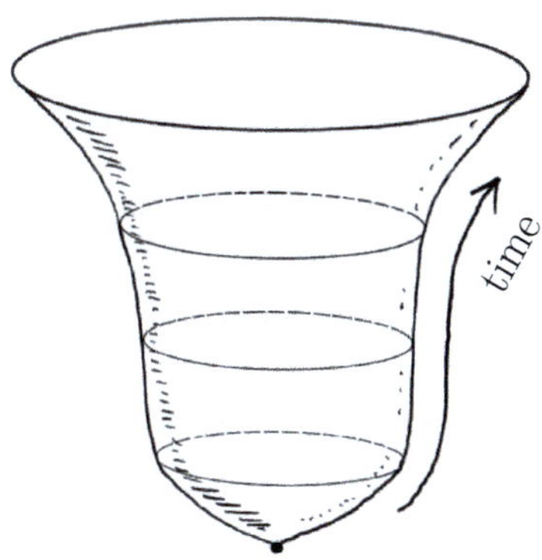

Fig. 18.8 The evolution of **Universe 9** which has $k = 1$ and vacuum energy $\Lambda > \Lambda_E$.

Example 18.5

Consider Universe 9 for the case where $\Lambda = (1+\varepsilon)^2 \Lambda_{\rm E}$, where ε is assumed small. Since, for the Einstein Universe we have $a_0 = (4\pi\rho_{\rm d})^{-\frac{1}{2}}$ and $a_0 = \Lambda_{\rm E}^{-\frac{1}{2}}$, we can write for the dust that

$$\frac{8\pi\rho(t)}{3} = \frac{2}{3\Lambda_{\rm E}^{\frac{1}{2}} a(t)^3}. \tag{18.29}$$

Substituting $\Lambda/(1+\varepsilon)^2$ for $\Lambda_{\rm E}$ and setting $k = 1$, we write the first Friedmann equation in the form

$$\dot{a}^2 + 1 = V_{\rm eff}(a), \tag{18.30}$$

where we've defined an effective potential

$$V_{\rm eff}(a) = \frac{2(1+\varepsilon)}{3\Lambda^{\frac{1}{2}} a} + \frac{\Lambda}{3}a^2. \tag{18.31}$$

From the last example we see that when a is small, the radius of Universe 9 varies as $\dot{a} \propto 1/a^{\frac{1}{2}}$. We examined this behaviour in the Einstein–de Sitter model (Universe 3) and found that it implies that the system expands from an initial singularity as $a \propto t^{\frac{2}{3}}$. However, as a becomes larger the rate of expansion slows, reaching a minimum when $\dot{a}$ is a minimum. This is the case when[4]

$$a = a_{\rm min} = \frac{(1+\varepsilon)^{\frac{1}{3}}}{\Lambda^{\frac{1}{2}}}. \tag{18.32}$$

The universe then **coasts** for a period, before speeding up again.

Example 18.6

Expanding the effective potential $V(a)$ about the minimum, we have

$$\dot{a}^2 \approx -1 + V(a_{\rm min}) + \frac{1}{2}(a - a_{\rm min})^2 \left.\frac{\partial^2 V}{\partial a^2}\right|_{a=a_{\rm min}}$$

$$= -1 + (1+\varepsilon)^{\frac{2}{3}} + \left[\Lambda^{\frac{1}{2}} a - (1+\varepsilon)^{\frac{1}{3}}\right]^2. \tag{18.33}$$

This equation can be integrated. The result (a standard integral that can be looked up) is

$$a(t) = \frac{(1+\varepsilon)^{\frac{1}{3}}}{\Lambda^{\frac{1}{2}}} \left\{ 1 + \left[1 - (1+\varepsilon)^{-\frac{2}{3}}\right]^{\frac{1}{2}} \sinh\left[\Lambda^{\frac{1}{2}}(t - t_{\rm m})\right] \right\}, \tag{18.34}$$

where $t_{\rm m}$ is the time at which $\dot{a}$ reaches its minimum. In the limit of small ε, we have

$$a \approx a_{\rm min}\left[1 + \frac{\varepsilon}{3}\sinh\Lambda^{\frac{1}{2}}(t - t_{\rm m})\right]. \tag{18.35}$$

As a result, a is approximately $a_{\rm min}$ until

$$\varepsilon \sinh \Lambda^{\frac{1}{2}}(t - t_{\rm m}) \approx 1, \tag{18.36}$$

which is to say, for a time

$$(t - t_{\rm m}) \approx \frac{\ln\left(\frac{1}{\varepsilon}\right)}{\Lambda^{\frac{1}{2}}}. \tag{18.37}$$

This coasting period can therefore be made arbitrarily long with a sufficiently small ε. Eventually, the expansion starts again with $(\dot{a}/a)^2 \to \Lambda/3$ as we had for the de Sitter model (Universe 1).

[4]The algebraic steps are examined in Exercise 18.4.

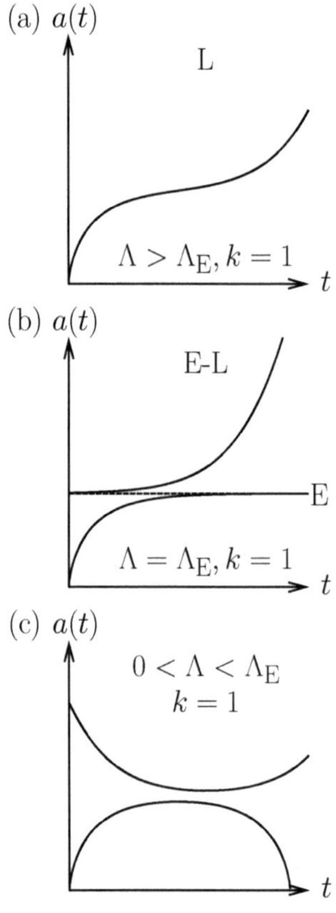

(a) $a(t)$

L

$\Lambda > \Lambda_{\rm E}, k = 1$

(b) $a(t)$

E-L

E

$\Lambda = \Lambda_{\rm E}, k = 1$

(c) $a(t)$

$0 < \Lambda < \Lambda_{\rm E}$
$k = 1$

Fig. 18.9   The radius function $a(t)$ for the some $k = 1$ universes. (a) **Universe 9** (the Lemaître model) with $\Lambda > \Lambda_{\rm E}$ showing its famous coasting period; (b) **Universe 8** (the Einstein static universe with $\Lambda_{\rm E}$) and **Universe 10** (The Eddington–Lemaître model), in its two forms, distinguished by different initial conditions; (c) **Universe 11**, with $0 < \Lambda < \Lambda_{\rm E}$ for two distinct initial conditions.

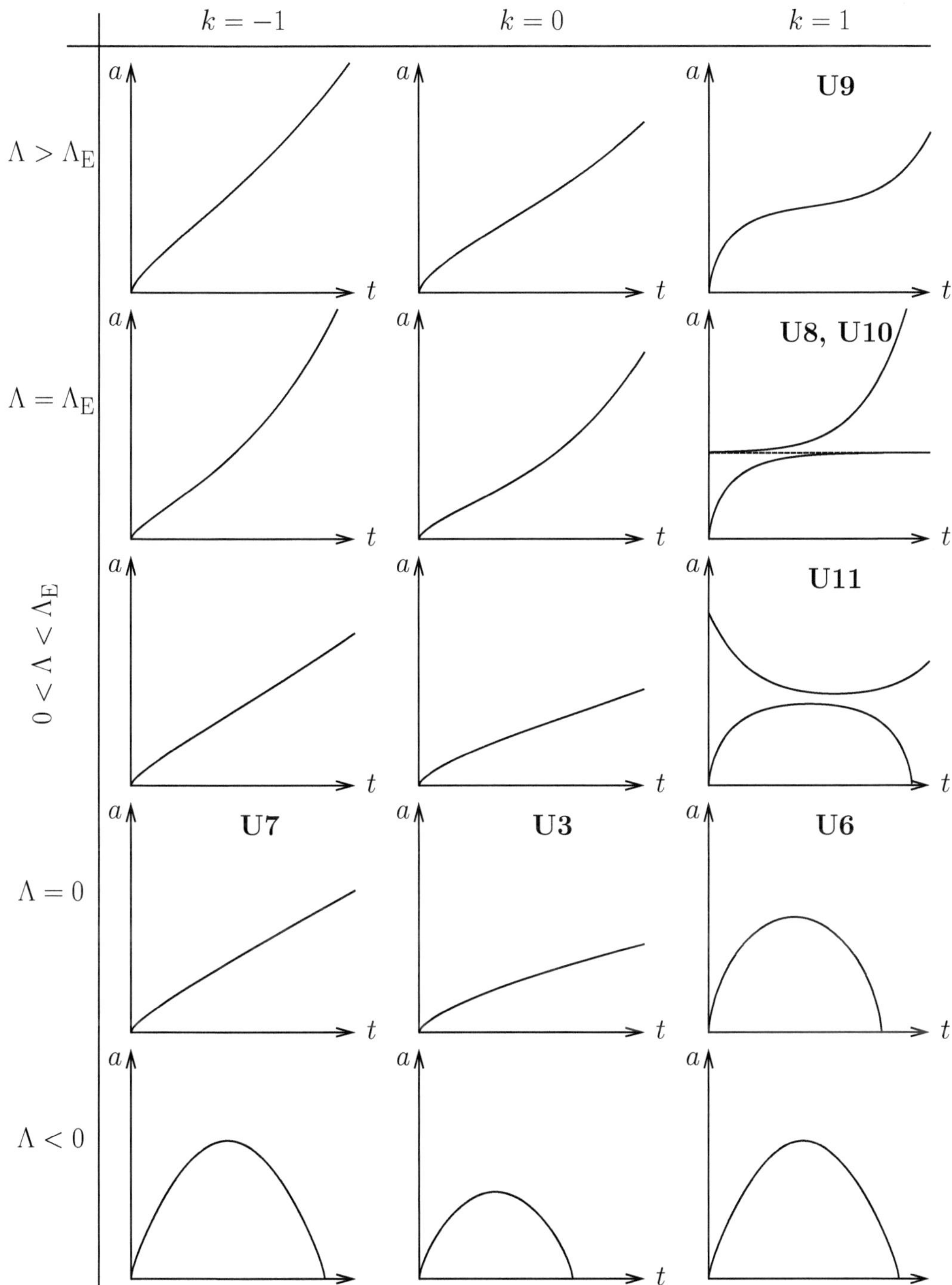

Fig. 18.10 The origins and fates of some Robertson–Walker spacetimes. These were computed for dust and vacuum energy universes using eqn 18.11.

[5]Arthur Eddington (1882–1944) was an early champion of general relativity, writing several articles describing it for an English-speaking audience, who were cut off from many developments in Germany owing to World War I.

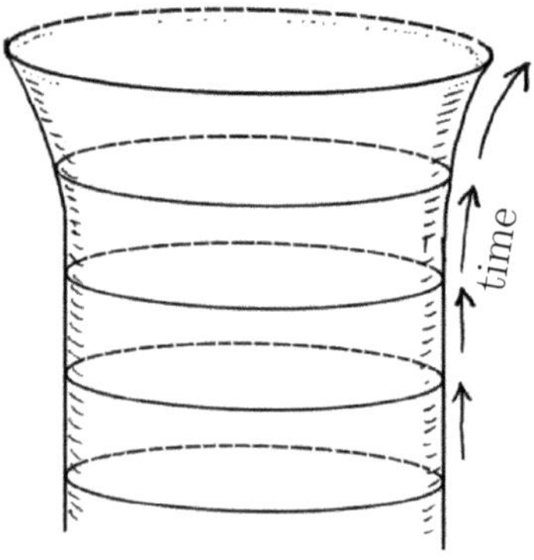

Fig. 18.11 Universe 10, the Eddington–Lemaître model, starts similarly to the Einstein static universe and begins to expand after a certain time.

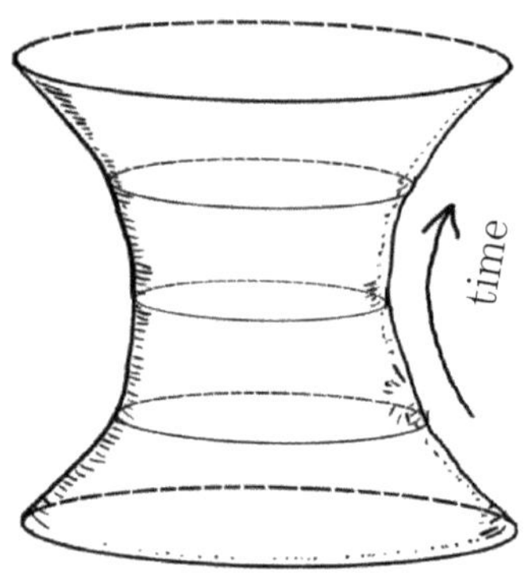

Fig. 18.12 Universe 11, showing an initial contraction, followed by a re-expansion.

The behaviour of the radius function in Universe 9 is shown in Fig. 18.9(a).

As a result of studying the Lemaître model, Arthur Eddington[5] proposed a model without a Big-Bang singularity. Instead the model starts by coasting at a value of a close to the static Einstein value until, at some point, it starts expanding. This is **Universe 10**: the **Eddington–Lemaître model** shown in Fig. 18.9(b) and Fig. 18.11. Actually, there is another variant of this model: we could start from a Big-Bang singularity and asymptotically approach the Einstein Universe as shown in Fig. 18.9(b).

Finally, what if we have $0 < \Lambda < \Lambda_E$? This is the content of **Universe 11**. In this case, there are two possibilities, depending on the initial conditions. If we start from an initial Big-Bang singularity then there will not be enough vacuum energy to prevent the matter collapsing back in itself and we have a Big Crunch singularity. Perhaps more interestingly, if we start at a very large value of a, then the universe contracts, reaches a minimum radius, and then expands again. It has $a \propto e^{H(-t)}$ as $t \to -\infty$ and $a \propto e^{Ht}$ as $t \to \infty$. This is shown in Fig. 18.12 with a radius function as shown in Fig. 18.9(c).

We can summarize some of our progress by examining the evolution of several universes in Fig. 18.10.

18.4   A brief history of model universes

We have presented several model universes generated by varying the content and geometry of the models described by the Friedmann equations. This was not the order in which they were discussed historically. The discovery that our Universe is expanding was established through observations made by a number of people in the decades following 1910. Perhaps most famous of the astronomers in this period is Edwin Hubble, who in 1929 established that the speed of recession of a galaxy v_d is proportional to its distance from us. This is Hubble's law that $v_d = H \times$ distance.

Somewhat earlier, back in 1917, Einstein had considered filling a spatially spherical universe with uniform matter and realized that, owing to the gravitational attraction of the matter, the resulting model could not be static without the incorporation into the theory of an extra energy contribution Λ, that he called the cosmological term. In the same year, de Sitter worked out that, for an isotropic model, the cosmological-constant term could provide a repulsion that exactly cancelled the gravitational attraction. This gives us the Einstein model (Universe 8). The model ceased to be considered seriously after it became established that the physical Universe was expanding. Eddington in 1930 showed that this universe was unstable to perturbation and, in 1931, Einstein recommended the cosmological term be dropped from general relativity.

In 1917, de Sitter had also observed that he could model an expanding universe if he included only the Λ term. Alexander Friedmann derived his results for evolving isotropic and homogenous universes in 1922, which also suggested the Hubble-law expansion $v_{\mathrm{d}} = H \times \text{distance}$. (In 1923, Hermann Weyl pointed out that test particles in the de Sitter model would separate at a rate consistent with this law.) In 1927, the Catholic priest and physicist George Lemaître independently derived Friedmann's homogeneous, isotropic models and proposed the Eddington–Lemaître model (Universe 10). This suggestion of a model that started at the Einstein solution and expanded was promoted by Eddington. However, Lemaître would in 1931 suggest that the universe could have expanded from what he termed a 'Primeval Atom' (his Universe 9), an idea later described rather dismissively by Fred Hoyle[6] as the 'Big Bang theory'. Robertson and Walker would independently show in 1935 that the line element Lemaître was using represents the most general geometry compatible with homogeneity and isotropy.

There was a significant problem with models involving the expansion of the universe from an initial point in spacetime. The age of rocks (and hence the Earth) had been estimated from the radioactive properties of uranium to be at least 3×10^9 years, but this didn't seem to allow long enough for the universe to evolve into its current state.

[6]Fred Hoyle (1915–2001) rejected the Big Bang theory and supported an alternative steady-state picture, in which there is no cosmological constant, but where a creation field (or C-field) is responsible for matter generation as the Universe expands, allowing the density to remain constant in time. The C-field has negative energy density (allowing for conservation of energy as new matter is created) and so exerts a negative pressure. The inability of the steady-state model to account for the cosmic microwave background, discovered in 1964, is one of the reasons why it was abandoned by cosmologists. The term 'Big Bang' was coined in a talk given by Hoyle on BBC radio in 1949. Hoyle denied that he intended to be insulting; rather that he was emphasizing the difference between the models for a general audience.

Example 18.7

Let's start by assuming that the acceleration of our Universe has always been negative (i.e. the rate of expansion has always been slowing down). If we consider the evolution of $a(t)$ then its tangent evaluated now $\dot{a}(t_{\mathrm{now}})$ cuts the axis at an earlier time t_1 where

$$a(t_{\mathrm{now}}) = \dot{a}(t_{\mathrm{now}})(t_{\mathrm{now}} - t_1). \tag{18.38}$$

This is shown in Fig. 18.13. Rearranging, we find

$$t_{\mathrm{now}} - t_1 = \frac{a(t_{\mathrm{now}})}{\dot{a}(t_{\mathrm{now}})} = H(t_{\mathrm{now}})^{-1}. \tag{18.39}$$

Before the 1950s the quantity $H(t_{\mathrm{now}})^{-1}$ was estimated to be 1.8×10^9 years. If the universe does start at, or around, $a(t = 0) = 0$, then its age must be less than $H(t_{\mathrm{now}})^{-1}$.

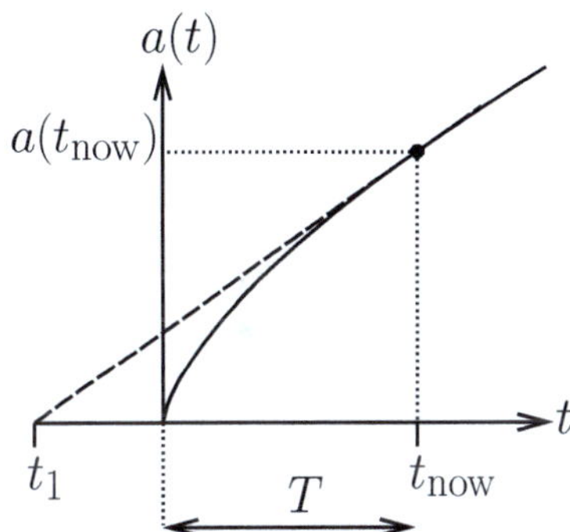

Fig. 18.13 Estimation of the age T of the Universe using $a(t)$. Assuming the Universe has been monotonically decelerating, we must have $T < t_{\mathrm{now}} - t_1$.

If the ages of the rocks are as claimed on the basis of the nuclear physics, then there doesn't seem to have been enough time for the Universe to evolve before the creation of rocks. This is known as the **time-scale problem**. One solution was the introduction into the Lemaître model of large positive values of Λ that cause a large acceleration. It is also solved by the Eddington–Lemaître model, since this starts from the Einstein solution before expansion. However, in that model the assumption that the stars condense at the moment that the expansion starts is problematic, since it is not clear how this can happen within a static and homogeneous model.

By the 1930s, nuclear physics was also offering explanations of the genesis of the elements (nucleogenesis) which seemed consistent with Big

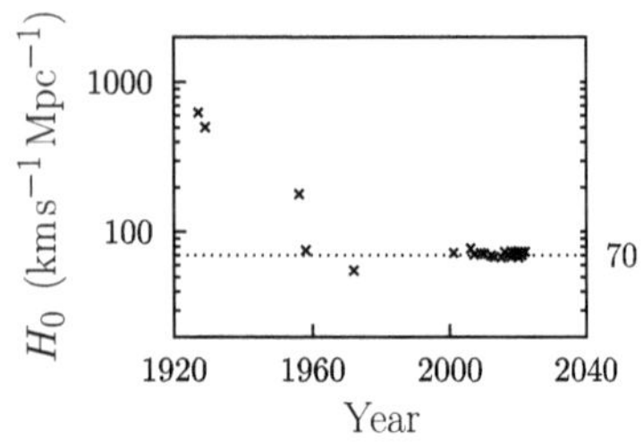

Fig. 18.14 Selected measurements of H_0.

> ⤷ **The evolution of the very early Universe is discussed in Chapter 41. See Chapter 15 for a brief description of the anisotropy of the CMB.**

[7] Recent estimates give h to be close to 0.7 (see Fig. 18.14).

> ⤷ **In Section 49.8 we will discuss our current understanding, based on experimental evidence, on the question of which type of Universe we are actually living in.**

Bang models, since as the scale factor a tends to zero, the temperature should go as $T \propto a^{-1}$ allowing, in the early moments after the Big Bang, the extreme conditions needed to fuse hydrogen to make the heavier elements. This led Lemaître to promote the idea of a **hot Big Bang**.

By the 1950s new measurements of Hubble's constant H_0 (see Fig. 18.14 for a timeline) suggested an increased lifetime of the Universe of 1.3×10^{10} years, effectively resolving the timescale problem without the need for a large cosmological constant, so that Big Bang models with $\Lambda = 0$ could now be seriously considered. This had an aesthetic appeal, owing to the slightly ad-hoc nature of the introduction of the cosmological constant. The $\Lambda = 0$ models became known as the **standard models** of cosmology. These are, for $k = 0$, Universe 3 (Einstein–de Sitter), for $k = 1$, Universe 6 and for $k = -1$, Universe 7. They are all Big Bang models.

Compelling evidence for a Big-Bang cosmology came from the measurement of the cosmic microwave background (CMB), which is the faint electromagnetic radiation remaining from recombination events that occurred during the early stages of the Universe. The existence of the CMB had originally been predicted in 1948 by Ralph Alpher and Robert Herman, following work by George Gamow. Alpher and Herman predicted that high-energy radiation from the very early Universe should have been shifted into the microwave region of the electromagnetic spectrum, and that this should result in an effective black-body temperature of space of $T = 5$ K. The CMB was accidentally discovered by Arno Penzias and Robert Wilson in 1964, and has since been subject to very precise measurements, that determine its temperatures to be 2.73 K, with faint anisotropies.

Which of the three standard models our Universe most resembles is determined by how the mass-energy content of the Universe compares with the critical density (discussed in the last chapter) of $\rho_c = 3H_0^2/8\pi$. Expressing the Hubble constant[7] as $H_0 = h \times 100$ km s^{-1} Mpc^{-1}, the critical density is $\rho_c = 1.88 \times 10^{-26}h^2$ kg m^{-3}. Observations suggests that the density of observable matter in the Universe is actually only a small fraction of this, at around 10^{-28} kg m^{-3}. However, observations also suggest that the properties of the Universe are consistent with a flat, $k = 0$ model. This discrepancy is known as the **missing mass problem** and remains one of the key questions about our knowledge of the Universe.

So far we have been quite cavalier in accepting the presence of the Big-Bang singularity. We are now at the point where we must engage with it, and the break down of physical laws that it brings. This is the goal of the next chapter.

Chapter summary

- The Robertson–Walker spaces filled with a perfect fluid provide a range of models, many of which start with a Big Bang.
- The Einstein Universe is static, but unstable; the Einstein–de Sitter model expands without bound and the Lemaître model has a coasting period.
- The standard models of cosmology have zero cosmological constant Λ, and all begin with a Big-Bang singularity.

Exercises

(18.1) Let's examine a Robertson–Walker universe with no matter or vacuum energy in it, by letting $\rho = p = \Lambda = 0$.

(a) Set $k = -1$ and show that the resulting metric line element can be written as

$$\mathrm{d}s^2 = -\mathrm{d}t^2 + t^2\left[\mathrm{d}\chi^2 + \sinh^2\chi(\mathrm{d}\theta^2 + \sin^2\theta\mathrm{d}\phi^2)\right]. \tag{18.40}$$

(b) Using the transformations

$$r = t\sinh\chi, \quad T = t\cosh\chi \tag{18.41}$$

show that this is a familiar spacetime in disguise.

(18.2) Verify by substitution that (a) eqn 18.13 and (b) eqn 18.14 are solutions to the Friedmann equations.

(18.3) Verify eqns 18.25 and 18.28.

(18.4) Verify eqns 18.32 and 18.33.

There is a neat geometrical interpretation of the **de Sitter universes**, *which are those that are driven only by a non-zero cosmological constant* Λ. *(Universe 1 is the $k = 0$ version. There are also $k = \pm 1$ versions.) The construction involves embedding a hyperboloid*

$$-T^2 + W^2 + X^2 + Y^2 + Z^2 = \alpha^2, \tag{18.42}$$

with α a constant, in a (4+1)-dimensional Minkowski space with line element $\mathrm{d}s^2 = -\mathrm{d}T^2 + \mathrm{d}X^2 + \mathrm{d}Y^2 + \mathrm{d}Z^2 + \mathrm{d}W^2$. The hyperboloid is shown in Fig. 18.15, with two of the dimensions suppressed (i.e. the circular cuts should be regarded as 3-spheres in five dimensions). This seems very abstract, by the idea of covering the hyperboloid with different coordinate system to see the features of the physics, can be demonstrated on the two-dimensional surface of a hyperboloid described by $-T^2 + X^2 + Y^2 = 1$, embedded in (2+1)-dimensional Minkowski space, as demonstrated by the following exercises.

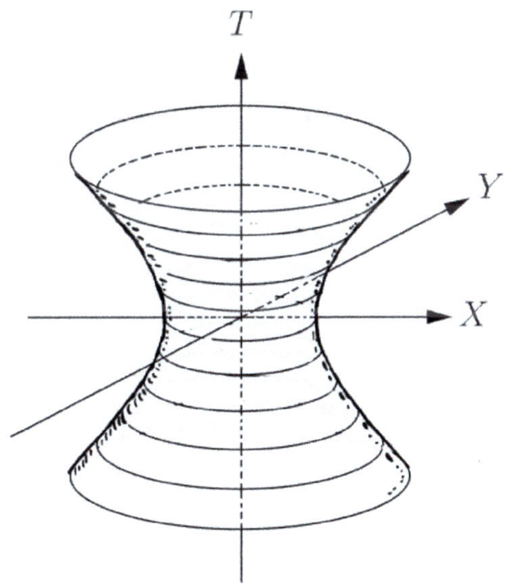

Fig. 18.15 de Sitter spacetime represented as a hyperboloid.

(18.5) As a warm-up exercise, use the method from Appendix D to embed the hyperbola $-T^2 + X^2 = 1$ in (1+1)-dimensional Minkowski space with line element $ds^2 = -dT^2 + dX^2$.

(18.6) Embed the hyperbolic surface $-T^2 + X^2 + Y^2 = 1$ shown in Fig. 18.15 in Minkowski space with line element $ds^2 = -dT^2 + dX^2 + dY^2$ by eliminating the variable Y, to show

(a) the line element can be written in terms of variables v and ψ as

$$ds^2 = -\frac{dv^2}{1 + v^2} + v^2 d\psi^2; \qquad (18.43)$$

(b) the latter expression can be reduced to the hyperbolic line element

$$ds^2 = -d\chi^2 + \sinh^2 \chi d\psi^2, \qquad (18.44)$$

where you should determine the relationship between coordinates (χ, ψ) and (T, X, Y).

(c) Compare the latter with the other form of the line element from the chapter, which was

$$ds^2 = -d\chi^2 + \cosh^2 \chi d\theta^2. \qquad (18.45)$$

At constant timelike variable χ there is a close resemblance between (i) eqn 18.44 and the line element of a spherical universe with $a(t) = \sinh t$ and (ii) eqn 18.45 and the line element of a hyperbolic Universe with $a(t) = \cosh t$.

(d) Confirm this is the case by considering the Friedmann equations.

Notice how that, by choosing coordinates that split the space and time up in different ways, we are able to use de Sitter spacetime to represent cosmological-constant-driven models with different spatial curvatures. This works because the different cross sections of the hyperboloid represented by each coordinate system possess different spatial curvatures.

(18.7) Start again with a hyperbolic surface $-T^2 + X^2 + Y^2 = 1$ embedded in (2+1)-dimensional Minkowski space in the form that results from eliminating Y:

$$ds^2 = -dT^2 + dX^2 + \frac{(TdT - XdX)^2}{(1 + T^2 - X^2)}. \qquad (18.46)$$

Now define variables

$$T = \sinh t + \frac{x^2}{2} e^t,$$

$$X = \cosh t - \frac{x^2}{2} e^t, \qquad (18.47)$$

implying $Y = x e^t$.

(a) Show that the line element becomes

$$ds^2 = -dt^2 + e^{2t} dx^2, \qquad (18.48)$$

which is the line element for the spatially expanding de Sitter Universe (Universe 1) in (1+1) dimensions.

(b) Show that these coordinates only cover half of the hyperboloid.

(c) What kind of fixed-time slices do these coordinates make on the hyperboloid?

(d) Find a transformation that turns eqn 18.48 into the line element

$$ds^2 = \frac{1}{u^2} \left(-du^2 + dx^2 \right). \qquad (18.49)$$

The latter is a type of Poincaré line element for Minkowski space, demonstrating that the Poincaré line element describes a hyperbolic spacetime.

(18.8) Using the same method as in Exercise 18.7, show that a description of (1+1)-dimensional de Sitter spacetime using coordinates

$$T = (1 - x)^{\frac{1}{2}} \sinh t,$$

$$X = (1 - x^2)^{\frac{1}{2}} \cosh t, \qquad (18.50)$$

gives the Schwarzschild-like static line element

$$ds^2 = -(1 - x^2) dt^2 + \frac{dx^2}{1 - x^2}. \qquad (18.51)$$

How much of the hyperboloid is covered by these coordinates?

The key feature of this coordinate system is that it is static. Since a cosmological constant is time-independent, this is not really a surprise.

Causality, infinity, and horizons

19

I cannot help it; — in spite of myself, infinity torments me
Alfred de Musset (1810–1857)

Space is almost infinite. As a matter of fact, we think it is infinite.
Dan Quayle (1947–)

> ↱ This is a mathematical chapter dealing with the treatment of infinities. Readers eager to explore more of the mechanics of general relativity can skip it on a first reading.

We have now seen several examples of the universes predicted by Einstein's equation. In this chapter, we ask if there's a general way to understand the spacetime structure of these universes and the consequences of this on the observations that are made by observers. We shall call this the spacetime's **causal structure**.

In order to discuss causal structure, we need to be able identify what is in the future and what is in the past. Technically, this involves being able to *consistently* classify vectors into those that point forward in time and those that point backwards, just as we can do in flat Minkowski spacetime [a single point of which is shown in Fig. 19.1(a)]. A spacetime is said to be **time-orientable** if this is possible. [Examples of orientable and non-orientable spacetimes are shown in Fig. 19.1(b,c).] In practice, we might try to work out which direction coincides with an increase of entropy and assign that as the forward time direction.

The idea that no signal can propagate faster than light sets strong limitations on matters related to **causality**. If some event causes another to occur then these two events must have a time ordering in which the cause precedes the effect. In addition, the spacetime interval between the events must be null or timelike. As a result, a particle can only influence points that appear inside its local forward light cone (and it cannot therefore itself appear outside this light cone), and can only be influenced by events in its local backward light cone. In a curved spacetime where light cones tip over due to the curvature, we saw in Chapter 5 that light cones can be drawn at each point along a world line, with the resulting light-cone structure telling us about the limitations imposed by causality. The only way that causality could be violated is if a timelike curve were to close, forming a timelike loop as shown in Fig. 19.2. In this case, a particle could always remain in its local forward light cone, but meet itself again at some event in its history. That is to say, we would have the science fiction conceit that (i) an observer's future could become their past, and (ii) the observer could conceivably influence their past, causing all manner of contradictions (e.g. killing one's

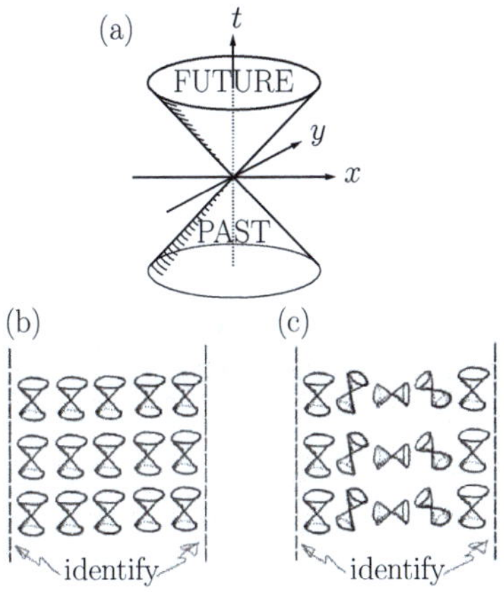

Fig. 19.1 (a) A light cone in Minkowski space distinguishes past from future. (b) In an orientable spacetime (here think of wrapping this sheet of spacetime around into a cylinder by identifying the left and right sides), we can uniquely identify future-pointing and past-pointing directions. (c) In a non-orientable spacetime, there is no way of uniquely determining a future direction and a past direction.

[1] We shall meet some, for example: in the interior of certain types of black hole.

Fig. 19.2 Closed timelike curves.

[2] Roger Penrose (1931–). His influence on modern general relativity is difficult to overstate, including contributions to the theory of singularities and the physics of black holes for which he was awarded the Nobel Prize in Physics in 2020.

[3] Conformal transformations are those that locally preserve angles.

> Topology, the study of those properties of shapes or spaces that are preserved despite deformations, is introduced in Appendix C.

[4] These are sometimes known as Penrose–Carter diagrams, also crediting Brandon Carter (1942–). Carter also contributed to the proof of the no-hair theorem described in Chapter 29, and named the **anthropic principle**. The latter was explained by Penrose as follows: 'The argument can be used to explain why the conditions happen to be just right for the existence of (intelligent) life on the Earth at the present time. For if they were not just right, then we should not have found ourselves to be here now, but somewhere else, at some other appropriate time.'

ancestors). Fortunately, there are very few regions of spacetime that can support timelike loops;[1] instead, timelike curves are generally open, joining events far in the past to events far in the future, allowing us to distinguish past and future owing to the orientation of the light cones along the curve.

In addition to causality, when we investigate which events can possibly occur, we are also forced to understand what we mean by **infinity**: the limiting behaviour over large intervals of spacetime. What does it mean for a signal to come from very far, or infinitely far, away? If we wait long enough, can we potentially receive signals from the entire universe? Do we have a unlimited realm of influence or, even if we wait forever, are there regions of the universe we cannot, even in principle, contact with a light signal? In this and future chapters, it will become clear that the answers to these questions depend on the large-scale structure of spacetime and, in general, our domain of influence as observers is limited. Just as the **horizon** on the skyline marks the boundary between those parts of the world we can see and those we cannot, we shall identify horizons that divide events in spacetime into those that observers can access and those they cannot.

One thing we have largely neglected so far in our discussion of the model universes is the presence and nature of infinities. In this chapter, we introduce a toolkit for understanding infinity that was developed in the 1960s, largely by Roger Penrose.[2] This involves classifying different types of infinity according to whether they represent the limiting behaviour of null, timelike, or spacelike curves. In order to understand infinities, we use coordinate transformations to map infinities back to finite coordinates. We then compare the structure of the resulting metrics using the idea of **conformal structure**.[3] This allows us to remove some of the extraneous geometrical baggage in our descriptions and simply analyse the **topology** of the universes via their conformal structure. In this spirit, the conformal structure can be presented using a **Penrose diagram**, which is a cartoon of the infinities in a space that presents their causal consequences.[4] Although the origin of the mathematical tricks used to generate Penrose diagrams can sometimes look obscure, the idea is simple: a Penrose diagram represents all of spacetime in a two-dimensional sketch. Its boundaries are the infinities and light travels along lines 45° to the vertical, making the light-cone structure look like that of flat space. Let's make a start investigating infinities in spacetime.

19.1 Penrose diagrams

The first tool we shall need is a method of comparing model universes. We are interested in the causal structure of our models, encoded in the arrangement of light cones. We therefore seek a method to compare metrics that preserves this light-cone structure, and hence also the causal structure of the universe. This is where conformal structure comes in.

Two metrics g and $\hat{g}$ are said to be **conformally related** if

$$\hat{g}(x) = \Omega(x)^2 g(x), \qquad (19.1)$$

for some differentiable function of spacetime coordinates $\Omega(x)$, known as a conformal factor. Equation 19.1 allows us to write

$$\frac{\hat{g}(x,y)}{\hat{g}(v,w)} = \frac{g(x,y)}{g(v,w)}, \qquad (19.2)$$

for vectors x, y, v and w. With these definitions, both (i) the angles between vectors,[5] and (ii) the ratios of the magnitudes of vectors, are identical for two conformally related metrics.[6] A null vector in a metric therefore remains null in a conformally related metric, so the light-cone structure of the spacetime described by both metrics is identical. We conclude that using this idea of conformally related metrics to compare the structure of spacetimes allows us to investigate the causal structure and infinities in our cosmological models.

Example 19.1

We have met hyperbolic spacetimes several times. In the hyperbolic (sometimes called Lobachevskian) geometry, Euclid's parallel postulate fails and the internal angles of triangles sum to less than π. The hyperbolic plane is represented using the Poincaré disc in Fig. 19.3, a mathematical image inspired by the famous woodcut by M. C. Escher (1898–1972). Notice how the straight lines of this geometry all meet the circular boundary at right angles, and each black triangle is in an equivalent environment to every other. The representation shown in this figure is a conformal one, which ensures that all of the angles in the triangles are maintained throughout the picture.

Before ploughing into the details of conformal transformations, let's start to think intuitively about spacetime, and begin with flat Minkowski spacetime. This, as you will recall, involves three space coordinates, which we will simplify by considering a single radial coordinate r, and one time coordinate t, allowing us to represent Minkowski space on a two-dimensional plane (strictly a half-plane, as the radial coordinate $r > 0$), as shown in Fig. 19.4. In such a diagram, the horizontal lines represent spacelike surfaces; vertical lines represent timelike surfaces; diagonal lines lying at 45° to these directions (shown as dotted lines in Fig. 19.4) are null surfaces, on which photons can travel.

Buzz Lightyear may have said 'To Infinity, and beyond!', but where exactly is infinity? Standing at the origin in our Minkowski diagram, we can identify five distinct types of infinity. These are shown in Fig. 19.5(a), and we will go through them carefully in turn.

- Standing at the origin, we can think about the far, far future. What will happen as t goes to infinity at our present location? This is a future we can potentially influence, as it falls within our light cone. We will label this point in our infinite future by the symbol i^+ and call it the **future timelike infinity**.

[5] The **cosine angle** between vectors x and y is defined in terms of metric as

$$\frac{g(x,y)}{|g(x,x)g(y,y)|^{\frac{1}{2}}}. \qquad (19.3)$$

[6] See exercises.

Fig. 19.3 A conformal representation of the hyperbolic plane.

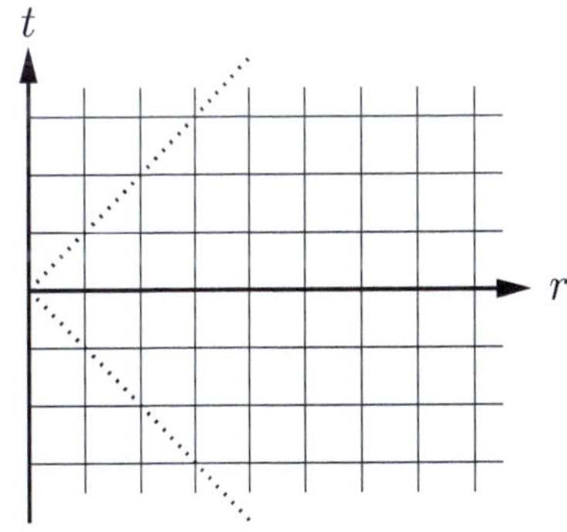

Fig. 19.4 A simple representation of Minkowski spacetime with a radial coordinate r and a time coordinate t. Horizontal lines are spacelike surfaces. Vertical lines are timelike surfaces. The diagonal lines (dotted) are null surfaces, on which photons can travel.

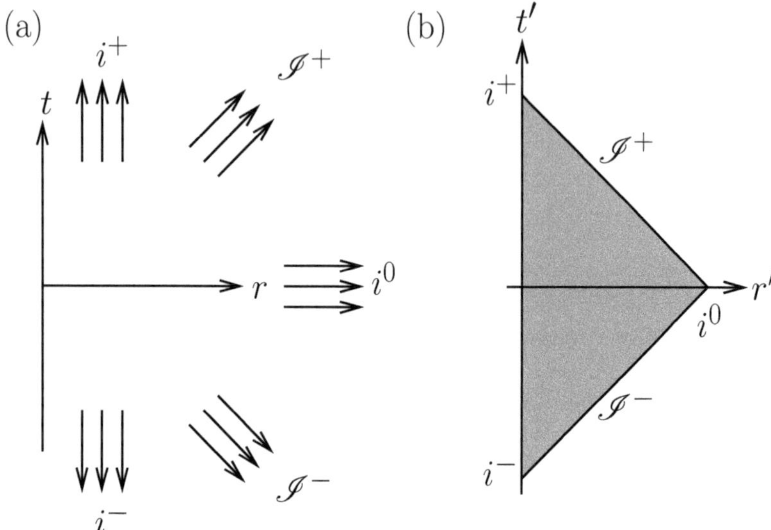

Fig. 19.5 (a) The different types of infinity in Minkowski spacetime. (b) The same information, now plotted on a Penrose diagram. All the points in Minkowski space are mapped into the shaded region in the Penrose diagram.

- Similarly, we can ask where we came from in the far, far past. In flat Minkowski spacetime, there is no Big Bang. The Universe was always here and so out there in the distant past, shown by the downward arrows in Fig. 19.5(a), is a point we will label i^-, the **past timelike infinity**.

- We can also identify a region far away from us in a spatial sense, out at infinite distance. Thus, our arrows point somewhere over to the right. This region has no causal connection to us, since light would never reach us from this distance, and so we will label this point i^0 and call it **spacelike infinity**.

- What if we send light rays out from around the origin? They will travel on lines at 45° to the horizontal and will end up in the region indicated by the arrows travelling towards a surface labelled[7] $\mathscr{I}^+$, known as the **future null infinity**.

- We can also receive light signals from a surface $\mathscr{I}^-$ called the **past null infinity**.

Pointing in the general direction of these five different infinities is not a very grown-up way of dealing with the mathematics. What we want to do is to find a way of mapping infinite Minkowski space into a finite region,[8] and this will result in the Penrose diagram shown in Fig. 19.5(b). The Penrose diagram [Fig. 19.5(b)] has exactly the same information as the Minkowski space picture [Fig. 19.5(a)], but the points at infinity have been mapped to particular points (i^+, i^0 and i^-) or lines ($\mathscr{I}^+$ and $\mathscr{I}^-$) in the Penrose diagram. The whole of Minkowski space has been squeezed into the shaded region in Fig. 19.5(b), which means that the horizontal and vertical lines in Minkowski space are now mapped into curves in the Penrose diagram. The boundary of the shaded region represents the 'dangerous' infinities of the Minkowski spacetime, which are now 'tamed' (well, at least made finite) in the Penrose diagram.

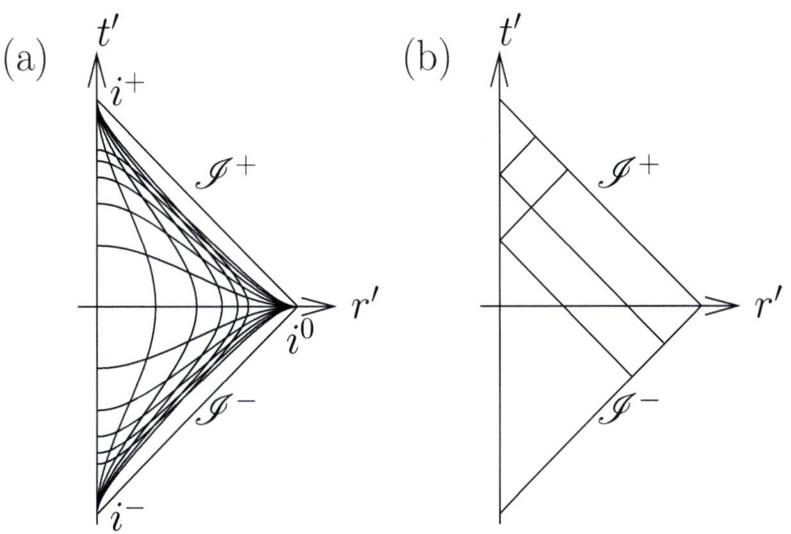

Fig. 19.7 (a) The Penrose diagram for Minkowski space is criss-crossed by timelike and spacelike surfaces. (b) The null surfaces, showing how rays of light travel, remain straight lines and connect $\mathscr{I}^-$ and $\mathscr{I}^+$.

Let's see how particular features in Minkowski space are mapped into the Penrose diagram.

- Vertical lines in Minkowski space (the timelike surfaces) are mapped to the curves shown in Fig. 19.6(a). Each curve starts at i^- and ends at i^+. Since all massive particles fall along timelike geodesics, we note that every particle's world line must also start at i^- and end at i^+.

- Horizontal lines in Minkowski space (the spacelike surfaces) are mapped to the curves shown in Fig. 19.6(a). All spacelike geodesics start and end at i^0 (remember that r is a radial coordinate, and so the spacelike geodesic starts at i^0, goes through the origin and heads back out to i^0).

- Collecting these two results, the horizontal and vertical lines of Minkowski space (Fig. 19.4) become the curves of Fig. 19.7(a).

- Light rays from any point in Minkowski space are lines at $\pm 45°$, and our mapping will preserve this feature in the Penrose diagram. Consider first the forward light cone of a point on the vertical axis ($r' = 0$). Outgoing, future-directed rays meet the surface $\mathscr{I}^+$ as shown in Fig 19.7(b). Incoming, future-directed rays meet the vertical axis. However, since the diagram is symmetric about

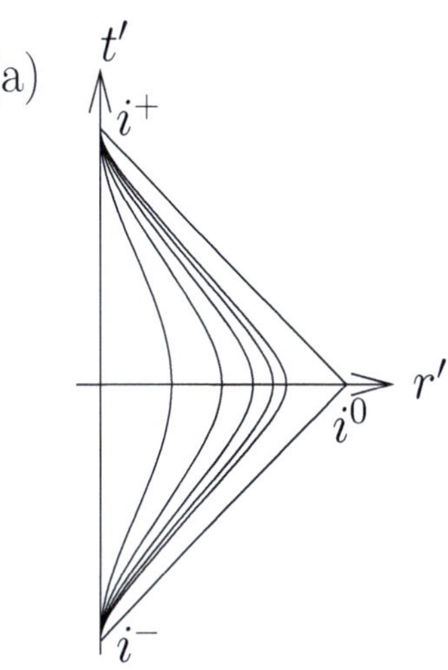

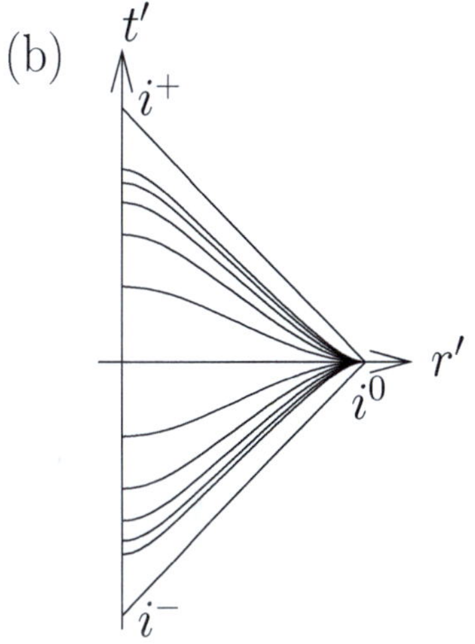

Fig. 19.6 The Penrose diagram for Minkowski space showing (a) timelike surfaces and (b) spacelike surfaces. (Surfaces shown are for r' and $t' = 0.5, 1, 1.5, 2$ and 2.5.)

the vertical axis when considered in three dimensions, the forward light cone from any point meets some part of the boundary at $\mathscr{I}^+$. Similarly, all rays forming the backward light cone meet the surface $\mathscr{I}^-$ and, therefore, *all* light rays must start at past null infinity $\mathscr{I}^-$ and end up at future null infinity $\mathscr{I}^+$. Thus, we conclude that future-directed null geodesics all start on $\mathscr{I}^-$ and end on $\mathscr{I}^+$.

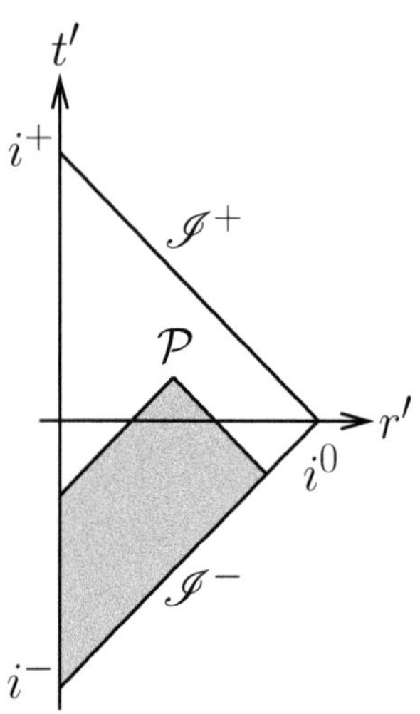

Fig. 19.8 The Penrose diagram for Minkowski space, showing the backward light cone from point $\mathcal{P}$.

Example 19.2

Now consider the state of affairs shown in Fig. 19.8. The shaded region shows the parts of spacetime from which light signals can reach point $\mathcal{P}$. An observer at $\mathcal{P}$ can only have knowledge of events that have occurred in the shaded region, but has no knowledge of events outside this region. Since all timelike geodesics must end at i^+, this implies that in Minkowski space an observer on a timelike geodesics eventually reaches a point where light signals can have reached them from any point in spacetime. It is also the case that the backward light cone from $\mathcal{P}$ always includes i^- and so there are no particles in the Minkowski Universe from whom the observer at $\mathcal{P}$ can't receive light signals if she waits long enough. The observer in Minkowski space can therefore potentially see every particle in the universe.

Now let's try and work out some of the details of the conformal mapping. As a first step, we express Minkowski space in light-cone coordinates.

Example 19.3

Minkowski space has the usual metric $\boldsymbol{\eta}$ with line element

$$\mathrm{d}s^2 = -\mathrm{d}t^2 + \mathrm{d}x^2 + \mathrm{d}y^2 + \mathrm{d}z^2. \tag{19.4}$$

In spherical polar coordinates, this becomes

$$\mathrm{d}s^2 = -\mathrm{d}t^2 + \mathrm{d}r^2 + r^2(\mathrm{d}\theta^2 + \sin^2\theta\mathrm{d}\phi^2). \tag{19.5}$$

Since we're interested in the light-cone structure, we re-express the metric line element in **light-cone coordinates**

$$u = t - r, \quad v = t + r, \tag{19.6}$$

where $r^2 = x^2 + y^2 + z^2$. In terms of these coordinates, outgoing light rays (which have r increasing and t increasing) are lines of constant u, while incoming light rays (r decreasing for increasing time t) are lines of constant v. If we express the Minkowski metric in light-cone coordinates, we obtain

$$\mathrm{d}s^2 = -\mathrm{d}u\mathrm{d}v + \frac{1}{4}(v - u)^2(\mathrm{d}\theta^2 + \sin^2\theta\mathrm{d}\phi^2). \tag{19.7}$$

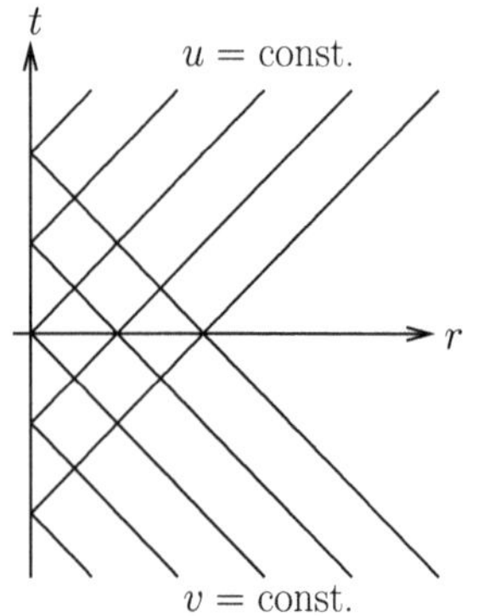

Fig. 19.9 Minkowski spacetime in light-cone coordinates. Each point represents a sphere of radius r.

The part of the metric $(\mathrm{d}\theta^2 + \sin^2\theta\mathrm{d}\phi^2)$ is simply that of a 2-sphere. If we suppress two dimensions, as we did when discussing Minkowski spacetime, then we can draw the two-dimensional plot of this spacetime shown in Fig. 19.9, as long as we interpret each point as representing a 2-sphere of radius r.

Minkowski space is covered by light-cone coordinates $-\infty \leq u, v \leq \infty$, but we want to avoid letting our coordinates go to infinity. Our Penrose diagram will need to be constructed to deal with the structure of spacetime at infinity; the idea is that we would like to be able to grab hold of the metric at infinity and bring it back to finite coordinates where it can be studied. One way of doing this would be to use a function like $\tan^{-1} x$, shown in Fig. 19.10. This function allows us to map the whole of the x axis (including $\pm\infty$) to the coordinate range $-\pi/2 \leq y \leq \pi/2$. We attempt this strategy in the next example.

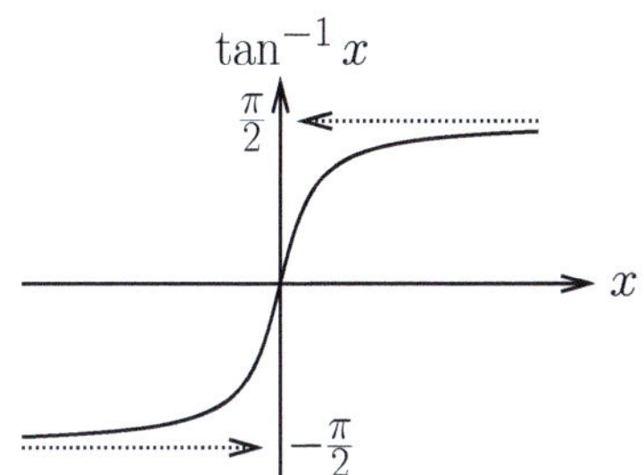

Fig. 19.10 The function $y = \tan^{-1} x$ can be used to map infinities in x back to an angle-like variable y.

Example 19.4

To examine an infinity in the Minkowski metric $\boldsymbol{\eta}$, we could try the simple tactic of using a new coordinate $V = 1/v$, since this maps infinite v to $V = 0$. However, if we try substituting this we end up with a line element for Minkowski space of

$$\mathrm{d}s^2 = \frac{1}{V^2}\mathrm{d}u\,\mathrm{d}V + \frac{1}{4}\left(\frac{1}{V} - u\right)^2 (\mathrm{d}\theta^2 + \sin^2\theta\mathrm{d}\phi^2). \tag{19.8}$$

This metric is still the Minkowski metric $\boldsymbol{\eta}$, but in these coordinates it is singular at $V = 0$, where the line element blows up. This first attempt has failed.

Undeterred, we make a sidestep and consider a new, **unphysical metric** $\bar{\boldsymbol{g}}$ such that

$$\boldsymbol{\eta} = \frac{1}{V^2}\bar{\boldsymbol{g}}, \tag{19.9}$$

which is to say that the two metrics $\boldsymbol{\eta}$ and $\bar{\boldsymbol{g}}$ are conformally related (i.e. $\bar{\boldsymbol{g}} = \Omega^2\boldsymbol{\eta}$ with a conformal factor $\Omega(x) = V$). As a result, the line element of the unphysical metric is given by $\mathrm{d}\bar{s}^2 = V^2\mathrm{d}s^2$ or

$$\mathrm{d}\bar{s}^2 = \mathrm{d}u\,\mathrm{d}V + \frac{1}{4}(1 - uV)^2(\mathrm{d}\theta^2 + \sin^2\theta\mathrm{d}\phi^2), \tag{19.10}$$

which is well behaved at $V = 0$.[9] Since the unphysical metric $\bar{\boldsymbol{g}}$ and the Minkowski metric $\boldsymbol{\eta}$ are related by a conformal transformation, they share the same light-cone structure and so we can equally well study $\bar{\boldsymbol{g}}$ as a more convenient alternative to $\boldsymbol{\eta}$.

Our strategy is now clear: (i) use coordinates that map infinities to finite points and then (ii) identify an unphysical, conformally related metric that does not have any problems with its coordinates. We can then study the unphysical metric since the conformal relationship guarantees it has the same causal structure.

[9]Technically, what we have done here it to extend Minkowski space by adding points represented by $V = 0$. Unlike $\eta_{\mu\nu}$, the unphysical metric components $\bar{g}_{\mu\nu}$ can be extended to $V = 0$.

Example 19.5

The same strategy employed in the last example can be enacted more cleverly by defining $\bar{\boldsymbol{g}} = \Omega^2\boldsymbol{\eta}$ with

$$\Omega^2 = \frac{4}{(1 + v^2)(1 + u^2)}. \tag{19.11}$$

Now define p and q by

$$v = \tan p, \quad u = \tan q \tag{19.12}$$

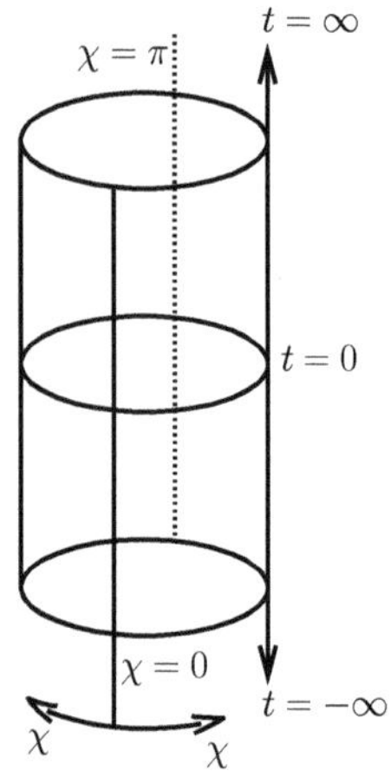

Fig. 19.11 The Einstein static Universe represented on a cylinder to demonstrate its topology. Each circle represents a 3-sphere.

[10]This is achieved technically by embedding the cylinder $x^2 + y^2 = 1$ in (2+1)-dimensional Minkowski spacetime. The full spacetime can be embedded as the cylinder $x^2+y^2+z^2+w^2 = 1$ in (4+1)-dimensional Minkowski spacetime (see Appendix D).

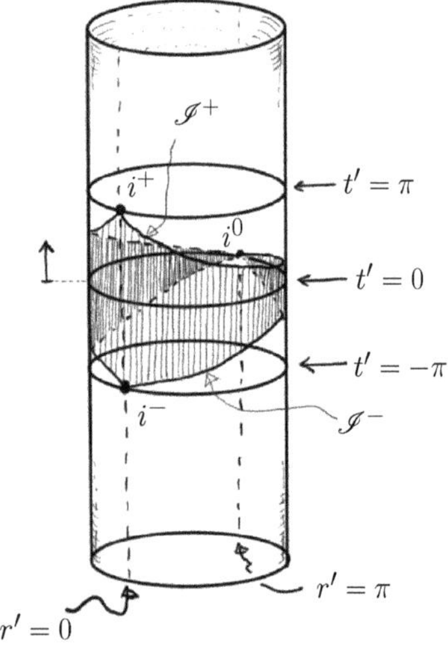

Fig. 19.12 The causal structure of Minkowski space mapped onto a limited region of a larger space. The larger space is the Einstein static Universe.

where
$$-\pi/2 < p < \pi/2, \quad -\pi/2 < q < \pi/2, \quad p \geq q. \tag{19.13}$$
Using $\sec^2 \theta = 1 + \tan^2 \theta$, the function Ω^2 becomes $\Omega^2 = 4/(\sec^2 p \sec^2 q)$, and the metric can be rewritten as
$$ds^2 = \sec^2 p \sec^2 q \left[-dp\,dq + \frac{1}{4}\sin^2(p - q)(d\theta^2 + \sin^2 \theta d\phi^2) \right]$$
$$= \frac{1}{\Omega^2} \left[-4dp\,dq + \sin^2(p - q)(d\theta^2 + \sin^2 \theta d\phi^2) \right]. \tag{19.14}$$
We therefore identify the line element $d\bar{s}$ of the unphysical metric $\bar{g}$ as being given by
$$d\bar{s}^2 = -4dp\,dq + \sin^2(p - q)(d\theta^2 + \sin^2 \theta d\phi^2). \tag{19.15}$$
This is the metric whose causal structure can be investigated. The line element $d\bar{s}^2$ can be reduced to a more familiar form by defining
$$t' = p + q, \quad r' = p - q, \tag{19.16}$$
where
$$-\pi < t' + r' < \pi, \quad -\pi < t' - r' < \pi, \quad r' \geq 0. \tag{19.17}$$
With this choice of coordinates, the metric becomes
$$d\bar{s}^2 = -(dt')^2 + (dr')^2 + \sin^2 r'(d\theta^2 + \sin^2 \theta d\phi^2). \tag{19.18}$$
This is the line element of Universe 8 (the Einstein static Universe, a closed $k = 1$ cosmology) which has a metric line element
$$ds^2 = -dt^2 + d\chi^2 + \sin^2 \chi(d\theta^2 + \sin^2 \theta d\phi^2), \tag{19.19}$$
and has spatial hypersurfaces that are all spherical. Specifically, these are 3-spheres covered by the coordinates
$$0 \leq \chi \leq \pi, \quad 0 \leq \theta \leq \pi, \quad -\pi < \phi \leq \pi. \tag{19.20}$$
Since the Einstein model is static, the spheres are unchanging in time. Note that eqn 19.18 is defined over a different set of coordinates to eqn 19.19.

It will turn out that the Einstein universe will be a very useful model against which to compare our other model spacetimes. Let's therefore attempt to draw Einstein spacetime with its unchanging 3-spheres. As you'll appreciate, it's a challenge to draw anything more complicated than a representation of a three-dimensional object on a two-dimensional page. Nonetheless, owing to the spherical symmetry of the model, we can suppress two of the spatial dimensions and just consider circles (or 1-spheres, if you like) as a function of time. The overall shape or topology of the spacetime manifold now comprises circles of constant radius (representing the 3-spheres) repeated as a function of time. We say that the topology of the manifold is cylindrical. The space can be represented[10] in the diagram shown in Fig. 19.11, where θ and ϕ have been suppressed. We notice the awkward way that the χ variable increases in two directions, reflecting the fact that it varies between 0 and π.

Although it has involved a certain amount of abstraction, in our cylindrical representation of the Einstein Universe, we have a structure with which we can now compare other spacetimes.

The last example shows how the whole of Minkowski space can be mapped to a *limited* region of the Einstein static metric. This allows us to draw a representation of Minkowski space on the cylinder in Fig. 19.11. The result is Fig. 19.12. The fact that it only covers a limited region means that *the boundary of Minkowski space represents the infinities of this space*. Strictly, it represents the **conformal structure of infinity**. The structure in Fig. 19.12 can be pared down to its bare essentials, and it is this that leads to the Penrose diagram that we saw in Fig. 19.5(b) which depicts the boundary in Fig. 19.12 plotted in the (t', r') plane, not including the superfluous half where r' increases anticlockwise around the cylinder. This approach also allows us to identify some different types of infinity that we encountered earlier.

Example 19.6

Referring to the labelled points in Fig. 19.12 we see that, in terms of the variables (t', r'), we have labelled

$$i^0 = (0, \pi), \quad i^+ = (\pi, 0), \quad i^- = (-\pi, 0). \tag{19.21}$$

It follows that the line labelled $\mathscr{I}^+$ is the line $t+r = \pi$ and $\mathscr{I}^-$ is the line $t-r = -\pi$. These lines, shown in Fig. 19.12 (which actually represents surfaces in spacetime, owing to our having suppressed two dimensions) are null. Since future-directed light rays have $t' - r' = 2q = $ const., then *all* future-directed light rays terminate on $\mathscr{I}^+$. Similarly, the line $\mathscr{I}^-$ also represents a null surface and all inward or past-directed light cones terminate on it (i.e. rays for which $t' + r' = 2p = $ const).

The definitions and rules introduced so far are applied in the next section. Suitably expanded and generalized, can be used as the foundation of a set of **global techniques** that can also be used to make very powerful and general arguments about singularities.[11]

[11]We discuss an example in Chapter 50. See Penrose (1973) for the full story.

19.2 The de Sitter spacetime

Now that we have a picture of Minkowski space and its conformal structure, let's examine Universe 1 (de Sitter spacetime). This is the spacetime with flat and expanding space described by line element

$$ds = -dt^2 + e^{2Ht} \left[dr^2 + r^2 \left(d\theta^2 + \sin^2\theta d\phi^2 \right) \right]. \tag{19.22}$$

With a little algebra it can be shown[12] that the line element for Universe 1 can be rewritten in terms of a new timelike variable t' as

$$ds^2 = \alpha^2 \sec^2 t' d\bar{s}^2, \tag{19.23}$$

[12]See exercises.

where $d\bar{s}^2$ is the line element for the Einstein Universe [with coordinates (t', r, θ, ϕ)], where $-\frac{\pi}{2} < t' < \frac{\pi}{2}$, $\alpha^2 = 1/H^2$, and the other coordinates take the same values as in the Einstein static Universe. This means that de Sitter space is conformal to the Einstein static Universe, but only for times $-\frac{\pi}{2} < t' < \frac{\pi}{2}$.

The resulting Penrose diagram is shown in Fig. 19.13. Strikingly, there are no unique timelike ($i^\pm$) or spacelike (i^0) infinities on which world lines terminate, as we had for Minkowski spacetime. Instead, there is a past null infinity surface $\mathscr{I}^-$ and a future null infinity surface $\mathscr{I}^+$. These are spacelike: they link up all of χ at a single instant ($t' = \pm\pi/2$). As a result of this difference in conformal structure, causality in de Sitter spacetime is rather different to that in Minkowski spacetime and provides the possibility of some new features: **particle horizons** and **event horizons**. Since, in de Sitter spacetime, all timelike geodesics start at $\mathscr{I}^-$ and end at $\mathscr{I}^+$, the past light cone of an observer at a point p does not, generally, capture all points in the past. This means that there are events, and even whole world lines of events, that an observer never sees. An example is shown in Fig. 19.14: some free particles fall

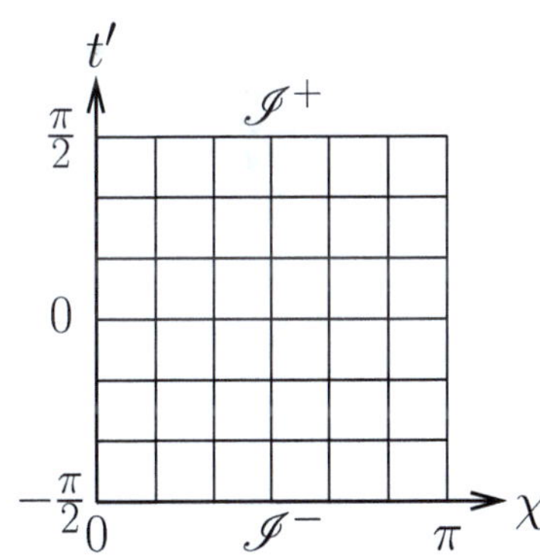

Fig. 19.13 Penrose diagram for de Sitter spacetime.

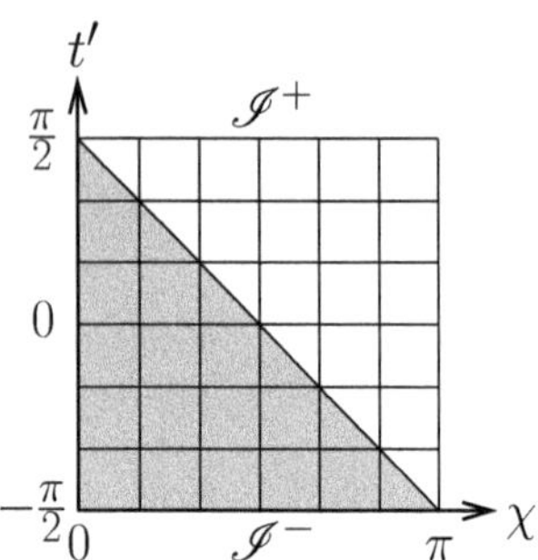

Fig. 19.15 The observer whose world line coincides with the origin of the radial coordinate $\chi = 0$ at $t' = \pi/2$ has a light cone that sweeps out the whole χ axis.

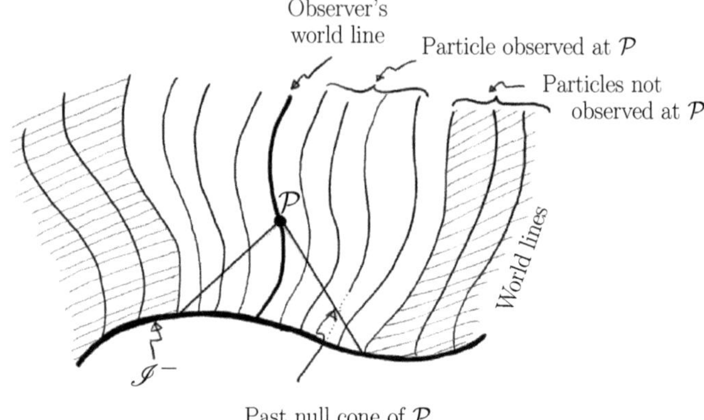

Fig. 19.14 The particle horizon for an observer at point $\mathcal{P}$, in a spacetime with a spacelike past null infinity $\mathscr{I}^-$. There exist particles whose world lines do not intersect the past null cone of $\mathcal{P}$ and so are not yet visible to the observer.

[13] It is perhaps tempting to think that an observer at $\chi = \pi/2$ can have seen events from the entire χ axis at the earlier time $t' = 0$. However, there cannot be a difference since each value of χ is equivalent. The problem here is that χ is a radius-like variable so it only makes sense to consider the light cone over increasing *or* decreasing χ when considering the question of the furthest a light ray can travel in a given time.

along world lines that lie outside the backward light cone of the observer at $\mathcal{P}$. This leads to a division between those geodesics that can be seen by the observer at $\mathcal{P}$ and those that can't. This division is known as the **particle horizon** at $\mathcal{P}$. As the observer moves along their world line, they are able to see more and more particle geodesics, so this horizon is observer dependent. Moreover, when the observer in de Sitter space finally meets the future null infinity $\mathscr{I}^+$, their past light cone does not sweep out all of spacetime. [As shown in Fig. 19.15, the backward light cone of an observer whose world line coincides with the origin of the radial coordinate χ does sweep over all values of χ at future infinity ($t' = -\pi/2$) for this spacetime, allowing the observer to see, at different times, signals from all of the events occurring in the shaded region.[13]] There are, therefore, events that can never been seen by the observer. The surface formed from the backward light cone on the future infinity $\mathscr{I}^+$ is known as the **future event horizon**. By contrast, since the observer in Minkowski space can see all of spacetime when they reach i^+, there is no future event horizon in Minkowski spacetime.

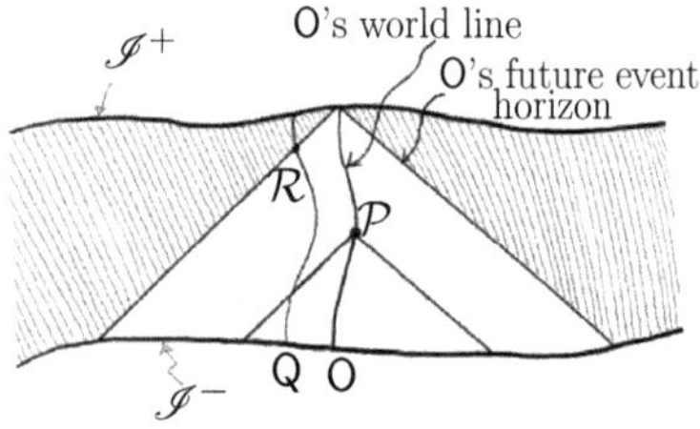

Fig. 19.16 The future event horizons for an observer travelling along a world line O. The shaded region lie's outside O's future event horizon and events in this region will never be seen by O.

Example 19.7

Consider the Penrose diagram in Fig. 19.16. The observer at point $\mathcal{P}$ on world line O can see signals emitted from part of a world line Q originating at $\mathscr{I}^-$. In other words, since part of curve Q is inside the particle horizon at $\mathcal{P}$, a particle travelling along world line Q is sometimes visible to the observer at $\mathcal{P}$. Owing to the existence of future event horizons in this spacetime, there is a point $\mathcal{R}$, where the world line Q meets the observer's future event horizon. The observer on $\mathcal{P}$ has no possibility of seeing events any further in the future along Q than $\mathcal{R}$. Imagine travelling along the observer's world line O. They can only see the event $\mathcal{R}$ when they (the observer) are at infinity (since this is the first time that the event at $\mathcal{R}$ lies on their past light cone).

Just as we have a future event horizon, we can also define a **past event horizon**.[14] This is the surface traced out by the forward light cone of a point on the past null infinity $\mathscr{I}^-$ as also shown in Fig. 19.17. This surface divides up all of the events that can be influenced or contacted by an observer from those that can't (with events outside the past event horizon unable to be influenced by the observer).

[14]Sit at a point at *future* infinity and draw a backward light cone get the *future* event horizon; sit at *past* infinity and cast the forward light cone to get the *past* event horizon.

Example 19.8

Consider the Penrose diagram shown in Fig. 19.17. Region I contains events that can be both seen and influenced by the observer; region II contains events that can be influenced, but never seen, and region III contains events that can be seen, but never influenced.

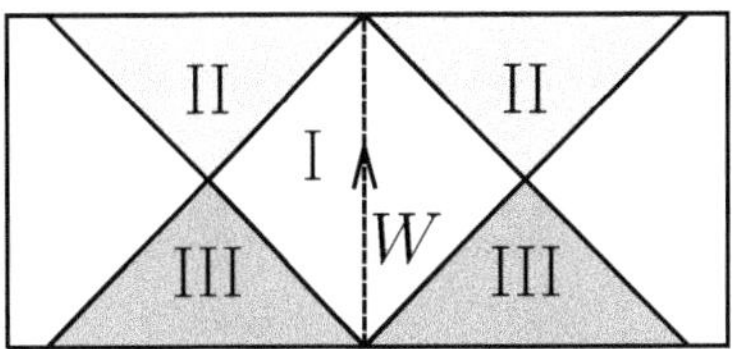

Fig. 19.17 Past and future event horizons for a world line W, dividing up spacetime.

19.3 Big-Bang singularities

Finally, we turn to the conformal structure of the universes based on the Robertson–Walker geometries. The Robertson–Walker spaces are defined a metric with line element[15]

$$\mathrm{d}s^2 = -\mathrm{d}t^2 + a(t)^2 \left[\mathrm{d}\chi^2 + \Sigma(\chi)^2(\mathrm{d}\theta^2 + \sin^2\theta\mathrm{d}\phi^2) \right]. \tag{19.24}$$

These solutions have a new feature: the initial Big-Bang singularity when $a(t=0)=0$. We should therefore expect that they should cover a limited region of the cylinder covered by the Einstein static Universe, and that there must be a boundary representing the Big Bang.

To investigate the conformal structure we start by using a new conformal time coordinate τ defined by

$$\mathrm{d}t = a(t)\mathrm{d}\tau. \tag{19.25}$$

The idea is that we then have $\mathrm{d}s^2 = a(t)^2\mathrm{d}\bar{s}^2$, with

$$\mathrm{d}\bar{s}^2 = -\mathrm{d}\tau^2 + \mathrm{d}\chi^2 + \Sigma(\chi)^2(\mathrm{d}\theta^2 + \sin^2\theta\mathrm{d}\phi^2). \tag{19.26}$$

This line-element is in a form that can be compared to the Einstein static Universe.

[15]The function $\Sigma(\chi)$ is given by

$$\begin{aligned} \Sigma(\chi) &= \sin\chi, & (k=+1), \\ \Sigma(\chi) &= \chi, & (k=0), \\ \Sigma(\chi) &= \sinh\chi, & (k=-1). \end{aligned}$$

Example 19.9

We can show that the Robertson–Walker spaces with $p = \Lambda = 0$ are conformal to the Einstein static Universe. We first set $\chi = r'$ and $\tau = t'$, where the primed coordinates are those from Example 19.5. For $k=1$ the line element is then conformal to the Einstein Universe but, since τ is defined for $0 < \tau < \infty$, we find that the space is mapped into the Einstein Universe for $t' = 0$ (when $\tau = 0$) to $t' = \pi$ (when $\tau = \infty$). For $k=0$ we have flat space, so, using the same coordinates and the procedure from Example 19.5, we again obtain a diamond-shaped region on the cylindrical surface, but this time with $0 < t < \pi$. For $k = -1$ things are more tricky and we must make a coordinate transformation[16] with the result that the metric becomes conformal to the region

$$-\tfrac{\pi}{2} \le t' + r' \le \tfrac{\pi}{2}, \quad -\tfrac{\pi}{2} \le t' - r' \le \tfrac{\pi}{2}. \tag{19.27}$$

[16]The transformation turns out to be

$$\begin{aligned} t' &= \tan^{-1}\left[\tanh\tfrac{1}{2}(t+\chi)\right] \\ &\quad + \tan^{-1}\left[\tanh\tfrac{1}{2}(t-\chi)\right], \\ r' &= \tan^{-1}\left[\tanh\tfrac{1}{2}(t+\chi)\right] \\ &\quad + \tan^{-1}\left[\tanh\tfrac{1}{2}(t-\chi)\right]. \end{aligned}$$

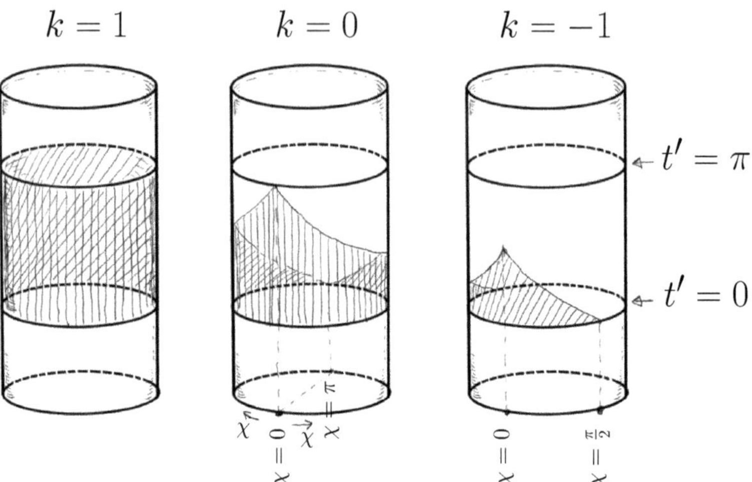

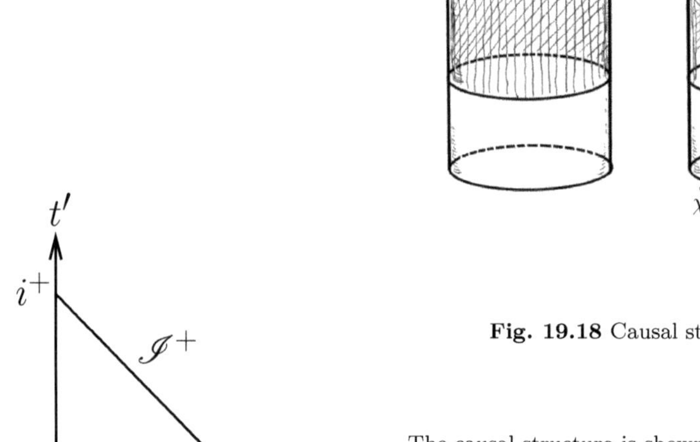

Fig. 19.18 Causal structure of the Robertson–Walker spacetimes.

The causal structure is shown on the Einstein cylinder in Fig. 19.18.

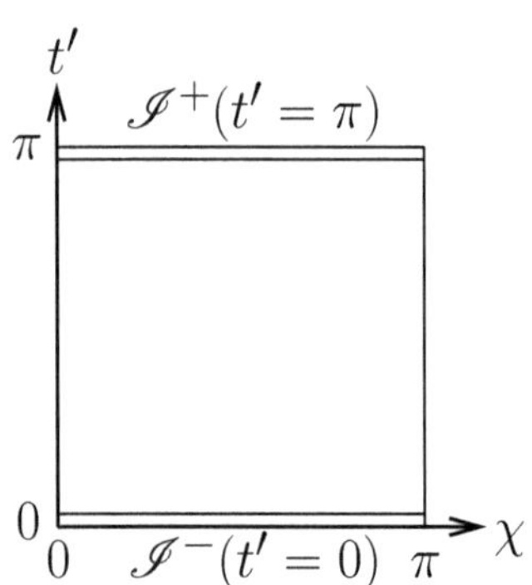

Fig. 19.19 Penrose diagram for the $k = 0$ and $k = -1$ Robertson–Walker spacetimes. The spacelike singularity is shown as a double line.

With the causal structure of the Robertson–Walker spaces determined, we draw their Penrose diagrams. As predicted, these feature a singularity. In terms of their conformal structure, the singularity is not represented as a point, but as a surface. Specifically, the Robertson–Walker spacetimes have a **singular surface** at $t' = 0$ representing the Big Bang, from which all world lines must originate. This surface is spacelike, so we say that it represents a **spacelike singularity**.

For the flat $k = 0$ and hyperbolic $k = -1$ spaces, the Penrose diagram (Fig. 19.19) resembles the upper half of the Penrose diagram for Minkowski space, but starting from the Big-Bang singularity. Timelike geodesics start at any point on the surface $\mathscr{I}^-$, but all terminate at the point i^+. There is therefore a past event horizon for an observer on a timelike geodesic, but there is not a future event horizon: when they reach i^+ the observer can see every event in the spacetime. (There is also a particle horizon in this spacetime.) Like massive particles, light rays also start on $\mathscr{I}^-$, but these terminate on $\mathscr{I}^+$.

The situation is different for an spherical (closed) $k = 1$ space: this is because the radial variable is constrained to lie in the range $0 < \chi < \pi$. In this case, the universe expands from an initial singularity (the Big Bang) and then contracts towards a second singularity (the Big Crunch). There are therefore two singular surfaces in the Penrose diagram, as shown in Fig. 19.20. All world lines start on the line $\mathscr{I}^-$ and terminate on $\mathscr{I}^+$. Like the conformal structure of the de Sitter spacetime, the $k = 1$ space allows past and future event horizons.

Fig. 19.20 Penrose diagram for the $k = 1$ Robertson–Walker spacetime. The spacelike singularities are depicted by double lines.

Example 19.10

Consider Universe 6 ($k = 1, \Lambda = 0$) from Chapter 18, with its Big Bang, expansion, contraction and Big Crunch, and causal structure depicted in Fig. 19.20. Since this universe is closed with $\Sigma(\chi) = \sin\chi$, the χ axis is periodic. We therefore identify $\chi = 0$ and $\chi = \pi$ in the Penrose diagram, so that light rays meeting the edges at $\chi = 0$ and $\chi = \pi$ wrap around the diagram. As shown in Fig. 19.21(a), this allows us (the observers at the origin) to see signals from all points on the χ axis at time $t' = \pi/2$. If there were always a star at each point on the χ grid, we would see all of the stars in the universe at this time. As shown in Fig. 19.22(a), this corresponds to radial light rays travelling to reach us by traversing the two halves of the unit circle that represents the angle-like radial coordinate χ.

During the expansion of this universe, we therefore see more and more stars [Fig. 19.21(a)] until, at the time $t' = \pi/2$ when the universe has its maximum radius $a(t' = \pi/2)$, we see all of the stars: half in front of us, and half behind. As the universe subsequently contracts, our backward light cone continues to get larger, and we start to see duplicates of the more distant stars, as light travelling over more than half of the circle begins to reach us. [An example of a duplicate is shown in Fig. 19.21(b) and Fig. 19.22(b)]. As the contraction of the universe continues, we come to see duplicates of more and more of the stars. We expect eventually to be able to see the back of our own heads (as they appeared shortly after the Big Bang!), when radial light rays from the base of our own world lines at $t' = 0$ reach us after traversing the entire universe. However, the observation of these rays coincides with the Big Crunch, and so we are crushed to death at exactly this moment (see Exercise 19.8). Moreover, since nothing can travel faster than light, there is no possibility of any particle to traverse the whole of the closed universe.

We can also analyse the story in terms of the red- and blueshifts that the observer measures. Initially, all of the stars seen will be redshifted since $a(t_{\text{em}}) < a(t_{\text{ob}})$. During the collapsing period, there will times when starlight traversing the most direct route will be blueshifted, while light from the duplicate image, traversing the longer route, will be redshifted. During the final stages of the collapse, most of the light received will be blueshifted since $a(t_{\text{em}}) > a(t_{\text{ob}})$.

We have now seen how to describe the incorporation of singularities like the Big Bang into the conformal structure of spacetime. Singularities can now be understood in the same way that we understand infinities: *they represent the boundaries of spacetime*. They are different from infinities in that they could exist at a finite interval from an observer. (By definition this is not the case for an infinity!) Singularities are still potentially a cause for concern since, as they are regions where the laws of physics break down, you never know what will come out of a singularity. However, the toolkit offered by conformal structure allows us to tame them geometrically and Penrose diagrams give us a convenient way to visualize them. We shall encounter Penrose diagrams again in Chapter 25 when we examine black holes and where, once again, their usefulness in incorporating a singularity will be important.

Finally, we are finishing Part III on cosmology without saying too much about which universe we are actually in, and what evidence might exist for determining that. This is because this is a book on general relativity as a theory, rather than a primer on cosmology. But we will return to this important issue at the end of the book in Chapter 50.

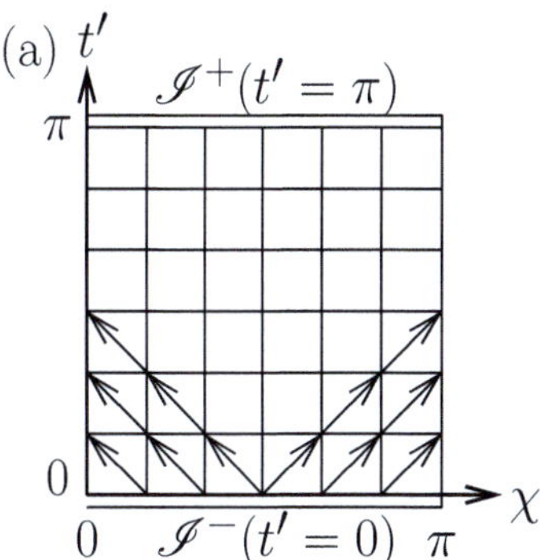

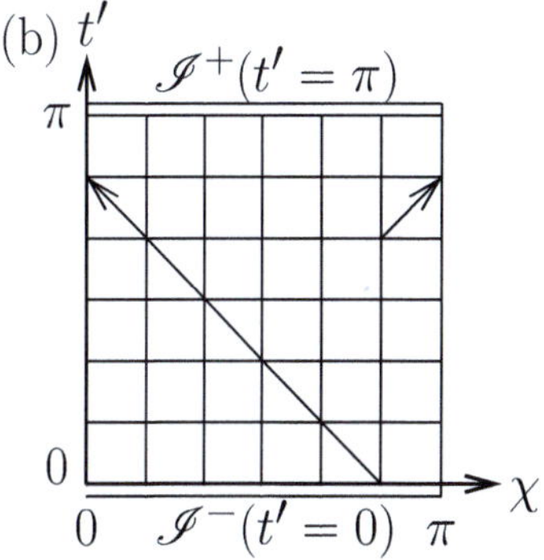

Fig. 19.21 (a) In Universe 6, the observer at $\chi = 0$ can see events from each point on the χ grid at $t' = \pi/2$. This is because $\chi = 0$ is identified with $\chi = \pi$. (b) An example of the two images of a single star seen by the observer.

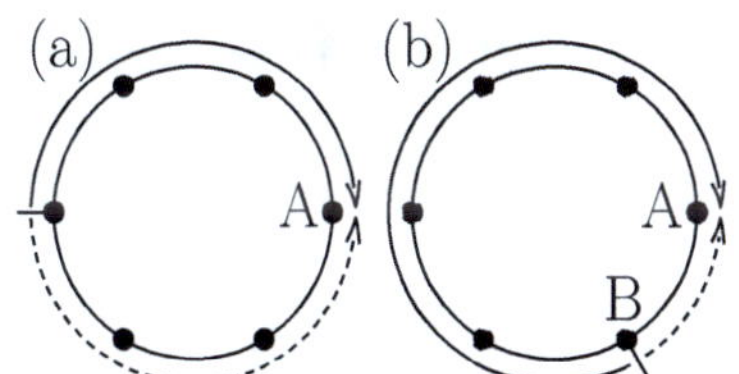

Fig. 19.22 (a) For the closed Universe 6 at the point of maximum expansion, light can reach us (at point A) from stars in the whole spherical universe, with half going one way (appearing in front of us, say) and half going the other way (appearing behind us). (b) As the universe contracts, light from B can reach us via the direct route (anticlockwise, as shown) and the indirect route (clockwise), so we see two images of B. (Don't confuse the variable χ with the angle around the unit circle: they differ by a factor of 2.)

Chapter summary

- The conformal structure of a model universe allows us to treat its infinities as points or surfaces at the edge of spacetime.
- Penrose diagrams allow us to visualize the conformal structure of a spacetime, revealing particle and event horizons.
- Singularities can be incorporated into Penrose diagrams: they form boundaries to spacetime that can be separated from an observer by a finite interval.

Exercises

(19.1) In Minkowski spacetime in (1+1) dimensions, light-cone coordinates are defined by the transformation

$$u = t - x,$$
$$v = t + x. \tag{19.28}$$

(a) Confirm that the metric line element, expressed in these coordinates, becomes

$$ds^2 = -du\,dv. \tag{19.29}$$

(b) How are proper lengths and proper times computed in these coordinates?

(19.2) Consider a particle moving at constant velocity along the x axis such that $x = \beta t$.

(a) Transform to light-cone coordinates ($u = t - x, v = t + x$) and show that the light-cone velocity is

$$\frac{du}{dv} = \frac{1-\beta}{1+\beta}. \tag{19.30}$$

(b) Investigate the light-cone velocities of particles with different values of β.

(19.3) The cosine angle between vectors $\boldsymbol{u}$ and $\boldsymbol{v}$ is defined as

$$\frac{g(\boldsymbol{u},\boldsymbol{v})}{[g(\boldsymbol{u},\boldsymbol{u})g(\boldsymbol{v},\boldsymbol{v})]^{\frac{1}{2}}}, \tag{19.31}$$

(a) Show that a conformal transformation $g_{\mu\nu} \to f(x)g_{\mu\nu}$ preserves all angles.
(b) Show that all null curves remain null under this transformation.

(19.4) Consider the de Sitter line element in Cartesian coordinates

$$ds^2 = -d\hat{t}^2 + e^{\frac{2\hat{t}}{\alpha}}\left(d\hat{x}^2 + d\hat{y}^2 + d\hat{z}^2\right), \tag{19.32}$$

where $\alpha = H^{-1}$. In the exercises in Chapter 18, we claimed that de Sitter spacetime can be represented on a hyperboloid embedded in five-dimensional Minkowski space, given by an equation

$$-v^2 + w^2 + x^2 + y^2 + z^2 = \alpha^2. \tag{19.33}$$

(a) By making the transformations

$$\hat{t} = \alpha \ln\frac{w+v}{\alpha}, \quad \hat{x} = \frac{\alpha x}{w+v},$$
$$\hat{y} = \frac{\alpha y}{w+v}, \quad \hat{z} = \frac{\alpha z}{w+v}, \tag{19.34}$$

show that ds^2 can be recast as the Minkowski line element

$$ds^2 = -dv^2 + dw^2 + dx^2 + dy^2 + dz^2. \tag{19.35}$$

(b) Now consider a set of coordinates on the hyperboloid given by

$$v = \alpha \sinh(Ht),$$
$$w = \alpha \cosh(Ht)\cos\chi,$$
$$x = \alpha \cosh(Ht)\sin\chi\cos\theta, \tag{19.36}$$
$$y = \alpha \cosh(Ht)\sin\chi\sin\theta\cos\phi,$$
$$z = \alpha \cosh(Ht)\sin\chi\sin\theta\sin\phi.$$

Show that the de Sitter Universe takes on the metric line element

$$ds^2 = -dt^2 + \alpha^2 \cosh^2(\alpha^{-1}t) \tag{19.37}$$
$$\times \left[d\chi^2 + \sin^2\phi(d\theta^2 + \sin^2\theta d\phi^2)\right].$$

(c) Using the result of the previous problem, use the transformation

$$\tan\left(\frac{t'}{2} + \frac{\pi}{4}\right) = e^{t/\alpha}, \qquad (19.38)$$

where $-\pi/2 < t' < \pi/2$, to show that de Sitter spacetime can be recast as in eqn 19.23.

(19.5) The action for the electromagnetic field is invariant with respect to conformal transformations. Consider a Robertson–Walker spacetime for which we make a transformation $dt = a(t)d\tau$ leading to the conformal transformation $ds^2 = a(t)^2 d\bar{s}^2$.
(a) Consider a solution to the electromagnetic wave equation of the form $A_\mu \propto e^{iB\tau}$, with B a constant, in the spacetime with line element $d\bar{s}$. What form will this solution take in the spacetime with line element ds in terms of the variable t?
(b) Explain why this implies that the instantaneous frequency of radiation $\omega(t)$ obeys $\omega(t)a(t) =$ const.
This equation is another demonstration of the cosmological redshift that occurs for light, emitted at time t_1 and observed at time t_2, for an expanding universe with $a(t_2) > a(t_1)$.

(19.6) What is the relationship between coordinates (t, x) and (t', x') in the Penrose diagram for Minkowski space? Use this to justify the behaviour of the geodesics in Fig. 19.6.

(19.7) Consider the Penrose diagram for Minkowski space and those geodesics starting at the origin moving radially outwards.

(a) Draw the null geodesic followed by a light pulse.
(b) Draw the timelike geodesic followed by a particle with a speed $\beta = 0.3$.
(c) Draw the spacelike geodesic for $\beta = 1.1$.

(19.8) Consider the story told in Example 19.10. We shall analyse this using Sidney Coleman's method (explained further in Coleman, 2022).
(a) Show that the Friedmann equation predicts that during the evolution, we have

$$dt = \frac{da}{(C/a - 1)^{\frac{1}{2}}}. \qquad (19.39)$$

(b) Define the comoving distance η via $dt = a(t)d\eta$ and show that a radial light ray travels a distance given by η.
This comoving distance η is identical to the conformal time coordinate in eqn 19.25.
(c) Show further that

$$\cos \eta = 1 - \frac{2a}{C}, \qquad (19.40)$$

solves the equation of motion.
(d) Finally, show that, as parametrized by η, we have

$$a(\eta) = \frac{C}{2}(1 - \cos \eta),$$
$$t(\eta) = \frac{C}{2}(\eta - \sin \eta). \qquad (19.41)$$

(e) Use this parametrization to verify the story in the example.

Part IV

Orbits, stars, and black holes

In Part IV of the book, we turn to the questions of some of the strongly gravitating objects in the Universe and the motions that they promote.

- In Chapter 20, we review the machinery of Newtonian gravity and the methods for determining the possible motions in a Newtonian gravitating potential.

- In Chapter 21, we discuss the Schwarzschild metric, which describes a stationary, spherically symmetric distribution of mass. We then determine the possible motions allowed by such a metric in Chapters 22, 23 and 24.

- In Chapters 25–29, we examine static black holes, objects that gravitate so strongly that not even light can escape their interior. Black holes likely contain a singularity in spacetime, which are understood using tools from Chapters 26 and 27. In Chapter 28, we discuss black hole thermodynamics and the radiation that can emerge from black holes. Finally, in Chapter 29, we discuss the rich structures that result if we allow our black holes to carry charge or to rotate in space.

20 Newtonian orbits

↱ **The material in this chapter provides some background to the Newtonian theory of gravitation. Readers eager to explore orbital motion in general relativity right away can skip to the next chapter.**

[1]We shall temporarily restore factors of G in this chapter to make contact with familiar-looking equations.

[2]The dot denotes a derivative with respect to t in this chapter.

[3]The particle is at $\vec{r} = r\vec{e}_{\hat{r}}$ and note that in polar coordinates

$$\frac{\partial \vec{e}_{\hat{r}}}{\partial r} = 0, \quad \frac{\partial \vec{e}_{\hat{r}}}{\partial \theta} = \vec{e}_{\hat{\theta}},$$

$$\frac{\partial \vec{e}_{\hat{\theta}}}{\partial r} = 0, \quad \frac{\partial \vec{e}_{\hat{\theta}}}{\partial \theta} = -\vec{e}_{\hat{r}},$$

where, as advertised in Example 10.4, we use orthonormal coordinates in this problem (denoted by hats on indices). Hence,

$$\vec{v} = \dot{\vec{r}} = \dot{r}\vec{e}_{\hat{r}} + r\dot{\vec{e}}_{\hat{r}} = \dot{r}\vec{e}_{\hat{r}} + r\dot{\theta}\vec{e}_{\hat{\theta}}.$$

The moon gravitates towards the earth and by the force of gravity is continually drawn off from a rectilinear motion and retained in its orbit.
Isaac Newton

'But the Solar System!' I protested.
'What the deuce is it to me?' (Sherlock Holmes) interrupted impatiently: 'you say that we go round the sun. If we went round the moon it would not make a penny-worth of difference to me and my work.'
Arthur Conan Doyle (1859–1930) *A Study in Scarlet*

General relativity is well known to provide corrections to Newtonian gravitation. In this chapter, we take a step back, and present a set of methods to find and describe the trajectories allowed by the $\Phi(r) \propto -1/r$ potential of *Newtonian* gravitation. On solving the problem we shall find that there is a restricted class of possible trajectories. It turns out that the same methods employed here can be used to deal with the more varied trajectories allowed in general relativity.

In Newtonian gravitation, the potential energy due to the gravitational interaction between two particles with masses m and M, separated by a distance r, is given by[1] $U(r) = -GMm/r$. We take the mass M to be fixed at the origin and the mass m to be separated from M by the 3-vector $\vec{r}$, and moving with momentum $\vec{p} = m\vec{v}$. The Newtonian force on the mass m, given by $\vec{F}(r) = -GMm\vec{r}/r^3$, acts radially, and therefore does not give rise to any torque $\vec{\tau}$. This is because $\vec{\tau} = \vec{r} \times \vec{F}$ and so, for a central force such as Newtonian gravitation, $\vec{\tau} \propto \vec{r} \times \vec{r} = 0$. Since the angular momentum of the moving mass $\vec{L} = \vec{r} \times \vec{p}$ is related to the torque via[2] $\vec{\tau} = \dot{\vec{L}}$, the angular momentum for any central force is a constant of the motion. This constant angular momentum can be used, along with the constant energy of the system E, to classify the possible trajectories of the moving particle.

As a further result of the conservation of angular momentum, the vector $\vec{L}$ (which is, by definition, perpendicular to $\vec{r}$ and $\vec{p}$) remains fixed. This means that the path of a particle in Newtonian gravitation can be taken as being confined to the equatorial plane of a set of three-dimensional spherical coordinates. Taking $\vec{L}$ to be pointing along the z-direction, we can therefore follow the paths using the two-dimensional cylindrical coordinates (r, θ). In these polar coordinates, we have[3] velocity components $v_{\hat{r}} = \dot{r}$ and $v_{\hat{\theta}} = r\dot{\theta}$ and hence a squared velocity

$v^2 = (\dot{r}^2 + r^2\dot{\theta}^2)$. The system has a Lagrangian

$$L = \frac{1}{2}m(\dot{r}^2 + r^2\dot{\theta}^2) + \frac{GMm}{r}. \tag{20.1}$$

Feeding this into the Euler–Lagrange equation, we obtain the two equations of motion[4]

$$\ddot{r} - r\dot{\theta}^2 = -\frac{GM}{r^2},$$
$$r\ddot{\theta} + 2\dot{r}\dot{\theta} = 0. \tag{20.5}$$

We will meet some solutions to these equations in this chapter.

[4]The components of acceleration can be derived using

$$\vec{a} = \dot{\vec{v}} = \ddot{r}\vec{e}_{\hat{r}} + \dot{r}\dot{\vec{e}}_{\hat{r}} + \dot{r}\dot{\theta}\vec{e}_{\hat{\theta}} + r\ddot{\theta}\vec{e}_{\hat{\theta}} + r\dot{\theta}\dot{\vec{e}}_{\hat{\theta}}$$
$$= a_{\hat{r}}\vec{e}_{\hat{r}} + a_{\hat{\theta}}\vec{e}_{\hat{\theta}}, \tag{20.2}$$

where

$$a_{\hat{r}} = \ddot{r} - r\dot{\theta}^2, \tag{20.3}$$
$$a_{\hat{\theta}} = r\ddot{\theta} + 2\dot{r}\dot{\theta}. \tag{20.4}$$

This means that eqn 20.5 is a statement of $\vec{F} = m\vec{a}$ with m divided out.

20.1 Kepler's laws

An **orbit** is the closed, periodic trajectory that each planet in our Solar System is, to a good approximation, observed to follow. Before Isaac Newton provided a solution to the problem of the description of orbits, Johannes Kepler[5] had formulated three laws of planetary motion resulting from his analysis of the observed trajectories of planets in our solar system. These laws are useful in understanding what the allowed orbits are. Each can be proven using the Newtonian approach, and we shall do that here. The laws are given below.

Kepler's first law: Planetary orbits follow an elliptical trajectory, with the Sun at a focus of the ellipse.

Kepler's second law: The line from Sun to planet sweeps out equal areas in equal times.

Kepler's third law: The square of the period of an orbit is proportional to the cube of the semi-major axis of the ellipse that describes its orbit.

We can immediately show how Kepler's second law is a consequence of the conservation of angular momentum.

[5]Johannes Kepler (1571–1630) was plagued by ill health, complaining of myopia, multiple vision, sores, stomach and gall bladder problems, piles, rashes, mange, worms, and the delusion that he was a dog. At one point he had to defend his mother against charges of witchcraft. His work on orbits was motivated by his attempts to describe the planetary orbits in terms of plantonic solids, using data from Tycho Brahe's (1546–1601) naked-eye observations. The problem of computing the orbits of a single particle under Newtonian gravity continues to be called the **Kepler problem**.

Example 20.1

Since there is no component of the Newtonian force in the θ direction, we have, from eqn 20.4, that

$$r\ddot{\theta} + 2\dot{r}\dot{\theta} = 0. \tag{20.6}$$

Integrating, we find

$$r^2\dot{\theta} = C, \tag{20.7}$$

where C is a constant. As shown in Fig. 20.1, C can be interpreted geometrically as twice the rate at which the radius vector sweeps out area A. We therefore have

$$\frac{dA}{dt} = \frac{1}{2}r^2\frac{d\theta}{dt} = \text{const.} \tag{20.8}$$

This is Kepler's second law (i.e. constant $\dot{A}$), proven from the equations of motion. As a bonus, we notice that the angular momentum is given by

$$\vec{L} = rmv_{\hat{\theta}}\vec{e}_{\hat{z}} = mr^2\dot{\theta}\vec{e}_{\hat{z}}. \tag{20.9}$$

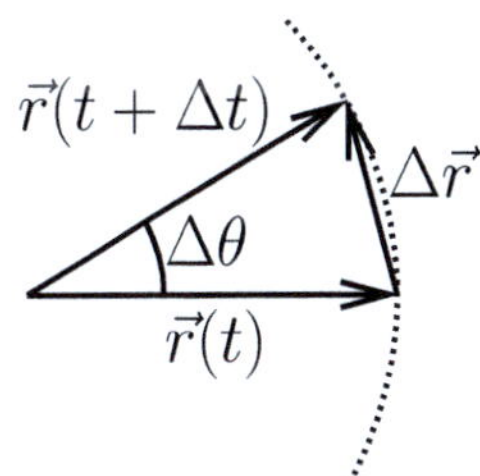

Fig. 20.1 A small displacement to the radius vector $\vec{r}(t)$ changes the vector by an amount $|\Delta\vec{r}| = 2r\sin\Delta\theta/2 \approx r\Delta\theta$. The area of the triangle in the figure is then $\Delta A = \frac{1}{2}r^2\Delta\theta$, and so area is swept out at a rate $\dot{A} = r^2\dot{\theta}/2(= C/2)$.

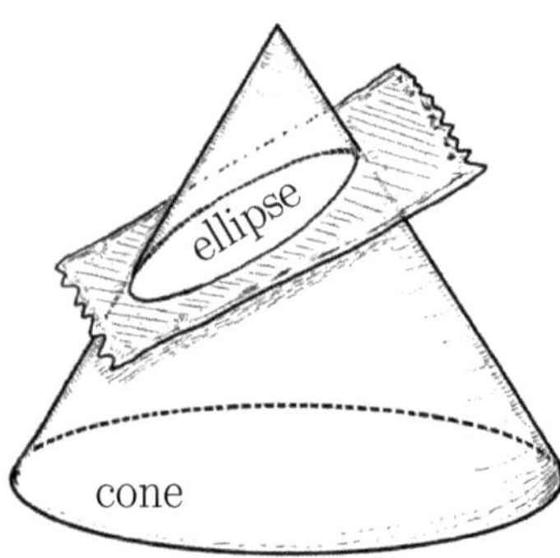

Fig. 20.2 The ellipse is a conic section, achieved via a slice made at an angle to the base. A slice made parallel to the base is a circle.

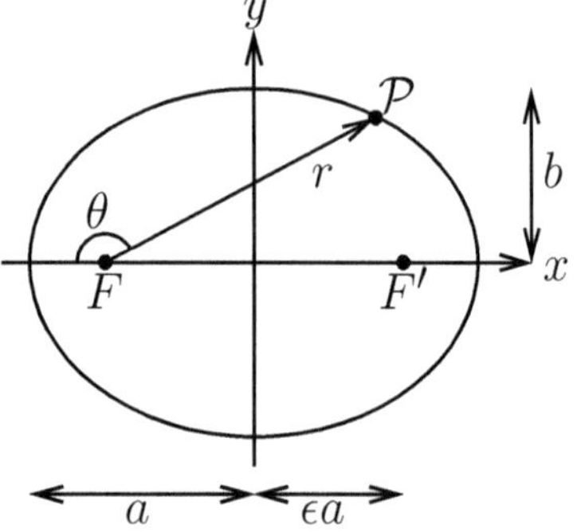

Fig. 20.3 The anatomy of an ellipse, showing the foci F and F', semi-major axis a, the semi-minor axis b and the eccentricity ϵ. The point $\mathcal{P}$ is a distance r from F. A circular orbit is a special case where $a = b$, $\epsilon = 0$ and F coincides with F'.

[6]More specifically, since $a_{\hat{r}} = -GM/r^2$, we can use eqn 20.16 and $C = |L|/m$ to show that we have

$$\frac{a}{b^2} = \frac{GMm^2}{L^2}.$$

Later, we call this quantity u_0.

Taking the area to be a vector $\vec{A}$ parallel to $\vec{e}_{\hat{z}}$, we have

$$\frac{\mathrm{d}\vec{A}}{\mathrm{d}t} = \frac{1}{2}r^2\frac{\mathrm{d}\theta}{\mathrm{d}t}\vec{e}_{\hat{z}} = \frac{\vec{L}}{2m}. \tag{20.10}$$

Since $\vec{L}$ is a constant, $\dot{\vec{A}}$ is also constant. Kepler's second law therefore reflects the conservation of angular momentum, which itself follows for any central gravitational force.

20.2 Anatomy of an orbit

Kepler's first law states that orbits follow elliptical paths. The description of orbits enjoys a rich range of terminology and here we review the necessary vocabulary.

An ellipse is one of a class of two-dimensional figures known as the **conic sections** that also includes the circle, the parabola and the hyperbola. As the name suggests, these are figures that are generated by taking slices of cones. The slice through a cone needed to generate an ellipse is shown in Fig. 20.2 and a typical ellipse is shown in Fig. 20.3. An ellipse is described by an equation in polar coordinates

$$\frac{1}{r} = \frac{a}{b^2}(1 + \epsilon\cos\theta), \quad 0 \le \epsilon < 1, \tag{20.11}$$

where a and b are, respectively, the **semi-major** and **semi-minor** axes of the ellipse and ϵ is known as the **eccentricity**. [The angle θ is the one that the radial vector $\vec{r}$ (linking the planet's position $\mathcal{P}$ and the focus) makes to the radial vector corresponding to the planet being at its position of closest approach to the focus, as shown in Fig. 20.3.]

With an equation for the elliptical trajectory available, we can show that it is compatible with a $1/r^2$ central force.

Example 20.2

We shall investigate the nature of the acceleration of a particle following an elliptical trajectory. Differentiate the equation for an ellipse to find

$$\frac{\dot{r}}{r^2} = \frac{\epsilon a}{b^2}\sin\theta\dot{\theta}. \tag{20.12}$$

We have $r^2\dot{\theta} = C$, where C is a constant equal to $|\vec{L}|/m$. Substituting this gives

$$\dot{r} = \frac{C\epsilon a}{b^2}\sin\theta, \tag{20.13}$$

and then

$$\ddot{r} = \frac{C\epsilon a}{b^2}\cos\theta\dot{\theta}. \tag{20.14}$$

Using the radial acceleration $a_{\hat{r}} = \ddot{r} - r\dot{\theta}^2$ we find

$$a_{\hat{r}} = \frac{C^2}{r^2}\left(\frac{\epsilon a\cos\theta}{b^2} - \frac{1}{r}\right). \tag{20.15}$$

Referring back to the equation for an ellipse we see that the part in the bracket is equal to[6] $-a/b^2$ and so we can write

$$a_{\hat{r}} = -\frac{C^2 a}{b^2}\frac{1}{r^2}. \tag{20.16}$$

Since acceleration is proportional to force, we conclude that an elliptical orbit is compatible with a $1/r^2$ force law.

The expression for radial acceleration allows us to prove Kepler's third law.

Example 20.3

The area of an ellipse is given by $A = \pi ab$. Using the notation from the previous example, the period of the orbit is $T = \pi ab/(C/2)$. Substituting into eqn 20.16 we have

$$a_{\hat{r}} = -\frac{4\pi^2 a^3}{T^2}\frac{1}{r^2}. \tag{20.17}$$

We also know that $a_{\hat{r}} = -GM/r^2$ and upon equating these expressions we have

$$T^2 = \frac{4\pi^2 a^3}{GM}, \tag{20.18}$$

which shows that the square of the period is proportional to the cube of a, the semi-major axis of the ellipse.

For an object in orbit around the Sun, the **perihelion** is the position of closest approach to the Sun. The **aphelion** is the position in the orbit furthest from the Sun (Fig. 20.4). These words are often incorrectly applied to the orbits of objects around bodies other than the Sun. In fact, for (i) orbits around the Earth the points are called the **perigee** and **apogee**; (ii) for orbits around a star we call them the **periastron** and **apastron** and (iii), most generally, for orbits around any centre of mass, they are called the **periapsis** and **apoapsis**.

We can exploit the properties of the aphelion to relate the total energy of the orbiting planet to the dimensions of the ellipse.

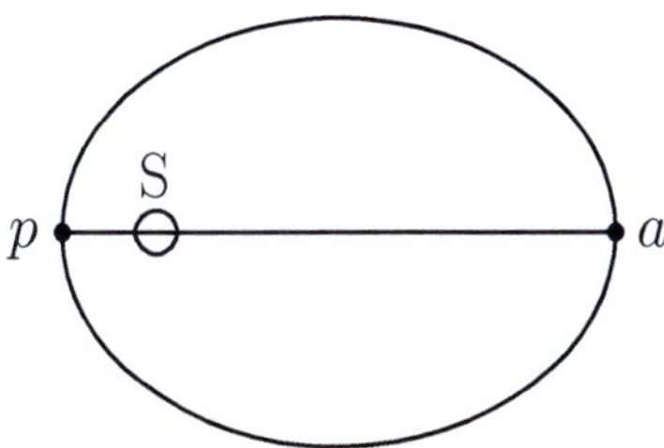

Fig. 20.4 The perihelion p and aphelion a of an orbit around the Sun S at one of the foci of the elliptical orbit.

Example 20.4

At the aphelion we have $r = a(1 + \epsilon)$. Also, since at this point $\theta = \pi$ and so, from eqn 20.11, $b^2 = a^2(1 - \epsilon^2)$. The potential energy at the aphelion for a planet in orbit around the Sun is

$$U = -\frac{GMm}{a(1 + \epsilon)}. \tag{20.19}$$

At this point $v_{\hat{r}} = 0$ and so the kinetic energy is

$$K = \frac{1}{2}mv_{\hat{\theta}}^2 = \frac{1}{2}mr^2\dot{\theta}^2 = \frac{mC^2}{2r^2}, \tag{20.20}$$

where we've used eqn 20.7 in the final step. Substituting for r we find

$$K = \frac{mC^2}{2a^2(1 + \epsilon)^2}, \tag{20.21}$$

or, substituting[7] for C, we obtain

$$K = \frac{GMm(1 - \epsilon)}{2a(1 + \epsilon)}. \tag{20.23}$$

This allows us to sum

$$E = K + U = \frac{GMm(1 - \epsilon)}{2a(1 + \epsilon)} - \frac{GMm}{a(1 + \epsilon)}, \tag{20.24}$$

or

$$E = -\frac{GMm}{2a}, \tag{20.25}$$

giving us an expression for the total energy of the orbit.

---

[7]Write

$$C^2 = \frac{4\pi^2 a^2 b^2}{T^2} = \frac{4\pi^2 a^4(1 - \epsilon^2)}{T^2},$$

and use Kepler's third law

$$T^2 = \frac{4\pi^2 a^3}{GM}.$$

Since $|L| = Cm$, we can use this to derive the useful result that

$$L^2 = GMm^2 a(1 - \epsilon^2), \tag{20.22}$$

for an elliptical orbit.

20.3   Effective potentials

The method of **effective potentials** is especially useful in understanding which trajectories are possible. We shall use it in the relativistic case and so we introduce it here. If the particle of mass m is moving with velocity $\vec{v}$ in the field of the stationary mass M then the energy of the two-particle system is written as

$$E = \frac{1}{2}m\left(\dot{r}^2 + r^2\dot{\theta}^2\right) - \frac{GMm}{r}. \tag{20.26}$$

The angular momentum has a magnitude $|L| = m\dot{\theta}r^2$, which allows us to substitute for $\dot{\theta}$ and then drop the angular contribution into an **effective potential energy** function $U_{\text{eff}}(r)$. Specifically, we write

$$E = \frac{1}{2}m\dot{r}^2 + U_{\text{eff}}(r), \tag{20.27}$$

where the effective potential energy is given by

$$U_{\text{eff}}(r) = \frac{L^2}{2mr^2} - \frac{GMm}{r}. \tag{20.28}$$

Given an initial angular momentum of a particle in this field, we can write the effective potential. As we shall see, it is this effective potential that can be used to classify and understand all of the trajectories that are possible for particles in the Newtonian potential. The potential energy function $U_{\text{eff}}(r)$ has the two contributions shown in Fig. 20.5. There is (i) a repulsive contribution whose strength depends on the angular momentum,[8] going as $1/r^2$; and (ii) an attractive contribution, given by the usual $-1/r$ potential. The difference in power law and sign means that the resultant potential can have a minimum. The limits of the effective potential for small and large r are instructive too: (i) the effect of gravitation dies off at large distances as $1/r$ irrespective of the angular momentum; (ii) the potential gets very large at small r owing to the angular momentum term. This means that any particle with non-zero angular momentum is scattered by the potential. (There is no option to spiral into the origin, for example). The only trajectory that can end up at the origin has $|L| = 0$, which amounts to a **radial plunge** into the source of the potential.

The values of r for which $E = U_{\text{eff}}(r)$ set some limits on the motion. We have this condition when $\dot{r} = 0$. This doesn't mean that the particle has stopped (because we can still have $\dot{\theta} \neq 0$), rather, the particle is turning at constant r, as it does at the points nearest and furthest from the focus of the ellipse. If $\dot{r} = 0$ at all times, then the trajectory must be circular (because there is never a change in r) with a radius r_0 that corresponds to the minimum[9] of U_{eff}. This occurs when $\partial U_{\text{eff}}(r)/\partial r|_{r=r_0} = 0$ or

$$\frac{L^2}{mr_0} = GMm \quad \text{(circular orbit)}. \tag{20.29}$$

The circular orbit has a negative total energy $E = -GMm/2r_0$, so that the orbiting particle is in a bound state that requires energy to

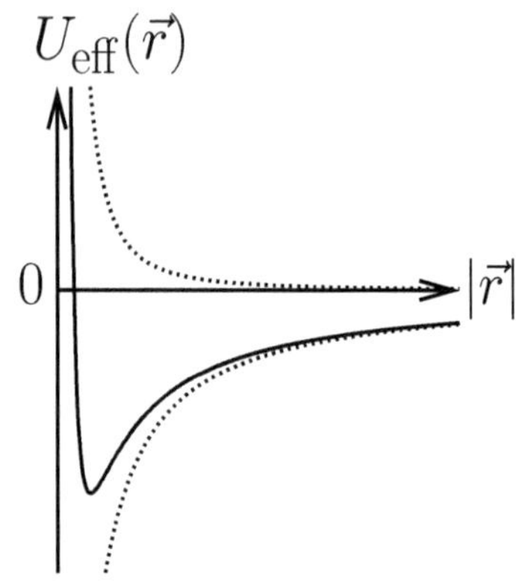

Fig. 20.5 The two contributions to the Newtonian effective potential energy (the repulsive angular momentum barrier and attractive $-1/r$ contribution, shown by dotted lines) give the curve shown with the solid line.

[8]This term is sometimes called the angular momentum barrier.

[9]It lies at the minimum because the system's energy is determined only by U_{eff}.

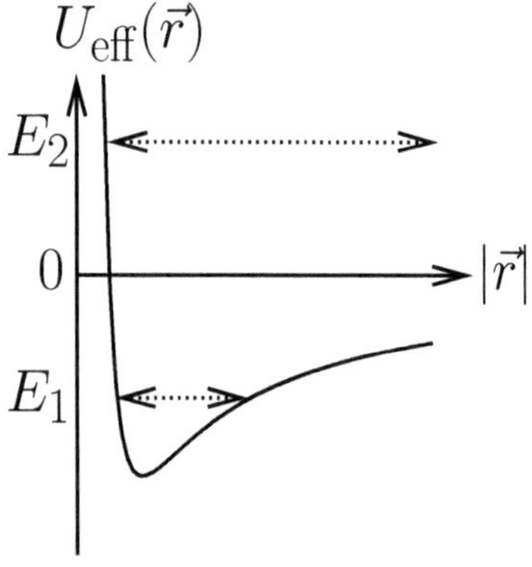

Fig. 20.6 Energies for an elliptical orbit with energy $E_1 < 0$ and an unbound trajectory ($E_2 > 0$).

be inputted in order to escape. There are other possible bound orbits with $E < 0$, where the particle can be thought of as being confined by the effective potential, bouncing back and forth between the two values where $E = U_{\text{eff}}$.[10] These are, of course, the ellipses that Kepler's first law describes. We can immediately gain some insight by looking at the negative values of $U_{\text{eff}}(r)$, since $E = U_{\text{eff}}$ for the elliptical paths at the perihelion and aphelion, where $\dot{r} = 0$. As shown in Fig. 20.6, lines of negative constant total energy E intersect $U_{\text{eff}}(r)$ at two points when $\dot{r} = 0$, providing the length of semi-major and semi-minor axes.

In contrast, an unbounded trajectory (i.e. not an orbit) with a positive energy has a unique distance of closest approach to the gravitating mass. At this position $\dot{r} = 0$ and so the distance of closest approach occurs when the energy of the system $E = U_{\text{eff}}(r)$. We can therefore use the graph of the effective potential energy (Fig. 20.6) to describe all of the trajectories. We imagine the particle moving horizontally as its value of r varies; it is deflected whenever it meets the curve U_{eff}. Remember that the form of U_{eff} is determined by the (constant) angular momentum of the particle. We shall use this graphical method again when we look into the trajectories allowed by general relativity.

[10]The fact that the potential has a minimum implies the motion for particles bound in the potential is stable.

↷ This graphical method is discussed in the context of general relativity in Chapter 23.

20.4 Allowed trajectories

We can now solve the problem once and for all, by determining all of the possible motions in the Newtonian potential. We have seen that orbits are expected in a Newtonian potential and that these have a negative total energy E. If the energy is positive, then we expect the trajectory to be unbounded. We shall now solve the equations of motion for the trajectories of the particles and then use the sign of the total energy to classify the orbits. The technique of choice employs the variable $u = 1/r$, which we shall also use in solving the relativistic problem.

Example 20.5

What is the equation of motion for the trajectories? It's convenient to use $u = 1/r$ and write

$$\frac{\mathrm{d}r}{\mathrm{d}t} = -\frac{1}{u^2}\frac{\mathrm{d}u}{\mathrm{d}t} = -\frac{L}{m}\frac{\mathrm{d}u}{\mathrm{d}\theta}, \tag{20.30}$$

and

$$\frac{\mathrm{d}^2 r}{\mathrm{d}t^2} = -\frac{L}{m}\frac{\mathrm{d}^2 u}{\mathrm{d}\theta^2}\frac{\mathrm{d}\theta}{\mathrm{d}t} = -\frac{L^2}{m^2}u^2\frac{\mathrm{d}^2 u}{\mathrm{d}\theta^2}. \tag{20.31}$$

Newton's second law becomes

$$F_{\hat{r}} = -GMmu^2 = m\left[\frac{\mathrm{d}^2 r}{\mathrm{d}t^2} - r\left(\frac{\mathrm{d}\theta}{\mathrm{d}t}\right)^2\right]$$

$$= -\frac{L^2}{m}u^2\left(\frac{\mathrm{d}^2 u}{\mathrm{d}\theta^2} + u\right). \tag{20.32}$$

We end up with a differential equation for the trajectories

$$\frac{\mathrm{d}^2 u}{\mathrm{d}\theta^2} + u = \frac{GMm^2}{L^2}. \tag{20.33}$$

The solution to this equation is a function $u(\theta)$ that gives the trajectory.

Rewriting the differential equation from the last example, we use the variable $u_0 = \frac{GMm^2}{L^2}$, and end up with an equation of motion for the variable u which is

$$\frac{\mathrm{d}^2}{\mathrm{d}\theta^2}(u - u_0) = -(u - u_0). \tag{20.34}$$

The general solution of this equation, that gives access to all of the allowed trajectories, is

$$u - u_0 = B\cos\theta, \tag{20.35}$$

where B is a constant.

Now for some interpretation. If we can write the expression for the trajectories in terms of the total energy E, then we can classify orbits. If the energy is negative we have a bound state or, in other words, an orbit. If the energy is positive, the trajectory is not bounded.

We shall rewrite eqn 20.35 in terms of the values of several physical quantities evaluated at the perihelion (i.e. when $\theta = 0$). At this point, the radius is $r = r_1$ (and so $u = u_1$) the kinetic energy and potential energy are K_1 and U_1 respectively and angular momentum is L. At the perihelion, the radius is perpendicular to the velocity and so the angular momentum is $L = mv_{\hat{\theta}}r_1$, and so $K_1 = L^2/2m$. This allows us to say

$$u_0 = \frac{GMm^2}{L^2} = -\frac{U_1}{2K_1}u_1. \tag{20.36}$$

Since when $u = u_1$ we have $\theta = 0$, eqn 20.35 gives us an expression $u_1 - u_0 = B$, which, together with the total energy $E = K_1 + U_1$, allows us to write and equation linking the constant total energy to the parameters u_0 and B

$$E = \frac{u_1}{K_1}(B - u_0). \tag{20.37}$$

Using this expression, we can classify the three general types of trajectory that are possible, via the sign of E.

- When $B = u_0$, we have $E = 0$ and the trajectory is the **parabola** $u = u_0(1 + \cos\theta)$. This allows r to go to infinity when $\theta = \pi$ so the trajectory is, only just, unbounded.[11]

- When $B > u_0$, we have E positive and the trajectory is a **hyperbola**. This allows r to go to infinity when $\theta = \cos^{-1}(-u_0/B)$. This is an unbounded trajectory.

- Finally, in order to have E negative (resulting in a bounded orbit), we must have $B < u_0$. Comparing $u - u_0 = B\cos\theta$ to eqn 20.11 for the ellipse, we see that it is identical if we take $u_0 = a/b^2$ and the eccentricity of the ellipse is given by $B/u_0 = \epsilon$. The equation $u - u_0 = B\cos\theta$ with $B < u_0$ then describes an ellipse with limiting radii $r = (u_0 + B)^{-1}$ and $r = (u_0 - B)^{-1}$. (The case of the circular orbit corresponds to $B = 0$, as it must.)

---

[11] The parabola corresponds to the case $\epsilon = 1$ in eqn 20.11.

Example 20.6

Another useful route to the finding the possible trajectories is to start with the equation for the total energy

$$E = \frac{1}{2}m\dot{r}^2 + \frac{1}{2}mr^2\dot{\theta}^2 - \frac{GMm}{r}, \tag{20.38}$$

and divide through by $L^2/2m$. Writing $u' = \partial u/\partial\theta$ we obtain, after a little algebra, that

$$(u')^2 + u^2 - \frac{2GMm^2}{L^2}u - \frac{2Em}{L^2} = 0. \tag{20.39}$$

Spotting the presence of $u_0 = 1/r_0 = GMm^2/L^2$ we choose to rewrite this quadratic equation as

$$(u')^2 + (u - u_0)^2 - u_0^2\epsilon^2 = 0, \tag{20.40}$$

where $u_0^2(1 - \epsilon^2) = -2Em/L^2$. Try a solution $u = c + d\cos\theta$, and find $c = u_0$ and $d = \pm u_0\epsilon$, so that we have

$$u = u_0(1 \pm \epsilon\cos\theta). \tag{20.41}$$

With our definition of θ the equation for the ellipse takes the positive sign and we have $u = u_0(1 + \epsilon\cos\theta)$.

20.5   The *why?* of orbits

What guarantees that there are orbits at all? That is, why should any trajectory close and hence be periodic? Let's temporarily examine non-relativistic orbits more generally, taking the central potential energy to be $U(r)$, which is not necessarily the Newtonian potential $U(r) \propto -1/r$. (In any case, the angular momentum $\vec{L}$ is still conserved owing to the lack of torque from a central field.) We can rewrite eqn 20.27 for the total energy as

$$\frac{\mathrm{d}r}{\mathrm{d}t} = \left\{\frac{2}{m}\left[E - U(r)\right] - \frac{L^2}{m^2r^2}\right\}^{\frac{1}{2}}, \tag{20.42}$$

from which we find the time taken for the motion between two values of r to be

$$t = \int_{r_1}^{r_2} \frac{\mathrm{d}r}{\left\{\frac{2}{m}\left[E - U(r)\right] - \frac{L^2}{m^2r^2}\right\}^{\frac{1}{2}}}, \tag{20.43}$$

or, since $\mathrm{d}\theta = L\mathrm{d}t/mr^2$, we have

$$\theta = \int_{r_1}^{r_2} \frac{\mathrm{d}r\, L/r^2}{\left\{2m\left[E - U(r)\right] - \frac{L^2}{r^2}\right\}^{\frac{1}{2}}}. \tag{20.44}$$

If the motion has two limiting radii, $r_{\min}$ and $r_{\max}$ then, during the time in which r varies from $r_{\min}$ to $r_{\max}$ and back, the radius vector turns through an angle

$$\Delta\theta = 2\int_{r_{\min}}^{r_{\max}} \frac{\mathrm{d}r\, L/r^2}{\left\{2m\left[E - U(r)\right] - \frac{L^2}{r^2}\right\}^{\frac{1}{2}}}. \tag{20.45}$$

[12]The amount of precession per orbit is given by

$$\delta\theta = \Delta\theta - 2\pi. \qquad (20.46)$$

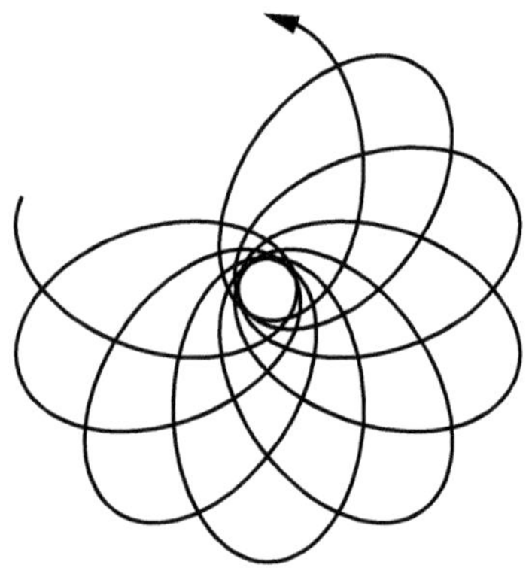

Fig. 20.7 The precession of an orbit. After n periods we must make m complete revolutions in order for the orbit to close.

[13]Alternatively, the integral is carried out on page 30 of the book by Zee.

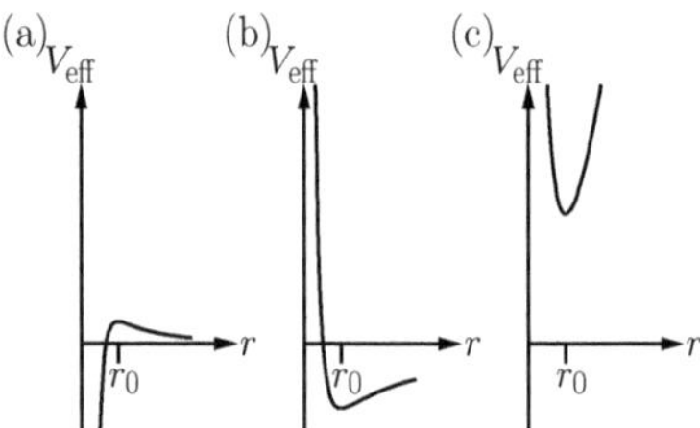

Fig. 20.8 The effective potential for three cases: (a) $n < -3$, (b) $-3 < n < -1$, (c) $n > -3$. Stable circular orbits at $r = r_0$ are only possible if $n > -3$.

For this motion to form a **closed orbit** we must have $\Delta\theta = 2\pi q/n$ with q and n integers. That is to say, after n periods of motion between $r_{\min}$ and $r_{\max}$, the radius vector has made q complete revolutions in θ and we're back where we started. This allows for the **precession** of the orbits, as shown in Fig. 20.7.[12]

Example 20.7

We can demonstrate that the orbit closes for the Newtonian case, that is, when $U(r) = -\frac{\alpha}{r}$. Rewrite the integral as

$$\Delta\theta = -2\frac{\partial}{\partial L}\int_{r_{\min}}^{r_{\max}} dr \left\{ 2mE + \frac{2m\alpha}{r} - \frac{L^2}{r^2} \right\}^{\frac{1}{2}}. \qquad (20.47)$$

This can be rewritten as an elliptical integral whose result can be looked up:[13] it turns out to yield $\Delta\theta(= 2\pi q/n) = 2\pi$. This implies that $q = n$ and there is no precession of any Newtonian orbit.

Another way of looking at the closing of trajectories is covered in the following example.

Example 20.8

For a central field of force $F = -Ar^n$, the effective potential is

$$V_{\text{eff}} = \frac{Ar^{n+1}}{n+1} + \frac{L^2}{2mr^2}. \qquad (20.48)$$

This gives a stable circular orbit at $r = r_0$ only if $n > -3$. (The form of the effective potential is shown in Fig. 20.8 for three distinct cases.) Near a circular orbit at $r = r_0$, we can write the effective potential as a Taylor expansion

$$V_{\text{eff}} = V(r_0) + \frac{1}{2}(r - r_0)^2 \left(\frac{d^2 V_{\text{eff}}}{dr^2}\right)_{r=r_0} + \cdots, \qquad (20.49)$$

where $d^2 V_{\text{eff}}/dr^2$ evaluated at $r = r_0$ is given by $(n+3)L^2/(mr_0^4)$. Small oscillations for a particle of mass m near the bottom of the well, i.e. orbiting close to $r = r_0$, are therefore governed by

$$m\ddot{r} + \frac{(n+3)L^2}{mr_0^4}(r - r_0) = 0, \qquad (20.50)$$

which is simple harmonic motion with period τ given by

$$\tau = \frac{2\pi m r_0^2}{(n+3)^{1/2}L}. \qquad (20.51)$$

The orbital period $T = 2\pi/\omega = 2\pi m r_0^2/L$ and so these two periods are related by

$$\tau = \frac{T}{(n+3)^{1/2}}. \qquad (20.52)$$

- For the $n = -2$ (Newtonian) case, $\tau = T$ and so the precession leads to elliptical orbits. Another stable case occurs for $n = 1$ (simple harmonic motion) where $\tau = T/2$.

- For general $n > -3$, excepting these special cases, the precession τ is incommensurate with the orbital motion. For this reason, any departures from $n = -2$ in some imagined small departure from Newtonian gravity should be easy to determine from observations since it will lead to precession of elliptical orbits.

The closing of trajectories is further demonstrated for Newtonian orbits in the next example.

Example 20.9

We start by proving a geometric result. The **Laplace–Runge–Lenz vector**[14] is defined as

$$\vec{\mathcal{L}} = (\vec{p} \times \vec{L}) - GMm^2 \frac{\vec{r}}{r}. \tag{20.53}$$

Taking the time derivative of this vector, we find

$$
\begin{aligned}
\dot{\vec{\mathcal{L}}} &= (\vec{F} \times \vec{L}) + (\vec{p} \times \dot{\vec{L}}) - GMm^2 \frac{\dot{\vec{r}}}{r} + GMm^2 \frac{\vec{r}}{r^2} \dot{r} && \text{(time derivative)} \\
&= \vec{F} \times \vec{r} \times \vec{p} - GMm^2 \frac{\dot{\vec{r}}}{r} + GMm^2 \frac{\vec{r}}{r^2} \dot{r} && \text{(using } \tau = \dot{\vec{L}} = 0) \\
&= \vec{r}(\vec{F} \cdot \vec{p}) - \vec{p}(\vec{F} \cdot \vec{r}) - GMm^2 \frac{\dot{\vec{r}}}{r} + GMm^2 \frac{\vec{r}}{r^2} \dot{r} && \text{(triple vector product)} \\
&= -GMm^2 \frac{\vec{r}}{r^2} \dot{r} + GMm^2 \frac{\dot{\vec{r}}}{r} - GMm^2 \frac{\dot{\vec{r}}}{r} + GMm^2 \frac{\vec{r}}{r^2} \dot{r} \\
&= 0.
\end{aligned}
$$

Now consider the orbit at the perihelion (and aphelion). Here we have that the momentum $\vec{p}$ is perpendicular to the semi-major axis of the ellipse. Since $\vec{L}$ points out of the plane of the orbit we can take $\vec{\mathcal{L}}$ to point from the focus to the perihelion. Since the vector $\vec{\mathcal{L}}$ is constant, it always points to the perihelion and so the perihelion cannot move. The orbit never precesses and must therefore close.[15]

With this review of Newtonian orbits under our belts, we can examine some of the richness afforded by general relativity's correction to the Newtonian picture. In the next chapter, we meet a metric field analogous to the spherically symmetric $1/r$ potential of the Newtonian case. This is the celebrated Schwarzschild metric.

Chapter summary

- Orbits in a Newtonian potential are elliptical. They can be understood by identifying an effective potential.

- The variable $u = 1/r$ allows us to rewrite the problem and solve the equations of motion. We make use of conserved energy and angular momentum.

- Newtonian orbits never precess.

[14]Pierre-Simon Laplace (1749–1827), Carl Runge (1856–1927) and Wilhelm Lenz (1888–1957). The vector is also known as the Laplace vector, the Runge–Lenz vector and the Lenz vector. In fact, the quantity seems to predate all of these people. Herbert Goldstein's two short articles [Am. J. Phys. **43**, 737 (1975) and **44**, 1123 (1976)] outline its interesting history, where it is suggested that priority actually belongs to Jakob Hermann (1678–1733) and Johann Bernoulli (1667–1748). As discussed in the book by Gutzwiller, Wolfgang Pauli used the properties of $\vec{\mathcal{L}}$ to compute the quantum-mechanical spectrum of the hydrogen atom exactly in 1926.

[15]The components of the vector $\vec{\mathcal{L}}$, along with the energy E, and the components of the angular momentum $\vec{L}$ give us seven quantities. In the Kepler problem, the length of $\vec{\mathcal{L}}$ is constant and we also have $\vec{\mathcal{L}} \cdot \vec{L} = 0$, giving us five independent constants of the motion. In general, a mechanical systems with d degrees of freedom can have, at most, $2d - 1$ constants of the motion. [This is because there are $2d$ initial conditions (the components of position and velocity) and the initial time cannot be determined by a constant of the motion.] The Kepler problem has $d = 3$, so we have the maximum number of constants of the motion possible. For this reason, the Kepler problem is sometimes called 'maximally superintegrable'.

Exercises

(20.1) Fill in the algebra leading to eqn 20.39.

(20.2) Two particles, each of mass m, move under the influence of their mutual gravitational attraction $-\frac{Gm^2}{r^2}$. Initially, the particles are a large distance apart and approach each other with velocities $\vec{v}$ and $-\vec{v}$ along parallel paths a distance b apart.

(a) What is the angular momentum of the system during the motion?

(b) Write down the energy of the system at the start of the motion and at the point in the motion when the particles are closest to each other.

(c) Show that the least distance d between the particles in their subsequent motion is given by

$$d = -\frac{Gm}{2v^2} + \sqrt{\left(\frac{Gm}{2v^2}\right)^2 + b^2}.$$

(20.3) A particle with mass m is placed a distance x from the centre of a thin ring of radius a, along the line through the centre of the ring and perpendicular to its plane.

(a) Assuming the ring has a total mass M, which is uniformly distributed along the ring, show that the total gravitational potential energy of the ring and particle is given by

$$U = -\frac{GMm}{(a^2 + x^2)^{\frac{1}{2}}}. \qquad (20.54)$$

(b) Find the magnitude and direction of the force on the particle. Comment on this result in the case that the distance between the particle and ring is very large compared to the ring's radius.

Now consider a mass Ω uniformly distributed over a disc of radius L. A particle of mass m is placed a distance x from the centre of the disc, along the line through its centre and perpendicular to its plane.

(c) Using the result from part (a), or otherwise, find the gravitational potential energy of the disc and particle system.

(20.4) (a) Show that an equation of motion

$$\frac{\mathrm{d}p_\alpha}{\mathrm{d}\tau} = -m\frac{\partial \Phi}{\partial x^\alpha}, \qquad (20.55)$$

where Φ is the time-independent Newtonian potential, is incompatible with the rule from special relativity that $\boldsymbol{a} \cdot \boldsymbol{u} = 0$ (where $\boldsymbol{a}$ is acceleration and $\boldsymbol{u}$ is velocity).

(b) Show that the equation of motion

$$\frac{\mathrm{d}p^\alpha}{\mathrm{d}\tau} = -m\left(\eta^{\alpha\beta} + u^\alpha u^\beta\right)\frac{\partial \Phi}{\partial x^\beta}, \qquad (20.56)$$

does not suffer from the problem in (a).

The latter can be regarded as an alternative theory of gravity in which the gravitational field exists in Minkowski spacetime. The projection operator $(\eta + \boldsymbol{u} \otimes \boldsymbol{u})$ picks out the part perpendicular to $\boldsymbol{u}$ (i.e. perpendicular to the world line, ensuring the orthogonality of this part and $\boldsymbol{u}$).

(20.5) Consider the equation of motion from Exercise 20.4(b). By taking the limit of zero mass, such that $\lambda = \tau/m$ remains constant as m and τ go to zero, show that $p^\alpha e^{\Phi(r)}$ (for $\alpha = 0$–4) remains constant along the world line of a photon.

This approach (based on that of Thorne and Blandford) shows that light cannot be deflected by the Sun in this theory, in contradiction to experiment. Gravity cannot, therefore, simply be incorporated into special relativity as another field in Minkowski spacetime.

The Schwarzschild geometry

21

As you see the war treated me kindly enough, in spite of heavy gunfire, to allow me to get away from it all and take this walk in the land of your ideas.
Karl Schwarzschild (1873–1916) *in a 1915 letter to Einstein*

After Einstein wrote down the field equation of general relativity he did not expect it to admit exact solutions owing to its complexity. He himself used an approximate solution in his 1915 article about the perihelion of Mercury. It therefore came as something of a surprise when Einstein received a letter from Karl Schwarzschild at the end of 1915 detailing a rather simple exact solution.[1] Schwarzschild was, despite being over forty years old, then serving as a soldier in World War I and carried out his work while at the Russian front.

The **Schwarzschild solution** is a solution of the Einstein field equation for the geometry *outside* a spherically symmetric, gravitating mass distribution. (We will find that the solution *inside* a mass distribution is different.) The Schwarzschild solution gives us a useful metric tensor that we will use to describe the spacetime outside stars and black holes, along with the motion of objects orbiting these bodies. First, the answer: the Schwarzschild metric line element for the space outside a static, spherically symmetric gravitating body of mass M at the origin of a set of coordinates (t, r, θ, ϕ) is written as[2]

$$ds^2 = -\left(1 - \frac{2M}{r}\right) dt^2 + \left(1 - \frac{2M}{r}\right)^{-1} dr^2 + r^2 \left(d\theta^2 + \sin^2\theta d\phi^2\right).$$

$$(21.1)$$

Notable features of this line element include: (i) it is static: none of the components[3] $g_{\mu\nu}$ of the metric field depend on time; (ii) it is asymptotically flat: as $r \to \infty$ it looks like the Minkowski metric; and (iii) it appears badly behaved at the origin, and also when $r = 2M$, where the second term becomes singular. One additional reason why this line element is so special is that **Birkhoff's theorem**[4] tells us that any spherically symmetric solution to the Einstein equation outside a gravitating object (they don't need to be static, for example) will be identical to Schwarzschild's static solution.

In this chapter, we will examine where the Schwarzschild solution comes from, and how we justify its form. Our results will be useful in the later chapters in this part of the book, where we will apply the

[1] In 1915, Schwarzschild started suffering from pemphigus, a rare autoimmune skin disease that likely led to his death in 1916. English speakers should consider saying 'shvarts-shilt' rather than the commonly heard 'shworts-child'. The name means black shield, rather than black child, in any case.

[2] Our choice of units in this part of the book is $G = c = 1$. To obtain real-world units in the metric substitute $M \to GM/c^2$ and $t \to ct$.

[3] The non-zero components of the metric tensor are

$$g_{tt} = -\left(1 - \frac{2M}{r}\right),$$

$$g_{rr} = \left(1 - \frac{2M}{r}\right)^{-1},$$

$$g_{\theta\theta} = r^2,$$

$$g_{\phi\phi} = r^2 \sin^2\theta.$$

[4] George David Birkhoff (1884–1944). Birkhoff's theorem says that any spherically symmetric solution of the vacuum field equations must be static and asymptotically flat.

[5] We'll postpone examining the light cone structure of this metric field until then.

In non-natural units, the Schwarzschild metric is given by

$$
\begin{aligned}
ds^2 = & - \left(1 - \frac{2GM}{c^2 r}\right) c^2 dt^2 \\
& + \left(1 - \frac{2GM}{c^2 r}\right)^{-1} dr^2 \\
& + r^2 \left(d\theta^2 + \sin^2\theta d\phi^2\right).
\end{aligned} \quad (21.2)
$$

Note that, unlike the coordinate systems we met in the previous part of the book, Schwarzschild coordinates are not comoving, since $g_{tt,r} \neq 0$. Physically, this is because the Schwarzschild coordinates privilege the rest frame of the spherically symmetric mass distribution that acts as the source of the curvature.

[6] Recall our slogan that coordinates have no intrinsic significance in the metric field, so one radius-like variable is just as good as another.

[7] This also provides us with an operational definition of when a field can be characterized as being weak (i.e. restoring factors, we want $2\Phi/c^2 \ll 1$).

metric and use the curvature tensor to examine motion such as orbits and objects like black holes.[5]

21.1 Justifying the solution

We seek a solution to the Einstein equation for a static, spherically symmetric mass distribution centred on the origin of a set of spherical coordinates (t, r, θ, ϕ). Although, as we have discussed, the names of these components have no intrinsic metric significance, it will be helpful to give them some temporarily in an effort to make sensible assumptions. We start by assuming that the metric field $\boldsymbol{g}$ is static. This means that intervals between events are time independent and so components obey $dg_{\alpha\beta}/dt = 0$. We assume that the spherical symmetry means that world lines of constant r, θ and ϕ are orthogonal to $t = (\text{const.})$ hypersurfaces. We also assume asymptotic flatness, which is to say that as $r \to \infty$, we must have flat spacetime (that is, the gravitational forces vanish at infinity).

A good place to start is therefore with spherically symmetric, flat spacetime, which obeys the above assumptions and has a metric line element

$$
ds^2 = -dt^2 + dr^2 + r^2 d\Omega^2, \quad (21.3)
$$

where $d\Omega^2 = d\theta^2 + \sin^2\theta d\phi^2$. The simplest way to proceed is then to guess that these components assume different values close to the gravitating object, subject to obeying the constraints of the previous paragraph. Owing to spherical symmetry, the new components can only depend on the radius coordinate r and so we write an expression with three functions

$$
ds^2 = -e^{2\Phi(r)} dt^2 + e^{2\Lambda(r)} dr^2 + R(r)^2 d\Omega^2. \quad (21.4)
$$

We can immediately now restrict the number of variables from three to two. This is because we can simply reinterpret $R(r)$ as the r coordinate and rescale the other functions.[6] So we have a general line element

$$
ds^2 = -e^{2\Phi(r)} dt^2 + e^{2\Lambda(r)} dr^2 + r^2 d\Omega^2, \quad (21.5)
$$

where we constrain $\Phi(\infty) = \Lambda(\infty) = 0$.

Expressions for $\Phi(r)$ and $\Lambda(r)$ will come from linking the line element in eqn 21.5 to the physics of gravitating objects, which we do below. However, we can immediately gain an insight into the function $\Phi(r)$ by comparison with the weak-field metric, which has a line element, expressed in spherical coordinates, of

$$
ds^2 = -(1 + 2\Phi) dt^2 + dr^2 + r^2 d\Omega^2, \quad (21.6)
$$

where Φ is the Newtonian gravitational potential. In eqn 21.5, we observe that in the limit of small Φ we have that $-e^{2\Phi} \approx -(1 + 2\Phi)$, suggesting that this is the same gravitational potential Φ that features in the weak-field metric.[7]

Now that we have, in eqn 21.5, a candidate metric, we can feed it through the Einstein field equation $\boldsymbol{G} = 8\pi\boldsymbol{T}$. This allows us to discover that the metric does indeed solve the problem and also how to provide expressions for Φ and Λ in terms of physically meaningful quantities.

21.2 Components of the Riemann tensor

We start by finding the left-hand side of the Einstein field equation. This involves finding the curvature tensor $\boldsymbol{R}$ for the metric field in eqn 21.5 and then generating the Einstein tensor $\boldsymbol{G}$. The components of $\boldsymbol{R}$ can be generated in a number of ways. To maximize simplicity, we will work in the orthonormal frame.[8]

Example 21.1

Written in its general form, the spherically symmetric metric line element looks like

$$\mathrm{d}s^2 = -\mathrm{e}^{2\Phi}\mathrm{d}t^2 + \mathrm{e}^{2\Lambda}\mathrm{d}r^2 + r^2\left(\mathrm{d}\theta^2 + \sin^2\theta\mathrm{d}\phi^2\right), \tag{21.7}$$

where Φ and Λ are functions of r only. Feeding this into the equations to find the components of the Riemann tensor, we obtain[9]

$$\begin{aligned}
R^{\hat{t}\hat{r}}{}_{\hat{t}\hat{r}} = E, \qquad R^{\hat{t}\hat{\theta}}{}_{\hat{t}\hat{\theta}} = \bar{E}, \qquad R^{\hat{t}\hat{\phi}}{}_{\hat{t}\hat{\phi}} = \bar{E}, \\
R^{\hat{\theta}\hat{\phi}}{}_{\hat{\theta}\hat{\phi}} = F, \qquad R^{\hat{\phi}\hat{r}}{}_{\hat{r}\hat{\phi}} = -\bar{F}, \qquad R^{\hat{r}\hat{\theta}}{}_{\hat{r}\hat{\theta}} = \bar{F},
\end{aligned} \tag{21.8}$$

where

$$\begin{aligned}
E &= -\,\mathrm{e}^{-2\Lambda}\left(\Phi'' + \Phi'^2 - \Phi'\Lambda'\right), \\
\bar{E} &= -\frac{1}{r}\mathrm{e}^{-2\Lambda}\Phi', \\
F &= \frac{1}{r^2}\left(1 - \mathrm{e}^{-2\Lambda}\right), \\
\bar{F} &= \frac{1}{r}\mathrm{e}^{-2\Lambda}\Lambda',
\end{aligned} \tag{21.9}$$

and we have used the dash notation for derivatives with respect to r.

These components of the Riemann tensor $\boldsymbol{R}$ give us the non-zero components of the Einstein tensor[10]

$$\begin{aligned}
G^{\hat{t}}{}_{\hat{t}} &= -G_{\hat{t}\hat{t}} = -(F + 2\bar{F}), \\
G^{\hat{r}}{}_{\hat{r}} &= G_{\hat{r}\hat{r}} = -(F + 2\bar{E}), \\
G^{\hat{\theta}}{}_{\hat{\theta}} &= G^{\hat{\phi}}{}_{\hat{\phi}} = -(E + \bar{E} + \bar{F}).
\end{aligned} \tag{21.10}$$

Note that $\boldsymbol{G}$ will vanish outside a mass distribution, although we can have curvature (i.e. non-zero $\boldsymbol{R}$) this region. The tensor $\boldsymbol{G}$ is zero here because the Einstein equation forces $\boldsymbol{G}$ to be proportional to the energy-momentum tensor $\boldsymbol{T}$, which itself vanishes in the vacuum.

---

[8]Remember that one advantage of this is that components can be raised and lowered with the Minkowski tensor $\boldsymbol{\eta}$. The vielbein components for observers in the stationary orthonormal frame are

$$\begin{aligned}
(\boldsymbol{e}_t)^{\hat{t}} &= \mathrm{e}^{\Phi}, \\
(\boldsymbol{e}_r)^{\hat{r}} &= \mathrm{e}^{\Lambda}, \\
(\boldsymbol{e}_\theta)^{\hat{\theta}} &= r, \\
(\boldsymbol{e}_\phi)^{\hat{\phi}} &= r\sin\theta.
\end{aligned}$$

In the Schwarzschild case, we can safely call the observer in the this frame stationary, since we can say they are stationary relative to the mass at the origin.

[9]See the exercises at the end of Chapter 11 and also Chapter 36.

[10]The simplest way to compute these is to use the useful rules that

$$\begin{aligned}
G^0{}_0 &= -(R^{12}{}_{12} + R^{23}{}_{23} + R^{31}{}_{31}), \\
G^1{}_1 &= -(R^{02}{}_{02} + R^{03}{}_{03} + R^{23}{}_{23}), \\
G^0{}_1 &= R^{02}{}_{12} + R^{03}{}_{13}, \\
G^1{}_2 &= R^{10}{}_{20} + R^{13}{}_{23},
\end{aligned}$$

where other components can be found using cyclic permutations. You're invited to prove these rules in Exercise 21.3.

This gives us the left-hand (geometrical) side of the Einstein equation. In the next section, we look at the right-hand (physical) side.

21.3   A gravitating object

Now for the right-hand side of the Einstein equation. We make a spherically symmetric distribution of static perfect fluid so that, in the orthonormal frame, the inside of the mass distribution is described by $T^{\hat\mu\hat\nu} = \mathrm{diag}(\rho, p, p, p)$, where ρ is the mass density and p is the pressure.[11] We'll take the total mass of the distribution giving rise to the metric to be M, and stipulate that the mass distribution stretches out to some maximum radius $r = R$. Outside this radius there is no matter and so the components of $\boldsymbol{T}$ must vanish locally.

The function $\Lambda(r)$ can be determined by considering the 00th component of the Einstein equation. We have, from the previous section, that the component $G_{\hat t\hat t} = 8\pi T_{\hat t\hat t}$ can be written as

$$\frac{1}{r^2}\frac{\mathrm{d}}{\mathrm{d}r}\left[r\left(1 - \mathrm{e}^{-2\Lambda}\right)\right] = 8\pi\rho. \tag{21.12}$$

We will see very shortly that it makes sense to call the quantity in the square brackets $2m(r)$, which is to say[12]

$$r\left(1 - \mathrm{e}^{-2\Lambda}\right) = 2m(r). \tag{21.14}$$

Returning to the Einstein equation 21.12, this definition of $m(r)$ gives us

$$\frac{2}{r^2}\frac{\mathrm{d}m(r)}{\mathrm{d}r} = 8\pi\rho, \tag{21.15}$$

whose solution is

$$m(r) = \int_0^r \mathrm{d}r\, 4\pi r^2 \rho + m(0). \tag{21.16}$$

Since ρ is a density, this expression makes physical sense for a spherically symmetrical object if we interpret $m(r)$ to be the mass contained in a sphere of radius r.

In order to find $\Phi(r)$, we need only consider the $\hat r\hat r$ component of the Einstein equation, which gives us

$$-\frac{1}{r^2} + \frac{1}{r^2}\mathrm{e}^{-2\Lambda} + \frac{2}{r}\mathrm{e}^{-2\Lambda}\frac{\mathrm{d}\Phi}{\mathrm{d}r} = G_{\hat r\hat r} = 8\pi p. \tag{21.17}$$

Substituting for Λ with $m(r)$ (from eqn 21.13), this expression becomes[13]

$$\frac{\mathrm{d}\Phi}{\mathrm{d}r} = \frac{m(r) + 4\pi r^3 p}{r\left[r - 2m(r)\right]}. \tag{21.19}$$

This looks complicated, but if we concentrate on the field outside of the mass distribution, were $p = 0$, then we can solve eqn 21.19 straightforwardly. In the region $r > R$, we deduce from eqn 21.16 that the mass function $m(r > R)$ must be constant. This constant must, of course, be the total mass M and so we exchange $m(r)$ for M, set $p = 0$ and write an expression valid for outside the mass distribution of

$$\frac{\mathrm{d}\Phi}{\mathrm{d}r} = \frac{M}{r\left(r - 2M\right)}. \tag{21.20}$$

[11]That is, in the orthonormal frame with basis vectors $\boldsymbol{e}_{\hat t}, \boldsymbol{e}_{\hat r}, \boldsymbol{e}_{\hat\theta}, \boldsymbol{e}_{\hat\phi}$ we have

$$T^{\hat\mu\hat\nu} = \begin{pmatrix} \rho & 0 & 0 & 0 \\ 0 & p & 0 & 0 \\ 0 & 0 & p & 0 \\ 0 & 0 & 0 & p \end{pmatrix}. \tag{21.11}$$

[12]This implies that

$$\mathrm{e}^{2\Lambda} = \left(1 - \frac{2m(r)}{r}\right)^{-1}. \tag{21.13}$$

[13]In the Newtonian limit, the potential becomes

$$\frac{\mathrm{d}\Phi}{\mathrm{d}r} = \frac{m}{r^2}, \tag{21.18}$$

as we might expect.

The solution to the latter expression is[14]

$$\Phi(r) = \tfrac{1}{2} \ln\left(1 - \tfrac{2M}{r}\right) \quad \text{(outside the mass).} \tag{21.22}$$

The resulting expressions for $e^{2\Lambda}$ and $e^{2\Phi}$ give us all we need to justify the form of the Schwarzschild line element outside a gravitating body, which is[15]

$$ds^2 = -\left(1 - \frac{2M}{r}\right)dt^2 + \left(1 - \frac{2M}{r}\right)^{-1} dr^2 + r^2\left(d\theta^2 + \sin^2\theta\, d\phi^2\right). \tag{21.23}$$

Example 21.2

Substituting the more familiar variables M and r, we find

$$\Phi = \tfrac{1}{2}\ln\left(1 - \tfrac{2M}{r}\right), \quad \Phi' = \frac{M}{r(r-2M)}, \quad \Phi'' = \frac{-2M(r-M)}{r^2(r-2M)^2},$$
$$\Lambda = \tfrac{1}{2}\ln\left(\frac{r}{r-2M}\right), \quad \Lambda' = \frac{-M}{r(r-2M)}, \tag{21.24}$$

and also

$$E = \frac{2M}{r^3}, \quad \bar{E} = \frac{-M}{r^3},$$
$$F = \frac{2M}{r^3}, \quad \bar{F} = \frac{-M}{r^3}. \tag{21.25}$$

In terms of the familiar variables M and r, the useful components of the Riemann curvature tensor are[16]

$$R_{\hat{t}\hat{r}\hat{t}\hat{r}} = -\frac{2M}{r^3}, \quad R_{\hat{t}\hat{\theta}\hat{t}\hat{\theta}} = \frac{M}{r^3}, \quad R_{\hat{t}\hat{\phi}\hat{t}\hat{\phi}} = \frac{M}{r^3},$$
$$R_{\hat{\theta}\hat{\phi}\hat{\theta}\hat{\phi}} = \frac{2M}{r^3}, \quad R_{\hat{r}\hat{\phi}\hat{r}\hat{\phi}} = -\frac{M}{r^3}, \quad R_{\hat{r}\hat{\theta}\hat{r}\hat{\theta}} = -\frac{M}{r^3}. \tag{21.26}$$

We also observe that components of $\boldsymbol{G}$ vanish as they must, since we're in a region with $\rho = p = 0$ and so $0 = 8\pi\boldsymbol{T} = \boldsymbol{G}$.

Example 21.3

The arguments in this section can also be used to derive the **Tolman-Oppenheimer-Volkov** (TOV) equations that describe the structure of a static, spherical perfect fluid, and which therefore provides us with a relativistic model of a star. Starting with the stress-energy tensor for a perfect fluid, we can use $\boldsymbol{\nabla} \cdot \boldsymbol{T} = 0$ to show[17] that in the Schwarzschild geometry

$$\frac{\partial p}{\partial r} + (p + \rho)\frac{\partial \Phi}{\partial r} = 0. \tag{21.27}$$

This equation can be combined with eqn 21.19 to give the first TOV equation

$$\frac{\partial p}{\partial r} = \frac{[p(r) + \rho(r)]\left[m(r) + 4\pi r^3 p(r)\right]}{r^2\left[1 - \frac{2m(r)}{r}\right]}. \tag{21.28}$$

The second TOV equation gives us a quantity to be interpreted as the mass

$$m(r) = 4\pi \int_0^r dr\, r^2 \rho(r), \tag{21.29}$$

where we have fixed the integration constant $m(0) = 0$. The solution to these equations requires a choice of equation of state (that is, a link between p and ρ). They are generally integrated numerically from the origin outwards until $p(r = R) = 0$, which we take to give the surface of the star, with $m(R)$ giving us the stellar mass.

---

[14]This gives us

$$e^{2\Phi(r)} = \left(1 - \frac{2M}{r}\right). \tag{21.21}$$

The connection coefficients for the Schwarzschild metric are

$$\Gamma^t{}_{rt} = \Phi',$$
$$\Gamma^r{}_{tt} = \Phi' e^{2\Phi} e^{-2\Lambda},$$
$$\Gamma^r{}_{rr} = \Lambda',$$
$$\Gamma^r{}_{\theta\theta} = -re^{-2\Lambda},$$
$$\Gamma^r{}_{\phi\phi} = -re^{-2\Lambda}\sin^2\theta,$$
$$\Gamma^\theta{}_{r\theta} = \tfrac{1}{r},$$
$$\Gamma^\theta{}_{\phi\phi} = -\sin\theta\cos\theta,$$
$$\Gamma^\phi{}_{r\phi} = \tfrac{1}{r},$$
$$\Gamma^\phi{}_{\theta\phi} = \frac{\cos\theta}{\sin\theta},$$

where a dash denotes a derivative with respect to the r coordinate.
(These are computed in Exercise 9.7).

[15]In a letter to Schwarzschild in 1916, Einstein wrote "I had not expected that one could formulate the exact solution of the problem so simply. The analytical treatment of the problem seems to me to be excellent."

[16]Achieved by lowering indices using the Minkowski tensor (i.e. lowering a $\hat{t}$ gives a minus sign, lowering everything else has no effect).

Richard C. Tolman (1881–1948).
J. Robert Oppenheimer (1904–1967).
George Volkoff (1914–2000).
Oppenheimer and Volkoff used Tolman's work as a basis that led to their prediction of the existence of neutron stars.

[17]See Exercise 39.5.

[18]As examined in the exercises, the Newtonian prediction for hydrostatic equilibrium is

$$4\pi r^2 \mathrm{d}p = -\frac{m\mathrm{d}m}{r^2}. \qquad (21.31)$$

It will be useful in later chapters to have in mind a simple picture of **stellar evolution**. Stars are formed from gas clouds that collapse under gravity to eventually achieve equilibrium, where gravitational collapse is balanced against outward radiation pressure resulting from nuclear fusion of hydrogen in the star's core. The hydrogen in the core eventually becomes exhausted, causing many stars (i.e. those of around one solar mass $M_\odot$), to fuse helium inside their core and hydrogen outside, leading to expansion into a **red giant**. Once the nuclear fuel is exhausted, the star collapses under its own gravity, typically forming a **white dwarf**, with outer layers of mass thrown off as planetary nebula. The white dwarf achieves equilibrium with gravitation collapse now balanced by the outward pressure caused by the electronic matter making up the dwarf being **degenerate**. The latter implies the density of electronic matter is such that the quantum energy levels are completely occupied by electrons, causing them to stack up in energy owing to the Pauli principle. This leads to a large outward pressure reflecting the difficulty in rearranging particles between energy levels while preventing multiple occupancy. For stars with masses $\gtrsim 1.5M_\odot$, the white dwarf cannot achieve this equilibrium owing to the size of the inward gravitational force, and a **neutron star** or **black hole** can be formed. The former involves electrons and protons fusing to form neutrons, with equilibrium now achieved from the outward pressure of the degenerate neutron matter. Pulsars were discovered by Jocelyn Bell Burnell (1943–) in 1967, and were later identified as rapidly rotating neutron stars. Black holes are discussed in Chapter 25.

[19]We will often invoke the observer at infinity, whose proper time $\tau = t$.

To interpret the first equation it can be rewritten as

$$4\pi r^2 \mathrm{d}p(r) = -\frac{m(r)\mathrm{d}m}{r^2}\left[1 + \frac{p(r)}{\rho(r)}\right]\left[1 + \frac{4\pi r^3 p(r)}{m(r)}\right]\left[1 - \frac{2m(r)}{r}\right]^{-1}. \qquad (21.30)$$

The first term on the right is the Newtonian prediction[18] with the remaining terms giving the relativistic corrections. In low-mass stars, the largest contribution to ρ is from baryons (i.e. nuclear matter), which don't contribute significantly to the pressure (which turns out to be provided by electrons). This means $p/\rho \approx 0$ and $p/m \approx 0$, so that the relativistic corrections are not large. In larger stars, the pressure and energy density increase the right-hand side of the equation, steepening the pressure gradient. This ultimately causes the radius of large stars to be reduced compared to the prediction of Newtonian physics.

21.4   The meaning of the coordinates

In general relativity, coordinates have no *intrinsic* metric significance. However, we *can* relate the coordinates to the quantities of interest in describing stars and orbits in this specific case. This is our next task.

Example 21.4

Let's consider a circle in the equatorial plane ($\theta = \pi/2$) at an instant in time ($\mathrm{d}t = 0$). We then have

$$\mathrm{d}s^2 = r^2 \mathrm{d}\phi^2. \qquad (21.32)$$

Taking a square root and integrating, we find the circumference of the circle is

$$C = \int_{\phi=0}^{2\pi} r\mathrm{d}\phi = 2\pi r. \qquad (21.33)$$

We can therefore call r a radius, in that it supplies the correct scaling to work out the circumference of circles. We must be careful, since if we attempt to compute the distance between two events along a radial line, we obtain

$$\Delta s = \int_{r_\mathrm{A}}^{r_\mathrm{B}} \left(1 - \frac{2M}{r}\right)^{-\frac{1}{2}} \mathrm{d}r, \qquad (21.34)$$

which will be greater than $r_\mathrm{B} - r_\mathrm{A}$.

We conclude that r does not give the distance from the origin in the Schwarzschild geometry. Since it does give the radius of circles it is sometimes known as the **circumferential radial coordinate**.

Next, we turn to the time. We find that the proper time between two events is

$$\Delta\tau = \left(1 - \frac{2M}{r}\right)\Delta t. \qquad (21.35)$$

For $r \to \infty$ we have $\Delta\tau = \Delta t$. We conclude that the t coordinate is the time between events as measured by a clock at infinity.[19]

We now have a suitable metric field we can use it to examine the motion that is possible in the curved spacetime that it describes. In the next chapter, we look at general properties of motion in the Schwarzschild geometry before, in Chapter 23, turning to the question of orbits.

Chapter summary

- The Schwarzschild geometry refers to static, spherically symmetric spacetime.
- The metric field in the Schwarzschild geometry is given by the Schwarzschild line element

$$\mathrm{d}s^2 = -\left(1 - \frac{2M}{r}\right)\mathrm{d}t^2 + \left(1 - \frac{2M}{r}\right)^{-1}\mathrm{d}r^2 + r^2\left(\mathrm{d}\theta^2 + \sin^2\theta\mathrm{d}\phi^2\right). \tag{21.36}$$

- The Schwarzschild metric is static, asymptotically flat and is badly behaved at $r = 0$ and $r = 2M$.
- Birkhoff's theorem says that any spherically symmetric solution to Einstein's equation is identical to the Schwarzschild solution.

Exercises

(21.1) Use the vielbein components to express the components of the energy-momentum tensor $\boldsymbol{T}$ for a perfect fluid in a (t, r, θ, ϕ) coordinate system.

(21.2) (a) Verify eqns 21.10 using the method suggested in the text.
(b) Using eqns 21.25, write the components of $\boldsymbol{G}$ in the orthonormal frame using the familiar polar coordinates and show that each component vanishes.

(21.3) Using the symmetries of $\boldsymbol{R}$, prove the useful rule in Sidenote 10.

(21.4) (a) Confirm that eqn 21.19 follows from eqn 21.17.
(b) Show that eqn 21.19 implies eqn 21.18 in the Newtonian limit.

(21.5) (a) By computing the relevant determinant g, compute the area of a spherical surface of fixed r and t in the Schwarzschild geometry.
(b) Compute a circumference at fixed r, t and θ.

(21.6) (a) Show that in Newtonian gravity, hydrostatic equilibrium requires that

$$4\pi r^2 \mathrm{d}p = -\frac{GM\mathrm{d}m}{r^2}, \tag{21.37}$$

where $\mathrm{d}m = 4\pi r^2 \rho \mathrm{d}r$.
Hint: Generalize the usual derivation of Pascal's law of hydrostatics in a uniform gravitational field for the case of a non-uniform field.

(21.7) *Justification of Birkhoff's theorem.* The most general spherically symmetric line element has the form

$$\mathrm{d}s^2 = -A(r,t)\mathrm{d}t^2 + B(r,t)\mathrm{d}r\mathrm{d}t + 2C(r,t)\mathrm{d}r^2 + r^2\left(\mathrm{d}\theta^2 + \sin^2\theta\mathrm{d}\phi^2\right). \tag{21.38}$$

(a) Show that a transformation $t \to t + h(r,t)$, where $h(r,t)$ is some function, can be used to eliminate the cross term proportional to $\mathrm{d}r\mathrm{d}t$, such that we have

$$\mathrm{d}s^2 = -\mathrm{e}^{\Phi(r,t)}\mathrm{d}t^2 + \mathrm{e}^{\Lambda(r,t)}\mathrm{d}r^2 + r^2\left(\mathrm{d}\theta^2 + \sin^2\theta\mathrm{d}\phi^2\right). \tag{21.39}$$

In Exercise 36.8, we compute the curvature properties of this spacetime. It follows from these that the components of the Einstein tensor are

$$G_{\hat{t}\hat{t}} = r^{-2}(1 - \mathrm{e}^{-2\Lambda}) + 2(\Lambda_{,r}/r)\mathrm{e}^{-2\Lambda},$$
$$G_{\hat{r}\hat{t}} = 2(\Lambda_{,t}/r)\mathrm{e}^{-(\Lambda+\Phi)},$$
$$G_{\hat{r}\hat{r}} = 2(\Phi_{,r}/r)\mathrm{e}^{-2\Lambda} + r^{-2}(\mathrm{e}^{-2\Lambda} - 1),$$
$$G_{\hat{\theta}\hat{\theta}} = G_{\hat{\phi}\hat{\phi}}, =$$
$$(\Phi_{,rr} + \Phi_{,r}^2 - \Phi_{,r}\Lambda_{,r} + \Phi_{,r}/r - \Lambda_{,r}/r)\mathrm{e}^{-2\Lambda}$$
$$- (\Lambda_{,tt} + \Lambda_{,t}^2 - \Phi_{,t}\Lambda_{,t})\mathrm{e}^{-2\Phi}. \tag{21.40}$$

(b) Show that if we are in a vacuum, then $\Lambda(r,t)$ is independent of time.

(c) Use the remaining components of the Einstein equation to show that

$$\Lambda = -\frac{1}{2}\ln\left|1 - \frac{2M}{r}\right|. \tag{21.41}$$

(d) Show further that

$$\Phi(r,t) = -\Lambda(r) + f(t), \tag{21.42}$$

where $f(t)$ is a function of time.

(e) Use this to prove Birkhoff's theorem, which says that the Schwarzschild geometry is the most general, asymptotically flat, spherically symmetric solution of the Einstein equation.

(21.8) Consider filling a universe described by the line element in eqn 21.5 with energy density ζ, such that we have $T_{\hat{t}\hat{t}} = 8\pi\zeta$ and $T_{\hat{r}\hat{r}} = -8\pi\zeta$.

(a) Use the results from this chapter to show that this energy density is consistent with $\Phi = -\Lambda$.

(b) Show that the line element

$$ds^2 = -\left(1 - H^2 r^2\right) dt^2 + \left(1 - H^2 r^2\right)^{-1} dr^2 + r^2 d\Omega^2, \tag{21.43}$$

is a solution to the Einstein equation. Here $d\Omega^2 = d\theta^2 + \sin^2\theta d\phi^2$ and H is a constant you should determine.

This solution is actually the de Sitter model (Universe 1 from Chapter 15) again, expressed in a different set of coordinates. The link is made in Exercise 18.8.

Motion in the Schwarzschild geometry

22

We are merely the stars' tennis-balls, struck and bandied
Which way to please them
John Webster (1580–1625) *The Duchess of Malfi*

Thanne longen folk to goon on pilgrimages
Geoffrey Chaucer (c.1343–1400)
Prologue to the Cantebury Tales

We found in the last chapter that the Schwarzschild line element is given by

$$\mathrm{d}s^2 = -\left(1 - \frac{2M}{r}\right)\mathrm{d}t^2 + \left(1 - \frac{2M}{r}\right)^{-1}\mathrm{d}r^2 + r^2\left(\mathrm{d}\theta^2 + \sin^2\theta\mathrm{d}\phi^2\right).$$

(22.2)

This equation represents the metric field outside a spherically symmetric object, such as a star, of mass M. In this chapter and the following ones, we shall examine the free-falling motion of particles in this metric field, where the particles follow the geodesics of this geometry. We will find that general relativity allows a richer range of possible motions than are found in the Newtonian problem, including, in addition to the interesting trajectories of massive particles, effects on the motion of photons: the particles of light.

At first glance the problem of relativistic motion looks like a formidable one involving deriving solutions to the geodesic equation,[1] with terms provided by the connection coefficients found from eqn 22.2. However, just as problems involving Newtonian orbits are greatly simplified by knowing about conserved quantities, we shall see how an analysis of conserved quantities, along with simple facts about the velocity $\boldsymbol{u}$, allow us to solve a variety of key problems. In particular, we make a lot of use of the identity[2] $\boldsymbol{u} \cdot \boldsymbol{u} = -1$ for massive particles and $\boldsymbol{u} \cdot \boldsymbol{u} = 0$ for massless ones. In addition, just as energy and angular momentum are conserved in the Newtonian potential, we will see that something very much like these is also conserved in the relativistic case, as we might naively expect. We turn first, therefore, to the method for extracting conserved quantities.

In this chapter, we'll need to use the components of the Schwarzschild metric

$$g_{tt} = -\left(1 - \frac{2M}{r}\right),$$
$$g_{rr} = \left(1 - \frac{2M}{r}\right)^{-1},$$
$$g_{\theta\theta} = r^2,$$
$$g_{\phi\phi} = r^2 \sin^2\theta.$$

(22.1)

[1]This is the differential equation

$$\frac{\mathrm{d}^2 x^\mu}{\mathrm{d}\tau^2} + \Gamma^\mu{}_{\alpha\beta}\frac{\mathrm{d}x^\alpha}{\mathrm{d}\tau}\frac{\mathrm{d}x^\beta}{\mathrm{d}\tau} = 0,$$

with the connection coefficients determined from the metric.

[2]Reminder: A massive particle's world line is parametrized by proper time so that, in a coordinate system with $x^\mu = (t, r, \theta, \phi)$, the velocity $\boldsymbol{u}$ has components $u^\mu = (u^t, u^r, u^\theta, u^\phi)$, i.e.

$$u^\mu = \left(\frac{\mathrm{d}t}{\mathrm{d}\tau}, \frac{\mathrm{d}r}{\mathrm{d}\tau}, \frac{\mathrm{d}\theta}{\mathrm{d}\tau}, \frac{\mathrm{d}\phi}{\mathrm{d}\tau}\right).$$

22.1 Constants of the motion

One of the most important factors in analysing motion in mechanics is the identification of constants of the motion. In geometrical problems, exemplified by the physics of gravitation, conserved quantities can be identified by finding a set of fields known as the **Killing vector fields** $\boldsymbol{\xi}$ of the metric. These are very simply identified by inspection, by noting which coordinates do not feature in the components of the metric. For now we will give (but not prove) a simple recipe for identifying conserved quantities.

To find a conserved quantity from a metric field:

Step 1: Look at the metric components and identify those coordinates on which none of the metric components depend.

Step 2: If the variable x^α does not feature, we have a **Killing vector** $\boldsymbol{\xi} = \boldsymbol{e}_\alpha$.

Step 3: Particles travelling along a geodesic have a velocity $\boldsymbol{u} = u^\mu \boldsymbol{e}_\mu$, which is the tangent to a geodesic. The quantity

$$\boldsymbol{\xi} \cdot \boldsymbol{u} = g_{\mu\nu}\xi^\mu u^\nu \tag{22.3}$$

is **conserved** along the geodesic.

To identify the Killing vectors for the Schwarzschild geometry, we spot (**step 1**) that the metric components are all independent of the coordinates t and ϕ. Ordering the coordinates (t, r, θ, ϕ) we identify two corresponding Killing vectors (**step 2**). First, from the independence of t we extract a Killing vector $\boldsymbol{\xi}$ via

$$\boldsymbol{\xi} = \boldsymbol{e}_t, \quad \xi^\mu = (1,0,0,0), \quad \text{(time independence)}, \tag{22.4}$$

and from the independence of ϕ we identify a Killing vector $\boldsymbol{\eta}$ as we have

$$\boldsymbol{\eta} = \boldsymbol{e}_\phi, \quad \eta^\mu = (0,0,0,1), \quad \text{(rotational symmetry)}. \tag{22.5}$$

We now use the fact that $\boldsymbol{\xi} \cdot \boldsymbol{u}$ is a constant along a geodesic (**step 3**) to identify the conserved quantities. The first is identified as[3]

$$\boldsymbol{\xi} \cdot \boldsymbol{u} = u^\mu \boldsymbol{e}_\mu \cdot \boldsymbol{e}_t = u^\mu g_{\mu t} = u_t, \tag{22.6}$$

or equivalently

$$\boldsymbol{\xi} \cdot \boldsymbol{u} = g_{tt} u^t = -\left(1 - \frac{2M}{r}\right) u^t = -\left(1 - \frac{2M}{r}\right) \frac{\mathrm{d}t}{\mathrm{d}\tau}, \tag{22.7}$$

which is conventionally named $-\tilde{E}$, which is to say we have a conserved quantity

$$\tilde{E} = -\boldsymbol{\xi} \cdot \boldsymbol{u} = \left(1 - \frac{2M}{r}\right) u^t. \tag{22.8}$$

This quantity $\tilde{E}$ tells us about the energy per unit rest mass.[4] In general, it can be interpreted, for timelike geodesics, as total energy per unit rest mass, measured by a static observer at infinity.

Killing vectors are discussed in detail in Chapter 33 in the context of geometry.

[3] Remember the very useful fact that $g_{\mu\nu} = \boldsymbol{e}_\mu \cdot \boldsymbol{e}_\nu$.

[4] In general, we can compute the energy of a particle with momentum $\boldsymbol{p}$, measured by an observer with velocity $\boldsymbol{u}$ by computing $E = -\boldsymbol{p} \cdot \boldsymbol{u}$. Notice that an observer at rest at infinity in the field of a spherical star has, in the rest frame of the star (which is also the observer's rest frame), a velocity $\boldsymbol{u}_{\text{obs}}$ with components $u^\mu = (1,0,0,0)$, so that $\boldsymbol{u}_{\text{obs}} = \boldsymbol{\xi}$. (This follows because spacetime is flat at infinity. In general, u^t for an observer is determined in their rest frame using the condition $\boldsymbol{u} \cdot \boldsymbol{u} = -1$.) Consider a particle with momentum $\boldsymbol{p}$ travelling along a geodesic in the field of a spherical mass distribution. Its energy, measured locally at infinity by an observer at rest, is therefore $E_\infty = -\boldsymbol{p} \cdot \boldsymbol{\xi}$. We will use this result below.

The other conserved quantity, called $\tilde{L}$ is also straightforward to pick out as $\boldsymbol{\eta} \cdot \boldsymbol{u} = u_\phi$, or, from $u_\phi = g_{\phi\phi} u^\phi$

$$\tilde{L} = \boldsymbol{\eta} \cdot \boldsymbol{u} = r^2 \sin^2 \theta u^\phi. \tag{22.9}$$

We interpret this as angular momentum per unit rest mass. Note that for equatorial motion (i.e. with $\theta = \pi/2$) this becomes $\tilde{L} = r^2 \dot{\phi}$, which looks just like the expression for angular momentum per unit mass in cylindrical coordinates.

So far we have considered only the geometry, and not what sort of particle is in free fall. This information is supplied via the square of the velocity vector $\boldsymbol{u} \cdot \boldsymbol{u}$. Massive particles move along timelike geodesics and so we fix the magnitude of the velocity vector (which is tangent to the geodesic) by saying

$$\boldsymbol{u} \cdot \boldsymbol{u} = g_{\mu\nu} \frac{\mathrm{d}x^\mu}{\mathrm{d}\tau} \frac{\mathrm{d}x^\nu}{\mathrm{d}\tau} = -1, \quad \text{(massive particles)}, \tag{22.10}$$

where the proper time τ has been employed as an affine parameter marking off the trajectory of the particle. Photons travel along null geodesics, whose velocity tangent vector has the property

$$\boldsymbol{u} \cdot \boldsymbol{u} = g_{\mu\nu} \frac{\mathrm{d}x^\mu}{\mathrm{d}\lambda} \frac{\mathrm{d}x^\nu}{\mathrm{d}\lambda} = 0, \quad \text{(massless photons)}, \tag{22.11}$$

where λ is an affine parameter for the photon world line.

22.2 Gravitational redshift

With these preliminaries, we can derive some results. The first is a result for photons: the gravitational redshift caused by the Schwarzschild geometry.[5] To analyse this, we consider the motion of a photon with null momentum $\boldsymbol{p}$ propagating along a radial line. By symmetry, this is a geodesic. The photon has the property[6] that $\boldsymbol{p} \cdot \boldsymbol{\xi}$ is conserved along the geodesic, which implies that, in this geometry, $g_{tt} p^t = (1 - 2M/r) p^t(r)$ is conserved as r is varied (rather than the flat-space energy p^t, as we might naively have expected). Remember, however, that an observer at r does not measure p^t; they measure the local value $p^{\hat{t}} = \hbar\omega$.

We can find out how the photon's measured frequency ω changes as the photon moves radially outwards from a massive star. We consider two static observers, one at radius $r = R$ and one at $r = \infty$ (Fig. 22.1). Being observers, their world lines have tangents that are timelike velocity vectors. The energy of a photon with momentum $\boldsymbol{p}$, measured by an observer[7] with velocity $\boldsymbol{u}_{\mathrm{obs}}$ is

$$\hbar\omega = -\boldsymbol{p} \cdot \boldsymbol{u}_{\mathrm{obs}}. \tag{22.12}$$

If the observer is at rest with respect to the star, then $u^i_{\mathrm{obs}} = 0$ and we can determine the timelike component of $\boldsymbol{u}_{\mathrm{obs}}$ using the velocity identity

$$g_{tt} \left(u^t_{\mathrm{obs}} \right)^2 = -1. \tag{22.13}$$

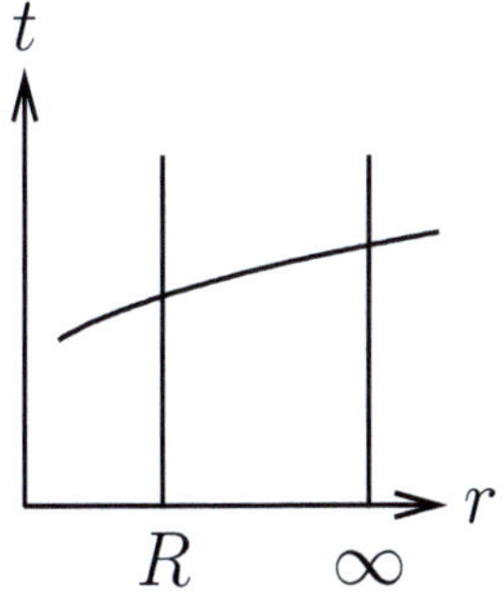

Fig. 22.1 A photon world line meets an observer at R and one at infinity.

[5] Here we revisit our second cause of frequency shifts discussed earlier in Chapter 6. (The others are the Doppler effect and the cosmological redshift.)

[6] This follows since we shall assume $\boldsymbol{p}$ to be parallel to $\boldsymbol{u}$ for light and $\boldsymbol{p}$ is therefore tangent to the geodesics. (For a more careful discussion of photon momentum, see Section 24.7.)

[7] An alternative rule would be to use the vielbein components to say that the measured energy $p^{\hat{t}}$ is related to p^t via

$$p^{\hat{t}}(r) = (e_\mu)^{\hat{t}}(r) p^\mu(r).$$

This alternative approach is examined in the exercises. Here, instead, since we are trying to relate measurements made by local observers in two places, we use the conservation laws encoded in the Killing vector $\boldsymbol{\xi}$ (i.e. that $\boldsymbol{u} \cdot \boldsymbol{\xi}$ is conserved along a geodesic). Our strategy is therefore to incorporate the energy Killing vector $\boldsymbol{\xi}$ into the description of the stationary observer's velocity.

With the Schwarzschild metric component g_{tt}, this gives a velocity component at a radius r of

$$u^t_{\text{obs}}(r) = \left(1 - \frac{2M}{r}\right)^{-\frac{1}{2}}, \qquad (22.14)$$

or in terms of one of the Killing vectors,[8] $\boldsymbol{u}_{\text{obs}}(r) = (1 - 2M/r)^{-\frac{1}{2}}\boldsymbol{\xi}$, which is the key equation. This expression means we can write eqn 22.12 in terms of the Killing vector $\boldsymbol{\xi}$. Specifically, the energy measured by the stationary observer sat at a radius R is

$$\hbar\omega_R = \left(1 - \frac{2M}{R}\right)^{-\frac{1}{2}} (-\boldsymbol{\xi} \cdot \boldsymbol{p})_R, \qquad (22.15)$$

where subscript R implies the quantity is evaluated at this radius. For the energy measured by the observer at infinity, the factor $(1 - 2M/r)^{-\frac{1}{2}} \to 1$ and so $\hbar\omega_\infty = (-\boldsymbol{\xi} \cdot \boldsymbol{p})_\infty$. However, since the photon is moving along a geodesic, the quantity $\boldsymbol{\xi} \cdot \boldsymbol{p}$ is conserved in the motion, and so $(-\boldsymbol{\xi} \cdot \boldsymbol{p})_R = (-\boldsymbol{\xi} \cdot \boldsymbol{p})_\infty$ and we must therefore have

$$\omega_\infty = \omega_R \left(1 - \frac{2M}{R}\right)^{\frac{1}{2}}. \qquad (22.16)$$

The frequency at infinity is less than the frequency at the point R (see Fig. 22.2). One can rationalize this result by saying that the photon's energy is lowered through its climbing of the gravitational potential.

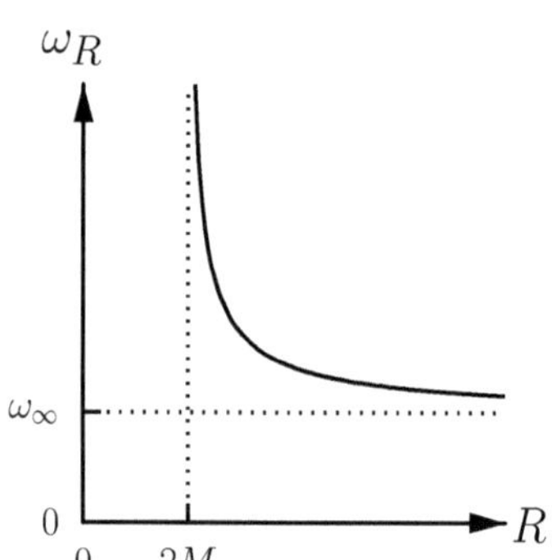

Fig. 22.2 A plot of ω_R against R for $R > 2M$.

22.3 Motion in Schwarzschild spacetime

We now turn to the motion of massive particles in the Schwarzschild geometry.[9] It might be assumed that we need to solve the full geodesic equation to understand the motion of particles and photons. Fortunately for us, the only ingredients needed are the constants of the motion given by the Killing vectors, and the velocity identity $\boldsymbol{u} \cdot \boldsymbol{u} = -1$.

Example 22.1

For simplicity, we assume motion in the equatorial plane of the geometry, which is to say that we fix $\theta = \pi/2$ and so $u^\theta = 0$. To analyse motion of a particle we start by using the velocity identity for the particle $\boldsymbol{u} \cdot \boldsymbol{u} = g_{\mu\nu}u^\mu u^\nu = -1$, to write

$$-\left(1 - \frac{2M}{r}\right)(u^t)^2 + \left(1 - \frac{2M}{r}\right)^{-1}(u^r)^2 + r^2(u^\phi)^2 = -1. \qquad (22.18)$$

This can be rewritten using the conserved quantities $\tilde{E}$ and $\tilde{L}$ as

$$-\left(1 - \frac{2M}{r}\right)^{-1}\tilde{E}^2 + \left(1 - \frac{2M}{r}\right)^{-1}\left(\frac{dr}{d\tau}\right)^2 + \frac{\tilde{L}^2}{r^2} = -1. \qquad (22.19)$$

A little massaging of the above expression gives

$$\frac{\tilde{E}^2 - 1}{2} = \frac{1}{2}\left(\frac{dr}{d\tau}\right)^2 + \frac{1}{2}\left[\left(1 - \frac{2M}{r}\right)\left(1 + \frac{\tilde{L}^2}{r^2}\right) - 1\right]. \qquad (22.20)$$

We can summarize the important results that we shall use to compute trajectories:

■ For massive particles, the particle velocity $\boldsymbol{u}$ is tangent to the world line and has the property $\boldsymbol{u} \cdot \boldsymbol{u} = -1$.

■ For photons we have $\boldsymbol{u} \cdot \boldsymbol{u} = 0$. The interval $ds = 0$ on the world line provides a useful constraint.

■ We compute conservation laws using the Killing rule, that says that if the metric components are independent of variable x^α, the component u_α is conserved along a geodesic.

■ To access the coordinate velocity dx^i/dt we use the trick

$$\frac{dx^i}{dt} = \frac{dx^i}{d\tau}\frac{d\tau}{dt} = \frac{1}{u^0}\frac{dx^i}{d\tau}. \qquad (22.17)$$

Now we introduce a new constant energy-like variable,[10] $\mathcal{E} = (\tilde{E}^2 - 1)/2$, and define an effective potential for motion in the Schwarzschild geometry of

$$V_{\text{eff}}(r) = \frac{1}{2}\left[\left(1 - \frac{2M}{r}\right)\left(1 + \frac{\tilde{L}^2}{r^2}\right) - 1\right] = -\frac{M}{r} + \frac{\tilde{L}^2}{2r^2} - \frac{M\tilde{L}^2}{r^3}, \qquad (22.21)$$

and we end up with the familiar equation $\mathcal{E} = \frac{1}{2}\left(\frac{dr}{d\tau}\right)^2 + V_{\text{eff}}(r)$.

The result of the last example is that the motion of a massive particle obeys the effective-potential equation

$$\mathcal{E} = \frac{1}{2}\left(\frac{dr}{d\tau}\right)^2 + V_{\text{eff}}(r), \qquad (22.22)$$

with $\mathcal{E} = (\tilde{E}^2 - 1)/2$ and

$$V_{\text{eff}}(r) = -\frac{M}{r} + \frac{\tilde{L}^2}{2r^2} - \frac{M\tilde{L}^2}{r^3}. \qquad (22.23)$$

This is similar to the equation for motion in an effective potential that we saw[11] in Chapter 20 for Newtonian motion. We conclude that motion takes place in an effective potential V_{eff}. The potential for the Schwarzschild geometry is shown in Fig. 22.3. Compared to the Newtonian potential, this one has more structure: specifically, with increasing r, an initial increase in V_{eff} to a maximum, followed by behaviour that looks similar to its Newtonian cousin. This new structure leads to the richness of the new trajectories allowed by relativity.

Example 22.2

We can restore factors of c and G to find

$$V_{\text{eff}} = \frac{1}{c^2}\left(-\frac{GM}{r} + \frac{\tilde{L}^2}{2r^2} - \frac{GM\tilde{L}^2}{c^2 r^3}\right). \qquad (22.24)$$

Now define the non-relativistic, Newtonian energy E_{N} via

$$\tilde{E} = \frac{mc^2 + E_{\text{N}}}{mc^2}. \qquad (22.25)$$

This allows us to write

$$E_{\text{N}} = \frac{m}{2}\left(\frac{dr}{d\tau}\right)^2 + \frac{L^2}{2mr^2} - \frac{GMm}{r} - \frac{GML^2}{mc^2 r^3}, \qquad (22.26)$$

where $L = m\tilde{L}$. We conclude that the usual Newtonian energy equation is augmented at order $1/c^2$ by the relativistic, $1/r^3$ term.

We are now in the position to be able to provide a description of the motion. If we are interested in what would be measured by an observer, we will need access to the vielbein components for the Schwarzschild geometry in order to shift into the orthonormal frame of the observer. The vielbein for a stationary observer is given in the margin. It is worth noting that for this observer at infinity, measurements of the $\hat{t}$ and $\hat{r}$ components of a vector will give the coordinate values as both of the relevant vielbein components are unity at infinity.

[10] Particles coming in from infinity have $\tilde{E} > 1$ and so have effective energy $\mathcal{E} > 0$.

[11] The equations of motion differ by a factor of m. The Newtonian version has an effective potential energy U_{eff} and so an effective potential

$$V_{\text{eff}}(r) = \frac{U_{\text{eff}}(r)}{m} = -\frac{M}{r} + \frac{\tilde{L}^2}{2r^2}.$$

The relativistic version has an extra term with a $1/r^3$ dependence.

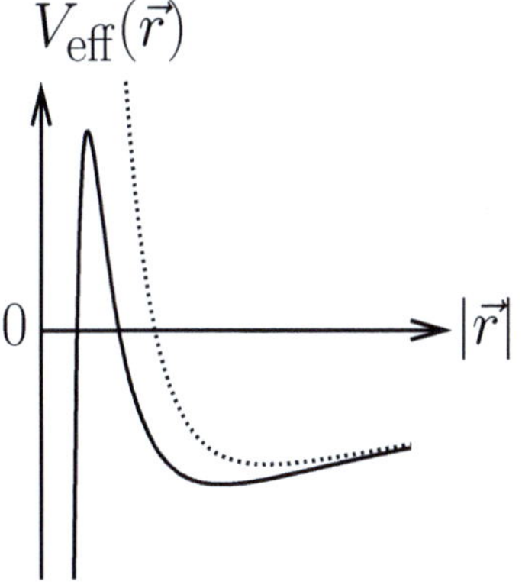

Fig. 22.3 The relativistic effective potential for a particle (solid line, with $\tilde{L}/M = 4.3$), compared to the Newtonian one (dotted line).

The vielbein components for an observer in the orthonormal frame (where the observer is stationary with respect to the coordinate frame) has components

$$(e_t)^{\hat{t}} = \left(1 - \frac{2M}{r}\right)^{\frac{1}{2}},$$

$$(e_r)^{\hat{r}} = \left(1 - \frac{2M}{r}\right)^{-\frac{1}{2}},$$

$$(e_\theta)^{\hat{\theta}} = r,$$

$$(e_\phi)^{\hat{\phi}} = r\sin\theta.$$

It is important to remember that if we want to shift into the frame of a moving observer we will need a different vielbein. An example we have seen before that we will use several times is the freely falling observer's frame.

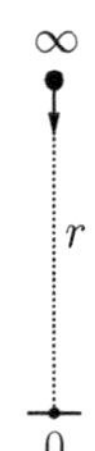

Fig. 22.4 The radial plunge.

[12]At infinity, spacetime is flat and the particle is at rest. Initially, the components of the velocity must therefore be $u^\mu(r=\infty) = \frac{dx^\mu}{d\tau} = (1,0,0,0)$. Recall also that the energy

$$\tilde{E} = \left(1 - \frac{2M}{r}\right) u^t$$

is constant along a geodesic. For $r \to \infty$ we have $\tilde{E} = 1$. Also recall that the (constant) effective energy is

$$\mathcal{E} = \frac{\tilde{E}^2 - 1}{2} = 0.$$

[13]There is no mechanism to pick up angular velocity, hence the zeros for the θ and ϕ components of velocity.

[14]The falling observer's local frame is flat, but moving with respect to the coordinate frame, so differs from the conventional orthonormal frame we often use. To find this frame, note that we always take $\boldsymbol{e}_{\hat{t}} = \boldsymbol{u}$. Then, for convenience choose $(\boldsymbol{e}_{\hat{\theta}})^\mu = (0,0,1,0)$ and $(\boldsymbol{e}_{\hat{\phi}})^\mu = (0,0,0,1)$. This means that the one remaining vielbein component is

$$(\boldsymbol{e}_{\hat{r}})^\mu = \begin{pmatrix} \dfrac{-\sqrt{2M/r}}{1 - 2M/r} \\ 1 \\ 0 \\ 0 \end{pmatrix}. \tag{22.30}$$

[15]In this section, it is necessary to fix the value of r at some value of the proper time τ in order to fix the integration constant.

[16]A useful quantity to note for computations is the coordinate velocity

$$\frac{dr}{dt} = -\left(1 - \frac{2M}{r}\right)\left(\frac{2M}{r}\right)^{\frac{1}{2}}. \tag{22.34}$$

The magnitude of this quantity is the coordinate escape velocity of a particle, as discussed in the exercises.

22.4 Example: the radial plunge

Let's consider a massive particle that starts at rest at infinity and plunges towards a star along a radial line (Fig. 22.4). This line is a geodesic.

Example 22.3

We have immediately that the angular momentum $\tilde{L} = 0$ since the plunge is radial. 'At rest at infinity' implies that at a great distance from the gravitating object we have[12] $u^t = dt/d\tau = 1$ and so, for the duration of the freefall, $\tilde{E} = 1$ and $\mathcal{E} = 0$. The fact that $\tilde{E} = 1$ allows us to pick out, at later times, that

$$u^t = \frac{dt}{d\tau} = \left(1 - \frac{2M}{r}\right)^{-1}. \tag{22.27}$$

The effective energy $\mathcal{E}$ in eqn 22.22 for a radial plunge with $\tilde{L} = 0$ and $\mathcal{E} = 0$ gives us an equation of motion

$$0 = \frac{1}{2}\left(\frac{dr}{d\tau}\right)^2 - \frac{M}{r}, \tag{22.28}$$

which provides us with an expression for the radial velocity $u^r = dr/d\tau = -(2M/r)^{\frac{1}{2}}$, where the sign is chosen so that the particle is falling towards the gravitating star. We conclude that the velocity during the fall is given by a vector $\boldsymbol{u}$ with components in the coordinate frame of[13]

$$u^\mu = \left(\left(1 - \frac{2M}{r}\right)^{-1}, -\left(\frac{2M}{r}\right)^{\frac{1}{2}}, 0, 0\right). \tag{22.29}$$

The vielbein components for this observer's frame can also be computed.[14]

Let's consider how much proper time and coordinate time elapses for the falling particle.

Example 22.4

First the proper time. We rewrite eqn 22.28 as an integral

$$\int_r^0 r^{\frac{1}{2}}\,dr = -\int_\tau^{\tau_*} (2M)^{\frac{1}{2}}\,d\tau, \tag{22.31}$$

where the negative square root has been chosen for an inward-travelling particle and τ_* fixes the value of the proper time[15] at the end of the plunge, when $r = 0$. This can be integrated with the result that

$$r(\tau) = \left(\frac{3}{2}\right)^{\frac{2}{3}} (2M)^{\frac{1}{3}} (\tau_* - \tau)^{\frac{2}{3}}. \tag{22.32}$$

We can then say that the proper time for a falling astronaut is given by

$$\tau(r) = \tau_* - \left(\frac{2}{3}\right)\left(\frac{1}{2M}\right)^{\frac{1}{2}} r^{\frac{3}{2}}. \tag{22.33}$$

The variation of τ with the Schwarzschild radius coordinate r is shown in Fig. 22.5. Notice that, from an arbitrary starting value of r it takes a finite amount of proper time to reach any value of r, as we might expect.

Next, the amount of elapsed coordinate time t. Since $(\boldsymbol{e}_t)^{\hat{t}} = 1$ at infinity, this is also the time measured by an observer at infinity watching the particle plunge. We have[16]

$$\frac{dt}{dr} = \frac{dt}{d\tau}\frac{d\tau}{dr} = \frac{u^t}{u^r} = -\left(\frac{2M}{r}\right)^{-\frac{1}{2}}\left(1 - \frac{2M}{r}\right)^{-1}. \tag{22.35}$$

This equation can be integrated to give

$$t = t_* + 2M \left[-\frac{2}{3}\left(\frac{r}{2M}\right)^{\frac{3}{2}} - 2\left(\frac{r}{2M}\right)^{\frac{1}{2}} + \ln \left| \frac{\left(\frac{r}{2M}\right)^{\frac{1}{2}} + 1}{\left(\frac{r}{2M}\right)^{\frac{1}{2}} - 1} \right| \right], \qquad (22.36)$$

where t_* gives the time coordinate when $r = 0$. This function is graphed in Fig. 22.6.

There are some curious facts about the results from the last example: the most notable being that, from the point of view of the observer at infinity, it takes an *infinite* amount of coordinate time for an in-falling astronaut to reach $r = 2M$ (but a finite proper time). We will return to this point in Chapter 25.

Finally, let's examine the coordinate velocity of the particle that plunges radially in a Schwarzschild metric.

Example 22.5

We won't assume the particle starts from rest this time. The velocity identity gives us

$$-1 = \boldsymbol{u} \cdot \boldsymbol{u}$$

$$= -\left(1 - \frac{2M}{r}\right)(u^t)^2 + \left(1 - \frac{2M}{r}\right)^{-1}(u^r)^2$$

$$= \left[-\left(1 - \frac{2M}{r}\right) + \left(1 - \frac{2M}{r}\right)^{-1}\left(\frac{dr}{dt}\right)^2\right](u^t)^2$$

$$= \left[-\left(1 - \frac{2M}{r}\right) + \left(1 - \frac{2M}{r}\right)^{-1}\left(\frac{dr}{dt}\right)^2\right](u_t)^2\left(1 - \frac{2M}{r}\right)^{-2}, \qquad (22.37)$$

where, in the final line we've written the equation in terms of the constant of the motion $\boldsymbol{u} \cdot \boldsymbol{\xi} = u_t$. Solving for $(dr/dt)^2$ we find

$$\left(\frac{dr}{dt}\right)^2 = \left(1 - \frac{2M}{r}\right)^2\left[1 - \left(1 - \frac{2M}{r}\right)\frac{1}{(u_t)^2}\right]. \qquad (22.38)$$

This coordinate velocity vanishes as the particle approaches $r = 2M$ (Fig. 22.7).

What does a stationary observer determine as the coordinate velocity? We won't assume that they are at infinity here and so we shift to the orthonormal frame using the vielbein. Remember that for 1-forms like $\boldsymbol{dt}$ and $\boldsymbol{dr}$ we have $\boldsymbol{\omega}^{\hat{\mu}} = (e_\mu)^{\hat{\mu}}\boldsymbol{\omega}^\mu$. We find

$$\boldsymbol{d\hat{t}} = \left(1 - \frac{2M}{r}\right)^{\frac{1}{2}}\boldsymbol{dt},$$

$$\boldsymbol{d\hat{r}} = \left(1 - \frac{2M}{r}\right)^{-\frac{1}{2}}\boldsymbol{dr}. \qquad (22.39)$$

Noting that the derivative $dr/dt \equiv \boldsymbol{dr}/\boldsymbol{dt}$, we have that the stationary observer observes a plunge with a velocity

$$\frac{d\hat{r}}{d\hat{t}} = \left(1 - \frac{2M}{r}\right)^{-1}\frac{dr}{dt}. \qquad (22.40)$$

Irrespective of u_t, and hence the initial velocity, this locally measured velocity approaches unity or, in real units the speed of light, as $r \to 2M$ (Fig. 22.7).

Using only the velocity identity and the conservation of energy and angular momentum, we have shown a number of curious relativistic effects in a particle's motion. We'll pick up these points in a few chapters' time. In the next chapter, we turn to orbits.

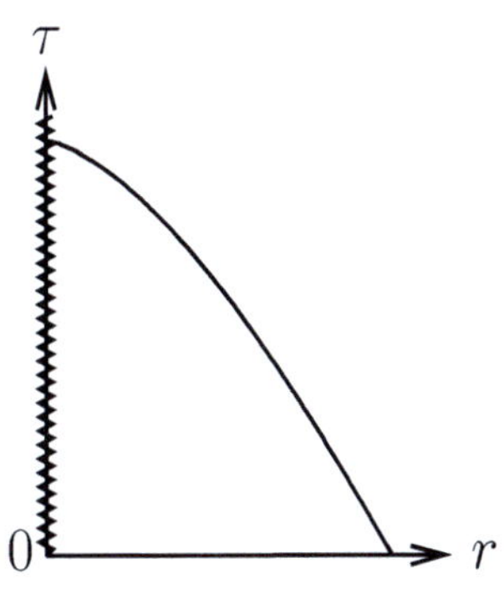

Fig. 22.5 The evolution of the coordinate r with the proper time τ for the radial plunge.

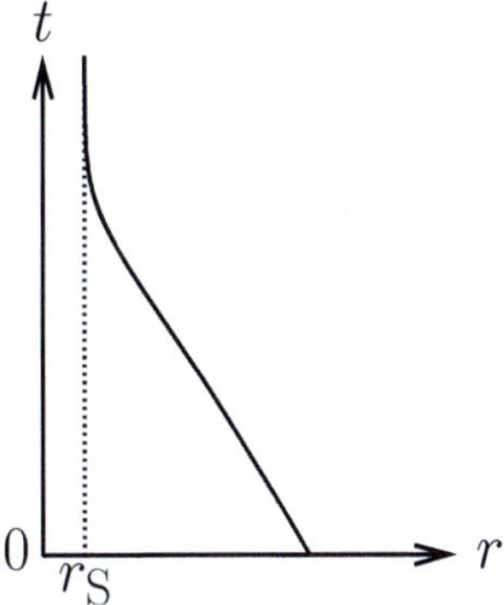

Fig. 22.6 The evolution of the coordinate r with the coordinate time t for the radial plunge. The dotted line shows $r_S = 2M$.

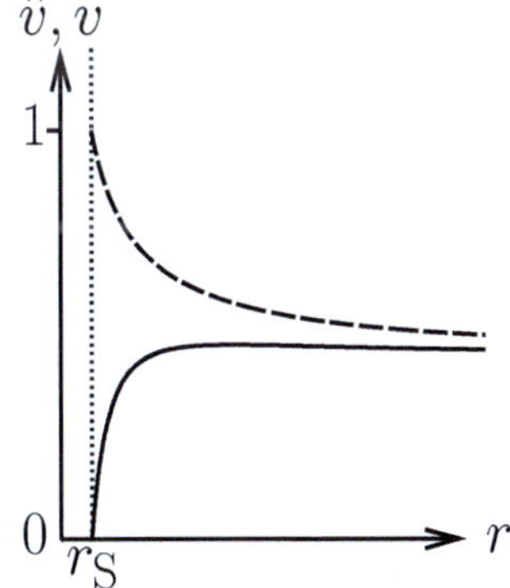

Fig. 22.7 Example evolution of the coordinate velocity $v = dr/dt$ with radial coordinate r for the radial plunge (solid line). The dashed line shows the coordinate velocity $\hat{v} = d\hat{r}/d\hat{t}$ measured by a stationary observer.

Chapter summary

- Motion in the Schwarzschild geometry can often be computed using the velocity $\boldsymbol{u}$, subject to the constraint that $\boldsymbol{u} \cdot \boldsymbol{u} = -1$. The constants of the motion $\tilde{E}$ and $\tilde{L}$ follow from the independence of the metric components from coordinates t and ϕ.

- The effective potential method allows motion to be analysed in terms of a relativistic V_{eff}.

- A particle that plunges from rest at infinity has $\mathcal{E} = 0$ and a velocity $\boldsymbol{u}$ with components

$$u^\mu = \left(\left(1 - \frac{2M}{r}\right)^{-1}, -\left(\frac{2M}{r}\right)^{\frac{1}{2}}, 0, 0 \right). \qquad (22.41)$$

Exercises

(22.1) Consider the acceleration $\boldsymbol{a} = \boldsymbol{\nabla}_{\boldsymbol{u}} \boldsymbol{u}$. By the geodesic equation, the components of the acceleration $a^\mu = u^\alpha u^\mu{}_{;\alpha}$ vanish in free fall.
(a) What are the components of the acceleration for a particle at rest in a static gravitational field with metric components $g_{\mu\nu}$?
(b) What are the components of the acceleration of a particle at rest at a fixed value of r, in a Schwarzschild coordinate system.
(c) What is the proper acceleration α (i.e. the acceleration measured in the frame of the particle)?

(22.2) *Although we have used a variety of shortcuts to compute the equations of motion, the same information is available from the geodesic equation.*
(a) Compute the components of the geodesic equation for the Schwarzschild geometry with line element given in the form of eqn 22.2.
If you've done this previously for the line element expressed in different variables, you could simply re-express the equations in terms of (t, r, θ, ϕ).
(b) How do these expressions simplify for motion in the plane with $\theta = \pi/2$?
(c) Show that the equation of motion for the r coordinate is consistent with eqn 22.22.

(22.3) Derive eqn 22.16 using the vielbein components, and the fact that p_t is conserved on the geodesic.

(22.4) Verify eqn 22.26 in the Newtonian limit.

(22.5) A stationary observer on the surface of a spherical mass M of radius R launches a projectile at its escape velocity. It does not experience any force during its motion and so it follows a geodesic.
(a) What is the escape velocity of the projectile measured in the observer's rest frame?
(b) What is the energy of the projectile as measured by the observer?

(22.6) *Gravitational clock effect:* Consider the Schwarzschild geometry describing the gravitational field of the Earth, which is rotating on its axis with angular velocity ω.
(a) For an observer at rest on the Earth's surface at the equator, show that the proper time interval is given approximately by

$$d\tau_1 \approx \left(1 - \frac{M}{r} - \frac{r^2\omega^2}{2}\right) dt. \qquad (22.42)$$

(b) Now consider a second observer who flies eastwards around the equator with velocity v at a height h, whose measures a proper time interval $d\tau_2$. Show that

$$d\tau_1 - d\tau_2 \approx \left[-\frac{Mh}{r^2} + \frac{(2r\omega + v)v}{2}\right] dt. \qquad (22.43)$$

(c) Define $\Delta = (d\tau_1 - d\tau_2)/d\tau_1$ and estimate this quantity for an eastward flight around the Earth.

(d) Constrast the value of Δ for a similar flight in a westward direction.

This experiment was carried out in the early 1970s by Hafele and Keating, whose results were consistent with predictions in this problem [which follows the method in Ryder (2009)].

(22.7) Two observers start on the same radial line in the Schwarzschild geometry, one at rest at r_1 and another at rest at $r_2 > r_1$. At $t = 0$, the observer at r_2 begins to fall freely. What is the relative velocity of the observers as they meet?

Hint: Recall from Chapter 2 that γ, corresponding to the instantaneous relative velocity $v_{\rm rel}$, can be found by taking the dot product of the two observers' velocity vectors: $-\gamma(v_{\rm rel}) = \boldsymbol{u} \cdot \boldsymbol{v}$.

23 Orbits in the Schwarzschild geometry

The Schwarzschild line element is given by

$$\mathrm{d}s^2 = -\left(1 - \frac{2M}{r}\right)\mathrm{d}t^2$$
$$+ \left(1 - \frac{2M}{r}\right)^{-1}\mathrm{d}r^2$$
$$+ r^2\left(\mathrm{d}\theta^2 + \sin^2\theta\,\mathrm{d}\phi^2\right).$$

The conserved quantities are

$$\tilde{E} = \left(1 - \frac{2M}{r}\right)u^t,$$

and

$$\tilde{L} = r^2\sin^2\theta\,u^\phi.$$

[1] See Exercise 22.2.

I am about to make my last voyage, a great leap in the dark
Thomas Hobbes (1588–1679)

The description of orbits given in Chapter 20 proved a crowning triumph of Newtonian mechanics. One of the first successes of general relativity was to provide predictions of orbits that, although similar to Kepler and Newton's ellipses, actually provided closer agreement with observation. We now turn to the possible orbits in the Schwarzschild geometry. These are a special class of geodesic, where a particle executes periodic motion by freely falling along a closed path in space. As in the last chapter, we will make use of the Schwarzschild line element and the conserved quantities $\tilde{E}$ and $\tilde{L}$. Previously, we have confined our attention to those plunging motions with $\tilde{L} = 0$. We now relax that constraint in order to allow angular motion. Our discussion of the resulting stable orbits will yield the famous prediction of the precession of the perihelion of an orbit in the Schwarzschild geometry.

23.1 Orbits for massive particles

Orbits made by massive particles are on geodesics parametrized by their proper time τ. For simplicity, we restrict our attention to particles moving in an equatorial plane, so that we can set $\theta = \pi/2$ and so $u^\theta = \mathrm{d}\theta/\mathrm{d}\tau = 0$. There is no loss of generality here since, as in the Newtonian case, orbiting particles are confined to a plane, as we shall now prove.

Example 23.1

If we consider the geodesic equation found by varying the Schwarzschild line element with respect to the angle[1] θ, we obtain

$$\frac{\mathrm{d}}{\mathrm{d}\tau}\left(r^2\frac{\mathrm{d}\theta}{\mathrm{d}\tau}\right) = r^2\sin\theta\cos\theta\left(\frac{\mathrm{d}\phi}{\mathrm{d}\tau}\right)^2. \tag{23.1}$$

This equation of motion is solved by $\theta = \pi/2$ for all τ. At this angle the right-hand side of the equation is zero and so the quantity $r^2\mathrm{d}\theta/\mathrm{d}\tau$ is conserved. Since we initially fix $\frac{\mathrm{d}}{\mathrm{d}\tau}\theta = 0$, the particle never acquires any acceleration, and the motion is confined to the plane. Just as in the Newtonian case, the confinement to the plane can be attributed to conservation of angular momentum. That is, if we rotate our coordinates such that we initially have $\theta = \frac{\pi}{2}$, then the equation of motion for u^ϕ gives $\frac{\mathrm{d}}{\mathrm{d}\tau}\tilde{L} = 0$ on the whole geodesic.

Our strategy will be to understand the possible orbits by analysing an effective potential, much as we did both in the Newtonian case and also in the last chapter. Using the effective energy variable $\mathcal{E} = (\tilde{E}^2 - 1)/2$,

$$\mathcal{E} = \frac{1}{2}\left(\frac{\mathrm{d}r}{\mathrm{d}\tau}\right)^2 + V_{\text{eff}}(r). \tag{23.2}$$

Using the results from Example 22.1 we have an effective potential for particles in the Schwarzschild geometry, given by

$$V_{\text{eff}}(r) = -\frac{M}{r} + \frac{\tilde{L}^2}{2r^2} - \frac{M\tilde{L}^2}{r^3}. \tag{23.3}$$

As we noted in the last chapter, this setup is designed to look as much like the Newtonian version as possible.[2] Here, the extra attractive term $-M\tilde{L}^2/r^3$ is the source of the richness of relativistic trajectories. This term is largest for large M and $\tilde{L}$ and small values of r.

An example of the effective potential V_{eff} is shown in Fig. 23.1, where we see its two characteristic extrema: (i) a minimum at a radius r_+, which resembles the minimum in the Newtonian effective potential; and (ii) a maximum is a smaller radius r_-, which has no analogue in Newtonian physics. These turning points in the relativistic potential, occurring for $V'_{\text{eff}}(r_\pm) = 0$, are found at

$$r_\pm = \frac{\tilde{L}^2}{2M}\left[1 \pm \left(1 - \frac{12M^2}{\tilde{L}^2}\right)^{\frac{1}{2}}\right]. \tag{23.4}$$

This equation implies that the nature of the extrema depends on the ratio $\tilde{L}/M$. That is, a combination of the angular momentum of the particle in motion and the mass of the gravitating object determines the effective potential, just as in Newtonian physics.

Example 23.2

In Fig. 23.2, we show some examples of motion in the potential determined by one choice of M and (non-zero) $\tilde{L}$. In these plots, we represent $V_{\text{eff}}(r)$ and the particle's value of $\mathcal{E}$. Where $\mathcal{E} = V_{\text{eff}}$ and the lines meet, the particle has no velocity in the radial direction, although it is important to note that the particle has a non-zero value of $\tilde{L}$ (and therefore u^ϕ) and so is still in motion.

• Figure 23.2(a) shows a circular orbit, with a particle sat at the minimum of V_{eff} at a (constant) radius r_+. The particle has effective energy $\mathcal{E} = V_{\text{eff}}(r_+)$, which takes a negative value, corresponding to the particle being bound in a potential.

• Figure 23.2(b) shows a particle in-falling from infinity (so therefore having a positive effective energy[3] $\mathcal{E}$), but colliding with the peak in the potential, which scatters it back off to infinity.

• Figure 23.2(c) shows the motion of a particle with a still higher energy. This one is not scattered since its effective energy is too high for it to hit the peak in the potential $V_{\text{eff}}(r_-)$. Instead, this particle spirals towards the centre of the gravitating mass, where it presumable crashes into the mass distribution at some small radius. This trajectory is a consequence of the relativistic potential being distinct from the Newtonian version, in that the latter has an energetic barrier at small r. This means that a Newtonian particle with positive energy, as long as it has some angular momentum, will always be deflected by the potential (assuming it doesn't crash into the source of the potential first!).

---

[2] Recall the Newtonian effective potential energy is given by

$$U_{\text{eff}}(r) = \frac{L^2}{2mr^2} - \frac{GMm}{r}.$$

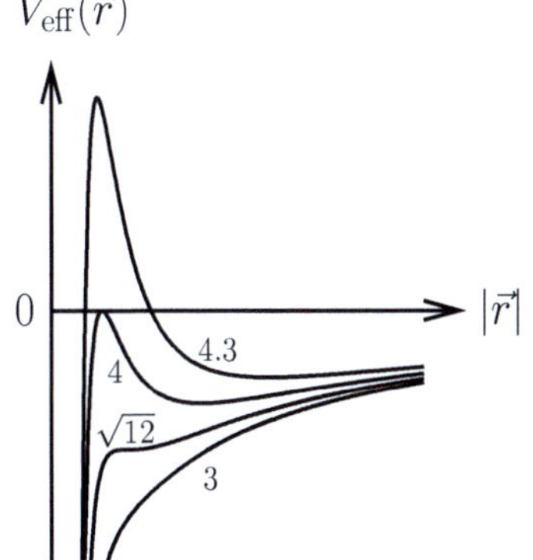

Fig. 23.1 The relativistic effective potential for given values of $\tilde{L}/M$.

(a) $V_{\text{eff}}(\vec{r})$

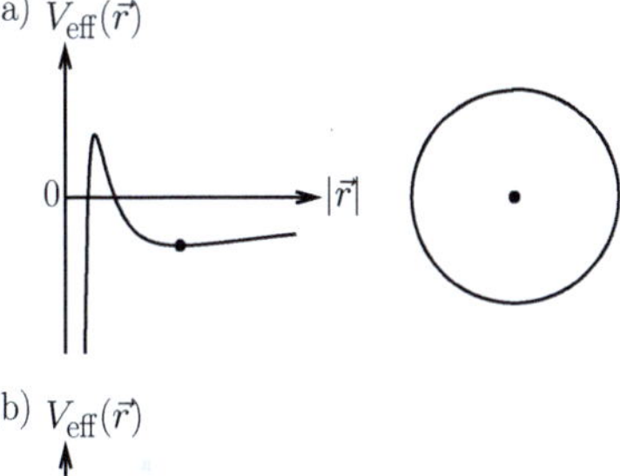

(b) $V_{\text{eff}}(\vec{r})$

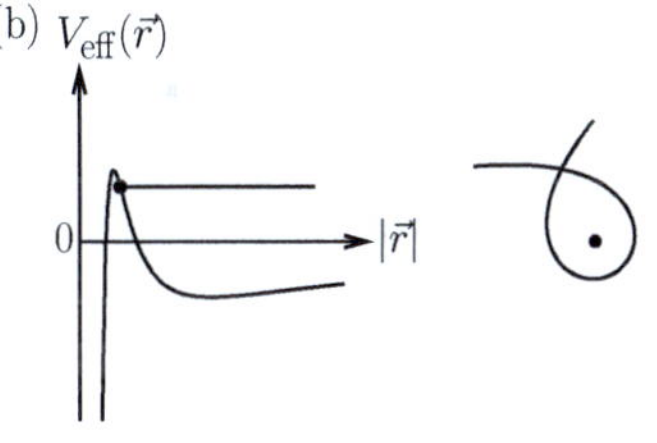

(c) $V_{\text{eff}}(\vec{r})$

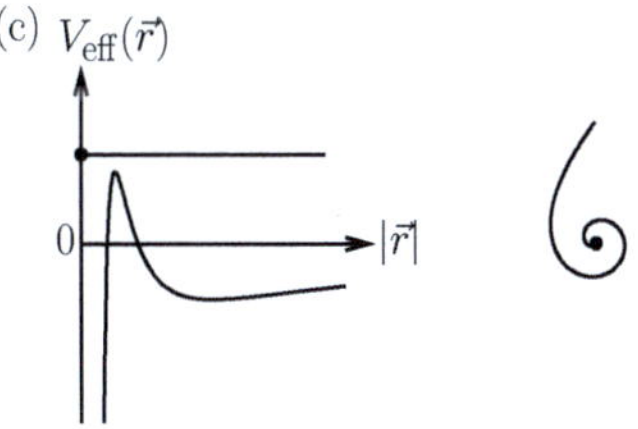

Fig. 23.2 Some allowed trajectories in the relativistic potential (computed using eqn 23.15). (a) Circular; (b) an encounter with the star leading to scattering; (c) a spiral into the star.

[3] Recall that particles arriving from infinity have $\tilde{E} > 1$ so $\mathcal{E} = (\tilde{E}^2 - 1)/2 > 0$.

One thing that we can spot from eqn 23.4 and Fig. 23.1 is that if $\tilde{L}/M < \sqrt{12}\ (= 3.464...)$ then there is only one extremum, a point of inflection. For the corresponding motions at low $\tilde{L}/M$ there is not enough angular momentum to stabilize an orbit. The stable orbits are only possible for $\tilde{L}/M \geq \sqrt{12}$ and, as we shall discover below, these stable orbits are almost, but not quite, elliptical. The difference between these orbits and their Newtonian analogues is that the motion can precess, such that the near-ellipses traced by the trajectories rotate.

It is also the case that if $\tilde{L}/M = 4$, then the maximum of V_{eff} occurs at $r_- = 4M$ where $V_{\text{eff}} = 0$, which is significant as it means that particles coming in from infinity, which have $\mathcal{E} > 0$, can never be scattered by such a potential.[4] If the angular momentum is greater than $4M$ then there are possible trajectories that can scatter particles coming in from infinity. Less angular momentum than this implies a spiral into the origin for $\mathcal{E} > 0$ particles.

[4]These particles will simply spiral into the origin.

23.2   Stable circular orbits

In the Newtonian case, we saw how the minimum of the effective potential gave us the circular orbits. We can repeat the computation for the relativistic case and show the properties of circular orbits here too.[5] Stable circular orbits occur when $r = r_+$ in eqn 23.4. These depend on $\tilde{L}/M$, with the minimum angular momentum occurring for $\tilde{L}/M = \sqrt{12}$, for which we have $r_+ = 6M$. This represents a minimum radius for stable circular orbits in the Schwarzschild geometry.

[5]In addition to the minimum, it's notable that the maximum of V_{eff} also represents a possible circular orbit. Being a maximum, it is manifestly unstable. The analogous trajectory will be considered for photons in the next chapter.

Example 23.3

From Chapter 20 we know that for a circular orbit we have that the effective potential is a minimum and that the effective energy $\mathcal{E} = (\tilde{E}^2 - 1)/2$ equals the value of the effective potential $V_{\text{eff}}(r_+)$. Rearranging eqn 23.2, this amounts to the condition

$$\tilde{E}^2 = \left(1 - \frac{2M}{r}\right)\left(1 + \frac{\tilde{L}^2}{r_+^2}\right). \tag{23.5}$$

This equation, along with eqn 23.4, can be usefully combined to find the ratio $\tilde{L}/\tilde{E}$ for a circular orbit. From eqn 23.4 we have

$$\left(2Mr_+ - \tilde{L}^2\right)^2 = \tilde{L}^4\left(1 - \frac{2M^2}{\tilde{L}^2}\right), \tag{23.6}$$

from which, on rearranging, we obtain

$$\frac{\tilde{L}^2}{r_+^2} = \frac{M}{r_+ - 3M}. \tag{23.7}$$

Combining with eqn 23.5, we find

$$\tilde{E}^2 = \left(1 - \frac{2M}{r_+}\right)\left(\frac{r_+ - 2M}{r_+ - 3M}\right), \tag{23.8}$$

from which we can write

$$\left(\frac{\tilde{L}}{\tilde{E}}\right)^2 = Mr_+^2\left(1 - \frac{2M}{r_+}\right)^{-2}. \tag{23.9}$$

The result of the previous example is useful, since it shows that the angular velocity of a particle in an equatorial orbit is given by[6]

$$\frac{\mathrm{d}\phi}{\mathrm{d}t} = \frac{\mathrm{d}\phi/\mathrm{d}\tau}{\mathrm{d}t/\mathrm{d}\tau} = \frac{1}{r^2}\left(1 - \frac{2M}{r}\right)\left(\frac{\tilde{L}}{\tilde{E}}\right). \tag{23.10}$$

Plugging the result for $\tilde{L}/\tilde{E}$ into eqn 23.10 leads to relativistic version of Kepler's law given by

$$\left(\frac{\mathrm{d}\phi}{\mathrm{d}t}\right)^2 = \frac{M}{r^3}. \tag{23.11}$$

This is, of course, the same as Kepler's original version of his law, and so we conclude that Kepler's third law for circular orbits is not altered by relativity.

23.3 Precession of the perihelion

The power of general relativity lies partially in explaining effects we find in observations that could not occur in Newtonian gravity. Historically, one of the most important was the precession of the perihelion of mercury. This is one of the three **classical solar-system tests of general relativity**. Below we shall derive the equation for the trajectory of a particle in the relativistic effective potential. The general idea is illustrated in the following example.

Example 23.4

The precessing orbit is shown in Fig. 23.3 (for a vastly exaggerated angle of precession compared to what we will compute below). The particle oscillates between two points on the effective potential curve, hitting maximum and minimum distances from the origin. As it does so, the trajectories resemble ellipses whose axes precess in the plane of motion.

Since our trajectories are confined to the plane $\theta = \pi/2$, then in order to find the shape of planetary orbits we must write an equation for the coordinate r in terms of ϕ (or vice-versa). Since we are not solving the full geodesic equations, but instead exploiting (i) the square of $\boldsymbol{u}$ and (ii) the conserved quantities $\tilde{E}$ and $\tilde{L}$, the way to do this is to find $\mathrm{d}r/\mathrm{d}\tau$ and $\mathrm{d}\phi/\mathrm{d}\tau$ and take their ratio to find $\mathrm{d}r/\mathrm{d}\phi$.

Starting with the velocity identity $\boldsymbol{u}\cdot\boldsymbol{u} = -1$, for fixed θ we have

$$g^{tt}(u_t)^2 + g^{\phi\phi}(u_\phi)^2 + g^{rr}(u_r)^2 = -1. \tag{23.12}$$

Noting that $g^{rr}(u_r)^2 = g_{rr}(u^r)^2$ for this diagonal metric,[7] we have

$$(u^r)^2 = \left(\frac{\mathrm{d}r}{\mathrm{d}\tau}\right)^2 = \frac{-1 - g^{tt}\tilde{E}^2 - g^{\phi\phi}\tilde{L}^2}{g_{rr}}, \tag{23.13}$$

[6] We make use of

$$u^\mu = \left(\frac{\mathrm{d}t}{\mathrm{d}\tau}, \frac{\mathrm{d}r}{\mathrm{d}\tau}, \frac{\mathrm{d}\theta}{\mathrm{d}\tau}, \frac{\mathrm{d}\phi}{\mathrm{d}\tau}\right),$$

and the equations for conserved quantities

$$\tilde{E} = \left(1 - \frac{2M}{r}\right)u^t,$$

and

$$\tilde{L} = r^2 u^\phi.$$

The three classical solar system tests of relativity are:
- Precession of the perihelion of mercury;
- Bending of light by the sun;
- Gravitational redshift.

We met gravitational redshift in Chapter 6. In Chapter 9, we also met a fourth test: Shapiro time delay.

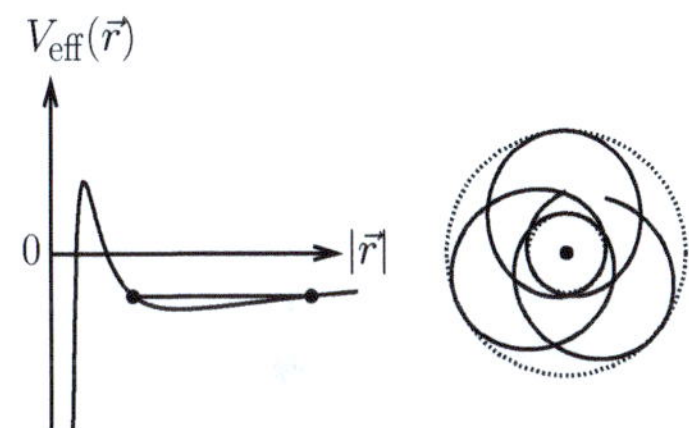

Fig. 23.3 The precessing orbit in the relativistic potential calculated using eqn 23.15. The motion is bounded at the points shown on the left, represented by the dotted lines on the right.

The components of the inverse Schwarzschild metric are

$$g^{tt} = -\left(1 - \frac{2M}{r}\right)^{-1},$$

$$g^{rr} = \left(1 - \frac{2M}{r}\right),$$

$$g^{\theta\theta} = r^{-2},$$

$$g^{\phi\phi} = (r\sin\theta)^{-2}.$$

[7] That is, $g^{rr}(u_r)^2 = g^{rr}(g_{rr}u^r)^2$ and $g_{rr}g^{rr} = 1$.

and so

$$\frac{\mathrm{d}r}{\mathrm{d}\tau} = \pm\left\{\left[-1 + \tilde{E}^2\left(1 - \frac{2M}{r}\right)^{-1} - \frac{\tilde{L}^2}{r^2}\right]\left(1 - \frac{2M}{r}\right)\right\}^{\frac{1}{2}}. \quad (23.14)$$

Since we also know from $u^\phi = g^{\phi\phi}\tilde{L}$ that $\frac{\mathrm{d}\phi}{\mathrm{d}\tau} = \frac{\tilde{L}}{r^2}$, we are able to isolate the equation that gives us the trajectories in the Schwarzschild geometry through the multiplication $\frac{\mathrm{d}r}{\mathrm{d}\tau}\frac{\mathrm{d}\tau}{\mathrm{d}\phi}$, which gives

$$\frac{\mathrm{d}r}{\mathrm{d}\phi} = \pm\frac{r^2}{\tilde{L}}\left\{\left[-1 + \tilde{E}^2\left(1 - \frac{2M}{r}\right)^{-1} - \frac{\tilde{L}^2}{r^2}\right]\left(1 - \frac{2M}{r}\right)\right\}^{\frac{1}{2}}. $$
$$(23.15)$$

This is our basic equation that tells us the paths in Schwarzschild coordinates that particles can take.[8] We can analyse this equation using some of the same techniques that we used in Chapter 20, where we introduced the variable[9] $u = 1/r$.

Example 23.5

Here we introduce the scaled variable $u = M/r$ and find

$$\left(\frac{\mathrm{d}u}{\mathrm{d}\phi}\right)^2 = \frac{\tilde{E}^2 - (1 - 2u)\left[1 + \left(\tilde{L}/M\right)^2 u^2\right]}{\left(\tilde{L}/M\right)^2}. \quad (23.17)$$

This equation can be rewritten in a form compatible with the Newtonian version. Specifically, it can be massaged into the expression

$$\left(\frac{\mathrm{d}u}{\mathrm{d}\phi}\right)^2 + (u - u_0)^2 = 6u_0(u - u_0)^2 + 2(u - u_0)^3 + \frac{M^2}{\tilde{L}^2}(\tilde{E}^2 - \tilde{E}_0{}^2). \quad (23.18)$$

Expanding out and comparing with eqn 23.17, we find that

$$6u_0 = 1 \pm \left(1 - \frac{12M^2}{\tilde{L}^2}\right)^{\frac{1}{2}}, \quad (23.19)$$

and

$$E_0^2 = (1 - 2u_0)\left(1 + \frac{\tilde{L}^2}{M^2}u_0^2\right). \quad (23.20)$$

These expressions are consistent with the intuitions we built up about which orbits are possible. In order to compare more closely to the Newtonian version, we write $\epsilon u_0^2 = \frac{M^2}{\tilde{L}^2}(\tilde{E}^2 - \tilde{E}_0{}^2)$ and we have

$$\left(\frac{\mathrm{d}u}{\mathrm{d}\phi}\right)^2 + (u - u_0)^2 - \epsilon^2 u_0^2 = 6u_0(u - u_0)^2 + 2(u - u_0)^3. \quad (23.21)$$

The terms on the right are relativistic corrections to the Newtonian expression on the left (see Sidenote 8). The Newtonian solution is $u = u_0(1 + \epsilon\cos\phi)$. The first-order correction term is $6u_0(u-u_0)^2$, which is of order $O(u_0^3\epsilon^2) = O(M^3\epsilon^2/r_0^3)$. In contrast, the second-order term is $2(u - u_0)^3$, which is of order $O(u_0^3\epsilon^3) = O(M^3\epsilon^3/r_0^3)$. For small eccentricities, the second-order correction can be ignored.

[8] The Newtonian version of this equation is

$$\frac{\mathrm{d}r}{\mathrm{d}\theta} =$$

$$\pm\frac{r^2}{L}\left\{2m\left[E + \frac{GMm}{r}\right] - \frac{L^2}{r^2}\right\}^{\frac{1}{2}},$$
$$(23.16)$$

or, using $u = 1/r$,

$$(u')^2 + u^2 - \frac{2GMm^2}{L^2}u - \frac{2Em}{L^2} = 0.$$

Recall that we defined $u_0 = GMm^2/L^2$ and $u_0(1 - \epsilon^2) = -2Em/L^2$ and wrote

$$(u')^2 + (u - u_0)^2 - u_0^2\epsilon^2 = 0.$$

[9] Don't confuse $u = 1/r$ with the components of the velocity $\boldsymbol{u}$, despite their sharing a symbol.

From the last example we conclude that the orbits are almost Newtonian ellipses with $u = u_0(1 + \epsilon \cos \phi)$, but are slightly perturbed by the corrections terms. Considering only the larger, first-order corrections, we must solve

$$(u')^2 = (1 - 6u_0)(u - u_0)^2 = u_0^2 \epsilon^2. \tag{23.22}$$

To solve the equation, define $\psi = (1 - 6u_0)^{\frac{1}{2}} \phi$ and $\mu = (u - u_0)$. In terms of these variables, we have an equation of motion

$$\left(\frac{du}{d\psi}\right)^2 + \mu^2 = \frac{u_0^2 \epsilon^2}{(1 - 6u_0)}. \tag{23.23}$$

The solution to this equation of motion is periodic in ψ. This means that for each orbit $\Delta\psi = 2\pi$. However, for the angle ϕ in which we are interested, $\Delta\phi \neq 2\pi$. Instead, from the definition of ψ, we can infer that an orbit corresponds to a change in ϕ of

$$\Delta\phi = \frac{2\pi}{\left(1 - \frac{6M}{r_0}\right)^{\frac{1}{2}}} \approx 2\pi\left(1 + \frac{3M}{r_0}\right). \tag{23.24}$$

The geometry for this effect is shown in Fig. 23.4. It implies that the change in the position of the perihelion can be written as $\Delta\phi = 2\pi + \delta\phi$ where for each orbit

$$\frac{\delta\phi}{2\pi} = \frac{3M}{r_0}. \tag{23.25}$$

This provides the basis of our estimate for the precession of the perihelion of mercury.

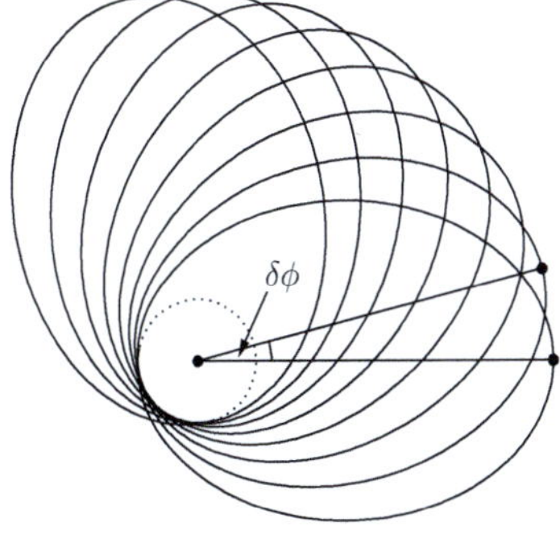
Fig. 23.4 The geometry for the precession of the perihelion.

Example 23.6

The orbit of Mercury has a period of 88 days ($\approx 7.6 \times 10^6$ s). If we take $r_0 = 5.8 \times 10^{10}$ m and $M_\odot = 2 \times 10^{30}$ kg then, restoring constants, we compute

$$\delta\phi = 6\pi \frac{GM_\odot}{c^2 r_0} \approx 4.8 \times 10^{-7} \text{ rad per orbit.} \tag{23.26}$$

This is equivalent to $\approx 2 \times 10^{-4}$ radians per century, or about 41 seconds of arc per century. A more accurate computation (also using general relativity, but with less severe approximations) predicts 43.0 seconds. The observed additional precession is 43.1 ± 0.5 seconds. Mercury's orbit actually precesses by around 575 arcseconds per century. Most of this is due to the tug from other bodies in the Solar System, and a precession of 532 arcseconds per century could be deduced by including the effects of the other planets. The oblateness of the Sun gives an additional contribution of about 0.03 arcseconds per century. But until the advent of general relativity, the remaining 43 arcseconds per century could not be accounted for, however much people tried to tweak the existing (Newtonian) models.[10]

The reason for precession is partly an effect of the curvature of space. This causes the length of the orbiting trajectory to be shorter than our flat-space intuition tells us, so that we can think of there not being enough path length to complete a precession-free Newtonian orbit. However, the larger share of the effect is also due to the curvature effects on the time-coordinate part of the Schwarzschild geometry, so this really is a *spacetime* effect.[11]

[10] Others had attempted to resolve the puzzle by tweaking Venus' mass up by 10%, which was way more than was believable, or by positing a vast swarm of asteroids close to Mercury, which had never been observed. As reported in Abraham Pais' biography, Einstein was strongly affected by the remarkable agreement between the prediction of his theory and the astronomical observations. This instantly resolved a mystery which had stubbornly resisted solution since the 1850s. Einstein wrote to Paul Ehrenfest after his discovery, saying that "For a few days, I was beside myself with joyous excitement," and told another colleague that the discovery had given him heart palpitations. He also said that when he saw the agreement between his calculations and the unexplained observations, he had the feeling that something actually snapped in him.

[11] One lesson here is that although it is possible to think of curvature in terms of a rubber sheet model, this only tells us about spatial curvature and neglects half of the story. Indeed, the time part is the only part that matters for stationary particles. This point is examined in the exercises.

Chapter summary

- Orbits of massive particles in the Schwarzschild geometry can be computed using the conservation of $\tilde{E}$ and $\tilde{L}$ and the identity $\boldsymbol{u}^2 = -1$.
- The precession of the perihelion of Mercury is one of the classical tests of general relativity. The theory predicts results in excellent agreement with experiment.

Exercises

(23.1) Consider a planet in a circular orbit, with radius r, around a spherically symmetric star in the $\theta = \pi/2$ plane of a set of Schwarzschild coordinates.
(a) If the planet has a constant coordinate velocity $v = r\mathrm{d}\phi/\mathrm{d}t$, use the line element to find the proper time taken to complete an orbit. Give your answer in terms of v and r.
(b) Using the conservation laws in this chapter, show that the proper time to complete a circular orbit is given more generally as $\Delta\tau = 2\pi r \left(\frac{r}{M} - 3\right)^{\frac{1}{2}}$, and show that this is compatible with the answer in (a).

(23.2) Consider circular motion in a de Sitter geometry with metric

$$\mathrm{d}s^2 = -\mathrm{d}t^2 + a(t)^2 \left(\mathrm{d}r^2 + r^2\mathrm{d}\phi^2\right). \qquad (23.27)$$

Calculate the proper time taken to complete an orbit, assuming constant velocity $v = a(t)R\mathrm{d}\phi/\mathrm{d}t$ and a factor $a(t) = e^t$.

(23.3) For a free particle at a constant radial coordinate in the Schwarzschild geometry, how does the t variable vary with proper time τ?

(23.4) A particle moves at a fixed radius in a Schwarzschild geometry at constant angular speed, such that it follows a world line given by

$$t = C\tau \qquad r = r_0$$
$$\theta = \pi/2 \qquad \phi = \omega\tau,$$

where τ is the proper time and r_0, C and ω are constants.
(a) Use the constraint on the velocity to find an expression for C.
(b) Compute the components of the acceleration

of the particle.
(c) Under what circumstances does the acceleration vanish?
There are more complex (solved) problems of this sort in Blennow and Ohlsson, from which the preceeding ones have been adapted.

(23.5) The precession of the perihelion results from two curvature effects: g_{tt} affects the rate of clock ticks and g_{rr} leads to a curvature of space. This latter effect can be isolated.
(a) The spatial slice of the Schwarzschild spacetime with $t = $ const. and $\theta = \pi/2$ has line element

$$\mathrm{d}s^2 = \frac{\mathrm{d}r^2}{1 - \frac{2M}{r}} + r^2\mathrm{d}\phi^2. \qquad (23.28)$$

Show that this slice can be embedded in flat spacetime with cylindrical coordinates and

$$z(r)^2 = \pm\left[8M(r - 2M)\right]. \qquad (23.29)$$

*Hint: This is one of the cases we meet in Appendix D. The resulting surface is called **Flamm's paraboloid** [after Ludwig Flamm (1885–1964)]. The particle orbits, not in flat space, but instead on the surface represented by the paraboloid. This resembles, at least approximately, a cone that is tangent to the paraboloid as shown in Fig. 23.5. The distance from the origin in this approximation is R. For a circular orbit therefore, the apparent flat-space trajectory seems to have length $2\pi R$, but this is too long when curvature is taken into account. This is because the real radius of the orbit is $2\pi r = 2\pi R\cos\alpha$. We must therefore cut out a slice of angle δ to turn the circle of radius R into the circle of radius r.*

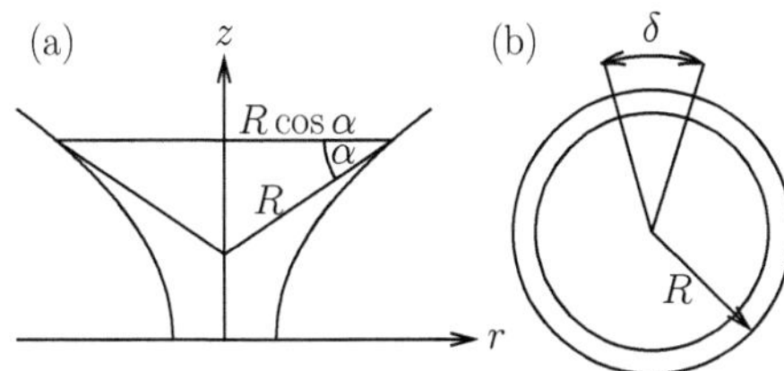

Fig. 23.5 (a) The flat-space radius R and the real radius $R\cos\alpha$. (b) A wedge of angular size δ must be cut out of the large circle so that its circumference matches the smaller one.

(b) Using the fact that the tangent to the paraboloid is $\tan\alpha = dz/dr$, show that

$$\delta \approx \frac{2\pi M}{r}.$$
(23.30)

(c) If the orbit is elliptical, argue that the perihelion must advance by an amount δ on every orbit.

(d) How much of the perihelion shift is accounted for by the spatial curvature?

See the book by Moore for more details of this approach.

(23.6) Consider the flat-space (but curved spacetime) metric

$$ds^2 = -\left(1 - \frac{2M}{r}\right)dt^2 + dr^2$$
$$+ r^2\left(d\theta^2 + \sin^2\theta\, d\phi^2\right).$$
(23.31)

This can be used to calculate the contribution to the perihelion shift from the time component of the metric.

(a) Using the usual velocity identity, find an expression for $dr/d\phi$ for motion in the equatorial plane in this spacetime.

(b) Show that this leads to an equation of motion

$$\frac{d^2 u}{d\phi^2} + u = \frac{M\tilde{E}}{\tilde{L}^2\left(1 - 2Mu\right)^2}.$$
(23.32)

(c) Expand this in the limit $r \gg 2M$ to show that for a circular orbit, with radius r_c we have

$$\left[1 - 4\left(\frac{M\tilde{E}}{\tilde{L}}\right)^2\right] u_c \approx \frac{M\tilde{E}^2}{\tilde{L}^2},$$
(23.33)

and use this to derive an approximate expression for $u_c = 1/r_c$ in terms of the constants in the problem.

(d) By expanding in terms of a perturbation to the circular orbit $u(\phi) = u_c + u_c w(\phi)$, find an equation for the perturbation $w(\phi)$.

(e) Hence, show that the perihelion shift for this metric accounts for $2/3$ of the total perihelion shift for the Schwarzschild metric.

Hint: You should obtain a harmonic oscillator equation for $w(\phi)$ with characteristic frequency ω_0. Argue that the distance of closest approach is found when $\omega_0\phi = 2\pi n$, where n is an integer, and use this to determine the shift $\Delta\phi$ between successive closest approaches. This method can be used to compute the perihelion shift for the Schwarzschild metric and is discussed in many books.

(23.7) Write a simple program to compute trajectories using eqn 23.15 for the case $\tilde{L}/M = 4.3$.

Hint: Consider how to choose the sign in front of the equation, as this changes depending on whether the orbiting particle is heading towards or away from the point of closest approach.

24

Photons in the Schwarzschild geometry

[1] In cylindrical polar coordinates, straight lines have the equation

$$u = \frac{1}{r} = -\frac{m\cos\theta - \sin\theta}{c}, \quad (24.1)$$

where m is the gradient and c the intercept on the y-axis.

[2] Johann Georg von Soldner (1776–1833). Other neglected predictions of gravitational effects on light rays were made by Henry Cavendish (1731–1810), John Michell (1724–1793) and Newton himself. Michell in particular has been described as one of the greatest unsung scientific heroes of all time, and was also the first person to suggest the existence of black holes, to explain the existence of double stars in terms of gravitation, and to apply statistics to cosmology.

[3] Since in the weak-field limit we write $\sqrt{-g_{00}} \approx 1 + 2\Phi$, we see that this coordinate speed c' varies with the gravitational potential Φ.

Music is the arithmetic of sounds as optics is the geometry of light.
Claude Debussy (1862–1918)

In Newtonian gravitation, as we have described it, the photon does not interact with the gravitational field and so the paths of all photons in free space are straight lines.[1] There had, however, been predictions of the bending of starlight around massive objects, notably by Johann Soldner[2] who used Newton's corpuscular theory of light to predict that starlight passing close to the Sun would cause an apparent shift in the position of stars of around 0.8 arc seconds. In this chapter, we see what general relativity has to say about the gravitational interaction of light and matter.

Example 24.1

So why should light rays be affected by gravity? To get an idea, we consider a static, diagonal metric. For the world line of a light ray, we have $0 = c^2 g_{00}\mathrm{d}t^2 + g_{ij}\mathrm{d}x^i\mathrm{d}x^j$, or

$$-c^2 g_{00} = g_{ij}\frac{\mathrm{d}x^i}{\mathrm{d}t}\frac{\mathrm{d}x^j}{\mathrm{d}t} = c'^2. \quad (24.2)$$

That is to say that the speed of light c' determined using the coordinate time varies.[3] Although the world line of a light ray is a null geodesic in spacetime, it is not necessarily a geodesic in space. In fact, $\sqrt{-g_{00}}$ looks a lot like a refractive index and it can be shown that the spatial part of the trajectory obeys a version of **Fermat's principle of least time** that says

$$\delta \int \left(\frac{g_{ij}\mathrm{d}x^i\mathrm{d}x^j}{-g_{00}}\right)^{\frac{1}{2}} = 0. \quad (24.3)$$

This provides at least some justification for the bending of light in a static gravitational field.

24.1 Photon trajectories

In general relativity, the curvature of spacetime due to the presence of massive objects causes a photon trajectory to be deflected compared to its path in flat spacetime. One immediate complication with understanding the motion of photons is the impossibility of defining a proper time τ. Instead we must make use of an affine parameter λ that marks off regular intervals along the world line of the photon. Key to computing the influence of gravitation on light is the observation that photons travel along null geodesics. These are paths whose velocity tangent vector $\boldsymbol{u}$ has the property

$$\boldsymbol{u} \cdot \boldsymbol{u} = g_{\mu\nu} \frac{\mathrm{d}x^{\mu}}{\mathrm{d}\lambda} \frac{\mathrm{d}x^{\nu}}{\mathrm{d}\lambda} = 0 \quad \text{(photons)}, \tag{24.4}$$

where λ is an affine parameter. We can write the null condition for photons in the Schwarzschild geometry in order to derive an equation of motion. Written out in terms of conserved quantities[4] $\tilde{E}$ and $\tilde{L}$, the null condition becomes

$$-\left(1 - \frac{2M}{r}\right)^{-1} \tilde{E}^2 + \left(1 - \frac{2M}{r}\right)^{-1} \left(\frac{\mathrm{d}r}{\mathrm{d}\lambda}\right)^2 + \frac{\tilde{L}^2}{r^2} = 0. \tag{24.5}$$

Multiplying by $(1 - 2M/r)$, dividing by $\tilde{L}^2$ and following the algebra through, we end up with an effective energy equation for photons, just as we had for massive particles. We cast this in the form $\mathcal{E} = \mathcal{T} + W_{\mathrm{eff}}$, where $\mathcal{T}$ is the kinetic-energy-like contribution.

The effective energy equation for photons is

$$\frac{1}{b^2} = \frac{1}{\tilde{L}^2}\left(\frac{\mathrm{d}r}{\mathrm{d}\lambda}\right)^2 + W_{\mathrm{eff}}(r), \tag{24.6}$$

where $b(\equiv \mathcal{E}^{-1/2}) = \tilde{L}/\tilde{E}$ and there is an effective potential W_{eff} for photons of

$$W_{\mathrm{eff}}(r) = \frac{1}{r^2}\left(1 - \frac{2M}{r}\right). \tag{24.7}$$

So the quantity $1/b^2$ plays the role analogous to total effective energy $\mathcal{E}$, while the role of kinetic energy $\mathcal{T}$ is taken by $(\mathrm{d}r/\mathrm{d}\lambda)^2/\tilde{L}^2$.

[4]The conserved quantities followed from geometric considerations in the last chapter and so apply for photons as well as for massive particles, although we cannot parametrize the world lines of photons with the proper time.

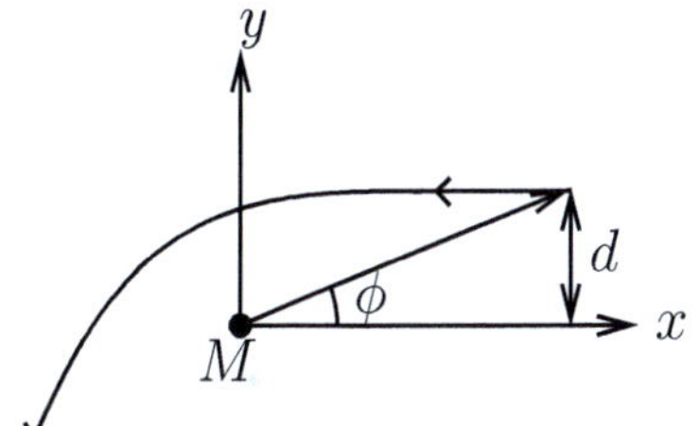

Fig. 24.1 The geometry defining the impact parameter.

Example 24.2

In order to interpret the physical meaning of b, consider the geometry in Fig. 24.1. A photon is initially moving parallel to the x-axis, a perpendicular distance d away from it, such that $y = d$ at infinity. This defines the **impact parameter** d. Far from any source of curvature the photon moves in a straight line, so for $r \gg 2M$ we have

$$b = \left|\frac{\tilde{L}}{\tilde{E}}\right| = \frac{r^2\mathrm{d}\phi/\mathrm{d}\lambda}{\mathrm{d}t/\mathrm{d}\lambda} = r^2\frac{\mathrm{d}\phi}{\mathrm{d}t}. \tag{24.8}$$

At large r we also have $\phi = \sin^{-1}(d/r) \approx d/r$ and $\mathrm{d}r/\mathrm{d}t \approx -1$ so that

$$\frac{\mathrm{d}\phi}{\mathrm{d}t} = \frac{\mathrm{d}\phi}{\mathrm{d}r}\frac{\mathrm{d}r}{\mathrm{d}t} = \frac{d}{r^2}. \tag{24.9}$$

We conclude that $b = d$. That is, b is the impact parameter for a light ray that reaches infinity.

The definition of b makes sense physically in that the impact parameter is a measure of the angular momentum: it evaluates how far from the origin a trajectory is, but decreases with increasing energy, since there is less scattering of an energetic particle by the gravitational potential compared to one with less energy.

The most important quantity in understanding the trajectories of light rays is the effective potential W_{eff} which is shown in Fig. 24.2. Unlike the analogous case for massive particles, the effective potential is independent of $\tilde{L}$ for photons. Like the relativistic V_{eff}, the potential W_{eff} has a maximum; this occurs at $r_0 = 3M$ where $W_{\text{eff}}(r_0) = \frac{1}{(27M^2)}$. When an incoming photon has a large impact parameter, such that $1/b^2 < W(r_0) = 1/27M^2$, then the photon will always be deflected by the potential. However, if an incoming photon has a small enough impact parameter such that $1/b^2 > W(r_0)$ then the photon will spiral in towards $r = 0$ where it is destroyed. Also unlike both the relativistic and Newtonian effective potentials for massive particles, W_{eff} has no local minimum, so we should not expect any stable circular orbits for photons. However, at the maximum at r_0 circular orbits of light are possible, with impact parameter $b^2 = 27M^2$, in a region of space called the photon sphere. Since this solution lies at the maximum of W_{eff} these trajectories are unstable.

Using the same techniques as we employed for massive particles, we can come up with an equation of motion for the photons, again using the variable $u = M/r$.

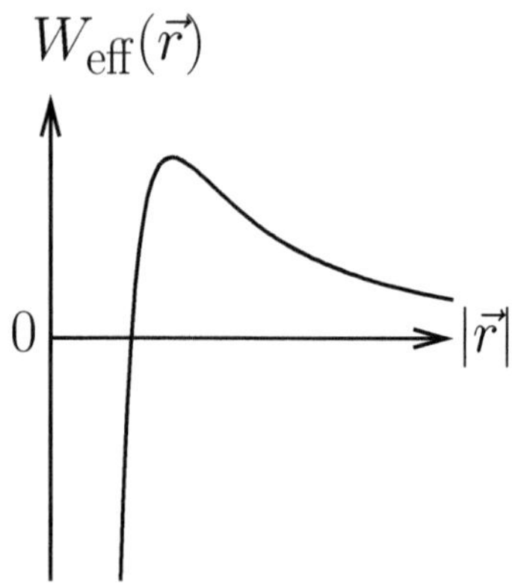

Fig. 24.2 The relativistic effective potential for photons.

Example 24.3

Let's examine light rays in the $\theta = \pi/2$ plane so we have that $u^\theta = 0$. The equation providing the trajectory (the change in radius with ϕ) is given by

$$\frac{dr}{d\phi} = \frac{dr}{d\lambda}\frac{d\lambda}{d\phi} = \frac{u^r}{u^\phi}. \tag{24.10}$$

Since all light rays are null, we have $\boldsymbol{u} \cdot \boldsymbol{u} = 0$ and so

$$g_{tt}(u^t)^2 + g_{rr}(u^r)^2 + g_{\phi\phi}(u^\phi)^2 = 0, \tag{24.11}$$

or, dividing through by $(u^\phi)^2$, we find

$$g_{tt}\frac{(u^t)^2}{(u^\phi)^2} + g_{rr}\left(\frac{dr}{d\phi}\right)^2 + g_{\phi\phi} = 0. \tag{24.12}$$

The first term can be rewritten[5] using $(u_t/u_\phi)^2 = 1/b^2$. On inserting the metric components, we have

$$\left(\frac{dr}{d\phi}\right)^2 = \left(1 - \frac{2M}{r}\right)\left[\frac{1}{b^2}\left(1 - \frac{2M}{r}\right)^{-1}r^4 - r^2\right]. \tag{24.13}$$

Now introduce the variable $u = M/r$ and find

$$\frac{M^2}{u^4}\left(\frac{du}{d\phi}\right)^2 = (1 - 2u)\left[\frac{1}{b^2}(1 - 2u)^{-1}\left(\frac{M}{u}\right)^4 - \left(\frac{M}{u}\right)^2\right], \tag{24.14}$$

or

$$\left(\frac{du}{d\phi}\right)^2 = \frac{M^2}{b^2} - u^2 + 2u^3. \tag{24.15}$$

Differentiating this latter equation, we find an equation of motion

$$\left(\frac{d^2u}{d\phi^2}\right) + u = 3u^2. \tag{24.16}$$

The Newtonian analogue of this equation can be extracted from eqn 20.33 by setting $m = 0$, yielding $u'' + u = 0$. The right-hand side of eqn 24.16 therefore provides the relativistic correction.

---

[5] This step involves a little rearrangement of the indices

$$g_{tt}\frac{(u^t)^2}{(u^\phi)^2} = \frac{g^{tt}(u_t)^2}{(g^{\phi\phi})^2(u_\phi)^2}.$$

The equation of motion allows us to evaluate how light rays are deflected by gravitation. In fact, this effect is the second of our classical solar-system tests of general relativity. The geometry for light deflection is shown in Fig. 24.3.

Example 24.4

Consider the equation of motion. In the limit $M/b \ll 1$, we expect the straight line solution

$$u_0 = \frac{M}{b} \sin \phi, \tag{24.17}$$

which solves the Newtonian equation $u_0'' + u_0 = 0$. We can expand the equation of motion in terms of perturbations to this straight line as $u(\phi) \approx u_0(\phi) + u_1(\phi)$. Doing this in eqn 24.16, we find

$$u_1'' + u_1 \approx 3u_0^2 = 3\left(\frac{M}{b}\right)^2 \sin^2 \phi \tag{24.18}$$

$$= \frac{3}{2}\left(\frac{M}{b}\right)^2 (1 - \cos 2\phi), \tag{24.19}$$

where we have retained only the largest term on the right (i.e. u_0) in the first of these expressions. The solution to this equation is given by

$$u_1 = \frac{1}{2}\left(\frac{M}{b}\right)^2 (3 + \cos 2\phi), \tag{24.20}$$

so that we have a trajectory given by

$$u \approx \left(\frac{M}{b}\right)\sin \phi + \frac{1}{2}\left(\frac{M}{b}\right)^2 (3 + \cos 2\phi). \tag{24.21}$$

A deflection angle can be found by calculating the two angles for which $u = 1/r = 0$. We find $u = 0$ for

$$2 \sin \phi \approx -\left(\frac{M}{b}\right)(3 + 2 \cos \phi). \tag{24.22}$$

Angles that approximately satisfy this condition are found at $\phi \approx -2(M/b)$ and $\phi \approx \pi + 2(M/b)$. Our conclusion is that the total deflection angle $\Delta\phi$ for light is therefore $4M/b$.

Example 24.5

For light that just grazes the Sun,[6] we have $b = 6.9 \times 10^8$ m and we compute a deflection (with constants restored) of

$$\Delta\phi = \frac{4GM}{c^2 b} \approx 8.6 \times 10^{-6} \text{ rad}, \tag{24.23}$$

which is equivalent to a prediction of 1.8 arc seconds (around twice Soldner's prediction based on Newtonian corpuscular theory). This was the prediction of general relativity that was tested by Arthur Eddington's 1919 observation of the shift in stellar positions during a solar eclipse, when starlight had passed close to the Sun would become temporarily visible.[7] The expedition gave measurements that agreed with the prediction within estimated experimental uncertainties. This made Einstein an internationally celebrated figure. Some later analyses of the Eddington results have questioned the systematic errors involved in the measurements. Later high-precision measurements using radio sources have confirmed that the measured shifts are in excellent agreement with the predictions of general relativity (and the picture shown in Fig. 24.4 is a common way of illustrating the effect).

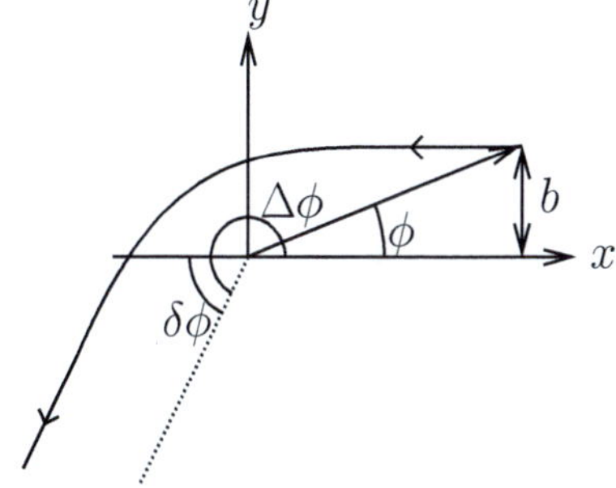

Fig. 24.3 The geometry for the deflection of a photon.

[6]The solar radius is usually taken to be $\approx 696{,}000$ km.

[7]For an account of the history, see J. Earman and C. Glymour, Historical Studies in the Physical Sciences **11**, 175 (1980).

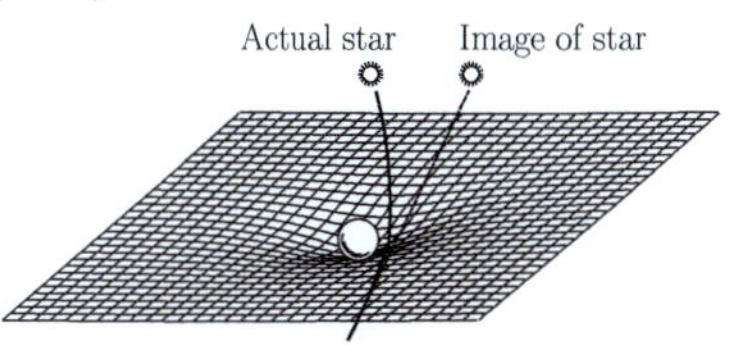

Fig. 24.4 An illustration of how curved space bends starlight around the Sun.

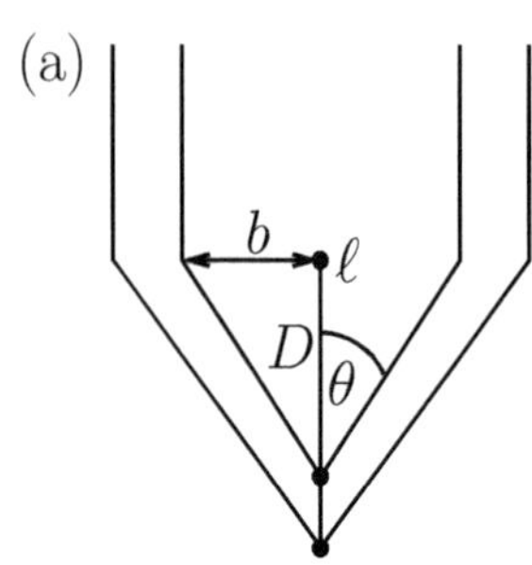

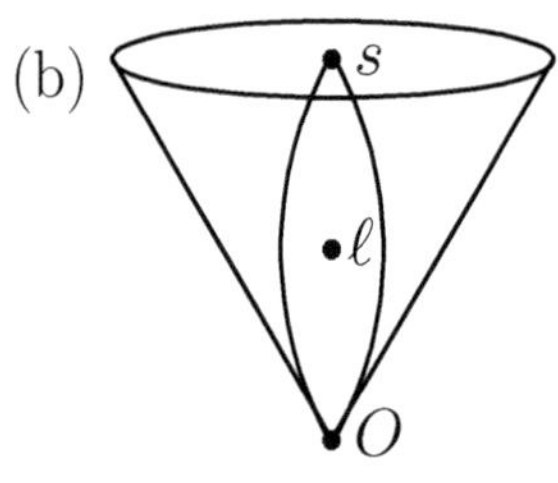

Fig. 24.5 (a) Light bent by a gravitating mass is not focussed to a point. (b) An 'Einstein ring' is formed by the deflection of light originating from a source s around a mass at ℓ. The trajectory of the light is such that an observer at O will trace the rays back via straight lines to form the circular image in the same plane as s.

[8] The consequences of this are examined in the exercises.

The deflection of light rays due to gravity allows us to treat gravitating masses a little like lenses. However, as discussed in the next example, the analogy is not an exact one.

Example 24.6

Consider the setup in Fig. 24.5(a) where light coming from infinity is bent by a gravitating mass at ℓ to reach the observer. Working in the small-angle approximation, light is detected a distance D from ℓ where $D = b/\theta$. The deflection angle θ for a gravitating mass M is $4M/b$ and so we obtain

$$D = b^2/4M. \tag{24.24}$$

This shows that the light rays are focussed down to different points depending on how far they are from the axis. They are not, therefore, focussed down to a point. This is examined further in the exercises.

In the case that the source, lens, and observer are all in a line, the image takes the form of a ring,[8] known as the **Einstein ring**, as demonstrated in Fig. 24.5(b).

24.2   Looking around

The passage of light rays in a gravitational field determines what observers can see. To investigate this we must consider the momentum of photons. Recall that owing to photons being massless, the components of the photon momentum vector in flat spacetime are related to the velocity components of the photon via $p^\mu = E u^\mu$, where E is the photon energy. Our first task is to update this for the curved spacetime we experience in the Schwarzschild geometry.

Example 24.7

A momentum vector $\boldsymbol{p}$ for a particle of mass m has components in the coordinate frame of

$$p^\mu = m\frac{\mathrm{d}x^\mu}{\mathrm{d}\tau} = m\frac{\mathrm{d}x^\mu}{\mathrm{d}t}\frac{\mathrm{d}t}{\mathrm{d}\tau} = m\frac{\mathrm{d}x^\mu}{\mathrm{d}t}u^t. \tag{24.25}$$

For motion in the Schwarzschild geometry, we have

$$\tilde{E} = \left(1 - \frac{2M}{r}\right)u^t, \tag{24.26}$$

and so we can write

$$p^\mu = \left(1 - \frac{2M}{r}\right)^{-1}m\tilde{E}\frac{\mathrm{d}x^\mu}{\mathrm{d}t}. \tag{24.27}$$

The quantity $m\tilde{E}$ is the energy of the particle at infinity. If we write this quantity as ε we have an expression

$$p^\mu = \left(1 - \frac{2M}{r}\right)^{-1}\varepsilon\frac{\mathrm{d}x^\mu}{\mathrm{d}t}. \tag{24.28}$$

This is well defined in the limit $m \to 0$ and so is suitable as a definition of the momentum components for the photon. The price we have paid is the inclusion of the coordinate t. In a coordinate system (t, r, θ, ϕ), the momentum components become

$$p^t = \varepsilon \left(1 - \frac{2M}{r}\right)^{-1}, \quad p^r = \varepsilon \left(1 - \frac{2M}{r}\right)^{-1} \frac{\mathrm{d}r}{\mathrm{d}t},$$
$$p^\theta = 0, \quad p^\phi = \varepsilon \left(1 - \frac{2M}{r}\right)^{-1} \frac{\mathrm{d}\phi}{\mathrm{d}t}. \tag{24.29}$$

We also have $b = r^2(1 - 2M/r)^{-1}\mathrm{d}\phi/\mathrm{d}t$ and, from the null condition for photons, we can write

$$0 = -\left(1 - \frac{2M}{r}\right) + \left(1 - \frac{2M}{r}\right)^{-1}\left(\frac{\mathrm{d}r}{\mathrm{d}t}\right)^2 + r^2\left(\frac{\mathrm{d}\phi}{\mathrm{d}t}\right)^2, \tag{24.30}$$

or

$$1 = \left(1 - \frac{2M}{r}\right)^{-2}\left(\frac{\mathrm{d}r}{\mathrm{d}t}\right)^2 + \left(1 - \frac{2M}{r}\right)\frac{b^2}{r^2}, \tag{24.31}$$

which gives us an expression for $\mathrm{d}r/\mathrm{d}t$. Collecting these facts, the final result for the momentum components of light is

$$p^t = \varepsilon \left(1 - \frac{2M}{r}\right)^{-1}, \quad p^r = \pm\varepsilon \left[1 - \frac{b^2}{r^2}\left(1 - \frac{2M}{r}\right)\right]^{\frac{1}{2}},$$
$$p^\theta = 0, \quad p^\phi = \varepsilon \frac{b}{r^2}. \tag{24.32}$$

The geometry for photons emitted from a point at some radius r is shown in Fig 24.6. The most striking thing here is that photons emitted beyond a particular angle ψ will have $(1/b^2) > W(r_0)$ (see Fig. 24.7) and will therefore spiral into the origin, where they are destroyed. The function $\sin\psi$ equals the sideways component of the coordinate velocity of the photon. Referring to the figure, we see that, in the observer's local frame, this can be written as

$$\sin\psi = \frac{u^{\hat{\phi}}}{u^{\hat{t}}} = \frac{p^{\hat{\phi}}}{p^{\hat{t}}}, \tag{24.33}$$

where, in the second step, we've used $p^{\hat{\mu}} = Eu^{\hat{\mu}}$ in the local frame. In order to use this, we need the conventional vielbein components for an orthonormal frame[9] to shift components via

$$\sin\psi = \frac{(\boldsymbol{e}_\phi)^{\hat{\phi}} p^\phi}{(\boldsymbol{e}_t)^{\hat{t}} p^t}. \tag{24.34}$$

Plugging in from the previous example we have

$$\sin\psi = \frac{b}{r}\left(1 - \frac{2M}{r}\right)^{\frac{1}{2}}. \tag{24.35}$$

The geometry is such that if the angle ψ is greater than a critical angle ψ_c then $1/b^2 > W(r_0)$ and the photon will spiral into the origin. This critical angle occurs when $1/b^2 = W(r_0)$, or $b = \sqrt{27}M$. [When the equality holds we expect the photons to (unstably) orbit the central mass.] Substituting, we conclude that those photons with angles greater than $\sin\psi_\mathrm{c} = (\sqrt{27}M/r)(1 - 2M/r)^{\frac{1}{2}}$ are condemned to fall into the central mass, where they are destroyed. Photons with angles $\psi < \psi_\mathrm{c}$

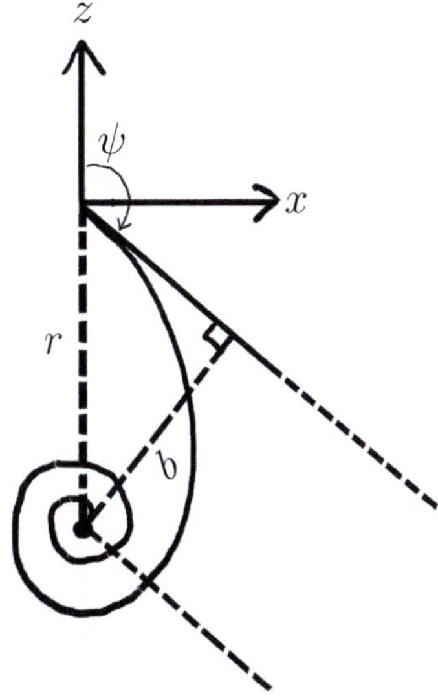

Fig. 24.6 The geometry for photons emitted from a point at radius r.

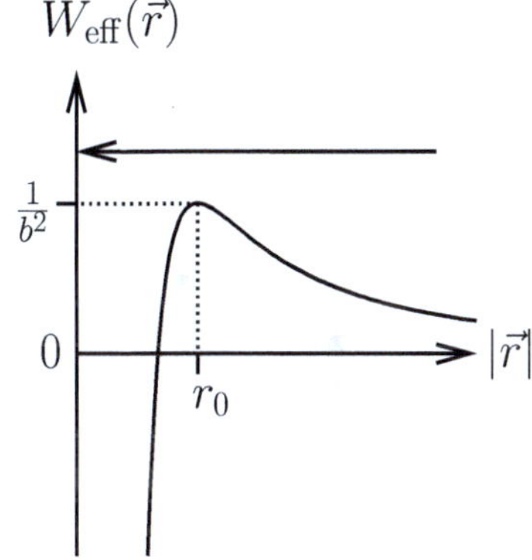

Fig. 24.7 A photon with $1/b^2 > W_\mathrm{eff}(r_0)$ will spiral into the origin of the coordinate system.

[9]The vielbein components for $\theta = \pi/2$ are

$$(\boldsymbol{e}_t)^{\hat{t}} = \left(1 - \frac{2M}{r}\right)^{\frac{1}{2}},$$
$$(\boldsymbol{e}_r)^{\hat{r}} = \left(1 - \frac{2M}{r}\right)^{-\frac{1}{2}},$$
$$(\boldsymbol{e}_\theta)^{\hat{\theta}} = r,$$
$$(\boldsymbol{e}_\phi)^{\hat{\phi}} = r.$$

escape, but as the radius r decreases, more and more of the photons are captured since the critical angle becomes smaller. The critical angle vanishes for $r = 2M$ and all photons are captured.

The Schwarzschild geometry is invariant with respect to time reversal. That is, if we play a film of the trajectories in reverse, the geometry is unchanged and so this same physics largely applies to the photons received by an observer at position r. The exception is the behaviour of those photons that were destroyed by spiralling into the origin: when we play the film backwards, these photons are not emitted and that part of space at angles $\psi > \psi_c$ is black. We will see the consequence of this in the next chapter, whose subject is black holes.

Chapter summary

- Photons are deflected by gravitating objects in general relativity.
- An effective-energy equation allows deflection to be calculated.
- The deflection of light by gravitating stars is one of the classical tests of general relativity. The theory predicts results in excellent agreement with experiment.

It is useful at this stage to summarize the equations for motion in the Schwarzschild geometry for massive particles and for photons, where we assume the motion takes place in the equatorial plane.

	Massive particles
Useful constants	$\tilde{E} = \left(1 - \frac{2M}{r}\right) \frac{dt}{d\tau}$ $\tilde{L} = r^2 \frac{d\phi}{d\tau}$
$\frac{dr}{d\tau}$	$\pm \left[\tilde{E}^2 - \left(1 - \frac{2M}{r}\right)\left(1 + \frac{\tilde{L}}{r^2}\right) \right]^{\frac{1}{2}}$
$\frac{dr}{dt}$	$\pm \left(1 - \frac{2M}{r}\right)\left[1 - \frac{1}{\tilde{E}^2}\left(1 - \frac{2M}{r}\right)\left(1 + \frac{\tilde{L}^2}{r^2}\right) \right]^{\frac{1}{2}}$
$\frac{d\phi}{dt}$	$\frac{\tilde{L}}{r^2 \tilde{E}}\left(1 - \frac{2M}{r}\right)$

	Photons
Useful constant	$b = \frac{\tilde{L}}{\tilde{E}} = r^2 \left(1 - \frac{2M}{r}\right)^{-1} \frac{d\phi}{dt}$
$\frac{dr}{dt}$	$\pm \left(1 - \frac{2M}{r}\right)\left[1 - \left(1 - \frac{2M}{r}\right)\frac{b^2}{r^2} \right]^{\frac{1}{2}}$
$\frac{d\phi}{dt}$	$\frac{b}{r^2}\left(1 - \frac{2M}{r}\right)$

Exercises

(24.1) Confirm eqns 24.6 and 24.7.

(24.2) Confirm that eqn 24.21 solves eqn 24.19.

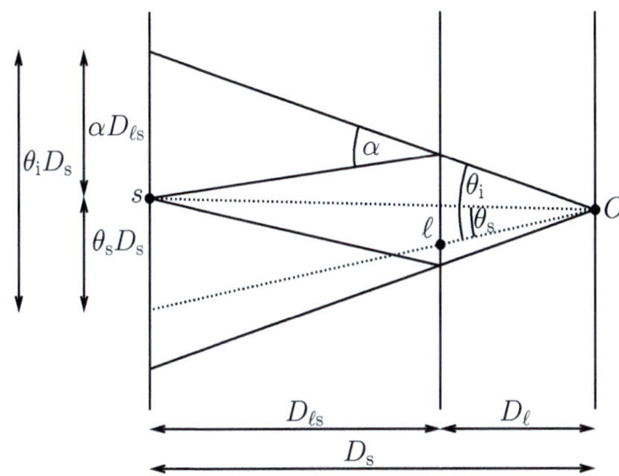

Fig. 24.8 Ray diagram for Exercise 24.3, showing a source s, a lensing mass ℓ and an observer O.

(24.3) Consider the optical diagram in Fig. 24.8.
(a) Using the small-angle approximation, show that

$$\theta_{\rm i} = \theta_{\rm s} + \frac{\theta_{\rm E}^2}{\theta_{\rm i}}, \qquad (24.36)$$

where we have defined the **Einstein angle** $\theta_{\rm E}^2 = 4M D_{\ell s}/D_\ell D_{\rm s}$.

(b) Find an expression of the angular size of the Einstein ring for the case that the source is at infinity and the source, lensing mass and observer are collinear.

(24.4) A light ray travels along a radial line in a spacetime described by the Robertson–Walker line element

$$\mathrm{d}s^2 = -\mathrm{d}t^2 + a(t)^2 \left(\frac{\mathrm{d}r^2}{1 - kr^2} + r^2 \mathrm{d}\Omega^2 \right), \quad (24.37)$$

where $\mathrm{d}\Omega^2 = \mathrm{d}\theta^2 + \sin^2\theta \mathrm{d}\phi^2$.
(a) Show that, when considering trajectories along the ray, the metric can be rewritten as

$$\mathrm{d}s^2 = -\mathrm{d}t^2 + a(t)^2 \mathrm{d}\chi^2, \qquad (24.38)$$

where χ is a function of r only.
(b) Show that the velocity component u^χ for this metric can be written as

$$u^\chi = \frac{C}{a(t)^2}, \qquad (24.39)$$

where C is a constant.
(c) Show further that the affine parameter evolves along the light ray according to

$$\lambda = \int \frac{1}{C} \frac{a(t)^2 \mathrm{d}r}{(1 - kr^2)^{\frac{1}{2}}}. \qquad (24.40)$$

The knowledge of how the affine parameter varies along a geodesic is often useful. See Exercise 26.4 for an example.

25 Black holes

[1]John Michell, and subsequently Pierre-Simon Laplace (1749–1827), had suggested the possibility of a dark star whose gravitation would be such that not even light would have the necessary velocity to escape the star's surface. The modern concept of a black hole was investigated by Robert Oppenheimer (1904–1967) and coworkers in the late 1930s, who proposed that the gravitational collapse of neutron stars could potentially cause them to become black holes (although the term 'black hole' was not used at this point; it was popularized by John Wheeler in the 1960s). The key paper on gravitational collapse, co-authored by Oppenheimer and Hartland Snyder (1913–1962) was published in 1939 on the eve of World War II, a conflict in which Oppenheimer would, of course, play a significant role.

[2]The radius r_S corresponds to a spherical shell of singularities in (3+1)-dimensional spacetime. Since we usually draw space on a two-dimensional page, it is sometimes called a a 'ring-like' singularity. The Schwarzschild singularity is sometimes also called the 'Hadamard catastrophe', as this is what Einstein jokingly called it, referring to the Jacques Hadamard's (1865–1963) suggestion that it could be regarded realistically.

[3]That is, frequency goes to zero since an oscillation takes infinite time.

> *When black clouds envelop stars which shone bright,*
>   *They can no longer pour forth their light.*
> Boethius (c.480–c.524/6) *The Consolation of Philosophy*

A black hole[1] is a part of spacetime where gravitational effects are so severe that light cannot escape from the inside of the region. It is likely that a black hole is the state of matter that, if massive enough, a star can adopt at the end of its life. Spacetime with a non-rotating black hole at its centre is spherically symmetric and must therefore, by Birkhoff's theorem, be described by the Schwarzschild metric.

The Schwarzschild geometry for the spacetime outside a gravitating object with mass M is represented by the line element

$$\mathrm{d}s^2 = -\left(1 - \frac{2M}{r}\right)\mathrm{d}t^2 + \left(1 - \frac{2M}{r}\right)^{-1}\mathrm{d}r^2 + r^2\left(\mathrm{d}\theta^2 + \sin^2\theta\,\mathrm{d}\phi^2\right).$$

(25.1)

It is notable that the Schwarzschild line element looks to be very badly behaved at two coordinate values: $r = 0$ and also at $r = r_S = 2M$, where r_S is known as the **Schwarzschild radius**.[2] On the three-dimensional spherical surface defined by r_S, trouble comes from two places: (i) the vanishing of the first term tells us that clocks at the Schwarzschild radius measure no time, which causes light signals to be infinitely redshifted;[3] (ii) the divergence of the second term means that $\partial s/\partial r$ is infinite: a small change in r leads to a divergent change in the invariant interval $\mathrm{d}s$. When Schwarzschild's solution was first discussed, it was expected that since the singularity at $r = 0$ was hidden deep inside objects like stars, its effects should be unobservable. The singularity at r_S originally had a similar status: if we restore units and plug numbers in, we find

$$r_S = \frac{2GM}{c^2} \approx \frac{3M}{M_\odot}\ \mathrm{km},$$

(25.2)

where $M_\odot \approx 2 \times 10^{30}$ kg is the solar mass. For objects with masses of order a solar mass, it appeared that this badly behaved point was buried well within the innards of the star. However, the discovery of incredibly massive, dense objects with very strong gravitational fields has forced relativists to confront the details of the singular coordinate r_S head on. Specifically, the mass distribution for a black hole is actually confined to an infinitely dense point at the origin, with the result that the region $0 < r < 2M$ lies *outside* of the mass that forms the hole.

It is the behaviour of massive particles and of light close to the radius $r_S = 2M$ that leads to many of the most notable features of a black hole. We shall see that although the metric at $r_S = 2M$ has several interesting features, it does not lead to any bad behaviour in the local physical fields. The curvature of spacetime, for example, varies smoothly over this region. An astronaut travelling towards the origin would not notice anything strange in their trajectory as they passed this point, with their watch continuing to tick regularly as they cross the surface $r = r_S$ in a finite interval of proper time. (Indeed, we saw in Chapter 22 that any point in the geometry can be reached, via a radial plunge, in a finite interval of proper time.) In fact, the threat of an infinity at r_S is actually the consequence of our choice of coordinates in the Schwarzschild geometry and much of our task of the next chapter will be to identify a set of coordinates that elucidates the behaviour of light and particles near black holes. In this chapter, our task is to examine the behaviour of geodesics that pass close to $r = 2M$ using the Schwarzschild geometry.[4]

Example 25.1

What is the evidence for the existence of black holes? As they cannot be directly seen (since, of course, light cannot escape their vicinity), the evidence for them is necessarily somewhat indirect. The observational evidence falls into four main categories.

- *Stars close to the centre of the Milky Way* have proper motions that indicate that they are whirling around a very compact radio source, believed to be a supermassive black hole of mass around $4 \times 10^6 M_\odot$, named Sagittarius A*. The accretion disc around this was imaged by the Event Horizon Telescope in 2022 (see Fig. 25.1).
- *Binary star systems that are strong sources of X-rays.* The mass of an unseen partner in a **binary star system** can be determined by its influence on the visible partner and can be inferred to be too massive to be a white dwarf or a neutron star. The X-ray emission occurs owing to the mass at the centre of the unseen part of the binary accreting an energetic disc of massive material from its partner. This matter orbits around the mass (at a distance larger than the event horizon, see below). Owing to the large velocities of the matter in the accretion disc, (which are caused by the large amount of angular momentum involved in the formation of the disc), the material in the disc has a temperature high enough to radiate strongly in the X-ray region of the electromagnetic spectrum.

    The most famous example is Cygnus-X1, an X-ray source found in the Cygnus constellation. The object is partnered with the blue supergiant HDE 226868 in a binary system. The X-ray emission is consistent with a source with diameter smaller than several hundred kilometres. The mass at the centre of Cygnus-X1 is estimated to be $\approx 15 M_\odot$, which is far too large to be a neutron star.
- *Anomalous behaviour close to the centre of a galaxy*, that can be explained by an enormous, massive, but unseen, object. Such objects can also be sources of large amounts of energy. **Quasars**[5] are very luminous (and therefore hugely energetic) objects that lie at the centre of galaxies. They are thought to comprise a supermassive black hole (i.e. black holes of millions to billions of $M_\odot$) surrounded by an accretion disc. The most distant quasar known is ULAS J1342+0928, which is expected to comprise the oldest known supermassive black hole, with a mass estimated at $8 \times 10^8 M_\odot$.
- *Gravitational wave observations that can be explained by the collision of black holes.* We will return to these when we discuss gravitational waves in Chapter 46.

[4]It's perhaps worth stressing that the picture of a black hole as an entity relentlessly sucking in all of the matter in the Universe is not an accurate one. Its exterior geometry is no different to that of any other star. As long as an observer has access to some means of propulsion, they can always escape a black hole (although, as we shall see, this is as long as they never get closer than $r = r_S$).

Fig. 25.1 An image of the supermassive black hole, Sagittarius A*, at the centre of the Milky Way, captured by an array of eight radio observatories distributed around Earth, networked together as the Event Horizon Telescope (EHT). The image shows a dark central region (called a shadow) surrounded by a bright ring-like structure. A mass of $4 \times 10^6 M_\odot$ is shoehorned into a region of space of diameter of only 0.3 AU. (Courtesy ESO.)

[5]This curious word is a contraction of quasi-stellar radio sources. As the name suggests, they were originally thought to be stars. In spite of this, their large energies, large redshifts and the jets that they can emit are good evidence that they cannot be stars.

25.1   The surface $r = 2M$

Let's consider the regions close to the spherical surface defined by $r_S = 2M$. First, we look at the geometry outside this surface, where $r > 2M$. Light cones in the Schwarzschild geometry, like all light cones, have $ds^2 = 0$, so we have from eqn 25.1 that, for constant θ and ϕ, they are described by

$$\frac{dt}{dr} = \pm \left(1 - \frac{2M}{r}\right)^{-1}. \tag{25.3}$$

The light cones have slope ± 1 for large r where spacetime becomes (asymptotically) flat. As we approach the surface r_S from $r > 2M$ (denoted $r \to 2M^+$ below) the slope of the light cone approaches $\pm\infty$ (meaning that the null surfaces of the cone point vertically in a standard r vs. t plot), closing up (or eclipsing) as shown in Fig. 25.2. As always, massive particles cannot travel faster than light, so are confined to the timelike part of the cone. The consequence of the narrowing of the light cones as $r \to 2M^+$ is that the trajectories of particles become more vertical in the r-t plane, which is to say that r cannot change as much for a given time interval. Therefore, as the particle approaches $r_S = 2M$ it seems to take longer and longer in coordinate time t to make a change in r. In fact, this is the effect we saw in Chapter 22 where it takes infinite amount of coordinate time to reach the surface at r_S during a radial plunge.[6]

Let's now examine the geometry of spacetime just inside the surface at r_S. Introducing the coordinate ε via $r = 2M - \varepsilon$, the line element can be written as

$$ds^2 = \frac{\varepsilon}{2M - \varepsilon}dt^2 - \frac{2M - \varepsilon}{\varepsilon}d\varepsilon^2 + (2M - \varepsilon)^2 d\Omega^2. \tag{25.4}$$

We see that if we fix t, θ and ϕ at constant values, the interval is $ds^2 = -(2M - \varepsilon)d\varepsilon^2/\varepsilon$. Remarkably, despite our having fixed things so that $dt = 0$, this interval has $ds^2 < 0$, which is to say that it is timelike! This implies that ε (the 'radial' coordinate inside the surface) is timelike, rather than spacelike as we expect for spatial coordinates. A timelike ε is doomed to constantly increase (like time in flat spacetime), causing the radial coordinate r to decrease until eventually we meet the singularity at $r = 0$. By the same token, the 'time' coordinate t in eqn 25.4 has become spacelike[7] inside the surface $r_S = 2M$.

A physical consequence of this topsy-turvy state of affairs can be seen by considering an astronaut at radius $r < 2M$ who sends out a photon. As shown in Fig. 25.2, the light cones inside $r = 2M$ tip over, spitting particles towards the singularity at $r = 0$. That is to say that the photon must go forward in time, which means r decreases and the photon has no option but to fall towards the origin. This means that for any observer within the event horizon, the future points towards $r = 0$ or, to sloganize: *the future is inwards.*[8] Photons inside the surface $r = 2M$ are therefore **trapped** and, as a consequence, so are all massive particles. With the impossibility of photons escaping the r_S surface, it is impossible

[6]As we saw in Chapter 22, coordinate time and the proper time of the observer travelling towards the surface are very different.

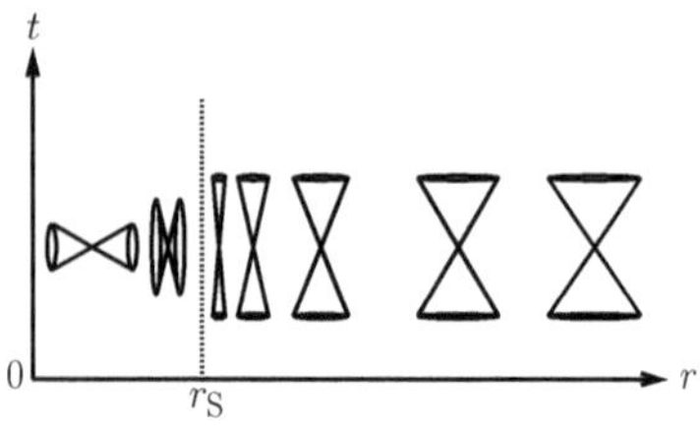

Fig. 25.2 Starting at large r, light cones eclipse as r decreases on the approach to a black hole, before turning over for $r < r_S$.

[7]This is a vivid manifestation of the lack of any metric significance of coordinates in general relativity: a variable called t has no more claim to tell the time than one called u or χ.

[8]This section should therefore really have been called 'Inside and after the surface $r = 2M$'.

for anything inside r_S to be seen. In the same way that we cannot see beyond the Earth's horizon, we call the surface at r_S the **event horizon** of the black hole.

Taking the above considerations into account, we can plot the light cones in Schwarzschild coordinates, which are shown in Fig. 25.3. The light cones have the alarming feature that null trajectories appear singular at r_S with incoming trajectories shooting off to infinite t before returning. It will turn out that this is not, in fact, real physical behaviour and in the next chapter we shall use an alternative coordinate description to understand the illusory nature of this feature.

25.2 The tortoise coordinate

The investigation of black holes in general relativity is, in large part, an exercise in finding the right coordinates with which to describe them. In evaluating the coordinate time t measured by some observer, we will often have cause to integrate equations like eqn 25.3. Integrals of the form

$$\Delta t = \int \frac{\mathrm{d}r}{1 - \frac{2M}{r}}, \tag{25.5}$$

lead to logarithmic contributions to the coordinate-time interval. To understand these logarithms, an interesting coordinate that we can employ is the **tortoise coordinate** r^*, defined by

$$\frac{\mathrm{d}r^*}{\mathrm{d}r} = \left(1 - \frac{2M}{r}\right)^{-1}, \tag{25.6}$$

so that we find (Fig. 25.4)

$$r^* = r + 2M \ln \left| \frac{r}{2M} - 1 \right| + \text{const.} \tag{25.7}$$

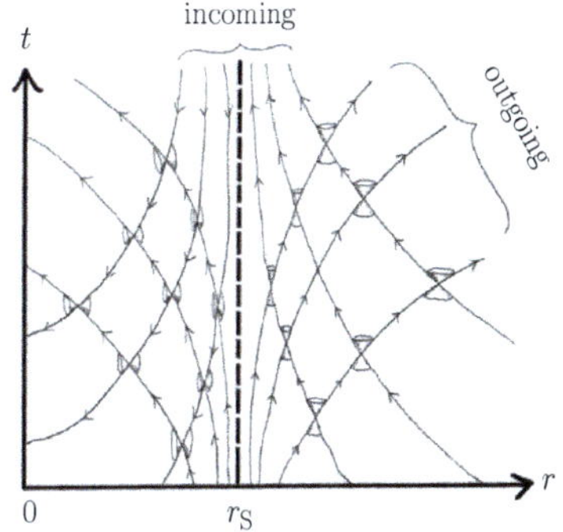

Fig. 25.3 Light cones in the Schwarzschild geometry.

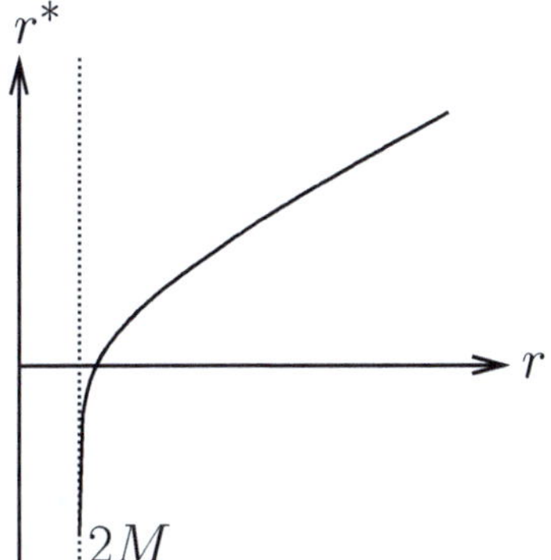

Fig. 25.4 A plot of eqn 25.7.

Example 25.2

This coordinate is named in honour of the story of Achilles[9] who races the tortoise and, despite moving at a high speed, apparently can never overtake it. This is because when Achilles catches up with the tortoise, the tortoise has moved to a new location. Achilles catches up again, but again, by the time he does, the tortoise has moved a bit further along the race track.

In our coordinates, we imagine a rather contrived race between Achilles, whose separation from the tortoise is given by $d = r - r_S$, so that when $r = r_S = 2M$, Achilles expects to pass the tortoise. We track progress in the race using the coordinate r^* which one can think of as a curious sort of race timer, that starts at positive values of r^* and monotonically decreases, with the possibility of taking all values $-\infty \le r^* \le \infty$.[10] We start with r^* taking a large, positive value, corresponding to the athletes being well separated (i.e. large $d = r - r_S$). As we time-evolve the race, we allow the timer r^* to decrease, and Achilles closes in on the tortoise, with r approaching r_S. As the race coordinate r^* becomes large and negative and r gets close to r_S, we see from eqn 25.6 that r changes more and more slowly with the race time r^*, since the race velocity $\mathrm{d}r/\mathrm{d}r^* \to 0$. We continue to time-evolve the race, making the r^* coordinate more and more negative. However, Achilles never reaches the tortoise.

[9]Achilles was a hero of the Trojan war and has a starring role in Homer's Iliad. Achilles is played by Brad Pitt in the film *Troy*, and even appears in the title of the Led Zeppelin track 'Achilles Last Stand'. The philosopher Zeno of Elea (Fifth century BC) picked Achilles for his paradox because, with regards to land speed, the slowly moving tortoise would be expected to be no match for the supremely athletic Achilles.

[10]Since coordinates have no intrinsic metric significance, we don't worry about exactly what this means in terms of the workings of the clock.

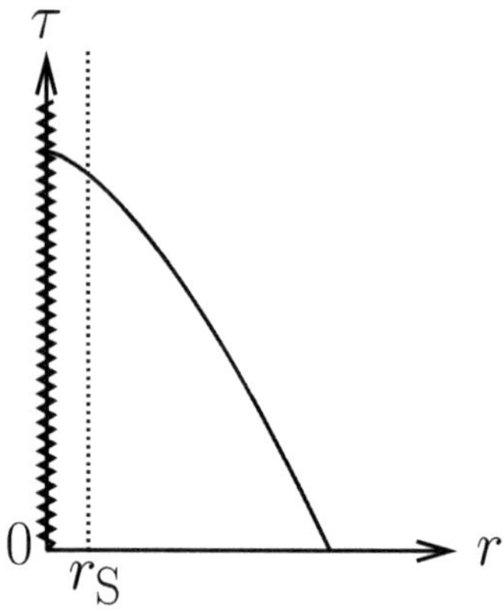

Fig. 25.5   Motion of an astronaut falling into a black hole as a function of proper time.

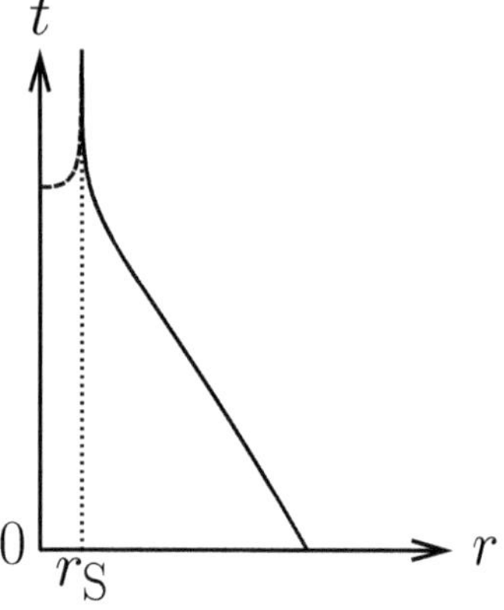

Fig. 25.6   Motion of an astronaut falling into a black hole as a function of Schwarzschild coordinate time.

[11] The equation of motion for the separation of two point is given, in the orthonormal frame, by

$$\frac{\mathrm{d}^2 \chi^{\hat{i}}}{\mathrm{d}\tau^2} = -R^{\hat{i}}{}_{\hat{\tau}\hat{j}\hat{\tau}}\chi^{\hat{j}}.$$

The components of the Riemann tensor are

$$R_{\hat{\tau}\hat{r}\hat{\tau}\hat{r}} = -\frac{2M}{r^3},$$

$$R_{\hat{\theta}\hat{\phi}\hat{\theta}\hat{\phi}} = +\frac{2M}{r^3},$$

$$R_{\hat{\tau}\hat{\theta}\hat{\tau}\hat{\theta}} = R_{\hat{\tau}\hat{\phi}\hat{\tau}\hat{\phi}} = +\frac{M}{r^3},$$

$$R_{\hat{r}\hat{\theta}\hat{r}\hat{\theta}} = R_{\hat{r}\hat{\phi}\hat{r}\hat{\phi}} = -\frac{M}{r^3}.$$

Notice that the curvature has no notable behaviour at $r = 2M$.

The tortoise coordinate is useful as it can be used to prevent the eclipse of the light cones that we saw occurring in the Schwarzschild coordinates as $r \to 2M^+$ owing to the property that $\mathrm{d}t/\mathrm{d}r = \pm(1 - 2M/r)^{-1} \to \infty$. The solution is to note that the light cones can be described by setting

$$t = \pm r^* + \text{const.,} \tag{25.8}$$

where r^* is the tortoise coordinate. If we then rewrite the metric we obtain

$$\mathrm{d}s^2 = \left(1 - \frac{2M}{r}\right)\left[-\mathrm{d}t^2 + (\mathrm{d}r^*)^2\right] + r^2\mathrm{d}\Omega^2, \tag{25.9}$$

with r a function of r^*. This solves the problem of the light cones closing at $r = 2M$ and it also prevents the metric having any bad behaviour around $r = 2M$. Nonetheless, it is not an easy set of coordinates to work with, since the point $r = 2M$ now occurs at infinite r^*. We return to the problem of finding better coordinates in the next chapter.

25.3   Death of an astronaut

We observe an astronaut undergoing a radial plunge towards a black hole. She has a torch that emits light pulses. We saw in the last chapter that, from her point of view, the proper time during a radial plunge changes with the r coordinate via

$$\tau = \tau_* + \frac{2}{3}\frac{r^{\frac{3}{2}}}{r_S^{\frac{1}{2}}}. \tag{25.10}$$

The proper time for the astronaut to fall through the horizon is finite, and is shown by a solid line in Fig. 25.5. The astronaut notices nothing special about her clock as she passes the point of no return at r_S.

However, from our point of view as observers at spatial infinity, our spacetime is flat and our proper time coincides with the coordinate time t. Using the result from the last chapter, the coordinate time during a radial plunge evolves according to

$$t = t_* + r_S\left[-\frac{2}{3}\left(\frac{r}{r_S}\right)^{\frac{3}{2}} - 2\left(\frac{r}{r_S}\right)^{\frac{1}{2}} + \ln\left|\frac{\left(\frac{r}{r_S}\right)^{\frac{1}{2}} + 1}{\left(\frac{r}{r_S}\right)^{\frac{1}{2}} - 1}\right|\right]. \tag{25.11}$$

This quantity diverges as $r \to r_S$ and so, we see the astronaut taking an infinite amount of time to cross the boundary. This is shown in Fig. 25.6.

Example 25.3

What does the astronaut feel as she falls?[11] We evaluated the forces that act on an astronaut in the Schwarzschild geometry in the orthonormal frame in Chapter 11. These are

$$\frac{\mathrm{d}^2\chi^{\hat{r}}}{\mathrm{d}\tau^2} = \frac{2M}{r^3}\chi^{\hat{r}}, \quad \frac{\mathrm{d}^2\chi^{\hat{\theta}}}{\mathrm{d}\tau^2} = -\frac{M}{r^3}\chi^{\hat{\theta}}, \quad \frac{\mathrm{d}^2\chi^{\hat{\phi}}}{\mathrm{d}\tau^2} = -\frac{M}{r^3}\chi^{\hat{\phi}}. \tag{25.12}$$

The astronaut is stretched out like spaghetti in the $\hat{r}$ direction, and compressed in the $\hat{\theta}$ and $\hat{\phi}$ directions. Notice that there are no special forces that the astronaut feels as she passes r_S. Nevertheless, the astronaut is doomed to be stretched out (and compressed) to death by the forces whose magnitudes all diverge as $r \to 0$.

The last example suggests that, whatever the status of the bad behaviour at r_S, we should take the singular behaviour of the metric at $r = 0$ very seriously. This is indeed the case as the point $r = 0$ does represent a **physical singularity**. Any body that meets the point $r = 0$ must be destroyed by the infinite forces at this point. Meeting this point is the fate of anything that finds itself within the event horizon of a black hole.

What about the light pulses that the astronaut emits? Light rays travel along null geodesics and so, assuming the pulses are emitted radially, the interval between two events on the world line of the photons emitted by the astronaut are given by

$$0 = -\left(1 - \frac{r_\mathrm{S}}{r}\right)\mathrm{d}t^2 + \frac{\mathrm{d}r^2}{1 - \frac{r_\mathrm{S}}{r}}, \tag{25.13}$$

or

$$\frac{\mathrm{d}t}{\mathrm{d}r} = \pm\frac{1}{1 - \frac{r_\mathrm{S}}{r}}. \tag{25.14}$$

The photons of interest are those travelling radially outwards, so we choose the $+$ sign (i.e. r increases as t increases). The journey time for a photon emitted at (t_1, r_1) and detected at (t_2, r_2) is

$$\begin{aligned}
t_2 - t_1 &= \int_{r_1}^{r_2} \frac{\mathrm{d}r}{1 - \frac{r_\mathrm{S}}{r}} \\
&= (r_2 - r_1) + r_\mathrm{S}\ln\left(\frac{r_2 - r_\mathrm{S}}{r_1 - r_\mathrm{S}}\right).
\end{aligned} \tag{25.15}$$

The coordinate-time interval is therefore corrected from the usual (flat space) value $(r_2 - r_1)$, with the journey time increased by the logarithmic correction (Fig. 25.7). The logarithm gets larger and larger as the point of emission r_1 gets closer to r_S, causing the interval to diverge. As seen by us distant observers, the light pulses from the astronaut become less and less frequent. If we are able to see the astronaut during her descent then she will appear to never quite reach the horizon owing to the coordinate time interval getting larger and larger as we saw above. In addition to the light pulses emitted from the astronaut becoming less frequent, they will also become dimmer as light is severely redshifted by the gravitational effect.[12]

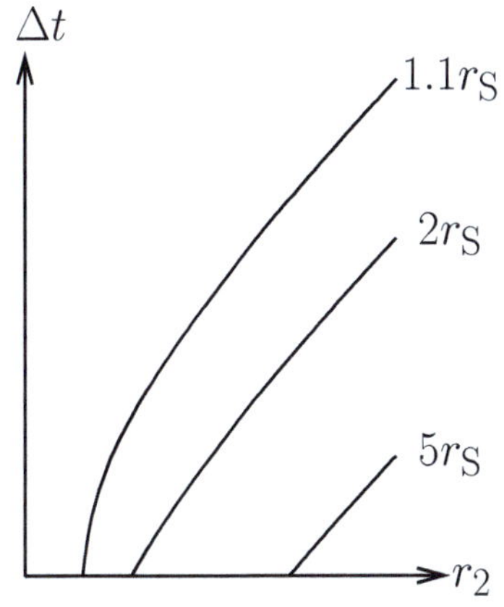

Fig. 25.7 Equation 25.15, giving an interval $\Delta t = t_2 - t_1$, plotted as a function of r_2 for given values of r_1.

[12]Since this observer is falling, we cannot simply use the result from eqn 22.16 which assumes a stationary observer. In fact, we will examine just how severe the redshift in the next chapter after we have identified a more suitable set of coordinates to describe the situation.

25.4 Looking around near a black hole

What does the astronaut see as she plunges towards the hole? Recall from the previous chapter that there is a critical angle for photons emerging from a point at radius r in the Schwarzschild geometry, given by

$$\sin \psi_{\mathrm{c}} = \frac{\sqrt{27}M}{r}\left(1 - \frac{2M}{r}\right)^{\frac{1}{2}}.$$ (25.16)

At angles greater than ψ_{c}, the photons spiral into the origin and are captured. Notice how the angle decreases to zero for $r = 2M$: we cannot see any light signal from this point as all photons are captured.

The Schwarzschild metric is invariant with respect to time reversal, which means that this argument also applies to photons arriving at the location r where the astronaut finds herself. This affects what the astronaut observes: photons from $r \leq 2M$ cannot reach her since photons do not emerge from a black hole. For photons from elsewhere, ψ_{c} represents a limit to what the astronaut outside the hole can see, with incoming photons from angles less than ψ_{c} being the only ones that she is able to detect. The rest of her field of vision is black: this is the astronaut 'seeing' the black hole.

Example 25.4

For example, when $r = 3M$ we have $\sin \psi_{\mathrm{c}} = 1$ and $\psi_{\mathrm{c}} = \pi/2$. The black hole then occupies exactly half of the sky.[13] As the astronaut gets closer to the black hole, the hole takes up more and more of the sky, until, very close to the horizon, light from the rest of the universe is only experienced through a small cone.

The photons that do reach the astronaut from other stars have their energies increased (or **blueshifted**) compared to the energies of these photons at infinity. This can be seen by evaluating[14] the energy measured by a stationary observer at radius r, which is

$$E_{\mathrm{obs}} = p^{\hat{t}} = (\boldsymbol{e}_t)^{\hat{t}} \, p^t,$$ (25.17)

with $(\boldsymbol{e}_t)^{\hat{t}} = \left(1 - \frac{2M}{r}\right)^{\frac{1}{2}}$. Recall that p_t is conserved along the geodesics, so[15] $p^t(r) = \hbar\omega_\infty(1 - 2M/r)^{-1}$. This yields

$$E_{\mathrm{obs}} = \hbar\omega_\infty\left(1 - \frac{2M}{r}\right)^{-\frac{1}{2}},$$ (25.18)

which is simply eqn 22.16 for the gravitational redshift rearranged. The final thing the astronaut sees is a tiny chink of blue light. Bon voyage.

25.5   Gravitational collapse

How are black holes created? When a main-sequence (i.e. a fairly average-sized) star exhausts its nuclear fuel, it generally expands to form a red giant, and subsequently collapses under its own gravitational attraction to form a white dwarf star. However, when a star of mass $\gtrsim 1.4M_\odot$ uses up its nuclear fuel,[16] then things can proceed differently.

---

[13]Interestingly, since when the equality holds we expect the photons to orbit the central mass, photons emitted from an observer at $r = 3M$ and ψ_{c} can be observed by that same observer after completing an orbit. This implies that the astronaut can see the back of her head.

[14]The use of the vielbein components here is equivalent to the usual rule that $E = -\boldsymbol{p} \cdot \boldsymbol{u}_{\mathrm{obs}}$ that we used to work out the gravitational redshift in Chapter 22. Recall also that our conventional vielbein corresponds to a stationary observer's local rest frame.

[15]That is, $p_t(r) = (1 - 2M/r)p^t(r)$ is conserved. Therefore, $p^t(\infty) = \hbar\omega_\infty = (1 - 2M/r)p^t(r)$, so $p^t(r) = \hbar\omega_\infty(1 - 2M/r)^{-1}$.

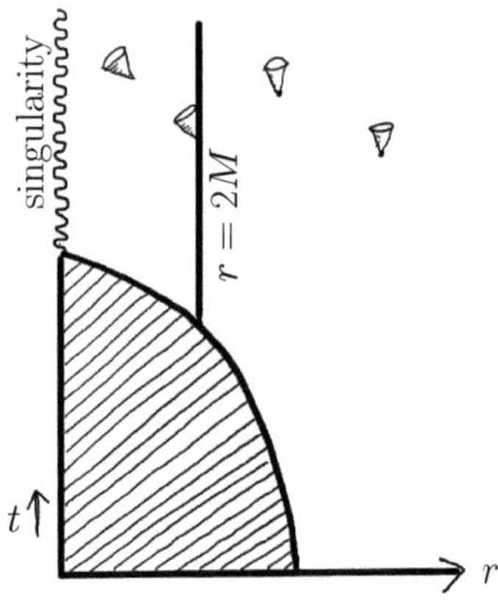

Fig. 25.8 Gravitational collapse with two spatial dimensions suppressed. Each point represents a 2-sphere.

[16]Subrahmanyan Chandrasekhar (1910–1995) computed that a white dwarf with a mass $\gtrsim 1.4M_\odot$ (the **Chandrasekhar limit**) would be unstable to further gravitational collapse. This results in a neutron star or, if the star is still more massive, a black hole. The most massive known neutron star is $\approx 2.4M_\odot$; the least massive black hole is thought to be $\approx 4M_\odot$.

If the star remains massive enough (i.e. it does not shed its outer layers of mass) it does not achieve the equilibrium state of a stable white dwarf. Instead, it will continue to collapse. If massive enough it contracts through the Schwarzschild radius $r = 2M$ as shown in Figs. 25.8 and 25.9. It then continues to collapse further until it is compressed to infinite density. That is, its mass is compressed to a point at the origin, with the result being a singularity in spacetime at $r = 0$.

Can an observer ever hope to observe this singularity? Not at all. We saw in the last section that regular light signals from an in-falling astronaut become less frequent (and also highly redshifted) as the astronaut nears r_S. This is also true of photons from the surface of a collapsing star. The image of the shrinking star appears to slow and darken. It freezes, becoming completely black in appearance at its outer surface approaches $r_S = 2M$ from above. The light cones of the photons emitted from the surface of the star are shown in Figs. 25.8 and 25.9. Light emitted as the surface of the star passes through r_S has an ingoing wavefront that falls into the singularity at $r_S = 0$ and an outgoing wavefront that remains at the radius $r_S = 2M$ for ever. Light emitted as the surface collapses further is drawn in towards the black hole, no matter whether it was directed towards larger or smaller r. (As explained in the next chapter, this curious situation leads to the formation of a so-called closed trapped surface.) At this point, the star itself is something of an irrelevance for the outside world which can no longer interact with it in any way. It is as if the star has now entered a Universe of its own. What matters for the outside Universe is now only the event horizon and the exterior geometry. By the same token, the singularity is hidden from the outside world.

Recall from Chapter 19 that we can map the presence of infinities and singularities in a spacetime using a Penrose diagram to depict the conformal structure. The conformal structure of spacetime that follows from the gravitational collapse that results in a black hole is shown in the Penrose diagram in Fig. 25.10. The black-hole singularity at $r = 0$ is shown by the double line, along with the event horizon H at $r_S = 2M$. The black-hole singularity is a spacelike surface (just as we had for the $t = 0$ point in the Robertson–Walker Penrose diagram in Chapter 19.) The reason for this is the swapped roles of time and space inside the event horizon. This means we describe $r = 0$ as a spacelike singularity.

For travellers on timelike geodesics that end up at i^+ all of spacetime is visible, except the region for which $r < 2M$. For those unlucky enough to have fallen through the event horizon, all light rays (which, remember, travel at $45°$ in these diagrams) will meet the singularity. There is a future horizon at the singularity: the observer in the region $r < 2M$ cannot access photons from the whole of the space. The Penrose diagram in Fig. 25.11 also shows the surface of a star as it collapses, passing though r_S and eventually meeting the singularity.

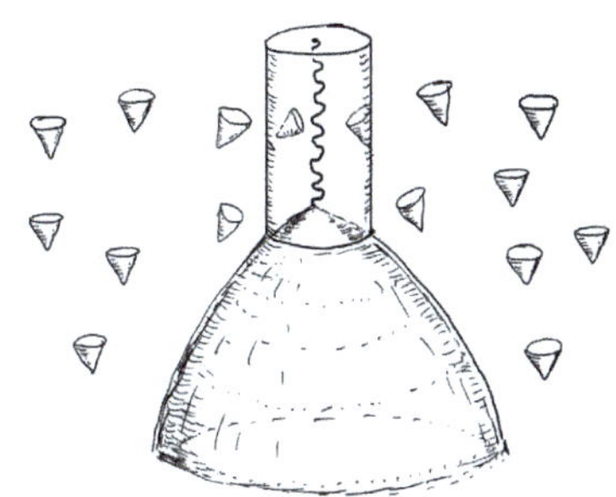

Fig. 25.9 Gravitational collapse with one spatial dimension suppressed.

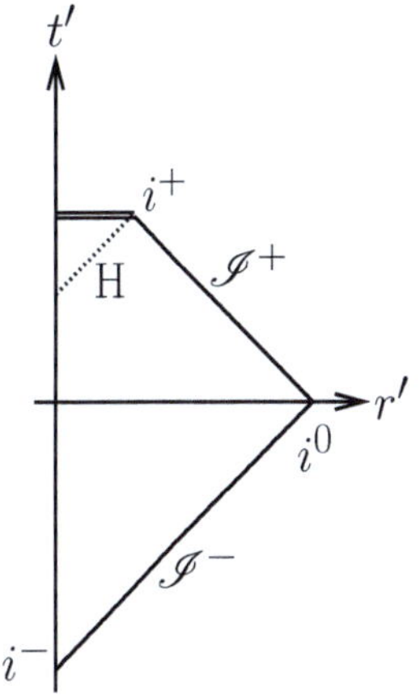

Fig. 25.10 Penrose diagram showing the structure of spacetime containing a black hole with event horizon H.

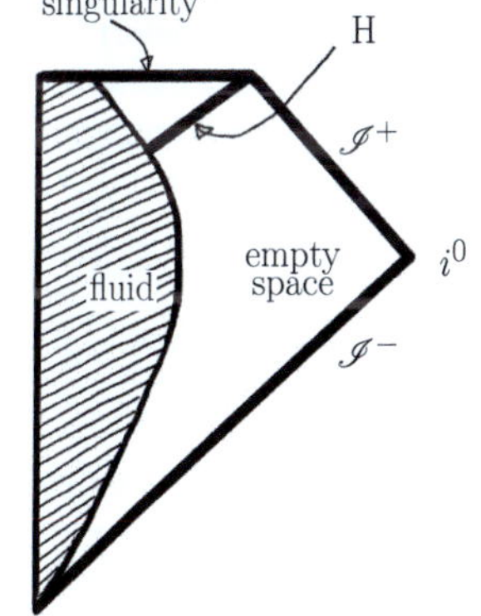

Fig. 25.11 Penrose diagram showing the gravitational collapse of a star (hatched area) to form a black hole.

Chapter summary

- Black holes are regions of spacetime where gravitational curvature doesn't allow light to escape. The spacetime of spherically symmetric black holes is described by the Schwarzschild metric.

- The Schwarzschild radius $r_\mathrm{S} = 2M$ represents the event horizon for the black hole. Inside the event horizon, the light cone structure means that all matter will inevitably meet the singularity in spacetime at $r = 0$.

- An astronaut will not notice anything special when passing the point $r = r_\mathrm{S}$ since this does not represent a physical singularity in spacetime. They will experience huge radially stretching gravitational forces as they approach $r = 0$.

- Although it takes an astronaut a finite proper time to meet the singularity, the coordinate time diverges.

Exercises

(25.1) Consider the Schwarzschild metric in the form

$$\mathrm{d}s^2 = -(1 - v^2)\mathrm{d}t^2 + (1 - v^2)^{-1}\mathrm{d}r^2 + r^2\mathrm{d}\Omega^2, \tag{25.19}$$

where $v^2 = 2M/r$ and $\mathrm{d}\Omega^2 = \mathrm{d}\theta^2 + \sin^2\theta\,\mathrm{d}\phi$.

(a) Compute the metric that results from a coordinate transformation $\mathrm{d}t = \mathrm{d}T - R(r)\mathrm{d}r$, where $R(r)$ is a function of r only.

(b) What function $R(r)$ is required to give $g_{rr} = 1$?

(c) Using the function from (b), show that the metric becomes

$$\mathrm{d}s^2 = -\mathrm{d}T^2 + \left[\mathrm{d}r + \left(\frac{2M}{r}\right)^{\frac{1}{2}}\mathrm{d}T\right]^2 + r^2\mathrm{d}\Omega^2. \tag{25.20}$$

(d) Comment on the form of the metric describing a hypersurface of constant T.

(e) What are the velocities $\mathrm{d}r/\mathrm{d}\tau$ and $\mathrm{d}r/\mathrm{d}T$ for a radial plunge in these coordinates.

Hint: Make use of the radial plunge results from Chapter 22.

(f) How do light cones behave in these coordinates?

(g) Show that the coordinate speed for a falling observer is always less than the coordinate speed of light.

*The coordinates used in eqn 25.20 are known as Painlevé–Gullstand coordinates or **global rain coordinates** and demonstrate that there is no physical singularity at $r = 2M$ (see the next chapter). The line element in these coordinates was independently proposed in 1922 by Paul Painlevé (twice Prime Minister of the French Third Republic) and Allvar Gullstrand (winner of the Nobel Prize in Physiology in 1911). Both were concerned that the solution to the Einstein equation they had discovered showed that relativity allowed infinite numbers of solutions, and so was incomplete. Lemaître showed in 1933 that these newly discovered solutions were simply the results of a coordinate transformation of the Schwarzschild line element, as the problem demonstrates. Incidentally, Gullstand had also blocked Einstein from receiving the Nobel Prize in Physics for his formulation special relativity in 1905, as he believed that theory to be incorrect.*

(25.2) *We shall derive, and justify the name of, the global rain coordinates from the previous problem using an array of in-falling clocks. See the books by Taylor, Wheeler, and Bertschinger, and by Moore (whose approach we follow here) for more details.*

The synchronized clocks are dropped at a steady rate from rest at infinity (where their readings coincide with the Schwarzschild coordinate time t), and fall radially into the black hole. Observers travel with the falling clocks and when they coincide with an event, they read off the time $\breve{t}$ and record the value of r, θ and ϕ. To compute the metric line element in these coordinates we note that $\breve{t} = \breve{t}(r, t)$ and so

$$\mathrm{d}\breve{t} = \left(\frac{\partial \breve{t}}{\partial t}\right)\mathrm{d}t + \left(\frac{\partial \breve{t}}{\partial r}\right)\mathrm{d}r. \tag{25.21}$$

(a) Argue that, from the setup, we must have

$$\left(\frac{\partial \breve{t}}{\partial t}\right) = 1. \tag{25.22}$$

Next, we want to work out $\left(\frac{\partial \breve{t}}{\partial r}\right)$, which we interpret as the difference in $\breve{t}$ measured by two clocks, separated by a distance $\mathrm{d}r$ evaluated a coordinate time t.

(b) Consider a pair of events separated by $\mathrm{d}r$ that occur at the same t. Show that for the observers attached to the clocks at these events

$$\frac{\mathrm{d}r}{\mathrm{d}\tau} = -\left(\frac{2M}{r}\right)^{\frac{1}{2}}, \tag{25.23}$$

and

$$\frac{\mathrm{d}r}{\mathrm{d}t} = -\left(1 - \frac{2M}{r}\right)\left(\frac{2M}{r}\right)^{\frac{1}{2}}. \tag{25.24}$$

(c) Call the time interval in t between the clocks being dropped from infinity $\mathrm{d}t_{\mathrm{d}}$. Show that

$$\mathrm{d}t_{\mathrm{d}} = \frac{\mathrm{d}r}{(1 - 2M/r)(2M/r)^{\frac{1}{2}}}. \tag{25.25}$$

(d) Show that the proper time measured for a clock that falls radially from r_{A} to r_{B} is given by

$$\Delta\tau = \frac{2}{3}\frac{\left(r_{\mathrm{A}}^{\frac{3}{2}} - r_{\mathrm{B}}^{\frac{3}{2}}\right)}{(2M)^{\frac{1}{2}}}. \tag{25.26}$$

(e) Explain why $\mathrm{d}\breve{t} = \mathrm{d}t_{\mathrm{d}} + \mathrm{d}\tau$.

(f) Show that

$$\frac{\partial \breve{t}}{\partial r} = \left(\frac{2M}{r}\right)^{\frac{1}{2}}\left(1 - \frac{2M}{r}\right)^{-1}. \tag{25.27}$$

(25.3) Using the result of the previous problem, show that the metric for the global rain coordinates is

$$\begin{aligned}\mathrm{d}s^2 = &-\left(1 - \frac{2M}{r}\right)\mathrm{d}\breve{t}^2 + 2\left(\frac{2M}{r}\right)^{\frac{1}{2}}\mathrm{d}\breve{t}\mathrm{d}r \\ &+ \mathrm{d}r^2 + r^2\left(\mathrm{d}\theta^2 + \sin^2\theta \mathrm{d}\phi^2\right).\end{aligned} \tag{25.28}$$

How can this metric be used to argue that objects following timelike geodesics inside the event horizon must move inwards?

(25.4) A particle in the Schwarzschild geometry around a black hole starts in the $\theta = \pi/2$ plane at radius r, moving purely tangentially with a local velocity of magnitude v_0, as measured by a stationary observer.

(a) Using the fact that the relative velocity v for observers with velocities $\boldsymbol{u}$ and $\boldsymbol{v}$ is determined by $\boldsymbol{u} \cdot \boldsymbol{v} = -\gamma(v)$, determine the components of the particle's velocity $\boldsymbol{v}$ in Schwarzschild coordinates at r.

(b) Assuming the particle moves freely, what are its values of the constants of the motion $\tilde{E}$ and $\tilde{L}$ in terms of v_0?

(c) If $r = 4M$, will a value of $v_0 = 1/\sqrt{2}$ be sufficient to prevent the particle falling into the hole? *See Blennow and Ohlsson for a more complete discussion of this problem.*

26

Black-hole singularities

> ↷ **This chapter and the next one deal with the technicalities of finding good coordinates suitable to deal with black-hole physics. Those wanting to see more of the physical consequences of black holes can skip to Chapter 28.**

[1] Although closed trapped surfaces were conceived as part of the discussion of singularities, the existence of a trapped surface doesn't necessarily imply that there is a singularity present. However, when a trapped surface is formed, it does at least suggest the presence of a singularity.

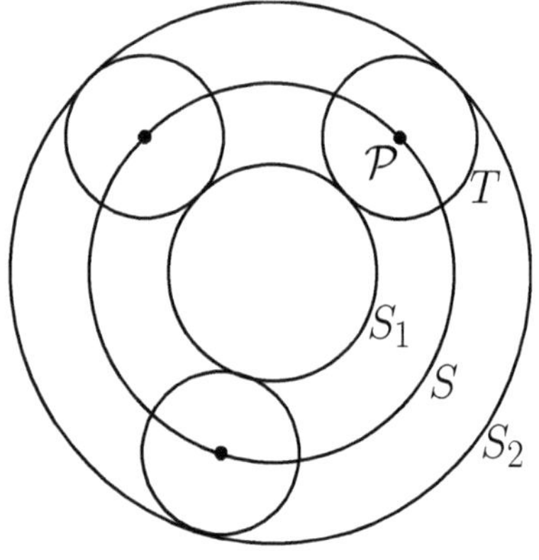

Fig. 26.1 A construction to illustrate closed trapped surfaces, which occur when the area of both S_1 and S_2 are smaller than that of S.

If the doors of perception were cleansed every thing would appear to man as it is, Infinite. For man has closed himself up, till he sees all things thro' narrow chinks of his cavern.
William Blake (1757–1827)

In 1965, Roger Penrose described the geometry of the inner region of a black hole's event horizon using the concept of a **closed trapped surface**. The argument goes like this. The 2-sphere S encloses a large spherical distribution of mass. Let a point $\mathcal{P}$ on S emit light pulses. At some later time the pulse will form a sphere T around $\mathcal{P}$. If all points on S emit pulses then the incoming and outgoing wavefronts will have envelopes forming two 2-spheres, S_1 and S_2, respectively, as shown in Fig. 26.1. Normally the area of the sphere S_1 (formed from all of the incoming parts) must be less than that of S, while S_2, formed from the outgoing parts, will be larger. However, when the amount of mass enclosed by S is sufficiently large, the areas of S_1 and S_2 will *both be less than S*. The surface S is then known as a **trapped surface** and all outward-going light rays will converge towards the centre of the mass distribution. Clearly something significant has caused the trapped surface: here it is a singularity at the centre of a black hole.

This argument illustrates some of the strange behaviour that can occur near a physical singularity in spacetime.[1] In this chapter, we look in more detail at the singularities occurring in the physics of black holes. As we shall see, the identification of genuine singularities is rather difficult, hence the need for physical arguments like the one above.

26.1 Singularities

One difficulty in dealing with the metric field of general relativity is the occurrence of singularities. A singularity is, roughly speaking, a point where a mathematical object is undefined, or fails to be well-behaved in some way. An example might be curvature blowing up or some other pathological behaviour in the metric line element. Near a true, **physical singularity** in the curvature of the metric, it is impossible to find an area small enough that space looks locally flat. (If in doubt consider the surface of a cone close to its apex). We therefore lose the smoothness property of spacetime, putting our description in terms of fields at risk. Physically, such a singularity is a point where our laws of physics break down.

Some singularities are more harmful that others. The less harmful variety, known as a **coordinate singularity** simply reflect our choice of coordinates. These are still a problem since they affect our ability to work with the geometry near the singular point, but they can effectively be eliminated. An instance of a coordinate singularity is given in the next example.[2]

Example 26.1

Two-dimensional cylindrical coordinates have line element

$$ds^2 = dr^2 + r^2 d\theta^2. \tag{26.1}$$

These coordinates become singular at the point $r = 0$. Here, changes in θ become meaningless, all values of θ apparently labelling a single point at $r = 0$. Put more vividly: there is a breakdown in the meaning of our coordinates. However, expressed in two-dimensional Cartesian coordinates, we have the usual line element $ds^2 = dx^2 + dy^2$. The Cartesian coordinates are well behaved around the origin, showing that the problem we identified from the cylindrical coordinate description does not reflect a problem with the spacetime.[3]

The Schwarzschild metric[4] with which we are concerned in this chapter is singular in two places: at $r = 0$ and at $r_S = 2M$. The singularity at $r = 0$ is a genuine physical singularity reflecting an infinite curvature of spacetime. (In this case, it turns this can be demonstrated by evaluating a scalar[5] $R_{\mu\nu\alpha\beta}R^{\mu\nu\alpha\beta}$ built from the components of the Riemann tensor, which diverges there.) In contrast, the singularity at the event horizon at $r_S = 2M$ is a coordinate singularity, reflecting a breakdown in the Schwarzschild-coordinate description of the spacetime. The point of this section is to examine how to remove the apparent singularity so that we can describe the physical behaviour close to the event horizon at r_S.

To understand coordinate singularities it's useful to have access to another conceptual tool. Some metric spaces are **geodesically complete**. This means that a geodesic, described by a function $x^\mu(\lambda)$, can be continued out to arbitrary points in space by varying the affine parameter λ. In contrast, some spaces are geodesically incomplete: an attempt to extend the geodesics hits a badly defined point. This is shown in Fig. 26.2. In general relativity, this idea allows a more robust definition of a singularity than the rough one offered so far: a singularity occurs in a spacetime if there are incomplete geodesics.

Example 26.2

Consider a spacetime with line element

$$ds^2 = -\frac{1}{t^4} dt^2 + dx^2, \tag{26.3}$$

which appears to be badly behaved at $t = 0$ and $t = \infty$. First, there appears to be a singularity at $t = 0$. However, if we make the coordinate transformation $y = \frac{1}{t}$, we find that

$$ds^2 = -dy^2 + dx^2, \tag{26.4}$$

$$ds^2 = -\left(1 - \frac{2M}{r}\right) dt^2$$
$$+ \left(1 - \frac{2M}{r}\right)^{-1} dr^2 + r^2 d\Omega^2, \tag{26.2}$$

where $d\Omega^2 = d\theta^2 + \sin^2\theta \, d\phi^2$. The components of the Riemann tensor in the orthonormal frame are

$$R_{\hat{r}\hat{t}\hat{r}\hat{t}} = -\frac{2M}{r^3},$$
$$R_{\hat{\theta}\hat{\phi}\hat{\theta}\hat{\phi}} = +\frac{2M}{r^3},$$
$$R_{\hat{r}\hat{\theta}\hat{r}\hat{\theta}} = R_{\hat{r}\hat{\phi}\hat{r}\hat{\phi}} = +\frac{M}{r^3},$$
$$R_{\hat{r}\hat{\theta}\hat{r}\hat{\theta}} = R_{\hat{r}\hat{\phi}\hat{r}\hat{\phi}} = -\frac{M}{r^3}.$$

Notice that the curvature has no unusual behaviour at $r = 2M$, but that the components do blow up at $r = 0$.

(a) $\qquad\qquad$ (b)

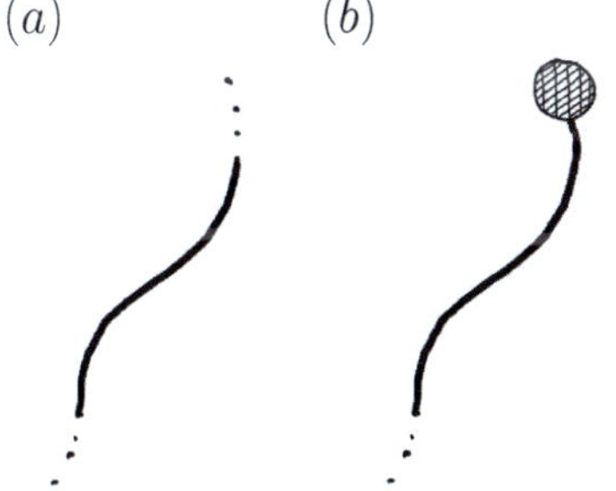

Fig. 26.2 (a) A geodesically complete space where all geodesics can be extended to infinity. (b) A geodesically incomplete space. The hatched area represents a singularity, through which the geodesics can't be extended.

§

[6]We can write $y(\lambda) = mx(\lambda) + c$ and use the parametrization $x = \lambda$ so $t = (m\lambda + c)^{-1}$. Make λ arbitrarily large and $t \to 0$. The geodesics are complete around here: we can make λ as large as we like and we still return a valid geodesic. One the other hand, as we send λ towards $-c/m$ we have a problem. Do this from above ($\lambda = -c/m + \varepsilon$, for $\varepsilon \to 0^+$) and $t \to \infty$; do it from below and $t \to -\infty$. The geodesics are therefore not complete when t becomes very large. They must stop at positive infinity, preventing us setting $\lambda < -c/m$.

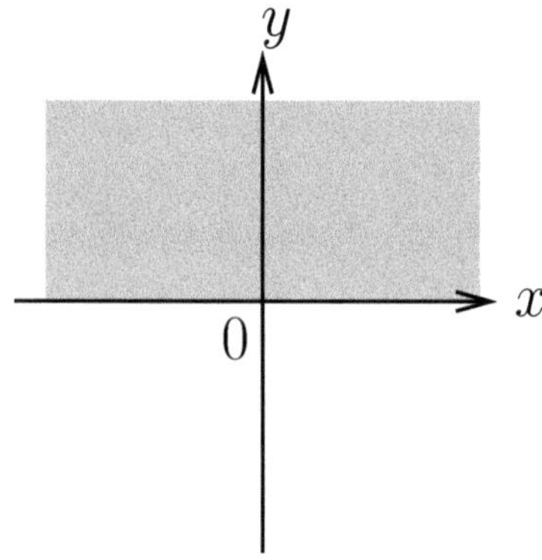

Fig. 26.3 The part of Minkowski space covered by the coordinate system in eqn 26.3. To cover the whole of Minkowski spacetime we need to add on the part below $y = 0$.

[7]We already did this in setting up light-cone coordinates in Minkowski space. These coordinates rely on the definitions

$$u = t - x,$$
$$v = t + x.$$

The line element for Minkowski spacetime can be written as

$$\mathrm{d}s^2 = -\mathrm{d}u\mathrm{d}v.$$

Notice that this is not a diagonal metric. Null geodesics are defined by $\mathrm{d}s^2 = 0$. Outgoing null geodesics, for which x increases as t increases, have $u = \text{const.}$ and $\mathrm{d}u = 0$; incoming null geodesics have $v = \text{const.}$ and $\mathrm{d}v = 0$. The null geodesics can therefore be used to make a grid covering flat space.

[8]See Chapter 5 for a picture of the light-cone structure for this metric.

that is, the metric in eqn 26.3 is Minkowski space in disguise. The apparent singularity at $t = 0$ corresponds to the point $y \to \infty$ in the Minkowski space. The cause of the problem was that all points at infinity in the Minkowski space were being mapped to $t = 0$ in the original coordinate system of eqn 26.3, and were therefore not given unique labels.

We can also carry out an analysis using geodesics, which are straight lines in this flat space.[6] The metric in eqn 26.3 is geodesically complete as $y \to \infty$, or as $t \to 0$, confirming that $t = 0$ is merely a coordinate singularity. However, we are not out of the woods since, in the original coordinates, the space is not geodesically complete for large t. Specifically, because of the form of the metric, the geodesics stop as $t \to \infty$. In Minkowski coordinates, $t \to \infty$ is equivalent to the point $y = 0$, implying that those original coordinates cannot access the parts of the space corresponding to negative y.

The cause of this singularity is that the metric in eqn 26.3 does not cover all of Minkowski space; only that portion with $y > 0$, since $y \to 0$ for $t \to \infty$ as shown in Fig. 26.3. However, we can extend the spacetime *beyond* $t = \infty$. This is achieved by considering the transformed spacetime in eqn 26.4 and *adding* the portion $y \leq 0$ to the spacetime. We see that we were misled by the original coordinate description of this well-behaved spacetime. There is nothing singular about it beyond the coordinate singularity.

The previous example gives us a clue of how to identify, interpret and, where possible, eliminate coordinate singularities. In many applications, such as the black holes examined in this chapter, coordinate singularities are transformed away with an inspired choice of coordinates. Although this often looks like magic, it is well motivated by a sensible method. In the next example, we show how a tricky spacetime with an apparent singularity can be understood with a change of coordinates that aims to use null geodesics as a coordinate system. We will sketch out a generalizable strategy to remove the singular-looking parts of a spacetime. The key is to examine the incoming and outgoing null geodesics and use these as a grid.[7]

Example 26.3

Consider Rindler spacetime[8], which has line element

$$\mathrm{d}s^2 = -x^2\mathrm{d}t^2 + \mathrm{d}x^2. \tag{26.5}$$

This appears to have a singularity at $x = 0$. At this point, the determinant of the matrix representing the metric $g(= -x^2)$ vanishes, making the inverse components, $g^{\mu\nu}$, singular. In two dimensions, null geodesics divide up into incoming and outgoing. Within each class they do not cross, so make a natural grid that can be used as a coordinate system. The null geodesics in Rindler spacetime may be found by considering $\mathrm{d}s^2 = 0$, or

$$\left(\frac{\mathrm{d}t}{\mathrm{d}x}\right)^2 = \frac{1}{x^2}, \tag{26.6}$$

so that, along the geodesics,

$$t = \pm \ln x + \text{const.}, \tag{26.7}$$

where c is a constant. The plus sign gives the outgoing geodesics, the minus sign the incoming ones. The equations for the light cones, by analogy with the version for Minkowski space, allow us to define a set of **null coordinates** (u, v)

$$u = t - \ln x,$$
$$v = t + \ln x. \tag{26.8}$$

Using the null coordinates, we write the metric as

$$\mathrm{d}s^2 = -\mathrm{e}^{v-u}\,\mathrm{d}u\,\mathrm{d}v. \tag{26.9}$$

This looks rather like the Minkowski metric expressed in light-cone coordinates $\mathrm{d}s^2 = -\mathrm{d}u\,\mathrm{d}v$ (albeit with a prefactor). However, it's notable that the coordinates u and v only extend over the regions of $x > 0$, so we need to re-parametrize the spacetime in order to make it geodesically complete by introducing new variables $U(u)$ and $V(v)$. Given the resemblance to Minkowski space, it is possible to guess a set of coordinates that will do this. The answer is to transform u and v into

$$U = -\mathrm{e}^{-u}, \quad V = \mathrm{e}^{v}. \tag{26.10}$$

The coordinates U and V are allowed to take all values from $-\infty \le U, V \le \infty$. The metric line element then becomes

$$\mathrm{d}s^2 = -\mathrm{d}U\,\mathrm{d}V, \tag{26.11}$$

which we recognize as the interval for Minkowski space expressed in light-cone coordinates. This space is geodesically complete and free from singularities. To confirm this, we could make a final transformation

$$T = \tfrac{(U+V)}{2}, \quad X = \tfrac{(V-U)}{2}, \tag{26.12}$$

and conclude that[9]

$$\mathrm{d}s^2 = -\mathrm{d}T^2 + \mathrm{d}X^2. \tag{26.14}$$

We were dealing with flat space all along!

So what was the reason for the coordinate singularity in Rinder spacetime? Rinder spacetime covers only a wedge of Minkowski space: that part with $X > |T|$, as shown in Fig. 26.4. The transformation to U and V allowed us to break through the coordinate barrier and extend our coordinates over all of the spacetime.

26.2 Eddington–Finkelstein coordinates

After taming Rindler space, we need to do the same for the space described by the Schwarzschild metric. Our aim is to find a new set of coordinates that shows that the singularity at $r_\mathrm{S} = 2M$ can be eliminated, just as we eliminated coordinate singularities in Rindler space. Taking tentative steps, we start with a warm-up exercise.

Example 26.4

Consider the following two-dimensional metric line element (for a sort of baby-Schwarzschild metric)

$$\mathrm{d}s^2 = -\left(1 - \frac{2M}{r}\right)\mathrm{d}t^2 + \left(1 - \frac{2M}{r}\right)^{-1}\mathrm{d}r^2, \tag{26.15}$$

and repeat the argument from Rindler space. The null geodesics, found from setting $\mathrm{d}s^2 = 0$, are written as

$$\left(\frac{\mathrm{d}t}{\mathrm{d}r}\right)^2 = \left(\frac{1}{1 - \frac{2M}{r}}\right)^2. \tag{26.16}$$

Integrating, we find that the radial null geodesics satisfy

$$t = \pm r^*, \tag{26.17}$$

[9]The original coordinates in terms of the final coordinates are

$$x = (X^2 - T^2)^{\frac{1}{2}},$$

$$t = \tanh^{-1}\left(\frac{T}{X}\right). \tag{26.13}$$

It would have been quite mysterious to have picked these coordinates out of nowhere.

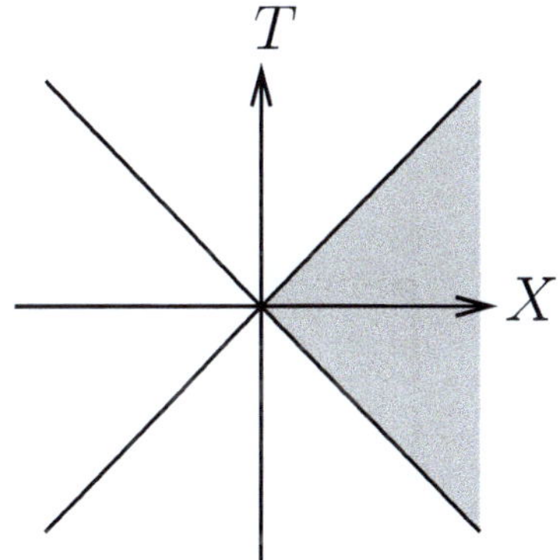

Fig. 26.4 The part of Minkowski space (T, X) covered by the Rindler metric coordinates.

[10] Recall from Chapter 25 that

$$r^* = r + 2M \ln \left| \frac{r}{2M} - 1 \right| + \text{const.}$$

We will set the constant to zero here.

[11] David Finkelstein (1929–2016). In addition to his work on quantum gravity, Finkelstein also worked on ball lightning and provided a detailed analysis of Albrecht Dürer's Melencolia I. He was also responsible for naming the sine-Gordon model: a joke, based on the Klein–Gordon model, that he later regretted.

[12] As in Schwarzschild coordinates we have the conserved quantity $\tilde{L}$. Notice also that there's no dependence on V in the metric and so in place of a conserved quantity $\tilde{E} = -u_t$, we have the conserved quantity $\tilde{E} = -u_V$.

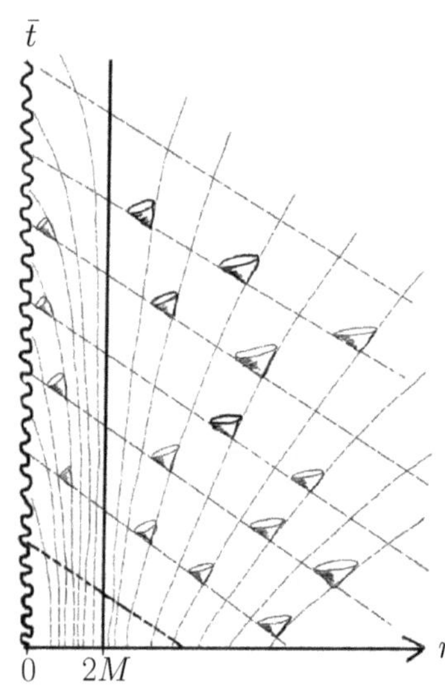

Fig. 26.5 Light cones in incoming Eddington–Finkelstein coordinates. Here we set $\bar{t} = V - r$ and plot the $\bar{t}$-r plane.

where r^* is the tortoise coordinate.[10] Define null coordinates by

$$\begin{aligned} U &= t - r^*, \\ V &= t + r^*. \end{aligned} \tag{26.18}$$

Outgoing, null geodesics are given by $U = \text{const}$. Incoming, null geodesics are given by $V = \text{const}$. In these coordinates, the metric becomes

$$\mathrm{d}s^2 = -\left(1 - \frac{2M}{r}\right) \mathrm{d}U \mathrm{d}V, \tag{26.19}$$

where r is defined implicitly by

$$r + 2M \ln \left| \frac{r}{2M} - 1 \right| = r^* = \frac{(V - U)}{2}. \tag{26.20}$$

Equation 26.19 has the form of a function multiplied by the Minkowski light-cone metric, just as we had in the last example. Although we could push through to complete the process, we pause at this point to examine a useful set of halfway-house coordinates.

Inspired by the argument above, one set of coordinates that are often used in the study of black holes are **incoming Eddington–Finkelstein coordinates**.[11] These coordinates are particularly useful for describing things that fall into back holes. They are a hybrid set of coordinates that use r and (the incoming null coordinate) V in place of r and t. That is, we substitute for the time coordinate t in the Schwarzschild metric with

$$t = V - r^*. \tag{26.21}$$

The Schwarzschild metric line element becomes

$$\mathrm{d}s^2 = -\left(1 - \frac{2M}{r}\right) \mathrm{d}V^2 + 2\mathrm{d}V\mathrm{d}r + r^2(\mathrm{d}\theta^2 + \sin^2\theta\,\mathrm{d}\phi^2). \tag{26.22}$$

This metric is not diagonal: it has the cross term $2\mathrm{d}V\mathrm{d}r$. There is no component $\mathrm{d}r^2$, but we shouldn't forget the presence of r in the calculations.[12]

Example 26.5

What do the light cones look like using these coordinates? To see them, set $\mathrm{d}s^2 = 0$ and find

$$-\left(1 - \frac{2M}{r}\right) \mathrm{d}V^2 + 2\mathrm{d}V\mathrm{d}r = 0. \tag{26.23}$$

Incoming rays have $V = \text{const}$ (as we should have expected). The other solution is

$$-\left(1 - \frac{2M}{r}\right) \mathrm{d}V + 2\mathrm{d}r = 0, \tag{26.24}$$

which we integrate to find that outgoing rays follow

$$V - 2\left(r + 2M \ln \left| \frac{r}{2M} - 1 \right|\right) = \text{const.} \tag{26.25}$$

The Eddington–Finkelstein coordinates therefore use one incoming null coordinate and one tortoise-like coordinate to form a grid for the light cones. Incoming Eddington–Finkelstein coordinates have the advantage that incoming light rays are simply continuous straight lines that head towards the singularity at $r = 0$, as shown in Fig. 26.5. They show no unusual behaviour passing through r_S. The price we pay is that outgoing rays still look singular at r_S. The 'outgoing' rays reach infinity if they start outside the horizon, but fall into the $r = 0$ singularity if they start inside the horizon. Notice from eqn 26.23 that if $r = 2M$ then we have $V = r = \text{const}$. The light rays are then trapped at the horizon, neither falling into the hole, nor emerging from it.

Having a well-behaved incoming coordinate V allows us to continuously describe things falling through the horizon at r_S.

Example 26.6

Consider an astronaut standing on the surface of a collapsing star. This is shown in incoming Eddington–Finkelstein coordinates in Fig. 26.6. The surface of the star shrinks radially and, because of the choice of coordinates, looks intuitively as expected. If the astronaut sends out regular light pulses, these follow the lines of constant $V - 2r^*$ as shown. We see that the interval between light pulses being received at a large distance from the hole increases. Once the astronaut has fallen through r_S the light pulses do not emerge from the hole.

We can also define **outgoing Eddington–Finkelstein coordinates**, by using r and U in place of r and t (i.e. we set $t = U + r^*$). The Schwarzschild metric in these coordinates becomes

$$\mathrm{d}s^2 = -\left(1 - \frac{2M}{r}\right)\mathrm{d}U^2 + 2\mathrm{d}U\mathrm{d}r + r^2(\mathrm{d}\theta^2 + \sin^2\theta\,\mathrm{d}\phi^2). \tag{26.26}$$

By evaluating the light cones, we see that one solution to $\mathrm{d}s^2 = 0$ has $U = \mathrm{const}$ and the other has

$$\frac{\mathrm{d}U}{\mathrm{d}r} = -\frac{2}{1 - \frac{2M}{r}}. \tag{26.27}$$

Comparing with the incoming version, we see that (as is easily guessed) outgoing Eddington–Finkelstein coordinates have continuous null geodesics for outgoing light rays and singular ones for incoming rays.

One interesting use of Eddington–Finkelstein coordinates is to investigate the radio pulses sent out by our astronaut colleague from Chapter 25 as she falls into the black hole.[13]

Example 26.7

Using outgoing Eddington–Finkelstein coordinates, the outgoing radial photons travel along lines of constant U, as shown in Fig. 26.7. The astronaut, whose proper time is τ, undergoes a radial plunge with velocity $\boldsymbol{u}$ with components $u^\mu = (u^U, u^r, 0, 0)$. The astronaut emits regular light pulses which are received by an observer at infinity, whose proper time is t.

Let's start by finding the velocity components $\frac{\mathrm{d}U}{\mathrm{d}\tau} = u^U$ and $\frac{\mathrm{d}r}{\mathrm{d}\tau} = u^r$. Notice how the metric doesn't depend on U, in exactly the manner in which it doesn't depend on t, so we have constant of the motion $u_t = u_U = -\tilde{E}$. Notice also that the cross term in the Eddington–Finkelstein coordinates means we need to be slightly careful in dealing with up and down components. We have, expressing the (U, r) part of the metric in matrix form,

$$g_{\mu\nu} = \begin{pmatrix} -\left(1 - \frac{2M}{r}\right) & -1 \\ -1 & 0 \end{pmatrix}, \tag{26.28}$$

whose inverse is

$$g^{\mu\nu} = \begin{pmatrix} 0 & -1 \\ -1 & \left(1 - \frac{2M}{r}\right) \end{pmatrix}. \tag{26.29}$$

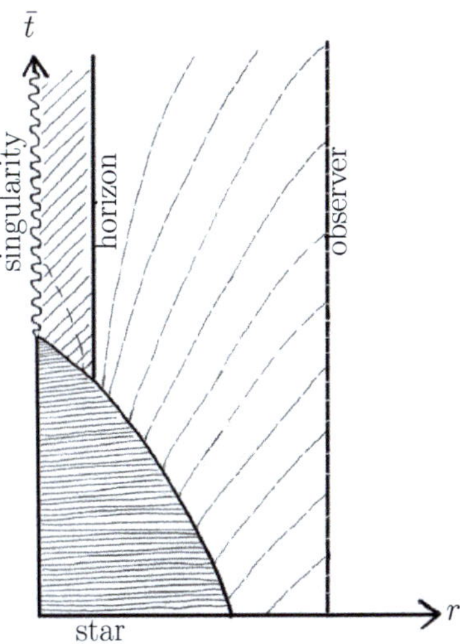

Fig. 26.6 Light pulses, shown in incoming Eddington–Finkelstein coordinates, sent outwards by an astronaut stood on the surface of a collapsing star.

[13]In this example, we follow the approach used in the book of problems by Lightman *et al.*

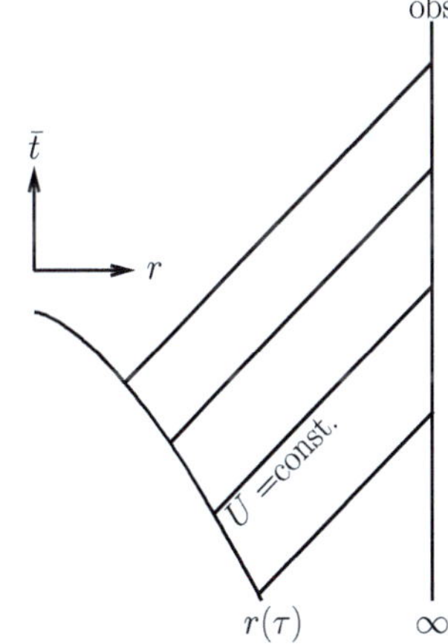

Fig. 26.7 An astronaut falling in the Schwarzschild geometry, emitting light pulses. These are shown in outgoing Eddington–Finkelstein coordinates where we set $\bar{t} = U + r$ and view the $\bar{t}$-r plane.

Normalization tells is that $\boldsymbol{u} \cdot \boldsymbol{u} = -1$, so we have

$$g^{rr}(u_r)^2 + 2g^{rU}u_U u_r + g^{UU}u_U u_U = -1, \tag{26.30}$$

from which, on substituting $\tilde{E} = -u_U$, we obtain

$$\left(1 - \frac{2M}{r}\right)(u_r)^2 + 2\tilde{E}u_r + 1 = 0, \tag{26.31}$$

and so

$$u_r = \frac{-\tilde{E} - \left[\tilde{E}^2 - \left(1 - \frac{2M}{r}\right)\right]^{\frac{1}{2}}}{1 - \frac{2M}{r}}. \tag{26.32}$$

We then have

$$u^U = g^{UU}u_U + g^{Ur}u_r = 0 + \frac{\tilde{E} + \left(\tilde{E}^2 - 1 + \frac{2M}{r}\right)^{\frac{1}{2}}}{1 - \frac{2M}{r}}, \tag{26.33}$$

and also

$$u^r = g^{Ur}u_U + g^{rr}U_r$$
$$= \tilde{E} + \left[-\tilde{E} - \left(\tilde{E}^2 - 1 + \frac{2M}{r}\right)^{\frac{1}{2}}\right] = -\left(\tilde{E}^2 - 1 + \frac{2M}{r}\right)^{\frac{1}{2}}. \tag{26.34}$$

This completes our analysis of the velocity in these coordinates.

Now consider the light pulses in Fig. 26.7. The astronaut sends (by her reckoning) regular pulses of light at intervals $(\Delta\tau)_{\rm em}$ to the observer at infinity who detects them separated by intervals $(\Delta t)_\infty$. We are interested in the ratio of wavelengths[14]

$$\frac{\lambda_\infty}{\lambda_{\rm em}} = \frac{(\Delta t)_\infty}{(\Delta\tau)_{\rm em}} = \frac{(\Delta U)_\infty}{(\Delta\tau)_{\rm em}}. \tag{26.35}$$

Now we receive the payoff for using these coordinates: since U is constant along the null geodesics then the change in U between pulses must be the same at the astronaut's position and at infinity (or $\Delta U_{\rm em} = \Delta U_\infty$, see Fig. 26.7). We therefore have

$$\frac{\lambda_\infty}{\lambda_{\rm em}} = \frac{(\Delta U)_{\rm em}}{(\Delta\tau)_{\rm em}} = \left.\frac{{\rm d}U}{{\rm d}\tau}\right|_{\rm em}. \tag{26.36}$$

The ratio is equal to the velocity component u^U of the emitter. This is good progress, but is rather difficult to interpret. However, we can massage this into something more enlightening by using

$$\frac{u^r}{u^U} = \frac{{\rm d}r}{{\rm d}U} = \frac{-\left(\tilde{E}^2 - 1 + \frac{2M}{r}\right)^{\frac{1}{2}}\left(1 - \frac{2M}{r}\right)}{\tilde{E} + \left(\tilde{E}^2 - 1 + \frac{2M}{r}\right)^{\frac{1}{2}}}. \tag{26.37}$$

This latter expression can be expanded near $r = 2M$ to yield the shift in the pulses emitted close to the horizon. We obtain

$${\rm d}U \approx -2\left(1 - \frac{2M}{r}\right)^{-1}{\rm d}r \approx -2\left(\frac{r}{2M} - 1\right)^{-1}{\rm d}r, \tag{26.38}$$

so that

$$U \approx -4M\ln\left(\frac{r}{2M} - 1\right) + {\rm const}, \tag{26.39}$$

or

$$1 - \frac{2M}{r} \approx {\rm e}^{-\frac{U}{4M}}. \tag{26.40}$$

From which we predict from eqn 26.33, that in the limit $r \to 2M$ we have

$$u^U \propto {\rm e}^{U/4M}. \tag{26.41}$$

Finally, we note that for an observer at large r we have $U \approx t + {\rm const}$ and so

$$\frac{\omega_\infty}{\omega_{\rm em}} = \frac{\lambda_{\rm em}}{\lambda_\infty} \approx {\rm e}^{-\frac{t}{4M}}. \tag{26.42}$$

We conclude that the light at infinity is redshifted: the detected frequency ω_∞ drops to zero exponentially with a time constant $4M$, that depends on the mass of the black hole.

[14]Since $t = U + r^*$ and the observations are made at the same position, we have $\Delta t_\infty = \Delta U_\infty$.

So, from the point of view of the observer at infinity, not only does it take an infinite interval in coordinate time for the astronaut to fall through the event horizon, the light that comes from the observer is redshifted, eventually by an infinite amount. We, while sat safely at infinity, would see the astronaut simultaneously slowing and fading as she approaches the Schwarzschild radius.

Although useful, Eddington–Finkelstein coordinates do not provide a solution to the central problem of this chapter: finding a coordinate system to describe the whole of the Schwarzschild geometry. That solution is the subject of the next chapter.

Chapter summary

- The Schwarzschild black hole has a physical singularity located at $r = 0$, whose presence gives rise to closed trapped surfaces. It has a coordinate singularity at r_S that can be removed by a change in coordinates.
- Eddington–Finkelstein coordinates remove the coordinate singularity for either incoming or outgoing rays, but not for both.

Exercises

(26.1) Show that for the Schwarzschild metric we have

$$R^{\mu\nu\alpha\beta} R_{\mu\nu\alpha\beta} = \frac{12M^2}{r^6}. \qquad (26.43)$$

Use this to discuss the nature of the singularities at (i) $r = 2M$ and (ii) $r = 0$.

(26.2) Show that the Eddington–Finkelstein metric has determinant $g = -r^4 \sin^2 \theta$.

(26.3) Verify eqn 26.29.

(26.4) In example 26.3, we guessed that $U(u) = e^{-u}$ and $V(v) = e^v$. The guesswork isn't necessary: there's a method we can use by considering how the affine parameter λ must vary along the null geodesics.

(a) Show that we can write a constant of the motion

$$\tilde{E} = x^2 \frac{dt}{d\lambda}. \qquad (26.44)$$

(b) Setting u constant, show that this leads to

$$\lambda = C + \left(\frac{e^{-u}}{2\tilde{E}} \right) e^v, \qquad (26.45)$$

with C a constant, showing that $\lambda_{\mathrm{out}} = e^v$ is an affine parameter along the outgoing geodesics.

(c) Show that a similar calculation yields $\lambda_{\mathrm{in}} = -e^{-u}$ for the incoming geodesics.

We can then reparametrize the null coordinates u and v using the affine coordinates in the hope of simplifying them.

27

Kruskal–Szekeres coordinates

[1]Martin Kruskal (1925–2006) made several contributions across a wide range of subjects in mathematical physics and was especially noted for his work on solitons.

[2]In addition to George Szekeres' (1911–2005) contribution to relativity was his work with Esther Klein on an important problem in combinatorial geometry. This was dubbed the 'Happy-Ending problem' by Paul Erdős after the collaboration resulted in Szekeres and Klein's marriage.

[3]The term $r^2\mathrm{d}\Omega^2$ is unchanged by our various manipulations and so we drop it in the next example, before restoring it in the solution.

And to add a new awareness to what the sky possesses in its size and formlessness, there is involved the quality of decay. For all the wonder of these everlasting stars, eternal spheres, and what not, they are not everlasting, they are not eternal; they burn out like candles.
Thomas Hardy (1840–1928) *Two on a Tower*

In this chapter, we introduce the solution to the main problem of the last chapter: defining a set of coordinates that are continuous over the event horizon and intuitively useful. These are the **Kruskal**[1]–**Szekeres**[2] **coordinates**.

27.1 Enter the Kruskal metric

The problem with Eddington–Finkelstein coordinates is that they represent two coordinate systems, neither satisfactory. For the outgoing version, incoming light rays are discontinuous at r_S (and vice-versa). Recall that, in the last chapter, we paused our programme of removing the singularity at $r_S = 2M$ at the point of having massaged the Schwarzschild metric into the form

$$\mathrm{d}s^2 = -\left(1 - \frac{2M}{r}\right)\mathrm{d}U\mathrm{d}V + r^2\mathrm{d}\Omega^2. \tag{27.1}$$

where[3] $\mathrm{d}\Omega^2 = \mathrm{d}\theta^2 + \sin^2\theta\,\mathrm{d}\phi^2$. We will now complete the process of removing this singularity.

Example 27.1

Use the fact that $r^* = r + 2M \ln|r/2M - 1| = (V - U)/2$ and rewrite the two-dimensional version of the metric as

$$\mathrm{d}s^2 = -\left(\frac{2M}{r}\mathrm{e}^{-\frac{r}{2M}}\right)\mathrm{e}^{\frac{(V-U)}{4M}}\mathrm{d}U\mathrm{d}V. \tag{27.2}$$

The message here is that we have factorized the metric into a function of r that is non-singular as $r \to 2M$, by (i) an exponential function of U and V and (ii) the Minkowski light-cone metric $-\mathrm{d}U\mathrm{d}V$. Following exactly the same path as we did in the Rindler case in the last chapter, we transform away the exponential dependence on U and V using a change in coordinates

$$u = -\,\mathrm{e}^{-U/4M},$$

$$v = \mathrm{e}^{V/4M}. \tag{27.3}$$

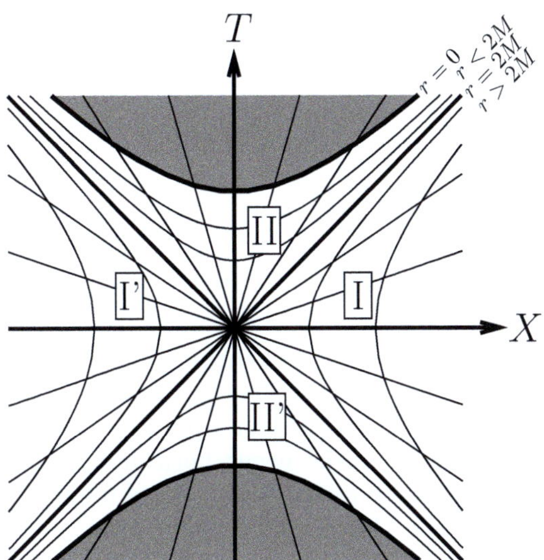

Fig. 27.1 Kruskal spacetime. The thick lines show the physical singularities at $r = 0$. Regions I and II are the parts through which a collapsing star moves. Regions I' and II' are described in Section 27.2.

This gives a metric

$$\mathrm{d}s^2 = -\frac{32M^3\mathrm{e}^{-r/2M}}{r}\,\mathrm{d}u\mathrm{d}u. \tag{27.4}$$

Once again, we can make the (now familiar) final transformation

$$T =(u + v)/2,$$
$$X =(v - u)/2. \tag{27.5}$$

From the argument in the last example we obtain[4] the **Kruskal metric**

$$\mathrm{d}s^2 = \frac{32M^3\mathrm{e}^{-r/2M}}{r}(-\mathrm{d}T^2 + \mathrm{d}X^2) + r^2(\mathrm{d}\theta^2 + \sin^2\theta\mathrm{d}\phi^2). \tag{27.6}$$

The Kruskal metric is not singular at $r = 2M$, demonstrating that this point is indeed a coordinate singularity, as we had previously claimed. The metric's structure is represented in Fig. 27.1 in the X-T plane with two coordinates suppressed, in a so-called **Kruskal diagram**. (Suppressing two coordinates means that each point in the Kruskal diagram represents a two-dimensional sphere of radius r.) By construction, all null geodesics are $45°$ lines, not just half of them as we had in Eddington–Finkelstein coordinates. Therefore, the light-cone structure is that of Minkowski space with future light cones heading towards positive T, and past light cones heading towards negative T. Inward-directed rays head towards small X, outward-directed rays head towards large X. Note that the metric becomes flat, asymptotically, in the limit $X \to \infty$.

Let's look at some of the features of the Kruskal diagram.

[4]The relation between old and new coordinates is

$$\left(1 - \frac{r}{2M}\right)\mathrm{e}^{r/2M} =T^2 - X^2,$$
$$\frac{t}{4M} =\tanh^{-1}\left(\frac{T}{X}\right).$$

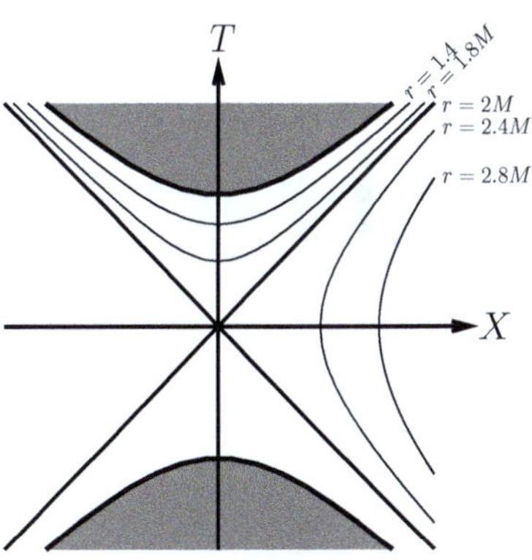

Fig. 27.2 Kruskal spacetime with lines of constant r plotted. They obey $T^2 - X^2 = (1 - r/2M)\mathrm{e}^{r/2M}$.

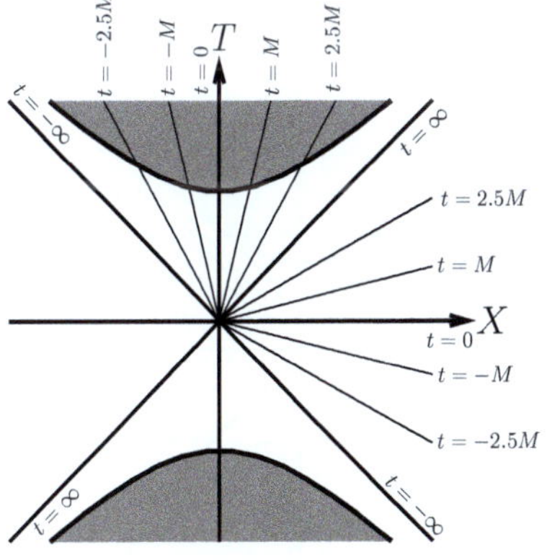

Fig. 27.3 Kruskal spacetime with (straight) lines of constant t plotted. They obey $T/X = \tanh(t/4M)$ in region I and $X/T = \tanh(t/4M)$ in region II.

[5] If you're tempted to give up at this point, remember the motivation: these coordinates ultimately make things as simple as possible, not least as they remove the coordinate singularity evident in the Schwarzschild coordinates at r_S. In the words of J. L. Synge, "If the problem of spherical symmetry had been attacked originally in this way, it would never have occurred to anyone to think of a singularity here."

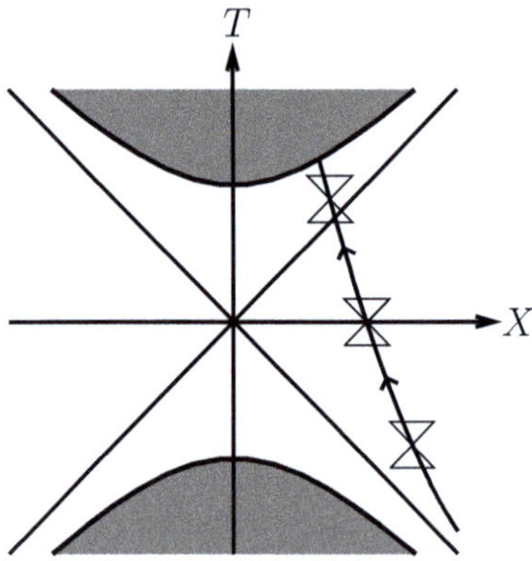

Fig. 27.4 Kruskal diagram showing an in-falling timelike world line.

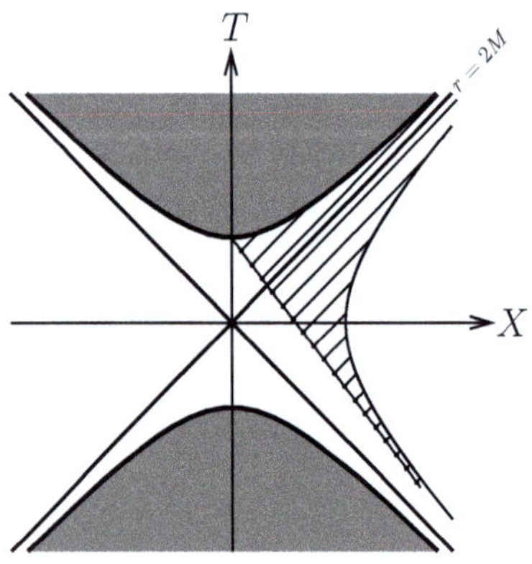

Fig. 27.5 Kruskal diagram showing light pulses emitted by an astronaut falling rapidly into a black hole, received by an observer at a fixed radius.

- In the Kruskal diagram, lines of constant r are curves of constant $X^2 - T^2 \left[= -(1 - r/2M)\mathrm{e}^{r/2M}\right]$ and are therefore hyperbolae, as shown in Fig. 27.2. One notable hyperbola is found for $r = 2M$, for which $X^2 - T^2 = 0$, resulting in the lines $X = \pm T$, which cut the plane into four regions, labelled I, II, I' and II' on the diagram in Fig. 27.1. Another notable hyperbola is that for $r = 0$, which is a physical singularity shown conventionally by thick (or sometimes jagged) lines which represent $X^2 - T^2 = -1$. There are two $r = 0$ singularities: one in the past (at the bottom of the diagram) and one in the future (at the top).

- The spacetime covered by Schwarzschild coordinates $r > 0$ and $-\infty < t < \infty$ is represented by that half of the Kruskal diagram with $T > -X$, comprising regions I and II. Schwarzschild coordinates outside of the event horizon ($2M < r < \infty$ and $-\infty < t < \infty$) cover region I with $X > 0$ and $-X < T < X$. In contrast, region II covers that part within the event horizon ($0 < r < 2M$ and $-\infty < t < \infty$) where $T > 0$ and $-T < X < T$.

- Spacelike hypersurfaces of constant t are represented by lines of constant T/X and so are straight lines, as shown in Fig. 27.3. The line corresponding to the surface $t = 0$ has $T = 0$ for $r > 2M$, but $X = 0$ for $r < 2M$. The coordinate $t = \infty$ is represented by $X = T$ and the coordinate $t = -\infty$ is represented by $X = -T$. In each quadrant, the t coordinate advances between $t = -\infty$ and ∞.

Clearly, these coordinates take a little getting used to.[5] However, the fact that the light rays all travel at $45°$ to the vertical simplifies things nicely. Let's look at an example of using the coordinates.

Example 27.2

The path of the in-falling astronaut is shown in a Kruskal diagram in Fig. 27.4. The world line shows the astronaut passing from region I, through the horizon into region II, and then falling towards the singularity at $r = 0$. Remember that light rays travel at $45°$, so on passing the $r = 2M$ horizon, all light rays emitted along the astronaut's path will end up hitting the $r = 0$ singularity or, if you like, they fall into the hole.

An illustration of the light signals emitted by the in-falling victim is shown in Fig. 27.5. The observer in region I remains at a fixed value of r (presumably maintained by firing a rocket, or some such) and their world line is therefore a hyperbola. The astronaut falls into the hole (rather fast, as shown, although the angle is less than $45°$ to maintain a subluminal speed). As the astronaut falls they regularly emit light pulses. Since the observer follows a hyperbolic path on the Kruskal diagram, we see that the observer receives the signals less and less frequently. The last ones are received after long intervals of coordinate time, with the ones emitted just above the horizon not reaching the observer until $t \to \infty$. Pulses emitted after the astronaut has fallen into the hole are seen to hit the singularity at $r = 0$, as indeed the astronaut does, where they are destroyed.

It is immediately apparent that Schwarzschild coordinates only cover regions I and II of the Kruskal diagram, and so the other half-plane (with $T < -X$) doesn't yet seem of much physical significance. However, it is possible to interpret the whole of the X-T plane as a meaningful coordinate patch. Taking all of the Kruskal coordinates to describe spacetime involves the extension of Schwarzschild space into the $T < -X$ half plane (the previously mysterious regions I' and II' in Fig. 27.1), and is therefore known as the **Kruskal extension**. Before exploring this feature, let's try to make ourselves a little more comfortable with the Kruskal coordinates in the next example.

Example 27.3

We can understand the physical basis of the (still admittedly rather mysterious) Kruskal coordinates with an analogy made by Wolfgang Rindler. We consider the acceleration of a particle in Minkowski space using coordinates (T, X, Y, Z). The particle's world line is $X^2 - T^2 = x^2$, which corresponds to a proper acceleration in the X direction of $1/x$. [This problem is discussed in Chapter 2. Some example hyperbolic world lines are shown in Fig. 27.6(a).] Now consider a change of variables to a set of coordinates (t, x, y, z) where

$$T = x \sinh t, \quad X = x \cosh t, \quad Y = y, \quad Z = z. \tag{27.7}$$

This transformation itself obeys the world-line equation for an accelerating particle and so (t, x, y, z) represent an accelerating set of coordinates.[6] Lines of constant x, when graphed in region I of the (T, X, Y, Z) coordinate system on a Minkowski diagram in the T-X plane, give hyperbolic curves [Fig. 27.6(b)].[7] So, it is as if the coordinates (t, x, y, z) are the coordinates used inside a rocket that accelerates in the X-direction. (By the same token, region I' corresponds to a rocket accelerating in the $-X$ direction.)

In order to describe regions II and II' in the same way, we make the transformation

$$T = x \cosh t, \quad X = x \sinh t, \quad Y = y, \quad Z = z, \tag{27.8}$$

which obeys $X^2 - T^2 = -x^2$, showing an exchange in roles of the T and X coordinates in these regions. The resulting accelerating world lines in regions I and II are shown in Fig. 27.6(b).

Using the (t, x, y, z) coordinate system in regions I and I' causes the Minkowski metric to change to the Rindler metric

$$\mathrm{d}s^2 = -x^2\mathrm{d}t^2 + \mathrm{d}x^2 + \mathrm{d}y^2 + \mathrm{d}z^2, \tag{27.9}$$

while in regions II and II' we have

$$\mathrm{d}s^2 = x^2\mathrm{d}t^2 - \mathrm{d}x^2 + \mathrm{d}y^2 + \mathrm{d}z^2. \tag{27.10}$$

If we then make the change of variables $x^2 = (2r - 1)$ in regions I and I' and $-x^2 = (2r - 1)$ in regions II and II', then we can describe the whole of Minkowski space with the accelerating coordinates and a metric

$$\mathrm{d}s^2 = -(2r - 1)\mathrm{d}t^2 + (2r - 1)^{-1}\mathrm{d}x^2 + \mathrm{d}y^2 + \mathrm{d}z^2. \tag{27.11}$$

This previous equation rather closely resembles the Schwarzschild metric line element!

So the accelerating coordinate system (used as a natural choice inside a rocket) can be likened to the Schwarzschild system, while Minkowski space is analogous to the Kruskal system. We therefore identify the accelerating coordinate system (t, x, y, z) with Schwarzschild coordinates and the Minkowski space [with coordinates (T, X, Y, Z)] with Kruskal space and we have the analogy:

$$\begin{pmatrix} \text{Minkowski} \\ \text{coordinates} \end{pmatrix} \longleftrightarrow \begin{pmatrix} \text{Accelerating Minkowski} \\ \text{coordinates} \end{pmatrix}$$
$$\begin{pmatrix} \text{Kruskal} \\ \text{coordinates} \end{pmatrix} \longleftrightarrow \begin{pmatrix} \text{Schwarzschild} \\ \text{coordinates} \end{pmatrix}. \tag{27.12}$$

We can use the analogy to say that, in region I, analogous to the accelerating rocket system in a Minkowski coordinate system, is a Schwarzschild system held outside $r = r_{\mathrm{S}}$ represented in the Kruskal system.

---

[6]Specifically, each choice of x represents the path of an accelerating particle in the (T, X, Y, Z) system, with acceleration $1/x$. Region I of the T-X plane can also be used to represent the acceleration of a rigid body. Measured in the (T, X, Y, Z) system, we see from the horizontal bars in Fig. 27.6(a) that if the front of a rocket has some acceleration $a = 1/x$, then the back must have greater acceleration leading to a increasing length contraction.

[7]Since we also have $X/T = \coth t$, lines of constant t are straight in the T-X plane.

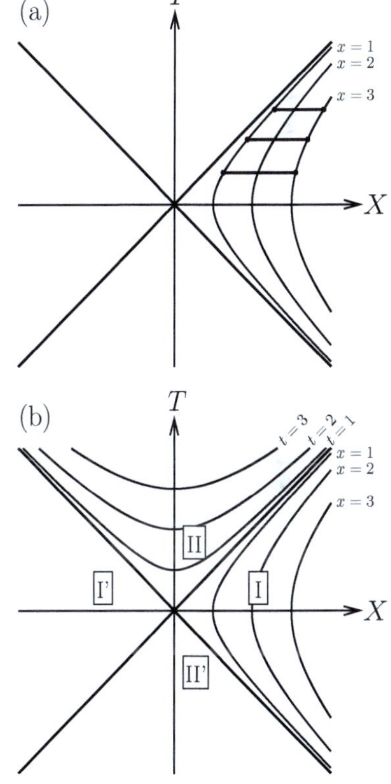

Fig. 27.6 (a) Hyperbolae in region I showing accelerating particles in Minkowski space. Horizontal lines linking two coordinates values of x at fixed T show how the back of a rigid object appears to accelerate faster than the front. (b) The accelerating coordinate system in regions I (constant x) and II (constant t). The resulting picture resembles the Kruskal diagram.

The last example shows that rather than regarding the Kruskal coordinates as a complicated and contrived system, we should perhaps think of them as being as natural a system as the Minkowski coordinates are in flat space. The Schwarzschild coordinates are then like the 'difficult' coordinates of an accelerating frame.

27.2 Wormholes

Let's now return to the idea of taking the Kruskal system to cover all of spacetime, previously named the Kruskal extension. Intriguingly, the Kruskal extension describes the spacetime of a **wormhole**. Large positive values of X in region I corresponds to flat spacetime and so, in the same way, large, negative X in region I' also corresponds to a flat spacetime. Region I' therefore looks very much like the spacetime for a spherical mass, outside the Schwarzschild radius. It is therefore interpreted as *another universe* outside of a black hole. Since time increases up the page, region II' is that part of spacetime that some past-directed world lines will fall into from region I' (just as future directed light rays can fall from region I into the black hole in region II). This implies that the primed universe (regions I' and I") is time-reversed compared to the universe comprising regions I and II, which is to say that the coordinate t advances in the opposite sense. Region II' has its own $r = 0$ singularity and so any particle in region II' must have been 'created' at this past singularity, just as any particle in region II must be destroyed at the future singularity. From this point of view, the past singularity in region II' might be characterized as a **white hole** from which all particles and light in region II' must originate.

The two singular $r = 0$ lines in regions II and II' form a **throat** in spacetime separating the two asymptotically flat universes in regions I and I'. A spacelike hypersurface at $T = 0$ is shown in Fig. 27.7 where the throat, of radius $2M$ at this point is shown.

Is it possible to travel between the two universes? That is, can a traveller traverse the throat in Fig. 27.7? The short answer is no: there are no null or timelike geodesics linking regions I and I', and so all photons and massive particles crossing the line $r = 2M$ are doomed to fall towards a singularity. This can be seen in Fig. 27.1, where we can see that no non-spacelike (i.e. $\leq 45°$ to the vertical) curve can get from region I to region I'. (Notice how the Kruskal coordinates, now proving their worth, allow this to be seen relatively simply.) Despite the impossibility of a trip between universes, there are some intriguing features connected with the wormhole that we examine in the next example.

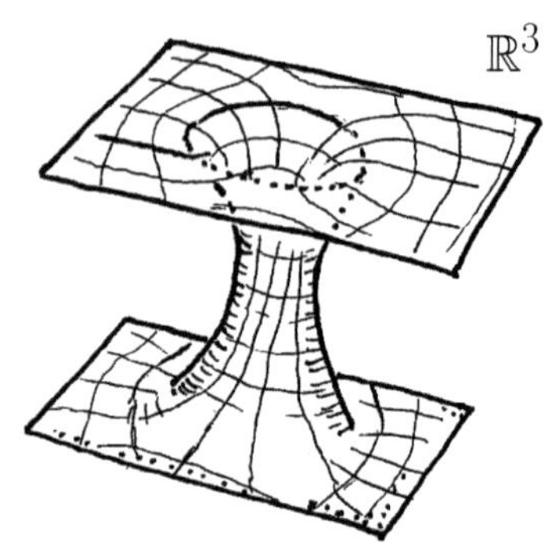

Fig. 27.7 A timelike slice of Kruskal spacetime taken at $T = 0$. This shows the spacetime throat between the universe I (top) and universe I' (bottom).

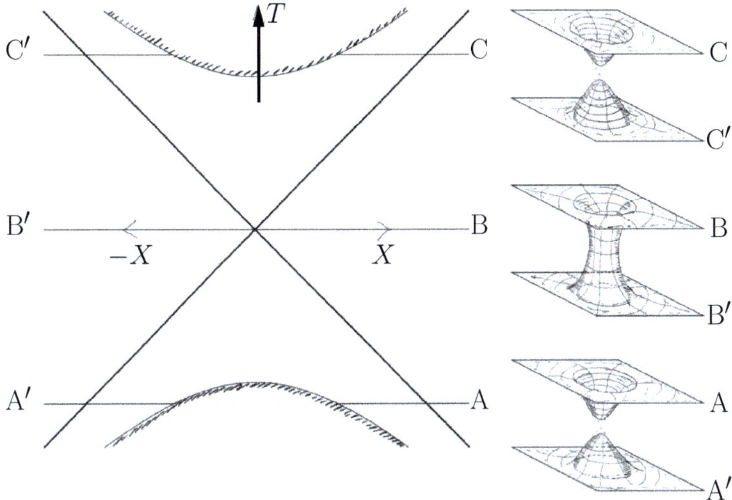

Fig. 27.8 Evolution of timelike slices of Kruskal spacetime with T.

Example 27.4

We usually regard the Schwarzschild geometry as static, with increments δt in time t causing no change in the geometry. This enables us to identify spacelike hypersurfaces, such as that shown in Fig. 27.7 for $T = 0$, showing a throat linking the two asymptotically flat universes. However, the idea of the static metric only makes sense in regions I and I'. In regions II and II', increments in the coordinate t are actually increments in space and so spacelike hypersurfaces are not really static. Instead, they can be thought of as evolving in the increasing T direction (i.e. the future, according to light cones). From this point of view we can trace the evolution of an arbitrary spacelike slice[8] as a function of T.

Consider, for example, the spacelike slice A'-A show in Fig. 27.8. We see (right) that the two universes are unlinked. Each slice curves towards the singularity at $r = 0$ represented by a cusp, which occurs where the slice hits the singularity. At a larger value of T, when the slice just touches the singularity, the universes must meet at a pinch point and then, as T increases, the universes are linked by a throat that avoids $r = 0$. The throat takes its maximum size of $r = 2M$ at $T = 0$ (B'-B) before contracting. The universes then reach a pinch point again before separating (C'-C). Although it might look temptingly like a particle or signal could be sent from region I, through the throat of the wormhole, to region I', it cannot. From the point of view of the spacelike surfaces, the particle cannot move fast enough to traverse the throat. The spacelike slices of the spacetime evolve so fast that the universes pinch off before the signal (or traveller) has reached the middle of the throat. The signal is doomed to hit the $r = 0$ singularity where it will be destroyed.

[8]There is no meaningful, special spacelike slice we can take. We simply select an arbitrary one to show the evolution.

27.3 Another Penrose diagram

The Penrose diagram for the Kruskal extension is shown in Fig. 27.9(a). It shows regions I, II, I' and II'. There are two spacelike singularities, one

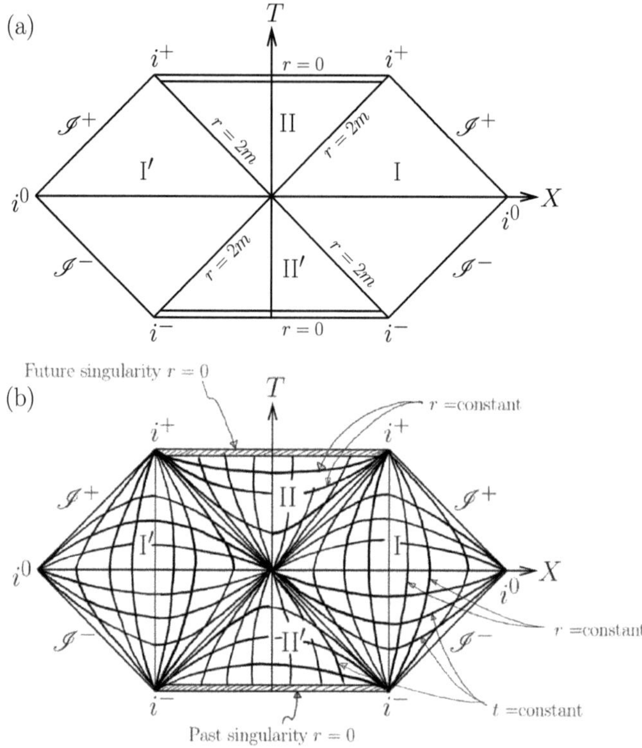

Fig. 27.9 (a) Penrose diagram of the Kruskal extension. (b) The same, but showing lines of constant r and t.

in the past and one in the future. Notice how these exist at the edges of the diagram or, if you prefer, the edges of spacetime. Unlike the case of the Robertson–Walker spacetimes in Chapter 19, not all world lines meet the singularities. In fact there are future and past null surfaces along with future timelike (i^+), past timelike (i^-) and spacelike (i^0) infinities. The existence of these infinities in this case is due to the fact that the spacetime is asymptotically flat, and so we expect to recover something like the Minkowski conformal structure far at large distances from $r = 0$. We also see how the singularity in the past is visible to observers in region I. This is known as a **naked singularity**, and can influence event in the future. Naked singularities are not a feature of normal physical processes and so its presence here is likely the result of the artificial extension we have made.

In Fig. 27.9(b), we have added to the Penrose diagram some lines of constant t (which run side to side in region I and up and down in region II) and lines of constant r (which run up and down in region I and side to side in region II). We can now justify the presence of trapped surfaces that we used to motivate the last chapter.

Example 27.5

In region I, a light pulse is emitted at point $\mathcal{P}$. At a later instant, defined by the future light cone (found by moving up to the next line running side to side), the outgoing wavefront s has a larger radius (further to the right in the diagram). The ingoing wavefront q has a smaller radius (further to the left). This contrasts with the situation in region II. Here a light pulse is emitted at point $\mathcal{Q}$. At a later instant (again defined by the future light cone: move up to the next line running side to side) both outgoing s and incoming q wavefronts are at a smaller value of r. This is because the next line up that runs side to side is a line of constant r and is smaller, since r decreases towards $r = 0$, which is the singularity at the top of the diagram. We conclude that the areas of the spherical wavefronts decrease with time. These are trapped surfaces and warn us that there the wavefronts will meet a singularity in the future, as indeed they shall at $r = 0$.

After two rather geometry-heavy chapters discussing the spacetime near a Schwarzschild black hole, we turn to the physics of the thermodynamics of the hole in the next chapter.

Chapter summary

- Kruskal–Szekeres coordinates successfully remove the singularity at r_S. They suggest that the coordinates can be extended to cover apparently unphysical regions of spacetime.
- Kruskal coordinates and their link to Schwarzschild coordinates can be understood by analogy to the relation between Minkowski coordinates and accelerating Minkowski coordinates.
- The conformal structure of the system reveals how a physical singularity can be thought of as living at the edge of spacetime.

Exercises

(27.1) Verify eqn 27.2.

(27.2) Show that the Kruskal coordinates are related to the Schwarzschild ones via

$$T^2 - X^2 = \left(1 - \frac{r}{2M}\right) e^{r/2M},$$

$$\tanh^{-1}\left(\frac{T}{X}\right) = \frac{t}{4M}. \tag{27.13}$$

(27.3) An accelerating particle in Minkowski space with coordinates (T, X) follows a world line $X^2 - T^2 =$

g^{-2} with

$$T(\tau) = \frac{1}{g} \sinh g\tau,$$

$$X(\tau) = \frac{1}{g} \cosh g\tau. \tag{27.14}$$

Show that accelerating observers follow world lines with constant x in Rindler spacetime, which is described by a line element $ds^2 = -x^2 dt^2 + dx^2$.

(27.4) Consider Rindler spacetime again with the metric given in the last question. Compute directly the components of the acceleration $\boldsymbol{a} = \boldsymbol{\nabla}_{\!u}\boldsymbol{u}$ for a particle held at constant $x = x_0$.

Hint: The relevant connection coefficients are given in the answers to Exercise 9.4.

(27.5) *We saw in this chapter how light from the event horizon was exponentially redshifted. We also saw how Schwarzschild coordinates can be thought of as accelerating Minkowski coordinates. We can link these ideas in very general terms, as we demonstrate here.*

An observer moves in an inertial frame with trajectory $X = vT$.

(a) Show that the observer's world line can be parametrized using $T(\tau) = \gamma\tau$ and $X(\tau) = \gamma v\tau$.

Now consider a light wave propagating with amplitude $A = e^{i\Omega(T-X)}$, with angular frequency Ω.

(b) By using the parametrization in terms of τ, show that the frequency measured by the observer

will be

$$\Omega' = \Omega \left(\frac{1-v}{1+v} \right)^{\frac{1}{2}}. \qquad (27.15)$$

This is just the expected result for relativistic Doppler shift.

(c) Now consider a uniformly accelerated observer with

$$T(\tau) = \tfrac{1}{g}\sinh g\tau, \quad X(\tau) = \tfrac{1}{g}\cosh g\tau. \qquad (27.16)$$

Using the same procedure as above, show that the observer now measures the frequency of the wave as being exponentially redshifted according to

$$\omega(\tau) = \Omega e^{-g\tau}. \qquad (27.17)$$

We conclude the frequency decreases exponentially with proper time for the accelerated observer.

Hawking radiation

28

All nature wears one universal grin
Henry Fielding (1707–1754)

In 1974, Stephen Hawking[1] showed that black holes are not simply holes in the Universe that consume matter. In fact, they continuously emit radiation just as a thermodynamic black body does. This effect is properly explained using quantum field theory: in particular, the notion that the ground state of a quantum field, known as the vacuum due to the absence of particles, allows **vacuum fluctuations**. The idea here is that it is possible to produce particle-antiparticle pairs that exist fleetingly before annihilating. The energy borrowed from the vacuum to produce the particles ΔE and the lifetime of the pair Δt are constrained by an uncertainty relation $\Delta E \Delta t \approx \hbar$. The reason for this is that these **virtual particles** exist **off mass shell**, which is to say that for a particle with momentum $\boldsymbol{p}$ and mass m, we no longer have the invariant relation $\boldsymbol{p} \cdot \boldsymbol{p} = -m^2$. As a result, the energy of a particle is not determined by its usual dispersion relationship. Quantum uncertainty sets a time limit $\Delta t \approx \Delta E/\hbar$ on how long this can be accommodated, leading to the finite lifetime of the virtual particle and antiparticle pair.

[1]Stephen Hawking's advanced-level book *The Large Scale Structure of Space-Time*, cowritten with George F. R. Ellis (1939–), is warmly recommended for its clarity, insight, and beautiful diagrams. Hawking is now celebrated to the extent that he was voted the 25th greatest-ever Briton in a poll conducted by the BBC in 2002, one place below Queen Elizabeth II. Of the other scientists mentioned in this book, Newton polled 6th, Faraday 22nd and Maxwell 91st. The poll was noted for several surprising results, including the placing of the actor Michael Crawford at seventeen, one place above Queen Victoria.

28.1 Hawking radiation

The mechanism for black holes to radiate relies on a particle-antiparticle pair being produced just outside the event horizon of the hole, as shown in Fig. 28.1. One particle falls into the hole and is destroyed; the other propagates freely out to infinity and therefore constitutes the radiation. The idea is that if one particle from the pair crosses the event horizon to the region $r < 2M$ then the status of the pair can change. They were previously virtual particles forced to recombine within a time $\Delta t \approx \hbar/E$. However, within the event horizon the changing roles of time and space can allow the (previously virtual) in-falling particle to propagate freely, no longer subject to the constraint on its lifetime. This also frees the outgoing particle. The resulting stream of real, outgoing particles from such processes transport energy from the hole and form **Hawking radiation**. This radiation has yet to be measured experimentally.

The key to understanding why this can occur is to examine the energy-momentum of the virtual particle. For a particle to propagate with a momentum vector $\boldsymbol{p}$ the observed energy $E_{\mathrm{obs}} = -\boldsymbol{p} \cdot \boldsymbol{u}$, as measured by

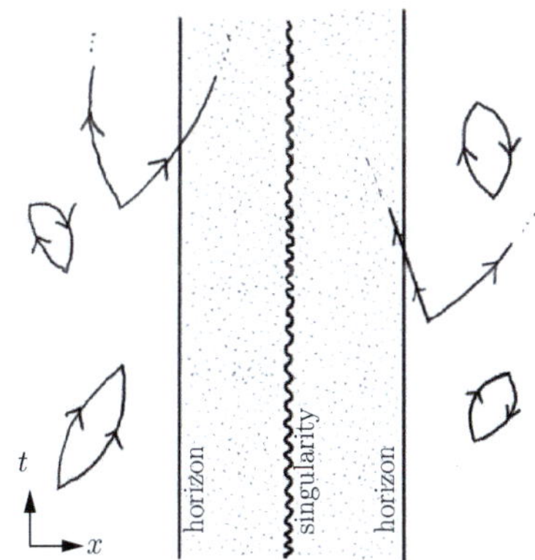

Fig. 28.1 Particle-antiparticle fluctuations near the event horizon of a black hole.

[2]It is often claimed that antiparticles have negative energy. In fact, all particles and antiparticles that have been observed have positive energy. The confusion arises as the antiparticles come from considering the two solutions to the dispersion relation equation $E = \pm(\vec{p}^{\,2} + m^2)^{\frac{1}{2}}$, which seems to allow negative energy states. In order to make physical sense, we must reinterpret the negative antiparticle solution with four-momentum $p^\mu = (-E, \vec{p})$. This, which describes a negative energy particle with outgoing momentum, is reinterpreted as a positive energy particle with incoming momentum $p^\mu = (E, -\vec{p})$, that is, we switch the sign of the momentum, so the antiparticle's direction is reversed, in order to guarantee positive energy.

[3]The link between the sign of energy and the passage of time can be seen in quantum mechanics by noting that quantum particles propagate with their wavefunction carrying a phase $e^{-iEt/\hbar}$. From the point of view of the wavefunction, a positive-energy particle propagating forward in time is equivalent to a negative energy one propagating backward in time.

[4]Such a path is timelike owing to the change in status of time and space inside the event horizon.

[5]Recall the effective energy equation for a photon is

$$\frac{1}{b^2} = \frac{1}{\tilde{L}^2}\left(\frac{dr}{d\lambda}\right)^2 + \frac{1}{r^2}\left(1 - \frac{2M}{r}\right),$$

with $b = \tilde{L}/\tilde{E}$. For $\tilde{L} \to 0$, we have that

$$\tilde{E} = \pm\left(\frac{dr}{d\lambda}\right) = \pm p^r.$$

[6]The observer has velocity with components $u^\mu = (1, 0, 0, 0)$; the Killing vector $\boldsymbol{\xi}$ has components $\xi^\mu = (1, 0, 0, 0)$. Recall that $\tilde{E} = -\boldsymbol{\xi} \cdot \boldsymbol{u}$ is the energy per unit mass (hence a tilde on the E). Therefore, $E = -\boldsymbol{\xi} \cdot \boldsymbol{p}$ gives us the energy (no tilde) since $\boldsymbol{p} = m\boldsymbol{u}$, where m is the rest mass.

[7]That is, $\boldsymbol{\xi} \cdot \boldsymbol{\xi} = \boldsymbol{e}_t \cdot \boldsymbol{e}_t = g_{tt} = -(1 - 2M/r)$ changes sign on either side of the horizon at $r = 2M$.

[8]Of course, the observer is at infinity where they make measurements locally, and so they are not measuring this latter quantity.

a *local* observer with velocity $\boldsymbol{u}$, must be positive.[2] If we do not make this stipulation, particles could travel backwards in time along timelike geodesics, which could allow violations of causality.[3] The positivity of energy will be important in demonstrating the possibility of propagation inside the event horizon.

Example 28.1

We start by considering photons. A photon is identical to an antiphoton, so we can consider one member of a pair of virtual photons, with momentum $\boldsymbol{p}$, produced by a vacuum fluctuation just outside $r = 2M$. We will show how it can propagate freely at radii $r < 2M$.

To do this, we put ourselves in the (unenviable) position of an observer inside the horizon, undergoing a radial plunge that follows a curve with tangent vector $\boldsymbol{u}$. To make things as simple as possible we choose a plunging path with $u^0 = 0$, so that the only non-zero component of $\boldsymbol{u}$ is[4] u^r. From the identity $\boldsymbol{u} \cdot \boldsymbol{u} = -1$, we have, for $r < 2M$, that

$$u^r = -\left(\frac{2M}{r} - 1\right)^{\frac{1}{2}}. \tag{28.1}$$

The condition for a photon trajectory to be allowed is that the photon has a positive energy, as measured by us locally. This means that we require that its energy obeys $E_{\text{obs}} = -\boldsymbol{p} \cdot \boldsymbol{u} > 0$. Let's consider a photon that, like us, is on a $\tilde{L} = 0$ radial plunge and has[5] $\tilde{E} = \pm p^r$. Its energy, measured by us, is then

$$E_{\text{obs}} = -\boldsymbol{p} \cdot \boldsymbol{u} = -g_{rr}p^r u^r = -\left(\frac{2M}{r} - 1\right)^{-\frac{1}{2}} p^r. \tag{28.2}$$

This is only positive if $p^r < 0$, which is always true for an in-falling photon. The key here is that this puts no constraint on $E = p^t$, the timelike component of the photon momentum in flat spacetime. This effectively frees the photon from the constraints of the uncertainty relation: it can propagate freely inside the event horizon. The other member of the particle-antiparticle pair, an outgoing photon, is also then free to propagate.

We conclude that virtual photons can propagate inside the event horizon, effectively removing the constraint that the particle-antiparticle pair has a limited lifespan.

In order to understand how this leads to outgoing radiation, and hence energy being transported out of the hole, we now consider the fate of a massive particle and antiparticle pair, and the energy measured by an observer at rest at infinity. Since this observer's velocity[6] is identical to the Killing vector $\boldsymbol{\xi}$, the observer measures the energy $E_{\text{obs}}^\infty = -\boldsymbol{p} \cdot \boldsymbol{\xi}$, which is a quantity conserved along the particle's geodesic. However, the status of the Killing vector $\boldsymbol{\xi} = \boldsymbol{e}_t$ changes from timelike outside the horizon to spacelike within it.[7] This allows the possibility that the quantity $-\boldsymbol{p} \cdot \boldsymbol{\xi}$ is *negative* inside the event horizon without it being unphysical.[8] Therefore, if a particle-antiparticle pair is produced close to the event horizon, to ensure the positivity of $-\boldsymbol{p} \cdot \boldsymbol{\xi}$ for the particle measured by the observer at infinity, we require that the particle with negative $-\boldsymbol{p} \cdot \boldsymbol{\xi}$ falls into the hole, while one with positive $-\boldsymbol{p} \cdot \boldsymbol{\xi}$ (of equal magnitude) travels out to infinity, where it is measured, locally, by the observer. This guarantees that the observer at infinity measures a positive energy particle and that, for the observer at infinity, energy has been conserved. This might seem to contradict the idea that we do not interpret any particles/antiparticles in Nature as carrying negative energy. However, it is really a symptom of the geometrical property of the quantity $-\boldsymbol{p}\cdot\boldsymbol{\xi}$, unique to black holes, rather than an admission of negative energy particles as measured locally. As we saw in the last example, measured locally, the energy of propagating particles and antiparticles is necessarily positive.

This curious state of affairs allows the outgoing particle produced in the pair to propagate outside the horizon and to carry its energy out to infinity. We compute how much energy in the next example.

Example 28.2

We, the observers, will watch a single vacuum fluctuation that creates a photon pair just outside the horizon. We work in a freely falling reference frame that is momentarily at rest at $r = 2M + \varepsilon$, that subsequently falls inwards following a trajectory with $\tilde{L} = 0$ and[9]

$$\tilde{E} = \left(1 - \frac{2M}{2M + \varepsilon}\right)^{\frac{1}{2}} \approx \left(\frac{\varepsilon}{2M}\right)^{\frac{1}{2}}. \tag{28.4}$$

The proper time for the frame to fall to the horizon is given (rearranging eqn 28.3) by

$$\Delta\tau = -\int_{2M+\varepsilon}^{2M} \frac{\mathrm{d}r}{\left(\frac{2M}{r} - \frac{2M}{2M+\varepsilon}\right)^{\frac{1}{2}}} \approx 2(2M\varepsilon)^{\frac{1}{2}}. \tag{28.5}$$

The photon must, at this stage of the process, exist subject to the Heisenberg uncertainty relation $\Delta E \Delta\tau \approx \hbar$, which means it has an energy, as measured in our local inertial frame, of

$$E_{\text{obs}} \approx \frac{\hbar}{2(2M\varepsilon)^{\frac{1}{2}}}. \tag{28.6}$$

Once the in-falling particle has crossed the horizon, the outgoing photon disappears to infinity taking its energy with it. Let's find this energy at infinity. The conserved quantity on the photon's trajectory is $\boldsymbol{p}\cdot\boldsymbol{\xi} = p_t$, which is also minus the energy E_{obs}^{∞}, measured by the observer at infinity. This can be linked with the energy measured by us locally in the frame just outside the horizon. We have that the energy in the local frame $E_{\text{obs}} = -\boldsymbol{p}\cdot\boldsymbol{u}$ with the only non-zero component of our instantaneous velocity[10] $-u_t = \tilde{E} \approx (\varepsilon/2M)^{\frac{1}{2}}$. We therefore have

$$E_{\text{obs}} = -g^{tt} p_t u_t = -g^{tt} E_{\text{obs}}^{\infty} \tilde{E}, \tag{28.7}$$

where g^{tt} is evaluated at $r = 2M + \varepsilon$. For a Schwarzschild black hole

$$g^{tt} = -\left[1 - \frac{2M}{2M + \varepsilon}\right]^{-1} = -\left(1 + \frac{2M}{\varepsilon}\right) \approx -\frac{2M}{\varepsilon}. \tag{28.8}$$

This gives us, plugging into eqn 28.7,

$$\frac{\hbar}{2(2M\varepsilon)^{\frac{1}{2}}} = \left(\frac{2M}{\varepsilon}\right) E_{\text{obs}}^{\infty} \left(\frac{\varepsilon}{2M}\right)^{\frac{1}{2}}, \tag{28.9}$$

or

$$E_{\text{obs}}^{\infty} = \frac{\hbar}{4M}. \tag{28.10}$$

In conclusion, the photon takes an amount of energy out to infinity which is inversely proportional to the mass of the black hole.

[9] Recall that

$$\frac{\tilde{E}^2 - 1}{2} = \frac{1}{2}(u^r)^2 + V_{\text{eff}}(r),$$

and so in this case, where $\tilde{L} = 0$, we have

$$\left(\frac{\mathrm{d}r}{\mathrm{d}\tau}\right)^2 = \tilde{E}^2 - 1 + \frac{2M}{r}. \tag{28.3}$$

If $u^r = 0$ at $r = 2M + \varepsilon$ we obtain the quoted result.

[10] Recall that

$$\tilde{E} = -\boldsymbol{\xi}\cdot\boldsymbol{u} = -u_t.$$

[11] The area A can be computed from the Schwarzschild metric (see exercises).

[12] We have assumed here that the black hole is not consuming energy from the outside Universe as it emits Hawking radiation. However, our Universe is aglow with the left-over radiation from the Big Bang. The cosmic microwave background (CMB, Chapter 15) with a black-body temperature of $T_{\mathrm{CMB}} = 2.73$ K, and this radiation continually feeds any black hole. (This is in addition to the effect of nearby dust or gas, which dominates the growth of black holes that have been observed in the Universe.) Black holes with temperatures $T < T_{\mathrm{CMB}}$ therefore cannot actually shrink due to Hawking radiation, even in principle, but must continue to grow, at least until that point in time when the CMB and other sources of cosmic radiation sufficiently cool down to temperatures below that of the black hole. This is compatible with the second law of thermodynamics, which tells us that heat flows from a hotter object to a colder one, and not the other way round. To get a sense of how small a black hole would have to be such that its temperature would exceed that of the present-day CMB, we can use eqn 28.12 to calculate the Hawking temperature of black hole with one solar mass, $M_\odot = 1.99 \times 10^{30}$ kg. This yields a temperature of just 62 nK $\ll T_{\mathrm{CMB}}$. In fact, using eqn 28.12 to calculate the upper bound on the black hole's mass such that its temperature would at least match that of the CMB, i.e. setting $T = T_{\mathrm{CMB}}$, we obtain $M_{\mathrm{CMB}} = 4.5 \times 10^{23}$ kg $= 2.3 \times 10^{-8} M_\odot$. Any black hole with a mass larger than M_{CMB} will thus continue to grow for quite some time yet, at least in the present-day universe.

[13] A **Killing horizon** is a sort of generalized geometrical upgrade of the idea of an event horizon. It is a null hypersurface defined by the vanishing of the norm of a Killing vector field $\boldsymbol{\xi}$. This bit of formalism allows a physical definition of the surface gravity κ of a Killing horizon: it is the acceleration, exerted at infinity, needed to keep an object at the Killing horizon. See Poisson for proof of this and justification of eqn 28.13. The logic here is to choose $\boldsymbol{\xi}$ to give a suitable Killing horizon on which to evaluate the surface gravity.

Although we have been dealing with a vastly simplified picture, the rigorous quantum field theory computation gives a result that is (perhaps surprisingly) fairly close to the estimate from the previous example. We examine the consequences of this in the next section.

28.2   Black-hole thermodynamics

The rigorous result is that the typical outgoing photon energy measured at infinity is actually $E_{\mathrm{obs}}^\infty = \hbar/8\pi M$, but that they can be detected with exactly the range of energies that one would expect for a thermodynamic black body. As a result, black holes are black bodies with a characteristic temperature

$$k_{\mathrm{B}} T \approx \frac{\hbar}{8\pi M}. \tag{28.11}$$

Since the hole is emitting radiation, it must be losing energy and consequently mass. This is consistent with the negativity of $-\boldsymbol{p} \cdot \boldsymbol{\xi}$ for the member of the particle-antiparticle pair that falls into the hole. Let's look into this.

Example 28.3

The temperature of a black hole is proportional to M^{-1}. For any black body, the rate of emission of energies varies as AT^4, where A is the area of the emitter, in this case the area of the black hole's event horizon.[11] For the Schwarzschild black hole, the area is $A = 16\pi M^2$. The luminosity therefore goes as M^{-2}, and is the result of a decrease in the mass of the hole, and so we can write

$$\frac{\mathrm{d}M}{\mathrm{d}t} \propto -\frac{1}{M^2}. \tag{28.12}$$

Solving this we see that the decrease in mass of the hole with time leads us to conclude that the lifetime τ of a black hole must vary as[12] $\tau \sim M^3$.

The last example demonstrates that black-hole radiation causes an isolated black hole to effectively evaporate, with its mass decreasing and temperature increasing until, presumably, it disappears. We return to the consequence of this later.

We have the result that, through Hawking's proposed mechanism, a black hole radiates such that it is a black body with a temperature T. In fact, this is only the start of a complete set of thermodynamic laws that govern the mechanics of black holes. These follows from the fact that the temperature T is equivalent to another property of the black hole: the **surface gravity** κ. This is defined geometrically as the quantity

$$\xi^\mu \xi^\nu{}_{;\mu} = \kappa \xi^\nu, \tag{28.13}$$

evaluated at the event horizon. Here $\boldsymbol{\xi}$ is a suitably chosen linear combination of Killing vectors.[13]

Example 28.4

For the Schwarzschild solution we have a Killing vector in Schwarzschild coordinates (t, r, θ, ϕ) of $\xi^\mu = (1, 0, 0, 0)$. Using the incoming Eddington–Finkelstein coordinates (V, r, θ, ϕ) examined in Chapter 26, we find that this quantity has components $\xi^\nu = (1, 0, 0, 0)$.[14] On carrying out the necessary computation using eqn 28.13, we obtain

$$-\frac{\partial}{\partial r}\left(-1 + \frac{2M}{r}\right) - \frac{M}{r^2} = \kappa, \tag{28.15}$$

giving, on the surface $r = 2M$, a surface gravity

$$\kappa = \frac{1}{4M}. \tag{28.16}$$

Since the temperature is given by $k_B T = \hbar/8\pi M$, we can relate the temperature and the surface gravity by

$$k_B T = \frac{\hbar \kappa}{2\pi}. \tag{28.17}$$

So surface gravity and temperature are proportional.

[14]Since we need to take a derivative, it turns out that we need to compute the components of the corresponding 1-form to obtain

$$\xi_\nu = \left(-1 + \frac{2M}{r}, 1, 0, 0\right). \tag{28.14}$$

This computation is discussed in the Exercises. We can see that this Killing field is null on the event horizon, giving us the Killing horizon on which we evaluate the surface gravity.

The last example shows that surface gravity and temperature are effectively equivalent for a black hole.

In classical thermodynamics,[15] the conjugate variable of temperature is entropy S. The property of the black hole analogous to the entropy is the area A of the event horizon. To see this we start with the first law of thermodynamics, which says that, in the absence of any work being done, we have $dU = TdS$. Taking the mass M of the hole as a measure of its internal energy, we have for a Schwarzschild black hole that the area $A = 16\pi M^2$ and so we can write

$$dM = \frac{dA}{32\pi M} = \frac{\hbar}{8\pi k_B M} d\left(\frac{A k_B}{4\hbar}\right). \tag{28.18}$$

[15]See Chapter 39 for further discussion of thermodynamics.

Since $T = \hbar/8\pi k_B M$ is the temperature, we can use the resemblance to $dU = TdS$ to link entropy S and area A via

$$S = \frac{A k_B}{4\hbar}. \tag{28.19}$$

We now generalize by stating the laws of black hole mechanics. For a black hole of mass M, angular momentum L, charge q, horizon area A and surface gravity κ, we have the four laws:[16]

0th law The surface gravity κ is constant over the horizon.
1st law For two stationary black holes that differ only by small variations in M, L and q, we have

$$dM = \frac{\kappa}{8\pi} dA + \Omega dL + \Phi dq, \tag{28.20}$$

where Ω and Φ are, respectively, the angular velocity and electric potential at the horizon.
2nd law The area A of the horizon never decreases, which is to say that $dA \geq 0$, as a function of cosmic time.
3rd law It is impossible to reduce the surface gravity to zero in a finite number of steps.

[16]The laws of thermodynamics are:

0th law The temperature is the same in interacting systems at equilibrium,
1st law $dU = TdS - pdV$,
2nd law $dS(t) \geq 0$,
3rd law It is impossible to reduce T to zero in a finite number of steps.

[17]We have the correspondences with conventional thermodynamics (assuming a sign convention for the work term $\mathrm{d}W = -p\mathrm{d}V$):

$$\begin{aligned}
M &\equiv U, \\
\kappa &\equiv T, \\
A &\equiv S, \\
\Omega \mathrm{d}L &\equiv -\mathrm{d}W, \\
\Phi \mathrm{d}q &\equiv -\mathrm{d}W.
\end{aligned} \qquad (28.21)$$

[18]For more on black hole thermodynamics, see the excellent review articles by Page (2005) and Carlip (2014) in the Further Reading in Appendix A.

[19]Hawking offered two quotations for inclusion on the cover of the spoof newspaper *The Framley Examiner*: 'This is absolute rubbish' and 'Laughed briefly'. The first was used.

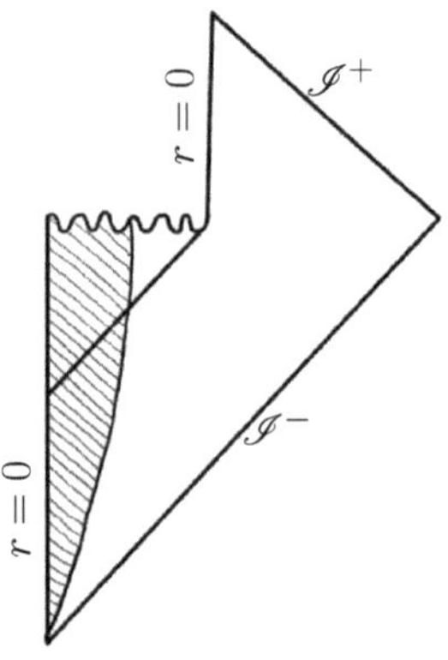

Fig. 28.2 Penrose diagram showing spacetime featuring a completely evaporating black hole.

These four laws bear an intended resemblance to the four laws of thermodynamics.[17] One quite dramatic consequence of the second law is that two colliding black holes must give rise to a hole that has an horizon area that is greater than or equal to the areas of the two holes we started with.

The second law demonstrates that when matter falls into a black hole, even though we lose the entropy of this matter (we no longer have access to its thermodynamic degrees of freedom), the increase in the black hole's mass M leads to an increase in area A and hence entropy S which means the entropy of the Universe has a net increase. We might wonder whether the second law of thermodynamics is still compatible with the decrease in mass, and hence area, that occurs when the black hole radiates; this process clearly violates $\mathrm{d}A \geq 0$, the formulation of the second law we have given above. For this reason, we need to invoke a **generalized second law** which applies to the physical situation more holistically than just considering only the black hole. This new law states that the total entropy of a black hole *and its environment* cannot decrease. Once the emitted radiation from a black hole has thermalized to the temperature of the external heat bath (one can perhaps assume the rest of the Universe acts as some kind of thermal reservoir) then the law implies that the entropy resulting from the radiation should at least compensate the decrease in entropy of the shrinking black hole. This result can be proved rigorously, albeit by making certain assumptions.[18]

We might also wonder about the status of **information**. Since objects cannot be seen after passing into the event horizon of a black hole, the information about them seems to be lost. As the black hole evaporates, it seems that this loss might be permanent. This is a worry since, in quantum mechanics, wavefunctions evolve unitarily so that the state of the system at a later time can always be deduced from that at an earlier time, and vice versa. Irreversible destruction of information about a physical system appears to break this property. This has been deemed so potentially disturbing that it has led some to propose that the end point of an evaporating black hole might not be its disappearance. There have been proposals that the information is permanently stored in a nugget, left over at the end of the evaporation, or that the death of a black hole is accompanied by an explosion that returns the information to the Universe. Hawking's own opinion was that the loss was permanent.[19]

Example 28.5

As the black hole radiates its mass reduces and the hole effectively evaporates. Whether this can continue until the black hole ceases to exist is another open question. This is because, as the black hole becomes small, a classical theory such as general relativity should not be expected to describe the physics and so we expect to need to employ quantum gravity (Chapter 49), which does not yet exist as a complete theory. Nevertheless, if we accept that the black hole will vanish, spacetime will be described by the Penrose diagram shown in Fig. 28.2.

Before moving on, it's worth pausing to recall from Chapter 26 that Schwarzschild coordinates can be though of as being equivalent to accelerating Minkowski coordinates. This suggests that some of the subtle behaviour that occurs near a black hole might have an analogy in a system of accelerating particles. This is the basis of the **Unruh effect**[20] which predicts that owing to quantum fluctuations in the vacuum, an observer accelerating through empty space with no other heat sources present will still observe black body radiation (i.e. the observer will think they are moving through a thermal bath), whereas an unaccelerated (inertial) observer will not observe any temperature. The **Unruh temperature** of this perceived blackbody radiation is given by $\hbar a/(2\pi c k_{\mathrm{B}})$, where a is the acceleration of the observer.

[20] W. G. 'Bill' Unruh (1945–). Speaking about his effect, Unruh quipped that it shows that you could cook your steak just by accelerating it. He also admitted that an acceleration of 10^{24} cm s^{-1} would be required to achieve a temperature of 300 °C, which one could also obtain from a good charcoal grill.

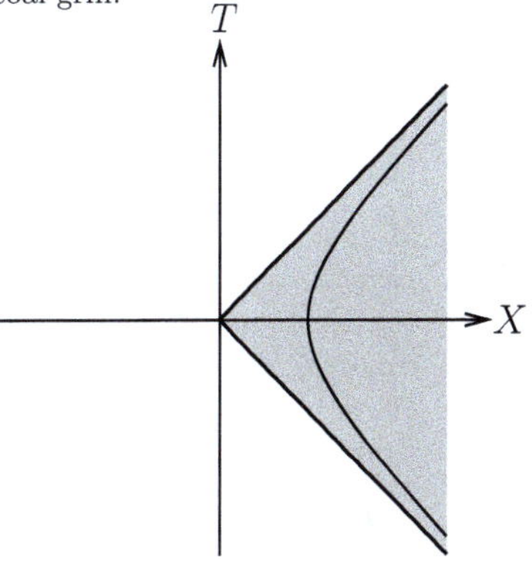

Fig. 28.3 An accelerated observer follows a hyperbolic path in Minkowski spacetime. At elevated values of acceleration they are close to a horizon that results from Rindler coordinates only covering a wedge of Minkowski spacetime.

Example 28.6

Close to the event horizon of a Schwarzschild black hole, the spacetime can be expressed in terms of Rindler coordinates which has line element

$$\mathrm{d}s^2 = -x^2\mathrm{d}t^2 + \mathrm{d}x^2. \qquad (28.22)$$

Accelerated observers in Minkowski spacetime[21] have world lines that coincide with lines of constant x in the Rindler spacetime. As discussed in Chapter 26, and shown in Fig. 28.3, Rindler coordinates only cover that part of Minkowski spacetime with $X > |T|$. Nothing at $X < |T|$ can ever catch up with the accelerated observer, so the line $T = \pm X$ is an effective event horizon for the accelerated observer. With this setup we can rerun the argument for Hawking radiation. A particle-antiparticle pair is created at the event horizon $T = X$. One of the pair passes through the horizon into the region $X < T$ where it is lost to the observer, leaving the other free to escape into the allowed region $X > |T|$. This free particle is then responsible for the Unruh radiation. The temperature of the radiation for an observer with acceleration a can be estimated via dimensional analysis to be $k_{\mathrm{B}}T \sim \hbar a/c$.

[21] As in our previous discussions of Rindler spacetime, we describe Minkowski spacetime with coordinates (T, X) and line element $\mathrm{d}s^2 = -\mathrm{d}T^2 + \mathrm{d}X^2$.

Chapter summary

- Black holes radiate Hawking radiation through a process where a particle-antiparticle pair is created at the event horizon. Energy is transported out to infinity.

- Black holes are thermodynamic black bodies with temperatures $T \propto M^{-1}$, where M is the mass of the hole.

- A set of laws of black hole mechanics can be formulated that closely resemble the laws of thermodynamics. Surface gravity is analogous to temperature and the area of the event horizon is analogous to entropy.

Exercises

(28.1) Compute the area of the event horizon of a Schwarzschild black hole.

(28.2) Show that the heat capacity C of a Schwarzschild black hole is given by

$$C = T\frac{\mathrm{d}S}{\mathrm{d}T} = -\frac{\hbar^2}{16\pi k_{\mathrm{B}}T^2}. \qquad (28.23)$$

Explain why the minus sign makes physical sense.

(28.3) *We shall carry out the computation sketched in Example 28.4*
(a) Using incoming Eddington–Finkelstein coordinates, show that the vector with components $\xi^\mu = (1,0,0,0)$ corresponds to the 1-form with components

$$\xi_\mu = \left(-1 + \frac{2M}{r}, 1, 0, 0\right). \qquad (28.24)$$

(b) All Killing vectors obey Killing's equation $\xi_{\mu;\nu} + \xi_{\nu;\mu} = 0$. Use this to show that an alternative expression for the surface gravity is

$$-\xi^\mu \xi_\mu{}^{;\sigma} = \kappa\xi^\sigma. \qquad (28.25)$$

(c) Hence, show that the differential equation we need to solve is

$$\xi_{V;r} = -\kappa. \qquad (28.26)$$

(d) The necessary connection coefficients are $\Gamma^V{}_{rV} = 0$ and $\Gamma^r{}_{rV} = -M/r^2$. Use these and the rule

$$X_{\alpha;\mu} = X_{\alpha,\mu} - \Gamma^\sigma{}_{\alpha\mu}X_\sigma, \qquad (28.27)$$

to obtain eqn 28.15.

(28.4) (a) Show that near the event horizon, the Schwarzschild metric takes the form

$$\mathrm{d}s^2 = -\frac{(r-2M)}{2M}\mathrm{d}t^2 + \frac{2M}{r-2M}\mathrm{d}r^2 + 4M^2\mathrm{d}\Omega^2, \qquad (28.28)$$

with $\mathrm{d}\Omega^2 = \mathrm{d}\theta + \sin^2\theta\mathrm{d}\phi^2$.
(b) By changing variables to $\rho^2 = 8M(r-2M)$, show that we obtain a line element

$$\mathrm{d}s^2 = -\frac{\rho^2}{16M^2}\mathrm{d}t^2 + \mathrm{d}\rho^2 + 4M^2\mathrm{d}\Omega^2. \qquad (28.29)$$

Readers with a background in quantum field theory will be aware that to compute thermodynamic effects in a field theory we must transform from evolution in time to evolution in Euclidean time, by making a Wick rotation. This takes the time evolution from coordinate time to a periodic imaginary time t_{E} which in turn corresponds to a temperature T via $\mathrm{i}t_{\mathrm{E}} = \hbar/k_{\mathrm{B}}T$. See the book by Zee for further discussion.
(c) Carry out the Wick rotation by setting $t = -\mathrm{i}t_{\mathrm{E}}$ and make a change of variable $t_{\mathrm{E}} = 4M\psi$ to obtain the line element

$$\mathrm{d}s^2 \approx \mathrm{d}\rho^2 + \rho^2\mathrm{d}\psi^2 + 4M^2\mathrm{d}\Omega^2. \qquad (28.30)$$

For fixed Ω this is simply flat space in cylindrical coordinates (ρ, ψ). The final term puts a 2-sphere of radius $2M$ at every point in spacetime.
(d) Interpreting ψ as an angle, identify the period of t_{E} and determine the Hawking temperature near the horizon of the black hole.

Charged and rotating black holes

29

The finite is annihilated in the presence of the infinite, and becomes a pure nothing. So our spirit before God, so our justice before divine justice.
Blaise Pascal (1623–1662)

The Schwarzschild black hole is only the simplest of a more general class of objects featuring in solutions to the Einstein equation describing the spacetime around a distribution of mass. In this chapter, we examine the consequence of (i) the mass carrying electrical charge and (ii) the mass rotating. We shall assume that gravitational collapse of the mass distribution exposes the region of space close to the origin, so that a black hole is created. In both cases, we shall encounter singularities along with a rich seam of new physics.

Owing to the presence of an event horizon, a black hole is necessarily hidden from the view of external observers. The only properties of the black hole that we observers should expect to be able to access are the mass, charge and angular momentum. Further, owing to the singularity, it is likely that these three properties alone describe the metric field near the hole. As a result, the properties of the black-hole spacetime are remarkably independent of the history of the spacetime, such as the matter content that has fallen into the hole. This state of affairs was described by the slogan 'black holes have no hair' by John Wheeler[1] and consequently elevated to a mathematical **no-hair theorem**, using hair as a metaphor for all other information in the spacetime. As a consequence of this, we expect our discussion to be widely applicable to black holes in the Universe, however they have arisen.

29.1 Charged black holes

In most astrophysical contexts, it is unlikely that an electrically charged black hole could be stabilized, since we would expect that it would attract matter with the opposite electrical charge and eventually become charge neutral. As we shall see, the consequence of allowing a black hole to have a charge is rather interesting.[2]

Suspending our disbelief, we therefore consider an electrically charged distribution of mass M with total charge q. The presence of the charge means that the solution to the Einstein equation must allow for the right-

[1]Wheeler reported that his student Jacob Bekenstein (1947–2015) had invented the phrase. The justification of the theorem was due to Stephen Hawking, Brandon Carter and Werner Israel (1931–2022).

[2]The physics is described by the Reissner–Nordström line element which we will now introduce. Hans Reissner (1874–1967), an aeronautical engineer, first discovered this solution. It was also found by: (i) Gunnar Nordström (1881–1923), described as the 'Finnish Einstein', who ultimately developed an alternative theory of gravitation to general relativity; (ii) Hermann Weyl (1885–1955), the genius who made significant advances across large numbers of topics in mathematical physics including general relativity; and (iii) George Barker Jeffery (1891–1957), who was famed in Britain as a translator of the original papers on relativity into the English language.

[3]The possibility of this configuration realizing a black hole might lead one to wonder if the metric describes the spacetime close to an electron, which, after all, is an infinitely dense, charged object. This turns out not to be the case since the electronic *spin* is not taken into account in our description.

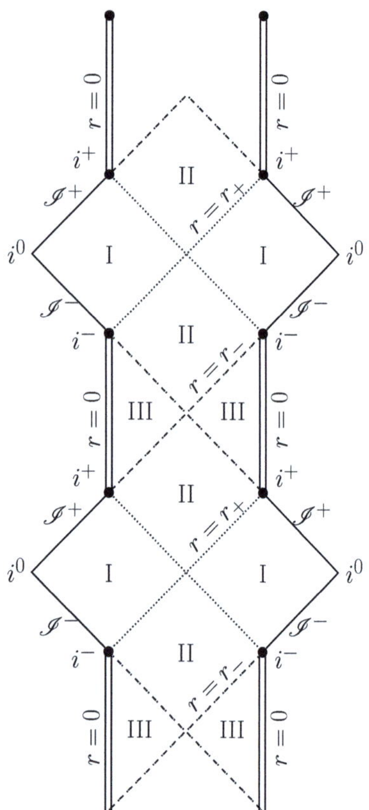

Fig. 29.1   Penrose diagram for a charged black hole. Dotted lines are $r = r_+$; dashed lines are $r = r_-$; doubled lines are $r = 0$.

[4]Recall that the singularity at $t = 0$ in the Robertson–Walker Penrose diagram is spacelike, as is the $r = 0$ singularity in the Schwarzschild black hole, owing to the swap between the roles of space and time inside the event horizon. This is the first timelike singularity that we have encountered. It is depicted by a double vertical line in the Penrose diagram.

hand side to feature electromagnetic energy density. This results in an upgraded metric field that, in the limit $q \to 0$, gives the Schwarzschild line element.[3]

The **Reissner–Nordström line element** describes the metric field outside of the electrically charged, spherically symmetric mass distribution. It is given by

$$\mathrm{d}s^2 = -\left(1 - \frac{2M}{r} + \frac{q^2}{r^2}\right)\mathrm{d}t^2 + \left(1 - \frac{2M}{r} + \frac{q^2}{r^2}\right)^{-1}\mathrm{d}r^2$$
$$+ r^2(\mathrm{d}\theta^2 + \sin^2\theta\,\mathrm{d}\phi^2), \tag{29.1}$$

If $q^2 > M^2$ the metric is non-singular everywhere, except for $r = 0$. If $q^2 \le M^2$ there are coordinate singularities at radii r_+ and r_-, where $r_\pm = M \pm (M^2 - q^2)^{\frac{1}{2}}$.

Using these singularities, we identify three regions of the spacetime. Region I takes the large values of the radial coordinate: $r_+ < r < \infty$; region II has $r_- < r < r_+$; and region III is the inner region with $0 < r < r_-$. The first and third regions are static; region II is not. (Note that if $q^2 = M^2$ then region II does not exist.)

Maximally extended Penrose diagrams for the charged black hole are shown in Figs. 29.1 and 29.2, where regions I, II and III are indicated. Region I is asymptotically flat and therefore features null, spacelike and timelike infinities. There are an infinite number of these regions I in the Penrose diagram, all connected by regions II and III (we shall see why shortly). In region I, surfaces of constant t are spacelike and surfaces of constant r are timelike, in the usual manner. In region II, surfaces of constant r are spacelike and surfaces of constant t timelike, just as we have in the region II of a Schwarzschild black hole. Each point in region II therefore represents a closed trapped surface.

In region III, all lines of constant r are timelike and lines of constant t are spacelike, just as in region I (Fig. 29.2). Region III has a physical singularity at $r = 0$. but, owing to the different roles of space and time compared to the Schwarzschild case, it is not spacelike but timelike.[4] The singularity in region III is not only timelike but is also *repulsive*, since no timelike geodesic intersects it and so freely falling massive particles will not collide with it. The singularity can then be avoided by a timelike curve that crosses the surface r_+ from region I into region II and then into region III. A timelike curve can then pass into another region II and into a further (asymptotically flat) region I. The traveller following this curve will have tunnelled into a new Universe through the wormhole caused by the presence of the charge! This is a one-way process, so the traveller will not return, since there is no future-directed timelike curve that allows the traveller back to the original region-I universe.

What would observers see in this spacetime? As in the Schwarzschild case, an observer close to infinity would see an astronaut infinitely redshifted as she approaches the surface r_+. From the geometry in Fig. 29.1, an observer crossing surface r_- into region III would potentially be able to see the whole history of a region I in a finite time. Light from objects in region I would appear infinitely blueshifted as the astronaut approached i^+.

29.2 Kerr black holes

We have done a lot using the Schwarzschild solution, but now there's some bad news: most astronomical objects rotate and so are not described by the Schwarzschild geometry, which assumes spherical symmetry (which is a symmetry that is broken by rotation). The **Kerr geometry**[5] is the only known exact solution to the Einstein equation that is able to represent a stationary, axially symmetric, asymptotically flat spacetime outside of a rotating distribution of mass.

Example 29.1

So far, most of our geometries have been expressible in terms of a diagonal metric.[6] The (3+1)-dimensional metric describing the rotating Kerr geometry is stationary (i.e. none of the metric components depend on t) and *axisymmetric*, rather than spherically symmetric, such that we expect that there is no dependence of the components on ϕ. The most general form of such a stationary, axisymmetric line element is often written in terms of an angular velocity ω as

$$\mathrm{d}s^2 = -\mathrm{e}^{2\nu}\mathrm{d}t^2 + \mathrm{e}^{2\psi}\left(\mathrm{d}\phi - \omega\mathrm{d}t\right)^2 + \mathrm{e}^{2\mu_2}(\mathrm{d}x^2)^2 + \mathrm{e}^{2\mu_3}(\mathrm{d}x^3)^2, \tag{29.2}$$

where ν, ψ, ω, μ_2, and μ_3 are functions of variables x^2 and x^3. Notice that the second term implies that the resulting metric will not be diagonal, but will include a cross term proportional to $\omega\mathrm{d}\phi\mathrm{d}t$. We shall see later that the cross term leads to lots of the notable properties of this geometry.

The Kerr geometry is usually written in yet another new coordinate system: the so-called **Boyer–Lindquist coordinates**[7] (t, r, θ, ϕ). Although these seem to resemble ordinary spherical polars, they are slightly unusual. In particular, although t and ϕ are the same as their counterparts in standard spherical polars; θ and r are not. These latter two coordinates are related to Cartesian coordinates via

$$x = \left(r^2 + a^2\right)^{\frac{1}{2}}\sin\theta\cos\phi,$$
$$y = \left(r^2 + a^2\right)^{\frac{1}{2}}\sin\theta\sin\phi. \tag{29.3}$$

Notice that these coordinates are a little like spherical polars with the radius adjusted. In terms of these coordinates, a surface of constant r is no longer a sphere, but rather is an ellipsoid. (Setting $r = 0$ does not, therefore, result in a point, so the physical singularity for a Kerr black hole is more complex than in the cases examined so far.)

In Boyer–Lindquist coordinates, we can express the metric field for the Kerr geometry as follows:

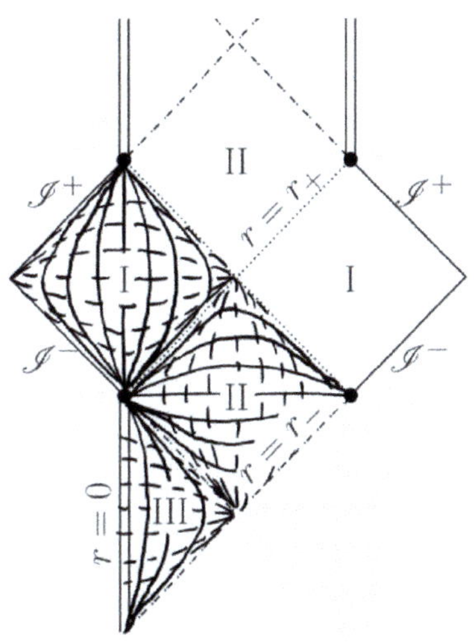

Fig. 29.2 Penrose diagram for the charged hole, showing surfaces of constant r (solid lines) and surfaces of constant t (dashed lines).

[5]Roy Kerr (1934–) is, in addition to his mathematical achievements, noted as a bridge player, having represented New Zealand in the 1970s. The generalization of the metric for a charged, rotating mass is called the Kerr–Newman geometry, named also after Ezra 'Ted' Newman (1929–2021).

[6]In fact, the Cotton–Darboux theorem guarantees that the metric $\boldsymbol{g} = g^{\mu\nu}\boldsymbol{e}_\mu \otimes \boldsymbol{e}_\nu$, with $\mu, \nu = 0, 1, 2$ in a three-dimensional space, can *always* be brought into diagonal form by a local coordinate transformation. See Chandrasekhar for the proof.

[7]Robert H. Boyer (1932–1966) and Richard W. Lindquist. Their paper on these coordinates was published in 1967, posthumously for Boyer, who was tragically killed in the infamous University of Texas tower shooting.

The Kerr metric has components

$$g_{tt} = -\left(1 - \frac{2Mr}{\rho^2}\right)$$

$$= -\frac{\Delta - a^2 \sin^2\theta}{\rho^2},$$

$$g_{t\phi} = -\frac{2Mar\sin^2\theta}{\rho^2},$$

$$g_{\phi\phi} = \frac{(r^2 + a^2)^2 - a^2\Delta\sin^2\theta}{\rho^2}\sin^2\theta,$$

$$g_{rr} = \frac{\rho^2}{\Delta},$$

$$g_{\theta\theta} = \rho^2. \tag{29.5}$$

We also have Killing vectors $\boldsymbol{\xi}$ and $\boldsymbol{\eta}$ with components

$$\xi^\mu = (1, 0, 0, 0),$$

$$\eta^\mu = (0, 0, 0, 1).$$

We therefore have conserved quantities

$$\tilde{E} = -\boldsymbol{\xi}\cdot\boldsymbol{u} = -u_t,$$

and

$$\tilde{L} = \boldsymbol{\eta}\cdot\boldsymbol{u} = u_\phi.$$

Note that, in the spirit of eqn 29.2, the line element can be rewritten as

$$ds^2 = -\frac{\rho^2\Delta}{\Sigma^2}dt^2 + \frac{\Sigma^2\sin^2\theta}{\rho^2}(d\phi - \omega dt)^2$$

$$+ \frac{\rho^2}{\Delta^2}dr^2 + \rho^2 d\theta^2, \tag{29.6}$$

where $\Sigma^2 = (r^2 + a^2)^2 - a^2\Delta\sin^2\theta$ and $\omega = 2Mar/\Sigma^2$.

$$ds^2 = -\left(1 - \frac{2Mr}{\rho^2}\right)dt^2 - \frac{4Mar\sin^2\theta}{\rho^2}dtd\phi$$

$$+ \frac{(r^2 + a^2)^2 - a^2\Delta\sin^2\theta}{\rho^2}\sin^2\theta d\phi^2$$

$$+ \frac{\rho^2}{\Delta}dr^2 + \rho^2 d\theta^2, \tag{29.4}$$

where $\Delta = r^2 - 2Mr + a^2$, and $\rho^2 = r^2 + a^2\cos^2\theta$.

Here M is the mass of the distribution and $J = Ma$ is its angular momentum as measured by an observer at infinity. The parameter a, which has units of length, is known as the **Kerr parameter**. It is a measure of the rotation.

The Kerr geometry is invariant under the simultaneous transformation $t \to -t$ and $\phi \to -\phi$, just as we should expect for any axially rotating object (i.e. if we reverse the direction of time, the mass rotates in the opposite direction). In the limit of large r we recover Minkowski space, confirming that the line element is asymptotically flat. When $a \to 0$, we recover the Schwarzschild geometry. By inspection of the Killing vectors, we have, just as for the Schwarzschild geometry, conservation of u_t and u_ϕ (although the presence of cross terms in the metric complicates the computation of u^t and u^ϕ).

The anatomy of this spacetime is very rich in features and is shown in cartoon form in Fig. 29.3. There are two sets of singularities in this geometry: (i) $\rho = 0$ and (ii) $\Delta = 0$. The former is a class of physical singularity, while the latter are coordinate singularities. Substituting into eqn 29.3, we find that $\rho = 0$ corresponds to a **ring** of Cartesian coordinates with $x^2 + y^2 = a^2$ and $z = 0$ (or, equivalently, $\theta = \pi/2$). These are physical singularities at the heart of the Kerr black hole where the curvature of spacetime is infinite.

When the parameters satisfy $0 \le a^2 \le M^2$ (the case we will examine exclusively in this section) we have a situation where three regions of interest are separated by singular surfaces $r_\pm$ (Fig. 29.3), not unlike the Reissner–Nordström black hole. Here the coordinate singularities occur where Δ vanishes at values of r given by

$$r_\pm = M \pm (M^2 - a^2)^{\frac{1}{2}}. \tag{29.7}$$

These equations describe spherical surfaces, both of which behave as event horizons. In Fig. 29.3, they are labelled as the inner (r_-) and outer (r_+) event horizons. Note that when $a = 0$ we have that the outer event horizon has $r_+ = 2M$, while the inner event horizon r_- vanishes, and so we recover the Schwarzschild case. The presence of the square root in eqn 29.7 shows that the event horizons only exists for $a \le M$, and so the hole has a maximum angular momentum $J = M^2$. Black holes with this maximum angular momentum are known as **extremal Kerr black holes**.

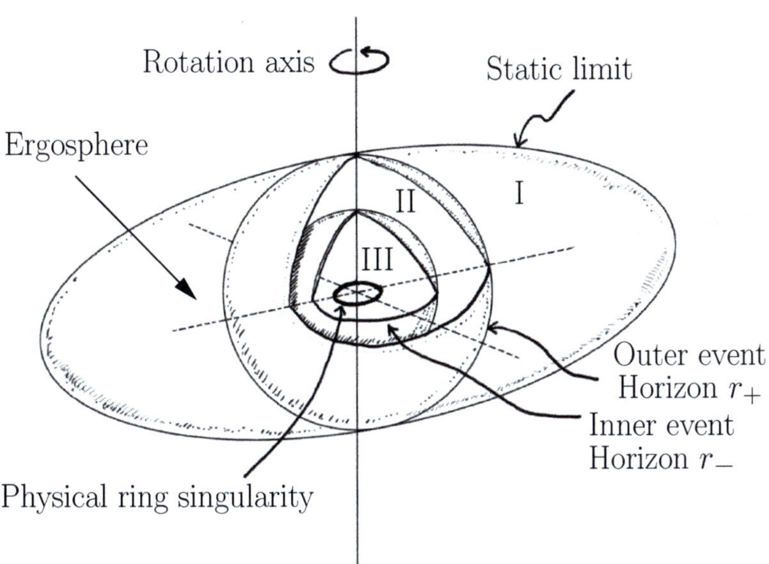

Fig. 29.3 The singular surfaces in the Kerr geometry.

Example 29.2

To compute the area of the r_+ horizon in the Kerr black hole, we set $dt = dr = 0$ and substitute $r = r_+$ and $\rho(r_+)$ to obtain

$$
ds^2 = \rho_+^2\, d\theta^2 + \left(r_+^2 + a^2 + \frac{2Mr_+ a^2 \sin^2 \theta}{\rho_+^2} \right) \sin^2 \theta\, d\phi^2
$$

$$
= \rho_+^2\, d\theta^2 + \left(\frac{2Mr_+}{\rho_+} \right)^2 \sin^2 \theta\, d\phi^2, \tag{29.8}
$$

where $\rho_+^2 = r_+^2 + a^2 \cos^2 \theta$ and r_+ is defined by $r_+^2 + a^2 = 2Mr_+$. The area of the Kerr horizon is then

$$
A = \int_0^{2\pi} d\phi \int_0^{\pi} d\theta\, g = 8\pi M r_+, \tag{29.9}
$$

where g is the determinant of the matrix of metric components. The integral is examined further in the exercises.

The Penrose diagram for the Kerr black hole is shown in Fig. 29.4. In this case, the maximal extension of the spacetime allows us to avoid the ring of physical singularities, since they do not occur at $r = 0$. As a result, we can extend spacetime into a region of negative r, in order to show the surfaces on which null lines (i.e. light rays) terminate. As in the Reissner–Nordström case, we can characterize the three regions: an asymptotically flat region I, region II containing closed trapped surfaces and region III containing the singularities. The Penrose diagram shows that a traveller can fall from region I, through the event horizon at r_+, to find themselves in region II, where the closed, trapped surfaces occur. The traveller is forced into region III which contains the singularities.

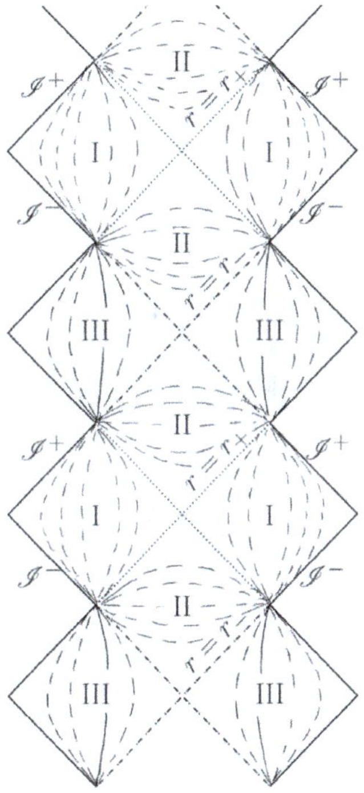

Fig. 29.4 Penrose diagram for the Kerr geometry, showing lines of constant r (dashed). The solid curved lines (in region III) show $r = 0$, the dotted lines shows $r = r_+$ and the dot-dashed line is $r = r_-$.

The singularities are timelike, and so can be avoided by timelike cures, just as we saw earlier. Also notable in region III is the existence of closed, timelike curves which violate causality. As in the Reissner–Nordström case, it is possible to tunnel through a wormhole to a new region II and then to a new region I.

To understand these three regions of the Kerr geometry in more detail, we shall drop some test particles into the spacetime and make use of the conserved quantities in this geometry. As mentioned above, the conserved quantities are

$$-\tilde{E} = u_t = g_{tt}u^t + g_{t\phi}u^\phi,$$
$$\tilde{L} = u_\phi = g_{\phi t}u^t + g_{\phi\phi}u^\phi. \tag{29.10}$$

Example 29.3

Consider a particle dropped into the geometry with no angular momentum, so that $u_\phi = 0$ initially. We will compute $d\phi/dt$. We have

$$\frac{d\phi}{dt} = \frac{d\phi}{d\tau}\frac{d\tau}{dt} = \frac{u^\phi}{u^t}. \tag{29.11}$$

Recall that the cross terms in the metric give us[8]

$$u^\phi = g^{\phi\phi}u_\phi + g^{\phi t}u_t,$$
$$u^t = g^{tt}u_t + g^{t\phi}u_\phi, \tag{29.14}$$

but since $u_\phi = 0$ we have

$$\frac{d\phi}{dt} = \frac{g^{\phi t}}{g^{tt}} \neq 0. \tag{29.15}$$

The particle therefore picks up some non-zero coordinate velocity in the ϕ-direction.

[8]Some algebra (see exercises) gives

$$g^{tt} = -\frac{(r^2 + a^2)^2 - a^2\Delta\sin^2\theta}{\rho^2\Delta},$$
$$g^{t\phi} = g^{\phi t} = -\frac{2Mra}{\rho^2\Delta},$$
$$g^{\phi\phi} = \frac{\Delta - a^2\sin^2\theta}{\rho^2\Delta\sin^2\theta}, \tag{29.12}$$

and therefore we find

$$\frac{g^{\phi t}}{g^{tt}} = \frac{2Mra}{(r^2 + a^2)^2 - a^2\Delta\sin^2\theta}. \tag{29.13}$$

From the previous example, we have the remarkable result that a particle dropped in to the Kerr geometry with no angular momentum gains some by virtue of the gravitational effect of the mass. This effect is often called **inertial frame dragging**, reflecting the fact that the system is dragged by the geometry in the direction of the mass's rotation. We see that the effect occurs because of the presence of a cross term $g^{\phi t}$ in the metric.

While we are considering massive particles, we can compute the effective energy equation governing the possible orbits.

Example 29.4

Consider the paths of massive particles in the Kerr metric confined to the $\theta = \pi/2$ plane. We obtain coordinate velocity components

$$u^t = \frac{1}{\Delta}\left[\left(r^2 + a^2 + \frac{2Ma^2}{r}\right)\tilde{E} - \frac{2Ma}{r}\tilde{L}\right],$$
$$u^\phi = \frac{1}{\Delta}\left[\left(1 - \frac{2M}{r}\right)\tilde{L} + \frac{2Ma}{r}\tilde{E}\right]. \tag{29.16}$$

Since we also have $u^\theta = 0$ and $\mathbf{u}^2 = -1$ we can solve for u^r to yield up a result similar in form to the one we obtained in the Schwarzschild geometry. This is

$$\frac{\tilde{E}^2 - 1}{2} = \frac{1}{2}\left(\frac{dr}{d\tau}\right)^2 + V_{\text{eff}}(r, \tilde{E}, \tilde{L}), \tag{29.17}$$

where in the Kerr geometry, we have an effective potential[9]

$$V_{\text{eff}}(r, \tilde{E}, \tilde{L}) = -\frac{M}{r} + \frac{\tilde{L}^2 - a^2(\tilde{E}^2 - 1)}{2r^2} - \frac{M(\tilde{L} - a\tilde{E})^2}{r^3}.$$ (29.18)

Next, we consider dropping a photon into the equatorial plane of the geometry, defined by $\theta = \pi/2$.

Example 29.5

Consider a photon initially travelling in the $\pm\phi$ direction, so that only dt and dϕ are non-zero. For photons, we can compute the motion using the null condition

$$ds^2 = g_{tt}dt^2 + 2g_{t\phi}dtd\phi + g_{\phi\phi}d\phi^2 = 0,$$ (29.19)

leading to an expression for the coordinate velocity in the ϕ direction of

$$\frac{d\phi}{dt} = -\frac{g_{t\phi}}{g_{\phi\phi}} \pm \left[\left(\frac{g_{t\phi}}{g_{\phi\phi}}\right)^2 - \frac{g_{tt}}{g_{\phi\phi}}\right]^{\frac{1}{2}}.$$ (29.20)

In the special case of the above example that[10] $g_{tt} = 0$, we have two solutions

$$\frac{d\phi}{dt} = -\frac{2g_{t\phi}}{g_{\phi\phi}}, \quad \frac{d\phi}{dt} = 0.$$ (29.21)

The first solution applies for photons sent in the same direction that the hole is spinning. The second applies for the photons sent against the direction of the hole's rotation. This last result is truly remarkable: in a region where $g_{tt} = 0$, a photon sent in opposite direction to that of the rotation of the mass initially cannot propagate. Since this applies for photons, which provide a universal speed limit via the light-cone structure, it applies for all particles, even those with large values of angular momentum. The effect can be further illustrated by considering a stationary observer, who has velocity with components $u^\mu = (u^t, 0, 0, 0)$. Since $\boldsymbol{u}^2 = -1$ we must have $g_{tt}(u^t)^2 = -1$, which can't be satisfied if g_{tt} vanishes, and so the observer cannot remain stationary and must start to rotate in the direction of the hole.

The surface on which $g_{tt} = 0$ lies outside the surface r_+ that defines the black hole's event horizon. It is called the **stationary limit**, since inside this surface it is impossible for particles to remain at rest. Setting $g_{tt} = 0$, we see that we obtain an ellipsoidal surface

$$r_0 = M + \left(M^2 - a^2\cos^2\theta\right)^{\frac{1}{2}}.$$ (29.22)

Notice how the ellipsoid coincides with r_+ at the poles. The region $r_+ \leq r \leq r_0$ is known as the **ergosphere**.[11] It is part of region I and is shown in Figs. 29.3 and 29.5.

More insight into this spacetime can be found by further examining the behaviour of light in each of the regions.

(a)

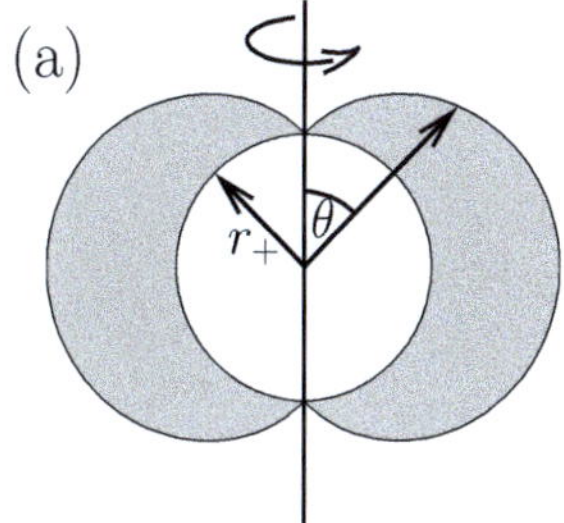

(b)

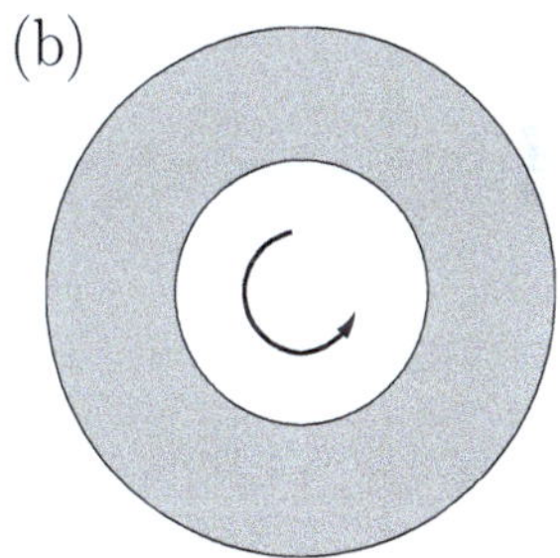

Fig. 29.5 The ergosphere for a near-extremal Kerr black hole, shown (a) from the side (i.e. observer at $\theta = \pi/2$, looking towards the origin); and (b) from the top (observer at $\theta = 0$).

[11]Note that $g_{tt} < 0$ outside the ergosphere, but $g_{tt} > 0$ inside. This is the same as the state of affairs for the Schwarzschild metric either side of the Schwarzschild radius. We shall use this fact at the end of the chapter.

Example 29.6

The effective energy equation for photons orbiting in the plane $\theta = \pi/2$ looks like

$$\frac{1}{\tilde{L}^2}\left(\frac{\mathrm{d}r}{\mathrm{d}\lambda}\right)^2 = \frac{1}{b^2} - W_{\text{eff}}(r,b,\sigma), \tag{29.23}$$

where, $\sigma = \text{sign}(\tilde{L})$ and[12]

$$W_{\text{eff}}(r,b,\sigma) = \frac{1}{r^2}\left[1 - \left(\frac{a}{b}\right)^2 - \frac{2M}{r}\left(1 - \sigma\frac{a}{b}\right)^2\right]. \tag{29.24}$$

The slightly peculiar factor of σ arises here telling us that the potential depends on the sign of the angular momentum of the photon. This is because the potential depends on whether the photon's angular momentum is in the direction of orbit of the black hole, or in the opposite direction. This is related to the frame-dragging that occurs for the Kerr black hole.

Finally, the light-cone structure of the black hole is shown in Fig. 29.6, where each of the regions can be identified and where the photon trajectories depend on the sense of rotation.

29.3 Interacting with the Kerr geometry

Roger Penrose suggested a method for extracting energy from a Kerr black hole. A particle starts at infinity and falls into the ergosphere of a black hole, where it decays into two particles. One of the particles falls through the event horizon and will eventually be consumed by the singularities; the other particle escapes. If the escaping particle carries away more energy to infinity than the original particle had, then we have removed energy from the black hole.

Locally, energy-momentum must be conserved, so we have that the incoming particle's 4-momentum $\boldsymbol{p}_{\text{in}}$ can be written as

$$\boldsymbol{p}_{\text{in}} = \boldsymbol{p}_{\text{out}} + \boldsymbol{p}_{\text{bh}}, \tag{29.25}$$

where the outgoing particle has 4-momentum $\boldsymbol{p}_{\text{out}}$ and the particle that falls into the hole has momentum $\boldsymbol{p}_{\text{bh}}$.

As discussed in the last chapter, a particle's conserved energy, as measured by an observer at infinity, can be obtained from minus the dot product of the timelike Killing vector $\boldsymbol{\xi}$ and its momentum. As a result of a dot product of the last equation with $\boldsymbol{\xi}$ we find

$$E_{\text{out}}^{\infty} = E_{\text{in}}^{\infty} - E_{\text{bh}}^{\infty}. \tag{29.26}$$

So far so good: this is simply a form of conservation of energy which, under normal circumstances, implies that $E_{\text{out}}^{\infty} \leq E_{\text{in}}^{\infty}$ for positive $E_{\text{bh}}^{\infty} = -\boldsymbol{p}_{\text{bh}} \cdot \boldsymbol{\xi}$. The unusual feature comes from considering the special situation inside the ergosphere where the Killing vector $\boldsymbol{\xi}$ becomes

$$\boldsymbol{\xi} \cdot \boldsymbol{\xi} = \boldsymbol{e}_t \cdot \boldsymbol{e}_t = g_{tt} = -\left(1 - \frac{2Mr}{\rho^2}\right) > 0, \tag{29.27}$$

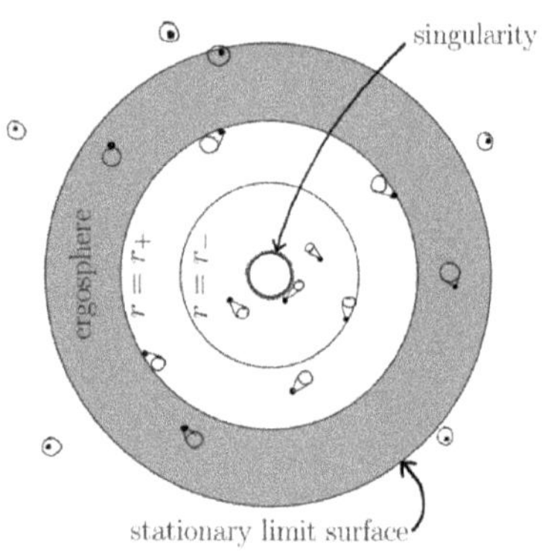

Fig. 29.6 The light-cone structure of the Kerr geometry.

and so the Killing vector is now spacelike. Just as in the argument for Hawking radiation in the last chapter, the dot product $-\boldsymbol{p}_{\mathrm{bh}} \cdot \boldsymbol{\xi}$ is now no longer constrained to be positive. Therefore, if the decay has $E_{\mathrm{bh}}^{\infty} = -\boldsymbol{p}_{\mathrm{bh}} \cdot \boldsymbol{\xi} < 0$ we have have $E_{\mathrm{out}}^{\infty} > E_{\mathrm{in}}^{\infty}$ and energy can be extracted from the black hole. The consequence is that negative E_{bh} will consume some of the mass of the black hole. It also removes some angular momentum, as demonstrated in the next example.

Example 29.7
Consider a rotating observer who remains at fixed r and θ in a Kerr black hole. (They cannot help but be rotated in ϕ.) They have a velocity

$$u^{\mu} = u^t \left(1, 0, 0, \Omega\right), \tag{29.28}$$

where Ω is the angular speed. This velocity can be written in vector form using the Killing vectors as

$$\boldsymbol{u}_{\mathrm{obs}} = u^t \left(\boldsymbol{\xi} + \Omega \boldsymbol{\eta}\right). \tag{29.29}$$

The observer measures a positive energy for the particle, and so using $E = -\boldsymbol{u}_{\mathrm{obs}} \cdot \boldsymbol{p}$, we find

$$-\left(\boldsymbol{\xi} + \Omega \boldsymbol{\eta}\right) \cdot \boldsymbol{p}_{\mathrm{bh}} \geq 0, \tag{29.30}$$

for any value of Ω that allows $\boldsymbol{u} \cdot \boldsymbol{u} = -1$. Using $E = -\boldsymbol{\xi} \cdot \boldsymbol{p}$ and $L = \boldsymbol{\eta} \cdot \boldsymbol{p}$, we have

$$E_{\mathrm{bh}} \geq \Omega L_{\mathrm{bh}}, \tag{29.31}$$

where L_{bh} is the angular momentum of the particle that falls into the black hole. Since Ω is positive (since the observer must rotate in the same direction as the black hole), then L_{bh} must be negative if we are to have $E_{\mathrm{bf}} < 0$ and extract energy from the hole. As a result, the angular momentum of the hole is reduced.

The reader will by now have built up a healthy respect for the Kerr solution. However, like many results in mathematical physics, it's rather hard to imagine how anyone could have come up with it! Kerr's announcement of the solution in 1963 did not give a derivation. A paper by Kerr and Schild from 1965 does so, but warns "the calculations giving these results are by no means simple", while Landau and Lifshitz (1975) say that "there is no constructive analytic derivation of the metric that is adequate in its physical ideas, and even a check of this solution of Einstein's equations involves cumbersome calculations". However, later work has elucidated the physical motivation, with Chandrasekhar describing the derivation he gives in his (excellent, but very advanced) book as "really very simple" with an "adequate base of physical and mathematical motivations". It is perhaps sobering that even Chandrasekhar provides a warning regarding perturbations to the Kerr solution, saying "the reductions that are necessary to go from one step to another are often very elaborate and, on occasion, may require as many as ten, twenty, or even fifty pages". He goes on to say that a copy of his derivations "(in some 600 legal-size pages and six additional notebooks)" are available via the Joseph Regenstein Library, University of Chicago, "in case some reader may wish to undertake a careful scrutiny of the entire development".

Chapter summary

- Charged black holes are unlikely to be realized in Nature. They are described by the Reissner–Nordstöm coordinates and feature three different regions of spacetime.

- The singularity at the centre of a charged black hole is repulsive. A traveller can use the wormhole created by the charge to tunnel into a new Universe.

- Rotating black holes are described by the Kerr geometry given in Boyer–Lindquist coordinates. There is a ring of physical singularities and an ellipsoidal event horizon at r_0.

- In the Kerr geometry, the region between the coordinate singularity at r_+ and the surface r_0 where g_{tt} vanishes is the ergosphere, in which it is impossible for a particle to remain stationary.

Exercises

(29.1) Verify that the coordinate singularities for the Reissner–Nordström line element occur at $r_\pm = M \pm (M^2 - q^2)^{\frac{1}{2}}$.

(29.2) (a) Show that g_{tt} for the Kerr solution can also be rewritten as

$$g_{tt} = -\frac{\Delta - a^2 \sin^2 \theta}{\rho^2}. \qquad (29.32)$$

(b) Show further that $g_{\phi\phi}$ can be rewritten as

$$g_{\phi\phi} = \left(\frac{2Mra^2 \sin^2 \theta}{\rho^2} + r^2 + a^2 \right) \sin^2 \theta. \qquad (29.33)$$

(29.3) Verify eqn 29.6.

(29.4) Show that

$$ds^2 = \rho_+ d\theta^2$$

$$+ \left(r_+^2 + a^2 + \frac{2Mr_+ a^2 \sin^2 \theta}{\rho_+^2} \right) \sin^2 \theta d\phi^2$$

$$= \rho_+^2 d\theta^2 + \left(\frac{2Mr_+}{\rho_+} \right)^2 \sin^2 \theta d\phi^2. \qquad (29.34)$$

(29.5) By evaluating the determinant g and integrating, show that

$$A = \int_0^{2\pi} d\phi \int_0^{\pi} d\theta g = 8\pi M r_+. \qquad (29.35)$$

(29.6) Verify eqns 29.12 by inverting the matrix

$$\begin{pmatrix} g_{tt} & g_{t\phi} \\ g_{\phi t} & g_{\phi\phi} \end{pmatrix}. \qquad (29.36)$$

(29.7) (a) Verify eqn 29.18.
(b) Verify eqn 29.24.

(29.8) What is the interval in proper time required to orbit a black hole described by the Kerr metric at fixed $\theta = \pi/2$, constant $r = R$ and coordinate speed $v = rd\phi/dt$.

Hint: An analogous problem appears in Chapter 23 in Exercise 23.1.

Part V

Geometry

In Part V of the book, we take a closer look at the mathematics of differential geometry and curvature. These ideas will be put to work in the final part of the book.

- In Chapter 30, we look at how curved lines and surfaces can be described mathematically using classical differential calculus.
- Then, in Chapters 31 and 32, we proceed to upgrade our basic mathematical machinery to provide a modern treatment of tensors.
- In Chapters 33 and 34, we meet the derivatives that are naturally adapted to a coordinate-free description of a manifold.
- In Chapter 35, we once again encounter curvature, but now from a thoroughly modern point of view. This enables us, in Chapter 36, to compute the curvature from a metric using an incredibly efficient procedure invented by Élie Cartan.
- In Chapter 37, we discuss how to deal with areas and volumes in coordinate-free language.
- In Chapter 38, we see how this new approach gives insights into how we understand integrals in physics.
- Appendix C contains some further description of the mathematical background behind this part of the book and can be consulted for interest or reference.

For many readers with a physics background, the material in this part of the book might seem rather unfamiliar. It's worth keeping in mind that Einstein was also initially unfamiliar with differential geometry when he started working on general relativity. He benefitted from the help of his friend Marcel Grossmann (1878–1936) for an introduction to many of the ideas.

<table><tr><td>

30

</td><td>

Classical curvature

</td></tr></table>

> ↷We call the approach outlined in this chapter **classical curvature** to distinguish it from the more modern, coordinate-free approach outlined in the following chapters. As a result, this chapter can be skipped if so desired.

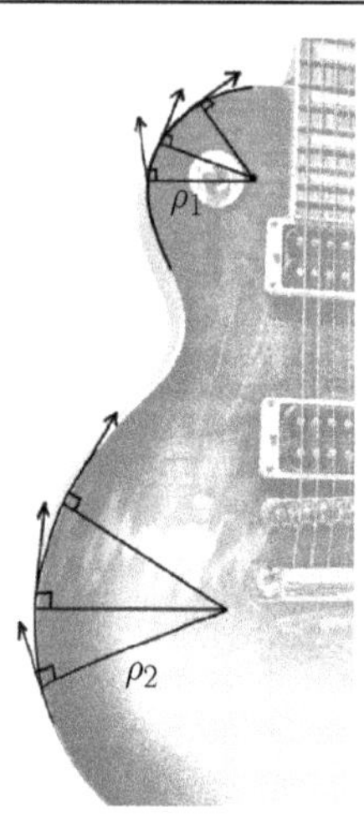

Fig. 30.1 The outline of this guitar traces out a path in space. The curvature of the upper region with radius ρ_1 is greater than that with radius ρ_2.

[1] Augustin-Louis Cauchy (1789–1857).

This paper will no doubt be found interesting by those who take an interest in it.
John Dalton (1766–1844)

The study of curvature has a rich history in mathematics. Our knowledge of the curvature of a line was developed by applying calculus to geometry, thereby founding the study of **differential geometry**. In this chapter, we follow some of the key developments made by mathematicians such as Leonhard Euler and Augustin-Louis Cauchy that culminate in the Frenet–Serret equations, which represent an important milestone in our understanding of the curvature of paths in space. Our knowledge of the curvature of planes is largely based around the work of two mathematicians: Carl Friedrich Gauss and Bernhard Riemann. Gauss showed how the curvature of a two-dimensional plane can be described by measurements made by beings entirely confined to that plane. Riemann generalized the result to higher dimensions, and it is Riemann's toolkit of insights that are incorporated into general relativity.

30.1 Curvature of a line

How do we measure the **curvature** of a path in two dimensions? The curved edge of the object shown in Fig. 30.1 has lots of features, but it's reasonable to say that the part at the top has a larger curvature than the part at the bottom, since its tangent vectors turn more rapidly as we traverse that part of the curve. If this isn't obvious, imagine driving in a car traversing the two indicated bends of the curve at a constant speed and consider how much you would need to turn the steering wheel in each case.

In order to quantify curvature, let's identify a figure, such as the circle, against which curvature can be compared. The circle, after all, involves a line element constantly changing its direction about the central point, so seems to have a suitably curvy quality. A small circle has greater curvature than a circle of larger radius, while a straight line has no curvature. This suggests we can use the reciprocal of a circle's radius as the measure of curvature. Let's see how we might do this systematically, by following Augustin-Louis Cauchy,[1] and evaluate the curvature at a specific point $\mathcal{P}$ on the curve.

A circle's centre can always be found by seeing where two normals to the circle intersect. We will use this fact to identify the **osculating circle**,[2] which will provide our measure of curvature. Referring to Fig. 30.2 where we determine the curvature of the curve at point $\mathcal{P}$. We define the **centre of curvature** $\mathcal{C}$ as the point where two closely spaced normals to the curve intersect, allowing us to draw the osculating circle, centred at $\mathcal{C}$ and intersecting $\mathcal{P}$. The **radius of curvature** ρ is then the distance from $\mathcal{P}$ to $\mathcal{C}$ and we define the curvature κ to be $\kappa = 1/\rho$. Referring back to Fig. 30.1, we see how this does the job: the feature at the bottom has an osculating circle with a smaller radius and, therefore, a larger curvature κ, as we had suggested. The curvature of a curve at a point $\mathcal{P}$ may, therefore, be thought of as the inverse radius of a circle that locally follows the curve at $\mathcal{P}$. We could repeat this procedure as we move along the curve, assessing the curvature at different points.

Example 30.1

Recall that a curve is parametrized by some parameter λ that varies along the curve. We shall choose to use length parametrization[3] so that we set λ equal to the distance s measured along the curve (Fig. 30.3). An alternative way to find the curvature is to see how the directions of the normals, or equivalently,[4] the tangent vectors to the curve, rotate as we traverse the curve. Specifically, we work out the tangent vectors at two ends of a short section of curve of length ds and measure the angle $d\theta$ between the two tangents. Extracting the tangent vectors from the figure, normalizing them, and arranging them in the unit circle, as shown in Fig. 30.4, provides us with a method for calculating the curvature

$$\kappa(s) = \lim_{\Delta s \to 0} \frac{\left(\begin{array}{c} \text{Angle } \Delta\theta \text{ swept out on} \\ \text{the unit circle by normals or tangents} \end{array} \right)}{\left(\begin{array}{c} \text{Corresponding arc length } \Delta s \\ \text{on actual curve} \end{array} \right)}. \tag{30.1}$$

Putting this together we have

$$\kappa(s) = \frac{1}{\rho(s)} = \frac{d\theta(s)}{ds}. \tag{30.2}$$

To see that this makes sense, consider a circle of radius ρ shown in Fig. 30.4. The tangents (and normals) sweep out angle $\Delta\theta$ corresponding to the arc length $\rho\Delta\theta$. It's fairly easy to see that the curvature is $\kappa = 1/\rho$.

A slightly more coordinate-friendly method of assessing the curvature of a line is to measure the departure y of the curve from a tangent line as a function of x. The curvature κ is then written as

$$\kappa = \frac{1}{\rho} = \frac{d^2y}{dx^2}. \tag{30.3}$$

Example 30.2

How does this latter method work? Consider the tangent to the curve at point $\mathcal{P}$ with coordinate x_p. If x is the coordinate measured along the tangent relative to x_p, then $y(x)$ is the height of the curve above the tangent. For small x we can expand

$$y(x) = y(x_p) + \left.\frac{dy}{dx}\right|_{x_p} (x - x_p) + \left.\frac{d^2y}{dx^2}\right|_{x_p} \frac{(x - x_p)^2}{2} + \dots \tag{30.4}$$

[2]Outside of mathematics, osculate is a verb derived from the Latin verb *oscu-lare*, meaning to kiss (*Da mihi osculum* $\equiv$ give me a kiss).

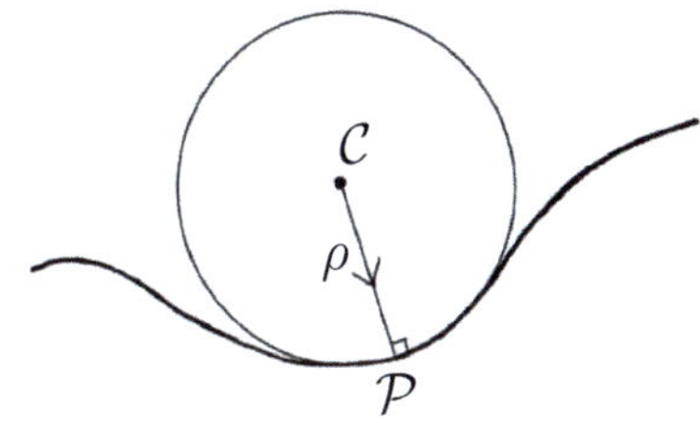

Fig. 30.2 The osculating circle for a point $\mathcal{P}$ on a curve. It is centred on $\mathcal{C}$ and has radius ρ.

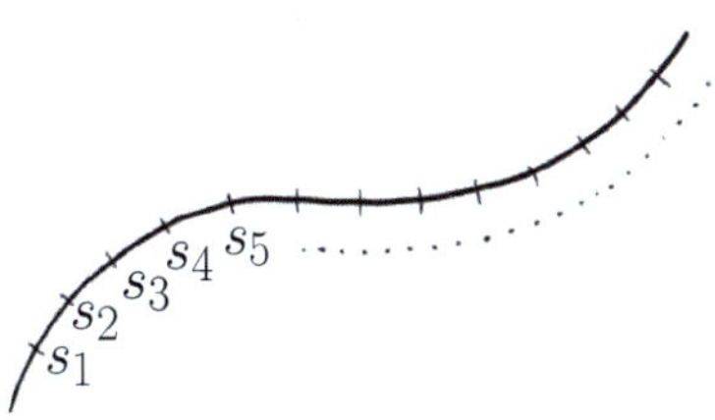

Fig. 30.3 Length parametrization of the curve uses the distance s along the curve as a parameter.

[3]See Chapter 9.

[4]The normals to a curve are simply the tangents rotated by 90°.

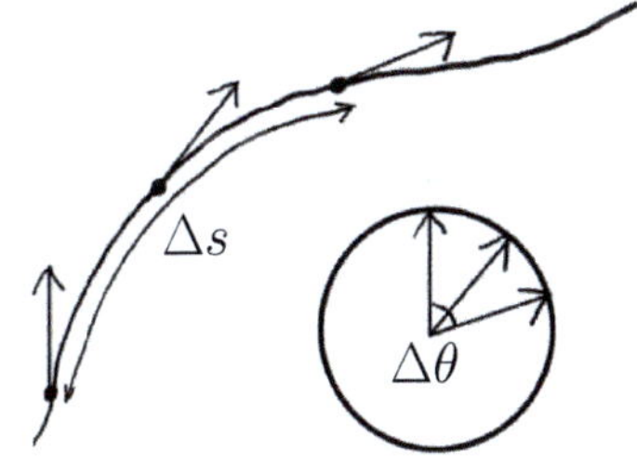

Fig. 30.4 Top: a curve with unit tangent vectors evolving over a distance Δs. Bottom: arranging the tangent vectors in the unit circle provides a measure of $\Delta\theta$.

As the tangent is defined by $\mathrm{d}y/\mathrm{d}x = 0$, taking $x_p = 0$ we have $y(x) = \kappa x^2/2$.

Next, imagine a circle of radius ρ which is tangent to the curve at p. Its height is $y = \rho(1 - \cos\theta)$. For small θ, we have $\theta = x/\rho$ and so $y = x^2/2\rho$ and we conclude that $\kappa = 1/\rho$ as we had before.[5]

[5]Notice that, because we measure curvature in terms of infinitesimals, in order to extract the curvature κ in the coordinate version we compare the curvature to a parabola (i.e. the height of the circle close to x_p is $y = x^2/2\rho$) which has $\mathrm{d}^2y/\mathrm{d}x^2 = $ (const.). We shall return to this when we look at the curvature of two-dimensional surfaces.

[6]This is a temporary measure for this chapter.

[7]This use of the displacement $\boldsymbol{X}$ relies on the one-dimensional curve being embedded in flat three-dimensional space. We shall see in the next chapter that it is not a strategy that we shall attempt in curved spacetime.

30.2   Curvature with vectors

A still more formal way of examining the curvature properties of a line is to invoke differential geometry. We will work in flat three-dimensional space, denoting 3-vectors by bold letters.[6] As usual, we parametrize a curve using the distance parameter s. We say that the curve is defined by vectors $\boldsymbol{X}(s)$ (i.e. insert a value of s giving the position on the curve and return a vector taking us from the origin to that point on the curve).[7] This curve has a tangent

$$\boldsymbol{t} = \frac{\mathrm{d}\boldsymbol{X}}{\mathrm{d}s}. \tag{30.5}$$

The tangent vector $\boldsymbol{t}$ can be expanded in terms of coordinates x^μ by using basis vectors defined along the curve and writing $\boldsymbol{t} = t^\mu \boldsymbol{e}_\mu$. Then we have

$$\boldsymbol{t} = \frac{\mathrm{d}\boldsymbol{X}}{\mathrm{d}s} = \frac{\mathrm{d}x^\mu}{\mathrm{d}s}\frac{\partial \boldsymbol{X}}{\partial x^\mu} \equiv t^\mu \boldsymbol{e}_\mu. \tag{30.6}$$

The tangent vector's components are $t^\mu = \frac{\mathrm{d}x^\mu}{\mathrm{d}s}$ and the basis vectors on the curve are $\boldsymbol{e}_\mu = \frac{\partial \boldsymbol{X}}{\partial x^\mu}$.

If s parametrizes length along the curve then $\mathrm{d}s^2 = \mathrm{d}\boldsymbol{X} \cdot \mathrm{d}\boldsymbol{X}$, and so we must have $\boldsymbol{t} \cdot \boldsymbol{t} = 1$, which is to say that the tangents are normalized to unity. Differentiation of $\boldsymbol{t} \cdot \boldsymbol{t} = 1$ implies that $\boldsymbol{t} \cdot \frac{\mathrm{d}\boldsymbol{t}}{\mathrm{d}s} = 0$. We therefore define

$$\frac{\mathrm{d}\boldsymbol{t}}{\mathrm{d}s} = \kappa\boldsymbol{p}, \tag{30.7}$$

where $\boldsymbol{p}$, known as the **principal normal vector**, is orthogonal to $\boldsymbol{t}$, and we also fix $\boldsymbol{p} \cdot \boldsymbol{p} = 1$. As we prove in the next example, the constant κ is the curvature of the curve that we defined above. The vectors $\boldsymbol{t}$ and $\boldsymbol{p}$ span a two-dimensional surface known as the **osculating plane**, as shown in Fig. 30.5. The osculating plane is carried along as we traverse the curve. If the curve is confined to a plane (a so-called plane curve), the osculating plane will always be the same plane, although the directions of $\boldsymbol{p}$ and $\boldsymbol{t}$ will change.

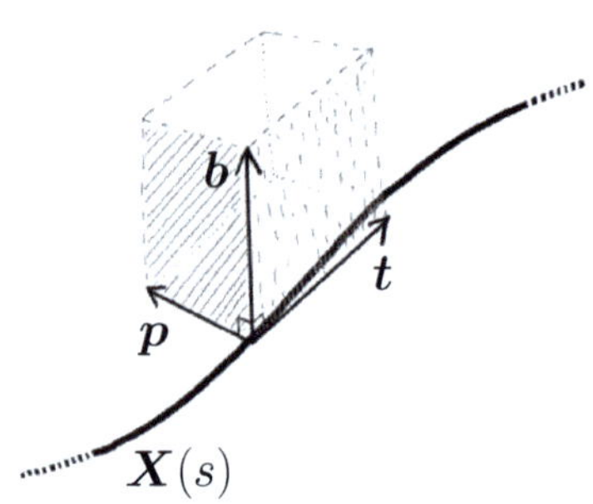

Fig. 30.5 The triplet formed from tangent $\boldsymbol{t}$, normal $\boldsymbol{p}$ and binormal $\boldsymbol{b} = \boldsymbol{t} \times \boldsymbol{p}$. The osculating plane is defined by $\boldsymbol{t}$ and $\boldsymbol{p}$.

Example 30.3

We pause to check that κ in the previous equation is indeed the curvature of the curve. Let $\Delta\theta$ be the angle between tangent $\boldsymbol{t}(s)$ at $\boldsymbol{X}(s)$ and tangent $\boldsymbol{t}(s + \Delta s)$ at the point $\boldsymbol{X}(s+\Delta s)$. If the tangent is normalized to unity, the vector $|\boldsymbol{t}(s+\Delta s)-\boldsymbol{t}(s)|$ is the base of an isosceles triangle with unit sides. Hence,

$$|\boldsymbol{t}(s + \Delta s) - \boldsymbol{t}(s)| = 2\sin\left(\frac{\Delta\theta}{2}\right) = \Delta\theta + O(\Delta\theta^3). \tag{30.8}$$

Then, noting that $|\boldsymbol{p}| = 1$ and writing[8] $\boldsymbol{t} = \mathrm{d}\boldsymbol{t}/\mathrm{d}s$, we have

$$|\kappa| = |\dot{\boldsymbol{t}}| = \left| \lim_{\Delta s \to 0} \frac{\boldsymbol{t}(s + \Delta s) - \boldsymbol{t}(s)}{\Delta s} \right|$$

$$= \lim_{\Delta s \to 0} \frac{\Delta\theta + O(\Delta\theta^3)}{\Delta s}. \tag{30.9}$$

This leads to

$$|\kappa| = \lim_{\Delta s \to 0} \frac{\Delta\theta}{\Delta s} = \frac{\mathrm{d}\theta}{\mathrm{d}s}, \tag{30.10}$$

demonstrating that κ is the curvature of the curve, just as we had before.

To describe curves in three dimensions (i.e. curves that are not plane curves) it would be useful to have an orthogonal triplet of vectors that spans the space. To make such a triplet, we also define the **binormal vector**, another unit vector ($\boldsymbol{b} \cdot \boldsymbol{b} = 1$), as

$$\boldsymbol{b}(s) = \boldsymbol{t}(s) \times \boldsymbol{p}(s), \tag{30.12}$$

Example 30.4

This definition fixes the direction of $\dot{\boldsymbol{b}} \equiv \mathrm{d}\boldsymbol{b}/\mathrm{d}s$. Since we define $\boldsymbol{b} \cdot \boldsymbol{b} = 1$, we must have $\boldsymbol{b} \cdot \dot{\boldsymbol{b}} = 0$. Noting also that $\boldsymbol{b} \cdot \boldsymbol{t} = 0$, we have

$$\dot{\boldsymbol{b}} \cdot \boldsymbol{t} + \boldsymbol{b} \cdot \dot{\boldsymbol{t}} = 0$$

$$\dot{\boldsymbol{b}} \cdot \boldsymbol{t} = -\boldsymbol{b} \cdot \dot{\boldsymbol{t}}$$

$$= -\kappa \boldsymbol{b} \cdot \boldsymbol{p} = 0. \tag{30.13}$$

We conclude that $\dot{\boldsymbol{b}}$ is perpendicular to $\boldsymbol{b}$ and $\boldsymbol{t}$, so must lie along $\boldsymbol{p}$.

Motivated by the previous example, we define

$$\frac{\mathrm{d}\boldsymbol{b}}{\mathrm{d}s} = -\tau\boldsymbol{p}, \tag{30.14}$$

where τ is known as the **torsion** of the curve. Since the three vectors $\boldsymbol{t}$, $\boldsymbol{p}$ and $\boldsymbol{b}$ form an orthogonal triplet, we can also write $\boldsymbol{p} = \boldsymbol{b} \times \boldsymbol{t}$. Differentiating this latter equation, we find that

$$\frac{\mathrm{d}\boldsymbol{p}}{\mathrm{d}s} = \frac{\mathrm{d}\boldsymbol{b}}{\mathrm{d}s} \times \boldsymbol{t} + \boldsymbol{b} \times \frac{\mathrm{d}\boldsymbol{t}}{\mathrm{d}s} = \tau\boldsymbol{b} - \kappa\boldsymbol{t}. \tag{30.15}$$

We can think of the vector $\dot{\boldsymbol{b}}$ as measuring the rate of change of the osculating plane as we traverse the curve. This in turn is a measure of the rate at which the curve deviates from being a curve confined to a plane.

We now have a useful triplet to describe curves in three dimensions in terms of how the three quantities vary along the curve. Everything goes together to form[9] the **Frenet–Serret equations**. They are written in matrix form as

$$\frac{\mathrm{d}}{\mathrm{d}s} \begin{pmatrix} \boldsymbol{t} \\ \boldsymbol{p} \\ \boldsymbol{b} \end{pmatrix} = \begin{pmatrix} 0 & \kappa & 0 \\ -\kappa & 0 & \tau \\ 0 & -\tau & 0 \end{pmatrix} \begin{pmatrix} \boldsymbol{t} \\ \boldsymbol{p} \\ \boldsymbol{b} \end{pmatrix}. \tag{30.16}$$

These equations summarize the curvature of paths in three-dimensions and are the key results, so far, from our investigation of classical curvature. They show that we only need two parameters: κ and τ.

---

[8] In this chapter, from here on, we will denote $\mathrm{d}/\mathrm{d}s$ by a dot.

By cyclic permutation, we have

$$\boldsymbol{b} = \boldsymbol{t} \times \boldsymbol{p},$$
$$\boldsymbol{t} = \boldsymbol{p} \times \boldsymbol{b},$$
$$\boldsymbol{p} = \boldsymbol{b} \times \boldsymbol{t}. \tag{30.11}$$

[9] Jean Frédéric Frenet (1816–1900), Joseph Alfred Serret (1819–1885). The equations were obtained by Serret in 1851, but can be found in Frenet's 1847 thesis, an abstract of which was published in 1852. As a result of their slightly complicated birth, they are sometimes called the Serret–Frenet equations.

30.3 Two-dimensional surfaces

So far we have considered the curvature of lines in three dimensions. We now upgrade the discussion to deal with the curvature of two-dimensional surfaces embedded in flat three-dimensional space. For one-dimensional surfaces (i.e. curves) we started by comparing the behaviour near closely spaced points to that of points on a circle (and then, subsequently, on a parabola). For the two-dimensional surface, we might imagine a sphere would be the measure of curvature. However, we can't say that the curvature will be the same in two different directions, since this is generally not true. In fact, we must evaluate two curvature constants, κ_1 and κ_2, to characterize the curvature of a plane. It seems like a tough problem to determine these two constants, since we don't know in which directions to evaluate the curvature. The task is dramatically simplified by a discovery by Leonhard Euler.[10] If the two curvature constants aren't equal then there will be some direction where the curvature with be a minimum and another where it is a maximum. Euler's discovery was that these directions are *perpendicular*. As a result, just as the curvature of a line is measured by the radius of a circle, the curvature of the two-dimensional plane is measured using the major and minor axes of an ellipse.

We start with the coordinate approach, where we measure curvature by comparing the height of a curved surface above a plane which forms the tangent to the surface at some nearby point (Fig. 30.6). Taking our lead from our analysis of the curve, in two dimensions we have that the height z of the curved surface, written in terms of coordinates x^i in the tangent plane, is

$$z = \frac{1}{2}K_{ij}x^i x^j, \tag{30.17}$$

where K_{ij} are the components of a 2×2 real symmetric tensor. This tensor is the key to understanding the classical curvature of curved surfaces since it defines the ellipse whose major and minor axes are the eigenvalues of K_{ij}. The inverse of these eigenvalues will give the two components of curvature, as we shall see below.

[10]Leonhard Euler (1707–1783). "The study of Euler's works will remain the best school for the different fields of mathematics, and nothing else can replace it." So said Johann Carl Friedrich Gauss (1777–1855), another giant of mathematics whose work we will discuss shortly.

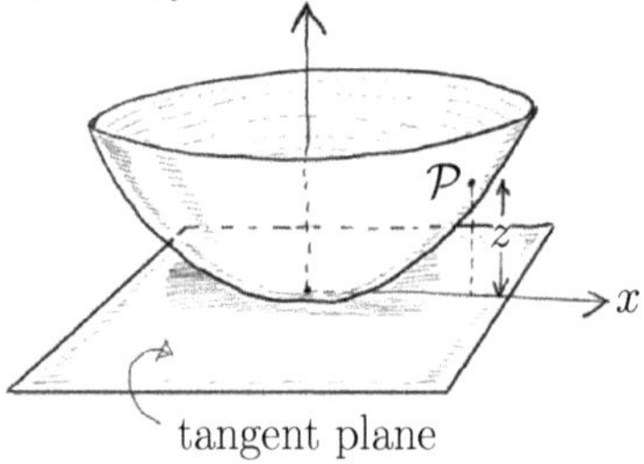

Fig. 30.6 A curved, two-dimensional surface is embedded in three dimensions. We measure the curvature of the point using its tangent plane. The height of the curve above the tangent plane is z.

Example 30.5

Consider a surface embedded in three-dimensional space. The departure z of the curved surface from the flat one can be expanded as

$$z = \frac{1}{2}ax^2 + bxy + \frac{1}{2}cy^2, \tag{30.18}$$

or, equivalently,

$$z = \frac{1}{2}\begin{pmatrix} x & y \end{pmatrix}\begin{pmatrix} a & b \\ b & c \end{pmatrix}\begin{pmatrix} x \\ y \end{pmatrix}. \tag{30.19}$$

It is always possible make a coordinate transformation, effectively by rotating the axes

$$\begin{pmatrix} x \\ y \end{pmatrix} = \begin{pmatrix} \cos\alpha & \sin\alpha \\ -\sin\alpha & \cos\alpha \end{pmatrix}\begin{pmatrix} \xi \\ \eta \end{pmatrix}. \tag{30.20}$$

We choose the angle α to diagonalize the matrix, such that we have

$$z = \frac{1}{2} \begin{pmatrix} \xi & \eta \end{pmatrix} \begin{pmatrix} \kappa_1 & 0 \\ 0 & \kappa_2 \end{pmatrix} \begin{pmatrix} \xi \\ \eta \end{pmatrix}, \tag{30.21}$$

or equivalently

$$z = \frac{1}{2}\kappa_1 \xi^2 + \frac{1}{2}\kappa_2 \eta^2. \tag{30.22}$$

The new variables, ξ and η are the coordinates on the **principal axes** of the ellipse. In this context, the eigenvalues κ_i are called the **principal curvatures** and can be written in terms of radii of curvature ρ_i as $\kappa_1 = 1/\rho_1$ and $\kappa_2 = 1/\rho_2$. The principal axes give us the largest (major) and smallest (minor) widths of the ellipse, which provide our measure of curvature.

Let's apply this approach to a spherical surface of radius a. For the surface balanced on a plane we can write $z = -\left(a^2 + x^2 + y^2\right)^{\frac{1}{2}} + a$. Near the point of contact, we have $z \approx \frac{1}{2a^2}(x^2 + y^2)$ and so we have two principal curvatures of $1/a$, both with radius of curvature a, just as we would expect.

In order for the previous description to work, we have been forced to embed the two-dimensional surface in a three-dimensional space. However, the beings inhabiting a space are generally confined to that space, unable to jump into higher dimensions, i.e. to leave their space and observe it from afar. Gauss thought this was a serious problem in the description of curvature and set out to find a satisfactory solution.

Gauss wanted[11] to confine measurements to the surface itself and come up with a complete description of curvature based only on the measurements that the inhabitants of a curved space could make. He succeeded. He found just such a measure of curvature known now, appropriately, as the **Gaussian curvature** and given the symbol K. Specifically, Gauss proved what is now called the *theorema egregium* or, 'remarkable theorem',[12] that says K can be determined purely by making measurements confined within the surface. *Gaussian curvature* is therefore an **intrinsic** measure of curvature, depending only on distances that are measured on the surface, not on the way it is embedded in a space.

Example 30.6

Here's a look at Gauss' remarkable result.[13] In two dimensions, the flat, Euclidean plane is described by a line element

$$ds^2 = \left(d\xi^1\right)^2 + \left(d\xi^2\right)^2. \tag{30.23}$$

Gauss assumed that in a sufficiently small (that is, infinitesimal) patch of a curved space, it would always be possible to describe the space using such a coordinate system. However, a general two-dimensional curved space will not be able to be described by the (flat-space) ξ^i coordinates over a finite neighbourhood. Instead, there will be a perfectly good coordinate system (x^1, x^2) that does cover the curved space. In this latter coordinate system, the line element is written as (just as we've had previously)

$$ds^2 = g_{ij}dx^i dx^j. \tag{30.24}$$

As usual, the values of g_{ij} depend on the particular coordinate system chosen (although it will turn out that they also encode the intrinsic properties of the space).

[11] Gauss' own version can be found in his 1827 paper *Disquisitiones generales circa superficies curvas* (General Investigations of Curved Surfaces). A translation of the paper along with a running commentary given in terms of modern mathematical notation can be found in Volume II of Michael Spivak's *A Comprehensive Introduction to Differential Geometry* and comes highly recommended.

[12] Modern English usage of the word egregious, which formerly meant 're-markable', usually implies that something is remarkably bad. H. W. Fowler (1858–1933) notes that it is especially applied to the nouns "ass, coxcomb, liar, imposter, folly, blunder, waste", and that "Reversion to the original sense ... is mere pedantry."

[13] See the excellent book by Needham (2021) for more details and several proofs. Needham makes sense of the theorem which, in Gauss' original form (discussed here), admittedly looks like it was plucked magically from the air.

Gauss sought to define a measure of curvature in terms of a function of the components of the metric of g_{ij} and their derivatives that depends only on the intrinsic properties of the space and not on the particular coordinate system chosen. The formidable equation that Gauss discovered is (employing the comma notation for derivatives)

$$
\begin{aligned}
K(x^1, x^2) = &\frac{1}{2g} \left[2g_{12,12} - g_{11,22} - g_{22,11} \right] \\
&- \frac{g_{22}}{4g^2} \left[g_{11,1}(2g_{12,2} - g_{22,1}) - g_{11,2}^2 \right] \\
&+ \frac{g_{12}}{4g^2} \left[g_{11,1}g_{22,2} - 2g_{11,2}g_{22,1} + (2g_{12,1} - g_{11,2})(2g_{12,2}g_{22,1}) \right] \\
&- \frac{g_{11}}{4g^2} \left[g_{22,2}(2g_{12,1} - g_{11,2}) - g_{22,1}^2 \right],
\end{aligned}
\tag{30.25}
$$

where g is the determinant

$$
g(x_1, x_2) = g_{11}g_{22} - g_{12}^2.
\tag{30.26}
$$

The significance of this is that someone in possession of the components of the metric also has a failsafe means of calculating the curvature of the surface that cannot be the consequence merely of a particular choice of coordinates. Gauss was, understandably, elated to have discovered this wonderful result.

As a concrete example, we have for the sphere that $g_{\theta\theta} = a^2$ and $g_{\phi\phi} = a^2 \sin^2 \theta$. We find that

$$
K(\theta, \phi) = \frac{1}{a^2},
\tag{30.27}
$$

which is to say that the curvature function is $1/(\text{radius})^2$, which is only given in terms of a coordinate of the space itself, its radius. It is also heartening that the flat, Euclidean plane gives

$$
K(\xi^1, \xi^2) = 0.
\tag{30.28}
$$

We will have more to say about Gaussian curvature but, happily won't have to make use of Gauss' remarkable equation as there are several alternative ways of extracting K.

30.4 Gauss' equation

In one dimension, we considered lengths of lines, so it would seem natural to look at areas in two dimensions. In this spirit, we can define the curvature of a surface at a point $\mathcal{P}$ as

$$
K(\mathcal{P}) = \frac{d\Omega}{dA} = \lim_{\Delta A \to 0} \frac{\left(\begin{array}{c} \text{dimensionless area swept out} \\ \text{on unit sphere by normals} \end{array} \right)}{\left(\begin{array}{c} \text{corresponding area } \Delta A \\ \text{on actual surface} \end{array} \right)}.
\tag{30.29}
$$

Turning to the description in terms of vectors, we define a two-dimensional surface embedded in flat three-dimensional space by a vector-valued function $\boldsymbol{X}(x^1, x^2)$, where the coordinates x_1 and x_2 parametrize the surface [i.e. the coordinate system (x^1, x^2) lies within the surface[14]]. That is, given a point with coordinates in the surface of x^1 and x^2, the vector-valued function $\boldsymbol{X}$ returns the vector from the origin to the surface at that point. The two-dimensional surface has two basis vectors, which are given by

$$
e_1 = \frac{\partial \boldsymbol{X}}{\partial x^1}, \quad e_2 = \frac{\partial \boldsymbol{X}}{\partial x^2}.
\tag{30.30}
$$

[14]This makes it similar to s in the one-dimensional case.

Example 30.7

Let's now use this expression to derive the Gaussian curvature using these two basis vectors. The normal to the surface is $\boldsymbol{n} = \boldsymbol{e}_1 \times \boldsymbol{e}_2$ and so the unit normal[15] is

$$\hat{\boldsymbol{n}} = \frac{(\boldsymbol{e}_1 \times \boldsymbol{e}_2)}{|\boldsymbol{e}_1 \times \boldsymbol{e}_2|}. \tag{30.31}$$

An element of area on the surface is given by

$$\mathrm{d}A = |\boldsymbol{e}_1 \times \boldsymbol{e}_2|\,\mathrm{d}x^1\mathrm{d}x^2 = \left|\frac{\partial \boldsymbol{X}}{\partial x^1} \times \frac{\partial \boldsymbol{X}}{\partial x^2}\right|\partial x^1\partial x^2. \tag{30.32}$$

For the area swept out by the unit normals, we note that it will be proportional to

$$\mathrm{d}\Omega \propto \left|\frac{\partial \hat{\boldsymbol{n}}}{\partial x^1} \times \frac{\partial \hat{\boldsymbol{n}}}{\partial x^2}\right|\mathrm{d}x^1\mathrm{d}x^2. \tag{30.33}$$

This makes it parallel to $\hat{\boldsymbol{n}}$ itself. To pick out the sign, we dot it with $\hat{\boldsymbol{n}}$ to obtain

$$\mathrm{d}\Omega = \hat{\boldsymbol{n}} \cdot \left(\frac{\partial \hat{\boldsymbol{n}}}{\partial x^1} \times \frac{\partial \hat{\boldsymbol{n}}}{\partial x^2}\right)\mathrm{d}x^1\mathrm{d}x^2. \tag{30.34}$$

The Gaussian curvature is then given by

$$K = \frac{\mathrm{d}\Omega}{\mathrm{d}A} = \frac{\hat{\boldsymbol{n}} \cdot \left(\frac{\partial \hat{\boldsymbol{n}}}{\partial x^1} \times \frac{\partial \hat{\boldsymbol{n}}}{\partial x^2}\right)}{|\boldsymbol{e}_1 \times \boldsymbol{e}_2|}. \tag{30.35}$$

We can now compute the curvature of a surface from knowledge of its basis vectors.

[15] It is the unit normal that is the useful quantity in this context, so we divide by the magnitude $|\boldsymbol{n}| = |\boldsymbol{e}_1 \times \boldsymbol{e}_2|$. There are several cases below where we differentiate $\hat{\boldsymbol{n}}$. However, although $|\boldsymbol{n}|$ can depend on the coordinates, we don't differentiate this normalization factor. Rather we scale $\boldsymbol{n}$ by a factor $1/|\boldsymbol{n}|$ at some position to make it a unit vector, and then treat the factor as a constant.

Although this is all very interesting, an approach even more closely based on modern differential geometry will be more useful to us. Consider the 3-vector $\boldsymbol{e}_\nu$ and its derivative $\frac{\partial \boldsymbol{e}_\nu}{\partial x^\mu}$. In general, we can choose to write the components of the derivative in the form

$$\frac{\partial \boldsymbol{e}_\nu}{\partial x^\mu} = \Gamma^\lambda{}_{\mu\nu}\boldsymbol{e}_\lambda + K_{\mu\nu}\hat{\boldsymbol{n}}. \tag{30.36}$$

This has a part expressed in terms on vectors confined to the surface (the first term on the right) and a part expressed in terms of the bit sticking out of the surface (the second term). Since $\hat{\boldsymbol{n}} \cdot \hat{\boldsymbol{n}} = 1$ and $\hat{\boldsymbol{n}} \cdot \boldsymbol{e}_\nu = 0$, we also find an expression for the matrix $K_{\mu\nu}$

$$K_{\mu\nu} = \frac{\partial \boldsymbol{e}_\nu}{\partial x^\mu} \cdot \hat{\boldsymbol{n}}, \tag{30.37}$$

which is known[16] as **Gauss' equation**. The matrix $K_{\mu\nu}$ is designed to look like the Gaussian curvature function $K(x_1, x_2)$, but it is not yet clear how. We look into this in the next section.

[16] Gauss came up with a lot of equations, and so this name doesn't uniquely identify it!

Example 30.8

Let's compute the components of $K_{\mu\nu}$ for a cylinder. The points on a cylinder's surface are described by vectors, with Cartesian components given by

$$\boldsymbol{X}(\theta, z) = \begin{pmatrix} a\cos\theta \\ a\sin\theta \\ z \end{pmatrix}. \tag{30.38}$$

[17]For later use, we also have metric components $g_{\mu\nu} = \boldsymbol{e}_\mu \cdot \boldsymbol{e}_\nu$ of $g_{\theta\theta} = a^2$ and $g_{zz} = 1$.

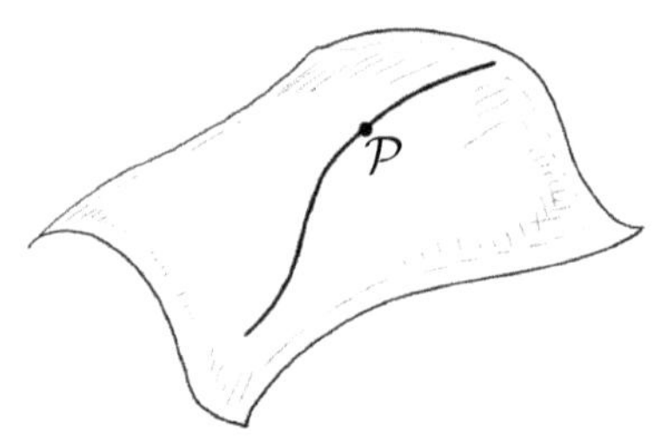

Fig. 30.7 A curve, passing through point $\mathcal{P}$, embedded in a two-dimensional surface.

[18]Some terminology in classical geometry includes the **first fundamental form**, defined as

$$F_1 = g_{\mu\nu} dX^\mu dX^\nu. \qquad (30.45)$$

This is simply ds^2 that we've written before. The **second fundamental form** is found by considering $\kappa_n = K_{\mu\nu}t^\mu t^\nu = K_{\mu\nu}(dX^\mu/ds)(dX^\nu/ds)$, and removing the path elements ds to write

$$F_2 = K_{\mu\nu} dX^\mu dX^\nu. \qquad (30.46)$$

We won't have cause to use this terminology.

The basis vectors and unit normal are found to be[17]

$$\boldsymbol{e}_\theta = \begin{pmatrix} -a\sin\theta \\ a\cos\theta \\ 0 \end{pmatrix}, \quad \boldsymbol{e}_z = \begin{pmatrix} 0 \\ 0 \\ 1 \end{pmatrix}, \quad \hat{\boldsymbol{n}} = \begin{pmatrix} \cos\theta \\ \sin\theta \\ 0 \end{pmatrix}. \qquad (30.39)$$

Differentiating, the only non-zero component results from

$$\frac{\partial \boldsymbol{e}_\theta}{\partial \theta} = \begin{pmatrix} -a\cos\theta \\ -a\sin\theta \\ 0 \end{pmatrix}, \qquad (30.40)$$

and dotting this with $\boldsymbol{n}$ yields $-a$. We conclude $K_{\theta\theta} = -a$ and all other components of $K_{\mu\nu}$ vanish.

30.5 Intrinsic and extrinsic curvature

We can now find out how the symmetric matrix $K_{\mu\nu} = \frac{d\boldsymbol{e}_\nu}{dx^\mu} \cdot \hat{\boldsymbol{n}}$, that we met in the last section, relates to the Gaussian curvature. Consider a curve lying in the surface that passes through a point $\mathcal{P}$ (Fig. 30.7), where it has tangent vector $\boldsymbol{t}$. As usual, we parametrize the curve with the arc-length parameter s, but we note for later that we also have access to the coordinates x^μ that lie within the surface. Differentiating the tangent vector along the curve, we have

$$\frac{d\boldsymbol{t}}{ds} = \frac{dt^\mu}{ds}\boldsymbol{e}_\mu + t^\mu \frac{d\boldsymbol{e}_\mu}{ds}. \qquad (30.41)$$

We expand this derivative in terms of two constants, κ_g and κ_n, as

$$\frac{d\boldsymbol{t}}{ds} = \frac{dt^\mu}{ds}\boldsymbol{e}_\mu + t^\mu \frac{d\boldsymbol{e}_\mu}{ds} = \kappa_g \boldsymbol{e} + \kappa_n \hat{\boldsymbol{n}}, \qquad (30.42)$$

where $\boldsymbol{e}$ is some unit vector formed from a linear combination of $\boldsymbol{e}_1$ and $\boldsymbol{e}_2$, and is, as such, perpendicular to $\hat{\boldsymbol{n}}$.

Now consider the derivative $\frac{d}{ds}$. When represented in terms of the coordinates in the surface, this is

$$\frac{d}{ds} = \frac{dx^\nu}{ds}\frac{\partial}{\partial x^\nu} = t^\nu \frac{\partial}{\partial x^\nu}. \qquad (30.43)$$

Therefore, a term $t^\mu \frac{d\boldsymbol{e}_\mu}{ds}$ can be written as $t^\mu t^\nu \frac{\partial \boldsymbol{e}_\mu}{\partial x^\nu}$. As a result, when we project eqn 30.42 along $\hat{\boldsymbol{n}}$ we can extract

$$\frac{d\boldsymbol{t}}{ds} \cdot \hat{\boldsymbol{n}} = t^\mu t^\nu \frac{\partial \boldsymbol{e}_\mu}{\partial x^\nu} \cdot \hat{\boldsymbol{n}} = \kappa_n, \qquad (30.44)$$

where we've used that $\hat{\boldsymbol{n}} \cdot \boldsymbol{e} = 0$ and $\hat{\boldsymbol{n}} \cdot \hat{\boldsymbol{n}} = 1$. Finally, recall from Gauss' equation $\frac{\partial \boldsymbol{e}_\mu}{\partial x^\nu} \cdot \hat{\boldsymbol{n}} = K_{\nu\mu} = K_{\mu\nu}$, which says that the part of $\partial \boldsymbol{e}_\mu / \partial x^\nu$ pointing along $\hat{\boldsymbol{n}}$ is $K_{\mu\nu}$ and so[18]

$$\kappa_n = t^\mu t^\nu K_{\mu\nu}. \qquad (30.47)$$

To relate this expression to curvature, we will follow Gauss and look not just at this one curve, but at all of the curves passing through $\mathcal{P}$. For each, we calculate κ_n and then find the extremal value, i.e. the largest and smallest values of κ_n.

Example 30.9

To do this, we extremize the quantity $K_{\mu\nu}t^\mu t^\nu$ subject to the constraint that the tangent vectors are properly normalized to unity, written[19] as $g_{\mu\nu}t^\mu t^\nu = 1$. We can carry out this procedure using a Lagrange multiplier k. So, following the usual procedure, we calculate

$$\frac{\mathrm{d}}{\mathrm{d}t^\mu}\left[K_{\mu\nu}t^\mu t^\nu - k(g_{\mu\nu}t^\mu t^\nu - 1)\right] = 0. \tag{30.48}$$

The result is

$$K_{\mu\nu}t^\nu - kg_{\mu\nu}t^\nu = 0. \tag{30.49}$$

Multiply through by the metric components to find

$$g^{\lambda\mu}K_{\mu\nu}t^\nu - k\delta^\lambda{}_\nu t^\nu = 0. \tag{30.50}$$

This is an eigenvalue equation for the matrix $\underline{A}$ with components[20] $A^\lambda{}_\nu = g^{\lambda\mu}K_{\mu\nu}$. It is the eigenvalues k_1 and k_2 of $\underline{A}$ that will be important. To see how, we multiply through by $t_\lambda = g_{\lambda\sigma}t^\sigma$ and contract the index λ to obtain

$$t^\mu K_{\mu\nu}t^\nu - kt_\nu t^\nu = 0, \tag{30.54}$$

and so

$$k = K_{\mu\nu}t^\mu t^\nu = \kappa_n^{\text{extremal}}. \tag{30.55}$$

We conclude that the two eigenvalues k_1 and k_2 of the matrix $\underline{A}$ with components $g^{\lambda\mu}K_{\mu\nu}$ give the extremal values of the curvature $\kappa_n^{\text{extremal}}$.

The approach from the last example gives us access to two *invariants* of the matrix $\underline{A}$ that characterize the curvature: (i) the determinant $\det\underline{A} = \det\underline{K}/\det\underline{g} = k_1 k_2$ and (ii) the trace $\text{Tr}\underline{A} = \text{Tr}(\underline{g}^{-1}\underline{K}) = k_1+k_2$. It turns out that the Gaussian curvature or **intrinsic curvature** that we previously called K is given by the determinant $k_1 k_2$, while the trace $(k_1 + k_2)$ gives a quantity called the **extrinsic curvature**. What do these terms mean? A good example is a cylinder (see Fig. 30.8) which has extrinsic curvature by virtue of the way in which the two-dimensional surface is embedded in three-dimensional space. However, it is not intrinsically curved because that surface can easily be unwrapped and placed on a flat surface (as shown in the figure).

Unlike the cylinder (see Fig. 30.8), a sphere cannot be made from a flat piece of paper (at least, not without cutting the paper) and so possesses both intrinsic and extrinsic curvature. The cylinder, on the other other hand, has extrinsic curvature but zero intrinsic curvature. It is therefore intrinsic curvature, that cannot be removed by unwrapping the surface without cutting the paper.[21]

Example 30.10

In this example, we evaluate the intrinsic and extrinsic curvature of two surfaces embedded in three-dimensional space.

(a) Consider a parabolic surface. This is described by coordinates $(x^1, x^2) = (x, y)$ with

$$\boldsymbol{X} = \begin{pmatrix} x \\ y \\ \frac{1}{2}ux^2 + \frac{1}{2}vy^2 \end{pmatrix}. \tag{30.56}$$

[19]Note how the metric is needed here as the surface is, in general, curved.

[20]Example: For the cylinder in Example 30.8 we write a matrix with components $K_{\mu\nu}$

$$\underline{K} = \begin{pmatrix} -a & 0 \\ 0 & 0 \end{pmatrix}, \tag{30.51}$$

and a matrix with components $g^{\lambda\mu}$

$$\underline{g} = \begin{pmatrix} 1/a^2 & 0 \\ 0 & 1 \end{pmatrix}, \tag{30.52}$$

which gives the matrix

$$\underline{A} = \begin{pmatrix} -1/a & 0 \\ 0 & 0 \end{pmatrix}. \tag{30.53}$$

Fig. 30.8 A cylinder (the tin can) looks curved, but it is not intrinsically curved but only curved extrinsically (by the way it is embedded in three-dimensional space). The cylindrical surface can be unwound and placed on a flat surface without tearing or distorting. For a cylinder of radius a, the eigenvalues of the matrix $\underline{A}$ are $k_1 = -1/a$ and $k_2 = 0$, so we have zero intrinsic curvature $K = k_1 k_2 = 0$, even though the extrinsic curvature is non-zero $(k_1 + k_2 = -1/a)$.

[21]For one-dimensional curves, a notion of intrinsic curvature doesn't really make sense, since the curve can always be flattened out; however curved around your shoelaces are, you can always unthread them and lay them out in straight lines. Thus, one-dimensional curves only possess extrinsic curvature.

The metric is then described by the matrix

$$g_{\mu\nu} = \begin{pmatrix} 1 + u^2 x^2 & uvxy \\ uvxy & 1 + v^2 y^2 \end{pmatrix}. \tag{30.57}$$

We have the basis vectors and unit normal

$$\boldsymbol{e}_x = \begin{pmatrix} 1 \\ 0 \\ ux \end{pmatrix}, \quad \boldsymbol{e}_y = \begin{pmatrix} 0 \\ 1 \\ vy \end{pmatrix}, \quad \hat{\boldsymbol{n}} = \frac{1}{|\boldsymbol{n}|} \begin{pmatrix} -ux \\ -vy \\ 1 \end{pmatrix}, \tag{30.58}$$

where $|\boldsymbol{n}| = (1 + u^2 x^2 + v^2 y^2)^{\frac{1}{2}}$. Taking derivatives yields

$$\frac{\partial \boldsymbol{e}_x}{\partial x} = \begin{pmatrix} 0 \\ 0 \\ u \end{pmatrix}, \quad \frac{\partial \boldsymbol{e}_y}{\partial y} = \begin{pmatrix} 0 \\ 0 \\ v \end{pmatrix}, \tag{30.59}$$

with other derivatives vanishing. Let's concentrate on the curvature evaluated at the origin, where we have that $g_{uu} = g_{vv} = 1$, $g_{uv} = g_{vu} = 0$, $K_{11} = u$ and $K_{22} = v$. We have that the intrinsic curvature is $K = uv$ and the extrinsic curvature is $u + v$.

(b) Next, consider the spherical surface with coordinates $(x^1, x^2) = (\theta, \phi)$. The surface is written as[22]

$$\boldsymbol{X} = \begin{pmatrix} a \sin\theta \cos\phi \\ a \sin\theta \sin\phi \\ a \cos\theta \end{pmatrix}. \tag{30.62}$$

The basis vectors and unit normal follow as

$$\boldsymbol{e}_\theta = \begin{pmatrix} a \cos\theta \cos\phi \\ a \cos\theta \sin\phi \\ -a \sin\theta \end{pmatrix}, \quad \boldsymbol{e}_\phi = \begin{pmatrix} -a \sin\theta \sin\phi \\ a \sin\theta \cos\phi \\ 0 \end{pmatrix}, \quad \hat{\boldsymbol{n}} = \begin{pmatrix} \sin\theta \cos\phi \\ \sin\theta \sin\phi \\ \cos\theta \end{pmatrix}. \tag{30.63}$$

Taking derivatives we obtain

$$\frac{\partial \boldsymbol{e}_\theta}{\partial \theta} = -a\hat{\boldsymbol{n}}, \quad \frac{\partial \boldsymbol{e}_\theta}{\partial \phi} = \frac{\partial \boldsymbol{e}_\phi}{\partial \theta} = a \cot\theta\, \boldsymbol{e}_\phi, \quad \frac{\partial \boldsymbol{e}_\phi}{\partial \phi} = -a \sin\theta \cos\theta\, \boldsymbol{e}_1 - a \sin^2\theta\, \hat{\boldsymbol{n}}. \tag{30.64}$$

We have non-zero components $K_{11} = -a$ and $K_{22} = -a \sin^2\theta$. We find an intrinsic curvature $\det\boldsymbol{K} / \det\boldsymbol{g} = 1/a^2$ and an extrinsic curvature $\mathrm{Tr}(\boldsymbol{g}^{-1}\boldsymbol{K}) = -2/a$.

30.6 Riemann's project

Bernhard Riemann aimed to generalize Gauss' remarkable result so that any observer confined to an n-dimensional space would be able to evaluate the intrinsic curvature.[23] The success of his approach laid the foundations for general relativity.

Broadly speaking, Riemann's method was to ask: when does the introduction of a new coordinate system change a metric $\boldsymbol{g}$ with components $g_{\mu\nu}$, defined in n-dimensional space, into some other metric $\boldsymbol{a}$ with components $a_{\mu\nu}$? Riemann argued that the metric $\boldsymbol{g}$ is determined by $n(n+1)/2$ functions[24] but that a new coordinate system can be defined using n functions. As a result, the new metric must be determined by $n(n+1)/2 - n = n(n-1)/2$ functions. Riemann then claimed that there is a quadratic function[25] Q that uniquely assigns a number to a two-dimensional space W and that (i) for a two-dimensional manifold $-3Q(W)$ is the Gaussian curvature; (ii) that for higher dimensional manifolds, $-3Q(W)$ describes the Gaussian curvature of the two-dimensional subspace W of the manifold; and (iii) there are $n(n-1)/2$ independent

[22]Recall that the metric tensor $\boldsymbol{g}$ has components

$$g_{\mu\nu} = \begin{pmatrix} a^2 & 0 \\ 0 & a^2 \sin^2\theta \end{pmatrix}, \tag{30.60}$$

so that $\det\boldsymbol{g} = a^4 \sin^2\theta$. The inverse metric has components

$$g^{\mu\nu} = \begin{pmatrix} \frac{1}{a^2} & 0 \\ 0 & \frac{1}{a^2 \sin^2\theta} \end{pmatrix}. \tag{30.61}$$

[23]Riemann sets out his aims in his inaugural lecture delivered at the University of Göttingen in 1853. The lecture was intended to be accessible to the entire faculty and so contains minimal mathematical detail. It's immense significance is explained in Spivak, Vol. II.

[24]Why? Since $g_{\mu\nu} = g_{\nu\mu}$, then the independent components (i) are the diagonal elements (of which there are n) and (ii) half of the off-diagonal elements (and there are $(n^2 - n)$ off-diagonal elements). Adding these we have $n + (n^2 - n)/2 = n(n+1)/2$, as claimed.

[25]The function Q is a function of $2n$ variables which take the form of two vectors $\boldsymbol{X}$ and $\boldsymbol{Y}$. We assume these two vectors span the two-dimensional space W and simply write $Q(W)$ for simplicity here.

two-dimensional subspaces for an n-dimensional vector space; if $Q(W)$ is known for each of these, then the metric is completely determined. This all seems a little obscure, but was shortly to be reformulated into something more familiar.

Some eight years later, in 1861, Riemann submitted a paper to the Paris Academy as an entry to a competition aiming to provide the answer to a question in the problem of heat conduction.[26] The paper contains the first instance of (what we would now call) the Riemann curvature tensor $\boldsymbol{R}(\ ,\ ,\ ,\ )$ in terms of the components of the metric. Riemann's paper asked what conditions make a space flat and gave the answer that what is needed is that the components of $\boldsymbol{R}$ should vanish. The number $Q(W)$ turns out to be proportional to the output of the (0,4) version of the tensor $\boldsymbol{R}$ when its slots are filled by linearly independent vectors $\boldsymbol{X}$ and $\boldsymbol{Y}$ that span the two-dimensional subspace W.[27] The significance of this is that *the curvature of space determines the metric*.

We shall not examine the details of Riemann's method in any more detail.[28] The important point is that Riemann's curvature tensor, which can be written in terms of the components of the metric only, not only contains the Gaussian curvature as a special case in two dimensions but generalizes a notion of curvature to higher dimensional cases. The promised link in two dimensions between the Gaussian curvature K and the Riemann tensor is

$$K = \frac{R_{1212}}{g}, \tag{30.65}$$

where $g = \det g_{\mu\nu}$, the determinant of the metric tensor $\boldsymbol{g}$.

[26] Riemann, whose health was poor at the time, didn't win the prize owing to the lack of detail in his paper. Nobody else won the prize and it was withdrawn in 1868.

[27] The explicit equation is $3Q(\boldsymbol{X},\boldsymbol{Y}) = -R(\boldsymbol{X},\boldsymbol{Y},\boldsymbol{X},\boldsymbol{Y})$, where $\boldsymbol{R}$ is the $(0,4)$ version of the Riemann tensor, with components $R_{\mu\nu\alpha\beta}$.

[28] The interested reader should consult the masterful discussion in Spivak, Vol. II.

Example 30.11

We can now justify the link between the Riemann tensor and the Gaussian curvature. The simplest route is to use eqn 30.65 to write $4gR_{1212} = 4g^2 K$. We then make use of the relationship from eqn 11.22 between the coordinates of $\boldsymbol{R}$ and the metric

$$R_{1212} = \frac{1}{2}\left(g_{12,21} - g_{11,22} + g_{21,12} - g_{22,11}\right) + g_{\sigma\rho}\left(\Gamma^{\sigma}{}_{21}\Gamma^{\rho}{}_{12} - \Gamma^{\sigma}{}_{22}\Gamma^{\rho}{}_{11}\right). \tag{30.66}$$

We also need the expression for the connection coefficients

$$\Gamma^{\rho}{}_{\mu\sigma} = \frac{g^{\rho\lambda}}{2}\left(\frac{\partial g_{\lambda\mu}}{\partial x^{\sigma}} + \frac{\partial g_{\lambda\sigma}}{\partial x^{\mu}} - \frac{\partial g_{\mu\sigma}}{\partial x^{\lambda}}\right), \tag{30.67}$$

so that, for example,

$$\Gamma^{\rho}{}_{12} = g^{\rho 1}\left(g_{11,2} + g_{12,1} - g_{12,1}\right) + g^{\rho 2}\left(g_{21,1} + g_{22,1} - g_{12,2}\right). \tag{30.68}$$

If we also use the result that

$$g^{11} = g_{22}/g, \quad g^{12} = g^{21} = -g_{12}/g = g_{21}/g, \quad g^{22} = g_{22}/g, \tag{30.69}$$

then, put all together with eqn 30.25 allows one to verify the claim. Try it and see!

This completes our review of classical curvature. In the next chapter, we shall begin to examine the more modern approach to geometry that most naturally fits with the physics of general relativity.

Chapter summary

- The curvature of a line can be described in differential geometry using the Frenet–Serret equations.
- Gaussian curvature allows the measurement of the intrinsic curvature of a two-dimensional surface, using measurements made entirely within that surface.
- Riemann's result upgrades the Gaussian curvature to higher dimensions.

Exercises

(30.1) A mechanical definition of the curvature of a curve is possible if we imagine a unit mass traversing the curve (confined to a plane) at unit speed. The curvature is the force perpendicular to the curve required to maintain this motion. Prove that this definition is equivalent to that used in this chapter.

(30.2) Another method to compute Gaussian curvature is to use parallel transport to carry a vector around a loop on a surface. We then define

$$\left(\begin{array}{c}\text{Gaussian}\\\text{curvature}\end{array}\right) = \frac{\left(\begin{array}{c}\text{angle vector}\\\text{turns through}\end{array}\right)}{\left(\begin{array}{c}\text{area}\\\text{of loop}\end{array}\right)}. \tag{30.70}$$

Show that for a loop on the surface of a 2-sphere

$$(\text{curvature}) = \frac{1}{a^2}. \tag{30.71}$$

(30.3) *We first met the Bertand–Diquet–Puiseux theorem in Exercise 3.3. Here we use it again.*
Another way of describing Gaussian curvature is to use the (normalized) difference between the circumference of a circle in the plane and a circle on the surface. The circumference of the circular locus of points a distance ϵ from a point differs from the Euclidean value $2\pi\epsilon$ by a correction factor. The curvature is defined as

$$K = \lim_{\epsilon \to 0} \frac{6}{\epsilon^2}\left(1 - \frac{\text{circumference}}{2\pi\epsilon}\right). \tag{30.72}$$

Use this theorem to work out whether a space described by the following metric is flat

$$ds^2 = dr^2 + \sin^2 r d\theta^2. \tag{30.73}$$

(30.4) (a) Consider a matrix form of eqn 30.49. Show that

$$\det(\underline{g}^{-1}\underline{K} - k\underline{I}) = 0. \tag{30.74}$$

(b) By expanding the previous equation in terms of a trace, show that

$$k^2 - \mathrm{Tr}(\underline{g}^{-1}\underline{K})k + \det(\underline{g}^{-1}\underline{K}) = 0. \tag{30.75}$$

(c) Use this to argue that

$$\det(\underline{g}^{-1}\underline{K}) = \frac{\det \underline{K}}{\det \underline{g}}, \tag{30.76}$$

gives the product of eigenvalues, and

$$\mathrm{Tr}(\underline{g}^{-1}\underline{K}), \tag{30.77}$$

represents the sum of the eigenvalues.

(30.5) (a) Consider a curve embedded in a surface. Differentiate the equation $\boldsymbol{t} \cdot \hat{\boldsymbol{n}} = 0$ along the curve and show that

$$\boldsymbol{t} \cdot \frac{d\hat{\boldsymbol{n}}}{ds} = -\hat{\boldsymbol{n}} \cdot \frac{d\boldsymbol{t}}{ds} = -K_{\mu\nu}t^{\mu}t^{\nu}. \tag{30.78}$$

(b) The previous equation can be used to derive a relationship between $\hat{\boldsymbol{n}}$ and $K_{\mu\nu}$. Show that

$$g_{\mu\sigma}\frac{\partial \hat{n}^{\sigma}}{\partial x^{\nu}} + K_{\mu\nu} = 0. \tag{30.79}$$

This is known as **Weingarten's equation**, which is usually written as

$$K_{\mu\nu} = -\boldsymbol{e}_{\mu} \cdot \frac{\partial \hat{\boldsymbol{n}}}{\partial x^{\nu}}, \tag{30.80}$$

and gives us another method of extracting $K_{\mu\nu}$.

(c) Using Weingarten's equation in the form

$$\frac{\partial \hat{\boldsymbol{n}}}{\partial x^\nu} = -K^\mu{}_\nu \boldsymbol{e}_\mu, \qquad (30.81)$$

show

$$K = \frac{\det \boldsymbol{K}}{\det \boldsymbol{g}}. \qquad (30.82)$$

as we had before.

(30.6) (a) Verify the result for the parabolic surface in Example 30.10 using Gauss's equation (eqn 30.37) to compute the components $K_{\mu\nu}$ for a general point (i.e. not the origin).

(b) Show that the same result is found using Weingarten's equation from the previous problem.

(c) Verify the result is obtained for the intrinsic curvature K using eqn 30.35.

(30.7) *Consider a three-dimensional spacelike hypersurface Σ with a timelike unit-normal $\boldsymbol{n}$. Construct a (1,1) projection operator $\boldsymbol{P}$ with components*

$$P^\mu{}_\nu = \delta^\mu{}_\nu + n^\mu n_\nu. \qquad (30.83)$$

(a) Evaluate $P^\mu{}_\alpha P^\alpha{}_\nu$.

(b) Show that the operator $P^\mu{}_\nu$ acts on vector $\boldsymbol{X}$ to project a vector $\bar{\boldsymbol{X}}$ that is tangent to the surface.

(c) Define an **induced metric h** via

$$h_{\alpha\beta} = P^\mu{}_\alpha P^\nu{}_\beta g_{\mu\nu}. \qquad (30.84)$$

Show that (i) $h_{\alpha\beta} = g_{\alpha\beta} + n_\alpha n_\beta$ and (ii) $h_{\alpha\beta}\bar{X}^\alpha \bar{Y}^\beta = g_{\alpha\beta}\bar{X}^\alpha \bar{Y}^\beta$.

(30.8) *Now define an **induced covariant derivative** $\boldsymbol{D}$, whose components are given via*

$$(\boldsymbol{D}_\mu \boldsymbol{X})^\nu = P^\nu{}_\alpha P^\beta{}_\mu (\boldsymbol{\nabla}_\beta \boldsymbol{X})^\alpha, \qquad (30.85)$$

where, now, the vector field $\boldsymbol{X}$ is restricted to be tangent to Σ. The double projection, as opposed to the single projection $P^\beta{}_\mu (\boldsymbol{\nabla}_\beta \boldsymbol{X})^\alpha$, is designed to output a derivative as a vector that is within Σ, cutting off the normal part.

(a) By generalizing this definition, show that $(\boldsymbol{D}_\mu \boldsymbol{h})_{\alpha\beta} = 0$.

(b) Show that, since $\boldsymbol{X}$ is tangent to Σ, we have $P^\mu{}_\nu X^\nu = X^\mu$.

(c) Starting from the fact that $n_\mu X^\mu = 0$, show further that

$$n_\beta P^\mu{}_\alpha (\boldsymbol{\nabla}_\mu \boldsymbol{X})^\beta = -X^\nu P^\beta{}_\nu P^\mu{}_\alpha (\boldsymbol{\nabla}_\mu \boldsymbol{n})_\beta. \qquad (30.86)$$

*This motivates the definition of the **extrinsic curvature tensor $\boldsymbol{K}$** with components*

$$K_{\alpha\beta} = -P^\mu{}_\alpha P^\nu{}_\beta (\boldsymbol{\nabla}_\mu \boldsymbol{n})_\nu. \qquad (30.87)$$

(d) Show finally that, for vector field $\boldsymbol{Y}$ that is also tangent to Σ, we have

$$\boldsymbol{\nabla_Y X} = \boldsymbol{D_Y X} + K(\boldsymbol{X}, \boldsymbol{Y})\boldsymbol{n}. \qquad (30.88)$$

Hint: The first term is clearly the part tangent to Σ. To obtain the second term, consider the normal component of $P^\mu{}_\alpha (\boldsymbol{\nabla}_\mu \boldsymbol{X})^\beta$.
This formalism is useful in that we can use it to describe the Riemann tensor in terms of quantities projected into the hyperspace Σ. These are the **Gauss equation**

$$P^\mu{}_\alpha P^\nu{}_\beta P^\sigma{}_\gamma P^\rho{}_\delta R_{\mu\nu\sigma\rho}$$
$$=^{(3)} R_{\alpha\beta\gamma\delta} - K_{\alpha\delta}K_{\beta\gamma} + K_{\alpha\gamma}K_{\beta\delta}, \qquad (30.89)$$

and the **Codazzi equation**

$$P^\mu{}_\alpha P^\nu{}_\beta P^\sigma{}_\gamma n^\rho R_{\mu\nu\sigma\rho} = D_\beta K_{\alpha\gamma} - D_\alpha K_{\beta\gamma}. \qquad (30.90)$$

31 A reintroduction to geometry

[1]That is to say, we leave out the structure given by the metric, its vector-space structure and also its topological structure. We shall gradually reintroduce these features in the following chapters.

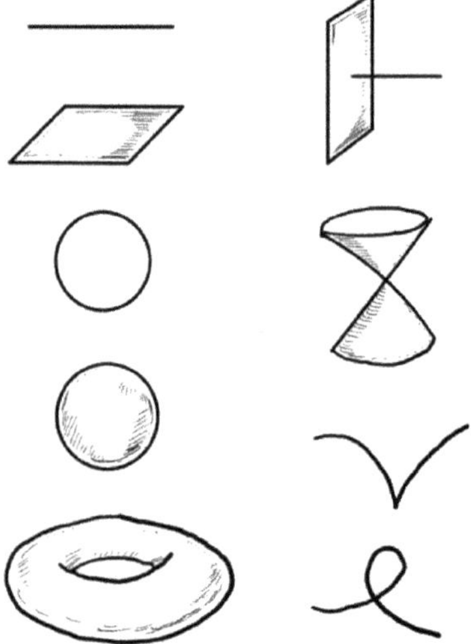

Fig. 31.1 Left: some manifolds, smooth enough that the region around any point looks locally flat; right: some non-manifolds with points whose neighbourhood does not look smooth at any level of magnification.

It is no matter what you teach them first, any more than what leg you shall put into your breeches first.
Samuel Johnson (1709–1784)

In this chapter, we once again meet the geometry of vectors and 1-forms. Our goal is to more thoroughly define an approach that is free from the shackles of coordinates. This will hinge on the observation that each vector is equivalent to a derivative.

Although not essential for our purposes here, it is worth keeping the idea in mind that the arena in which we work in this chapter is a **manifold**. This is a space, often called $\mathcal{M}$, that is smooth: locally resembling the smoothness of $\mathbb{R}^n$, the usual Euclidean (i.e. flat) n-dimensional space. It is this smoothness that characterizes a manifold, since we leave out the rest of the rich structure[1] of $\mathbb{R}^n$. In this smooth space, there are points called things like $\mathcal{P}$, $\mathcal{Q}$, $\mathcal{A}$ and $\mathcal{B}$. We can cover patches of this space with coordinates, so that $\mathcal{P}$ can be described by a set of n coordinates $(x^1...x^n)$, although this won't always be necessary. Curves are well-defined objects in our manifolds, parametrized by some quantity λ that varies monotonically along the curve. Differentiation, which measures changes along curves, is also a well-defined operation.

Example 31.1

It is useful to consider which spaces are smooth enough to qualify as a manifold. Examples of spaces that are manifolds include all of the n-dimensional Euclidean spaces called $\mathbb{R}^n$, which include $\mathbb{R}^1$ (a line), $\mathbb{R}^2$ (a plane) etc. (These clearly are locally identical to $\mathbb{R}^n$, since they *are* the spaces $\mathbb{R}^n$.) Other good examples which are smooth enough to look locally flat are (i) the one-dimensional circle, called S^1; (ii) the two-dimensional surface of a sphere, called S^2; and (iii) the two-dimensional surface of a torus. Some of these are shown on the left-hand side of Fig. 31.1. Perhaps more illuminating are spaces that *aren't* smooth enough to qualify as manifolds. Examples of these include (i) a line that juts out of a plane or (ii) a double cone; (iii) a line with a kink; (iv) a line that crosses itself. All of these structures have a point which, even when blown up to a very large size, will never look locally flat. These latter examples are shown on the right-hand side of Fig. 31.1.

Our first task in this chapter will be to identify vectors and 1-forms. These don't actually live in the manifold $\mathcal{M}$ itself, but in related manifolds. Specifically, a vector defined at a point $\mathcal{P}$ lives a manifold called a tangent space, while 1-forms live in a dual space. For now, we will keep things as simple as possible and put these subtleties aside, retaining only the notion of points in a smooth manifold $\mathcal{M}$ that, locally, looks like flat Euclidean space. So forget, temporarily, about the metric, and also the covariant derivatives and curvature tensors that the metric field can generate, as we look at how a smooth space, with a minimum of structure, can host a powerful geometry that will allow us many insights into the physics of relativity.

31.1 Old notions of vectors and gradients

We want to define vectors and describe surfaces in space. Let's review the old technology for doing this.

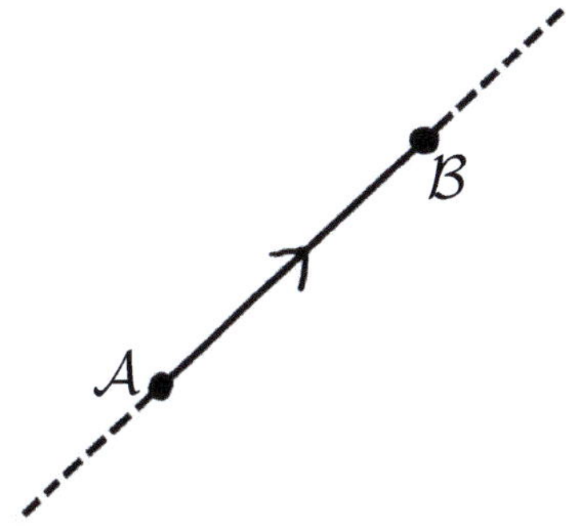

Fig. 31.2 A vector stretching between points $\mathcal{A}$ and $\mathcal{B}$.

Example 31.2

The first time we meet vectors we are taught to picture them as a directed straight line linking two points. For example, the line from point $\mathcal{A}$ to point $\mathcal{B}$ (Fig. 31.2). We then write a vector in terms of coordinates as

$$\boldsymbol{u} = u^\alpha \boldsymbol{e}_\alpha = u^0 \boldsymbol{e}_0 + u^1 \boldsymbol{e}_1 + u^2 \boldsymbol{e}_2 + u^3 \boldsymbol{e}_3 + \dots \tag{31.1}$$

or $\boldsymbol{u} = u^\alpha \boldsymbol{e}_\alpha$, where the $\boldsymbol{e}_\alpha$ make an appropriate basis, spanning the space in which we're working. Often we use an orthonormal basis such that $\boldsymbol{e}_\alpha \cdot \boldsymbol{e}_\beta = \delta_{\alpha\beta}$.

If we are presented with a curved surface defined, in three spatial dimensions, by a function such as $z = f(x, y)$, a particularly useful object is the **gradient vector** of the surface. We find this by rewriting f as a function $h(x, y, z) = 0$, whose gradient vector is

$$\boldsymbol{\nabla} h(x, y, z) = \frac{\partial h}{\partial x} \boldsymbol{e}_x + \frac{\partial h}{\partial y} \boldsymbol{e}_y + \frac{\partial h}{\partial z} \boldsymbol{e}_z. \tag{31.2}$$

More generally we could write

$$\boldsymbol{\nabla} h = \frac{\partial h}{\partial x^\alpha} \boldsymbol{e}_\alpha. \tag{31.3}$$

This defines[2] a vector pointing normal to the **tangent plane** of the surface. If we combine a vector $\boldsymbol{v}$ and the gradient vector using the dot product, we obtain a useful object: the **directional derivative** of h, often denoted $\partial_{\boldsymbol{v}} h$, given by

$$\partial_{\boldsymbol{v}} h = \boldsymbol{v} \cdot \boldsymbol{\nabla} h = v^\alpha \frac{\partial h}{\partial x^\alpha}, \tag{31.4}$$

which tells us the change in the function h along the direction of the vector $\boldsymbol{v}$. That is, the value of h at the tip of the vector $\boldsymbol{v}$, minus the value of h at the base of $\boldsymbol{v}$. The output of the directional derivative is a number.

[2]Interpreting this as a vector immediately looks wrong from our tensor conventions of summing over one up and down index, since it appears that we have components $h_{,\alpha} \boldsymbol{e}_\alpha$.

For our purposes this old technology simply will not do. However, these ideas, when taken together with some new ones, allow us to tighten up the definitions and come up with a more useful geometry. For example, we will see how having 1-forms allows us to abandon this rather forced (and non-covariant) notion of the gradient outputting a vector normal to a surface.

31.2   Vectors and vector fields

We often think of a vector as a directed line $\mathcal{B} - \mathcal{A}$, joining two points $\mathcal{A}$ and $\mathcal{B}$ as in Fig. 31.2. Defining a vector in terms of two points is cumbersome. We would prefer to have the concept of a vector at a single point $\mathcal{P}$. Let's parametrize a straight line with a parameter λ by writing $\mathcal{P}(\lambda) = \mathcal{A} + \lambda(\mathcal{B} - \mathcal{A})$. This allows us to extract the vector as the difference between tip and base via

$$\frac{\mathrm{d}\mathcal{P}(\lambda)}{\mathrm{d}\lambda} = \mathcal{B} - \mathcal{A}. \tag{31.5}$$

This idea of a vector as equivalent to a derivative is the key one in this chapter.

Advancing beyond straight lines, we can define a curve in terms of a another parametrized path[3] $\mathcal{P}(\lambda)$, as we have in Fig. 31.3. The derivative formulation of vectors in the previous equation allows us to define a **tangent vector** to this curve via

$$\boldsymbol{v} = \frac{\mathrm{d}\mathcal{P}(\lambda)}{\mathrm{d}\lambda}. \tag{31.6}$$

An interpretation of this expression, due to Élie Cartan, is that the vector represents the movement of the point $\mathcal{P}$ with the derivative evaluating the difference between the point at the tip of the tangent vector, and the point at the base of this vector. Of course, we don't only have access to the parameter λ. We often describe curves in the manner shown in Fig. 31.4 using a coordinate system like x^α. Using these coordinates, the parametrized path is written as $x^\alpha(\lambda)$, and we use the chain rule to say that our vector is expressed as

$$\boldsymbol{v} = \frac{\mathrm{d}\mathcal{P}}{\mathrm{d}\lambda} = \frac{\mathrm{d}x^\alpha}{\mathrm{d}\lambda}\frac{\partial\mathcal{P}}{\partial x^\alpha}. \tag{31.7}$$

The next step is to realize that, instead of interpreting a vector as the movement of the point $\mathcal{P}$, we could strip off the point to make a more general **vector operator** along the curve

$$\boldsymbol{v}[\ ] = \frac{\mathrm{d}}{\mathrm{d}\lambda} = \frac{\mathrm{d}x^\alpha}{\mathrm{d}\lambda}\frac{\partial}{\partial x^\alpha}, \tag{31.8}$$

where the brackets in $\boldsymbol{v}[\ ]$ indicate that we need to provide a point on the curve as an input to the vector operator. In other words, the tangent vector $\boldsymbol{v}$ at a point is identified with a derivative $\mathrm{d}/\mathrm{d}\lambda$ at that point.[4] Therefore, we don't need to view the vector as an arrow representing the movement of a point, but rather as an object that is attached to a particular point on the curve. The vector will vary as you evaluate it at different points along the curve.

Finally, we return to our old notion of a vector, written as $\boldsymbol{v} = v^\alpha \boldsymbol{e}_\alpha$. Comparison with our new formulation of a vector in eqn 31.8 allows us to identify components and basis vectors as follows:

$$\frac{\mathrm{d}x^\alpha}{\mathrm{d}\lambda} = v^\alpha, \quad \frac{\partial}{\partial x^\alpha} = \boldsymbol{e}_\alpha, \tag{31.9}$$

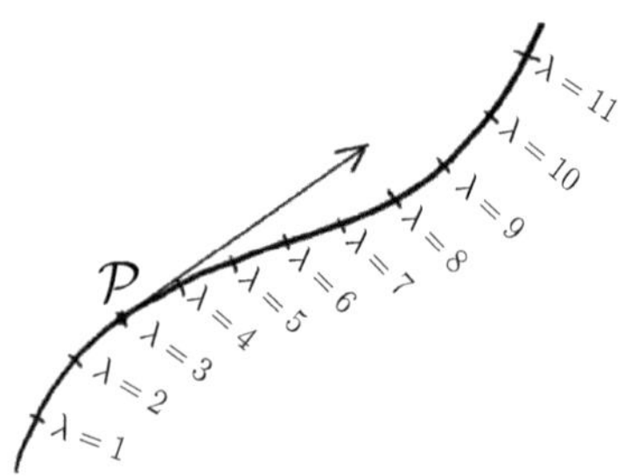

Fig. 31.3 A vector as a derivative of a parametrized curve.

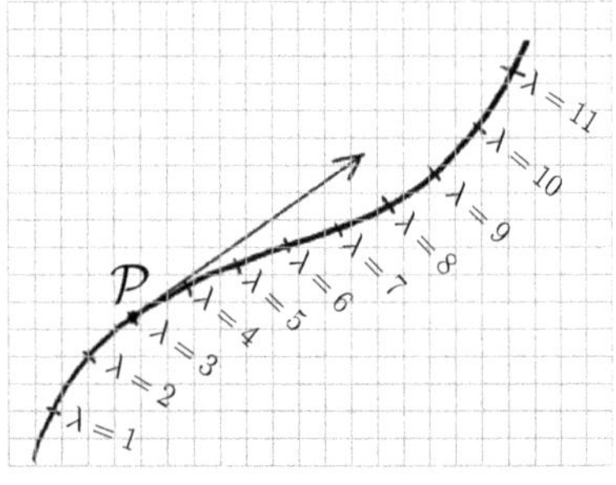

Fig. 31.4 A vector as a derivative in a coordinate system. We use the relationship between the point and the coordinates and also between the coordinates and the parametrization of the curve. These are linked by the chain rule.

[4]Or, if you prefer, the tangent vector $\boldsymbol{v}$ *is a derivative* d/dλ.

where we note that $\boldsymbol{e}_\alpha$ could be written as $\boldsymbol{e}_\alpha[\ ]$, since this is the part in which we input points on the curve.[5] Mixing this notation, allows us to write our vector as

$$\boldsymbol{v}[\ ] \equiv \frac{\mathrm{d}}{\mathrm{d}\lambda} = v^\alpha \frac{\partial}{\partial x^\alpha} = \partial_{\boldsymbol{v}}, \tag{31.10}$$

where we've introduced the directional derivative operator $\partial_{\boldsymbol{v}}$ in the last step.

We therefore have a new definition of a vector as identified with a directional derivative operator. This is sloganized as *every vector can be identified with a derivative*. We summarize our progress in the next box.

A vector $\boldsymbol{v}$ that is tangent to a curve $\mathcal{P}(\lambda)$ can be written as a derivative operator

$$\boldsymbol{v}[\ ] = \frac{\mathrm{d}}{\mathrm{d}\lambda} = \frac{\mathrm{d}x^\alpha}{\mathrm{d}\lambda}\frac{\partial}{\partial x^\alpha} = v^\alpha \boldsymbol{e}_\alpha, \tag{31.11}$$

where λ parametrizes the curve. The vector's components are $v^\alpha = \frac{\mathrm{d}x^\alpha}{\mathrm{d}\lambda}$ and the basis vectors are $\boldsymbol{e}_\alpha = \frac{\partial}{\partial x^\alpha}$.

The derivative in the definition of the vector could continue to be fed points like $\mathcal{P}$ on the curve as it was in eqn 31.6. However, its real value is that we can input functions into this vector operator. We shall write the action of a vector field on a function[6] $f(x)$ as

$$\begin{aligned} \boldsymbol{v}[f(x)] &= \frac{\mathrm{d}f(x)}{\mathrm{d}\lambda} = \frac{\mathrm{d}x^\alpha}{\mathrm{d}\lambda}\frac{\partial f(x)}{\partial x^\alpha} \\ &= v^\alpha \frac{\partial f(x)}{\partial x^\alpha} = \partial_{\boldsymbol{v}} f(x). \end{aligned} \tag{31.12}$$

This object evaluates the change in the function $f(x)$ along the direction of the vector $\boldsymbol{v}$, that is, the directional derivative.[7] So to recap:

Every vector is equivalent to a differential operator that inputs a function and outputs the directional derivative with respect to the vector.

This idea that **every vector is a derivative** links the concepts of geometry and analysis, as shown in Fig. 31.5. This helps explain the power of these techniques in the study of general relativity.

We can show that tangent vectors defined via derivatives evaluated at a point $\mathcal{P}$, give a **vector space**. This relies on the notion that lots of curves can pass through a given point $\mathcal{P}$, allowing us to compare the different tangent vectors to each curve evaluated at $\mathcal{P}$ (Fig. 31.6).

Example 31.3

A vector $\boldsymbol{u}$, tangent to the curve parametrized by μ, is written in coordinate space

$$\boldsymbol{u} = \frac{\mathrm{d}}{\mathrm{d}\mu} = \frac{\mathrm{d}x^\alpha}{\mathrm{d}\mu}\frac{\partial}{\partial x^\alpha}. \tag{31.13}$$

Consider adding two different vectors at the same point $\mathcal{P}$ (Fig. 31.6). In the derivative language, these vectors are tangent to two different curves at $\mathcal{P}$: a curve $x(\lambda)$ and a curve $x(\mu)$. We write a linear combination, in terms of constants a and b, as

$$a\boldsymbol{v} + b\boldsymbol{u} = a\frac{\mathrm{d}}{\mathrm{d}\lambda} + b\frac{\mathrm{d}}{\mathrm{d}\mu} = \left(a\frac{\mathrm{d}x^\alpha}{\mathrm{d}\lambda} + b\frac{\mathrm{d}x^\alpha}{\mathrm{d}\mu}\right)\frac{\partial}{\partial x^\alpha}, \tag{31.14}$$

[5] In other words, the components tell us how the coordinates x^α change with the curve's parameter λ; the basis vectors represent the rate of change with respect to our choice of coordinates.

[6] We continue to use square brackets for this, and round brackets for the vector's slot that inputs a 1-form.

[7] It might sometimes help to continue to think of this as the value of the function at the tip of the vector, minus the value at the base of the vector.

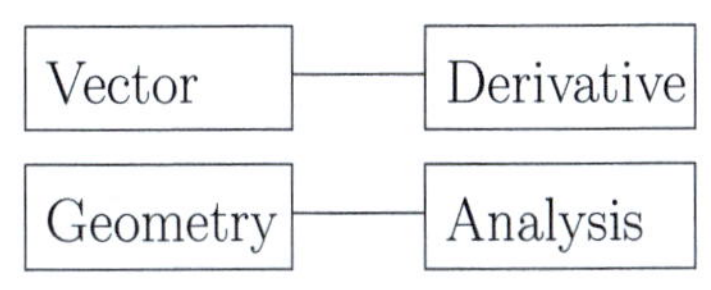

Fig. 31.5 Links between geometry and analysis.

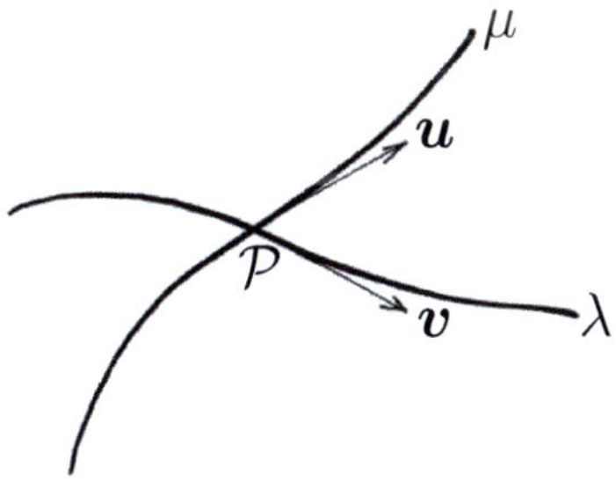

Fig. 31.6 Two of the many curves passing through point $\mathcal{P}$ parametrized by λ and μ, respectively. Their tangents at $\mathcal{P}$ are shown.

where we've expanded both vectors in terms of our choice of coordinates x^α in the second step, so that they're referred to the same set of basis vectors $\boldsymbol{e}_\alpha = \partial/\partial x^\alpha$. The previous equation defines a new vector $\boldsymbol{w}$, with components $\left(a\frac{\mathrm{d}x^\alpha}{\mathrm{d}\lambda} + b\frac{\mathrm{d}x^\alpha}{\mathrm{d}\mu}\right)$. This new vector must be tangent to some other curve through $\mathcal{P}$ with, say, a parameter ϕ, so we write

$$\boldsymbol{w} = \frac{\mathrm{d}}{\mathrm{d}\phi} = \left(a\frac{\mathrm{d}x^\alpha}{\mathrm{d}\lambda} + b\frac{\mathrm{d}x^\alpha}{\mathrm{d}\mu}\right)\frac{\partial}{\partial x^\alpha} = w^\alpha \boldsymbol{e}_\alpha. \tag{31.15}$$

We conclude that

$$\frac{\mathrm{d}}{\mathrm{d}\phi} = a\frac{\mathrm{d}}{\mathrm{d}\lambda} + b\frac{\mathrm{d}}{\mathrm{d}\mu}, \tag{31.16}$$

or $\boldsymbol{w} = a\boldsymbol{v} + b\boldsymbol{u}$. This means that the directional derivatives do indeed form a vector space[8] at $\mathcal{P}$. Why? It is because linear combinations of a set of vectors ($\mathrm{d}/\mathrm{d}\lambda$ and $\mathrm{d}/\mathrm{d}\mu$ here) at $\mathcal{P}$ can express the directional derivatives of other curves at this point.

To summarize, our new formulation of vectors as derivative operators has been based on evaluating tangent vectors along a curve at a point such as $\mathcal{P}$. The vector $\boldsymbol{v} = \mathrm{d}/\mathrm{d}\lambda|_{\mathcal{P}}$ is tangent to the curve parametrized by λ at point[9] $\mathcal{P}$. Many other curves pass through point $\mathcal{P}$ and we have seen how the vectors defined at this point make a vector space. We can express an arbitrary vector at $\mathcal{P}$ as a superposition of other vectors at $\mathcal{P}$.

Now let's imagine there being a large family of curves that fill the space, but don't intersect, as shown in Fig. 31.7. Each curve is parametrized by parameter λ. (However, a point such as $\lambda = 2$ should be expected to label points different distances along each curve.) Such a set of space-filling curves that never intersect each other is called a **congruence**.[10] The existence of a congruence means that at any point in space there is a curve with a unique derivative. Since derivatives are equivalent to tangent vectors, this gives us the notion of a **vector field**: at any point in a space we can identify a unique vector. At point $\mathcal{Q}$, for example, we call that vector $\mathrm{d}/\mathrm{d}\lambda|_{\mathcal{Q}}$. It is the tangent to that curve from the congruence that exists at point $\mathcal{Q}$. Put the other way around, we can pick a point and the vector field will input this point and output a vector. The vector-field machine works out what the vector is by computing the tangent to that curve found in the congruence at that particular point.

In this formulation, it only makes sense to compare different vectors defined at a single point, such as $\mathcal{Q}$. There is no obvious way to compare vectors defined at another point $\mathcal{P}$ with those at $\mathcal{Q}$. In order to do this, we would need a way of moving vectors around, such that they can be compared at a single point. The method of moving vectors would require the extra information on how vectors change as they move throughout the space, in order to make the comparison a meaningful one. In the previous chapters, we saw that this requires the notion of the connection that led to the covariant derivative. The connection allowed us to formulate a notion of parallelism: how to move a vector through space such that the only changes would be due to how that space varied.[11] For now, we will continue to compare objects defined at only a single point.

---

[8]Roughly speaking, a vector space is characterized by linear combinations of vectors and has the property of *closure* (so that any linear combination of vectors belonging to the vector space can only produce a vector which is itself a member of the vector space). The defining features of a vector space include the existence of a zero vector and the existence of an inverse of any non-zero vector, and also properties such as associativity, distributivity and commutativity. There must be a basis of linearly independent vectors that spans the space, the number of which are equal to the number of dimensions of the space.

[9]We write this $\mathrm{d}/\mathrm{d}\lambda|_{\mathcal{P}}$ to remove any ambiguity about which point we mean. Note that we don't write this $\mathrm{d}\mathcal{P}/\mathrm{d}\lambda$ since we don't any longer need to input the point $\mathcal{P}$ into the derivative and generally, we won't. The vector will most often operate on a function to deliver the function's directional derivative along the vector.

[10]These are important in physics. For example, we shall see in Chapter 39 how the congruence represents the streamlines of a fluid.

Fig. 31.7 A congruence of curves fills the space but the curves never intersect. Their tangent vectors form a vector field.

[11]In the following chapters, we will see that there is another derivative that can also be used to compare vector fields at different points: this is the so-called Lie derivative.

31.3 Linear slot machines again

Next, we shall expand the menu of objects to which we have access by considering 1-forms. These are members of a family of objects called 'differential forms' or simply 'forms' that will occupy us for the remainder of this part of the book. Forms were proposed (or discovered) by Elié Cartan in around 1900. While vectors have become an essential tool whose properties are taught to all undergraduate physicists, forms do not currently share this status. This is arguably a mistake, since they are no more complicated than vectors, and allow access to a far more simple and clear path to understanding geometry than is often presented.

We have, from our progress in this chapter, the notion of vector fields as derivatives that act on functions at a particular point in space. Quite separately from this, each vector has another possible input: vectors have a slot $v(\ )$ that accepts a 1-form and returns a scalar. Calling a typical 1-form $\tilde{\sigma}$, we have $v(\tilde{\sigma}) \equiv \langle \tilde{\sigma}, v \rangle = \text{(number)}$.[12]

Vectors, 1-forms and numbers live, in a sense, in different spaces. We mentioned how vectors live in a tangent space and, by the same token, 1-forms live in a **dual space**. In addition, scalar numbers live on the real line $\mathbb{R}^1$, giving a situation shown in Fig. 31.8. The important thing about the inner product summarized in the last equation is that it allows us to map between these spaces.

A 1-form maps a vector onto a number; a vector maps a 1-form onto a number.

This mapping onto a number (i.e. a point on the real line) is what we mean when we say that a vector is **dual** to a 1-form. An important property of the mapping/inner product operation is linearity, which is to say

$$\langle (a\tilde{\sigma} + b\tilde{\rho}), v \rangle = a\langle \tilde{\sigma}, v \rangle + b\langle \tilde{\rho}, v \rangle \tag{31.18}$$

and similarly, $\langle \tilde{\sigma}, (nv + mu) \rangle = n\langle \tilde{\sigma}, v \rangle + m\langle \tilde{\sigma}, u \rangle$.

We saw in Chapter 4 that while a vector can be thought of as a directed arrow, a 1-form can the thought of an infinite set of equally spaced surfaces (Fig. 31.9). In this picture, the inner product of a vector and a 1-form outputs a number that corresponds to the number of surfaces that the vector pierces.

Example 31.4

Consider a 1-form $\tilde{A}$. Working in Euclidean space we find the number of surfaces pierced by a unit vector in the x-direction is $\langle \tilde{A}, e_1 \rangle = A_1$. This implies that the spacing of the surfaces along this direction is $1/(A_1)$. Multiplying the form by some factor F causes the density of surfaces to be increased by the factor F.

Just as a vector can be written in terms of components and basis vectors, a 1-form $\tilde{\sigma}$ is written in terms of components σ_α and basis 1-forms ω^α as $\tilde{\sigma} = \sigma_\alpha \omega^\alpha$. To allow the mapping between vectors, 1-forms and

[12]We shall maintain the symmetry of the operation, so that the 1-form has a slot in which we can insert a vector $\tilde{\sigma}(v) = \text{(number)}$, where, in all cases we shall examine, $v(\tilde{\sigma}) = \tilde{\sigma}(v)$. This is just how we define the inner product between a vector and a 1-form. This is sometimes denoted by a dot, and when the dot would be confusing (as it usually denotes the scalar product of two vectors), more often by angle brackets $\tilde{\sigma} \cdot v \equiv \langle \tilde{\sigma}, v \rangle$. In summary,

$$\tilde{\sigma} \cdot v \equiv \langle \tilde{\sigma}, v \rangle = v(\tilde{\sigma}) = \tilde{\sigma}(v)$$
$$= \text{(number)}. \tag{31.17}$$

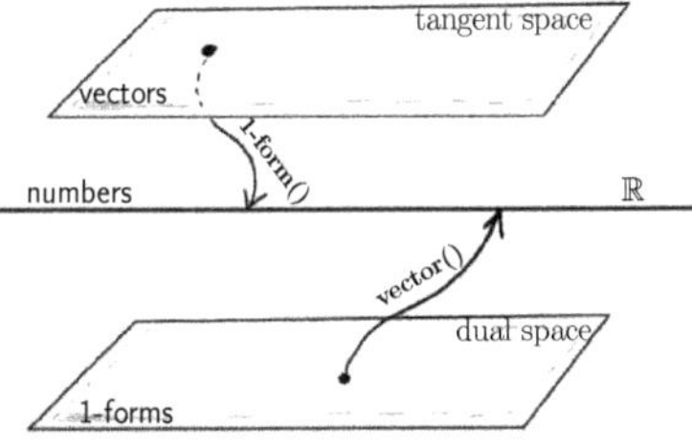

Fig. 31.8 Vectors live in a tangent space, 1-forms in a dual space and numbers along the real line $\mathbb{R}^1$. A vector maps a 1-form onto a number; a 1-form maps a vector on to a number.

Fig. 31.9 The 1-form $\tilde{A}$ represented as a set of repeating surfaces. The inner product $\langle \tilde{A}, X \rangle$ tells us how many surfaces are pierced by vector X.

numbers, we define an inner product between basis 1-forms and basis vectors as

$$\langle \boldsymbol{\omega}^\beta, \boldsymbol{e}_\alpha \rangle = \delta^\beta{}_\alpha. \tag{31.19}$$

In the new philosophy of having a vector correspond to a derivative, we have basis vectors $\boldsymbol{e}_\mu = \frac{\partial}{\partial x^\mu}$, where x^μ are a set of coordinates. We might also ask ourselves whether the 1-form corresponds to another operation? Of course it does: it is the operation that is dual to the operation of taking a derivative. Specifically, each basis 1-form corresponds to a **differential**[13]

$$\boldsymbol{\omega}^\alpha = \boldsymbol{dx}^\alpha. \tag{31.21}$$

This works in our coordinate basis since

$$\langle \boldsymbol{\omega}^\mu, \boldsymbol{e}_\nu \rangle = \left\langle \boldsymbol{dx}^\mu, \frac{\partial}{\partial x^\nu} \right\rangle = \frac{\partial x^\mu}{\partial x^\nu} = \delta^\mu{}_\nu. \tag{31.22}$$

The simplest example of 1-form $\tilde{\boldsymbol{\sigma}}$ is the differential of a function $f(x)$, which we write

$$\tilde{\boldsymbol{\sigma}} = \boldsymbol{df} = \frac{\partial f}{\partial x^\alpha} \boldsymbol{dx}^\alpha = \frac{\partial f}{\partial x^\alpha} \boldsymbol{\omega}^\alpha.$$

The identification of the 1-form as a differential allows us to reconcile the two sets of slots we have given the vector $\boldsymbol{u}$: square brackets that accept a function and round ones that accept a 1-form. We have that to operate on a function $f(x)$ with the vector $\boldsymbol{u}[\ ]$, to make $\boldsymbol{u}[f]$, we can equivalently take an inner product between the vector $\boldsymbol{u}$ and the 1-form $\boldsymbol{df}$, written as $\boldsymbol{u}(\boldsymbol{df})$ or, more helpfully $\langle \boldsymbol{df}, \boldsymbol{u} \rangle$. This is to say that we write[14]

$$\boldsymbol{u}[f] \equiv \langle \boldsymbol{df}, \boldsymbol{u} \rangle = \left\langle \frac{\partial f}{\partial x^\alpha} \boldsymbol{\omega}^\alpha, u^\beta \boldsymbol{e}_\beta \right\rangle$$

$$= u^\beta \frac{\partial f}{\partial x^\alpha} \langle \boldsymbol{\omega}^\alpha, \boldsymbol{e}_\beta \rangle$$

$$= u^\beta \frac{\partial f}{\partial x^\alpha} \delta^\alpha{}_\beta = u^\alpha \frac{\partial f}{\partial x^\alpha}. \tag{31.23}$$

We conclude that the inner product gives the same directional derivative we had before (telling us how the function changes along the vector $\boldsymbol{u}$) and so, in terms of the directional derivative operator, we write

$$\boldsymbol{u}[f] = \partial_{\boldsymbol{u}} f = \langle \boldsymbol{df}, \boldsymbol{u} \rangle. \tag{31.24}$$

Finally, we note that we have been dealing with vectors and 1-forms defined at the same point in space $\mathcal{P}$. Just as we define a vector field we also define a 1-form field. That is to say, at every point in space we have access to a unique 1-form. This can be combined with the vector we obtain at this point from a vector field to make a number. The numbers evaluated at each point in space themselves form a scalar field.

---

[13] Recall that the differential $\boldsymbol{d}f$ of a function $f(x, y)$ is written as

$$\boldsymbol{d}f = \left(\frac{\partial f}{\partial x} \right)_y \boldsymbol{d}x + \left(\frac{\partial f}{\partial y} \right)_x \boldsymbol{d}y. \tag{31.20}$$

We use a bold $\boldsymbol{d}$ here to denote the differential operation. The reason for this will be explained in a few Chapters' time.

[14] As a result of this manipulation we are able to effectively retire the vector's square bracket slot and rely on the combination of the vector and 1-form to make numbers via tensor operations.

31.4   Tensors again

Tensors can be regarded as generalized slot machines.[15] They are machines in which we insert combinations of vectors and 1-forms in order to return numbers. Generally, a valence (n, m) **tensor** has slots for the insertion of n lots of 1-forms and m lots of vectors, in order to return a number. We have the defining rule for a (n, m) tensor that

$$\boldsymbol{T}\left(\tilde{\boldsymbol{\sigma}}^1, ..., \tilde{\boldsymbol{\sigma}}^n, \boldsymbol{v}^1, ..., \boldsymbol{v}^m\right) = \text{(number)}. \tag{31.25}$$

Tensors can be built using a **tensor product**. The tensor product[16] $\otimes$ of two vectors gives us a $(2, 0)$ tensor

$$\boldsymbol{T}(\ ,\ ) = \boldsymbol{u} \otimes \boldsymbol{v} = \boldsymbol{u}(\ ) \otimes \boldsymbol{v}(\ ). \tag{31.26}$$

This object has two slots into which 1-forms can be inserted. To find the components of the tensor (which are a set of numbers) insert basis 1-forms into each of the slots.

Example 31.5

How do we extract the components of the (2,0) tensor $\boldsymbol{T} = \boldsymbol{u} \otimes \boldsymbol{v}$? We just insert the basis 1-forms into the slots. Here's how it goes.

$$\begin{aligned} T^{\mu\nu} = \boldsymbol{T}(\boldsymbol{\omega}^\alpha, \boldsymbol{\omega}^\beta) &= \boldsymbol{u}(\boldsymbol{\omega}^\alpha) \otimes \boldsymbol{v}(\boldsymbol{\omega}^\beta) \\ &= \langle \boldsymbol{\omega}^\alpha, \boldsymbol{u}\rangle\langle \boldsymbol{\omega}^\beta, \boldsymbol{v}\rangle \\ &= u^\alpha v^\beta. \end{aligned} \tag{31.27}$$

We conclude $T^{\alpha\beta} = u^\alpha v^\beta$. We can reconstruct the tensor from the components using

$$\boldsymbol{T} = u^\alpha v^\beta \boldsymbol{e}_\alpha \otimes \boldsymbol{e}_\beta. \tag{31.28}$$

which is a shorthand for a version written with the slots:

$$\boldsymbol{T}(\ ,) = u^\alpha v^\beta \boldsymbol{e}_\alpha(\ ) \otimes \boldsymbol{e}_\beta(\ ). \tag{31.29}$$

We can use the tensor product to build objects out of 1-forms too.

Example 31.6

Consider a set of basis 1-forms $\boldsymbol{dx}^\mu = \boldsymbol{\omega}^\mu$. The metric is a written as a $(0,2)$ tensor

$$\boldsymbol{g} = g_{\mu\nu} \boldsymbol{dx}^\mu \otimes \boldsymbol{dx}^\nu = g_{\mu\nu} \boldsymbol{\omega}^\mu \otimes \boldsymbol{\omega}^\nu. \tag{31.30}$$

This is sometimes called $\boldsymbol{ds}^2$. With this the metric becomes a two-slot, $(0,2)$ object called $\boldsymbol{g}(\ ,\ )$, into which we insert vectors to return a number.

An arbitrary tensor (n, m) tensor can be built from n vectors and m 1-forms

$$\boldsymbol{T} = \boldsymbol{u}^1 \otimes ... \otimes \boldsymbol{u}^n \otimes \tilde{\boldsymbol{\sigma}}^1 \otimes ... \otimes \tilde{\boldsymbol{\sigma}}^m. \tag{31.31}$$

[15]Tensor calculus was developed mainly by Gregorio Ricci-Curbastro (1853–1925). His most famous work on the calculus of tensors was co-authored with his former student, Tullio Levi-Civita (1873–1941), and signed Gregorio Ricci. (Curiously, all of Ricci's other works feature his full name, Ricci-Curbastro.) This work *Méthods de calcul différential absolu et leurs applications* from 1901 greatly simplified the presentation of Riemannian geometry and gave us the version of geometry that Einstein used as the basis for his general relativity.

[16]Recall that $\otimes$ simply tells us to retain the order of the slots.

To find the tensor's components by inserting n basis 1-forms and m basis vectors:

$$T^{\mu_1\cdots\mu_n}{}_{\nu_1\dots\nu_m} = \boldsymbol{u}^1(\boldsymbol{\omega}^{\mu_1})\dots\boldsymbol{u}^n(\boldsymbol{\omega}^{\mu_n})\tilde{\boldsymbol{\sigma}}^1(\boldsymbol{e}_{\nu_1})\dots\tilde{\boldsymbol{\sigma}}^m(\boldsymbol{e}_{\nu_m}). \qquad (31.32)$$

If we know the components, we can recreate the tensor by expanding

$$\boldsymbol{T} = T^{\mu_1\cdots\mu_n}{}_{\nu_1\dots\nu_m}\left(\boldsymbol{e}_{\mu_1}\otimes\dots\otimes\boldsymbol{e}_{\mu_n}\otimes\boldsymbol{\omega}^{\nu_1}\otimes\dots\otimes\boldsymbol{\omega}^{\nu_m}\right). \qquad (31.33)$$

We will frequently encounter tensors built using tensor products from a selection of vectors and 1-forms. One useful rule is that, although you can't alter the order of tensors or vectors amongst themselves, the 1-form parts and vector parts commute.[17]

It should be no surprise that in the same way that we defined vector fields and 1-form fields, we can also define tensor fields. That is, at each point in space, we have a unique tensor that can be combined with vectors and 1-forms found at that same point to make a number.

31.5   Examples of tensor operations

Once we have our tensors we can apply a range of tools to manipulate them. Some of the common operations are described in the example below.[18]

Example 31.7

Consider, for example, a tensor $\boldsymbol{P}$ of rank (0,3), which we can write

$$\boldsymbol{P}(\;,\;,\;) = P_{\alpha\beta\gamma}\boldsymbol{\omega}^\alpha\otimes\boldsymbol{\omega}^\beta\otimes\boldsymbol{\omega}^\gamma. \qquad (31.34)$$

In words: $\boldsymbol{P}$ has three slots which all take vectors. If we fill the slots with vectors we output a scalar given by

$$\begin{aligned}\boldsymbol{P}(\boldsymbol{u},\boldsymbol{v},\boldsymbol{w}) &=P_{\alpha\beta\gamma}\boldsymbol{\omega}^\alpha(\boldsymbol{u})\boldsymbol{\omega}^\beta(\boldsymbol{v})\boldsymbol{\omega}^\gamma(\boldsymbol{w})\\ &=P_{\alpha\beta\gamma}u^\alpha v^\beta w^\gamma.\end{aligned} \qquad (31.35)$$

Here is a list of things one can do with a tensor.

• Take the **gradient**. The gradient operation $\boldsymbol{\nabla}$ adds an extra (0,1) slot to the tensor. Therefore, $\boldsymbol{\nabla}\boldsymbol{P}$ has four slots that accept vectors. We interpret this as meaning

$$\boldsymbol{\nabla}P(\boldsymbol{u},\boldsymbol{v},\boldsymbol{w},\boldsymbol{z}) \approx \left(\begin{array}{c} \boldsymbol{P}(\boldsymbol{u},\boldsymbol{v},\boldsymbol{w})\text{ at tip of }\boldsymbol{z}\\ -\boldsymbol{P}(\boldsymbol{u},\boldsymbol{v},\boldsymbol{w})\text{ at base of }\boldsymbol{z}\end{array}\right). \qquad (31.36)$$

In flat space, we write $\boldsymbol{\nabla}P(\boldsymbol{u},\boldsymbol{v},\boldsymbol{w},\boldsymbol{z})$ in component form as

$$\frac{\partial P_{\alpha\beta\gamma}}{\partial x^\mu}u^\alpha v^\beta w^\gamma z^\mu = P_{\alpha\beta\gamma,\mu}u^\alpha v^\beta w^\gamma z^\mu. \qquad (31.37)$$

• **Contraction** reduces the rank of a tensor by two. We do this by inserting a basis vector and basis 1-form and sum over corresponding components. Let's do this to the (1,3) tensor $\boldsymbol{S}$, defined as

$$\boldsymbol{S} = S^\mu{}_{\alpha\beta\gamma}\boldsymbol{e}_\mu\otimes\boldsymbol{\omega}^\alpha\otimes\boldsymbol{\omega}^\beta\otimes\boldsymbol{\omega}^\gamma, \qquad (31.38)$$

to make a (0,2) tensor $\boldsymbol{Q}$ with components $Q_{\alpha\gamma}$. We can write an expression for a scalar given by

$$\boldsymbol{Q}(\boldsymbol{u},\boldsymbol{v}) = \boldsymbol{S}(\boldsymbol{\omega}^\mu,\boldsymbol{u},\boldsymbol{e}_\mu,\boldsymbol{v}), \qquad (31.39)$$

---

[17] The tensor product sign tells us to maintain the order of vectors parts and, separately, the 1-form parts. However, it doesn't matter if you list the basis vectors first or the basis 1-forms first, as long as you know which is which.

[18] Our presentation here follows Misner, Thorne and Wheeler, which can be consulted for further details.

where a sum over the repeated index is assumed. In components, this is written as

$$\boldsymbol{Q}(\boldsymbol{u}, \boldsymbol{v}) = Q_{\alpha\beta} u^\alpha v^\gamma = S^\mu{}_{\alpha\mu\gamma} u^\alpha v^\gamma. \tag{31.40}$$

The summing over a common up and down index is often called 'contracting an index'.

• **Divergence** Contract the gradient with a slot on the tensor. Let's consider our (1,3) tensor $\boldsymbol{S}$ again. We'll assume we're taking a divergence with respect to the first slot

$$\begin{aligned} \boldsymbol{\nabla} \cdot \boldsymbol{S}(\boldsymbol{u}, \boldsymbol{v}, \boldsymbol{w}) &= \boldsymbol{\nabla} \boldsymbol{S}(\boldsymbol{\omega}^\mu, \boldsymbol{u}, \boldsymbol{v}, \boldsymbol{w}, \boldsymbol{e}_\mu) \\ &= \frac{\partial S^\mu{}_{\alpha\beta\gamma}}{\partial x^\mu} u^\alpha v^\beta w^\gamma = S^\mu{}_{\alpha\beta\gamma,\mu} u^\alpha v^\beta w^\gamma, \end{aligned} \tag{31.41}$$

where the component version applies in flat space.

• **Transpose** This swaps the contents of two slots

$$\boldsymbol{P}(\boldsymbol{u}, \boldsymbol{v}, \boldsymbol{w}) \to \boldsymbol{P}(\boldsymbol{v}, \boldsymbol{u}, \boldsymbol{w}). \tag{31.42}$$

What this does to a tensor depends on the properties of that tensor. A tensor is **symmetric** if it is unaffected by a the transpose of two slots.[19] If it is unaffected by the swap of any two slots it is **totally symmetric**

$$\boldsymbol{P}(\boldsymbol{u}, \boldsymbol{v}, \boldsymbol{w}) = \boldsymbol{P}(\boldsymbol{v}, \boldsymbol{u}, \boldsymbol{w}) = \boldsymbol{P}(\boldsymbol{w}, \boldsymbol{u}, \boldsymbol{v}) = \ldots \tag{31.43}$$

A tensor is antisymmetric if the sign is reversed when two slots are swapped.[20] It is **totally antisymmetric** if the sign is reversed if any two slots are swapped.

$$\boldsymbol{P}(\boldsymbol{u}, \boldsymbol{v}, \boldsymbol{w}) = -\boldsymbol{P}(\boldsymbol{v}, \boldsymbol{u}, \boldsymbol{w}) = +\boldsymbol{P}(\boldsymbol{w}, \boldsymbol{u}, \boldsymbol{v}) = \ldots \tag{31.44}$$

• **Symmetrization** In components, the **symmetrization** of a (2,0) tensor $\boldsymbol{T}$ involves the following action on the components:

$$T^{(\mu\nu)} = \frac{1}{2!} \left(T^{\mu\nu} + T^{\nu\mu} \right). \tag{31.45}$$

A symmetric tensor then has the property $T^{(\mu\nu)} = T^{\mu\nu}$. An antisymmetric tensor has the property $A^{(\mu\nu)} = 0$. Symmetrization therefore extracts the symmetric part of a tensor. For a (n, m) tensor we have the rule

$$T^{(\alpha\ldots\kappa)}{}_{\mu\ldots\sigma} = \frac{1}{n!} \left(\begin{array}{c} \text{Sum over all permutations} \\ \text{of the } n \text{ indices } \alpha\ldots\kappa \end{array} \right). \tag{31.46}$$

• **Antisymmetrization** In components the **antisymmetrization** of a tensor involved the following action on the components:

$$T^{[\mu\nu]} = \frac{1}{2!} \left(T^{\mu\nu} - T^{\nu\mu} \right). \tag{31.47}$$

A symmetric tensor has the property $T^{[\mu\nu]} = 0$. An antisymmetric tensor has the property $A^{[\mu\nu]} = A^{\mu\nu}$. Antisymmetrization therefore extracts the antisymmetric part of a tensor. For a (n, m) tensor we have the rule for antisymmetrization that

$$T^{[\alpha\ldots\kappa]}{}_{\mu\ldots\sigma} = \frac{1}{n!} \left(\begin{array}{c} \text{Alternating sum over all} \\ \text{permutations of the } n \text{ indices } \alpha\ldots\kappa \end{array} \right). \tag{31.48}$$

• The **wedge product** is an antisymmetrized tensor product. Take vectors $\boldsymbol{u}$ and $\boldsymbol{v}$ and build the **bivector** $\boldsymbol{b}$ using the wedge product

$$\boldsymbol{b} = \boldsymbol{u} \wedge \boldsymbol{v} = \boldsymbol{u} \otimes \boldsymbol{v} - \boldsymbol{v} \otimes \boldsymbol{u}. \tag{31.49}$$

The components of the bivector are

$$\begin{aligned} b^{\mu\nu} = \boldsymbol{b}(\boldsymbol{\omega}^\mu, \boldsymbol{\omega}^\nu) &= \boldsymbol{u}[\boldsymbol{\omega}^\mu] \otimes \boldsymbol{v}[\boldsymbol{\omega}^\nu] - \boldsymbol{v}[\boldsymbol{\omega}^\mu] \otimes \boldsymbol{u}[\boldsymbol{\omega}^\nu] \\ &= u^\mu v^\nu - v^\mu u^\nu. \end{aligned} \tag{31.50}$$

[19] If a $(2,0)$ tensor is symmetric then the components obey $T^{\mu\nu} = T^{\nu\mu}$.

[20] If a $(2,0)$ tensor is antisymmetric then the components obey $A^{\mu\nu} = -A^{\nu\mu}$.

We shall use these tricks in the chapters that follow.

We now have a set of vectors, 1-forms and tensors with which to describe geometry. They can be written in a coordinate-independent form (i.e. $\boldsymbol{v}, \tilde{\boldsymbol{\sigma}}, \boldsymbol{T}$) or we can write them in terms of a set of coordinates by invoking basis vectors and components. By combining the objects using the relevant slots we can map between them and numbers. Each of these objects is useful in describing physical quantities commonly found in Nature.

In the next chapter, we will treat a special class of tensor that are created from a special combination of 1-forms. These are called differential forms and are of particular importance owing to their use in describing curvature.

Chapter summary

- Vectors can be identified with derivatives. The tangent vector at a point parametrized by λ on a curve $\mathcal{P}(\lambda)$ is $\boldsymbol{v} = \mathrm{d}/\mathrm{d}\lambda$. A vector field allows us to find a vector at each point.

- A vector can be combined with a 1-form $\tilde{\boldsymbol{\sigma}}$ to make a number. A 1-form field gives a 1-form at every point in space. It only makes sense to combine vectors and 1-forms defined at the same point.

- Tensors generalize the notion of vectors and 1-forms. Tensors can be manipulated in a large number of ways to produce tensors of different valences.

Exercises

(31.1) The basis vectors and basis 1-forms for a coordinate frame are given by $\partial/\partial x^\mu$ and $\boldsymbol{dx}^\alpha$, respectively. Evaluate:
(a) $\langle \boldsymbol{dx}^1, \partial/\partial x^1 \rangle$,
(b) $\langle \boldsymbol{dx}^1, \partial/\partial x^0 \rangle$,
(c) $\partial/\partial x^\alpha \cdot \partial/\partial x^\beta$,
(d) $\boldsymbol{dx}^\alpha \cdot \boldsymbol{dx}^\beta$.

(31.2) Find a 1-form $\tilde{\boldsymbol{W}}$ that acts on a displacement $\vec{X}$ to give the work done by a force $\vec{f}$ in Euclidean space.

(31.3) Expand (a) $T^{(\alpha\beta\gamma)}{}_{\delta\epsilon\zeta}$; (b) $T^{\alpha\beta\gamma}{}_{[\delta\epsilon\zeta]}$.

(31.4) Consider the contraction $A^{\mu\nu}T_{\mu\nu}$, where $T_{\mu\nu}$ are the components of some general tensor.
(a) Show that if $A^{\mu\nu}$ is symmetric, only the symmetric part of $T_{\mu\nu}$ contributes to the contraction.

(b) Show that if $A^{\mu\nu}$ is antisymmetric, only the asymmetric part of $T_{\mu\nu}$ contributes to the contraction.

(31.5) (a) Expand $F_{[\mu\nu;\lambda]}$ if $F_{\mu\nu}$ is an antisymmetric tensor.
(b) Show that

$$F^{\mu\nu}\eta_{\mu\alpha}\eta_{\nu\beta}u^\alpha u^\beta = 0, \qquad (31.51)$$

where $F^{\mu\nu}$ are the components of an antisymmetric tensor.

(31.6) By removing the point ξ^μ, show that the expression

$$\frac{\partial^2 \xi^\mu}{\partial x^\lambda \partial x^\sigma} = \Gamma^\nu{}_{\lambda\sigma}\frac{\partial \xi^\mu}{\partial x^\nu}, \qquad (31.52)$$

is equivalent to

$$\frac{\partial \boldsymbol{e}_\sigma}{\partial x^\lambda} = \Gamma^\nu{}_{\lambda\sigma}\boldsymbol{e}_\nu. \tag{31.53}$$

(31.7) Roger Penrose has invented a diagrammatic notation for tensor equations that provides an entertaining way of visualizing complicated expressions. Tensors are represented by shapes. The rules are that a valence (n, m) tensor has n lines emerging from its top and m lines from the bottom. The position of the page allows one to keep track of indices. A thick bar denotes antisymmetrization and a wiggly bar denotes symmetrization (although the numerical factors associated with these operations are not conventionally included, so can be drawn on the diagrams).

(a) Using these rules and referring to Fig. 31.10, explain why diagram (i) represents the component expression $A^{\mu\nu}{}_{\alpha\beta\gamma} + 2A^{\mu\nu}{}_{\beta\gamma\alpha}$.

(b) What expression does diagram (ii) represent? What about (iii)?

(c) If covariant derivatives are represented by a large circle around the tensor, suggest which equations of electromagnetism are represented by diagrams (iv) and (v).

(d) Which expression involving the Riemann tensor is represented by diagram (vi)?

See Penrose (2004) for more details.

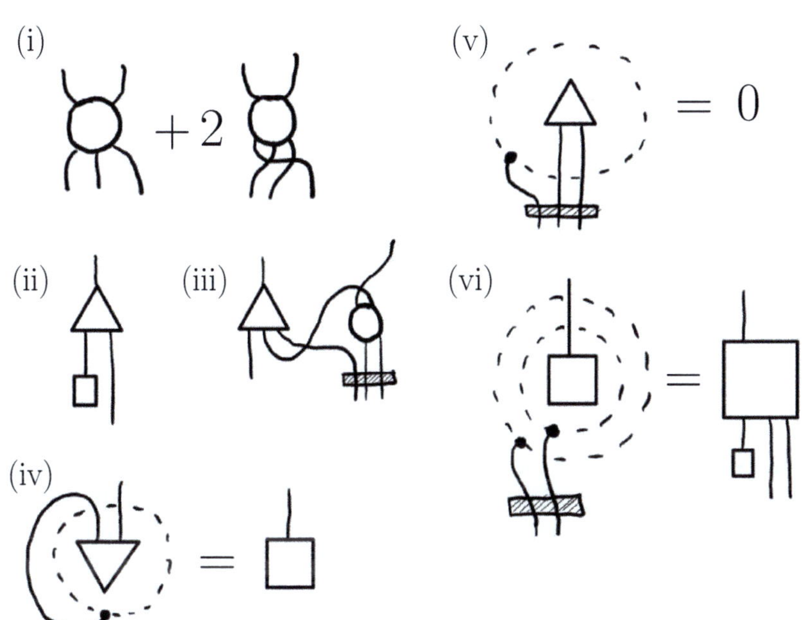

Fig. 31.10 Tensor diagrams from Exercise 31.7.

32

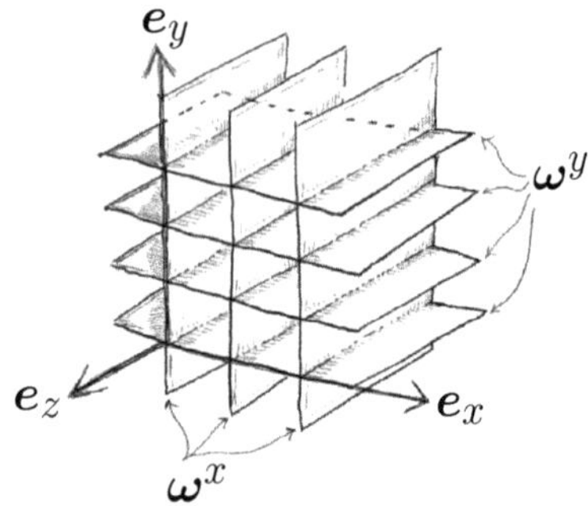

Fig. 32.1 An example 2-form $\tilde{\beta}(\ ,\ )$ made from $\boldsymbol{\omega}^x \wedge \boldsymbol{\omega}^y$.

[1] There's an analogy here between the vectors and 1-forms that we have described, with their characteristic relationship $\langle \boldsymbol{\omega}^\mu, \boldsymbol{e}_\mu \rangle = \delta^\mu{}_\nu$, and the idea of lattice vectors $\boldsymbol{a}_i$ and reciprocal lattice vectors $\boldsymbol{A}_j$ in solids, which have an analogous flat-space relationship $\boldsymbol{A}_i \cdot \boldsymbol{a}_j = 2\pi\delta_{ij}$. We can use reciprocal lattice vectors to label sets of lattice planes, just as we label repeating planes using 1-forms. See the book by Pauli, Section 10.

[2] As we saw in the last chapter, we can also use the wedge product on vectors like $\boldsymbol{u}$ and $\boldsymbol{v}$ to make bivectors

$$\boldsymbol{u} \wedge \boldsymbol{v} = \boldsymbol{u} \otimes \boldsymbol{v} - \boldsymbol{v} \otimes \boldsymbol{u}. \qquad (32.1)$$

Bivectors are antisymmetric (2,0) objects with two slots, each of which takes a 1-form.

Differential forms

Good God, there's two of them!
Audience heckle (on seeing the comic Bernie Winters take the stage at the Glasgow Empire, after his brother Mike had been poorly received)

In the last chapter, we reacquainted ourselves with vectors and 1-forms. With certain qualifications, the former can be thought of as arrows; the latter as a set of regularly repeating surfaces. In this chapter, we start with the 1-form and use it to generate a family of objects known as differential forms or *p*-forms. These have several applications in general relativity: a particularly useful example is that curvature can be described very efficiently as an object known as a **2-form**. We will see how to encode information about areas and volumes in forms, and this will lead us, in Chapter 38, to the profound idea that *every* integral can be reinterpreted as an integral over a form.

32.1   2-forms

1-forms are the simplest of a family of tensor objects called **differential forms**. This family of forms can be built from 1-forms. Recall the geometric interpretation of a 1-form as an infinite number of equally spaced surfaces.[1] By combining the surfaces of 1-forms to build more complicated structures, we can generate all of the other differential forms.

Let's look at an example. Working in Cartesian coordinates, the surfaces of the basis 1-form $\boldsymbol{\omega}^\mu$ are perpendicular to the direction of the basis vector $\boldsymbol{e}_\mu$. By combining the surfaces $\boldsymbol{\omega}^x$ and $\boldsymbol{\omega}^y$ as shown in Fig. 32.1 we can create the structure of tubes shown in the figure. This structure corresponds to an antisymmetric $(0,2)$ tensor $\tilde{\beta}(\ ,\ )$, which is our first example of a **2-form**. To achieve the combination of the surfaces symbolically, we use the **wedge product** to combine the 1-forms.[2]

The **wedge product** of two 1-forms, $\tilde{\sigma}$ and $\tilde{\tau}$, is the antisymmetrized tensor product

$$\tilde{\sigma} \wedge \tilde{\tau} = \tilde{\sigma} \otimes \tilde{\tau} - \tilde{\tau} \otimes \tilde{\sigma}. \qquad (32.2)$$

The wedge product to two 1-forms is a 2-form. We notice immediately that with this definition

$$\tilde{\sigma} \wedge \tilde{\sigma} = 0, \qquad (32.3)$$

where $\tilde{\sigma}$ is any 1-form.

With this notation, our example 2-form $\tilde{\boldsymbol{\beta}}(\ ,)$ can be generated by saying

$$\tilde{\boldsymbol{\beta}}(\ ,) = \boldsymbol{\omega}^x \wedge \boldsymbol{\omega}^y. \tag{32.4}$$

The wedge-product operation does not generate an arbitrary $(0,2)$ tensor; rather, it is **antisymmetric**.[3] The antisymmetry is interpreted geometrically in terms of the tubes in Fig. 32.1 having a handedness. To visualize this, we could picture the 2-form circulating in a particular direction in each tube as shown in Fig. 32.2.

Let's now consider a more general 2-form $\tilde{\boldsymbol{B}}(\ ,) = \tilde{\boldsymbol{\sigma}} \wedge \tilde{\boldsymbol{\tau}}$, built from two arbitrary 1-forms $\tilde{\boldsymbol{\sigma}}$ and $\tilde{\boldsymbol{\tau}}$. Any 2-form can be written in terms of its components $B_{\mu\nu}$ as

$$\tilde{\boldsymbol{B}}(\ ,) = \frac{1}{2}B_{\mu\nu}\boldsymbol{\omega}^\mu \wedge \boldsymbol{\omega}^\nu, \tag{32.5}$$

where $\boldsymbol{\omega}^\mu$ are basis 1-forms. The asymmetry of the 2-form means that $B_{\mu\nu} = -B_{\nu\mu}$. The factor $1/2$ is a convention, that we justify in the next example.[4]

Example 32.1

The motivation for the $1/2$ factor is seen by expanding the 2-form

$$\begin{aligned}\tilde{\boldsymbol{B}} &= \frac{1}{2}B_{\mu\nu}\boldsymbol{\omega}^\mu \wedge \boldsymbol{\omega}^\nu \\ &= \frac{1}{2}B_{\mu\nu}\left(\boldsymbol{\omega}^\mu \otimes \boldsymbol{\omega}^\nu - \boldsymbol{\omega}^\nu \otimes \boldsymbol{\omega}^\mu\right).\end{aligned} \tag{32.8}$$

Now insert basis vectors $\boldsymbol{e}_\alpha$ and $\boldsymbol{e}_\beta$ to extract the components of the 2-form

$$\begin{aligned}\tilde{\boldsymbol{B}}(\boldsymbol{e}_\alpha, \boldsymbol{e}_\beta) &= \frac{B_{\mu\nu}}{2}\left(\delta^\mu{}_\alpha \delta^\nu{}_\beta - \delta^\nu{}_\alpha \delta^\mu{}_\beta\right) \\ &= \frac{1}{2}\left(B_{\alpha\beta} - B_{\beta\alpha}\right) = B_{\alpha\beta},\end{aligned} \tag{32.9}$$

where, in the final step we use asymmetry to say $B_{\alpha\beta} = -B_{\beta\alpha}$.

Just as with the tensors from the last chapter, it is possible to use the slot-machine structure of forms to extract their components.

Example 32.2

We can extract the components of the 2-form $\tilde{\boldsymbol{B}} = \tilde{\boldsymbol{\sigma}} \wedge \tilde{\boldsymbol{\tau}}$ in terms of the components of $\tilde{\boldsymbol{\sigma}}$ and $\tilde{\boldsymbol{\tau}}$ by inserting basis vectors $\boldsymbol{e}_\mu$ into the slots in the expression $\tilde{\boldsymbol{B}}(\ ) = \tilde{\boldsymbol{\sigma}}(\ )\wedge\tilde{\boldsymbol{\tau}}(\ )$ to find

$$\begin{aligned}B_{\mu\nu} &= \tilde{\boldsymbol{B}}(\boldsymbol{e}_\mu, \boldsymbol{e}_\nu) \\ &= \tilde{\boldsymbol{\sigma}}(\boldsymbol{e}_\mu) \otimes \tilde{\boldsymbol{\tau}}(\boldsymbol{e}_\nu) - \tilde{\boldsymbol{\tau}}(\boldsymbol{e}_\mu) \otimes \tilde{\boldsymbol{\sigma}}(\boldsymbol{e}_\nu) \\ &= \sigma_\mu \tau_\nu - \tau_\mu \sigma_\nu.\end{aligned} \tag{32.10}$$

[3]Recall from the previous chapter that a $(0,2)$ tensor is antisymmetric if $\tilde{\boldsymbol{B}}(\boldsymbol{x},\boldsymbol{y}) = -\tilde{\boldsymbol{B}}(\boldsymbol{y},\boldsymbol{x})$, where $\boldsymbol{x}$ and $\boldsymbol{y}$ are vectors.

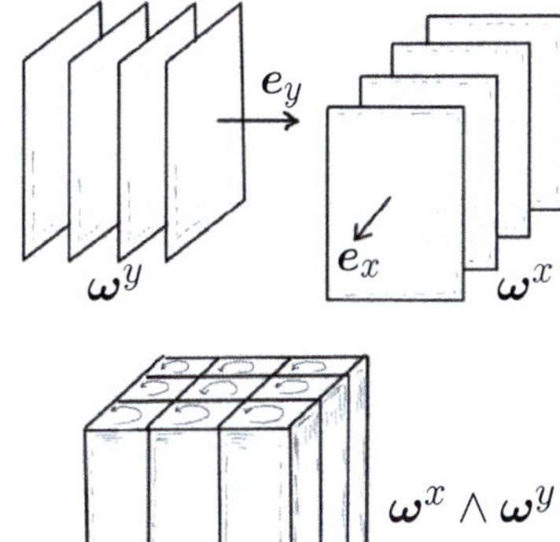

Fig. 32.2 The 1-forms $\boldsymbol{\omega}^x$ and $\boldsymbol{\omega}^y$ can be thought of as sets of parallel surfaces. Their combination gives the tubular structure described by $\boldsymbol{\omega}^x\wedge\boldsymbol{\omega}^y$, with a particular handedness, as depicted by the arrows in the tubes.

[4]As shown in the example, the factor $1/2$ only arises when we're summing over indices, as in eqn 32.5. If we choose to specify specific components, rather than sum over them, no factor explicit factor is needed. For example, if a 2-form in Euclidean 3-space has only one non-zero component, we write

$$\tilde{\boldsymbol{B}} = \gamma(\boldsymbol{\omega}^2 \wedge \boldsymbol{\omega}^3). \tag{32.6}$$

Computing components we find

$$\begin{aligned}B_{\mu\nu} &= \gamma[\boldsymbol{\omega}^2(\boldsymbol{e}_\mu) \otimes \boldsymbol{\omega}^3(\boldsymbol{e}_\nu) \\ &\quad - \boldsymbol{\omega}^3(\boldsymbol{e}_\mu) \otimes \boldsymbol{\omega}^2(\boldsymbol{e}_\nu)] \\ &= \gamma[\delta^2_\mu \delta^3_\nu - \delta^3_\mu \delta^2_\nu],\end{aligned} \tag{32.7}$$

so $B_{23} = \gamma$ and $B_{32} = -\gamma$.

[5]Recall from the last chapter that the spacing of the surfaces in the Cartesian μ-direction is $1/\sigma_\mu$.

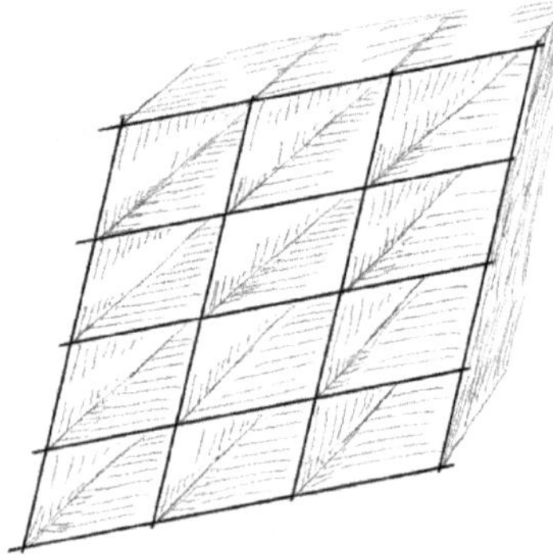

Fig. 32.3 An arbitrary 2-form.

[6]We define a 0-form to be a function such as $f(x)$.

[7]For example in three-dimensional space

$$\tilde{B} = \frac{1}{2} B_{\mu\nu} \left(\omega^\mu \wedge \omega^\nu \right)$$

$$= \frac{1}{2} \left(B_{12} \omega^1 \wedge \omega^2 + B_{21} \omega^2 \wedge \omega^1 \right.$$

$$+ B_{13} \omega^1 \wedge \omega^3 + B_{31} \omega^3 \wedge \omega^1$$

$$\left. + B_{23} \omega^2 \wedge \omega^3 + B_{32} \omega^3 \wedge \omega^2 \right).$$

Using (i) $B_{\mu\nu} = -B_{\nu\mu}$ and (ii) $\omega^\mu \wedge \omega^\nu = -\omega^\nu \wedge \omega^\mu$, we can simplify this to read

$$\tilde{B} = \frac{1}{2} B_{\mu\nu} \left(\omega^\mu \wedge \omega^\nu \right)$$

$$= \left(B_{12} \omega^1 \wedge \omega^2 + B_{13} \omega^1 \wedge \omega^3 \right.$$

$$\left. + B_{23} \omega^2 \wedge \omega^3 \right)$$

$$= B_{|\mu\nu|} \omega^\mu \wedge \omega^\nu. \tag{32.14}$$

Example 32.3

Let's try inserting a single vector $\boldsymbol{u}$ into the first slot of a 2-form. (Remember that the tensor products $\otimes$ are intended to maintain the order of the linear slots.) We find

$$\tilde{B}(\boldsymbol{u},\ ) = \tilde{\boldsymbol{\sigma}} \wedge \tilde{\boldsymbol{\tau}}(\boldsymbol{u},\ ) = \tilde{\boldsymbol{\sigma}}(\boldsymbol{u}) \otimes \tilde{\boldsymbol{\tau}}(\ ) - \tilde{\boldsymbol{\tau}}(\boldsymbol{u}) \otimes \tilde{\boldsymbol{\sigma}}(\ ). \tag{32.11}$$

The result is a 1-form: a $(0,1)$ tensor object with a single available slot that accepts a vector.

An arbitrary 1-form $\tilde{\boldsymbol{\sigma}} = \sigma_\mu \boldsymbol{\omega}^\mu$ is represented by a series of surfaces oriented according to, and with spacing determined by, the components[5]. The product of two such arbitrary 1-forms resembles the example shown in Fig. 32.3 made of intersecting surfaces giving a series of (parallelepiped) tubes. As in our simple example earlier, the tubes come with a sense of direction since there is a difference between $\tilde{\boldsymbol{\sigma}} \wedge \tilde{\boldsymbol{\tau}}$ and $\tilde{\boldsymbol{\tau}} \wedge \tilde{\boldsymbol{\sigma}}$. We think of the 2-form field as circulating in each of the cells.

32.2 p-forms

We've seen that the wedge product of two 1-forms is a 2-form. In fact, we can generalize this notion and say that, if $\tilde{\boldsymbol{\mu}}$ is a p-form and $\tilde{\boldsymbol{\nu}}$ is a q-form, then we can use the wedge product to make a $(p+q)$-form $\tilde{\boldsymbol{\mu}} \wedge \tilde{\boldsymbol{\nu}}$, with the property that

$$\tilde{\boldsymbol{\mu}} \wedge \tilde{\boldsymbol{\nu}} = (-1)^{pq} \tilde{\boldsymbol{\nu}} \wedge \tilde{\boldsymbol{\mu}}. \tag{32.12}$$

Using this idea, we can generate p-forms by making more and more wedge products.[6]

A p-form $\tilde{\boldsymbol{\alpha}}$ can be written as

$$\tilde{\boldsymbol{\alpha}} = \frac{1}{p!} \alpha_{i_1, i_2, \ldots, i_p} \boldsymbol{\omega}^{i_1} \wedge \boldsymbol{\omega}^{i_2} \wedge \ldots \wedge \boldsymbol{\omega}^{i_p}$$

$$= \alpha_{|i_1, i_2, \ldots, i_p|} \boldsymbol{\omega}^{i_1} \wedge \boldsymbol{\omega}^{i_2} \wedge \ldots \wedge \boldsymbol{\omega}^{i_p}. \tag{32.13}$$

Here, the vertical bars around a list of i_n mean we fix $i_1 < i_2 < \ldots < i_p$ and don't therefore need the $1/p!$ normalization in the first.[7]

Example 32.4

Let's make a 3-form from the 1-forms $\tilde{\boldsymbol{\sigma}}$, $\tilde{\boldsymbol{\tau}}$ and $\tilde{\boldsymbol{\kappa}}$. We have

$$\tilde{\boldsymbol{\sigma}} \wedge (\tilde{\boldsymbol{\tau}} \wedge \tilde{\boldsymbol{\kappa}}) = (\tilde{\boldsymbol{\sigma}} \wedge \tilde{\boldsymbol{\tau}}) \wedge \tilde{\boldsymbol{\kappa}} = \tilde{\boldsymbol{\sigma}} \otimes \tilde{\boldsymbol{\tau}} \otimes \tilde{\boldsymbol{\kappa}} - \tilde{\boldsymbol{\sigma}} \otimes \tilde{\boldsymbol{\kappa}} \otimes \tilde{\boldsymbol{\tau}}$$

$$+ \tilde{\boldsymbol{\kappa}} \otimes \tilde{\boldsymbol{\sigma}} \otimes \tilde{\boldsymbol{\tau}} - \tilde{\boldsymbol{\kappa}} \otimes \tilde{\boldsymbol{\tau}} \otimes \tilde{\boldsymbol{\sigma}}$$

$$+ \tilde{\boldsymbol{\tau}} \otimes \tilde{\boldsymbol{\kappa}} \otimes \tilde{\boldsymbol{\sigma}} - \tilde{\boldsymbol{\tau}} \otimes \tilde{\boldsymbol{\sigma}} \otimes \tilde{\boldsymbol{\kappa}}. \tag{32.15}$$

Geometrically, we can build this object by stepping up from the equally spaced planes of the 1-form, via the tube-like structure of the 2-form to the cellular structure of the 3-form.

A simple example of a p-form that we can build in n-dimensional space involves multiplying a scalar function $\alpha_{|i_1,i_2...i_p|} = f(x^1, x^2, ..., x^p)$ by wedge products of the basis 1-forms or, equivalently, the differentials of coordinates to build a p-form tensor $\tilde{\beta}$

$$\tilde{\beta} = f(x^1, x^2, ..., x^p)\boldsymbol{dx}^{i_1} \wedge \boldsymbol{dx}^{i_2} \wedge ... \wedge \boldsymbol{dx}^{i_p}. \tag{32.16}$$

All p-forms are linear and can therefore be added to other p-forms to make the arbitrary p-form of your choosing.

Example 32.5

In three-dimensional space with basis 1-forms $\boldsymbol{\omega}^1 = \boldsymbol{dx}$, $\boldsymbol{\omega}^2 = \boldsymbol{dy}$ and $\boldsymbol{\omega}^3 = \boldsymbol{dz}$, we can investigate some p-form objects that we make from functions of the coordinates (x, y, z) and the three basis forms:

$$
\begin{aligned}
\text{0-form} \quad & \tilde{\alpha} = f(x, y, z), \\
\text{1-form} \quad & \tilde{\beta} = A(x, y, z)\boldsymbol{dx} + B(x, y, z)\boldsymbol{dy} + C(x, y, z)\boldsymbol{dz}, \\
\text{2-form} \quad & \tilde{\mu} = D(x, y, z)\boldsymbol{dx} \wedge \boldsymbol{dy} + E(x, y, z)\boldsymbol{dy} \wedge \boldsymbol{dz} + F(x, y, z)\boldsymbol{dz} \wedge \boldsymbol{dx}, \\
\text{3-form} \quad & \tilde{\nu} = G(x, y, z)\boldsymbol{dx} \wedge \boldsymbol{dy} \wedge \boldsymbol{dz}.
\end{aligned}
\tag{32.17}
$$

Notice how having two identical 1-forms in any wedge product causes it to vanish (e.g. $\boldsymbol{dx} \wedge \boldsymbol{dy} \wedge \boldsymbol{dx} = 0$). This constrains the possible p-forms to those above. It is not possible, for example, for a 4-form to exist in three spatial dimensions.

32.3 p-vectors

Consider vectors $\boldsymbol{u}$ and $\boldsymbol{v}$ in Euclidean 2-space. We can build a parallelogram in Fig. 32.4 from these vectors. Do this by starting at the origin and drawing $\boldsymbol{u}$ and then $\boldsymbol{v}$, returning to the origin and drawing $\boldsymbol{v}$ and then $\boldsymbol{u}$. The area of this parallelogram is given by the computation of the determinant

$$(\text{Area}) = \begin{vmatrix} u^x & u^y \\ v^x & v^y \end{vmatrix} = u^x v^y - v^x u^y = \varepsilon_{ij} u^i v^j, \tag{32.18}$$

where ε_{ij} is the two-dimensional Levi-Civita symbol.[8] The signed area is antisymmetric in the components of $\boldsymbol{u}$ and $\boldsymbol{v}$, which can be thought of as giving a handedness to the area. The area of the parallelogram is related to the **bivector b** defined as the antisymmetric (2,0) tensor[9]

$$\boldsymbol{b}(\ ,\ ) = \boldsymbol{u} \wedge \boldsymbol{v} = \boldsymbol{u} \otimes \boldsymbol{v} - \boldsymbol{v} \otimes \boldsymbol{u}. \tag{32.19}$$

In fact, it is fairly easy to see that the area is of the parallelogram is equal to the component b^{xy}, accessible by inserting basis 1-forms $\boldsymbol{\omega}^x$ and $\boldsymbol{\omega}^y$ into the slots of the bivector [i.e. $\boldsymbol{b}(\boldsymbol{\omega}^x, \boldsymbol{\omega}^y)$]. We will return to this in Chapter 37.

For now, we note that the existence of p-vectors, formed from the wedge products of vectors with components

$$
\begin{aligned}
\boldsymbol{v} &= \frac{1}{p!} v^{i_1, i_2 ... i_p} \boldsymbol{e}_{i_1} \wedge \boldsymbol{e}_{i_2} \wedge ... \wedge \boldsymbol{e}_{i_p} \\
&= v^{|i_1, i_2 ... i_p|} \boldsymbol{e}_{i_1} \wedge \boldsymbol{e}_{i_2} \wedge ... \wedge \boldsymbol{e}_{i_p}.
\end{aligned}
\tag{32.20}
$$

[8] In two dimensions, $\varepsilon_{12} = 1$, $\varepsilon_{21} = -1$, and $\varepsilon_{11} = \varepsilon_{22} = 0$. In general, $\varepsilon_{i_1...i_n} = (-1)^P$, where P is the parity of the permutation of $1, 2, 3...n$ in the indices (i.e. how many pairwise swaps need to be made to a string of numbers to work it back to the order $1...n$), and $\varepsilon_{i_1...i_n} = 0$ if any of the indices are repeated.

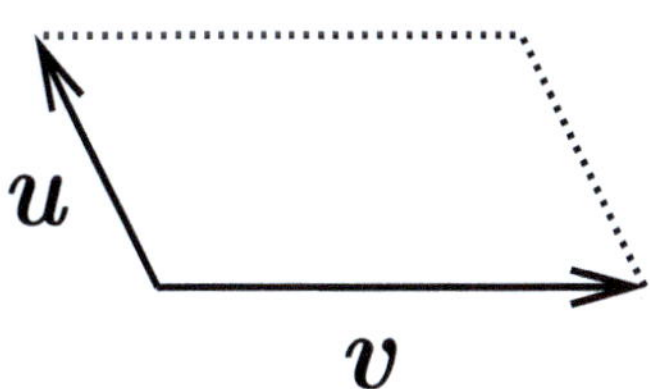

Fig. 32.4 The parallelogram formed by vectors $\boldsymbol{u}$ and $\boldsymbol{v}$.

[9] As for the case of forms, we can write the components of a bivector as

$$\boldsymbol{b} = \frac{1}{2} b^{\mu\nu} \boldsymbol{e}_\mu \wedge \boldsymbol{e}_\nu.$$

The existence of p-vectors allows us to define a set of generalized inner products of p-forms, such that p-vectors and p-forms can be mapped on to numbers. In order to make the inner product of a p-form $\tilde{\alpha} = \alpha_{|i_1\ldots i_p|}\omega^{i_1}\wedge\ldots\wedge\omega^{i_p}$ and a p-vector $v = v^{|j_1\ldots j_p|}e_{j_1}\wedge\ldots\wedge e_{j_p}$, we have the rule[10]

$$\langle\tilde{\alpha}, v\rangle = \alpha_{|i_1\ldots i_p|}v^{|j_1\ldots j_p|}\langle\omega^{i_1}\wedge\ldots\wedge\omega^{i_p}, e_{j_1}\wedge\ldots\wedge e_{j_p}\rangle$$
$$= \alpha_{|i_1\ldots i_p|}v^{|j_1\ldots j_p|}\delta^{i_1\ldots i_p}_{j_1\ldots j_p}$$
$$= \alpha_{|i_1\ldots i_p|}v^{i_1\ldots i_p}. \tag{32.21}$$

[10]The symbol $\delta^{i_1\ldots i_p}_{j_1\ldots j_p}$ is defined as

$$\delta^{i_1\ldots i_p}_{j_1\ldots j_p} =$$

$$\begin{cases} +1 & \text{if } (i_1\ldots i_p) \text{ is an even} \\ & \text{permutation of } (j_1\ldots j_p), \\ -1 & \text{if } (i_1\ldots i_p) \text{ is an odd} \\ & \text{permutation of } (j_1\ldots j_p), \\ 0 & \text{any two } i\text{s are the same,} \\ 0 & \text{any two } j\text{s are the same,} \\ 0 & \text{the } i\text{s and } j\text{s are} \\ & \text{different integers.} \end{cases}$$

Example 32.6

Work in three spatial dimensions, with components arranged in the order $(x^1, x^2, x^3) = (x, y, z)$. Combine a 2-form $\tilde{F}$ and a bivector b to make a number thus

$$\langle\tilde{F}, b\rangle = F_{xy}b^{xy} + F_{xz}b^{xz} + F_{yz}b^{yz}. \tag{32.22}$$

This has a pleasing resemblance to the dot product of vectors, or the inner product of a vector and a 1-form. The geometrical interpretation of this is that the scalar outputted by the inner product represents the number of tubes of the 2-form $\tilde{F}$ contained in the parallelogram representing the bivector b (Fig. 32.5).

It's also useful to note that if the bivector b is formed from vectors c and d via $b = c\wedge d$ then we also have that

$$\langle\tilde{F}, c\wedge d\rangle = \tilde{F}(c, d). \tag{32.23}$$

This is proved in the exercises.

To add to our list of known tensor objects, we now have p-forms and q-vectors, which are examples of antisymmetric tensors. In the next chapter, our study of differential geometry starts in earnest, as we start to take derivatives.

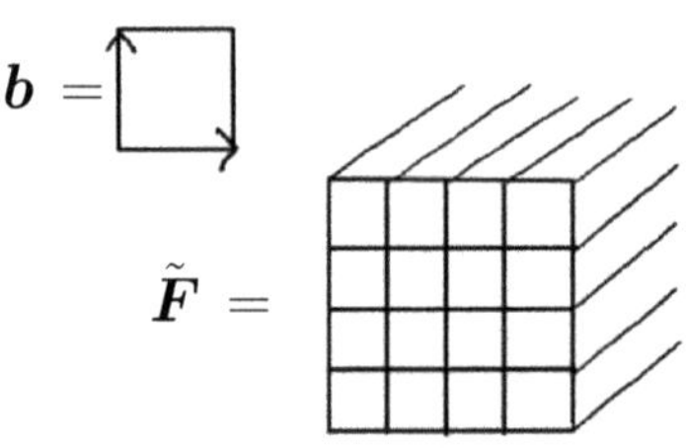

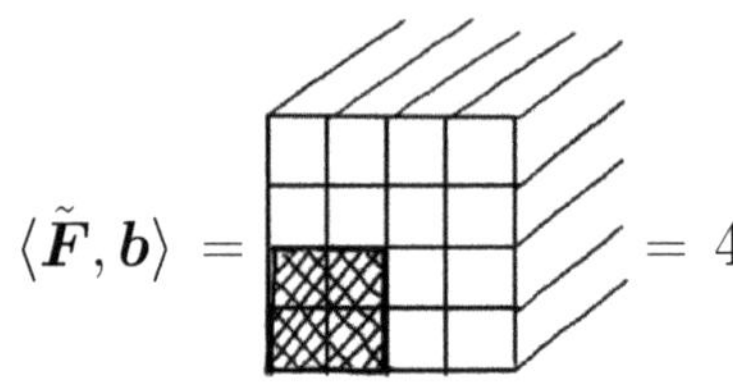

Fig. 32.5 The bivector b and 2-form $\tilde{F}$ are used to for the inner product $\langle\tilde{F}, b\rangle$. This outputs a number equal to the number of tubes of $\tilde{F}$ contained within the parallelogram b, which in this case is 4.

Chapter summary

- Differential forms can be built from 1-forms using the wedge product. A p-form is an antisymmetric $(0, p)$ tensor.

- 2-forms can be represented as a tube-like structure with a handedness.

- p-vectors can be built from vectors using the wedge product. A p-vector and a p-form can be combined to make a number.

Exercises

(32.1) Verify eqn 32.23 by computing components.

(32.2) *Fermi transport*: Consider a sphere, and vectors $\boldsymbol{a}$, $\boldsymbol{b}$ and $\boldsymbol{c}$ that are orthogonal to each other and to the velocity vector $\boldsymbol{u}$ of the sphere (i.e. the vector field that is tangent to the sphere's world line). The three vectors determine the orientation of the sphere along its world line and can be used to measure the rotation of the sphere. We can write down the equations of motion that tells us there is no rotation. The extent to which this equation is satisfied measures the rotation. The equations of motion, known as the **Fermi transport** equations,

are

$$\boldsymbol{\nabla}_u \boldsymbol{a} = -\left(\boldsymbol{A} \wedge \boldsymbol{u}\right)\left(\,,\tilde{\boldsymbol{a}}\right),$$
$$\boldsymbol{\nabla}_u \boldsymbol{b} = -\left(\boldsymbol{A} \wedge \boldsymbol{u}\right)\left(\,,\tilde{\boldsymbol{b}}\right),$$
$$\boldsymbol{\nabla}_u \boldsymbol{c} = -\left(\boldsymbol{A} \wedge \boldsymbol{u}\right)\left(\,,\tilde{\boldsymbol{c}}\right), \qquad (32.24)$$

where $\boldsymbol{A} = \boldsymbol{\nabla}_u \boldsymbol{u}$ is the acceleration.

(a) Write one of these equations in components.
(b) What happens for transport along a geodesic world line?
(c) Show that if $\boldsymbol{a}$, $\boldsymbol{b}$ and $\boldsymbol{c}$ are initially orthogonal, they stay orthogonal.
(d) Show that a vector transported by these equations remains orthogonal to the four-velocity.
(e) Compare Fermi transport with parallel transport.

33 Exterior and Lie derivatives

More ways of killing a cat than choking her with cream
Charles Kingsley (1819–1875)

Derivatives are undeniably important in physics. We would like a way of taking derivatives of tensor fields that does not rely on coordinates. We naturally reach the idea of a derivative if we think in terms of a tensors changing as we move around a space. More precisely, we work with the notion of a tensor field which, at each point in space, provides us with a tensor. In working out derivatives, we are faced with the problem that there is no obvious way to compare tensor fields evaluated at two different points in a space. One way around this is to use the metric to define a notion of parallelism. This requires a connection: a method of connecting different points, so that tensors can be moved in a parallel manner around the space. We examine this connection in more detail in Chapter 34. If we don't have this parallelism (or we don't have a metric) then how can we proceed? In this chapter, we examine two ways of dealing with more primitive notions of rates of change of tensors. The first type of derivative applies only to fields of forms and is called the **exterior derivative**. We shall later use this derivative to describe the curvature of spacetime. The second type of derivative applies to all tensors and relies on being able to carry the tensor field of interest across the space using a separate vector field. This is called a **Lie derivative** and is the subject of the second part of this chapter. The Lie derivative is especially useful in cosmology for describing changes as we travel along with the cosmological fluid. However, perhaps its most useful feature is in identifying conserved quantities via **Killing vectors**, which we describe at the end of the chapter.

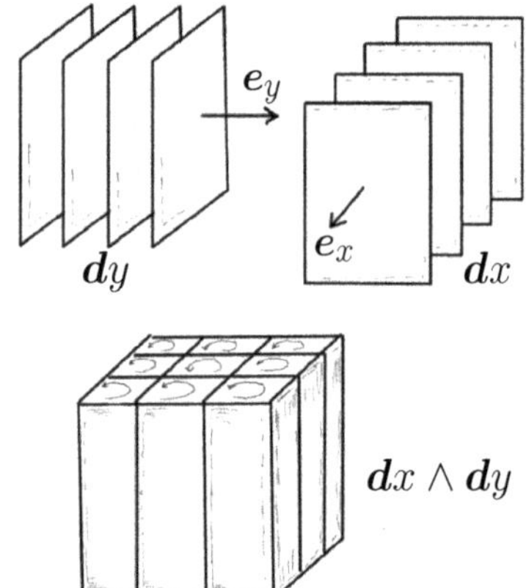

Fig. 33.1 The action of the exterior derivative $\boldsymbol{d}$, on the 1-form $f(x)\boldsymbol{dy}$ is to convert it into $\partial f(x)/\partial x(\boldsymbol{dx} \wedge \boldsymbol{dy})$.

33.1 Exterior calculus

The **exterior derivative** is a differential operation that can be applied to forms. In measuring a rate of change of a p-form, exterior differentiation converts the p-form into a $(p + 1)$-form. This has a pictorial interpretation which is clearest for the action of the exterior derivative on a 1-form and is shown in Fig. 33.1. Recall that a 1-form can be thought of as a set of parallel surfaces in space, while a 2-form is a tubular structure. By measuring how fast a component of the 1-form changes along a some direction, the exterior derivative builds us a 2-form by adding surfaces normal to this direction.

To use the exterior derivative we define an operator $\boldsymbol{d}$ that acts on forms. The simplest 1-form is the differential $\boldsymbol{df}$. This 1-form can be thought of as originating from the action of the **exterior derivative operator** $\boldsymbol{d}$ on the 0-form $f(x)$. In general, the operator $\boldsymbol{d}$ can be thought of as a $(0,1)$ object that creates the $(p+1)$-form $\boldsymbol{d\tilde{A}}$ from a p-form $\boldsymbol{\tilde{A}}$. It obeys the rules listed in the margin.

The rules feel rather austere. We get a clearer idea of what the operator does in practice if we consider its action on an arbitrary p-form

$$\boldsymbol{\tilde{A}} = \frac{1}{p!} A_{\alpha,\beta..\sigma} \boldsymbol{dx}^\alpha \wedge \boldsymbol{dx}^\beta \wedge ... \wedge \boldsymbol{dx}^\sigma. \tag{33.6}$$

To be consistent with the rules, we must have

$$\boldsymbol{d\tilde{A}} = \frac{1}{p!} dA_{\alpha\beta...\sigma} \wedge \boldsymbol{dx}^\alpha \wedge \boldsymbol{dx}^\beta \wedge ... \wedge \boldsymbol{dx}^\sigma, \tag{33.7}$$

where

$$\boldsymbol{d}A_{\alpha\beta...\sigma} = \frac{\partial A_{\alpha\beta...\sigma}}{\partial x^\mu} \boldsymbol{dx}^\mu. \tag{33.8}$$

As a result, we have the following:[1]

The exterior derivative of a p-form $\boldsymbol{\alpha}$ is given by

$$\boldsymbol{d\tilde{A}} = \frac{1}{p!} \frac{\partial A_{\alpha\beta...\sigma}}{\partial x^\mu} \boldsymbol{dx}^\mu \wedge \boldsymbol{dx}^\alpha \wedge \boldsymbol{dx}^\beta \wedge ... \wedge \boldsymbol{dx}^\sigma. \tag{33.10}$$

As with all calculus, the only way to really see what's going on is to examine some examples.

Example 33.1

We work in $(3+1)$-dimensional spacetime with spherical coordinates $x^\mu = (t, \chi, \theta, \phi)$. We act with $\boldsymbol{d}$ on a 1-form $\boldsymbol{\tilde{A}} = A_\alpha(t,\chi,\theta,\phi)\boldsymbol{dx}^\alpha$, where A_α are functions of the coordinates. We will obtain

$$\boldsymbol{d\tilde{A}} = \frac{\partial A_\alpha}{\partial x^\mu} \boldsymbol{dx}^\mu \wedge \boldsymbol{dx}^\alpha. \tag{33.11}$$

Example 0: Consider a 1-form $\boldsymbol{\tilde{A}} = \boldsymbol{dt}$. Acting with $\boldsymbol{d}$ we obtain

$$\boldsymbol{d\tilde{A}} = \boldsymbol{ddt} = 0. \tag{33.12}$$

where we've used $\boldsymbol{dd} = 0$ (rule IV in the box).
Example 1: Consider a 1-form $\boldsymbol{\tilde{A}} = a(t)\boldsymbol{d\chi}$, where $a(t)$ is a scalar function of t only. Acting with $\boldsymbol{d}$ we obtain

$$\boldsymbol{d\tilde{A}} = \frac{\partial a(t)}{\partial t} \boldsymbol{dt} \wedge \boldsymbol{d\chi}, \tag{33.13}$$

where we've used the fact that $\partial a(t)/\partial x^i = 0$, for $i \neq t$.
Example 2: Consider a 1-form $\boldsymbol{\tilde{A}} = a(t)\sin\theta\boldsymbol{d\theta}$. Taking the exterior derivative we obtain

$$\boldsymbol{d\tilde{A}} = \frac{\partial a(t)}{\partial t} \sin\theta \boldsymbol{dt} \wedge \boldsymbol{d\theta} + a(t)\cos\theta \boldsymbol{d\theta} \wedge \boldsymbol{d\theta}$$

$$= \frac{\partial a(t)}{\partial t} \sin\theta \boldsymbol{dt} \wedge \boldsymbol{d\theta}. \tag{33.14}$$

where we've used the fact that $\boldsymbol{dx}^\mu \wedge \boldsymbol{dx}^\mu = 0$.

The exterior derivative operator $\boldsymbol{d}$ has the following properties when acting on the p-form $\boldsymbol{\tilde{\alpha}}$ and q-form $\boldsymbol{\tilde{\beta}}$:

I: It is linear

$$\boldsymbol{d}(\boldsymbol{\tilde{\alpha}} + \boldsymbol{\tilde{\beta}}) = \boldsymbol{d\tilde{\alpha}} + \boldsymbol{d\tilde{\beta}}. \tag{33.1}$$

II: For the 0-form f (i.e. a scalar function), the 1-form $\boldsymbol{df}$ is defined by the usual relationship

$$\langle \boldsymbol{df}, \boldsymbol{v} \rangle = \boldsymbol{v}[f] = \partial_{\boldsymbol{v}} f, \tag{33.2}$$

where $\boldsymbol{v}$ is a vector.
In coordinates, the 1-form is given by

$$\boldsymbol{df} = \frac{\partial f}{\partial x^k} \boldsymbol{dx}^k. \tag{33.3}$$

III: The action of $\boldsymbol{d}$ on a wedge product is given by

$$\boldsymbol{d}\left(\boldsymbol{\tilde{\alpha}} \wedge \boldsymbol{\tilde{\beta}}\right) = \boldsymbol{d\tilde{\alpha}} \wedge \boldsymbol{\tilde{\beta}} + (-1)^p \boldsymbol{\tilde{\alpha}} \wedge \boldsymbol{d\tilde{\beta}}. \tag{33.4}$$

IV: The exterior derivative of an exterior derivative is zero

$$\boldsymbol{d}^2 = \boldsymbol{dd} = 0. \tag{33.5}$$

[1]Alternatively, we could write

$$\boldsymbol{d\tilde{A}} =$$
$$\frac{\partial A_{|\alpha\beta...\sigma|}}{\partial x^\mu} \boldsymbol{dx}^\mu \wedge \boldsymbol{dx}^\alpha \wedge \boldsymbol{dx}^\beta \wedge ... \wedge \boldsymbol{dx}^\sigma \tag{33.9}$$

where we don't require the factor $1/p!$

Example 3: Consider a 1-form $\tilde{\boldsymbol{A}} = a(t)\sin\chi\sin\theta\boldsymbol{d\phi}$. This time we have the exterior derivative

$$\boldsymbol{d\tilde{A}} = \frac{\partial a(t)}{\partial t}\sin\theta\boldsymbol{dt}\wedge\boldsymbol{d\phi} + a(t)\cos\chi\sin\theta\boldsymbol{d\chi}\wedge\boldsymbol{d\phi} + a(t)\sin\chi\cos\theta\boldsymbol{d\theta}\wedge\boldsymbol{d\phi}. \tag{33.15}$$

All of the 1-forms chosen in this example will be met again when we examine the metric of the expanding Universe.

The rule $\boldsymbol{dd} = 0$ puts constraints on the sorts of objects that can be accommodated in $(3+1)$-dimensional spacetime, as we shall see in the next example.

Example 33.2

We work in $(3+1)$-dimensional spacetime. Starting with a function f, we can construct a 1-form $\tilde{\boldsymbol{\alpha}}$ using the exterior derivative

$$\tilde{\boldsymbol{\alpha}} = \boldsymbol{df}. \tag{33.16}$$

The 1-form looks like a set of evenly spaced planes. If we try to obtain a 2-form by acting on $\tilde{\boldsymbol{\alpha}}$ with the exterior derivative we obtain

$$\boldsymbol{d}\tilde{\boldsymbol{\alpha}} = \boldsymbol{dd}f = 0. \tag{33.17}$$

If we start with a 1-form $\tilde{\boldsymbol{A}}$, we can construct a 2-form by acting on the 1-form with $\boldsymbol{d}$, which gives

$$\tilde{\boldsymbol{F}} = \boldsymbol{d\tilde{A}}. \tag{33.18}$$

The 2-form looks like a set of tubes in space, with the 2-form field circulating in each of the tubes. If we try to obtain a 3-form from the 2-form using the exterior derivative we obtain

$$\boldsymbol{d\tilde{F}} = \boldsymbol{dd\tilde{A}} = 0. \tag{33.19}$$

Now start with a 2-form $\tilde{\boldsymbol{\nu}}$ and construct a 3-form by acting on $\tilde{\boldsymbol{\nu}}$ with $\boldsymbol{d}$, so that we have

$$\tilde{\boldsymbol{\mu}} = \boldsymbol{d\tilde{\nu}}. \tag{33.20}$$

The 3-form $\tilde{\boldsymbol{\mu}}$ looks like a set of cubes with the 3-form field circulating inside. If we try to obtain a 4-form from the 3-form using the exterior derivative we obtain

$$\boldsymbol{d\tilde{\mu}} = \boldsymbol{dd\tilde{\nu}} = 0. \tag{33.21}$$

Continuing, we can start with a 3-form and make a 4-form using $\boldsymbol{d}$. A 5-form, however, cannot be accommodated in our four-dimensional space.

[2]Marius Sophus Lie (1842–1899). His name is pronounced 'Lee' (as in Bruce Lee, to rhyme with ski). Continuous transformation groups are now called Lie groups in honour of his greatest work in mathematics, and are very widely used in quantum mechanics. Élie Cartan was a doctoral student of Lie.

We now turn to the second sort of derivative for coordinate-free geometry: the **Lie derivative**.[2] This derivative is most simply applied using the notion of a commutator, so we first pause to explore this in the next section.

33.2 Commutators

Adopting the idea of a vector as an operator, we define the **commutator** $[\boldsymbol{u}, \boldsymbol{v}]$ of two vector fields $\boldsymbol{u} = \mathrm{d}/\mathrm{d}\lambda$ and $\boldsymbol{v} = \mathrm{d}/\mathrm{d}\mu$ as

$$\begin{aligned} [\boldsymbol{u}, \boldsymbol{v}] &= \boldsymbol{uv} - \boldsymbol{vu} \\ &= \frac{\mathrm{d}}{\mathrm{d}\lambda}\frac{\mathrm{d}}{\mathrm{d}\mu} - \frac{\mathrm{d}}{\mathrm{d}\mu}\frac{\mathrm{d}}{\mathrm{d}\lambda}. \end{aligned} \tag{33.22}$$

The brackets in the notation for commutators are also known as **Lie brackets**. Since vector fields act on functions, we can also write

$$[\boldsymbol{u}, \boldsymbol{v}][f] = \boldsymbol{u}\boldsymbol{v}[f] - \boldsymbol{v}\boldsymbol{u}[f]$$
$$= \frac{\mathrm{d}}{\mathrm{d}\lambda}\frac{\mathrm{d}f}{\mathrm{d}\mu} - \frac{\mathrm{d}}{\mathrm{d}\mu}\frac{\mathrm{d}f}{\mathrm{d}\lambda}. \tag{33.23}$$

Example 33.3
Worked through in coordinates on a function f, the commutator becomes

$$[\boldsymbol{u}, \boldsymbol{v}][f] = u^\alpha \frac{\partial}{\partial x^\alpha}\left(v^\beta \frac{\partial f}{\partial x^\beta}\right) - v^\alpha \frac{\partial}{\partial x^\alpha}\left(u^\beta \frac{\partial f}{\partial x^\beta}\right)$$
$$= \left(u^\alpha \frac{\partial v^\beta}{\partial x^\alpha} - v^\alpha \frac{\partial u^\beta}{\partial x^\alpha}\right)\frac{\partial f}{\partial x^\beta}. \tag{33.24}$$

We could write the right-hand side of this equation out as another vector field $w^\beta \partial/\partial x^\beta$, acting on the function f. The components w^β are given by $w^\beta = \left(u^\alpha \frac{\partial v^\beta}{\partial x^\alpha} - v^\alpha \frac{\partial u^\beta}{\partial x^\alpha}\right)$, which will, in general, give is a set of non-zero numbers. We conclude that the commutator generates another vector field, whose components do not, in general, vanish.

Another conclusion that can be drawn from the previous example, through the removal of the function is that the commutator, written in terms of coordinates, is

$$[\boldsymbol{u}, \boldsymbol{v}] = \left(u^\alpha \frac{\partial v^\beta}{\partial x^\alpha} - v^\alpha \frac{\partial u^\beta}{\partial x^\alpha}\right)\frac{\partial}{\partial x^\beta}. \tag{33.25}$$

This expression, built from directional derivatives, gives rise to a simple pictorial view of the commutator. The first term is an instruction to evaluate the difference $\boldsymbol{\Delta}_2$ in the vector field $\boldsymbol{v}$ evaluated at the tip and base of the vector $\boldsymbol{u}$, as shown in Fig. 33.2(b). The second term tells us to evaluate the difference $\boldsymbol{\Delta}_1$ in $\boldsymbol{u}$ evaluated at the tip and base of the vector $\boldsymbol{v}$, as shown in Fig. 33.2(a). The commutator $(\boldsymbol{\Delta}_2 - \boldsymbol{\Delta}_1)$ therefore tells us to move along the vector $\boldsymbol{v}$ and then along the vector $\boldsymbol{u}$ and compare this to the final position if we move along $\boldsymbol{u}$ and then $\boldsymbol{v}$. In pictures, this is a measurement of the vector that closes the figures in Fig. 33.2(c). If this were Euclidean space with constant vector fields, then the parallelogram that is made is a closed figure and the commutator is zero. However, in general, we cannot guarantee that vector fields make closed figures when transported along each other's lengths. The amount by which the figure fails to close is measured by the vector that is outputted by the commutator, as shown in Fig. 33.2(c). So, in short, the commutator measures the amount by which our attempt to make a parallelogram from vectors $\boldsymbol{u}$ and $\boldsymbol{v}$ fails to close.

In Chapter 10, we discussed the case that the vectors in the Lie brackets are the basis vectors $\boldsymbol{e}_\mu$ of a space. Since basis vectors should, in coordinates, be expressible simply as partial derivatives $\partial/\partial x^\beta$, then we expect the part in the brackets to vanish. If it does not, it implies that

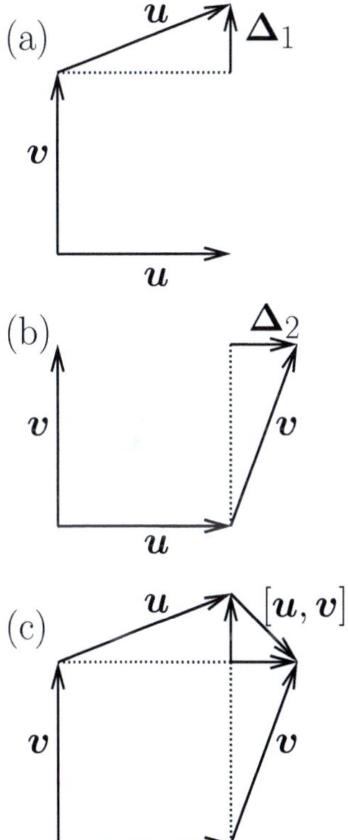

Fig. 33.2 (a) The sequence $\boldsymbol{vu}$ involves following $\boldsymbol{v}$ then $\boldsymbol{u}$. This vector $\boldsymbol{u}$ at the tip of $\boldsymbol{v}$ will generally be different to the one at the base of $\boldsymbol{v}$. (b) The sequence $\boldsymbol{uv}$ involves following $\boldsymbol{u}$ and then $\boldsymbol{v}$. (c) The commutator $[\boldsymbol{u}, \boldsymbol{v}]$.

[3]This is just what we discussed in Chapters 3 and 10.

[4]Although this might all seem somewhat contrived, the process of being carried along by a vector field is an example of a **diffeomorphism**, which is the generalization of an active coordinate transformation to the manifold. Diffeomorphisms allow us to describe the smoothness of a spacetime and are important in the mathematical description of relativity. See Appendix C for more details. In terms of physics, the Lie derivative is so important to us because, in cosmology, we consider that we are all being carried along the current of the cosmological fluid of matter that fills the Universe. Fluids themselves are examined in detail in Chapter 39.

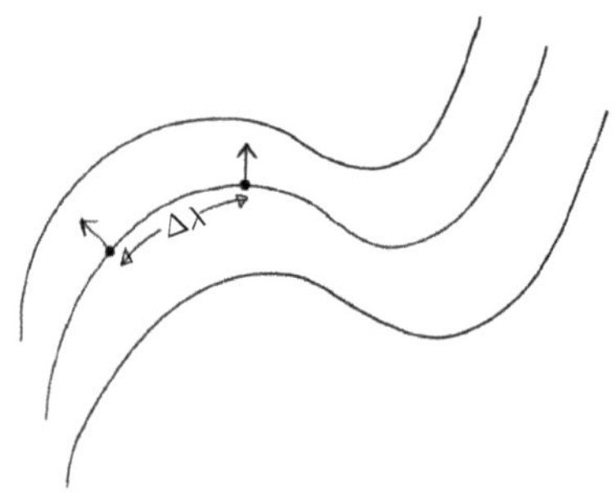

Fig. 33.3 Moving along a congruence of curves by an amount $\Delta\lambda$ and comparing vectors from the field $\boldsymbol{v}$. The congruence is formed from the streamlines of a vector field $\boldsymbol{u}$. We take Lie derivatives with respect to this second field.

[5]The pound sign $\pounds$ means 'L' for Lie here. It's use in denoting the currency stems from the origin of the English word pound, which comes from the Latin *libra pondo* meaning a 'pound by weight'.

[6]In Chapter 31, we used given curves to provide a tangent vector fields. Now we turn the tables and take a vector field to generate a series of curves to which the field is tangent.

it is *impossible* to express the basis in terms of derivatives of coordinates. Such bases are **non-coordinate bases**. In contrast, the set of basis vectors $\{e_\mu\}$ is a **coordinate basis** if and only if $[e_\alpha, e_\beta] = 0$.[3] In the formulation of general relativity, many expressions are simplest when referred to a coordinate basis. However, we make and interpret measurements locally in orthonormal frames, which are generally non-coordinate frames.

33.3 Lie derivatives of vectors

Previously, we saw that the exterior derivative measures the rate of change of a 1-form in the direction of the normal to the 1-form's surface. When generalized, this derivative applied only to p-forms. We now turn to an alternative, primitive, method of taking derivatives. Recall that a problem we face is that it's difficult to know how to compare tensor fields at two different points in space. Lie differentiation presents a solution to this problem by using a second vector field to provide a means of moving the original vectors around.

Imagine being carried along by the current in a river.[4] In order to evaluate the change of some vector field $\boldsymbol{v}$, we can extract a vector at some position, allow ourselves to be carried along by the river current, and then compare the vector we originally extracted to another one at the new position at which we find ourselves. This is the essence of the Lie derivative. The current of the river can itself be represented by a second vector field $\boldsymbol{u}$ representing the velocity field of the river fluid. The vectors from the field $\boldsymbol{u}$ are tangent to a congruence of curves that we call the **streamlines** of the vector field $\boldsymbol{u}$. These are shown in Fig. 33.3. We need to know both $\boldsymbol{v}$ and $\boldsymbol{u}$ to take the Lie derivative, which will turn out simply to be given by the commutator of the two fields $[\boldsymbol{u}, \boldsymbol{v}]$.

Lie differentiation doesn't just apply to vectors like $\boldsymbol{v}$; it can be applied to any tensor $\boldsymbol{Z}$. We write the Lie derivative of $\boldsymbol{Z}$ as[5]

$$\pounds_{\boldsymbol{u}} \boldsymbol{Z} = \left(\begin{array}{c} \text{Lie derivative of tensor } \boldsymbol{Z} \\ \text{carried along vector field } \boldsymbol{u} \end{array} \right). \tag{33.26}$$

We can think of the Lie derivative as being a derivative of $\boldsymbol{Z}$ with respect to $\boldsymbol{u}$, where the field $\boldsymbol{u}$ is used to provide a yardstick against which to measure changes in $\boldsymbol{Z}$. Importantly, if there is no difference in a tensor after being carried along by the current, we have $\pounds_{\boldsymbol{u}} \boldsymbol{Z} = 0$ and we say that the tensor field has been **Lie dragged**. Geometrically, a Lie-dragged vector that joins two streamlines at some point will continue to join them after being carried along by the current, as shown in Fig. 33.4.

Next, we show that the Lie derivative of a vector field $\boldsymbol{v}$ obeys $\pounds_{\boldsymbol{u}} \boldsymbol{v} = [\boldsymbol{u}, \boldsymbol{v}]$. First, we need a suitable set of streamlines along which to carry the field. In a fluid, a streamline is a curve whose tangent at some point $\mathcal{P}$ gives the velocity $\boldsymbol{u}$ of the element of fluid at $\mathcal{P}$. So we can think of the vector field $\boldsymbol{u}(x)$ as supplying the streamlines.[6]

Example 33.4

More formally, we have a congruence when there is a unique curve $c(\lambda)$ through each point $\mathcal{P}$ of space such that: (i) $c(\lambda = 0) = \mathcal{P}$, and (ii) the tangent vector at a point $c(\lambda)$ is $\boldsymbol{u}|_{c(\lambda)}$. This curve therefore has a tangent vector that is always given by the vector field $\boldsymbol{u}$. The curve is called an **integral curve** of the vector field $\boldsymbol{u}$. It is not obvious that such a curve will exist for some given field $\boldsymbol{u}$, so we pause to consider it. The components of the vector field $\boldsymbol{u}$ at $\mathcal{P}$ are $u^\mu(\mathcal{P})$ and change as a function of this position. If we work in a coordinate system the curve has coordinates $x^\mu(\lambda)$. Saying that the components $u^\mu(x^1(\lambda), x^2(\lambda), ..x^n(\lambda))$ are tangent to the curve with parameter λ amounts to the set of differential equations

$$\frac{\mathrm{d}x^\mu(\lambda)}{\mathrm{d}\lambda} = u^\mu(x^1(\lambda), x^2(\lambda), ..., x^n(\lambda)). \tag{33.27}$$

This set of first-order differential equations is guaranteed[7] to have unique solutions in the region near $\mathcal{P}$. This result guarantees the existence of the required congruence of integral curves.

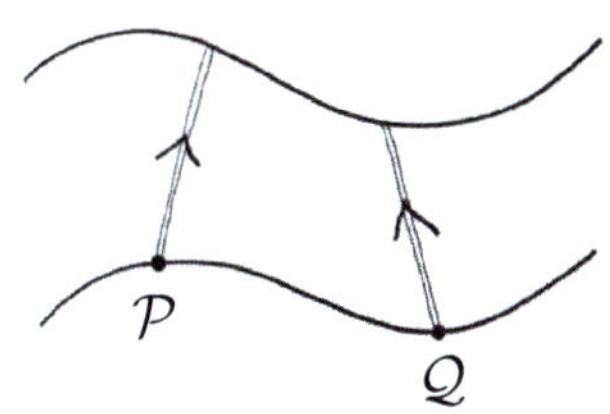

Fig. 33.4 A Lie dragged vector continues to link two streamlines after being carried along the congruence.

[7]This is a result which is proved in many of standard references. An example is Choquet-Bruhat, DeWitt-Morette and Dillard-Bleick *Analysis, Manifolds and Physics*.

With all of the ingredients in place, we're now ready to compute the derivative. We take the vector field at $\mathcal{P}$, called $\boldsymbol{v}(\mathcal{P})$, we carry this vector a distance $\Delta\lambda$ along an integral curve of the vector field $\boldsymbol{u}$ to a point $\mathcal{Q}$. This turns $\boldsymbol{v}(\mathcal{P})$ into the vector $\boldsymbol{v}'(\mathcal{Q})$. We compare this transported vector to the vector $\boldsymbol{v}(\mathcal{Q})$ found at $\mathcal{Q}$. The definition of the Lie derivative is then

$$\pounds_{\boldsymbol{u}}\boldsymbol{v}(\mathcal{P}) = \lim_{\Delta\lambda\to0}\frac{1}{\Delta\lambda}\left\{\boldsymbol{v}(\mathcal{Q}) - \boldsymbol{v}'(\mathcal{Q})\right\}. \tag{33.28}$$

In the next example, we carry out this computation in a coordinate system.

Example 33.5

In terms of a coordinate system, we have a curve parametrized by λ with a tangent vector with components $u^\mu = \frac{\mathrm{d}x^\mu}{\mathrm{d}\lambda}$. The point $\mathcal{P}$ has a coordinate x^μ. The coordinates of the nearby point $\mathcal{Q}$ are $x^{\mu'}$ and these can be found using a coordinate transformation that translates us from $\mathcal{P}$ to $\mathcal{Q}$, along the vector $\boldsymbol{u}$

$$x^{\mu'} = x^\mu + \mathrm{d}x^\mu = x^\mu + u^\mu \mathrm{d}\lambda. \tag{33.29}$$

To find the Lie derivative, we need to compare the field at a point $\mathcal{Q}$ (with components v^β) to the field carried from $\mathcal{P}$ to $\mathcal{Q}$ (with components $v^{\beta'}$)

$$(\pounds_{\boldsymbol{u}}\boldsymbol{v}(\mathcal{P}))^\beta = \frac{v^\beta(\mathcal{Q}) - v^{\beta'}(\mathcal{Q})}{\mathrm{d}\lambda}. \tag{33.30}$$

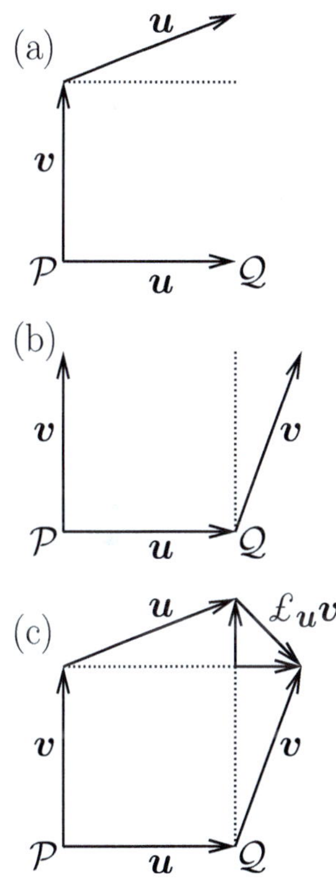

Fig. 33.5 (a) The effect of transporting the vector $\boldsymbol{v}$ from $\mathcal{P}$ to $\mathcal{Q}$ is measured in terms of the change in $\boldsymbol{u}$. (b) The vector $\boldsymbol{v}$ evaluated at $\mathcal{P}$ and $\mathcal{Q}$. (c) The difference between (a) and (b) gives the Lie derivative: a vector that closes the circuit.

[8]Remember that components must transform as $A^{\mu'} = \frac{\partial x^{\mu'}}{\partial x^\nu}A^\nu$.

Step I: We first work out the components of the transported vector. We do this by transforming the vector into the coordinate frame at $\mathcal{Q}$ using the tensor transformation law[8] and then substituting eqn 33.29, thus

$$\begin{aligned} v^{\beta'}(\mathcal{Q}) &= \frac{\partial x^{\beta'}}{\partial x^\alpha}v^\alpha(\mathcal{P}) \\ &= \left(\delta^\beta{}_\alpha + \frac{\partial u^\beta}{\partial x^\alpha}\mathrm{d}\lambda\right)v^\alpha(\mathcal{P}) \\ &= v^\beta(\mathcal{P}) + \frac{\partial u^\beta}{\partial x^\alpha}v^\alpha(\mathcal{P})\mathrm{d}\lambda. \end{aligned} \tag{33.31}$$

[9]Note that if we have a connection, $\Gamma^\beta{}_{\alpha\gamma}$, this expression can be rewritten in terms of covariant derivatives in an equivalent form

$$(\pounds_u v(\mathcal{P}))^\beta = u^\alpha \left(\frac{\partial v^\beta}{\partial x^\alpha} + \Gamma^\beta{}_{\alpha\gamma} v^\gamma \right)$$
$$- v^\alpha \left(\frac{\partial u^\beta}{\partial x^\alpha} + \Gamma^\beta{}_{\alpha\gamma} u^\gamma \right)$$
$$= u^\alpha v^\beta{}_{;\alpha} - v^\alpha u^\beta{}_{;\alpha}. \tag{33.33}$$

This is strictly optional: the connection is not needed to define the Lie derivative.

We see that the effect of carrying the vector v is measured by $\frac{\partial u^\beta}{\partial x^\alpha} v^\alpha(\mathcal{P})$, which is the difference in the velocity field u measured at the base and tip of the v, as shown in Fig. 33.5(a)

Step II: Second, we compute the vector v at $\mathcal{Q}$ with components $v^\beta(\mathcal{Q})$. This is done by seeing how the field v itself changes between $\mathcal{P}$ and $\mathcal{Q}$ as we move along u

$$v^\beta(\mathcal{Q}) = v^\beta(\mathcal{P}) + \frac{\partial v^\beta}{\partial x^\alpha} u^\alpha(\mathcal{P}) d\lambda. \tag{33.32}$$

This is shown geometrically in Fig. 33.5(b).

Step III: The previous two expressions involve the field evaluated at the same point, so can be combined. The Lie derivative is, therefore[9]

$$(\pounds_u v(\mathcal{P}))^\beta = \frac{v^\beta(\mathcal{Q}) - v^{\beta\prime}(\mathcal{Q})}{d\lambda}.$$
$$= u^\alpha(\mathcal{P}) \frac{\partial v^\beta}{\partial x^\alpha} - v^\alpha(\mathcal{P}) \frac{\partial u^\beta}{\partial x^\alpha}. \tag{33.34}$$

Comparing with eqn 33.25, we see that this is given simply by the Lie bracket of the fields u and v.

The result from the last example implies that, for a vector field, we have

$$\pounds_u v = [u, v]. \tag{33.35}$$

We use this to formalize our set of rules for finding the Lie derivative of a field.

The Lie derivative of a scalar function with respect to the field u is defined to be the directional derivative along u:

$$\pounds_u f = u[f] = \partial_u f. \tag{33.36}$$

The Lie derivative of a vector field $v = d/d\mu$ with respect to $u = d/d\lambda$ is

$$\pounds_u v = \frac{d}{d\lambda} v - \frac{d}{d\mu} u = [u, v] = \left(u^\alpha v^\beta{}_{,\alpha} - v^\alpha u^\beta{}_{,\alpha} \right) \frac{\partial}{\partial x^\beta}. \tag{33.37}$$

Notice from the geometrical interpretation of the Lie derivative that $\pounds_u v$ evaluated at point $\mathcal{P}$ does not just depend on the value of v at $\mathcal{P}$. It also depends on the value of v at surrounding points. This is unlike the ordinary partial derivative, which only depends on the point $\mathcal{P}$ in question. As a result, the Lie derivative is not powerful enough to become the derivative for use describing the curvature in manifolds. This is the reason why, ultimately, we must introduce a connection and a covariant derivative.

Example 33.6

Choose a coordinate system where u is a coordinate basis vector $e_1 = \partial/\partial x^1$, then we have

$$(\pounds_{e_1} W)^\mu = \delta^\nu{}_1 \frac{\partial}{\partial x^\nu} W^\mu - W^\nu \frac{\partial}{\partial x^\nu} \delta^\mu{}_\nu = \frac{\partial W^\mu}{\partial x^1}. \tag{33.38}$$

This suggests a sense in which we can interpret the Lie derivative as a coordinate-free version of the partial derivative.

33.4 Lie derivatives of tensors

We advertised the Lie derivative as working for all tensors. Another object we need to know how to differentiate in this way is therefore a 1-form. To find this, consider the contraction $\langle \tilde{\sigma}, w \rangle$ and take the Lie derivative with respect to the vector field v. To do this, we simply use the Leibniz rule[10] $\mathcal{L}_v \langle \tilde{\sigma}, w \rangle = \langle \mathcal{L}_v \tilde{\sigma}, w \rangle + \langle \tilde{\sigma}, \mathcal{L}_v w \rangle$. Since, for fields, $\langle \tilde{\sigma}, w \rangle = f$, where f is some scalar field, we have $\mathcal{L}_v \langle \tilde{\sigma}, w \rangle = \partial_v f$ and a result

$$\langle \mathcal{L}_v \tilde{\sigma}, w \rangle = \partial_v f - \langle \tilde{\sigma}, \mathcal{L}_v w \rangle. \tag{33.41}$$

[10]Using the linearity of the derivative operation, we have for arbitrary tensors, that

$$\mathcal{L}_v(S \otimes T) = \mathcal{L}_v S \otimes T + S \otimes \mathcal{L}_v T. \tag{33.39}$$

If we contract this, we obtain a Leibniz rule

$$\mathcal{L}_v \langle S, T \rangle = \langle \mathcal{L}_v S, T \rangle + \langle S, \mathcal{L}_v T \rangle. \tag{33.40}$$

Example 33.7

From the previous expression we can, with some relabelling of indices, deduce the components of the Lie derivative of a 1-form

$$(\mathcal{L}_v \tilde{\sigma})_\mu w^\mu = v^\nu \frac{\partial}{\partial x^\nu}(\sigma_\mu w^\mu) - \sigma_\mu \left(v^\nu \frac{\partial}{\partial x^\nu} w^\mu - w^\nu \frac{\partial}{\partial x^\nu} v^\mu \right)$$

$$= \left(v^\nu \frac{\partial}{\partial x^\nu} \sigma_\mu + \sigma_\nu \frac{\partial}{\partial x^\mu} v^\nu \right) w^\mu. \tag{33.42}$$

Since the w^μ are arbitrary, we conclude

$$(\mathcal{L}_v \tilde{\sigma})_\mu = v^\nu \frac{\partial}{\partial x^\nu} \sigma_\mu + \sigma_\nu \frac{\partial}{\partial x^\mu} v^\nu. \tag{33.43}$$

If we want to work out Lie derivatives of larger tensor objects, we again use the Leibniz rule and find a more general expression

$$\mathcal{L}_v T(\tilde{\sigma}, ...; U....) = (\mathcal{L}_v T)(\tilde{\sigma}, ...; U....) + T(\mathcal{L}_v \tilde{\sigma}, ...; U....) + ...$$
$$+ T(\tilde{\sigma}, ...; \mathcal{L}_v U....) + ... \tag{33.44}$$

Perhaps most useful here is the general expression in component form

$$(\mathcal{L}_v T)^{\alpha\beta...\kappa}_{\nu\xi...\omega} = v^\mu \left(\frac{\partial T^{\alpha\beta...\kappa}_{\nu\xi...\omega}}{\partial x^\mu} \right) - T^{\mu\beta...\kappa}_{\nu\xi...\omega} \frac{\partial v^\alpha}{\partial x^\mu} - T^{\alpha\mu...\kappa}_{\nu\xi...\omega} \frac{\partial v^\beta}{\partial x^\mu} ... - T^{\alpha\beta...\mu}_{\nu\xi...\omega} \frac{\partial v^\kappa}{\partial x^\mu}$$

$$+ T^{\alpha\beta...\kappa}_{\mu\xi...\omega} \frac{\partial v^\mu}{\partial x^\nu} + T^{\alpha\beta...\kappa}_{\nu\mu...\omega} \frac{\partial v^\mu}{\partial x^\xi} ... + T^{\alpha\beta...\kappa}_{\nu\xi...\mu} \frac{\partial v^\mu}{\partial x^\omega}. \tag{33.45}$$

Notice the pleasing way in which the indices behave, generalizing the rules for the up and down components.

Example 33.8

A very useful result, which follows from eqn 33.45, is that the components of the Lie derivative of the metric 2-form are given by

$$(\mathcal{L}_v g)_{\alpha\beta} = v^\gamma \frac{\partial}{\partial x^\gamma} g_{\alpha\beta} + g_{\gamma\beta} \frac{\partial}{\partial x^\alpha} v^\gamma + g_{\alpha\gamma} \frac{\partial}{\partial x^\beta} v^\gamma. \tag{33.46}$$

> ⤷ **Section 33.4 gives the technical recipe for applying the Lie derivative to tensors. It can be skipped if you're happy to take the useful expression in eqn 33.46 on trust.**

After all of this setting up, we should mention some uses of the Lie derivative. The Lie derivative is the natural derivative for expressing the invariance of a tensor under a change in position along a curve. The Lie derivative arises most often in relativity in cases where we imagine ourselves carried along a particular world line while making measurements. We often use comoving coordinates in which we float along with an element of fluid, as in cosmology, for example. Another use arises when examining geodesic deviation. The vector linking the two geodesics has zero Lie derivative along the velocity field of the geodesics.

33.5   Killing vectors

A particularly important class of vectors are those that represent conserved quantities. In mechanics and field theory, **Noether's theorem** tells us that wherever there's a **symmetry** then we have a conserved quantity. There is also a geometric method of extracting the conserved quantities from the metric that relies on the use of the Lie derivative and which is particularly useful in relativity.

> ⌢↷ **See Chapter 22 for a first example of the use of Killing vectors. Noether's theorem is discussed in Chapter 40.**

Geometrically, conserved quantities can be found by identifying *Killing vectors*.[11] We say that a geometry has an **isometry** if we can identify a vector field $\boldsymbol{\xi}$ with the property that if a set of points is displaced along the streamlines of $\boldsymbol{\xi}$, then all distance relationships are unchanged. This vector field $\boldsymbol{\xi}$ is then a **Killing vector field** for the geometry. Distance relationships are encoded by the metric tensor, and so we want the metric tensor to remain unchanged as we carry it along the streamlines of $\boldsymbol{\xi}$. This implies we have

$$\mathcal{L}_{\boldsymbol{\xi}}\boldsymbol{g} = 0, \tag{33.47}$$

and so, if the metric tensor is Lie dragged along the congruence formed by the field $\boldsymbol{\xi}$, then $\boldsymbol{\xi}$ is a Killing field.

When we have access to a connection, the condition $\mathcal{L}_{\boldsymbol{\xi}}\boldsymbol{g} = 0$ implies that $\boldsymbol{\xi}$ satisfies Killing's equation[12]

$$\xi_{\alpha;\beta} + \xi_{\beta;\alpha} = 0. \tag{33.48}$$

[11] Wilhelm Killing (1847–1923). Killing invented Lie algebras, independently of Sophus Lie, who was dismissive of Killing's work, claiming that he (Lie) had proven all that was valid in the subject, while all that was invalid was added by Killing. Killing was very modest about his own work, which was arguably intended to be more general than Lie's. In it, Killing made a number of unproven conjectures, which only later turned out to be true.

[12] Note that Killing's equation can also be written in the memorable shorthand form $\xi_{(\alpha;\beta)} = 0$.

Example 33.9

We pause to prove $\xi_{(\alpha;\beta)} = 0$ using the result in example 33.8. Write the defining equation in coordinates

$$(\mathcal{L}_{\boldsymbol{\xi}}\boldsymbol{g})_{\alpha\beta} = \xi^{\gamma}\frac{\partial}{\partial x^{\gamma}}g_{\alpha\beta} + g_{\gamma\beta}\frac{\partial}{\partial x^{\alpha}}\xi^{\gamma} + g_{\alpha\gamma}\frac{\partial}{\partial x^{\beta}}\xi^{\gamma} = 0. \tag{33.49}$$

Use the *commas go to semicolons* rule, to obtain

$$\xi^{\gamma}g_{\alpha\beta;\gamma} + g_{\gamma\beta}\xi^{\gamma}{}_{;\alpha} + g_{\alpha\gamma}\xi^{\gamma}{}_{;\beta} = 0. \tag{33.50}$$

Finally, note that for any affine connection[13] $g_{\alpha\beta;\gamma} = 0$, to obtain

$$\xi_{\alpha;\beta} + \xi_{\beta;\alpha} = 0. \tag{33.51}$$

---

[13] This is the compatibility property of the metric from Chapter 7. It is examined in more detail in the next chapter.

Often Killing vectors can be written down by inspection of the form of the metric. From Example 33.6, we see that if $\boldsymbol{\xi}$ were to be, say, the basis vector $\boldsymbol{e}_1$, then we have

$$(\mathcal{L}_{\boldsymbol{e}_1}\boldsymbol{g})_{\alpha\beta} = \frac{\partial}{\partial x^1}g_{\alpha\beta} = 0, \tag{33.52}$$

and so the metric components can't depend on coordinate x^1. Therefore, a metric that is independent of x^1 has a Killing vector $\boldsymbol{e}_1$. This is the most straightforward way to find Killing vectors: simply see which coordinates do not feature in any of the components of the metric and identify the corresponding basis vectors. Finally, we need to know how to find conserved quantities. Here's the rule:

The dot product $\boldsymbol{u} \cdot \boldsymbol{\xi}$ of a Killing field $\boldsymbol{\xi}$ and tangent $\boldsymbol{u}$ to a geodesic is a constant of the motion along that geodesic.

Example 33.10

This latter claim is easily proved by using the covariant derivative $\boldsymbol{\nabla}_{\boldsymbol{u}}$ to find the change along the geodesic of the dot product $\boldsymbol{\nabla}_{\boldsymbol{u}}(\boldsymbol{u} \cdot \boldsymbol{\xi})$. Using the Leibniz rule, we have

$$\boldsymbol{\nabla}_{\boldsymbol{u}}(\boldsymbol{u} \cdot \boldsymbol{\xi}) = (\boldsymbol{\nabla}_{\boldsymbol{u}}\boldsymbol{u}) \cdot \boldsymbol{\xi} + \boldsymbol{u} \cdot \boldsymbol{\nabla}_{\boldsymbol{u}}\boldsymbol{\xi}. \tag{33.53}$$

By definition, a geodesic parallel transports its own tangent vector, meaning $\boldsymbol{\nabla}_{\boldsymbol{u}}\boldsymbol{u} = 0$, and so the first term is zero. The second term is

$$\boldsymbol{u} \cdot \boldsymbol{\nabla}_{\boldsymbol{u}}\boldsymbol{\xi} = u^\alpha u^\beta \xi_{\alpha;\beta}. \tag{33.54}$$

Since we're contracting both indices against the components of $\boldsymbol{u}$ then this last term is symmetric in α and β, allowing us to write

$$\boldsymbol{\nabla}_{\boldsymbol{u}}(\boldsymbol{u} \cdot \boldsymbol{\xi}) = u^\alpha u^\beta \xi_{(\alpha;\beta)} = 0, \tag{33.55}$$

where the final equality follows from Killing's equation $\xi_{(\alpha;\beta)} = 0$. We conclude that the quantity $\boldsymbol{u} \cdot \boldsymbol{\xi}$ is unchanged along the geodesic.

Let's now use this toolkit to extract some Killing vectors.

Example 33.11

The Schwarzschild metric has a line element

$$\mathrm{d}s^2 = \left(1 - \frac{2M}{r}\right)\mathrm{d}t^2 + \left(1 - \frac{2M}{r}\right)^{-1}\mathrm{d}r^2 + r^2(\mathrm{d}\theta^2 + \sin^2\theta\,\mathrm{d}\phi^2). \tag{33.56}$$

The components have no dependence on t and ϕ and so we have Killing vectors

$$\boldsymbol{e}_t = \frac{\partial}{\partial t}, \quad \boldsymbol{e}_\phi = \frac{\partial}{\partial \phi}. \tag{33.57}$$

As we saw in Chapter 22, this leads to conserved quantities along geodesics of

$$\boldsymbol{e}_t \cdot \boldsymbol{u} = u_t, \quad \boldsymbol{e}_\phi \cdot \boldsymbol{u} = u_\phi. \tag{33.58}$$

where $\boldsymbol{u}$ is tangent to the geodesics.

Example 33.12

Consider Rindler spacetime with metric line element $ds^2 = -x^2 dt^2 + dx^2$. Labelling coordinates (t, x), we see that the coordinate t does not feature in any of the components of the metric. The vector $\boldsymbol{\xi} = \boldsymbol{e}_t$ is therefore the Killing vector. This implies that we have a conserved quantity along the geodesic of $\boldsymbol{u} \cdot \boldsymbol{\xi} = u_t$ or

$$\boldsymbol{u} \cdot \boldsymbol{\xi} = g_{\mu\nu} u^\mu \delta^t_\nu = g_{tt} u^t = -x^2 u^t. \tag{33.59}$$

Chapter summary

- The exterior derivative operator $\boldsymbol{d}$ inputs a p-form and outputs a $(p+1)$-form.
- The action of $\boldsymbol{d}$ on a p-form $\tilde{\boldsymbol{A}}$ is given by

$$d\tilde{\boldsymbol{A}} = \frac{1}{p!} \frac{\partial A_{\alpha\beta\ldots\sigma}}{\partial x^\mu} \boldsymbol{dx}^\mu \wedge \boldsymbol{dx}^\alpha \wedge \boldsymbol{dx}^\beta \wedge \ldots \wedge \boldsymbol{dx}^\sigma. \tag{33.60}$$

- The Lie derivative is a method of taking derivatives of a tensor field. It requires an additional vector field $\boldsymbol{u}$.
- The Lie derivative of a vector $\boldsymbol{v}$ with respect to $\boldsymbol{u}$ is $\mathcal{L}_{\boldsymbol{u}} \boldsymbol{v} = [\boldsymbol{u}, \boldsymbol{v}]$.
- Killing vectors allow us access to conserved quantities in geometrical theories.
- Along a geodesic with a tangent vector field $\boldsymbol{u}$, the quantity $\boldsymbol{u} \cdot \boldsymbol{\xi}$ is conserved, where $\boldsymbol{\xi}$ is a Killing vector.

Exercises

(33.1) Consider the amount by which the figure fails to close in Example 33.3. We shall write the distance between initial and final points as

$$\mathcal{P}_4 - \mathcal{P}_3 = [\boldsymbol{u}(\mathcal{P}_0) + \boldsymbol{v}(\mathcal{P}_1)] - [\boldsymbol{u}(\mathcal{P}_2) + \boldsymbol{v}(\mathcal{P}_0)], \tag{33.61}$$

where $\boldsymbol{u}(\mathcal{P}_i)$ is the vector field $\boldsymbol{u}$ evaluated at point $\mathcal{P}_i$. Show that this expression yields the commutator of the vector fields $\boldsymbol{u}$ and $\boldsymbol{v}$, evaluated at $\mathcal{P}_0$.

(33.2) (a) Determine the components of the Lie derivative of the $(1,1)$ tensor $\boldsymbol{A}$, written $(\mathcal{L}_{\boldsymbol{u}} \boldsymbol{A})^\mu_\nu$.
(b) Contraction can be thought of as multiplication by the Kronecker delta $\delta^\mu_{\ \nu} = \langle \boldsymbol{\omega}^\mu, \boldsymbol{e}_\nu \rangle$. By

taking a Lie derivative of $\delta^\mu_{\ \nu}$, show that Lie differentiation commutes with contraction.

(33.3) Show that

$$\mathcal{L}_{\boldsymbol{u}} \mathcal{L}_{\boldsymbol{v}} - \mathcal{L}_{\boldsymbol{v}} \mathcal{L}_{\boldsymbol{u}} = \mathcal{L}_{[\boldsymbol{u}, \boldsymbol{v}]}, \tag{33.62}$$

by acting with this combination on (i) a scalar function and (ii) a vector field.
Hint: For (ii), use the Jacobi identity $[\boldsymbol{x}, [\boldsymbol{y}, \boldsymbol{z}]] + [\boldsymbol{z}, [\boldsymbol{x}, \boldsymbol{y}]] + [\boldsymbol{y}, [\boldsymbol{z}, \boldsymbol{x}]] = 0$.

(33.4) Using the results of the last exercise, show that the commutator of two Killing fields is a Killing field.

Geometry of the connection

34

Difficult you call it, Sir? I wish it were impossible.
Samuel Johnson (1709–1784) on hearing a famous violinist

To understand the physics of relativity we need the notion of a derivative to encode rates of change. So far in this part of the book, we have encountered two derivatives: the exterior derivative and the Lie derivative. The exterior derivative is designed to work on forms only. The Lie derivative, which works on all tensors, depends on the behaviour of a field at two points, which is unlike the partial derivative in ordinary calculus. Neither of these derivatives is particularly satisfactory to describe the physics of gravitation. In order to have a satisfactory derivative, we need to add a little more structure to our primitive spaces (or manifolds). The solution is found by adding the notion of **parallelism** to our toolkit.[1] In order to compare two vectors at different points in space, we must have the ability to parallel transport one to the other's location. Being able to keep a vector parallel requires that we must be able to set up basis vectors and 1-forms at each point and, somehow, to **connect** them. This then allows us to form the **covariant derivative** that compares a vector at a point to its parallel transported counterpart from another point. It is this derivative that is the most useful for formulating general relativity.

We therefore define[2] a **connection**, with symbol ∇. This is not a tensor itself, but can be applied to any tensor field and, like the conventional derivative, only takes in the properties of space at a single point. Like the exterior derivative d, the connection can be thought of as a $(0, 1)$ object. In fact, the exterior derivative d and the connection ∇ are identical in their action on scalars.[3] As we shall see, the introduction of ∇, which could be viewed as an upgrade of the exterior derivative d, allows access to an efficient means of extracting the all-important curvature of a spacetime.

The connection can be used to measure the change in a quantity on being transported along a vector $\boldsymbol{u}$. The connection directed along $\boldsymbol{u}$ is $\boldsymbol{u} \cdot \nabla$, which is the covariant derivative and given the symbol $\nabla_{\boldsymbol{u}}$. The covariant derivative tells us to move along the integral curves of the field $\boldsymbol{u}$, comparing tensors as we go, using parallel transport to account for any changes in coordinate system at different points in space. Rules for the workings of the connection are given in the margin.

↷ This chapter introduces the connection ∇ and its properties, proving a number of useful results on the way (many of which have been used previously in the book). However, it can be skipped on a first reading.

[1]We shall see in more detail in this chapter how parallelism relies on having a metric. This fundamentally links the covariant derivative and the metric. Since general relativity of the physics of the metric field, this explains why the covariant derivative features so heavily in the description of the mathematics of relativity.

[2]The abstract treatment of the covariant derivative $\nabla_{\boldsymbol{u}}$ was developed by Jean-Louis Koszul (1921–2018). The operator ∇ wasn't employed in anger until around 1954 in a paper by Katsumi Nomizu (1924–2008) who calls it t rather than ∇.

[3]When a connection is present, the action of d and ∇ on vectors, or indeed on any $(n, 0)$ tensor, will be identical. However, d and ∇ are not identical in their action on forms.

In order for the connection to only use the information at a single point, we require linearity in the direction along which we point it:

$$\nabla_{a\boldsymbol{u}+b\boldsymbol{v}}\boldsymbol{w} = a\nabla_{\boldsymbol{u}}\boldsymbol{w}+b\nabla_{\boldsymbol{v}}\boldsymbol{w}. \quad (34.1)$$

We also want it to be linear in the argument on which it operates

$$\nabla_{\boldsymbol{u}}(\alpha\boldsymbol{c}+\beta\boldsymbol{d}) = \alpha\nabla_{\boldsymbol{u}}\boldsymbol{c}+\beta\nabla_{\boldsymbol{u}}\boldsymbol{d}. \quad (34.2)$$

Finally, we want the connection to obey a Leibniz product rule

$$\nabla(\boldsymbol{S}\otimes\boldsymbol{T}) = (\nabla\boldsymbol{S}\otimes\boldsymbol{T})+(\boldsymbol{S}\otimes\nabla\boldsymbol{T}). \quad (34.3)$$

[4]The (1,2) torsion tensor $\boldsymbol{\tau}$ along with the (1,3) Riemann curvature tensor $\boldsymbol{R}$ are two independent characteristic features of a connection. The torsion-free ($\boldsymbol{\tau}=0$) spaces we consider in general relativity all share the property of symmetry of the connection

$$\Gamma^{\mu}{}_{\alpha\beta} = \Gamma^{\mu}{}_{\beta\alpha},$$

when expressed in a coordinate basis.

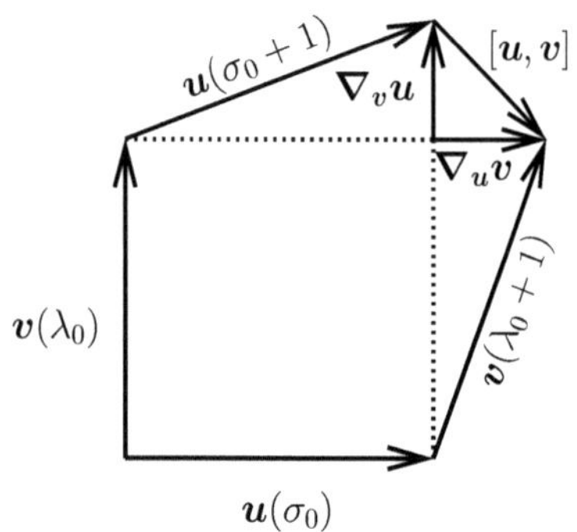

Fig. 34.1 After the vectors $\boldsymbol{u}$ and $\boldsymbol{v}$ have been transported, the quadrilateral fails to close.

[5]The symbol $\nabla_{\boldsymbol{u}}\boldsymbol{v}$ tells us to transport the vector $\boldsymbol{v}$ along the vector field $\boldsymbol{u}$ and measure the difference in $\boldsymbol{v}$. We could, of course, transport the vector field $\boldsymbol{u}$ along the vector $\boldsymbol{v}$ by considering $\nabla_{\boldsymbol{v}}\boldsymbol{u}$.

34.1 Covariant derivative in pictures

In the sorts of manifolds with which we work in general relativity, there is a neat way of visualizing the properties of the covariant derivative.

Although general relativity embraces a geometrical description of gravity, it does so by singling out curvature as the fundamental geometrical object associated with a connection, to the detriment of a quantity called **torsion**. The torsion of two vector fields $\boldsymbol{u}$ and $\boldsymbol{v}$ is a (1,2) tensor $\boldsymbol{\tau}$ defined by

$$\boldsymbol{\tau}(\boldsymbol{u},\boldsymbol{v}) = \nabla_{\boldsymbol{u}}\boldsymbol{v} - \nabla_{\boldsymbol{v}}\boldsymbol{u} - [\boldsymbol{u},\boldsymbol{v}]. \quad (34.4)$$

It's certainly worth noting that alternative theories of gravity can be formulated in terms of torsion, which potentially provide a richer phenomenology than the conventional curvature-based theory. However, in the spaces that we conventionally consider in general relativity, the torsion vanishes,[4] so we always have the property that

$$\nabla_{\boldsymbol{u}}\boldsymbol{v} - \nabla_{\boldsymbol{v}}\boldsymbol{u} = [\boldsymbol{u},\boldsymbol{v}]. \quad (34.5)$$

As we see in the next example, this condition allows us to relate the covariant and Lie derivatives.

Example 34.1

When the torsion vanishes, the covariant derivatives of vectors are related to Lie derivatives via

$$\mathcal{L}_{\boldsymbol{u}}\boldsymbol{v} = [\boldsymbol{u},\boldsymbol{v}] = \nabla_{\boldsymbol{u}}\boldsymbol{v} - \nabla_{\boldsymbol{v}}\boldsymbol{u},$$
$$(\mathcal{L}_{\boldsymbol{u}}\boldsymbol{v})^{\alpha} = v^{\alpha}{}_{;\beta}u^{\beta} - u^{\alpha}{}_{;\beta}v^{\beta}, \quad (34.6)$$

where the second expression is given in terms of components and semicolon notation.

Since the vanishing of torsion allows us to link the connection and the Lie derivative, we can also interpret the connection visually, in terms of whether a figure fails to close, in the same way that we interpreted the Lie derivative in the last chapter.

Example 34.2

As in the last chapter, we evaluate the vector that closes the figure $\boldsymbol{v}$-$\boldsymbol{u}$-$\boldsymbol{v}$-$\boldsymbol{u}$. As we've seen, this is given by the commutator $[\boldsymbol{v},\boldsymbol{u}] = \boldsymbol{v}\boldsymbol{u} - \boldsymbol{u}\boldsymbol{v}$. Consider Fig. 34.1, where we use the parameters λ and σ to parametrize the vector fields $\boldsymbol{v}$ and $\boldsymbol{u}$ respectively. The figure is formed from the vector field $\boldsymbol{v}$ evaluated at parameter point λ_0 and $\lambda_0 + 1$, which gives us two different vectors. (They are different vectors, because we've evaluated the vector field at two different points.) We also evaluate the field $\boldsymbol{u}$ at points σ_0 and $\sigma_0 + 1$ and use these four vectors to form the figure. We see that the figure does not close. Using the fact that[5]

$$v(\lambda_0 + 1) - v(\lambda_0) \approx \nabla_{\boldsymbol{u}}\boldsymbol{v},$$
$$u(\sigma_0 + 1) - u(\sigma_0) \approx \nabla_{\boldsymbol{v}}\boldsymbol{u}, \quad (34.7)$$

we see that the amount by which it fails to close is given by $\nabla_{\boldsymbol{u}}\boldsymbol{v} - \nabla_{\boldsymbol{v}}\boldsymbol{u}$. It follows that $\nabla_{\boldsymbol{u}}\boldsymbol{v} - \nabla_{\boldsymbol{v}}\boldsymbol{u} = [\boldsymbol{u},\boldsymbol{v}]$.

For cases where the commutator $[\boldsymbol{u}, \boldsymbol{v}]$ vanishes (i.e. when the quadrilateral closes), we have the useful property of the **symmetry of the covariant derivative** which says

$$\nabla_{\boldsymbol{u}} \boldsymbol{v} = \nabla_{\boldsymbol{v}} \boldsymbol{u}. \tag{34.8}$$

34.2 Connection and exterior derivative

How do we understand the connection symbol ∇? That is to say, what is the meaning of a covariant derivative without the specification of a direction along which to take the derivative? The connection ∇ is much like the exterior derivative $\boldsymbol{d}$, and is defined on a scalar function via

$$\nabla f = \boldsymbol{d} f. \tag{34.9}$$

The resulting quantity is a 1-form. We can give it a direction using $\langle \boldsymbol{d} f, \boldsymbol{u} \rangle = \partial_{\boldsymbol{u}} f$, or equivalently

$$\boldsymbol{u} \cdot \nabla f = \nabla_{\boldsymbol{u}} f \equiv u^{\mu} \frac{\partial f}{\partial x^{\mu}} \equiv \partial_{\boldsymbol{u}} f, \tag{34.10}$$

where we've written the results in several forms of notation.[6]

Previously, we confined the use of the differential operator $\boldsymbol{d}$ to forms [i.e. antisymmetric $(0, n)$ objects] and, in this spirit, we used it above on a function, which is a 0-form. We now extend the use of this tool to vectors, which are $(1, 0)$ objects. When acting on vectors, the action of $\boldsymbol{d}$ and ∇ are identical. We interpret the action of $\boldsymbol{d}$ and ∇ on a vector as saying that the object is vector-valued [that is, there's a 1 in the first slot of the valency $(1, 0)$], but that it is also a 0-form [that is, the 0 part of $(1, 0)$]. Crucial here is having a connection available. In its absence, the action of $\boldsymbol{d}$ on a vector is undefined.

In component notation, the vector $\boldsymbol{v}$ is written as $v^{\mu} \boldsymbol{e}_{\mu}$. Acting on this with $\boldsymbol{d}$ gives rise to (i) a contribution from $\boldsymbol{d} v^{\mu}$, which is simply the action of $\boldsymbol{d}$ on a scalar function, and (ii) a contribution from the action on the vector-valued 0-form $\boldsymbol{e}_{\mu}$. The covariant derivative of a vector $\boldsymbol{v} = v^{\mu} \boldsymbol{e}_{\mu}$ can then be written as

$$\nabla \boldsymbol{v} \equiv \boldsymbol{d} \boldsymbol{v} = \boldsymbol{d} v^{\mu} \otimes \boldsymbol{e}_{\mu} + v^{\mu} \boldsymbol{d} \boldsymbol{e}_{\mu}. \tag{34.12}$$

The quantity $\boldsymbol{d} \boldsymbol{v}$ is known as a **vector-valued 1-form**.[7] Note that it depends on $\boldsymbol{d} \boldsymbol{e}_{\mu}$, which tells us how the basis vectors change in space. This quantity gives the nature of the connection. We therefore define[8] **connection coefficients** in terms of the action of $\boldsymbol{d}$ (or ∇) on the basis vectors $\boldsymbol{e}_{\mu}$ as follows:

$$\nabla \boldsymbol{e}_{\mu} \equiv \boldsymbol{d} \boldsymbol{e}_{\mu} = \Gamma^{\alpha}{}_{\nu\mu} \left(\boldsymbol{\omega}^{\nu} \otimes \boldsymbol{e}_{\alpha} \right). \tag{34.14}$$

We can relate this more geometric definition of the connection to our previous notion of connection coefficients (see eqn 7.6) in the following example.

↷ This section introduces some formal ideas that will be taken up in **Chapter 36** to formulate a highly efficient method for extracting curvature from a metric.

[6]This doesn't exhaust the notational possibilities. For example, an equivalent description in a more geometrical notation is

$$\nabla_{\boldsymbol{u}} f \equiv \langle \boldsymbol{d} f, \boldsymbol{u} \rangle \equiv \boldsymbol{u}[f]. \tag{34.11}$$

[7]Why not simply call this a $(1,1)$ tensor? It's more subtle than that. A general $(1,1)$ tensor may be written as

$$\boldsymbol{S} = S^{\mu}{}_{\nu} \boldsymbol{\omega}^{\nu} \otimes \boldsymbol{e}_{\mu}. \tag{34.13}$$

However, the vector-valued 1-form features the term $\boldsymbol{d} \boldsymbol{e}_{\mu}$, which contains information about how the basis vectors themselves change in space. This relies on the connection, and so the vector-valued 1-form requires more structure than is needed to define a standard tensor like $\boldsymbol{S}$.

[8]Previously (Chapter 9), we found the connection coefficients by considering derivatives of the metric, which has been notable by its absence in this part of the book. We shall see at the end of the chapter how parallelism, that gives us the covariant derivative (and therefore these coefficients), relies on having a metric.

Example 34.3

Noting that $\langle \boldsymbol{\nabla} A, e_\beta \rangle = \boldsymbol{\nabla}_\beta A$, we have

$$
\begin{aligned}
\boldsymbol{\nabla}_\beta e_\mu &= \Gamma^\alpha{}_{\nu\mu}\langle(\omega^\nu \otimes e_\alpha), e_\beta\rangle \\
&= \Gamma^\alpha{}_{\nu\mu}\omega^\nu(e_\beta) \otimes e_\alpha && \text{(inserting } e_\beta \text{ in the first slot)} \\
&= \Gamma^\alpha{}_{\nu\mu}\delta^\nu{}_\beta e_\alpha = \Gamma^\alpha{}_{\beta\mu} e_\alpha && \text{(using } \omega^\nu(e_\beta) = \delta^\nu{}_\beta),
\end{aligned}
$$

or, picking out components by taking an inner product with ω^σ

$$
\langle \omega^\sigma, \boldsymbol{\nabla}_\beta e_\mu \rangle = \Gamma^\sigma{}_{\beta\mu}. \tag{34.15}
$$

Without worrying about the direction of the derivative, we're now able to write down the connection's action on the vector field $\boldsymbol{v}$.

Example 34.4

We can write

$$
\begin{aligned}
\boldsymbol{\nabla} \boldsymbol{v} &= \boldsymbol{\nabla} v^\mu \otimes e_\mu + v^\mu \boldsymbol{\nabla} e_\mu && \text{(chain rule)} \\
&= \boldsymbol{d} v^\mu \otimes e_\mu + v^\alpha \Gamma^\mu{}_{\beta\alpha}(\omega^\beta \otimes e_\mu) && \text{(eqn 34.14)} \\
&= \frac{\partial v^\mu}{\partial x^\beta}(\boldsymbol{d}x^\beta \otimes e_\mu) + v^\alpha \Gamma^\mu{}_{\beta\alpha}(\omega^\beta \otimes e_\mu) && \text{(action of } \boldsymbol{d} \text{ on } v_\mu).
\end{aligned} \tag{34.16}
$$

Using the last example and noting that $\boldsymbol{d}x^\beta \equiv \omega^\beta$, we have the final expression,

$$
\boldsymbol{\nabla} \boldsymbol{v} = \left(\frac{\partial v^\mu}{\partial x^\beta} + v^\alpha \Gamma^\mu{}_{\beta\alpha} \right) (\omega^\beta \otimes e_\mu). \tag{34.17}
$$

↪ **See Chapter 7 for a reminder of the action of the covariant derivative on vectors.**

Note that this can also be written in semicolon notation as $\boldsymbol{\nabla}\boldsymbol{v} = v^\mu{}_{;\beta}(\omega^\beta \otimes e_\mu)$. That is to say that the components of $\boldsymbol{\nabla}\boldsymbol{v}$ are the covariant derivative components $v^\mu{}_{;\beta}$. We conclude that from the linear slot machine point of view, the connection symbol $\boldsymbol{\nabla}$ acting on a vector $\boldsymbol{v}$ results in the 2-slotted object

$$
\boldsymbol{\nabla} \boldsymbol{v}(\ ,\ ) = v^\mu{}_{;\nu}\omega^\nu(\ ) \otimes e_\mu(\ ). \tag{34.18}
$$

This is the **gradient** of $\boldsymbol{v}$. We summarize the properties of the gradient in the next example.

Example 34.5

The **gradient** of $\boldsymbol{v}$ is neither a vector nor a 1-form, but a vector-valued 1-form, sometimes written as $\boldsymbol{d}\boldsymbol{v}$. You must insert both a vector and a 1-form to get a number. It has components

$$
\boldsymbol{\nabla} \boldsymbol{v}(e_\alpha, \omega^\mu) = (\boldsymbol{\nabla}\boldsymbol{v})^\mu{}_\alpha = \left(\frac{\partial v^\mu}{\partial x^\alpha} + v^\lambda \Gamma^\mu{}_{\alpha\lambda} \right). \tag{34.19}
$$

If we insert $\boldsymbol{v} = e_\gamma$ into this last equation we effectively extract the components of $\boldsymbol{\nabla}$

$$
(\boldsymbol{\nabla} e_\gamma)^\mu{}_\alpha = \delta^\lambda{}_\gamma \Gamma^\mu{}_{\alpha\lambda} = \Gamma^\mu{}_{\alpha\gamma}. \tag{34.20}
$$

That is, the components of the connection ∇ can be thought of as the connection coefficients.

The quantity $\nabla v(u,\,) = \nabla_u v$ is a **vector**. It is the covariant derivative of v along u with components

$$\langle \omega^\mu, \nabla_u v \rangle = (\nabla_u v)^\mu = u^\alpha \left(\frac{\partial v^\mu}{\partial x^\alpha} + v^\lambda \Gamma^\mu{}_{\alpha\lambda} \right). \tag{34.21}$$

To get a **number** from ∇v, insert a 1-form $\tilde{\sigma}$ and a vector u:

$$\nabla v(u, \tilde{\sigma}) = \langle \tilde{\sigma}, \nabla_u v \rangle = (\text{number}). \tag{34.22}$$

The number that is outputted evaluates the number of times the vector $\nabla_u v$ pierces the surfaces of the 1-form $\tilde{\sigma}$.

Finally, we shall derive the transformation properties of the connection coefficients.

Example 34.6

Relating primed and unprimed frames, we write

$$\Gamma^{\alpha'}{}_{\beta'\gamma'} = \langle \omega^{\alpha'}, \nabla_{e_{\beta'}} e_{\gamma'} \rangle$$

$$= \left\langle \Lambda^{\alpha'}{}_\alpha \omega^\alpha, \nabla_{\Lambda^\beta{}_{\beta'} e_\beta} \left(\Lambda^\gamma{}_{\gamma'} e_\gamma \right) \right\rangle. \tag{34.23}$$

Using linearity and the rule that $\nabla_{au} = a\nabla_u$ we have

$$\Gamma^{\alpha'}{}_{\beta'\gamma'} = \Lambda^{\alpha'}{}_\alpha \Lambda^\beta{}_{\beta'} \left\langle \omega^\alpha, \nabla_{e_\beta} \left(\Lambda^\gamma{}_{\gamma'} e_\gamma \right) \right\rangle. \tag{34.24}$$

Now using the Leibniz rule and the rule that[9] $\nabla_{e_\beta} f = e_\beta [f]$ that we find

$$\Gamma^{\alpha'}{}_{\beta'\gamma'} = \Lambda^{\alpha'}{}_\alpha \Lambda^\beta{}_{\beta'} \left(\left\langle \omega^\alpha, e_\beta \left[\Lambda^\gamma{}_{\gamma'} \right] e_\gamma \right\rangle + \Lambda^\gamma{}_{\gamma'} \Gamma^\alpha{}_{\beta\gamma} \right)$$

$$= \Lambda^{\alpha'}{}_\alpha \Lambda^\beta{}_{\beta'} e_\beta \left[\Lambda^\gamma{}_{\gamma'} \right] \delta^\alpha{}_\gamma + \Lambda^{\alpha'}{}_\alpha \Lambda^\beta{}_{\beta'} \Lambda^\gamma{}_{\gamma'} \Gamma^\alpha{}_{\beta\gamma}$$

$$= \Lambda^{\alpha'}{}_\alpha e_{\beta'} \left[\Lambda^\alpha{}_{\gamma'} \right] + \Lambda^{\alpha'}{}_\alpha \Lambda^\beta{}_{\beta'} \Lambda^\gamma{}_{\gamma'} \Gamma^\alpha{}_{\beta\gamma}. \tag{34.25}$$

The second term is the usual tensor transformation law for the components of a tensor. However, it's the first term that ruins the tensor transformation properties. We can see how if we use a coordinate frame (and recall that $\Lambda^\alpha{}_\beta = \partial x^\alpha / \partial x^\beta$ and $e_\beta [f] = \partial f / \partial x^\beta$), where we find

$$\Gamma^{\alpha'}{}_{\beta'\gamma'} = \frac{\partial x^{\alpha'}}{\partial x^\alpha} \frac{\partial^2 x^\alpha}{\partial x^{\beta'} \partial x^{\gamma'}} + \frac{\partial x^{\alpha'}}{\partial x^\alpha} \frac{\partial x^\beta}{\partial x^{\beta'}} \frac{\partial x^\gamma}{\partial x^{\gamma'}} \Gamma^\alpha{}_{\beta\gamma}. \tag{34.26}$$

[9] Recall that the square brackets here mean $\partial_{e_\beta} f$ or, in a coordinate basis $\partial f / \partial x^\beta$ and that this quantity is a number.

This provides the proof of the expression in eqn 7.11 from Chapter 7.

34.3 Covariant derivative of tensors

The rule for $(1,0)$ objects that $dv \equiv \nabla v$ carries over to all $(n,0)$ tensors, that is, all tensor valued 0-forms, where we have

$$dS = \nabla S, \tag{34.27}$$

where S is a $(n,0)$ tensor. It might seem that we can simply make d equivalent to ∇ in its action on all objects. We cannot. This is

This section gives the technical arguments leading to the recipe for taking covariant derivatives of tensors. These are results we have used previously in the book without derivation. You can skip to the grey box at the end of the section if you're willing to take them on trust.

because the action of $\boldsymbol{d}$ on a p-form produces a $(p+1)$-form and not the object whose components are the covariant derivatives. This can be traced back to the fact that the action of $\boldsymbol{d}$ on p-forms with $p \geq 1$ involves the antisymmetric wedge product $\wedge$ and not the tensor product $\otimes$. This presents a problem since the action of the connection operator on a tensor does not yield an antisymmetric object, but simply a tensor built from ordinary tensor products. In fact, for a general (n, m) tensor $\boldsymbol{S}$, we define

$$\boldsymbol{\nabla S} = S^{i_1,\dots i_n}{}_{j_1\dots j_m;k}\left(\boldsymbol{e}_{i_1} \otimes \dots \otimes \boldsymbol{e}_{i_n} \otimes \boldsymbol{\omega}^{j_1} \otimes \dots \otimes \boldsymbol{\omega}^{j_m} \otimes \boldsymbol{\omega}^{k}\right), \quad (34.28)$$

which does not involve the wedge product.[10] The point here is that we need a different approach to evaluate the covariant derivative of a p-form.

[10] We explain how to access the components $S^{i_1,\dots i_n}{}_{j_1\dots j_m;k}$ in the next sections.

Example 34.7

Let's derive the action of the covariant derivative $\boldsymbol{\nabla}_\mu$ on 1-forms. We use the same method we used in the last chapter. This involves combining the 1-form with a vector via the inner product to make a number. The derivative of the combined vector and 1-form may be found using the Leibniz product rule for tensors.

This product rule for the connection is $\boldsymbol{\nabla}(\boldsymbol{S} \otimes \boldsymbol{T}) = \boldsymbol{\nabla S} \otimes \boldsymbol{T} + (\boldsymbol{S} \otimes \boldsymbol{\nabla T})$. Take $\boldsymbol{S} = \boldsymbol{\omega}^\nu$ and $\boldsymbol{T} = \boldsymbol{e}_\mu$ and replace the $\otimes$ operation with $\langle\,,\rangle$. This is permissible as the inner product preserves the linear structure enforced by tensor product $\otimes$. The derivative of $\langle\boldsymbol{\omega}^\nu, \boldsymbol{e}_\mu\rangle = \delta^\nu{}_\mu$ is zero, so we obtain

$$\begin{aligned}
0 = \langle\boldsymbol{\nabla}(\boldsymbol{\omega}^\nu, \boldsymbol{e}_\mu)\rangle &= \langle\boldsymbol{\nabla\omega}^\nu, \boldsymbol{e}_\mu\rangle + \langle\boldsymbol{\omega}^\nu, \boldsymbol{\nabla e}_\mu\rangle \\
&= \boldsymbol{\nabla\omega}^\nu(\boldsymbol{e}_\mu, \ ) + \Gamma^\alpha{}_{\beta\mu}\boldsymbol{\omega}^\beta(\ ) \otimes \boldsymbol{e}_\alpha(\boldsymbol{\omega}^\nu) \\
&= \boldsymbol{\nabla\omega}^\nu(\boldsymbol{e}_\mu, \ ) + \Gamma^\alpha{}_{\beta\mu}\boldsymbol{\omega}^\beta(\ )\delta^\nu{}_\alpha, \quad (34.29)
\end{aligned}$$

where, in the second line, we used $\boldsymbol{\nabla e}_\mu = \Gamma^\alpha{}_{\beta\mu}\boldsymbol{\omega}^\beta \otimes \boldsymbol{e}_\alpha$.

From the last example we conclude that[11]

$$\boldsymbol{\nabla\omega}^\nu(\boldsymbol{e}_\mu, \ ) = -\Gamma^\nu{}_{\beta\mu}\boldsymbol{\omega}^\beta. \quad (34.32)$$

We can use this to calculate the covariant derivative of the 1-form

$$\begin{aligned}
\boldsymbol{\nabla}_\beta\tilde{\sigma} &= (\boldsymbol{\nabla}_\beta\sigma_\nu)\boldsymbol{\omega}^\nu + \sigma_\nu\boldsymbol{\nabla}_\beta\boldsymbol{\omega}^\nu \\
&= \frac{\partial\sigma_\nu}{\partial x^\beta}\boldsymbol{\omega}^\nu - \sigma_\nu\Gamma^\nu{}_{\beta\mu}\boldsymbol{\omega}^\mu. \quad (34.33)
\end{aligned}$$

The result, on relabelling indices, is an equation for the action of the covariant derivative operator on a 1-form[12]

$$\boldsymbol{\nabla}_\mu\tilde{\sigma} = \left(\frac{\partial\sigma_\nu}{\partial x^\mu} - \sigma_\lambda\Gamma^\lambda{}_{\mu\nu}\right)\boldsymbol{\omega}^\nu. \quad (34.34)$$

We can also write an equation for the components in semicolon notation

$$\sigma_{\nu;\mu} = \sigma_{\nu,\mu} - \sigma_\lambda\Gamma^\lambda{}_{\mu\nu}, \quad (34.35)$$

or an equation for the 1-form that results if the direction of the derivative is given by a vector $\boldsymbol{u}$,

$$(\boldsymbol{\nabla}_{\boldsymbol{u}}\tilde{\sigma})_\alpha = u^\mu\left(\frac{\partial\sigma_\alpha}{\partial x^\mu} - \sigma_\lambda\Gamma^\lambda{}_{\mu\alpha}\right). \quad (34.36)$$

[11] By extension, we have

$$\boldsymbol{\nabla\omega}^\nu = -\Gamma^\nu{}_{\beta\mu}\boldsymbol{\omega}^\nu \otimes \boldsymbol{\omega}^\beta. \quad (34.30)$$

This should be contrasted with a result we shall establish in Chapter 36 that

$$\boldsymbol{d\omega}^\nu = \Gamma^\nu{}_{\beta\mu}\boldsymbol{\omega}^\nu \wedge \boldsymbol{\omega}^\beta. \quad (34.31)$$

[12] Notice how this expression has minus sign in front of the connection coefficients, in contrast to the vector version.

↰ This provides the proof of the expressions in eqns 7.40 and 7.34 from Chapter 7.

As shown in the next example, we can also express the result of applying the exterior derivative to 1-forms in a coordinate frame in terms of the action of the covariant derivative.

Example 34.8

A 2-form is obtained by acting on the 1-form $\tilde{\boldsymbol{A}} = A_\mu \boldsymbol{dx}^\mu$ with the exterior derivative operator $\boldsymbol{d}$. This is written

$$\boldsymbol{d}\tilde{\boldsymbol{A}} = \frac{\partial A_\mu}{\partial x^\alpha} \boldsymbol{dx}^\alpha \wedge \boldsymbol{dx}^\mu. \tag{34.37}$$

To point the derivative along the $\boldsymbol{e}_\beta$ direction, we contract, which is equivalent to filling the first slot of the 2-form

$$\begin{aligned}
\langle \boldsymbol{d}\tilde{\boldsymbol{A}}, \boldsymbol{e}_\beta \rangle = \boldsymbol{d}\tilde{\boldsymbol{A}}(\boldsymbol{e}_\beta, \ ) &= \frac{\partial A_\mu}{\partial x^\alpha} [\boldsymbol{dx}^\alpha \wedge \boldsymbol{dx}^\mu](\boldsymbol{e}_\beta, \ ) \\
&= \frac{\partial A_\mu}{\partial x^\alpha} [\boldsymbol{dx}^\alpha \otimes \boldsymbol{dx}^\mu(\boldsymbol{e}_\beta, \ ) - \boldsymbol{dx}^\mu \otimes \boldsymbol{dx}^\alpha(\boldsymbol{e}_\beta, \ )] \\
&= \frac{\partial A_\mu}{\partial x^\alpha} \left(\delta^\alpha{}_\beta \boldsymbol{dx}^\mu - \delta^\mu{}_\beta \boldsymbol{dx}^\alpha \right) \\
&= \frac{\partial A_\mu}{\partial x^\beta} \boldsymbol{dx}^\mu - \frac{\partial A_\beta}{\partial x^\alpha} \boldsymbol{dx}^\alpha \\
&= \left(\frac{\partial A_\mu}{\partial x^\beta} - \frac{\partial A_\beta}{\partial x^\mu} \right) \boldsymbol{dx}^\mu.
\end{aligned} \tag{34.38}$$

This is a 1-form. Extract the γ component by contracting with $\boldsymbol{e}_\gamma$ to find

$$\boldsymbol{d}\tilde{\boldsymbol{A}}(\boldsymbol{e}_\beta, \boldsymbol{e}_\gamma) = \left(\frac{\partial A_\gamma}{\partial x^\beta} - \frac{\partial A_\beta}{\partial x^\gamma} \right). \tag{34.39}$$

Compare this to the coordinate-frame equation

$$\begin{aligned}
A_{\gamma;\beta} - A_{\beta;\gamma} &= \left(\frac{\partial A_\gamma}{\partial x^\beta} - \Gamma^\sigma{}_{\beta\gamma} A_\sigma \right) - \left(\frac{\partial A_\beta}{\partial x^\gamma} - \Gamma^\sigma{}_{\gamma\beta} A_\sigma \right) \\
&= \frac{\partial A_\gamma}{\partial x^\beta} - \frac{\partial A_\beta}{\partial x^\gamma}.
\end{aligned} \tag{34.40}$$

We see that we can write the components of the 2-form $\boldsymbol{d}\tilde{\boldsymbol{A}} = \frac{1}{2}(\boldsymbol{d}\tilde{\boldsymbol{A}})_{\beta\gamma} \boldsymbol{dx}^\beta \wedge \boldsymbol{dx}^\gamma$ as

$$(\boldsymbol{d}\tilde{\boldsymbol{A}})_{\beta\gamma} = (A_{\gamma;\beta} - A_{\beta;\gamma}) = -2A_{[\beta;\gamma]}, \tag{34.41}$$

where the square bracket notation for antisymmetrization has been used in the last line.[13]

We now extend the action of the covariant derivative from vectors and 1-forms to tensors in general. Compare the vector $\boldsymbol{v} = v^\mu \boldsymbol{e}_\mu$ to the $(2,0)$ tensor

$$\boldsymbol{T} = T^{\mu\nu} \boldsymbol{e}_\mu \otimes \boldsymbol{e}_\nu. \tag{34.44}$$

The vector $\boldsymbol{v}$ is made up of components v^μ and the basis vectors $\boldsymbol{e}_\mu$ and we saw that, when acted on by the derivative operator $\boldsymbol{\nabla}_\alpha$ (i.e. the covariant derivative in the direction $\boldsymbol{e}_\alpha$), we obtain a contribution from the components $\partial v^\mu / \partial x^\alpha$, added to a contribution from the basis vectors of $\Gamma^\mu{}_{\alpha\beta} v^\beta$. The tensor $\boldsymbol{T}$ is very similar, only with an additional contribution from a second basis vector. We therefore obtain an additional connection Γ for this basis vector. The rule for the derivatives of a $(2,0)$ tensor are therefore, in component form,

$$(\boldsymbol{\nabla}_\alpha \boldsymbol{T})^{\mu\nu} = \frac{\partial T^{\mu\nu}}{\partial x^\alpha} + \Gamma^\mu{}_{\alpha\beta} T^{\beta\nu} + \Gamma^\nu{}_{\alpha\beta} T^{\mu\beta}. \tag{34.45}$$

[13]This expression is the basis of the useful equation for the 1-form $\tilde{\alpha}$ that says

$$\langle \boldsymbol{d}\tilde{\alpha}, \boldsymbol{u} \wedge \boldsymbol{v} \rangle = \langle \boldsymbol{\nabla}_{\boldsymbol{u}}\tilde{\alpha}, \boldsymbol{v} \rangle - \langle \boldsymbol{\nabla}_{\boldsymbol{v}}\tilde{\alpha}, \boldsymbol{u} \rangle. \tag{34.42}$$

This can be shown by writing the inner product in components:

$$(\alpha_{|\gamma;\beta|} - \alpha_{|\beta;\gamma|})(u^\beta v^\gamma - u^\gamma v^\beta). \tag{34.43}$$

The details are left as an exercise, and the expression is used in Chapter 43.

⤳ **Eqn 34.45 was originally claimed in eqn 12.37 in Chapter 12.**

Notice how the connection-coefficients are contracted against each tensor index in turn.

Since basis 1-forms make a $-\Gamma$ contribution, we similarly expect a contribution of one of these for each basis 1-form. So, for the $(0,2)$ tensor $\boldsymbol{\xi}$ we have

$$(\boldsymbol{\nabla}_\alpha\boldsymbol{\xi})_{\mu\nu} = \frac{\partial\xi_{\mu\nu}}{\partial x^\alpha} - \Gamma^\lambda{}_{\alpha\mu}\xi_{\lambda\nu} - \Gamma^\lambda{}_{\alpha\nu}\xi_{\mu\lambda}. \tag{34.46}$$

Again, we must contract the connection coefficients against each down index.

Generalizing, the routine for a general (m,n) tensor $\boldsymbol{A}$ is simply to add additional Γs for each up index and subtract them for each down index:

$$\begin{aligned}
(\boldsymbol{\nabla}_\alpha\boldsymbol{A})^{\mu\nu\xi\ldots\omega}_{\beta\gamma\delta\ldots\kappa} =& \frac{\partial A^{\mu\nu\xi\ldots\omega}_{\beta\gamma\delta\ldots\kappa}}{\partial x^\alpha} + \Gamma^\mu{}_{\alpha\lambda}A^{\lambda\nu\xi\ldots\omega}_{\beta\gamma\delta\ldots\kappa} + \Gamma^\nu{}_{\alpha\lambda}A^{\mu\lambda\xi\ldots\omega}_{\beta\gamma\delta\ldots\kappa} \\
& + \Gamma(\text{terms for all up indices}) \\
& - \Gamma^\lambda{}_{\alpha\beta}A^{\mu\nu\xi\ldots\omega}_{\lambda\gamma\delta\ldots\kappa} - \Gamma^\lambda{}_{\alpha\gamma}A^{\mu\nu\xi\ldots\omega}_{\beta\lambda\delta\ldots\kappa} \\
& - \Gamma(\text{terms for all down indices}).
\end{aligned} \tag{34.47}$$

This completes the description of the connection and covariant derivative. Of course, the reason we need all of this formalism is that $\boldsymbol{\nabla}_{\boldsymbol{v}}$ is the derivative that is most useful in describing the curvature of spacetime encoded into the metric field. We make this link between the covariant derivative to the metric in the next and final section of this chapter.

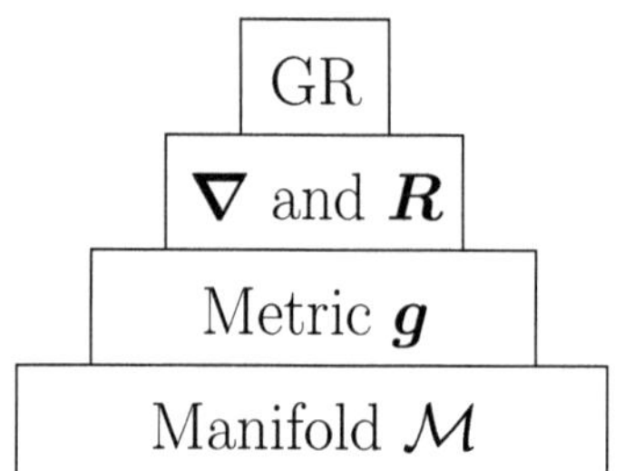

Fig. 34.2 A manifold with a metric provides enough structure to define an affine connection, and then curvature, and then the Einstein tensor. These are all required for general relativity.

[14]The Riemann tensor will be revisited in Chapter 35.

[15]This statement is the final word on what the spacetime of general relativity is, from the mathematical point of view. For those interested in the history, Jean le Rond d'Alembert (1717–1783) was probably the first person to express time as the fourth dimension and thereby invent the notion of spacetime. This advance was attributed to Lagrange by E. T. Bell, but there is little evidence for this. See R. G. Van Oss, *Historia Mathematica* **10**, 455 (1983) for a discussion.

34.4   The metric revisited

The central pillar of our geometric theory of Nature is the metric, the fundamental field of the theory of general relativity. Both the connection $\boldsymbol{\nabla}$ and the Riemann tensor[14] $\boldsymbol{R}$ can be thought of as structures that derive from the metric $\boldsymbol{g}$ defined on the manifold $\mathcal{M}$, and these structures underpin general relativity (this idea is described schematically in Fig. 34.2). To specify these essential ingredients we write $(\mathcal{M},\boldsymbol{g})$. A mathematical statement of general relativity is then as follows:[15]

Spacetime is the manifold $\mathcal{M}$ on which there is a Lorentz metric $\boldsymbol{g}$. The curvature of spacetime, described by $\boldsymbol{g}$, is related to the distribution of matter in spacetime by the Einstein equation.

As we've seen previously in this book, the metric tensor is defined mathematically to be a non-singular $(0,2)$ tensor with the property $\boldsymbol{g}\left(\boldsymbol{e}_\mu,\boldsymbol{e}_\nu\right) = g_{\mu\nu}$. The definition implies that the metric tensor can be built from the tensor product of basis 1-forms: $\boldsymbol{g} = g_{\mu\nu}\boldsymbol{\omega}^\mu \otimes \boldsymbol{\omega}^\nu$, or, introducing some alternative notation, we can write things in terms of a **line element tensor**, $\boldsymbol{ds}^2 = g_{\mu\nu}\boldsymbol{dx}^\mu \otimes \boldsymbol{dx}^\nu$. Written in slot machine form, we have

$$\boldsymbol{ds}^2(\ ,\ ) \equiv \boldsymbol{g}(\ ,\ ) = g_{\mu\nu}\boldsymbol{dx}^\mu(\ ) \otimes \boldsymbol{dx}^\nu(\ ). \tag{34.48}$$

With the metric in our toolkit, we can show how possessing a connection ∇ allows us access to the notion of parallelism and parallel transport. One possible version of parallel transport might be that the tangent vector $\boldsymbol{u}$ to a geodesic, parametrized by λ, is moved along the curve and simply returns a vector proportional to that tangent vector[16]

$$\nabla_{\boldsymbol{u}}\boldsymbol{u} \equiv \frac{D\boldsymbol{u}}{d\lambda} \propto \boldsymbol{u}. \tag{34.49}$$

A connection that obeys this, very general, statement of parallelism is called a **non-affine connection**. Such connections are characterized by parametrizations where λ does *not* mark of regular intervals along the curve, in Fig 34.3 (bottom). When a connection is non-affine the vector $\boldsymbol{u}$ can get longer or shorter as it moves around, but always remains parallel to itself.

In contrast, an **affine connection**, where λ marks off regular intervals along the path, is characterized by the much more restrictive statement of parallelism that, for a geodesic parametrized by λ, we have[17]

$$\nabla_{\boldsymbol{u}}\boldsymbol{u} = \frac{D\boldsymbol{u}}{d\lambda} = 0, \tag{34.50}$$

which we recognize as the geodesic equation. When this condition holds we have the state of affairs we have previously described as parallel transport. The tangent vector is transported along the geodesic remaining parallel to itself and, crucially, its length remains constant (Fig 34.3, top). Tangent vectors of constant length are something we certainly want from our physical theory, since we require, for example, that the velocity vector $\boldsymbol{u}$, which is tangent to the world line of a massive particle, has a constant magnitude such that $\boldsymbol{u} \cdot \boldsymbol{u} = -1$.

It is the *metric* that guarantees that the connection is affine. As a consequence, we must now permanently fix the connection and the metric together. This link between connection and metric is made with the compatibility condition[18]

$$\nabla\boldsymbol{g} = 0. \tag{34.51}$$

This equation implies that the covariant derivative $\nabla_{\boldsymbol{u}}\boldsymbol{g} = 0$ when taken along a curve with tangent vector $\boldsymbol{u}$. As we show in the example below, this condition is enough to ensure that the length of *any* vector is a constant when it is parallel transported.

[16]As in Part II, we again use the $D/d\lambda$ notation here to denote a covariant derivative taken along the curve parametrized by λ in a spacetime with a connection.

[17]In Exercise 34.7, we show that if a parametrization λ that obeys this equation, then an alternative parametrization $s = a\lambda + b$, with a and b constants, also obeys the equation.

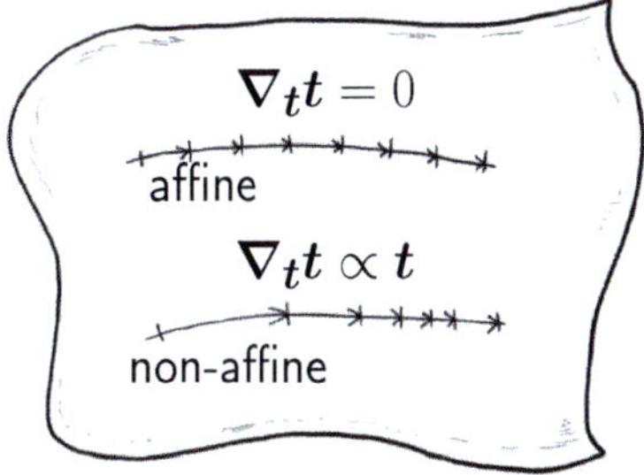

Fig. 34.3 Affine and non-affine connections.

[18]We note that there is an approach to gravity where the connection and the metric are treated as independent variables, with respect to which the Einstein–Hilbert action is varied (see Chapter 40). This is known as Palatini gravity. However, if the action is given by eqn 40.36, formed from the (not necessarily metric-compatible) connection, then the two versions of gravity are equivalent.

Example 34.9

The vectors $\boldsymbol{v}$ and $\boldsymbol{w}$ are parallel transported along the curve whose tangent is $\boldsymbol{u}$. That is

$$\begin{aligned} \nabla_{\boldsymbol{u}}\boldsymbol{v} = 0, \qquad & \nabla_{\boldsymbol{u}}\boldsymbol{w} = 0, \\ u^\alpha v^\beta{}_{;\alpha} = 0, \qquad & u^\alpha w^\gamma{}_{;\alpha} = 0. \end{aligned} \tag{34.52}$$

If the lengths of the vectors are unchanged by this operation then the inner product $\boldsymbol{g}(\boldsymbol{v}, \boldsymbol{w})$ cannot change on parallel transportation. As a result we have

$$\nabla_{\boldsymbol{u}}\boldsymbol{g}(\boldsymbol{v}, \boldsymbol{w}) = 0, \tag{34.53}$$

or, written in components

$$u^\alpha (g_{\beta\gamma} v^\beta w^\gamma)_{;\alpha} = 0. \tag{34.54}$$

Using the Leibniz rule:

$$u^\alpha g_{\beta\gamma;\alpha} v^\beta w^\gamma + u^\alpha g_{\beta\gamma} v^\beta{}_{;\alpha} w^\gamma + u^\alpha g_{\beta\gamma} v^\beta w^\gamma{}_{;\alpha} = 0. \tag{34.55}$$

The parallel transport conditions in eqn 34.52 kill the latter two terms on the left, and we obtain

$$t^\alpha v^\beta w^\gamma g_{\beta\gamma;\alpha} = 0. \tag{34.56}$$

This holds if, and only if,

$$g_{\beta\gamma;\alpha} = 0. \tag{34.57}$$

The components of $\boldsymbol{\nabla} g$ are all zero therefore, and so this object vanishes.

As a special case, we also see that for a geodesic with tangent $\boldsymbol{u}$, then fixing $g_{\beta\gamma;\alpha} = 0$ and $\boldsymbol{g}(\boldsymbol{u}, \boldsymbol{u}) = -1$, ensures the geodesic equation $\boldsymbol{\nabla}_{\boldsymbol{u}} \boldsymbol{u} \equiv u^\nu u^\mu{}_{;\nu} = 0$ is obeyed, as required.[19]

[19] In Exercise 34.8, we show that fixing $g_{\beta\gamma;\alpha} = 0$ and $\boldsymbol{g}(\boldsymbol{u}, \boldsymbol{u}) = -1$ along a curve that is not a geodesic means that the velocity $\boldsymbol{u}$ is always perpendicular to the non-zero acceleration $\boldsymbol{a} = \boldsymbol{\nabla}_{\boldsymbol{u}} \boldsymbol{u}$.

This makes the crucial link between the connection and the metric. The condition $\boldsymbol{\nabla} g = 0$ (along with the vanishing of torsion) is enough the fix the action of $\boldsymbol{\nabla}$ as the operator giving an affine connection. The reason that the covariant derivative is the suitable derivative for relativity is because it is the derivative that is built on the foundation of parallelism, which involved having access to a metric, and general relativity is the field theory of the metric field.

On a more practical note, metric components are useful in raising and lowering indices. The semicolon notation provides the means for doing this in expressions involving the covariant derivative, as we examine below.

Example 34.10

We act on the components of the covariant derivative with the metric components. We have

$$\begin{aligned} g_{\mu\nu} v^\mu{}_{;\alpha} &= g_{\mu\nu} \left(v^\mu{}_{,\alpha} + \Gamma^\mu{}_{\alpha\lambda} v^\lambda \right) \\ &= v_{\nu,\alpha} + \Gamma_{\nu\alpha\lambda} v^\lambda \\ &= v_{\nu,\alpha} + g^{\lambda\gamma} \Gamma_{\nu\alpha\lambda} v_\gamma. \end{aligned} \tag{34.58}$$

This last equation causes us to pause since Γ is not a tensor in its latter two down components, and so we cannot simply raise an index as we might be tempted to. This isn't a problem if we simply follow eqn 34.35 and write

$$g_{\mu\nu} v^\mu{}_{;\alpha} = v_{\nu;\alpha} = v_{\nu,\alpha} - \Gamma^\gamma{}_{\alpha\nu} v_\gamma, \tag{34.59}$$

which gives us a way to interpret the final line in eqn 34.58.

Chapter summary

- The connection $\boldsymbol{\nabla}$ allows a derivative to be defined that is satisfactory to describe tensor fields in the curved spacetimes of general relativity.
- The components of $\boldsymbol{\nabla}$ are the connection coefficients $\Gamma^\mu{}_{\alpha\beta}$.
- An affine connection is guaranteed by the condition $\boldsymbol{\nabla} g = 0$.

Exercises

(34.1) Use the symmetry of the covariant derivative to show

$$\nabla_{u+n} v = \nabla_u v + \nabla_n v. \qquad (34.60)$$

(34.2) (a) Using the definition $e_\alpha \cdot e_\beta = g_{\alpha\beta}$, along with the action of ∇ on basis vectors, to compute the derivative $\nabla(e_\alpha \cdot e_\beta)$, and use this to show

$$g_{\alpha\beta,\sigma} = \Gamma^\mu{}_{\sigma\alpha} g_{\mu\beta} + \Gamma^\mu{}_{\sigma\beta} g_{\mu\alpha}. \qquad (34.61)$$

(b) Use the result of (a) to prove the expression

$$\Gamma_{\alpha\beta\sigma} = \frac{1}{2}\left(g_{\alpha\beta,\sigma} + g_{\alpha\sigma,\beta} - g_{\beta\sigma,\alpha}\right), \qquad (34.62)$$

that links the connection coefficients to the derivatives of the components of g.

(34.3) (a) For an arbitrary matrix $\underline{M}$ there is an identity

$$\frac{\partial}{\partial x^\lambda} \ln \det \underline{M}(x) = \mathrm{Tr}\left\{\underline{M}^{-1}(x)\frac{\partial}{\partial x^\lambda}\underline{M}(x)\right\}. \qquad (34.63)$$

Prove this by considering a variation in $\ln(\det \underline{M})$ owing to a variation δx^λ that gives

$$\delta \ln \det \underline{M} = \ln \det(\underline{M}+\delta\underline{M}) - \ln \det \underline{M}. \qquad (34.64)$$

(b) Use the identity

$$\Gamma^\mu{}_{\lambda\mu} = \frac{1}{2}g^{\mu\rho}\frac{\partial g_{\rho\mu}}{\partial x^\lambda}, \qquad (34.65)$$

along with the expression in part (a) to show

$$\Gamma^\mu{}_{\lambda\mu} = \frac{1}{\sqrt{|g|}}\frac{\partial}{\partial x^\lambda}\sqrt{|g|}. \qquad (34.66)$$

(c) Use this to show further that the divergence is given by

$$v^\mu{}_{;\mu} = \frac{1}{\sqrt{|g|}}\frac{\partial}{\partial x^\mu}\left(\sqrt{|g|}v^\mu\right). \qquad (34.67)$$

(34.4) Using components, take the covariant derivative ∇_u of an inner product of vector and 1-form $\langle \tilde{\sigma}, v \rangle = f$, where f is a function, and use this to prove the rule for the components of the covariant derivative of a 1-form.

(34.5) (a) If λ is an affine parameter such that $u^\alpha = \frac{dx^\alpha}{d\lambda}$, show that

$$(\nabla_u u) \cdot e_1 = \frac{du_1}{d\lambda} - u^\alpha u_\sigma \Gamma^\sigma{}_{\alpha 1}. \qquad (34.68)$$

(b) Using this, show that if the metric functions $g_{\mu\nu}$ are independent of a coordinate x^1 then u_1 is a constant along the particle's geodesic world line. *This amounts to the rule for finding Killing vectors that we discussed in the last chapter.*

(34.6) *The exterior derivative of a 1-form*
Consider the number formed by filling the slots of a 2-form as follows: $d\tilde{A}(u, v)$, where $\tilde{A}$ is a 1-form and u and v are vectors.
(a) Show that when there is a connection, we can express this quantity as

$$d\tilde{A}(u, v) = \nabla_u \tilde{A}(v) - \nabla_v \tilde{A}(u). \qquad (34.69)$$

(b) Show that this can also be expressed as

$$d\tilde{A}(u, v) = \nabla_u \langle \tilde{A}, v \rangle - \nabla_v \langle \tilde{A}, u \rangle - \tilde{A}([u, v]). \qquad (34.70)$$

This suggests another route to eqn 34.69. Since we know that the action of the exterior and the covariant derivative on a scalar are identical, we could have started by considering the candidate quantity $\nabla_u \langle \tilde{A}, v \rangle$ and then antisymmetrized it. We then expand the resulting candidate equation for $d\tilde{A}(u, v)$ of $\nabla_u \langle \tilde{A}, v \rangle - \nabla_v \langle \tilde{A}, u \rangle$ and obtain

$$\nabla_u \tilde{A}(v) - \nabla_v \tilde{A}(u) + \tilde{A}([u, v]). \qquad (34.71)$$

Since there should be no dependence on the choice of vectors u and v, we would then need to subtract the part of the resulting expression that depends on $[u, v]$, since this encodes the Lie derivative of the vectors.
(c) Starting from the expression from part (b) that we have now 'derived', show that

$$d\tilde{A} = \partial_\mu A_\nu (dx^\mu \wedge dx^\nu). \qquad (34.72)$$

Hint: Assume the arbitrary vectors u and v in (b) are constant in space.

(34.7) Consider a path parametrized by an affine parameter λ. Define a new parametrization $s = f(\lambda)$. Show that this makes no difference to the geodesic equation if s and λ are linearly related.

(34.8) Consider a non-geodesic curve in a spacetime with an affine connection. Show that the compatibility condition implies that if the magnitude of the tangent vector $\boldsymbol{u}$ is a constant along the curve, then $\boldsymbol{u} \cdot (\boldsymbol{\nabla}_{\boldsymbol{u}} \boldsymbol{u}) = 0$.

Physically, this means that an affine parametrization that keeps the magnitude of the velocity vector constant for any curve, has a $\boldsymbol{u}$ that is perpendicular to the acceleration $\boldsymbol{a} = (\boldsymbol{\nabla}_{\boldsymbol{u}} \boldsymbol{u})$, just as we have in the flat spacetime of special relativity. See Needham (2021) for a discussion of the material in this problem.

Riemann curvature revisited

35

After Riemann had made known his discoveries, mathematicians busied themselves with working out his system of geometrical ideas formally; chief among these were Christoffel, Ricci, and Levi-Civita. Riemann ... clearly left the real development of his ideas in the hands of some subsequent scientist whose genius as a physicist could rise to equal flights with his own as a mathematician. After a lapse of seventy years this mission has been fulfilled by Einstein.
Hermann Weyl (1885–1955) *Space - Time - Matter*

You wake up in a space station and feel a force keeping you in bed. If you can't tell whether this is because there's a gravitational field present or because the station is accelerating, then there's a simple experiment to carry out. Allow two particles to fall freely, starting them in motion on parallel paths, and later measure their relative acceleration. If this acceleration is non-zero it suggests the presence of a genuine curvature in spacetime. This experiment, a measurement of geodesic deviation, is a true test of spacetime curvature and hence of real gravitational fields, as it allows access to the Riemann tensor $\boldsymbol{R}$. It is this tensor that provides the key to assessing whether spacetime is curved.

In this chapter, we revisit the Riemann tensor and investigate the geometrical method of calculating $\boldsymbol{R}$. The tools we need are the Lie and covariant derivatives from the previous two chapters. The discussion here will lead us to an operator equation for the curvature tensor that is very useful in actually computing the curvature for different spacetimes.[1]

> ↷ This chapter revisits the Riemann curvature tensor using the geometrical tools we have built in this part of the book, allowing us to rederive some key results in a more elegant manner and derive some new ones. This material provides an insight into the role curvature plays in the physics of the Universe.

[1]These computations are the subject of the next chapter.

35.1 Geodesic deviation (slight return)

In (3+1) dimensions, the Riemann tensor $\boldsymbol{R}(\ ,\ ,\ ,)$ is a (1,3) object that encodes the curvature of spacetime. If we insert three carefully chosen vectors into the latter slots, $\boldsymbol{R}$ returns a vector whose physical interpretation is that it encodes the relative acceleration of neighbouring geodesics. The properties of the tensor can be derived by considering the deviation of two freely falling particles following separate geodesics, much as in the thought experiment described above. This is our first task in this chapter.

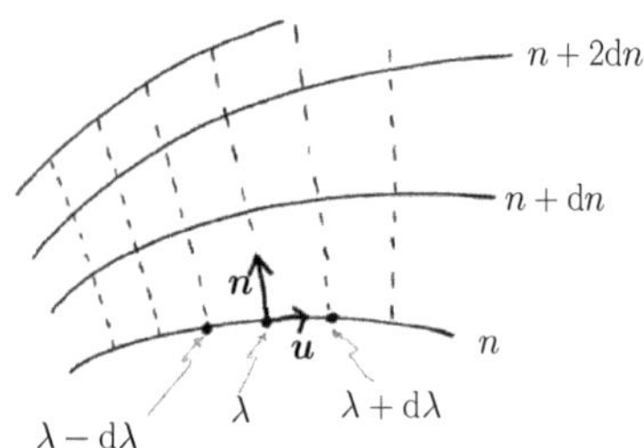

Fig. 35.1 Geodesics for different freely falling particles used to compute geodesic deviation.

^{2}This expression is sometimes called the 'geodesic deviation equation', and sometimes called the 'Jacobi equation', reflecting Jacobi's discovery of it in 1837.

The separation of the geodesics is described by the vector $\boldsymbol{n}$. This important vector can be understood with reference to Fig. 35.1. A typical geodesic, labelled n, is parametrized with affine parameter λ and has velocity given by the value of the tangent vector field $\boldsymbol{u}$ evaluated at a point on the geodesic. There will be lots of other geodesics in the vicinity of a particular point λ_0. The closest is labelled $n + \mathrm{d}n$, the next $n + 2\mathrm{d}n$ and so on. This gives us access to the vector $\boldsymbol{n} = \mathrm{d}/\mathrm{d}n$ that we interpret as the providing a measure of the relative separation of neighbouring geodesics.

Now that we have the separation vector $\boldsymbol{n}$ we need to evaluate it as the particles fall along their geodesic world lines. The tangents to the particle's geodesics are also provided by the velocity field $\boldsymbol{u}$ evaluated at the points of interest on the relevant geodesic. (Recall that the geodesics form a congruence of curves for the vector field $\boldsymbol{u}$.) The falling of the particles can then be viewed in terms of the vector $\boldsymbol{n}$ being carried along the streamlines of the velocity field $\boldsymbol{u}$. The change in $\boldsymbol{n}$ is therefore described by the Lie derivative $\mathcal{L}_{\boldsymbol{u}}\boldsymbol{n}$.

Now recall the notion of a field being **Lie dragged**. This occurs if the vector that stretches between two curves of a congruence still stretches between them after being carried along the congruence. This is exactly the situation we require here to describe the defining property of $\boldsymbol{n}$. From the definition of $\boldsymbol{n}$ being the vector that links the two particles at all points as they fall, we must have the condition that, although $\boldsymbol{n}$ can change as it the particles fall, its Lie derivative must vanish: $\mathcal{L}_{\boldsymbol{u}}\boldsymbol{n} = 0$. This encodes the condition that the vector $\boldsymbol{n}$ still stretches between the particles after being transported through $\boldsymbol{u}$. We are now almost ready to derive the most important equation of this chapter. This is the equation that says that relative acceleration of the geodesics is related to the Riemann tensor by^2

$$\frac{D^2\boldsymbol{n}}{d\lambda^2} + \boldsymbol{R}(\ ,\boldsymbol{u},\boldsymbol{n},\boldsymbol{u}) = 0. \tag{35.1}$$

In components, this reads

$$\frac{D^2 n^\mu}{d\lambda^2} + R^\mu{}_{\nu\alpha\beta} u^\nu n^\alpha u^\beta = 0. \tag{35.2}$$

We shall provide the geometric derivation of this equation and this will also give us the Riemann tensor.

Example 35.1

Before we get there, it's very useful, as a warm up, to first consider the behaviour of two particles falling in a gravitational field according to Newtonian gravitation. Specifically, let's consider the vector $\boldsymbol{n}$, with components n^k, that connects the trajectories of the two particles. We'll calculate the relative acceleration $\ddot{n}^k$. This is useful as it closely mirrors the full relativistic derivation that we shall discuss afterwards. The vector of interest is $\boldsymbol{n} \equiv \mathrm{d}/\mathrm{d}n = (\mathrm{d}x^k/\mathrm{d}n)(\partial/\partial x^k)$. The component n^k is therefore $\mathrm{d}x^k/\mathrm{d}n$. We therefore want to find

$$\frac{\partial^2 n^k}{\partial t^2} = \frac{\partial^2}{\partial t^2}\frac{\mathrm{d}x^k}{\mathrm{d}n} = \frac{\mathrm{d}}{\mathrm{d}n}\frac{\partial^2 x^k}{\partial t^2}. \tag{35.3}$$

The acceleration of a particle at coordinate x^k is given by Newton's force law

$$\frac{\partial^2 x^k}{\partial t^2} + \eta^{ik}\frac{\partial \Phi}{\partial x^i} = 0, \tag{35.4}$$

where Φ is the Newtonian potential, and so we have

$$
\begin{aligned}
\frac{\partial^2 n^k}{\partial t^2} &= \frac{\mathrm{d}}{\mathrm{d}n}\frac{\partial^2 x^k}{\partial t^2} && \text{(from eqn 35.3)}\\
&= -\eta^{ik}\frac{\mathrm{d}}{\mathrm{d}n}\frac{\partial \Phi}{\partial x^i} && \text{(substituting eqn 35.4)}\\
&= -\eta^{ik}\frac{\mathrm{d}x^j}{\mathrm{d}n}\frac{\partial}{\partial x^j}\frac{\partial \Phi}{\partial x^i} && \text{(chain rule)}\\
&= -\eta^{ik}\frac{\partial^2 \Phi}{\partial x^j \partial x^i}n^j = -R^k{}_j n^j && \text{(tidying).}
\end{aligned}
$$

This gives us an equation of motion $\ddot{n}^k + R^k{}_j n^j = 0$, and an expression for the components[3] $R^k{}_j$ of a tensor that can be used to determine the acceleration of the separation of the trajectories.

[3] This expression reads

$$R^k{}_j = \eta^{ik}\frac{\partial^2 \Phi}{\partial x^j \partial x^i}. \tag{35.5}$$

Let's now consider the full, geometrical version of the problem considered in the previous example. This follows exactly the same pattern: we simply want to calculate the acceleration of n. That is to say, we want to find the double derivative of n with respect to an affine parameter: $\mathrm{D}^2/\mathrm{d}\lambda^2$, which is equivalent to $\nabla_u\nabla_u n$. From above, we recall that the relative separation vector n is Lie dragged, which is to say that

$$\mathscr{L}_u n = [u, n] = 0. \tag{35.6}$$

We also know that in a torsion-free system $\nabla_u n - \nabla_n u = [u, n]$, so the Lie dragging is equivalently described by the equation

$$\nabla_u n = \nabla_n u, \tag{35.7}$$

which is the symmetry of the covariant derivative, from the last chapter. We now have the tools at our disposal to find $\nabla_u\nabla_u n$.

Example 35.2

Consider a geodesic with tangent vector u. As usual it is described by the geodesic equation $\nabla_u u = 0$. We take the covariant derivative of this expression along the n direction

$$\nabla_n\nabla_u u = 0. \tag{35.8}$$

Next, we use the commutator $[\nabla_n, \nabla_u] = \nabla_n\nabla_u - \nabla_u\nabla_n$, to write

$$(\nabla_u\nabla_n + [\nabla_n, \nabla_u])\,u = 0. \tag{35.9}$$

Finally, use the symmetry of the covariant derivative $\nabla_n u = \nabla_u n$ to say

$$\nabla_u\nabla_u n + [\nabla_n, \nabla_u]\,u = 0, \tag{35.10}$$

or, parametrizing the geodesic with affine parameter λ,

$$\frac{\mathrm{D}^2 n}{\mathrm{d}\lambda^2} + [\nabla_n, \nabla_u]\,u = 0. \tag{35.11}$$

Although this looks like it solves the problem via an operator $[\nabla_n, \nabla_u]$, it does not quite. We discuss this below.

The operator $[\boldsymbol{\nabla}_a, \boldsymbol{\nabla}_b]$ isn't quite the one we need to compute the Riemann tensor $\boldsymbol{R}$ to evaluate the curvature of general spaces. We saw above that $[\boldsymbol{u}, \boldsymbol{n}] = 0$, which implies that a loop formed by vectors $\boldsymbol{u}$ and $\boldsymbol{n}$ closes. However, a loop constructed by travelling along arbitrary vectors $\boldsymbol{a}$ then $\boldsymbol{b}$ won't necessarily meet up with one where we travel along $\boldsymbol{b}$ and then along $\boldsymbol{a}$. The distance between the end points is measured by the Lie bracket $[\boldsymbol{a}, \boldsymbol{b}]$. As a result we actually need the slightly upgraded operator $[\boldsymbol{\nabla}_a, \boldsymbol{\nabla}_b] - \boldsymbol{\nabla}_{[a,b]}$ to correctly generalize the operator so that we capture the Riemann tensor. In a coordinate frame, we have $[\boldsymbol{e}_\mu, \boldsymbol{e}_\nu] = 0$, simplifying the curvature operator. We shall often work in coordinate frames, making this correction a negligible detail.

The result of this argument is that the curvature tensor $\boldsymbol{R}$ can be formed into a **Riemann curvature operator** $\hat{R}$ defined by

$$\boldsymbol{R}(\ , \boldsymbol{c}, \boldsymbol{a}, \boldsymbol{b}) = \hat{R}(\boldsymbol{a}, \boldsymbol{b})\boldsymbol{c} = \left([\boldsymbol{\nabla}_a, \boldsymbol{\nabla}_b] - \boldsymbol{\nabla}_{[a,b]}\right)\boldsymbol{c}. \tag{35.12}$$

The operator needs two vectors (here, $\boldsymbol{a}$ and $\boldsymbol{b}$) in order to build it. It can then act on a vector (here, $\boldsymbol{c}$) to output another vector. In fact, as shown in the exercises, in a coordinate frame where the loop formed by the basis vectors closes, we end up with the memorable component equation[4]

$$[\boldsymbol{\nabla}_\mu, \boldsymbol{\nabla}_\nu]\, c^\alpha = c^\alpha{}_{;\nu\mu} - c^\alpha{}_{;\mu\nu} = R^\alpha{}_{\beta\mu\nu}c^\beta, \tag{35.13}$$

where c^α are the components of a vector.

The Riemann curvature operator turns out to be very useful and we shall work with it in the next chapter where we calculate the tensor for a variety of spacetimes. For now, we conclude that the Riemann curvature results from the double derivative of the vector field $\boldsymbol{u}$ via

$$\boldsymbol{\nabla}_u \boldsymbol{\nabla}_u \boldsymbol{n} + \hat{R}(\boldsymbol{n}, \boldsymbol{u})\boldsymbol{u} = 0. \tag{35.14}$$

This is simply eqn 35.1 rewritten in terms of covariant derivatives and the Riemann operator. One of the most important things to note about the Riemann tensor and operator is that the curvature is represented by a double covariant derivative. This point will be important in the next chapter when we come to finding an efficient means of computing $\boldsymbol{R}$.

35.2   Components of the curvature tensor

Since the curvature tensor $\boldsymbol{R}$ is a $(1,3)$ tensor, inserting the basis vectors gives us an expression for the components $R^\alpha{}_{\beta\gamma\delta}$, which can be interpreted in terms of the connection coefficients $\Gamma^\alpha{}_{\mu\nu}$. To extract the components we write

$$R^\alpha{}_{\beta\gamma\delta} = \boldsymbol{R}(\boldsymbol{\omega}^\alpha, \boldsymbol{e}_\beta, \boldsymbol{e}_\gamma, \boldsymbol{e}_\delta) = \langle \boldsymbol{\omega}^\alpha, \hat{R}(\boldsymbol{e}_\gamma, \boldsymbol{e}_\delta)\boldsymbol{e}_\beta \rangle, \tag{35.15}$$

where we've used the curvature operator in the final expression. Let's now show that this gives the expression for components of $\boldsymbol{R}$ that we derived in Chapter 11.

---

[4]We define the second derivative $\boldsymbol{\nabla}_\mu \boldsymbol{\nabla}_\nu \boldsymbol{c}$ to be $\boldsymbol{\nabla}_\mu (\boldsymbol{\nabla}_\nu \boldsymbol{c})$ with components $c^\alpha{}_{;\nu\mu} = \left(c^\alpha{}_{;\nu}\right)_{;\mu}$.

Example 35.3

We work in a coordinate basis, so that $\nabla_{[e_\alpha, e_\beta]} = 0$. We therefore have a curvature operator

$$\hat{R}(e_\gamma, e_\delta)e_\beta = [\nabla_\gamma, \nabla_\delta]\, e_\beta$$
$$= \nabla_\gamma \nabla_\delta e_\beta - \nabla_\delta \nabla_\gamma e_\beta. \tag{35.16}$$

Recall that we defined[5] $\nabla_\mu e_\nu = \Gamma^\alpha{}_{\mu\nu} e_\alpha$. Acting on e_ν with a term like $\nabla_\alpha \nabla_\beta$ we obtain[6]

$$\nabla_\alpha \nabla_\beta e_\nu = \nabla_\alpha \left(\Gamma^\sigma{}_{\beta\nu} e_\sigma \right)$$
$$= (\nabla_\alpha \Gamma^\sigma{}_{\beta\nu}) e_\sigma + \Gamma^\sigma{}_{\beta\nu} (\nabla_\alpha e_\sigma)$$
$$= \frac{\partial \Gamma^\sigma{}_{\beta\nu}}{\partial x^\alpha} e_\sigma + \Gamma^\sigma{}_{\beta\nu} \Gamma^\lambda{}_{\alpha\sigma} e_\lambda. \tag{35.19}$$

Putting this together via $R^\alpha{}_{\beta\gamma\delta} = \langle \omega^\alpha, \hat{R}(e_\gamma, e_\delta)e_\beta \rangle$ we have

$$R^\alpha{}_{\beta\gamma\delta} = \frac{\partial \Gamma^\alpha{}_{\delta\beta}}{\partial x^\gamma} - \frac{\partial \Gamma^\alpha{}_{\gamma\beta}}{\partial x^\delta} + \Gamma^\alpha{}_{\gamma\mu} \Gamma^\mu{}_{\delta\beta} - \Gamma^\alpha{}_{\delta\mu} \Gamma^\mu{}_{\gamma\beta}, \tag{35.20}$$

just as we found in Chapter 11. We conclude that the Riemann tensor defined geometrically in terms of geodesic deviation is identical to the one we introduced back in Chapter 11.

The geodesic deviation equation gives us a tool to expand on the discussion of the local orthonormal frames from Chapter 10. There we introduced the freely falling frame as one example of a local inertial frame (LIF). In the next example, we shall find a useful set of coordinates that describe another sort of LIF, known as **Riemann normal coordinates.**

Example 35.4

This idea is most straightforwardly described in flat space. Erect a set of orthonormal axes[7] e_α at the origin and then send straight lines out in every direction. To get to any point $\mathcal{P}$ in the neighbourhood of the origin we need only pick a straight line and then travel along it through a distance λ until we reach $\mathcal{P}$. The straight lines can be characterized by their (normalized, unit) gradient vectors u [Fig. 35.2(a)]. Any point can therefore be reached by translating through a vector λu. The coordinates of this point ζ^α, can be written as

$$\zeta^\alpha e_\alpha = \lambda u^\alpha e_\alpha, \tag{35.21}$$

or $\zeta^\alpha = \lambda u^\alpha$. As a concrete example, the point $(x, y) = (2, 1)$ [Fig. 35.2(b)] lies along the line defined by the tangent $u = \frac{1}{\sqrt{5}}(2e_x + e_y)$, so that we have

$$\zeta^\alpha e_\alpha = \frac{\lambda}{\sqrt{5}}(2e_x + e_y). \tag{35.22}$$

If we simply parametrize the line using the distance, then the point $(2,1)$ is found at $\lambda = \sqrt{5}$, so the coordinates are $\zeta^x = 2$, $\zeta^y = 1$. If we choose our parametrization differently, these would be scaled accordingly.

The idea of Riemann normal coordinates is to generalize this procedure to curved space. So we start by erecting a set of orthonormal axes e_α at the origin. Since the generalization of a straight line is a geodesic, we send out geodesics in every direction from the origin [Fig. 35.2(c)]. The geodesics can be timelike or spacelike. They are characterized by their (unit) tangent vectors u and are parametrized by affine parameter λ that tells us where we are on the curve. Select an event at $\mathcal{P}$ and reach it by travelling a distance λ along the geodesic with gradient u. The coordinates of the point $\mathcal{P}$ are given the Riemann normal coordinates ζ^μ such that

$$\zeta^\alpha e_\alpha = \lambda u^\alpha e_\alpha, \tag{35.23}$$

[5]Alternatively, recall the definition from Chapter 34 that

$$\langle \omega^\alpha, \nabla_\mu e_\nu \rangle = \Gamma^\alpha{}_{\mu\nu}. \tag{35.17}$$

[6]Reminder: Since $\Gamma^\sigma{}_{\nu\beta}$ is a function, we have

$$\nabla_\alpha \Gamma^\sigma{}_{\nu\beta} = \frac{\partial \Gamma^\sigma{}_{\nu\beta}}{\partial x^\alpha}. \tag{35.18}$$

(a)

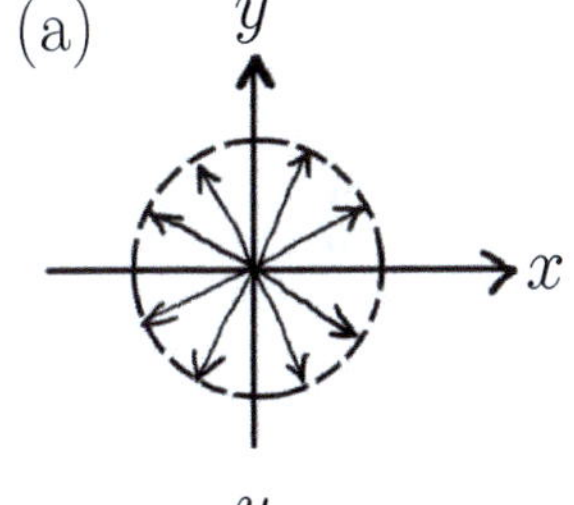

(b)

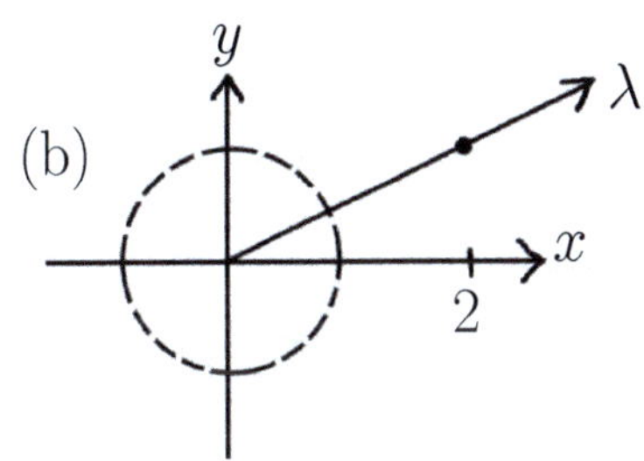

(c)

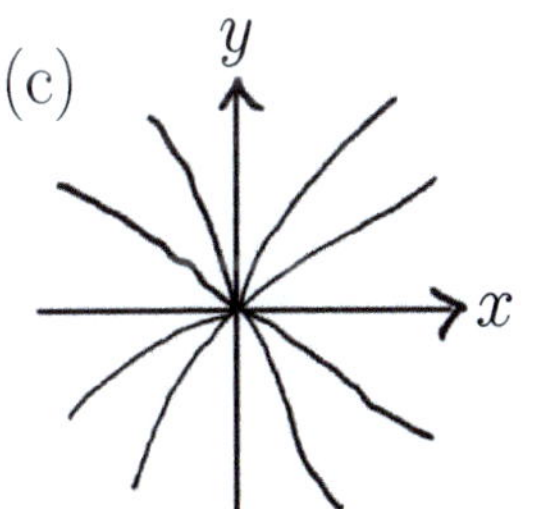

Fig. 35.2 (a) Unit tangents to geodesics in a flat plane. (b) The point $(x, y) = (2, 1)$ reached following a unit tangent by a distance given by the parameter λ. (c) Geodesics in a curved spacetime.

[7]We won't give the indices hats in this example to save on clutter.

so that we have $\zeta^\alpha = \lambda u^\alpha$.

In terms of the Riemann normal coordinates, the spacetime near the origin $\mathcal{O}$ can be shown[8] to have have metric components

$$g_{\mu\nu} = \eta_{\mu\nu} - \frac{1}{3}R_{\mu\alpha\nu\beta}(\mathcal{O})\zeta^\alpha\zeta^\beta + O(\zeta^3). \tag{35.24}$$

At the origin we have $\zeta^\alpha = 0$, and so $g_{\mu\nu} = \eta_{\mu\nu}$. Since there are no linear terms in ζ^μ in eqn 35.24 we have $g_{\mu\nu,\alpha}(\mathcal{O}) = 0$, implying that the connection coefficients vanish. We have therefore described a LIF: flat with vanishing connection coefficients.

35.3   Parallel transport again

Using the geometrical concepts of the previous few chapters, we can return to the parallel transport method that we employed back in Chapter 11 to compute the components of the Riemann tensor. There we carried out a rather messy computation that we can simplify enormously with our new machinery. As we saw in Chapter 11, the Riemann tensor measures the curvature via the angle through which a vector changes when it is parallel transported around a closed path. Here we shall use a coordinate-free approach employing the covariant derivative to assess the change $\delta\boldsymbol{A}$ in a vector $\boldsymbol{A}$, when transported around the loop shown in Fig. 35.3.[9]

We carry a vector $\boldsymbol{A}$ along a vector $\boldsymbol{u}$ using the covariant derivative $\nabla_{\boldsymbol{u}}\boldsymbol{A}$, which computes the value of $\boldsymbol{A}$ at the tip of $\boldsymbol{u}$ minus the value at the base. There are two contributions to this directional derivative: the total change in the vector field $\boldsymbol{A}$ between the start and end points, and the correction due to the change in the basis vectors. The latter is achieved through subtracting off the vector $\boldsymbol{A}$ parallel transported along the path. Since we are going to close the loop, the first of these contributions is zero (since the field $\boldsymbol{A}$ is single valued) and we are simply left with the parallel transport correction. As a result, computing the covariant derivative of the vector field $\boldsymbol{A}$ around a closed loop, outputs (minus) the result of parallel transporting the vector $\boldsymbol{A}$ around the loop. Referring to Fig. 35.3, we shall show in the next example, that the change in the vector field $\delta\boldsymbol{A}$ on traversing the loop is given by

$$\delta\boldsymbol{A} + \hat{\boldsymbol{R}}(\boldsymbol{u}, \boldsymbol{v})\boldsymbol{A}\Delta a\Delta b = 0, \tag{35.25}$$

where $\hat{\boldsymbol{R}}$ is the curvature operator.

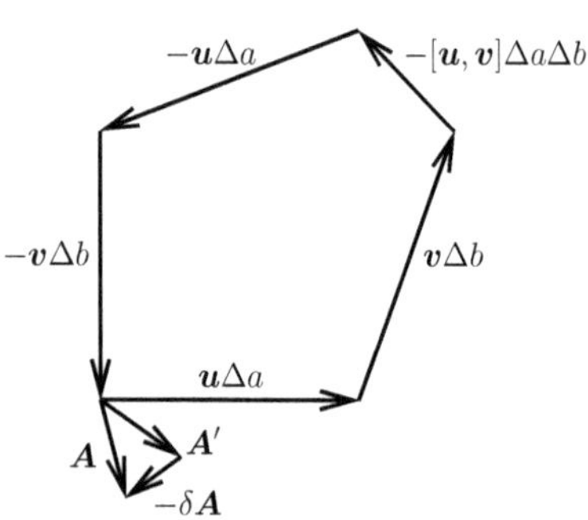

Fig. 35.3 Parallel transport of a vector around a loop gives access to the Riemann tensor. The vector is $\boldsymbol{A}$ originally and changes to $\boldsymbol{A}' = \boldsymbol{A} + \delta\boldsymbol{A}$ after transport.

Example 35.5

In terms of the covariant derivative, the change $-\delta\boldsymbol{A}$ is given by traversing the loop in Fig. 35.3 in an anticlockwise direction, which we reorder into contributions:

$$
\begin{aligned}
-\delta\boldsymbol{A} = \quad &+\nabla_{\boldsymbol{v}}\boldsymbol{A}\Delta b && (\text{moving along path labelled } \boldsymbol{v}\Delta b) \\
&-\nabla_{\boldsymbol{v}}\boldsymbol{A}\Delta b && (\text{moving along } -\boldsymbol{v}\Delta b) \\
&-\nabla_{\boldsymbol{u}}\boldsymbol{A}\Delta a && (\text{moving along } -\boldsymbol{u}\Delta a) \\
&+\nabla_{\boldsymbol{u}}\boldsymbol{A}\Delta a && (\text{moving along } \boldsymbol{u}\Delta a) \\
&-\nabla_{[\boldsymbol{u}, \boldsymbol{v}]}\boldsymbol{A}\Delta a\Delta b && (\text{moving along } -[\boldsymbol{u}, \boldsymbol{v}]\Delta a\Delta b).
\end{aligned}
\tag{35.26}
$$

The key here is to recognize that the movement along $\boldsymbol{v}\Delta b$ is displaced from the movement along $-\boldsymbol{v}\Delta b$ by a distance $\boldsymbol{u}\Delta a$. We can therefore make the replacement

$$\boldsymbol{\nabla}_{\boldsymbol{v}}A\Delta b\big|_{0} - \boldsymbol{\nabla}_{\boldsymbol{v}}A\Delta b\big|_{\boldsymbol{u}\Delta a} = \boldsymbol{\nabla}_{\boldsymbol{u}}\boldsymbol{\nabla}_{\boldsymbol{v}}A\Delta a\Delta b. \tag{35.27}$$

Making an analogous replacement for the displacements along $\boldsymbol{u}\Delta a$ we end up with

$$\begin{aligned}-\delta\boldsymbol{A} &=(\boldsymbol{\nabla}_{\boldsymbol{u}}\boldsymbol{\nabla}_{\boldsymbol{v}} - \boldsymbol{\nabla}_{\boldsymbol{v}}\boldsymbol{\nabla}_{\boldsymbol{u}} - \boldsymbol{\nabla}_{[\boldsymbol{u},\boldsymbol{v}]})A\Delta a\Delta b\\ &=\hat{\boldsymbol{R}}(\boldsymbol{u},\boldsymbol{v})A\Delta a\Delta b.\end{aligned} \tag{35.28}$$

The Riemann operator is, therefore,

$$\begin{aligned}\hat{\boldsymbol{R}}(\boldsymbol{u},\boldsymbol{v}) &=\boldsymbol{\nabla}_{\boldsymbol{u}}\boldsymbol{\nabla}_{\boldsymbol{v}} - \boldsymbol{\nabla}_{\boldsymbol{v}}\boldsymbol{\nabla}_{\boldsymbol{u}} - \boldsymbol{\nabla}_{[\boldsymbol{u},\boldsymbol{v}]})\\ &=[\boldsymbol{\nabla}_{\boldsymbol{u}},\boldsymbol{\nabla}_{\boldsymbol{v}}] - \boldsymbol{\nabla}_{[\boldsymbol{u},\boldsymbol{v}]}.\end{aligned} \tag{35.29}$$

This is, of course, exactly what we had in the previous section.

This general discussion of parallel transport gives us yet another insight into what curvature is. If we take two vectors that are orthogonal at a point and parallel transport the pair, then they must remain orthogonal, by definition of parallel transport. This seems like a way to construct an orthogonal Cartesian grid across all spacetime. However, it is just an illusion. Parallel transport is defined along a particular curve, and we have just seen how transporting a vector around a loop of curves leads, in curved space, to the vector rotating. This means that the vector field created through the parallel transport of a vector defined at some point $\mathcal{P}$ is not single valued. (This is because the vector rotates, giving two vectors when we carry out a loop starting at $\mathcal{P}$: the original one and the rotated one.) We can therefore conclude that another description of curvature is the property that prevents us using parallel transport to set up a Cartesian grid across all spacetime.

Finally, in our discussion of the properties of $\boldsymbol{R}$, we consider how the symmetries affect the number of independent components that the Riemann tensor carries.[10]

Example 35.6

How many independent components does $R_{\alpha\beta\gamma\delta}$ have in n dimensions?
Solution: The number of components is

$$\begin{pmatrix}\text{Number of ways of}\\ \text{choosing } \alpha\beta\gamma\delta \text{ subject}\\ \text{to pair symmetries}\end{pmatrix} - \begin{pmatrix}\text{Number of constraints}\\ \text{due to cyclic symmetry}\end{pmatrix}. \tag{35.32}$$

To understand the first term, we note that the antisymmetry of $(\alpha\beta)$ and $(\gamma\delta)$ means there are $M = \frac{n}{2}(n-1)$ ways of choosing the pairs $(\alpha\beta)$ and M of choosing $(\gamma\delta)$. The symmetry with respect to exchanging the pairs means there are $M(M+1)/2$ independent choices of pairs or pairs (i.e. a M-dimensional symmetric matrix A_{ij} has M diagonal components and $(M-1)/2$ independent off-diagonal ones). We conclude that $\frac{M}{2}(M+1)$ ways of choosing $\alpha\beta\gamma\delta$, when we take all of the pair symmetries into account, where $M = n(n-1)/2$.
To understand the second term, we start with the cyclic symmetry which is encoded in the expression

$$R_{\alpha\beta\gamma\delta} + R_{\alpha\delta\beta\gamma} + R_{\alpha\gamma\delta\beta} = 0. \tag{35.33}$$

---

[10] Recall the discussion of Chapter 11, where we saw that the Riemann tensor obeyed the pair-symmetry equations

$$\begin{aligned}R_{\alpha\beta\gamma\delta} &= - R_{\beta\alpha\gamma\delta},\\ R_{\alpha\beta\gamma\delta} &= - R_{\alpha\beta\delta\gamma},\\ R_{\alpha\beta\gamma\delta} &= + R_{\gamma\delta\alpha\beta},\end{aligned} \tag{35.30}$$

along with a cyclic identity

$$R_{\alpha\beta\gamma\delta} + R_{\alpha\delta\beta\gamma} + R_{\alpha\gamma\delta\beta} = 0. \tag{35.31}$$

This means we need all four indices to be distinct. The number of constraints is then the number of combinations of four objects can be taken from a collection of n objects, which is

$$\frac{n!}{(n-4)!4!},\tag{35.34}$$

or

$$\frac{M}{2}(M+1) - \frac{n!}{(n-4)!4!} = \frac{n^2(n^2-1)}{12}.\tag{35.35}$$

In $n=4$ dimensions, we have twenty independent components.

After having derived the properties of the Riemann curvature tensor and operator, we shall, in the next chapter, finally calculate its components for a variety of spacetimes. Before we get there we pause to consider the Ricci tensor, which incorporates part of the Riemann tensor into Einstein's equation.

35.4   The meaning of the Ricci tensor

The Ricci tensor is the result of the only contraction of the Riemann tensor that does not vanish. In four spacetime dimensions, it has ten of the twenty independent components of the Riemann tensor. Physically, it can be thought of as that part of the Riemann curvature that causes volumes of matter (i.e. sources of curvature) to shrink. As a consequence of the Einstein equation, the Ricci tensor vanishes in free space.[11] In contrast, if we remove the Ricci tensor part from the Riemann tensor, we obtain the **Weyl curvature tensor** C, defined as having components

$$C_{\mu\nu\alpha\beta} = R_{\mu\nu\alpha\beta} - \frac{1}{2}R_{\mu\alpha}g_{\nu\beta} + \frac{1}{2}R_{\mu\beta}g_{\nu\alpha} + \frac{1}{2}R_{\nu\alpha}g_{\mu\beta} - \frac{1}{2}R_{\nu\beta}g_{\mu\alpha}$$
$$+ \frac{1}{6}R\left(g_{\mu\alpha}g_{\nu\beta} - g_{\mu\beta}g_{\nu\alpha}\right).\tag{35.36}$$

The Weyl tensor contains the other ten independent components of R and does not generally vanish in the vacuum, in contrast to the Ricci tensor.[12] An important physical effect of the Weyl curvature tensor C is in its role to distort geodesics.

Example 35.7

Since Ricci curvature can sometimes also distort timelike geodesics, it's best to compare the two sorts of curvature in terms of their action on light rays (or null geodesics). The trace-reversed part of the Ricci tensor has components

$$G_{\mu\nu} = R_{\mu\nu} - \frac{1}{4}Rg_{\mu\nu}.\tag{35.37}$$

This part of the curvature causes light rays from a source to focus, just like a positive focusing lens. The part of the curvature captured by the Weyl curvature has the effect of an astigmatic lens, which focuses positively in one plane and defocuses in a perpendicular plane.

[11]Spacetimes with **constant curvature** have Ricci tenors with components $R_{\mu\nu} = Cg_{\mu\nu}$, with C constant. Spacetimes where $R_{\mu\nu} = 0$ are called **Ricci flat**. Of course, this latter property does not mean that all of the components of the Riemann tensor necessarily vanish; rather that the positive parts of curvature cancel out the negative parts when the Ricci tensor is computed. In this sense, the Ricci tensor is a sort of average of the Riemann tensor.

[12]The Weyl curvature tensor is invariant with respect to conformal transformation. Its components vanish in Minkowski space (as might be expected) and also in the Robertson–Walker spacetimes. These spaces are therefore sometimes called **conformally flat**. If the early Universe resembles a Robertson–Walker spacetime then we expect the Weyl curvature to be very small, or zero. As matter clumps together and black holes are formed, the Weyl curvature must increase, diverging at the black hole singularities. However, the idea that at the initial Big-Bang singularity the Weyl curvature is constrained to be small (or zero) is known as the **Weyl curvature hypothesis**, and is a strong constraint on cosmological models.

Roger Penrose invites us to think about this[13] by imagining the action of a transparent non-refracting Sun on rays from distant stars that lie behind it, as shown in Fig. 35.4. Without any gravitational field the light from each star should form a circular image, as shown by the circles drawn with dashed lines. In the presence of the gravitational field from our hypothetical transparent Sun, those rays that pass through the Sun are mostly affected by the Ricci part of the curvature, causing them to be magnified by the positive focusing effect to form a (larger) circular image (shown by the circles drawn with solid lines). On the other hand, stars whose rays pass beyond the rim of the Sun will only experience the Weyl effect, which causes their images to be distorted into ellipses (as shown in the bottom left-hand corner of Fig. 35.4).

We can make these ideas a little more mathematical by considering geodesic deviation.

Example 35.8

Recall the geodesic equation in component form

$$\frac{D^2 n^\mu}{d\lambda^2} = \left(-R^\mu{}_{\nu\alpha\beta} u^\nu u^\beta\right) n^\alpha. \tag{35.38}$$

Notice how the velocity components are contracted against the indices in the second and fourth positions but that we trace over the other two components to make the Ricci tensor.

Consider the matrix $K^\mu{}_\alpha = -R^\mu{}_{\nu\alpha\beta} u^\nu u^\beta$. If we choose $\boldsymbol{n}$ to be an eigenvector of this matrix, then $\boldsymbol{K}\boldsymbol{n} = \lambda\boldsymbol{n}$, outputting a parallel vector that is scaled by some amount. If we take the eigenvectors of $\boldsymbol{K}$ to describe a volume in spacetime, then the trace over $\boldsymbol{K}$ gives us the rate of change of the volume.[14] We have then that

$$K^\mu{}_\mu = -R^\mu{}_{\nu\mu\beta} u^\nu u^\beta = -R_{\nu\beta} u^\nu u^\beta, \tag{35.40}$$

where $R_{\nu\beta}$ are the components of the Ricci tensor. The rate of change of the volume element δV is then

$$\frac{D^2 \delta V}{d\lambda^2} = (-R_{\nu\beta} u^\nu u^\beta)\delta V. \tag{35.41}$$

This demonstrates that the Ricci tensor describes the evolution of volumes. The contraction that creates it, which involves tracing over the components of the Riemann tensor, amounts to constructing the response of the volume from the individual lengths that make it up.

Returning the to full geodesic equation $D^2 n^\mu/d\lambda^2 = K^\mu{}_\nu n^\nu$, we could also consider the fate of a small sphere as we move along a geodesic. We choose geodesics that follow the edges of the sphere to be initially parallel to the geodesic whose tangent is $\boldsymbol{u}$. Then geodesic deviation causes the sphere to deform into an ellipsoid, whose axes are the eigenvalues of $\boldsymbol{K}$.

This chapter has considered the geometric origin of the Riemann tensor, and we have already said that this tensor derives from the fundamental field of general relativity, the metric field. In the following chapter, we consider the most efficient method to extract the Riemann tensor from a metric.

[13]This is discussed further in Penrose (2004), Section 28.8.

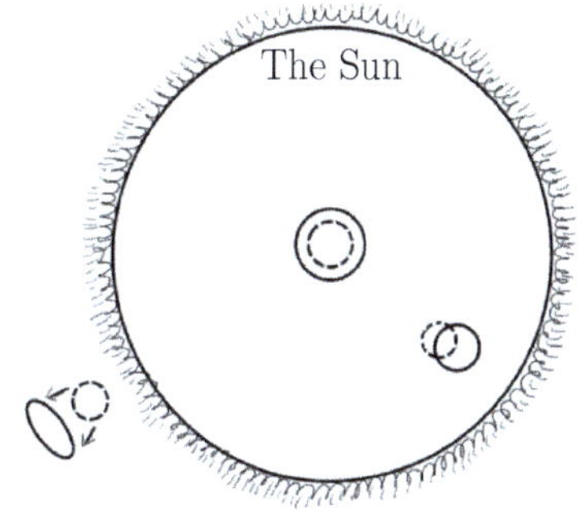

Fig. 35.4 The magnification and distortion of the images of distant stars due to a transparent Sun. The stars actual sizes are shown by the dashed circles; their images are shown by solid lines.

[14]To see this consider a two-dimensional area $A = \delta x \delta y$ where δx and δy are the eigenfunctions of a linear operator $D^2/d\lambda^2 = \hat{K}$. Operating on A we obtain

$$\begin{aligned} D^2 A/d\lambda^2 &= \hat{K}A \\ &= (\lambda_x \delta x)\delta y + \delta x (\lambda_y \delta y) \\ &= (\lambda_x + \lambda_y)A = (\text{Tr}\hat{K})A. \end{aligned} \tag{35.39}$$

Chapter summary

- Geodesic deviation can be calculated geometrically by considering the condition that the Lie derivative $\pounds_{\boldsymbol{u}}\boldsymbol{n}$ vanishes.

- Instead of the Riemann tensor, we can use the Riemann curvature operator $\hat{\boldsymbol{R}}$, defined by

$$\hat{\boldsymbol{R}}(\boldsymbol{a},\boldsymbol{b})\boldsymbol{c} = \left([\boldsymbol{\nabla}_a,\boldsymbol{\nabla}_b] - \boldsymbol{\nabla}_{[a,b]}\right)\boldsymbol{c}. \tag{35.42}$$

- The Ricci tensor represents the part of the Riemann tensor that has a focussing effect on volumes. The Weyl tensor represents the parts left over, which have a distorting effect on light rays.

Exercises

(35.1) Show that, in terms of coordinates, $\boldsymbol{\nabla}_{[\boldsymbol{X},\boldsymbol{Y}]}\boldsymbol{Z}$ can be represented as

$$\left(X^\gamma Y^\delta_{\;;\gamma} - Y^\gamma X^\delta_{\;;\gamma}\right) Z^\alpha_{\;;\delta}. \tag{35.43}$$

(35.2) (a) Show that, in a coordinate frame,

$$R^\mu_{\;\nu\alpha\beta}v^\nu = v^\mu_{\;;\beta\alpha} - v^\mu_{\;;\alpha\beta}. \tag{35.44}$$

(b) Show further that

$$v^{\mu;\alpha}_{\quad\mu} - v^\mu_{\;;\mu}^{\;\;\alpha} = R^\alpha_{\;\mu}v^\mu. \tag{35.45}$$

(35.3) Consider the flat space equation

$$-\partial^2 A^\mu + \partial^\mu \partial_\nu A^\nu = J^\mu. \tag{35.46}$$

(a) Write the equation in index comma notation.
(b) Show that using the *comma goes to semicolon rule* results in an ambiguity.
(c) Use the result of Exercise 35.2 to show that the two possible forms of the equation can be related using the component of the Ricci tensor $R_{\mu\nu}$. *There is no definite rule on how to avoid this ambiguity. Misner, Thorne, and Wheeler give a set of rules of thumb to choose the most physically reasonable.*

(35.4) *We are now used to setting up a local orthonormal frame at some point in curved spacetime. Here, following the approach of Poisson and Will, we shall set up Riemann normal coordinates ζ^μ, which provide a local inertial frame where, in addition to the property $g_{\hat\mu\hat\nu}(\mathcal{O}) = \eta_{\hat\mu\hat\nu}$ at the origin $\mathcal{O}$, the connection coefficients are also guaranteed to vanish. To do this, we shall work with the basis vectors $\boldsymbol{e}_\mu$ of an orthonormal frame defined at the origin. (We won't follow our usual convention of gives these directions hats in this case, to save on clutter.)*

Consider the setup for the Riemann normal coordinates $\zeta^\mu = \lambda u^\mu$, where $\boldsymbol{u}$ is tangent to a geodesic and λ measures the distance from $\mathcal{O}$. The components here are those of an orthonormal frame that we set up at the origin $\mathcal{O}$ (only). If we vary the directions of the tangents we obtain more geodesics that originate from $\mathcal{O}$. These are linked by deviation vectors $\boldsymbol{n}_\nu$ (one for each component ν of the tangent) with components

$$n^\mu_\nu = \frac{\partial\zeta^\mu}{\partial u^\nu} = s\delta^\mu_\nu. \tag{35.47}$$

For definiteness, we'll treat the geodesics as spacelike.
(a) Show that at any point on the geodesic $\Gamma^\mu_{\;\alpha\beta}u^\alpha u^\beta = 0$. Why does this imply that the connection coefficients vanish at the origin?
(b) Since the connection coefficients vanish at $\mathcal{O}$ we can expand

$$\Gamma^\mu_{\;\alpha\beta} = \frac{\partial\Gamma^\mu_{\;\alpha\beta}}{\partial\zeta^\sigma}\zeta^\sigma + \dots \tag{35.48}$$

Use this to show that the first derivative of the deviation vectors is

$$\left(\frac{D\boldsymbol{n}_\nu}{\mathrm{d}\lambda}\right)^\mu = \delta^\mu_\nu + \lambda^2 \frac{\partial \Gamma^\mu_{\ \alpha\nu}(\mathcal{O})}{\partial \zeta^\sigma} u^\alpha u^\sigma + O(s^3).$$

(35.49)

(c) Show that the second derivative is

$$\left(\frac{D^2\boldsymbol{n}_\nu}{\mathrm{d}\lambda^2}\right)^\mu = 3\lambda \frac{\partial \Gamma^\mu_{\ \alpha\nu}(\mathcal{O})}{\partial \zeta^\sigma} u^\alpha u^\sigma + O(s^2).$$ (35.50)

(d) Show that the geodesic deviation equation is then

$$\left[3\frac{\partial \Gamma^\mu_{\ \alpha\nu}(\mathcal{O})}{\partial \zeta^\sigma} + R^\mu_{\ \alpha\nu\sigma}(\mathcal{O})\right] u^\alpha u^\sigma + O(s^2) = 0.$$

(35.51)

(e) Show that this implies that, at the origin, we have

$$\frac{\partial \Gamma^\mu_{\ \alpha\nu}(\mathcal{O})}{\partial \zeta^\sigma} + \frac{\partial \Gamma^\mu_{\ \sigma\nu}(\mathcal{O})}{\partial \zeta^\alpha} = -\frac{1}{3}\left(R^\mu_{\ \alpha\nu\sigma} + R^\mu_{\ \sigma\nu\alpha}\right).$$

(35.52)

(f) By permuting indices, use the previous expression to obtain

$$\frac{\partial \Gamma^\mu_{\ \nu\sigma}(\mathcal{O})}{\partial \zeta^\alpha} = -\frac{1}{3}\left(R^\mu_{\ \nu\sigma\alpha} + R^\mu_{\ \sigma\nu\alpha}\right).$$ (35.53)

(g) Use the expression from (f) to show

$$\frac{\partial^2 g_{\mu\nu}(\mathcal{O})}{\partial \zeta^\alpha \partial \zeta^\sigma} = -\frac{1}{3}\left(R_{\mu\sigma\nu\alpha} + R_{\mu\alpha\nu\sigma}\right).$$ (35.54)

Hint: We saw in Exercise 34.2 that taking derivatives of $g_{\mu\nu} = \boldsymbol{e}_\mu \cdot \boldsymbol{e}_\nu$ yields the useful identity

$$g_{\mu\nu,\sigma} = g_{\mu\beta}\Gamma^\beta_{\ \sigma\nu} + g_{\nu\beta}\Gamma^\beta_{\ \sigma\mu}.$$ (35.55)

Take the derivative of this to compute the expression.

(h) Finally, expand the metric in the Riemann normal coordinates, to recover eqn 35.24.

(35.5) Consider a two-dimensional surface spanned by basis vectors $\boldsymbol{A}$ and $\boldsymbol{B}$, where $[\boldsymbol{A}, \boldsymbol{B}] = 0$. Given that the torsion vanishes and that $\nabla_{\boldsymbol{A}}\boldsymbol{A} = -\boldsymbol{B}$, $\nabla_{\boldsymbol{B}}\boldsymbol{B} = \boldsymbol{B}$ and $\nabla_{\boldsymbol{B}}\boldsymbol{A} = \boldsymbol{A}$, show that the Riemann tensor vanishes.

(35.6) Consider the case where we have the basis vectors $\boldsymbol{X}$ and $\boldsymbol{Y}$, with the properties $[\boldsymbol{X}, \boldsymbol{Y}] = 0$ and

$$\nabla_{\boldsymbol{X}}\boldsymbol{X} = 0, \qquad \nabla_{\boldsymbol{X}}\boldsymbol{Y} = \boldsymbol{X} + \boldsymbol{Y},$$
$$\nabla_{\boldsymbol{Y}}\boldsymbol{X} = \boldsymbol{X} - \boldsymbol{Y}, \qquad \nabla_{\boldsymbol{Y}}\boldsymbol{Y} = 0.$$

(a) Compute the connection coefficients.
(b) Compute the components of the Riemann tensor.

36

Cartan's method

[1] Recall that we put hats on indices (e.g. $A_{\hat{\mu}}$) to remind ourselves of the use of the orthonormal frame.

[2] We can expand this using vielbein by writing $\boldsymbol{g} = g_{\alpha\beta} \boldsymbol{dx}^{\alpha} \otimes \boldsymbol{dx}^{\beta}$ with components

$$g_{\alpha\beta} = \eta_{\hat{\mu}\hat{\nu}}(\boldsymbol{e}_{\alpha})^{\hat{\mu}}(\boldsymbol{e}_{\beta})^{\hat{\nu}}, \qquad (36.1)$$

and then write orthonormal basis 1-forms $\boldsymbol{\omega}^{\hat{\mu}} = (\boldsymbol{e}_{\alpha})^{\hat{\mu}} \boldsymbol{dx}^{\alpha}$. Putting these expressions together yields eqn 36.2.

The finest collection of frames I ever saw
Sir Humphery Davy (1778–1829) *when asked what he thought of the Paris art galleries*

The calculation of curvature tensors is, generally speaking, a tedious business. However, using the geometrical techniques we have built up in the last few chapters, we can come up with a more efficient means of extracting the tensor $\boldsymbol{R}$ from the metric field tensor $\boldsymbol{g}$. The key, discovered by Élie Cartan, is to make use of forms. Cartan's method is especially effective because we often choose to work in the orthonormal frame[1] with metric[2]

$$\boldsymbol{ds}^2 = \boldsymbol{g} = \eta_{\hat{\mu}\hat{\nu}} \boldsymbol{\omega}^{\hat{\mu}} \otimes \boldsymbol{\omega}^{\hat{\nu}}, \qquad (36.2)$$

where $\eta_{\hat{\mu}\hat{\nu}}$ are the components of the Minkowski metric tensor. That is to say, in a typical coordinate frame (t, χ, θ, ϕ) with diagonal metric, we cast the metric field in the form

$$\boldsymbol{ds}^2 = \boldsymbol{g} = -(\boldsymbol{\omega}^{\hat{t}})^2 + (\boldsymbol{\omega}^{\hat{\chi}})^2 + (\boldsymbol{\omega}^{\hat{\theta}})^2 + (\boldsymbol{\omega}^{\hat{\phi}})^2. \qquad (36.3)$$

In this chapter, we show how a routine based on taking exterior derivatives of the basis 1-forms $\boldsymbol{\omega}^{\hat{\mu}}$ does the job of extracting the curvature.

> ↷The first two sections of this chapter build up the logic behind Cartan's method and provide derivations of the key results. Once this ground is covered, we are left with a turn-the-handle method for computing curvature, so will not need to use many of the methods. So, if you find these sections a little hard-going, consider turning to the start of Section 36.3 (and the rules given in the grey box just before this) and have a look at the method itself. We follow Misner, Thorne, and Wheeler closely in our discussion of the method, which should be consulted for further discussion.

36.1 Connection 1-forms

In the last chapter, we identified a curvature operator $\hat{\boldsymbol{R}}(\boldsymbol{a}, \boldsymbol{b}) = [\boldsymbol{\nabla}_{\boldsymbol{a}}, \boldsymbol{\nabla}_{\boldsymbol{b}}] - \boldsymbol{\nabla}_{[\boldsymbol{a},\boldsymbol{b}]}$. This is a complicated way of saying that the curvature tensor relies on the behaviour of second covariant derivatives of a vector field. We might wonder if there's another way of framing this second derivative, using the simple connection symbol $\boldsymbol{\nabla}$. The key is to exploit the similarity between $\boldsymbol{\nabla}$ and the exterior derivative $\boldsymbol{d}$ in their action on vectors.

Originally, we didn't know how to use $\boldsymbol{d}$ on anything other than forms. In Chapter 34, we defined the connection $\boldsymbol{\nabla}$ and the covariant derivative $\boldsymbol{\nabla}_{\boldsymbol{v}} = \boldsymbol{v} \cdot \boldsymbol{\nabla}$. Strip the vector field $\boldsymbol{v}$ from $\boldsymbol{\nabla}$ and its action on a scalar function $\boldsymbol{\nabla}f$ is defined to be equivalent to that of the exterior derivative $\boldsymbol{d}f$. We then asked how to use $\boldsymbol{d}$ on vector fields and came up with the answer that, for vectors, $\boldsymbol{d} \equiv \boldsymbol{\nabla}$. We defined the resulting object $\boldsymbol{dv}$ to be a vector-valued 1-form. The exterior derivative d acts on the vector field $v = v^\mu e_\mu$ one part at a time

$$dv = (dv^\mu) \otimes e_\mu + v^\mu de_\mu. \tag{36.4}$$

Let's now slightly diverge from our path in Chapter 34, which next involved writing connection coefficients. Instead, knowing that we are working on a manifold with a connection, we expand the vector-valued 1-form de_μ, which tells us how the basis vectors change in space. We write this in the (slightly complicated-looking) manner

$$de_\mu = \omega^\nu{}_\mu \otimes e_\nu. \tag{36.5}$$

The objects $\omega^\nu{}_\mu$ are 1-forms that express the change of the basis vectors in space: they therefore express the connection itself, and we call them **connection 1-forms**. The right-hand side of eqn 36.5 tells us that, at a particular point in space, the way in which the basis vectors change can be given by an expansion of tensor products of the basis vectors and connection 1-forms $\omega^\nu{}_\mu$.

We can relate the connection 1-forms to connection coefficients, as they describe the same property of the space. We write

$$\Gamma^\nu{}_{\alpha\mu}e_\nu = \nabla_\alpha e_\mu = \langle \nabla e_\mu, e_\alpha \rangle \equiv \langle de_\mu, e_\alpha \rangle = \langle (\omega^\nu{}_\mu \otimes e_\nu), e_\alpha \rangle. \tag{36.6}$$

We spot that, in order for this to be consistent, the connection 1-forms can be written in terms of connection coefficients as follows:

$$\omega^\nu{}_\mu = \Gamma^\nu{}_{\lambda\mu}\omega^\lambda. \tag{36.7}$$

We check this in the next example.

Example 36.1

Checking this previous equation makes sense, we have

$$
\begin{aligned}
\langle (\omega^\nu{}_\mu \otimes e_\nu), e_\alpha \rangle &= \Gamma^\nu{}_{\lambda\mu}\langle (\omega^\lambda \otimes e_\nu), e_\alpha \rangle && \text{(from eqn 36.7)} \\
&= \Gamma^\nu{}_{\lambda\mu}\omega^\lambda(e_\alpha) \otimes e_\nu && \text{(inserting } e_\alpha \text{ in the first slot)} \\
&= \Gamma^\nu{}_{\alpha\mu}e_\nu && \text{(using } \omega^\lambda(e_\alpha) = \delta^\lambda{}_\alpha\text{).}
\end{aligned}
$$

This is exactly what we started with, validating the definitions above.

We now have a 1-form $\omega^\mu{}_\nu$ that expresses the connection.[3] Next, we need to determine the curvature.

Roughly speaking we defined curvature in the last chapter in terms of a double derivative of a field v by using the operator ∇. Here we shall take the double exterior derivative of v using d. The first derivative results in a vector-valued 1-form, and is given by

$$dv = e_\mu \otimes (dv^\mu + \omega^\mu{}_\nu v^\nu). \tag{36.8}$$

[3] In fact, eqn 36.6 provides us with an interpretation of $\omega^\mu{}_\nu$. The object $\nabla_\alpha e_\mu$ tells us the rate of change of the basis vector e_μ along the α direction, and, by the last step in eqn 36.6, that this is equal to $\omega^\nu{}_\mu(e_\alpha) \otimes e_\nu$. We can therefore interpret $\omega^\nu{}_\mu(e_\alpha)$ as encoding the rate at which e_μ rotates towards e_ν as we move along a curve whose tangent vector is given by e_α.

[4] It's important to note that, because the derivative of a vector results in a (1,1) object built from the tensor product of a vector and a 1-form (and not the antisymmetric wedge product between two 1-forms), they generally have non-zero exterior derivatives. In short, vectors allow the existence of the peculiar-looking double exterior derivative and it is this that gives us access to the curvature.

Differentiate again[4] and we obtain a vector-valued 2-form, given by

$$
\begin{aligned}
\boldsymbol{d}^2\boldsymbol{v} =&\boldsymbol{d}\boldsymbol{e}_\mu \wedge \left(\boldsymbol{d}v^\mu + \boldsymbol{\omega}^\mu{}_\nu v^\nu\right) + \boldsymbol{e}_\mu \otimes \left(\boldsymbol{d}^2 v^\mu + \boldsymbol{d}\boldsymbol{\omega}^\mu{}_\nu v^\nu - \boldsymbol{\omega}^\mu{}_\nu \wedge \boldsymbol{d}v^\nu\right) \\
=&\boldsymbol{e}_\mu \otimes \left(\boldsymbol{\omega}^\mu{}_\nu \wedge \boldsymbol{d}v^\nu + \boldsymbol{\omega}^\mu{}_\alpha \wedge \boldsymbol{\omega}^\alpha{}_\nu v^\nu + \boldsymbol{d}^2 v^\mu + \boldsymbol{d}\boldsymbol{\omega}^\mu{}_\nu v^\nu \right. \\
& \left. - \boldsymbol{\omega}^\mu{}_\nu \wedge \boldsymbol{d}v^\nu\right).
\end{aligned}
\tag{36.9}
$$

Noting that since the component v^μ is itself merely a scalar function and so $\boldsymbol{d}^2 v^\mu = 0$ (for the usual reason that $\boldsymbol{d}^2 = 0$ when acting on any function), we conclude that

$$
\boldsymbol{d}^2\boldsymbol{v} = \boldsymbol{e}_\mu \otimes \left(\boldsymbol{d}\boldsymbol{\omega}^\mu{}_\nu + \boldsymbol{\omega}^\mu{}_\alpha \wedge \boldsymbol{\omega}^\alpha{}_\nu\right) v^\nu,
\tag{36.10}
$$

or

$$
\boldsymbol{d}^2\boldsymbol{v} = \left(\boldsymbol{e}_\mu \otimes \mathcal{R}^\mu{}_\nu\right) v^\nu,
\tag{36.11}
$$

where we define the **curvature 2-form** $\mathcal{R}^\mu{}_\nu$ by

$$
\mathcal{R}^\mu{}_\nu = \boldsymbol{d}\boldsymbol{\omega}^\mu{}_\nu + \boldsymbol{\omega}^\mu{}_\alpha \wedge \boldsymbol{\omega}^\alpha{}_\nu.
\tag{36.12}
$$

Equation 36.12 is the most important one of this chapter. It is built from two terms, the first is the exterior derivative of the connection 1-forms, expressing how their rate of change in space; the second is a wedge product of the connection 1-forms, expressing their tube-like structure in space. Both of these aspects are needed to describe curvature. We also write, for economy, a curvature 2-form operator $\mathcal{R}(\ )$, which is a (2,2) object designed for the input of a vector[5]

$$
\mathcal{R}(\ ) = \boldsymbol{\omega}^\nu(\ ) \otimes \boldsymbol{e}_\mu \otimes \mathcal{R}^\mu{}_\nu.
\tag{36.13}
$$

[5] We interpret this as

$$
\begin{aligned}
\mathcal{R}(\boldsymbol{v}) =&\boldsymbol{\omega}^\nu(\boldsymbol{v}) \otimes \boldsymbol{e}_\mu \otimes \mathcal{R}^\mu{}_\nu \\
=&\boldsymbol{\omega}^\nu(\boldsymbol{e}_\alpha)\boldsymbol{e}_\mu \otimes \mathcal{R}^\mu{}_\nu v^\alpha \\
=& \left(\boldsymbol{e}_\mu \otimes \mathcal{R}^\mu{}_\alpha\right) v^\alpha,
\end{aligned}
$$

which is what we had in eqn 36.11.

This allows us to write the memorable equation for the second exterior derivative of a vector in terms of the curvature 2-form operator as

$$
\boldsymbol{d}^2\boldsymbol{v} = \mathcal{R}(\boldsymbol{v}).
\tag{36.14}
$$

Example 36.2

Let's prove this is equivalent to the previously defined curvature tensor by contracting the 2-form $\mathcal{R}^\mu{}_\nu = \boldsymbol{d}\boldsymbol{\omega}^\mu{}_\nu + \boldsymbol{\omega}^\mu{}_\alpha \wedge \boldsymbol{\omega}^\alpha{}_\nu$ with two vectors. First, consider the 1-form

$$
\boldsymbol{\omega}^\mu{}_\nu = \Gamma^\mu{}_{\sigma\nu}\boldsymbol{\omega}^\sigma.
\tag{36.15}
$$

Take its exterior derivative

$$
\boldsymbol{d}\boldsymbol{\omega}^\mu{}_\nu = \frac{\partial \Gamma^\mu{}_{\sigma\nu}}{\partial x^\lambda}\boldsymbol{\omega}^\lambda \wedge \boldsymbol{\omega}^\sigma.
\tag{36.16}
$$

Contract this with vectors $\boldsymbol{e}_\alpha$ and $\boldsymbol{e}_\beta$ as follows:

$$
\begin{aligned}
\boldsymbol{d}\boldsymbol{\omega}^\mu{}_\nu(\boldsymbol{e}_\alpha, \boldsymbol{e}_\beta) =&\frac{\partial \Gamma^\mu{}_{\sigma\nu}}{\partial x^\lambda}\boldsymbol{\omega}^\lambda \wedge \boldsymbol{\omega}^\sigma(\boldsymbol{e}_\alpha, \boldsymbol{e}_\beta) \\
=&\frac{1}{2}\frac{\partial \Gamma^\mu{}_{\sigma\nu}}{\partial x^\lambda}\left[\boldsymbol{\omega}^\lambda \otimes \boldsymbol{\omega}^\sigma(\boldsymbol{e}_\alpha, \boldsymbol{e}_\beta) - \boldsymbol{\omega}^\sigma \otimes \boldsymbol{\omega}^\lambda(\boldsymbol{e}_\alpha, \boldsymbol{e}_\beta)\right] \\
=&\frac{1}{2}\frac{\partial \Gamma^\mu{}_{\sigma\nu}}{\partial x^\lambda}\left(\delta^\lambda_\alpha\delta^\sigma_\beta - \delta^\sigma_\alpha\delta^\lambda_\beta\right) \\
=&\frac{1}{2}\frac{\partial \Gamma^\mu{}_{\beta\nu}}{\partial x^\alpha} - \frac{\partial \Gamma^\mu{}_{\alpha\nu}}{\partial x^\beta}.
\end{aligned}
\tag{36.17}
$$

Now contract the second term $\omega^\mu{}_\sigma \wedge \omega^\sigma{}_\nu$ in the same way

$$\omega^\mu{}_\sigma \wedge \omega^\sigma{}_\nu (e_\alpha, e_\beta) = \frac{1}{2} \left[(\omega^\mu{}_\sigma \otimes \omega^\sigma{}_\nu - \omega^\sigma{}_\nu \otimes \omega^\mu{}_\sigma) (e_\alpha, e_\beta) \right]$$

$$= \frac{1}{2} \left[\Gamma^\mu{}_{\lambda\sigma} \Gamma^\sigma{}_{\rho\nu} \omega^\lambda(e_\alpha) \otimes \omega^\rho(e_\beta) - \Gamma^\sigma{}_{\rho\nu} \Gamma^\mu{}_{\lambda\sigma} \omega^\rho(e_\alpha) \otimes \omega^\lambda(e_\beta) \right]$$

$$= \left(\Gamma^\mu{}_{\alpha\sigma} \Gamma^\sigma{}_{\beta\nu} - \Gamma^\sigma{}_{\alpha\nu} \Gamma^\mu{}_{\beta\sigma} \right). \tag{36.18}$$

We conclude that

$$\mathcal{R}^\mu{}_\nu (e_\alpha, e_\beta) = \frac{1}{2} \left(\frac{\partial \Gamma^\mu{}_{\beta\nu}}{\partial x^\alpha} - \frac{\partial \Gamma^\mu{}_{\alpha\nu}}{\partial x^\beta} + \Gamma^\mu{}_{\alpha\sigma} \Gamma^\sigma{}_{\beta\nu} - \Gamma^\mu{}_{\beta\sigma} \Gamma^\sigma{}_{\alpha\nu} \right)$$

$$= \frac{1}{2} R^\mu{}_{\nu\alpha\beta}. \tag{36.19}$$

From the last example we see that we can write the curvature 2-form $\mathcal{R}^\mu{}_\nu$ in terms of the components of the Riemann tensor $\boldsymbol{R}$:

$$\mathcal{R}^\mu{}_\nu = R^\mu{}_{\nu|\alpha\beta|} \omega^\alpha \wedge \omega^\beta, \tag{36.20}$$

where the restriction on the sum removes the factor of $1/2$ that we had above.[6] We now have an expression for the curvature 2-form, but lack a simple method for finding the connection 1-forms $\omega^\mu{}_\alpha$ from the metric field $\boldsymbol{g}$. We now turn to this matter, since it's our reason for pursuing this formalism.

36.2 Two rules

The previous section resulted in a number of equivalent expressions designed to compute the curvature of a spacetime. They rely on using the connection 1-forms of the spacetime. So how do we find the connection 1-forms $\omega^\mu{}_\nu$? The good news is that, given the metric, they can usually be found by guesswork! More specifically, we express our $(0,2)$ metric field $\boldsymbol{g}$ in terms of basis 1-forms $\boldsymbol{g} = g_{\mu\nu} \omega^\mu \otimes \omega^\nu$ and use the exterior derivatives of the basis 1-forms to guess the connection 1-forms.

What enables this strategy are two very restrictive rules that we can derive by simply considering the action of a Leibniz product rule on a dot product

$$d(\boldsymbol{u} \cdot \boldsymbol{v}) = (d\boldsymbol{u}) \cdot \boldsymbol{v} + \boldsymbol{u} \cdot (d\boldsymbol{v}). \tag{36.22}$$

We're familiar with the important rule $\langle \omega^\mu, e_\nu \rangle = \delta^\mu{}_\nu$. The conceptual trick here, due to Cartan, is to take a field of vectors and a field of 1-forms and consider the contraction of a basis vector e_μ and a basis 1-form ω^μ, at a specific point $\mathcal{P}$. In this spirit, we write that the derivative of a point $d\mathcal{P}$ corresponds to

$$d\mathcal{P} = e_\mu \otimes \omega^\mu. \tag{36.23}$$

What does this mean and how does this work? Cartan's view was that a vector could be seen as the movement of a point and so the most primitive version of a vector would be the movement of a point $d\mathcal{P}$.

[6]The $|\alpha\beta|$ notation here should be interpreted as only allowing components in the order $\alpha\beta$ and not $\beta\alpha$. In the interests of writing memorable equations, we can also write things in terms of the $\mathcal{R}(\ )$ operator, but defining the $(2,0)$ $\mathcal{R}^{\mu\nu}$, such that $\mathcal{R}(\ ) = \frac{1}{2}(e_\mu \wedge e_\nu)\mathcal{R}^{\mu\nu}$. Then we can write the curvature operator in terms of the components of the Riemann tensor. The curvature operator $\mathcal{R}$ is given by

$$\mathcal{R}(\ ) = \frac{1}{4} (e_\mu \wedge e_\nu) R^{\mu\nu}{}_{\alpha\beta} \left(\omega^\alpha \wedge \omega^\beta \right). \tag{36.21}$$

This is not yet quite a vector. The object $d\mathcal{P}(\ ,\ )$ should be interpreted as a $(1,1)$ quantity. This means that it's a vector that hasn't yet been given a direction, and so is a sort of proto-vector that doesn't point anywhere. In fact, to point it along a direction we insert a vector $\boldsymbol{v}$ into one of its slots, to find $\langle d\mathcal{P}, \boldsymbol{v}\rangle = \boldsymbol{v}$. For this to work, we must therefore have that[7] $d\mathcal{P} = \boldsymbol{e}_\mu \otimes \boldsymbol{\omega}^\mu$.

[7]That is to say

$$\boldsymbol{e}_\mu \otimes \boldsymbol{\omega}^\mu(\boldsymbol{v}) = \boldsymbol{e}_\mu v^\mu$$
$$= \boldsymbol{v}. \tag{36.24}$$

Next, we apply the product rule to the exterior derivative of $d\mathcal{P} = \boldsymbol{e}_\mu \otimes \boldsymbol{\omega}^\mu$. Since $\mathcal{P}$ acts like a function, we must have $\boldsymbol{d}^2\mathcal{P} = 0$, and so

$$\begin{aligned} 0 =\boldsymbol{d}^2\mathcal{P} &= d\boldsymbol{e}_\mu \wedge \boldsymbol{\omega}^\mu + \boldsymbol{e}_\mu \otimes \boldsymbol{d}\boldsymbol{\omega}^\mu \\ &= \boldsymbol{e}_\mu \otimes (\boldsymbol{\omega}^\mu{}_\nu \wedge \boldsymbol{\omega}^\nu + \boldsymbol{d}\boldsymbol{\omega}^\mu). \end{aligned} \tag{36.25}$$

[8]This is also sometimes known as Cartan's first structural equation. Note that combining it with $\boldsymbol{\omega}^\nu{}_\mu = \Gamma^\nu{}_{\lambda\mu}\boldsymbol{\omega}^\lambda$ we obtain

$$\boldsymbol{d}\boldsymbol{\omega}^\mu = \Gamma^\mu{}_{\beta\nu}\boldsymbol{\omega}^\nu \wedge \boldsymbol{\omega}^\beta, \tag{36.26}$$

as we claimed in Chapter 34. We can compare this with

$$\boldsymbol{d}\boldsymbol{e}_\mu = \Gamma^\nu{}_{\beta\mu}\boldsymbol{\omega}^\beta \otimes \boldsymbol{e}_\nu, \tag{36.27}$$

which follows from eqn 36.5, and (also from Chapter 34)

$$\boldsymbol{\nabla}\boldsymbol{\omega}^\mu = -\Gamma^\mu{}_{\beta\nu}\boldsymbol{\omega}^\nu \otimes \boldsymbol{\omega}^\beta. \tag{36.28}$$

This gives us the first of our two rules: a restrictive equation known as[8] the **symmetry condition** (rule 1)

$$\boldsymbol{d}\boldsymbol{\omega}^\mu + \boldsymbol{\omega}^\mu{}_\nu \wedge \boldsymbol{\omega}^\nu = 0. \tag{36.29}$$

There is second simplification we can make to restrict the connection 1-forms. This one is based on the definition $g_{\mu\nu} = \boldsymbol{e}_\mu \cdot \boldsymbol{e}_\nu$. Insert $\boldsymbol{u} = \boldsymbol{e}_\mu$ and $\boldsymbol{v} = \boldsymbol{e}_\nu$ into the product rule (eqn 36.22) to find

$$\begin{aligned} dg_{\mu\nu} = d(\boldsymbol{e}_\mu \cdot \boldsymbol{e}_\nu) &= d\boldsymbol{e}_\mu \cdot \boldsymbol{e}_\nu + \boldsymbol{e}_\mu \cdot d\boldsymbol{e}_\nu \\ &= (\boldsymbol{\omega}^\sigma{}_\mu \otimes \boldsymbol{e}_\sigma) \cdot \boldsymbol{e}_\nu + \boldsymbol{e}_\mu \cdot (\boldsymbol{\omega}^\sigma{}_\nu \otimes \boldsymbol{e}_\sigma). \end{aligned} \tag{36.30}$$

Again using the fact that a dot is an instruction to combine vectors according to the rule $\boldsymbol{e}_\mu \cdot \boldsymbol{e}_\nu = g_{\mu\nu}$, we obtain

$$dg_{\mu\nu} = \boldsymbol{\omega}^\sigma{}_\mu g_{\sigma\nu} + g_{\mu\sigma}\boldsymbol{\omega}^\sigma{}_\nu. \tag{36.31}$$

Finally, we lower indices with the metric components and tidy up to obtain our second rule, the **compatibility condition** (rule 2)

$$dg_{\mu\nu} = \boldsymbol{\omega}_{\mu\nu} + \boldsymbol{\omega}_{\nu\mu}. \tag{36.32}$$

The compatibility condition is actually equivalent to the condition $\boldsymbol{\nabla}\boldsymbol{g} = 0$ seen in Chapter 34.

With these two key equations available to help us turn the metric into the connection 1-forms, we make a final simplifying move. We agree to work in an orthonormal basis, which means $g_{\mu\nu} = \eta_{\hat{\mu}\hat{\nu}}$ and $\boldsymbol{d}\eta_{\hat{\mu}\hat{\nu}} = 0$, so, by the compatibility condition, we have connections $\boldsymbol{\omega}_{\hat{\mu}\hat{\nu}} = -\boldsymbol{\omega}_{\hat{\nu}\hat{\mu}}$.

Having covered a lot of technical ground, let's collect together the key ideas, in the order they will be used:

Idea 1: Cartan's two rules for finding curvature 1-forms

$$0 = d\boldsymbol{\omega}^{\mu} + \boldsymbol{\omega}^{\mu}{}_{\nu} \wedge \boldsymbol{\omega}^{\nu},$$
$$d\eta_{\hat{\mu}\hat{\nu}} = 0 = \boldsymbol{\omega}_{\hat{\mu}\hat{\nu}} + \boldsymbol{\omega}_{\hat{\nu}\hat{\mu}}. \tag{36.33}$$

Idea 2: The curvature 2-form in terms of connection 1-forms

$$\mathcal{R}^{\mu}{}_{\nu} = d\boldsymbol{\omega}^{\mu}{}_{\nu} + \boldsymbol{\omega}^{\mu}{}_{\alpha} \wedge \boldsymbol{\omega}^{\alpha}{}_{\nu}. \tag{36.34}$$

Idea 3: The components of the Riemann tensor

$$\mathcal{R}^{\mu}{}_{\nu} = R^{\mu}{}_{\nu|\alpha\beta|}\boldsymbol{\omega}^{\alpha} \wedge \boldsymbol{\omega}^{\beta}. \tag{36.35}$$

As advertised, we shall find in the examples below that Idea 1 and some (rather uninspired) guesswork solves the problem of finding the connection 1-forms $\boldsymbol{\omega}^{\hat{\mu}}{}_{\hat{\nu}}$. We then immediately have access to the curvature 2-form and can extract the components of the Riemann tensor. Although, at this stage, this no doubt looks rather involved, in practice, this makes working out Riemann tensor components a fairly simple job.

36.3 Le repère mobile

Cartan intended his method to be used in an orthonormal frame of reference. Although we have been referring regularly to *the* orthonormal frame, we should perhaps have spoken of orthonormal *frames*. The idea, after all, is that, at a particular point in spacetime, we erect a local frame of reference and normalize its orthogonal axes. This gives us an orthonormal frame at each point in spacetime, which we might helpfully think of as a field of frames. The relationship between the orthonormal frames found at each point is encoded in the connection, expressed in Cartan's method as the 1-form field $\boldsymbol{\omega}^{\mu}{}_{\nu}(x)$. Cartan himself imagined moving through space watching the orthonormal frame picked out at each point seemingly change as a function of position. He called this changing frame of reference *Le repère mobile* ($\approx$ the moving frame).[9] Whichever way we think of it, the field of orthonormal frames is the starting point for using Cartan's method.[10] Once we have the orthonormal frame(s), we can find the connection 1-forms.

In what follows, we shall be working in local orthonormal frames for our computations. These are frames in which the Minkowski metric raises and lowers indices, but where the connection coefficients do not vanish. As we've stressed previously, there is a simple prescription for obtaining the orthonormal frame: we choose the one appropriate for an observer at rest in the coordinate frame at a particular point.

In general, the metric can be written as

$$\boldsymbol{g} = g_{\mu\nu}\boldsymbol{\omega}^{\mu} \otimes \boldsymbol{\omega}^{\nu}. \tag{36.36}$$

[9] It is somehow appropriate that the French reflexive verb *se repérer* means *to find one's way around*, something we are needing to do all the time in curved spaces!

[10] In reviewing this procedure here, we are just upgrading the procedure of normalizing the components of a diagonal metric, discussed in Chapter 10.

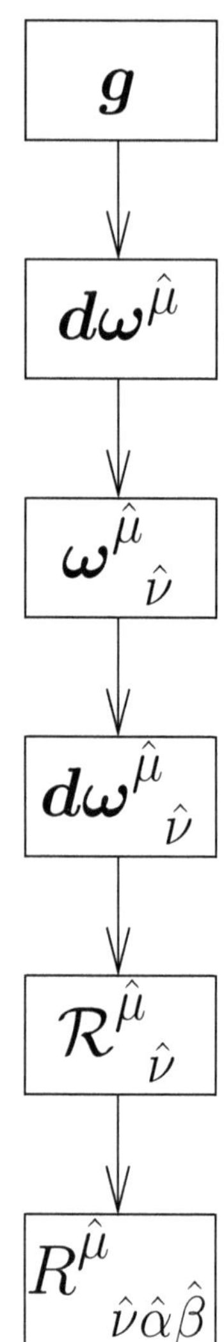

Fig. 36.1 The stages in Cartan's recipe for computing the Riemann tensor.

Should you want them, you can also extract connection coefficients using $\omega^{\hat{\nu}}{}_{\hat{\mu}} = \Gamma^{\hat{\nu}}{}_{\hat{\lambda}\hat{\mu}}\,\omega^{\hat{\lambda}}$. However, you should remember that, not only are these not required to access the curvature, but that they cannot simply be transformed between frames as tensors and that, importantly here, in non-coordinate frames we lose the symmetry in the lower indices.

From here we write $\boldsymbol{\omega}^\mu \otimes \boldsymbol{\omega}^\mu = (\boldsymbol{\omega}^\mu)^2$, so we have, for the usual case of a diagonal metric in (3+1)-dimensional spacetime,

$$\boldsymbol{g} = g_{tt}(\boldsymbol{dt})^2 + g_{rr}(\boldsymbol{dr})^2 + g_{\theta\theta}(\boldsymbol{d\theta})^2 + g_{\phi\phi}(\boldsymbol{d\phi})^2 \tag{36.37}$$

where we've arbitrarily chosen to use spherical polars. The orthonormal frame has a metric

$$\boldsymbol{g} = -(\boldsymbol{\omega}^{\hat{t}})^2 + (\boldsymbol{\omega}^{\hat{r}})^2 + (\boldsymbol{\omega}^{\hat{\theta}})^2 + (\boldsymbol{\omega}^{\hat{\phi}})^2. \tag{36.38}$$

Immediately, therefore, we must have

$$\boldsymbol{\omega}^{\hat{t}} = \sqrt{-g_{tt}}\,\boldsymbol{dt}, \quad \boldsymbol{\omega}^{\hat{r}} = \sqrt{g_{rr}}\,\boldsymbol{dr}, \quad \boldsymbol{\omega}^{\hat{\theta}} = \sqrt{g_{\theta\theta}}\,\boldsymbol{d\theta}, \quad \boldsymbol{\omega}^{\hat{\phi}} = \sqrt{g_{\phi\phi}}\,\boldsymbol{d\phi}. \tag{36.39}$$

This corresponds to the orthonormal frame with basis vectors

$$\boldsymbol{e}_{\hat{t}} = \frac{1}{\sqrt{-g_{tt}}}\frac{\partial}{\partial t}, \quad \boldsymbol{e}_{\hat{r}} = \frac{1}{\sqrt{g_{rr}}}\frac{\partial}{\partial r}, \quad \boldsymbol{e}_{\hat{\theta}} = \frac{1}{\sqrt{g_{\theta\theta}}}\frac{\partial}{\partial\theta}, \quad \boldsymbol{e}_{\hat{\phi}} = \frac{1}{\sqrt{g_{\phi\phi}}}\frac{\partial}{\partial\phi}. \tag{36.40}$$

These latter two equations allow us to read off the non-zero vielbein components

$$\begin{aligned}
(\boldsymbol{e}_t)^{\hat{t}} &= \sqrt{-g_{tt}}, & (\boldsymbol{e}_r)^{\hat{r}} &= \sqrt{g_{rr}}, & (\boldsymbol{e}_\theta)^{\hat{\theta}} &= \sqrt{g_{\theta\theta}}, & (\boldsymbol{e}_\phi)^{\hat{\phi}} &= \sqrt{g_{\phi\phi}}, \\
(\boldsymbol{e}_{\hat{t}})^{t} &= \frac{1}{\sqrt{-g_{tt}}}, & (\boldsymbol{e}_{\hat{r}})^{r} &= \frac{1}{\sqrt{g_{rr}}}, & (\boldsymbol{e}_{\hat{\theta}})^{\theta} &= \frac{1}{\sqrt{g_{\theta\theta}}}, & (\boldsymbol{e}_{\hat{\phi}})^{\phi} &= \frac{1}{\sqrt{g_{\phi\phi}}}.
\end{aligned} \tag{36.41}$$

With this, we are finally ready to try some example computations.

36.4 Example computations

Here's the recipe for how to calculate the Riemann tensor components using these ideas (see also Fig. 36.1).

Step I: Identify the orthonormal basis from the metric, by comparing the given metric to

$$ds^2 = \boldsymbol{g} = -(\boldsymbol{\omega}^{\hat{t}})^2 + (\boldsymbol{\omega}^{\hat{x}})^2 + (\boldsymbol{\omega}^{\hat{\theta}})^2 + (\boldsymbol{\omega}^{\hat{\phi}})^2. \tag{36.42}$$

In this basis, we have $dg_{\mu\nu} = d\eta_{\hat{\mu}\hat{\nu}} = 0$, so, by the compatibility condition $\omega_{\hat{\mu}\hat{\nu}} = -\omega_{\hat{\nu}\hat{\mu}}$.

Step II: Take the exterior derivatives of the basis 1-forms. We have Idea 1 (the symmetry condition $\boldsymbol{d\omega}^{\hat{\mu}} = -\boldsymbol{\omega}^{\hat{\mu}}{}_{\hat{\nu}} \wedge \boldsymbol{\omega}^{\hat{\nu}}$), and so it is an easy job to simply guess the values of $\boldsymbol{\omega}^{\hat{\mu}}{}_{\hat{\nu}}$ from the results of the derivatives.

Step III: Take the exterior derivatives of the connection 1-forms $\boldsymbol{\omega}^{\hat{\mu}}{}_{\hat{\nu}}$.

Step IV: Assemble the curvature 2-form $\mathcal{R}^{\hat{\mu}}{}_{\hat{\nu}}$ using Idea 2.

Step V: Extract components $R^{\hat{\mu}}{}_{\hat{\nu}\hat{\alpha}\hat{\beta}}$ as needed using Idea 3. Remember that you're currently working in the orthonormal frame, so it might be necessary to transform out of it using the vielbein components.

Let's discuss some examples. First we warm up with the (fairly trivial) case of flat two-dimensional space. Of course we expect the curvature tensor to vanish if the space is truly flat. We won't include time in these first few computations, meaning that indices are raised and lowered in the orthonormal frame using the metric components $\eta^{\mu\nu} = \mathrm{diag}(1,1)$, so that up and down become equivalent since this is Euclidean space.

Example 36.3

Flat space in polar coordinates has a metric line element $\boldsymbol{ds}^2 = \boldsymbol{dr}^2 + r^2\boldsymbol{d\theta}^2$. We can identify an orthonormal basis with basis 1-forms

$$\boldsymbol{\omega}^{\hat{r}} = \boldsymbol{dr}, \quad \boldsymbol{\omega}^{\hat{\theta}} = r\boldsymbol{d\theta}. \tag{36.45}$$

The exterior derivatives of each of these are

$$\boldsymbol{d\omega}^{\hat{r}} = 0, \quad \boldsymbol{d\omega}^{\hat{\theta}} = \boldsymbol{dr} \wedge \boldsymbol{d\theta} = -\frac{\boldsymbol{\omega}^{\hat{\theta}}}{r} \wedge \boldsymbol{\omega}^{\hat{r}}\ . \tag{36.46}$$

Using $\boldsymbol{d\omega}^{\hat{\theta}} = -\boldsymbol{\omega}^{\hat{\theta}}{}_{\hat{r}} \wedge \boldsymbol{\omega}^{\hat{r}}$, we extract a non-zero connection 1-form

$$\boldsymbol{\omega}^{\hat{\theta}}{}_{\hat{r}} = \frac{\boldsymbol{\omega}^{\hat{\theta}}}{r} = \boldsymbol{d\theta}. \tag{36.47}$$

Taking the exterior derivative of this connection 1-form,[11] we find

$$\boldsymbol{d\omega}^{\hat{\theta}}{}_{\hat{r}} = \boldsymbol{dd\theta} = 0. \tag{36.48}$$

As a result $\mathcal{R}^{\hat{\mu}}{}_{\hat{\nu}} = 0$ for all components. This means that the Riemann tensor vanishes confirming (as we certainly knew) that this space is flat.

We have basis vectors

$$\boldsymbol{e}_{\hat{r}} = \frac{\partial}{\partial r}, \quad \boldsymbol{e}_{\hat{\theta}} = \frac{1}{r}\frac{\partial}{\partial\theta}, \tag{36.43}$$

and vielbein components

$$\begin{aligned}(\boldsymbol{e}_r)^{\hat{r}} &= 1, & (\boldsymbol{e}_\theta)^{\hat{\theta}} &= r, \\ (\boldsymbol{e}_{\hat{r}})^r &= 1, & (\boldsymbol{e}_{\hat{\theta}})^\theta &= 1/r.\end{aligned} \tag{36.44}$$

[11]Warning: To avoid making a mistake in taking this derivative, it's best to consider the version written in terms of the 1-form $\boldsymbol{dx}^\mu$ rather than $\boldsymbol{\omega}^\mu$. Note also that since $\boldsymbol{\omega}_{\hat{i}\hat{j}} = -\boldsymbol{\omega}_{\hat{j}\hat{i}}$ (see Exercises) we have $\boldsymbol{\omega}^{\hat{\theta}}{}_{\hat{r}} = -\boldsymbol{\omega}^{\hat{r}}{}_{\hat{\theta}} = \boldsymbol{\omega}^{\hat{\theta}}/r$, so that $\boldsymbol{\omega}^{\hat{\theta}}{}_{\hat{r}} \wedge \boldsymbol{\omega}^{\hat{r}}{}_{\hat{\theta}} = 0$.

Next, we look at the space on the surface of the 2-sphere. Here we expect to see the effect of curvature.

Example 36.4

Spherical space has a metric line element $\boldsymbol{ds}^2 = a^2\boldsymbol{d\theta}^2 + a^2\sin^2\theta\boldsymbol{d\phi}^2$, with a constant. We identify orthonormal basis 1-forms

$$\boldsymbol{\omega}^{\hat{\theta}} = a\boldsymbol{d\theta}, \quad \boldsymbol{\omega}^{\hat{\phi}} = a\sin\theta\boldsymbol{d\phi}. \tag{36.51}$$

We find exterior derivatives

$$\boldsymbol{d\omega}^{\hat{\theta}} = 0, \tag{36.52}$$

and

$$\boldsymbol{d\omega}^{\hat{\phi}} = a\cos\boldsymbol{d\theta} \wedge \boldsymbol{d\phi} = -\frac{\cos\theta}{a\sin\theta}\boldsymbol{\omega}^{\hat{\phi}} \wedge \boldsymbol{\omega}^{\hat{\theta}}. \tag{36.53}$$

We read off that a non-zero connection 1-form is given by

$$\boldsymbol{\omega}^{\hat{\phi}}{}_{\hat{\theta}} = \frac{\cos\theta}{a\sin\theta}\boldsymbol{\omega}^{\hat{\phi}} = \cos\theta\boldsymbol{d\phi}, \tag{36.54}$$

and, using $\boldsymbol{\omega}_{\hat{i}\hat{j}} = -\boldsymbol{\omega}_{\hat{j}\hat{i}}$, get for free that[12]

$$\boldsymbol{\omega}^{\hat{\theta}}{}_{\hat{\phi}} = -\frac{\cos\theta}{a\sin\theta}\boldsymbol{\omega}^{\hat{\phi}} = -\cos\theta\boldsymbol{d\phi}. \tag{36.56}$$

Taking the exterior derivative of the non-zero connection 1-form yields

$$\boldsymbol{d\omega}^{\hat{\theta}}{}_{\hat{\phi}} = \sin\theta\boldsymbol{d\theta} \wedge \boldsymbol{d\phi} = \frac{1}{a^2}\boldsymbol{\omega}^{\hat{\theta}} \wedge \boldsymbol{\omega}^{\hat{\phi}}. \tag{36.57}$$

From this we can construct the non-zero components of the curvature 2-form. We find

$$\mathcal{R}^{\hat{\theta}}{}_{\hat{\phi}} = \frac{1}{a^2}\boldsymbol{\omega}^{\hat{\theta}} \wedge \boldsymbol{\omega}^{\hat{\phi}}. \tag{36.58}$$

Basis vectors:

$$\boldsymbol{e}_{\hat{\theta}} = \frac{1}{a}\frac{\partial}{\partial\theta}, \quad \boldsymbol{e}_{\hat{\phi}} = \frac{1}{a\sin\theta}\frac{\partial}{\partial\phi}, \tag{36.49}$$

and vielbein components:

$$\begin{aligned}(\boldsymbol{e}_\theta)^{\hat{\theta}} &= a, & (\boldsymbol{e}_\phi)^{\hat{\phi}} &= a\sin\theta, \\ (\boldsymbol{e}_{\hat{\theta}})^\theta &= \frac{1}{a}, & (\boldsymbol{e}_{\hat{\phi}})^\phi &= \frac{1}{a\sin\theta}.\end{aligned} \tag{36.50}$$

[12]We could at this point extract connection coefficients

$$\Gamma^{\hat{\theta}}{}_{\hat{\phi}\hat{\phi}} = -\frac{1}{a}\cot\theta,$$

$$\Gamma^{\hat{\phi}}{}_{\hat{\phi}\hat{\theta}} = \frac{1}{a}\cot\theta. \tag{36.55}$$

[13]Here the symbol $|\alpha\beta|$ orders components in the order they appear in the wedge product. That is, $|\hat\alpha\hat\beta| = \hat\theta\hat\phi$.

[14]The easiest way to proceed here is to manipulate indices using the symmetries of $\boldsymbol{R}$. In this case,

$$\frac{1}{a^2} = R^{\hat\theta}{}_{\hat\phi\hat\theta\hat\phi} = R_{\hat\theta\hat\phi\hat\theta\hat\phi}. \tag{36.62}$$

Recall that a swap within the two pairs of indices pick up a minus sign, so

$$R_{\hat\theta\hat\phi\hat\theta\hat\phi} = R_{\hat\phi\hat\theta\hat\phi\hat\theta}, \tag{36.63}$$

and finally, since up and down are equivalent for spatial indices

$$R^{\hat\phi}{}_{\hat\theta\hat\phi\hat\theta} = \frac{1}{a^2}. \tag{36.64}$$

[15]We called this Universe 1 in Chapter 15.

Basis vectors

$$e_{\hat t} = \frac{\partial}{\partial t},$$

$$e_{\hat k} = \frac{1}{a(t)}\frac{\partial}{\partial x^k}.$$

Vielbein components

$$(e_t)^{\hat t} = 1, \quad (e_k)^{\hat k} = a(t),$$
$$(e_{\hat t})^t = 1, \quad (e_{\hat k})^k = \frac{1}{a(t)}.$$

[16]Useful here is to recall that raising a lowering time indices in the orthonormal frame picks up a minus sign, so we have

$$\omega^{\hat k}{}_{\hat t} = \omega_{\hat k\hat t} = -\omega_{\hat t\hat k} = \omega^{\hat t}{}_{\hat k}. \tag{36.75}$$

Using[13]

$$\mathcal{R}^{\hat\theta}{}_{\hat\phi} = R^{\hat\theta}{}_{\hat\phi|\hat\theta\hat\phi|}\,\boldsymbol{\omega}^{\hat\theta} \wedge \boldsymbol{\omega}^{\hat\phi}, \tag{36.59}$$

we read off

$$R^{\hat\theta}{}_{\hat\phi\hat\theta\hat\phi} = \frac{1}{a^2}. \tag{36.60}$$

Note that in the coordinate frame

$$R^{\theta}{}_{\phi\theta\phi} = R^{\hat\theta}{}_{\hat\phi\hat\theta\hat\phi}(e_{\hat\theta})^{\theta}(e_{\hat\phi})^{\hat\phi}(e_\theta)^{\hat\theta}(e_\phi)^{\hat\phi}$$

$$= \frac{1}{a^2}a^2\sin^2\theta = \sin^2\theta. \tag{36.61}$$

We also find[14]

$$R^{\hat\phi}{}_{\hat\theta\hat\phi\hat\theta} = \frac{1}{a^2}, \tag{36.65}$$

which, in the coordinate frame, becomes

$$R^{\phi}{}_{\theta\phi\theta} = \frac{1}{a^2}(e_\theta)^{\hat\theta}(e_\theta)^{\hat\theta} = 1. \tag{36.66}$$

The components of the Ricci tensor calculated in the orthonormal frame, are

$$R_{\hat\phi\hat\phi} = R_{\hat\theta\hat\theta} = \frac{1}{a^2}, \tag{36.67}$$

and so, finally, the Ricci scalar is

$$R = \frac{2}{a^2}. \tag{36.68}$$

Note that, being a scalar, R is the same if we choose to evaluate it in the coordinate frame using the Ricci tensor components $R_{\mu\nu}$.

What can be done on the surface of a sphere can be done on other surfaces and there are several examples in the exercises at the end of the chapter. Let's now include time and find the curvature of some non-trivial spacetimes. First is an expanding Universe model from Chapter 18 where space is flat, but spacetime is not.

Example 36.5

The de Sitter model of the Universe[15] has a metric with line element

$$\boldsymbol{ds}^2 = -\boldsymbol{dt}^2 + a(t)^2\left(\boldsymbol{dx}^2 + \boldsymbol{dy}^2 + \boldsymbol{dz}^2\right). \tag{36.69}$$

We have orthonormal basis 1-forms

$$\boldsymbol{\omega}^{\hat t} = \boldsymbol{dt}, \quad \boldsymbol{\omega}^{\hat k} = a(t)\boldsymbol{dx}^k. \tag{36.70}$$

First we note

$$\boldsymbol{d\omega}^{\hat t} = 0, \tag{36.71}$$

and we use $\boldsymbol{d\omega}^{\hat t} = -\boldsymbol{\omega}^{\hat t}{}_{\hat k} \wedge \boldsymbol{\omega}^{\hat k}$, from which we guess that $\boldsymbol{\omega}^{\hat t}{}_{\hat k} \propto \boldsymbol{\omega}^{\hat k}$. Then

$$\boldsymbol{d\omega}^{\hat k} = \frac{\dot a}{a}\boldsymbol{\omega}^{\hat t} \wedge \boldsymbol{\omega}^{\hat k}. \tag{36.72}$$

Writing

$$\boldsymbol{d\omega}^{\hat k} = -\boldsymbol{\omega}^{\hat k}{}_{\hat t} \wedge \boldsymbol{\omega}^{\hat t} - \boldsymbol{\omega}^{\hat k}{}_{\hat l} \wedge \boldsymbol{\omega}^{\hat l}, \tag{36.73}$$

we guess $\boldsymbol{\omega}^{\hat k}{}_{\hat t} = \frac{\dot a}{a}\boldsymbol{\omega}^{\hat k}$ and $\boldsymbol{\omega}^{\hat k}{}_{\hat l} = 0$. Now to compute the curvature 2-form

$$\mathcal{R}^{\hat t}{}_{\hat k} = \boldsymbol{d\omega}^{\hat t}{}_{\hat k} = \boldsymbol{d}\left(\dot a\,\boldsymbol{dx}^k\right) = \frac{\ddot a}{a}\boldsymbol{\omega}^{\hat t} \wedge \boldsymbol{\omega}^{\hat k}. \tag{36.74}$$

We must also not forget that there are other components in this case. We have[16]

$$\mathcal{R}^{\hat k}{}_{\hat l} = \boldsymbol{d\omega}^{\hat k}{}_{\hat l} + \boldsymbol{\omega}^{\hat k}{}_{\hat t} \wedge \boldsymbol{\omega}^{\hat t}{}_{\hat l}$$

$$= 0 + \left(\frac{\dot a}{a}\right)^2 \boldsymbol{\omega}^{\hat k} \wedge \boldsymbol{\omega}^{\hat l}. \tag{36.76}$$

We obtain the resulting Riemann tensor components[17]

$$R^{\hat{t}}{}_{\hat{k}\hat{t}\hat{k}} = \frac{\ddot{a}}{a}, \qquad R^{\hat{k}}{}_{\hat{l}\hat{k}\hat{l}} = \left(\frac{\dot{a}}{a}\right)^2, \tag{36.77}$$

where we have temporarily suspended the Einstein summation convention. The Ricci tensor has $\hat{0}\hat{0}$th component

$$R_{\hat{t}\hat{t}} = \sum_{k=1}^{3} R^{k}{}_{\hat{t}\hat{k}\hat{t}} = -3\frac{\ddot{a}}{a}. \tag{36.78}$$

To find the $\hat{i}\hat{i}$th components we have $R_{\hat{i}\hat{i}} = R^{\hat{t}}{}_{\hat{i}\hat{t}\hat{i}} + \sum_{j=1}^{3} R^{\hat{j}}{}_{\hat{i}\hat{j}\hat{i}}$. However, the component with $\hat{i} = \hat{j}$ is $R^{\hat{i}}{}_{\hat{i}\hat{i}\hat{i}} = 0$ because of the symmetries of the Riemann tensor. As a result we have

$$R_{\hat{i}\hat{i}} = \frac{\ddot{a}}{a} + 2\left(\frac{\dot{a}}{a}\right)^2. \tag{36.79}$$

The Ricci scalar is therefore

$$R = 3\frac{\ddot{a}}{a} + 3\left[\frac{\ddot{a}}{a} + 2\left(\frac{\dot{a}}{a}\right)^2\right] = 6\left(\frac{\ddot{a}}{a} + \frac{\dot{a}^2}{a^2}\right). \tag{36.80}$$

In this case, it's meaningful to extract the components of the Einstein tensor $G_{\mu\nu} = R_{\mu\nu} - \frac{1}{2}g_{\mu\nu}R$. These are

$$G_{\hat{t}\hat{t}} = 3\left(\frac{\dot{a}}{a}\right)^2, \qquad G_{\hat{i}\hat{i}} = -2\frac{\ddot{a}}{a} - \frac{\dot{a}}{a}. \tag{36.81}$$

Finally, we can compute the curvature properties of the Schwarzschild metric from Chapter 21.

Example 36.6

Written in its general form, the spherically symmetric metric line element looks like

$$\boldsymbol{ds}^2 = -\mathrm{e}^{2\Phi}\boldsymbol{dt}^2 + \mathrm{e}^{2\Lambda}\boldsymbol{dr}^2 + r^2\left(\boldsymbol{d\theta}^2 + \sin^2\theta\boldsymbol{d\phi}^2\right), \tag{36.82}$$

where Φ and Λ are functions of r only. We identify orthonormal basis 1-forms

$$\boldsymbol{\omega}^{\hat{t}} = \mathrm{e}^{\Phi}\boldsymbol{dt}, \quad \boldsymbol{\omega}^{\hat{r}} = \mathrm{e}^{\Lambda}\boldsymbol{dr}, \quad \boldsymbol{\omega}^{\hat{\theta}} = r\boldsymbol{d\theta}, \quad \boldsymbol{\omega}^{\hat{\phi}} = r\sin\theta\boldsymbol{d\phi}. \tag{36.83}$$

The connections are

$$\boldsymbol{\omega}^{\hat{t}}{}_{\hat{r}} = \Phi'\mathrm{e}^{\Phi}\mathrm{e}^{-\Lambda}\boldsymbol{dt}, \quad \boldsymbol{\omega}^{\hat{\theta}}{}_{\hat{r}} = \mathrm{e}^{-\Lambda}\boldsymbol{d\theta},$$
$$\boldsymbol{\omega}^{\hat{\phi}}{}_{\hat{r}} = \sin\theta\,\mathrm{e}^{-\Lambda}\boldsymbol{d\phi}, \quad \boldsymbol{\omega}^{\hat{\phi}}{}_{\hat{\theta}} = \cos\theta\boldsymbol{d\phi}, \tag{36.84}$$

where dashes denote derivatives with respect to r. The exterior derivatives of the connections can be computed and give

$$\boldsymbol{d\omega}^{\hat{t}}{}_{\hat{r}} = -\left[\Phi'' + (\Phi')^2 - \Phi'\Lambda'\right]\mathrm{e}^{-2\Lambda}\boldsymbol{\omega}^{\hat{t}} \wedge \boldsymbol{\omega}^{\hat{\theta}},$$

$$\boldsymbol{d\omega}^{\hat{\theta}}{}_{\hat{r}} = \frac{\Lambda'\mathrm{e}^{-2\Lambda}}{r}\boldsymbol{\omega}^{\hat{\theta}} \wedge \boldsymbol{\omega}^{\hat{r}},$$

$$\boldsymbol{d\omega}^{\hat{\phi}}{}_{\hat{r}} = \left(-\frac{\mathrm{e}^{-\Lambda}\cos\theta}{r^2\sin\theta} + \frac{\Lambda\mathrm{e}^{-2\Lambda}}{r}\right)\boldsymbol{\omega}^{\hat{\phi}} \wedge \boldsymbol{\omega}^{\hat{\theta}},$$

$$\boldsymbol{d\omega}^{\hat{\phi}}{}_{\hat{\theta}} = \frac{1}{r^2}\boldsymbol{\omega}^{\hat{\phi}} \wedge \boldsymbol{\omega}^{\hat{\theta}}. \tag{36.85}$$

Next, we extract the non-zero components of the curvature 2-form

$$\mathcal{R}^{\hat{t}}{}_{\hat{r}} = \boldsymbol{d\omega}^{\hat{t}}{}_{\hat{r}},$$
$$\mathcal{R}^{\hat{t}}{}_{\hat{\theta}} = \boldsymbol{\omega}^{\hat{t}}{}_{\hat{r}} \wedge \boldsymbol{\omega}^{\hat{r}}{}_{\hat{\theta}},$$
$$\mathcal{R}^{\hat{t}}{}_{\hat{\phi}} = \boldsymbol{\omega}^{\hat{t}}{}_{\hat{r}} \wedge \boldsymbol{\omega}^{\hat{r}}{}_{\hat{\phi}},$$
$$\mathcal{R}^{\hat{\theta}}{}_{\hat{\phi}} = \boldsymbol{d\omega}^{\hat{\theta}}{}_{\hat{\phi}} + \boldsymbol{\omega}^{\hat{\theta}}{}_{\hat{r}} \wedge \boldsymbol{\omega}^{\hat{r}}{}_{\hat{\phi}},$$
$$\mathcal{R}^{\hat{\phi}}{}_{\hat{r}} = \boldsymbol{d\omega}^{\hat{\phi}}{}_{\hat{r}} + \boldsymbol{\omega}^{\hat{\phi}}{}_{\hat{\theta}} \wedge \boldsymbol{\omega}^{\hat{\theta}}{}_{\hat{r}},$$
$$\mathcal{R}^{\hat{r}}{}_{\hat{\theta}} = \boldsymbol{d\omega}^{\hat{r}}{}_{\hat{\theta}} + \boldsymbol{\omega}^{\hat{r}}{}_{\hat{\phi}} \wedge \boldsymbol{\omega}^{\hat{\phi}}{}_{\hat{\theta}}. \tag{36.86}$$

[17] We can therefore say $R^{\hat{k}}{}_{\hat{t}\hat{k}\hat{t}} = R_{\hat{k}\hat{t}\hat{k}\hat{t}} = R_{\hat{t}\hat{k}\hat{t}\hat{k}} = -R^{\hat{t}}{}_{\hat{k}\hat{t}\hat{k}}$.

Basis vectors

$$\boldsymbol{e}_{\hat{t}} = \frac{1}{\mathrm{e}^{-\Phi}}\frac{\partial}{\partial t},$$
$$\boldsymbol{e}_{\hat{r}} = \frac{1}{\mathrm{e}^{\Lambda}}\frac{\partial}{\partial r},$$
$$\boldsymbol{e}_{\hat{\theta}} = \frac{1}{r}\frac{\partial}{\partial\theta},$$
$$\boldsymbol{e}_{\hat{\phi}} = \frac{1}{r\sin\theta}\frac{\partial}{\partial\phi}.$$

Vielbein components

$$(\boldsymbol{e}_t)^{\hat{t}} = \mathrm{e}^{-\phi}, \qquad (\boldsymbol{e}_r)^{\hat{r}} = \mathrm{e}^{\Lambda},$$
$$(\boldsymbol{e}_\theta)^{\hat{\theta}} = r, \qquad (\boldsymbol{e}_\phi)^{\hat{\phi}} = r\sin\theta,$$
$$(\boldsymbol{e}_{\hat{t}})^t = \mathrm{e}^{\Phi}, \qquad (\boldsymbol{e}_{\hat{r}})^r = \mathrm{e}^{-\Lambda},$$
$$(\boldsymbol{e}_{\hat{\theta}})^\theta = \frac{1}{r}, \qquad (\boldsymbol{e}_{\hat{\phi}})^\phi = \frac{1}{r\sin\theta}.$$

[18]We write the expressions in this form for simplicity, where the index is raised by the appropriate components of the Minkowski tensor, since we're working in the orthonormal frame.

We find[18]

$$\mathcal{R}^{\hat{t}\hat{r}} = E\boldsymbol{\omega}^{\hat{t}} \wedge \boldsymbol{\omega}^{\hat{r}}, \quad \mathcal{R}^{\hat{t}\hat{\theta}} = \bar{E}\boldsymbol{\omega}^{\hat{t}} \wedge \boldsymbol{\omega}^{\hat{\theta}}, \quad \mathcal{R}^{\hat{t}\hat{\phi}} = \bar{E}\boldsymbol{\omega}^{\hat{t}} \wedge \boldsymbol{\omega}^{\hat{\phi}},$$
$$\mathcal{R}^{\hat{\theta}\hat{\phi}} = F\boldsymbol{\omega}^{\hat{\theta}} \wedge \boldsymbol{\omega}^{\hat{\phi}}, \quad \mathcal{R}^{\hat{\phi}\hat{r}} = \bar{F}\boldsymbol{\omega}^{\hat{\phi}} \wedge \boldsymbol{\omega}^{\hat{r}}, \quad \mathcal{R}^{\hat{r}\hat{\theta}} = \bar{F}\boldsymbol{\omega}^{\hat{r}} \wedge \boldsymbol{\omega}^{\hat{\theta}}, \tag{36.87}$$

where

$$E = -\,\mathrm{e}^{-2\Lambda}\left(\Phi'' + \Phi'^2 - \Phi'\Lambda'\right),$$
$$\bar{E} = -\,\frac{1}{r}\mathrm{e}^{-2\Lambda}\Phi',$$
$$F = \frac{1}{r^2}\left(1 - \mathrm{e}^{-2\Lambda}\right),$$
$$\bar{F} = \frac{1}{r}\mathrm{e}^{-2\Lambda}\Lambda'. \tag{36.88}$$

[19]Apply the rule

$$\mathcal{R}^{\hat{\mu}\hat{\nu}} = R^{\hat{\mu}\hat{\nu}}{}_{|\hat{\alpha}\hat{\beta}|}\,\tilde{\boldsymbol{\omega}}^{\alpha} \wedge \tilde{\boldsymbol{\omega}}^{\beta}.$$

The only non-trivial component is

$$\mathcal{R}^{\hat{\phi}\hat{r}} = -\bar{F}\tilde{\boldsymbol{\omega}}^{\hat{r}} \wedge \tilde{\boldsymbol{\omega}}^{\hat{\phi}} = R^{\hat{\phi}\hat{r}}{}_{\hat{r}\hat{\phi}}\tilde{\boldsymbol{\omega}}^{\hat{r}} \wedge \tilde{\boldsymbol{\omega}}^{\hat{\phi}},$$

where we are forced to rearrange the components in order to obey the instruction $|\hat{\alpha}\hat{\beta}|$.

[20]The simplest way to do this is to use the rule from Chapter 21 that

$$G^0{}_0 = -\left(R^{12}{}_{12} + R^{23}{}_{23} + R^{31}{}_{31}\right),$$
$$G^1{}_1 = -\left(R^{02}{}_{02} + R^{03}{}_{03} + R^{23}{}_{23}\right),$$
$$G^0{}_1 = R^{02}{}_{12} + R^{03}{}_{13},$$
$$G^1{}_2 = R^{10}{}_{20} + R^{13}{}_{23},$$

where other components can be found using cyclic permutations.

We can extract the components of the Riemann tensor[19]

$$R^{\hat{t}\hat{r}}{}_{\hat{t}\hat{r}} = E, \quad R^{\hat{t}\hat{\theta}}{}_{\hat{t}\hat{\theta}} = \bar{E}, \quad R^{\hat{t}\hat{\phi}}{}_{\hat{t}\hat{\phi}} = \bar{E},$$
$$R^{\hat{\theta}\hat{\phi}}{}_{\hat{\theta}\hat{\phi}} = F, \quad R^{\hat{\phi}\hat{r}}{}_{\hat{r}\hat{\phi}} = -\bar{F}, \quad R^{\hat{r}\hat{\theta}}{}_{\hat{r}\hat{\theta}} = \bar{F}. \tag{36.89}$$

This gives us the non-zero components of the Einstein equation[20]

$$G^{\hat{t}}{}_{\hat{t}} = -G_{\hat{t}\hat{t}} = -\left(F + 2\bar{F}\right),$$
$$G^{\hat{r}}{}_{\hat{r}} = G_{\hat{r}\hat{r}} = -\left(F + 2\bar{E}\right),$$
$$G^{\hat{\theta}}{}_{\hat{\theta}} = G^{\hat{\phi}}{}_{\hat{\phi}} = -\left(E + \bar{E} + \bar{F}\right). \tag{36.90}$$

There are several more examples to try in the exercises.

We now have an efficient method to extract curvature of a spacetime from its metric. In the final chapters of this part of the book, we discuss some different aspects of geometric techniques.

Chapter summary

- To compute the Riemann tensor, work in the orthonormal frame and compute the curvature 1-forms, which have the properties

$$\boldsymbol{d\omega}^{\mu} + \boldsymbol{\omega}^{\mu}{}_{\nu} \wedge \boldsymbol{\omega}^{\nu} = 0, \tag{36.91}$$

and $\boldsymbol{\omega}_{\hat{\mu}\hat{\nu}} = -\boldsymbol{\omega}_{\hat{\nu}\hat{\mu}}$.

- The curvature 2-form is given by

$$\mathcal{R}^{\mu}{}_{\nu} = \boldsymbol{d\omega}^{\mu}{}_{\nu} + \boldsymbol{\omega}^{\mu}{}_{\alpha} \wedge \boldsymbol{\omega}^{\alpha}{}_{\nu}. \tag{36.92}$$

- The components of the Riemann tensor are related to the curvature 2-form via

$$\mathcal{R}^{\mu}{}_{\nu} = R^{\mu}{}_{\nu|\alpha\beta|}\boldsymbol{\omega}^{\alpha} \wedge \boldsymbol{\omega}^{\beta}. \tag{36.93}$$

Exercises

(36.1) Show that, when working in the orthonormal frame,
(a) $\boldsymbol{\omega}^{\hat{t}}{}_{\hat{i}} = \boldsymbol{\omega}^{\hat{i}}{}_{\hat{t}}$.
(b) $\boldsymbol{\omega}^{\hat{i}}{}_{\hat{j}} = -\boldsymbol{\omega}^{\hat{j}}{}_{\hat{i}}$.

(36.2) Read off the connection coefficients in an orthonormal frame for flat space described using cylindrical polar coordinates.

(36.3) A surface has a metric with line element
$$ds^2 = d\sigma^2 + r(\sigma)^2 d\phi^2. \qquad (36.94)$$

Compute $\mathcal{R}^{\hat{\sigma}}{}_{\hat{\phi}}$.

(36.4) The parabolic surface has a metric with line element
$$ds^2 = (1 + a^2 r^2)dr^2 + r^2 d\theta^2. \qquad (36.95)$$

Compute (i) the components of $\boldsymbol{R}$, (ii) the components of the Ricci tensor, and (iii) the Ricci scalar.

(36.5) The torus metric has a line element
$$ds^2 = (c + a\cos v)^2 du^2 + a^2 dv^2. \qquad (36.96)$$

Compute: (a) the components of $\boldsymbol{R}$, (b) the components of the Ricci tensor, and (c) the Ricci scalar.

(36.6) The Poincaré half-plane has line element
$$ds^2 = \frac{dx^2}{r^2} + \frac{dr^2}{r^2}. \qquad (36.97)$$

Compute (a) the components of $\boldsymbol{R}$, (b) the components of the Ricci tensor, and (c) the Ricci scalar.

(36.7) Consider the expanding universe of the Friedman model, with metric line element
$$ds^2 = -dt^2 + a(t)^2 \left[d\chi^2 + \sin^2 \chi (d\theta^2 + \sin^2 \theta d\phi^2) \right]. \qquad (36.98)$$
Compute (a) $\mathcal{R}^{\hat{\mu}}{}_{\hat{\nu}}$, (b) the Ricci scalar and, (c) the components of the Einstein tensor $\boldsymbol{G}$ in the orthonormal frame.

(36.8) Consider a time-evolving star with a spherically symmetric, time-dependent metric line element
$$ds^2 = -e^{2\Phi} dT^2 + e^{2\Lambda} dR^2 + r^2 (d\theta^2 + \sin^2 \theta d\phi^2), \qquad (36.99)$$
where $\Phi(R,T)$, $\Lambda(R,T)$ and $r(R,T)$ are all functions of the coordinates R and T.
(a) Show that
$$\boldsymbol{\omega}^{\hat{R}}{}_{\hat{T}} = \Phi' e^{\Phi} e^{-\Lambda} dT + \dot{\Lambda} e^{\Lambda} e^{-\Phi} dR, \qquad (36.100)$$

where dashes denote derivatives with respect to R and dots denote derivatives with respect to T.
(b) Show that the curvature 2-forms are given by
$$
\begin{aligned}
\mathcal{R}^{\hat{T}}{}_{\hat{R}} &= E\boldsymbol{\omega}^{\hat{T}} \wedge \boldsymbol{\omega}^{\hat{R}}, \\
\mathcal{R}^{\hat{T}}{}_{\hat{\theta}} &= \bar{E}\boldsymbol{\omega}^{\hat{T}} \wedge \boldsymbol{\omega}^{\hat{\theta}} + H\boldsymbol{\omega}^{\hat{R}} \wedge \boldsymbol{\omega}^{\hat{\theta}}, \\
\mathcal{R}^{\hat{T}}{}_{\hat{\phi}} &= \bar{E}\boldsymbol{\omega}^{\hat{T}} \wedge \boldsymbol{\omega}^{\hat{\phi}} + H\boldsymbol{\omega}^{\hat{R}} \wedge \boldsymbol{\omega}^{\hat{\phi}}, \\
\mathcal{R}^{\hat{\theta}}{}_{\hat{\phi}} &= F\boldsymbol{\omega}^{\hat{\theta}} \wedge \boldsymbol{\omega}^{\hat{\phi}}, \\
\mathcal{R}^{\hat{R}}{}_{\hat{\theta}} &= \bar{F}\boldsymbol{\omega}^{\hat{R}} \wedge \boldsymbol{\omega}^{\hat{\theta}} - H\boldsymbol{\omega}^{\hat{T}} \wedge \boldsymbol{\omega}^{\hat{\theta}}, \\
\mathcal{R}^{\hat{R}}{}_{\hat{\phi}} &= \bar{F}\boldsymbol{\omega}^{\hat{R}} \wedge \boldsymbol{\omega}^{\hat{\phi}} - H\boldsymbol{\omega}^{\hat{T}} \wedge \boldsymbol{\omega}^{\hat{\phi}}, \qquad (36.101)
\end{aligned}
$$

where
$$
\begin{aligned}
E &= e^{-2\Phi}(\ddot{\Lambda} + \dot{\Lambda}^2 - \dot{\Lambda}\dot{\Phi}) - e^{-2\Lambda}(\Phi'' + \Phi'^2 - \Phi'\Lambda'), \\
\bar{E} &= \frac{1}{r}e^{-2\Phi}(\ddot{r} - \dot{r}\dot{\Phi}) - \frac{1}{r}e^{-2\Lambda}r'\Phi', \\
H &= \frac{1}{r}e^{-\Phi}e^{-\Lambda}(\dot{r}' - \dot{r}\Phi' - r'\dot{\Lambda}), \\
F &= \frac{1}{r^2}(1 - (r')^2 e^{-2\Lambda} + \dot{r}^2 e^{-2\Phi}), \\
\bar{F} &= \frac{1}{r}e^{-2\Phi}\dot{r}\dot{\Lambda} + \frac{1}{r}e^{-2\Lambda}(r'\Lambda' - r''). \qquad (36.102)
\end{aligned}
$$

See Misner, Thorne, and Wheeler Chapters 14, 26, and 32 for further details, including the use of this metric in studying stellar pulsations and gravitational collapse.

37

Duality and the volume form

> The material in this chapter is useful in understanding the geometrical interpretation of electromagnetism (Chapter 42) and the Bianchi identity (Chapter 43).

[1] In Chapter 32 we saw that an antisymmetric tensor, formed from N vectors wedged together is known as a N-vector. It is an antisymmetric $(N, 0)$ tensor.

[2] Sir William V. D. Hodge (1903–1975). The influential mathematician Sir Michael Atiyah was one of his doctoral students. Hodge's three-volume work *Methods of Algebraic Geometry*, cowritten with Daniel Pedoe (1910–1998), freely used the component notation referred to by Cartan in the quotation above.

The utility of the absolute differential calculus of Ricci and Levi-Civita must be tempered by an avoidance of excessively formal calculations, where the debauch of indices disguises an often very simple geometrical reality.
Élie Cartan (1869–1951)

It turns out to be very useful to be able to **map** between different tensors that encode the same information. By this we mean that we start with one sort of tensor and uniquely obtain another that expresses related physical content. In this section, we shall see how to map between a p-form, which is an $(0, p)$ antisymmetric tensor that can built from 1-forms using the wedge product, and an antisymmetric q-vector $(q, 0)$, built from vectors, again combined using the wedge product.[1] This mapping is known as a **duality**, and can be carried out using an operation known as the **Hodge star**.[2] Our rather formal discussion in this chapter provides a general method of mapping between two different sorts of tensor. The payoff from this formalism will be a new insight into how volumes can be encoded in tensor algebra.

37.1 Motivation: 2-forms and flux

Let's consider a 2-form in three-dimensional Euclidean space with components

$$\tilde{\boldsymbol{F}}(\,,) = F^1(\boldsymbol{dy} \wedge \boldsymbol{dz}) + F^2(\boldsymbol{dz} \wedge \boldsymbol{dx}) + F^3(\boldsymbol{dx} \wedge \boldsymbol{dy}). \qquad (37.1)$$

By virtue of working in space with $n = 3$ dimensions, this object has three components. Now let's extract the components and use them to form a vector $\boldsymbol{u}$, which we shall write

$$\boldsymbol{u}(\,) = F^1\boldsymbol{e}_x + F^2\boldsymbol{e}_y + F^3\boldsymbol{e}_z. \qquad (37.2)$$

What is the relationship between $\tilde{\boldsymbol{F}}$ and $\boldsymbol{u}$? We shall see shortly that these objects are *dual* to each other and can be related via the Hodge star operation, written as $\star\tilde{\boldsymbol{F}} = \boldsymbol{u}$. Physically, we can see how they are related by using the notion of flux.

We shall consider a fluid flowing throughout a volume with velocity $\boldsymbol{u}$. The **flux** of the fluid is the amount of fluid flowing through an area

A in unit time. If the area is a parallelogram with sides given by vectors $\boldsymbol{a}$ and $\boldsymbol{b}$, then shall show in Example 37.1 that

$$(\text{Flux of } \boldsymbol{u} \text{ through } A) = \tilde{\boldsymbol{F}}(\boldsymbol{a}, \boldsymbol{b}). \tag{37.3}$$

From our discussion in Example 32.6 of Chapter 32, it follows that we can determine the flux by computing $\tilde{\boldsymbol{F}}(\boldsymbol{a}, \boldsymbol{b}) = \langle \tilde{\boldsymbol{F}}, \boldsymbol{a} \wedge \boldsymbol{b} \rangle$, that is, the inner product of the 2-form $\boldsymbol{F}$ and the bivector $\boldsymbol{a} \wedge \boldsymbol{b}$. However, in the next example, we show that the same result can be found using two related objects: the vector $\boldsymbol{u}$ constructed from $\tilde{\boldsymbol{F}}$ and a 1-form $\tilde{\boldsymbol{S}}$ constructed from the bivector $\boldsymbol{a} \wedge \boldsymbol{b}$.

Example 37.1

Suppose the only non-zero component of $\boldsymbol{u}$ is F^3. Then, in three dimensions, the amount of fluid per unit time passing through A will be given by the projection of $\boldsymbol{u}$ onto the area: $F^3 \boldsymbol{e}_z \cdot (\boldsymbol{a} \times \boldsymbol{b})$, where we have used the conventional cross product for three-dimensional space to express the area spanned by the $\boldsymbol{a}$ and $\boldsymbol{b}$. Expanding components we find this is equal to $F^3(a^x b^y - b^x a^y) = F^3 \boldsymbol{dx}(\boldsymbol{a}) \wedge \boldsymbol{dy}(\boldsymbol{b})$. We now let F^1 and F^2 take non-zero values and repeat the argument for these components to find that the total flux is given by

$$F^1 \boldsymbol{dy}(\boldsymbol{a}) \wedge \boldsymbol{dz}(\boldsymbol{b}) + F^2 \boldsymbol{dz}(\boldsymbol{a}) \wedge \boldsymbol{dx}(\boldsymbol{b}) + F^3 \boldsymbol{dx}(\boldsymbol{a}) \wedge \boldsymbol{dy}(\boldsymbol{b}) = \tilde{\boldsymbol{F}}(\boldsymbol{a}, \boldsymbol{b}). \tag{37.4}$$

This proves eqn 37.3.

It's useful to note that the bivector $\boldsymbol{a} \wedge \boldsymbol{b}$ has non-zero components $(\boldsymbol{a} \wedge \boldsymbol{b})^{yx} = a^y b^z - b^y a^z$, $(\boldsymbol{a} \wedge \boldsymbol{b})^{zx} = a^z b^x - b^z a^x$ and $(\boldsymbol{a} \wedge \boldsymbol{b})^{xy} = a^x b^y - b^x a^y$. We notice that if these were to be arranged to make the components of a 1-form, we would have

$$\tilde{\boldsymbol{S}} = (a^y b^z - b^y a^z)\boldsymbol{\omega}^x + (a^z b^x - b^z a^x)\boldsymbol{\omega}^y + (a^x b^y - b^x a^y)\boldsymbol{\omega}^z. \tag{37.5}$$

The flux can then also be expressed as $\langle \tilde{\boldsymbol{S}}, \boldsymbol{u} \rangle$.

This example demonstrates that, in addition to the 2-form $\tilde{\boldsymbol{F}}$, there is a vector $\boldsymbol{u}$ that encodes the same information via the physical notion of a flux. The vector $\boldsymbol{u}$ is the dual of $\tilde{\boldsymbol{F}}$. Similarly, the bivector $\boldsymbol{a} \wedge \boldsymbol{b}$ has a dual 1-form $\tilde{\boldsymbol{S}}$. We can obtain the dual of an object using the Hodge star operation; the object outputted by this operation depends on the dimensionality of the space in which we're working.[3] We often work in $n = 3$-dimensional space, in which the Hodge star takes a q-form and outputs a $(3 - q)$-vector. As a result, the Hodge star of a 2-form is a vector. Similarly, in three dimensions the Hodge star converts a p-vector into a $(3 - p)$-form, so that[4] the bivector is dual to a 1-form.

In the next section, we describe the procedure required to compute a dual and its components. Although fairly straightforward, this is a slightly tedious business when $n \neq 3$, and so this section can be skipped on a first reading. (Some useful rules are collected in the margin at the end of the next section.)

[3] In general, the Hodge star takes a q-form and outputs an $(n - q)$-vector. It also takes a p-vector and outputs an $(n - p)$ form.

[4] This property of three-dimensional Euclidean space is what allows our conventional vector calculus to work as neatly as it does. In particular, it explains the existence of the cross product $\boldsymbol{a} \times \boldsymbol{b}$ which is the vector that is dual to a bivector $\boldsymbol{a} \wedge \boldsymbol{b}$.

37.2 Hodge star operation

The duality transformation, also known as the Hodge star operation, is carried out using a special tensor called the **volume form**. This

[5]Recall that the orthonormal basis is an example of a local frame in which measurements can be made. For (3+1)-dimensional spacetime it has metric tensor

$$g = -(\boldsymbol{\omega}^{\hat{0}}) \otimes (\boldsymbol{\omega}^{\hat{0}}) + (\boldsymbol{\omega}^{\hat{1}}) \otimes (\boldsymbol{\omega}^{\hat{1}}) + (\boldsymbol{\omega}^{\hat{2}}) \otimes (\boldsymbol{\omega}^{\hat{2}}) + (\boldsymbol{\omega}^{\hat{3}}) \otimes (\boldsymbol{\omega}^{\hat{3}}),$$

which has signature $(-+++)$.

[6]It's a candidate at this stage as we don't yet have a means of determining the ordering of the basis 1-forms. This gives rise to a sign ambiguity that we resolve in the next Example.

[7]We also have the rule, if we have M indices, that

$$\varepsilon_{ij...k}\varepsilon^{lm...r} = M!\delta^{lm...r}_{ij...k}. \tag{37.8}$$

is another example of a p-form, this one built from the basis forms of the space in which we're working. Working in the four-dimensional orthonormal basis[5] with basis 1-forms $\boldsymbol{\omega}^{\hat{0}}, ..., \boldsymbol{\omega}^{\hat{3}}$. We then construct a candidate volume 4-form $\tilde{\boldsymbol{\omega}}$ via[6]

$$\tilde{\boldsymbol{\omega}} = \boldsymbol{\omega}^{\hat{\mu}} \wedge \boldsymbol{\omega}^{\hat{\nu}} \wedge \boldsymbol{\omega}^{\hat{\alpha}} \wedge \boldsymbol{\omega}^{\hat{\beta}}. \tag{37.6}$$

In the orthonormal basis, the volume form has components $\omega_{\hat{\mu}\hat{\nu}\hat{\alpha}\hat{\beta}} = \varepsilon_{\mu\nu\alpha\beta}$, where $\varepsilon_{\mu\nu\alpha\beta}$ is the four-dimensional **Levi-Civita symbol**.

Example 37.2

We can treat the Levi-Civita symbol $\varepsilon_{\mu\nu\alpha\beta}$ as being the components of a **Levi-Civita tensor** $\boldsymbol{\varepsilon}(\ ,\ ,\ ,\ )$. The tensor is antisymmetric in all of its slots, so it changes sign when any two vectors exchange their slots. In Minkowski spacetime, we fix the sign by saying

$$\boldsymbol{\varepsilon}(\boldsymbol{e}_0, \boldsymbol{e}_1, \boldsymbol{e}_2, \boldsymbol{e}_3) = \varepsilon_{0123} = 1. \tag{37.7}$$

The components are then[7]

$$\varepsilon_{\mu\nu\alpha\beta} = \begin{cases} 0 \text{ unless } \mu, \nu, \alpha, \beta \text{ are all different,} \\ +1 \text{ for even permutations of } 0, 1, 2, 3, \\ -1 \text{ for odd permutation.} \end{cases} \tag{37.9}$$

It's important to note that, if we're treating $\varepsilon_{\mu\nu\alpha\beta}$ as the components of a tensor, as we do here, then in Minkowski spacetime the all-up version has the property that

$$\varepsilon^{0123} = -\varepsilon_{0123} = -1. \tag{37.10}$$

In terms of components we can then, using the bar notation of the previous chapter, fix the sign by writing the volume 4-form as

$$\tilde{\boldsymbol{\omega}} = \frac{1}{4!}\varepsilon_{\mu\nu\alpha\beta}\boldsymbol{\omega}^{\hat{\mu}} \wedge \boldsymbol{\omega}^{\hat{\nu}} \wedge \boldsymbol{\omega}^{\hat{\alpha}} \wedge \boldsymbol{\omega}^{\hat{\beta}} = \varepsilon_{|\mu\nu\alpha\beta|}\boldsymbol{\omega}^{\hat{\mu}} \wedge \boldsymbol{\omega}^{\hat{\nu}} \wedge \boldsymbol{\omega}^{\hat{\alpha}} \wedge \boldsymbol{\omega}^{\hat{\beta}}$$
$$= \boldsymbol{\omega}^{\hat{0}} \wedge \boldsymbol{\omega}^{\hat{1}} \wedge \boldsymbol{\omega}^{\hat{2}} \wedge \boldsymbol{\omega}^{\hat{3}}. \tag{37.11}$$

If, instead of an orthonormal basis, we consider a general metric space, we can use an arbitrary coordinate system with basis 1-forms $\boldsymbol{dx}^0, ..., \boldsymbol{dx}^n$. In that case, we must include a factor of the determinant of the metric in our definition of the volume form thus:

$$\tilde{\boldsymbol{\omega}} = \sqrt{(-1)^s g}\, \boldsymbol{dx}^0 \wedge \boldsymbol{dx}^1 \wedge ... \wedge \boldsymbol{dx}^n, \tag{37.12}$$

where g is the determinant of the metric and s is the number of minus signs in the signature of the metric. The $(-1)^s$ factor ensures that we do not attempt to take the square root of a negative number.

Example 37.3

[8]This metric has $s = 0$.

Consider a flat two-dimensional plane expressed in polar coordinates with line element[8] $\mathrm{d}s^2 = \mathrm{d}r^2 + r^2\mathrm{d}\theta^2$. Expressed in orthonormal coordinates $(x^1, x^2) = (r, \theta)$, we have basis 1-forms

$$\boldsymbol{\omega}^{\hat{r}} = \boldsymbol{dr}, \quad \boldsymbol{\omega}^{\hat{\theta}} = r\boldsymbol{d\theta}. \tag{37.13}$$

The volume form is given by

$$\tilde{\boldsymbol{\omega}} = \boldsymbol{\omega}^{\hat{r}} \wedge \boldsymbol{\omega}^{\hat{\theta}} = r\boldsymbol{dr} \wedge \boldsymbol{d\theta}. \tag{37.14}$$

This volume form has components given by

$$\omega_{12} = \tilde{\boldsymbol{\omega}}(\boldsymbol{e}_{\hat{r}}, \boldsymbol{e}_{\hat{\theta}}) = \left[\boldsymbol{\omega}^{\hat{r}}(\boldsymbol{e}_{\hat{r}}) \otimes \boldsymbol{\omega}^{\hat{\theta}}(\boldsymbol{e}_{\hat{\theta}}) - \boldsymbol{\omega}^{\hat{\theta}}(\boldsymbol{e}_{\hat{r}}) \otimes \boldsymbol{\omega}^{\hat{r}}(\boldsymbol{e}_{\hat{\theta}})\right] = 1,$$

$$\omega_{21} = \tilde{\boldsymbol{\omega}}(\boldsymbol{e}_{\hat{\theta}}, \boldsymbol{e}_{\hat{r}}) = \left[\boldsymbol{\omega}^{\hat{r}}(\boldsymbol{e}_{\hat{\theta}}) \otimes \boldsymbol{\omega}^{\hat{\theta}}(\boldsymbol{e}_{\hat{r}}) - \boldsymbol{\omega}^{\hat{\theta}}(\boldsymbol{e}_{\hat{\theta}}) \otimes \boldsymbol{\omega}^{\hat{r}}(\boldsymbol{e}_{\hat{r}})\right] = -1, \tag{37.15}$$

or

$$\omega_{ij} = \varepsilon_{ij}. \tag{37.16}$$

If, instead, we choose to work with in a coordinate basis with basis 1-forms $\boldsymbol{\omega}^r = \boldsymbol{dr}$ and $\boldsymbol{\omega}^\theta = \boldsymbol{d\theta}$ we must multiply $\boldsymbol{dr} \wedge \boldsymbol{d\theta}$ a factor of $\sqrt{g} = \sqrt{\det g_{ij}} = r$ and obtain the same result for the volume form $\tilde{\boldsymbol{\omega}}$. Note, however, that in the coordinate basis, the volume form has components $\sqrt{g}\varepsilon_{ij}$, or

$$\omega_{ij} = r\varepsilon_{ij}. \tag{37.17}$$

With the volume form $\tilde{\boldsymbol{\omega}}$ in hand, we can define the duality transformation using the Hodge star operation. In n-dimensional space, this operation takes a $(q, 0)$ tensor and outputs a p-form, where $p = n - q$. Consider a $(q, 0)$ antisymmetric tensor $\boldsymbol{T}$ with components $T^{\alpha \dots \kappa} = T^{[\alpha \dots \kappa]}$. The p-form $\tilde{\boldsymbol{A}}$ that is *dual* to $\boldsymbol{T}$ is written as

$$\tilde{\boldsymbol{A}} = \star \boldsymbol{T} = \frac{A_{\mu \dots \omega}}{p!} \boldsymbol{\omega}^\mu \wedge \dots \wedge \boldsymbol{\omega}^\omega, \tag{37.18}$$

where the $\boldsymbol{\omega}^\mu$ are the basis 1-forms, $p = n - q$ and the dual tensor's components are given by

$$A_{\mu \dots \omega} = \frac{1}{q!} \omega_{\alpha \dots \kappa \mu \dots \omega} T^{\alpha \dots \kappa}. \tag{37.19}$$

Despite the fussy definition, the practice of taking the dual turns out to be rather simple, especially in the usual $(x^0, \dots, x^n)$ Minkowski space. Here the volume form is given by $\boldsymbol{dx}^0 \wedge \dots \wedge \boldsymbol{dx}^n$, and has components given by the Levi-Civita symbol $\varepsilon_{0 \dots n}$.

Example 37.4

In three-dimensional Euclidean space in Cartesian coordinates, we have the volume 3-form $\tilde{\boldsymbol{\omega}} = \frac{1}{3!}\varepsilon_{\mu\nu\sigma}\boldsymbol{dx}^\mu \wedge \boldsymbol{dx}^\nu \wedge \boldsymbol{dx}^\sigma$ with components $\varepsilon_{\mu\nu\sigma}$, where $\varepsilon_{123} = 1$.

Consider a bivector $\boldsymbol{T} = \boldsymbol{e}_2 \wedge \boldsymbol{e}_3$, which has components $T^{\alpha\beta} = \boldsymbol{T}(\boldsymbol{\omega}^\alpha, \boldsymbol{\omega}^\beta)$ given by

$$T^{\alpha\beta} = \boldsymbol{e}_2(\boldsymbol{\omega}^\alpha) \otimes \boldsymbol{e}_3(\boldsymbol{\omega}^\beta) - \boldsymbol{e}_3(\boldsymbol{\omega}^\alpha) \otimes \boldsymbol{e}_2(\boldsymbol{\omega}^\beta)$$
$$= \delta_2^\alpha \delta_3^\beta - \delta_3^\alpha \delta_2^\beta. \tag{37.20}$$

This $q = 2$ vector is dual to an object with valence $(0, p) = (0, 3 - 2) = (0, 1)$, that is, a 1-form $\tilde{\boldsymbol{A}} = \star \boldsymbol{T}$. This 1-form has components

$$A_\mu = \frac{1}{q!}\varepsilon_{\alpha\beta\mu}T^{\alpha\beta}$$
$$= \frac{1}{2}\varepsilon_{\alpha\beta\mu}\left(\delta_2^\alpha \delta_3^\beta - \delta_3^\alpha \delta_2^\beta\right) = \frac{1}{2}\left(\varepsilon_{23\mu} - \varepsilon_{32\mu}\right). \tag{37.21}$$

We conclude

$$A_1 = \frac{1}{2}\left(\varepsilon_{231} - \varepsilon_{321}\right) = 1, \tag{37.22}$$

with other components vanishing. We have, therefore, that $\tilde{A} = \omega^1 = dx^1$. In brief, $\star(e_y \wedge e_x) = dx$ in three dimensions.

[9] This fussy sign convention is the one used in most general relativity texts and so we employ it here. It is useful since it allows us to raise the components of ω with the metric as we expect to be able to do with the components of any tensor in a metric space.

In order to define the inverse map that inputs the p-form and outputs a q-tensor, we need the components $\omega^{\alpha\cdots\kappa}$ of a n-vector version of the volume n-form $\tilde{\omega}$. In a general metric space, these components are determined by the equation

$$\omega^{\alpha\cdots\kappa}\omega_{\alpha\cdots\kappa} = (-1)^s n!, \tag{37.23}$$

where, once again, s is the number of negative signs in the metric's signature.[9] This guarantees the normalization condition that $\omega^{123\cdots n} = (-1)^s/\omega_{123\cdots n}$.

Example 37.5

For an orthonormal basis with $s = 1$ we have that the components of the volume tensor are

$$\omega^{\alpha\beta\cdots\kappa} = \varepsilon^{\alpha\beta\cdots\kappa}, \tag{37.24}$$

and so $\omega^{123\cdots n}\omega_{123\cdots n} = \varepsilon^{\alpha\beta\cdots\kappa}\varepsilon_{\alpha\beta\cdots\kappa} = -n!$

In the usual $(3+1)$-dimensional metric spacetime with $s = 1$, this gives us the relations

$$\omega_{\mu\nu\alpha\beta} = \sqrt{-g}\,\varepsilon_{\mu\nu\alpha\beta}, \quad \omega^{\mu\nu\alpha\beta} = \frac{1}{\sqrt{-g}}\varepsilon^{\mu\nu\alpha\beta}, \tag{37.25}$$

where $\varepsilon^{0123} = -\varepsilon_{0123} = -1$.

In n-dimensional space, the dual $(q,0)$ tensor S of the p-form $\tilde{B}$ is given by

$$S = \star\tilde{B} = \frac{S^{\alpha\cdots\kappa}}{q!}e_\alpha \wedge \cdots \wedge e_\kappa, \tag{37.26}$$

where $q = n - p$ and the components are

$$S^{\alpha\cdots\kappa} = \frac{1}{p!}\omega^{\mu\cdots\omega\alpha\cdots\kappa}B_{\mu\cdots\omega}. \tag{37.27}$$

Example 37.6

In flat $n = 3$-dimensional space, we have the volume 3-form $\tilde{\omega} = dx^1 \wedge dx^2 \wedge dx^3$, with components $\varepsilon_{\mu\nu\sigma}$. The corresponding up version has components $\varepsilon^{\mu\nu\sigma}$, with $\varepsilon^{123} = 1$. Consider the $p = 2$-form $\tilde{A} = dx^2 \wedge dx^3$, which has components

$$\begin{aligned}A_{\mu\nu} &= \tilde{A}(e_\mu, e_\nu) = dx^2(e_\mu) \otimes dx^3(e_\nu) - dx^3(e_\mu) \otimes dx^2(e_\nu) \\ &= \delta_\mu^2\delta_\nu^3 - \delta_\mu^3\delta_\nu^2.\end{aligned} \tag{37.28}$$

This $(p = 2)$-form is dual to an object with valence $(q,0) = (3 - 2, 0) = (1,0)$, that is, a vector. The components of the vector $S = \star\tilde{A}$ are

$$S^\alpha = \frac{1}{p!}\omega^{\mu\nu i}A_{\mu\nu} = \frac{1}{2!}\varepsilon^{\mu\nu\alpha}(\delta_\mu^2\delta_\nu^3 - \delta_\mu^3\delta_\nu^2). \tag{37.29}$$

We conclude

$$S^1 = \frac{1}{2}\left(\varepsilon^{231} - \varepsilon^{321}\right) = 1, \tag{37.30}$$

with other components vanishing. We have, therefore, $\boldsymbol{S} = \boldsymbol{e}_1$. In brief, $\star(\boldsymbol{dy} \wedge \boldsymbol{dz}) = \boldsymbol{e}_x$ in three dimensions.

Most of our work will be in Minkowski space, similar to the next example.

Example 37.7

In $n = 4$-dimensional Minkowski space, we have the volume form $\tilde{\boldsymbol{\omega}} = \boldsymbol{dt} \wedge \boldsymbol{dx} \wedge \boldsymbol{dy} \wedge \boldsymbol{dz}$, with components $\varepsilon_{\mu\nu\alpha\beta}$. A $p = 2$-form $\tilde{\boldsymbol{M}} = B\boldsymbol{dy} \wedge \boldsymbol{dz}$ has components

$$M_{\mu\nu} = \tilde{\boldsymbol{M}}(\boldsymbol{e}_\mu, \boldsymbol{e}_\nu) = B\left(\delta_\mu^2 \delta_\nu^3 - \delta_\mu^3 \delta_\nu^2\right), \tag{37.35}$$

which is to say that $M_{23} = -M_{32} = B$ and all other components vanish. The dual to $\tilde{\boldsymbol{M}}$ is a $(2,0)$ tensor $\star\tilde{\boldsymbol{M}} = \boldsymbol{M}$ whose components are

$$\begin{aligned}
M^{\alpha\beta} &= \frac{1}{p!}\varepsilon^{\mu\nu\alpha\beta} M_{\mu\nu} \\
&= \frac{B}{2}\varepsilon^{\mu\nu\alpha\beta}(\delta_\mu^2 \delta_\nu^3 - \delta_\mu^3 \delta_\nu^2) \\
&= \frac{B}{2}\left(\varepsilon^{23\alpha\beta} - \varepsilon^{32\alpha\beta}\right) = B\varepsilon^{23\alpha\beta}.
\end{aligned} \tag{37.36}$$

We conclude that the only non-zero components are $M^{01} = -M^{10}$. So we have the bivector

$$\boldsymbol{M} = B\boldsymbol{e}_x \wedge \boldsymbol{e}_t. \tag{37.37}$$

We can also see how $\star(\boldsymbol{dy} \wedge \boldsymbol{dz}) = \boldsymbol{e}_x \wedge \boldsymbol{e}_t$.

Some useful examples of commonly encountered duals are collected in the margin.

Example 37.8

Most people are already familiar with a vector that results from taking a dual: the cross product. Consider the 2-form $\tilde{\boldsymbol{u}} \wedge \tilde{\boldsymbol{v}}$ in Euclidean $n = 3$-dimensional space. It has components $w_{ij} = u_i v_j - u_j v_i$. Its dual is a $q = 1$ vector $\boldsymbol{t}$, whose first component is

$$\begin{aligned}
t^1 &= \frac{1}{p!}\varepsilon^{\mu\nu 1}\left(u_\mu v_\nu - v_\mu u_\nu\right) \\
&= \frac{1}{2!}\varepsilon^{231}\left(u_2 v_3 - v_2 u_3\right) + \frac{1}{2!}\varepsilon^{321}\left(u_3 v_2 - v_3 u_2\right) \\
&= u_2 v_3 - u_3 v_2.
\end{aligned} \tag{37.38}$$

We also obtain

$$\begin{aligned}
t^2 &= u_1 v_3 - v_1 u_3, \\
t^3 &= u_1 v_2 - v_1 u_2.
\end{aligned} \tag{37.39}$$

Since, in ordinary Euclidean 3-space, there is no difference between and up and down index, we have recreated the cross product vector

$$\boldsymbol{t} = \boldsymbol{u} \times \boldsymbol{v}, \tag{37.40}$$

where $\boldsymbol{u}$ and $\boldsymbol{v}$ are the vector versions of the corresponding 1-forms.

Some useful special cases:
In $n = 3$, Euclidean space, where $\varepsilon_{123} = \varepsilon^{123}$ we have

$$\begin{aligned}
\star\boldsymbol{e}_x &= \boldsymbol{dy} \wedge \boldsymbol{dz}, \\
\star\boldsymbol{e}_y &= \boldsymbol{dz} \wedge \boldsymbol{dx}, \\
\star\boldsymbol{e}_z &= \boldsymbol{dx} \wedge \boldsymbol{dy},
\end{aligned} \tag{37.31}$$

and

$$\begin{aligned}
\star(\boldsymbol{dy} \wedge \boldsymbol{dz}) &= \boldsymbol{e}_x, \\
\star(\boldsymbol{dz} \wedge \boldsymbol{dx}) &= \boldsymbol{e}_y, \\
\star(\boldsymbol{dx} \wedge \boldsymbol{dy}) &= \boldsymbol{e}_z,
\end{aligned} \tag{37.32}$$

and therefore $\star\star = 1$.
In $n = 4$ Minkowski space, where $\varepsilon^{0123} = -\varepsilon_{0123} = -1$, we have

$$\begin{aligned}
\star(\boldsymbol{dx} \wedge \boldsymbol{dt}) &= \boldsymbol{e}_y \wedge \boldsymbol{e}_z, \\
\star(\boldsymbol{dy} \wedge \boldsymbol{dt}) &= \boldsymbol{e}_z \wedge \boldsymbol{e}_x, \\
\star(\boldsymbol{dz} \wedge \boldsymbol{dt}) &= \boldsymbol{e}_x \wedge \boldsymbol{e}_y, \\
\star(\boldsymbol{dy} \wedge \boldsymbol{dz}) &= \boldsymbol{e}_x \wedge \boldsymbol{e}_t, \\
\star(\boldsymbol{dz} \wedge \boldsymbol{dx}) &= \boldsymbol{e}_y \wedge \boldsymbol{e}_t, \\
\star(\boldsymbol{dx} \wedge \boldsymbol{dy}) &= \boldsymbol{e}_z \wedge \boldsymbol{e}_t,
\end{aligned} \tag{37.33}$$

and

$$\begin{aligned}
\star(\boldsymbol{e}_y \wedge \boldsymbol{e}_z) &= -\boldsymbol{dx} \wedge \boldsymbol{dt}, \\
\star(\boldsymbol{e}_z \wedge \boldsymbol{e}_x) &= -\boldsymbol{dy} \wedge \boldsymbol{dt}, \\
\star(\boldsymbol{e}_x \wedge \boldsymbol{e}_y) &= -\boldsymbol{dz} \wedge \boldsymbol{dt}, \\
\star(\boldsymbol{e}_x \wedge \boldsymbol{e}_t) &= -\boldsymbol{dy} \wedge \boldsymbol{dz}, \\
\star(\boldsymbol{e}_y \wedge \boldsymbol{e}_t) &= -\boldsymbol{dz} \wedge \boldsymbol{dx}, \\
\star(\boldsymbol{e}_z \wedge \boldsymbol{e}_t) &= -\boldsymbol{dx} \wedge \boldsymbol{dy},
\end{aligned} \tag{37.34}$$

and therefore $\star\star = -1$.

As discussed the exercises, we have for the case of n dimensions for a p-form $\tilde{\boldsymbol{B}}$ and q-vector $\boldsymbol{T}$ that

$$\begin{aligned} \star\star\tilde{\boldsymbol{B}} &= (-1)^{s+p(n-p)}\tilde{\boldsymbol{B}}, \\ \star\star\boldsymbol{T} &= (-1)^{s+q(n-q)}\boldsymbol{T}. \end{aligned} \tag{37.41}$$

That is, a dual of a dual is, up to a sign, the tensor we started with.

37.3   Volume forms

After that interlude into the abstract mathematics of the mappings between forms and vectors, we turn to the business of calculating areas and volumes in four-dimensional space. This can be done quite simply using the volume tensor discussed in the previous section.

In Chapter 32, we considered $n = 2$-dimensional Euclidean space and the case shown in Fig. 37.1, where the area of the parallelogram is given by

$$(\text{Area}) = \begin{vmatrix} u^x & u^y \\ v^x & v^y \end{vmatrix} = u^x v^y - v^x u^y = \varepsilon_{\mu\nu} u^\mu v^\nu. \tag{37.42}$$

There are two interesting observations to make about this. The first is that the area results from inserting vectors $\boldsymbol{u}$ and $\boldsymbol{v}$ into a two-dimensional volume tensor $\tilde{\boldsymbol{\omega}}(\ ,\ ) = \boldsymbol{dx} \wedge \boldsymbol{dy}$, which has components $\omega_{ij} = \varepsilon_{ij}$. That is

$$(\text{Area}) = \tilde{\boldsymbol{\omega}}(\boldsymbol{u}, \boldsymbol{v}). \tag{37.43}$$

The second is that, since the dual of a bivector is a number for $n = 2$, we can link the area (a number) to the $(2,0)$ bivector $\boldsymbol{b} = \boldsymbol{u} \wedge \boldsymbol{v}$

$$(\text{Area}) = \star\boldsymbol{b} = \star(\boldsymbol{u} \wedge \boldsymbol{v}). \tag{37.44}$$

We demonstrate this latter feature in the next example.

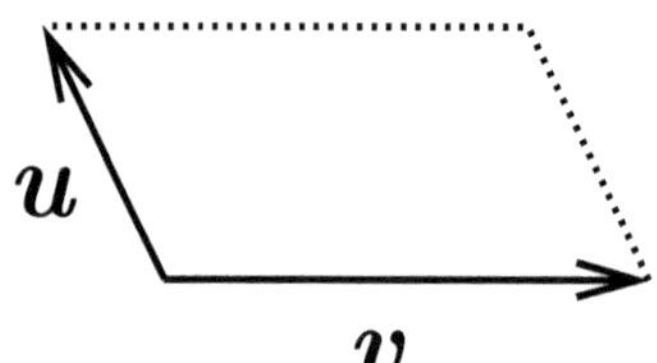

Fig. 37.1   The parallelogram formed by vectors $\boldsymbol{u}$ and $\boldsymbol{v}$.

Example 37.9

We are working in (Euclidean) $n = 2$ space and so a dual of a $(2,0)$ tensor gives us a 0-form (or number). The bivector has components $b^{\mu\nu} = u^\mu v^\nu - v^\mu u^\nu$, and so the dual A has components

$$A = \frac{1}{2}\varepsilon_{\mu\nu}\left(u^\mu v^\nu - v^\mu u^\nu\right) = \varepsilon_{\mu\nu} u^\mu v^\nu, \tag{37.45}$$

which is the expression for the area we had above.

We conclude that the volume of the parallelogram in $n = 2$ space (also known as the area) is the dual of the bivector formed from its sides.

In the same way, a 4-volume $\mathcal{V}$ in $(3+1)$-dimensional spacetime can be determined by filling in slots in a volume 4-form. As might be expected $\tilde{\boldsymbol{\omega}}(\boldsymbol{A}, \boldsymbol{B}, \boldsymbol{C}, \boldsymbol{D})$ outputs the 4-volume $\mathcal{V}$ of a four-dimensional parallelepiped with sides formed from the 4-vectors $\boldsymbol{A}$, $\boldsymbol{B}$, $\boldsymbol{C}$ and $\boldsymbol{D}$. In terms of components in Minkowski spacetime, this is written as

$$\mathcal{V} = \varepsilon_{\mu\nu\alpha\beta} A^\mu B^\nu C^\alpha D^\beta = \det \begin{vmatrix} A^0 & A^1 & A^2 & A^3 \\ B^0 & B^1 & B^2 & B^3 \\ C^0 & C^1 & C^2 & C^3 \\ D^0 & D^1 & D^2 & D^3 \end{vmatrix}. \tag{37.46}$$

Example 37.10

Recall that in a coordinate representation, the volume 4-form for a metric space is

$$\tilde{\boldsymbol{\omega}} = \sqrt{-g}\, \boldsymbol{dx}^0 \wedge \boldsymbol{dx}^1 \wedge \boldsymbol{dx}^2 \wedge \boldsymbol{dx}^3. \tag{37.47}$$

For an infinitesimal box with sides described by vectors $\mathrm{d}x^0 \boldsymbol{e}_0$, $\mathrm{d}x^1 \boldsymbol{e}_1$, $\mathrm{d}x^2 \boldsymbol{e}_2$ and $\mathrm{d}x^3 \boldsymbol{e}_3$, we have an infinitesimal 4-volume

$$\mathrm{d}\mathcal{V} = \sqrt{-g}\, \mathrm{d}x^0 \mathrm{d}x^1 \mathrm{d}x^2 \mathrm{d}x^3. \tag{37.48}$$

This means that an infinitesimal volume element (useful for integration) should be written as $\mathrm{d}\mathcal{V} = \sqrt{-g}\, \mathrm{d}^4 x$, where $\mathrm{d}^4 x$ is the usual Cartesian flat-space volume element $\mathrm{d}x^0 \mathrm{d}x^1 \mathrm{d}x^2 \mathrm{d}x^3$.

We also see that we have access to the same volume $\mathcal{V}$ by taking the dual of the tetravector $(\boldsymbol{A} \wedge \boldsymbol{B} \wedge \boldsymbol{C} \wedge \boldsymbol{D})$, which is to say

$$\mathcal{V} = \star(\boldsymbol{A} \wedge \boldsymbol{B} \wedge \boldsymbol{C} \wedge \boldsymbol{D}). \tag{37.49}$$

Another useful object is the 3-volume.[10] Still working in (3+1)-dimensional spacetime, this quantity can be represented using a dual of a trivector by defining a **3-volume 1-form** $\tilde{\boldsymbol{\sigma}}$. Noting that the dual of a (3,0) tensor $\boldsymbol{A} \wedge \boldsymbol{B} \wedge \boldsymbol{C}$ is a 1-form in four-dimensional spacetime, we define $\tilde{\boldsymbol{\sigma}}$ to be given by

$$\tilde{\boldsymbol{\sigma}} = -\star(\boldsymbol{A} \wedge \boldsymbol{B} \wedge \boldsymbol{C}). \tag{37.50}$$

Why define it like this? It's because this fits nicely with our use of the volume 4-form. In fact, filling in three slots of the volume 4-form gives us the 3-volume 1-form:

$$\tilde{\boldsymbol{\sigma}}(\ ) = \tilde{\boldsymbol{\omega}}(\ , \boldsymbol{A}, \boldsymbol{B}, \boldsymbol{C}), \tag{37.51}$$

with components in Minkowski space of

$$\sigma_\mu = \tilde{\boldsymbol{\omega}}(\boldsymbol{e}_\mu, \boldsymbol{A}, \boldsymbol{B}, \boldsymbol{C}) = \varepsilon_{\mu\alpha\beta\gamma} A^\alpha B^\beta C^\gamma. \tag{37.52}$$

We see how this can be used in the next example.

[10] The 3-volume of an object in (3+1)-dimensional space can be thought of as a spacelike hypersurface. That is, a slice of four-dimensional spacetime at a constant time. The normal to a spacelike hypersurface is timelike. We can also identify 3-volumes that are timelike hypersurfaces, which are slices of spacetime taken at a constant value of one spatial coordinate. These are defined by their normals being spacelike.

Example 37.11

Consider Minkowski spacetime. A cuboid box has sides parallel to the spacelike basis vectors $\boldsymbol{e}_i$, with lengths δx, δy and δz. The (spacelike) 3-volume of the box in its rest frame is given by

$$\sigma_0 = \tilde{\boldsymbol{\omega}}(\boldsymbol{e}_0, \delta x^1 \boldsymbol{e}_1, \delta x^2 \boldsymbol{e}_2, \delta x^3 \boldsymbol{e}_3) = \varepsilon_{0123} \delta x^1 \delta x^2 \delta x^3 = \delta x \delta y \delta z, \tag{37.53}$$

which is the expected answer for a 3-volume of a box. All of the other components of $\tilde{\boldsymbol{\sigma}}$ vanish.

[11] Recall that an observer with velocity u measures an energy $E = -\tilde{p}(u)$ for a particle with momentum p a number density of particles $n = -\tilde{J}(u)$. We can add that they measure a 3-volume $V = \tilde{\sigma}(u)$.

[12] One way to think of this sign is that it reflects the fact that if a surface moves in some direction and passes through dust particles, then in the rest frame of the surface, the flux of dust particles is directed towards the surface. Since we define the positive sense of flux through a surface as directed being outwards, this gives us a minus sign. Another way to think about this is to characterize the three-dimensional hypersurfaces using the direction of their normal, similar to the way we write $d\vec{S} = \vec{n}dS$ for a two-dimensional surface. From our definitions the signs are accounted for if we have that spacelike hypersurfaces have outward-directed (timelike) normals, while timelike hypersurfaces have inward directed ones (spacelike) normals. This becomes important when choosing the orientation of a surface over which to integrate in later chapters.

[13] Recall for an observer at rest in a spacetime with a diagonal metric we have $u^0 = (-g_{tt})^{-\frac{1}{2}}$, showing how the timelike component of the metric is effectively divided out of the determinant.

In the local rest frame of an observer, their velocity vector u is taken to be equal to $e_{\hat{0}}$. As a result, we can say that the 3-volume V of a parallelepiped with sides A, B and C measured by an observer with velocity u is[11]

$$V = \tilde{\omega}(u, A, B, C) = \tilde{\sigma}(u). \tag{37.54}$$

While the 0th component of $\tilde{\sigma}$ gives the three-volume of a spacelike surface, the other components tell us about the 3-volume of timelike surfaces.

Example 37.12

Again in Minkowski space, consider that spatial surface of our cuboid box with lengths $\delta y e_2$ and $\delta z e_3$. When moving at a velocity u, in a proper time $\delta\tau$ the surface sweeps out a volume whose third edge is $u\delta\tau$. We then write a 3-volume 1-form

$$\tilde{\sigma}(\ ) = \tilde{\omega}(\ , \delta y e_2, \delta z e_3, u\delta\tau). \tag{37.55}$$

In the rest frame of the box, we then have $u = e_0$ and so the components of $\tilde{\sigma}(\ )$ in this frame are given by

$$\begin{aligned}
\sigma_0 &= \tilde{\omega}(e_0, \delta y e_2, \delta z e_3, e_0 \delta\tau) = 0, \\
\sigma_1 &= \tilde{\omega}(e_1, \delta y e_2, \delta z e_3, e_0 \delta\tau) = \varepsilon_{1230} \delta y \delta z \delta\tau, \\
\sigma_2 &= \tilde{\omega}(e_2, \delta y e_2, \delta z e_3, e_0 \delta\tau) = 0, \\
\sigma_3 &= \tilde{\omega}(e_3, \delta y e_2, \delta z e_3, e_0 \delta\tau) = 0.
\end{aligned} \tag{37.56}$$

We conclude that $\sigma_1 = -\delta y \delta z \delta\tau$, and the other components vanish. The minus sign might be slightly unexpected since we have chosen the surface using the usual right-handed conventions, but it follows from the fact that the surface is at rest in the frame in which we've worked out the components.[12]

The volume form is often used to compute volume elements as examined in the next example.

Example 37.13

Working in a coordinate frame we have

$$\tilde{\omega} = \sqrt{-g}\, dx^0 \wedge dx^1 \wedge dx^2 \wedge dx^3. \tag{37.57}$$

For an infinitesimal box with sides described in its rest frame by spacelike vectors $dx^1 e_1$, $dx^2 e_2$ and $dx^3 e_3$ we have a 3-volume 1-form integration element

$$d\tilde{\sigma}(\ ) = \sqrt{-g}\, \varepsilon_{0123} dx^0(\ ) dx^1 dx^2 dx^3. \tag{37.58}$$

To turn this into a volume, insert the velocity and we find (noting $\varepsilon_{0123} = 1$)

$$\begin{aligned}
dV = d\tilde{\sigma}(u) &= \sqrt{-g} dx^0(u) dx^1 dx^2 dx^3 \\
&= \sqrt{-g}\, u^0 dx^1 dx^2 dx^3,
\end{aligned} \tag{37.59}$$

leading us to conclude that the invariant proper 3-volume element is[13] $dV = \sqrt{-g}\, u^0 dx^1 dx^2 dx^3$.

The components $d\sigma_\mu$ of the infinitesimal 3-volume 1-form will always be contracted against the components of a vector in expressions like $J^\mu d\sigma_\mu = d\tilde{\sigma}(J)$.

Example 37.14

If the integration surface is the infinitesimal box in 3-space from above, we have a contribution

$$J^0 \mathrm{d}\sigma_0 = \sqrt{-g}\, J^0 \varepsilon_{0123} \mathrm{d}x^1 \mathrm{d}x^2 \mathrm{d}x^3 = \sqrt{-g}\, J^0 \mathrm{d}x \mathrm{d}y \mathrm{d}z, \qquad (37.60)$$

which computes the amount of the timelike (i.e. charge-like) component of vector $\boldsymbol{J}$ in the box. A different choice of the three-dimensional integration surface might lead to a non-zero contribution like

$$J^1 \mathrm{d}\sigma_1 = \sqrt{-g}\, J^1 \varepsilon_{1230} \mathrm{d}x^2 \mathrm{d}x^3 \mathrm{d}x^0 = -\sqrt{-g}\, J^1 \mathrm{d}y \mathrm{d}z \mathrm{d}t. \qquad (37.61)$$

which computes the flux of the J^1 component of the current though the face of a surface with sides parallel to $\boldsymbol{e}_2$ and $\boldsymbol{e}_3$ in coordinate time $\mathrm{d}t$.

The material in this chapter will be used most extensively in Chapters 42 and 43. In the next chapter, we continue to examine the mathematics of forms and turn to the question of how they give rise to a new insight into what an integral is.

Chapter summary

- The Hodge star operation allows us to map between forms and antisymmetric tensors built from vectors.
- Duality allows us to compute volumes using the volume tensor $\tilde{\omega}$. This is a 4-form in our usual $(3+1)$-dimensional spacetime.
- The 3-volume of a box with sides described by vectors $\boldsymbol{A}$, $\boldsymbol{B}$ and $\boldsymbol{C}$, measured by an observer with velocity $\boldsymbol{u}$, is given by $\tilde{\sigma}(\boldsymbol{u}) = \tilde{\omega}(\boldsymbol{u}, \boldsymbol{A}, \boldsymbol{B}, \boldsymbol{C})$.

Exercises

(37.1) Consider the volume 1-form $\tilde{\sigma}$ describing the 3-volume of a box. Insert this into the (2,0) energy-momentum tensor $\boldsymbol{T}$ and we have that momentum $\boldsymbol{p}$ is given via

$$\boldsymbol{T}(\ , \tilde{\sigma}) = \begin{pmatrix} \text{momentum crossing from} \\ \text{negative to positive side of box} \end{pmatrix}. \qquad (37.62)$$

An observer with velocity $\boldsymbol{u}$ carrying the box measures its 3-volume as V.
(a) Show that the volume 1-form $\tilde{\sigma}$ is related to the velocity 1-form by $\tilde{\sigma} = -V\tilde{u}$.

Use this result to find expressions for: (b) the momentum and (c) the total energy, in terms of V and the components of $\boldsymbol{T}$ and $\tilde{u}$.

(37.2) The observer with velocity vector $\boldsymbol{u}$ carries a box with sides $\boldsymbol{A}$, $\boldsymbol{B}$ and $\boldsymbol{C}$ which has permeable walls, allowing a particle current $\boldsymbol{J}$ to enter it.
(a) Explain why the observer can resolve the current into the form

$$\boldsymbol{J} = n\boldsymbol{u} + a\boldsymbol{A} + b\boldsymbol{B} + \gamma\boldsymbol{C}, \qquad (37.63)$$

where n is the number density of particles measured by the observer.

(b) Show that the number of particles N inside the box is

$$N = \tilde{\omega}(\boldsymbol{J}, \boldsymbol{A}, \boldsymbol{B}, \boldsymbol{C}). \tag{37.64}$$

(c) Now consider a second observer with velocity $\boldsymbol{v}$ carrying a box with sides $\boldsymbol{A}'$, $\boldsymbol{B}'$ and $\boldsymbol{C}'$. What condition is necessary to ensure the boxes capture the same number of particles?

(37.3) (a) In a three-dimensional coordinate system with a metric, show that the first component of the curl of a vector $\boldsymbol{v}$ can be written as

$$\frac{1}{\sqrt{g}} \left(\frac{\partial V_3}{\partial x^2} - \frac{\partial V_2}{\partial x^3} \right). \tag{37.65}$$

(b) Show further that the curl is given, in an orthonormal frame in three dimensions by

$$(\boldsymbol{\nabla} \times \boldsymbol{v})^{\hat{1}} =$$
$$\left(\frac{g_{11}}{g} \right)^{\frac{1}{2}} \left[\frac{\partial}{\partial x^3} \left(g_{22}^{\frac{1}{2}} v^{\hat{2}} \right) - \frac{\partial}{\partial x^2} \left(g_{33}^{\frac{1}{2}} v^{\hat{3}} \right) \right], \tag{37.66}$$

and cyclic permutations.

(37.4) Show that

$$\begin{aligned} \star\star\tilde{\boldsymbol{B}} &= (-1)^{s+p(n-p)} \tilde{\boldsymbol{B}}, \\ \star\star\boldsymbol{T} &= (-1)^{s+q(n-q)} \boldsymbol{T}. \end{aligned} \tag{37.67}$$

Hint: This proof, given in the 1980 book by Schutz, relies on manipulating the indices of the Levi-Civita symbol. If in doubt, prove the first identity for $n = 4$ and $p = 2$ or similar, and then generalize. Once the first identity is proven, the second identity follows using very similar arguments.

(37.5) The Lie derivative gives access to a general method for computing the divergence θ_v of a field $\boldsymbol{v}$. The divergence arises when we take the Lie derivative of the volume tensor $\tilde{\omega} = \varepsilon_{|\alpha\beta\gamma\delta|} \boldsymbol{dx}^\alpha \wedge \boldsymbol{dx}^\beta \wedge \boldsymbol{dx}^\gamma \wedge \boldsymbol{dx}^\delta$. We define the divergence via

$$\mathcal{L}_v\tilde{\omega} = -\theta_v\tilde{\omega}. \tag{37.68}$$

(a) Using $\boldsymbol{d}\tilde{\omega} = 0$ show that

$$\boldsymbol{\nabla}_\mu\tilde{\omega} = 0. \tag{37.69}$$

(b) Using the previous result, show further that

$$\begin{aligned} -(\mathcal{L}_v\tilde{\omega})_{\mu\nu\alpha\beta} &= \varepsilon_{\sigma\nu\alpha\beta}v^\sigma{}_{;\mu} + \varepsilon_{\mu\sigma\alpha\beta}v^\sigma{}_{;\nu} \\ &\quad + \varepsilon_{\mu\nu\sigma\beta}v^\sigma{}_{;\alpha} + \varepsilon_{\mu\nu\alpha\sigma}v^\sigma{}_{;\beta}. \end{aligned} \tag{37.70}$$

(c) As a result, prove that the divergence is calculated using the connection through the familiar rule

$$\theta_v = -\boldsymbol{\nabla} \cdot \boldsymbol{v} = -v^\mu{}_{;\mu}. \tag{37.71}$$

(37.6) Consider a uniform distribution of dust. We have a congruence of curves formed from the world lines of dust particles with a corresponding current (tangent) field $\boldsymbol{J}$. Consider again an observer with velocity $\boldsymbol{u}$ carrying a permeable box of volume V spanned by vectors $\boldsymbol{A}$, $\boldsymbol{B}$ and $\boldsymbol{C}$.

(a) Explain why

$$\mathcal{L}_{\boldsymbol{J}}\boldsymbol{J} = \mathcal{L}_{\boldsymbol{J}}\boldsymbol{A} = \mathcal{L}_{\boldsymbol{J}}\boldsymbol{B} = \mathcal{L}_{\boldsymbol{J}}\boldsymbol{C} = 0. \tag{37.72}$$

(b) Show that for $\chi = \tilde{\omega}(\boldsymbol{J}, \boldsymbol{A}, \boldsymbol{B}, \boldsymbol{C})$, where $\tilde{\omega}$ is the volume 4-form, we have

$$\mathcal{L}_{\boldsymbol{J}}\chi = -\theta_{\boldsymbol{J}}\chi. \tag{37.73}$$

(c) Why is the case that, if the number of particles in the box is constant (as we would expect for a uniform distribution), then

$$\mathcal{L}_{\boldsymbol{J}}\chi = 0? \tag{37.74}$$

(d) Show finally that

$$\boldsymbol{\nabla} \cdot \boldsymbol{J} = 0. \tag{37.75}$$

See the book by Ludvigsen for a discussion of the material in this problem.

Forms, chains, and Stokes' theorem

38

Only by taking infinitesimally small units for observation (the differential of history, that is, the individual tendencies of men) and attaining to the art of integrating them (that is, finding the sum of these infinitesimals) can we hope to arrive at the laws of history.
Leo Tolstoy (1828–1910) *War and Peace*

> ⤵ This Chapter represents the logical endpoint for the material in Part V of the book. However, it lies outside most of our remaining topics, so can be skipped on a first reading. Stokes' theorem is used in Chapter 43.

38.1 Integration

This chapter is going to focus on integrals, which were originally formulated in calculus as a type of sum. In fact, the integral symbol $\int$, introduced by Leibniz, is a 'long s' to indicate the Latin word *summa* (or *ſumma* in old-fashioned writing) meaning sum. The fundamental theorem of calculus states that differentiation and integration are the inverses of each other, so that

$$\int_a^b \frac{\mathrm{d}f}{\mathrm{d}x}\,\mathrm{d}x = f(b) - f(a). \tag{38.1}$$

This equation has a rather intriguing feature: we are integrating the derivative $\mathrm{d}f/\mathrm{d}x$ over a region ($a \leq x \leq b$) and the answer depends only on the function f evaluated at the boundary of the region. Moreover, there is an interesting property of the signs on the right-hand side of eqn 38.1: $f(b)$ is included with a $+$ sign, but $f(a)$ is included with a $-$ sign. You can see where this comes from by thinking of the integral as a sum, slicing up the interval, so that we could write

$$\int_a^b \frac{\mathrm{d}f}{\mathrm{d}x}\,\mathrm{d}x \approx \Delta x \sum_{i=1}^{N-1} \frac{f(x_{i+1}) - f(x_i)}{\Delta x}, \tag{38.2}$$

where Δx is the width of each slice, so that adjacent terms cancel and all that is left is $f(x_N) - f(x_1) \equiv f(b) - f(a)$, the difference between the value of f at the boundary of the region of integration.

This result can be generalized to three dimensions, so that[1]

$$\int \vec{\nabla} f \,\mathrm{d}V = \int f \,\mathrm{d}\vec{S}, \tag{38.3}$$

---

[1] In components, eqn 38.3 becomes

$$\int \frac{\partial f}{\partial x^i}\,\mathrm{d}V = \int f \,\mathrm{d}S^i.$$

where the equivalent of the sign change seen on the right-hand side of eqn 38.1 is accomplished by the surface vector $\mathrm{d}\vec{S}$ changing orientation on either side of the surface. Then putting $f = a^j$ (where $j = x,\ y$ or z) into eqn 38.3 and then summing over j yields

$$\int \vec{\nabla} \cdot \vec{a}\,\mathrm{d}V = \int \vec{a} \cdot \mathrm{d}\vec{S}, \tag{38.4}$$

which is the divergence theorem, familiar from vector calculus (and used extensively in electromagnetism and fluid dynamics). This famous result preserves the intuition we have gained from eqn 38.1, namely that the integral of the derivative of a quantity in the interior of a space is equal to some kind of sum of that quantity on the boundary of that space.

We leapt from one dimension to three dimensions. What about the result in the middle? This is a little more complicated and a two-dimensional version of this argument yields the result that

$$\int \vec{\nabla} f \times \mathrm{d}\vec{S} = -\oint f\,\mathrm{d}\vec{\ell}, \tag{38.5}$$

where the right-hand side is an integral round a contour. This equation follows from the idea that all the gradients of f inside the surface sum to nothing, leaving only a circulating term around the boundary. The ith component of this equation can be written out as $\varepsilon_{ikm} \int \partial^k f\,\mathrm{d}S^m = -\oint f\,\mathrm{d}\ell_i$, so that substituting in $f = a^j$ and then setting $i = j$ yields Stokes' theorem[2]

$$\int \vec{\nabla} \times \vec{a} \cdot \mathrm{d}\vec{S} = \oint \vec{a} \cdot \mathrm{d}\vec{\ell}. \tag{38.6}$$

This equation is also familiar from vector calculus, electromagnetism, and fluid mechanics, but for the purposes of this chapter it is important to notice something about the structure of this equation. It is complicated by the vector product symbol, but once again we have emerged with the intuition that an integral of some kind of gradient of a function inside the interior of a space is related to the integral of the function on the boundary. We are left with the feeling that these various results (eqns 38.1, 38.4, and 38.6) are expressing the same underlying truth, but they all look a bit different and it is hard to see how to unify them.

Can we cut through this dense thicket of different theorems for different dimensionalities? It turns out that we can, using the mathematical apparatus that we have been developing in the last few chapters. Where we will end up is the following generalized Stokes' theorem:

$$\int_C d\tilde{\Omega} = \int_{\partial C} \tilde{\Omega}. \tag{38.7}$$

This is an integral over a space C of the exterior derivative of an n-form $\tilde{\Omega}$ is equal to the integral over the boundary of a space C, written as ∂C, of the n-form itself. Notice that it expresses exactly the intuition we have gleaned from our previous examples that were specific to particular dimensions. This new equation works in all directions and has buried

[2] Sir George G. S. Stokes, 1st Baronet (1919–1903) made numerous, important contributions to mathematical physics in the 19th century. He was the first person to simultaneously be Lucasian Professor of Mathematics, President of the Royal Society, and Member of Parliament for Cambridge University. Newton had also held all three posts, but not at the same time.

inside it all the other vector theorems we have just discussed (and with which we suspect most readers will be familiar).

To make progress, and to reach our generalized Stokes' theorem (eqn 38.7), we will need to upgrade our notation. Consider a typical line integral, such as the one used to compute the work done by a force $\boldsymbol{F}$ in a three-dimensional Euclidean space expressed in Cartesian coordinates,[3]

$$I_1 = \int_C \boldsymbol{F} \cdot \mathrm{d}\boldsymbol{r} = \int_C F_x \mathrm{d}x + F_y \mathrm{d}y + F_z \mathrm{d}z. \tag{38.8}$$

This involves the integral of something over a one-dimensional path C. From our experience in the previous chapters, we recognize that the 'something' in this case resembles a 1-form, which can be rewritten in geometric notation as $\tilde{\alpha} = F_x \boldsymbol{dx} + F_y \boldsymbol{dy} + F_z \boldsymbol{dz}$. We shall write this integral as

$$I_1 = \int_C \tilde{\alpha}. \tag{38.9}$$

Admittedly, it looks as if we are making a terrible *faux pas* in writing an integral without including a term d(something), violating a rule that will have been drummed into most of us by successive teachers of mathematics. Nevertheless, this notation is correct as we understand this to be part of the differential form, as we will discuss below.

Next, consider a surface integral. Recalling from Chapters 32 and 37 that areas can be written as bivectors, we can write a typical surface integral of a function as

$$I_2 = \int_S \boldsymbol{G} \cdot \mathrm{d}\boldsymbol{S} = \int_S (G_x \boldsymbol{dy} \wedge \boldsymbol{dz} + G_y \boldsymbol{dz} \wedge \boldsymbol{dx} + G_z \boldsymbol{dx} \wedge \boldsymbol{dy}), \tag{38.10}$$

where the elements of the surface S have been recast using the wedge product. Written in this way, the integral involves the integral of a 2-form $\tilde{\beta} = G_x \boldsymbol{dy} \wedge \boldsymbol{dz} + G_y \boldsymbol{dz} \wedge \boldsymbol{dx} + G_z \boldsymbol{dx} \wedge \boldsymbol{dy}$ over a surface S:

$$I_2 = \int_S \tilde{\beta}. \tag{38.11}$$

In the remainder of this chapter, we generalize these observations and see how *all* multiple integrals can be understood as an integral of a form. We will start by investigating how this might work and why forms are the natural content of the integrand of an integral. We then go on to investigate the surfaces over which we integrate: these are members of a family of objects called **chains** that are dual to forms, which is to say we have an inner product-like relationship

$$\langle \text{form}, \text{chain} \rangle \text{ “} = \text{” } \left(\begin{array}{c} \text{multiple} \\ \text{integral} \end{array} \right) = (\text{number}). \tag{38.12}$$

The rest of this chapter discusses this idea, leading eventually to our generalized version of Stokes' theorem, which is one of the milestones of differential geometry.

[3] We shall restrict our attention to flat space and spacetime in this chapter.

38.2   Integrating over forms

A 0-form is a function. A 1-form looks like a series of planes. A 2-form looks like a set of crossed surfaces forming a series of parallel tubes. A 3-form looks like an array of cells. These geometric objects can be used to provide the basic integrand of (function) × (integration cells) over which to integrate.

A definite integral outputs a number. A multiple integral is carried out over a generalized *surface*: a one-dimensional line, two-dimensional surface, three-dimensional volume or even a higher dimensional surface. Geometrically, it is the combination of a differential form and a surface that gives rise to the number. Specifically, the line integral evaluates the number of times the line crosses the surfaces of the 1-form. That is, for a 1-form $\tilde{\boldsymbol{Y}}$, we have

$$\int_{\text{line}} \tilde{\boldsymbol{Y}} = \left(\begin{array}{c} \text{number of} \\ \text{surfaces cut} \end{array} \right). \tag{38.13}$$

In the same way, the surface integral counts the number of tubes of the 2-form that cut the 2-surface. That is, for a 2-form $\tilde{\boldsymbol{X}}$, we have

$$\int_{\text{surface}} \tilde{\boldsymbol{X}} = \left(\begin{array}{c} \text{number of} \\ \text{tubes cut} \end{array} \right). \tag{38.14}$$

The volume integral evaluates the number of cells of the 3-form contained by a 3-volume, or, for a 3-form $\tilde{\boldsymbol{W}}$, we have

$$\int_{\text{volume}} \tilde{\boldsymbol{W}} = \left(\begin{array}{c} \text{number of} \\ \text{cells contained} \end{array} \right). \tag{38.15}$$

If we accept that integrals are indeed all carried out over forms, then we have a neat explanation for a potentially puzzling aspect of multiple integrals: the existence of the **Jacobian**.[4] Recall how multiple integration is often introduced. We split a surface up into an array of infinitesimal cells and evaluate a function over the surface by summing its value on each of the cells. If we use a Cartesian coordinate system, a typical volume integral then looks like

$$\int_{V} f(x, y, z) \, \mathrm{d}x \mathrm{d}y \mathrm{d}z. \tag{38.16}$$

[4]We met this object back in Section 5.4. Here we will see how the Jacobian arises naturally from the algebra of forms.

However, when we generalize to different coordinate systems, we need to include a factor of the **Jacobian determinant** J, which guarantees that the cells defined by the new coordinates correctly mesh, allowing the volume to be covered without any gaps. Given two coordinate systems, with coordinates $(x^0, ..., x^n)$ and $(y^0, ...y^n)$, the Jacobian determinant is defined as

$$J = \frac{\partial(x^0, ..., x^n)}{\partial(y^0, ..., y^n)} = \begin{vmatrix} \frac{\partial x^0}{\partial y^0} & \cdots & \frac{\partial x^0}{\partial y^n} \\ \vdots & \ddots & \vdots \\ \frac{\partial x^n}{\partial y^0} & \cdots & \frac{\partial x^n}{\partial y^n} \end{vmatrix}. \tag{38.17}$$

The determinant is an antisymmetric object and can be formed using[5] the antisymmetric symbol $\varepsilon_{\mu\alpha\beta\gamma}$.

So how does the Jacobian factor arise? Let's investigate what happens to an element of surface, written as a 2-form $d\alpha^1 \times d\alpha^2$, when a change of coordinates is made.

Example 38.1

Suppose we have a mesh given in terms of coordinates α^1 and α^2 and we want to write it in terms of coordinates x^1 and x^2. We write the basis 1-forms in terms of the new coordinates as follows:

$$d\alpha^1 = \frac{\partial\alpha^1}{\partial x^1}dx^1 + \frac{\partial\alpha^1}{\partial x^2}dx^2,$$

$$d\alpha^2 = \frac{\partial\alpha^2}{\partial x^1}dx^1 + \frac{\partial\alpha^2}{\partial x^2}dx^2. \tag{38.19}$$

In terms of the new coordinates, the areal element is

$$
\begin{aligned}
d\alpha^1 \wedge d\alpha^2 &= \left(\frac{\partial\alpha^1}{\partial x^1}dx^1 + \frac{\partial\alpha^1}{\partial x^2}dx^2\right) \wedge \left(\frac{\partial\alpha^2}{\partial x^1}dx^1 + \frac{\partial\alpha^2}{\partial x^2}dx^2\right) \\
&= \frac{\partial\alpha^1}{\partial x^1}\frac{\partial\alpha^2}{\partial x^2}(dx^1 \wedge dx^2) + \frac{\partial\alpha^1}{\partial x^2}\frac{\partial\alpha^2}{\partial x^1}(dx^2 \wedge dx^1) \\
&= \left(\frac{\partial\alpha^1}{\partial x^1}\frac{\partial\alpha^2}{\partial x^2} - \frac{\partial\alpha^1}{\partial x^2}\frac{\partial\alpha^2}{\partial x^1}\right)(dx^1 \wedge dx^2) \\
&= \varepsilon_{\mu\nu}\frac{\partial\alpha^\mu}{\partial x^1}\frac{\partial\alpha^\nu}{\partial x^2}dx^1 \wedge dx^2 \\
&= \frac{\partial(\alpha^1, \alpha^2)}{\partial(x^1, x^2)}dx^1 \wedge dx^2. \tag{38.20}
\end{aligned}
$$

Where we recognize the Jacobian determinant J in the final line.

We conclude that the Jacobian determinant is generated naturally by the algebra of forms.

38.3 Anatomy of an integral

We saw in the last section how a surface and a form are combined to give a number. This is reminiscent of how a 1-form $\tilde{\alpha}$ and its dual, a vector v, are combined to make a number. This, of course, is done using an inner product $\langle\tilde{\alpha}, v\rangle$. By examining the anatomy of a multiple integral, we shall see how a form can be combined in an inner product with its dual in this context: the generalized surface. This inner product gives us our integral.

A general multiple integral can be written as[6]

$$\mathcal{I} = \int_{\mathcal{S}(\lambda^1, ..., \lambda^n)} f(x^1, ..., x^n)\frac{\partial(x^1, ..., x^n)}{\partial(\lambda^1, ..., \lambda^n)}d\lambda^1, ..., d\lambda^n. \tag{38.21}$$

This says that there's a surface $\mathcal{S}$ parametrized by some variables λ^i. We integrate a function $f(x^1, ..., x^n)$ by summing the function multiplied by a surface element $dx^1...dx^n$ written in the function's coordinates. The

[5] Specifically, we can also write J as a sum

$$J = \varepsilon_{i_0 i_1 ... i_n}\frac{\partial x^{i_0}}{\partial y^0}\frac{\partial x^{i_1}}{\partial y^1}...\frac{\partial x^{i_n}}{\partial y^n}. \tag{38.18}$$

[6] We label coordinates here starting from 1 (rather than zero, as we have been doing for (3+1)-dimensional spacetime).

Jacobian guarantees that the surface elements (given in terms of x^i) mesh to cover the surface (given in terms of λ^i). The key points of this section are that:

• The function and surface element together can be represented as an n-form $\tilde{\alpha}$;

• The surface can be represented as a n-vector $\boldsymbol{S}$;

• The integrand, complete with Jacobian, can be recreated by taking the inner product of the n-form $\tilde{\alpha}$ and n-vector $\boldsymbol{S}$;

• Geometrically, the integral is an inner product that counts the number of n-cells representing $\tilde{\alpha}$, contained in the n-parallelepiped that represents $\boldsymbol{S}$.

Let's break this up into parts. We first look at the function and surface element that is contained in the form $\tilde{\alpha}$, which looks like

$$\tilde{\alpha} = f(x^1, ..., x^n)\, \boldsymbol{dx}^1 \wedge ... \wedge \boldsymbol{dx}^n. \tag{38.22}$$

This clearly falls apart into (i) a function and (ii) a surface element built from forms. The form in the surface element has the necessary properties to naturally give the Jacobian matrix required for our choice of coordinates. In the next example, we remind ourselves how to represent a 4-volume and 3-volume in terms of forms.

Example 38.2

Working in (3+1)-dimensional flat space with coordinates (t, x, y, z), we can interpret the 4-volume element as a volume 4-form[7]

$$\tilde{\omega}(\ ,\ ,\ ,\ ) = \frac{1}{4!}\varepsilon_{\mu\nu\alpha\beta}\, \boldsymbol{dx}^\mu \wedge \boldsymbol{dx}^\nu \wedge \boldsymbol{dx}^\alpha \wedge \boldsymbol{dx}^\beta$$

$$= \varepsilon_{|\mu\nu\alpha\beta|}\, \boldsymbol{dx}^\mu \wedge \boldsymbol{dx}^\nu \wedge \boldsymbol{dx}^\alpha \wedge \boldsymbol{dx}^\beta$$

$$= \varepsilon_{0123}\, \boldsymbol{dt} \wedge \boldsymbol{dx} \wedge \boldsymbol{dy} \wedge \boldsymbol{dz},$$

$$= \boldsymbol{dt} \wedge \boldsymbol{dx} \wedge \boldsymbol{dy} \wedge \boldsymbol{dz}, \tag{38.23}$$

where we've used the $|\alpha\beta\gamma\delta|$ notation.[8] (One message here is that we can leave the sum unrestricted and include the factor 4!, or restrict the sum.)

In the same way, the 3-volume element can be provided by a 3-form[9]

$$\boldsymbol{d\tilde{\Sigma}}_\mu(\ ,\ ,\ ) = \frac{1}{3!}\varepsilon_{\mu\alpha\beta\gamma}\, \boldsymbol{dx}^\alpha \wedge \boldsymbol{dx}^\beta \wedge \boldsymbol{dx}^\gamma$$

$$= \varepsilon_{\mu|\alpha\beta\gamma|}\, \boldsymbol{dx}^\alpha \wedge \boldsymbol{dx}^\beta \wedge \boldsymbol{dx}^\gamma, \tag{38.24}$$

In many applications, we shall integrate 3-form objects written as $J^\mu \boldsymbol{d\tilde{\Sigma}}_\mu$ over some surface. In terms of forms, this is to be interpreted as

$$J^\mu \boldsymbol{d\tilde{\Sigma}}_\mu = \frac{1}{3!} J^\mu \varepsilon_{\mu\alpha\beta\gamma}\, \boldsymbol{dx}^\alpha \wedge \boldsymbol{dx}^\beta \wedge \boldsymbol{dx}^\gamma$$

$$= J^\mu \varepsilon_{\mu|\alpha\beta\gamma|}\, \boldsymbol{dx}^\alpha \wedge \boldsymbol{dx}^\beta \wedge \boldsymbol{dx}^\gamma. \tag{38.25}$$

With the integrand identified as a form, we now attempt to describe the generalized surface as an n-vector. This is possible since, just as n-forms are built from wedge products of 1-forms, the generalized surface over

[7] This is the volume 4-form we met in Chapter 37. This 4-form is dual to the number 1, i.e. $\tilde{\omega} = \star(1)$.

[8] Recall that this notation forces $\alpha < \beta < \gamma < \delta$.

[9] The 3-form $\boldsymbol{d\tilde{\Sigma}}_\mu$ is dual to a basis vector $\boldsymbol{e}_\mu$, i.e. $\boldsymbol{d\tilde{\Sigma}}_\mu = \star\boldsymbol{e}_\mu$. If we fill the three slots of the tensor $\boldsymbol{d\tilde{\Sigma}}_\mu$, we obtain the components of the 3-volume 1-form $\boldsymbol{d\tilde{\sigma}}(\boldsymbol{e}_\mu)$ from the last chapter.

which we integrate can be built from wedge products of vectors. Let's consider doing a general multiple integral

$$I = \int_{\mathcal{S}} \tilde{\alpha}, \tag{38.26}$$

where the form $\tilde{\alpha}$ is integrated over the n-dimensional surface $\mathcal{S}$. To see how to construct an element of generalized n-dimensional surface $\mathcal{S}$, consider the vicinity of a points $\mathcal{P}$ on a surface $\mathcal{P}(\lambda^1, \lambda^2, ..., \lambda^n)$, where λ^i are the coordinates parametrizing the surface. A small displacement in the λ^i direction is written as $\left(\partial\mathcal{P}/\partial\lambda^i\right)\Delta\lambda^i$, which is formed from a vector $\partial\mathcal{P}/\partial\lambda^i$ and infinitesimal length element $\Delta\lambda^i$. In building a surface, we can take the wedge products of the vector parts to form a generalized n-dimensional parallelepiped. Therefore, tangent to the surface $\mathcal{S}$ at point $\mathcal{P}$ is an infinitesimal parallelepiped built from vectors

$$\frac{\partial\mathcal{P}}{\partial\lambda^1}\Delta\lambda^1 \wedge \frac{\partial\mathcal{P}}{\partial\lambda^2}\Delta\lambda^2 \wedge ... \wedge \frac{\partial\mathcal{P}}{\partial\lambda^n}\Delta\lambda^n$$

$$= \left(\frac{\partial\mathcal{P}}{\partial\lambda^1} \wedge \frac{\partial\mathcal{P}}{\partial\lambda^2} \wedge ... \wedge \frac{\partial\mathcal{P}}{\partial\lambda^n}\right)\Delta\lambda^1\Delta\lambda^2...\Delta\lambda^n. \tag{38.27}$$

To perform the integral, we evaluate how many infinitesimal cells represented by the integration form $\tilde{\alpha}$ are cut by the surface. Recall that the machine that works this out is the inner product,[10] since this tells us how a n-form and n-vector mesh together.[11] We therefore interpret the integral symbol as follows:

$$\int_{\mathcal{S}} \tilde{\alpha} = \int \int ... \int \left\langle \tilde{\alpha}, \frac{\partial\mathcal{P}}{\partial\lambda^1} \wedge \frac{\partial\mathcal{P}}{\partial\lambda^2} \wedge ... \wedge \frac{\partial\mathcal{P}}{\partial\lambda^n} \right\rangle \mathrm{d}\lambda^1\mathrm{d}\lambda^2...\mathrm{d}\lambda^n. \tag{38.30}$$

Let's see how this generates the Jacobian. Consider a basis forms $dx^1, ..., dx^n$ such that the form is written as

$$\alpha = \alpha(x^1, ..., x^n)dx^1 \wedge dx^2 \wedge ... \wedge dx^n, \tag{38.31}$$

then we have, in the inner product,

$$\left\langle dx^1 \wedge dx^2 \wedge ... \wedge dx^n, \frac{\partial\mathcal{P}}{\partial\lambda^1} \wedge \frac{\partial\mathcal{P}}{\partial\lambda^2} \wedge ... \wedge \frac{\partial\mathcal{P}}{\partial\lambda^n} \right\rangle = \frac{\partial(x^1, x^2, ..., x^n)}{\partial(\lambda^1, \lambda^2, ..., \lambda^n)}. \tag{38.32}$$

We check this gives the expected answer for the simplest possible case in the next example.

Example 38.3

Consider a line integral $\int_{\text{curve}} df$. This can be done by inspection: the answer is simply the difference in the values of the function f at the start and end of the curve.[12] We parametrize the curve with $0 \leq \lambda \leq 1$ and we have

$$\int_{\text{curve}} df = \int_0^1 \left\langle df, \frac{\partial\mathcal{P}}{\partial\lambda} \right\rangle \mathrm{d}\lambda$$

$$= \int_0^1 \frac{\mathrm{d}f}{\mathrm{d}\lambda}\mathrm{d}\lambda = f[\mathcal{P}(1)] - f[\mathcal{P}(0)]. \tag{38.33}$$

[10] Consider a 2-form $\tilde{F}$ and a bivector $u \wedge v$. The contraction

$$\langle \tilde{F}, u \wedge v \rangle, \tag{38.28}$$

tells us to evaluate how much many tubes of $\tilde{F}$ are cut by the bivector $u \wedge v$. Looking ahead, we put the material in this section to work in Chapter 43 by identifying an infinitesimal surface $\Delta x e_x \wedge \Delta y e_y$ and evaluating the integral

$$\int_{\Delta x e_x \wedge \Delta y e_y} \tilde{F}$$

$$= \int \left\langle \tilde{F}, e_x \wedge e_y \right\rangle \Delta x \Delta y$$

$$= \int \tilde{F}(e_x, e_y)\Delta x \Delta y$$

$$= \int F_{xy}\Delta x \Delta y. \tag{38.29}$$

See Chapter 43 for the details.

[11] Remember the rule that $\langle \tilde{\alpha}, v \rangle = \alpha_{|i_1...i_n|}v^{i_1...i_n}$.

[12] This integral is independent of the path taken by the curve, showing that the form df corresponds to a conservative force field. We can generalize that the integral of an exact 1-form df is path independent, and given by the change in the function f along the curve.

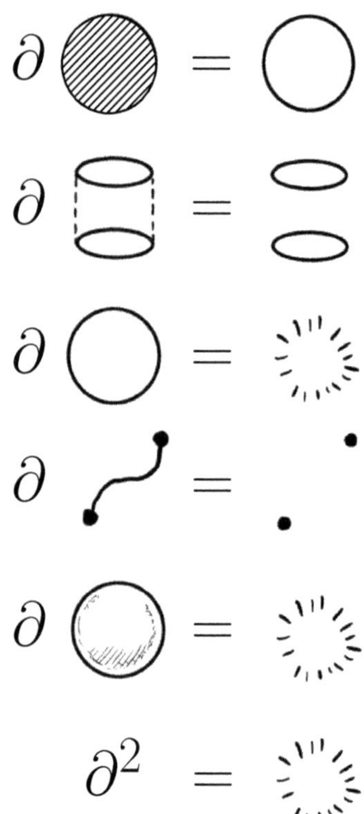

Fig. 38.1 Boundaries for several of objects. In some cases, for spaces with no boundary, we generate the empty set (shown schematically as the space vanishing 'in a puff of logic', in Douglas Adams' memorable phrase). This figure illustrates some of the examples given in Example 38.4.

[13]The concept of chains comes from algebraic topology, where the chain group is an algebraic structure composed by constructing linear combinations of 'simplicial complexes', the idea being that a topological space can be constructed by gluing together little triangles or tetrahedra, or other high-dimensional basic objects (the simplicial complexes). For more details, see e.g. Nakahara (1990), but this algebraic treatment is not important for our purposes here and we can just think of chains simply as topological spaces.

We recover the expected result.

We conclude that an integral is an inner product between a form and a surface. Since it outputs a number, the surfaces must be dual to the forms. In the next section, we examine the properties of the surfaces in more detail.

38.4   Boundaries and chains

Here we examine the properties of the generalized surfaces we integrate over. First a definition: the **boundary** ∂M of a region of M, consists of those points of M that do not lie in the interior.

Example 38.4
The boundary of the closed unit disc is the unit circle.
The boundary of a finite cylinder is two circles.
The boundary of a line is its two end points.
The boundary of the unit sphere is empty (i.e. there isn't one).

We regard ∂ as an operator that extracts the boundary from a space M. Some examples of the use of this operator are shown in Fig. 38.1.

With boundaries at our disposal we turn to the main topic of this section. The generalized surfaces over which we integrate form a family of objects we call **chains**.[13] As described above, forms are **dual** to chains, in that we combine them to output a number. We define and denote our chains as follows:

$$
\begin{aligned}
C_0 \quad & \text{0-chain} = \text{point}, \\
C_1 \quad & \text{1-chain} = \text{line}, \\
C_2 \quad & \text{2-chain} = \text{area (or 2-volume)}, \\
C_3 \quad & \text{3-chain} = \text{3-volume}, \\
C_n \quad & n\text{-chain} = n\text{-volume}.
\end{aligned}
\tag{38.34}
$$

The boundary operator ∂ acts on the chains, mapping a chain C_n on to a chain C_{n-1}. That is, it inputs an n-chain and outputs an $(n-1)$-chain. We have, therefore, that

$$\partial C_n = C_{n-1}. \tag{38.35}$$

Some chains have no boundaries. A closed line like a circle has no boundary. We call these boundary-free chains **closed chains** or **cycles**, often denoted Z_n with the property $\partial Z_n = 0$. Some chains are boundaries to higher dimensional chains. A closed surface is boundary to a volume; a closed line is boundary to a surface; the boundary of a line is two points. In short, the boundary of an $n+1$-chain is an n-chain. If B_n is the boundary of an $(n+1)$-chain we have

$$B_n = \partial C_{n+1}. \tag{38.36}$$

However, boundaries themselves are closed: they *are* boundaries, but don't themselves have boundaries. So any boundary has

$$\partial B_n = 0. \tag{38.37}$$

Combining the two previous equations we see that $\partial\partial C_{n+1} = 0$. Since there was nothing special about C_{n+1} we conclude that

$$\partial\partial = \partial^2 = 0, \tag{38.38}$$

or, in words, **the boundary of a boundary is zero**.[14]

38.5 Stokes' theorem

An integral involves the inner product of an n-chain and an n-form. The pinnacle of this description of calculus in terms of chains and forms is **Stokes' theorem**.[15] In order to understand the content of the theorem, we need to bring in one final (but familiar) piece of technology: the exterior derivative d. The key to Stokes' theorem is to note that just as the boundary operator ∂ takes a $(n+1)$-chain and outputs an n-chain, the exterior derivative operator d takes a n form and outputs a $(n+1)$-form.

Example 38.5

Recall that in three dimensions with basis 1-forms dx, dy and dz, we have the following forms:

$$
\begin{aligned}
\text{0-form} \quad & \tilde{Z} = f, \\
\text{1-form} \quad & \tilde{Y} = A\,dx + B\,dy + C\,dz, \\
\text{2-form} \quad & \tilde{X} = a\,dx \wedge dy + b\,dy \wedge dz + c\,dz \wedge dx, \\
\text{3-form} \quad & \tilde{W} = F\,dx \wedge dy \wedge dz.
\end{aligned}
\tag{38.39}
$$

We can operate with the exterior derivative operator on the 1-form to make

$$
\begin{aligned}
d\tilde{Y} &= \frac{\partial A}{\partial y}\,dy \wedge dx + \frac{\partial A}{\partial z}\,dz \wedge dx \\
&+ \frac{\partial B}{\partial x}\,dx \wedge dy + \frac{\partial B}{\partial z}\,dz \wedge dy \\
&+ \frac{\partial C}{\partial x}\,dx \wedge dz + \frac{\partial C}{\partial y}\,dy \wedge dz \\
&= \left(\frac{\partial C}{\partial y} - \frac{\partial B}{\partial z}\right) dy \wedge dz + \left(\frac{\partial A}{\partial z} - \frac{\partial C}{\partial x}\right) dz \wedge dx \\
&+ \left(\frac{\partial B}{\partial x} - \frac{\partial A}{\partial y}\right) dx \wedge dy.
\end{aligned}
\tag{38.40}
$$

However, if we operate a second time, we find $dd\tilde{Y} = 0$. This property that two operations of the exterior derivative operator yields zero is a general one. Trying again by starting on the 2-form, we find

$$d\tilde{X} = \left(\frac{\partial a}{\partial z} + \frac{\partial b}{\partial x} + \frac{\partial c}{\partial y}\right) dx \wedge dy \wedge dz, \tag{38.41}$$

but then $dd\tilde{X} = 0$.

---

[14] A chain which is a boundary is closed. Is a closed chain always the boundary of another chain? In a simply connected space: yes; in general: no. Note that a space is described as being 'simply connected' if any path between two points can be continuously deformed into any other path joining those points. This relies on there being no holes in the space, for example.

[15] As pointed out in the book by Needham (2021), it is rather inappropriate that this theorem is named after Stokes alone. Vladimir Arnold jokingly calls it the **Newton–Leibniz–Gauss–Green–Ostragradskii–Stokes–Poincaré formula**. In view of this difficulty, Penrose has promoted calling it the **fundamental theorem of exterior calculus**. The version that we describe was first introduced by Cartan in a lecture course he gave in 1937. The history of the theorem is sketched out in V. J. Katz, *The History of Stokes' Theorem*, Mathematics Magazine, **52**, 146 (1979)

Each of the results from the last example are summarized with the expression, familiar from Chapter 33 that

$$dd = 0. \tag{38.42}$$

This equation is the dual equation of $\partial^2 = 0$.[16]

With the operators d and ∂ in hand, we can write down Stokes' theorem. The content of the theorem reflects the ability to use the operators to equate the inner product of an $(n+1)$-chain C and $(n+1)$-form $d\tilde{\Omega}$, with that of an n-chain ∂C and n-form $\tilde{\Omega}$.

Stokes' theorem says that if $\tilde{\Omega}$ is a n-form and C is a $(n+1)$-chain, then we have

$$\int_{\partial C} \tilde{\Omega} = \int_C d\tilde{\Omega}, \tag{38.43}$$

where the orientation of the surface ∂C must be chosen[17] such that if the orientation for C is

$$\frac{\partial \mathcal{P}}{\partial \lambda^1} \wedge \frac{\partial \mathcal{P}}{\partial \lambda^2} \wedge \dots \wedge \frac{\partial \mathcal{P}}{\partial \lambda^n}, \tag{38.44}$$

then the orientation for ∂C is

$$\frac{\partial \mathcal{P}}{\partial \lambda^2} \wedge \dots \wedge \frac{\partial \mathcal{P}}{\partial \lambda^n}. \tag{38.45}$$

We shall not prove Stokes' theorem here,[18] instead, we put the theorem to work in the examples below. We warm up with the computation of some volume elements.

Example 38.6

Working in a coordinate frame in three-dimensional space with coordinate system (x, y, z) and a metric g, we want to integrate a function $f(x, y, x)$ over a 3-volume V. The volume 3-form[19] for the space is given by

$$\tilde{\omega}(\ ,\ ,\ ) = \sqrt{g}\, dx(\ ) \wedge dy(\ ) \wedge dz(\ ). \tag{38.46}$$

The chain $C_3 = V$ tells us the size of the volume over which to integrate, using an integration volume element dV. The latter is a box with sides $dx e_x$, $dy e_y$ and $dz e_z$, such that we have $dV = \tilde{\omega}(dx e_x, dy e_y, dz e_z) = \sqrt{g}\, dx dy dz$. A volume integral is then written as

$$\int_{C_3} f\tilde{\omega} = \int_{C_3} f\sqrt{g}\, dx \wedge dy \wedge dz = \int_V f(x, y, z)\sqrt{g}\, dx dy dz. \tag{38.47}$$

In flat space, of course, we have $g = 1$.

Example 38.7

Let's integrate a function $f(x, y, z)$ over an area S in flat space.[20] One way to construct the integration surface element is to identify an element spanned by vectors $d\boldsymbol{u}$ and $d\boldsymbol{v}$ with unit outward normal $\hat{\boldsymbol{n}}$ and write as

$$\tilde{\omega}(d\boldsymbol{u}, d\boldsymbol{v}, \hat{\boldsymbol{n}}) = \hat{\boldsymbol{n}} \cdot d\boldsymbol{u} \times d\boldsymbol{v} = \hat{\boldsymbol{n}} \cdot d\boldsymbol{S}, \tag{38.48}$$

---

[16]An n-form $\tilde{\boldsymbol{A}}$ is **closed** if $d\tilde{\boldsymbol{A}} = 0$. An n-form is called exact if it is the derivative of an $(n-1)$-form. All exact forms are clearly closed. Are all closed forms exact? That is, is the property that the exterior derivative gives zero enough to guarantee that we can write the form as the exterior derivative of another object. The **Poincaré lemma** says that if $d\tilde{\boldsymbol{A}} = 0$ throughout a simply connected region of space, then $\tilde{\boldsymbol{A}} = d\tilde{\boldsymbol{B}}$ for some $\tilde{\boldsymbol{B}}$ and so the form is exact. A space that is not simply connected is discussed in the exercises.

[17]This fussy definition really just means we need to define a consistent positive sense of pointing away from the volume C.

[18]Stokes' theorem is proved in Needham (2021) and, at a more advanced level, in Spivak's *Calculus on Manifolds*, both of which are highly recommended. Of Stokes' theorem, Spivak notes: '1. It is trivial. 2. It is trivial because the terms appearing in it have been properly defined. 3. It has significant consequences.'

[19]Recall that in the coordinate frame, we need a factor $[(-1)^s \det g_{\mu\nu}]^{\frac{1}{2}}$, where s is the number of minuses in the metric signature. We'll assume here that, since we're looking at three-dimensional space, $s = 0$ and write $\det g_{\mu\nu} = g$.

[20]In curved space, or if we weren't using Cartesian coordinates, we would need to include the extra factor $\sqrt{g}$ once again in the definition of the volume element.

where $\mathrm{d}\boldsymbol{S} = \mathrm{d}\boldsymbol{u} \times \mathrm{d}\boldsymbol{v}$. We spot that the expression in eqn 38.48 is equivalent to the 2-form

$$\tilde{\boldsymbol{\Sigma}}(\ ,\ ) = \hat{n}^x \boldsymbol{dy} \wedge \boldsymbol{dz} + \hat{n}^y \boldsymbol{dz} \wedge \boldsymbol{dx} + \hat{n}^z \boldsymbol{dx} \wedge \boldsymbol{dy}, \tag{38.49}$$

when its slots are filled with $\mathrm{d}\boldsymbol{u}$ and $\mathrm{d}\boldsymbol{v}$. We can therefore perform the surface integral

$$\int_{C_2} f\tilde{\boldsymbol{\Sigma}} = \int_S f(x,y,z)\hat{\boldsymbol{n}} \cdot \mathrm{d}\boldsymbol{S}. \tag{38.50}$$

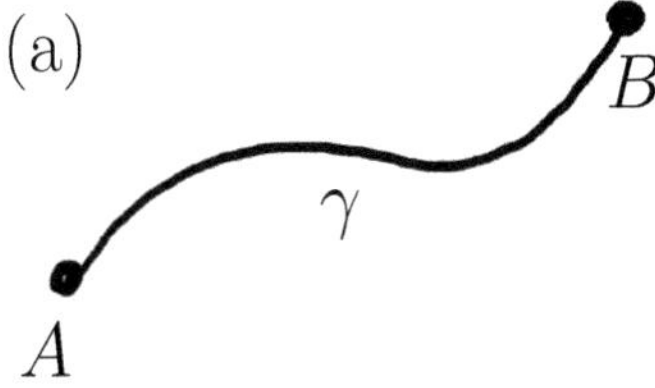

Example 38.8

We can now use Stokes' theorem to see how the flat-space integral theorems of vector calculus can be described in the new language by considering a function f (i.e. a 0-form) integrated over the geometries shown in Fig. 38.2. In Fig. 38.2(a), we have

$$\int_\gamma \boldsymbol{d}f = f[\mathcal{P}(B)] - f[\mathcal{P}(A)], \tag{38.51}$$

as we saw in Example 12. Similarly, for Fig. 38.2(b) we have

$$\int_S \boldsymbol{d}f \wedge \boldsymbol{dx}^i = \int_\Gamma f\boldsymbol{dx}^i, \tag{38.52}$$

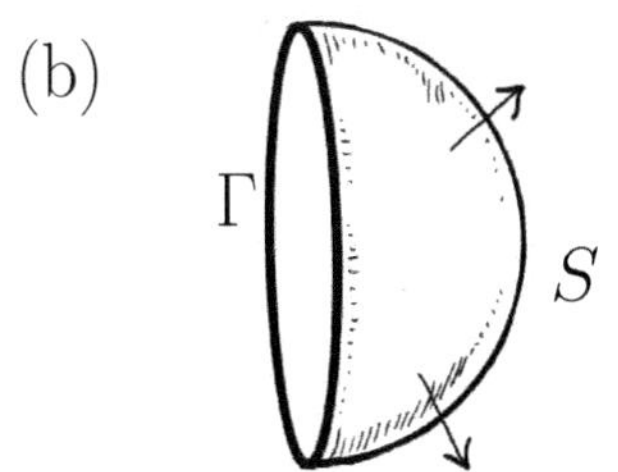

and for the volume V in Fig. 38.2(c) we obtain

$$\int_V \boldsymbol{d}f \wedge \boldsymbol{dx}^i \wedge \boldsymbol{dx}^j = \int_\Sigma f\boldsymbol{dx}^i \wedge \boldsymbol{dx}^j. \tag{38.53}$$

To investigate this final geometry in more detail, consider the 2-form

$$\tilde{\boldsymbol{A}} = A_x \boldsymbol{dy} \wedge \boldsymbol{dz} + A_y \boldsymbol{dz} \wedge \boldsymbol{dx} + A_z \boldsymbol{dx} \wedge \boldsymbol{dy}, \tag{38.54}$$

and the 3-chain $C_3 = V$, a volume in flat space, with boundary $\partial V = \Sigma$. We then have, by Stokes' theorem

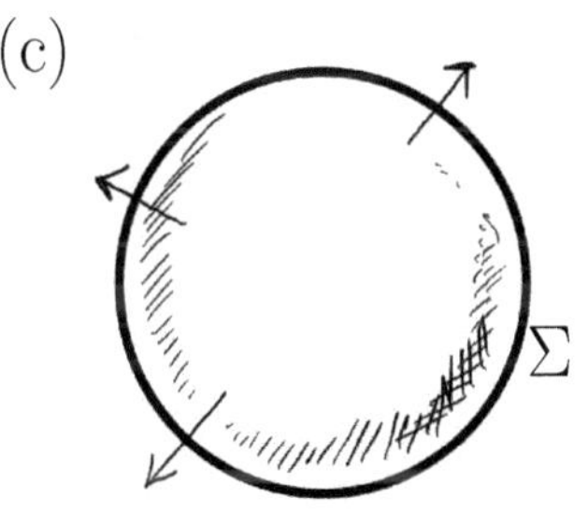

$$\int_\Sigma A_x \boldsymbol{dy} \wedge \boldsymbol{dz} + A_y \boldsymbol{dz} \wedge \boldsymbol{dx} + A_z \boldsymbol{dx} \wedge \boldsymbol{dy}$$
$$= \int_V \left(\frac{\partial A_x}{\partial x} + \frac{\partial A_y}{\partial y} + \frac{\partial A_z}{\partial z}\right) \boldsymbol{dx} \wedge \boldsymbol{dy} \wedge \boldsymbol{dz}. \tag{38.55}$$

In ordinary Cartesian coordinates, this is equivalent to

$$\int_\Sigma A_x \mathrm{d}y\mathrm{d}z + A_y \mathrm{d}z\mathrm{d}x + A_z \mathrm{d}x\mathrm{d}y = \int_V \vec{\nabla} \cdot \vec{A}\,\mathrm{d}x\mathrm{d}y\mathrm{d}z, \tag{38.56}$$

or, in terms of the element $\mathrm{d}\vec{\Sigma}$ of the surface Σ and element $\mathrm{d}V$ of the volume V,

$$\int_\Sigma \vec{A} \cdot \mathrm{d}\vec{\Sigma} = \int_V \vec{\nabla} \cdot \vec{A}\,\mathrm{d}V, \tag{38.57}$$

which is the divergence theorem.

Fig. 38.2 Geometries for the theorems of integral calculus.

Stokes' theorem is an essential part of our geometrical description of the conservation of charge in electromagnetism, and also of the constraints on curvature caused by the geometry of spacetime. These topics will be introduced as we take a closer look at the physics of fields, which forms the final part of the book.

Chapter summary

- All integrals can be expressed as an integral over a form. The integral expresses the inner product of the n-form and the n-vector that represents the surface over which the integral is carried out.
- The general surfaces over which we integrate are chains. The boundaries of chains are found with the boundary operator.
- $\partial\partial = 0$ expresses the fact that a boundary of a boundary is zero.
- Stokes' theorem says that

$$\int_{\partial C} \tilde{\Omega} = \int_{C} d\tilde{\Omega}. \tag{38.58}$$

Exercises

(38.1) Consider the 1-form $\tilde{\boldsymbol{A}} = A_i dx^i$ in flat, three-dimensional space. Use Stokes' theorem to prove the Kelvin–Stokes theorem

$$\int_{S} (\vec{\nabla} \times \vec{A}) \cdot d\vec{S} = \oint_{\partial S} \vec{A} \cdot d\vec{l}. \tag{38.59}$$

(38.2) The ball S^2 has no boundary.
(a) Prove that if a 2-form $\tilde{\Omega}$ is exact on S^2, we must have

$$\int_{S^2} \tilde{\Omega} = 0. \tag{38.60}$$

(b) Show that the 2-form $\tilde{\boldsymbol{G}} = x^1 \boldsymbol{dx}^2 \wedge \boldsymbol{dx}^3$ defined in three-dimensional space, is not exact on S^2.

(38.3) Consider the vortex 1-form field

$$\tilde{\phi} = \frac{A}{r^2} (-y\boldsymbol{dx} + x\boldsymbol{dy}), \tag{38.61}$$

where $r^2 = x^2 + y^2$ and A is a constant. This form is not exact since $\tilde{\phi} \neq \boldsymbol{df}$, for any function f. Show, however, that the vortex form is closed, with $\boldsymbol{d}\tilde{\phi} = 0$.
This is an example of a closed form that is not exact. Although we appear to be working in Euclidean 2-space (which is simply connected) where we expect closed forms to be exact, the field is actually singular at the origin. As a result, we must remove this troublesome point. We are not therefore really dealing with a simply connected space and so Poincaré's lemma does not apply.

(38.4) (a) By computing connection coefficients in the orthonormal frame, show that the geodesic equation on the surface of a unit sphere can be written as

$$\begin{pmatrix} \ddot{x}^{\hat{1}} \\ \ddot{x}^{\hat{2}} \end{pmatrix} = -\dot{\phi}\cos\theta \begin{pmatrix} 0 & -1 \\ 1 & 0 \end{pmatrix} \begin{pmatrix} \dot{x}^{\hat{1}} \\ \dot{x}^{\hat{2}} \end{pmatrix}, \tag{38.62}$$

where $(x^1, x^2) = (\theta, \phi)$ and dots denote derivatives with respect to the proper time τ.
This equation can be solved by a velocity of the form $\boldsymbol{u}(t) = e^{\boldsymbol{A}(t)}\boldsymbol{u}(0)$ where, for our chosen basis,

$$\underline{\boldsymbol{A}}(t) = -\int_0^t d\tau \cos\theta(\tau)\dot{\phi}(\tau) \begin{pmatrix} 0 & -1 \\ 1 & 0 \end{pmatrix}. \tag{38.63}$$

(b) Use Stokes' theorem to verify Green's theorem in the plane

$$\oint_\gamma (L dx + M dy) = \int_S dx dy \left(\frac{\partial M}{\partial x} - \frac{\partial L}{\partial y} \right). \tag{38.64}$$

(c) Use Green's theorem in the plane to show that if the motion on the sphere forms a loop after a time T, then the integral $\mathcal{I}$ gives

$$\mathcal{I} = \int_0^T d\tau \cos\theta(\tau)\dot{\phi}(\tau) = -\mathcal{A}, \tag{38.65}$$

where $\mathcal{A}$ is the area enclosed by the loop.
(d) Hence, show that if the vector $\boldsymbol{u}$ is parallel

transported around a closed loop, the result can be described by the operation $\underline{\boldsymbol{R}}\boldsymbol{u}$, where $\underline{\boldsymbol{R}}$ is a matrix corresponding to rotation by an angle $\mathcal{A}$.

See Blennow and Ohlsson, whose approach we have followed here, for a more complete discussion of this problem. The result of this problem is a useful one in magnetism for computing geometric phases.

Part VI

Classical and quantum fields

Having sharpened our geometrical tools in Part V, we now apply these ideas to a geometrical description of the physics of general relativity. Our goal is to understand the physics of relativity from both the point of view of differential geometry and also from that of classical field theory. 'Classical' here means pre-quantum, but including relativity. The extension to quantum physics is our focus towards the end of this part of the book.

- In Chapter 39, we examine the physics of fluids, including their equations of motion and thermodynamics.

- In Chapter 40, we approach the Einstein equations from the point of view of classical field theory where using the idea of a Lagrangian density we can see how general relativity fits into the framework of field theory. We apply this to the problem of the early Universe in Chapter 41.

- In Chapter 42, we formulate a geometrical description of electro-magnetism. This is developed in Chapter 43, where we use differential geometry to understand the conservation of charge.

- In Chapter 44, we examine the link between relativity and the very influential idea of a gauge in a field theory.

- In Chapter 45, we examine gravitation in the domain where it is usually encountered: the limit of small field. This enables us to discuss gravitational radiation in Chapter 46.

- In Chapters 47–49, we investigate how quantum mechanics might be linked with gravitation, and how having extra dimensions available might enable this.

- Finally, in Chapter 50, we show the inevitability of a Big Bang from our geometrical viewpoint and its reliance on smooth spacetimes.

<table>
<tr><td>

39

</td><td>

Fluids as dry water

</td></tr>
</table>

> ⤳ **The material in this chapter fills in some background on how fluids are treated in physics. It can be skipped on a first reading.**

[1] This approach is adopted in Vol. II, Lecture 40 of the *Feynman Lectures in Physics*, R.P. Feynman, R. Leighton and M. Sands. In the lecture after this, Feynman examines some of the consequences of including viscosity, and the interested reader is advised to look there in the first instance.

[2] We shall distinguish the mass density ρ_0 of the fluid and the energy density ρc^2, which includes a contribution from $\rho_0 c^2$ and also sources of internal energy.

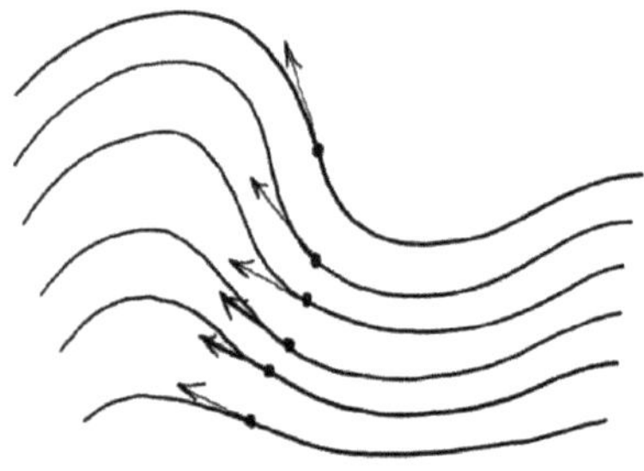

Fig. 39.1 Flow lines form a congruence of curves. The tangents give the velocity field.

He who drinks a tumbler of water in London has literally in his stomach more animated beings than there are men, women and children on the face of the globe.
Sydney Smith (1771–1845)

A particularly satisfying classical field theory is that of the fluid. In this chapter, we start by examining a non-relativistic fluid in flat space-time, and formulate mass conservation and an equation of motion, and then look at the consequences of energy conservation. We then turn to the problem of the relativistic fluid. At the risk of being perverse, we shall ignore exactly that feature that gives fluids their characteristic wetness: viscosity. This omission is made because the fluids we describe in the context of relativity are well described as **inviscid** (i.e. viscosity-free). This approach[1] also has the attractive property of making the subject as straightforward as possible.

A **fluid** is matter that continuously deforms if it is subject to a shear stress. In modelling a fluid, we identify elements of the fluid, initially at some coordinate $x^\mu = (t, \vec{x})$, that we expect to move around. This movement is called a **flow**. To describe a fluid, we specify its mass density,[2] described by a scalar field $\rho_0(t, \vec{x})$ and its velocity, which is described by a vector field $\vec{v}(t, \vec{x})$.

The velocity is best thought of as being a tangent field. The tangents in question are tangents to a congruence of curves. As usual, the congruence fills all of the space, so that there is a single tangent at a particular point. This tangent is identical to the velocity vector field evaluated at that point (Fig. 39.1). The congruent curves are known as the **streamlines** of the fluid. (They are also known as the **flow lines**, and they do indeed describe the flow of the fluid.) Defined in this way, the flow lines tell us about the *velocities* of the elements of the fluid, rather than their trajectories. Flow lines are only identical to the trajectories of the fluid elements in the case of **steady flow** of the fluid.

We want to ensure that the mass of the fluid is **locally conserved**, which is to say that (in the absence of sources and sinks of mass), the change of fluid mass in a particular volume is accounted for by the flow of mass into or out of that volume. We define the fluid's mass current vector $\boldsymbol{J}$, which has components $J^\mu = (\rho_0, \rho_0 \vec{v})$. Conservation of mass is then written as $\boldsymbol{\nabla} \cdot \boldsymbol{J} = 0$, or in components

$$\frac{\partial \rho_0}{\partial t} + \vec{\nabla} \cdot (\rho_0 \vec{v}) = 0. \tag{39.1}$$

Finally, a useful quantity known as the **vorticity** of the fluid is defined as

$$\vec{\omega} = \vec{\nabla} \times \vec{v}. \tag{39.2}$$

A fluid with $\vec{\omega} = 0$ is called **irrotational**.

39.1 Euler's equation

We would like to identify an equation of motion for the fluid. We shall do this non-relativistically by using Newton's second law. If a volume of fluid has a **pressure** $p(t, \vec{x})$ then we have access to the local force on an element of fluid

$$\vec{F} = -\oint \mathrm{d}\vec{S}\, p(t, \vec{x}) = -\int \mathrm{d}^3x\, \vec{\nabla}p, \tag{39.3}$$

where $\mathrm{d}\vec{S}$ is a directed element of area of the fluid element, and d^3x is an element of volume (Fig. 39.2). The equation effectively defines the pressure of the fluid in terms of it providing an inward-directed force (hence the minus sign) on an element.[3] The product of mass and acceleration of the fluid is

$$\int \mathrm{d}^3x\, \rho_0 \frac{\mathrm{d}\vec{v}}{\mathrm{d}t}. \tag{39.4}$$

Equating these two expressions[4] gives an equation of motion

$$\rho_0 \frac{\mathrm{d}\vec{v}}{\mathrm{d}t} = -\vec{\nabla}p. \tag{39.6}$$

At this point, we need to pause and consider the details of the frames of reference we are using. The velocity field is a function of time and position. The total time derivative $\mathrm{d}/\mathrm{d}t$ we have assumed does not fix the space coordinates like a partial derivative does.[5] The total derivative therefore corresponds to the rate of change of the velocity as observed by an observer who travels along with fluid, so that their spatial coordinate is carried along with the element of fluid under consideration. This is the same situation we had in Chapter 15 where we described the observer as **comoving**. We conclude that the comoving observer probes the fluid using the total derivative $\mathrm{d}/\mathrm{d}t$.

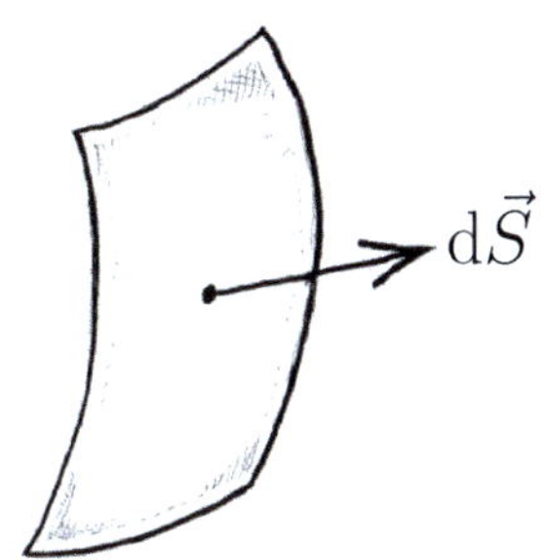

Fig. 39.2 An outwardly directed element of fluid surface at pressure p used to define the force.

[3]The surface integral in eqn 39.3 is taken over the surface of the element of fluid, on which the pressure of the rest of the fluid acts. From the volume integral we see immediately that if there is no gradient in the pressure field, there is no net force on the element, which makes sense.

[4]That is, we find that

$$\int \mathrm{d}^3x \left(\rho_0 \frac{\mathrm{d}\vec{v}}{\mathrm{d}t} + \vec{\nabla}p \right) = 0, \tag{39.5}$$

and so the integrand must vanish.

[5]The partial derivative with respect to time should be interpreted as reading

$$\left(\frac{\partial}{\partial t} \right)_{\text{fixed position}}.$$

Example 39.1

The total derivative describes the rate of change measured by an observer moving locally with the fluid. However, we typically have access to the partial derivative: the rate of change of the fluid from our fixed frame of reference, relative to which the fluid is moving. We can relate these quantities. Starting with the differential for the ith component of a vector field $\vec{A}(t, \vec{x})$, we find

$$\mathrm{d}A^i = \frac{\partial A^i}{\partial t}\mathrm{d}t + \frac{\partial A^i}{\partial x^j}\mathrm{d}x^j$$

$$\frac{\mathrm{d}A^i}{\mathrm{d}t} = \frac{\partial A^i}{\partial t} + \frac{\partial A^i}{\partial x^j}\frac{\mathrm{d}x^j}{\mathrm{d}t}$$

$$= \frac{\partial A^i}{\partial t} + v^j \frac{\partial A^i}{\partial x^j} = \left(\frac{\partial}{\partial t} + v^j \frac{\partial}{\partial x^j} \right) A^i, \tag{39.7}$$

where $v^j = \partial x^j / \partial t$ is a component of the fluid's velocity. This expression provides a useful link between total derivative (appropriate for the comoving observer) and partial derivative (appropriate for the stationary observer).

Using the previous example, we have the useful equation in vector notation, that

$$\frac{\mathrm{d}\vec{A}}{\mathrm{d}t} = \frac{\partial \vec{A}}{\partial t} + \left(\vec{v} \cdot \vec{\nabla}\right) \vec{A}. \tag{39.8}$$

This allows us to define the **convective derivative** operator

$$\frac{\mathrm{d}}{\mathrm{d}t} = \frac{\partial}{\partial t} + \vec{v} \cdot \vec{\nabla}. \tag{39.9}$$

Using this operator, we can convert our equations between comoving and stationary frames. Notice how the right-hand side of this equation has two contributions: (i) the rate of change of the vector field at a fixed spatial coordinate, added to (ii) the directional derivative (i.e. the derivative directed along the velocity vector $\vec{v}$). Another way of describing the directional derivative is that $\vec{v} \cdot \vec{\nabla}$ represents a spatial derivative *taken along the flow lines.*

Example 39.2

Our mass conservation equation $\boldsymbol{\nabla} \cdot \boldsymbol{J} = 0$ was previously written in terms of partial derivatives and so corresponds to the viewpoint of a stationary observer. We can use the convective derivative equation to shift this into the comoving frame. First expand the conservation equation, and then use the convective derivative to yield

$$\frac{\partial \rho_0}{\partial t} + \left(\vec{\nabla}\rho_0\right) \cdot \vec{v} + \rho_0 \vec{\nabla} \cdot \vec{v} = 0 \quad \text{(expanding)}$$

$$\frac{\mathrm{d}\rho_0}{\mathrm{d}t} = -\rho_0 \vec{\nabla} \cdot \vec{v} \quad \text{(using convective derivative).}$$

In the comoving frame, the observer who follows the fluid is able to identify a fixed mass of fluid M, which should not vary as a function of time. The mass of fluid instantaneously fills some volume V and has a density ρ_0 with $M = \rho_0 V$. We have, therefore, that

$$\frac{\mathrm{d}M}{\mathrm{d}t} = \frac{\mathrm{d}\rho_0}{\mathrm{d}t}V + \rho_0 \frac{\mathrm{d}V}{\mathrm{d}t} = 0. \tag{39.10}$$

Combining the previous two expressions allows us to conclude that the divergence of the velocity is given in terms of the fluid's volume by[6]

$$\vec{\nabla} \cdot \vec{v} = \frac{1}{V}\frac{\mathrm{d}V}{\mathrm{d}t}. \tag{39.11}$$

[6]We summarize two useful rules for non-relativistic fluid dynamics:
(i) The derivative rule: we use $\mathrm{d}/\mathrm{d}t$ in the comoving frame and $\partial/\partial t + \vec{v} \cdot \vec{\nabla}$ in the stationary frame.
(ii) The divergence rule: the velocity's divergence (a quantity evaluated in the stationary frame) is given in the comoving frame in terms of the rate of change of the volume of a fluid element, *viz.*

$$\vec{\nabla} \cdot \vec{v} = \frac{1}{V}\frac{\mathrm{d}V}{\mathrm{d}t}.$$

Expanding the equation of motion (eqn 39.6) in terms of the convective derivative, results in **Euler's equation** for the motion of the fluid for the stationary observer, which reads

$$\frac{\partial \vec{v}}{\partial t} + \left(\vec{v} \cdot \vec{\nabla}\right) \vec{v} = -\frac{1}{\rho_0}\vec{\nabla}p. \tag{39.12}$$

If gravity is also present then this provides an additional acceleration $\vec{g}$ and we write

$$\frac{\partial \vec{v}}{\partial t} + \left(\vec{v} \cdot \vec{\nabla}\right) \vec{v} = -\frac{1}{\rho_0}\vec{\nabla}p + \vec{g}. \tag{39.13}$$

Example 39.3

Hydrostatics refers to an equilibrium state of affairs where we have a vanishing velocity field, so we then have a hydrostatic equation

$$\vec{\nabla}p = \rho_0\vec{g} \quad \text{(hydrostatic fluid).} \tag{39.14}$$

The existence of a hydrostatic fluid allows us to derive[7] **Pascal's law**. Expanding the last expression in Cartesian coordinates and assuming gravity acts in the $-z$-direction, we have

$$\frac{\partial p}{\partial x} = \frac{\partial p}{\partial y} = 0, \quad \frac{\partial p}{\partial z} = -\rho_0 g, \tag{39.15}$$

which we integrate, to give

$$p(z) = -\rho_0 g z + \text{const}, \tag{39.16}$$

which is Pascal's law, giving the variation of pressure with depth in a hydrostatic fluid.

[7]Blaise Pascal (1623–1662), mathematician, physicist, philosopher, theologian and inventor, was described by Roberto Rossellini as 'a boring man who never made love in his life'. Rossellini made a well-regarded film about Pascal's life for French television in 1972.

The gravitational field can be written in terms of a gravitational potential $\vec{g} = -\vec{\nabla}\Phi$. The gravitational potential fits more comfortably into a description of fluids in terms of their energy.

Example 39.4

Consider the cross product of the velocity and the vorticity ($\vec{\omega} = \vec{\nabla} \times \vec{v}$)

$$\vec{v} \times \vec{\omega} = \vec{v} \times (\vec{\nabla} \times \vec{v})$$

$$= \frac{1}{2}\vec{\nabla}v^2 - (\vec{v} \cdot \vec{\nabla})\vec{v}, \tag{39.17}$$

where we have used the rule for a double cross product in the second line. This expression can be used to remove the directional derivative term from Euler's equation (eqn 39.12). We also take the opportunity to replace $\vec{g}$ with $-\vec{\nabla}\Phi$, with the result

$$\frac{\partial \vec{v}}{\partial t} + \vec{\nabla}\left(\frac{1}{2}v^2 + \Phi\right) + \frac{\vec{\nabla}p}{\rho_0} - \vec{v} \times \vec{\omega} = 0. \tag{39.18}$$

This is a restatement of the Euler equation, but is also the most general form of what is known as **Bernoulli's equation**.[8] We shall see how this equation arises from energetic considerations in the next section.

[8]Daniel Bernoulli (1700–1782) was the son of Johann Bernoulli (famed for his developments in calculus) and the nephew of Jacob Bernoulli (who founded the study of probability). Daniel's work on fluids was published as *Hydrodynamica*. Parts of his work were plagiarized by his father and published in Johann's book *Hydraulica*. Such behaviour is not uncommon in the story of the Bernoulli family.

The Euler equation gives us an equation of motion for the fluid. However, there's lots of additional insight we can gain into motion if we consider the energetics of the fluid. These considerations are our next topic.

39.2 Energy and Bernoulli's equation

The thermodynamics of the fluid at an equilibrium temperature T may be investigated using the first law of thermodynamics

$$\mathrm{d}U = T\mathrm{d}S - p\mathrm{d}V, \tag{39.19}$$

where $U(S,V)$ is the internal energy and S is the entropy.[9] Of course,

[9]Writing $U(S,V)$ implies that entropy and volume are the **natural variables** of U. These are the quantities that appear as differentials in the first law.

the first law simply expresses energy conservation. We want to work with the first law in units where the internal energy and the entropy are given per unit mass (these quantities are denoted by lower-case letters). Unit mass implies that volume V can be replaced with $1/\rho_0$ and we have a unit-mass version of the first law, written as

$$\mathrm{d}u = T\mathrm{d}s - p\,\mathrm{d}\left(\frac{1}{\rho_0}\right). \tag{39.20}$$

The natural variables of u are s and ρ_0. In situations where the natural variables of a problem are s and p, it is often useful to describe the physics using the **enthalpy** $H(S,p) = U + pV$ and so we have, in unit mass form,

$$\mathrm{d}h = T\mathrm{d}s + \frac{1}{\rho_0}\mathrm{d}p, \tag{39.21}$$

where h is enthalpy per unit mass. This implies we can write the enthalpy as $h = u + p/\rho_0$.

In a **perfect fluid**, there is no dissipation, and so we must have no entropy production, i.e. $\mathrm{d}s = 0$. In this case,

$$\begin{aligned} \mathrm{d}u &= -p\,\mathrm{d}\left(\tfrac{1}{\rho_0}\right), \\ \mathrm{d}h &= \tfrac{1}{\rho_0}\mathrm{d}p \end{aligned} \qquad \left(\begin{array}{c}\text{perfect}\\\text{fluid}\end{array}\right). \tag{39.22}$$

From the former expression for the first law $\mathrm{d}u = -p\,\mathrm{d}(1/\rho_0)$, we find an expression for the pressure of a perfect fluid, which is

$$p = \rho_0^2 \frac{\mathrm{d}u}{\mathrm{d}\rho_0} \qquad \left(\begin{array}{c}\text{perfect}\\\text{fluid}\end{array}\right). \tag{39.23}$$

Example 39.5

Evaluated in the static frame, we note that $\partial s/\partial t = 0$ for the perfect fluid. However, the entropy is constant when following the flow (in the comoving frame with $\mathrm{d}/\mathrm{d}t$) as well as in the static frame (where we use the convective derivative $\partial/\partial t + \vec{v}\cdot\vec{\nabla}$), with the consequence that we can write

$$\frac{\mathrm{d}s}{\mathrm{d}t} = (\vec{v}\cdot\vec{\nabla})s = 0 \qquad \left(\begin{array}{c}\text{perfect}\\\text{fluid}\end{array}\right). \tag{39.24}$$

This latter equation is useful in simplifying thermodynamic descriptions of a perfect fluid.

For the special case of **steady flow**, we have the defining property $\frac{\partial}{\partial t}(\text{anything}) = 0$. As a result, we can use a simplifying expression

$$\frac{\mathrm{d}}{\mathrm{d}t} = (\vec{v}\cdot\vec{\nabla}) \quad \text{(steady flow)}. \tag{39.25}$$

This simply says that when the flow is steady, the rate of change of quantities with time for the comoving observer is given by the spatial derivative along the flow lines.

For the perfect fluid in a state of steady flow, eqn (39.25) implies two thermodynamic equations for the enthalpy, one for the comoving frame and one for the stationary frame. These are, respectively,

$$\frac{dh}{dt} = \frac{1}{\rho_0}\frac{dp}{dt}, \qquad \rho_0(\vec{v}\cdot\vec{\nabla})h = (\vec{v}\cdot\vec{\nabla})p \qquad \left(\begin{array}{c}\text{perfect fluid,}\\ \text{steady flow}\end{array}\right). \tag{39.26}$$

We can use the latter of these to obtain a more memorable version of Bernoulli's equation (eqn 39.18) for steady flow in a perfect fluid.

Example 39.6

Dot Bernoulli's equation, in the form of eqn 39.18, with the velocity vector to obtain

$$\vec{v}\cdot\frac{\partial\vec{v}}{\partial t} + \vec{v}\cdot\vec{\nabla}\left(\frac{1}{2}v^2 + \Phi\right) + \vec{v}\cdot\frac{\vec{\nabla}p}{\rho_0} - \vec{v}\cdot(\vec{v}\times\vec{\omega}) = 0. \tag{39.27}$$

Spot that $\vec{v}\cdot\vec{v}\times\omega = 0$ and then consider the case for steady flow by setting $\partial/\partial t = 0$ to obtain

$$\begin{aligned} 0 &= \vec{v}\cdot\vec{\nabla}\left(\frac{1}{2}v^2 + \Phi\right) + \vec{v}\cdot\vec{\nabla}h \\ &= \vec{v}\cdot\vec{\nabla}\left(\frac{1}{2}v^2 + \Phi + h\right), \end{aligned} \tag{39.28}$$

where we have used the second of eqn 39.26 to substitute the pressure with the enthalpy for the case of a perfect fluid. This resulting expression requires a little unpacking at this stage.

From the last example, we have a result given in terms of the operator $\vec{v}\cdot\vec{\nabla}$ appropriate for the stationary observer. Since this is steady flow, we can switch to the comoving frame using the rule that $\vec{v}\cdot\vec{\nabla} \to \frac{d}{dt}$ as we move along the flow lines. This gives an alternative form for Bernoulli's equation for steady flow of a perfect fluid in the comoving frame as

$$\frac{d}{dt}\left(\tfrac{1}{2}v^2 + \Phi + h\right) = 0 \qquad \left(\begin{array}{c}\text{perfect fluid,}\\ \text{steady flow}\end{array}\right). \tag{39.29}$$

Notice how the terms in the bracket represent kinetic energy, gravitational energy and enthalpy. We conclude that for steady flow, Bernoulli's equation, which we introduced as a restatement of Euler's equation, simply expresses conservation of energy. It applies for the comoving observer, which is to say that conservation of energy in the form

$$\tfrac{1}{2}v^2 + \Phi + h = \text{const.} \qquad \left(\begin{array}{c}\text{perfect fluid,}\\ \text{steady flow}\end{array}\right), \tag{39.30}$$

applies along the flow lines. Since $h = u + p/\rho_0$, we can rewrite the conservation of energy equation with pressure included explicitly. That is, along the flow lines we have that Bernoulli's equation reads

$$\tfrac{1}{2}v^2 + \Phi + u + \frac{p}{\rho_0} = \text{const.} \qquad \left(\begin{array}{c}\text{perfect fluid,}\\ \text{steady flow}\end{array}\right). \tag{39.31}$$

To complete our look at the energetics of the fluid we consider an irrotational flow, where $\vec{\nabla} \times \vec{v} = 0$. In this case we can introduce a **velocity potential** ϕ via

$$\vec{v} = \vec{\nabla}\phi, \tag{39.32}$$

which has the property $\vec{\nabla} \times \vec{\nabla}\phi = 0$. Let's see how this works in Bernoulli's equation.

Example 39.7

From the first law for enthalpy we have $\mathrm{d}h = T\mathrm{d}s + \mathrm{d}p/\rho_0$. Taking derivatives with respect to the spatial coordinates x^i we have

$$\vec{\nabla}h = T\vec{\nabla}s + \frac{1}{\rho_0}\vec{\nabla}p. \tag{39.33}$$

If the flow is isentropic the middle term falls out and we have $\vec{\nabla}h = \vec{\nabla}p/\rho_0$. As a result, we now write an expression for unsteady, isentropic, irrotational flow of a perfect fluid. Once again, start with Bernoulli's equation

$$0 = \frac{\partial \vec{v}}{\partial t} + \vec{\nabla}\left(\frac{1}{2}v^2 + \Phi\right) + \frac{\vec{\nabla}p}{\rho_0} - (\vec{v} \times \vec{\omega}), \tag{39.34}$$

and then use $\vec{v} = \vec{\nabla}\phi$, $\vec{\omega} = 0$ and $\vec{\nabla}h = \vec{\nabla}p/\rho_0$ to say

$$0 = \vec{\nabla}\frac{\partial \phi}{\partial t} + \vec{\nabla}\left(\frac{1}{2}v^2 + \Phi\right) + \vec{\nabla}h \tag{39.35}$$

$$= \vec{\nabla}\left(\frac{1}{2}v^2 + \Phi + h + \frac{\partial \phi}{\partial t}\right). \tag{39.36}$$

To guarantee that the right-hand side of this last equation vanishes, we part in the bracket must be constant in space.

The result of the last example is Bernoulli's equation for *unsteady* flow of an irrotational, perfect fluid. Notice how the operator on the bracket was $\vec{\nabla}$ and not $\vec{v}\cdot\vec{\nabla}$. This implies that the equation does not just apply along the flow lines (identified with $\vec{v}\cdot\vec{\nabla}$) but *everywhere*. That is to say, throughout the fluid, we have

$$\left(\tfrac{1}{2}v^2 + \Phi + h + \tfrac{\partial \phi}{\partial t}\right) = \text{const.} \quad \left(\begin{array}{c}\text{irrotational perfect fluid,}\\ \text{unsteady flow}\end{array}\right). \tag{39.37}$$

This differs from the steady flow version through the time derivative of the velocity potential and in its realm of applicability.

39.3   Energy-momentum tensor

We have already seen the importance of the energy-momentum tensor. Here we consider its form in the non-relativistic fluid. The main message here is that we can solve many matter-field problems using the key equation

$$\vec{\nabla}\cdot\boldsymbol{T} = 0, \tag{39.38}$$

which expresses conservation of energy-momentum.[10]

---

[10]See Chapter 12 for a discussion.

Recall that the energy-momentum tensor for the relativistic fluid was given by

$$\boldsymbol{T} = (\rho + p)\,\boldsymbol{v} \otimes \boldsymbol{v} + p\boldsymbol{g}, \tag{39.39}$$

where 4-vectors are used for velocity $\boldsymbol{v}$ and $\boldsymbol{g}$ is the metric. Here ρ is the energy density of the fluid (as measured in its local rest frame). In general, it will receive contributions from the mass density ρ_0 and also the internal energy density $\rho_0 u$, where u is the specific internal energy of the fluid.[11] In the case of flat spacetime, we have $\boldsymbol{g} = \boldsymbol{\eta}$ and so, if fluid has a *non-relativistic* velocity $\boldsymbol{v}$, we have $\rho \gg p/c^2$ and

$$\boldsymbol{T} = \rho\boldsymbol{v} \otimes \boldsymbol{v} + p\boldsymbol{\eta}, \tag{39.40}$$

or, in components,

$$T^{\mu\nu} = \rho v^\mu v^\nu + p\eta^{\mu\nu}. \tag{39.41}$$

[11] Restoring factors of c, we have an energy density $\rho c^2 = \rho_0 c^2 + \rho_0 u$. In natural units, we write $\rho = \rho_0(1 + u)$.

Example 39.8

The last expression for $\boldsymbol{T}$ can be plugged into the key equation $\boldsymbol{\nabla} \cdot \boldsymbol{T} = 0$ to yield

$$\frac{\partial T^{\mu\nu}}{\partial x^\mu} = \frac{\partial \rho}{\partial x^\mu} v^\mu v^\nu + \rho \frac{\partial v^\mu}{\partial x^\mu} v^\nu + \rho v^\mu \frac{\partial v^\nu}{\partial x^\mu} + \frac{\partial p}{\partial x^\mu}\eta^{\mu\nu} = 0. \tag{39.42}$$

Next, we use the non-relativistic expression for the components of velocity: $v^\mu = (1, \vec{v})$, to write

$$\frac{\partial T^{\mu\nu}}{\partial x^\mu} = \frac{\partial \rho}{\partial t} v^\nu + \vec{v}\cdot\vec{\nabla}\rho v^\nu + \rho\vec{\nabla}\cdot\vec{v}v^\nu + \rho\frac{\partial v^\nu}{\partial t} + \rho\vec{v}\cdot\vec{\nabla}v^\nu - \frac{\partial p}{\partial t}\delta^\nu{}_t + (\vec{\nabla}p)^i\delta^\nu{}_i = 0. \tag{39.43}$$

When $\nu = t$ we have[12]

$$\frac{\partial p}{\partial t} = \frac{\partial \rho}{\partial t} + \vec{v}\cdot\vec{\nabla}\rho + \rho\vec{\nabla}\cdot\vec{v}$$

$$= \frac{\partial \rho}{\partial t} + \vec{\nabla}\cdot(\rho\vec{v}). \tag{39.44}$$

[12] The untidy justification presented here can be made more respectable by projecting out parts of the divergence using the velocity vector. That method is presented in the next section.

This is a statement of conservation of energy. From the first law we see the left-hand side gives $\partial p/\partial t = \rho_0 \partial h/\partial t$, telling us about sources and sinks of energy. The right-hand side expresses the local flow of energy density $\boldsymbol{\nabla} \cdot \boldsymbol{J}$. In the absence of energy sources, we have $\boldsymbol{\nabla} \cdot \boldsymbol{J} = 0$.

Considering next the spatial components of eqn 39.43 (by setting $\nu = i$) , we find

$$\frac{\partial \rho}{\partial t} v^i + \left(\vec{v}\cdot\vec{\nabla}\rho\right)v^i + \left(\rho\vec{\nabla}\cdot\vec{v}\right)v^i + \rho\frac{\partial v^i}{\partial t} + \rho\vec{v}\cdot\vec{\nabla}v^i + (\vec{\nabla}p)^i = 0. \tag{39.45}$$

If we assume that the internal energy density contribution to ρ is small in the non-relativistic limit, then the first three terms express local conservation of mass-density. Setting them to zero (i.e. assuming conservation of mass), the final three terms yield

$$\rho\frac{\partial \vec{v}}{\partial t} + \rho\vec{v}\cdot\vec{\nabla}\vec{v} = -\vec{\nabla}p, \tag{39.46}$$

which is Euler's equation.

We see how the conservation of energy-momentum approach yields up both (i) the conservation of mass-energy and (ii) the equations of motion. We shall find the same idea applies for relativistic fluids, allowing us to extract a wealth of information from $\boldsymbol{\nabla} \cdot \boldsymbol{T} = 0$.

39.4 Relativistic fluids

We shall apply the same methodology that we applied to the non-relativistic fluid to its relativistic counterpart. In particular, we require relativistic upgrades of the appropriate derivatives, expressions for energy and for the energy-momentum tensor.

Mathematically, relativistic fluids are described by an energy density and a set of flow lines. The flow lines are the integral curves of the velocity field[13] $\boldsymbol{u}$. The (3+1)-dimensional derivative along the flow lines is given by the covariant derivative $\boldsymbol{u} \cdot \boldsymbol{\nabla} = \boldsymbol{\nabla_u}$. In general, the flow lines are not geodesics described by $\boldsymbol{\nabla_u u} = 0$. We shall see that the equation of motion (a relativistic Euler equation) tells us directly how much the flow lines depart from being geodesics. The departure is caused by gradients in pressure.

The relativistic fluid in which we're interested is unlikely to be very fast-moving water, or similar. Rather, we are interested in the cosmological fluid filling spacetime, or the energetic matter inside stars. This is made up, principally of massive particles we'll call[14] **baryons**. We also need to be slightly more careful about our definitions of thermodynamic variables. Let's define a set of objects $n, \rho, p, T \ s$. These are scalar fields and so vary as a function of position in spacetime. In the rest frame of an element of fluid, they correspond to the results of measurements of the following physical quantities:[15]

- The number density of baryons is n. If the average rest mass per baryon is $\bar{m}_{\mathrm{b}}$, then the rest-mass density of baryons is $\rho_0 = \bar{m}_{\mathrm{b}} n$.
- The total energy density is ρ. This includes the mass-energy of baryons ρ_0 and internal energy. If the specific internal energy is u then we have a total energy density[16]

$$\rho = \rho_0(1 + u). \tag{39.48}$$

- The isotropic pressure is p.
- The temperature is T.
- The entropy per baryon is s.

In non-relativistic fluid dynamics, we had two useful rules, (i) the derivative rule: that we use $\mathrm{d}/\mathrm{d}t$ in the comoving frame and $\partial/\partial t + \vec{v} \cdot \vec{\nabla}$ otherwise, and (ii) the divergence rule $\vec{\nabla} \cdot \vec{v} = 1/V(\mathrm{d}V/\mathrm{d}t)$. These have memorable relativistic counterparts: (i) a comoving observer measures rates of change using their proper time[17] $\mathrm{D}/\mathrm{d}\tau$, while the natural upgrade for the convective derivative is the covariant derivative $\boldsymbol{\nabla_u}$; (ii) the derivative rule becomes $\boldsymbol{\nabla} \cdot \boldsymbol{u} = (1/V)\mathrm{d}V/\mathrm{d}\tau$.

<hr>

Example 39.9

Let's examine the rules in more detail. For rule (i), we consider a particle that moves along a world line parametrized by proper time τ which has a tangent vector $\boldsymbol{u}$. In the comoving frame in flat space, we recall that the rate of change of a vector quantity like $\boldsymbol{A}$ is

$$\frac{\mathrm{D}\boldsymbol{A}}{\mathrm{d}\tau} = \boldsymbol{\nabla_u A} = u^\alpha \left(\frac{\partial A^\beta}{\partial x^\alpha}\right) \boldsymbol{e}_\beta, \tag{39.50}$$

---

[13]Put the other way, the flow lines of a fluid have timelike tangent vectors $\boldsymbol{w} = \mathrm{d}/\mathrm{d}\tau$, where τ is the proper time. When suitably normalized, according to $\boldsymbol{u} = \boldsymbol{w}/\sqrt{[-\boldsymbol{g}(\boldsymbol{w}, \boldsymbol{w})]}$ these become the velocity field, with its characteristic property $\boldsymbol{u}^2 = -1$.

[14]Baryon derives from the Greek word for heavy and describes those composite particles comprising ≥ 3 quarks. This includes protons and neutrons and excludes free electrons and neutrinos. The idea here is that the term accounts for the most common and most massive particles that make up the stuff in the Universe.

[15]As the quantities are scalars, their values are frame independent, depending only on the point $\mathcal{P}$ in spacetime at which they are evaluated. However, we specify that they are defined in terms of what would be measured in the local frame of a comoving observer, at rest with respect to an element of the fluid.

[16]This means that the Lagrangian density for the relativistic fluid is

$$\mathcal{L} = -\rho = -\rho_0(1 + u). \tag{39.47}$$

This is used in Chapter 40.

[17]This is simply $\mathrm{d}/\mathrm{d}\tau$ when acting on a scalar field.

To summarize:
(i) comoving observers measure $\mathrm{D}/\mathrm{d}\tau$, and we use $\boldsymbol{\nabla_u}$ in a general coordinate frame.
(ii) The derivative rule is

$$\frac{1}{V}\frac{\mathrm{d}V}{\mathrm{d}\tau} = \boldsymbol{\nabla} \cdot \boldsymbol{u}, \tag{39.49}$$

where V is the 3-volume.

where we've used the fact that $\Gamma^{\mu}{}_{\alpha\beta} = 0$ for flat space. Expanding components we have

$$\frac{\mathrm{D}\boldsymbol{A}}{\mathrm{d}\tau} = \boldsymbol{\nabla}_{\boldsymbol{u}}\boldsymbol{A} = \left(u^t\frac{\partial}{\partial t} + u^i\vec{\nabla}_i\right)\boldsymbol{A}, \tag{39.51}$$

which, recalling that $\boldsymbol{u}$ has components $(\gamma(u), \gamma(u)\vec{u})$, illustrates how the covariant derivative, with its directional property in spacetime, supplies the natural relativistic analogue of our previous expression for the convective derivative.

For rule (ii), we seek an upgrade of the rule to convert between the divergence of velocity and fluid volume whose form, we recall, in non-relativistic Euclidean space was $\frac{1}{V}\frac{\mathrm{d}V}{\mathrm{d}t} = \vec{\nabla}\cdot\vec{v}$. To justify our proposed upgrade, we note that non-relativistically $u^t = \frac{\mathrm{d}t}{\mathrm{d}\tau} = 1$ and so, motivated by rule (i), we can suggest the compatible expression[18]

$$\frac{1}{V}\frac{\mathrm{d}V}{\mathrm{d}\tau} = \boldsymbol{\nabla}\cdot\boldsymbol{u}, \tag{39.52}$$

where $\boldsymbol{u}$ is the 4-velocity of the fluid. Since our expressions are valid covariant statements in flat spacetime, they must, by the principle of covariance, be the case in curved spacetime. With this in hand, we can check the conservation equation.

[18] Recall that $\vec{\nabla}\cdot\vec{v} = v^i{}_{,i}$ and $\boldsymbol{\nabla}\cdot\boldsymbol{u} = u^{\mu}{}_{;\mu}$.

As we've seen previously, the relativistic energy-momentum tensor $\boldsymbol{T}$ for a perfect fluid is given by

$$\boldsymbol{T} = (\rho + p)\boldsymbol{u}\otimes\boldsymbol{u} + p\boldsymbol{g}, \quad T^{\mu\nu} = (\rho + p)\,u^{\mu}u^{\nu} + pg^{\mu\nu}. \tag{39.53}$$

In the fluid's rest frame, we have $T^{00} = \rho$, $T^{j0} = 0$ and $T^{ij} = p\delta^{ik}$. The conservation equation takes the form

$$\boldsymbol{\nabla}\cdot\boldsymbol{T} = 0, \quad T^{\mu\nu}{}_{;\nu} = 0, \tag{39.54}$$

where we've generalized to the version appropriate to curved spacetime using the semicolon notation.

Example 39.10

As in Chapter 12, take the divergence of $\boldsymbol{T}$ and use the Leibniz rule to find the following:

$$(\rho + p)_{,\beta}u^{\alpha}u^{\beta} + (\rho + p)\left(u^{\alpha}{}_{;\beta}u^{\beta} + u^{\alpha}u^{\beta}{}_{;\beta}\right) + p_{,\beta}g^{\alpha\beta} = 0. \tag{39.55}$$

This expression is very useful in what follows.

With the above concepts in place, we can discuss the key ideas that govern relativistic fluid dynamics.

I: Baryon conservation is perhaps the most primitive conservation law we can specify. It is written in the comoving frame as[19]

$$\frac{\mathrm{d}(nV)}{\mathrm{d}\tau} = 0. \tag{39.57}$$

In general curved spacetime, this is written in terms of the covariant derivative as

$$\boldsymbol{\nabla}\cdot(n\boldsymbol{u}) = 0, \quad (nu^{\alpha})_{;\alpha} = 0, \tag{39.58}$$

where the semicolon notation has been employed in the component version on the right.

[19] Of course, we could equivalently write this as the conservation of rest-mass by multiplying by m_{b} and writing

$$\frac{\mathrm{d}(\rho_0 V)}{\mathrm{d}\tau} = 0. \tag{39.56}$$

[20] Also recall that $\boldsymbol{\nabla}_{\boldsymbol{u}} \equiv \boldsymbol{u} \cdot \boldsymbol{\nabla}$.

[21] It's also useful to note that we also have

$$\boldsymbol{\nabla} \cdot \boldsymbol{u} = -\frac{1}{\rho_0}\frac{\mathrm{d}\rho_0}{\mathrm{d}t}. \qquad (39.61)$$

Example 39.11

We prove this using our rules. The comoving rest mass conservation equation becomes

$$0 = \frac{\mathrm{d}n}{\mathrm{d}\tau} + \frac{n}{V}\frac{\mathrm{d}V}{\mathrm{d}\tau}. \qquad (39.59)$$

Consider the terms on the right, using rule (i) on the first term and rule (ii) on the second to get[20]

$$\begin{aligned} 0 =& \boldsymbol{\nabla}_{\boldsymbol{u}} n + n(\boldsymbol{\nabla} \cdot \boldsymbol{u}) \\ =& \boldsymbol{u} \cdot \boldsymbol{\nabla} n + n(\boldsymbol{\nabla} \cdot \boldsymbol{u}) = \boldsymbol{\nabla} \cdot (n\boldsymbol{u}). \end{aligned} \qquad (39.60)$$

The vanishing of the final expression, whose coordinate version is $(nu^\alpha)_{;\alpha}$ is what we wanted to prove.[21]

II: Entropy conservation The amount of entropy in a volume V is $S = nsV$. In a perfect fluid, we assume that there is no heat flow between flow lines, which implies

$$\frac{\mathrm{d}(nsV)}{\mathrm{d}\tau} \geq 0. \qquad (39.62)$$

This can be combined with the conservation of baryons $\mathrm{d}(nV)/\mathrm{d}\tau$ to allow us to conclude that $\frac{\mathrm{d}s}{\mathrm{d}\tau} \geq 0$, which is the second law of thermodynamics. In many cases, we consider a perfect fluid, where we have $\frac{\mathrm{d}s}{\mathrm{d}\tau} = 0$.

III: Energy conservation is expressed via $\boldsymbol{\nabla} \cdot \boldsymbol{T} = 0$. As in the non-relativistic example, this powerful equation actually outputs too much in the case that we substitute the expression for the energy-density of the relativistic fluid. That is to say it outputs the energy conservation and also the equation of motion. To restrict the result to energy conservation we use a contracted version of this equation

$$u_\alpha T^{\alpha\beta}{}_{;\beta} = 0. \qquad (39.63)$$

This is clearly true and the projection along the velocity direction is enough to extract the energy conservation.

Example 39.12

[22] Useful here is the ever-trusty condition $|u|^2 = u^\alpha u_\alpha = -1$.

We project eqn 39.55, describing $T^{\alpha\beta}{}_{;\beta}$ for the perfect fluid, along u_α and[22] find

$$\begin{aligned} 0 =& (\rho + p)_{,\beta}|u|^2 u^\beta + (\rho + p)\left(u^\alpha{}_{;\beta}u^\beta u_\alpha + |u|^2 u^\beta{}_{;\beta}\right) + p_{,\beta}u^\beta 0 \\ =& -(\rho + p)_{,\beta}u^\beta + (\rho + p)\left(u^\alpha{}_{;\beta}u^\beta u_\alpha - u^\beta{}_{;\beta}\right) + p_{,\beta}u^\beta \\ =& -\rho_{,\beta}u^\beta + (\rho + p)\left(u^\alpha{}_{;\beta}u^\beta u_\alpha - u^\beta{}_{;\beta}\right). \end{aligned} \qquad (39.64)$$

To break this down further, concentrate on the term $u^\alpha{}_{;\beta}u^\beta u_\alpha$. Translating back into vector notation, this is rewritten as

$$u^\alpha{}_{;\beta}u^\beta u_\alpha \equiv \boldsymbol{u} \cdot \boldsymbol{\nabla}_{\boldsymbol{u}}\boldsymbol{u}. \qquad (39.65)$$

This is useful since we can spot that in the same way as for vectors in flat space, where we have $\boldsymbol{v} \cdot \frac{\mathrm{d}\boldsymbol{v}}{\mathrm{d}x} = \frac{1}{2}\frac{\mathrm{d}\boldsymbol{v}^2}{\mathrm{d}x}$, we have here that

$$\boldsymbol{u} \cdot \nabla_{\boldsymbol{u}}\boldsymbol{u} = \frac{1}{2}\nabla_{\boldsymbol{u}}(\boldsymbol{u}^2) = 0, \tag{39.66}$$

which again relies on $\boldsymbol{u}^2 = -1$. This allows us to say

$$-\rho_{,\beta}u^\beta - (\rho + p)u^\beta{}_{;\beta} = 0. \tag{39.67}$$

Written in coordinate-free notation, the conclusion from the last example, for the projection of the equation $\nabla \cdot \boldsymbol{T} = 0$ along the velocity direction, is

$$\nabla_{\boldsymbol{u}}\rho + (\rho + p)\nabla \cdot \boldsymbol{u} = 0. \tag{39.68}$$

This does not look much like conservation of energy, but it can be rewritten in a more enlightening form as shown in the next example.

Example 39.13
Using our rule (ii) $\nabla \cdot \boldsymbol{u} = (1/V)\mathrm{d}V/\mathrm{d}\tau$ we write

$$\frac{\mathrm{d}\rho}{\mathrm{d}\tau} + \frac{(\rho + p)}{V}\frac{\mathrm{d}V}{\mathrm{d}\tau} = 0. \tag{39.69}$$

Then, using $\mathrm{d}(nV)/\mathrm{d}\tau = 0$ we have

$$\frac{\mathrm{d}\rho}{\mathrm{d}\tau} - \frac{(\rho + p)}{n}\frac{\mathrm{d}n}{\mathrm{d}\tau} = 0. \tag{39.70}$$

We shall see very shortly that this expression does indeed describe energy conservation.

Let's link the result of the previous example to energy conservation. In relativity, the first law of thermodynamics may be expressed as[23]

$$\mathrm{d}\left(\begin{array}{c}\text{Energy in a volume element} \\ \text{with fixed number } N \text{ of baryons}\end{array}\right) = -p\mathrm{d}V + T\mathrm{d}S. \tag{39.71}$$

We evaluate

$$\mathrm{d}\left(\frac{\rho N}{n}\right) = -p\mathrm{d}\left(\frac{N}{n}\right) + T\mathrm{d}\left(Ns\right), \tag{39.72}$$

where $V = N/n$. We note that the factors of N are common to all terms, so taking the derivatives with respect to n and rearranging, we obtain the **first law of thermodynamics** for our relativistic variables as

$$\mathrm{d}\rho = \frac{p + \rho}{n}\mathrm{d}n + nT\mathrm{d}s. \tag{39.73}$$

Setting $\mathrm{d}s = 0$ and taking derivatives with respect to the proper time τ we obtain eqn 39.70.

IV: Momentum conservation As in the non-relativistic case, this comes from a projection of $\nabla \cdot \boldsymbol{T} = 0$. This time we project out the part of the divergence perpendicular to the velocity $\boldsymbol{u}$ using the (0,2) projection-operator tensor[24] $\boldsymbol{P}$ which has components $P_{\alpha\mu} = g_{\alpha\mu} + u_\alpha u_\mu$

[23]Interesting here is that the differential in this expression can be interpreted as the exterior derivative $\boldsymbol{d}$. There is no assumption made in the first law about following the flow of the fluid.

[24]To find the part of a vector $\boldsymbol{v}$ perpendicular to a velocity vector $\boldsymbol{u}$ we write

$$\boldsymbol{v} - \frac{(\boldsymbol{u} \cdot \boldsymbol{v})}{(\boldsymbol{u} \cdot \boldsymbol{u})}\boldsymbol{u}$$
$$= \boldsymbol{v} + (\boldsymbol{u} \cdot \boldsymbol{v})\boldsymbol{u}. \tag{39.74}$$

Comparing the action of $\boldsymbol{P}$ we find

$$P_{\alpha\mu}v^\mu = v_\alpha + (u_\mu v^\mu)u_\alpha, \tag{39.75}$$

which is the down-index version of the simple prescription above.

and write

$$P_{\alpha\mu}T^{\mu\nu}{}_{;\nu} = 0. \tag{39.76}$$

This expresses momentum conservation and, as we shall see, results in the equation of motion for the fluid: the relativistic analogue of the Euler equation. The equation of motion is given as an equation for the flow lines.

[25]This can be seen by expanding the $(2,0)$ version of $\boldsymbol{P}$ in the product $\boldsymbol{P}\cdot(\boldsymbol{\nabla}p)$ in components. Explicitly,

$$\left(g^{\alpha\beta} + u^\alpha u^\beta\right)\frac{\partial p}{\partial x^\beta}$$
$$= g^{\alpha\beta}\frac{\partial p}{\partial x^\beta} + u^\alpha\left(\boldsymbol{u}\cdot\boldsymbol{\nabla}p\right),$$

which is the up version of the right-hand side of eqn 39.78, which can also be written as $\boldsymbol{P}(\boldsymbol{\nabla}p, \ )$. You can then quickly check that $\boldsymbol{P}(\boldsymbol{\nabla}p, \tilde{\boldsymbol{u}}) = 0$, which guarantees that $\boldsymbol{u}\cdot\boldsymbol{\nabla}_{\boldsymbol{u}}\boldsymbol{u} = 0$, as usual for an accelerated world line.

[26]The non-relativistic Euler equation is

$$\rho\left[\frac{\partial\vec{u}}{\partial t} + \left(\vec{u}\cdot\vec{\nabla}\right)\vec{u}\right] = -\vec{\nabla}p.$$

Some equations of state for fluids that we consider in this book include the following:
vacuum energy: $p = -\rho$,
radiation: $p = \rho/3$,
dust: $p = 0$,
spatially homogeneous scalar field:

$$p = \frac{\dot{\phi}/2 + V(\phi)}{\dot{\phi}/2 - V(\phi)}\rho.$$

Example 39.14

Applying the projection operator, we obtain

$$\begin{aligned}
0 &= P_{\alpha\mu}T^{\mu\nu}{}_{;\nu}\\
&= 0 + P_{\alpha\mu}(\rho + p)u^\mu{}_{;\nu}u^\nu + p_{,\nu}P_{\alpha\mu}\\
&= (\rho + p)u_{\alpha;\nu}u^\nu + p_{,\alpha} + p_{,\nu}u^\nu u_\alpha.
\end{aligned} \tag{39.77}$$

In vector notation, this looks like

$$(\rho + p)\boldsymbol{\nabla}_{\boldsymbol{u}}\tilde{\boldsymbol{u}} = -\left[\boldsymbol{\nabla}p + (\boldsymbol{\nabla}_{\boldsymbol{u}}p)\tilde{\boldsymbol{u}}\right]. \tag{39.78}$$

This may also be written, in vector notation, using the $(2,0)$ version of $\boldsymbol{P}$, as[25]

$$(\rho + p)\boldsymbol{\nabla}_{\boldsymbol{u}}\boldsymbol{u} + \boldsymbol{P}\cdot\boldsymbol{\nabla}p = 0. \tag{39.79}$$

The result is that the relativistic Euler equation is written as

$$(\rho + p)\boldsymbol{\nabla}_{\boldsymbol{u}}\boldsymbol{u} = -\boldsymbol{P}\cdot\boldsymbol{\nabla}p. \tag{39.80}$$

Comparing this to the non-relativistic version[26] we remember that the non-relativistic limit requires $\rho \gg p/c^2$. Our rule (i) for swapping convective for covariant derivatives is seen on the left-hand side.

Important here is a comparison with the geodesic equation $\boldsymbol{\nabla}_{\boldsymbol{u}}\boldsymbol{u} = 0$. We see from eqn 39.80 how in the relativistic fluid, it is pressure gradients $\boldsymbol{\nabla}p$ perpendicular to the velocity that causes the flow lines to deviate from describing geodesics. A fluid with a uniform pressure simply describes particles that all follow geodesics. A special case of this is the pressureless **dust** that we often use to describe the cosmological fluid. So, once more with emphasis: *dust is a pressureless fluid where all particles follow geodesics.*

V: Equation of state The thermodynamics of a fluid is described by an equation of state. The nature of the equation of state depends on the circumstances and the resulting natural variables for the problem at hand.

Example 39.15

An example equation of state for the relativistic fluid might be written as

$$\rho = \rho(n, s), \tag{39.81}$$

that is, we specify the total energy density as a function of baryon density and specific entropy. With this assignment, we can use the first law (eqn 39.73) to gives us an expression for pressure in terms of energy. Taking the derivative with respect to n at constant s leads to

$$p(n, s) = n \left(\frac{\partial \rho}{\partial n} \right)_s - \rho. \tag{39.82}$$

An alternative might we to specify a free energy function a such as

$$a = a(n, T), \tag{39.83}$$

where the natural variables are n and T. Again, we can derive an expression for pressure. The pressure (and remember that it is gradients in this quantity that cause flow lines to depart from being geodesics) is generally a first derivative of the free energy. In this case

$$p(n, T) = n^2 \left(\frac{\partial a}{\partial n} \right)_T. \tag{39.84}$$

This is the form used in Chapter 40.

This completes our review of fluids. One final comment to make is that for the comoving observer, the natural derivative is the Lie derivative described in Chapter 33. This is examined in the exercises.

Chapter summary

- Fluid mechanics relies on the velocity field of the fluid or, equivalently its integral curves.
- Euler's equation expresses the equations of motion of the fluid; Bernoulli's equation describes the conservation of energy. Both of these are contained in the expression $\boldsymbol{\nabla} \cdot \boldsymbol{T} = 0$.
- In relativistic fluid mechanics, Euler's equation expresses the deviation from geodesic behaviour caused by gradients in the pressure perpendicular to the local velocity.

Exercises

(39.1) Use Bernoulli's theorem to work out how fast fluid flows from a pipe at the bottom of a reservoir. Give your answer in terms of the heights of the top of the fluid y_2 and the outlet pipe y_1.

(39.2) In special relativity, we write the components of the velocity vector as $u^\mu = (u^0, \vec{u}) = (\gamma, \gamma \vec{v})$, where $\gamma = (1 - |\vec{v}|^2)^{-\frac{1}{2}}$. The flat space, relativistic version of Bernoulli's equation for steady flow

is given by

$$\frac{dB}{d\tau} = \gamma \vec{v} \cdot \vec{\nabla} B = 0, \tag{39.85}$$

where $B = \gamma(\rho + p)/\rho_0$. We shall prove this.
(a) Start by showing that $T^{0\mu}{}_{,\mu} = 0$ leads to

$$\frac{d}{d\tau} [\gamma(\rho + p)] + (p + \rho)\gamma \vec{\nabla} \cdot \vec{u} = 0. \tag{39.86}$$

(b) Next, use the conservation of rest mass equation, $\boldsymbol{\nabla} \cdot \boldsymbol{u} = -\frac{1}{\rho_0} \frac{d\rho_0}{d\tau}$, to prove eqn 39.85.

(39.3) *The speed of sound in a static, perfect fluid.* Consider a density wave in flat spacetime, defined by

$$
\begin{aligned}
n &= n_* + \delta n, \quad \rho = \rho_* + \delta\rho, \\
p &= p_* + \delta p, \quad s = s_* + \delta s, \\
&\text{and} \qquad \vec{v} = \delta\vec{v},
\end{aligned}
\tag{39.87}
$$

where the subscript * denotes the constant, equilibrium values.

(a) Show that the baryon conservation law yields

$$
\frac{\partial}{\partial t}\delta n + n_* \vec{\nabla}\cdot\delta\vec{v} = 0.
\tag{39.88}
$$

(b) Show further that

$$
(\rho_* + p_*)\frac{\partial}{\partial t}\delta\vec{v} + \vec{\nabla}\delta p = 0.
\tag{39.89}
$$

(c) Explain why we must also have

$$
n_*\delta\rho = (\rho_* + p_*)\delta n.
\tag{39.90}
$$

We define the speed of sound as

$$
c_{\mathrm s}^2 = \left(\frac{\partial p}{\partial\rho}\right)_s.
\tag{39.91}
$$

(d) Use the expressions above to show that the equation of motion for δn is the wave equation

$$
\frac{1}{c_{\mathrm s}^2}\frac{\partial^2}{\partial t^2}\delta n = \vec{\nabla}^2\delta n.
\tag{39.92}
$$

Hint: The velocity has components $u^\mu = (1,0)$ and $(\delta u)^\mu = (\gamma, \gamma\delta\vec{v})$. However, since $\delta\vec{v}$ is small, we have $\gamma \approx 1$ and so $(\delta u)^\mu = (1, \delta\vec{v})$.

(39.4) Define the mass current of point particles as

$$
J_{\mathrm m}^\alpha = \int d\tau\,\frac{m}{\sqrt{-g}}\frac{dz^\alpha}{d\tau}\delta^{(4)}[x - z(\tau)].
\tag{39.93}
$$

Use the useful identity $\nabla_\mu A^\mu = \frac{1}{\sqrt{-g}}\frac{\partial}{\partial x^\mu}(\sqrt{-g}A^\mu)$ (proved in Exercise 34.3) to show that the mass current is divergenceless.

Hint: Use the distributional identity

$$
u^\alpha\partial_\alpha\delta^{(4)}[x - z(\tau)] = -\frac{d}{d\tau}\delta^{(4)}[x - z(\tau)].
\tag{39.94}
$$

(39.5) Using the connection coefficients for the Schwarzschild geometry given in Chapter 21, show that $\nabla\cdot T = 0$ implies that, for a perfect fluid,

$$
\frac{\partial p}{\partial r} + (p + \rho)\frac{\partial\Phi}{\partial r} = 0.
\tag{39.95}
$$

Hint: Remember to represent the components of T in spherical polar coordinates. These can be transformed from the orthonormal-frame version using the appropriate vielbein.

(39.6) In a non-relativistic fluid, conservation of mass is expressed as

$$
\frac{\partial\rho_0}{\partial t} + \vec{\nabla}\cdot(\rho_0\vec{v}) = 0.
\tag{39.96}
$$

The comoving observer is carried along by the fluid's 3-velocity field $\vec{v}$ and measures the change in the mass field $\rho_0 V$. We shall upgrade this into the 3-volume V into the language of geometry. We first define a volume element 3-form $\tilde{\omega} = dx \wedge dy \wedge dz$.

(a) Let's find the Lie derivative $\pounds_{\vec{v}}(\rho_0\tilde{\omega})$. Show, using components, that

$$
\pounds_{\vec{v}}(\rho_0\tilde{\omega}) = \left[\vec{\nabla}\cdot(\rho_0\vec{v})\right]\tilde{\omega}.
\tag{39.97}
$$

(b) For the Lie derivatives of p-forms, there is a useful rule that says

$$
\pounds_{\boldsymbol v}\tilde{\omega} = \boldsymbol{d}\langle\tilde{\omega}, \boldsymbol{v}\rangle + \langle\boldsymbol{d}\tilde{\omega}, \boldsymbol{v}\rangle,
\tag{39.98}
$$

where $\boldsymbol{v}$ is a vector field. [This is proved in Schutz (1980), whose discussion we follow here.] Use this rule to confirm the result in part (a).

(c) Comparing the result proved in the first parts of the question to the conservation of mass equation, show that

$$
\left(\frac{\partial}{\partial t} + \pounds_{\vec{v}}\right)(\rho_0\tilde{\omega}) = 0.
\tag{39.99}
$$

See Schutz (1980) for further details.

(39.7) We can write the non-relativistic Euler equation in flat space as

$$
\frac{\partial}{\partial t}v_i + v^j\frac{\partial}{\partial x^j}v_i + \frac{1}{\rho_0}\frac{\partial}{\partial x^i}p + \frac{\partial}{\partial x^i}\Phi = 0.
\tag{39.100}
$$

Show that, with $\tilde{v} = v_i dx^i$, this can be rewritten in terms of Lie and exterior derivatives as

$$
\left(\frac{\partial}{\partial t} + \pounds_{\vec{v}}\right)\tilde{v} + \frac{1}{\rho_0}dp + \boldsymbol{d}(\Phi - v^2/2) = 0.
\tag{39.101}
$$

(39.8) If $\vec{v}$ is a uniform velocity field for a fluid in Euclidean space, find a 2-form $\tilde{f}$ that outputs the flux of the fluid through a parallelogram formed by two displacements $\vec{a}$ and $\vec{b}$.

(39.9) *We can use fluid mechanics to compute the behaviour of the cosmological fluid in the Einstein–de Sitter universe (Universe 3 in Chapter 18), whose dynamics we can assume is Newtonian.*
(a) Show that for a Hubble-law expansion with $\vec{v} = (\dot{a}/a)\vec{r}$ we expect a non-relativistic Euler equation of

$$-\vec{\nabla}\Phi = \frac{\partial \vec{v}}{\partial t} + (\vec{v} \cdot \vec{\nabla})\vec{v} = \left(\frac{\ddot{a}}{a}\right)\vec{r}, \qquad (39.102)$$

where Φ is the potential.
(b) Show that this is consistent with (i) the Friedmann equations, and (ii) a Poisson equation with potential $\Phi = 2\pi\rho r^2/3$.
Now consider a perturbation

$$\begin{aligned} \rho &\to \rho + \delta\rho, \quad \vec{v} \to \vec{v} + \delta\vec{v}, \\ \Phi &\to \Phi + \delta\Phi. \end{aligned} \qquad (39.103)$$

(c) Show that the linearized mass-continuity equation becomes

$$0 = \left[\frac{\partial}{\partial t} + (\vec{v} \cdot \vec{\nabla})\right]\frac{\delta\rho}{\rho} + \vec{\nabla} \cdot \delta\vec{v}. \qquad (39.104)$$

(d) Show that the linearized equation of motion becomes

$$0 = \left[\frac{\partial}{\partial t} + (\vec{v} \cdot \vec{\nabla})\right]\delta\vec{v} + \frac{\dot{a}}{a}\delta\vec{v} = -\vec{\nabla}\delta\Phi. \qquad (39.105)$$

(e) We can explicitly make a change to comoving coordinates with the transformation

$$t = t', \quad x^i = a(t')x^{i'}. \qquad (39.106)$$

Show that this is consistent with the expected rule

$$\frac{\partial}{\partial t} + \vec{v} \cdot \vec{\nabla} = \frac{\partial}{\partial t'}. \qquad (39.107)$$

(f) In terms of these new coordinates, verify that eqns 39.104 and 39.105 can be combined to give

$$\left[\frac{\partial^2}{\partial t'^2} + \frac{2\dot{a}}{a}\frac{\partial}{\partial t'}\right]\frac{\delta\rho}{\rho} = \frac{1}{a^2}\vec{\nabla}'^2\delta\Phi. \qquad (39.108)$$

(g) By invoking the properties of the Einstein–de Sitter universe, show that this in turn reduces to

$$\left[\frac{\partial^2}{\partial t'^2} + \frac{4}{3t'}\frac{\partial}{\partial t'} - \frac{2}{3t'^2}\right]\frac{\delta\rho}{\rho} = 0. \qquad (39.109)$$

(h) Demonstrate that the equation has two solutions: one that decays as $1/t'$ and another that grows as $t'^{2/3}$.
This is important as it shows that any perturbation to this universe either decays or grows very slowly.

40 Lagrangian field theory

> ⤳ The material in this chapter is useful for understanding inflation (Chapter 41) and electromagnetism (Chapter 42). It also forms the basis for string theory (Chapter 49).

[1] David Hilbert (1862–1943) was one of the most influential mathematicians of the nineteenth and twentieth centuries. An example of his influence was the collection of mathematical problems, presented in 1900, that largely set the agenda for research in the twentieth century. The Einstein–Hilbert action was proposed by Hilbert in 1915.

[2] Einstein developed general relativity between 1907 and 1915. He arrived at his field equations in late 1915. The history is presented in Pais' *Subtle is the Lord* (2005), and also in Cheng's *Einstein's Physics* (2013). As pointed out by Ohanian and Ruffini, Einstein gave us general relativity early: we really only deserved it around twenty years later when the relativistic gauge-field techniques discussed in this part of the book had matured.

The reader will find no figures in this work. The methods which I set forth do not require either constructions or geometrical or mechanical reasonings: but only algebraic operations, subject to a regular and uniform rule of procedure. Those who love Analysis will see with pleasure that Mechanics has become a branch of it, and will be grateful to me for having thus extended its domain.

Joseph Lagrange (1736–1813) *Mécanique Analytique*

A classical field can be thought of as a machine whose input is a position x in spacetime and whose output is the amplitude of the field at that point. General relativity is a classical field theory and so, like all field theories, we should therefore be able to deal with it using the machinery of Lagrangians. This approach has many advantages, not least that it represents a systematic means of deriving information about the fields and their interactions. Previously, for example, we have had to make educated guesses about the shape and content of the energy-momentum tensor $\boldsymbol{T}$. This is rather unsatisfactory given the importance of $\boldsymbol{T}$ in supplying the right-hand side of the Einstein field equation. In contrast to this ad-hoc approach, a Lagrangian field theory has an inbuilt, turn-the-handle recipe for deriving the energy-momentum tensor. We need only supply the matter fields and the Lagrangian-field-theory machine outputs a suitable field $\boldsymbol{T}(x)$ containing all of the information about the energy content of the fields we have inputted. In this chapter, we rederive the Einstein equation systematically by following the Lagrangian method. When general relativity was being formulated, **David Hilbert**[1] searched for a Lagrangian formulation that would describe gravitating systems. He found it (and it is now named in his honour) but only after Einstein had found his field equations via a less systematic method, narrowly beating Hilbert to the solution.[2]

Before starting, it is worth a reminder of the fields with which we are concerned. General relativity couples the geometry of spacetime to the **matter fields** of the Universe. One class of fields is therefore the matter fields that describes the distribution of both massive and massless matter throughout spacetime. (This class of matter fields also include the 1-form field $\tilde{\boldsymbol{A}}(x)$ that describes electromagnetism.) These matter fields go together to form $\boldsymbol{T}(x)$. The other class has a single member: it is the field that governs the curvature of spacetime. This is the (0,2) metric field $g(x)$. A Lagrangian field theory version of general relativity must couple these fields.

Let's now build a general Lagrangian formulation of classical field theory. We do this in such a way as to maximize the resemblance of field theory to the Lagrangian formulation of classical particle mechanics, that we have already met. Our procedure will involve writing down a Lagrangian that describes the metric field, the matter fields and their interactions. The Lagrangian can then be fed through a set of Euler–Lagrange equations to output a generalized form of the Einstein field equations.

40.1 Matter fields

Every particle in the Universe is an excitation of a matter field. This statement is the basis of **quantum field theory** (QFT), which describes the properties of matter fields on a quantum level. It does this by taking a classical description of the matter fields and quantizing it. That is, it turns the fields into quantum-mechanical operators that have commutation relations with each other. The quantum-field operators act on quantum states, creating or annihilating excitations.[3]

General relativity deals with classical fields, without the step of quantization described above. However, we still start by following the same route as is followed in QFT, where we write down a description of the matter in the Universe in terms of classical matter fields. We might wonder what the fields represent. We've said that a field can be thought of as a machine that inputs a position in spacetime and outputs an amplitude; in a classical field theory this amplitude typically describes the mass-energy density of matter.

[3]See Lancaster and Blundell, *Quantum Field Theory for the Gifted Amateur* (2014), for a full account.

> Chapter 49 contains a discussion of approaches to quantum gravity.

Example 40.1

We shall meet several kinds of classical field in this chapter.

• A **scalar field** inputs a position in spacetime x and outputs a scalar amplitude that describes the distribution of matter. Specifically, the real scalar field $\phi(x)$ gives the probability amplitude for finding matter at point $\vec{x}$ at a time t, such that the probability of finding a particle in a region $d\Sigma$ centred on $\vec{x}$ is $\phi(x)^2 d\Sigma$, where $d\Sigma$ is an element of 3-volume. This field is quantized in QFT to describe excitations in the field, which are massive, spinless particles.

• A **vector field** or **1-form field** inputs a position in spacetime and outputs a vector-valued quantity or 1-form valued quantity, respectively. An important example is the electromagnetic 1-form gauge field $\tilde{A}(x)$ that can be used to determine the electric and magnetic fields at a point in spacetime. It is quantized in QFT into photons, the massless particle excitations of light.

• In general relativity, we most often deal with the mass density field $\rho(x)$ for a **perfect fluid**. This tells us the density of fluid at a point x in spacetime, such that the mass of fluid in a volume $d\Sigma$ centred at $\vec{x}$ at a particular time is $\rho(x)d\Sigma$. This is not a field that features in most standard treatments of QFT.

> Gauge fields are described in Chapter 44. Chapter 42 contains an introduction to the case of the electromagnetic field.

Knowing a field locally (that is, at our position in spacetime x) gives us access to the distribution of matter described by the field close to us (that is, the fields are local). If we also have the field's equation of motion, we shall be able to make predictions about the behaviour of the field at other points in spacetime, and the consequences of its interactions with other fields. The first step towards finding this equation is to write down the **action** that describes the system, and it is to this we now turn.

40.2 Action and equations of motion

We have met the action of a system in particle mechanics. Mathematically the action is a **functional** $S[\ ]$: a machine that inputs a function $\phi(x)$ (or possibly several functions), and outputs a number $S[\phi]$ with units of (energy) × (time). Classical particle mechanics involves writing the action S of a system in terms of the coordinates q_i of the particles involved and their time derivatives $\dot{q}_i$. The action is found using the Lagrangian $L(q_i, \dot{q}_i)$ which is a function that we integrate with respect to time to obtain the action itself. Specifically, for N particles we have

$$S[q_1, q_2, ..., q_N] = \int dt\, L(q_1, q_2, ..., q_N, \dot{q}_1, \dot{q}_2, ..., \dot{q}_N). \tag{40.1}$$

In particle mechanics, the action is simply the difference between the kinetic and potential energies of the system.

We now start upgrading this scheme to work on fields. In flat, three-dimensional space, the Lagrangian is related to the **Lagrangian density** $\mathcal{L}$ by

$$L = \int d^3x\, \mathcal{L}, \tag{40.2}$$

so that

$$S = \int d^4x\, \mathcal{L}. \tag{40.3}$$

The Lagrangian is so important that we often use a shorthand and call a Lagrangian or action that describes fields and their interaction a **field theory**.

Classical field theory involves writing the action S of a system in terms of the matter field ϕ involved and its derivatives with respect to space and time.[4] This allows us to identify a Lagrangian density for the matter fields $\mathcal{L}_m(\phi, \partial_\mu \phi)$: a function of the field and its derivatives that we integrate over all space and time to obtain the action. In flat space, we have, therefore, that

$$S[\phi] = \int d^4x\, \mathcal{L}_m(\phi, \partial_\mu \phi). \tag{40.5}$$

The Lagrangian density of matter fields is analogous to the Lagrangian in particle mechanics, it reflects the difference in kinetic and potential energy densities of a field. In order to describe the matter fields in the Universe, we need to be able to write their Lagrangian densities in terms of the fields and their derivatives.[5]

---

[4] Some often-used notation is that, in flat space, partial derivatives with respect to a coordinate x^μ are often written as

$$\partial_\mu \phi \equiv \frac{\partial \phi}{\partial x^\mu}. \tag{40.4}$$

In comma notation, we would write $\partial_\mu \phi \equiv \phi_{,\mu}$. The advantage of this latter notation is that in curved space, commas can be upgraded to semicolons to signify a covariant partial derivative.

[5] Lagrangian density is usually shortened to 'Lagrangian' to save effort, which we'll do from here.

Example 40.2

Consider waves on a string of mass m and length ℓ. We define the mass density $\rho = m/\ell$, tension $\mathcal{T}$ and displacement $\psi(x,t)$ from the equilibrium (see Fig. 40.1). The kinetic energy T can then be written as $T = \frac{1}{2}\int_0^\ell \mathrm{d}x\, \rho(\partial\psi/\partial t)^2$ and the potential energy $V = \frac{1}{2}\int_0^\ell \mathrm{d}x\, \mathcal{T}(\partial\psi/\partial x)^2$. The action is then

$$S[\psi(x,t)] = \int \mathrm{d}t\,(T - V) = \int \mathrm{d}t\,\mathrm{d}x\, \mathcal{L}_\mathrm{m}\left(\psi, \frac{\partial\psi}{\partial t}, \frac{\partial\psi}{\partial x}\right), \qquad (40.6)$$

where

$$\mathcal{L}_\mathrm{m}\left(\psi, \frac{\partial\psi}{\partial t}, \frac{\partial\psi}{\partial x}\right) = \frac{\rho}{2}\left(\frac{\partial\psi}{\partial t}\right)^2 - \frac{\mathcal{T}}{2}\left(\frac{\partial\psi}{\partial x}\right)^2, \qquad (40.7)$$

is the Lagrangian density.

In most cases we assume a form for the Lagrangian such as those discussed in the next example.

Example 40.3

A **free scalar field** has the Lagrangian

$$\mathcal{L}_\mathrm{m} = -\frac{1}{2}(\partial_\mu\phi)^2 - \frac{m^2\phi^2}{2}, \qquad (40.8)$$

where m is a constant and

$$\begin{aligned}(\partial_\mu\phi)^2 &= (\partial_\mu\phi)\cdot(\partial_\mu\phi)\\ &= g^{\mu\nu}(\partial_\mu\phi)(\partial_\nu\phi).\end{aligned} \qquad (40.9)$$

In Minkowski space, this yields

$$\mathcal{L}_\mathrm{m} = \frac{1}{2}\left(\frac{\partial\phi}{\partial t}\right)^2 - \frac{1}{2}\left(\vec{\nabla}\phi\right)^2 - \frac{m^2}{2}\phi^2. \qquad (40.10)$$

This can be interpreted as the difference between a kinetic energy density T and a potential energy density V. The first term is the kinetic energy, proportional to the time derivative of the field. The second is the potential energy cost for a field changing in space. The third term represents the energy cost of the field existing in spacetime at all.

It is important to note that, if there is more than one field present, we must write the action for all of the fields in terms of the sum of their Lagrangians along with a Lagrangian for any interactions between the fields. So if we have two fields ϕ and ψ then the action is written

$$S = \int \mathrm{d}^4x\, \mathcal{L} = \int \mathrm{d}^4x\, [\mathcal{L}_\mathrm{m}(\phi, \phi_{,\mu}) + \mathcal{L}_\mathrm{m}(\psi, \psi_{,\mu}) + \mathcal{L}_\mathrm{int}(\phi, \psi)], \qquad (40.11)$$

where $\mathcal{L}_\mathrm{int}$ encodes the interaction between the fields. This can also be written as a sum of actions

$$S = S[\phi] + S[\psi] + S_\mathrm{int}[\phi, \psi]. \qquad (40.12)$$

As for particles, the equations of motion of a field are found by extremizing the total action by varying the fields and setting the variation in the action to zero. Extremizing the action S amounts to the condition[6] $\delta S = \delta\int \mathrm{d}^4x\, \mathcal{L} = \int \mathrm{d}^4x\, \delta\mathcal{L} = 0$, where the variation in the Lagrangian can be calculated by using the well-known rule for differentials.[7]

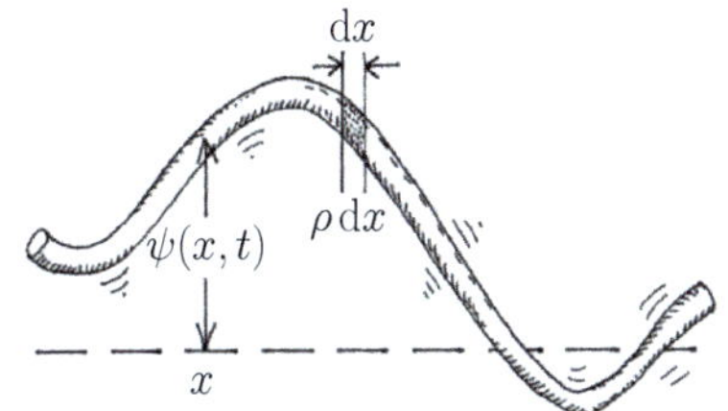

Fig. 40.1 Displacements of a string. The displacement from equilibrium is $\psi(x,t)$ and the equation of motion can be derived by considering an element of the string of length $\mathrm{d}x$ and mass $\rho\,\mathrm{d}x$. The figure shows a short section in the middle of the string which is assumed to be tethered at either end so that $\psi(0,t) = \psi(\ell,t)$.

[6]This only works assuming that the integration measure doesn't change when we vary the matter field. This latter point is where the complication comes in gravitation, since we vary action with respect to the components of the metric field, which in turn changes the integration measure.

[7]For a function $f(x,y)$ we write

$$\delta f = \left(\frac{\partial f}{\partial x}\right)_y \delta x + \left(\frac{\partial f}{\partial y}\right)_x \delta y. \qquad (40.13)$$

Example 40.4

We vary the action S by changing the fields within a $(3+1)$-dimensional region $\mathcal{D}$ of flat spacetime, subject to the condition that the variation of the fields $\delta\phi$ vanishes at the boundary of $\mathcal{D}$. The variation δS yields

$$\delta S = \int \mathrm{d}^4 x \left[\frac{\partial \mathcal{L}_{\mathrm{m}}}{\partial \phi} \delta\phi + \frac{\partial \mathcal{L}_{\mathrm{m}}}{\partial \phi_{,\mu}} \delta(\phi_{,\mu}) \right] = 0. \tag{40.14}$$

However, we note that $\delta(\phi_{,\mu}) = (\delta\phi)_{,\mu}$, so the product rule says that this second term becomes,

$$\int \mathrm{d}^4 x \left[\left(\frac{\partial \mathcal{L}_{\mathrm{m}}}{\partial \phi_{,\mu}} \delta\phi \right)_{,\mu} - \left(\frac{\partial \mathcal{L}_{\mathrm{m}}}{\partial \phi_{,\mu}} \right)_{,\mu} \delta\phi \right]. \tag{40.15}$$

The first term in eqn 40.15 is a total derivative and can, using the divergence theorem, be rewritten as

$$\int \mathrm{d}^4 x \left(\frac{\partial \mathcal{L}_{\mathrm{m}}}{\partial \phi_{,\mu}} \delta\phi \right)_{,\mu} = \int_{\partial \mathcal{D}} \mathrm{d}\sigma \left(\frac{\partial \mathcal{L}_{\mathrm{m}}}{\partial \phi_{,\mu}} \delta\phi \right), \tag{40.16}$$

where $\mathrm{d}\sigma$ labels elements of the (three-dimensional) surface on the boundary $\partial\mathcal{D}$ of our region $\mathcal{D}$. This integral must be zero since $\delta\phi$ vanishes at the boundary $\partial\mathcal{D}$. Collecting the remaining terms from eqns 40.14 and 40.15 we see that

$$\int \mathrm{d}^4 x \left[\frac{\partial \mathcal{L}_{\mathrm{m}}}{\partial \phi} - \left(\frac{\partial \mathcal{L}_{\mathrm{m}}}{\partial \phi_{,\mu}} \right)_{,\mu} \right] \delta\phi = 0. \tag{40.17}$$

This integral is guaranteed to vanish if the integrand does, which yields the Euler–Lagrange equation for fields.

[8]Recall that for a particle the Euler–Lagrange equation is

$$\frac{\partial L}{\partial q} = \frac{\mathrm{d}}{\mathrm{d}t} \left(\frac{\partial L}{\partial \dot{q}} \right). \tag{40.18}$$

Note that using the alternative notation in the field case, we have

$$\frac{\partial \mathcal{L}_{\mathrm{m}}}{\partial \phi} = \partial_\mu \frac{\partial \mathcal{L}_{\mathrm{m}}}{\partial(\partial_\mu \phi)}. \tag{40.19}$$

In Minkowski space, this leads to the Euler–Lagrange equations[8]

$$\frac{\partial \mathcal{L}_{\mathrm{m}}}{\partial \phi} = \left(\frac{\partial \mathcal{L}_{\mathrm{m}}}{\partial \phi_{,\mu}} \right)_{,\mu}. \tag{40.20}$$

This closely resembles the version for particles we've used previously.

Example 40.5

For the string in Example 40.2 we then have

$$0 = \frac{\partial \mathcal{L}_{\mathrm{m}}}{\partial \psi} - \frac{\partial}{\partial x} \frac{\partial \mathcal{L}_{\mathrm{m}}}{\partial(\partial \psi/\partial x)} - \frac{\partial}{\partial t} \frac{\partial \mathcal{L}_{\mathrm{m}}}{\partial(\partial \psi/\partial t)}$$

$$= 0 + \mathcal{T} \frac{\partial^2 \psi}{\partial x^2} - \rho \frac{\partial^2 \psi}{\partial t^2}, \tag{40.21}$$

and so the wave equation $(\partial^2 \psi/\partial x^2) = (1/v^2)(\partial^2 \psi/\partial t^2)$ with $v = \sqrt{\mathcal{T}/\rho}$ emerges almost effortlessly.[9] We say that the field ψ hosts wave-like excitations.

[9]Note for later that a general **wave equation** can be written (in units where the wave speed is normalized to unity) as

$$-\frac{\partial^2 \psi}{\partial t^2} + \vec{\nabla}^2 \psi = 0, \tag{40.22}$$

or as $\partial^2 \psi = 0$ for short. In Chapter 46, we shall see how excitations in the metric field obey this equation, giving us gravitational waves.

Example 40.6

For the free scalar field in Minkowski space from Example 40.3 we find

$$\frac{\partial \mathcal{L}_{\mathrm{m}}}{\partial \phi} = -m^2 \phi, \qquad \frac{\partial \mathcal{L}_{\mathrm{m}}}{\partial \phi_{,\mu}} = -\partial^\mu \phi, \tag{40.23}$$

leading to the equation of motion

$$(-\partial^2 + m^2)\phi = 0, \tag{40.24}$$

where $\partial^2 \phi = \partial_\mu \partial^\mu = g^{\mu\nu} \partial_\mu \partial_\nu = -\frac{\partial^2 \phi}{\partial t^2} + \frac{\partial^2 \phi}{\partial x^2} + \frac{\partial^2 \phi}{\partial y^2} + \frac{\partial^2 \phi}{\partial x^2}$. The equation of motion is known as the **Klein–Gordon equation**.

40.3 Fields in curved spacetime

When spacetime is curved, the usual flat-space volume element $\mathrm{d}^4 x$ is replaced by the invariant element[10] $\sqrt{-g}\,\mathrm{d}^4 x$. For a scalar matter field ϕ, the action functional is given by

$$S[\phi] = \int \mathrm{d}^4 x \sqrt{-g}\,\mathcal{L}_{\mathrm{m}}\left(\phi, \frac{\partial\phi}{\partial x^\mu}\right). \tag{40.25}$$

When spacetime is curved the partial derivative ∂_μ is upgraded to the covariant derivative $\boldsymbol{\nabla}_\mu$. For a scalar we note that $\partial_\mu \phi = \boldsymbol{\nabla}_\mu \phi$ and so our previous equation for the action remains valid with $\mathcal{L}_{\mathrm{m}}(\phi, \phi_{,\mu}) \to \mathcal{L}_{\mathrm{m}}(\phi, \phi_{;\mu})$. We might guess that with the inclusion of the covariant derivative, the Euler–Lagrange equations for scalar fields should look like

$$\frac{\partial\mathcal{L}_{\mathrm{m}}}{\partial\phi} = \boldsymbol{\nabla}_\mu \frac{\partial\mathcal{L}_{\mathrm{m}}}{\partial\left(\boldsymbol{\nabla}_\mu \phi\right)}. \tag{40.26}$$

It does.

When applied to vector or tensor fields, the Euler–Lagrange equation outputs the equations of motion for each component of a field. This is because the Lagrangian machinery works in terms of components. In generalizing the above equation, we therefore need a notation that picks out the components of the covariant derivative. This is the useful semicolon notation,[11] which allows us to write the Euler–Lagrange equation for scalar fields as

$$\frac{\partial\mathcal{L}_{\mathrm{m}}}{\partial\phi} = \left(\frac{\partial\mathcal{L}_{\mathrm{m}}}{\partial\phi_{;\mu}}\right)_{;\mu}. \tag{40.27}$$

We therefore infer that the equations of motion for the components of a vector field $\boldsymbol{A}$ are given by the Euler–Lagrange equation

$$\frac{\partial\mathcal{L}_{\mathrm{m}}}{\partial A^\nu} = \left(\frac{\partial\mathcal{L}_{\mathrm{m}}}{\partial A^\nu_{\;;\mu}}\right)_{;\mu}. \tag{40.28}$$

Dealing with the equations of motion of matter fields is fairly straightforward in curved spacetime. The complication compared to Minkowski space being that we must be careful to include the components of the metric field in our manipulations.[12]

Example 40.7

For the scalar field in curved spacetime we have the Lagrangian density

$$\mathcal{L}_{\mathrm{m}} = -\frac{1}{2}\phi_{;\mu}\phi_{;\nu}g^{\mu\nu} - \frac{m^2}{2}\phi^2. \tag{40.29}$$

We have $\partial\mathcal{L}_{\mathrm{m}}/\partial\phi = -m^2\phi$. Differentiating $\mathcal{L}$ with respect to $\boldsymbol{\nabla}_\mu \phi = \phi_{;\mu}$ gives us

$$\left(\frac{\partial\mathcal{L}_{\mathrm{m}}}{\partial\phi_{;\mu}}\right) = -g^{\mu\nu}\phi_{;\nu}, \tag{40.30}$$

from which we conclude[13]

$$\left(\frac{\partial\mathcal{L}_{\mathrm{m}}}{\partial\phi_{;\mu}}\right)_{;\mu} = -g^{\mu\nu}\phi_{;\nu\mu}. \tag{40.32}$$

[10] The volume 4-form, written in the coordinate frame, is

$$\tilde{\boldsymbol{\omega}} = \sqrt{-g}\,\boldsymbol{dt}(\ )\wedge\boldsymbol{dx}(\ )\wedge\boldsymbol{dy}(\ )\wedge\boldsymbol{dz}(\ ).$$

Inserting Cartesian infinitesimal vectors $\mathrm{d}t\boldsymbol{e}_t$, $\mathrm{d}x\boldsymbol{e}_x$, $\mathrm{d}y\boldsymbol{e}_y$ and $\mathrm{d}z\boldsymbol{e}_z$, yields the volume element $\sqrt{-g}\,\mathrm{d}t\mathrm{d}x\mathrm{d}y\mathrm{d}z$.

[11] Remember $A^\nu_{\;;\mu} = \partial_\mu A^\nu + \Gamma^\nu_{\;\mu\beta}A^\beta$ and $A_{\nu;\mu} = \partial_\mu A_\nu - \Gamma^\beta_{\;\mu\nu}A_\beta$.

[12] Here it is helpful to remember that the partial derivative ∂_μ is *naturally lowered*, which is to say that the components are in the down position. This is also true of the covariant derivative $\boldsymbol{\nabla}_\mu$. We therefore make products of these derivatives using the all-up components of the (0,2) metric $g^{\mu\nu}$. It is also worth noting (since we usually deal with the all-down components of the metric $\boldsymbol{g}$) that it is only in the case that the metric is diagonal that we have $g_{\mu\nu} = 1/g^{\mu\nu}$.

[13] The semicolon notation is especially useful here. Explicitly we have that

$$\phi_{;\mu\nu} = (\partial_\nu\phi)_{;\mu}$$
$$= \partial_\mu\partial_\nu\phi - \Gamma^\sigma_{\;\mu\nu}\partial_\sigma\phi. \tag{40.31}$$

The equation of motion is, therefore

$$g^{\mu\nu}\phi_{;\nu\mu} - m^2\phi = 0. \tag{40.33}$$

We therefore have a systematic means of dealing with matter fields in curved spacetime.

40.4 Motivating the Einstein equation

How do we deal with the geometry of spacetime, expressed via $\boldsymbol{g}(x)$ in the Lagrangian formalism? This is the problem that was solved by David Hilbert, with the solution (now called) the **Einstein–Hilbert action** S_{EH}. The idea is that we need to add this new term S_{EH} that reflects the $\boldsymbol{g}(x)$ field to the action S_{m} of the system of matter fields. The total action is then

$$S = S_{\mathrm{EH}} + \alpha S_{\mathrm{m}}, \tag{40.34}$$

where α is a constant[14] and S_{m} contains the contributions from all of the matter fields in the Universe. The equation of motion (a.k.a. the Einstein equation) then results from the variation of the total action, found by setting $\delta S = 0$. The variation of the matter fields should be expected to supply the right-hand side of the Einstein equation reflecting the mass-energy in the Universe, while the variation of the Einstein–Hilbert action should give us the geometrical left-hand side.

A complication here is that, unfortunately, we can't simply guess a total Lagrangian and feed this into the Euler–Lagrange equations to extract the Einstein equation. This is because the field that we must vary, subject to the condition that $\delta S = 0$, is the metric field $\boldsymbol{g}(x)$. The variation therefore has the effect of changing the intervals between points in spacetime itself and, in particular, changing the volume element $\sqrt{-g}\,\mathrm{d}^4x$ in the action interval. We must therefore vary the total action with respect to the components of the field $g_{\mu\nu}$ of the metric from first principles, such that δS vanishes. It is also important that we vary the action with respect to $g_{\mu\nu}$ and not the all-up components $g^{\mu\nu}$. (This is because the gravitational field is most directly represented by the (0,2) tensor field $\boldsymbol{g}(\ ,)$ in which we input vectors.) We therefore seek

$$\delta S = \int \mathrm{d}^4x\, \frac{\partial(\sqrt{-g}\mathcal{L})}{\partial g_{\mu\nu}} \delta g_{\mu\nu} = 0. \tag{40.35}$$

After all of these preliminaries, we are now ready to meet Hilbert's famous result.[15] The Lagrangian density for geometry is given by the Ricci scalar field $R(x)$, such that the Einstein–Hilbert action is written as

$$S_{\mathrm{EH}} = \int \mathrm{d}^4x\, \sqrt{-g(x)}R(x). \tag{40.36}$$

This is added to the matter action $\alpha S_{\mathrm{m}} = \int \mathrm{d}^4x\, \sqrt{-g(x)}\alpha\mathcal{L}_{\mathrm{m}}$, where α is the constant discussed above. Varying this total Lagrangian with

[14]The constant is used here to ensure that however we have defined the matter fields, when combined with the Einstein–Hilbert action they correctly produce the 8π term in the Einstein equation $\boldsymbol{G} = 8\pi\boldsymbol{T}$.

[15]We state this without proof, but the inclusion of the Ricci scalar field $R(x)$ is perhaps not completely unexpected. By the principle of general covariance, we would be looking for a field which transforms properly, and we might expect something fairly simple and fundamental to appear there. As we shall see though, the proof of the pudding is that it produces the right answer.

respect to the components of the metric results in the Einstein equations describing the dynamics that link matter fields and geometrical fields.

In order to vary the geometry Lagrangian in eqn 40.36, we shall need three new mathematical tricks: two are rules for the variation of the metric tensor and a third is for varying the Riemann tensor. The first comes from our need to vary the action with respect to the components $g_{\mu\nu}$. It turns out that this forces us to also consider the result of varying components $g^{\mu\nu}$. The rule we need is that there's an extra minus sign relating the variation of the up and down versions of the vector

$$\delta g^{\sigma\rho} = -g^{\sigma\mu}\delta g_{\mu\nu}g^{\nu\rho} \quad \text{(Rule 1)}. \tag{40.37}$$

We prove this below.

Example 40.8

The identity can be proven by considering the variation of the matrix equation $\underline{M}\underline{M}^{-1} = I$ to give

$$(\delta\underline{M})\underline{M}^{-1} + \underline{M}\delta\underline{M}^{-1} = 0. \tag{40.38}$$

From this we derive the matrix equation

$$\delta\underline{M}^{-1} = -\underline{M}^{-1}(\delta\underline{M})\underline{M}^{-1}. \tag{40.39}$$

Note the important minus sign! Representing the components of the metric as a matrix with $\underline{M}_{\mu\nu} = g_{\mu\nu}$ and $(\underline{M}^{-1})_{\mu\nu} = g^{\mu\nu}$, we must then have

$$\delta g^{\alpha\delta} = -g^{\alpha\beta}\delta g_{\beta\gamma}g^{\gamma\delta}, \tag{40.40}$$

Conclusion: There's an extra minus sign when you vary the up-index version of the metric.

The second rule we need is for the variation of a determinant

$$\delta\sqrt{-g} = \tfrac{1}{2}\sqrt{-g}\,g^{\mu\nu}\delta g_{\mu\nu} \quad \text{(Rule 2)}. \tag{40.41}$$

Example 40.9

To prove this we use the identity[16] $\ln\det\underline{M} = \text{Tr}\ln\underline{M}$. Differentiate the identity to find

$$\begin{aligned} \delta\det\underline{M} &= \det\underline{M}\,\delta(\text{Tr}\ln\underline{M}) \\ &= \det\underline{M}\,\text{Tr}(\underline{M}^{-1}\delta\underline{M}). \end{aligned} \tag{40.42}$$

Plugging in the metric leads to the equation claimed above.

The third and final rule we shall need is the **Palatini identity,**[17] which says that the variation of components of the Ricci tensor $R_{\alpha\beta}$ can be written in terms of covariant derivatives of the variation of the connection coefficients. The identity reads

$$\delta R_{\alpha\beta} = \nabla_\mu\left(\delta\Gamma^\mu{}_{\alpha\beta}\right) - \nabla_\beta\left(\delta\Gamma^\mu{}_{\alpha\mu}\right) \quad \text{(Rule 3)}. \tag{40.43}$$

With these three identities under our belts, we can vary the Einstein–Hilbert action. There's nothing for it now but to plunge into the algebra.

[16] The proof of this identity, which can be found in many sources, is left as an exercise. Alternatively, one can follow Exercise 34.3.

[17] Attilio Palatini (1889–1949) showed that the variations of the connection coefficients are the coordinate components of a tensor. We won't reproduce the derivation here, but note that the Palatini identity in eqn 40.43 is proved in Zee (2013).

Example 40.10

Written out in terms of components of the Ricci tensor,[18] we have that the Einstein–Hilbert action comprises three objects

$$S_{\mathrm{EH}} = \int \mathrm{d}^4 x \, \sqrt{-g} \, g^{\sigma\rho} R_{\sigma\rho}. \tag{40.44}$$

Varying the action gets us three terms, one for each object

$$\delta S = \int \mathrm{d}^4 x \, \left[\sqrt{-g} \, g^{\sigma\rho} \delta R_{\sigma\rho} + \sqrt{-g} \, \delta g^{\sigma\rho} R_{\sigma\rho} + \delta(\sqrt{-g}) \, R \right]$$

$$= \delta I_1 + \delta I_2 + \delta I_3 = 0. \tag{40.45}$$

We deal with these in reverse order. For δI_3 we need Rule 2 from eqn 40.37, which yields

$$\delta I_3 = \int \mathrm{d}^4 x \, \delta(\sqrt{-g}) R$$

$$= \frac{1}{2} \int \mathrm{d}^4 x \, \sqrt{-g} \, g^{\mu\nu} \delta g_{\mu\nu} R$$

$$= \int \mathrm{d}^4 x \, \sqrt{-g} \left(\frac{1}{2} g^{\mu\nu} R \right) \delta g_{\mu\nu}. \tag{40.46}$$

For δI_2 we need Rule 1 (eqn 40.41) for lowering the indices of the components of the metric. Applying this, we obtain

$$\delta I_2 = \int \mathrm{d}^4 x \sqrt{-g} R_{\sigma\rho} \left(-g^{\sigma\mu} \delta g_{\mu\nu} g^{\nu\rho} \right)$$

$$= - \int \mathrm{d}^4 x \sqrt{-g} R^{\mu\nu} \delta g_{\mu\nu}. \tag{40.47}$$

The result in terms of $\delta g_{\mu\nu}$ and $R^{\mu\nu}$ confirms that writing δI_2 in terms of $R_{\sigma\rho}$ and not $R^{\sigma\rho}$ was the correct choice: the latter would not have had indices consistent with the term from δI_3.

The most problematical part is the final integral for δI_1. Fortunately, this only contributes a surface term which vanishes. The key to seeing this is to use the Palatini identity (Rule 3) which says

$$\delta R_{\nu\beta} = \delta R^{\mu}{}_{\nu\mu\beta} = \boldsymbol{\nabla}_\mu \delta \Gamma^{\mu}{}_{\nu\beta} - \boldsymbol{\nabla}_\beta \delta \Gamma^{\mu}{}_{\nu\mu}. \tag{40.48}$$

Integrating this

$$\delta I_2 = \int \mathrm{d}^4 x \sqrt{-g} g^{\nu\beta} \left(\boldsymbol{\nabla}_\mu \delta \Gamma^{\mu}{}_{\nu\beta} - \boldsymbol{\nabla}_\beta \delta \Gamma^{\mu}{}_{\nu\mu} \right)$$

$$= - \int \mathrm{d}^4 x \sqrt{-g} \, \boldsymbol{\nabla}_\beta \left(g^{\nu\beta} \delta \Gamma^{\mu}{}_{\nu\mu} - g^{\nu\mu} \delta \Gamma^{\beta}{}_{\nu\mu} \right), \tag{40.49}$$

using $\boldsymbol{\nabla}_\alpha g^{\mu\nu} = 0$. Finally, we use the divergence theorem to say that the derivative may be written as a surface integral over the boundary. This vanishes because the variation vanishes on the boundary by definition and so this term does not contribute. The result is that the contribution to the equations of motion is

$$\delta I_2 + \delta I_3 = - \int \mathrm{d}^4 x \sqrt{-g} \left(R^{\mu\nu} - \frac{1}{2} g^{\mu\nu} R \right) \delta g_{\mu\nu}. \tag{40.50}$$

The variation of the action should vanish, so we must have that $\delta S = \delta I_2 + \delta I_3 = 0$. The way to guarantee this is that the integrand must vanish.

The conclusion of this lengthy set of manipulations is that the components of the Einstein tensor $G^{\mu\nu} = \left(R^{\mu\nu} - \frac{1}{2} g^{\mu\nu} R \right)$ emerge as the contribution to the (Einstein) equation of motion and that they must vanish. This gives the Einstein equation in the slightly stripped-down form $R^{\mu\nu} - \frac{1}{2} g^{\mu\nu} R = 0$. This simplified form arises because we have not yet included the matter fields in the action. The matter fields must therefore spit out the energy-momentum tensor, and so we now turn to examine how the energy-momentum tensor can be derived from the action.

40.5 Energy-momentum tensor

Often the energy-momentum tensor is determined ad hoc, using all of the physical insight we can muster. However, field theory tells us that there should be a way to simply and mechanically extract it from the Lagrangian. There is, as we see in the next example.

Example 40.11

We can gain access to the energy-momentum tensor if we vary the action with respect to the components of the metric. The action is given by

$$S = S_{\mathrm{EH}} + \alpha S_{\mathrm{m}} = \int \mathrm{d}^4 x \sqrt{-g}\left(\mathcal{L}_{\mathrm{EH}} + \alpha \mathcal{L}_{\mathrm{m}}\right). \tag{40.51}$$

If we vary with respect to the components of the metric, we obtain

$$\int \mathrm{d}^4 x \left[\frac{\partial\left(\sqrt{-g}\mathcal{L}_{\mathrm{EH}}\right)}{\partial g_{\mu\nu}} + \alpha \frac{\partial\left(\sqrt{-g}\mathcal{L}_{\mathrm{m}}\right)}{\partial g_{\mu\nu}}\right]\delta g_{\mu\nu} = 0. \tag{40.52}$$

We have from the previous section that

$$\int \mathrm{d}^4 x \frac{\partial\left(\sqrt{-g}\mathcal{L}_{\mathrm{EH}}\right)}{\partial g_{\mu\nu}}\delta g_{\mu\nu} = \int \mathrm{d}^4 x \sqrt{-g}G^{\mu\nu}\delta g_{\mu\nu}, \tag{40.53}$$

where we have used the components of the Einstein tensor $G^{\mu\nu} = R^{\mu\nu} - \frac{1}{2}g^{\mu\nu}R$. Comparing eqn 40.52 with the Einstein equation, we make the suggestion

$$T^{\mu\nu} = \frac{2}{\sqrt{-g}}\frac{\delta\left(\sqrt{-g}\mathcal{L}_{\mathrm{m}}\right)}{\delta g_{\mu\nu}}. \tag{40.54}$$

We immediately see that this is a good choice, as equating the integrand of eqn 40.52 to zero leads to an expression

$$\sqrt{-g}G^{\mu\nu} - \alpha\frac{\sqrt{-g}T^{\mu\nu}}{2} = 0. \tag{40.55}$$

We choose $\alpha = 16\pi$, which leads to the Einstein equation $G_{\mu\nu} = 8\pi T_{\mu\nu}$, which is what the variation must produce. We conclude that the action

$$S = \int \mathrm{d}^4 x \sqrt{-g}\left[R(x) + 16\pi\mathcal{L}_{\mathrm{m}}\right], \tag{40.56}$$

when varied with respect to the components of the metric, leads to the Einstein equation.

Expanding eqn (40.54), we have a final prescription for calculating the energy-momentum tensor from the matter action:

$$T^{\mu\nu} = 2\frac{\delta\mathcal{L}_{\mathrm{m}}}{\delta g_{\mu\nu}} + g^{\mu\nu}\mathcal{L}_{\mathrm{m}}. \tag{40.57}$$

We use this below. However, because the derivative operator ∂_μ is naturally an object with a down index, the action for matter fields is most simply written in terms of a derivative with respect to $g^{\mu\nu}$. It's then easiest to use an analogous expression for the all-down components of $\boldsymbol{T}$, which is[19]

$$T_{\mu\nu} = -2\frac{\partial\mathcal{L}_{\mathrm{m}}}{\partial g^{\mu\nu}} + g_{\mu\nu}\mathcal{L}_{\mathrm{m}}. \tag{40.59}$$

Let's see some examples.

[19]To prove this, use the chain rule to say

$$\begin{aligned}T^{\mu\nu} &= 2\frac{\partial\mathcal{L}_{\mathrm{m}}}{\partial g^{\alpha\beta}}\frac{\partial g^{\alpha\beta}}{\partial g_{\mu\nu}} + g^{\mu\nu}\mathcal{L}_{\mathrm{m}}\\ &= -2g^{\alpha\mu}g^{\nu\beta}\frac{\partial\mathcal{L}_{\mathrm{m}}}{\partial g^{\alpha\beta}} + g^{\mu\nu}\mathcal{L}_{\mathrm{m}}.\end{aligned}$$

Then lower indices in the usual way

$$\begin{aligned}T_{\gamma\sigma} &= T^{\mu\nu}g_{\mu\gamma}g_{\nu\sigma}\\ &= -2g^{\alpha\mu}g^{\nu\beta}g_{\mu\gamma}g_{\nu\sigma}\frac{\partial\mathcal{L}_{\mathrm{m}}}{\partial g^{\alpha\beta}}\\ &\quad + g^{\mu\nu}g_{\mu\gamma}g_{\nu\sigma}\mathcal{L}_{\mathrm{m}}\\ &= -2\delta^\alpha{}_\gamma\delta^\beta{}_\sigma\frac{\partial\mathcal{L}_{\mathrm{m}}}{\partial g^{\alpha\beta}}\\ &\quad + g_{\gamma\sigma}\mathcal{L}_{\mathrm{m}}, \tag{40.58}\end{aligned}$$

where we contracted on μ and ν in the final step.

Example 40.12

The **scalar field Lagrangian** is given by

$$\mathcal{L}_{\mathrm{m}} = -\frac{1}{2}g^{\mu\nu}\phi_{;\mu}\phi_{;\nu} - \frac{m^2}{2}\phi^2. \tag{40.60}$$

We have

$$\frac{\partial\mathcal{L}}{\partial g^{\mu\nu}} = -\frac{1}{2}\phi_{;\mu}\phi_{;\nu}, \tag{40.61}$$

leading to the energy-momentum tensor

$$T_{\mu\nu} = \phi_{;\mu}\phi_{;\nu} - g_{\mu\nu}\left(\frac{1}{2}g^{\alpha\beta}\phi_{;\alpha}\phi_{;\beta} + \frac{m^2}{2}\phi^2\right). \tag{40.62}$$

In flat space, this becomes

$$T_{\mu\nu} = \partial_\mu\phi\partial_\nu\phi - \frac{1}{2}\eta_{\mu\nu}(\partial_\mu\phi)^2 - \eta_{\mu\nu}\frac{m^2}{2}\phi^2, \tag{40.63}$$

which has timelike component $T_{00} = \frac{1}{2}(\partial_0\phi)^2 + \frac{1}{2}(\vec{\nabla}\phi)^2 + \frac{m^2}{2}\phi^2$, which simply reflects the sum of kinetic energy and potential energy.
Electromagnetism is described by the Lagrangian

$$\mathcal{L}_{\mathrm{m}} = -\frac{1}{4}F_{\alpha\beta}F_{\gamma\delta}g^{\alpha\gamma}g^{\beta\delta}. \tag{40.64}$$

We have

$$\frac{\partial\mathcal{L}}{\partial g^{\alpha\gamma}} = -\frac{1}{2}F_{\alpha\beta}F_{\gamma\delta}g^{\beta\delta}. \tag{40.65}$$

The energy-momentum tensor is, therefore,

$$T_{\alpha\gamma} = \left(F_{\alpha\beta}F_{\gamma\delta}g^{\beta\delta} - \frac{1}{4}g_{\alpha\gamma}F_{\sigma\rho}F_{\mu\nu}g^{\sigma\mu}g^{\rho\nu}\right). \tag{40.66}$$

The previous example shows how we can extract the energy-momentum tensor for the fields in the Universe. An exception is gravitation itself. In the same way that general relativity removes the notion of a gravitational force, it also prevents us from identifying an energy-moment tensor for a gravitational field.

40.6 Noether's theorem

One of the most profound statements in field theory is **Noether's theorem**.[20] It says that if a field theory has a **symmetry** then there is a corresponding **conservation equation** corresponding to a **conserved quantity**. Noether's theorem elegantly reveals these conservation equations and will consider how it can be used on the metric field.

Consider the differential of the Lagrangian in Minkowski space

$$\delta\mathcal{L} = \frac{\partial\mathcal{L}}{\partial\phi}\delta\phi + \frac{\partial\mathcal{L}}{\partial(\partial_\mu\phi)}\delta(\partial_\mu\phi) + \frac{\partial\mathcal{L}}{\partial\eta_{\mu\nu}}\delta\eta_{\mu\nu}. \tag{40.67}$$

If we make a continuous transformation of the variables in the Lagrangian and find that the action doesn't change[21] we say that the transformation represents a symmetry. In order to examine the symmetries corresponding to the geometric features of the theory, we are

---

Electromagnetism is discussed in detail in Chapter 42.

[20]Amalie Emmy Noether (1882–1935).

[21]Strictly, if it doesn't change up to a surface term.

going to pull this Lagrangian along a vector $\boldsymbol{\xi}$ and compare the result with our original Lagrangian.

The mathematical tool for comparing a field after it has been transported by a vector is, of course, the Lie derivative. We therefore upgrade the differential $\delta\mathcal{L}$ to a Lie derivative

$$\delta\mathcal{L} = \pounds_{\boldsymbol{\xi}}\mathcal{L} = \xi^\alpha \partial_\alpha \mathcal{L}, \tag{40.68}$$

where the final equality holds because $\mathcal{L}$ is a scalar function. Exchanging δ for the Lie derivative, eqn 40.67 becomes

$$\xi^\alpha \partial_\alpha \mathcal{L} = \frac{\partial\mathcal{L}}{\partial\phi}\pounds_{\boldsymbol{\xi}}\phi + \frac{\partial\mathcal{L}}{\partial(\partial_\mu\phi)}\pounds_{\boldsymbol{\xi}}(\partial_\mu\phi) + \frac{\partial\mathcal{L}}{\partial\eta_{\mu\nu}}\pounds_{\boldsymbol{\xi}}\eta_{\mu\nu}. \tag{40.69}$$

We have seen that in examining geometry (encoded in the metric field), symmetries are described in terms of isometries in the metric, which give rise to Killing fields. So we set $\boldsymbol{\xi}$ to be a Killing field, defined by $\pounds_{\boldsymbol{\xi}}\eta_{\alpha\beta} = 0$, which wipes out the final term in eqn 40.69. We can then use the Euler–Lagrange equation[22] to substitute for $\partial\mathcal{L}/\partial\phi$. Rearranging then gives

$$\partial_\alpha\left[\frac{\partial\mathcal{L}}{\partial(\partial_\alpha\phi)}\pounds_{\boldsymbol{\xi}}\phi - \xi^\alpha\mathcal{L}\right] = 0. \tag{40.71}$$

This is a conservation equation[23] that tells us the quantity in the bracket is conserved. This is Noether's theorem: a symmetry (identified via the Killing vector $\boldsymbol{\xi}$) has given rise to a conserved Noether current $\boldsymbol{J}_{\mathrm{N}}$ with components

$$J_{\mathrm{N}}^\alpha = \frac{\partial\mathcal{L}}{\partial(\partial_\alpha\phi)}\pounds_{\boldsymbol{\xi}}\phi - \xi^\alpha\mathcal{L}. \tag{40.72}$$

In particular, we note that for Minkowski space that we're considering, translation vectors in each of the Cartesian directions are Killing vectors.

Example 40.13

Translations in the x-direction leave the metric unchanged and so we have a Killing vector

$$\boldsymbol{\xi} = \boldsymbol{e}_x = \frac{\partial}{\partial x}. \tag{40.73}$$

This gives a Noether current with components

$$J_{\mathrm{N}}^\alpha = \frac{\partial\mathcal{L}}{\partial(\partial_\alpha\phi)}\frac{\partial\phi}{\partial x} - \delta^\alpha{}_x\mathcal{L}. \tag{40.74}$$

We can group the four Noether currents that result from translations into a single tensor $\boldsymbol{S}$. We call this quantity the **canonical energy-momentum tensor**. It has components

$$S^{\alpha\beta} = \frac{\partial\mathcal{L}}{\partial(\partial_\alpha\phi)}\partial^\beta\phi - g^{\alpha\beta}\mathcal{L}, \tag{40.75}$$

so that $\partial_\alpha S^{\alpha\beta} = 0$.

The tensors $\boldsymbol{S}$ and the energy-momentum tensor $\boldsymbol{T}$ are the same for scalar fields, but differ for more complicated matter fields. Generally, $\boldsymbol{S}$ isn't symmetrical and doesn't generalize for curved spacetimes, so is generally unsuitable for incorporation into Einstein's equation.[24]

[22] The Euler–Lagrange equation is

$$\frac{\partial\mathcal{L}}{\partial\phi} = \partial_\alpha\frac{\partial\mathcal{L}}{\partial(\partial_\alpha\phi)}. \tag{40.70}$$

[23] This is an equation of continuity, telling us that the divergence of something is zero. In some cases, we may have to worry about the global boundary conditions to show that there is no effect of a surface term. We will not worry about that here.

[24] The need for a symmetrical energy-momentum tensor is discussed in the exercises.

40.7   The perfect fluid

Fluids were examined in detail in Chapter 39. Here we show how the perfect fluid's equation of motion and energy-momentum tensor can be derived directly from a Lagrangian. The perfect fluid has no viscosity and moves adiabatically. In field theory, a fluid is described by two features: (i) its mass-density field $\rho_0(x)$ and (ii) a congruence of curves called flow lines (or stream lines) with timelike tangent vectors $\boldsymbol{w} = \frac{\partial}{\partial \tau}$. When normalized, the tangent $\boldsymbol{w}$ becomes the velocity $\boldsymbol{v}$ field of the fluid. That is to say

$$\boldsymbol{v} = \frac{\boldsymbol{w}}{[-g(\boldsymbol{w}, \boldsymbol{w})]^{\frac{1}{2}}}, \tag{40.76}$$

so that $\boldsymbol{v}^2 = g(\boldsymbol{v}, \boldsymbol{v}) = -1$, as we require for a velocity field. The mass current is $\boldsymbol{J} = \rho_0 \boldsymbol{v}$ and we impose the condition that $\boldsymbol{\nabla} \cdot \boldsymbol{J} = 0$ or equivalently $J^\alpha{}_{;\alpha} = 0$, which is to say that the current is conserved. The fluid has a Lagrangian

$$\mathcal{L}_{\mathrm{m}} = -\rho = -\rho_0(1 + u), \tag{40.77}$$

where u is the specific internal energy, which is a function of ρ_0 [that is $u = u(\rho_0)$].[25] The quantity $\rho = \rho_0(1 + u)$ the total energy density of the fluid.

To derive the equations of motion we need to vary the action, which involves varying the flow lines. This changes $\boldsymbol{v}$, so while we vary the flow lines, we imagine adjusting ρ_0 to ensure that the current $\boldsymbol{J}$ is conserved throughout. This enables us to continue using the conservation law. The variation of the flow lines can be followed by defining $\boldsymbol{K}$ as the displacement vector of a point on a flow-line as we vary the flow. How much the flow has changed can then be measured by carrying the flow pattern along the vector field $\boldsymbol{K}$ and comparing the tangent fields before and after. That is to say, the effect of the variation is contained in the variation in $\boldsymbol{w}$, which is measured by a Lie derivative

$$\delta \boldsymbol{w} = \pounds_{\boldsymbol{K}} \boldsymbol{w}. \tag{40.79}$$

What follows is a long computation that we split into several steps.[26]

[25]Under these conditions then, in terms of these quantities, the pressure is

$$p = \rho_0^2 \frac{\partial u}{\partial \rho_0}. \tag{40.78}$$

[26]We follow the approach of Hawking and Ellis (1973), filling in some intermediate steps that are left out of their discussion.

Example 40.14

Step 0: Write the variation in S in terms of a differential $\delta\rho_0$ in the usual way, by saying

$$\delta S = -\int \mathrm{d}^4 x \sqrt{-g}\, \frac{\mathrm{d}\mathcal{L}}{\mathrm{d}\rho_0} \delta\rho_0$$

$$= -\int \mathrm{d}^4 x \sqrt{-g} \left(1 + \frac{\mathrm{d}(\rho_0 u)}{\mathrm{d}\rho_0}\right) \delta\rho_0 = 0. \tag{40.80}$$

Step I: In order to use the expression for δS, we need an expression for $\delta\rho_0$. One can be found from our conservation of current equation. Since $\delta(J^\alpha{}_{;\alpha}) = 0 = (\delta J^\alpha)_{;\alpha}$ we have

$$(\delta J^\alpha)_{;\alpha} = [(\delta\rho_0)v^\alpha + \rho_0 \delta v^\alpha]_{;\alpha}$$

$$= (\delta\rho_0)_{;\alpha} v^\alpha + \delta\rho_0 v^\alpha{}_{;\alpha} + (\rho_0)_{;\alpha} \delta v^\alpha + \rho_0 (\delta v^\alpha)_{;\alpha} = 0. \tag{40.81}$$

Step II: In order to use the previous equation, we next need an expression for $\delta \boldsymbol{v}$. This can be extracted from $\delta \boldsymbol{w}$, which was given by the Lie derivative. So since $\boldsymbol{v} = [-\boldsymbol{g}(\boldsymbol{w}, \boldsymbol{w})]^{-\frac{1}{2}} \boldsymbol{w}$ we have, using the chain rule,

$$\delta v^\alpha = \frac{1}{2} [-\boldsymbol{g}(\boldsymbol{w}, \boldsymbol{w})]^{-\frac{3}{2}} [\delta \boldsymbol{g}(\boldsymbol{w}, \boldsymbol{w})] w^\alpha + [-\boldsymbol{g}(\boldsymbol{w}, \boldsymbol{w})]^{-\frac{1}{2}} \delta w^\alpha. \tag{40.82}$$

This is simplified by considering the term

$$\begin{aligned} \delta \boldsymbol{g}(\boldsymbol{w}, \boldsymbol{w}) &= 2 \frac{\partial \boldsymbol{g}(\boldsymbol{w}, \boldsymbol{w})}{\partial w^\gamma} \delta w^\gamma \\ &= 2 g_{\beta\gamma} w^\beta \delta w^\gamma. \end{aligned} \tag{40.83}$$

We note further that, by definition of the Lie derivative, we have that

$$\delta w^\alpha = (\pounds_{\boldsymbol{K}} \boldsymbol{w})^\alpha = w^\alpha_{\;;\beta} K^\beta - K^\alpha_{\;;\beta} w^\beta. \tag{40.84}$$

Putting this together gives the result of Step II: an expression for δv^α, which is

$$\begin{aligned} \delta v^\alpha =& \frac{1}{2} [-\boldsymbol{g}(\boldsymbol{w}, \boldsymbol{w})]^{-\frac{3}{2}} 2 g_{\beta\gamma} w^\beta (w^\gamma_{\;;\sigma} K^\sigma - K^\gamma_{\;;\sigma} w^\sigma) w^\alpha \\ &+ [-\boldsymbol{g}(\boldsymbol{w}, \boldsymbol{w})]^{-\frac{1}{2}} (w^\alpha_{\;;\beta} K^\beta - K^\alpha_{\;;\beta} w^\alpha) \\ =& v^\alpha_{\;;\beta} K^\beta - K^\alpha_{\;;\beta} v^\beta - v^\alpha v^\beta K_{\beta;\gamma} v^\gamma, \end{aligned} \tag{40.85}$$

where, in the final line, we've written everything in terms of components of $\boldsymbol{v}$ and the displacement field $\boldsymbol{K}$.

Step III: Substitute for δv^α in eqn 40.81 (which, remember, only describes the conservation law $J^\alpha_{\;;\alpha} = 0$). We find that the number of terms seems to increase rather alarmingly! Specifically, we find

$$\begin{aligned} 0 =& (\delta \rho_0)_{;\alpha} v^\alpha + \delta \rho_0 v^\alpha_{\;;\alpha} + (\rho_0)_{;\alpha} \delta v^\alpha + \rho_0 (\delta v^\alpha)_{;\alpha} \\ =& (\delta \rho_0)_{;\alpha} v^\alpha + \delta \rho_0 v^\alpha_{\;;\alpha} + (\rho_0)_{;\alpha} \left(v^\alpha_{\;;\beta} K^\beta - K^\alpha_{\;;\beta} v^\beta - v^\alpha v^\beta K_{\beta;\gamma} v^\gamma \right) \\ &+ \rho_0 \left(v^\alpha_{\;;\beta} K^\beta - K^\alpha_{\;;\beta} v^\beta - v^\alpha v^\beta K_{\beta;\gamma} v^\gamma \right)_{;\alpha} \\ =& (\delta \rho_0)_{;\alpha} v^\alpha + \delta \rho_0 v^\alpha_{\;;\alpha} + (\rho_0)_{;\alpha} \left(v^\alpha_{\;;\beta} K^\beta - K^\alpha_{\;;\beta} v^\beta - v^\alpha v^\beta K_{\beta;\gamma} v^\gamma \right) \\ &+ \rho_0 \left(v^\alpha_{\;;\beta\alpha} K^\beta + v^\alpha_{\;;\beta} K^\beta_{\;;\alpha} - K^\alpha_{\;;\beta\alpha} v^\beta - K^\alpha_{\;;\beta} v^\beta_{\;;\alpha} \right. \\ &\left. - v^\alpha_{\;;\alpha} v^\beta K_{\beta;\gamma} v^\gamma - v^\alpha v^\beta_{\;;\alpha} K_{\beta;\gamma} v^\gamma - v^\alpha v^\beta K_{\beta;\gamma\alpha} v^\gamma - v^\alpha v^\beta K_{\beta;\gamma} v^\gamma_{\;;\alpha} \right). \tag{40.86} \end{aligned}$$

This looks truly formidable, but can be simplified dramatically if we *integrate along the flow lines*. The key here is that covariant derivatives (identified in the above using the semicolon notation) are directional derivatives along the flow lines. Integrating along the flow lines can then be thought of as undoing these covariant derivatives. Along the flow lines we have local mass conservation[27] written as $J^\alpha_{\;;\alpha} = (\rho_0 v^\alpha)_{;\alpha} = 0$. It turns out that if we collect terms which contain a factor $v^\alpha_{\;;\alpha}$ and demand that these vanish, this provides the relationship between $\delta \rho_0$ and v^α, ρ_0 and K^α that we need. The instruction to integrate along the flow lines is especially important as it allows us to integrate those terms involving a covariant derivative by parts, effectively changing the positions of the semicolons. This allows us to massage the indices in such a way as to invoke $(\rho_0 v^\alpha)_{;\alpha} = 0$.[28] The terms contributing a factor $v^\alpha_{\;;\alpha}$ are

$$\int_{\text{flow lines}} \left(\delta \rho_0 v^\alpha_{\;;\alpha} + \rho_0 v^\alpha_{\;;\beta\alpha} K^\beta - \rho_0 v^\alpha_{\;;\alpha} v^\beta K_{\beta;\gamma} v^\gamma \right) = 0. \tag{40.88}$$

It's not obvious that the second term is one of the ones we want, until we integrate it by parts to get $-\int (\rho_0 K^\beta)_{;\beta} v^\alpha_{\;;\alpha}$.

We have now collected all terms in the integrand with a factor $v^\alpha_{\;;\alpha}$, and since the divergence of v^α does not vanish (unless the density is constant) this integrand must do. The result is an equation for $\delta \rho_0$:

$$\delta \rho_0 = (\rho_0 K^\beta)_{;\beta} + \rho_0 K_{\beta;\gamma} v^\beta v^\gamma. \tag{40.89}$$

We can plug this into eqn 40.80, which expresses the variation of the action.

[27]In general, as we shall show below, they don't obey the geodesic equation and so $v^\beta (v^\alpha)_{;\beta} \neq 0$. This is due to the possibility of pressure gradients perpendicular to the flow lines in the fluid.

[28]As an example of integration by parts, consider

$$\int v^\alpha_{\;;\beta\alpha} K^\beta = \left[v^\alpha_{\;;\beta} K_\beta \right]_{-\infty}^{\infty} - \int v^\alpha_{\;;\alpha} K^\beta_{\;;\beta}, \tag{40.87}$$

where the integrals are carried out along the flow lines. The first term on the right vanishes as we demand that all fields die off to zero at infinity. This leave the second term. We see how the position of the $;\beta$ has changed and we have picked up a minus sign.

Step IV: The variation in the Lagrangian is therefore

$$\delta S = -\int d^4x \sqrt{-g}\,\left\{\left[(\rho_0 K^\beta)_{;\beta} + \rho_0 K_{\beta;\gamma}v^\beta v^\gamma\right]\left[1 + \frac{d(\rho_0 u)}{d\rho_0}\right]\right\} = 0. \qquad (40.90)$$

Step V: In the final step, we integrate this equation by parts, moving the semicolons from the factors of K^β onto the other terms, in such a way that they have a common factor of K^β. The result is

$$\delta S = \int d^4x \sqrt{-g}\,\left\{\rho_0\left[1 + \frac{d(\rho_0 u)}{d\rho_0}\right]v^\alpha{}_{;\beta}v^\beta + \rho_0\left[\frac{d(\rho_0 u)}{d\rho_0}\right]_{,\gamma}(g^{\gamma\alpha} + v^\gamma v^\alpha)\right\}K_\alpha = 0.$$

$$(40.91)$$

Since K_α does not vanish, the integrand must do, and we have our result for the variation of the fluid action. Finally, we interpret $\dot{v}^\alpha = v^\alpha{}_{;\beta}v^\beta$ as the acceleration of the flow lines and recall the pressure of the fluid is given by $p = \rho_0^2 \partial u/\partial \rho_0$.

We conclude that the equation of motion for the fluid is given by

$$(\rho + p)\dot{v}^\alpha = -p_{,\beta}(g^{\alpha\beta} + v^\beta v^\alpha), \qquad (40.92)$$

where the energy density is written as $\rho = \rho_0(1+u)$.[29] This the equation of motion for the fluid, Euler's equation, written for curved spacetime. As promised, the geodesic equation $\dot{v}^\alpha = 0$ is not obeyed: the flow lines accelerate in response to transverse pressure gradients.

Next, we extract the energy-momentum tensor using out prescription from earlier (eqn 40.57) that the up components are given by $T^{\mu\nu} = 2\frac{\delta\mathcal{L}_m}{\delta g_{\mu\nu}} + g^{\mu\nu}\mathcal{L}_m$.

Example 40.15

Starting with the matter Lagrangian $\mathcal{L} = -\rho_0(1+u)$, we have that

$$\left(\frac{\partial\mathcal{L}_m}{\partial g_{\alpha\beta}}\right)_{\rho_0} = -\left(\frac{\partial\mathcal{L}_m}{\partial\rho_0}\right)_{g_{\alpha\beta}}\left(\frac{\partial\rho_0}{\partial g_{\alpha\beta}}\right)_{\mathcal{L}} = \left(1 + u + \rho_0\frac{du}{d\rho_0}\right)\frac{\partial\rho_0}{\partial g_{\alpha\beta}}. \qquad (40.93)$$

We need access to the final term. This is achieved with a trick: the divergence of the current can be written in terms of the determinant of the metric matrix as[30]

$$(J^\alpha)_{;\alpha} = \frac{1}{(-g)^{\frac{1}{2}}}\frac{\partial}{\partial x^\alpha}\left[(-g)^{\frac{1}{2}}J^\alpha\right] = 0. \qquad (40.94)$$

This is helpful as it tells us that the quantity $(-g)^{\frac{1}{2}}J^\alpha$ is unchanged as the metric is varied. Dotting two currents together we have

$$\rho_0^2 = g^{-1}\left[(-g)^{\frac{1}{2}}J^\alpha(-g)^{\frac{1}{2}}J^\beta)g_{\alpha\beta}\right], \qquad (40.95)$$

and so

$$2\rho_0\delta\rho_0 = \left(J^\alpha J^\beta - J^\gamma J_\gamma g^{\alpha\beta}\right)\delta g_{\alpha\beta}$$

$$= \rho_0^2(v^\alpha v^\beta - v^\gamma v_\gamma g^{\alpha\beta})\delta g_{\alpha\beta}, \qquad (40.96)$$

giving

$$2\delta\rho_0 = \rho_0(v^\alpha v^\beta + g^{\alpha\beta})\delta g_{\alpha\beta}. \qquad (40.97)$$

Using $\delta\rho_0 = \frac{\partial\rho_0}{\partial g_{\alpha\beta}}\delta g_{\alpha\beta}$, we extract the result we were after

$$\frac{\partial\rho_0}{\partial g_{\alpha\beta}} = \frac{1}{2}\rho_0(v^\alpha v^\beta + g^{\alpha\beta}). \qquad (40.98)$$

The first term in the energy-momentum tensor becomes

$$2\frac{\delta\mathcal{L}_m}{\delta g_{\alpha\beta}} = \left(1 + u + \rho_0\frac{du}{d\rho_0}\right)\rho_0(v^\alpha v^\beta + g^{\alpha\beta}). \qquad (40.99)$$

We add to this the term

$$g^{\alpha\beta}\mathcal{L}_m = -\rho_0(1+u)g^{\alpha\beta}. \qquad (40.100)$$

Putting this all together, we conclude that the components of the energy-momentum tensor are given by

$$T^{\alpha\beta} = \left[\rho_0(1+u) + \rho_0^2\frac{\mathrm{d}u}{\mathrm{d}\rho_0}\right]v^\alpha v^\beta + \rho_0^2\frac{\mathrm{d}u}{\mathrm{d}\rho_0}g^{\alpha\beta}. \qquad (40.101)$$

We may reinterpret this using the symbols used above, to find the components of the energy-momentum tensor for the perfect fluid are given by

$$T^{\alpha\beta} = (\rho + p)v^\alpha v^\beta + pg^{\alpha\beta}. \qquad (40.102)$$

This completes our review of the Lagrangian formulation. We shall use it in the next two chapters, where we use it in thinking about symmetry breaking in cosmology (Chapter 41) and then meet one of the most successful field theories: electromagnetism (Chapter 42).

Chapter summary

- Equations of motion for matter fields can be found using

$$\frac{\partial\mathcal{L}_\mathrm{m}}{\partial A^\nu} = \left(\frac{\partial\mathcal{L}_\mathrm{m}}{\partial A^\nu_{\;;\mu}}\right)_{;\mu}. \qquad (40.103)$$

- The Einstein–Hilbert action is given by

$$S_\mathrm{EH} = \int \mathrm{d}^4x\,\sqrt{-g(x)}R(x). \qquad (40.104)$$

- The components of the energy-momentum tensor can be found using the expression

$$T^{\mu\nu} = 2\frac{\delta\mathcal{L}_\mathrm{m}}{\delta g_{\mu\nu}} + g^{\mu\nu}\mathcal{L}_\mathrm{m}. \qquad (40.105)$$

Exercises

(40.1) Consider the Lagrangian for a particle of unit mass: $L(x^\mu, \dot{x}^\mu) = \frac{1}{2}g_{\mu\nu}\dot{x}^\mu\dot{x}^\nu$, where dots denote a derivative with respect to proper time τ.

(a) Confirm that the equation of motion for the particle is the geodesic law.

Recall from Chapter 2 that we define the canonical momentum as

$$p_\mu = \frac{\partial L}{\partial\dot{x}^\mu}, \qquad (40.106)$$

and the Hamiltonian as the function

$$H(x^\mu, p_\mu) = p_\mu\dot{x}^\mu - L(x^\mu, \dot{x}^\mu). \qquad (40.107)$$

(b) Show that the Hamiltonian derived from our example Lagrangian is

$$H = \frac{1}{2}g^{\mu\nu}p_\mu p_\nu. \qquad (40.108)$$

The Hamiltonian allows access to another route to finding the equations of motion in terms of p_μ, and

x^μ *via a pair of first-order derivatives*

$$\frac{\partial H}{\partial p_\mu} = \dot{x}^\mu, \qquad \frac{\partial H}{\partial x^\mu} = -\dot{p}_\mu. \tag{40.109}$$

These are **Hamilton's equations**.

(c) Use Hamilton's equations to compute the geodesic equation again.

(d) What happens to the equation of motion if a metric coefficient is independent of one of the coordinates?

(40.2) Assuming a Lagrangian of the same form as used in the last question, a particle in a Schwarzschild geometry has

$$L = \frac{1}{2}\left[-\left(1 - \frac{2M}{r}\right)\dot{t}^2 + \left(1 - \frac{2M}{r}\right)^{-1}\dot{r}^2 \right.$$
$$\left. + r^2\dot{\theta}^2 + r^2\sin^2\theta\,\dot{\phi}^2 \right]. \tag{40.110}$$

(a) Compute the momenta and show the Hamiltonian $H = L$.

(b) Use Hamilton's equations to show that p_t and p_ϕ are constants of the motion.

(40.3) Consider the Lagrangian density

$$\mathcal{L} = \frac{1}{2}\dot{\phi}^2 - \frac{1}{2}(\vec{\nabla}\phi)^2 - \frac{1}{2}m^2\phi^2, \tag{40.111}$$

restricted to Minkowski space.

By analogy with the state of affairs in the last question, we define the **momentum density** *as*

$$\pi = \frac{\partial \mathcal{L}}{\partial \dot{\phi}}. \tag{40.112}$$

(a) Show that the momentum density for our Lagrangian is given by

$$\pi = \dot{\phi}. \tag{40.113}$$

Again, by analogy, we define the **Hamiltonian density** *as*

$$\mathcal{H} = \pi\dot{\phi} - \mathcal{L}. \tag{40.114}$$

(b) Show that, for our example, we find

$$\mathcal{H} = \frac{1}{2}\dot{\phi}^2 + \frac{1}{2}(\vec{\nabla}\phi)^2 + \frac{1}{2}m^2\phi^2. \tag{40.115}$$

(c) We also define the momentum-like quantities

$$\Pi^\mu = \frac{\partial \mathcal{L}}{\partial(\partial_\mu\phi)}. \tag{40.116}$$

Show that $\Pi^\mu = \partial^\mu\phi$ and $\Pi^0 = \pi = \dot{\phi}$.

(40.4) *The symmetry of* **T** *is necessary for angular momentum to be conserved. We define an antisymmetric four-dimensional angular-momentum tensor for a particle of* $J^{ik} = x^i p^k - x^k p^i$. *In flat space and Cartesian coordinates, the continuous version is then*

$$J^{ik} = \int \mathrm{d}^3x \left(x^i T^{k0} - x^k T^{i0} \right). \tag{40.117}$$

We then define a tensor with components $M^{\mu\nu\sigma} = x^\mu T^{\nu\sigma} - x^\nu T^{\mu\sigma}$. *Conservation of angular momentum in flat (3+1)-dimensional spacetime is then expressed via*

$$\frac{\partial M^{\mu\nu\sigma}}{\partial x^\sigma} = \frac{\partial}{\partial x^\sigma}\left(x^\mu T^{\nu\sigma} - x^\nu T^{\mu\sigma}\right) = 0. \tag{40.118}$$

Show that for this to be true, **T** must be symmetric.

(40.5) Show that eqn 40.88 is consistent with local mass conservation, as described in eqn 40.86.

Inflation

41

> *Before she had drunk half the bottle, she found her head*
> *pressing against the ceiling, and had to stoop to save her neck*
> *from being broken. She hastily put down the bottle, remark-*
> *ing, 'That's quite enough—I hope I sha'n't grow any more.'*
> Lewis Carroll (1832–1898) *Alice's Adventures in Wonderland*

Our story so far is that the Universe can be described, to a good approximation, as a Robertson–Walker spacetime that started in an initial
Big-Bang singularity. However, there are some problems with an explanation of the current state of our Universe using this standard model of
cosmology. We mentioned in Chapter 18 that observations suggest the
Universe appears very close to being a spatially flat, or $k = 0$, spacetime.
This can be expressed using the form of the Friedmann equation that
reads

$$\frac{k}{\dot{a}^2} = \Omega - 1. \tag{41.1}$$

The implication is either that $k = 0$ for all time, or that the current
density ratio $\Omega = \rho/\rho_c \approx 1$. Our observations provide an estimate of
$\Omega \approx 0.3$, which isn't unity, but isn't massively far away either. In the
case that $k = 0$, we might ask how and why the Universe was set up
to be flat. If $\Omega \approx 1$ now, but was not so in the past, this would seems
quite a coincidence, given the need for the Universe to be very **finely
tuned** in the past to conspire to give this result only now.[1] So what is
the reason that we find $\Omega \approx 1$? This collection of questions is known as
the **flatness problem**.

A further problem is that the Universe is also observed to be surprisingly homogeneous at earlier times.[2] This is especially surprising in the
context of a hot Big Bang model in which the matter in our Universe
cools very rapidly as it expands.

Example 41.1

A rough picture of the earliest stages of the Universe can be summarized as follows.
Evolution occurs via several **phase transitions**, the phenomena that form the basis
of this chapter.

• At a time $t \approx 10^{-43}$ s after the Big Bang, the temperature scale of the Universe is
$T \approx 10^{32}$ K and the corresponding energy scale is $E \approx 10^{19}$ GeV. Under these con-
dition, the strong and electroweak interactions are unified into a single interaction.
On cooling below this scale (at time $t \approx 10^{-36}$ s, energy 10^{15} GeV or temperature
10^{27} K) a phase transition[3] takes place causing the strong and electroweak interac-
tions to separate.

---

[1] The implications of the apparent
finely tuned nature of the Universe have
generated a lot of debate. For more on
this issue, see the book *A Fortunate
Universe* by Lewis and Barnes (listed
in the Further Reading in Appendix A)
or the article on this subject in the
Stanford Encyclopedia of Philosophy
(https://plato.stanford.edu/).

[2] The Cosmic Background Explorer
(COBE) satellite showed that the cosmic microwave background (CMB) has
a near-perfect black-body spectrum
with average temperature $T = 2.73$ K,
with only faint anisotropies at a level of
1 part in 10^5.

[3] This is known as the 'GUT phase
transition', referencing putative Grand
Unified (GUT) theories, that describe
the combination of the strong and electroweak interactions.

[4] By 'not causally connected' we mean that photons from one region have not had time to travel to another region, much less establish thermal equilibrium.

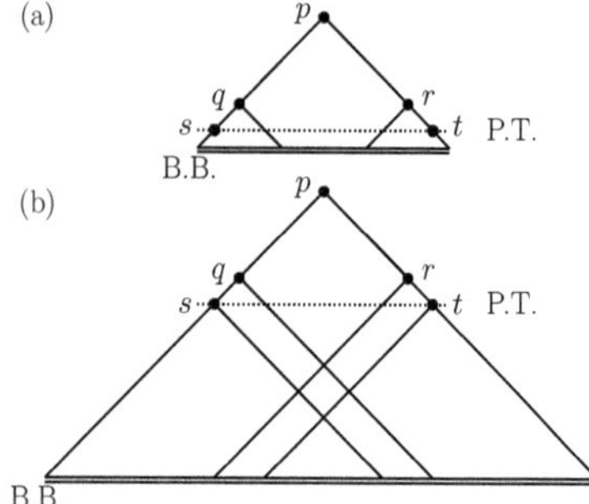

Fig. 41.1 (a) Penrose diagram showing the horizon problem. The dotted line represents a phase transition (P.T.) and B.B. labels the Big-Bang singularity. Events q and r do not have any shared history, and neither do events s and t that take place at the instant of the phase transition. (b) Inflation pushes back the Big Bang, so that the pasts of events q and r overlap for a period after the Big Bang. Events s and t that take place at the phase transition, then also have a shared history.

[5] Inflation was originally suggested by Alan Guth (1947–) who examined the case of a scalar field in a spatially flat Universe as we do in this chapter.

[6] See Penrose (2004) for a discussion.

[7] L. D. Landau (1908–1968), perhaps best known for foundational work in condensed matter physics, also made contributions to general relativity, including the independent computation of the Chandrasekhar limit (the maximum mass of a stable white dwarf) in 1932.

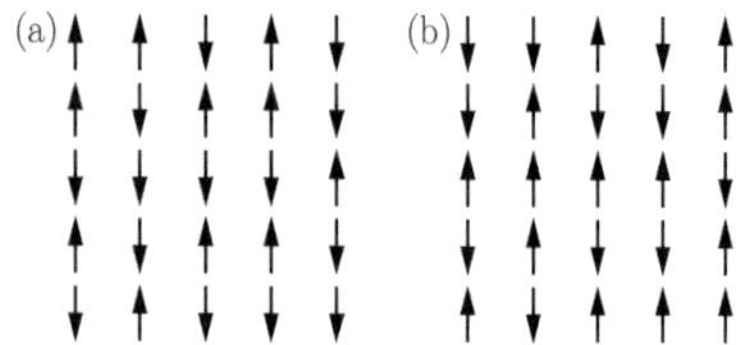

Fig. 41.2 The magnet at $T > T_c$. The magnetization, or average moment, is zero in (a) and, after rotation of each spin through 180°, is still zero in (b).

• At $t \approx 10^{-11}$ s (or $T \approx 10^{15}$ K, $E \approx 10^2$ GeV) another phase transition occurs causing the Higgs scalar field to take on a finite value. The result is that the electromagnetic and weak interactions separate. The Higgs mechanism causes some of the leptons to become massive. This is known as the electroweak symmetry-breaking transition.

• After $t \approx 10^{-5}$ s the Universe is cool enough ($T \approx 10^{12}$ K, $E \approx 100$ MeV) for a further phase transition to occur. This one allows the strong interaction to increase in strength, leading to quark confinement.

The expansion of the early Universe implies that regions of space that cannot have been causally connected[4] in the early stages of the cooling and expansion of the Universe are now found to be at almost exactly the same temperature. This is known as the **horizon problem**. Figure 41.1(a) shows a Penrose diagram illustrating the problem. An observer at p can see two events (q and r) by looking in different directions. These events are not causally connected, in that they don't have histories that overlap at any period after the Big Bang. It is therefore a problem if the regions of spacetime around the events are near-identical.

To explain these problems we might suggest that the Universe underwent a period of very rapid expansion in its early history, known as **inflation**.[5] This involves an acceleration of the scale factor $a(t)$ of the Universe. [This is in contrast to most of the results we have examined in which $a(t)$ is decelerating.] Inflation allows causally connected regions of spacetime to separate rapidly enough to give the homogeneous temperature distribution that is observed. As illustrated in Fig. 41.1(b), it effectively pushes back the Big Bang, allowing events taking place at a phase transition to have some region of shared history. It is notable that this does not necessarily solve the horizon problem, as although there is some communication in the past, events such as s and t in Fig. 41.1(b) do not share much of their history, so might well be rather different at the point of the phase transition.[6] Inflation also results in a spatially flat Universe.

Inflation can be understood using the machinery of field theory that we discussed in the last chapter. The key concept is one familiar from thermodynamics that has been invoked in the evolution of the early Universe: the symmetry-breaking phase transition.

41.1 Symmetry breaking

An intuitive and powerful description of a continuous phase transition was developed by Lev Landau.[7] Our discussion of Landau's theory of phase transitions begins with a simple observation about magnets. We imagine a magnet whose spin moments can point either up or down. At high temperatures each spin is equally likely to be found up or down. This system is shown in Fig. 41.2(a). Its magnetization, that is, the spatial average of the magnetic moment, is zero. The system has a symmetry: turn all of the spins through 180° [as in Fig. 41.2(b)] and the magnetization is still zero. Of course, each individual moment is pointing in the opposite direction, but the number pointing in the upwards direction is still half of the total and the magnetization is still zero. The symmetry here is a global one: we rotate *all* of the spins through the same angle, here 180°.

It is found experimentally that upon cooling the system through a critical temperature T_c, the system undergoes a phase transition and the magnetization M becomes non-zero as all of the spins line up along a single direction, as shown in Fig. 41.3(a). The direction along which the spins align could *either* be all in the up direction *or* the down direction. If we rotate each spin of the aligned system through 180° [Fig. 41.3(b)] M is obviously reversed. We say that the system has broken (or lowered) its symmetry in the ordered phase.

One puzzling feature of this description is the reason why the system chose to point all of the spins in one direction rather than the other. After all, there's nothing in the Hamiltonian which describes the system that distinguishes between up and down.[8] The result is that the ground state does not have the symmetry of the Hamiltonian describing the system. The original symmetry of the system appears to have *spontaneously* broken. The same thing happens in the Euler strut shown in Fig. 41.4. A weight is balanced on top of an elastic strut. If the weight is large enough the strut will buckle. The buckling can be *either* to the left *or* to the right. There is nothing in the underlying physics of the strut and weight that allows one to predict which way it will go.[9]

Since equilibrium in thermodynamic systems relies on both minimizing the internal energy U and maximizing the entropy S of a system, Landau considered the free energy $F = U - TS$. To find the equilibrium state of the system we need to minimize F. The free energy is a function of an **order parameter**, namely some field describing the system whose thermal average is zero in the $T > T_c$ unbroken-symmetry state and non-zero in the $T < T_c$ broken-symmetry state. For a magnet, the order parameter is simply the magnetization field M. With deliberate ignorance of the microscopic state of the system, Landau wrote $F(M)$ as a power series

$$F = F_0 + aM^2 + bM^4 + \dots, \tag{41.2}$$

where a and b (the latter assumed > 0) are parameters which are independent of M, but may in principle depend on the temperature. This free energy has the symmetry $M \to -M$, that is, reversing the magnetization doesn't affect the energy. If a and b are positive we have the free energy shown in Fig. 41.5(a), which is minimized at $M = 0$, which is the correct prediction for the high-temperature regime. If, however, the parameter a is negative then we have the free energy shown in Fig. 41.5(b), which has two minima at non-zero values of magnetization. These minima correspond to the spins all aligning up ($M = +M_0$) or all aligning down ($M = -M_0$). The previous minimum $M = 0$ is now at a position of metastable equilibrium.[10] If we take $a = a_0(T - T_c)$ (with a_0 a constant) and $b(> 0)$ to be T-independent, then clearly a is positive for $T > T_c$ and negative for $T < T_c$, and so we predict a phase transition at a temperature $T = T_c$.

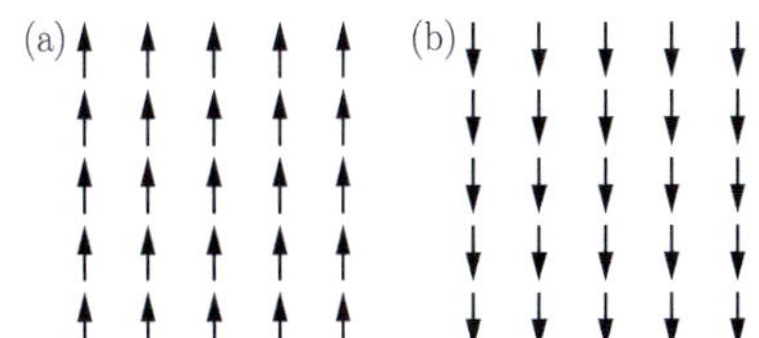

Fig. 41.3 For temperatures $T < T_c$ the spins align along a single direction. The magnetization in (a) is different in (b), where each spin has been rotated by 180°.

[8]For the magnet this Hamiltonian could be $\hat{H} = -J\sum_i \hat{S}_i^z \hat{S}_{i+1}^z$, where J is a coupling constant, $\hat{S}_i^z$ is the spin operator for the ith spin and the up direction is along z. This Hamiltonian has no bias either favouring the spins pointing up or favouring them pointing down, only for them pointing in the same direction.

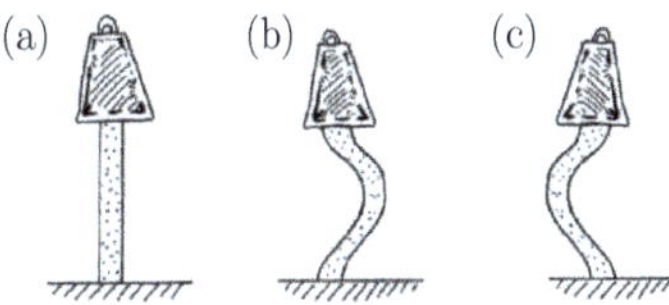

Fig. 41.4 The Euler strut. (a) This has a vertical axis of symmetry and can buckle either (b) one way or (c) the other, in each case breaking symmetry.

[9]Of course, in real life, some very small perturbation or fluctuation will have tipped the system towards its choice of ground state.

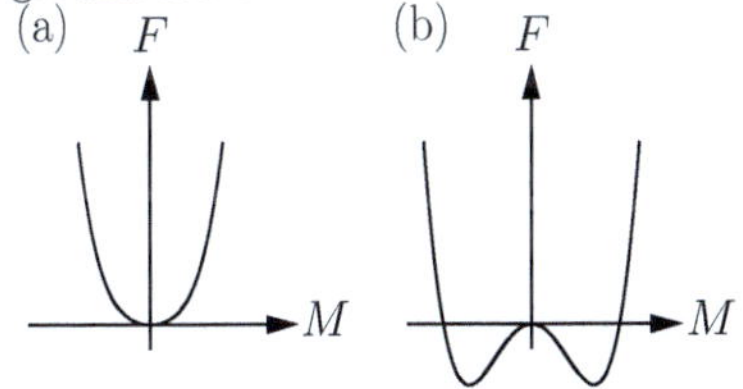

Fig. 41.5 The Landau free energy for (a) $T > T_c$ with a minimum at $M = 0$ and (b) for $T < T_c$ with minima at $\pm M_0$.

[10]This means that any slight perturbation of the system which has been prepared in a $M = 0$ state will drop it into one, or other, of the new minima that have emerged at $\pm M_0$.

Example 41.2

We can find the minima of F straightforwardly for $T < T_c$. We set

$$\frac{\partial F}{\partial M} = 2aM + 4bM^3 = 0, \tag{41.3}$$

from which we conclude that the minima occur at $M = M_0$ where

$$M_0^2 = -\frac{a}{2b}. \tag{41.4}$$

The square root gives a real M_0 since $a < 0$. There is also a maximum at $M = 0$.

For $T < T_c$ we have two minima at $M = \pm M_0$. The system will be propelled into one of these as its new ground state and the symmetry will be broken.

Let's now apply the arguments developed for magnets to our Lagrangian treatment of field theory. Here instead of following the ground state magnetization, we're interested in the equilibrium state value of $\phi(x)$, which is the order parameter for the field theory. We start with a scalar field theory

$$\mathcal{L} = -\frac{1}{2}(\partial_\mu \phi)^2 - U(\phi), \tag{41.5}$$

where we've split off all of the potential-energy-like terms and called them the potential energy density $U(\phi)$. For the so-called ϕ^4 theory we have that

$$U(\phi) = \frac{\mu^2}{2}\phi^2 + \frac{\lambda}{4!}\phi^4, \tag{41.6}$$

where μ and $\lambda (> 0)$ are parameters. This function resembles the free energy in our magnet example above. It admits the global symmetry $\phi(x) \rightarrow -\phi(x)$. Assuming μ^2 is positive we have a potential with a minimum at $\phi = 0$ as shown in Fig. 41.6(a). The minimum corresponds to a ground state, also known as the vacuum, of $\phi(x) = 0$.

What if we swap the sign in front of μ^2? In that case we have $U(\phi) = -\frac{\mu^2}{2}\phi^2 + \frac{\lambda}{4!}\phi^4$, and a potential that looks like Fig. 41.6(b). The minima[11] of the potential now occur at

$$\phi_0 = \pm\left(\frac{6\mu^2}{\lambda}\right)^{\frac{1}{2}}, \tag{41.7}$$

so we have the choice of two new vacua. Once the temperature has been lowered such that we have the characteristic broken-symmetry potential, we imagine the system lowering its energy by adopting one of the states ϕ_0 consistent with one of the minima. This can be thought of as the system *rolling down the potential* towards a minimum of the potential function $U(\phi_0) = U_0$.

This mechanism, involving a scalar field that breaks a symmetry, is the one responsible for the Higgs mechanism that leads to the generation of mass in the electroweak part of the Standard Model. Since there is good evidence for the existence of a scalar Higgs field, there is certainly motivation to examine the effect of scalar fields on the expansion of the Universe. This combination of a field and an expanding spacetime will provide a model of inflation.

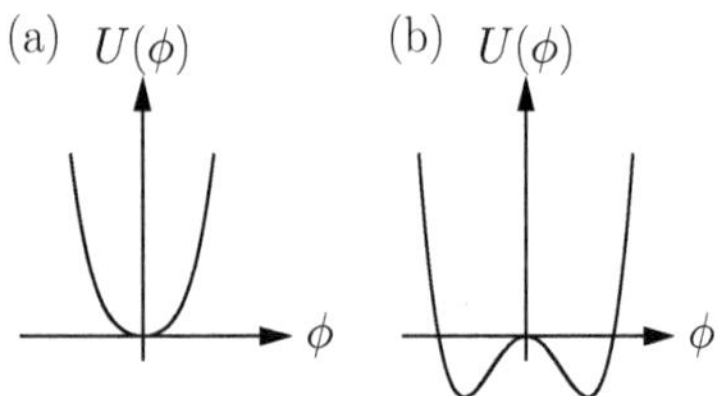

Fig. 41.6 Equation 41.6 for (a) $\mu^2 > 0$ with a minimum at $\phi = 0$ and (b) for $\mu^2 < 0$ with minima at $\pm\sqrt{6\mu^2/\lambda}$.

[11] From $\partial U/\partial \phi = 0$ we find

$$0 = -\mu^2 \phi + \frac{\lambda}{3!}\phi^3,$$

from which we deduce stationary points at $(0, \pm\sqrt{6\mu^2/\lambda})$. The second derivative,

$$\frac{\partial^2 U}{\partial \phi^2} = -\mu^2 + \frac{\lambda\phi^2}{2},$$

is negative (and equals $-\mu^2$) for $\phi = 0$ and is positive (and equals $+2\mu^2$) for the minima at $\phi_0 = \pm\sqrt{6\mu^2/\lambda}$.

41.2 Effective potentials

Our model for inflation involves the phase transition of a scalar field similar to that discussed above.[12] Let's consider a scalar field $\phi(x)$ in the Universe before the GUT phase transition. We'll use the ϕ^4 Lagrangian

$$\mathcal{L} = -\frac{1}{2}(\partial_\mu \phi)^2 - \frac{\mu^2(T)}{2}\phi^2 - \frac{\lambda}{4!}\phi^4, \tag{41.8}$$

so that the potential $U = \frac{\mu^2(T)}{2}\phi^2 + \frac{\lambda}{4!}\phi^4$. This field has an equation of motion

$$\phi_{;\mu\nu}g^{\mu\nu} = \frac{\partial U}{\partial \phi} = \mu^2(T)\phi^2 + \frac{\lambda}{3!}\phi^3. \tag{41.9}$$

In eqn 41.8, $T(t)$ is the time-dependent temperature which causes $\mu^2(T)$ to change sign as a function of time below a critical temperature T_c. At this point the field ϕ undergoes a phase transition. In the cosmology we're dealing with, we shall assume that the field ϕ is always uniform in space, so that $\vec{\nabla}\phi = 0$ at all times. Crucially, ϕ will vary as a function of time as the phase transition takes place.

We shall now show how the expansion of a Robertson–Walker spacetime causes a frictional term in the equation of motion for ϕ, that depends on the rate of expansion of spacetime. The energy-density and pressure of the ϕ field must also have an effect on the evolution of the Universe, determining the rate of expansion. Taken together, the resulting coupled equations of motion can lead to the rapid expansion of the Universe we're looking for. In the example below, we shall simplify the description for that appropriate to a flat Universe.[13]

[12]This field should not be identified with the Higgs field, but is rather some other field in the Universe. Confusingly, it is sometime called a Higgs field.

[13]This simplifying approximation is discussed in the next section.

Example 41.3

Consider the scalar field Lagrangian

$$\mathcal{L} = -\frac{1}{2}(\partial_\mu \phi)^2 - U(\phi), \tag{41.10}$$

in a spatially flat Robertson–Walker spacetime[14] with diagonal metric $g^{tt} = -1$ and $g^{ii} = 1/a(t)^2$. The equations of motion are

$$\phi_{;\mu\nu}g^{\mu\nu} = \frac{\partial U(\phi)}{\partial \phi}, \tag{41.11}$$

and since ϕ is a scalar we have $\phi_{;\mu} = \partial_\mu \phi$. We can then compute the second covariant derivative as

$$\phi_{;\mu\nu} = (\partial_\mu \phi)_{;\nu} = \partial_\nu \partial_\mu \phi - \Gamma^\sigma_{\ \nu\mu}\partial_\sigma \phi. \tag{41.12}$$

Since we shall trace over the tensor $\phi_{;\mu\nu}$, we need only compute the diagonal elements, and so we find

$$\phi_{;00} = \frac{\partial^2 \phi}{\partial t^2}, \quad \phi_{;ii} = \partial_i^2 \phi - \dot{a}a\frac{\partial \phi}{\partial t}, \tag{41.13}$$

and since we only deal with slow variations in space, we retain the time derivatives only. Tracing using our metric $g^{\mu\nu}$ we obtain

$$\frac{\partial^2 \phi}{\partial t^2} + 3\frac{1}{a}\frac{\partial a}{\partial t}\frac{\partial \phi}{\partial t} + \frac{\partial U}{\partial \phi} = 0. \tag{41.14}$$

[14]See Chapter 15 and Exercise 15.2. This spatially flat, expanding Universe has $\Gamma^0_{\ ij} = a\dot{a}\delta_{ij}$ and $\Gamma^i_{\ j0} = (\dot{a}/a)\delta^i_j$.

The second term of the latter equation supplies an effective drag factor $3H = 3\frac{\dot{a}}{a}$ that damps the motion of the ϕ field. Our theory has an energy-momentum tensor given by

$$T_{\mu\nu} = \partial_\mu\phi\partial_\nu\phi - g_{\mu\nu}\left[\frac{1}{2}g^{\alpha\beta}\partial_\alpha\phi\partial_\beta\phi + U(\phi)\right].\tag{41.15}$$

Again assuming spatially uniform ϕ, and working in the local frame of an observer, we can pick out an energy-density

$$\rho = \frac{1}{2}\dot{\phi}^2 + U(\phi),\tag{41.16}$$

and also a pressure

$$p = \frac{1}{2}\dot{\phi}^2 - U(\phi).\tag{41.17}$$

In the event that the fields are constant in time, we have $\rho = -p$, which, comparing to the case of a fluid[15] gives us an effective cosmological constant $\Lambda = 8\pi U(\phi)$. This is promising, since it implies that the non-zero scalar field can potentially cause the sort of rapid expansion we saw in the de Sitter spacetime of Universe 1 in Chapter 15. We also need to find the equation of motion for $a(t)$. The Friedmann equation for $k = 0$ tells us that

$$\left(\frac{\dot{a}}{a}\right)^2 = \frac{8\pi\rho}{3},\tag{41.19}$$

or, using our result for ρ for the scalar field

$$\left(\frac{\dot{a}}{a}\right)^2 = \frac{8\pi}{3}\left[\frac{1}{2}\dot{\phi}^2 + U(\phi)\right].\tag{41.20}$$

This completes the necessary ingredients to describe inflation.

The two most important equations from our analysis of the scalar field in an expanding universe are then: (i) the damped equation of motion for the ϕ field

$$\ddot{\phi} + 3\frac{\dot{a}}{a}\dot{\phi} + \frac{\partial U}{\partial\phi} = 0,\tag{41.21}$$

and (ii) the equation of motion for the expansion factor of a Universe filled with ϕ-field energy density

$$\left(\frac{\dot{a}}{a}\right)^2 = \frac{8\pi}{3}\left(\frac{1}{2}\dot{\phi} + U\right).\tag{41.22}$$

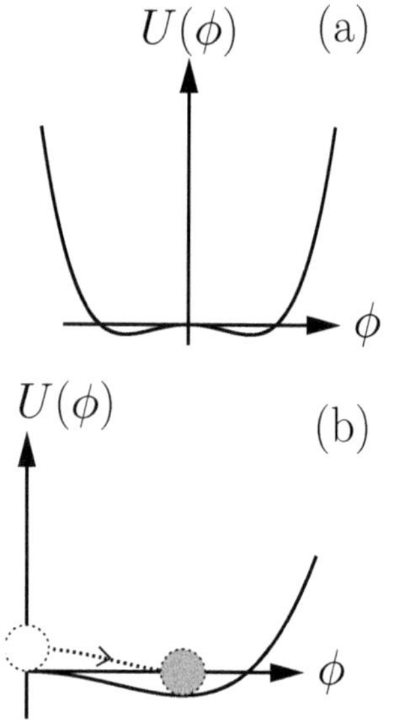

The key now is to look at the phase transition in the ϕ field and how ϕ makes its way to its new minimum as a function of time for $T < T_\mathrm{c}$. To make the description simple we make the **slow-rolling approximation**, that at temperatures below the transition, the ϕ field will change slowly in time. This requires a shallow potential $U(\phi)$, as shown in Fig. 41.7.

Slow rolling means that we have a negligible value of $\ddot{\phi}$ so that the equation of motion for the ϕ field (eqn 41.21) becomes $3(\dot{a}/a)\dot{\phi} = -\partial U/\partial\phi$. Near the transition the flat potential takes a near-constant value $U(\phi) \approx U(\phi_0) = U_0$, which implies that $\dot{\phi}$ also becomes small. What we're really interested in is the equation of motion for the expansion factor of the Universe $a(t)$, and so, considering eqn 41.22 with small $\dot{\phi}$, we deduce that

$$\frac{\dot{a}}{a} \approx \left(\frac{8\pi U_0}{3}\right)^{\frac{1}{2}}.\tag{41.23}$$

[15]For the fluid we had

$$T_{\mu\nu} = (\rho + p)u_\mu u_\nu + pg_{\mu\nu}.\tag{41.18}$$

Fig. 41.7 (a) The shallow potential below the transition temperature, required by the inflation model. (b) The slow-rolling approximation where the field ϕ slowly finds the potential minimum.

The solution to this latter equation is an exponential growth of $a(t)$ with

$$a(t) \propto \exp\left[\left(\frac{8\pi U_0}{3}\right)^{\frac{1}{2}} t\right].$$

(41.24)

The theory therefore predicts an inflationary period where, owing to the behaviour of the ϕ field close to its phase transition and the consequences on its energy density, the expansion factor $a(t)$ of the Universe increases exponentially.

41.3   Why flat?

In the above discussion, we have assumed the starting point of inflation was a $k = 0$ universe.[16] In this section, we shall see that the result of inflation is a flat Universe and, in fact, the end result of an inflationary expansion is *always* a flat space, no matter which curvature we start with. To understand this we first note that rapid expansion is always seen in models with a cosmological constant Λ and we can therefore compare our exponential expansion to these other analogous universes with cosmological-constant-driven expansion. These models predict expansions

$$a(t) \propto \begin{cases} \cosh Ht & (k = +1), \\ e^{Ht} & (k = 0), \\ \sinh Ht & (k = -1), \end{cases}$$

(41.25)

with $H^2 = \Lambda/3$. Owing to the behaviour of the hyperbolic functions, each of these cases evolve towards an exponential expansion. That is, they are eventually all consistent with the $\Omega = 1$ case that results in $k = 0$ flat space. So, even if Ω is not initially set close to 1, then exponential expansion forces Ω towards this value very rapidly.

The Friedmann equations tell us that in the absence of other sources of cosmological constant Λ, the acceleration of the Universe obeys

$$\frac{\ddot{a}}{a} = -4\pi\left(p + \frac{\rho}{3}\right),$$

(41.26)

so in order to achieve the positive acceleration needed for inflation we require $p < -\rho/3$. In terms of the estimates for p and ρ from our scalar field (eqns 41.16 and 41.17), this condition is equivalent to

$$\dot{\phi}^2 < U(\phi).$$

(41.27)

The condition for expansion provides a means for inflation to cease during the rolling of the field ϕ along the potential: that is, when the speed becomes large enough and/or the value of $U(\phi)$ is sufficiently small. One possibility is that ϕ reach the minimum at U_0 and then oscillates about this value as the damping term reduces the kinetic energy $\frac{1}{2}\dot{\phi}^2$ until the field is at rest. If $U_0 < 0$ at the minimum then inflation stops.[17] In practice, the presence of coupling to other fields will cause **reheating**, where

[16] Inflation occurs at the start of the expansion of the Universe so the effect of a particular value of k will be difficult to detect before the Universe has evolved sufficiently. Other values of k can be treated in the same spirit, but the differences in the predictions are small compared to the observables. Treating $k = 0$ is therefore a good approximation.

[17] The condition therefore also provides a means for inflation to continue indefinitely if the field comes to rest at the minimum and $U_0 > 0$.

the scalar field decays during the roll, producing other particle excitations. At the end of the period of inflation we are then left with a flat Universe filled with normal matter and radiation, allowing a standard-model expansion of the Universe afterwards.

Although the inflationary model has been hugely influential over the last few decades, it is not accepted by absolutely all cosmologists. Is it the correct model of the early Universe? It is probably fair to say that the jury is still out. However, cosmic inflation remains the reigning paradigm. As we've seen in this chapter, it solves some knotty problems in cosmology and it is likely to continue to remain our best picture of the Universe just after the Big Bang, unless of course some other revolutionary new idea or piece of contradictory observational evidence turns up. We discuss the current status of the field in Chapter 49.

Chapter summary

- To solve the flatness and horizon problems of cosmology, inflation theory proposes a rapid period of expansion in the early life of the Universe.

- The theory relies on the notion of a symmetry-breaking phase transition that takes place as the Universe cools.

- The slow-rolling approximation predicts an exponential expansion of the Universe for a limited period of time, leading to a spatially flat Universe at the end of the expansion.

Exercises

(41.1) (a) Compute the equation of motion for a scalar field Φ described by the Lagrangian in eqn 41.5 in a Robertson–Walker universe with $k \neq 0$.

Use the connection coefficients given in Chapter 16
(b) Verify that the equation of motion reduces to the expected form for $k = 0$.

The electromagnetic field

42

Electrical force is defined as something which causes motion of electrical charge; an electrical charge is something which exerts electric force.
Arthur Eddington (1882–1944) *The Nature of the Physical World*

O'er the wires the electric message came,
'He is no better; he is much the same.'
Alfred Austin (1835–1913) (attrib.) *On the Illness of the Prince of Wales*

In this chapter, we use the geometrical techniques from the previous part of the book to reformulate **electromagnetism**. In some ways, electromagnetism can be viewed as one of the simplest and most successful field theories in Nature. Einstein certainly thought so, and would return to electromagnetism regularly for inspiration throughout his formulation of relativity. In much the same way that in Chapter 0, where we characterized gravitation with John Wheeler's slogan

> spacetime tells matter how to move;
> matter tells spacetime how to curve,

we can similarly sloganize electromagnetism by saying

> electromagnetic fields tells electric charges how to move;
> electric charges tells electromagnetic fields where to go.

We shall return to the analogy between electromagnetism and gravitation repeatedly in the coming chapters. This chapter aims to set the scene. We start by working in flat, Minkowski spacetime.[1] We shall produce a Lagrangian description of electromagnetism in terms of the components of its field. We then formulate a description of electromagnetism using the coordinate-free, geometrical machinery of tensors.

[1] Remember, of course, that valid tensor equations in flat space are also valid in curved space, so this simplification does not limit our results significantly.

42.1 Electric charge in a field

A **charge** can be thought of as that property of a particle that couples to a field. How could a tensor field cause a charge to move? We have seen the answer for the case of the metric field where curvature causes the movement of massive particles. However, there are other ways that a field can cause a change in motion of a charge. Let's ask what the simplest method of coupling a particle and a field might be. This, it turns out, is how electromagnetism manifests itself in Nature.

The relativistic action for a free massive particle is given by

$$S_{\mathrm{m}} = -m \int \sqrt{-\mathrm{d}s^2},\tag{42.1}$$

where in flat, Minkowski space the interval $\mathrm{d}s^2 = \eta_{\mu\nu}\mathrm{d}x^\mu\mathrm{d}x^\nu$. We would like this particle to interact with another field. Perhaps the simplest coupling that obeys Lorentz symmetry couples the displacement of the particle $\mathrm{d}x^\mu$ to the components of a 1-form field $\tilde{\boldsymbol{A}}(x)$. This results in a contribution to the action of[2]

$$S_{\mathrm{em}} = q \int A_\sigma(x)\mathrm{d}x^\sigma,\tag{42.2}$$

where the charge q tells us the strength of the coupling between the field and the displacement. This action describes the interaction between the electromagnetic field[3] $\tilde{\boldsymbol{A}}(x)$ and a particle with electric charge q. The particle is assumed massive and so we choose to parametrize its path with an affine parameter τ. We then have a total action S for a particle in the field $\tilde{\boldsymbol{A}}(x)$ given by the sum of the free-particle action and the interaction of the particle and the electromagnetic field[4] $S = S_{\mathrm{m}} + S_{\mathrm{em}}$ or

$$S = \int \left[-m \left(-\eta_{\mu\nu}\frac{\mathrm{d}x^\mu}{\mathrm{d}\tau}\frac{\mathrm{d}x^\nu}{\mathrm{d}\tau} \right)^{\frac{1}{2}} + qA_\sigma\frac{\mathrm{d}x^\sigma}{\mathrm{d}\tau} \right] \mathrm{d}\tau.\tag{42.3}$$

This action can be used to derive the equations of motion for the particle.

[2] Recall that the Lagrangian description of fields relies on the components of tensors. We return to our coordinate-free, geometrical description below.

[3] We explain the link between the 1-form field $\tilde{\boldsymbol{A}}$ and the E- and B-fields of electromagnetism in the next section.

[4] This action does not include the dynamics of the electromagnetic fields themselves. We deal with this a little later.

Example 42.1

If the action is given by $S = \int L\mathrm{d}\tau$, then we can write the Lagrangian as $L = L_{\mathrm{m}} + L_{\mathrm{em}}$. We then utilize the Euler–Lagrange equations for each part in turn. The kinetic energy contribution of the free particle is

$$\frac{\partial L_{\mathrm{m}}}{\partial \left(\frac{\mathrm{d}x^\alpha}{\mathrm{d}\tau} \right)} = \frac{m}{2L_{\mathrm{m}}} \left(\eta_{\mu\nu}\delta^\mu{}_\alpha\frac{\mathrm{d}x^\nu}{\mathrm{d}\tau} + \eta_{\mu\nu}\frac{\mathrm{d}x^\mu}{\mathrm{d}\tau}\delta^\nu{}_\alpha \right)$$

$$= \frac{m}{L_{\mathrm{m}}}\eta_{\mu\alpha}\frac{\mathrm{d}x^\mu}{\mathrm{d}\tau}.\tag{42.4}$$

Using the usual routine for length parametrization we have

$$\frac{\mathrm{d}}{\mathrm{d}\tau}\frac{\partial L_{\mathrm{m}}}{\partial \left(\frac{\mathrm{d}x^\alpha}{\mathrm{d}\tau} \right)} = m\eta_{\mu\alpha}\frac{\mathrm{d}^2x^\mu}{\mathrm{d}\tau^2}.\tag{42.5}$$

Turning to the second, interaction, term

$$\frac{\partial L_{\mathrm{em}}}{\partial \left(\frac{\mathrm{d}x^\alpha}{\mathrm{d}\tau} \right)} = qA_\sigma\delta^\sigma{}_\alpha = qA_\alpha,\tag{42.6}$$

and so

$$\frac{\mathrm{d}}{\mathrm{d}\tau}\frac{\partial L_{\mathrm{em}}}{\partial \left(\frac{\mathrm{d}x^\alpha}{\mathrm{d}\tau} \right)} = q\frac{\mathrm{d}A_\alpha}{\mathrm{d}\tau} = q\frac{\mathrm{d}x^\beta}{\mathrm{d}\tau}\frac{\partial A_\alpha}{\partial x^\beta}.\tag{42.7}$$

Finally,

$$\frac{\partial L_{\text{em}}}{\partial x^\alpha} = q\frac{\partial A_\sigma}{\partial x^\alpha}\frac{dx^\sigma}{d\tau}. \tag{42.8}$$

Putting the parts together using the Euler–Lagrange equations gives

$$m\eta_{\mu\alpha}\frac{d^2 x^\mu}{d\tau^2} + q\frac{dx^\beta}{d\tau}\frac{\partial A_\alpha}{\partial x^\beta} = q\frac{\partial A_\sigma}{\partial x^\alpha}\frac{dx^\sigma}{d\tau}, \tag{42.9}$$

or

$$m\eta_{\mu\alpha}\frac{d^2 x^\mu}{d\tau^2} = q\left(\frac{\partial A_\beta}{\partial x^\alpha} - \frac{\partial A_\alpha}{\partial x^\beta}\right)\frac{dx^\beta}{d\tau}, \tag{42.10}$$

where we've reindexed the terms in the final line.

We can rewrite the final expression from the last example as[5]

$$m\eta_{\mu\alpha}\frac{d^2 x^\mu}{d\tau^2} = qF_{\alpha\beta}u^\beta, \tag{42.12}$$

where $u^\beta = dx^\beta/d\tau$ are the components of the velocity, and the interaction is encoded in the tensor components

$$F_{\alpha\beta} = A_{\beta,\alpha} - A_{\alpha,\beta}. \tag{42.13}$$

In fact, the $F_{\alpha\beta}$ are the components of the **Faraday tensor**.[6] This is a (0,2) tensor which can be written in a basis as

$$\boldsymbol{F}(\ ,\ ) = F_{\alpha\beta}\,\boldsymbol{dx}^\alpha \otimes \boldsymbol{dx}^\beta. \tag{42.14}$$

Note from the definition of the components that the tensor is antisymmetric and so the Faraday tensor can be written as the 2-form[7]

$$\tilde{\boldsymbol{F}}(\ ,\ ) = \frac{1}{2}F_{\alpha\beta}\,\boldsymbol{dx}^\alpha \wedge \boldsymbol{dx}^\beta. \tag{42.15}$$

We shall see that it is this tensor which gives us access to the electromagnetic fields. We can summarize the equation of motion symbolically as[8]

$$\frac{d\tilde{\boldsymbol{p}}}{d\tau} = q\tilde{\boldsymbol{F}}(\ ,\boldsymbol{u}), \tag{42.17}$$

where the momentum 1-form features on the left-hand side and the velocity $\boldsymbol{u}$ has been inserted into the second slot of the tensor $\tilde{\boldsymbol{F}}$ (corresponding to contracting against the second index of $F_{\alpha\beta}$ in eqn 42.17). Since this expression is a valid tensor expression in flat space, it must also be applicable to curved space and represents the answer to our question of how a charge couples to the electromagnetic field. Next, we turn to the dynamics of the 1-form field $\tilde{\boldsymbol{A}}$ and the 2-form $\tilde{\boldsymbol{F}}$, as these provide the equations of motion of the electromagnetic fields, also known as **Maxwell's equations**.[9]

42.2 Faraday tensor and Maxwell equations

The field $\tilde{\boldsymbol{A}}(x) = A_\mu(x)\boldsymbol{dx}^\mu$ is a 1-form field known as the electromagnetic **gauge field**. The four components $A_\mu = (A_0, A_i)$ may, in flat

[5]As examined in the Exercises, the spatial components of the right-hand side of this equation give the relativistic version of the Lorentz force law

$$\vec{F} = q(\vec{E} + \vec{v} \times \vec{B}). \tag{42.11}$$

[6]Michael Faraday (1791–1867).

[7]Remember that the factor $1/2$ arises because

$$\frac{1}{2}F_{\alpha\beta}\boldsymbol{dx}^\alpha \wedge \boldsymbol{dx}^\beta$$
$$= \frac{1}{2}(F_{\alpha\beta}\,\boldsymbol{dx}^\alpha \otimes \boldsymbol{dx}^\beta$$
$$\quad - F_{\alpha\beta}\,\boldsymbol{dx}^\beta \otimes \boldsymbol{dx}^\alpha)$$
$$= \frac{1}{2}(F_{\alpha\beta}\,\boldsymbol{dx}^\alpha \otimes \boldsymbol{dx}^\beta$$
$$\quad - F_{\beta\alpha}\,\boldsymbol{dx}^\alpha \otimes \boldsymbol{dx}^\beta)$$
$$= F_{\alpha\beta}\,\boldsymbol{dx}^\alpha \otimes \boldsymbol{dx}^\beta,$$

where the antisymmetry of the components is used to produce the last line.

[8]That is, we lower the index with $x_\alpha = \eta_{\alpha\mu}x^\mu$ and then see that the left-hand side can be written as

$$m\frac{d}{d\tau}\tilde{\boldsymbol{u}},$$

where $u_\alpha = dx_\alpha/d\tau$, before noting that $\tilde{\boldsymbol{p}} = m\tilde{\boldsymbol{u}}$. Note also that since $\tilde{\boldsymbol{F}}(\boldsymbol{u},\boldsymbol{u}) = 0$, which follows from the antisymmetry of the Faraday tensor, we have on the left-hand side an expression that effectively says $\boldsymbol{u} \cdot \boldsymbol{a} = 0$, where $\boldsymbol{a}$ is the acceleration, which is a condition required for our theory (see Chapter 1). Finally, the upgrade to curved space simply swaps commas for semicolons to give

$$\frac{D\tilde{\boldsymbol{p}}}{d\tau} \equiv m\boldsymbol{\nabla}_{\boldsymbol{u}}\tilde{\boldsymbol{u}} = q\tilde{\boldsymbol{F}}(\ ,\boldsymbol{u}). \tag{42.16}$$

The equation of motion tells us that the departure from geodesic free fall is given by the combination of the charge, Faraday tensor and particle velocity.

[9]James Clerk Maxwell (1831–1879).

[10]As usual, A^i are the components of the spatial 3-vector $\vec{A}$.

space, be mapped with metric $\boldsymbol{\eta}$ onto a (1,0) vector field $\boldsymbol{A}(x)$ with components $A^\mu = (A^0, A^i)$, where $A_0 = -A^0$ and[10] $A_i = A^i$. The timelike component A^0 is often called the **electrostatic potential** V.

We then define the 3-vector electric field $\vec{E}$ in terms of the components of the gauge field as

$$\vec{E} = -\vec{\nabla} A^0 - \frac{\partial \vec{A}}{\partial t}, \tag{42.18}$$

and the magnetic field $\vec{B}$ as

$$\vec{B} = \vec{\nabla} \times \vec{A}. \tag{42.19}$$

An important property of these equations is that there is some freedom in how the components A_μ are specified. In fact, if we make the **gauge transformation**

$$A_\mu(x) \to A_\mu(x) - \frac{\partial \chi(x)}{\partial x^\mu}, \tag{42.20}$$

where $\chi(x)$ is some arbitrary function, then the $\vec{E}$ and $\vec{B}$ fields are unchanged. This turns out to be of fundamental importance to the theory and is discussed further in the next section. Our task here is to derive the equations of motion of the electromagnetic field.

Since the components of the Faraday tensor are given by $F_{\mu\nu} = A_{\nu,\mu} - A_{\mu,\nu}$ then, in terms of the components of $\vec{E}$ and $\vec{B}$ fields, we can write the matrix as[11]

[11]Note that in flat space we can also transform these to the components of a (2,0) tensor $\boldsymbol{F} = F^{\mu\nu} \boldsymbol{e}_\mu \otimes \boldsymbol{e}_\nu = \frac{1}{2} F^{\mu\nu} \boldsymbol{e}_\mu \wedge \boldsymbol{e}_\nu$, where

$$F^{\mu\nu} = \begin{pmatrix} 0 & E^x & E^y & E^z \\ -E^x & 0 & B^z & -B^y \\ -E^y & -B^z & 0 & B^x \\ -E^z & B^y & -B^x & 0 \end{pmatrix}. \tag{42.21}$$

Notice how the components of the $\vec{E}$ field change their signs, but those of the $\vec{B}$ field do not.

[12]It is therefore $\boldsymbol{J}$ that tells the field lines where to go (and the field lines tell $\boldsymbol{J}$ how to evolve in spacetime).

[13]This is related to the Bianchi identity for gravitation that we have met in Chapter 13. The link will be discussed in detail in Chapter 43, but we shall show how the identity arises geometrically in electromagnetism at the end of this chapter.

[14]Set $\alpha = 1, \beta = 2$ and $\gamma = 3$.

[15]Set one of the indices equal to zero and this follows.

$$F_{\mu\nu} = \begin{pmatrix} 0 & -E^x & -E^y & -E^z \\ E^x & 0 & B^z & -B^y \\ E^y & -B^z & 0 & B^x \\ E^z & B^y & -B^x & 0 \end{pmatrix}. \tag{42.22}$$

The components of $\tilde{\boldsymbol{F}}$ have dynamics of their own, encoded in **Maxwell's equations of motion**. Recall that Maxwell's equations are written in natural units as

$$\vec{\nabla} \cdot \vec{E} = \rho, \qquad \vec{\nabla} \cdot \vec{B} = 0,$$
$$\vec{\nabla} \times \vec{E} = -\frac{\partial \vec{B}}{\partial t}, \qquad \vec{\nabla} \times \vec{B} = \vec{J} + \frac{\partial \vec{E}}{\partial t}. \tag{42.23}$$

We combine the charge density ρ and the current density $\vec{J}$ to create the current 4-vector $\boldsymbol{J}$ with components $J^\mu = (\rho, \vec{J})$. This vector represents the source of the electromagnetic fields (i.e. charges and currents) and also the bodies that are set in motion by the fields.[12]

Maxwell's equations can be re-expressed in terms of the components of the Faraday tensor. It can be checked that the[13] **Bianchi identity**

$$F_{\alpha\beta,\gamma} + F_{\gamma\alpha,\beta} + F_{\beta\gamma,\alpha} = 0, \tag{42.24}$$

yields the Maxwell equations[14] $\vec{\nabla} \cdot \vec{B} = 0$ and[15] $-\partial \vec{B}/\partial t = \vec{\nabla} \times \vec{E}$. The other two Maxwell equations (involving the components of the current $\boldsymbol{J}$) can be recreated using the equation

$$F^{\mu\nu}{}_{,\nu} = J^\mu. \tag{42.25}$$

Example 42.2

Returning to the Lagrangian point of view, we claimed in Chapter 40 that the Lagrangian density for the electromagnetic matter fields themselves was $\mathcal{L} = -\frac{1}{4}F_{\mu\nu}F^{\mu\nu}$. We can also include a generalized version of the coupling with Lagrangian $L = q\,\mathrm{d}x^\mu A_\mu$, which becomes a contribution to the Lagrangian density of $\mathcal{L}_{\mathrm{emf}} = J^\mu A_\mu$. We then have the action for the electromagnetic fields and their coupling to charges as given by

$$
\begin{aligned}
S_{\mathrm{em}} &= \int \mathrm{d}^4 x \left(-\frac{1}{4}F_{\mu\nu}F^{\mu\nu} + J^\mu A_\mu \right) \\
&= \int \mathrm{d}^4 x \left[-\frac{1}{4}\left(\partial_\mu A_\nu - \partial_\nu A_\mu\right)\left(\partial^\mu A^\nu - \partial^\nu A^\mu\right) + J^\mu A_\mu \right].
\end{aligned}
\tag{42.26}
$$

Feeding the Lagrangian density in the integrand through the Euler–Lagrange equations yields the Maxwell equations.

This is all very well, but in order to understand the Maxwell equations in the context of gravitation, we should examine the geometrical interpretation of electromagnetism. Indeed we shall shortly see how a geometrical viewpoint leads to the Maxwell equations.

Example 42.3

Before we get to a coordinate-free formulation, we can get an idea of how curvature changes electromagnetism.[16] For simplicity we set the currents J^μ to zero. In curved space with metric components $g_{\mu\nu}$, we produce the Faraday tensor using the commas-go-to-semicolons rule, to give

$$
F_{\mu\nu} = A_{\nu;\mu} - A_{\mu;\nu}\,(= 2A_{[\nu;\mu]}).
\tag{42.27}
$$

Since we must be careful to use the metric to raise indices, we rewrite the flat-space Lagrangian density $\mathcal{L}_{\mathrm{em}} = -\frac{1}{4}F^{\mu\nu}F_{\mu\nu}$, which becomes

$$
\mathcal{L}_{\mathrm{em}} = -\frac{1}{4}F_{\alpha\beta}F_{\gamma\delta}g^{\alpha\gamma}g^{\beta\delta}.
\tag{42.28}
$$

The equations of motion can then be found using the E-L equations. We have

$$
\frac{\partial \mathcal{L}_{\mathrm{m}}}{\partial A_{\alpha;\beta}} = 0,
\tag{42.29}
$$

and

$$
\frac{\partial \mathcal{L}_{\mathrm{m}}}{\partial A_{\alpha;\beta}} = (A_{\delta;\gamma} - A_{\gamma;\delta})g^{\alpha\gamma}g^{\beta\delta},
\tag{42.30}
$$

from which we find

$$
\begin{aligned}
\left(\frac{\partial \mathcal{L}_{\mathrm{m}}}{\partial A_{\alpha;\beta}} \right)_{;\beta} &= (A_{\delta;\gamma} - A_{\gamma;\delta})_{;\beta}\,g^{\alpha\gamma}g^{\beta\delta}, \\
&= F_{\gamma\delta;\beta}\,g^{\alpha\gamma}g^{\beta\delta} = 0,
\end{aligned}
\tag{42.31}
$$

where the last equality follows from the E-L equations. Since the $g^{\alpha\gamma}$ simply multiplies all of the terms in the sum, it can be dropped. The equation is then equivalent to

$$
F_{\gamma\delta;\beta}\,g^{\beta\delta} = 0.
\tag{42.32}
$$

This expression gives two of the Maxwell equations in curved space, in the absence of sources.[17]

[16]The electromagnetic field itself generates curvature via its energy-momentum tensor, as discussed below.

[17]The other two Maxwell equations are given by employing the *comma-goes-to semicolon* rule on eqn 42.24, to yield

$$
F_{\alpha\beta;\gamma} + F_{\gamma\alpha;\beta} + F_{\beta\gamma;\alpha} = 0. \tag{42.33}
$$

Central to the role of fields in general relativity is the energy-momentum tensor $\boldsymbol{T}(\ ,\ )$. We met a method in Chapter 40 that allows us to derive this object from the Lagrangian of the electromagnetic fields. In Minkowski spacetime, the (2,0) energy-momentum tensor $\boldsymbol{T}$ for the electromagnetic field has components

$$T^{\mu\nu} = F^{\mu\alpha}F^{\nu}{}_{\alpha} - \frac{1}{4}\eta^{\mu\nu}F_{\alpha\beta}F^{\alpha\beta}. \tag{42.34}$$

Example 42.4

The components of this tensor can be recast in terms of the E- and B-fields. We find that the energy density is

$$T^{00} = \frac{1}{2}(\vec{E}^2 + \vec{B}^2), \tag{42.35}$$

while the momentum density has components

$$T^{0j} = (\vec{E} \times \vec{B})^j. \tag{42.36}$$

These latter components are often given the name the **Poynting vector**.[18] Finally, the all-spatial components of $\boldsymbol{T}$ are

$$T^{jk} = -\left[(E^j E^k + B^j B^k) - \frac{1}{2}(\vec{E}^2 + \vec{B}^2)\delta^{jk}\right]. \tag{42.37}$$

[18]The Poynting vector, named after J. H. Poynting (1852–1914) who invented it, is defined in classical electromagnetism as the vector $\vec{S} = \vec{E} \times \vec{H}$ (equivalently $\vec{S} = \vec{E} \times \vec{B}/\mu_0$). Its magnitude gives the energy flux (the energy flow per unit time, per unit area) and its direction indicates the direction of the energy flow. Our expression for T^{0j} gives the jth component of $\vec{S}$, with the constant μ_0 disappearing in our units.

In curved spacetime, we can write the (0,2) version of the energy-momentum tensor $\boldsymbol{T}$ in terms of

$$T_{\mu\nu} = F_{\mu\alpha}F_{\nu\beta}g^{\alpha\beta} - \frac{1}{4}g_{\mu\nu}F_{\gamma\delta}F_{\lambda\rho}g^{\gamma\lambda}g^{\delta\rho}. \tag{42.38}$$

It is the tensor $\boldsymbol{T}(\ ,\ )$ with these components that couples to the curvature of spacetime.

42.3   Gauge freedom

The concept of gauges is important for describing several topics in the coming chapters, including gravitational waves. Gauges are explained in more detail in Chapter 44.

The electromagnetic 1-form field $\tilde{\boldsymbol{A}}(x)$ is an important one. In several forthcoming chapters, we shall have cause to invoke the principle of **gauge freedom**, by using eqn 42.20. The principle says that no physics changes if we make the change $A_\mu(x) \to A_\mu(x) - \chi_{,\mu}(x)$. That is to say, the gauge transformation cannot change the result of any measurement. The choice of the smooth function $\chi(x)$ is known as a choice of gauge.[19] A sensible choice of the function $\chi(x)$ can simplify a problem.

[19]It might be helpful to think of the choice of gauge as a choice of language, where we are able to communicate the same message no matter which language we choose.

Example 42.5

Consider how the expression $\partial_\nu F^{\mu\nu} = J^\mu$ can be rewritten in terms of the components A^μ as[20]

$$\partial^\mu(\partial_\nu A^\nu) - \partial^2 A^\mu = J^\mu. \tag{42.40}$$

It would simplify our use of this equation of motion if the first term on the left-hand side could be made to go away through a sensible choice of χ. To do this we write $A_\mu \to A'_\mu = A_\mu - \partial_\mu\chi$. What we want is then

$$\partial^\mu A'_\mu = \partial^\mu A_\mu - \partial^\mu\partial_\mu\chi = 0, \tag{42.41}$$

[20]The notation here makes use of

$$\partial^2 \equiv -\frac{\partial^2}{\partial t^2} + \frac{\partial^2}{\partial x^2} + \frac{\partial^2}{\partial y^2} + \frac{\partial^2}{\partial z^2}$$
$$= -\frac{\partial^2}{\partial t^2} + \vec{\nabla}^2. \tag{42.39}$$

which we can satisfy by setting $\partial^2 \chi = \partial^\mu A_\mu$. The equation of motion for the electromagnetic field then becomes

$$\partial^2 A^\mu = -J^\mu. \tag{42.42}$$

Note that in the absence of sources, this tells us $\partial^2 A^\mu = 0$, which is a wave equation[21] for the components A^ν, which is to say

$$-\frac{\partial^2 A^\mu}{\partial t^2} + \vec{\nabla}^2 A^\mu = 0. \tag{42.43}$$

The electromagnetic field propagates through space at the speed of light.

The choice of χ such that $\partial_\mu A^\mu = 0$ is known as[22] the **Lorenz gauge**. In fact, the Lorenz gauge does not exhaust the gauge freedom.

Example 42.6

This is because we can make a further shift $A'_\mu \to A''_\mu = A'_\mu - \partial_\mu \xi$ as long as $\partial^2 \xi = 0$ (so that both A'_μ and A''_μ satisfy the Lorenz condition). To make A''^μ unique, we shall choose

$$\partial_0 \xi = A'_0, \tag{42.44}$$

which implies $A''_0 = 0$. With this further choice, the Lorenz condition then reduces to $\vec{\nabla} \cdot \vec{A}'' = 0$. This choice is known as **Coulomb gauge**.[23]

Similar choices of gauge will be important when we come to examine gravitational waves in Chapter 46. For now, it's useful to note that we can write down the solutions to the equation of motion for the electromagnetic field: $\partial^2 A^\mu = -J^\mu$.

Example 42.7

Consider the electromagnetic field at position $\vec{x}$ that results from a source at $\vec{x}'$, as shown in Fig. 42.1. In the static case, we have

$$\vec{\nabla}^2 A^\mu = -J^\mu \quad \text{(static)}, \tag{42.45}$$

and so, for $A^\mu = (A^0, \vec{A})$ and $J^\mu = (\rho, \vec{J})$ we write

$$\vec{\nabla}^2 A^0 = -\rho, \quad \vec{\nabla}^2 \vec{A} = -\vec{J} \quad \text{(static)}. \tag{42.46}$$

We arrange the currents and charges in a three-volume $\mathcal{V}'$ as shown in Fig. 42.1. These latter equations then have solutions[24]

$$A^0(\vec{x}) = \frac{1}{4\pi} \int d\mathcal{V}' \frac{\rho(\vec{x}')}{|\vec{x}-\vec{x}'|}, \quad \vec{A}(\vec{x}) = \frac{1}{4\pi} \int d\mathcal{V}' \frac{\vec{J}(\vec{x}')}{|\vec{x}-\vec{x}'|} \quad \text{(static)}. \tag{42.47}$$

In words: The charges and currents at positions $\vec{x}'$ are added up to give the fields at $\vec{x}$. Since the electromagnetic field propagates at the speed of light, adding time dependence requires us to evaluate the charge and current distribution at the **retarded time**

$$t_{\rm r} = t - \frac{|\vec{x}-\vec{x}'|}{c}, \tag{42.48}$$

where we temporarily reinstate the speed of light c. This says that the field experienced at $x^\mu = (t, \vec{x})$ is determined by adding up the charges and currents at $\vec{x}'$ at a time $t_{\rm r}$, since it takes $|\vec{x}-\vec{x}'|/c$ seconds for the field to propagate to $\vec{x}$. The solutions are then given by the **retarded potentials**

$$A^0(t, \vec{x}) = \frac{1}{4\pi} \int d^3\mathcal{V}' \frac{\rho(t_{\rm r}, \vec{x}')}{|\vec{x}-\vec{x}'|}, \quad \vec{A}(t, \vec{x}) = \frac{1}{4\pi} \int d\mathcal{V}' \frac{\vec{J}(t_{\rm r}, \vec{x}')}{|\vec{x}-\vec{x}'|}. \tag{42.49}$$

[21] A wave equation has the form $\partial^2 \phi = -\frac{\partial^2 \phi}{\partial t^2} + \vec{\nabla}^2 \phi = 0$. Its solutions are plane waves of the form $\phi = C\mathrm{e}^{-\mathrm{i}(\omega t - \vec{k} \cdot \vec{x})}$.

[22] Actually it's more usually (and incorrectly) known as Lorentz gauge due to its misattribution to Hendrik Lorentz (1853–1928) rather than to the less famous Ludvig Lorenz (1829–1891) who used it first. See J. D. Jackson and L. B. Okun, Rev. Mod. Phys. **73**, 663 (2001) for details of the history. In gravitation, the analogous choice of gauge is sometimes known as the harmonic gauge or de Donder gauge [the latter named after Théophile de Donder (1872–1957)].

[23] Charles-Augustin de Coulomb (1736–1806).

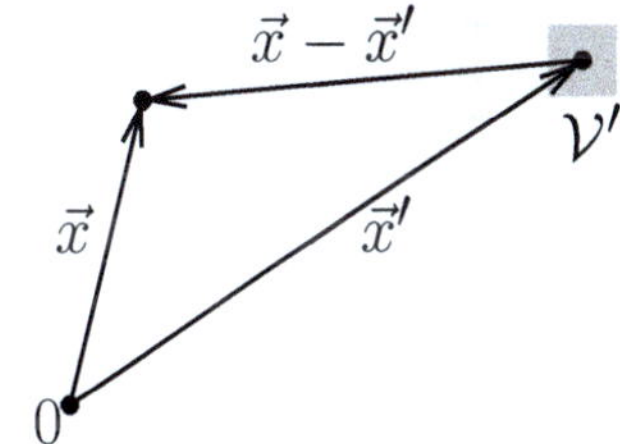

Fig. 42.1 The geometry for Example 42.7. The field at $\vec{x}$ results from a distribution of currents and charges with elements at positions $\vec{x}'$ in a three-volume $\mathcal{V}'$.

[24] See Exercise 42.7, where useful identities are given for filling in the algebra in this example.

or

$$A^\mu(t, \vec{x}) = \frac{1}{4\pi} \int \mathrm{d}\mathcal{V}' \frac{J^\mu(t_\mathrm{r}, \vec{x}')}{|\vec{x} - \vec{x}'|}. \tag{42.50}$$

We shall later use analogous potentials to describe the gravitational metric field generated by a distribution of masses and energies.

42.4   Geometrical electromagnetism

[25]Our coordinate-free expressions can be straightforwardly written in coordinates in flat or curved spacetime. We shall continue to assume flat spacetime in our discussion.

We now, finally, re-express the equations of electromagnetism in coordinate-free language.[25] We start with the electromagnetic 1-form gauge field $\tilde{\boldsymbol{A}}(x)$. We take its exterior derivative and find that we obtain the 2-form $\tilde{\boldsymbol{F}}$, or

$$\tilde{\boldsymbol{F}} = \boldsymbol{d}\tilde{\boldsymbol{A}}. \tag{42.51}$$

Example 42.8

We pause to prove the assertion above. We have $\tilde{\boldsymbol{A}} = A_\nu \boldsymbol{d}x^\nu$. The exterior derivative gives a 2-form

$$\begin{aligned}
\boldsymbol{d}\tilde{\boldsymbol{A}} &= \frac{\partial A_\nu}{\partial x^\mu} \boldsymbol{d}x^\mu \wedge \boldsymbol{d}x^\nu \\
&= \frac{1}{2} \frac{\partial A_\nu}{\partial x^\mu} \boldsymbol{d}x^\mu \wedge \boldsymbol{d}x^\mu - \frac{1}{2} \frac{\partial A_\nu}{\partial x^\mu} \boldsymbol{d}x^\nu \wedge \boldsymbol{d}x^\mu \\
&= \frac{1}{2} \left(\frac{\partial A_\nu}{\partial x^\mu} - \frac{\partial A_\mu}{\partial x^\nu} \right) \boldsymbol{d}x^\mu \wedge \boldsymbol{d}x^\nu.
\end{aligned} \tag{42.52}$$

We see that $\tilde{\boldsymbol{F}} = \boldsymbol{d}\tilde{\boldsymbol{A}}$, where $\tilde{\boldsymbol{F}} = \frac{1}{2} F_{\mu\nu} \boldsymbol{d}x^\mu \wedge \boldsymbol{d}x^\nu$ and

[26]We choose to write the indices for the components of $\vec{E}$ and $\vec{B}$ 3-vectors in the up position. These are not components of 4-vectors, so there is no implied difference between up and down versions.

$$F_{\mu\nu} = \left(\frac{\partial A_\nu}{\partial x^\mu} - \frac{\partial A_\mu}{\partial x^\nu} \right), \tag{42.53}$$

as we had identified originally.

Recall that the expression $\tilde{\boldsymbol{F}} = \boldsymbol{d}\tilde{\boldsymbol{A}}$ means that the 2-form $\tilde{\boldsymbol{F}}$ is exact and, therefore, closed, with the property that $\boldsymbol{d}\tilde{\boldsymbol{F}} = \boldsymbol{dd}\tilde{\boldsymbol{A}} = 0$.

Expressing electromagnetism in terms of the 2-form $\tilde{\boldsymbol{F}}$ allows us to illustrate the theory in a slightly different manner to the usual field lines of $\vec{E}$ and $\vec{B}$, by using the idea from Chapter 32 that a 2-form looks like a set of intersecting planes. Using the wedge product, we can identify the components of the Faraday 2-form with the components of the electric and magnetic 3-vectors[26] thus

$$\begin{aligned}
\tilde{\boldsymbol{F}} =\ & E^x \boldsymbol{d}x \wedge \boldsymbol{d}t + E^y \boldsymbol{d}y \wedge \boldsymbol{d}t + E^z \boldsymbol{d}z \wedge \boldsymbol{d}t \\
& + B^x \boldsymbol{d}y \wedge \boldsymbol{d}z + B^y \boldsymbol{d}z \wedge \boldsymbol{d}x + B^z \boldsymbol{d}x \wedge \boldsymbol{d}y.
\end{aligned} \tag{42.54}$$

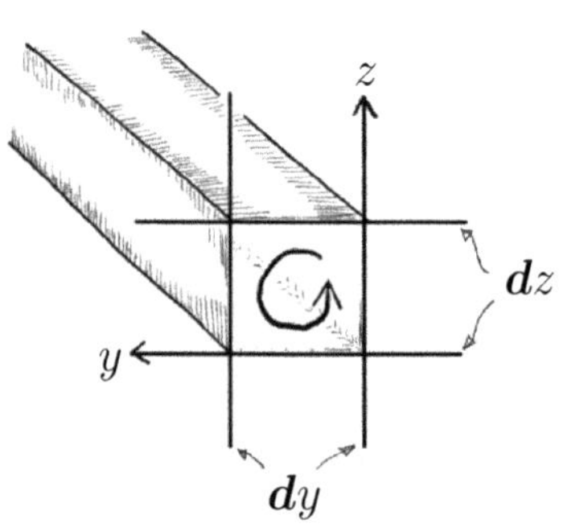

Fig. 42.2 A B-field in the geometrical language circulates in the tubes of a 2-form. The field $\tilde{\boldsymbol{F}} = B^x \boldsymbol{d}y \wedge \boldsymbol{d}z$ is represented here, with the field component B^x circulating in the tube formed by $\boldsymbol{d}y$ and $\boldsymbol{d}z$.

We can therefore picture the components E^i and B^i circulating in the tubes formed from the intersecting basis 1-forms $\boldsymbol{d}x^\mu$. An example is shown in Fig. 42.2.

Equation 42.54 allows us access to the components of the equations of motion via eqn 42.12. Specifically, we start with the momentum 1-form

$\tilde{p} = p_\mu dx^\mu$ and the velocity of the particle $\boldsymbol{u}$, which is tangent to its world line (parametrized by the proper time τ). The equation of motion is then $d\tilde{p}/d\tau = q\tilde{\boldsymbol{F}}(\ , \boldsymbol{u})$. That is, we fill one of the slots of the 2-form $\boldsymbol{F}$ with the velocity $\boldsymbol{u}$.

Example 42.9

A particle moves along the z-axis. In a region where there is a constant magnetic field in the x-direction, we have a Faraday tensor

$$\tilde{\boldsymbol{F}}(\ ,\ ) = B^x \boldsymbol{dy} \wedge \boldsymbol{dz}(\ ,\ ). \tag{42.55}$$

We find the equation of motion

$$\frac{d\tilde{\boldsymbol{p}}}{d\tau} = q\tilde{\boldsymbol{F}}(\ , \boldsymbol{u}) = qB^x \boldsymbol{dy} \wedge \boldsymbol{dz}(\ , \boldsymbol{u})$$
$$= qB^x \left[\boldsymbol{dy}(\ ) \otimes \boldsymbol{dz}(\boldsymbol{u}) - \boldsymbol{dz}(\ ) \otimes \boldsymbol{dy}(\boldsymbol{u}) \right]$$
$$= qB^x \left(\boldsymbol{dy}\, u^z - \boldsymbol{dz}\, u^y \right). \tag{42.56}$$

We extract equations in component form

$$\frac{dp_x}{d\tau} = 0,$$
$$\frac{dp_y}{d\tau} = qB^x u^z = qB^x \frac{dz}{d\tau},$$
$$\frac{dp_z}{d\tau} = -qB^x u^y = -qB^x \frac{dy}{d\tau}. \tag{42.57}$$

The solutions to these equations are the helical trajectories typical for a charge in a magnetic field, as shown in Fig. 42.3.

The geometrical version of the Faraday tensor is also helpful in understanding the gauge transformation in eqn 42.20 which can be rewritten as

$$A_\mu \boldsymbol{dx}^\mu \to A_\mu \boldsymbol{dx}^\mu - \frac{\partial \chi}{\partial x^\mu} \boldsymbol{dx}^\mu, \tag{42.58}$$

or

$$\tilde{\boldsymbol{A}} \to \tilde{\boldsymbol{A}} - \boldsymbol{d}\chi. \tag{42.59}$$

It is clear from this geometrical expression that a gauge transformation cannot have any effect on the Faraday 2-form $\tilde{\boldsymbol{F}}$, since

$$\tilde{\boldsymbol{F}} = \boldsymbol{d}\tilde{\boldsymbol{A}} \to \boldsymbol{d}\tilde{\boldsymbol{A}} - \boldsymbol{dd}\chi, \tag{42.60}$$

and, since $\boldsymbol{dd} = 0$, we see that $\tilde{\boldsymbol{F}} \to \tilde{\boldsymbol{F}}$ under a gauge transformation. Since a gauge transformation has no effect on the Faraday 2-form, it cannot affect the $\vec{E}$- and $\vec{B}$-fields that make up its components.

We now turn to the Maxwell equations. The first two Maxwell equations may be written in terms of the Faraday 2-form as

$$\boldsymbol{d}\tilde{\boldsymbol{F}} = 0, \tag{42.61}$$

which we know to be true by virtue of the definition $\tilde{\boldsymbol{F}} = \boldsymbol{d}\tilde{\boldsymbol{A}}$ and the fact that $\boldsymbol{dd} = 0$. To obtain the Maxwell equations in a familiar guise, we explicitly take an exterior derivative

$$\boldsymbol{d}\tilde{\boldsymbol{F}} = \frac{1}{2}\frac{\partial F_{\alpha\beta}}{\partial x^\sigma}\, \boldsymbol{dx}^\sigma \wedge \boldsymbol{dx}^\alpha \wedge \boldsymbol{dx}^\beta = 0. \tag{42.62}$$

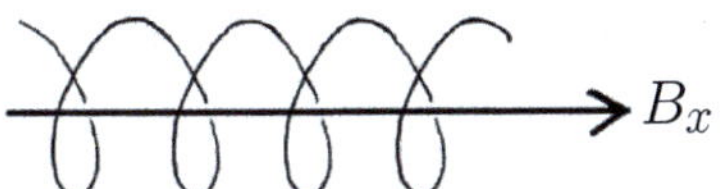

Fig. 42.3 Helical motion of a charged particle in a B-field.

Example 42.10

Doing this, we see

$$2d\tilde{\boldsymbol{F}} = \left(\frac{\partial B^x}{\partial x} + \frac{\partial B^y}{\partial y} + \frac{\partial B^z}{\partial z} \right) \boldsymbol{dx} \wedge \boldsymbol{dy} \wedge \boldsymbol{dz}$$
$$+ \left(\frac{\partial B^x}{\partial t} + \frac{\partial E^z}{\partial y} - \frac{\partial E^y}{\partial z} \right) \boldsymbol{dt} \wedge \boldsymbol{dy} \wedge \boldsymbol{dz}$$
$$+ \left(\frac{\partial B^y}{\partial t} + \frac{\partial E^x}{\partial z} - \frac{\partial E^z}{\partial x} \right) \boldsymbol{dt} \wedge \boldsymbol{dz} \wedge \boldsymbol{dx}$$
$$+ \left(\frac{\partial B^z}{\partial t} + \frac{\partial E^y}{\partial x} - \frac{\partial E^x}{\partial y} \right) \boldsymbol{dt} \wedge \boldsymbol{dx} \wedge \boldsymbol{dy} = 0. \tag{42.63}$$

Set each of the coefficients to zero, to guarantee that the expression always vanishes, and we extract

$$\vec{\nabla} \cdot \vec{B} = 0, \quad \vec{\nabla} \times \vec{E} = -\frac{\partial \vec{B}}{\partial t}. \tag{42.64}$$

Another way to express the sum of components in $\boldsymbol{d\tilde{F}} = 0$ is to note that, owing to the properties of the wedge product, any term with a repeated index in eqn 42.62 vanishes. As a result, we must have that the sum of the $\partial F_{\alpha\beta}/\partial x^\gamma$ with indices all different vanishes. We therefore write an equivalent version of the first two Maxwell equations as

$$\frac{\partial F_{\alpha\beta}}{\partial x^\gamma} + \frac{\partial F_{\gamma\alpha}}{\partial x^\beta} + \frac{\partial F_{\beta\gamma}}{\partial x^\alpha} = 0, \tag{42.65}$$

which is the Bianchi identity for electromagnetism that we gave in eqn 42.24. We'll examine its physical and geometrical significance further in the next chapter.

In order to recreate the other two Maxwell equations, we must find out how the current $\boldsymbol{J}$ is coupled to the electromagnetic fields. The way to couple the fields is not to use $\tilde{\boldsymbol{F}}$, but the 2-form that results from taking the dual of the (2,0) tensor $\boldsymbol{F}$, which is[27]

$$\star\boldsymbol{F} = \frac{1}{2}(\star\boldsymbol{F})_{\alpha\beta}\, \boldsymbol{dx}^\alpha \wedge \boldsymbol{dx}^\beta. \tag{42.66}$$

The object $\star\boldsymbol{F}$ is known as the **Maxwell tensor**.

[27]See Chapter 37 for the recipe for forming duals. An alternative method to the one used here is to take the dual of the (0,2) Faraday 2-form $\tilde{\boldsymbol{F}}$ to make the (2,0) Maxwell tensor $\star\boldsymbol{F}$. We would then need to lower the indices to construct the components of the 2-form.

Example 42.11

The dual of the (2,0) Faraday tensor $\boldsymbol{F} = \frac{1}{2}F^{\mu\nu}(\boldsymbol{e}_\mu \wedge \boldsymbol{e}_\nu)$ is a 2-form with components

$$(\star\boldsymbol{F})_{\alpha\beta} \equiv \frac{1}{2}\varepsilon_{\mu\nu\alpha\beta}F^{\mu\nu} = \varepsilon_{|\mu\nu|\alpha\beta}F^{\mu\nu}. \tag{42.67}$$

We obtain

$$(\star\boldsymbol{F})_{01} = \varepsilon_{2301}F^{23} = B^x,$$
$$(\star\boldsymbol{F})_{02} = \varepsilon_{1302}F^{13} = B^y,$$
$$(\star\boldsymbol{F})_{03} = \varepsilon_{1203}F^{12} = B^z,$$
$$(\star\boldsymbol{F})_{23} = \varepsilon_{0123}F^{01} = E^x,$$
$$(\star\boldsymbol{F})_{13} = \varepsilon_{0213}F^{02} = -E^y,$$
$$(\star\boldsymbol{F})_{12} = \varepsilon_{0312}F^{03} = E^z. \tag{42.68}$$

This 2-form has components[28]

$$(\star F)_{\mu\nu} = \begin{pmatrix} 0 & B^x & B^y & B^z \\ -B^x & 0 & E^z & -E^y \\ -B^y & -E^z & 0 & E^x \\ -B^z & E^y & -E^x & 0 \end{pmatrix}. \tag{42.69}$$

So the change from the 2-form $\tilde F$ to the 2-form $\star F$ can be achieved by making a transformation of components corresponding to $\vec E$ changing to $-\vec B$ and $\vec B$ changing to $\vec E$. It is also straightforward to show that $\star\star F = -F$.

[28]The corresponding (2,0) tensor has components

$$(\star F)^{\mu\nu} =$$
$$\begin{pmatrix} 0 & -B^x & -B^y & -B^z \\ B^x & 0 & E^z & -E^y \\ B^y & -E^z & 0 & E^x \\ B^z & E^y & -E^x & 0 \end{pmatrix}.$$

The Maxwell 2-form is written in terms of the 3-vector $\vec E$ and $\vec B$ fields as

$$\begin{aligned} \star F =\; & E^x\, dy \wedge dz + E^y\, dz \wedge dx + E^z\, dx \wedge dy \\ & + B^x\, dt \wedge dx + B^y\, dt \wedge dy + B^z\, dt \wedge dz. \end{aligned} \tag{42.70}$$

We shall see in the next chapter that the conservation of charge that is so fundamental to electromagnetism is expressed by the dual of the current $\star J$. It is this (0,3) tensor that couples to the 2-form $\star F$ to give the second pair of Maxwell's equations.

Example 42.12

Let's take the dual of the current vector J. In $n = 4$ dimensions, this results in a 3-form with components $J_{\alpha\beta\gamma} = \varepsilon_{\mu\alpha\beta\gamma} J^\mu$ given by

$$\begin{aligned} (\star J)_{123} &= \varepsilon_{0123} J^0 = \rho, \\ (\star J)_{023} &= \varepsilon_{1023} J^1 = -J^x, \\ (\star J)_{031} &= \varepsilon_{2031} J^2 = -J^y, \\ (\star J)_{012} &= \varepsilon_{3012} J^3 = -J^z. \end{aligned} \tag{42.71}$$

So we have the 3-form

$$\begin{aligned} \star J =\; & \rho\, dx \wedge dy \wedge dz - J^x\, dt \wedge dy \wedge dz \\ & - J^y\, dt \wedge dz \wedge dx - J^z\, dt \wedge dx \wedge dy. \end{aligned} \tag{42.72}$$

The final piece of the puzzle is to determine how to couple the 2-form $\star F$ to the 3-form $\star J$. The answer is to make another 3-form by taking the exterior derivative of the Maxwell 2-form[29] $\star F$.

[29]The only other 3-form available is $d\tilde F = dd\tilde A = 0$.

Example 42.13

Take the exterior derivative of the Maxwell 2-form to make a 3-form

$$\begin{aligned} d\star F =\; & \left(\frac{\partial E_x}{\partial x} + \frac{\partial E_y}{\partial y} + \frac{\partial E_z}{\partial z} \right) dx \wedge dy \wedge dz \\ & + \left(\frac{\partial E_x}{\partial t} + \frac{\partial B_z}{\partial y} - \frac{\partial B_y}{\partial z} \right) dt \wedge dy \wedge dz \\ & + \left(\frac{\partial E_y}{\partial t} + \frac{\partial B_x}{\partial z} - \frac{\partial B_z}{\partial x} \right) dt \wedge dz \wedge dx \\ & + \left(\frac{\partial E_z}{\partial t} + \frac{\partial B_y}{\partial x} - \frac{\partial B_x}{\partial y} \right) dt \wedge dx \wedge dy. \end{aligned} \tag{42.73}$$

Match each component with that of the 3-form $\star\boldsymbol{J}$

$$\star\boldsymbol{J} = \rho\, d\boldsymbol{x} \wedge d\boldsymbol{y} \wedge d\boldsymbol{z} - J^x d\boldsymbol{t} \wedge d\boldsymbol{y} \wedge d\boldsymbol{z}$$
$$- J^y d\boldsymbol{t} \wedge d\boldsymbol{z} \wedge d\boldsymbol{x} - J^z d\boldsymbol{t} \wedge d\boldsymbol{x} \wedge d\boldsymbol{y}. \tag{42.74}$$

This results in the remaining Maxwell equations.

We conclude that the second two Maxwell equations are given by

$$d\star\boldsymbol{F} = \star\boldsymbol{J}. \tag{42.75}$$

We have now built a geometrical version of the Maxwell equations. In the next chapter, we shall look a little more closely at the notion of charge conservation and see how this allows us to understand the Maxwell equations in their geometric form.

Chapter summary

- Interactions of changes with electromagnetic fields are described in flat space by the equation of motion

$$\frac{d\tilde{\boldsymbol{p}}}{d\tau} = q\tilde{\boldsymbol{F}}(\ ,\boldsymbol{u}). \tag{42.76}$$

- The Faraday tensor is the 2-form $\tilde{\boldsymbol{F}} = d\tilde{\boldsymbol{A}}$ where $\tilde{\boldsymbol{A}}$ is the gauge field.

- In coordinate free-notation, the Maxwell equations can be expressed as

$$d\tilde{\boldsymbol{F}} = 0, \quad d\star\boldsymbol{F} = \star\boldsymbol{J}. \tag{42.77}$$

- The (0,2) energy-momentum tensor $\boldsymbol{T}(\ ,\ )$ for the electromagnetic field has components

$$T_{\mu\nu} = F_{\mu\alpha}F_{\nu\beta}g^{\alpha\beta} - \frac{1}{4}g_{\mu\nu}F_{\gamma\delta}F_{\lambda\rho}g^{\gamma\lambda}g^{\delta\rho}. \tag{42.78}$$

Exercises

(42.1) Show that the equation of motion in eqn 42.12 is consistent with the Lorentz force law.

(42.2) Use the fact that under a Lorentz transformation

$$F^{\mu\nu} \to (F')^{\mu\nu} = \Lambda^{\mu}{}_{\alpha}\Lambda^{\nu}{}_{\beta}F^{\alpha\beta}, \tag{42.79}$$

to show how the components of $\vec{E}$ and $\vec{B}$ transform when a boost is made along the x axis.

(42.3) (a) Why in electronics are we free to set the zero of electric potential? Are we allowed to set zero volts differently in London and in Manchester?
(b) What is the effect on the equations for the $\vec{E}$-

and $\vec{B}$-fields of making the changes to the components $A^\mu = (V, \vec{A})$ of the electromagnetic field $\boldsymbol{A}(x)$ of

$$V \to V - \frac{\partial \chi}{\partial t},$$
$$\vec{A} \to \vec{A} + \vec{\nabla}\chi, \qquad (42.80)$$

where χ is a function of space and time coordinates?

(c) Consider Maxwell's equations given in terms of the field $\boldsymbol{A}$. What happens to the equations if we choose V and $\vec{A}$ such that

$$\vec{\nabla} \cdot \vec{A} + \frac{\partial V}{\partial t} = 0. \qquad (42.81)$$

(42.4) (a) In flat space, what $\vec{B}$-field results from a vector potential with components $\vec{A} = (0, Cx, 0)$, where C is a constant?

This is known as Landau gauge.

(b) What $\vec{B}$-field results from a vector potential $\vec{A} = \frac{1}{2}(-Cy, Cx, 0)$, where C is a constant?

(42.5) (a) Show that $F_{\alpha\beta,\gamma} + F_{\gamma\alpha,\beta} + F_{\beta\gamma,\alpha} = 0$, yields two Maxwell equations.

(b) Show that $F^{\mu\nu}{}_{,\nu} = J^\mu$ yields the other two.

(42.6) Derive the expressions linking the components of $\boldsymbol{T}$ to the $\vec{E}$ and $\vec{B}$ fields.

(42.7) By considering the spatial Laplacian of a component $\vec{\nabla}^2 A^\mu$, show that the retarded potential in eqn 42.50 satisfies the inhomogeneous wave equation $\partial^2 A^\mu(t, x^\mu) = -J^\mu(t, x^\mu)$.

Hint: Use the identities $\vec{\nabla}R = \hat{R}$, $\vec{\nabla}(1/R) = -\hat{R}/R^2$, $\vec{\nabla} \cdot (\hat{R}/R) = 1/R^2$ *and* $\vec{\nabla} \cdot (\hat{R}/R^2) = 4\pi\delta^{(3)}(\vec{R})$, *where* $R = |\vec{x} - \vec{x}'|$.

(42.8) We can show that eqn 42.50 transforms as a 4-vector in flat space. Consider field A^μ at point $\mathcal{P}$ at $(0, 0, 0, 0)$ resulting from light travelling from the current J^μ at points $\mathcal{Q}$ at $(-t, x, y, z)$. We have

$$A^\mu(\mathcal{P}) = \frac{1}{4\pi} \int \mathrm{d}\mathcal{V} \frac{J^\mu(\mathcal{Q})}{|\vec{r}|}, \qquad (42.82)$$

where $r^2 = x^2 + y^2 + z^2$, $|\vec{r}| = -t$ (and we drop the primed notation for the volume element $\mathrm{d}\mathcal{V}$ to avoid confusion with the primed coordinates examined in this problem!).

(a) In a primed frame moving at speed v along the x-direction relative to the unprimed frame, show that

$$r' = \gamma r(1 + v \cos\theta), \qquad (42.83)$$

where $\gamma = (1 - v)^{-\frac{1}{2}}$ and θ is the angle between $\vec{r}$ and the x axis.

(b) The three-volume $\mathrm{d}\mathcal{V}$ captures all of the events that intersect a spherical light wave launched backward in time from $\mathcal{P}$. Show that the volume element in the primed frame is given by $\mathrm{d}\mathcal{V}' = \mathrm{d}x'\mathrm{d}y'\mathrm{d}z' = \mathrm{d}x'\mathrm{d}y\mathrm{d}z$, with

$$\mathrm{d}x' = \gamma(\mathrm{d}x - v\mathrm{d}t). \qquad (42.84)$$

(c) In the previous equation, $\mathrm{d}t$ gives the difference in times for measuring the extremities of the interval $\mathrm{d}x$. Use this to show that

$$\mathrm{d}x' = \gamma\mathrm{d}x(1 + v\cos\theta), \qquad (42.85)$$

and, as a result, that the eqn 42.50 transforms as a 4-vector.

(42.9) Show that the solution to the equations of motion in eqn 42.57 are helices.

(42.10) We define **projection tensors** with components

$$P_\mathrm{L}^{\mu\nu} = \frac{p^\mu p^\nu}{p^2} \quad \text{and} \quad P_\mathrm{T}^{\mu\nu} = g^{\mu\nu} - \frac{p^\mu p^\nu}{p^2}, \qquad (42.86)$$

where p^μ are the components of a momentum vector. Verify that P_L and P_T are indeed projection operators by showing that $P^2 = P$.

(42.11) The Lagrangian for electromagnetism in vacuo is, in Minkowski space, $\mathcal{L} = -\frac{1}{4}F^{\mu\nu}F_{\mu\nu}$.

(a) Show that this can be rewritten as

$$\mathcal{L} = -\frac{1}{2}(\partial_\mu A_\nu \partial^\mu A^\nu - \partial_\mu A_\nu \partial^\nu A^\mu). \qquad (42.87)$$

(b) We define a generalized momentum density with components $\Pi^{\sigma\rho}$. Show that

$$\Pi^{\sigma\rho} \equiv \frac{\partial\mathcal{L}}{\partial(\partial_\sigma A_\rho)} = -F^{\sigma\rho}. \qquad (42.88)$$

(c) Hence, show that the canonical energy-momentum tensor

$$S^\mu{}_\nu = -\Pi^{\mu\sigma}\partial_\nu A_\sigma + \delta^\mu{}_\nu \mathcal{L}, \qquad (42.89)$$

can be written as

$$S^{\mu\nu} = F^{\mu\sigma}\partial^\nu A_\sigma - \frac{1}{4}\eta^{\mu\nu}F^{\alpha\beta}F_{\alpha\beta}. \qquad (42.90)$$

(42.12) The canonical energy-momentum tensor from the last question is not symmetric (and so cannot be used for general relativity). However, we can symmetrize it by adding an extra term $\partial_\lambda X^{\lambda\mu\nu}$, where $X^{\lambda\mu\nu} = -F^{\mu\lambda}A^\nu$. Show that a symmetrized energy-momentum tensor $T^{\mu\nu} = S^{\mu\nu} + \partial_\lambda X^{\lambda\mu\nu}$ can be written as

$$T^{\mu\nu} = F^{\mu\sigma}F^\nu{}_\sigma - \frac{1}{4}\eta^{\mu\nu}F^{\alpha\beta}F_{\alpha\beta}, \qquad (42.91)$$

as we had before.

(42.13) *We can attempt to formulate a scalar field theory for gravitation using the same method we used for electromagnetism. We follow the approach of Padmanabhan.*

Consider the action for the coupling between the scalar field $\phi(x^\mu)$ and matter

$$S = -\int d\tau \, (m + \lambda\phi) \left(-\eta_{\mu\nu} \frac{dx^\mu}{d\tau} \frac{dx^\nu}{d\tau} \right)^{\frac{1}{2}}. \tag{42.92}$$

(a) Verify that this is a sensible suggestion by restoring factors of c, and expanding the Lagrangian in the non-relativistic limit to $O(1/c^2)$.

(b) Show that the equation of motion for a particle is given by

$$\frac{du_\mu}{d\tau} = -\frac{\lambda}{(m + \lambda\phi)} \left(\phi_{,\mu} + u_\mu u^\sigma \phi_{,\sigma} \right). \tag{42.93}$$

(c) If ϕ describes gravitation then the equation of motion must obey the principle of equivalence. By rescaling the field ϕ, show that action must take the form

$$S = -\int d\tau \, (1 + \Phi) \left(-\eta_{\mu\nu} \frac{dx^\mu}{d\tau} \frac{dx^\nu}{d\tau} \right)^{\frac{1}{2}}. \tag{42.94}$$

The latter equation implies that the effect of gravitation can be accounted for by a change in the metric $\boldsymbol{\eta} \to \boldsymbol{g}$ with

$$g_{\mu\nu} = (1 + \Phi)^2 \eta_{\mu\nu}. \tag{42.95}$$

(42.14) *We can examine the equation of motion of the field Φ from the previous question.*

(a) Show that a Lagrangian density

$$\mathcal{L} = -\frac{1}{8\pi G} (\partial_\mu \Phi)^2 - \rho\Phi, \tag{42.96}$$

where ρ is the mass density of particles, leads to Poisson's equation for the gravitational field.

(b) How must this Lagrangian be altered to take into account gravitational coupling to energy in addition to mass?

Charge conservation and the Bianchi identity

43

For I have wings equipped to fly
Up to the high vault of the sky.
Once these are harnessed, your swift mind
Views earth with loathing, far behind.
Boethius (c.480–c.524/6) *The Consolation of Philosophy*

The conservation of charge, when described using geometry, leads to a mathematical expression known as the **Bianchi identity**. In electromagnetism, a Bianchi identity generates two of the Maxwell equations: the equations of motion of the electromagnetic fields. In the case of gravitation, the analogous equation (also called a Bianchi identity) encodes conservation of energy, and provides the constraint that tells us how matter couples to the curvature of spacetime. In this chapter, we follow a geometrical path, originally beaten by Misner, Thorne, and Wheeler, from charge conservation to the electromagnetic and then gravitational Bianchi identity.[1] The tools we need are mostly taken from Chapter 38 and the arguments in this chapter are mostly geometrical ones involving several computations. As a result, the chapter can be skipped on a first reading if you're happy to take the Bianchi identity on trust and without a physical or geometrical justification.

43.1 Conserving electric charge

Conservation of charge can be written as a divergence $\boldsymbol{\nabla} \cdot \boldsymbol{J} = 0$. In flat spacetime,[2] we can write this as the component equation[3]

$$J^{\mu}{}_{,\mu} = \frac{\partial J^{\mu}}{\partial x^{\mu}} = 0 \quad \text{(flat space)}. \tag{43.2}$$

This expresses the fact that charge is *locally* conserved, so no spontaneous generation of charges can take place in a given 4-volume $\mathcal{V}$. This is important in relativity. *Global* conservation of charge would mean that no charges can be created or destroyed, but would allow a charge to disappear at one end of the Universe and reappear at some arbitrary point in spacetime. This would allow superluminal transport of charges, and so is forbidden. We must therefore have the stronger condition of local charge conservation enshrined in eqn 43.2.

Let's unpack[4] eqn 43.2 a little more. We are free to integrate the conservation equation over the 4-volume $\mathcal{V}$ and then apply Stokes' theorem,

[1] Luigi Bianchi (1856–1928) rediscovered the identity for the Riemann tensor in 1902. This had (according to Tullio Levi-Civita) been discovered by Gregorio Ricci-Curbastro around 1889, who had supposedly forgotten about it. A contracted version (examined in the exercises) had been derived in 1880 by Aurel Voss (1845–1931).

[2] We work in flat spacetime for simplicity, but remember that our valid tensor equations also apply in curved space.

[3] In a more familiar form, we write this as the flat-space continuity equation

$$\frac{\partial \rho}{\partial t} + \vec{\nabla} \cdot \vec{J} = 0. \tag{43.1}$$

[4] As a warm up, consider the integral over a volume V in 3-space with a boundary S, of

$$\int_{V} \mathrm{d}V \left(\frac{\partial \rho}{\partial t} + \vec{\nabla} \cdot \vec{J} \right) = 0. \tag{43.3}$$

The first term gives us the rate of change of the total charge Q in the volume and, applying the divergence theorem, we obtain

$$\frac{\partial Q}{\partial t} + \int_{S} \mathrm{d}\vec{S} \cdot \vec{J} = 0. \tag{43.4}$$

Integrating this latter equation with respect to time, we have the result that the total change in the amount of charge in the volume V is equal to the total charge that flows through the surface S. This is local conservation of charge.

to yield

$$0 = \int_{\mathcal{V}} \frac{\partial J^\mu}{\partial x^\mu} d\mathcal{V} = \int_{\partial \mathcal{V}} J^\mu d\Sigma_\mu, \tag{43.5}$$

where $\partial \mathcal{V}$ is the boundary of the 4-volume and $d\Sigma_\mu$ is a 3-volume element. The right-hand side of eqn 43.5 is a statement that conservation means that the integral of currents across a boundary $\partial \mathcal{V}$ is zero. Charges enter and exit a given volume, but no new charge is generated in the volume $\mathcal{V}$. We shall now see how this fact is accounted for naturally in geometry.

The current of electric charge $\boldsymbol{J}$ is the source of the electromagnetic field tensor $\boldsymbol{F}$. The charges create the fields, the fields tell the charges how to move. The link between the source $\boldsymbol{J}$ and the field $\boldsymbol{F}$ have, hard-wired into them, the fact that the source is conserved. This follows from the geometrical law we met in Chapter 38 that *the boundary of a boundary is zero*. To see this, we need to work with a slightly different version of the conservation law. Rather than $\boldsymbol{\nabla} \cdot \boldsymbol{J} = 0$, we shall use the equivalent version $\boldsymbol{d} \star \boldsymbol{J} = 0$. We explain the origin and meaning of this expression below. We start by extracting the dual $\star \boldsymbol{J}$, which will be the flux in the new version of our conservation law.

Example 43.1

In flat (3+1)-dimensional spacetime, the vector $\boldsymbol{J}$ has components $J^\mu = (\rho, J^x, J^y, J^z)$ and so we have a dual 3-form $\star \boldsymbol{J}$, with components

$$(\star \boldsymbol{J})_{\nu\sigma\rho} = \varepsilon_{\mu\nu\sigma\rho} J^\mu. \tag{43.6}$$

For example, as we saw in the last chapter,

$$(\star \boldsymbol{J})_{012} \boldsymbol{dt} \wedge \boldsymbol{dx} \wedge \boldsymbol{dy} = \varepsilon_{3012} J^z \boldsymbol{dt} \wedge \boldsymbol{dx} \wedge \boldsymbol{dy}. \tag{43.7}$$

Explicitly, we have

$$\begin{aligned}
\star \boldsymbol{J} =& (\star \boldsymbol{J})_{123} \boldsymbol{dx}^1 \wedge \boldsymbol{dx}^2 \wedge \boldsymbol{dx}^3 + (\star \boldsymbol{J})_{023} \boldsymbol{dx}^0 \wedge \boldsymbol{dx}^2 \wedge \boldsymbol{dx}^3 \\
&+ (\star \boldsymbol{J})_{031} \boldsymbol{dx}^0 \wedge \boldsymbol{dx}^3 \wedge \boldsymbol{dx}^1 + (\star \boldsymbol{J})_{012} \boldsymbol{dx}^0 \wedge \boldsymbol{dx}^1 \wedge \boldsymbol{dx}^2 \\
=& \rho \boldsymbol{dx}^1 \wedge \boldsymbol{dx}^2 \wedge \boldsymbol{dx}^3 - J^1 \boldsymbol{dx}^0 \wedge \boldsymbol{dx}^2 \wedge \boldsymbol{dx}^3 \\
&+ J^2 \boldsymbol{dx}^0 \wedge \boldsymbol{dx}^1 \wedge \boldsymbol{dx}^3 - J^3 \boldsymbol{dx}^0 \wedge \boldsymbol{dx}^1 \wedge \boldsymbol{dx}^2. \tag{43.8}
\end{aligned}$$

It is simple to verify that the dual of the source field $\boldsymbol{J}$ can also be written as[5]

$$\star \boldsymbol{J} = J^\mu \boldsymbol{d\tilde{\Sigma}}_\mu = J^\mu \varepsilon_{\mu|\alpha\beta\gamma|} \boldsymbol{dx}^\alpha \wedge \boldsymbol{dx}^\beta \wedge \boldsymbol{dx}^\gamma, \tag{43.9}$$

which is the integrand in eqn 43.5. The exterior derivative of this quantity may be calculated straightforwardly, by following the usual rules

$$\begin{aligned}
\boldsymbol{d} \star \boldsymbol{J} =& \frac{\partial (\star \boldsymbol{J})_{123}}{\partial x^0} \boldsymbol{dx}^0 \wedge \boldsymbol{dx}^1 \wedge \boldsymbol{dx}^2 \wedge \boldsymbol{dx}^3 + \frac{\partial (\star \boldsymbol{J})_{023}}{\partial x^1} \boldsymbol{dx}^1 \wedge \boldsymbol{dx}^0 \wedge \boldsymbol{dx}^2 \wedge \boldsymbol{dx}^3 \\
&+ \frac{\partial (\star \boldsymbol{J})_{031}}{\partial x^2} \boldsymbol{dx}^2 \wedge \boldsymbol{dx}^0 \wedge \boldsymbol{dx}^3 \wedge \boldsymbol{dx}^1 + \frac{\partial \star \boldsymbol{J})_{012}}{\partial x^3} \boldsymbol{dx}^3 \wedge \boldsymbol{dx}^0 \wedge \boldsymbol{dx}^1 \wedge \boldsymbol{dx}^2. \tag{43.10}
\end{aligned}$$

In the same way that $\star \boldsymbol{J}$ represents $J^\mu d\Sigma_\mu$ in eqn 43.5, this latter equation for $\boldsymbol{d} \star \boldsymbol{J}$ expresses the integrand $J^\mu{}_{,\mu} d\mathcal{V}$. This is because the 4-volume element can be replaced with the 4-form $\varepsilon_{0123} \boldsymbol{dt} \wedge \boldsymbol{dx} \wedge \boldsymbol{dy} \wedge \boldsymbol{dz}$ and so we can write

$$\begin{aligned}
\boldsymbol{d} \star \boldsymbol{J} =& (\boldsymbol{\nabla} \cdot \boldsymbol{J}) \, \varepsilon_{|\alpha\beta\gamma\delta|} \boldsymbol{dx}^\alpha \wedge \boldsymbol{dx}^\beta \wedge \boldsymbol{dx}^\gamma \wedge \boldsymbol{dx}^\delta \\
=& \frac{1}{4!} (\boldsymbol{\nabla} \cdot \boldsymbol{J}) \, \varepsilon_{\alpha\beta\gamma\delta} \boldsymbol{dx}^\alpha \wedge \boldsymbol{dx}^\beta \wedge \boldsymbol{dx}^\gamma \wedge \boldsymbol{dx}^\delta. \tag{43.11}
\end{aligned}$$

---

[5] We saw in Chapter 38 that the 3-volume element can be written as the 3-form

$$\begin{aligned}
\boldsymbol{d\tilde{\Sigma}}_\mu =& \varepsilon_{\mu|\alpha\beta\gamma|} \boldsymbol{dx}^\alpha \wedge \boldsymbol{dx}^\beta \wedge \boldsymbol{dx}^\gamma, \\
=& \frac{1}{3!} \varepsilon_{\mu\alpha\beta\gamma} \boldsymbol{dx}^\alpha \wedge \boldsymbol{dx}^\beta \wedge \boldsymbol{dx}^\gamma.
\end{aligned}$$

We conclude that conservation of charge in eqn 43.5 may be written as

$$0 = \int_{\mathcal{V}} d\star \boldsymbol{J} = \int_{\partial\mathcal{V}} \star \boldsymbol{J}. \tag{43.12}$$

The conservation law is shown schematically in Fig. 43.1.

The point of this section is that the vanishing of these two integrals follows inevitably from (i) Maxwell's equations and (ii) the geometric fact that the boundary of a boundary is zero. Specifically, we can relate the dual of the source $\star\boldsymbol{J}$ to the dual of the electromagnetic field $\boldsymbol{F}$ using the Maxwell equation $\star\boldsymbol{J} = d\star\boldsymbol{F}$. All we need do now is write the conservation law in terms of an integral over $d\star\boldsymbol{F}$ rather than $\star\boldsymbol{J}$ and then apply Stokes theorem

$$0 = \int_{\partial\mathcal{V}} \star \boldsymbol{J} = \int_{\partial\mathcal{V}} d\star\boldsymbol{F} = \int_{\partial\partial\mathcal{V}} \star\boldsymbol{F}. \tag{43.13}$$

In this final equation we see that, since the boundary of the boundary $\partial\partial\mathcal{V}$ is zero, this integral *must* itself yield zero (since it's being computed over chain that vanishes). Working backwards then, we can view the conservation of the source field, encoded in $d\star\boldsymbol{J} = 0$ as following from the fact that $\partial\partial\mathcal{V} = 0$.[6]

43.2 Electromagnetic gauge field

The argument sketched in the last section can be applied to the electromagnetic gauge field $\tilde{\boldsymbol{A}}$ to extract a Bianchi identity for electromagnetism. We write the electromagnetic 2-form field in terms of the gauge 1-form field as $\tilde{\boldsymbol{F}} = d\tilde{\boldsymbol{A}}$. Consider how the expression $d\tilde{\boldsymbol{F}}$ is modified through the application of Stokes' theorem

$$\int_{\mathcal{V}} d\tilde{\boldsymbol{F}} = \int_{\partial\mathcal{V}} \tilde{\boldsymbol{F}}. \tag{43.14}$$

Now, inserting the gauge field, we see

$$\int_{\partial\mathcal{V}} \tilde{\boldsymbol{F}} = \int_{\partial\mathcal{V}} d\tilde{\boldsymbol{A}} = \int_{\partial\partial\mathcal{V}} \tilde{\boldsymbol{A}} = 0. \tag{43.15}$$

In words, the facts that (i) the boundary of a boundary is zero, and (ii) that $\tilde{\boldsymbol{F}}$ is exact, mean that it is inevitable[7] that $d\tilde{\boldsymbol{F}} = 0$. This is the first statement of the Bianchi identity for electromagnetism.

There are two ways to see the consequences of this, examined in the next examples.

Example 43.2

We examine the term $\int_{\partial\partial\mathcal{V}} \tilde{\boldsymbol{A}}$, integrating the 1-form $\tilde{\boldsymbol{A}} = A_\mu d\boldsymbol{x}^\mu$ over the boundary of a cube in flat spacetime. That is to say, we examine

$$\int_{\partial\partial\mathcal{V}} A_\mu d\boldsymbol{x}^\mu, \tag{43.16}$$

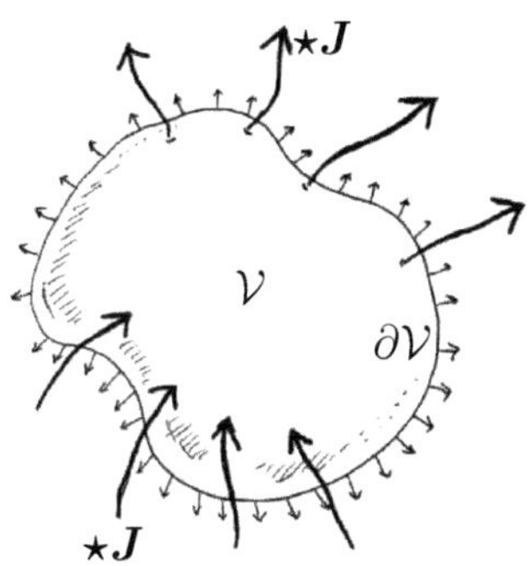

Fig. 43.1 Conservation of charge tells us that the rate of change of the amount of charge in volume $\mathcal{V}$ is equal to the flux $\star\boldsymbol{J}$ summed over the surface $\partial\mathcal{V}$.

[6]This explanation can be found in Cartan, although this was something Richard Feynman was not aware of, as shown by his remark in his Lectures of Gravitation that "I do not, off-hand, know the geometric significance of the Bianchi identity." The version we describe here is taken from Misner, Thorne, and Wheeler.

[7]Of course this was also mandated by $d\boldsymbol{d} = 0$ (via $d\boldsymbol{d}\tilde{\boldsymbol{A}} = d\tilde{\boldsymbol{F}} = 0$), and we can see here the sense in which this is dual to $\partial\partial = 0$.

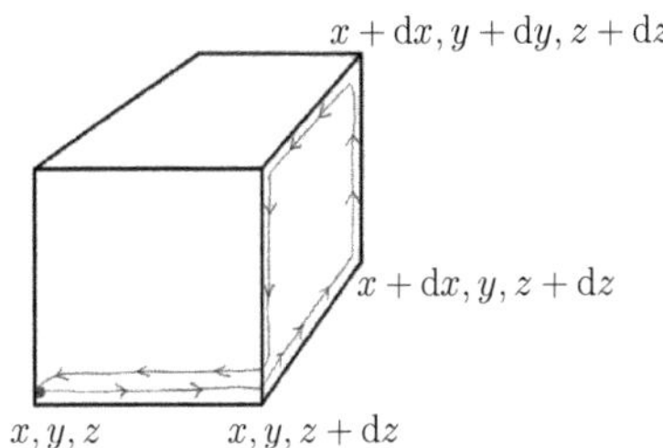

Fig. 43.2 A contribution to the integral over the boundary $\partial\partial\mathcal{V}$.

as we traverse each edge of the cube. Consider the path shown in Fig. 43.2. Labelling the corners of the square anticlockwise from x, y we find that in traversing a square we obtain a contribution

$$A_x(\mathcal{A})\mathrm{d}x + A_y(\mathcal{B})\mathrm{d}y - A_x(\mathcal{C})\mathrm{d}x - A_y(\mathcal{D})\mathrm{d}y$$
$$= [A_x(\mathcal{A}) - A_x(\mathcal{C})]\,\mathrm{d}x + [A_y(\mathcal{B}) - A_y(\mathcal{D})]\,\mathrm{d}y$$
$$= \left[A_x(\mathcal{A}) - A_x(\mathcal{A}) - \frac{\partial A_x}{\partial y}\mathrm{d}y\right]\mathrm{d}x + \left[A_y(\mathcal{B}) - A_y(\mathcal{B}) + \frac{\partial A_y}{\partial x}\mathrm{d}x\right]\mathrm{d}y$$
$$= \left(\frac{\partial A_y}{\partial x} - \frac{\partial A_x}{\partial y}\right)\mathrm{d}x\mathrm{d}y. \tag{43.17}$$

The contribution from the face shown is therefore given by

$$\int \left(\frac{\partial A_y}{\partial x} - \frac{\partial A_x}{\partial y}\right)\mathrm{d}x\mathrm{d}y = \int F_{xy}\mathrm{d}x\mathrm{d}y. \tag{43.18}$$

Repeating this procedure and adding to this the contribution from the opposite face, which is traversed in the other direction, we obtain

$$\int [F_{xy}(z + \mathrm{d}z) - F_{xy}(z)]\,\mathrm{d}x\mathrm{d}y = \int \frac{\partial F_{xy}}{\partial z}\mathrm{d}x\mathrm{d}y\mathrm{d}z. \tag{43.19}$$

Adding all of the contributions, we obtain

$$\int_{\partial\partial\mathcal{V}} A_\mu \boldsymbol{d}x^\mu = \int \left(\frac{\partial F_{xy}}{\partial z} + \frac{\partial F_{zx}}{\partial y} + \frac{\partial F_{yz}}{\partial x}\right)\mathrm{d}x\mathrm{d}y\mathrm{d}z. \tag{43.20}$$

The integrand on the right must be zero, since we known that $\int_{\partial\partial\mathcal{V}} \tilde{\boldsymbol{A}} = 0$. The cube can then be reoriented in the $(3+1)$ dimensions in which we're working, and the argument repeated for different coordinates.

We conclude that in flat spacetime the components of the Faraday tensor obey the constraining equation[8]

$$F_{\alpha\beta,\gamma} + F_{\gamma\alpha,\beta} + F_{\beta\gamma,\alpha} = 0. \tag{43.21}$$

This equation, a consequence of the fact that a boundary of a boundary is zero, is the Bianchi identity in component form. They gives us the two Maxwell's equations which, as we saw in the last chapter, are contained in $\boldsymbol{d}\tilde{\boldsymbol{F}} = 0$.

Example 43.3

Another way to approach this same problem, which proves useful when we consider gravity, is to consider $I = \int_{\partial\mathcal{V}} \tilde{\boldsymbol{F}} = 0$, that is, the surface integral of the 2-form $\tilde{\boldsymbol{F}}$. We can interpret this integral over a 2-form in the spirit of Chapter 38. We represent the surface by the bivector $\boldsymbol{u}\wedge\boldsymbol{v}$ and, to evaluate the integral, take the inner product of this bivector with the 2-form $\tilde{\boldsymbol{F}}$. Geometrically, this returns a number that tells us how many tubes of $\tilde{\boldsymbol{F}}$ are cut by the parallelogram $\boldsymbol{u}\wedge\boldsymbol{v}$. We have a contribution I_{xy} from the face shown in Fig. 43.2 given by

$$I_{xy} = \int_{\Delta x\boldsymbol{e}_x\wedge\Delta y\boldsymbol{e}_y} \tilde{\boldsymbol{F}} = \int \langle\tilde{\boldsymbol{F}}, \boldsymbol{e}_x\wedge\boldsymbol{e}_y\rangle\Delta x\Delta y. \tag{43.22}$$

Using the expression for the inner product of a bivector and a 2-form[9] we have

$$I_{xy} = \int F_{xy}\Delta x\Delta y. \tag{43.23}$$

Repeating the argument for the opposite face, as in the last example, and adding this (exercise) leads to the contribution

$$\int \frac{\partial F_{xy}}{\partial z}\Delta x\Delta y\Delta z. \tag{43.24}$$

Repeating and adding the contributions from the other faces gives us the Bianchi identity in component form once again.

---

[8] This flat-spacetime equation can be written using comma notation as $F_{[\alpha\beta,\gamma]} = 0$. The *comma-goes-to-semicolon* rule for curved spacetime then says $F_{[\alpha\beta;\gamma]} = 0$.

[9] The rule, from Chapter 32 is

$$\langle\tilde{\boldsymbol{F}}, \boldsymbol{e}_x\wedge\boldsymbol{e}_y\rangle = F_{|i_1 i_2|}(\boldsymbol{e}_x\wedge\boldsymbol{e}_y)^{i_1 i_2},$$

which, since the only non-vanishing component of the bivector is $(\boldsymbol{e}_x\wedge\boldsymbol{e}_y)^{xy} = 1$, yields F_{xy}. In fact, it's trivial to see this by simply saying $\langle\tilde{\boldsymbol{F}}, \boldsymbol{e}_x\wedge\boldsymbol{e}_y\rangle = \tilde{\boldsymbol{F}}(\boldsymbol{e}_x, \boldsymbol{e}_y)$.

43.3 Gravitational curvature

We saw in Chapter 13 that the Bianchi identity for gravitation was an essential part of justifying the Einstein equation. Just like its electromagnetic analogue, the gravitational Bianchi identity follows from considering the consequence of the boundary of a boundary being zero. In the electromagnetic case, we evaluated the effect of this on the 2-form field $\tilde{\boldsymbol{F}}$. For gravitation the analogous quantity is the (1,3) Riemann tensor $\boldsymbol{R}$. Instead of working with $\boldsymbol{R}$ directly, it is easiest to use Cartan's method (Chapter 36) to find the curvature 2-form, which allows us to act on vectors using the operator equation

$$\boldsymbol{d}^2 \boldsymbol{v} = \mathcal{R}(\boldsymbol{v}), \tag{43.25}$$

where the curvature operator is $\mathcal{R}(\ ) = \boldsymbol{\omega}^\nu(\ ) \otimes \boldsymbol{e}_\mu \otimes \mathcal{R}^\mu{}_\nu$ and $\mathcal{R}^\mu{}_\nu$ is the all-important curvature 2-form given in terms of the connection 1-forms[10] via

$$\mathcal{R}^\mu{}_\nu = \boldsymbol{d}\boldsymbol{\omega}^\mu{}_\nu + \boldsymbol{\omega}^\mu{}_\alpha \wedge \boldsymbol{\omega}^\alpha{}_\nu. \tag{43.27}$$

Why focus on $\mathcal{R}$? The answer is that, just as the electromagnetic Bianchi identity can be written as $\boldsymbol{d}\tilde{\boldsymbol{F}} = 0$, the gravitational version is

The **Bianchi identity for gravitation:**

$$\boldsymbol{d}\mathcal{R} = 0. \tag{43.28}$$

We prove this important result in three steps, each involving a computation.

Step I: The result of taking an exterior derivative of the curvature 2-form is

$$\boldsymbol{d}\mathcal{R}^\mu{}_\nu = \mathcal{R}^\mu{}_\alpha \wedge \boldsymbol{\omega}^\alpha{}_\nu - \boldsymbol{\omega}^\mu{}_\alpha \wedge \mathcal{R}^\alpha{}_\nu. \tag{43.29}$$

Example 43.4

Equation 43.29 can be proved as follows. We work in the orthonormal frame,[11] so that we have $\boldsymbol{\omega}_{\mu\nu} = -\boldsymbol{\omega}_{\nu\mu}$. Take the derivative of $\mathcal{R}^\mu{}_\nu$ to find

$$\begin{aligned}
\boldsymbol{d}\mathcal{R}^\mu{}_\nu &= \boldsymbol{d}\left(\boldsymbol{\omega}^\mu{}_\alpha \wedge \boldsymbol{\omega}^\alpha{}_\nu\right) \\
&= \boldsymbol{d}\boldsymbol{\omega}^\mu{}_\alpha \wedge \boldsymbol{\omega}^\alpha{}_\nu - \boldsymbol{\omega}^\mu{}_\alpha \wedge \boldsymbol{d}\boldsymbol{\omega}^\alpha{}_\nu \\
&= \left(\mathcal{R}^\mu{}_\alpha - \boldsymbol{\omega}^\mu{}_\beta \wedge \boldsymbol{\omega}^\beta{}_\alpha\right) \wedge \boldsymbol{\omega}^\alpha{}_\nu - \boldsymbol{\omega}^\mu{}_\alpha \wedge \left(\mathcal{R}^\alpha{}_\nu - \boldsymbol{\omega}^\alpha{}_\beta \wedge \boldsymbol{\omega}^\beta{}_\nu\right).
\end{aligned} \tag{43.30}$$

We find that the triple wedge products cancel and we have that

$$\boldsymbol{d}\mathcal{R}^\mu{}_\nu - \mathcal{R}^\mu{}_\alpha \wedge \boldsymbol{\omega}^\alpha{}_\nu + \boldsymbol{\omega}^\mu{}_\alpha \wedge \mathcal{R}^\alpha{}_\nu = 0. \tag{43.31}$$

Step II: We need the result of taking the exterior derivative of a bivector formed from the basis vectors.[12]

---

[10] We also have the equation that determines the connection 1-forms

$$\boldsymbol{d}\boldsymbol{\omega}^\mu + \boldsymbol{\omega}^\mu{}_\nu \wedge \boldsymbol{\omega}^\nu = 0. \tag{43.26}$$

[11] We won't put hats on the indices in this chapter, to prevent clutter.

[12] We need the rule that if $\tilde{\boldsymbol{p}}$ is a p-form and $\tilde{\boldsymbol{q}}$ is a 1-form then

$$\tilde{\boldsymbol{p}} \wedge \tilde{\boldsymbol{q}} = (-1)^{pq} \tilde{\boldsymbol{q}} \wedge \tilde{\boldsymbol{p}}. \tag{43.32}$$

We also have the rule that tensor products of 1-forms and 1-vectors commute, allowing us to write $\boldsymbol{e}_\mu \otimes \boldsymbol{\omega}^\nu = \boldsymbol{\omega}^\nu \otimes \boldsymbol{e}_\mu$.

Example 43.5

Compute as follows:

$$
\begin{aligned}
d(e_\mu \wedge e_\nu) &= d(e_\mu \otimes e_\nu - e_\nu \otimes e_\mu)\\
&= de_\mu \otimes e_\nu + e_\mu \otimes de_\nu - de_\nu \otimes e_\mu - e_\nu \otimes de_\mu\\
&= \omega^\alpha{}_\mu \otimes e_\alpha \otimes e_\nu + e_\mu \otimes \omega^\beta{}_\nu \otimes e_\beta - \omega^\gamma{}_\mu \otimes e_\gamma \otimes e_\mu - e_\nu \otimes \omega^\delta{}_\nu \otimes e_\delta\\
&= \omega^\alpha{}_\mu \otimes e_\alpha \otimes e_\nu + e_\mu \otimes \omega^\beta{}_\nu \otimes e_\beta - \omega^\beta{}_\mu \otimes e_\beta \otimes e_\mu - e_\nu \otimes \omega^\alpha{}_\nu \otimes e_\alpha\\
&= \omega^\alpha{}_\mu (e_\alpha \wedge e_\nu) + \omega^\beta{}_\nu (e_\mu \wedge e_\beta).
\end{aligned}
\tag{43.33}
$$

[13]Remember that, in an orthonormal frame, $\omega^{\mu\nu} = -\omega^{\nu\mu}$.

Step III: Working in the orthonormal frame, we define $\mathcal{R}^{\mu\nu} = \mathcal{R}^\mu{}_\alpha \eta^{\alpha\nu}$. We can then consider an alternative definition of the curvature operator

$$
\mathcal{R} = \frac{1}{2}\left(e_\mu \wedge e_\nu\right)\mathcal{R}^{\mu\nu},
\tag{43.34}
$$

where[13]

$$
\mathcal{R}^{\mu\nu} = d\omega^{\mu\nu} - \omega^\mu{}_\alpha \wedge \omega^{\nu\alpha}.
\tag{43.35}
$$

Using this, we can combine the results of steps I and II to prove our result.

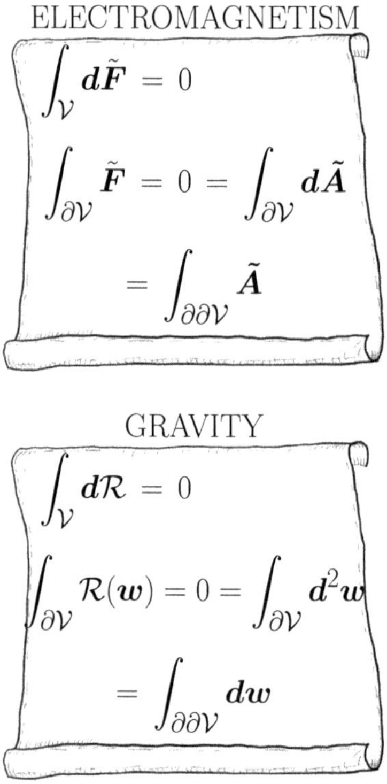

Example 43.6

Take the derivative of $\mathcal{R}$ to find

$$
\begin{aligned}
d\mathcal{R} &= \frac{1}{2} d\left(e_\mu \wedge e_\nu\right) \wedge \mathcal{R}^{\mu\nu} + \frac{1}{2}\left(e_\mu \wedge e_\nu\right) d\mathcal{R}^{\mu\nu}\\
&= \frac{1}{2}\left(e_\alpha \wedge e_\nu\right)\left(\omega^\alpha{}_\mu \wedge \mathcal{R}^{\mu\nu}\right) + \frac{1}{2}\left(e_\mu \wedge e_\beta\right)\left(\omega^\beta{}_\nu \wedge \mathcal{R}^{\mu\nu}\right)\\
&\quad + \frac{1}{2}\left(e_\mu \wedge e_\nu\right) d\mathcal{R}^{\mu\nu}\\
&= \frac{1}{2}\left(e_\mu \wedge e_\nu\right)\left(\omega^\mu{}_\alpha \wedge \mathcal{R}^{\alpha\nu} + \omega^\nu{}_\beta \wedge \mathcal{R}^{\mu\beta}\right.\\
&\quad \left. - \mathcal{R}^\mu{}_\beta \wedge \omega^{\nu\beta} + \omega^\mu{}_\alpha \wedge \mathcal{R}^{\nu\alpha}\right)\\
&= \frac{1}{2}\left(e_\mu \wedge e_\nu\right)\left(\omega^\nu{}_\beta \wedge \mathcal{R}^{\mu\beta} - \mathcal{R}^\mu{}_\beta \wedge \omega^{\nu\beta}\right)\\
&= \frac{1}{2}\left(e_\mu \wedge e_\nu\right)\left(\omega^\nu{}_\beta \wedge \mathcal{R}^{\mu\beta} - \omega^{\nu\beta} \wedge \mathcal{R}^\mu{}_\beta\right) = 0.
\end{aligned}
\tag{43.36}
$$

In the last line, we've used the fact that $\omega^\mu{}_\alpha \wedge \mathcal{R}^{\nu\alpha} = \omega^{\mu\alpha} \wedge \mathcal{R}^\nu{}_\alpha$, and that $\mathcal{R}^\mu{}_\nu$ is a 2-form and $\omega^\mu{}_\nu$ is a 1-form, so inverting the wedge product does not pick up a sign.

Although the computation was rather lengthy, we have established the Bianchi identity $d\mathcal{R} = 0$. This allows us to repeat the general argument above (and in Fig. 43.3) involving the boundary of a boundary. Since $d\mathcal{R} = 0$, we have that

$$
0 = \int_\mathcal{V} d\mathcal{R} = \int_{\partial\mathcal{V}} \mathcal{R}.
\tag{43.37}
$$

ELECTROMAGNETISM

$$
\int_\mathcal{V} d\tilde{F} = 0
$$

$$
\int_{\partial\mathcal{V}} \tilde{F} = 0 = \int_{\partial\mathcal{V}} d\tilde{A}
$$

$$
= \int_{\partial\partial\mathcal{V}} \tilde{A}
$$

GRAVITY

$$
\int_\mathcal{V} d\mathcal{R} = 0
$$

$$
\int_{\partial\mathcal{V}} \mathcal{R}(w) = 0 = \int_{\partial\mathcal{V}} d^2w
$$

$$
= \int_{\partial\partial\mathcal{V}} dw
$$

Fig. 43.3 Boundary of a boundary arguments for electromagnetism (top) and gravity (bottom).

Here, $\mathcal{R}$ is defined by the double derivative of a vector field $\boldsymbol{w}$ by writing $\boldsymbol{d}^2\boldsymbol{w} = \mathcal{R}(\boldsymbol{w})$. So we're now suggesting that $0 = \int_{\partial\mathcal{V}} \boldsymbol{d}^2\boldsymbol{w} = \int_{\partial\partial\mathcal{V}} \boldsymbol{dw}$. So we might expect to be able to derive a component version of the Bianchi identity by evaluating the vector-valued 1-form $\boldsymbol{dw} = \boldsymbol{\nabla}\boldsymbol{w}$ round a boundary, and setting it equal to zero. Equivalently, we can evaluate the surface integral of $\boldsymbol{d}^2\boldsymbol{w}$ and obtain the same answer. We shall do the latter.[14]

To evaluate our integral, we need the result of the inner product of the form $\boldsymbol{d}^2\boldsymbol{w}$ and a surface element, as examined in the next example.

Example 43.7

Let's consider $\langle \boldsymbol{d\tilde\alpha}, \boldsymbol{u} \wedge \boldsymbol{v} \rangle$, where $\tilde\alpha$ is a 1-form and $\boldsymbol{u}$ and $\boldsymbol{v}$ are vectors. This object counts the number of cells of the 2-form $\boldsymbol{d\tilde\alpha}$ in the parallelogram formed by the bivector $\boldsymbol{u} \wedge \boldsymbol{v}$.[15] It can be rewritten as[16]

$$\langle \boldsymbol{d\tilde\alpha}, \boldsymbol{u} \wedge \boldsymbol{v} \rangle = \langle \boldsymbol{d\tilde\alpha}, \boldsymbol{u} \otimes \boldsymbol{v} \rangle - \langle \boldsymbol{d\tilde\alpha}, \boldsymbol{v} \otimes \boldsymbol{u} \rangle$$
$$= \langle \boldsymbol{\nabla}_{\boldsymbol{u}}\tilde\alpha, \boldsymbol{v} \rangle - \langle \boldsymbol{\nabla}_{\boldsymbol{v}}\tilde\alpha, \boldsymbol{u} \rangle. \tag{43.38}$$

Expanding the actions of the covariant derivative on forms, we have[17]

$$\langle \boldsymbol{\nabla}_{\boldsymbol{u}}\tilde\alpha, \boldsymbol{v} \rangle - \langle \boldsymbol{\nabla}_{\boldsymbol{v}}\tilde\alpha, \boldsymbol{u} \rangle = \boldsymbol{\nabla}_{\boldsymbol{u}}\langle \tilde\alpha, \boldsymbol{v} \rangle - \langle \tilde\alpha, \boldsymbol{\nabla}_{\boldsymbol{u}}\boldsymbol{v} \rangle - \boldsymbol{\nabla}_{\boldsymbol{v}}\langle \tilde\alpha, \boldsymbol{u} \rangle + \langle \tilde\alpha, \boldsymbol{\nabla}_{\boldsymbol{v}}\boldsymbol{u} \rangle$$
$$= \boldsymbol{\nabla}_{\boldsymbol{u}}\langle \tilde\alpha, \boldsymbol{v} \rangle - \boldsymbol{\nabla}_{\boldsymbol{v}}\langle \tilde\alpha, \boldsymbol{u} \rangle - \langle \tilde\alpha, [\boldsymbol{u}, \boldsymbol{v}] \rangle. \tag{43.39}$$

This result generalizes to any tensor-valued 1-form $\boldsymbol{S}$, giving

$$\langle \boldsymbol{dS}, \boldsymbol{u} \wedge \boldsymbol{v} \rangle = \boldsymbol{\nabla}_{\boldsymbol{u}}\langle \boldsymbol{S}, \boldsymbol{v} \rangle - \boldsymbol{\nabla}_{\boldsymbol{v}}\langle \boldsymbol{S}, \boldsymbol{u} \rangle - \langle \boldsymbol{S}, [\boldsymbol{u}, \boldsymbol{v}] \rangle. \tag{43.40}$$

For the special case at hand, involving the vector-valued 1-form $\boldsymbol{dw}$, we use $\langle \boldsymbol{dw}, \boldsymbol{u} \rangle = \boldsymbol{\nabla}_{\boldsymbol{u}}\boldsymbol{w}$ and find

$$\langle \boldsymbol{d}^2\boldsymbol{w}, \boldsymbol{u} \wedge \boldsymbol{v} \rangle = \boldsymbol{\nabla}_{\boldsymbol{u}}\boldsymbol{\nabla}_{\boldsymbol{v}}\boldsymbol{w} - \boldsymbol{\nabla}_{\boldsymbol{v}}\boldsymbol{\nabla}_{\boldsymbol{u}}\boldsymbol{w} - \boldsymbol{\nabla}_{[\boldsymbol{u},\boldsymbol{v}]}\boldsymbol{w}$$
$$= \hat{\boldsymbol{R}}(\boldsymbol{u}, \boldsymbol{v})\boldsymbol{w}, \tag{43.41}$$

where, as before, $\hat{\boldsymbol{R}}$ is the covariant Riemann curvature operator $[\boldsymbol{\nabla}_{\boldsymbol{u}}, \boldsymbol{\nabla}_{\boldsymbol{v}}] - \boldsymbol{\nabla}_{[\boldsymbol{u},\boldsymbol{v}]}$.

In the electromagnetic case we examined in Example 43.3, we had $\langle \boldsymbol{F}, \boldsymbol{u} \wedge \boldsymbol{v} \rangle = \boldsymbol{F}(\boldsymbol{u}, \boldsymbol{v})$, allowing us to evaluate the surface integral over a closed surface. The Bianchi identity $\boldsymbol{d\tilde{F}} = 0$ allowed us to set this to zero. In the same way, we have the equation

$$\langle \boldsymbol{d}^2\boldsymbol{w}, \boldsymbol{u} \wedge \boldsymbol{v} \rangle = \hat{\boldsymbol{R}}(\boldsymbol{u}, \boldsymbol{v})\boldsymbol{w}, \tag{43.42}$$

which allows us to do the surface integral, which we know from $\boldsymbol{d}\mathcal{R} = 0$, must be zero. This provides a coordinate version of the Bianchi identity.

All that's left to do is to rerun the argument in Example 43.3. For simplicity we use Riemann normal coordinates.[18] The outward contribution from the the yz face of the cube at $x + \Delta x$ is given by

$$\hat{\boldsymbol{R}}(\boldsymbol{e}_y, \boldsymbol{e}_z)\boldsymbol{w}\Delta y\Delta z \tag{43.43}$$

In coordinates, this is

$$R^\alpha{}_{\beta yz}(\text{at } x + \Delta x)w^\beta \Delta y \Delta z \tag{43.44}$$

[14]The former approach is followed in Misner, Thorne, and Wheeler. You can see how taking $\boldsymbol{\nabla}\boldsymbol{w}$ around the boundary resembles the process in Example 35.5 for obtaining the components of $\boldsymbol{R}$. In this case, each loop comprises the face of the infinitesimal cube, and opposite faces are taken in different directions, leading to the derivatives. The details are left as an exercise.

[15]As mentioned in Chapter 32, this quantity is equivalent to the inner product $\langle \boldsymbol{d\tilde\alpha}, \boldsymbol{u} \wedge \boldsymbol{v} \rangle$.

[16]If in doubt, think of $\boldsymbol{d\tilde\alpha}$ as an object with two slots, which we'll fill with vectors $\boldsymbol{u}$ and $\boldsymbol{v}$ in order, via the object $\boldsymbol{u} \otimes \boldsymbol{v}$. Putting the vector $\boldsymbol{u}$ in the first $(\boldsymbol{d})$ slot gives $\boldsymbol{u} \cdot \boldsymbol{d} \equiv \boldsymbol{u} \cdot \boldsymbol{\nabla} = \boldsymbol{\nabla}_{\boldsymbol{u}}$. This was verified in Exercise 34.6.

[17]We note the recipe of how to take the covariant derivative of a 1-form like $\tilde\alpha$ is given by considering the inner product

$$\boldsymbol{\nabla}_{\boldsymbol{u}}\langle \tilde\alpha, \boldsymbol{w} \rangle = \langle \boldsymbol{\nabla}_{\boldsymbol{u}}\tilde\alpha, \boldsymbol{w} \rangle + \langle \tilde\alpha, \boldsymbol{\nabla}_{\boldsymbol{u}}\boldsymbol{w} \rangle.$$

[18]See Chapter 35 for a description of these. In short, they describe a local inertial frame, so the vanishing connection coefficients at the origin imply that we can treat ordinary derivatives as equivalent to covariant ones in this treatment.

The opposite face gives a similar contribution, except that the sign is reversed and we evaluate at point x rather than $x+\Delta x$. The combination yields

$$\frac{\partial R^{\alpha}{}_{\beta yz}}{\partial x} w^{\beta} \Delta x \Delta y \Delta z. \tag{43.45}$$

Repeating for all of the other faces, summing and setting equal to zero, we obtain the Bianchi identity for gravitation in component form

$$R^{\hat{\alpha}}{}_{\hat{\beta}\hat{y}\hat{z},\hat{x}} + R^{\hat{\alpha}}{}_{\hat{\beta}\hat{z}\hat{x},\hat{y}} + R^{\hat{\alpha}}{}_{\hat{\beta}\hat{x}\hat{y},\hat{z}} = 0, \tag{43.46}$$

where we've reinstated hats to stress that this is true in the local inertial frame. This valid tensor equation is also true in a general coordinate frame on using the, by now familiar, commas go to semicolons rule.[19] Finally, we can generalize the cube so that it's described by any set of coordinate axes. This gives a final expression for the Bianchi identity of[20]

$$R^{\alpha}{}_{\beta\mu\nu;\lambda} + R^{\alpha}{}_{\beta\nu\lambda;\mu} + R^{\alpha}{}_{\beta\lambda\mu;\nu} = 0. \tag{43.49}$$

This is the version of the equation we used in Chapter 13.

This chapter has given us a geometrical sense of where the Bianchi identity comes from, both in electromagnetism and in gravitation. There is one final link that we can make between these theories using the Bianchi identity. It is tempting to identify the Einstein field equation with the Maxwell equations, since these are the key equation of motion for both theories. However, a closer analogy can be made between the Maxwell equations and the Weyl tensor $\boldsymbol{C}$ that we met in Chapter 35.[21] This is examined in the next example.

Example 43.8

Since the Weyl tensor collects those parts of the Riemann tensor that do not appear in the Einstein field equation, it would seem that $\boldsymbol{C}$ represents that part of the spacetime curvature that is *not* generated locally by the matter distribution. The components of the Riemann tensor obey the Bianchi identity, which can be rewritten[22] in terms of the components of $\boldsymbol{C}$ as

$$C^{\mu\nu\alpha\beta}{}_{;\beta} = J^{\mu\nu\alpha}, \tag{43.51}$$

where $J^{\mu\nu\alpha} = R^{\alpha[\mu;\nu]} + \frac{1}{6}g^{\alpha[\nu}R^{;\mu]}$. Since eqn 43.51 looks so much like the Maxwell equations $F^{\alpha\beta}{}_{;\beta} = J^{\alpha}$, we can think of the Bianchi identity as a field equation for the Weyl tensor, telling us how matter determines curvature at other points in spacetime, just as currents determine electromagnetic field lines in other parts of spacetime.

In the next chapter, we turn to another important aspect of classical field theory that plays a major role in gravitation: the notion of a gauge field.

---

[19]That is, we have the curved space-time equation

$$R^{\alpha}{}_{\beta yz;x} + R^{\alpha}{}_{\beta zx;y} + R^{\alpha}{}_{\beta xy;z} = 0. \tag{43.47}$$

[20]In tensor notation, we can also write

$$R^{\alpha}{}_{\beta[\mu\nu;\lambda]} = 0. \tag{43.48}$$

[21]The components of $\boldsymbol{C}$ are given by

$$\begin{aligned}
C_{\mu\nu\alpha\beta} ={}& R_{\mu\nu\alpha\beta} - \frac{1}{2}R_{\mu\alpha}g_{\nu\beta} \\
&+ \frac{1}{2}R_{\mu\beta}g_{\nu\alpha} + \frac{1}{2}R_{\nu\alpha}g_{\mu\beta} \\
&- \frac{1}{2}R_{\nu\beta}g_{\mu\alpha} + \frac{1}{6}R(g_{\mu\alpha}g_{\nu\beta} \\
&- g_{\mu\beta}g_{\nu\alpha}).
\end{aligned} \tag{43.50}$$

[22]See Exercise 43.2.

Chapter summary

- Charge conservation can be written as

$$0 = \int_{\mathcal{V}} \boldsymbol{d} \star \boldsymbol{J} = \int_{\partial \mathcal{V}} \star \boldsymbol{J}. \qquad (43.52)$$

- The Bianchi identity for electromagnetism is a component version of the expression $\boldsymbol{d}\tilde{\boldsymbol{F}} = 0$. The flat-space component equation reads

$$F_{\alpha\beta,\gamma} + F_{\gamma\alpha,\beta} + F_{\beta\gamma,\alpha} = 0. \qquad (43.53)$$

- The gravitational Bianchi identity can be written as

$$\boldsymbol{d}\mathcal{R} = 0. \qquad (43.54)$$

In components, this yields a curved-space equation

$$R^{\alpha}{}_{\beta\mu\nu;\lambda} + R^{\alpha}{}_{\beta\nu\lambda;\mu} + R^{\alpha}{}_{\beta\lambda\mu;\nu} = 0. \qquad (43.55)$$

Exercises

(43.1) Verify that the Bianchi identity can also be written as

$$R^{\alpha}{}_{[\mu\nu;\lambda]} = 0. \qquad (43.56)$$

(43.2) Using the contracted Bianchi identity

$$R^{\alpha}{}_{\lambda;\alpha} = \frac{1}{2} R_{;\lambda}, \qquad (43.57)$$

that was derived in Chapter 13, verify that eqn 43.51 reduces to (a contraction of) the Bianchi identity.

(43.3) *Conservation of the source of gravitational curvature is written as* $\boldsymbol{\nabla} \cdot \boldsymbol{T} = 0$, *but it can also be written as*

$$\boldsymbol{d}\star\boldsymbol{T} = 0. \qquad (43.58)$$

Since the Einstein equations are simply $\boldsymbol{G} = 8\pi\boldsymbol{T}$, *the dual form is*

$$\star\boldsymbol{G} = 8\pi\star\boldsymbol{T}. \qquad (43.59)$$

We shall find expressions for $\star\boldsymbol{T}$ *and* $\star\boldsymbol{G}$.
Take the dual of the vector part of the $(1,1)$ version of $\boldsymbol{T} = T^{\mu}{}_{\nu}\boldsymbol{e}_{\mu} \otimes \boldsymbol{\omega}^{\nu}$ to show that

$$\star\boldsymbol{T} = T^{\mu\nu}\left(\boldsymbol{e}_{\mu} \otimes \mathrm{d}\boldsymbol{\Sigma}_{\nu}\right), \qquad (43.60)$$

and

$$\star\boldsymbol{G} = G^{\mu\nu}\left(\boldsymbol{e}_{\mu} \otimes \mathrm{d}\boldsymbol{\Sigma}_{\nu}\right), \qquad (43.61)$$

where $\mathrm{d}\boldsymbol{\Sigma}_{\nu}$ is the surface 3-form.

(43.4) *Recall the identity from Exercise 34.3, valid in a coordinate frame, that*

$$\Gamma^{\alpha}{}_{\beta\alpha} = \frac{\partial}{\partial x^{\beta}} \ln\left(\sqrt{-g}\right). \qquad (43.62)$$

Use this to show that $\boldsymbol{d}\star\boldsymbol{T} = 0$, as we claimed in the previous question.

(43.5) Show that we must have

$$\int_{\partial \mathcal{V}} \star\boldsymbol{G} = 0. \qquad (43.63)$$

See Misner, Thorne, and Wheeler for a discussion of the physical content of this equation.

44 Gauge fields

> ↷ Gauge theory is very helpful for understanding weak gravitational fields (Chapter 45) and gravitational waves (Chapter 46). It also forms the basis of high-dimensional theories (Chapter 48).

[1] 'Gauge' is an unfortunate term in this context, referring originally to the thickness of metal wires and rails, but we are stuck with it. In field theory, it derives from Hermann Weyl's use of the term in general relativity that we discuss in Chapter 45.

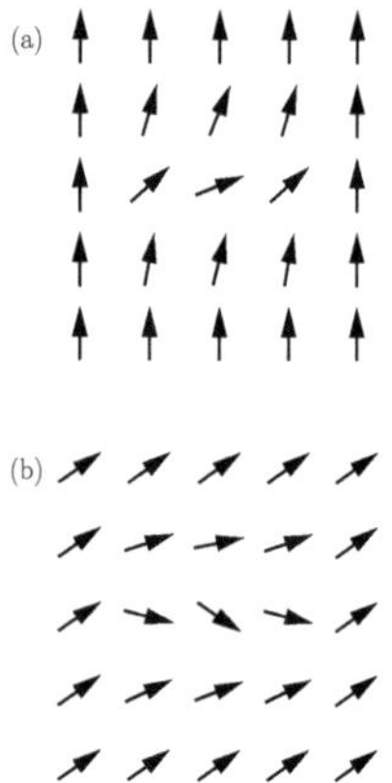

Fig. 44.1 (a) A representation of the field $\rho(x)e^{i\theta(x)}$, where the angle of each arrow to the vertical represents $\theta(x)$ at a particular point. (b) The effect of a global phase transformation $\theta(x) \to \theta(x) + \alpha$.

If you leave a thing alone you leave it to a torrent of change.
G. K. Chesterton (1874–1936) *Heretics*

Suppose you move through some distance in space and you measure a change in a vector. There are two possible reasons for the change: (i) the intrinsic change of the vector field with position and (ii) the change, with position, of the coordinate system you are using. In Chapter 7, these factors led to our definition of the covariant derivative, in words and coordinates respectively, as

$$\left(\begin{array}{c}\text{Covariant}\\\text{derivative}\end{array}\right) = \left(\begin{array}{c}\text{Change}\\\text{in vector}\end{array}\right) - \left(\begin{array}{c}\text{Change due to}\\\text{coordinate system}\end{array}\right)$$

$$(\boldsymbol{\nabla}_\alpha \boldsymbol{v})^\mu = \frac{\partial v^\mu}{\partial x^\alpha} + \Gamma^\mu{}_{\alpha\beta} v^\beta. \tag{44.1}$$

This effect of two contributions to the change in a vector with position is closely related to the physics of **phases** and **gauges**[1] in field theory. In particular, there is an illuminating analogy between the covariant derivative $\boldsymbol{\nabla}_\mu$ in relativity and the covariant derivative D_μ used to describe field theories with **local symmetries**. In this chapter, we explore this link.

44.1 Fibre bundles and gauge invariance

Fields depend on variables like x, the position in spacetime. We can think of the field ψ stretching out over all time and space, such that there is a value of $\psi(x)$ at each point in spacetime. A field theory is described by a Lagrangian $\mathcal{L}$. If we perform a transformation of the coordinates that feature in the theory and obtain the same value of $\mathcal{L}$, then the equations of motion that one derives from the Lagrangian are unchanged, and we say that the theory has a **symmetry**. In addition to their dependence on coordinate variables like x, fields can have other, **internal variables** such as a phase θ that causes the field carry around a factor $e^{i\theta}$. An example is a field, $\psi(x) = \rho(x)e^{i\theta(x)}$ which could be visualized as having an amplitude ρ at each point in space, as well as a phase θ, which could be represented in a diagram by an phasor arrow, as shown in Fig. 44.1(a). We can perform transformations on the phase itself, perhaps by making the change $\theta(x) \to \theta(x) + \alpha$, which amounts to a rotation of all of the arrows by an angle α as shown in Fig. 44.1(b). If the Lagrangian does not change under this transformation then we say that we have an **internal symmetry**.

Mathematically speaking, despite being a function of the position in spacetime x, the phase $\theta(x)$ does not itself live in (3+1)-dimensional spacetime. So the change in the phase does not correspond to any change in the spatial coordinates, which is to say that the rotation of the arrows is a rotation in the internal space of the field. To understand these internal variables, we make use of the mathematical notion of a **fibre bundle** over spacetime.[2] We imagine that floating above each point in spacetime is another space called a **fibre**, described by the internal variables. In the mathematical language, we call the usual (3+1)-dimensional spacetime the **base space** $\mathcal{M}$. The fibre bundle $\mathcal{B}$ is a structure formed from combining a fibre $\mathcal{V}$ at each point in $\mathcal{M}$ (or, in Roger Penrose's words, 'an $\mathcal{M}$ worth of $\mathcal{V}$'s'). The bundle $\mathcal{B}$ can then be thought of as comprising $\mathcal{M}$ with the structure floating above $\mathcal{M}$, as shown in Fig. 44.2.

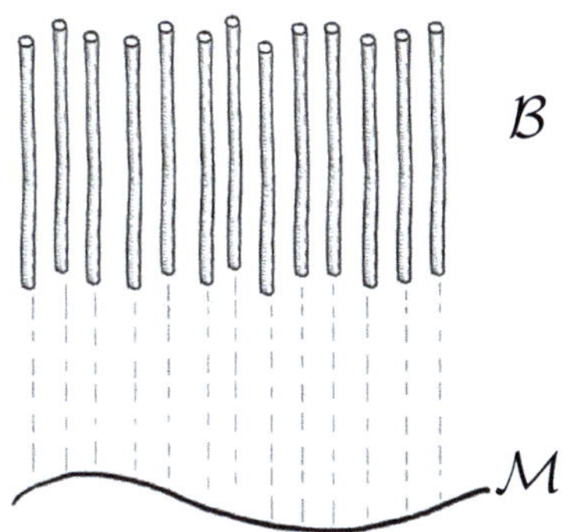

Fig. 44.2 A one-dimensional manifold $\mathcal{M}$ has a fibre $\mathcal{V}$ floating above each point. The combination of an $\mathcal{M}$ worth of $\mathcal{V}$s, makes the bundle $\mathcal{B}$.

[2] Fibre bundles are described more mathematically in Appendix C.

Example 44.1

• We have met such a structure before when we considered tangent vectors. A vector $\boldsymbol{v}$ that is tangent to a curve at a point $\mathcal{P}$ in a spacetime manifold $\mathcal{M}$, lives in a vector space known as a tangent space, denoted $\mathcal{T}_{\mathcal{P}}$ that can be thought of as floating above the point. The components of the vector can then be thought of as being stored in a fibre (that is, in the tangent space $\mathcal{T}_{\mathcal{P}}$) above $\mathcal{P}$. There is a different tangent space for each point in the spacetime. Combining all of the tangent spaces, we make a tangent bundle $\mathcal{T}\mathcal{M}$.

• Another example of bundles in physics is **isospin**. Many matter fields act as if they have an internal switch that allows us to change the identity of particle excitations at a particular point. This internal variable is isospin and can be treated in terms of fibres.

• A slightly different example is **Newtonian spacetime**, which is formed from a one-dimensional manifold R that corresponds to points in time. Floating above each point in time is a three-dimensional space that gives the configuration of the system in space at that time.[3]

[3] This structure is what general relativity replaces with our notion of metrics defined on a manifold.

In our example of a field ψ with a phase, it is the phase variable θ that is stored in a fibre above each point in spacetime. Since we want the field to be single valued, the angle θ must be the same as $\theta + 2\pi$, and we can visualize each of the fibres as a unit circle floating above a point in spacetime as shown (for two spatial dimensions) in Fig. 44.3. (This is, of course, equivalent to our picture of arrows in Fig. 44.1, but making the unit circles explicit is helpful for later applications.) The global rotation of the phases by an amount α, simply adds α to the values stored in each of the fibres. If in doubt, think of each unit circle fibre as a knob that can be turned. The global rotation corresponds to twisting all of the knobs by the same angle α.

A very special property of fields is revealed if we examine the transformation of the internal phase variable in detail. Let's confine ourselves to flat space and consider the so-called **complex scalar field theory**. This theory is built from complex-number-valued fields ψ that are de-

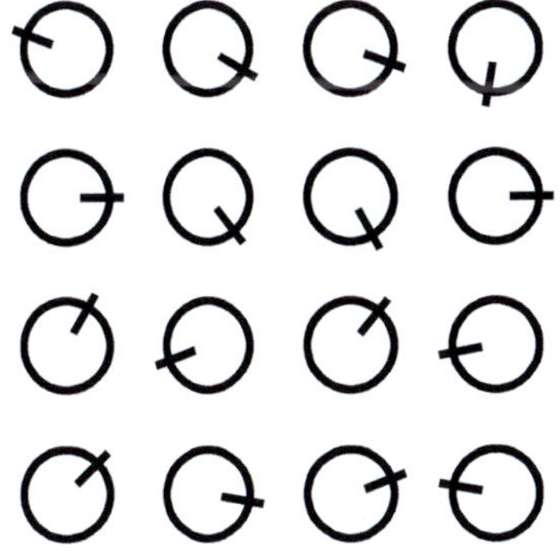

Fig. 44.3 View from above of a two-dimensional space with a circular fibre (or knob) floating above each point.

scribed by the matter-field Lagrangian

$$\mathcal{L} = -(\partial^{\mu}\psi)^{\dagger}(\partial_{\mu}\psi) - m^2\psi^{\dagger}\psi, \tag{44.2}$$

where m is a constant.[4] This theory has an internal symmetry known as a global phase symmetry, or global $U(1)$ symmetry. This is shown if we make the replacement

$$\psi(x) \to \psi(x)\mathrm{e}^{\mathrm{i}\alpha}, \tag{44.3}$$

since then the Lagrangian $\mathcal{L}$ (and hence the equation of motion) does not change.[5] The transformation in eqn 44.3 is a **global transformation** in that the phase changes by the same amount (i.e. an addition of α) at every point in spacetime.

We can now ask an interesting question: what if we attempt to change the phase by different amounts at each point in spacetime? This would imply we change the phases $\theta(x)$ by an amount $\alpha(x)$, that is, by an amount that depends in some arbitrary (but smoothly varying) manner on position in spacetime. This is known as a **local transformation**.

[4]To keep our expressions brief and manageable, we use the familiar ∂_{μ} notation, where

$$\partial_{\mu} = \frac{\partial}{\partial x^{\mu}}.$$

We interpret ∂^{μ} as

$$\partial^{\mu} = \eta^{\mu\nu}\partial_{\nu}.$$

[5]This is of course because $\psi^{\dagger} \to \psi^{\dagger}\mathrm{e}^{-\mathrm{i}\alpha}$ and so the angle α cancels in the combinations of the field and its hermitian conjugate in $\mathcal{L}$.

Example 44.2

Let's see how this unfolds. We have

$$\partial_{\mu}\psi(x) \to \partial_{\mu}\psi(x)\mathrm{e}^{\mathrm{i}\alpha(x)}$$
$$= \mathrm{e}^{\mathrm{i}\alpha(x)}\partial_{\mu}\psi(x) + \psi(x)\mathrm{e}^{\mathrm{i}\alpha(x)}\mathrm{i}\partial_{\mu}\alpha(x)$$
$$= \mathrm{e}^{\mathrm{i}\alpha(x)}\left\{\partial_{\mu} + \mathrm{i}\left[\partial_{\mu}\alpha(x)\right]\right\}\psi(x). \tag{44.4}$$

Similarly,

$$\partial^{\mu}\psi^{\dagger}(x) \to \mathrm{e}^{-\mathrm{i}\alpha(x)}\left\{\partial^{\mu} - \mathrm{i}\left[\partial^{\mu}\alpha(x)\right]\right\}\psi^{\dagger}(x). \tag{44.5}$$

The first term in the Lagrangian becomes

$$(\partial^{\mu}\psi^{\dagger})(\partial_{\mu}\psi) - \mathrm{i}\psi^{\dagger}(\partial^{\mu}\alpha)(\partial_{\mu}\psi) + \mathrm{i}\psi(\partial^{\mu}\psi^{\dagger})(\partial_{\mu}\alpha) + \psi^{\dagger}\psi(\partial^{\mu}\alpha)(\partial_{\mu}\alpha), \tag{44.6}$$

which is not what we started with.

The message is that, perhaps unsurprisingly, the theory is not invariant with respect to local phase transformations. However, we can fix things to guarantee local invariance, by adding another field into the mix. To that end we introduce a 1-form field $\tilde{\boldsymbol{A}}(x) = A_{\mu}(x)\boldsymbol{dx}^{\mu}$. This field, whose job is to cancel out the effect of the change in internal variable $\alpha(x)$ with position, is known as a **gauge field**. This enters into a new covariant field derivative D_{μ}, defined by

$$D_{\mu}\psi(x) = \partial_{\mu}\psi(x) + \mathrm{i}qA_{\mu}(x)\psi(x). \tag{44.7}$$

We define the gauge field such that when we change the phase by angle $\alpha(x)$, the components of the gauge field transform according to

$$A_{\mu}(x) \to A_{\mu}(x) - \frac{1}{q}\partial_{\mu}\alpha(x). \tag{44.8}$$

Let's see how things unfold now.

Example 44.3

If $\psi(x) \to \psi(x)e^{i\alpha(x)}$, then $\partial_\mu \psi \to e^{i\alpha}(\partial_\mu \psi) + i\psi e^{i\alpha}(\partial_\mu \alpha)$ and so

$$D_\mu \psi = (\partial_\mu + iqA_\mu)\psi \to e^{i\alpha}(\partial_\mu \psi) + i\psi e^{i\alpha}(\partial_\mu \alpha) + iqA_\mu \psi e^{i\alpha} - i\psi e^{i\alpha}(\partial_\mu \alpha)$$
$$= e^{i\alpha} D_\mu \psi. \tag{44.9}$$

This property makes the whole Lagrangian invariant if we replace ordinary derivatives by covariant ones

$$\mathcal{L} = -(D^\mu \psi)^\dagger (D_\mu \psi) - m^2 \psi^\dagger \psi, \tag{44.10}$$

since now with $D_\mu \psi \to e^{i\alpha} D_\mu \psi$, the first term is invariant.

The message is that if the phase $\alpha(x)$ is a function of x then, in order to guarantee local phase invariance, a new 1-form gauge field $\tilde{\boldsymbol{A}}(x)$ with components $A_\mu(x)$ is required to build the covariant field derivative. Furthermore, we recognize this field as akin to the electromagnetic gauge field introduced in Chapter 42, where we saw that one of its most intriguing features was that, without changing the Maxwell equations, it could be changed by an arbitrary amount

$$A_\mu(x) \to A_\mu(x) - \partial_\mu \chi(x), \tag{44.11}$$

which known as a gauge transformation. We see from eqn 44.8 that if we identify $\chi(x)$ with $\alpha(x)/q$, then the transformation demanded in eqn 44.8 is simply gauge invariance, confirming that $\boldsymbol{A}$ has the usual properties of a gauge field as defined in electromagnetism.

To summarize, the complex scalar field theory can be made to have a local phase symmetry [that is, an invariance of the Lagrangian under a transformation $\theta(x) \to \theta(x) + \alpha(x)$] by introducing a gauge field $\tilde{\boldsymbol{A}}(x)$ to build a covariant derivative with components $D_\mu = \partial_\mu + iqA_\mu$. The theory, made locally phase invariant by replacing $\partial_\mu \to D_\mu$ in the Lagrangian, is known as a gauge theory. The gauge field transforms via a gauge transformation $A_\mu \to A_\mu - \partial_\mu \alpha/q$. Gauge theories are the basis of particle physics as well as the physics of superconductivity and topological states of matter like the fractional quantum Hall fluid.[6]

In terms of our picture of the fibre bundle, the local transformation $\theta(x) \to \theta(x) + \alpha(x)$ corresponds to a shift in the value of the phase stored in the fibre floating above point x. Since the shifts are different on different fibres, the resulting bundle is often called a **strained bundle**, in the same way that a strained solid has atoms displaced by different amounts as a function of position. The derivative ∂_μ, **connects** the different fibres in that it forces us to consider differences between variables stored in different fibres. In order to take account of the strain in the bundle that we have produced, we say that we *introduce a strain into the connection* via the field $\tilde{\boldsymbol{A}}(x)$. This all sounds vaguely reminiscent of the connection coefficients in geometry. In fact, a further insight into gauge fields is revealed if we compare the covariant derivative from relativity $\boldsymbol{\nabla}_\mu$ with the covariant field derivative D_μ, which is our next subject for discussion.

[6]In the case we have examined, if we interpret q as the electric charge, $\tilde{\boldsymbol{A}}$ is the electromagnetic field.

44.2   Parallel transport and field strength

Let's revisit the covariant derivative ∇_μ and the parallel transport that it allows from the point of view of fields. Move from x to $x + \mathrm{d}x$ and a field ψ changes by an amount $\mathrm{d}\psi$. The way to make sense of the change is to note that $\psi(x)$ and $\psi(x + \mathrm{d}x)$ are measured in different coordinate systems. Thus, the apparently innocent equation

$$\mathrm{d}\psi = \psi(x + \mathrm{d}x) - \psi(x), \tag{44.12}$$

that describes the total change in the field in moving through a distance $\mathrm{d}x$, carries information about two coordinate systems.

We want to isolate properties that change due to changes in the internal space of the field as the field is moved around. To do this we make use of parallel transport. In the geometric description of Chapter 7, we made sense of the covariant derivative conceptually as

$$\left(\begin{array}{c} \text{Covariant} \\ \text{derivative of } \boldsymbol{v} \end{array} \right)_{\boldsymbol{u}} = \left(\begin{array}{c} \text{total change} \\ \text{in vector } \boldsymbol{v} \end{array} \right) - \left(\begin{array}{c} \text{change due to} \\ \text{coordinate system} \end{array} \right). \tag{44.13}$$

The condition for the parallel transport of a vector $\boldsymbol{v}$ along a path whose tangent vector is $\boldsymbol{u}$, is that $\nabla_{\boldsymbol{u}}\boldsymbol{v} = 0$, which is to say that the change in the vector components $(\partial v^\mu / \partial x^\nu)$ reflect only the change in the coordinate system (which changes by an amount measured by $-\Gamma^\mu{}_{\nu\alpha}v^\alpha$).

Example 44.4

We write $\nabla_{\boldsymbol{u}}\boldsymbol{v} = 0$ in components as

$$u^\nu \frac{\partial v^\mu}{\partial x^\nu} + \Gamma^\mu{}_{\nu\beta} v^\beta u^\nu = 0. \tag{44.14}$$

Cancelling u^ν, we see that the change in the vector components on parallel transport δv^μ is given by

$$\delta v^\mu = -\Gamma^\mu{}_{\nu\beta} v^\beta \mathrm{d}x^\nu. \tag{44.15}$$

The change in the components of the vector field are given by the connection coefficients, which encode how the coordinates change as we move around spacetime

---

In the same way that geometrical parallel transport allows us access to the change in the vector component δv^μ due to the change in underlying coordinates, parallel transport also allows us access to the change in the scalar field $\delta\psi$ due only to the effect of the change in coordinate system on the field's internal coordinates. For the complex scalar field theory described above, we write the change in parallel transport as[7]

$$\delta\psi(x) = -\mathrm{i}qA_\nu(x)\psi(x)\mathrm{d}x^\nu. \tag{44.16}$$

We therefore arrive at a general description of the covariant derivative as evaluating $\mathrm{d}\psi$, the total change in the scalar field on translation minus

[7]We read off that, for the geometric version in eqn 44.15, we need some extra indices to keep track of which component of the vector we are examining. However, we can see that there is an analogy if we take $\mathrm{i}qA_\nu(x)$ to be equivalent to $\Gamma^\mu{}_{\nu\beta}(x)$.

$\delta\psi$, the change due to the effects of the change in coordinate system on the internal variables of the field. In words and symbols

$$D\psi = \left(\begin{array}{c}\text{Change}\\\text{in field}\end{array}\right) - \left(\begin{array}{c}\text{Change due to}\\\text{coordinate system}\end{array}\right)$$
$$= \mathrm{d}\psi - \delta\psi$$
$$= \mathrm{d}\psi + iqA_\mu\psi\mathrm{d}x^\mu. \tag{44.17}$$

We then have a general covariant derivative fit for a description of internal variables of

$$\frac{D\psi}{\mathrm{d}x^\mu} \equiv D_\mu\psi = \partial_\mu\psi + iqA_\mu\psi. \tag{44.18}$$

Comparing this to the geometric version, we see how the components of the gauge field $\tilde{A}$ play the role of the connections $\Gamma^\mu{}_{\alpha\beta}$.

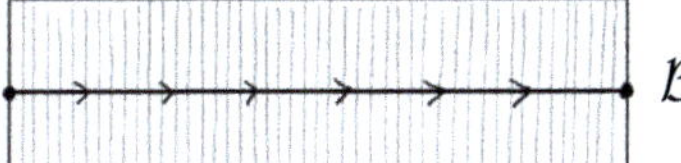

Example 44.5

Returning once more to the fibre bundle picture, the idea of parallel transport can be generalized. As usual, we picture the fibres floating above the base space. Instead of parallel transport, we now ask that a quantity found in a fibre that doesn't change as we move through spacetime, be connected by a **horizontal curve** as shown in Fig. 44.4. We therefore have a notion of **horizontal transport**, expressed mathematically by saying that for horizontal transport $D_\mu\psi = 0$, or $\partial_\mu\psi = -iqA_\mu$.

Fig. 44.4 The bundle $\mathcal{B}$ floating above the manifold $\mathcal{M}$. The analogue of parallel transport is horizontal transport through $\mathcal{B}$.

Example 44.6

One use of parallel transport in Chapter 11 was to parallel transport a vector around a loop. This resulted in a measure of the curvature of spacetime. Now that we have a general prescription for the covariant derivative of a gauge field ψ, we will try it out by transporting the field ψ around a closed infinitesimal quadrilateral loop $\mathcal{ABCDA}$ and see what happens. Since the field is single valued, the effect of the change in the field $\mathrm{d}\psi$ does not contribute, and we are be left with the change equivalent to the parallel transport of the field around the loop. Expanding to second order, the change in the field in moving from $\mathcal{A}$ to $\mathcal{B}$, along a line element Δx^μ is found by considering

$$\psi(x_\mathcal{B}) = \psi(x_\mathcal{A}) + D_\mu\psi(x_\mathcal{A})\Delta x^\mu + \frac{1}{2}D_\mu D_\nu\psi(x_\mathcal{A})\Delta x^\mu\Delta x^\nu$$
$$= \left(1 + \Delta x^\mu D_\mu + \frac{1}{2}\Delta x^\mu\Delta x^\nu D_\mu D_\nu\right)\psi(x_\mathcal{A}). \tag{44.19}$$

We agree to keep all terms to second order in going around the rest of the loop. The process is then repeated for the line section leading from points $\mathcal{B}$ to $\mathcal{C}$, which has length δx^μ. Denoting $\psi(x_\mathcal{B})$ as $\psi_\mathcal{B}$ we write

$$\psi_\mathcal{C} = \left(1 + \delta x^\mu D_\mu + \frac{1}{2}\delta x^\mu\delta x^\nu D_\mu D_\nu\right)\psi_\mathcal{B}$$
$$= \left[1 + (\delta x^\mu + \Delta x^\mu)D_\mu + \left(\frac{1}{2}\delta x^\mu\delta x^\nu + \delta x^\mu\Delta x^\nu + \frac{1}{2}\Delta x^\mu\Delta x^\nu\right)D_\mu D_\nu\right]\psi_\mathcal{A}.$$

Moving now from points $\mathcal{C}$ to $\mathcal{D}$, assumed to have separation Δx^{μ}, we have

$$\psi_{\mathcal{D}} = \left(1 - \Delta x^{\mu} D_{\mu} + \frac{1}{2}\Delta x^{\mu} \Delta x^{\nu} D_{\mu} D_{\nu}\right) \psi_{\mathcal{C}}$$

$$= \left[1 + \delta x^{\mu} D_{\mu} + \frac{1}{2}\delta x^{\mu} \delta x^{\nu} D_{\mu} D_{\nu} + (\delta x^{\mu} \Delta x^{\nu} - \Delta x^{\mu} \delta x^{\nu}) D_{\mu} D_{\nu}\right] \psi_{\mathcal{A}}$$

and finally back to $\mathcal{A}$, with the result that

$$\psi'_{\mathcal{A}} = (1 + \delta x^{\mu} \Delta x^{\nu} [D_{\mu}, D_{\nu}]) \psi_{\mathcal{A}}, \tag{44.20}$$

where the commutator $[D_{\mu}, D_{\nu}] = D_{\mu} D_{\nu} - D_{\mu} D_{\nu}$. Notice that the area of the loop is $\delta x^{\mu} \Delta x^{\nu}$ and so the change in the field is given by the action of $[D_{\mu}, D_{\nu}]$ on the field, multiplied by the area of the loop.

It is striking and significant that the parallel transport around an infinitesimal loop has resulted in a change determined by the action of a linear operator $[D_{\mu}, D_{\nu}]$ on the field ψ. This commutator is often given the name the **gauge field strength**, or

$$Q_{\mu\nu} = [D_{\mu}, D_{\nu}]. \tag{44.21}$$

Example 44.7

For our example of the complex scalar field, the gauge field strength is

$$Q_{\mu\nu} = [D_{\mu}, D_{\nu}] = [(\partial_{\mu} + iqA_{\mu}), (\partial_{\nu} + iqA_{\nu})]. \tag{44.22}$$

We obtain

$$[D_{\mu}, D_{\nu}] = iq(\partial_{\mu} A_{\nu} - \partial_{\nu} A_{\mu}) = iqF_{\mu\nu}, \tag{44.23}$$

where $F_{\mu\nu}$ are the components of the Faraday tensor $\tilde{\boldsymbol{F}}$.

We see from the last example that the commutator of the covariant derivative gives us access to the tensor $\tilde{\boldsymbol{F}}$ that tells us about the strength and dynamics of the gauge fields $\tilde{\boldsymbol{A}}$. In the fibre-bundle language, we can say that the $[D_{\mu}, D_{\nu}]$ is a **curvature operator** that tells us about the local strain in the bundle. This local strain is represented by the components of the curvature tensor $\tilde{\boldsymbol{F}}$.

What about the version for curved spacetime, rather than internal space? We saw in Chapter 35 that, when working in a coordinate system, parallel transport of a vector $\boldsymbol{Z}$ around an infinitesimal loop also resulted in the action of the commutator of the covariant derivative $\boldsymbol{\nabla}_{\mu}$. In direct analogy, we had the coordinate-frame expression

$$[\boldsymbol{\nabla}_{\mu}, \boldsymbol{\nabla}_{\nu}] Z^{\rho} = R^{\rho}{}_{\sigma\mu\nu} Z^{\sigma}. \tag{44.24}$$

It is clear that the Riemann tensor $\boldsymbol{R}$ is analogous to the Faraday tensor $\tilde{\boldsymbol{F}}$ in that it gives access to the local curvature. The difference is that the field strength $\boldsymbol{R}$ describes the curvature of spacetime, while the field strength $iq\tilde{\boldsymbol{F}}$ describes the curvature of the fibre bundle containing the internal variables. We summarize the analogy in the table below, where we also link in the terminology of the fibre bundle picture.

General relativity	Gauge theory	Fibre bundles
coordinate transformation	local phase transformation	strained bundle
connection coefficient $\Gamma^{\mu}{}_{\alpha\beta}$	gauge field A_{μ}	bundle connection
curvature $R^{\mu}{}_{\nu\alpha\beta}$	field strength $iqF_{\mu\nu}$	bundle curvature

In this chapter, we have shown one example of a gauge theory: the complex scalar field theory [also known as $U(1)$ theory]. In this case, the gauge field $\tilde{A}$ is exactly the gauge field that describes electromagnetism. There are many more examples of gauge field theories, of which the complex scalar field theory is the simplest. They all share the structure presented here with the field strength being provided by the commutator of the covariant derivative. In later chapters, we shall encounter gauges again. Most dramatically, we will see in Chapter 48 an attempt to add a dimension into spacetime in order to accommodate an internal parameter.

Chapter summary

- Internal variables carried by fields can be thought of as living in a fibre, suspended above a point in spacetime. The fibres are combined to make a fibre bundle.
- A field can be made invariant with respect to local changes of phase by introducing a gauge field $\tilde{A}(x)$. This leads to a covariant derivative $D_{\mu} = \partial_{\mu} + iqA_{\mu}$.
- Parallel transport of a field around a loop results in a contribution to the field proportional the action of the linear operator $[D_{\mu}, D_{\nu}] = iqF_{\mu\nu}$ on the field. This is analogous to the case of gravitation, where the Riemann tensor R takes the place of $\tilde{F}$.

Exercises

(44.1) *We saw in Chapter 43 how the Bianchi identity followed from the fact that the sum of curvature-induced rotations associated with the six faces of an elementary cube is zero. Following the approach of Ryder (1985), we can repeat this for a gauge field with field strength $[D_{\mu}, D_{\nu}] = -igG_{\mu\nu}$.*

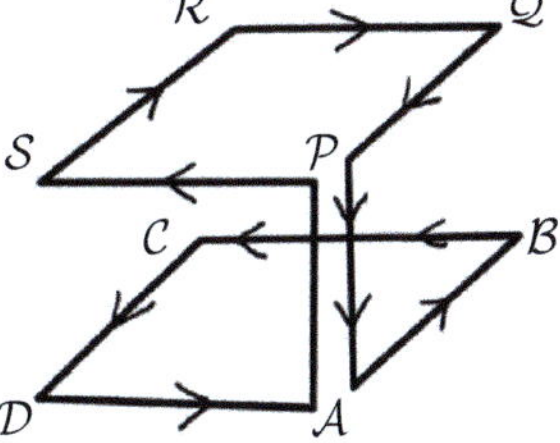

Fig. 44.5 A path around the faces of a cube.

Referring to Fig: 44.5, the path around the cube is

$$(\mathcal{ABCDAPSRQPA})$$
$$+(\mathcal{ADSPABQRCBA})$$
$$+(\mathcal{APQBADCRSDA}). \qquad (44.25)$$

We will evaluate the effect of traversing the first of these paths.

(a) Find the effect on a field ψ of traversing the path $\mathcal{ABCDA}$.

(b) Next, compute the effect of traversing the path $\mathcal{AP}$.

(c) Now the circuit $\mathcal{PSRQP}$.

(d) Then the return path $\mathcal{PA}$.

(e) Finally, combine all of the paths together to obtain the result to leading order in the interaction strength g.

(f) Show that if the procedure is repeated around each of the three paths, the result is that the field changes by a factor

$$1 - ig\Delta V^{\rho\mu\nu}(D_\rho G_{\mu\nu} + D_\mu G_{\nu\rho} + D_\nu G_{\rho\mu}), \quad (44.26)$$

where $V^{\rho\mu\nu}$ is the volume of the cube.

(g) Show that the previous expression is consistent with the Jacobi identity

$$[D_\rho,[D_\mu,D_\nu]]+[D_\mu,[D_\nu,D_\rho]]+[D_\nu,[D_\rho,D_\mu]] = 0. \qquad (44.27)$$

(44.2) Compute the energy-momentum tensor for complex scalar field theory in flat space.

Weak gravitational fields

45

We must touch his weaknesses with a delicate hand. There are some faults so nearly allied to excellence, that we scarce weed out the fault without eradicating the virtue.
Oliver Goldsmith (1728–1774) *The Good-Natured Man*

We have previously met applications of gravitation in circumstances in which gravity is at its strongest. This is exemplified by the physics of the black hole, where the extreme gravitational effects lead to a singularity. In this chapter, we discuss the opposite limit: **weak gravitational fields**. The subject is particularly important since this is the limit we experience most often in our interactions with Nature. Here we examine how best to treat gravitation in the limit that its effects are weak, and we shall build an approximate field theory that is notable in that it hosts wave-like excitations. As a warm up, we start by examining how the Einstein equations map onto Newton's theory of gravity in the limit of weak fields.[1]

[1]Some of the arguments presented in this chapter were trailed back in Section 14.1.

45.1 The Newtonian limit

The **Newtonian limit** is the limit of velocities which are small compared to c (as measured by a particular[2] observer). This corresponds to a velocity vector $\boldsymbol{u}$, with components (u^0, u^1, u^2, u^3), for which the timelike component u^0 much larger than the spacelike components u^i. In other words, slow-moving objects have velocities with components written approximately as $(1, \vec{u})$ with[3] $u^i \ll 1$. In order that gravitational effects do not cause masses to attain high velocities, this limit also implies that the gravitational field is limited in strength. Since it is mass-energy that causes spacetime to curve, a limitation of the strength of gravitational curvature constrains the energy-momentum tensor. Specifically, we demand that in the Newtonian limit, all stresses T^{ij} must be small compared to the density of mass energy T^{00}, so that[4]

$$\frac{|T^{ij}|}{T^{00}} \ll 1. \tag{45.2}$$

This provides a working definition of what we mean by weak gravitation.

In the weak-field limit, spacetime is almost flat. Perfectly flat spacetime is described by the Minkowski metric $\boldsymbol{\eta}$ and so the weak field causes a small perturbation to this underlying geometry. The deviation from the near-flatness is encoded in a small additive correction to

[2]Not all observers will agree that a particular velocity is in the Newtonian limit.

[3]That is to say, of course, that $u^i \ll c$. Our choice of units obscures the identification of when relativistic effects should be expected to be large. The relevant factor for velocities is v^2/c^2, which approaches unity in the relativistic limit. Using some dimensional analysis we can replace v^2 with GM/r, which has the same dimensions, and so we conclude that when the factor

$$\varepsilon = \frac{GM}{c^2 r} \tag{45.1}$$

is of order 1, relativistic effects are important. When $\varepsilon \ll 1$, we are in the Newtonian limit.

[4]See Section 14.1 for further justification of this point.

the Minkowski metric, such that the components of the full metric $\boldsymbol{g}$ are written as

$$g_{\alpha\beta} = \eta_{\alpha\beta} + h_{\alpha\beta}, \qquad (45.3)$$

with $|h_{\alpha\beta}| \ll 1$. The small additional components $h_{\alpha\beta}$ will act as a symmetric tensor field. To see this, consider a Lorentz transformation of coordinates $x^\mu = \Lambda^\mu{}_{\nu'} x^{\nu'}$. Equation 45.3 is altered when we try to transform the components of $\boldsymbol{g}$ from a frame with primed indices into a frame with unprimed indices using the Lorentz transformations. We have

$$\begin{aligned} g_{\alpha'\beta'} = \eta_{\alpha'\beta'} + h_{\alpha'\beta'} &= \Lambda^\mu{}_{\alpha'}\Lambda^\nu{}_{\beta'}(\eta_{\mu\nu} + h_{\mu\nu}) \\ &= \eta_{\alpha'\beta'} + \Lambda^\mu{}_{\alpha'}\Lambda^\nu{}_{\beta'}h_{\mu\nu}. \end{aligned} \qquad (45.4)$$

We conclude that the corrections $h_{\mu\nu}$ transform like the components of a tensor in *flat spacetime* with

$$h_{\alpha'\beta'} = \Lambda^\mu{}_{\alpha'}\Lambda^\nu{}_{\beta'}h_{\mu\nu}. \qquad (45.5)$$

It is convenient, therefore, to conceptualize weak gravitational fields as existing in a flat spacetime with Minkowski metric $\boldsymbol{\eta}$, but filled with a symmetric tensor field[5] $\boldsymbol{h}(x)$.

It should be stressed though that this picture is a fiction! It is not possible within general relativity to exactly describe a curved spacetime as flat Minkowski space with an additional field defined on it, hence the reason for Einstein's geometrical theory in the first place. So we should keep in mind that the spacetime is actually curved and it is only an approximation to treat it as flat. Indeed, $\boldsymbol{h}$ only looks like a tensor when we assume flat spacetime, and if we were to be forced to consider coordinate transformations other than the Lorentz transformations, then $\boldsymbol{h}$ would not necessarily transform as a tensor. Nonetheless, this picture of a flat spacetime with a field $\boldsymbol{h}(x)$ overlaid is a very useful one that allows us to compute the effects of gravitation using $\boldsymbol{h}(x)$, while raising and lowering indices using the components of the Minkowski metric. In addition, in order to match the boundary condition that the gravitation field dies away at infinity [i.e. the potential $\Phi(r \to \infty) = 0$], we will also demand that $h(r \to \infty) = 0$.

We have said several times how unique gravitation is in its treatment of the metric field. The weak-field approximation has the great advantage that it gives us an approximate theory that closely resembles all other classical theories of fields (in flat spacetime) in which classical fields fill all of spacetime. So now, instead of gravitation determining the metric field, gravitation is described by the field $\boldsymbol{h}$ in flat spacetime, and it is the dynamics of this $\boldsymbol{h}$-field that we seek to describe. This is a really useful simplification. The next step in our argument is to find a linear version of the Einstein equation that involves the $\boldsymbol{h}$-field, rather than the Einstein tensor $\boldsymbol{G}$, and this linearized approximation to general relativity, which is valid in the weak-field limit, will also set us up for exploring the properties of gravitational waves.

[5]This allows gravitation to be treated in the same way as the other classical field theories, such as electromagnetism, using the tools from Chapter 40.

45.2 Linearized theory of gravitation

General relativity is a nonlinear field theory. This may be traced back to the products of the connection coefficients that feature in the expression for the Riemann tensor.[6] In the limit of small interaction strength, many nonlinear field theories can be linearized by dropping nonlinear terms rendered small by the small size of the interaction strength. This is what we shall do here.

In the last section, we saw that the components of the metric may be written as $g_{\alpha\beta} = \eta_{\alpha\beta} + h_{\alpha\beta}$, where we agree to raise and lower indices with $\eta^{\mu\nu}$ and $\eta_{\mu\nu}$, respectively. With this in mind, the connection coefficients are given by

$$
\begin{aligned}
\Gamma^{\mu}{}_{\alpha\beta} &= \frac{1}{2}\eta^{\mu\nu}\left(h_{\alpha\nu,\beta} + h_{\beta\nu,\alpha} - h_{\alpha\beta,\nu}\right) \\
&= \frac{1}{2}\left(h_{\alpha}{}^{\mu}{}_{,\beta} + h_{\beta}{}^{\mu}{}_{,\alpha} - h_{\alpha\beta}{}^{,\mu}\right),
\end{aligned}
\tag{45.6}
$$

where we have extended the comma notation to take in up indices as well as down ones. We'll use this expression for the connections to compute the components of the Riemann tensor. However, since we are seeking a linearized theory, we ignore terms that are products of the Γs, leaving us with a simplified expression for the components of $\boldsymbol{R}$ that reads

$$
R_{\alpha\mu\beta\nu} \approx \eta_{\alpha\lambda}\left(\Gamma^{\lambda}{}_{\mu\beta,\nu} - \Gamma^{\lambda}{}_{\beta\mu,\nu}\right).
\tag{45.7}
$$

Example 45.1

In terms of the metric components, we obtain (after some reordering)

$$
R_{\alpha\mu\beta\nu} = \frac{1}{2}(h_{\mu\beta,\nu\alpha} + h_{\nu\alpha,\mu\beta} - h_{\mu\nu,\alpha\beta} - h_{\alpha\beta,\mu\nu}).
\tag{45.8}
$$

At the same level of approximation, the Ricci tensor is given by

$$
\begin{aligned}
R_{\mu\nu} &= \Gamma^{\alpha}{}_{\nu\mu,\alpha} - \Gamma^{\alpha}{}_{\alpha\mu,\nu} \\
&= \frac{1}{2}\left(h_{\mu}{}^{\alpha}{}_{,\nu\alpha} + h_{\nu}{}^{\alpha}{}_{,\mu\alpha} - h_{\mu\nu,\alpha}{}^{\alpha} - h_{,\mu\nu}\right),
\end{aligned}
\tag{45.9}
$$

where, in the final term we have defined the trace $h = h^{\alpha}{}_{\alpha} = \eta^{\alpha\beta}h_{\alpha\beta}$.

The tools from the last example allow us to work out the Einstein equation $G_{\mu\nu} = R_{\mu\nu} - \frac{1}{2}g_{\mu\nu}R = 8\pi T_{\mu\nu}$. We obtain the rather unmemorable expression

$$
h_{\mu\alpha,\nu}{}^{\alpha} + h_{\nu\alpha,\mu}{}^{\alpha} - h_{\mu\nu,\alpha}{}^{\alpha} - h_{,\mu\nu} - \eta_{\mu\nu}\left(h_{\alpha\beta}{}^{,\alpha\beta} - h_{,\beta}{}^{\beta}\right) = 16\pi T_{\mu\nu}.
\tag{45.10}
$$

To simplify this, we define the **bar notation** for trace-reversed[7] quantities

$$
\bar{h}_{\mu\nu} = h_{\mu\nu} - \frac{1}{2}\eta_{\mu\nu}h.
\tag{45.13}
$$

---

[6] Recall that

$$
\begin{aligned}
R^{\alpha}{}_{\beta\mu\nu} =\,& \Gamma^{\alpha}{}_{\nu\beta,\mu} - \Gamma^{\alpha}{}_{\mu\beta,\nu} \\
&+ \Gamma^{\alpha}{}_{\mu\sigma}\Gamma^{\sigma}{}_{\nu\beta} - \Gamma^{\alpha}{}_{\nu\sigma}\Gamma^{\sigma}{}_{\mu\beta}.
\end{aligned}
$$

Terms three and four on the right-hand side lead to nonlinearity.

[7] Recall that the name 'trace-reversed' comes from the fact that

$$
\bar{h} = \bar{h}^{\alpha}{}_{\alpha} = -h.
\tag{45.11}
$$

This means that $G_{\mu\nu} = \bar{R}_{\mu\nu}$. Note that $\bar{\bar{h}}_{\mu\nu} = h_{\mu\nu}$, so that

$$
h_{\mu\nu} = \bar{h}_{\mu\nu} - \frac{1}{2}\eta_{\mu\nu}\bar{h}.
\tag{45.12}
$$

Using the trace-reversed fields we find the Einstein tensor becomes

$$G_{\mu\nu} = -\frac{1}{2}\left[\bar{h}_{\mu\nu,\alpha}{}^{\alpha} + \eta_{\mu\nu}\bar{h}_{\alpha\beta,}{}^{\alpha\beta} - \bar{h}_{\mu\alpha,}{}^{\alpha}{}_{\nu} - \bar{h}_{\nu\alpha,}{}^{\alpha}{}_{\mu}\right], \tag{45.14}$$

so that we have another formidable-looking expression

$$-\bar{h}_{\mu\nu,\alpha}{}^{\alpha} - \eta_{\mu\nu}\bar{h}_{\alpha\beta,}{}^{\alpha\beta} + \bar{h}_{\mu\alpha,}{}^{\alpha}{}_{\nu} + \bar{h}_{\nu\alpha,}{}^{\alpha}{}_{\mu} = 16\pi T_{\mu\nu}. \tag{45.15}$$

We have succeeded in writing down a linearized version of the Einstein equation in terms of the h-field. It has the disadvantage of looking rather complicated. However, a drastic simplification is possible, by exploiting the notion of gauge freedom discussed in Chapters 42 and 44.

45.3 Exploiting gauges

In simplifying the Einstein equation, it would be very helpful to us if we were allowed to write

$$\bar{h}^{\mu\alpha}{}_{,\alpha} = 0, \tag{45.16}$$

since this would kill off several terms in eqn 45.15. It turns out that this is a condition we can indeed fix. The reason for this stems from the **gauge invariance** properties of the Einstein equation.[8]

In order to see the gauge transformation properties, we allow the coordinates to transform via an **infinitesimal coordinate transformation**[9]

$$x^{\mu'}(\mathcal{P}) = x^{\mu}(\mathcal{P}) + \xi^{\mu}(\mathcal{P}), \tag{45.17}$$

where $\mathcal{P}$ is some point and the shifts $\xi^{\mu}(\mathcal{P})$ are small enough that the components of h in the primed frame obey $|h_{\mu'\nu'}| \ll 1$. This tiny transformation is designed to make only tiny changes to the smoothly varying fields that exist in the almost-flat spacetime. However, there is one field where they cannot be ignored: the metric. This is because it is exactly the very small variations in the metric that describe gravitation in this limit. Recall that metric components have the transformation law

$$g_{\rho'\sigma'}[x^{\alpha'}(\mathcal{P})] = g_{\mu\nu}[x^{\alpha}(\mathcal{P})]\frac{\partial x^{\mu}}{\partial x^{\rho'}}\frac{\partial x^{\nu}}{\partial x^{\sigma'}}. \tag{45.18}$$

Substitute from eqns 45.3 and 45.17, and we see that

$$g_{\rho'\sigma'}(x^{\alpha'} = a^{\alpha}) = \eta_{\rho'\sigma'} + h_{\rho'\sigma'}(x^{\alpha'} = a^{\alpha}) - \xi_{\rho',\sigma'} - \xi_{\sigma',\rho'} + (\text{corrections}). \tag{45.19}$$

---

[8] Recall the situation in electromagnetism where we are free to pick the components of the electromagnetic field 1-form $\tilde{A}(x)$. Specifically, we can always make the transformation $A_{\mu} \to A_{\mu} - \chi_{,\mu}$ without making any changes to Maxwell's equations. We can therefore pick the function χ in such a way as to simplify the problem at hand.

[9] Unlike the transformation in Chapter 44: $\theta(x) \to \theta(x) + \alpha(x)$, where we were changing an internal variable at each point in spacetime, we must label the underlying points in the manifold with labels like $\mathcal{P}$, since the spacetime coordinates themselves are being transformed. The choice of length scale in the coordinates implied by the choice of $\xi^{\mu}(\mathcal{P})$ in eqn 45.17 was called a 'gauge' by Hermann Weyl, where the use of the term is motivated by the 'gauge' of wires and rails.

Example 45.2

Let's check this last expression. We have

$$x^{\mu} = x^{\mu'} - \xi^{\mu}, \tag{45.20}$$

and so

$$\frac{\partial x^{\mu}}{\partial x^{\rho'}} = \delta^{\mu'}{}_{\rho'} - \xi^{\mu}{}_{,\rho'}, \tag{45.21}$$

giving the combination

$$\frac{\partial x^\mu}{\partial x^{\rho'}}\frac{\partial x^\nu}{\partial x^{\sigma'}} = \delta^\mu{}_{\rho'}\delta^\nu{}_{\sigma'} - \xi^\mu{}_{,\rho'}\delta^\nu{}_{\sigma'} - \xi^\nu{}_{,\sigma'}\delta^\mu{}_{\rho'} + \cdots, \tag{45.22}$$

where we ignore higher order terms in the derivatives of the transformation components ξ^μ. We then have a right-hand side of eqn 45.18 given by

$$\begin{aligned}(\eta_{\mu\nu} + h_{\mu\nu})\frac{\partial x^\mu}{\partial x^{\rho'}}\frac{\partial x^\nu}{\partial x^{\sigma'}} &= \eta_{\rho',\sigma'} + h_{\rho'\sigma'} - \xi^\mu{}_{,\rho'}\eta_{\mu\sigma'} - \xi^\nu{}_{,\sigma'}\eta_{\rho'\nu} + \cdots \\ &= \eta_{\rho',\sigma'} + h_{\rho'\sigma'} - \xi_{\sigma'\rho'} - \xi_{\rho'\sigma'} + \cdots \end{aligned} \tag{45.23}$$

where we've retained the lowest order terms only.

We conclude that the components of $\boldsymbol{h}$ in new and old coordinates are[10]

$$h^{\mathrm{new}}_{\mu\nu} = h^{\mathrm{old}}_{\mu\nu} - \xi_{\mu,\nu} - \xi_{\nu,\mu}. \tag{45.24}$$

All other fields in the linearized theory are not changed appreciably by our infinitesimal coordinate transformations. We sketch this property for the Riemann tensor in the next example.

Example 45.3

We need to show $R^{\mathrm{new}}_{\mu\nu\alpha\beta} = R^{\mathrm{old}}_{\mu\nu\alpha\beta}$. Under the gauge transformation we have

$$h^{\mathrm{new}}_{\mu\nu} = h^{\mathrm{old}}_{\mu\nu} - 2\xi_{(\mu,\nu)}, \tag{45.25}$$

where we've used the useful bracket notation for symmetrization. We find

$$R^{\mathrm{new}}_{\alpha\mu\beta\nu} = R^{\mathrm{old}}_{\alpha\mu\beta\nu} - \xi_{\alpha[,\nu,\beta],\mu} + \xi_{\mu[,\nu,\beta],\alpha} - \xi_{\beta[,\mu,\alpha,],\nu} + \xi_{\nu[,\mu,\alpha],\beta}, \tag{45.26}$$

where here the bracket notation for antisymmetrization has been used. The added terms are all zero (since derivatives commute) and the Riemann tensor is left unchanged.

We are therefore free to make infinitesimal coordinate transformations without changing the physics. Using the analogy from Chapter 44, we note that our choice of coordinates in this context is the same as a choice of gauge. We therefore choose coordinates such that

$$\bar{h}^{\mu\alpha}{}_{,\alpha} = 0. \tag{45.27}$$

This is analogous to **Lorenz gauge** in electromagnetism, which is the condition $A^\mu{}_{,\mu} = 0$. This gauge cause the Einstein equation to reduce to the far friendlier expression

$$-\bar{h}_{\mu\nu,\alpha}{}^\alpha = 16\pi T_{\mu\nu}. \tag{45.28}$$

Written in another way, we have[11]

$$-\partial^2\bar{h}_{\mu\nu} = 16\pi T_{\mu\nu}, \tag{45.29}$$

where $\bar{h}_{\mu\nu} = h_{\mu\nu} - \frac{1}{2}\eta_{\mu\nu}h$.

We have therefore derived a linear field theory of gravitation. It has the form of a wave equation[12] in the absence of the **source term** $16\pi T_{\mu\nu}$.

[10]We can compare the situation in gravitation, where a coordinate change $x^{\mu'}(\mathcal{P}) \to x^\mu(\mathcal{P}) + \xi^\mu(\mathcal{P})$ mandates a change in the metric field components

$$h_{\mu\nu} \to h_{\mu\nu} - \xi_{\mu,\nu} - \xi_{\nu,\mu},$$

with that of electromagnetism, where a change in an internal variable $\theta(\mathcal{P}) \to \theta(\mathcal{P}) + \alpha(\mathcal{P})$ mandates a change in the electromagnetic field components

$$A_\mu \to A_\mu - \frac{1}{q}\alpha_{,\mu},$$

where q is the electromagnetic charge.

[11]We use the notation

$$\begin{aligned}\partial^2 &= -\frac{\partial^2}{\partial t^2} + \frac{\partial^2}{\partial x^2} + \frac{\partial^2}{\partial y^2} + \frac{\partial^2}{\partial z^2} \\ &= -\frac{\partial^2}{\partial t^2} + \vec{\nabla}^2.\end{aligned}$$

[12]Reminder: A wave equation has the form $\partial^2\phi = -\frac{\partial^2\phi}{\partial t^2} + \vec{\nabla}^2\phi = 0$. Its solutions are plane waves of the form $\phi = Ae^{-\mathrm{i}(\omega t - \vec{k}\cdot\vec{x})}$.

The analogous expression in electromagnetism is $-\partial^2 A^\mu = J^\mu$. The (retarded) potential in electromagnetism that solves this latter wave equation is

$$A^\mu(t, \vec{x}) = \frac{1}{4\pi} \int d^3 y \frac{J^\mu(t - |\vec{x} - \vec{y}|, \vec{y})}{|\vec{x} - \vec{y}|}. \qquad (45.30)$$

This suggests that the solution to the weak-field gravitation equation should be given by the retarded function

$$\bar{h}_{\mu\nu}(t, \vec{x}) = 4G \int d^3 y \frac{T_{\mu\nu}(t - |\vec{x} - \vec{y}|, \vec{y})}{|\vec{x} - \vec{y}|}, \qquad (45.31)$$

where the factor of G has been restored. We shall examine this further in the next two chapters.

Example 45.4

Let's check out eqn 45.31 in the limit of a non-relativistic stationary source (i.e. a static distribution of mass). This problem has no time dependence and so eqn 45.31 then reduces to

$$\bar{h}_{\mu\nu}(\vec{x}) = 4G \int d^3 y \frac{T_{\mu\nu}(\vec{y})}{|\vec{x} - \vec{y}|}, \qquad (45.32)$$

and in the weak-field limit the only significant term in $T_{\mu\nu}(\vec{y})$ is $T_{00}(\vec{y}) \approx \rho(\vec{y})$, where $\rho(\vec{y})$ is the density distribution of the source. Hence, we only have to worry about

$$\bar{h}_{00}(\vec{x}) = 4G \int d^3 y \frac{\rho(\vec{y})}{|\vec{x} - \vec{y}|}. \qquad (45.33)$$

However, this equation reminds us of the equation for the gravitational scalar potential $\Phi(\vec{x})$ which is

$$\Phi(\vec{x}) = -G \int d^3 y \frac{\rho(\vec{y})}{|\vec{x} - \vec{y}|}, \qquad (45.34)$$

from which we immediately deduce that the only non-negligible term in $\bar{h}_{\mu\nu}$ is

$$\bar{h}_{00}(\vec{x}) = -4\Phi(\vec{x}). \qquad (45.35)$$

This is exactly the same conclusion we came to in Example 14.2, and following the same argument we can quickly deduce that

$$h_{00} = h_{11} = h_{22} = h_{33} = -2\Phi, \qquad (45.36)$$

so that the metric can be written as

$$ds^2 = -(1 + 2\Phi)dt^2 + (1 - 2\Phi)(dx^2 + dy^2 + dz^2), \qquad (45.37)$$

which is, of course, the weak-field metric (eqn 5.22, derived in Section 14.1).

Example 45.5

The energy-momentum tensor for dust takes the form $T^{\mu\nu} = \rho u^\mu u^\nu$, and so we only keep the dominant term $T^{00} = \rho$ in the weak-field limit. The next-most important terms are $T^{0i} = \rho u^i$ and these terms are smaller than T^{00} by a factor[13] v/c. The terms $T^{ij} = \rho u^i u^j$ are smaller than T^{00} by a factor of $(v/c)^2$ and so can be safely ignored, but now let's consider the effect of the T^{0i} terms. We can now relate $\bar{h}_{0i}$ to a gravitational analogue of the vector potential $\vec{A}_{\mathrm{g}}$ by writing

$$A_{\mathrm{g}}^i(\vec{x}) = -G \int d^3 y \frac{J^i(\vec{y})}{|\vec{x} - \vec{y}|}, \qquad (45.38)$$

---

[13]Recall that, inserting the factors of c, that $T^{00} = \rho c^2$ and $T^{0i} = \rho c u^i$.

where $J^i(\vec{y}) = \rho(\vec{y})u^i(\vec{y})$ is the current density, in which case we deduce that (on lowering indices, and using $h_{0i} = \bar{h}_{0i}$)

$$\bar{h}_{0i}(\vec{x}) = -4A_{\mathrm{g}i}(\vec{x}). \tag{45.39}$$

Now, using our linearized equation of gravitation (eqn 45.29), and ignoring the smaller time-dependent term, we have

$$\nabla^2\Phi = 4\pi G\rho \quad \text{and} \quad \nabla^2\vec{A}_{\mathrm{g}} = 4\pi G\vec{J}, \tag{45.40}$$

which are the Poisson's equations for the gravitational scalar and vector potential. These are, of course, reminiscent of the analogous equations for electromagnetism, namely

$$\nabla^2 V = -\frac{\rho}{\epsilon_0} \quad \text{and} \quad \nabla^2\vec{A} = -\mu_0\vec{J}, \tag{45.41}$$

where the minus signs reflect the fact that like charges repel in electromagnetism but masses attract in gravity. Pursuing this analogy, we can define the **gravitoelectric field** $\vec{E}_{\mathrm{g}} = -\vec{\nabla}\Phi$ and the **gravitomagnetic field** $\vec{B}_{\mathrm{g}} = \vec{\nabla} \times \vec{A}_{\mathrm{g}}$ and, by analogy with electromagnetism, these will satisfy the **gravitational Maxwell equations**

$$\vec{\nabla} \cdot \vec{E}_{\mathrm{g}} = -4\pi G\rho, \quad \vec{\nabla} \cdot \vec{B}_{\mathrm{g}} = 0, \quad \vec{\nabla} \times \vec{E}_{\mathrm{g}} = 0, \quad \vec{\nabla} \times \vec{B}_{\mathrm{g}} = -4\pi G\vec{J}. \tag{45.42}$$

These are analogous to the electromagnetic Maxwell equations ignoring any time-dependence. This analogy between gravity and electromagnetism can be taken further and it can be shown that, just as a charge current in electromagnetism creates a magnetic field that exerts at $\vec{v} \times \vec{B}$ force, so a current of mass creates a gravitomagnetic field that exerts a $\vec{v} \times \vec{B}_{\mathrm{g}}$ force. This is the basis of the Lense–Thirring effect in which a spinning mass (in which you have rotational currents of mass) will drag objects around with the rotation.[14]

[14]See Exercise 45.9 for more details.

Weak-field gravitation is a very useful limit for a lot of the Universe we encounter for which gravity is a relatively small perturbation. After all, most of the Universe is pretty empty of matter. In the next chapter, we will solve eqn 45.29 in matter-free spacetime, so that the energy-momentum tensor $\boldsymbol{T} = 0$ and we then have $-\partial^2\bar{h}_{\mu\nu} = \left(\frac{\partial^2}{\partial t^2} - \vec{\nabla}^2\right)\bar{h}_{\mu\nu} = 0$. This is a wave equation and predicts the existence of gravitational waves!

Chapter summary

- Weak-field gravitation can be approximately described using a Minkowski metric with a small additive correction given by the weak-field metric $\boldsymbol{h}(x)$.

- In terms of the weak-field metric, the Einstein equation is

$$-\partial^2\bar{h}_{\mu\nu} = 16\pi T_{\mu\nu}, \tag{45.43}$$

where $\bar{h}_{\mu\nu} = h_{\mu\nu} - \frac{1}{2}\eta_{\mu\nu}h$. This is a wave equation in the absence of sources of mass-energy.

- Under an infinitesimal coordinate change $x^{\mu'} \to x^\mu + \xi^\mu$, we have

$$h_{\mu\nu} \to h_{\mu\nu} - 2\xi_{(\mu,\nu)}. \tag{45.44}$$

Exercises

(45.1) Verify (a) eqn 45.10;
(b) eqn 45.14;
(c) eqn 45.26.

(45.2) (a) Take a trace to show that

$$\partial^2 h = 16\pi T, \qquad (45.45)$$

where h is the trace of the tensor $\boldsymbol{h}$.
(b) Show that the weak-field, linearized Einstein equation can be rewritten as

$$-\partial^2 h_{\mu\nu} = 16\pi\left(T_{\mu\nu} - \frac{1}{2}\eta_{\mu\nu}T\right). \qquad (45.46)$$

(c) For a stationary source, explain why the solution to the Einstein equation

$$\vec{\nabla}^2 h_{\mu\nu} = -16\pi\left(T_{\mu\nu} - \frac{1}{2}\eta_{\mu\nu}T\right), \qquad (45.47)$$

take the form

$$h_{\mu\nu}(\vec{x}) = 2\int d\Sigma\, \frac{(2T_{\mu\nu} - \eta_{\mu\nu}T)}{|\vec{x} - \vec{x}'|}, \qquad (45.48)$$

where $d\Sigma$ is an element of 3-volume and the energy-momentum tensor is evaluated at position $\vec{x}'$.

(45.3) For a low-velocity perfect fluid in the weak-field limit, show the following:
(a) The **gravitoelectric energy density**, defined as $\rho_g = 2T_{tt} - \eta_{tt}T$, is given by

$$\rho_g \approx \rho + 3p. \qquad (45.49)$$

(b) The **gravitometric current density**, $\Pi_i = -T_{ti} + \frac{1}{2}\eta_{ti}T$, is given by

$$\Pi_i \approx (\rho + p)u_i, \qquad (45.50)$$

where u_i are velocity 1-form components.
(c) The **curvature energy density**, $\rho_c = 2T_{ii} - \eta_{ii}T$, is given by

$$\rho_c \approx \rho - p. \qquad (45.51)$$

This curvature energy density is the quantity in eqn 45.48 that causes the diagonal, spatial parts of $\boldsymbol{h}$ to be non-zero. These are the parts of $\boldsymbol{g}$ that give the deviations from a flat-spacetime metric.

(45.4) (a) Show that a particle moving in the weak, stationary field obeys the geodesic equation

$$\frac{d^2 x^i}{dt^2} \approx \eta^{ij}\left(-\partial_j\Phi + F_{jk}v^k\right), \qquad (45.52)$$

where $v^k = dx^k/dt$ and

$$\Phi = -h_{tt}/2,$$
$$F_{ij} = \partial_i h_{tj} - \partial_j h_{ti}. \qquad (45.53)$$

(b) Show that the source of Φ is given by evaluating

$$\Phi(\vec{x}) = -\int d^3x'\, \frac{\rho_g(\vec{x}')}{|\vec{x} - \vec{x}'|}, \qquad (45.54)$$

where ρ_g was defined in the previous question.

(45.5) (a) Show that in a Universe filled with only vacuum energy, the gravitoelectric energy density is negative, making gravity repulsive.
(b) Show further that in terms of the potential Φ, we obtain an equation $\vec{\nabla}^2\Phi = -\Lambda$, which is solved by a potential $\Phi = \alpha r^2$, where α is a constant that should be determined.

(45.6) *We shall demonstrate the precession of angular momentum in a gravitational field. We follow the approach of Ryder (2009).*
In the weak-field limit, the geometry outside a rotating object leads to a metric with components

$$g_{\mu\nu} = \begin{pmatrix} -(1 + 2\phi) & \zeta_1 & \zeta_2 & \zeta_3 \\ \zeta_1 & 1 - 2\phi & 0 & 0 \\ \zeta_2 & 0 & 1 - 2\phi & 0 \\ \zeta_3 & 0 & 0 & 1 - 2\phi \end{pmatrix},$$

$$(45.55)$$

where

$$\phi = -\frac{M}{r},$$
$$\zeta_i = \frac{2}{r^3}\varepsilon_{ijk}x^j J^k, \qquad (45.56)$$

where J^i are components of the angular momentum of the source, which has mass M.
(a) Assuming that, for application to the Earth, the quantities ϕ^2 and $\phi\zeta_i$ and $\zeta_i\zeta_j$ can be ignored, show that the components of $g^{\mu\nu}$ are the same as

those of $g_{\mu\nu}$ with the sign in front of ϕ reversed.
(b) Show that, to leading order, the connection coefficients are

$$\Gamma^0{}_{i0} = \nabla_i \phi,$$

$$\Gamma^k{}_{i0} = \frac{1}{2}\left(\zeta_{k,i} - \zeta_{i,k}\right),$$

$$\Gamma^k{}_{im} = \delta_{im}\eta^{lk}\nabla_l\phi - \delta_i^k\nabla_m\phi - \delta_m^k\nabla_i\phi,$$

$$\Gamma^0{}_{im} = -\frac{1}{2}\left(\zeta_{i,m} + \zeta_{m,i}\right). \tag{45.57}$$

(45.7) Define a (3+1)-dimensional **spin** by a 1-form with components S_μ which is orthogonal to the velocity.
(a) Show that $S_0 = -(\mathrm{d}x^j/\mathrm{d}t)S_i$.
(b) If such a spin is parallel transported along a geodesic, show further that

$$\frac{\mathrm{d}S_i}{\mathrm{d}t} = \left(-\Gamma^0{}_{0i}\frac{\mathrm{d}x^k}{\mathrm{d}t} - \Gamma^0{}_{ji}\frac{\mathrm{d}x^j}{\mathrm{d}t}\frac{\mathrm{d}x^k}{\mathrm{d}t}\right.$$

$$\left. + \Gamma^k{}_{0i} + \Gamma^k{}_{ji}\frac{\mathrm{d}x^j}{\mathrm{d}t}\right) S_k. \tag{45.58}$$

(c) Using the result of Exercise 45.6, show that the evolution of the spin in orbit around the Earth follows the equation of motion

$$\frac{\mathrm{d}S_i}{\mathrm{d}t} = [-(\nabla_i\phi)v^k + \frac{1}{2}(\zeta_{k,i} - \zeta_{i,k}) + (\delta_{im}\eta^{lk}\nabla_l\phi$$

$$- \delta_{ki}\nabla_m\phi - \delta_{km}\nabla\phi)v^m]S_k. \tag{45.59}$$

(45.8) Using the metric in Exercise 45.6, show that to order $v^2\vec{S}^2$ or $\phi\vec{S}^2$, the quantity

$$\vec{S}^2 + 2\phi\vec{S}^2 - (\vec{v}\cdot\vec{S})^2. \tag{45.60}$$

is a constant when parallel transported.

(45.9) To see the consequences of Exercise 45.7, we define a slightly updated version of the spin vector through the 3-vector equation

$$\vec{S} = (1-\phi)\vec{\Sigma} + \frac{1}{2}(\vec{v}\cdot\vec{\Sigma})\vec{v}. \tag{45.61}$$

Working to order $\vec{v}^2\vec{S}^2$ and $\phi\vec{S}^2$, and using the fact that the spin 4-vector is parallel transported, show that
(a) $\phi\vec{S}^2 = \phi\vec{\Sigma}^2$.
(b) $(\vec{v}\cdot\vec{S})^2 = (\vec{v}\cdot\vec{\Sigma})^2$,
and hence that
(c) $\vec{\Sigma}^2 = \text{const.}$

This new 3-vector has a constant length and so is suitable for demonstrating precession.
(d) Explain why

$$\frac{\mathrm{d}\phi}{\mathrm{d}t} = \vec{\nabla}\phi\cdot\vec{v}, \tag{45.62}$$

and $\mathrm{d}\vec{v}/\mathrm{d}t = -\vec{\nabla}\phi$.
At the level of approximation we're using, we have

$$\vec{\Sigma} = (1+\phi)\vec{S} - \frac{1}{2}(\vec{v}\cdot\vec{S})\vec{v}. \tag{45.63}$$

(e) Use the results from part (d) to show that, if we ignore terms of order $v^2\mathrm{d}\vec{S}/\mathrm{d}t$ or $\phi\mathrm{d}\vec{S}/\mathrm{d}t$, we have

$$\frac{\mathrm{d}\vec{\Sigma}}{\mathrm{d}t} = \frac{\mathrm{d}\vec{S}}{\mathrm{d}t} + (\vec{\nabla}\phi\cdot\vec{v})\vec{S} + \frac{1}{2}(\vec{v}\cdot\vec{S})\vec{\nabla}\phi + \frac{1}{2}(\vec{\nabla}\phi\cdot\vec{S})\vec{v}. \tag{45.64}$$

(f) Using the result of Exercise 45.7, show finally that

$$\frac{\mathrm{d}\vec{\Sigma}}{\mathrm{d}t} = \vec{\Omega}\times\vec{\Sigma}, \tag{45.65}$$

where

$$\vec{\Omega} = -\frac{1}{2}\vec{\nabla}\times\vec{\zeta} - \frac{3}{2}\vec{v}\times\vec{\nabla}\phi, \tag{45.66}$$

Hint: It is permissible to swap $\vec{S}$ for $\vec{\Sigma}$ in the final step to obtain the quoted equation of motion.
This is the equation of motion for the precession of a vector $\vec{\Sigma}$ around the direction of $\vec{\Omega}$, and implies that a gyroscope in orbit around the Earth will precess.
(g) Using $\phi = -M/r$ and $\zeta_i = (2/r^3)\varepsilon_{ikm}x^k J^m$, show that

$$\vec{\Omega} = \frac{1}{r^3}\left[\frac{3(\vec{J}\cdot\vec{r})\vec{r}}{r^2} - \vec{J}\right] + \frac{3M}{2r^3}\vec{r}\times\vec{v}. \tag{45.67}$$

*The first term in this equation depends on the angular momentum of the Earth $\vec{J}$ and gives rise to the **Lense–Thirring effect**. The second term gives the **de Sitter–Fokker effect**, usually called **geodetic motion**. Geodetic motion is a precession effect caused by the motion of the gyroscope around a geodesic and implies that on completing an orbit of a non-rotating mass, a gyroscope will precess. The Lense–Thirring precession is more remarkable as it arises purely from the rotation of the Earth. Notice how the field $\vec{\Omega}$ causing the precession has the form of an electromagnetic dipole.*

46

Gravitational waves

[1]LIGO stands for Laser Interferometer Gravitational-Wave Observatory.

Like as the waves make towards the pebbled shore,
So do our minutes hasten to their end
William Shakespeare (1564–1616) *Sonnet 60*

In this chapter, we discuss the waves that can propagate as excitations of the gravitational field. These waves were predicted by Einstein in 1906 on the basis of general relativity (they had been suggested previously by Henri Poincaré) and have been the subject of two relatively recent Nobel prizes: the 1993 award to Hulse and Taylor whose work on binary pulsars offered indirect evidence for the waves, and the 2017 prize for Weiss, Thorne and Barish. The latter was awarded in the wake of the direct observation (by the LIGO[1] collaboration) of gravitational waves that resulted from the merger of two black holes. We describe the LIGO experiment at the end of the chapter. We start, however, with a review of electromagnetic waves, before repeating the argument for waves in a weak gravitational field.

46.1 Waves in a gauge theory

We saw in Chapter 42 that, in flat spacetime with no sources, an equation of motion for the electromagnetic field can be written as[2]

$$-\partial_\nu(\partial_\mu A^\mu) + \partial^2 A_\nu = 0. \tag{46.1}$$

[2]In the presence of sources, the equation is

$$-\partial_\nu(\partial_\mu A^\mu) + \partial^2 A_\nu = -J_\nu.$$

This equation tells us that the electromagnetic field has dynamics of its own, independent of the presence of electric charges. These dynamics are wave-like excitations of the field.

As usual, we're free to make changes to $\tilde{A}$ subject to $A_\mu(x) \to A_\mu(x) - \partial_\mu\chi(x)$ since such changes in gauge do not alter the dynamics of the electromagnetic field, nor the things coupled to that field.

Example 46.1

As discussed in Chapter 42, we choose the Lorenz gauge, which means we define a new, but physically equivalent, gauge field with components A'_μ which obey the constraint $\partial_\mu A'^\mu = 0$, helpfully knocking out the first term in eqn 46.1. This leaves us with a simpler looking equation of motion

$$\partial^2 A'^\mu = 0, \tag{46.2}$$

whose solutions are plane waves of the form[3]

$$A^\mu = \mathrm{Re}\left[\epsilon^\mu(\boldsymbol{k})\mathrm{e}^{\mathrm{i}\boldsymbol{k}\cdot\boldsymbol{x}}\right], \tag{46.3}$$

where $\omega = |\vec{k}|$ and $\mathrm{Re}[\ ]$ reminds us to take the real part of a complex expression. However, we also saw in Chapter 42 that since the Lorenz gauge still leaves some freedom,[4] we could then impose the Coulomb gauge, upgrading to a new gauge field with components A''_μ where $A''_0 = 0$. With this further choice, the Lorenz condition then becomes $\vec{\nabla} \cdot \vec{A}'' = 0$, which further reduces the number of independent field components by one. This makes it clear that although the electromagnetic field has four components, the physics allows only *two* independent components.

The equations of motion[5] in the Lorenz gauge read $\partial^2 A^\mu = 0$, which, with $A^0 = 0$, has plane wave solutions $\vec{A} = \vec{\epsilon}\mathrm{e}^{\mathrm{i}\boldsymbol{k}\cdot\boldsymbol{x}}$. The equation encoding the Coulomb gauge condition, $\vec{\nabla} \cdot \vec{A} = 0$, leads to

$$\vec{k} \cdot \vec{A} = \vec{k} \cdot \vec{\epsilon} = 0, \tag{46.4}$$

which tells us that the direction of propagation of the wave is perpendicular to the polarization $\vec{\epsilon}$, i.e. the wave is *transverse*. For a wave propagating along z with null momentum components $k^\mu = (|\vec{k}|, 0, 0, |\vec{k}|)$, the components of the electromagnetic field must then be functions of the form

$$A^t = 0, \quad A^x = A^x(-\omega t + |\vec{k}|z), \quad A^y = A^y(-\omega t + |\vec{k}|z), \quad A^z = 0. \tag{46.5}$$

Comparing our plane wave, we spot solutions such as $A^j = \epsilon^j(\boldsymbol{k})\mathrm{e}^{-\mathrm{i}(\omega t - |\vec{k}|z)}$, for $j = x$ and y, with[6] $\omega/|\vec{k}| = 1$. Putting everything together, we see that two possible choices of basis polarization vectors are simply

$$\vec{\epsilon}_1(\boldsymbol{k}) = \begin{pmatrix} 1 \\ 0 \\ 0 \end{pmatrix}, \quad \vec{\epsilon}_2(\boldsymbol{k}) = \begin{pmatrix} 0 \\ 1 \\ 0 \end{pmatrix}, \tag{46.6}$$

corresponding to linear polarization along x or y, respectively. This explains, in classical terms, the propagation of light waves in the electromagnetic field.[7]

Now we repeat the argument for gravitation in the weak-field limit. Using the gravitational version of the Lorenz gauge $\partial^\nu \bar{h}_{\mu\nu} = 0$, we have a wave equation for gravity in the absence of sources, that says[8]

$$\partial^2 \bar{h}_{\mu\nu} = 0. \tag{46.8}$$

The good news is that this looks a lot like the wave equation for the electromagnetic field. It does however involve a second index, meaning that the polarizations of the fields that solve this wave equation have to be represented as square matrices, rather than simply as column vectors, as we had in the case of the electromagnetic field.

Let's consider an ansatz in the form of a plane gravitational wave. This can be written as

$$\bar{h}_{\mu\nu} = \mathrm{Re}\left[A_{\mu\nu}\mathrm{e}^{\mathrm{i}\boldsymbol{k}\cdot\boldsymbol{x}}\right], \tag{46.9}$$

where $\mathrm{Re}[\ ]$ again reminds us that we must take the real part of this plane wave solution to describe the physical amplitude of the wave.

Example 46.2

For this to work within the Lorenz gauge we require $\partial^\nu \bar{h}_{\mu\nu} = 0$, or

$$\bar{h}_{\mu\nu}{}^{,\nu} = \mathrm{i}k^\nu A_{\mu\nu}\mathrm{e}^{\mathrm{i}\boldsymbol{k}\cdot\boldsymbol{x}} = 0. \tag{46.10}$$

To satisfy the field equations we must also $\partial^2 \bar{h}_{\mu\nu} = 0$ (eqn 46.8), so that

$$-k_\sigma k^\sigma A_{\mu\nu}\mathrm{e}^{\mathrm{i}k\cdot x} = 0. \tag{46.11}$$

---

[3]The term $k_\mu x^\mu$ is written here as $\boldsymbol{k}\cdot\boldsymbol{x}$. We also drop the prime on the field A'^μ.

[4]Recall that this is because we can make a further shift $A'_\mu \to A''_\mu = A'_\mu - \partial_\mu\xi$ as long as $\partial^2\xi = 0$ (so that both A'_μ and A''_μ satisfy the Lorenz condition).

[5]We are now going to focus on A''_μ, so will drop the double prime from now on. We also suppress the $\mathrm{Re}[\ ]$ notation, which is assumed for the wave solutions.

[6]Of course, this implies that in SI units $\omega/|\vec{k}| = c$.

[7]The polarization vectors introduced here carry the information about the spin state of the photon. Further aspects of the quantum-mechanical treatment are the topic of the following chapter.

[8]Remember that when sources are present, the key equation is

$$-\partial^2 \bar{h}_{\mu\nu} = 16\pi T_{\mu\nu}. \tag{46.7}$$

We will return to this later when we put the sources back in, see eqn 46.39.

As a consequence of the last example, we have that the amplitude $A_{\mu\nu}$ and wavevector k^{μ} components obey two constraints

$$A_{\alpha\mu}k^{\mu} = 0, \quad \boldsymbol{k} \cdot \boldsymbol{k} = 0. \tag{46.12}$$

We conclude from the second expression that the gravitational wave is null, implying that the gravitational field propagates at the speed of light.[9] From the first condition we have that the wave's amplitude is orthogonal to its direction, making it a transverse plane wave. Analogous to the electromagnetic plane wave travelling along the z-direction, we have that the components of the gravitational field are given by functions

$$\bar{h}_{xx} = \bar{h}_{xx}(-\omega t + |k|z), \quad \bar{h}_{xy} = \bar{h}_{yx} = \bar{h}_{xy}(-\omega t + |k|z),$$
$$\bar{h}_{yy} = \bar{h}_{yy}(-\omega t + |k|z), \quad \bar{h}_{\mu z} = 0 \quad \text{for all } \mu. \tag{46.15}$$

Also by analogy with the plane wave, we would like to make a further choice of gauge to guarantee that $\bar{h}^{\mu 0} = 0$. It turns out that we can do exactly this, as we shall see in the next section.

46.2 Lorenz gauge for gravitational waves

We have already chosen the Lorenz gauge to guarantee the wave equation but, just as in the electromagnetic case, this doesn't exhaust the gauge freedom. Remembering that the gauge transformation we are considering is $x^{\mu} \to x^{\mu} + \xi^{\mu}$, we note by analogy with electromagnetism that we can still obey Lorenz gauge as long as we have[10]

$$\partial^2 \xi_{\alpha} = 0. \tag{46.16}$$

Gravitational waves are a little more complicated than their electromagnetic cousins, so we shall proceed step by step.

Example 46.3

We choose $\xi_{\alpha} = B_{\alpha}e^{i\boldsymbol{k}\cdot\boldsymbol{x}}$, so that the transformation is oscillatory in spacetime with amplitude B_{α}. The usual change in $\boldsymbol{h}$ resulting from a gauge transformation is given by

$$h'_{\alpha\beta} = h_{\alpha\beta} - \xi_{\alpha,\beta} - \xi_{\beta,\alpha}, \tag{46.17}$$

which means, for the trace-reversed components, that

$$\bar{h}'_{\alpha\beta} = \bar{h}_{\alpha\beta} - \xi_{\alpha,\beta} - \xi_{\beta,\alpha} + \eta_{\alpha\beta}\xi^{\mu}{}_{,\mu}. \tag{46.18}$$

Substituting the gauge choice gives a condition on the solution that

$$A'_{\alpha\beta} = A_{\alpha\beta} - i(B_{\alpha}k_{\beta} + B_{\beta}k_{\alpha} - \eta_{\alpha\beta}B^{\mu}k_{\mu}). \tag{46.19}$$

This satisfies our previous constraint $k^{\alpha}A'_{\alpha\beta} = 0$, as long as $A_{\alpha\beta}$ does too.

The amplitude of the transformation B_{α} is then chosen in such a way as to impose two additional (highly simplifying) constraints on the amplitude components $A_{\mu\nu}$ of our wave-like solution. These are

$$A^{\alpha}{}_{\alpha} = 0, \quad A_{\alpha\beta}u^{\beta} = 0, \tag{46.20}$$

[9]The following argument will make this statement a little more convincing. Let's consider a gravitational plane wave in $\bar{h}_{\mu\nu}$, which are constant on a surface on which its phase $\boldsymbol{k} \cdot \boldsymbol{x} = k_{\mu}x^{\mu}$ is constant. A photon moving in the direction of the null vector $\boldsymbol{k}$ travels on the curve

$$x^{\mu}(\lambda) = k^{\mu}\lambda + l^{\mu}, \tag{46.13}$$

where l^{μ} are the components of a constant vector and λ parametrizes the curve. Dotting the equation for the curve with k_{μ} and noting $\boldsymbol{k} \cdot \boldsymbol{k} = 0$, we find

$$k_{\mu}x^{\mu} = k_{\mu}l^{\mu} = \text{const.} \tag{46.14}$$

This implies that the photon wave and gravitational wave share the same surfaces on which their respective phases are constant, and in fact their respective phases can only differ by a constant scalar value. Thus, the two waves move essentially in lockstep. We can therefore conclude that the gravitational wave travels at the speed of light with $\vec{k}$ giving its direction of travel.

[10]The analogous expression for electromagnetism was $\partial^2 \xi = 0$. As you might expect, the gravitational case looks very similar but has an extra index.

where, in the second expression, u^β are the components of a fixed velocity. The first condition tells us that the wave is traceless, which means that, in this gauge, $h_{\mu\nu} = \bar{h}_{\mu\nu}$. The second says that the wave is orthogonal to a velocity vector $\boldsymbol{u}$. This state of affairs is known as **transverse-traceless gauge.**

Example 46.4

Choose a local inertial frame in which $\boldsymbol{u}$ has components $u^\mu = (1,0,0,0)$. We then have from $A_{\alpha\beta}u^\beta = 0$ (eqn 46.20) the condition that we wanted, that $A_{\alpha 0} = 0$. We arrange for the wave to travel along z, so we have $\boldsymbol{k}$ with components $k^\mu = (|\vec{k}|, 0, 0, |\vec{k}|)$. This means from $A_{\alpha\beta}k^\beta = 0$ (eqn 46.12) that $A_{\alpha z} = 0$ too. We therefore have a possibility of non-zero matrix elements for A_{xx}, A_{yy}, A_{xy} and A_{yx}. Since the wave is traceless, we must have $A_{xx} = -A_{yy}$. By symmetry $A_{xy} = A_{yx}$. We therefore have, in this frame, the amplitude components

$$A_{\alpha\beta} = \begin{pmatrix} 0 & 0 & 0 & 0 \\ 0 & A_{xx} & A_{xy} & 0 \\ 0 & A_{xy} & -A_{xx} & 0 \\ 0 & 0 & 0 & 0 \end{pmatrix}. \tag{46.21}$$

We then have a simplified solution to the wave equation $\bar{h}_{\alpha\beta} = A_{\alpha\beta}e^{i\boldsymbol{k}\cdot\boldsymbol{x}}$, with $\omega = |\vec{k}|$. Remembering that $g_{\mu\nu} = \eta_{\mu\nu} + h_{\mu\nu}$, and also that $\bar{h}_{\mu\nu} = h_{\mu\nu}$ with our choice of gauge, we deduce that this solution results in a metric line element

$$ds^2 = -\,dt^2 + \left(1 + A_{xx}e^{-i(\omega t - |k|z)}\right)dx^2 + 2A_{xy}e^{-i(\omega t - |k|z)}dxdy$$
$$+ \left(1 - A_{xx}e^{-i(\omega t - |k|z)}\right)dy^2 + dz^2. \tag{46.22}$$

The two independent solutions represented here can be disentangled as follows. If $A_{xy} = 0$, then our metric reduces to

$$ds^2 = -dt^2 + \left(1 + A_{xx}e^{-i(\omega t - |k|z)}\right)dx^2 + \left(1 - A_{xx}e^{-i(\omega t - |k|z)}\right)dy^2 + dz^2. \tag{46.23}$$

On the other hand, if $A_{xx} = 0$, our metric reduces to

$$ds^2 = -dt^2 + 2A_{xy}e^{-i(\omega t - |k|z)}dxdy + dz^2. \tag{46.24}$$

Each of these solutions represents a plane gravitational wave, and eqns 46.23 and 46.24 are related to each other by a 45° rotation.

The metric line element has some oscillatory terms in it, but what does this mean? Does this mean that individual masses suspended in space will bob up and down as a gravitational wave goes past, just like boats on the ocean do when ocean waves go past? It's a bit more complicated than that, as shown in the next example.

Example 46.5

Place a test mass at rest at the origin. Its velocity is then $\dot{x}^\mu = (1,0,0,0)$. The geodesic equation, eqn 8.24, is $\ddot{x}^\mu + \Gamma^\mu_{\alpha\beta}\dot{x}^\alpha\dot{x}^\beta = 0$, so that

$$\ddot{x}^i + \Gamma^i_{00}\dot{x}^0\dot{x}^0 = \ddot{x}^i + \Gamma^i_{00} = 0. \tag{46.25}$$

However, for a gravitational wave we must have $\Gamma^i_{00} = 0$ since $\Gamma^i_{ab} = \frac{1}{2}\eta^{ic}(\partial_a h_{cb} + \partial_b h_{ca} - \partial_c h_{ab}) = 0$ because $h_{0a} = 0$. This implies that $\ddot{x}^i = 0$ so that the particle doesn't move. Oh dear; this is not what we wanted!

However, the fact that the *coordinate* of our test mass does not change as the gravitational wave rolls past does not mean anything. By now, we have learned to be suspicious of coordinates which can be chosen in lots of different ways. We know the metric line element does oscillate, so for example if we put a first test mass at the origin and a second test mass displaced a distance L in the x-direction, then the distance between them should be

$$\int_0^L \sqrt{g_{xx}}\,\mathrm{d}x = \mathrm{Re}\left[L\sqrt{1 + A_{xx}\mathrm{e}^{-\mathrm{i}(\omega t - |k|z)}}\right] \approx L + \frac{LA_{xx}}{2}\cos(\omega t - |k|z).$$

$$(46.26)$$

Happily this does oscillate, and also gives a method of detecting gravitational waves: by measuring the distance between pairs of masses as a function of time.

We can therefore understand gravitational waves by assessing their influence on *groups* of tiny test masses. Let's therefore examine in more detail the geodesic deviation of a set of particles. Geodesic deviation is described by the equation

$$\frac{D^2\boldsymbol{n}}{\mathrm{d}\tau^2} + \boldsymbol{R}(\ ,\boldsymbol{u},\boldsymbol{n},\boldsymbol{u}) = 0, \qquad (46.27)$$

where $\boldsymbol{n}$ is the separation vector of the particles and the velocity $\boldsymbol{u}$ is tangent to the streamlines formed by the geodesics. We will work in the local frame of the particles, amounting to a choice of the components of $\boldsymbol{u}$ as $u^\mu = (1, 0, 0, 0)$, so all we need to do is compute the relevant components of $\boldsymbol{R}$ from the components of the $\boldsymbol{h}$ field.

Example 46.6

With our choice of $\boldsymbol{u}$ the geodesic deviation expression becomes the component equation

$$\frac{D^2 n^\mu}{\mathrm{d}\tau^2} = -R^\mu{}_{0\alpha 0}n^\alpha. \qquad (46.28)$$

For simplicity, let's choose $\boldsymbol{n}$ to initially have components $n^\mu = (0, a, 0, 0)$, implying that the two masses are separated by the spacelike interval a at the start of the motion. From the last chapter, we have that

$$R_{\alpha\beta\mu\nu} = \frac{1}{2}\left(h_{\alpha\nu,\beta\mu} - h_{\alpha\mu,\beta\nu} + h_{\beta\mu,\alpha\nu} - h_{\beta\nu,\alpha\mu}\right). \qquad (46.29)$$

Recalling also that we raise and lower indices in the weak-field limit using $\eta_{\mu\nu} = \mathrm{diag}(-1, 1, 1, 1)$, we find that the components of the Riemann tensor relevant to the geodesic equation then become

$$R^x{}_{0x0} = R_{x0x0} = -\frac{1}{2}\frac{\partial^2 h_{xx}}{\partial t^2},$$

$$R^y{}_{0x0} = R_{y0x0} = -\frac{1}{2}\frac{\partial^2 h_{xy}}{\partial t^2},$$

$$R^y{}_{0y0} = R_{y0y0} = -\frac{1}{2}\frac{\partial^2 h_{yy}}{\partial t^2} = -R^x{}_{0x0}. \qquad (46.30)$$

A further simplification is that, to first order in $h_{\mu\nu}$ we can make the replacement $\tau = t$. This means that the separation vector of the particles, originally separated along the x-direction by an interval a, obey the equations of motion

$$\frac{\partial^2 n^x}{\partial t^2} = \frac{1}{2}a\frac{\partial^2 h_{xx}}{\partial t^2}, \qquad \frac{\partial^2 n^y}{\partial t^2} = \frac{1}{2}a\frac{\partial^2 h_{xy}}{\partial t^2}. \qquad (46.31)$$

By the same token, two particles initially separated along y by a spacelike interval a obey

$$\frac{\partial^2 n^y}{\partial t^2} = -\frac{1}{2}a\frac{\partial^2 h_{xx}}{\partial t^2}, \quad \frac{\partial^2 n^x}{\partial t^2} = \frac{1}{2}a\frac{\partial^2 h_{xy}}{\partial t^2}. \tag{46.32}$$

On integrating these differential equations twice, we find that the separation n^i of particles can be described in terms of the components of the $\boldsymbol{h}$-field in transverse-traceless gauge by writing

$$n^i = \frac{1}{2}h^{\mathrm{TT}}_{ij}x^j, \tag{46.33}$$

where $h^{\mathrm{TT}}_{xx} = -h^{\mathrm{TT}}_{yy} = h_+$ and $h^{\mathrm{TT}}_{xy} = h^{\mathrm{TT}}_{yx} = h_\times$.

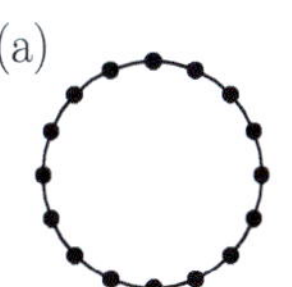

The equations from the previous example allow us to understand the polarization of the gravitational waves as causing motion of the particles arranged in a circle in Fig. 46.1(a). If the waves have $h_{xy} = 0$ and $h_{xx} \neq 0$, then the pattern of displacements corresponds to that shown in Fig. 46.1(b), with the masses moving along the x and y directions out of phase by 180°. This is sometimes called the $+$ polarization. If, instead, we have $h_{xx} = 0$ and $h_{xy} \neq 0$, then the pattern of displacements is that shown in Fig. 46.1(c). This is simply the pattern from Fig. 46.1(b) rotated by 45°, hence the name: $\times$ polarization. Notice how the two different polarizations are related by a 45° rotation[11] unlike the two electromagnetic linear polarizations, which are related by a 90° rotation. This is a consequence of the tensorial nature of the $(0,2)$ gravitational $\boldsymbol{h}$-field, as opposed to the 1-form field $\tilde{\boldsymbol{A}}$ that expresses electromagnetism.

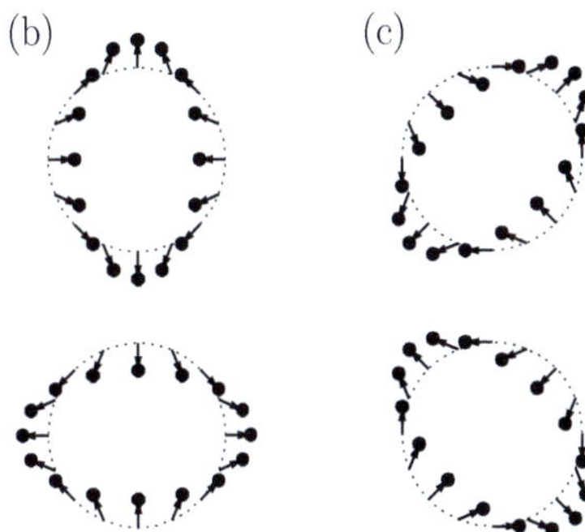

Fig. 46.1 Polarizations of gravitational waves. (a) A circle of masses. (b) The $+$ polarization. (c) The $\times$ polarization.

[11]As we found in Example 46.4.

Example 46.7

We can define the **tidal field** $\boldsymbol{\mathcal{E}}$ with components $\mathcal{E}_{ij} = R_{0i0j}$, which can be written in terms of the two possible polarizations of the waves as

$$\boldsymbol{\mathcal{E}} = -\frac{1}{2}\ddot{\boldsymbol{h}}^{\mathrm{TT}}$$

$$= -\frac{1}{2}\left(\ddot{h}_+ \boldsymbol{e}_+ + \ddot{h}_\times \boldsymbol{e}_\times\right), \tag{46.34}$$

where

$$\boldsymbol{e}_+ = \boldsymbol{e}_x \otimes \boldsymbol{e}_x - \boldsymbol{e}_y \otimes \boldsymbol{e}_y, \tag{46.35}$$

and

$$\boldsymbol{e}_\times = \boldsymbol{e}_x \otimes \boldsymbol{e}_y + \boldsymbol{e}_y \otimes \boldsymbol{e}_x, \tag{46.36}$$

are **polarization tensors**.

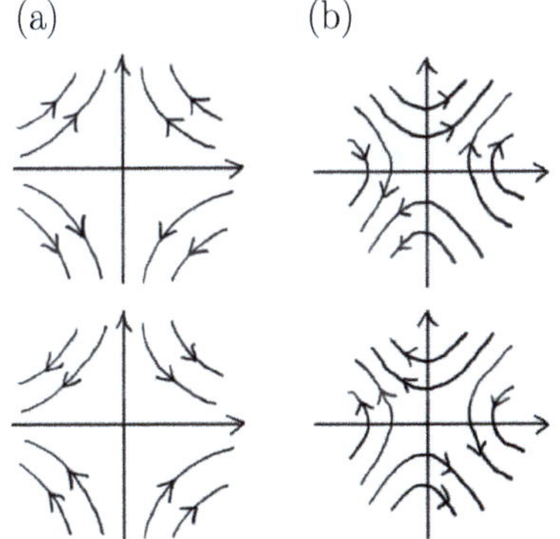

Fig. 46.2 Quadrupolar field lines representing the accelerations produced by a gravitational wave (a) with the $+$ polarization and (b) with the $\times$ polarization.

We can also depict the two polarizations by looking at the accelerations that are produced by the waves. This is shown in the diagrams in Fig. 46.2, where the field lines represent the acceleration field $D^2\boldsymbol{n}/\mathrm{d}\tau^2$. The four diagrams display the field at different points in the wave cycle for the two polarizations. These field diagrams are reminiscent of the magnetic field from a quadrupolar magnet and in fact these do indeed represent **quadrupolar** fields. This point deserves some further exploration, which we will do in the following example, first considering the case of sources of electromagnetic radiation before considering the analogous case of sources of gravitational radiation.

Example 46.8

● *(i) Electromagnetic waves:* In the study of electromagnetic radiation, the **multipole expansion** of a source is a useful technique. Consider some object sitting in empty space, and assume it is made up of various charges, perhaps both positive and negative. We are interested in the electromagnetic field at some point distant from this object due to the effect of the charges within the object and their individual motions. The multipole expansion involves writing the charge distribution of this bounded object at a particular instant in time as a sum of terms of increasing complexity. We start off by adding up all the charge in the object and shrinking it to a point; at sufficient distance, the object will after all look like a point charge. The first term is then the monopolar contribution, the next term will be the dipole term, then we have a quadrupolar contribution. The charge in the object might be in some complicated motion, so we will also have some dynamic current-carrying contributions such as the magnetic dipole moment, magnetic quadrupole moment, etc. (no magnetic monopolar term though, since this turns out to be identically zero).

To get electromagnetic radiation out of this object, the charges within need to jiggle around. The acceleration of those charges could then produce electromagnetic waves. For example, suppose we could get the total charge in our object (the monopolar term) to oscillate up and down, we could generate spherically symmetry electromagnetic waves. However, this can't happen because we are not allowed to vary the total charge in our bounded object, charge being a conserved quantity. What we can do is vary the dipolar term in an oscillatory fashion, and this is how transmitters (and aerials) work. We make one end of an object positive and the other end negative by driving a current in one direction, and then by reversing the current we reverse the polarity of the dipole moment, making the formerly positive end negative and the formerly negative end positive. This oscillating dipole produces dipole radiation. Higher multipoles of electromagnetic radiation are possible, but charge conservation forbids the monopole term.

● *(ii) Gravitational waves:* In the context of gravitation, we can consider an analogous multipole expansion of the mass distribution of an array of masses m_i at positions $\vec{x}_i$. (For simplicity, we will put the origin of our coordinates at the centre of mass of this distribution.) We are looking for contributions that aren't conserved, since these can vary and therefore act as the source of gravitational waves.

The monopole contribution is the total mass $\sum_i m_i$ (also known as the zeroth mass moment $\mathcal{I}_0$), which is constant owing to mass conservation. The dipole moment of the distribution (also known as the first mass moment $\mathcal{I}_1$), which is given by $\sum_i m_i \vec{x}_i$, but this is constant because of momentum conservation. The gravitational analogue of the magnetic moment is a first-order moment involving mass currents. It is given by the angular momentum $\vec{L} = \sum_i m_i (\vec{x}_i \times \dot{\vec{x}}_i)$. The first-order moment is therefore conserved because of angular momentum conservation. This implies that the lowest order contribution to gravitational radiation can only be from the next term: the quadrupolar field.

In expanding the metric close to the source in the weak-field regime, the 00 component can be expanded as a sum of the mass moments

$$g_{00} = -1 + a_0 \frac{\mathcal{I}_0}{r} + a_1 \frac{\mathcal{I}_1}{r^2} + a_2 \frac{\mathcal{I}_2}{r^3} + ..., \tag{46.37}$$

where $\mathcal{I}_\ell$ is the ℓth mass moment and a_i are a set of constants. Similarly, the $0j$ components of the metric can be expanded in terms of the current moments

$$g_{0j} = b_1 \frac{\mathcal{S}_1}{r^2} + b_2 \frac{\mathcal{S}_2}{r^3} + ..., \tag{46.38}$$

where $\mathcal{S}_\ell$ are current moments and b_i are constants. Since for a source of linear dimension L, we would expect on simple, dimensional grounds that $\mathcal{I}_\ell \propto ML^\ell$ and $\mathcal{S} \propto MvL^\ell$, where M is mass and v is velocity. We infer that the leading-order time-varying contribution to g_{00} is from the quadrupolar mass term $\mathcal{I}_2$, with the other terms contributing higher order corrections. This turns out to be the case.

46.3 Quadrupolar radiation

We will now derive a description of the quadrupolar radiation directly. We rewrite our wave equation with a source of radiation, so that

$$-\partial^2 \bar{h}_{\mu\nu} = 16\pi G T_{\mu\nu}, \tag{46.39}$$

where the factor of G has been restored. By analogy with the electromagnetic case,[12] we can write down the solution to this weak-field gravitation equation as

$$\bar{h}_{\mu\nu}(t, \vec{x}) = 4G \int \mathrm{d}^3 y \frac{T_{\mu\nu}(t - |\vec{x} - \vec{y}|, \vec{y})}{|\vec{x} - \vec{y}|}, \tag{46.41}$$

Assuming that the source is compact, so that it is concentrated in a small region Σ of linear size r a distance $R \gg r$ away, the spatial components $\bar{h}_{ij}$ are given by

$$\bar{h}_{ij}(t) \approx \frac{4G}{R} \int_\Sigma \mathrm{d}^3 y \, T_{ij}(t - R, \vec{y}). \tag{46.42}$$

The energy-momentum tensor is subject to a conservation law $\partial_\mu T^{\mu\nu} = 0$ (or equivalently $T^{\mu\nu}{}_{,\mu} = 0$) and including that allows[13] us to rewrite eqn 46.42 as

$$\bar{h}_{ij}(t) \approx \frac{2G}{R} \partial_0^2 \int_\Sigma \mathrm{d}^3 y \, y_i y_j T^{00}(t - R, \vec{y}). \tag{46.43}$$

The integral is just the second moment of the mass distribution, which is (in energy units) the moment of inertia tensor I_{ij}, so we can write this equation in the simplified form

$$\bar{h}_{ij}(t) \approx \frac{2G}{R} \ddot{I}_{ij}(t - R). \tag{46.44}$$

This is known as the **Einstein quadrupole formula**.[14] Note that the moment of inertia tensor $I_{ij} = \int_\Sigma \mathrm{d}^3 y \, \rho y_i y_j$ differs from the quadrupole moment $Q_{ij} = \int_\Sigma \mathrm{d}^3 y \, \rho \left(y_i y_j - \frac{1}{3} r^2 \delta_{ij} \right) = I_{ij} - \frac{1}{3} \mathrm{Tr} I$ solely by its trace. Since we are working in the transverse-traceless gauge, we are insensitive to the trace and hence we can think of the moment of inertia as a quadrupole moment.

[12] We are repeating the argument that led to eqn 45.31. In electromagnetism, the source of radiation is the four-current density J_μ (which includes both charge and current density) and hence

$$-\partial^2 A_\mu = \mu_0 J_\mu.$$

The solution at $(t, \vec{x})$ is given in terms of the retarded current density (due to the finite travel-time from a source at $\vec{y}$ to position $\vec{x}$) by

$$A_\mu(t, \vec{x}) = \frac{\mu_0}{4\pi} \int \mathrm{d}^3 y \frac{J_\mu(t - |\vec{x} - \vec{y}|, \vec{y})}{|\vec{x} - \vec{y}|}. \tag{46.40}$$

[13] This is worked out in detail in Exercise 46.3.

[14] Restoring the factors of c, one could write eqn 46.44 as

$$\bar{h}_{ij}(t) \approx \frac{2G}{c^4 R} \ddot{I}_{ij} \left(t - \frac{R}{c} \right), \tag{46.45}$$

if I_{ij} is in mass units.

Example 46.9

Let's put some numbers in to see how big an effect this could be. With a source at, say, $R = 100\,\mathrm{MPc}$ away from us, consisting of a pair of black holes, each of mass $M = 30 M_\odot$ orbiting each other at $f = \omega/(2\pi) = 10\,\mathrm{Hz}$ and separated by twice $a = 2000$ km and using $\ddot{I} \approx 4Ma^2\omega^2$ then

$$|\bar{h}| \approx \frac{2G}{c^4 R} 4Ma^2\omega^2 = \frac{32\pi^2 G M a^2 f^2}{c^4 R} \approx 5 \times 10^{-21}. \tag{46.46}$$

This will be a small effect!

The value of $\bar{h} \approx 10^{-21}$ is very small, and much tinier even than the Newtonian potential on the surface of the Earth, where $|h_{00}| = 2GM_\oplus/(R_\oplus c^2) \approx 10^{-9}$, and this highlights an issue with our approach so far. The gravitational waves that are detected in experiments to date have frequencies in the tens of Hz to a few kHz range and consequently have wavelengths (tens to thousands of km) which are short compared to a terrestrial scale. These ripples in spacetime are superimposed on a larger, more slowly varying background due to astrophysical objects (including the astrophysical object on which a gravitational detector might be mounted, i.e. the Earth!). We have described the waves using a metric $g_{\mu\nu} = \eta_{\mu\nu} + h_{\mu\nu}$, expanding around a flat spacetime, whereas we probably really should write something like

$$g_{\mu\nu} = g_{\mu\nu}^{\mathrm{b}} + h_{\mu\nu}, \tag{46.47}$$

where $g_{\mu\nu}^{\mathrm{b}}$ describes some background curved metric. However, even this may not be enough as it is not obvious which contributions should be background and which should be due to the gravitational waves. Moreover, as we will see, gravitational waves carry energy and momentum and so this is also going to act as a source of curvature of spacetime. Our linearized theory has ignored this effect, and so one way of including this is to say that our linearized Einstein tensor $G_{\mu\nu}^{(1)}$ is modified to

$$G_{\mu\nu}^{(1)} = 8\pi(T_{\mu\nu} + t_{\mu\nu}), \tag{46.48}$$

where $T_{\mu\nu}$ is due to matter and $t_{\mu\nu}$ is due to the effect of the gravitational field itself. Our exact Einstein equation is

$$G_{\mu\nu} = 8\pi T_{\mu\nu}, \tag{46.49}$$

where the exact Einstein tensor is written as

$$G_{\mu\nu} \equiv G_{\mu\nu}^{(1)} + G_{\mu\nu}^{(2)} + \cdots, \tag{46.50}$$

where the sum is over a linear approximation $G_{\mu\nu}^{(1)}$ (linear in $h_{\mu\nu}$), a quadratic approximation $G_{\mu\nu}^{(2)}$ (quadratic in $h_{\mu\nu}$), etc. If we truncate the series at the quadratic term, then these equations suggest that

$$t_{\mu\nu} = -\frac{1}{8\pi}G_{\mu\nu}^{(2)}. \tag{46.51}$$

It turns out that this quantity is not gauge invariant, and you probably wouldn't expect it to be.[15] The trick is to average this quantity over a region of spacetime that is spatially larger than the wavelength of a gravitational wave and temporally larger than the reciprocal a gravitational-wave frequency, thereby capturing the curvature of the background spacetime. Thus, we arrive at

$$t_{\mu\nu} = -\frac{c^4}{8\pi G}\langle G_{\mu\nu}^{(2)}\rangle, \tag{46.52}$$

where the angle brackets denote this averaging process and we have restored the factors of G and c.

[15]After all, the energy-momentum of the gravitational field has no real meaning locally since you can always transform to a freely falling frame and gravity disappears.

46.4 Radiated energy and power

The evaluation of an expression for $t_{\mu\nu}$ from eqn 46.52 is rather tedious, but in the transverse-traceless gauge it produces[16] the rather pleasing result

$$t_{\mu\nu} = \frac{c^4}{32\pi G}\langle \partial_\mu h_{\alpha\beta}^{\mathrm{TT}} \partial_\nu h_{\mathrm{TT}}^{\alpha\beta}\rangle. \tag{46.53}$$

In particular, the energy density is then given by[17]

$$t_{00} = \frac{c^4}{32\pi G}\langle \partial_0 h_{\alpha\beta}^{\mathrm{TT}} \partial_0 h_{\mathrm{TT}}^{\alpha\beta}\rangle = \frac{G}{8\pi c^6 R^2}\langle \dddot{I}_{ij}^{\mathrm{TT}} \dddot{I}_{ij}^{\mathrm{TT}}\rangle, \tag{46.54}$$

where we have used the Einstein quadrupole formula, eqn 46.44. We can work out the total flux of power from our source, namely the energy passing per second through a spherical surface of radius R using the fact the energy inside a volume V is $\int \mathrm{d}^3 x\, T^{00}$, so that the rate of change of energy is

$$\frac{\mathrm{d}}{\mathrm{d}t}\int \mathrm{d}^3 x\, T^{00} = \int \mathrm{d}^3 x\, \partial_t T^{00} = -c\int \mathrm{d}^3 x\, \partial_i T^{0i}, \tag{46.55}$$

where we have used $\partial_\mu T^{\mu\nu} = 0$ and $\partial_t = c\partial_0$. The energy carried by the gravitational wave is the negative of this (because energy lost inside the volume V is carried away by the gravitational waves), so together with Stokes' theorem we have the rate of change of energy emitted by quadrupolar waves as an integral over the surface S and hence

$$\frac{\mathrm{d}E}{\mathrm{d}t} = c\int_S T^{oi}\,\mathrm{d}\Sigma_i = cR^2 \int T^{0r}\,\mathrm{d}\Omega, \tag{46.56}$$

where the final result is an integral over solid angle (inside a sphere of radius R). In the present case, we are working in terms of $t_{\mu\nu}$ due to the effect of the gravitational field, and therefore we need t^{0r} in eqn 46.56. We can get this from eqn 46.53, but we also can use the fact that $\bar{h}_{ij}$ is a function of $t - R/c$ and hence

$$\partial_r \bar{h}_{ij} = -\partial_0 \bar{h}_{ij} = +\partial^0 \bar{h}_{ij}, \tag{46.57}$$

which means we can write $t^{0r} = t^{00}$. Thus, the rate of change of energy carried away by gravitational waves from their source is

$$\frac{\mathrm{d}E}{\mathrm{d}t} = cR^2 \int_\Omega t^{00}\,\mathrm{d}\Omega = \frac{G}{8\pi c^5}\int_\Omega \langle \dddot{I}_{ij}^{\mathrm{TT}} \dddot{I}_{ij}^{\mathrm{TT}}\rangle\,\mathrm{d}\Omega. \tag{46.58}$$

To evaluate this, we can use the identity[18]

$$M_{ij}^{\mathrm{TT}} M_{ij}^{\mathrm{TT}} = M_{ij}M_{ij} - 2n_a n_b M_{ia}M_{jb} + \frac{1}{2}n_a n_b n_c n_d M_{ab}M_{cd}. \tag{46.59}$$

The integral evaluates[19] to

$$\frac{\mathrm{d}E}{\mathrm{d}t} = \frac{G}{5c^5}\langle \dddot{I}_{ij} \dddot{I}_{ij}\rangle. \tag{46.60}$$

The prefactor $G/(5c^5)$ is extremely small and hence most potential sources of gravitational radiation produce only a very weak emitted power.[20]

[16] The calculation is rather tedious, but it is laid out in detail in Exercise 46.2 for any reader who wishes to follow it through.

[17] When we are just focussing on spatial vectors, the distinction between upstairs and downstairs indices becomes unnecessary, and following the lead of most authors in this field, we will write $\langle \dddot{I}_{ij}^{\mathrm{TT}} \dddot{I}_{ij}^{\mathrm{TT}}\rangle$ rather than $\langle \dddot{I}_{ij}^{\mathrm{TT}} \dddot{I}_{\mathrm{TT}}^{ij}\rangle$, which is less fussy. The Einstein summation convention still holds, so i and j are summed over.

[18] See Exercise 46.5(d).

[19] See Exercise 46.5(e). Note that this is the energy flux associated with the gravitational waves. The rate of change of energy inside the volume is *minus* this.

[20] Moreover, note that a spherically symmetric source (such as rotating star) has zero $\langle \dddot{I}_{ij} \dddot{I}_{ij}\rangle$ (its moment of inertia does not change with time) and will not emit gravitational waves.

[21] As shown in Exercise 46.4, the emitted power for most orbiting systems is tiny. To get some sizeable, and hence potentially detectable, emitted power, we need some pretty dramatic situations, such as the two closely spaced black holes whirling around each at a ferocious speed considered in this example. We could have chosen two neutron stars, but we wanted to be even more dramatic! Note that $L \propto \omega^6$ so high frequency signals contain much more power.

[22] The derivation is in Maggiore, Section 3.3.3. Note that this expression for the angular momentum implies that angular momentum will not be emitted if the source is axisymmetric.

Example 46.10

Returning to the numbers[21] in Example 46.9, we would estimate (using $\langle \dddot{I}_{ij} \dddot{I}_{ij} \rangle = 128 M^2 a^4 \omega^6$, as shown in eqn 46.106 from Exercise 46.6) that the gravitational luminosity L of our binary black-hole system would be

$$L = \frac{128 G}{5 c^5} M^2 a^4 \omega^6 \approx 10^{47} \ \text{W}, \tag{46.61}$$

although the power flux on Earth works out to be less than $1 \ \text{mW m}^{-2}$.

One can also show that not only is energy carried away by gravitational waves but also angular momentum. By analogy with eqn 46.56, the rate of change of angular momentum is given by

$$\frac{\mathrm{d}J_i}{\mathrm{d}t} = -\int_S \epsilon_{ijk} x^j T^{km} \, \mathrm{d}\Sigma_m, \tag{46.62}$$

and this can be used to show that the rate of change of angular momentum carried by the waves from the source is[22]

$$\frac{\mathrm{d}J^i}{\mathrm{d}t} = \frac{2G}{5c^5} \epsilon_{ijk} \langle \ddot{I}_{j\ell} \dddot{I}_{k\ell} \rangle. \tag{46.63}$$

For both eqn 46.60 and eqn 46.63 the quantities $\dddot{I}_{ij}$ are evaluated at the retarded time $t_{\mathrm{r}} = t - R/c$.

Example 46.11

An orbiting pair of black holes has a gravitational-wave luminosity L given by eqn 46.61 so that $L = \frac{128 G}{5 c^5} M^2 a^4 \omega^6$, but the two objects (assuming Newtonian mechanics holds) have a gravitational potential energy equal to $-GM^2/(2a)$ and a kinetic energy of $2 \times \frac{1}{2} M v^2 = GM^2/(4a)$ so that the total energy is $E = -GM^2/(4a) = -M a^2 \omega^2$ and

$$\omega = \sqrt{\frac{GM}{4a^3}}. \tag{46.64}$$

This expression for ω allows us to rearrange eqn 46.61 to give

$$L = \frac{\mathrm{d}E}{\mathrm{d}t} = \frac{128 G}{5 c^5} M^2 a^4 \omega^6 = \frac{2}{5} \frac{G^4 M^5}{a^5 c^5}. \tag{46.65}$$

It is useful to express E in terms of the orbital period $P = 2\pi/\omega$, and some rearrangement gives

$$E = -\left(\frac{G^2 M^5 \pi^2}{4} \right)^{\frac{1}{3}} P^{-\frac{2}{3}}, \tag{46.66}$$

and hence the rate of change of orbital period is

$$\frac{\mathrm{d}P}{\mathrm{d}t} = -\frac{3}{2} \frac{P}{E} \frac{\mathrm{d}E}{\mathrm{d}t}, \tag{46.67}$$

so that using eqn 46.65 we have

$$\frac{\mathrm{d}P}{\mathrm{d}t} = -\frac{3}{5} 4^{\frac{11}{3}} \pi^{\frac{8}{3}} \left(\frac{GM}{P} \right)^{\frac{5}{3}}. \tag{46.68}$$

Note that the two orbiting black holes are in a bound state so that the total energy E is negative. Therefore, when energy is lost via emission of gravitational waves the energy becomes more negative and hence $|E|$ is larger. This causes the orbital period P to decrease (hence the minus sign in eqn 46.68) and the orbiting becomes faster. This speed up of the orbital motion (called 'spin-up') was observed in a binary pulsar system in 1974 by Hulse and Taylor and was the first (indirect) discovery of gravitational waves (see Fig. 46.3). We receive radio emission only from one of the pulsars (which has a 59 ms period which can be measured very accurately) and the Doppler shift of the radio signal allows us to estimate the orbital period of about 7.75 hours and to measure that this orbit is very slowly speeding up, with a rate of decrease of orbital period around 76.5 microseconds per year.

As $|E|$ increases, the two black holes get closer together and a decreases. We can use eqn 46.63 to estimate the rate of change of angular momentum carried off by the gravitational waves. The only non-zero component is perpendicular to the orbital plane and yields[23]

$$\frac{\mathrm{d}J^z}{\mathrm{d}t} = \frac{128GM^2a^4\omega^5}{5c^5} = \frac{4G^{\frac{7}{2}}M^{\frac{9}{2}}}{5a^{\frac{7}{2}}c^5}. \tag{46.69}$$

The angular momentum of our orbiting pair of black holes is $J = 2Ma^2\omega = \sqrt{GMa^3} = -E\omega/2$. Also, eliminating a and ω we have $E = -G^2M^5/(4J^2)$ and hence $\mathrm{d}J/\mathrm{d}t = -(J/2E)\mathrm{d}E/\mathrm{d}t$ and therefore substituting in our expression for $\mathrm{d}E/\mathrm{d}t$ gives

$$\frac{\mathrm{d}J}{\mathrm{d}t} = -\frac{4G^{\frac{7}{2}}M^{\frac{9}{2}}}{5a^{\frac{7}{2}}c^5}, \tag{46.70}$$

in agreement with eqn 46.69 (the sign change being that the angular momentum lost by the orbiting pair is carried off by the gravitational waves).

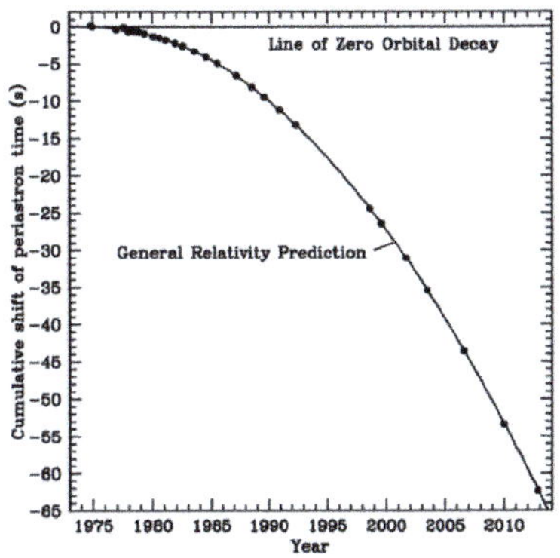

Fig. 46.3 The orbital decay of the Hulse–Taylor binary pulsar system PSR B1913+16, together with the prediction from general relativity. Our treatment has assumed circular orbits of two equal masses, but the curve here assumes the correct elliptical orbital parameters. This plot shows cumulative shifts in the time of periastron. [Figure reproduced from J. M. Weisberg and Y. Huang, Astrophysical Journal **829**, 55 (2016) doi:10.3847/0004-637X/829/1/55 ©American Astronomical Society.]

[23]See Exercise 46.7.

Our treatment of these astrophysical cases has assumed Newtonian dynamics and so therefore we would not expect it to apply in the final case of the inspiral of two compact objects as their orbital periods drop rapidly and their orbital velocities become relativistic. Our approach has also assumed a linearized approximation to general relativity and here we will encounter a big difference with electromagnetism. In the electromagnetic case, the force-mediating particle (the photon) and its associated wave (the electromagnetic wave) have no charge and so there is no nonlinear effect in free space. For gravity, our gravitational waves do transmit energy (i.e. mass) which is itself a source of gravity; thus, our theory is inherently nonlinear and we take a brief moment in the next section to examine the consequences of this.

46.5   An exact solution

So far we have only examined gravitational waves within the linear approximation. Will they survive in the exact, nonlinear theory of gravitation? We can show that they shall.

We argued in Section 46.2 for solutions as a function of $\phi = (-\omega t + kz)$ with $\omega = k^z = |k|$, since gravitational waves are described by null velocity vectors. Since we expect gravitational waves to be null, it is useful to work in light-cone coordinates $u = t - z$ and $v = t + z$ in which the Minkowski metric is written in terms of the line element as

$$\mathrm{d}s^2 = \eta_{\mu\nu}\mathrm{d}x^\mu\mathrm{d}x^\nu = -\mathrm{d}u\mathrm{d}v + \mathrm{d}x^2 + \mathrm{d}y^2. \tag{46.71}$$

We can therefore look for an exact, wavelike solution to the Einstein field equation with metric

$$ds^2 = (\eta_{\mu\nu} + h_{\mu\nu})\,dx^\mu dx^\nu = -dudv + F(u)^2 dx^2 + G(u)^2 dy^2, \quad (46.72)$$

where F and G are functions to be determined by the Einstein equation.

Example 46.12

Using the metric from this line element, we can generate the useful connections and Riemann components. These are

$$\Gamma^x{}_{xu} = \frac{1}{F}\frac{\partial F}{\partial u}, \quad \Gamma^y{}_{yu} = \frac{1}{G}\frac{\partial G}{\partial u}, \quad \Gamma^v{}_{xx} = \frac{2}{F}\frac{\partial F}{\partial u}, \quad \Gamma^v{}_{yy} = \frac{2}{G}\frac{\partial G}{\partial u}, \quad (46.73)$$

and

$$R^x{}_{uxu} = -\frac{1}{F}\frac{\partial^2 F}{\partial u^2}, \quad R^y{}_{uyu} = -\frac{1}{G}\frac{\partial^2 G}{\partial u^2}. \quad (46.74)$$

We therefore have the Einstein equation (for a vacuum) as

$$\frac{1}{F}\frac{\partial^2 F}{\partial u^2} + \frac{1}{G}\frac{\partial^2 G}{\partial u^2} = 0. \quad (46.75)$$

If $F(u) = 1 + \varepsilon(u)$, where the function $\varepsilon(u)$ is assumed small, then eqn 46.75 can be solved by $G(u) = 1 - \varepsilon(u)$, since we obtain

$$\frac{\varepsilon''(u)}{1+\varepsilon(u)} - \frac{\varepsilon''(u)}{1-\varepsilon(u)} \approx \varepsilon''(u)\left[1-\varepsilon(u)+...\right] - \varepsilon''(u)\left[1+\varepsilon(u)+...\right] \approx 0. \quad (46.76)$$

Since, in this case we have $h_{xx} = \varepsilon(u)$ and $h_{yy} = -\varepsilon(u)$, this is just the case shown in Fig. 46.1(b) of a $+$ polarized wave.

We conclude that the exact, nonlinear case therefore supports solutions similar to the ones found for the linear, weak-field theory.

46.6 The discovery of gravitational waves

Gravitational waves were for a hundred years, apart from the impressive but indirect evidence from the spin-up of the Hulse–Taylor binary pulsar, a theoretical construct. It was believed that they were there, but they had not been directly detected. That has all changed due to the extraordinary results that have been obtained from ground-based gravitational-wave observatories. These are designed to probe the relatively high-frequency portion of the gravitational-wave spectrum, from about 10 Hz to about 10 kHz. This spectrum is dominated by signals originating from stellar-mass compact sources, principally coalescing binary black hole and neutron star systems.

In ground-based observatories, the idea is to use the alternating motion of masses produced by a passing gravitational wave that we can detect using laser interferometry. The key here is to use the sensitivity of interference to optical path length to detect the oscillatory motion of the masses. We've seen how gravitational waves lead to quadrupolar oscillations of masses. To see the $+$ polarization for a wave propagating in the z-direction, for example, we would need to access the relative motion of at least two masses, such as one separated along x and one separated along y. In order to do this, we use the masses to form a Michelson[24] interferometer in the x-y plane, as shown in Fig. 46.4. Laser light is split by a beam-splitting mirror and travels along the two arms of the interferometer. The lengths of the arms are defined by mirrored masses, which reflect the light back to where the beams are recombined and detected via a photodetector as shown. The bad news is that the effect is small. Even for merging neutron stars or collapsing supernovae, we only expect fractional changes in the displacement of the mirrors in the experiment of 10^{-21}. In order to see an oscillation from a gravitational wave, a typical photon must remain in the system for at least half the period of the gravitational wave, which turns out to be of order milliseconds. As a result, we require interferometers with very long arms and exceedingly sensitive detection technologies. This is far from trivial. It took several decades to develop the technology to achieve this extraordinary level of sensitivity.

Example 46.13

The intensity at the photodetector is determined by the phase shift between the light combined from each arm of the interferometer. Take the arms to have lengths L_x and L_y. If the arm lengths vary by respective amounts δx and δy then the phase shift is $\Delta\phi = \omega_0(2\delta y - 2\delta x)$, where ω_0 is the laser frequency. Using eqn 46.33 we can rewrite this as

$$\Delta\phi(t) = (L_x + L_y)h_+(t), \tag{46.77}$$

allowing us to see the relationship between the h field and the phase shift. Assuming the arms are roughly the same length L we have that the intensity measured is

$$I \propto \Delta\phi(t) \approx 2Lh_+(t). \tag{46.78}$$

We conclude that the length of the arms must be maximized to have a chance of seeing the effect.

LIGO stands for the Laser Interferometry Gravitational Wave Observatory. It comprises two identical detectors: one interferometer in Livingston, Louisiana and one in Hanford, Washington. The arms of the interferometers are 4.2 km in length (for comparison, the Michelson–Morley experiment involved arms of length 1.3 m). However, these still cannot introduce a long-enough optical path difference to detect the oscillations from the tiny displacements caused by the perturbations to the metric caused by gravitational waves. As a result, Fabry–Perot cavities are also mounted along the arms, increasing the effective optical length of the arms up to ≈ 1200 km. The merger of two black holes provided the source of gravitational waves that were detected on 14 September, 2015 by both of the twin LIGO interferometers (Fig. 46.5). This particular gravitational-wave signal is thought to have been due to the inward spiral and merger of a pair of black holes, estimated to be around $36M_\odot$ and $29M_\odot$, and the subsequent 'ringdown' of the single resulting black hole of mass $62M_\odot$, with the remaining $3M_\odot c^2$ energy radiated as gravitational waves. This measurement and subsequent ones have provided

[24]Albert A. Michelson (1952–1931). The Michelson-Morley experiment was, of course, very important in the development of special relativity, although it didn't seem to be one of Einstein's main motivations. It did, however, represent the key test of a v^2/c^2 correction to the pre-relativistic theory, predicted by Lorentz on the strength of the ether theory using a locally defined time. The null result of the experiment led Lorentz to propose length contraction. See Cheng for an accessible account of the history.

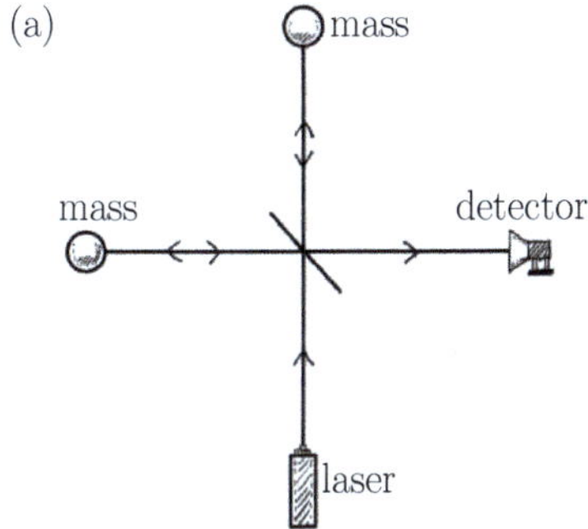

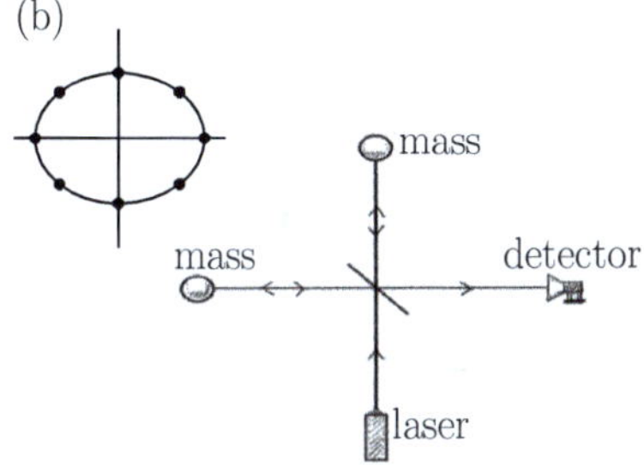

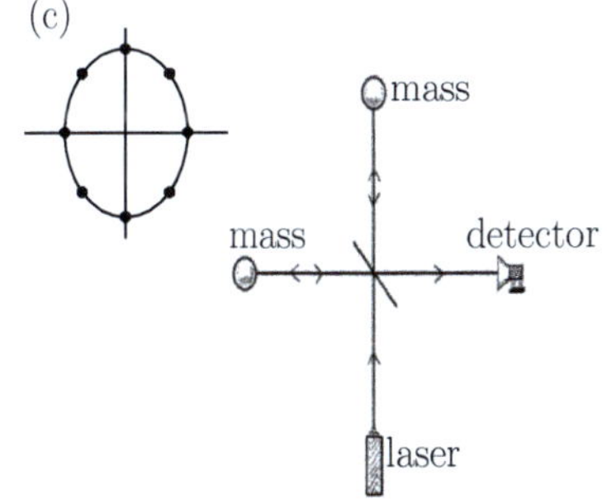

Fig. 46.4 (a) A schematic of a gravitational-wave detector. (b,c) The gravitational wave's effect on a test mass system and its corresponding effect on the arms of the interferometer.

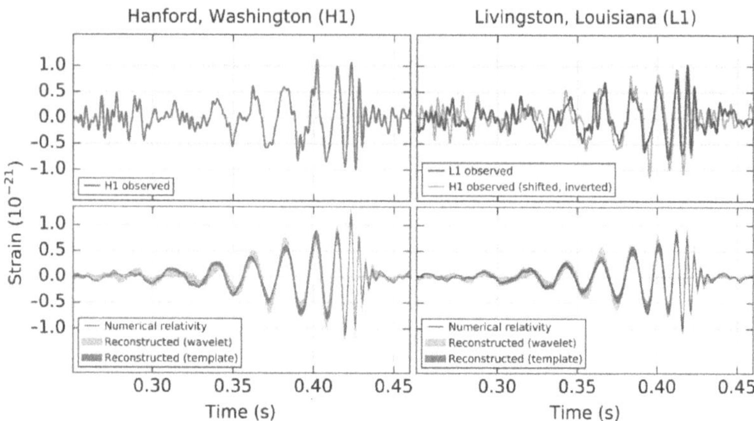

Fig. 46.5 The gravitational-wave event GW150914 observed by the LIGO Hanford (H1, left column panels) and Livingston (L1, right column panels) detectors. Times are shown relative to 14 September, 2015 at 09:50:45 UTC. [From B. P. Abbott *et al.* Phys. Rev. Lett. **116**, 061102 (2016), DOI: 10.1103/PhysRevLett.116.061102, published by the American Physical Society under the terms of the Creative Commons Attribution 3.0 License.]

The fascinating story of the LIGO work has been recounted in many places, but we particularly recommend the three Nobel lectures by Weiss, Barish, and Thorne, published in Rev. Mod. Phys. **90**, 040501 (2018), **90**, 040502 (2018), and **90**, 040503 (2018).

some of the most stringent experimental verifications of general relativity.

These results have stimulated the building of further gravitational wave observatories, including the Einstein Telescope and Cosmic Explorer which are planned to achieve an order of magnitude increase in sensitivity and would therefore be able to study the evolution of compact objects in the early Universe. However, other projects are aimed at building an observatory a long way from the ground. The proposed space-based Laser Interferometer Space Antenna (LISA) will be able to explore much lower frequency gravitational waves (from around 100 μHz to 100 mHz). It is expected to be capable of detecting at very high redshift the first seed black holes formed, as well as intermediate-mass and 'light' super-massive coalescing black hole systems in the 10^2–$10^7 M_\odot$ range. It should therefore be able to follow the evolution of black holes right from the early Universe.

To get to really low frequencies, the best technology for detecting gravitational waves is the use of pulsar timing arrays. These work in the nanohertz to microhertz frequency band and can be used to detect gravitational-wave remnants from the past mergers of supermassive black holes. The basic idea is that rather than using laser interferometers (as used in both LIGO and LISA), one measures the pulse arrival time at Earth from an array of millsecond pulsars (i.e. rapidly rotating neutron stars). These pulsars have extremely regular and stable periods and act as ideal timing sources. A gravitational wave emitted from some astrophysical source will pass the pulsar and the Earth and so this will produce two perturbations on the signal received from the pulsar: one from spacetime variations at the pulsar and the other from spacetime variations on Earth. Data from the pulsar arrays have to be accumulated over many years, so these experiments are not quick.

The stochastic background of remnant primordial gravitational waves produced during the Big Bang will dominate the gravitational-wave spectrum down to approximately 10^{-18} Hz and depending on the cosmological model then some of this background may lie in a region that could be probed by the technologies discussed above. This is an open question, but the modern era of gravitational-wave astronomy that has just begun looks like it is going to have an exciting and productive future in the coming decades.

We have seen in this chapter how general relativity predicts the existence of wave-like excitations in the gravitational field. In the next chapter, we examine gravitational waves from another point of view: that of *quantum* fields, where these waves are quantized into force-carrying particles known as gravitons.

Chapter summary

- A gravitational wave solution to the linearized Einstein equation has the form $\bar{h}_{\mu\nu} = \mathrm{Re}\left[A_{\mu\nu}e^{i\boldsymbol{k}\cdot\boldsymbol{x}}\right]$. The waves have null $\boldsymbol{k}$ and amplitude orthogonal to the direction of propagation.

- Transverse-traceless gauge can be used to simplify the solutions, leading to $+$ and $\times$ polarizations which are $45°$ out of phase.

- There are wave solutions containing nonlinear terms that solve the Einstein equations.

- The gravitational wave luminosity of a source is given by

$$L = \frac{\mathrm{d}E}{\mathrm{d}t} = \frac{G}{5c^5}\langle\dddot{I}_{ij}\,\dddot{I}_{ij}\rangle,$$

  and the radiation is quadrupolar.

- Gravitational wave astronomy is a rapidly growing area in astrophysics. The laser interferometers that make up the ground-based LIGO or the space-based LISA have, or will have, extraordinary sensitivity. Lower frequency gravitational waves are expected to be detected by pulsar timing arrays.

Exercises

(46.1) Verify the components in eqn 46.30.

(46.2) (a) Recall that the connection coefficients are

$$\Gamma^{\alpha}{}_{\mu\nu} = \frac{1}{2}g^{\alpha\lambda}(\partial_{\mu}g_{\nu\lambda} + \partial_{\nu}g_{\lambda\mu} - \partial_{\lambda}g_{\mu\nu}), \quad (46.79)$$

and the Ricci tensor is

$$R_{\mu\nu} = \partial_\alpha \Gamma^\alpha_{\ \nu\mu} - \partial_\nu \Gamma^\alpha_{\ \alpha\mu} + \Gamma^\beta_{\ \beta\alpha}\Gamma^\alpha_{\ \nu\mu} - \Gamma^\beta_{\ \nu\alpha}\Gamma^\alpha_{\ \beta\mu}.$$
$$(46.80)$$

In linearized gravity, we take $g_{\mu\nu} = \eta_{\mu\nu} + h_{\mu\nu}$ so that $g^{\mu\nu} = \eta^{\mu\nu} - h^{\mu\nu}$ (and hence $g_{\mu\alpha}g^{\alpha\nu} = \delta^\nu_\mu + O(h^2)$). The tensor field $h_{\mu\nu}$ is symmetric, so we are free to swap indices around since $h_{\mu\nu} = h_{\nu\mu}$. Hence, show that the quadratic contribution to the Ricci tensor is

$$R^{(2)}_{\mu\nu} = -\frac{1}{2}\partial_\alpha(h^{\alpha\lambda}[\partial_\nu h_{\mu\lambda} + \partial_\mu h_{\lambda\nu} - \partial_\lambda h_{\nu\mu}])$$
$$+ \frac{1}{2}\partial_\nu(h^{\alpha\lambda}[\partial_\alpha h_{\mu\lambda} + \partial_\mu h_{\lambda\alpha} - \partial_\lambda h_{\alpha\mu}])$$
$$+ \frac{1}{4}\eta^{\beta\lambda}\eta^{\alpha\xi}(\partial_\beta h_{\alpha\lambda} + \partial_\alpha h_{\lambda\beta} - \partial_\lambda h_{\beta\alpha})$$
$$\times (\partial_\nu h_{\mu\xi} + \partial_\mu h_{\xi\nu} - \partial_\xi h_{\nu\mu})$$
$$- \frac{1}{4}\eta^{\beta\lambda}\eta^{\alpha\xi}(\partial_\nu h_{\alpha\lambda} + \partial_\alpha h_{\lambda\nu} - \partial_\lambda h_{\nu\alpha})$$
$$\times (\partial_\beta h_{\mu\xi} + \partial_\mu h_{\xi\beta} - \partial_\xi h_{\beta\mu}).\qquad (46.81)$$

(b) Group the terms in eqn 46.81 and show that it reduces to

$$R^{(2)}_{\mu\nu} = \frac{1}{2}\left[\underbrace{\frac{1}{2}\partial_\mu h_{\alpha\beta}\partial_\nu h^{\alpha\beta}}_{1} + \underbrace{h^{\alpha\beta}\partial_\nu\partial_\mu h_{\alpha\beta}}_{2} \right.$$
$$+ \underbrace{(-h^{\alpha\beta}\partial_\nu\partial_\alpha h_{\mu\beta})}_{3} + \underbrace{(-h^{\alpha\beta}\partial_\beta\partial_\mu h_{\alpha\nu})}_{4}$$
$$+ \underbrace{h^{\alpha\beta}\partial_\alpha\partial_\beta h_{\mu\nu}}_{5} + \underbrace{\partial^\alpha h_\mu^{\ \beta}\partial_\alpha h_{\beta\nu}}_{6}$$
$$+ \underbrace{(-\partial^\alpha h_\mu^{\ \beta}\partial_\beta h_{\alpha\nu})}_{7} + \underbrace{(-\partial_\beta h^{\alpha\beta}\partial_\nu h_{\mu\alpha})}_{8}$$
$$+ \underbrace{\partial_\alpha h^{\alpha\beta}\partial_\beta h_{\mu\nu}}_{9} + \underbrace{(-\partial_\alpha h^{\alpha\beta}\partial_\mu h_{\beta\nu})}_{10}$$
$$+ \underbrace{(-\frac{1}{2}\partial_\alpha h_{\mu\nu}\partial^\alpha h)}_{11} + \underbrace{\frac{1}{2}\partial_\nu h_{\alpha\mu}\partial^\alpha h}_{12}$$
$$\left. + \underbrace{\frac{1}{2}\partial_\mu h_\nu^{\ \alpha}\partial_\alpha h}_{13} \right],\qquad (46.82)$$

where $h = h_\alpha^{\ \alpha}$.

(c) Show that the only two terms that survive in this expression are the first two, using

- Tracelessness ($h = 0$) [this annihilates terms 11, 12, and 13].

- The gauge condition $\partial_\mu h^{\mu\nu} = 0$ [this annihilates 4, 8, 9, and 10].

- Any divergence vanishes on the boundary (after averaging over a volume). [This results in 3, 5, and 7 vanishing once you also apply the gauge condition, and 6 similarly going once you use the field equations $\partial^2 h_{\alpha\mu} = 0$].

Show further that the Ricci scalar vanishes using these conditions.

(d) Hence, show that

$$\langle G^{(2)}_{\mu\nu}\rangle = -\frac{1}{4}\langle \partial_\mu h_{\alpha\beta}\partial_\nu h^{\alpha\beta}\rangle.\qquad (46.83)$$

(46.3) (a) Expand the conservation law $T^{\mu\nu}_{\ ,\mu} = 0$ and show that

$$\partial_0 T^{00} = -\partial_k T^{k0},\qquad (46.84)$$
$$\partial_0 T^{0i} = -\partial_k T^{ki}.\qquad (46.85)$$

(b) Define the tensor field $\mathfrak{I}^{ij}$ by

$$\mathfrak{I}^{ij} = x^i x^j T^{00}.\qquad (46.86)$$

This is related to the moment of inertia tensor I_{ij} via

$$I_{ij} = \int_\Sigma \mathrm{d}^3 y\, \mathfrak{I}_{ij}.\qquad (46.87)$$

Using the results in (a), show that $\dot{\mathfrak{I}}_{ij} = \partial_0 \mathfrak{I}_{ij}$ is given by

$$\dot{\mathfrak{I}}_{ij} = -\partial_k(x^i x^j T^{k0}) + x^j T^{i0} + x^i T^{j0}.\qquad (46.88)$$

Show further that

$$\ddot{\mathfrak{I}}_{ij} = -\partial_k(x^i x^j \partial_0 T^{k0} + x^j T^{ik} + x^i T^{jk}) + 2T^{ij}.$$
$$(46.89)$$

(c) Integrate eqn 46.89 over the region Σ and show that

$$\ddot{I}_{ij} = \int_\Sigma \mathrm{d}^3 y\, \ddot{\mathfrak{I}}_{ij} = 2\int_\Sigma \mathrm{d}^3 y\, T_{ij}.\qquad (46.90)$$

(d) Use eqn 46.90 to prove eqn 46.43, i.e. show that

$$\bar{h}_{ij}(t,\vec{x}) = \frac{2G}{R}\frac{\partial^2}{\partial t^2}\int_\Sigma \mathrm{d}^3 y\, [y_i y_j T^{00}(t - R, \vec{y})].$$
$$(46.91)$$

(46.4) In Newtonian gravitation, typical velocities and accelerations are $v^2 \approx Gm/r$ and $a \approx Gm/r^2$, respectively.

(a) Using the result of the previous problem, show that we expect

$$h \approx \frac{Gm}{R}v^2.\qquad (46.92)$$

(b) The second-order terms in the weak-field expansion suggest the gravitational energy-density t varies as $t \approx (\dot{h})^2/G$. Show that

$$t \approx \frac{1}{R}\frac{G^4 m^5}{r^5}.\qquad (46.93)$$

(c) Integrating over a sphere of radius R and restoring factors, show further that the power radiated via gravitational radiation is approximately

$$P \approx \frac{G^4 m^5}{r^5 c^5}. \qquad (46.94)$$

(d) Estimate the power radiated by (i) the solar system, (ii) a collapsing binary star, formed from two stellar-mass black holes, and (iii) a fist, shaken in anger.

(46.5) In Exercise 30.7, we considered a projection operator P_{ij} which projected a vector onto a unit spatial vector $\vec{n} = (n_1, n_2, n_3)$. We now want to find a traceless version of the same thing.
(a) Show that $P_{ij} = \delta_{ij} - n_i n_j$ acts as a projection operator on a vector $\vec{v}$ so that $n_i P_{ij} v_j = 0$. Show also that $\mathrm{Tr}P = P_{ii} = 2$ and $P_{ij} P_{jk} = P_{ik}$.
(b) The action of this projection operator on a tensor M_{ij} requires it to be used twice, so that

$$M'_{k\ell} = P_{ik} P_{j\ell} M_{ij}, \qquad (46.95)$$

and show that $M'_{k\ell} n_k = M'_{k\ell} n_\ell = 0$. However, it is not traceless, and you can show this by proving that $M'_{kk} = \mathrm{Tr}(PM)$.
(c) The solution is to use the **traceless transverse projection operator**

$$M^{\mathrm{TT}}_{k\ell} = (P_{ik} P_{j\ell} - \frac{1}{2} P_{k\ell} P_{ij}) M_{ij}, \qquad (46.96)$$

and show that this is traceless (i.e. $M^{\mathrm{TT}}_{kk} = 0$).
(d) If M itself is traceless and symmetric, show that

$$\begin{aligned} M_{ij}^{\mathrm{TT}} M_{\mathrm{TT}}^{ij} &= (P_{ik} P_{j\ell} P_{im} P_{jn} - P_{ik} P_{j\ell} P_{ij} P_{mn} \\ &\quad + \frac{1}{4} P_{ij} P_{ij} P_{k\ell} P_{mn}) M_{k\ell} M_{mn} \\ &= M_{ij} M_{ij} - 2 n_a n_b M_{ia} M_{jb} \\ &\quad + \frac{1}{2} n_a n_b n_c n_d M_{ab} M_{cd}. \end{aligned} \qquad (46.97)$$

(e) The integral over all solid angles $\int \mathrm{d}\Omega = 4\pi$. Show also that

$$\int n_i n_j \, \mathrm{d}\Omega = \frac{4\pi}{3} \delta_{ij}, \qquad (46.98)$$

and

$$\int n_i n_j n_k n_\ell \, \mathrm{d}\Omega = \frac{4\pi}{15} (\delta_{ij}\delta_{k\ell} + \delta_{ik}\delta_{j\ell} + \delta_{i\ell}\delta_{jk}). \qquad (46.99)$$

Hence, show that

$$\int \langle \dddot{I}^{\mathrm{TT}}_{ij} \dddot{I}^{\mathrm{TT}}_{ij} \rangle \, \mathrm{d}\Omega = \frac{24\pi}{15} \langle \dddot{I}_{ij} \dddot{I}_{ij} \rangle. \qquad (46.100)$$

This can be used to prove eqn 46.60.

(46.6) Consider two black holes of the same mass M in a circular orbit of radius a around their common centre of mass. At a time t the stars are at positions with Cartesian coordinates

$$(x, y, z) = (a \cos \omega t, a \sin \omega t, 0), \qquad (46.101)$$

and

$$(x, y, z) = (-a \cos \omega t, -a \sin \omega t, 0), \qquad (46.102)$$

respectively.
(a) Show that the angular frequency of the circular motion is given by $\omega = (GM/4a^3)^{\frac{1}{2}}$.
(b) Show that the moment of inertia tensor is given by

$$\begin{aligned} I^{ij} &= 2Ma^2 \begin{pmatrix} \cos^2 \omega t & \cos \omega t \sin \omega t & 0 \\ \cos \omega t \sin \omega t & \sin^2 \omega t & 0 \\ 0 & 0 & 0 \end{pmatrix} \\ &= Ma^2 \begin{pmatrix} 1 + \cos 2\omega t & \sin 2\omega t & 0 \\ \sin 2\omega t & 1 - \cos 2\omega t & 0 \\ 0 & 0 & 0 \end{pmatrix}. \end{aligned} \qquad (46.103)$$

(c) Show further that the oscillating gravitational field a distance R away is given by

$$\begin{aligned} \bar{h}_{\mu\nu}(t, R) &= -\frac{8MGa^2\omega^2}{R} \\ &\quad \times \begin{pmatrix} 0 & 0 & 0 & 0 \\ 0 & \cos 2\omega t_r & \sin 2\omega t_r & 0 \\ 0 & \sin 2\omega t_r & -\cos 2\omega t_r & 0 \\ 0 & 0 & 0 & 0 \end{pmatrix}, \end{aligned} \qquad (46.104)$$

where the retarded time is $t_r = t - R$. *This equation tells us that the oscillating field has a frequency twice that of the orbit of the binary system. It is in the form of eqn 46.21 and so also describes the gravitational radiation emitted in the z-direction.*
(d) Show finally that

$$\dddot{I}^{ij} = 8Ma^2\omega^3 \begin{pmatrix} \sin 2\omega t & -\cos 2\omega t & 0 \\ -\cos 2\omega t & -\sin 2\omega t & 0 \\ 0 & 0 & 0 \end{pmatrix}, \qquad (46.105)$$

and hence

$$\langle \dddot{I}^{ij} \dddot{I}_{ij} \rangle = 128M^2 a^4 \omega^6. \qquad (46.106)$$

(46.7) Using the results of the previous problem, derive eqn 46.69 using the expression given in eqn 46.63.

47 The properties of gravitons

[1] This is a question Richard Feynman (1918–1988) asked himself in his course on gravity. Our approach in this chapter follows Feynman's resulting *Lectures on Gravitation* (1995). This was also one of the early paths taken by several scientists (including Feynman) attempting to formulate a quantum theory of gravity. Although informative in many ways, it would ultimately prove unsuccessful. For an introduction to the history, see A. Ashketar *Quantum Gravity*, arXiv:gr-qc/0410054v2 (2004).

[2] We won't assume any familiarity with the techniques of quantum field theory in this chapter.

↻ Since this chapter discusses a hypothetical particle (i.e. the graviton) it can be skipped on a first reading.

Hideki Yukawa (1907–1981).

Though free to think and act, we are held together, like the stars in the firmament, with ties inseparable. These ties cannot be seen, but we can feel them. I cut myself in the finger, and it pains me: this finger is a part of me. I see a friend hurt, and it hurts me, too: my friend and I are one. And now I see stricken down an enemy, a lump of matter which, of all the lumps of matter in the universe, I care least for, and it still grieves me. Does this not prove that each of us is only part of a whole? Nikola Tesla (1856–1943)

Imagine a parallel world where civilization had formulated quantum field theory (QFT) but had no geometrical theory of gravitation. In seeking to describe gravity with the tools at hand, where would their reasoning take them?[1] We shall suggest in this chapter that gravitational excitations would be a natural place for them to start. This would lead to the idea of a **graviton**, a force-carrying particle derived from the quantization of gravitational waves.

In this chapter, we therefore pick up the discussion of the gravitational interactions between masses from the point of view of field theory. We shall work in the weak-field limit, where gravitation is described by a field $h(x)$ and indices are raised and lowered by the Minkowski metric η. Our plan is to work out as much as we can about gravity waves by drawing on the concepts of QFT,[2] where the waves are quantized into force-carrying graviton particles. Although we do not yet have a quantum field theory of gravitation, we can still make progress using the tools from field theory and, indeed, some candidate theories of gravitation predict the existence of gravitons. We shall see that, as in the previous two chapters, we can gain a certain amount of insight by comparing gravity waves to light waves, or in this quantum context, comparing photons to gravitons. The result will be a rather different way to think about gravitational interactions to that we have considered thus far.

47.1 Force-carrying particles

One of the most interesting things about particles is that they interact with each other. Hideki Yukawa's great insight was his suggestion that this interaction process itself involves particles. These **force-carrying particles** are slightly different to their more familiar cousins with whom we're already acquainted. Yukawa's idea centres around one key notion:

Particles interact by exchanging virtual, force-carrying particles.

Recall that a **virtual particle** is one defined as existing 'off mass-shell'.[3]

Interactions are described by field theories. The pattern we've seen in classical field theories (such as electromagnetism and gravitation) is (i) that sources of a field tell the field how to arrange itself; (ii) the field then acts on the sources, via virtual particles, telling them how to move. In quantum electrodynamics (or QED, which is the quantum upgrade of classical electromagnetism), the sources of the fields are electric charges and currents, arranged into a current 4-vector $\boldsymbol{J}$. The force-carrying particles are photons, which are the massless particle excitations of the electromagnetic field $\tilde{\boldsymbol{A}}(x)$. One explanation for the photon having zero mass is that a massless virtual particle is the only way to produce the Coulomb interaction potential $V(r) \propto 1/r$. A massive force-carrying particle would lead to a potential that falls off more rapidly with distance, as examined in the next example.

[3]See Chapter 28. We saw that the argument is that quantum mechanics allows us to violate this classical dispersion relation, as long as we don't do it for too long. By invoking energy-time uncertainty $\Delta E \Delta t \sim \hbar$, we can say that particles of energy E are allowed to exist off the mass-shell as long as they live for a short time $\Delta t \lesssim \hbar/E$. Virtual particles, therefore, must have a finite range since they can't live forever and they travel at finite velocity.

Example 47.1

The role of mass can be understood by considering the mathematical form of the interaction mediated by Yukawa's force-carrying particles. The Yukawa potential is written as $U(\vec{r}) \propto -\dfrac{e^{-\alpha m |\vec{r}|}}{4\pi |\vec{r}|}$, where m is the mass of a force-carrying particle and α is a parameter. This potential obeys the Green's function[4] equation

$$\left(\vec{\nabla}^2 - \alpha^2 m^2\right) U(\vec{r}) = \delta^{(3)}(\vec{r}). \tag{47.1}$$

This equation is a form of the Klein–Gordon equation (see Chapter 40), which is the equation of motion for massive, spinless ($S = 0$) particles. The effective potential representing the interaction mediated by such particles must also obey this equation of motion.

What is the analogous expression for electromagnetism? We know that the Coulomb potential $V(\vec{r}) \propto 1/|\vec{r}|$ must obey Poisson's equation, which is the name we give the Green's function equation

$$\vec{\nabla}^2 V(\vec{r}) = \delta^{(3)}(\vec{r}). \tag{47.2}$$

We see that everything is consistent if we set $m = 0$ for the case of electromagnetism (corresponding to a massless photon). Conversely, if $m \neq 0$, then the electromagnetic interaction would have an exponential contribution $e^{-\alpha m |\vec{r}|}$, which it does not.

[4]A **Green's function** G is the solution of a differential equation of the form $\hat{L} G = \delta$, where $\hat{L}$ is a linear operator and δ is a delta function.

A massless photon necessarily has only two polarization states, associated with the two polarizations of light. The photon also has a spin[5] $S = 1$. We would now like to identify the analogous properties of the **graviton**, the particle excitation of the metric field of gravity that mediates the gravitational force. The gravitational potential varies as $\Phi(r) \propto 1/r$ and so this constrains the mass of the virtual particle to be $m = 0$. What is the spin of the graviton? A spin $S = 1$ theory has the property that like charges repel and unlike charges attract.[6] This is incompatible with gravity, which is purely attractive. Even-integer spin exchange leads exclusively to forces of one sign (either purely attractive

[5]We summarize here the different spins and tensors associated with different field theories:

$$
\begin{array}{lll}
S = 0 & \phi(x) & \text{scalar} \\
S = 1 & A_\mu(x) & \text{vector/1-form} \\
S = 2 & h_{\mu\nu}(x) & \text{second-rank tensor.}
\end{array}
$$

[6]Or vice versa, if the coupling constant has opposite sign.

[7]See Exercise 42.13.

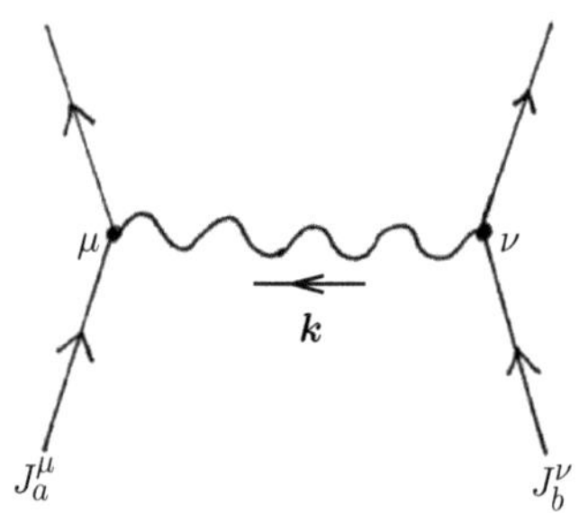

Fig. 47.1 The exchange of a virtual photon causes two currents to interact.

[8]Owing to the symmetry of the interaction, which is reflected in the diagram in Fig. 47.1, each particle actually plays both the role of the scattered particle and that of the source of the scattering potential. However, the argument presented here will get us to the right answer.

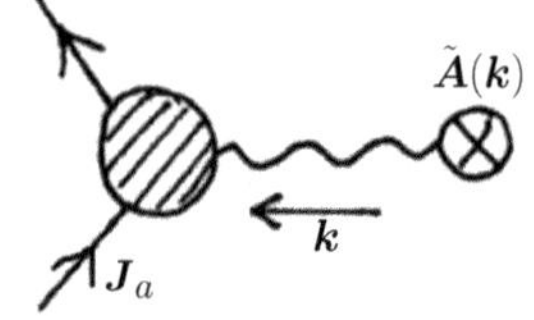

Fig. 47.2 One current acting as the source of the field with which the other interacts.

[9]We can see the motivation for this by combining eqn 47.3 with the interaction Lagrangian $\mathcal{L} = A_\mu(x) J^\mu(x)$, which leads us to predict an interaction with the form $J_a \frac{1}{k^2} J_b$. This is indeed the case.

or purely repulsive), so the graviton must be one of $S = 0, 2, 4, ...$ A scalar field $\phi(x)$ has $S = 0$. However, an $S = 0$ theory turns out to be too simple to capture gravitation[7] (it makes incorrect predictions for the energetics of gravitation for one thing). We shall therefore assume that the graviton is a $S = 2$ particle and, as a result, must be described by a symmetric tensor field. In order to extract some properties of the graviton, we first take a side step and examine the properties of the photon within quantum field theory. We shall then discuss gravitation by making exactly the same mathematical steps, substituting gravitation for electromagnetism. This will reveal the form of the graviton interaction and show the role of graviton polarization.

47.2   Photon propagation and polarization

We know that two electrons have interacted if, on approaching each other, their motion is altered by each other's presence. This is a rough description of scattering, where incoming particles change their momentum states owing to interactions. The probability of scattering is encoded in a quantum mechanical amplitude, and it is this that we shall compute in this section. A useful method of understanding and computing these amplitudes makes use of **Feynman diagrams**. These are momentum-space cartoons of the scattering processes that encode the equations involved in a perturbation expansion of the underlying quantum field theory. The simplest Feynman diagram for electromagnetic electron-electron interactions is shown in Fig. 47.1. Conceptually, the diagram can be understood in terms of the current representing the motion of one electron (called the b-particle for the sake of argument) being the source of the field $\tilde{A}(x)$ with which the current $J_a(x)$ representing the other electron (the a-particle) interacts, as shown in Fig. 47.2.[8]

In general, an electromagnetic field $\tilde{A}$ interacts with a current at a point x via a term in the Lagrangian $\mathcal{L} = J^\mu(x) A_\mu(x)$. As we said above, the source of the electromagnetic field in this case is the current of the b-particle with components $(J_b)^\mu$. The resulting $\tilde{A}$ field has components that can be described (after a suitable choice of gauge) by the equation of motion $\partial^2 A^\mu = -(J_b)^\mu$ or, equivalently, the momentum-space equation

$$A^\mu(\mathbf{k}) = \frac{1}{\mathbf{k}^2}(J_b)^\mu. \tag{47.3}$$

We can understand the interaction of this field with the other current by computing a **scattering amplitude** $\mathcal{A}$ for the current of b-electrons (J_b) to interact with a current of a-electrons (J_a). This has the component form[9]

$$i\mathcal{A} = (\text{Current})^\mu_a \left(\begin{array}{c} \text{Virtual-particle} \\ \text{propagator} \end{array} \right)_{\mu\nu} (\text{Current})^\nu_b. \tag{47.4}$$

The part in the middle, called the **propagator** tells us the probability amplitude for a virtual, force-carrying particle to interact with current

a and current b. This is the process shown in Fig. 47.1. The solid lines represent the currents; the wiggly line represents the photon propagator.

Working in momentum space, the flat-space photon propagator giving the amplitude describing a photon with wavevector $\boldsymbol{k}$ is given by a tensor $\boldsymbol{D}$ with components[10]

$$\tilde{D}_{\mu\nu}(k) = \frac{-i\eta_{\mu\nu}}{\boldsymbol{k}^2 + i\varepsilon}, \tag{47.5}$$

where $\boldsymbol{k}$ is the four-momentum of the photon. (The $i\varepsilon$ factor deals with the fact that $k^0 \neq |\vec{k}|$ for a virtual particle, but won't be important to us here and so will be dropped.) We therefore consider the amplitude

$$\mathcal{A} = -J_a^\mu \left(\frac{\eta_{\mu\nu}}{\boldsymbol{k}^2}\right) J_b^\nu. \tag{47.6}$$

We shall use this to discuss the nature of the interaction and the polarization states of the photon.

[10]The details of this equation won't be important to us, apart from the $1/\boldsymbol{k}^2$ part.

Example 47.2

If we work in a frame where $k^\mu = (k^0, 0, 0, k^3)$, then the amplitude looks like

$$\mathcal{A} = -\frac{\left(-J_a^0 J_b^0 + \vec{J}_a \cdot \vec{J}_b\right)}{-(k^0)^2 + (k^3)^2}. \tag{47.7}$$

We can immediately reduce the number of components using the momentum-space representation of the conservation of current $\boldsymbol{\nabla} \cdot \boldsymbol{J} = 0$, which implies that $k_\mu J^\mu = 0$. We then have $-k_0 J^0 + k_3 J^3 = 0$, which allows us to eliminate the component J^3 to yield

$$\mathcal{A} = \frac{J_a^0 J_b^0}{(k^3)^2} - \frac{J_a^1 J_b^1 + J_a^2 J_b^2}{-(k^0)^2 + (k^3)^2}. \tag{47.8}$$

Now for some interpretation. The first term is written in terms of $J^0 = \rho$, the electromagnetic charge density. If we (inverse) Fourier transform this quantity we obtain an instantaneously acting Coulomb potential, which is repulsive between like charges

$$\int \frac{\mathrm{d}^4 k}{(2\pi)^4}\, \mathrm{e}^{-ik\cdot x} \frac{J_a^0 J_b^0}{(k^3)^2} \propto \frac{q^2}{4\pi|\vec{r}|} \delta(t_a - t_b). \tag{47.9}$$

This only looks (unphysically) instantaneous because we've split up the propagator in a non-covariant manner. Moreover, this is the Coulomb term that dominates in the non-relativistic regime.

The part left over is retarded. That is, it depends on the finite time taken for the photon to propagate. For our case of photons propagating along the z- (or 3-) direction, we look at the amplitude of the second term and we see that there seem to be two sorts of photon: those that couple J^1 currents and those that couple J^2 currents. These are the two physical transverse photon polarizations. To see this, we decompose the term $(J_a^1 J_b^1 + J_a^2 J_b^2)$ into a different basis by writing

$$J_a^1 J_b^1 + J_a^2 J_b^2 = \frac{1}{\sqrt{2}}(J_a^1 + iJ_a^2)\frac{1}{\sqrt{2}}(J_b^1 + iJ_b^2)^\dagger + \frac{1}{\sqrt{2}}(J_a^1 - iJ_a^2)\frac{1}{\sqrt{2}}(J_b^1 - iJ_b^2)^\dagger. \tag{47.10}$$

This implies that two sorts of photons interact: the $J + iJ$ sort and the $J - iJ$ sort. These are indeed the two possible polarizations for the photons, although they are circularly polarized here (see below), compared to the linearly polarized states discussed in the last chapter.

If, in the last example, we use coordinates $J_a^1 = j \cos \phi$ and $J_a^2 = j \sin \phi$, we see that the two polarization can be represented as $je^{i\phi}$ and $je^{-i\phi}$. We conclude that these are circularly polarized photons and, using the angular momentum operator $\hat{L}_z = i\partial/\partial\phi$, they have spin ± 1 respectively.

This concludes a round-up of the properties of photons. Next, we turn to gravitons.

47.3 Graviton propagation and polarization

We examine the case of the graviton by following exactly the steps that we followed for the photon. By analogy with the electromagnetic case, we have that the gravitational field $\boldsymbol{h}(x)$, which exists by virtue of a mass distribution being present, has components

$$h_{\mu\nu}(\boldsymbol{k}) \propto \frac{1}{\boldsymbol{k}^2} T_{\mu\nu}. \tag{47.11}$$

The interaction of this field with another mass distribution can be examined by considering two distributions of mass-energy described by energy-momentum tensors $\boldsymbol{T}_a$ and $\boldsymbol{T}_b$ interacting via a propagator that reflects a massless gravity-carrying particle. Since $\boldsymbol{T}$ is a second-rank tensor, we need to deal with the extra indices, so the analogous scattering amplitude is given by

$$\mathcal{A} = -T_a^{\alpha\beta} \left(\frac{\eta_{\alpha\mu}\eta_{\beta\nu}}{\boldsymbol{k}^2 + i\varepsilon} \right) T_b^{\mu\nu}. \tag{47.12}$$

In the same way that the field $\boldsymbol{A}(x)$ interacts with current $\boldsymbol{J}$ via an interaction term $\mathcal{L} = A_\mu J^\mu$, we expect the interaction of the gravitational field $\boldsymbol{h}$ and the energy-momentum $\boldsymbol{T}$ to take the form

$$\mathcal{L} = h_{\mu\nu} T^{\mu\nu}, \tag{47.13}$$

where $h_{\mu\nu}$ are the components of the weak-field tensor $\boldsymbol{h}(x)$.

Newton's law can be expressed in terms of the instantaneous part of this interaction. Since we know that T^{00} represents ρ, the mass density, then we expect the part reflecting Newton's law to be

$$-\frac{T_a^{00} T_b^{00}}{(k^3)^2}, \tag{47.14}$$

where the sign ensures gravity is attractive. This equation is the (inverse) Fourier transform of the Newtonian potential energy. The retarded term then gives us the information on the graviton polarizations. We therefore expand out the amplitude and find

$$\begin{aligned} -T_a^{\alpha\beta} \frac{\eta_{\alpha\mu}\eta_{\beta\nu}}{\boldsymbol{k}^2} T_b^{\mu\nu} = {} & \frac{-1}{-(k^0)^2 + (k^3)^2} \big(T_a^{00} T_b^{00} - 2T_a^{03} T_b^{03} - 2T_a^{02} T_b^{02} \\ & - 2T_a^{01} T_b^{01} + 2T_a^{23} T_b^{23} + 2T_a^{31} T_b^{31} + 2T_a^{21} T_b^{21} \\ & + T_a^{33} T_b^{33} + T_a^{22} T_b^{22} + T_a^{11} T_b^{11} \big). \end{aligned} \tag{47.15}$$

One complication in dealing with a second-rank tensor $\boldsymbol{T}$ is that it carries around an invariant (i.e. scalar and therefore $S = 0$) part: its trace T. The consequence of this is that when we split up the amplitude $\mathcal{A}$ in terms of the components of $\boldsymbol{T}$, it can erroneously appear that the graviton has three polarizations, rather than the two it must have. The remedy is to use this trace part by adding to our amplitude $\mathcal{A} = T'_{\mu\nu}(1/\boldsymbol{k}^2)T^{\mu\nu}$ a multiple of the trace-part

$$\alpha T' \left(\frac{1}{\boldsymbol{k}^2} \right) T, \tag{47.16}$$

where T and T' denote traces. Here α is a constant that we are free to choose in order that we cancel off any illusory $S = 0$ part and leave only $S = 2$ gravitons. Let's set about decomposing the amplitude.

Example 47.3

As in the photon case, we use the momentum-space version of conservation of mass-energy, $\boldsymbol{\nabla} \cdot \boldsymbol{T} = 0$, to write

$$k_\mu T_b^{\mu\nu} = 0, \tag{47.17}$$

which gives us

$$k_0 T_b^{0\nu} = k_3 T_b^{3\nu}. \tag{47.18}$$

Using this to eliminate $T^{3\nu}$ we find that the instantaneous part becomes

$$-\frac{1}{(k^3)^2} \left[T_a^{00} T_b^{00} \left(1 - \frac{(k^0)^2}{(k^3)^2} \right) - 2T_a^{01} T_b^{01} - 2T_a^{02} T_b^{02} \right]. \tag{47.19}$$

The retarded term is then given by

$$\frac{-1}{-(k^0)^2 + (k^3)^2} \left(T_a^{11} T_b^{11} + T_a^{22} T_b^{22} + 2T_a^{21} T_b^{21} \right). \tag{47.20}$$

To remove the $S = 0$ contribution we add the term in eqn 47.16, which contributes a piece to the retarded term of

$$\alpha \frac{1}{-(k^0)^2 + (k^3)^2} (T_a^{11} + T_a^{22})(T_b^{11} + T_b^{22}). \tag{47.21}$$

Now we choose α so that there are only two terms in the $S = 2$ retarded part (reflecting the two polarizations). This is achieved by setting $\alpha = 1/2$ and so we obtain

$$\frac{-1}{-(k^0)^2 + (k^3)^2} \left[\frac{1}{2}(T_a^{11} - T_a^{22})(T_b^{11} - T_b^{22}) + 2T_a^{12} T_b^{12} \right]. \tag{47.22}$$

Since we can use the symmetry of $\boldsymbol{T}$ to rewrite $2T_b^{12} = (T_b^{12} + T_b^{21})$, we can write the retarded part of the amplitude as

$$\frac{-1}{-(k^0)^2 + (k^3)^2} \left[\frac{1}{2}(T_a^{11} - T_a^{22})(T_b^{11} - T_b^{22}) + \frac{1}{2}(T_a^{12} + T_a^{21})(T_b^{12} + T_b^{21}) \right]. \tag{47.23}$$

We conclude that there are two graviton polarizations: $(T_b^{11} - T_b^{22})$ and $(T_b^{12} + T_b^{21})$. The field required to generate these gravitons takes the form

$$h^{\mu\nu}(x) = A^{\mu\nu} e^{i\boldsymbol{k}\cdot\boldsymbol{x}}. \tag{47.24}$$

We can identify from the previous example that

$$A^{11} = \tfrac{1}{\sqrt{2}}, \quad -A^{22} = \tfrac{1}{\sqrt{2}}, \quad A^{12} = A^{21} = \tfrac{1}{\sqrt{2}}. \tag{47.25}$$

How do we know the expressions identified in the last example are the correct polarizations for an $S = 2$ field? For circularly polarized gravitons, we must be able to shift to a coordinate system where the phases behaves as $\mathrm{e}^{2i\theta}$ and $\mathrm{e}^{-2i\theta}$. As can be checked, it is possible to rewrite the polarization part of the retarded term as

$$\frac{1}{4}\left(T_a^{11} - T_a^{22} + 2\mathrm{i}T_a^{12}\right)\left(T_b^{11} - T_b^{22} - 2\mathrm{i}T_b^{12}\right)$$
$$+ \frac{1}{4}\left(T_a^{11} - T_a^{22} - 2\mathrm{i}T_a^{12}\right)\left(T_b^{11} - T_b^{22} + 2\mathrm{i}T_b^{12}\right). \tag{47.26}$$

Each bracket has the form $(xx - yy \pm 2\mathrm{i}xy)$, which is equivalent to $(x\pm\mathrm{i}y)(x\pm\mathrm{i}y)$. Since each of the bracketed terms in this last expression can be represented as a phase $\mathrm{e}^{\pm i\theta}$, their product has the required $\mathrm{e}^{\pm 2i\theta}$ phase.[11]

[11]This is examined further in the exercises.

Example 47.4

From the last example we see that the amplitude can be written as

$$\mathcal{A} = \frac{-(T_a)_{\mu\nu}T_b^{\mu\nu} + \frac{1}{2}(T_a)^{\mu}{}_{\mu}(T_b)^{\nu}{}_{\nu}}{k^2}. \tag{47.27}$$

We can then spot two things. This first is that the amplitude $\mathcal{A}$ can be rewritten in terms of the graviton propagator as $\mathcal{A} = T'^{\sigma\tau}D_{\sigma\tau\mu\nu}T^{\mu\nu}$ from which we find an expression for the propagator of

$$D_{\sigma\tau\mu\nu} = -\frac{1}{2}\frac{(\eta_{\mu\sigma}\eta_{\nu\tau} + \eta_{\mu\tau}\eta_{\nu\sigma} - \eta_{\mu\nu}\eta_{\sigma\tau})}{k^2}. \tag{47.28}$$

Since the amplitude $\mathcal{A}$ is also proportional to $h_{\mu\nu}T_a^{\mu\nu}$, we conclude that amplitude of gravitons emitted from a source can be written in terms of a field as

$$h_{\mu\nu}(\boldsymbol{k}) \propto \frac{1}{k^2}\left(T_{\mu\nu} - \frac{1}{2}\eta_{\mu\nu}T\right). \tag{47.29}$$

The part in brackets is, of course, familiar from the Einstein equation, so it is heartening to see it appear from this completely different approach. In fact, this equation is recognizable as the momentum-space version of the weak-field Einstein equation.

[12]The problem here is related to **renormalization**. Calculations of scattering processes using perturbation theory in QFT often lead to infinities. Unlike the case in some other QFTs, the infinities in gravitational perturbation theory, encountered at higher orders of scattering, cannot be removed by the set of techniques, known as renormalization, that have proved successful in removing infinities in theories like quantum electrodynamics or quantum chromodynamics. For example, if we compute the amplitude for gravitons to scatter from gravitons at energy E, perturbation theory predicts the amplitude is given by a series of the form $G\left[1 + GE^2 + (GE^2)^2 + ...\right]$, where G is the gravitational constant. Once the energy scale E reaches $G^{-\frac{1}{2}}$ this approach fails as the series diverges. (By analogy with the closely related Fermi theory of the weak interaction, we might expect some new physics to appear at this energy scale.) The infinities encountered in this approach are avoided to an extent in string theory, which follows a similar route to QFT and is described in Chapter 49. See Zee's *Quantum Field Theory in a Nutshell* (2003) for a discussion of the analogy between graviton-graviton scattering and the Fermi theory.

Although we might feel pleased with the progress made in this chapter, it's sobering to remember that nobody has yet quantized gravity consistently. The approach suggested here, which has perturbation theory at its root, has been attempted several times, but does not lead to a consistent theory.[12] A successful quantum field theory of gravity might still be expected to result in a prediction of the graviton excitations with properties something like those that we have discussed here. (However, there is nothing to guarantee this.) In Chapter 49, we shall look at some possible avenues for this project of finding quantum gravity. In order to get there, we shall need to make some more room in spacetime by considering the possibility of spacetime with more than $(3+1)$ dimensions, which is our next subject.

Chapter summary

- The graviton is a force-carrying particle with spin $S = 2$ and two polarizations.
- The graviton has spin $S = 2$ because its source is a second-rank tensor $T^{\mu\nu}$; the photon has spin $S = 1$ because its source is a first-rank tensor J^{μ}.
- In scattering theory, the gravitational interaction between masses can be written in terms of an amplitude as

$$\mathcal{A} = -T_a^{\alpha\beta} \left(\frac{\eta_{\alpha\mu}\eta_{\beta\nu}}{k^2 + i\varepsilon} \right) T_b^{\mu\nu}. \tag{47.30}$$

- The perturbative scattering approach fails to describe gravitation at higher orders of perturbation theory.

Exercises

(47.1) Verify that eqn 47.26 is equivalent to the retarded term in the graviton amplitude.

(47.2) The polarization vectors of a spin-1 particle change according to $(\epsilon'_\mu)^i = R^i{}_j(\theta)(\epsilon_\mu)^j$, where the rotation matrix is given by

$$R(\theta) = \begin{pmatrix} \cos\theta & \sin\theta & 0 \\ -\sin\theta & \cos\theta & 0 \\ 0 & 0 & 1 \end{pmatrix}. \tag{47.31}$$

Find combinations of the linear polarization vectors that obey the transformation law

$$R^i{}_j(\theta)(\epsilon_h)^j = e^{ih\theta}(\epsilon'_h)^i, \tag{47.32}$$

for helicities $h = 1, 0$ and -1.
For photons, only the $h = 1$ and $h = -1$ helicities are found in Nature.

(47.3) For gravitons, polarizations are expressed as (0,2) tensors ϵ with components ϵ_{ij} that transform as

$$(\epsilon'_\mu)_{ij} = R_i{}^m(\theta)R_j{}^n(\theta)(\epsilon_\mu)_{mn}, \tag{47.33}$$

where the rotation matrix is the same as in the last question. (Note the slightly awkward ordering of the components here. We take $R^i{}_j = R_i{}^j$

in any case.) For a graviton travelling along z, we found in the previous chapter that the only non-zero components are $\epsilon_{11} = -\epsilon_{22}$ and $\epsilon_{12} = \epsilon_{21}$.
(a) Show that under the transformation we obtain

$$\begin{aligned} (\epsilon')_{11} &= (\cos^2\theta - \sin^2\theta)\epsilon_{11} + 2\sin\theta\cos\theta\epsilon_{12}, \\ (\epsilon')_{12} &= -2\sin\theta\cos\theta\epsilon_{11} + (\cos^2\theta - \sin^2\theta)\epsilon_{12}. \end{aligned} \tag{47.34}$$

(b) Find linear combinations of the polarization tensors that yield the two polarization states of the graviton with $h = \pm 2$.

(47.4) (a) Show that the gravitational wave power L from a binary system of two masses, m_1 and m_2, separated by distance a, and in a circular orbit about their centre of mass, is given by

$$L = \frac{32G^4 m_1^2 m_2^2 (m_1 + m_2)}{5c^5 a^5}. \tag{47.35}$$

Estimate this quantity for the Earth–Sun system. Also, estimate the number of gravitons emitted per second.
(b) Estimate how long it would take you to emit a single graviton by frantically waving your arms around in the air.

48

Higher dimensional spacetime

[1] Theodor Kaluza (1885–1954). It is said that he taught himself to swim by reading a book, resulting in him successfully swimming on his first attempt.

[2] We follow the approach of Zee in this chapter. Kaluza's theory was rediscovered and developed by Oskar Klein (1894–1977) in 1926, who provided a quantum mechanical description of the theory. Gunnar Nordström had also independently developed a related theory before Kaluza.

[3] Max Planck (1858–1947) was an early champion of special relativity, extending the theory by formulating the relativistic action. Planck and Einstein were close friends who would meet to play music together. In his biography of Einstein, Abraham Pais notes Einstein's profound respect for Planck, both as a scientist and as a deeply principled human being.

[4] Remember the conceptual equation

$$\boxed{\partial g} + \boxed{\partial^2 g} \rightarrow \boxed{R}.$$

The idea that this can be achieved through a five dimensional cylinder-world has never occurred to me and would seem to be altogether new. I like your idea at first sight very much.
Albert Einstein, *letter to Theodor Kaluza* (1919)

General relativity presents us with a classical field theory of gravitation expressed using the tools of geometry. In Chapter 42, we met the classical field theory of electromagnetism expressed in similar geometric language. It's natural to ask whether gravitation and electromagnetism can be combined in such a way that they naturally arise as different facets of some master theory. This is the project of **unification** which, in a broader, modern sense, involves a combination of gravity and the standard model of particle physics. In this chapter, we examine an attempt, originally made by Theodor Kaluza[1] in 1919, to use the gauge structure of electromagnetism and gravity to combine these interactions. The solution, known as **Kaluza–Klein theory**,[2] involves adding an extra spatial dimension to spacetime.

Since, in this chapter, we shall be comparing theories in different numbers of dimensions, it will be helpful to make the action of our gravitation theory dimensionless. We do this by employing some dimensional analysis. We use units where $c = 1$ and also where Planck's constant[3] $\hbar = 1$. In such units, a mass has units of $1/(\text{length})$ or $1/L$. The Einstein–Hilbert action was previously written as

$$S_{\text{EH}} = \int \mathrm{d}^4 x \, \sqrt{-g} R(\boldsymbol{g}), \tag{48.1}$$

where we write the Ricci scalar as $R(\boldsymbol{g})$ to remind us that it is derived from the first and second derivatives of the components of the metric tensor.[4] The components of the metric tensor $\boldsymbol{g}$ are dimensionless. The Ricci scalar R involves two derivatives of the metric and therefore carries units of $1/L^2$. As a result, S_{EH} with its four contributions of length from $\mathrm{d}^4 x$, has units L^2. In order to make it dimensionless, we multiply by two powers of mass m_{P} with the result that

$$S_{\text{EH}} = \int \mathrm{d}^4 x \, \sqrt{-g} m_{\text{P}}^2 R(\boldsymbol{g}). \tag{48.2}$$

The mass we choose sets the scale of gravitational interactions and is known as the **Planck mass**. We shall discuss this quantity further in Chapter 49.

48.1 Gauge transformations in five dimensions

Kaluza's scheme for unifying electromagnetism and gravity can be understood by (once again) comparing the structure of the gauge transformations in electromagnetism and in gravitation. We saw in Chapter 42 that the gauge transformation in electromagnetism[5] can be written in terms of the components of the electromagnetic 1-form $\tilde{\boldsymbol{A}}$ as

$$A_\mu \to A_\mu - \chi_{,\mu}. \tag{48.3}$$

This transformation has no effect on the Faraday 2-form $\tilde{\boldsymbol{F}} = \boldsymbol{d}\tilde{\boldsymbol{A}}$ and hence on the underlying equations of motion of the electromagnetic fields.

From the point of view of Chapter 44, the field $\tilde{\boldsymbol{A}}$ was mandated by changes made in the internal phase variable $\theta(\mathcal{P}) \to \theta(\mathcal{P}) + \alpha(\mathcal{P})$, where $\mathcal{P}$ labels a point in space. This variable has its information stored in a bundle of fibres, one at each point $\mathcal{P}$, floating above the spacetime. In the weak-field limit of gravitation, we recall that the gauge structure is derived by considering invariance under a set of infinitesimal coordinate transformations $x^\mu(\mathcal{P}) \to x^\mu(\mathcal{P}) + \xi^\mu(\mathcal{P})$, which causes the components of the tensor $\boldsymbol{h}$ to change according to

$$h_{\mu\nu} \to h_{\mu\nu} - \xi_{\mu,\nu} - \xi_{\nu,\mu}. \tag{48.4}$$

The key to unification is to treat the internal variable θ as describing a coordinate in spacetime. That is, we regard a coordinate that was previously stored in a fibre as now describing a position in spacetime. This allows us to combine the electromagnetic and gravitational gauge transformations in a manner where they all derive from a single set of infinitesimal coordinate transformations, and reveals the structure needed to unify the two interactions.

Kaluza's inspired idea was therefore to add to our four-dimensional coordinates $x^\mu = (x^0, x^1, x^2, x^3)$ a fifth coordinate x^5 to form the five-dimensional coordinate set[6] $X^a = (x^0, x^1, x^2, x^3, x^5)$. We now demand invariance under the infinitesimal coordinate transformation

$$X^a(\mathcal{P}) \to X^a(\mathcal{P}) + \xi^a(\mathcal{P}). \tag{48.5}$$

In the new (4+1)-dimensional spacetime, we have a five-dimensional weak-field metric $\boldsymbol{H}$ with components

$$H_{ab} = \eta_{ab} + h_{ab}, \tag{48.6}$$

where the five-dimensional version of the Minkowski metric has components $\eta_{ab} = \mathrm{diag}(-1, 1, 1, 1, 1)$. We have again the gauge transformation property of the metric components that $h_{ab} \to h_{ab} - \xi_{a,b} - \xi_{b,a}$.

We now use the fifth dimension to accommodate the electromagnetic gauge freedom. To see this set the index $b = 5$ and we have

$$h_{\mu 5} \to h_{\mu 5} - \xi_{\mu,5} - \xi_{5,\mu}. \tag{48.7}$$

[5]Remember that this arose from demanding local phase invariance for the matter fields that fill spacetime.

[6]In what follows we always let μ run over values 0-3 and we let $a = 0, 1, 2, 3$ and 5. The reason for the introduction of x^5, rather than the more logical x^4, is historical: it was conventional to call the timelike component x^4 in the older literature, instead of the more modern choice of x^0.

[7] Remembering that Greek indices like μ run over 0-3 only.

We then (i) set $h_{\mu 5} = \ell A_\mu$, where ℓ is an arbitrary length; and (ii) assume ξ_μ is independent of x^5. The result is that eqn 48.7 becomes

$$\ell A_\mu \to \ell A_\mu - \xi_{5,\mu}, \tag{48.8}$$

which is identical to the electromagnetic gauge transformation in eqn 48.3 if we set $\chi = \xi_5/\ell$. In fact, if we take[7] $\xi_\mu = 0$ then the electromagnetic gauge transformation becomes

$$x^\mu \to x^\mu, \quad x^5 \to x^5 + \ell \chi(x^\mu). \tag{48.9}$$

In summary, the electromagnetic gauge transformation has been absorbed into the infinitesimal coordinate transformation. Specifically, we recall that the 1-form $\tilde{\boldsymbol{A}} = A_\mu \boldsymbol{dx}^\mu$ transforms according to

$$\tilde{\boldsymbol{A}} \to \tilde{\boldsymbol{A}} - \boldsymbol{d}\chi. \tag{48.10}$$

Since the coordinate x^5 transforms according to $x^5 \to x^5 + \ell\chi$ then we have $\boldsymbol{dx}^5 \to \boldsymbol{dx}^5 + \ell \boldsymbol{d}\chi$, and we can spot that the combination $(\boldsymbol{dx}^5 + \ell\tilde{\boldsymbol{A}})$ is gauge invariant. This quantity then, linking a spatial coordinate and the electromagnetic field, is key to unification.

From the point of view of the gauge structure, the extra dimension can be used to bring the electromagnetic gauge transformation down from the fibre bundle and into the heart of spacetime itself. Motivated by this, we shall see in the next section how the extra structure can be used to build a metric that incorporates electromagnetism and gravitation.

48.2   Unifying electromagnetism and gravitation

To simplify our notation a little, let's call the x^5 coordinate z. We write the action for the enlarged spacetime as

$$S = \int \mathrm{d}^4 x \mathrm{d} z \sqrt{-H}\, m_{\mathrm{K}}^3 R(\boldsymbol{H}), \tag{48.11}$$

where an extra mass factor m_{K} has been included since there are now 5 powers of length in the terms $\mathrm{d}^4 x \mathrm{d} z$. The mass m_{K} sets the scale for five-dimensional gravity, just as the Planck mass m_{P} did in four dimensions. The form of the metric is strongly constrained if we stipulate that it must be gauge invariant. Since the gauge transformation changes $z \to z + \ell\chi$ there is only one way a gauge invariant metric line element can be constructed. The line element must be

$$\mathrm{d}s^2 = g_{\mu\nu}\mathrm{d}x^\mu \mathrm{d}x^\nu + (\mathrm{d}z + \ell A_\mu \mathrm{d}x^\mu)^2. \tag{48.12}$$

[8] In these matrix equations, the usual $(3+1)$ dimensions described by Greek indices live in the top left block $M_{\mu\nu}$, with the new, fifth dimension in the bottom right component M_{55}. The off-diagonal components $M_{\mu 5}$ and $M_{5\mu}$ mix $(3+1)$ dimensions and the fifth dimension.

Expanding the bracket we can read off the components of the metric, which takes the form of a matrix[8]

$$H_{\mu\nu} = \begin{pmatrix} g_{\mu\nu} + \ell^2 A_\mu A_\nu & \ell A_\mu \\ \ell A_\mu & 1 \end{pmatrix}. \tag{48.13}$$

Example 48.1

The (4+1) metric can be used to generate a Ricci scalar $R(\boldsymbol{H})$, with the result that[9]

$$R(\boldsymbol{H}) = R(\boldsymbol{g}) - \frac{1}{4} F^{\mu\nu} F_{\mu\nu}. \tag{48.14}$$

This gives us contributions to the action from gravitation and from electromagnetism,

$$S = \int \mathrm{d}^4 x \mathrm{d}z \, \sqrt{-H} \left(R(\boldsymbol{g}) - \frac{1}{4} F^{\mu\nu} F_{\mu\nu} \right), \tag{48.15}$$

where $H = \det H_{ab}$ and with the electromagnetic part having the Lagrangian $\mathcal{L} = -\frac{1}{4} F^\mu F_{\mu\nu}$ that we met in Chapter 42.

We have seen that for the cost of adding an extra dimension to space, gravitation, and electromagnetism can be combined. But if this is a description of reality, where is this extra dimension and, since we don't appear to have detected it in our measurements, how can it be explored?

Taking our lead from the structure of our fibres in Chapter 44, we propose that the extra dimension is hidden from us by virtue of its being wound, or **compactified**, into a very small circle of radius a. As a result, $z \equiv x^5$ varies in the range $0 \le x^5 \le 2\pi a$. This curvature of space is permitted by general relativity and, since in our units energy has units $1/L$, we see that potentially enormous energies would be needed to experimentally resolve the dimension if it is small enough.[10] Put another way, in order to escape into this dimension, a particle would need (a huge) momentum of order $p \approx 1/a$. We therefore have the picture of spacetime in Fig. 48.1. It has, at each point x^μ in its (3+1)-dimensional subspace, a dimension resembling a tiny, circular knob. The electromagnetic gauge transformation $x^5 \to x^5 + \ell\chi(x^\mu)$ corresponds to a rotation of the knobs by different amounts at each point.

Example 48.2

Using this idea we can link the interaction scale m_K to the Planck mass m_P. In the absence of electromagnetic field, we have $H_{\mu\nu} = g_{\mu\nu}$, $H_{\mu 5} = 0$ and $H_{55} = 1$. The action then becomes

$$S = 2\pi a m_\mathrm{K}^3 \int \mathrm{d}^4 x \, \sqrt{-g} R(\boldsymbol{g}), \tag{48.16}$$

and so we recognize from eqn 48.2 that $m_\mathrm{P}^2 = 2\pi a m_\mathrm{K}^3$.

We can use the Kaluza–Klein metric $\boldsymbol{H}$ to work out the equations of motion for a particle in flat (3+1)-dimensional spacetime.

Example 48.3

We start from the action for a particle,[11] including the coordinate z, written as

$$S = -m \int \left[-\eta_{\mu\nu} \mathrm{d}x^\mu \mathrm{d}z^\nu + (\mathrm{d}z + \ell A_\mu \mathrm{d}x^\mu)^2 \right]^{\frac{1}{2}}. \tag{48.17}$$

[9]See the exercises and also the book by Zee (2013) for the details of how this is done.

[10]This theme of hidden, compactified dimensions is one we pick up in Chapter 49.

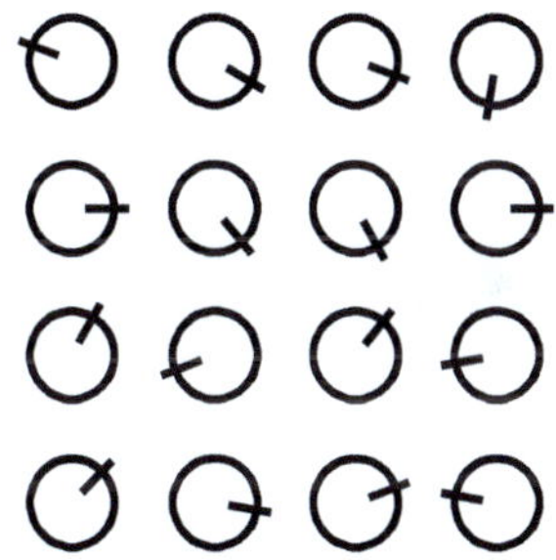

Fig. 48.1 Spacetime in the Kaluza–Klein theory. At every point in three-dimensional space an extra dimension can be found, wound up into a small circle of radius a.

[11]This is simply an extension of the usual action for a particle in an electromagnetic field of $S = -m \int \sqrt{-\eta_{\mu\nu} \mathrm{d}x^\mu \mathrm{d}x^\nu} + q \int A_\sigma \mathrm{d}x^\sigma$ from Chapter 42.

Using the Euler–Lagrange equations, we end up with two equations of motion. The first says that the momentum p^z in the z-direction is constant and given by

$$p^z = m\left(\frac{\mathrm{d}z}{\mathrm{d}\tau} + \ell A_\mu \frac{\mathrm{d}x^\mu}{\mathrm{d}\tau}\right). \tag{48.18}$$

The second equation of motion is

$$\frac{\mathrm{d}}{\mathrm{d}\tau}\left(-\eta_{\mu\nu}m\frac{\mathrm{d}x^\mu}{\mathrm{d}\tau} + (p^z\ell)A_\nu\right) = (p^z l)\frac{\partial A_\lambda}{\partial x^\nu}\frac{\mathrm{d}x^\lambda}{\mathrm{d}\tau}, \tag{48.19}$$

which becomes, on collecting the components of the Faraday tensor[12] $\boldsymbol{F}$ and doing some simplifying,

$$m\frac{\mathrm{d}^2x^\mu}{\mathrm{d}\tau^2} = (p^z\ell)F^\mu{}_\nu u^\nu, \tag{48.20}$$

where u^ν are components of the particle's velocity $\boldsymbol{u}$. Comparing with what we had in Chapter 42, we see that the electromagnetic charge is given by $q = p^z\ell$. In words: the momentum along the z-direction tells us q, the strength of the interaction between the particle and the electromagnetic field $\tilde{\boldsymbol{A}}$. Finally, since the wavefunction for a particle confined to a circle has the form $\psi(z) \propto e^{ip^z z}$ where, in order for the wavefunction to be single valued, we require a quantized $(p^z)_n = 2\pi n/2\pi a = n/a$, where n is an integer. This implies that electric charge in this picture must be quantized in units of

$$q = \frac{\ell}{a}, \tag{48.21}$$

that is, the ratio of length ℓ and the radius of the extra dimension a.

The unification that Kaluza–Klein theory achieves is very interesting, but ultimately we still lack a quantum theory of gravitation that combines gravity and the standard model of particle physics. The search for one is the subject of our next chapter. Before closing this chapter, we can use the tools we have developed to address a different extension of general relativity: what if, instead of extra dimensions, there are extra interactions?

[12] These are $F_{\mu\nu} = A_{\nu,\mu} - A_{\mu,\nu}$.

[13] In the last chapter, for example, we predicted with quantum field theory (QFT) that the amplitude for graviton-graviton scattering varies with energy E as $G(1 + GE^2 + ...)$, where G is the gravitational constant. Effective field theory says that such an approach is permissible as long as we confine ourselves to the realm of applicability of the theory. In this case, this is $E \ll G^{-\frac{1}{2}}$. As discussed in the next chapter, this limiting energy scale is that of the Planck mass m_P.

[14] Effective theories are described in more detail in Zee. This point of view has been influential in Condensed Matter and in Particle Physics, where it follows from analysis using renormalization group (RG) techniques. See our *Quantum Field Theory for the Gifted Amateur* (2014) for a description of RG.

Example 48.4

The field-theory approach allows us to treat general relativity as an **effective theory**. The idea here is that our observations are made at low energies and long-length scales (relative to some very small length scale ℓ, for example).[13] As a result, the terms in the gravitational action that determine the equations of motion that we can probe in our observations are those where the fields (such as the metric) vary most slowly. It might be that there are really higher order terms in the relativistic action where the fields vary more rapidly, while the ones we have identified represent an effective, low-energy approximation to a more complete theory of gravitation.[14] The higher order terms are those scalars that involve more derivatives of the metric, so will combine more multiples of the objects formed from the components of $\boldsymbol{R}$. We can use these to upgrade our Einstein–Hilbert action from $S_\mathrm{EH} = m_\mathrm{P}^2 \int \mathrm{d}^4x\sqrt{-g}R$ to

$$S'_\mathrm{EH} = m_\mathrm{P}^2 \int \mathrm{d}^4x\sqrt{-g}\left[R + \ell^2\left(\alpha R^2 + \beta R_{\mu\nu}R^{\mu\nu} + \gamma R_{\mu\nu\sigma\rho}R^{\mu\nu\sigma\rho}\right) + ...\right], \tag{48.22}$$

where ℓ is a length and α, β and γ are constants. The introduction of the length ℓ is required on dimensional grounds, since the terms in the brackets involve two more derivatives of the metric field than R does. This length allows us to fix the scale at which these extra interactions are important, just as an analogous quantity allowed to pick out the scale at which extra dimensions are important earlier in the chapter.

Fixing a probable value of ℓ will occupy us in the next chapter.

Chapter summary

- Kaluza–Klein theory combines electromagnetism and gravitation via their gauge structure by introducing an extra spatial dimension described by a coordinate x^5.
- A metric for the resulting $(4+1)$-dimensional spacetime incorporates the metric for $(3+1)$-dimensional spacetime along with the electromagnetic field $\tilde{\boldsymbol{A}}(x)$.
- The extra dimension is wound into a tiny circle. Escaping into this dimension costs enormous amounts of energy.

Exercises

(48.1) *We shall use the metric in eqn 48.13 to show eqn 48.14, using the method from Chapter 36. We follow the steps in the textbook by Zee, which should be consulted for further discussion.*
We absorb the factor of ℓ into A_μ so that the metric is

$$\boldsymbol{ds}^2 = g_{\mu\nu}\boldsymbol{\omega}^\mu \otimes \boldsymbol{\omega}^\nu + (\boldsymbol{dz} + A_\mu \boldsymbol{dx}^\mu)^2, \quad (48.23)$$

where Greek indices, as usual, range from 0 to 3. We have, in the orthonormal frame,

$$\boldsymbol{ds}^2 = \eta_{\hat{\alpha}\hat{\beta}}\boldsymbol{\omega}^{\hat{\alpha}} \otimes \boldsymbol{\omega}^{\hat{\beta}} + \boldsymbol{\omega}^{\hat{5}} \otimes \boldsymbol{\omega}^{\hat{5}}, \quad (48.24)$$

and

$$\boldsymbol{\omega}^{\hat{5}} = (\boldsymbol{dz} + A_{\hat{\alpha}}\boldsymbol{dx}^{\hat{\alpha}}). \quad (48.25)$$

(a) Evaluate $\boldsymbol{d\omega}^{\hat{5}}$ to show (using **idea 1** from Chapter 36) that

$$\boldsymbol{\omega}^{\hat{5}}{}_{\hat{\alpha}} = \frac{1}{2}F_{\hat{\alpha}\hat{\beta}}\boldsymbol{\omega}^{\hat{\beta}}. \quad (48.26)$$

(b) Show further that

$$\boldsymbol{\omega}^{\hat{\alpha}}{}_{\hat{\beta}} = \boldsymbol{\Omega}^{\hat{\alpha}}{}_{\hat{\beta}} - \frac{1}{2}F^{\hat{\alpha}}{}_{\hat{\beta}}\boldsymbol{\omega}^{\hat{5}}, \quad (48.27)$$

where $\boldsymbol{\Omega}^{\hat{\alpha}}{}_{\hat{\beta}}$ are the connection 1-forms for the usual four-dimensional metric.
(c) By using **Idea 2** from Chapter 36, verify that

$$\begin{aligned}\mathcal{R}^{\hat{\alpha}}{}_{\hat{\beta}} =&\,\boldsymbol{d\Omega}^{\hat{\alpha}}{}_{\hat{\beta}} - \frac{1}{2}F^{\hat{\alpha}}{}_{\hat{\beta},\hat{\gamma}}\boldsymbol{\omega}^{\hat{\gamma}} \wedge \boldsymbol{\omega}^{\hat{5}} - \frac{1}{2}F^{\hat{\alpha}}{}_{\hat{\beta}}\boldsymbol{d\omega}^{\hat{5}} \\ &+ \left(\boldsymbol{\Omega}^{\hat{\alpha}}{}_{\hat{\gamma}} - \frac{1}{2}F^{\hat{\alpha}}{}_{\hat{\gamma}}\boldsymbol{\omega}^{\hat{5}}\right) \wedge \left(\boldsymbol{\Omega}^{\hat{\gamma}}{}_{\hat{\beta}} - \frac{1}{2}F^{\hat{\gamma}}{}_{\hat{\beta}}\boldsymbol{\omega}^{\hat{5}}\right) \\ &- \left(\frac{1}{2}F^{\hat{\alpha}}{}_{\hat{\gamma}}\boldsymbol{\omega}^{\hat{\gamma}}\right) \wedge \left(\frac{1}{2}F_{\hat{\beta}\hat{\delta}}\boldsymbol{\omega}^{\hat{\delta}}\right). \end{aligned} \quad (48.28)$$

(d) Since we are only trying to compute the Ricci scalar, terms containing $\boldsymbol{\omega}^{\hat{\gamma}} \wedge \boldsymbol{\omega}^{\hat{5}}$ do not contribute. Use this fact to express the useful part of the curvature 2-form as

$$\begin{aligned}\mathcal{R}^{\hat{\alpha}}{}_{\hat{\beta}} =&\,\tilde{\mathcal{R}}^{\hat{\alpha}}{}_{\hat{\beta}} - \frac{1}{4}F^{\hat{\alpha}}{}_{\hat{\beta}}F_{\hat{\gamma}\hat{\delta}}\boldsymbol{\omega}^{\hat{\gamma}} \wedge \boldsymbol{\omega}^{\hat{\delta}} \\ &- \frac{1}{8}\left(F^{\hat{\alpha}}{}_{\hat{\gamma}}F_{\hat{\beta}\hat{\delta}} - F^{\hat{\alpha}}{}_{\hat{\delta}}F_{\hat{\beta}\hat{\gamma}}\right)\boldsymbol{\omega}^{\hat{\gamma}} \wedge \boldsymbol{\omega}^{\hat{\delta}} + ..., \end{aligned} \quad (48.29)$$

where $\tilde{\mathcal{R}}^{\hat{\alpha}}{}_{\hat{\beta}}$ is the four-dimensional part.
(e) Use this to show that the components of the Riemann tensor are given by

$$\begin{aligned}R^{\hat{\alpha}}{}_{\hat{\beta}\hat{\gamma}\hat{\delta}} =&\,\tilde{R}^{\hat{\alpha}}{}_{\hat{\beta}\hat{\gamma}\hat{\delta}} - \frac{1}{2}F^{\hat{\alpha}}{}_{\hat{\beta}}F_{\hat{\gamma}\hat{\delta}} \\ &- \frac{1}{4}\left(F^{\hat{\alpha}}{}_{\hat{\gamma}}F_{\hat{\beta}\hat{\delta}} - F^{\hat{\alpha}}{}_{\hat{\delta}}F_{\hat{\beta}\hat{\gamma}}\right), \end{aligned} \quad (48.30)$$

where $\tilde{R}^{\hat{\alpha}}{}_{\hat{\beta}\hat{\gamma}\hat{\delta}}$ is again the four-dimensional part.
(f) Turning now to the $\hat{5}$-components, verify that

$$\begin{aligned}\mathcal{R}^{\hat{5}}{}_{\hat{\alpha}} =&\,\frac{1}{2}F_{\hat{\alpha}\hat{\beta},\hat{\gamma}}\boldsymbol{\omega}^{\hat{\gamma}} \wedge \boldsymbol{\omega}^{\hat{\beta}} - \frac{1}{2}F_{\hat{\alpha}\hat{\beta}}\boldsymbol{\Omega}^{\hat{\alpha}}{}_{\hat{\gamma}} \wedge \boldsymbol{\omega}^{\hat{\gamma}} \\ &- \frac{1}{2}F_{\hat{\beta}\hat{\gamma}}\boldsymbol{\Omega}^{\hat{\beta}}{}_{\hat{\alpha}} \wedge \boldsymbol{\omega}^{\hat{\gamma}} \\ &- \frac{1}{4}F_{\hat{\beta}\hat{\gamma}}F^{\hat{\beta}}{}_{\alpha}\boldsymbol{\omega}^{\hat{\gamma}} \wedge \boldsymbol{\omega}^{\hat{5}}. \end{aligned} \quad (48.31)$$

(g) Use the previous results to compute

$$R^{\hat{5}}{}_{\hat{\alpha}\hat{5}\hat{\gamma}} = \frac{1}{4}F_{\hat{\beta}\hat{\gamma}}F^{\hat{\beta}}{}_{\hat{\alpha}}. \quad (48.32)$$

(h) Now compute the components of the Ricci tensor to find

$$R_{\hat{\beta}\hat{\delta}} = \tilde{R}_{\hat{\beta}\hat{\delta}} - \frac{1}{2}F^{\hat{\alpha}}{}_{\beta}F_{\hat{\alpha}\hat{\delta}},$$

$$R_{\hat{5}\hat{5}} = \frac{1}{4}F_{\hat{\gamma}\hat{\alpha}}F^{\hat{\gamma}\hat{\alpha}}. \tag{48.33}$$

(i) Finally, put these previous two equations together to show

$$R = \tilde{R} - \frac{1}{4}F^{\mu\nu}F_{\mu\nu}. \tag{48.34}$$

From classical to quantum gravity

49

What are the stars? ... They are bits of fire a few kilometres away. We could reach them if we wanted to. Or we could blot them out. The earth is the centre of the universe. The sun and the stars go round it.
George Orwell (1903–1950) *Nineteen Eighty-Four*

The fundamental interactions in Nature are gravitational, electromagnetic, weak and strong. In the last chapter, we met a scheme aimed at unifying electromagnetism and gravitation into a single master theory, using the concept of a gauge field. It relied on having access to an extra dimension in which to work. The project to combine the fundamental forces of nature into a single theory was one of the greatest triumphs of twentieth century physics. Quantum field theory (QFT) allowed (i) electromagnetism along with (ii) the weak and (iii) the strong interactions to be combined into a quantum gauge field theory known as the **Standard Model of particle physics**. The fundamental force that resists being incorporated into this scheme on the same basis as the others is gravitation.[1] Starting from the attempt of last chapter to accommodate gravitation through extra spatial dimensions, we shall introduce some more recent attempts that try to make sense of the place of gravitation in the quantum world.[2]

49.1 Extra dimensions

We saw in the last chapter the power of having extra spatial dimensions available in formulating theories. If these extra dimensions exist, we should ask ourselves why we haven't yet been able to detect them through their quantum-mechanical effects, given the ease with which we have detected the three spatial dimensions and single time dimension of conventional field theory. In the last chapter, we guessed that this was due to the large energies involved in their small scale, an intuition we confirm in the next example. This idea is that, in quantum mechanics, a particle can lower its kinetic energy if it spreads its wavefunction over a larger area. So if there are extra dimensions that a quantum particle can access, this leads to a different structure of energy levels compared to the case where the particle is confined to three spatial dimensions.

[1] If we accept the effective-field point of view from the last chapter, it is possible to incorporate gravitation into this scheme (even though, unlike the other forces, gravitation isn't renormalizable) as long as we accept it as a low-energy approximation that we expect to break down at sufficiently high energy.

[2] Before getting into the complicated story presented in this chapter, we note, following an argument by Sidney Coleman, that invoking a little quantum mechanics immediately allows us a quick route to gravitational redshift. In a uniform gravitational field, the free-particle Hamiltonian $H = p^0 = m$ picks up an extra potential energy term mgh, where $V = gh$ is equal to the gravitational potential, so we can write the effect of gravity by saying $H \to H + HV$. The time evolution operator for quantum states, determined by the Hamiltonian, then changes by the presence of the potential according to

$$\hat{U} = e^{-i\hat{H}t/\hbar} \to e^{-i\hat{H}(1+V)t/\hbar}. \quad (49.1)$$

Gravitation can thus be included in our equations by making the swap to the time parameter $t \to (1+V)t$. This tells us that clocks run slowest deep down in the potential, where h and therefore V are smallest.

Example 49.1

Consider a particle in a one-dimensional square well, extended to include an extra dimension. The resulting two-dimensional well is shown in Fig. 49.1(a). It has length L in the x direction. We suppose that the reason the extra (y) dimension is not apparent is that it has been compactified, or curled up into a circle. To describe this we can identify (x, y) and $(x, y + 2\pi a)$, and we have the situation shown in Fig. 49.1(b). The space is now a cylinder with circular cross section of circumference $2\pi a$. The Schrödinger equation for a particle confined in this space is

$$-\frac{\hbar^2}{2m}\left(\frac{\partial^2 \psi}{\partial x^2} + \frac{\partial^2 \psi}{\partial y^2}\right) = E\psi. \tag{49.2}$$

The solutions for the square well can be written down. They are

$$\psi = a_n \sin\left(\frac{n\pi x}{L}\right)\left[b_l \sin\left(\frac{ly}{a}\right) + c_l \cos\left(\frac{ly}{a}\right)\right], \tag{49.3}$$

where a_n, b_l and c_l are set of constants and n and l are integer quantum numbers. The resulting energies corresponding to these eigenfunctions are

$$E_{nl} = \frac{\hbar^2}{2m}\left[\left(\frac{n\pi}{L}\right)^2 + \left(\frac{l}{a}\right)^2\right]. \tag{49.4}$$

The quantum number l can vanish, and so if we set it to zero we recover the usual one-dimensional energy levels. New energy levels only arise for $l > 0$. If we assume the extra dimensions are curled up into a very small circle, then $a \ll L$ and the second term in eqn 49.4 is large compared to the first. The result is that the extra energy levels lie at $E \approx \hbar^2/2ma^2$, which is potentially a very large energy indeed. So, since a is assumed small, new energy-levels appear only at very large E.

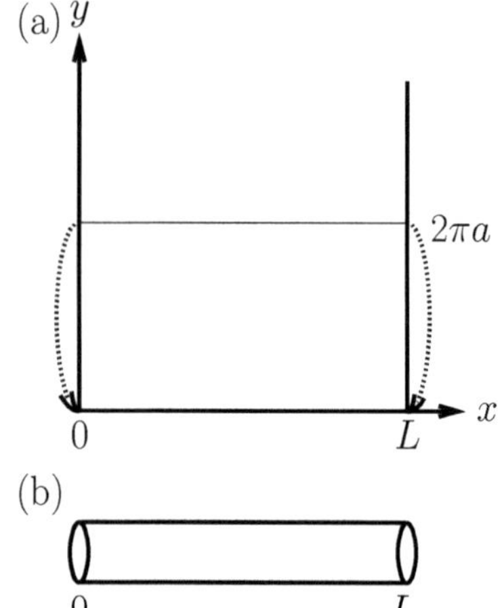

Fig. 49.1 (a) Two-dimensional square well. (b) The well compactified in the y direction.

The last example shows how extra dimensions, if compactified on a very small scale, might not necessarily give rise to measurable effects since they occur at such high energies compared to the capabilities of our measurements.

To get an idea of the length scales involved in these arguments, we can use a set of units originally formulated by Max Planck. René Descartes had originally proposed that there might be an underlying system of units that allowed the geometry of the Universe to be expressed in a natural manner. Around 250 years later, Planck proposed such a set of units, that combined three important constants of Nature, and so link gravity, relativity, and quantum mechanics. The interaction of these different effects should be expected to occur on this **Planck scale**.

Example 49.2

The idea is to use dimensional analysis to express the gravitational constant G, the speed of light c, and the Planck constant $\hbar$ in terms of a basic length, mass, and time scale ($\ell_{\mathrm{P}}, m_{\mathrm{P}}$ and t_{P}, respectively). From the dimensions of these constants we find

$$G = \frac{\ell_{\mathrm{P}}}{m_{\mathrm{P}} t_{\mathrm{P}}^2}, \quad c = \frac{\ell_{\mathrm{P}}}{t_{\mathrm{P}}}, \quad \hbar = \frac{m_{\mathrm{P}} \ell_{\mathrm{P}}^2}{t_{\mathrm{P}}}. \tag{49.5}$$

In order to express the constants of Nature that we measure, the Planck scales are defined as follows (in S.I. units)

$$\ell_{\mathrm{P}} = \left(\frac{G\hbar}{c^3}\right)^{\frac{1}{2}} = 1.616 \times 10^{-35} \text{ m},$$

$$t_{\mathrm{P}} = \left(\frac{G\hbar}{c^5}\right)^{\frac{1}{2}} = 5.391 \times 10^{-44} \text{ s},$$

$$m_{\mathrm{P}} = \left(\frac{\hbar c}{G}\right)^{\frac{1}{2}} = 2.176 \times 10^{-8} \text{ kg}. \tag{49.6}$$

The fact that the underlying Planck length scale is so small gives us reason to pause: the effects of quantum mechanics (via expressions involving $\hbar$) and gravity (involving G) together are likely only to become clear in processes sharing this underlying scale of units.[3]

In natural units, where $\hbar = c = 1$, length is inversely proportional to momentum and hence also to energy. Exploring a length scale of 10^{-35} m implies we explore energies[4] of 10^{19} GeV, which is well beyond the measurement capability of any existing accelerators.[5] As a result, what is going on at the Planck length scale remains mysterious. It might well be, for example, that we have the situation proposed in the previous chapter, where extra dimensions are compactified on length scales of ℓ_{P}, and they have hence escaped our notice. So if extra dimensions are at least possible from this point of view, it has been suggested that these might provide the necessary backdrop to formulate a unified theory that includes quantum fields and also gravitation.

If we accept that it might be plausible for extra spatial dimensions to exist, what would be their effect on gravity? In four dimensions [i.e. (3+1)-dimensional spacetime], the law of Newtonian gravitation says

$$\nabla^2 \Phi^{(4)} = 4\pi G \rho, \tag{49.7}$$

where $\Phi^{(D)}(x)$ is the gravitational potential for D-dimensional spacetime.[6] By definition, the mass density is always given by the mass divided by the number of spatial dimensions d, so if we increase the number of dimensions, the constant of gravitation will have to alter, so that we have

$$\nabla^2 \Phi^{(D)} = 4\pi G^{(D)} \rho. \tag{49.8}$$

Given the dimensionality of the mass density ρ, this latter equation also implies on dimensional grounds that the gravitational force falls off as $1/r^{d-1}$. That is, the force law differs in different numbers of dimensions.

Example 49.3

If there were five dimensions, with one compactified with radius a, we would have

$$\nabla^2 \Phi^{(5)} = 4\pi G^{(5)} \rho^{(5)} = 4\pi \frac{G^{(5)}}{2\pi a} \rho^{(4)}. \tag{49.10}$$

This would imply that the measured gravitation constant G is actually $G = \frac{G^{(5)}}{2\pi a}$. From this we can deduce that

$$\frac{G^{(5)}}{G} = 2\pi a = \ell_{\mathrm{c}}, \tag{49.11}$$

[3]It is notable that the Planck mass is much larger than that of any elemental particle and seems the least experimentally inaccessible of the Planck units. The Planck energy scale $E_{\mathrm{P}} = m_{\mathrm{P}} c^2$ that derives from this is 1.22×10^{28} eV (often quoted as 1.22×10^{19} GeV), which is seven or eight order of magnitudes larger than the highest energy cosmic ray yet observed. One can speculate whether or not this observation is significant (see Exercise 49.1).

[4]It is simplest to simply compute $E_{\mathrm{P}} = m_{\mathrm{P}} c^2$, as mentioned in the previous sidenote.

[5]The Large Hadron Collider at CERN, a multi-billion dollar project, accelerates proton beams up to nearly 7 TeV per beam. This is incredibly impressive, but a TeV is *only* 1000 GeV. We are a long way off 10^{19} GeV.

[6]We use D for the number of spacetime dimensions and $d = D - 1$ for the number of spatial dimensions.

The Planck length can be defined in other dimensions. Using some dimensional analysis one can show

$$\left(\ell_{\mathrm{P}}^{(D)}\right)^{D-2} = \frac{\hbar G^{(D)}}{c^3} = \ell_{\mathrm{P}}^2 \frac{G^{(D)}}{G}. \tag{49.9}$$

where ℓ_c is the characteristic length scale of the compacted dimension.

The generalization of the last example to D-dimensional spacetime is that

$$\frac{G^{(D)}}{G} = \ell_c^{(D-4)}. \tag{49.12}$$

The quantity on the right is the volume of the extra dimensions $\mathcal{V}_c$. This gives us an answer to how dimensionality changes the strength of gravitation.

49.2 String theory

The idea of extra dimensions is taken up in the formalism of **string theory**. This theory shares many of the techniques of QFT, but involves replacing the idea of fundamental particle-like excitations of fields with excitations of fundamental one-dimensional **strings**. In short, strings are one-dimensional objects with no internal structure, and particles are excitations of these strings. Strings are not built from parts and so we can't deal with some part of a string. Strings come in many forms: they can be open (meaning the ends do not join) or closed (where they form loops). Open strings can have free endpoints or fixed end points. The objects on which strings terminate to fix the ends are characterized by their dimensionality and are called D-branes.[7] An example of a string ending on a 2-brane is shown in Fig. 49.2.

First some good news: the use of strings instead of particles largely resolves one of the most persistent conceptual problems with QFT.

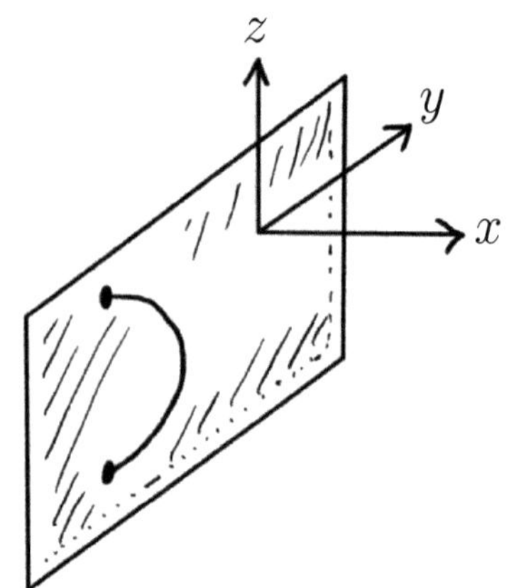

Fig. 49.2 String ending on a 2-brane.

[7]'D' here stands for Dirichlet, or Peter Gustav Lejeune Dirichlet (1805–1859). Have fixed string ends is equivalent to a Dirichlet boundary condition $\partial X^i/\partial\tau = 0$ at the end of the string.

[8]As a result of this striking feature, string theory has been viewed as the natural successor to QFT in attempting to find a consistent quantum gravity. It also has a natural description of a $S = 2$ massless excitation, i.e. a graviton, leading some to suggest that string theory might have gravitation built into it.

Example 49.4

A problem that bedevils QFT is the divergence of quantities and the consequent need for renormalization. A particle decay process in QFT is depicted in the Feynman diagram shown in Fig. 49.3(a), involving the interaction of fields (or particles) at a point. This is the source of the problem: the evaluation of the fields at the same point can cause contributions to perturbation theory to diverge, necessitating the removal of infinities using a number of sophisticated techniques, known as renormalization, whose validity have been the subject of persistent debate. Even if we trust renormalization, it fails to remove the infinities encountered in quantum gravity in the sorts of scattering processes described in Chapter 47. The QFT divergences are avoided to some extent in string theory, where particle processes resemble the example shown in Fig. 49.3(b). Here the particles, represented as closed strings, never meet at a point.[8]

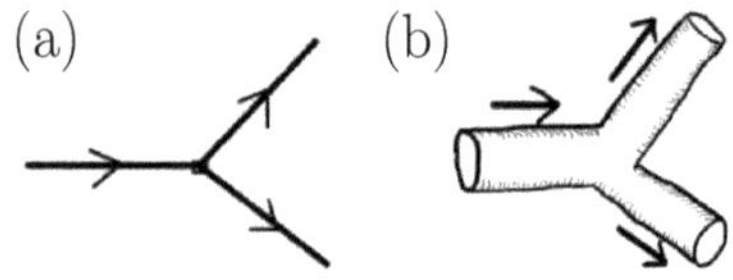

Fig. 49.3 (a) A particle decay process in quantum field theory. (b) A particle decay in string theory.

If we accept that strings might be of some use, then we should examine how to incorporate them into field theory. Fortunately, we can use many of the techniques developed in the course of this book to investigate strings. To get an idea how to do this, let's return to the relativistic description of particles and build a string theory from there.

A free particle of mass m has a world line found by extremizing an action

$$S = -m \int \mathrm{d}\tau \left(-g_{\mu\nu} \frac{\mathrm{d}x^\mu}{\mathrm{d}\tau} \frac{\mathrm{d}x^\nu}{\mathrm{d}\tau} \right)^{\frac{1}{2}}. \qquad (49.13)$$

Here the world line is parametrized by τ. This parametrization is usually the proper time for massive particles, but could be chosen to be another affine parameter.[9] We use a set of coordinates x^μ and capture the motion of the particle with which we can express the world line as a curve $x^\mu(\tau)$. The equation of motion derived from the particle's action is $\mathrm{d}p_\mu(\tau)/\mathrm{d}\tau = 0$, where the momentum $p_\mu = \partial L/\partial \dot{x}^\mu$.

The treatment of the string is very similar to that of the particle. The major difference is that since the string is a one-dimensional object, rather than a world line, it traces out a two-dimensional **world sheet**. Just as we parametrize the world line with a parameter τ, we parametrize the world sheet with two parameters τ and σ. The names are chosen to allow us to think of the τ parameter as telling us about the behaviour with respect to time and σ the coordinate telling us how far we are along the string.[10] For an open string σ ranges from 0 to some maximum σ_*. The world sheet is potentially curved in an interesting way and so we seek to embed it in Euclidean space, in the same way that we describe in Appendix D.

We shall choose to work in Euclidean space with coordinates $x^\mu = (t, x, y, z)$. In this space, positions on the string's world sheet have coordinates $X^\mu = (T, X, Y, Z)$, so that a position vector of a point on the string world sheet is written as $\boldsymbol{X}(\tau, \sigma)$ [i.e. input τ and σ labelling the part of the world sheet, return coordinates $x^\mu = (t, x, y, z) = (T, X, Y, Z) = X^\mu$ labelling the point on the world sheet]. The goal is to describe the world sheet of the string, some examples of which are shown in Fig. 49.4. To do this we embed the world sheet in Euclidean space, and then find the induced metric $\boldsymbol{\gamma}$ of the world sheet. As discussed in Appendix D, the metric $\boldsymbol{\gamma}$ has components given by

$$\gamma_{\alpha\beta} = \eta_{\mu\nu} \frac{\partial X^\mu}{\partial \xi^\alpha} \frac{\partial X^\nu}{\partial \xi^\beta}, \qquad (49.14)$$

and here we choose coordinates $\xi^1 = \tau$ and $\xi^2 = \sigma$. To simplify our expressions we define some notation

$$\dot{X}^\mu = \tfrac{\partial X^\mu}{\partial \tau}, \quad (X^\mu)' = \tfrac{\partial X^\mu}{\partial \sigma}. \qquad (49.15)$$

Using this notation, the induced metric can be written as a 2×2 matrix

$$\gamma_{\alpha\beta}(\tau, \sigma) = \begin{pmatrix} (\dot{\boldsymbol{X}})^2 & \dot{\boldsymbol{X}} \cdot \boldsymbol{X}' \\ \dot{\boldsymbol{X}} \cdot \boldsymbol{X}' & (\boldsymbol{X}')^2 \end{pmatrix}, \qquad (49.16)$$

where the products are things like $\dot{\boldsymbol{X}} \cdot \boldsymbol{X}' = \eta_{\mu\nu} \dot{X}^\mu (X')^\nu$, and so on. This provides us with a simple way to describe the world sheet.

Just as the action for a particle is proportional to the interval along the world line, the action for the string is proportional the effective area

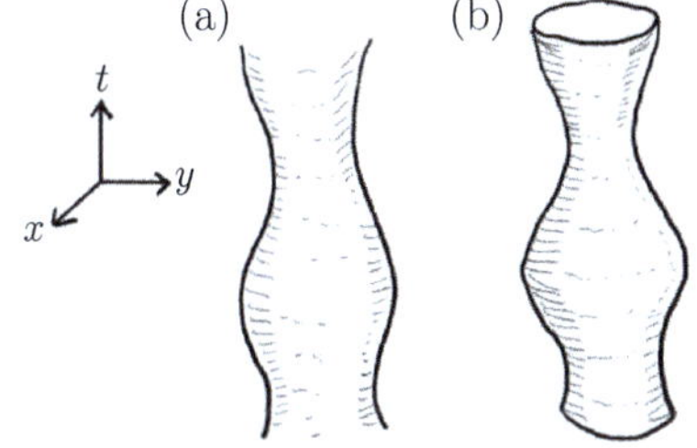

Fig. 49.4 (a) Open string world sheet. (b) Closed string world sheet.

of the world sheet. For the world sheet with a metric $\boldsymbol{\gamma}$, the area A is

$$A = \int \mathrm{d}\tau \mathrm{d}\sigma \sqrt{-\gamma}, \tag{49.17}$$

where γ is the determinant of the matrix $\gamma_{\alpha\beta}$. Using eqn 49.16 we can compute the determinant as $\gamma = \left(\dot{\boldsymbol{X}}\right)^2 (\boldsymbol{X}')^2 - (\boldsymbol{X} \cdot \boldsymbol{X}')^2$. Finally, the constant of proportionality that relates the area to the string action is the tension in the string T_0, which plays a role like the mass does in the particle theory.

With these ingredients we can write the resulting **Nambu–Goto string action** for a string theory as

$$S = -T_0 \int \mathrm{d}\tau \mathrm{d}\sigma \left[\left(\boldsymbol{X} \cdot \boldsymbol{X}'\right)^2 - \left(\dot{\boldsymbol{X}}\right)^2 (\boldsymbol{X}')^2 \right]^{\frac{1}{2}}. \tag{49.18}$$

This action corresponds to a Lagrangian density of

$$\mathcal{L} = -T_0 \left[\left(\boldsymbol{X} \cdot \boldsymbol{X}'\right)^2 - \left(\dot{\boldsymbol{X}}\right)^2 (\boldsymbol{X}')^2 \right]^{\frac{1}{2}}. \tag{49.19}$$

Given this Lagrangian, we can compute equations of motion.

Example 49.5

There are two variables in the problem and so we can identify two momentum-like variables.[11] We write

$$P_\mu^\tau = \frac{\partial \mathcal{L}}{\partial \dot{X}^\mu} = -\frac{T_0}{\mathcal{L}} \left[\left(\dot{\boldsymbol{X}} \cdot \boldsymbol{X}'\right) X'_\mu - (\boldsymbol{X}')^2 \dot{X}_\mu, \right], \tag{49.23}$$

and

$$P_\mu^\sigma = \frac{\partial \mathcal{L}}{\partial (X')^\mu} = -\frac{T_0}{\mathcal{L}} \left[\left(\dot{\boldsymbol{X}} \cdot \boldsymbol{X}'\right) \dot{X}_\mu - \dot{\boldsymbol{X}}^2 X'_\mu \right]. \tag{49.24}$$

Using these momenta, the equations of motion are

$$\frac{\partial P_\mu^\tau}{\partial \tau} + \frac{\partial P_\mu^\sigma}{\partial \sigma} = 0. \tag{49.25}$$

When written out, these look rather complicated, but they can be simplified, as we shall see in the next section.

49.3 Parametrizing the string

We can choose the parametrizations to simplify the equations of motion as much as possible, just as we did in Chapter 9. In this context, the choice of parametrization is known as choosing a gauge. First we make a choice of the timelike parameter τ. The most direct choice is to treat this as the coordinate time by fixing $\tau = t$, which is a choice known as **static gauge**. In this gauge, lines of constant τ represent static strings, i.e. an observer [with coordinates (t, x, y, z)] will see a static string at a

---

Yoichiro Nambu (1921–2015) and Tetsuo Goto (1931–1982).

[11] Recall from Chapter 40 that the Euler–Lagrange equation for a Lagrangian density $\mathcal{L}$ built from scalar fields is

$$\frac{\partial \mathcal{L}}{\partial \phi} = \left(\frac{\partial \mathcal{L}}{\partial \phi_{,\mu}} \right)_{,\mu}. \tag{49.20}$$

For a Lagrangian built from vector field we saw in Exercise 40.3 that we can identify momentum-like variables

$$\mathcal{P}_\mu^\nu = \frac{\partial \mathcal{L}}{\partial X^\mu_{,\nu}}, \tag{49.21}$$

where $\nu = \tau$ and σ and $\mu = t, x, y$ and z. We then write an Euler Lagrange equation

$$\frac{\partial \mathcal{L}}{\partial X^\mu} = \left(\frac{\partial \mathcal{L}}{\partial X^\mu_{,\nu}} \right)_{,\nu}$$
$$= \left(\mathcal{P}_\mu^\nu \right)_{,\nu}, \tag{49.22}$$

where we sum over $\nu = \tau$ and σ.

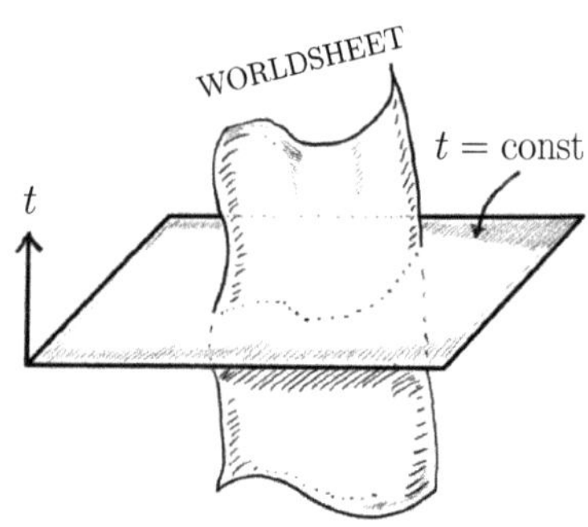

Fig. 49.5 Static gauge involves setting $\tau = t$. The plane $t = \mathrm{const.}$ then cuts the world sheet at $\tau = t$.

fixed time t, as shown in Fig. 49.5. In static gauge, the derivatives of the world sheet coordinates $X^\mu = (t, X^i)$ have components

$$\frac{\partial X^\mu}{\mathrm{d}\sigma} = \left(0, \frac{\partial \vec{X}}{\partial \sigma}\right), \quad \frac{\partial X^\mu}{\mathrm{d}\tau} = \left(1, \frac{\partial \vec{X}}{\partial t}\right), \tag{49.26}$$

where the 3-vector $\vec{X}$ has components X^i. Static gauge allows us to work in terms of the velocities (i.e. derivatives with respect to time $t = \tau$). The most useful of these is the **transverse velocity** that we discuss in the next example.

Example 49.6

Consider a static string at a fixed value of $t = \tau$. An element of string length can be written using an infinitesimal $\mathrm{d}s$, where

$$\mathrm{d}s = \left|\frac{\mathrm{d}\vec{X}}{\mathrm{d}\sigma}\right| \mathrm{d}\sigma. \tag{49.27}$$

Here, σ is the parameter along the length of the static string and the vector $\mathrm{d}\vec{X}/\mathrm{d}\sigma$ is tangent to the string. We can then deduce, using the previous equation, that

$$\frac{\partial \vec{X}}{\partial s} \cdot \frac{\partial \vec{X}}{\partial s} = \left(\frac{\partial \vec{X}}{\partial \sigma}\right)^2 \left(\frac{\partial \sigma}{\partial s}\right)^2 = 1, \tag{49.28}$$

so the quantity $\partial \vec{X}/\partial s$ is a unit vector. The vectors $\partial \vec{X}/\partial \sigma$ and $\partial \vec{X}/\partial s$ are parallel since $\partial \vec{X}/\partial s = (\partial \vec{X}/\partial \sigma)(\partial \sigma/\partial s)$, so $\partial \vec{X}/\partial s$ is a unit tangent vector to the string.

To project out the part of a vector $\boldsymbol{v}$ that is perpendicular to a unit vector $\boldsymbol{n}$ we write $v_\perp = \boldsymbol{v} - (\boldsymbol{v} \cdot \boldsymbol{n})\boldsymbol{n}$. With this simple rule we can use the string coordinate velocity $\vec{v} = \partial \vec{X}/\partial t$ and obtain the transverse velocity $\vec{v}_\perp$ thus

$$\vec{v}_\perp = \frac{\partial \vec{X}}{\partial t} - \left(\frac{\partial \vec{X}}{\partial t} \cdot \frac{\partial \vec{X}}{\partial s}\right) \frac{\partial \vec{X}}{\partial s}. \tag{49.29}$$

We then use this transverse velocity to simplify the string action. From the component eqns 49.26, we find

$$\dot{\boldsymbol{X}}^2 = -1 + \left(\frac{\partial \vec{X}}{\partial t}\right)^2, \quad (\boldsymbol{X}')^2 = \left(\frac{\partial \vec{X}}{\partial \sigma}\right)^2. \tag{49.30}$$

Putting everything together, the part in the square root of the string Lagrangian in eqn 49.19 becomes

$$(\boldsymbol{X} \cdot \boldsymbol{X}')^2 - \left(\dot{\boldsymbol{X}}\right)^2 (\boldsymbol{X}')^2 = \left(\frac{\partial \vec{X}}{\partial t} \cdot \frac{\partial \vec{X}}{\partial \sigma}\right)^2 + \left[1 - \left(\frac{\partial \vec{X}}{\partial t}\right)^2\right] \left(\frac{\partial \vec{X}}{\partial \sigma}\right)^2$$

$$= \left(\frac{\mathrm{d}s}{\mathrm{d}\sigma}\right)^2 \left[\left(\frac{\partial \vec{X}}{\partial t} \cdot \frac{\partial \vec{X}}{\partial s}\right)^2 + 1 - \left(\frac{\partial \vec{X}}{\partial t}\right)^2\right]$$

$$= \left(\frac{\mathrm{d}s}{\mathrm{d}\sigma}\right)^2 (1 - v_\perp^2). \tag{49.31}$$

Using the results from the last example, we can write down a simplified version of the string action

$$S = -T_0 \int \mathrm{d}\tau \mathrm{d}\sigma \left[\frac{\mathrm{d}s}{\mathrm{d}\sigma}(1 - v_\perp^2)^{\frac{1}{2}}\right]. \tag{49.32}$$

Although this is certainly an improvement, one final choice of σ parametrization gives us the simplified equation of motion that we're after.

Example 49.7

We shall attempt to fix the parametrization of σ such that the string velocity $\vec{v}$ is perpendicular to the string tangent, which would mean

$$\frac{\partial \vec{X}}{\partial t} \cdot \frac{\partial \vec{X}}{\partial \sigma} = 0. \tag{49.33}$$

This is useful as the transverse velocity becomes $\vec{v}_\perp = \frac{\partial \vec{X}}{\partial t}$, and, in terms of this $\vec{v}_\perp$, we have that the momenta from eqns 49.25 can be rewritten as

$$P^{\tau\mu} = T_0 \left(1 - v_\perp^2\right)^{-\frac{1}{2}} \frac{\mathrm{d}s}{\mathrm{d}\sigma} \frac{\partial X^\mu}{\partial t},$$

$$P^{\sigma\mu} = - T_0 \left(1 - v_\perp^2\right)^{\frac{1}{2}} \frac{\partial X^\mu}{\partial s}. \tag{49.34}$$

Plugging these latter equations into the Euler–Lagrange equation gives us the equation of motion

$$\frac{\partial^2 \vec{X}}{\partial t^2} = \left(1 - v_\perp^2\right)^{\frac{1}{2}} \frac{\mathrm{d}\sigma}{\mathrm{d}s} \frac{\partial}{\partial \sigma} \left[\left(1 - v_\perp^2\right)^{\frac{1}{2}} \frac{\mathrm{d}\sigma}{\mathrm{d}s} \frac{\partial \vec{X}}{\partial \sigma} \right]. \tag{49.35}$$

The choice of the parametrization σ that allows both the preceding chain of simplifications is to fix

$$\frac{\mathrm{d}s}{\mathrm{d}\sigma} (1 - v_\perp^2)^{-\frac{1}{2}} = 1. \tag{49.36}$$

This is a choice compatible with eqn 49.33, and also gives rise to the goal: a simplified equation of motion reading

$$\frac{\partial^2 \vec{X}}{\partial t^2} = \frac{\partial^2 \vec{X}}{\partial \sigma^2}. \tag{49.37}$$

This is a wave equation, whose solution (the dynamics of the string) can be written down: they are simply plane waves.

Our conclusion is that, with the right choice of parameters for the world sheet, the strings support wavelike excitations. These can be used to describe particle excitations. Further development of the theory would involve quantizing the string motion to extract the properties of the particles.[12] For example, the graviton can be identified as a vibrational state of a closed string.

49.4 Strings in relativity

Strings crop up in several places in general relativity. Most prominent are **superstrings**, which are strings with dimensions of the Planck length that possess **supersymmetry**.

Example 49.8

Observations suggest that most of the matter in the Universe is not seen: it is **dark matter**, which does not interact with light. In fact, it only seems to interact via its gravitational effect,[13] explaining why it has never been directly detected.[14] that are designed to One solution to this problem invokes supersymmetry. In supersymmetry, every boson particle in the Universe has a partner Fermi particle of the same mass (and vice versa). We can then identify a symmetry operator with the property that

$$\hat{Q}|\text{boson}\rangle = |\text{fermion}\rangle,$$

$$\hat{Q}|\text{fermion}\rangle = |\text{boson}\rangle. \tag{49.38}$$

[12]This is discussed in the book by Zwiebach (2009).

[13]The gravitational interaction between matter in a galaxy and dark matter leads to the formation of a dark matter halo surrounding each galaxy, inferred from the observed effect on the motion of stars orbiting the galaxy.

[14]All dark matter searches performed using detectors on Earth tacitly assume that candidate dark matter particles interact with ordinary matter by a non-gravitational force, e.g. electromagnetic coupling. Such experiments have not yet succeeded in detecting any such particles. But perhaps dark matter particles only interact gravitationally, explaining why they are so hard to find in particle physics experiments.

Acting on boson A with $\hat{Q}$ gives fermion A-ino. (Example: $\hat{Q}$ acting on a photon γ, gives a photonino $\tilde{\gamma}$.) Acting on fermion B gives boson sB. (Example: $\hat{Q}$ acting on a quark q gives a squark $\tilde{q}$.)

Perhaps these new particles generated by supersymmetry compose dark matter. If so, the darkness of dark matter follows from quantum mechanical selection rules that suppress the probability amplitudes for matter to interact with dark matter. There is currently no experimental evidence for the existence of these extra particles.

Superstring theory allows the incorporation of supersymmetric particles into the theory. A generalization of the original superstring models is called m-brane theory, or **M-theory**, where m is the number of dimensions. (The theory suggests that up to $m = 9$ branes can exist.) It is hoped that M-theory could allow a consistent theory to be developed that incorporates gravitation and the other three fundamental interactions. However, it is a well-known problem that M-theory has not yet been able to make falsifiable predictions that can be tested by experiment. We have seen how extra dimensions should lead to a modification of the dependence of the gravitational force with distance and so, at the Planck length, measurable differences might be detectable. Of course, we have no means of making measurements at the Planck length at present.

In a completely different context, there are suggestions that there exists another sort of string in the Universe, this type potentially spanning large distances. These **cosmic strings** are created in the early Universe on a microscopic length scale and stretched out during the subsequent expansion. They would have a gravitational effect owing to their tension meaning they could be detected via gravitational lensing (see Chapter 24). The lensing effect due to the string is predicted to cause a distant light source (e.g. a star) to show up as two images, owing the curvature of spacetime around the string.

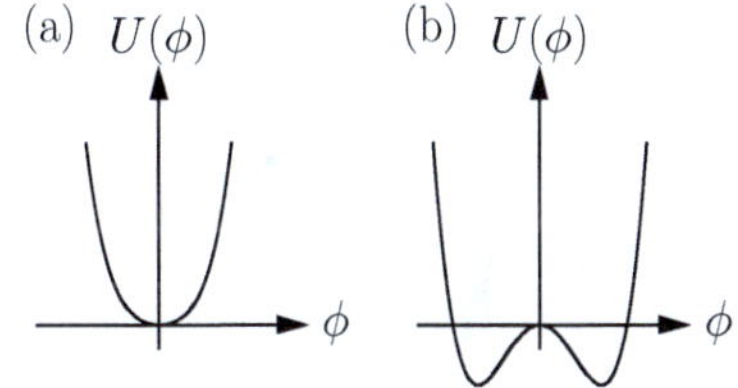

Fig. 49.6 The potential discussed in Chapter 41: (a) shows the high temperature potential; (b) shows the broken symmetry potential at low temperature.

Example 49.9

The origin of cosmic strings follows from the phase transition in the early Universe, that we discussed in Chapter 41. The most simple system showing a phase transition is that of the scalar field where, on cooling, the potential changes from that shown in Fig. 49.6(a) to that in 49.6(b), with the assumption that the system falls into one of these two minima, which occur at a field $\phi = \pm\phi_0$. However, in any phase transition, we have the possibility of forming **domains**. These result in the system breaking symmetry in different ways in different regions of space. That is, part of the system falls into one minimum in the potential, and part falls into another. For the scalar field we might have the situation shown in Fig. 49.7. On the left the system has fallen into the minimum in the broken symmetry potential at $-\phi_0$; on the right it has fallen into the minimum at ϕ_0. The space between these two domains involves a field that must smoothly evolve between the $-\phi_0$ and ϕ_0. These regions are called **defects**. A one-dimensional defect, known as a wall or a kink, is shown in Fig. 49.7.

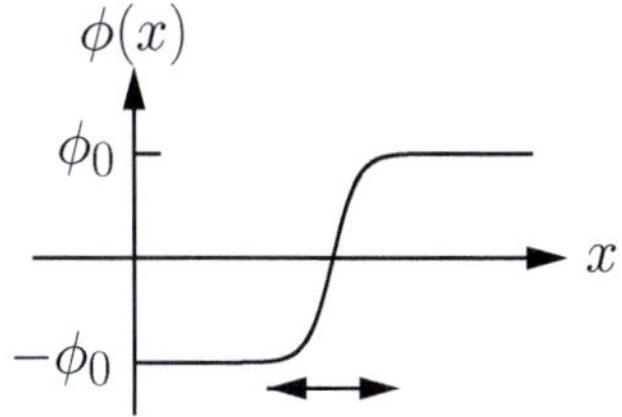

Fig. 49.7 A domain wall linking two regions where symmetry is broken in different ways.

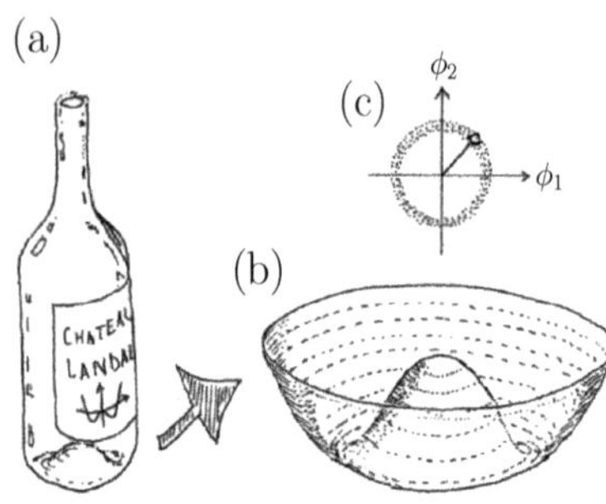

Fig. 49.8 The broken symmetry potential for the complex scalar field.

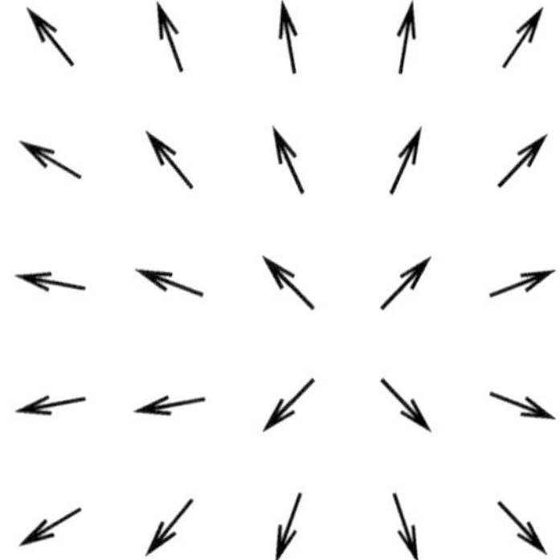

Fig. 49.9 The vortex field configuration.

[15]Further information on defects and vortices in field theory can be found in our *Quantum Field Theory for the Gifted Amateur*.

[16]The curvature around the cosmic string is examined further in the exercises.

A more interesting example takes place for the complex scalar field $\psi(x)$. The broken-symmetry potential-energy surface for this field is shown in Fig. 49.8. It is two dimensional, reflecting the two degrees of freedom that the complex field possesses (i.e. the real and imaginary parts). The potential resembles the punt at the bottom of a wine bottle, or a Mexican hat (it is sometimes called the Mexican-hat potential for this reason). There are an infinite number of minima in this potential that occur at the same radius $|\phi_0|$ in the complex plane, but at different values of the complex phase $\theta(x)$. In choosing a broken symmetry ground state, the complex field $\psi(x) = |\psi(x)|e^{i\theta(x)}$ picks a particular phase value $|\psi_0|e^{i\theta_0}$ by selecting the phase angle θ_0.

The defects in this potential are shown (in two spatial dimensions) in Fig. 49.9, and are known as **vortices**.[15] Like the domain wall, the vortex can be understood as the simplest way in which the system can have different spatial parts in different minima of the potential. The resulting vortex has the feature that the gradient in the phase $\nabla\theta(x)$ diverges at the centre of the vortex, giving rise to a singularity. To translate this picture into three dimensions, we imagine stacking vortices on top of each other, such that the singularities form a one-dimensional path. This curve through the vortex cores is the cosmic string.[16]

Finally, it has been suggested that Schwarzschild black holes might in fact be strings, whose physics is then amenable to a description using string theory. There are some hints that the properties of such a black hole coincide with those of a string, although we do not yet have conclusive evidence for this.

49.5   Superspace

String theory represents only one of many possible approaches that attempt to reconcile general relativity and quantum mechanics. A very different strategy is to accept that the uncertainty inherent in the quantum-mechanical description of particle mechanics means that spacetime, with its rigid (3+1)-dimensional structure, can itself only be a classical approximation. In fact, it approximates a more subtle and complex state of affairs that allows quantum states to be realized with certain probability amplitudes ψ. This is quite a radical approach, in that it implies that the successful unification of quantum mechanics and gravitation involves abandoning the picture of a structured spacetime describing events, as the basic arena of gravitation.

In order to build a more suitable foundation, we start with classical spacetime and note that we can deconstruct it by taking three-dimensional spacelike slices, or 3-surfaces $^{(3)}\mathcal{C}$. There is some freedom in how we do this, but this is subject to the constraint that we can rebuild spacetime by stacking up the slices in a well-defined manner. Next, in order to accommodate the probabilistic features of quantum mechanics, our collection of spacelike surfaces must be vastly increased to form a **superspace**. This superspace will include all possible 3-surfaces, which in a quantum theory, each occur with a particular probability amplitude $\psi(^{(3)}\mathcal{C})$.

Example 49.10

The quantum properties of a system are described in terms of a quantum amplitude, which is determined through the combination of the phases of interfering wavelike contributions. In Feynman's 'sum over histories' description of quantum mechanics,[17] the phase of each contribution is determined by the classical action $S\left(^{(3)}\mathcal{C}\right)$ of the corresponding configuration of 3-space $^{(3)}\mathcal{C}$, such that each contribution to the wavefunction take the form

$$\psi\left(^{(3)}\mathcal{C}\right) \sim \exp\left[\frac{\mathrm{i}S\left(^{(3)}\mathcal{C}\right)}{\hbar}\right]. \tag{49.39}$$

To obtain the quantum probability amplitudes, we must then sum all of the possible $\psi\left(^{(3)}\mathcal{C}\right)$ that are compatible with the problem we're considering.

We obtain constructive interference when the actions for two configurations match up. This implies that the dynamics of quantum gravity can be determined by computing the details of how wavefronts of constant action S propagate throughout the superspace. The equation governing this (entirely classical) propagation is known as the **Einstein–Hamilton–Jacobi equation**[18] and is given by

$$\frac{1}{\sqrt{\gamma}}\left(\frac{1}{2}\gamma_{ij}\gamma_{kl} - \gamma_{ik}\gamma_{jl}\right)\frac{\delta S}{\delta\gamma_{ij}}\frac{\delta S}{\delta\gamma_{kl}} + \sqrt{\gamma}\,^{(3)}R = 0, \tag{49.40}$$

where γ_{ij} are components of the three-dimensional metric describing a hypersurface and γ is its determinant. Here, $^{(3)}R$ is the Ricci scalar for the 3-geometry. Remarkably, this one equation contains the same information as the Einstein field equation.

The fluctuations that characterize the quantum world (e.g. the zero-point fluctuations in a quantum oscillator) are expected to occur in the metric field on scale of the Planck length. In the superspace approach, quantum fluctuations on this scale cause the probability amplitudes $\psi\left(^{(3)}\mathcal{C}_i\right)$ for a range of 3-surfaces $^{(3)}\mathcal{C}_i$ to take on appreciable values. This leads to a fundamental limit to how well the spacetime picture adopted elsewhere in general relativity approximates the fluctuating reality of the underlying quantum world.

Ultimately, the superspace approach, with its abstract geometry containing all of the 3-surfaces, and classical equation of motion for describing the phases corresponding to each, does not make any strong claims about the underlying structure of the interactions that allow quantum mechanics and gravitation to coexist and interact. This is, to its supporters, a positive aspect, since it represents a conservative and robust approach that relies on the metric field, instead of novel and unobserved features in Nature, such as those that are found in string theory.

49.6 Loop quantum gravity

An alternative approach to string theory that aims to combine quantum mechanics and gravitation is **loop quantum gravity** (LQG). This theory involves an attempt to quantize the geometry of spacetime itself. After all, lots of things in quantum mechanics become quantized, such as harmonic oscillator energy levels or angular momentum states, so why not spacetime geometry itself? The intuition is the following: imagine localizing a particle with precision L. Heisenberg uncertainty

[17] In brief, Feynman's approach to quantum mechanics involves considering every possible path a particle can take in getting between two points. We compute the classical action S_i for each trajectory and then build the quantum amplitude for the particle to travel between the two points by summing a factor $\mathrm{e}^{\mathrm{i}S_i/\hbar}$ for every possible trajectory, to give an amplitude $\mathcal{A} = \sum_i \mathrm{e}^{\mathrm{i}S_i/\hbar}$. In this way, the amplitude for a quantum mechanical process is built by a sum over all possible trajectories. This picture is described in more detail in our *Quantum Field Theory for the Gifted Amateur* (2014).

[18] The Hamilton–Jacobi equation in classical particle mechanics reads $H = -\frac{\partial S}{\partial t}$. It describes the evolution of the function S (equal to the classical action) resulting from a Hamiltonian function H. This is the only formulation of classical mechanics that represents particle motion in terms of the properties of a wave with a phase determined by S. There is no coincidence then, that Schrödinger's equation closely resembles the Hamilton–Jacobi equation. The version of the Hamilton–Jacobi equation in eqn 49.39, suitable for computations in general relativity, involves the evolution of the function S with respect to the 3-metric components, driven by a $\sqrt{\gamma}\,^{(3)}R$. More details of the Hamilton–Jacobi equation in classical particle mechanics can be found in Landau and Lifshitz (volume I, 1976).

would demand that $L\Delta p > \hbar$, and since $(\Delta p)^2 = \langle p^2 \rangle - \langle p \rangle^2$, this means $\langle p^2 \rangle > (\hbar/L)^2$. As we localize the particle in a smaller and smaller region, its momentum will go up and so will its energy E, and it will become ultra-relativistic so that $E \approx pc$ and hence E will exceed $\hbar c/L$. Energy E acts as gravitational mass via $E = Mc^2$ and a concentrated mass in this small region L will become a black hole with Schwarzschild radius $R = GM/c^2$. This horizon will reach L when $L = (G/c^2)(\hbar c/L)/c^2$, i.e. $L = \ell_P$ so that the particle is localized within a Planck length. Thus, we conclude that though spacetime might be smooth at length scales above the Planck length, things below the Planck length are 'hidden inside its own mini-black hole'.[19]

So how do we go about quantizing spacetime? LQG uses the quantum-mechanical intuition that it is the commutation relations between operators that give rise to quantization conditions. For example, the components of angular momentum $\hat{L}^i$ ($i = 1, 2, 3$) obey the commutation relation

$$\left[\hat{L}^i, \hat{L}^j \right] = \mathrm{i}\hbar \varepsilon^{ijk} \hat{L}^k, \tag{49.41}$$

and this leads to the quantization of angular momentum. Moreover, they lead to the interesting feature that you can know the (square of the) total angular momentum[20] and any one component of the angular momentum (such as $\hat{L}^3$), but not the other components of the angular momentum. In LQG, we try and do the same thing with an element of space, and we start with a very simple three-dimensional shape: the tetrahedron shown in Fig. 49.10(a). One way of describing this tetrahedron is by using four vectors [see Fig. 49.10(a)] which we will call $\vec{L}_a$ where $a = 1, 2, 3, 4$; the directions of these vectors are perpendicular to the four faces of the tetrahedron and the magnitudes are equal to the area of the four faces of the tetrahedron. These vectors satisfy the condition

$$\sum_{a=1}^{4} \vec{L}_a = 0, \tag{49.42}$$

and one can show that the volume V of the tetrahedron is given by $V^2 = \frac{2}{9}\vec{L}_1 \times \vec{L}_2 \cdot \vec{L}_3$.

Geometry itself can then be quantized by upgrading these vectors into operators and imposing commutation relations between their components. One possible quantization scheme for our tetrahedron would then be to write

$$\left[\hat{L}_a^i, \hat{L}_b^j \right] = \mathrm{i}\delta_{ab}\ell_0^2 \varepsilon^{ijk} \hat{L}_a^k. \tag{49.43}$$

Here ℓ_0 is a constant, which must have dimensions of length (since $\hat{L}_a^i$ is an operator whose eigenvalue has the dimensions of *area*), and it turns out that it should be a constant multiplied by the Planck length. Equation 49.43 is simply a postulate, but what would it imply if we accepted it? The first thing to note is that the area of the faces of this tetrahedron would behave analogously to angular momentum in ordinary quantum mechanics. Thus, the area of the ath face of a tetrahedron must

[19]This phrase is from Rovelli and Vidotto (2015), a highly readable introduction to LQG.

[20]The total angular momentum (squared) is related to the operator $\hat{L}^2 = (\hat{L}^1)^2 + (\hat{L}^2)^2 + (\hat{L}^3)^2$.

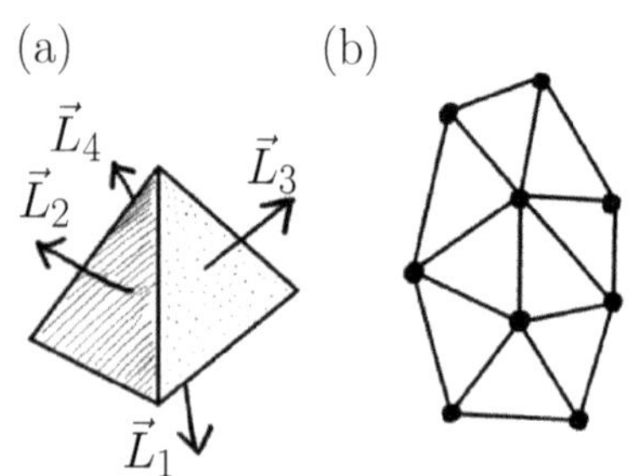

Fig. 49.10 (a) The quantization of a tetrahedral region of spacetime. (b) Overlapping volumes can be reduced to a graph in which closed paths over the nodes are the loops of the theory.

be quantized with eigenvalues

$$A_a = \ell_0^2 \sqrt{j_a(j_a + 1)}, \qquad (49.44)$$

with $j_a = 0, 1/2, 1, 3/2 \ldots$ The second thing one can conclude is that even if you know one Cartesian component of the area of one face of the tetrahedron, you won't know any of the other Cartesian components of the area of the other faces (for exactly the same reason that you can only know one component of the angular momentum). The normal vectors to each face of the tetrahedron are therefore known only partially and so these faces somehow blurrily shimmer with quantum uncertainty! We therefore deduce that geometry becomes fuzzy when you get down to the Planck scale; if your ambitions stretch to determining every aspect of the geometry of a shape at the smallest length scales, then you will be limited by fundamental quantum uncertainty. Thus, even though our argument has been formulated in terms of a tetrahedron, it would work if we had chosen some other shape and so we can't deduce that the smallest scale really does consist of a network of tetrahedra since we can't have precise information about microscopic quantum geometry. This then is the consequence of LQG: the Riemannian geometry that gives rise to gravitation must be replaced with quantum geometry that involves an inherent uncertainty at the Planck scale in at least some lengths, angles, and areas. In order to describe the curved spacetime of gravitation, we must consider a mesh of spacetime volumes such as the tetrahedron discussed above. These meshes can be reduced to graphs whose lines are analogous to lines of force. The loops in LQG are the closed paths linking nodes in these graphs [Fig. 49.10(b)].

Let's return to the tetrahedron, and imagine we know the eigenvalue j_a for each vector operator $\hat{\vec{L}}_a$, remembering that we also know that these vectors are subject to a closure property (eqn 49.42). The volume operator $\hat{V}$, defined via $\hat{V}^2 = \frac{2}{9}\hat{\vec{L}}_1 \times \hat{\vec{L}}_2 \cdot \hat{\vec{L}}_3$, commutes with the closure operator $\hat{C}$ (defined by $\hat{C} = \sum_{a=1}^{4} \hat{\vec{L}}_a$) and so it turns out that we can write states as $|j_1, j_2, j_3, j_4, v\rangle$, a common eigenstate of the four total area operators and the volume operator (with eigenvalue v). Thus, there is a fundamental 'quantum of space', so that the tetrahedron can grow or shrink only in discrete steps.

LQG remains a theory under construction and so it is not yet clear how well it describes our observations. Perhaps most seriously, we currently do not have a semiclassical limit of the theory that recovers general relativity. In addition to not reproducing the physical predictions of general relativity, LQG has not yet given rise to any prediction not made by the Standard Model. As a result, the jury is still out on this theory, as it is on all of the quantum approaches to gravitation.[21]

49.7 Anti-de Sitter spacetime

In studying quantum mechanics, we often use the idea of a particle confined to a box.[22] Clearly, confining the gravitational field to a box is

[21] Rovelli and Vidotto (2015) give much more detail and discussion of LQG in a highly engaging form. Readers interested in further alternative (and technical) approaches to quantum gravity can consult the vibrant literature on the subject. For example, for an introduction to twistor theory see the books by Penrose (2004) and by Zee (2013); for an introduction to Regge calculus see Misner, Thorne, and Wheeler (1973); for an introduction to spinors in relativity see Wald (1984) and also Misner, Thorne, and Wheeler (1973).

[22] Our discussion in this section follows that of Zee, which can be consulted for further details.

[23]We shall write d-dimensional AdS spacetime as AdS^d. The **holographic principle** was originally proposed by Gerard 't Hooft (1946–). It says that the description of a d-dimensional volume of space can be encoded on its $(d-1)$-dimensional boundary (similarly to how a three-dimensional image is captured in a two-dimensional interference pattern in optical holography). AdS^d spacetime represents a particularly vivid example of the holographic principle.

[24]CFT = conformal field theory, which is to say, a conformally invariant gauge field theory. AdS/CFT correspondence was originally proposed for spin theory in AdS space by Juan Maldacena (1968–). The idea is that some theories of quantum gravity are equivalent to quantum theories with no gravitational interaction in fewer dimensions. It has been suggested that the correspondence might also provide insight into research areas such as condensed matter physics in the future. See Năstase (2017) for further details.

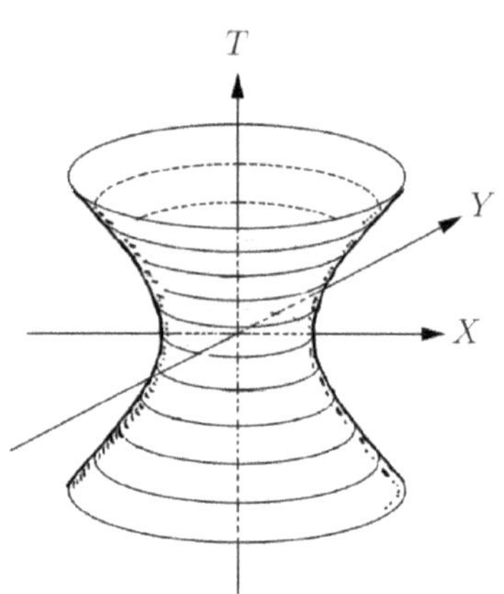

Fig. 49.11   de Sitter spacetime as a hyperboloid embedded in (4+1)-dimensional Minkowski spacetime.

[25]In Chapter 15, we described a spacetime of constant curvature as having a Riemann tensor determined by the Ricci scalar. Equivalently, we described it in Chapter 16 as having a Riemann tensor with components

$$R_{\mu\nu\alpha\beta} = K(g_{\mu\alpha}g_{\nu\beta} - g_{\mu\beta}g_{\nu\alpha}).$$
$$(49.48)$$

In this case, the constant $K = \alpha^{-2}$.

not something we can straightforwardly do in our own spacetime. It is possible to confine gravity in **Anti-de Sitter** (AdS) spacetime, whose geometry is related to the de Sitter spacetime we met in Chapter 15. A remarkable feature of d-dimensional AdS spacetime is that it possesses a spatial boundary made up of Minkowski spacetime with one fewer dimension.[23] AdS spacetime has caught the imaginations of many relativists, especially after the discovery that the physics of some gravitational theories in (4+1)-dimensional AdS spacetime (AdS^5) can be mapped onto (3+1)-dimensional Minkowski space. This is known as the AdS/CFT correspondence[24] and is an active area of research into quantum theories of gravity. AdS spacetime is not a quantum theory of gravity in itself, but might be an important ingredient, and we discuss it in this section.

Before we describe AdS space, let's revisit the view of de Sitter spacetime discussed the exercises for Chapter 18. We saw there that model universes driven by a non-zero cosmological constant Λ can be represented geometrically using this spacetime. de Sitter spacetime can be visualized as a hyperboloid defined by

$$-T^2 + X^2 + Y^2 + Z^2 + W^2 = \alpha^2, \qquad (49.45)$$

embedded in a flat five-dimensional space with metric

$$ds^2 = -dT^2 + dX^2 + dY^2 + dZ^2 + dW^2. \qquad (49.46)$$

The embedding is shown from Fig. 49.11 with two dimensions suppressed. Going through the embedding routine from Appendix D, the result of eliminating a dimension is a line element with the form

$$ds^2 = \left(\eta_{\mu\nu} - \frac{\eta_{\mu\alpha}\eta_{\nu\beta}X^\alpha X^\beta}{\eta_{\mu\nu}X^\mu X^\nu - \alpha^2} \right), \qquad (49.47)$$

where indices in the last equation run from 0 to 3. The topology of this spacetime is built from three-dimensional spheres S^3 that start at $T \to -\infty$ with infinite radius, shrink down to a minimum radius α, before start expanding again for $T \to \infty$. We call this topology $\mathbb{R}^1 \times S^3$ (i.e. a real line representing the time combined with 3-spheres at every instant). This is a spacetime of constant curvature[25] and consistent with a cosmological constant $\Lambda = R/4$ and Einstein tensor with components $G_{\mu\nu} = -\frac{1}{4}Rg_{\mu\nu}$, where $R > 0$.

We found in Chapter 18 that it is possible to represent models with different spatial curvatures by covering the hyperboloid with coordinates that make different cuts through the spacetime. This versatility of de Sitter spacetime follows from the high degree of symmetry that it possesses. In fact, another way of expressing its constant curvature is to say that de Sitter spacetime is an example of a **maximally symmetric** space. Here 'maximal symmetry' means that the space has the same number of symmetries as Euclidean space. A sphere also has this property and it is evident that the de Sitter spacetime can be thought of as a version of a higher dimensional sphere in Minkowski space, with the transformation $T^2 \to -T^2$ providing a means of swapping between a sphere in Euclidean space and de Sitter geometry in Minkowski spacetime.

Anti-de Sitter spacetime can be thought of as de Sitter spacetime with $R < 0$, corresponding to a negative cosmological constant $-\Lambda$.[26] It has topology $S^1 \times \mathbb{R}^3$ and can be represented as a hyperboloid

$$-T^2 + X^2 + Y^2 + Z^2 - U^2 = -\alpha^2. \tag{49.50}$$

This is to be embedded in a spacetime with line element

$$ds^2 = -\mathrm{d}T^2 + \mathrm{d}X^2 + \mathrm{d}Y^2 + \mathrm{d}Z^2 - \mathrm{d}U^2, \tag{49.51}$$

which is a (3+2)-dimensional Minkowski space with two timelike variables (i.e. variables that enter the metric with a minus sign). The embedding is shown in Fig. 49.12, with two dimensions suppressed. The spacetime resembling the de Sitter hyperboloid turned on its side. The embedding routine now results in a line element

$$ds^2 = \left(\eta_{\mu\nu} - \frac{\eta_{\mu\alpha}\eta_{\nu\beta}X^\alpha X^\beta}{\eta_{\mu\nu}X^\mu X^\nu + \alpha^2} \right), \tag{49.52}$$

with $\mu = 0$ to 3 again. In general, we can swap back and forth between results in de Sitter space and anti-de Sitter space by swapping $\alpha^2 \to -\alpha^2$.

The unusual shape of AdS spacetime allows for the existence of closed timelike loops. This is undesirable owing to the violation to causality that it allows. However, the tube-like shape of AdS spacetime can effectively be cut and unrolled with a good choice of coordinates. (We say the topology has been changed to $\mathbb{R}^4$ as a result.) The standard choice of coordinates that does this for three-dimensional AdS spacetime (AdS^3) is

$$T = (1+r^2)^{\frac{1}{2}}\cos t, \quad U = (1+r^2)^{\frac{1}{2}}\sin t, \quad X = r\cos\theta, \quad Y = r\sin\theta. \tag{49.53}$$

In terms of these coordinates, the line element becomes

$$ds^2 = -(1+r^2)\mathrm{d}t^2 + \frac{\mathrm{d}r^2}{1+r^2} + r^2\mathrm{d}\theta^2. \tag{49.54}$$

Example 49.11

We can produce a conformal version of the AdS line element using the ideas from Chapter 19. First, make the choice $r = \tan\psi$, then rewrite eqn 49.54 to say

$$ds^2 = \frac{1}{\cos^2\psi}\left(-\mathrm{d}t^2 + \mathrm{d}\psi^2 + \sin^2\psi\,\mathrm{d}\theta^2\right). \tag{49.55}$$

Here ψ is a latitude-like variable. As the radius-like coordinate r goes from 0 to ∞, the latitude ψ goes from 0 to $\pi/2$ (rather than π, as we might have expected). More colourfully, latitude in AdS spacetime goes from north pole to the equator, not to the south pole. With this observation, we have discovered the boundary AdS spacetime!

---

[26] It has a Riemann tensor with components

$$R_{\mu\nu\alpha\beta} = -\frac{1}{\alpha^2}(g_{\mu\alpha}g_{\nu\beta} - g_{\mu\beta}g_{\nu\alpha}). \tag{49.49}$$

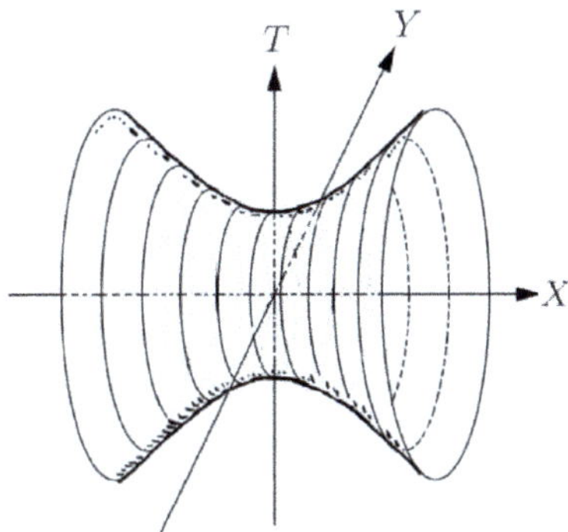

Fig. 49.12 Anti-de Sitter spacetime as a hyperboloid embedded in (3+2)-dimensional Minkowski spacetime.

[27] See exercises for a derivation. Notice the resemblance to what we previously called the Poincaré line element: $ds^2 = (dr^2 + dx^2)/r^2$.

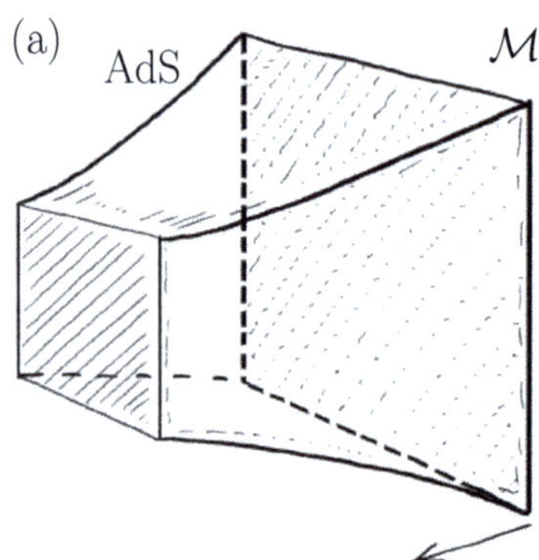

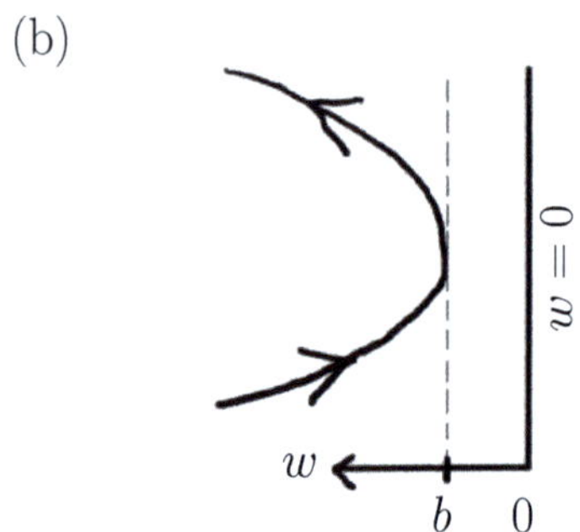

Fig. 49.13 (a) Anti-de Sitter spacetime with its boundary. (b) A massive particle bounces owing to the boundary in spacetime.

[28] This is of the form (kinetic energy) + (potential energy) = const.

[29] The latter was how it has been regarded for most of its lifetime. AdS spacetime was originally discussed in the 1920s by de Sitter (unhelpfully, both de Sitter and AdS were referred to as 'de Sitter spacetimes' for this reason). The spacetime was discovered independently by Tullio Levi-Civita.

[30] 'You may have enjoyed this course and decided that you would like to do your thesis research in general relativity. DON'T. Einstein spent the last thirty years of his life working on general relativity, and it led to nothing. And he was smarter than you.' So said Sidney Coleman (1937–2007), albeit in 1970.

To examine the consequence of the boundary seen in the last example, it is helpful to (once again!) recast AdS spacetime, this time in Poincaré coordinates[27] as

$$ds^2 = \frac{1}{w^2}\left(-dt^2 + dx^2 + dw^2\right). \tag{49.56}$$

The boundary now occurs at $w = 0$. We can see from this coordinate system how a slice made at a particular value of w is simply Minkowski space with one fewer spatial dimension. This idea is shown in Fig. 49.13(a), where AdS spacetime terminates on this flat boundary.

To see the physical influence of the boundary, we shall shoot photons and massive particles at it and see what happens.

Example 49.12

A light beam sent from a point w_0 to the boundary will, if reflected by a mirror $w = 0$ come back after a coordinate time $2w_0$. Things are different for a massive particle. A massive particle obeys the usual condition on its velocity $\boldsymbol{u} \cdot \boldsymbol{u} = -1$, which gives us the equation of motion

$$\dot{t}^2 - \dot{w}^2 = w^2, \tag{49.57}$$

where dots indicate a derivative with respect to proper time. Owing to the absence of the variable t in the metric components we have a Killing vector $\partial/\partial t$ and a conservation law $u_t = g_{tt}u^t = \text{const.}$, or

$$-\frac{1}{w^2}\frac{dt}{d\tau} = (\text{const.}). \tag{49.58}$$

Let's write this latter equation $\dot{t} = w^2/b$, with b a constant length determined by the initial conditions. Substituting the conservation law back into eqn 49.57 we obtain an equation of motion in terms of the coordinate time[28] of

$$\left(\frac{\partial w}{\partial t}\right)^2 + \frac{b^2}{w^2} = 1. \tag{49.59}$$

This describes the motion of a massive particle in a Newtonian potential with positive potential energy $V(w) = (b/w)^2$, which diverges as the boundary $w = 0$ is approached. As shown in Fig. 49.13(b), a particle set in motion in this potential never reaches the boundary. It must stop and turn back at the point $w = b$.

The last example hints at how the energy of a particle might be something that can be mapped to a characteristic value of $w = b$. In this way, the boundary of AdS space is able to encode the physics of the bulk.

Ultimately nobody yet knows whether AdS is an essential part of quantum gravity, or an intriguing curiosity.[29] Of course, the same could currently be said of each of the theories we have discussed in this chapter. It is perhaps not too optimistic to hope that, ultimately, experiment will provide the final word.[30]

49.8   Our current best guess

Which model best describes our Universe? In the last few decades, cosmology has gone from being a highly speculative field (in the 1950s people were arguing about whether the Universe had a beginning or not) to one which is now heavily constrained by very well tied down measurements. Its practitioners claim that we have now entered the era of **precision cosmology**. The cosmic background explorer (COBE) satellite revealed in 1992 that the cosmic microwave background (CMB) exhibits a beautiful blackbody spectrum with a temperature of 2.725 K, and this spectrum is pretty smooth across the sky (there are fluctuations but their amplitude is $\delta T/T \sim 10^{-5}$). In 2001, the Wilkinson Microwave Anisotropy Probe (WMAP) was launched to measure the angular spectrum of these fluctuations in greater detail, and 2009 saw the launch of the Planck satellite, which improved on these measurements even further. These data fit well with a Big-Bang cosmological model known as **ΛCDM**, where Λ refers to the cosmological constant and CDM refers to cold dark matter.

Let's unpack these terms. First, the presence of Λ in the model is consistent with the experimental observation that the expansion of the Universe is currently accelerating, which has been determined by measurements of type-Ia supernovae used as 'standardized candles'. An inflationary period in the Universe's history mandates that the Universe must be very close to its critical density, and the models show that Ω_0 (the ratio of the Universe's density to the critical density) is 0.999(2). From a variety of measurements, it is found that the baryonic[31] matter in the Universe has a density of $\Omega_B = 0.05$. Now we come to CDM, non-baryonic matter which resides in the halos of galaxies.[32] The CDM density comes out to be $\Omega_M = 0.26$, much larger than the baryonic density, but the sum of the two does not yield $\Omega_0 \approx 1$, so that $\Omega_\Lambda = 0.69$ has to make up the difference. The cosmological term has been termed dark energy; it is believed to have a very low density, but is completely uniform across all space[33] and so dominates the overall mass/energy of the Universe. It might be thought to be the energy of the quantum vacuum, though in 1968 Zel'dovich pointed out that though the energy of the quantum vacuum could contribute to Λ, it would result in an energy density fifty orders of magnitude larger than the critical density.

Here is our best guess of how all this fits together: at around 10^{-32} s after the Big Bang there is a period of inflation[34] in which the Universe expands exponentially, leading to a scale-invariant spectrum of gravitational waves (not yet detected in experiments, but may well be found in the coming years). This is terminated only when the scalar field potential energy is converted into particles. The resulting quark soup condenses into hadrons at $t \approx 10^{-5}$ s, with baryons and antibaryons in roughly equal number, and as abundant as thermal photons. However, at $t \sim 1\,$s, as the Universe cools below the temperature of the lightest baryon, most baryons and antibaryons annihilate each other, leaving a small excess[35] of baryons over antibaryons in the few[36] that remain. From about 0.01 s to 20 minutes, we have a period known as Big-Bang nucleosynthesis (BBN) when the lightest elements form, mostly ^{4}He, but also some deuterium (D), ^{3}Li, and ^{7}Li. The Universe cools further and the baryons fall into the gravitational potential wells produced by CDM

[31]Reminder: Baryons are particles like protons and neutrons which are composites of quarks, but this is a shorthand for the 'ordinary' matter in the Universe.

[32]Hot dark matter models were ruled out fairly quickly. CDM is cold, meaning that the dark matter particles are moving at speeds $\ll c$, and so become trapped in the gravitational potential wells of galaxies. They interact with gravity, but not with the strong or electromagnetic force (so we can't see them); they may or may not couple via the weak interaction. Since no-one has directly detected a CDM particle, we don't really know. The only reason we believe they are in halos around galaxies is the effect they have on the rotations of stars around galaxies via measurements that are known as **rotation curves**.

[33]This is unlike ordinary matter which is strongly clumped in stars and galaxies, with lots of regions of space empty of ordinary matter.

[34]Inflation is discussed in detail in Chapter 41.

[35]This is called **baryogenesis**. It is thought that non-equilibrium interactions that violate baryon number conservation, charge (C) conservation, and CP conservation, might allow the Universe to evolve a small net baryon abundance.

[36]Today there are only a few baryons per billion photons.

[37]There are also photons in the Universe, i.e. radiation. However, recall from Chapter 17 that cold matter density $\propto 1/a^3$, directly due to the volume expansion of the Universe, but radiation density $\propto 1/a^4$, which has an extra dependence on the scale factor given by $1/a$ due to cosmological redshift. Thus, the Universe becomes matter dominated, since the radiation density decreases more quickly.

[38]Thus, we have only a loose understanding of the physics occurring at energies corresponding to the era of baryogenesis, let alone the inflationary era.

[39]Or might not!

particles. At around 380,000 years after the Big Bang, the Universe is much cooler, so neutral atoms start to form (the nuclei find electrons to orbit around them) and the Universe becomes transparent to photons; the CMB dates from this era, giving us a snapshot of the Universe at this time. The expansion of the Universe is dominated[37] by the ordinary matter and dark matter. Both types of matter are 'cold', meaning that they are non-relativistic and pressure-less fluids. The Universe is now reasonably well described by the Einstein–de Sitter model, which we called 'Universe 3' back in Chapter 18. From about 5 Gyr ago, the expansion of the Universe starts to become dominated by the cosmological constant term, and so in other words dark energy takes over and the era of cosmic acceleration began. The Universe is now believed to be 13.80(2) Gyr old and its expansion is accelerating.

Can we believe this ΛCDM picture? One impressive feature is that the numerical values of the various constants (e.g. the Ω-values mentioned above) are pretty consistent between completely independent measurements based on (i) the gravity-driven acoustic oscillations in the CMB (which come from the surface of last scattering, determined at a time 380,000 years after the Big Bang) and on (ii) deuterium abundance (due to nuclear reactions that start to take place about one second after the Big Bang). The very early Universe involves some very high energies and temperatures[38] but the microphysics of the eras of (i) and (ii) above have been tested in the laboratory, so we can have some confidence that this picture might be right.

ΛCDM is therefore the current 'standard model' of modern cosmology and has survived various tests unscathed. However, it is worth saying that there are still some wrinkles. The abundance of ^{7}Li is not quite right, baryogenesis is not completely understood, and different measurements are giving slightly different values of the Hubble constant (they vary between about 67 and 74 km s^{-1} Mpc^{-1}), a problem which is called **Hubble tension**. These problems might[39] go away following more detailed measurements, or perhaps following better understanding of some of the systematic errors. More significant is the fact that the physics of inflation is not well tied down, and our current model of inflation is, at best, an approximation; where does inflation actually come from? Even worse, we have not been able to detect any candidate dark matter particles in experiments (despite intensive and patient searches) and so we don't really know what dark matter is. And as for dark energy, we have even less idea. Is dark energy perhaps simply a phantom effect, with the real reason for cosmic acceleration being a consequence of whatever theory succeeds general relativity?

However our understanding of the Universe evolves in the future, we can be reasonably confident that an unalterable part of the picture will be the existence of the Big-Bang singularity. Thus, in our final chapter, we will turn to the Big Bang and describe how this event fits naturally into our current best theory of gravity: general relativity.

Chapter summary

- Extra dimensions could exist if compactified on the Planck length scale. They would have an effect on the nature of gravity.
- String theory describes the dynamics of one-dimensional strings, whose excitations are quantum particles. The theory describes the world sheet of the string and leads to a wave-equation of motion.
- Superspace offers a different approach, where classical spacetime must be replaced with a quantum-mechanical structure.
- Another alternative approach to quantum gravity involving the quantization of spacetime is offered by loop quantum gravity.
- Anti-de Sitter spacetime has a boundary made up of Minkowski spacetime with one fewer spatial dimension.
- ΛCDM is the 'standard model' of modern cosmology and is supported by a great deal of experimental evidence. It might be correct.

Exercises

(49.1) Show that the wavelength of a particle with an energy equal to $m_{\mathrm{P}}c^2$ would be around the Planck length ℓ_{P}. Estimate the gravitational self-energy of such a particle, as well as its Compton wavelength and Schwarzschild radius. (Ignore factors of 2 and π.)

(49.2) By considering the string action, show that our string theory is invariant with respect to reparametrization.

(49.3) Verify eqn 49.9.

(49.4) Show that

$$\vec{v}_{\perp}^{2} = \left(\frac{\partial \vec{X}}{\partial t}\right)^{2} - \left(\frac{\partial \vec{X}}{\partial s} \cdot \frac{\partial \vec{X}}{\partial t}\right)^{2}. \quad (49.60)$$

(49.5) (a) Starting from the Lagrangian expressed within static gauge, compute the canonical momentum of the string

$$\vec{P}(t,\sigma) = \frac{\partial L}{\partial(\partial_t \vec{X})}. \quad (49.61)$$

(b) Use the definitions in Exercises 40.3 and 40.1, with the previous results, to compute the Hamiltonian

$$H = \int \mathrm{d}\sigma\, \mathcal{H}, \quad (49.62)$$

where $\mathcal{H}$ is the Hamiltonian density.

(49.6) Verify eqn 49.35.

(49.7) Consider a very narrow tube around a cosmic string. We will solve the Einstein equation for this situation, which we shall assume has an energy-momentum tensor with components $T_{\mu\nu} = \mathrm{diag}(\rho, 0, 0, -\rho)$, which is designed to be proportional to the Minkowski metric in the (t, z) plane. The region is described by the line element

$$\mathrm{d}s^2 = -\mathrm{d}t^2 + r_0^2(\mathrm{d}\theta^2 + \sin^2\theta \mathrm{d}\phi^2) + \mathrm{d}z^2, \quad (49.63)$$

with constant r_0.
If we accept that this looks similar to the flat-space metric $\mathrm{d}s^2 = -\mathrm{d}t^2 + \mathrm{d}r^2 + r^2\mathrm{d}\phi^2 + \mathrm{d}z^2$, then $r_0\theta$ in eqn 49.63 can be treated as a sort of radial variable.
(a) Compute the components of the Ricci tensor and the Ricci scalar for this spacetime and show that Einstein's equation is satisfied.
(b) If the outer surface of the string region occurs at θ_{m}, compute the cross-sectional area of the string and hence its mass per unit length.

(49.8) Following from the previous problem, now consider the region outside the string. The metric outside the string can be written as

$$ds^2 = -dt^2 + r_0^2 \left(\frac{\cos^2 \theta}{\cos^2 \theta_m} d\theta^2 + \sin^2 \theta d\phi^2 \right) + dz^2. \tag{49.64}$$

This has the property that it smoothly joins on to the interior metric from the last question at θ_m.
(a) By making a suitable transformation, show that this metric describes flat spacetime in cylindrical polar coordinates.
(b) Although the spacetime seems flat, it isn't really. What is the circumference of a large circle in this spacetime with $r' = a \gg r_0$?

(49.9) Consider a tetrahedron in three-dimensional space whose vertices are given by the four vectors $\vec{0}$, $\vec{a}$, $\vec{b}$, and $\vec{c}$. Derive expressions for the four area vectors $\vec{L}_1$, $\vec{L}_2$, $\vec{L}_3$, and $\vec{L}_4$, and show that they satisfy a closure property (eqn 49.42).

(49.10) Show that the volume of the tetrahedron in the previous exercise is given by

$$V = \frac{1}{6} |\vec{a} \times \vec{b} \cdot \vec{c}|. \tag{49.65}$$

Hence show that

$$V^2 = \frac{2}{9} |\vec{L}_1 \times \vec{L}_2 \cdot \vec{L}_3|. \tag{49.66}$$

(49.11) Show that de Sitter space solves an Einstein equation with components $G_{\mu\nu} = -(3/\alpha^2) g_{\mu\nu}$. How does this differ for anti-de Sitter space?

(49.12) Consider a defining equation for AdS^3 given by $(T^2 - X^2) + (W^2 - Y^2) = 1$. We solve this by writing

$$T^2 - X^2 = \frac{t^2 - x^2}{w^2}, \tag{49.67}$$

and

$$W^2 - Y^2 = \frac{x^2 - t^2}{w^2} + 1, \tag{49.68}$$

with $T = t/w$, $X = x/w$,

$$Y = \frac{1}{2w} \left(x^2 - t^2 + w^2 - 1 \right), \tag{49.69}$$

and

$$W = \frac{1}{2w} \left(x^2 - t^2 + w^2 + 1 \right). \tag{49.70}$$

Use these coordinates to confirm the Poincaré spacetime line element from eqn 49.56.

(49.13) Verify the expressions in Example 49.12.

The Big-Bang singularity

50

I have shown that all the realms of the universe
Are mortal, and the substance of the heavens
Had birth; and I have explained most of those things
That in heaven occur and must occur.
Lucretius (c.100 BC–c.50 BC) *On the Nature of Things*

The Robertson–Walker spaces, and many of the Friedmann universes that we have built from them, have one particularly notable feature: an initial Big-Bang singularity. In this chapter, we ask whether this is an artefact of our theory, or whether we should regard the Big Bang as a realistic event in our Universe. We shall show that, with a minimal set of assumptions, a singularity occurring at some time in the past is indeed a realistic, and possibly even an inevitable, prospect.[1]

Let's construct a spacelike hypersurface S in our spacetime at some fixed value of the cosmic time. We assume that it is a special sort of hypersurface known as a **Cauchy surface**. Such a surface has the property that the events on it completely determine some future surface, lying in what is known as the **domain of dependence** of S. If this is the case then it turns out there must exist a longest-timelike curve from S to some point on the domain of dependence.[2]

The existence of such a longest timelike curve doesn't seem especially scandalous, but it is the subject of the argument in this chapter that reveals, in general terms, the necessity for a Big-Bang singularity to have started our Universe. Let's go to work.

50.1 Facts about Euclidean geometry

Let's first review a few useful facts about Euclidean 3-space. (We will then apply similar ideas to curved $(3+1)$-dimensional space.) We shall deal with a class of geodesics that all meet points on the surface S. Although these, being geodesics, all represent the shortest distance between some point q and some specific point on the surface S, some will be shorter than others, depending on the point on S on which they end.

Consider a path γ between point p and surface S. If the path is the shortest one possible between p and S then γ must intersect S orthogonally. If it doesn't, as in the curve γ shown in Fig. 50.1, then there is a shorter path γ' that does meet the surface at right angles. We shall use the term 'orthogonal' to describe those geodesics that meet surfaces at right angles (they are also known as normal geodesics).

[1] In this chapter, we follow the argument in the form given by Geroch in *General Relativity, 1972 Lecture Notes* (2013). The techniques sketched here (sometimes called **global techniques** in their generalized form) are described in detail in Penrose (1973) and also in Wald (1984) and in Hawking and Ellis (1973).

[2] The necessity for this longest curve existing follows from a technicality: the fact that the collection of all timelike and null curves from a point on the domain of dependence to S is *compact*. Compactness is discussed in Appendix C, but can be thought of here as implying that no curves go off to infinity. As a result, length is a continuous function on this space of curves, and must achieve a maximum.

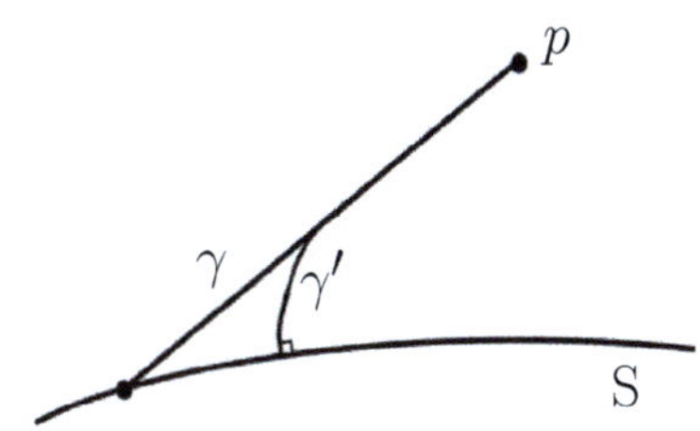

Fig. 50.1 Curve γ does not meet the surface S at a right angle. The curve γ' that does is shorter.

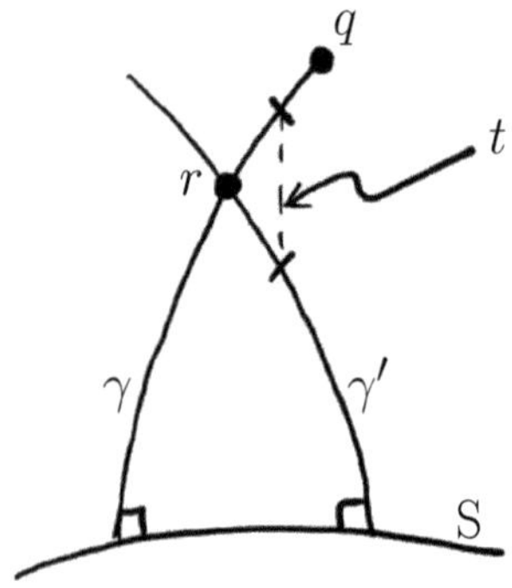

Fig. 50.2 The shortest point between q and the surface S involves travelling along γ, rounding off the corner to avoid the crossing point at r and then following γ' down to the surface S.

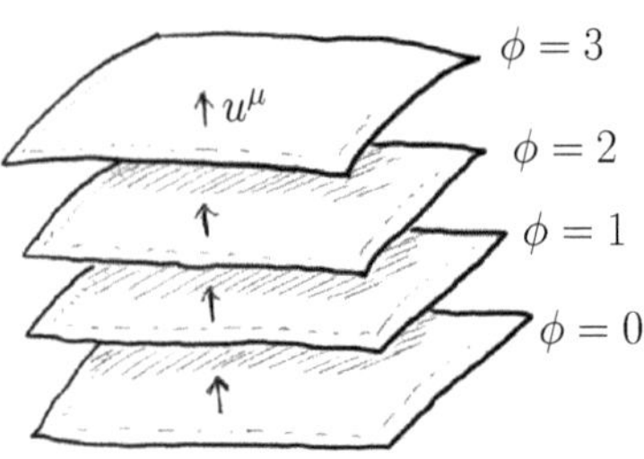

Fig. 50.3 The function ϕ measures the distance along the orthogonal geodesics from the surface S.

[3] This can be seen by writing $u_\mu = \partial\phi/\partial x^\mu$ and then writing $u_{\mu;\nu}$ as

$$\left(\frac{\partial\phi}{\partial x^\mu}\right)_{;\nu} = \frac{\partial^2\phi}{\partial x^\nu \partial x^\mu} - \Gamma^\alpha{}_{\nu\mu}\left(\frac{\partial\phi}{\partial x^\alpha}\right),$$

which is manifestly symmetric since $\Gamma^\alpha{}_{\mu\nu} = \Gamma^\alpha{}_{\nu\mu}$.

Now consider two orthogonal geodesics γ and γ' that cross at a point r (Fig. 50.2). In this case, γ can't be the shortest path from q to the surface S. This is because we can construct a shorter curve by rounding off the corner at the crossing point r and then following the other curve γ' down to S, as is shown in Fig. 50.2. This argument can be made a little more rigorous by recalling our discussion in Chapter 8 of whether the action for a particular trajectory is a minimum or not. We saw that the existence of a conjugate point along the trajectory guarantees that it does not represent the minimum. The crossing point of the two orthogonal geodesics in this example is just such a conjugate point.

We now turn to spacetime, where these facts will be used in a slightly modified form.

50.2   Orthogonal geodesics in spacetime

In curved spacetime, recall that the *longest lines* are the straightest. That is, owing to the sign of the metric the most proper time elapses along the lines with least acceleration. We therefore use the results argued for Euclidean space above, but with *longest* replacing *shortest*.

Consider again a surface S and its orthogonal geodesics in spacetime. We want a way of measuring the distance from S along all of the orthogonal geodesics. To do this, we define a function $\phi(x)$ that provides a measure of this distance for that geodesic that passes through point with coordinates x^μ, as shown in Fig. 50.3. We then define a velocity 1-form field $\tilde{u}(x)$ for the orthogonal geodesics with components $u_\mu = \nabla_\mu\phi$. We work in a spacetime with a metric, so we can raise indices and form a velocity vector field $u(x)$ with components u^μ. The velocity components have the usual property $u^\mu u_\mu = -1$. From the definition of $\tilde{u}(x)$ in terms of the scalar function $\phi(x)$, the components $u_{\nu;\mu}$ of the covariant derivative of $\tilde{u}$, are symmetric[3] with respect to exchange of μ and ν and so

$$(\nabla_u\tilde{u})_\nu = u^\mu u_{\nu;\mu} = u^\mu u_{\mu;\nu} = \frac{1}{2}(u^\mu u_\mu)_{;\nu} = 0, \qquad (50.1)$$

since $u^\mu u_\mu = -1$. Raising the ν index on $u^\mu u_{\nu;\mu} = 0$ we obtain $\nabla_u u = 0$. This implies that the velocity vector field $u(x)$ is the tangent field of *all* of the orthogonal geodesics. We shall study the **convergence** of this field c, defined as minus the divergence, or

$$c = -\nabla \cdot u = -u^\mu{}_{;\mu}. \qquad (50.2)$$

To make further progress, we need the result of the slightly tedious, but important, computation that follows in the next example.

Example 50.1

Consider the directional derivative of the scalar function c along our geodesics

$$\nabla_u c = u^\nu \nabla_\nu c = -u^\nu u^\alpha{}_{;\alpha\nu}. \qquad (50.3)$$

This can be linked to curvature by using the eqn 35.13 from Chapter 35 that says

$$u^{\alpha}{}_{;\nu\mu} - u^{\alpha}{}_{;\mu\nu} = R^{\alpha}{}_{\beta\mu\nu}u^{\beta}, \tag{50.4}$$

which can be massaged to read

$$u^{\alpha}{}_{;\nu\alpha} - u^{\alpha}{}_{;\alpha\nu} = R^{\alpha}{}_{\beta\alpha\nu}u^{\beta} = R_{\beta\nu}u^{\beta}, \tag{50.5}$$

where the components of the Ricci tensor have appeared in the final step. We have then, that

$$\nabla_{\boldsymbol{u}}c = -u^{\nu}u^{\alpha}{}_{;\nu\alpha} + R_{\beta\nu}u^{\nu}u^{\beta}. \tag{50.6}$$

This latter expression can be rewritten in a final form

$$\nabla_{\boldsymbol{u}}c = -(u^{\nu}u^{\alpha}{}_{;\nu})_{;\alpha} + (u^{\nu}{}_{;\alpha})(u^{\alpha}{}_{;\nu}) + R_{\beta\nu}u^{\nu}u^{\beta}. \tag{50.7}$$

The first term on the left is zero, since $\boldsymbol{u}$ is tangent to the geodesics, and so we have

$$\nabla_{\boldsymbol{u}}c = (u^{\nu}{}_{;\alpha})(u^{\alpha}{}_{;\nu}) + R_{\beta\nu}u^{\nu}u^{\beta}. \tag{50.8}$$

The result from the last example can be rewritten as

$$\nabla_{\boldsymbol{u}}c = (u^{\nu;\alpha})(u_{\alpha;\nu}) + R_{\beta\nu}u^{\nu}u^{\beta}. \tag{50.9}$$

Let's pause and interpret this equation. It is telling us about the change in divergence of the world lines of the orthogonal geodesics as we move along them. This depends on two terms: the second is related to the curvature of spacetime via the Ricci tensor, which tells us about the way that curvature causes volumes to shrink. The first term is given by the quantity $u_{\alpha;\nu}$, which is a symmetric matrix that we must effectively multiply by itself and then trace over the result. Although this seems a little abstract, the next example shows that this quantity obeys a neat inequality, which makes it very useful.

Example 50.2

We separate out the trace and the trace-free part of the quantity $u_{\alpha;\nu}$ by using the transverse projection operator[4] $P_{\alpha\beta} = g_{\alpha\beta} + u_{\alpha}u_{\beta}$. We write[5]

$$u_{\alpha;\nu} = -\frac{c}{3}P_{\alpha\nu} + \left(u_{\alpha;\nu} + \frac{c}{3}P_{\alpha\nu}\right). \tag{50.10}$$

This decomposition allows us to say

$$(u^{\nu;\alpha})(u_{\alpha;\nu}) = \left[-\frac{c}{3}P^{\nu\alpha} + \left(u^{\nu;\alpha} + \frac{c}{3}P^{\nu\alpha}\right)\right]\left[-\frac{c}{3}P_{\alpha\nu} + \left(u_{\alpha;\nu} + \frac{c}{3}P_{\alpha\nu}\right)\right]$$

$$= \frac{c^{2}}{3} + \left(u^{\nu;\alpha} + \frac{c}{3}P^{\nu\alpha}\right)\left(u_{\alpha;\nu} + \frac{c}{3}P_{\alpha\nu}\right). \tag{50.11}$$

Consider the second term on the right, which effectively instructs us to take the square of the traceless part and then trace over the result. One thing we can say about the number that results is that, if it is non-zero, it must be positive. This leads to the key inequality

$$(u^{\nu;\alpha})(u_{\alpha;\nu}) \geq \frac{c^{2}}{3}. \tag{50.12}$$

This is the important technical result we need.

[4]See Chapter 39 for a description of the transverse projection operator.

[5]Here we make use of the following facts (see exercises):
(i) both $P^{\alpha\beta}$ and $u_{\alpha;\beta}$ are symmetric.
(ii) The trace of the first term is non zero, since

$$\frac{c}{3}P_{\alpha\nu}g^{\alpha\nu} = c,$$

explaining why we call it the trace part.
(iii) The trace of the second term *is* zero

$$g^{\alpha\nu}\left(u_{\alpha;\nu} + \frac{c}{3}P_{\alpha\nu}\right) = -c + c,$$

explaining why we call it trace free.
(iv) The projection operator obeys

$$P^{\alpha\beta}P_{\alpha\beta} = 3,$$

(v) and, finally,

$$P^{\alpha\beta}u_{\alpha;\beta} = -c.$$

As a result, the trace-free part gives zero when contracted against $\boldsymbol{P}$.

Now consider the second term in eqn 50.9. Using a form of the Einstein equation we have, on multiplying by velocity components,

$$u^\mu u^\nu R_{\mu\nu} = 8\pi \left(T_{\mu\nu} - \frac{1}{2} g_{\mu\nu} T \right) u^\mu u^\nu. \tag{50.13}$$

The term on the right is essentially an energy density, as seen by an observer with a tangent vector to their world line of u. Remember from Chapter 13 that the weak energy condition says that this quantity must be positive. This means we can use eqn 50.12 to refine eqn 50.9 to read

$$\nabla_u c \geq \frac{c^2}{3}. \tag{50.14}$$

Physically, this says that gravity acts attractively, making world lines tend to converge along their length. This latter equation can also be solved. If the world lines are parametrized by a proper time τ then $\nabla_u \equiv D/d\tau$ and we need only solve the simple differential equation

$$\frac{dc}{d\tau} \geq \frac{c^2}{3}. \tag{50.15}$$

Taking the boundary conditions to be that, at the surface S, we have $\tau = 0$ and also $c(\tau = 0) = c_0$, then the result of integrating this equation is

$$c(\tau) \geq \frac{3}{3/c_0 - \tau}. \tag{50.16}$$

The solution tells us about the effect of gravity drawing the world lines together: the convergence becomes infinite by a proper time $\tau = 3/c_0$. An infinite convergence implies that the world lines have started to cross. This is not particularly surprising, since world lines are certainly allowed to cross, but it is interesting. It means the world lines form a **caustic**, which is an envelope around the congruence of world lines that forces them to cross at some point. This caustic is simply a consequence of the attractive nature of gravity.

So we have a situation where every orthogonal geodesic must cross another by the time it reaches an interval $\tau = 3/c_0$. Recall our argument about crossing: if one orthogonal geodesic crosses another, then the first cannot be the longest timelike curve from a point beyond S. But we also know (from our other geometric fact) that the longest curve must be an orthogonal geodesic. These contradictory statements can only be reconciled by the following:

If we go farther than $\tau = 3/c_0$ from S along any timelike curve to some point p, then *there is no longest timelike curve from p to S.*

This statement implies that sufficiently far from S, points in spacetime cannot be joined by an extremal timelike curve.

This all seems very abstract, as indeed it is, but the argument can be put straightforwardly: if we go a distance $3/c_0$ from S, then we reach a point where there is no possible longest timelike curve from S. However, recall from the start of this chapter that if S is a Cauchy surface, there *must* be a longest timelike curve to a future point. This contradiction has only one resolution: *it must simply not be possible to get more than* $3/c_0$ *away from S by following a timelike curve.*

50.3   Our Universe

After the lengthy and abstract argument of the last section, we are ready to apply the result to our Universe, which is likely very close to one of the Robertson–Walker spaces filled with perfect fluid, and hence is well-described by a Friedmann model.

Consider a small comoving volume of space of 3-volume $\mathcal{V}$ in a Friedmann model. Dimensionally speaking, the convergence c must scale as $-\dot{\mathcal{V}}/\mathcal{V}$. Since the volume $\mathcal{V}$ itself scales varies as $a(t)^3$, where $a(t)$ is the expansion factor, we conclude that the convergence of the Universe can be taken to be

$$c = -\frac{3\dot{a}(t)}{a(t)}. \tag{50.17}$$

Since $\dot{a}(t)$ and $a(t)$ are positive we see that, as a consequence of the expansion of the Universe, the world lines of dust particles in our Universe are diverging, rather than converging, as we assumed in the argument above. Rather disappointingly, therefore, we must conclude that the argument is inapplicable. Before losing hope, we note that if we play time backwards then the sign of $\dot{a}$ flips, while a does not, and so we can apply then the argument to the *past* of our Universe, even if it doesn't seem to work for the future.

So now we take S to be the current spacelike hypersurface of the Universe. Following the argument above, it follows that it is not possible to trace a timelike curve backwards in time further than an interval in proper time of $\tau = 3/c_0$. Why has this happened? We have defined our theory of cosmology to take place on the smooth manifold that describes the theory of fields. As a result, the only point that could be reached by following a timelike curve back in time for an interval of $3/c_0$ is one that does not live in the manifold at all. This point is the singularity for which we have been searching.

The key here is that a singularity cannot conform to the demand that the manifold is smooth everywhere. It is a point that we must cut out of the spacetime if we are to describe the spacetime manifold using a field theory like relativity (Fig. 50.4). If we follow a timelike curve backwards and we reach a gap in the manifold then the curve cannot continue. This then is our initial Big-Bang singularity. In a Universe whose character is well approximated by the state of affairs that we have described in this chapter, a singular point in the past is therefore an inevitability. We conclude that the Universe starts (and this book ends) not with a whimper, but with a Big Bang.

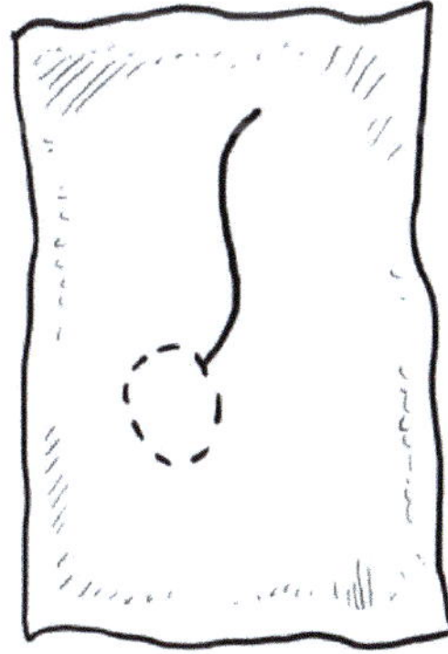

Fig. 50.4 We are forced to cut out a singular point from a manifold, which must be a smooth space.

Chapter summary

- A Cauchy surface S has a longest timelike curve from S to a point p on the domain of dependence.

- If a point p is further from S than $\tau = 3/c_0$ along any timelike curve then there is no timelike curve from S to p.

- A singularity in the spacetime manifold is the reason we cannot follow a timelike curve further back than $\tau = 3/c_0$ in a Robertson–Walker Universe.

- An initial, Big-Bang singularity is therefore expected for a Robertson–Walker Universe on general grounds.

Exercises

(50.1) The (0,2) projection tensor is given by

$$\boldsymbol{P}(\ ,) = \boldsymbol{g}(\ ,) + \tilde{\boldsymbol{u}}(\ ) \otimes \tilde{\boldsymbol{u}}(\ ), \qquad (50.18)$$

where $\boldsymbol{u}$ is a velocity vector.

(a) Write the tensor in component form.

(b) Show that if we insert a vector $\boldsymbol{v}$ into $\boldsymbol{P}$ we project $\boldsymbol{v}$ into the 3-surface that is orthogonal to $\boldsymbol{u}$.

(c) Evaluate $P^{\mu\nu} P_{\mu\nu}$.

(d) Evaluate the $P^{\alpha\beta} u_{\alpha;\beta}$ for the case that $\boldsymbol{u}$ is tangent to a geodesic.

(e) If $\boldsymbol{n}$ is a unit spacelike vector, show that $\boldsymbol{P} = \boldsymbol{g}(\ ,\ ) - \tilde{\boldsymbol{n}}(\ ) \otimes \tilde{\boldsymbol{n}}(\ )$ is the corresponding projection operator.

(50.2) *The conventional derivation of the Raychaudhuri equation uses a very similar argument to the one in Section 50.2. We follow the approach of Zee here, which can be referred to for further details.*
Consider a congruence of timelike geodesics, parametrized by proper time τ, with coordinates $x^\mu(\tau, \sigma^1, \sigma^2, \sigma^3)$ and tangent vectors $\boldsymbol{u} = (\partial x^\mu/\partial \tau)\boldsymbol{e}_\mu$. The vectors $\boldsymbol{W}_i = (\partial x^\mu/\partial \sigma^i)\boldsymbol{e}_\mu$ span a three-dimensional subspace which can be projected into using the operator $\boldsymbol{P}$ from the previous question.

(a) Show that the rate of change of $\boldsymbol{W}_i$ along the congruence is given by

$$(\boldsymbol{\nabla}_{\boldsymbol{u}} \boldsymbol{W}_i)^\mu = B^\mu{}_\nu W_i^\nu, \qquad (50.19)$$

where $B_{\mu\nu} = u_{\mu;\nu}$. The tensor $\boldsymbol{B}$ therefore measures the failure of the vector $\boldsymbol{W}_i$ to be parallel transported along the congruence.

(b) Show that $\boldsymbol{B}$ has the properties that $u^\mu B_{\mu\nu} = 0$ and $B_{\mu\nu} u^\nu = 0$, telling us that $\boldsymbol{B}$ also lives in the same three-space as $\boldsymbol{W}_i$.

The tensor $\boldsymbol{B}$ is conventionally split into: (i) a trace

$$\theta = P^{\mu\nu} B_{\mu\nu} = u^\mu{}_{;\mu}, \qquad (50.20)$$

which describes expansion; (ii) a symmetric, traceless part

$$\sigma_{\mu\nu} = \frac{1}{2}(B_{\mu\nu} + B_{\nu\mu}) - \frac{1}{3}\theta P_{\mu\nu}, \qquad (50.21)$$

which describes shear, and (iii) an antisymmetric part

$$\omega_{\mu\nu} = \frac{1}{2}(B_{\mu\nu} - B_{\nu\mu}), \qquad (50.22)$$

which describes rotation. We therefore have

$$B_{\mu\nu} = \sigma_{\mu\nu} + \frac{1}{3}\theta P_{\mu\nu} + \omega_{\mu\nu}. \qquad (50.23)$$

(c) Show that

$$(\boldsymbol{\nabla}_{\boldsymbol{u}} \boldsymbol{B})_{\mu\nu} = -B_{\mu\alpha} B^\alpha{}_\nu - R_{\beta\mu\alpha\nu} u^\beta u^\alpha, \qquad (50.24)$$

which is known as the **Raychaudhuri equation** and is very useful in proving singularity theorems. *Often the expansion parameter θ is useful to tell us if a congruence is expanding or contracting. Since $\mathrm{D}\theta/\mathrm{d}\tau = g^{\mu\nu}(\boldsymbol{\nabla}_{\boldsymbol{u}}\boldsymbol{B})_{\mu\nu}$, we can contract the result*

from (c) using $g^{\mu\nu}$.
(d) Show that

$$g^{\mu\nu}B_{\mu\alpha}B^{\alpha}{}_{\nu} = \sigma_{\mu\nu}\sigma^{\mu\nu} + \frac{1}{3}\theta^2 - \omega_{\mu\nu}\omega^{\mu\nu}. \quad (50.25)$$

(e) Use the previous result to obtain

$$\frac{\mathrm{D}\theta}{\mathrm{d}\tau} = -\frac{1}{3}\theta^2 - \sigma_{\mu\nu}\sigma^{\mu\nu} + \omega_{\mu\nu}\omega^{\mu\nu} - R_{\mu\nu}u^{\mu}u^{\nu}. \quad (50.26)$$

(50.3) In the case that $\omega_{\mu\nu} = 0$, the tensor $\sigma_{\mu\nu}$ is purely spatial, implying

$$\frac{\mathrm{D}\theta}{\mathrm{d}\tau} \leq -\frac{1}{3}\theta^2 - R_{\mu\nu}u^{\mu}u^{\nu}. \quad (50.27)$$

Show that if we adopt the **strong energy condition**

$$T_{\mu\nu}v^{\mu}v^{\nu} \geq \frac{1}{2}Tv^{\mu}v_{\mu}, \quad (50.28)$$

for all timelike vectors v, then all of the geodesics approach each other, i.e. gravitation has a focussing effect on the congruence.

A   Further reading

Books must follow sciences, and not sciences books
Francis Bacon (1561–1626)

In my situation as Chancellor of the University of Oxford, I have been much exposed to authors
Arthur Wellesley, Duke of Wellington (1769–1852)

[1] The books listed in the bibliography section of this Appendix are referred to throughout the text using the author's name where this is unambiguous. The publication date is also included in cases where several books by the same author feature in the bibliography.

There are many excellent books on general relativity, cosmology, geometry and related fields.[1] Like many introductory books this one contains a compilation of many arguments, explanations and examples formulated and presented by other authors. Our sources are discussed at the end of this appendix.

In learning most subjects, one usually benefits from having read ≥ 2 books and many of those mentioned in this chapter would provide a good supplement to this one. We like Geroch and Spivak's insightful explanations, Penrose (2004) and Hartle's approachability and Wald's precision. For learning general relativity, some recommended choices that share a similar approach to this book include: (i) Geroch (1978), Penrose (2004), Schutz (1985) and Hartle at an introductory level; (ii) d'Inverno, Guidry, Hobson/Efstathiou/Lasenby, and Zee at an intermediate level; and (iii) Geroch (2013, General Relativity), Hawking/Ellis, Landau/Lifshitz (1975), Misner/Thorne/Wheeler and Wald at an advanced level. For learning differential geometry we recommend: (i) Penrose (2004) and Schutz (1980) at an introductory level; Misner/Thorne/Wheeler at an intermediate level; and (iii) Geroch (1985 and 2013, Differential Geometry) and Spivak (2005) at a more advanced level (with the latter very suitable for readers with a background in mathematics). The problem books by Moore (at an elementary level) and by Lightman *et al.* and Blennow/Ohlsson (intermediate/advanced) are also warmly recommended. We have followed their approaches in some examples and exercises.

One thing to beware of in books on general relativity is the different sign conventions adopted. These change a number of the key equations. There are four conventions to watch out for (three of which are independent). We list the conventions of several books below.

[2] Our choice $s_1 = +$ is sometimes called *mostly plusses* or the *East Coast convention*. The choice $s_1 = -$ is called *mostly minuses/West Coast convention*.

- The sign s_1 in front of the line element of the metric[2] $\boldsymbol{g}$

$$s_1 \mathrm{d}s^2 = -\mathrm{d}x^0 + \mathrm{d}x^1 + \mathrm{d}x^2 + \mathrm{d}x^3. \tag{A.1}$$

- The sign s_2 in front of the components of the Riemann tensor $\boldsymbol{R}$

$$s_2 R^\mu{}_{\nu\alpha\beta} = \partial_\alpha \Gamma^\mu{}_{\beta\nu} - \partial_\beta \Gamma^\mu{}_{\alpha\nu} + \Gamma^\mu{}_{\alpha\sigma} \Gamma^\sigma{}_{\beta\nu} - \Gamma^\mu{}_{\beta\sigma} \Gamma^\sigma{}_{\alpha\nu},$$
$$s_2 \boldsymbol{R}(\boldsymbol{u}, \boldsymbol{v}) = \boldsymbol{\nabla}_{\boldsymbol{u}} \boldsymbol{\nabla}_{\boldsymbol{v}} - \boldsymbol{\nabla}_{\boldsymbol{v}} \boldsymbol{\nabla}_{\boldsymbol{u}} - \boldsymbol{\nabla}_{[\boldsymbol{u},\boldsymbol{v}]}. \tag{A.2}$$

- The sign s_3 in the Einstein equation

$$\boldsymbol{G} = s_3 8\pi \boldsymbol{T},$$
$$G_{\mu\nu} = R_{\mu\nu} - \frac{1}{2} g_{\mu\nu} R = s_3 8\pi T_{\mu\nu}. \tag{A.3}$$

- The sign $s_4 = s_3/s_2$ in the contraction

$$s_4 R_{\mu\nu} = R^\alpha{}_{\mu\alpha\nu}. \tag{A.4}$$

Here are some of the sign conventions used in well-known books.[3]

[3] References are given at the end of this appendix.

Reference	g	R	G
This book	+	+	+
Blennow and Ohlsson	−	+	+
Böhmer	+	+	+
Carlip (2005)	−	+	+
Choquet-Bruhat	+	+	+
d'Inverno	−	+	+
Einstein	−	+	−
Feynman (1995)	−	−	+
Foster and Nightingale	−	+	−
Geroch (2013, General Relativity)	+	+	+
Guidry	+	+	+
Hartle	+	+	+
Hawking and Ellis	+	+	+
Hobson, Efstathiou, and Lasenby	−	−	−
Hughston and Tod	−	−	−
Lambourne	−	+	−
Landau and Lifshitz (1975)	−	+	+
Lightman, Press, Price, and Teukolsky	+	+	+
Ludvigsen	−	−	−
Misner, Thorne, and Wheeler	+	+	+
Moore	+	+	+
Nakahara	+	+	+
Ohanian and Ruffini	−	+	−
Padmanabhan	+	+	+
Peacock	−	+	−
Penrose (2004)	−	−	−
Plebański and Krasiński	−	+	+
Poisson	+	+	+
Poisson and Will	+	+	+
Ryder (2009)	+	+	+
Schutz (1985)	+	+	+
Thorne and Blandford	+	+	+
Wald	+	+	+
Weinberg (1972)	+	−	−
Zee	+	+	+

For a biography of Einstein and a history of the formulation of general relativity we recommend Pais, which is a superb scientific biography and is unlikely to be surpassed. Cheng is helpful in introducing and discussing Einstein's work using modern notation. Einstein (1952) contains many of the key original papers.

Further reading by chapter:

Most of the topics covered in this book are also discussed in the standard references on general relativity. The further reading list given below is based on books that use a similar approach to us, and those whose presentation we've followed in some of our arguments. Some are at an introductory and some at a more advanced level.

Chapter 1: an accessible introduction to special relativity can be found in French and in Geroch (1978); a useful summary is given in Landau and Lifshitz (vol. II). Links between geometry and special relativity are discussed in Ellis/Williams. *Chapter 2*: vectors are discussed in Boas and in Penrose (2004); their use in relativity is covered in Hartle and in Schutz (1985). Chapter 3: coordinate transformations are introduced in French and in Schutz (1985). *Chapter 4*: 1-forms are introduced in Schutz (1980 and 1985), Misner/Thorne/Wheeler and in Guidry. Ludvigsen gives a geometrical introduction to the energy-momentum tensor. *Chapter 5*: metrics are introduced in Hartle and Zee. *Chapter 6*: the principles of relativity are discussed in all books on relativity, and in most detail by Weinberg (1972). A historical account can be found in Pais. See Einstein for a collection of the original papers, these are put in context by Cheng. *Chapter 7*: the covariant derivative and connection coefficients are discussed by Schutz (1985) and in Misner/Thorne/Wheeler. *Chapters 8 and 9*: the method used to extract connection coefficients can be found in Zee and in Hartle. *Chapter 10*: the importance of the vielbein is stressed in Hartle, whose approach and notation we follow. They are used extensively in Lightman *et al.* *Chapter 11*: an introduction to Riemann curvature is found in Hartle and in Misner/Thorne/Wheeler. *Chapter 12*: an intuitive introduction to the energy-momentum tensor is given in Hartle. Misner/Thorne/Wheeler provides lots of insight and useful diagrams. *Chapter 13*: the construction of the Einstein equation is justified in Schutz (1985). See Feynman (1995) for a rather different approach. Einstein's route to the field equation is described is Pais. *Chapter 15*: Cosmology is introduced in Lambourne. For a full account see Peacock and Weinberg (2006). *Chapter 16*: Robertson Walker spaces are introduced in Lambourne and described in detail in Misner/Thorne/Wheeler. More detail on hyperbolic spaces is covered in Penrose (2004) and in Needham (1997). *Chapters 17 and 18*: cosmological models are introduced in Penrose (2004) and, more systematically, in Lambourne. *Chapter 19*: conformal infinities and singularities are outlined in d'Inverno. Singularity theory is described at a more advanced level in Hawking/Ellis (whose presentation we follow in a highly simplified form) and in Penrose (1972). *Chapter 20*: Newtonian orbits are analysed in French and Ebbison. An advanced (but fascinating!) take is Gutzwiller. *Chapter 21*: the Schwarzschild geometry is introduced in Hartle, in Schutz (1985) and in Lambourne. Misner/Thorne/Wheeler gives a complete account. *Chapters 22, 23 and 24*: motion in the Schwarzschild geometry is discussed in Hartle (whose approach and notation we follow), in Moore and in Misner/Thorne/Wheeler. *Chapter 25*:

black holes are introduced in Blundell, and treated in all modern general relativity texts. See Hartle and Schutz (1985) for introductory treatments and Misner/Thorne/Wheeler for a more advanced discussion. Chandrasekhar gives a complete account (including a clear discussion of much of the material in this part of the book), albeit at a very advanced level. *Chapters 26 and 27*: black hole singularities are clearly explained in Wald, and we follow this approach. The analogy with accelerating Minkowski coordinates is discussed in Rindler. Wormholes are discussed in Misner/Thorne/Wheeler. *Chapter 28*: Hawking radiation is explained in Schutz (1985) and in Zee. For black hole thermodynamics, see Page (2005) and Carlip (2014). *Chapter 29*: charged and rotating black holes are introduced in Hartle. Our discussion of the Kerr metric follows Schutz (1985). Hawking/Ellis supplies additional insight. *Chapter 30*: classical curvature is introduced from a visual perspective in the wonderful book by Needham (2021), in Zee and (from a historical perspective) in Weinberg (1972). A full account is given in Lipschutz. Spivak (1999) gives translations of the key papers by Gauss and Riemann, along with Spivak's characteristically insightful commentary. *Chapters 31, 32 and 37*: modern geometry is introduced in Needham (2021), in Misner/Thorne/Wheeler and in Schutz (1980). We follow Misner/Thorne/Wheeler's presentation and notation in this part of the book. An introduction to the formal mathematics underlying this subject is given in Spivak (1971). See Spivak (2005) for the full story on all of the topics in this section. *Chapter 33*: an accessible introduction to the Lie derivative is found in Penrose (2004). The discussion in Schutz (1980) is also very accessible at an intermediate level. *Chapters 34 and 35*: the geometrical approach to the covariant derivative and the Riemann tensor is discussed in Penrose (2004) and Needham (2021) at an introductory level, and in Misner/Thorne/Wheeler at an advanced level. Spivak (2005) fills in the mathematical details. Hawking/Ellis provides lots of insight. *Chapter 36*: Cartan's method is explained in Needham (2021), and in more detail (with examples) in Misner/Thorne/Wheeler and in Nakahara. Some more applications can be found in Lightman *et al. Chapter 38*: chains are introduced very clearly in Ryder (1985). For the full story see Spivak (1971 and 2005). *Chapter 39*: a full account of fluid mechanics is given in Landau/Lifshitz (vol. VI). A introduction can be found in Feynman/Leighton/Sands (vol. II) and in Thorne/Blandford. *Chapter 40*: quantum field theory is described in Lancaster/Blundell. Some advanced topics are covered in Wald and in Padmanabhan. *Chapter 41*: inflation is discussed in Peacock. A more advanced discussion can be found in Padmanabhan. Fine tuning is considered in Lewis and Barnes. *Chapter 42*: the geometrical interpretation of electromagnetism is well described in Misner/Thorne/Wheeler. It's treatment as a field theory is discussed in Lancaster/Blundell. *Chapter 43*: the geometric view of the Bianchi identity is covered at an introductory level in Ryder (1985). See Misner/Thorne/Wheeler, whose approach we follow, for the full story. *Chapter 44*: gauge theory is discussed in Lancaster/Blundell and in Ryder (1985), whose approach we follow. A nice introduction is given in Penrose (2004). *Chapter 45*: the weak-field limit is discussed at an introductory level in Schutz (1985) and in more detail in Misner/Thorne/Wheeler. Feynman (1995) has a characteristically interesting take, as does Geroch (2013, *General Relativity*). We follow Ryder's (2009) discussion of the Lense–Thirring effect in the problems. *Chapter 46*: gravitational waves are introduced clearly in Schutz (1985), whose approach we follow. A complete and modern treatment can be found in Thorne/Blandford. *Chapter 47*: the properties of gravitons in a quantum field theory are discussed in similar terms in Feynman (1995). *Chapter 48*: Kaluza–Klein theory is introduced in Zee, whose discussion we follow. *Chapter 49*: string theory is introduced in Zwiebach. Loop quantum gravity is described in Rovelli/Vidotto. A short history of the latter field is given in the review by Ashketar. We follow Zee's discussion of particles in the AdS spacetime. *Chapter 50*: the argument we discuss is given in more detail in Geroch (2013, *General Relativity*). More detail on the methods can be found in Penrose (1973), Wald and also in Hawking/Ellis. *Appendix C*: good books on topological spaces include all of the lecture note volumes by Geroch (his course on Topology is the simplest), with more mathematical treatments available in Nakahara and in the book by Nash and Sen. See Penrose (2004) for a basic introduction to this material and Spivak for the full story. Geroch's book *Mathematical Physics* takes things further for the physicist. *Appendix D*: we follow Zee and Hartle's very clear discussions of embedding.

Bibliography

- V. I. Arnold, *Mathematical Methods of Classical Mechanics, 2nd edition*, Springer, New York (1989).

- A. Ashketar, *Quantum Gravity*, arXiv:gr-qc/0410054v2 (2004).

- M. Blennow and T. Ohlsson, *300 Problems in Special and General Relativity*, CUP, Cambridge (2022).

- K. M. Blundell, *Black Holes, a Very Short Introduction*, OUP, Oxford (2015).

- M. L. Boas, *Mathematical Methods in the Physical Sciences, 2nd edition*, Wiley, New York (1983).

- C. G. Böhmer, *Introduction to General Relativity and Cosmology*, World Scientific, London (2016).

- H. R. Brown, *Physical Relativity*, OUP, Oxford (2006).

- S. Carlip, Int. J. Mod. Phys. D **23**, 1430023 (2014) [arXiv:1410.1486].

- S. Carlip, *General Relativity, a Concise Introduction*, OUP, Oxford (2019).

- S. Carroll, *Spacetime and Geometry: An Introduction to General Relativity*, CUP, Cambridge (2019).

- S. Chandrasekhar, *The Mathematical Theory of Black Holes*, OUP, Oxford (1992).

- T.-P. Cheng, *Einstein's Physics*, OUP, Oxford (2013).

- Y. Choquet-Bruhat, C. DeWitt-Morette, and M. Dillard-Bleick, *Analysis, Manifolds and Physics*, North-Holland, Amsterdam (1977).

- Y. Choquet-Bruhat, *Introduction to General Relativity, Black Holes and Cosmology*, OUP, Oxford (2015).

- S. Coleman, *Sidney Coleman's Lectures on Relativity*, CUP, Cambridge, (2022).

- R. d'Inverno, *Introduction to Einstein's Relativity*, OUP, Oxford (1992).

- A. Einstein, *The Principle of Relativity*, Dover, New York (1952).

- G. F. R. Ellis and R. M. Williams, *Flat and Curved Space-Times, (2nd edition)*, OUP, Oxford (2000).

- R. P. Feynman, *Feynman Lectures on Gravitation*, Penguin, London (1995).

- R. P. Feynman, R. B. Leighton, and M. Sands, *The Feynman Lectures on Physics, Vol. II*, Pearson Addison Wesley, San Francisco (2006).

- J. Foster and D. J. Nightingale, *A Short Course in General Relativity, 3rd edition*, Springer, New York (2010).

- T. Frankel, *The Geometry of Physics, 2nd edition*, CUP, Cambridge (2004).

- A. P. French, *Special Relativity*, Chapman and Hall, London (1968).

- A. P. French and M. G. Ebbison, *Introduction to Classical Mechanics*, Chapman and Hall, London (1986).

- R. Geroch, *General Relativity from A to B*, University of Chicago Press, Chicago (1978).

- R. Geroch, *Differential Geometry, 1972 Lecture Notes*, Minkowski Institute Press, Montreal (2013).

- R. Geroch, *General Relativity, 1972 Lecture Notes*, Minkowski Institute Press, Montreal (2013).

- R. Geroch, *Geometrical Quantum Mechanics, 1974 Lecture Notes*, Minkowski Institute Press Montreal (2013).

- R. Geroch, *Topology, 1978 Lecture Notes*, Minkowski Institute Press, Montreal (2013).

- R. Geroch, *Mathematical Physics*, Chicago University Press, Chicago (1985).

- N. Gray, *A Student's Guide to General Relativity*, CUP, Cambridge (2019).

- Ø. Grøn and S. Hervik, *Einstein's General Theory of Relativity*, Springer, New York (2007).

- M. Guidry, *Modern General Relativity*, CUP, Cambridge (2019).

- M. C. Gutzwiller, *Chaos in Classical and Quantum Mechanics*, Springer-Verlag, New York (1990).

- J. B. Hartle, *Gravity: an Introduction to Einstein's General Relativity*, Pearson, Harlow (2014).

- S. W. Hawking and G. F. R. Ellis, *The Large Scale Structure of Space-time*, CUP, Cambridge (1973).

- M. P. Hobson, G. Efstathiou, and A. N. Lasenby, *General Relativity*, CUP, Cambridge (2006).

- L. P. Hughston and K. P. Tod, *An Introduction to General Relativity*, CUP, Cambridge (1990).

- R. J. A. Lambourne, *Relativity, Gravitation and Cosmology*, CUP, Cambridge (2010).

- T. Lancaster and S. J. Blundell, *Quantum Field Theory for the Gifted Amateur*, OUP, Oxford (2014).

- L. D. Landau and E. M. Lifshitz, *Mechanics* (volume I of Landau and Lifshitz), Pergamon, Oxford (1976).

- L. D. Landau and E. M. Lifshitz, *Classical Theory of Fields* (volume II of Landau and Lifshitz), Pergamon, Oxford (1975).

- L. D. Landau and E. M. Lifshitz, *Fluid Mechanics* (volume VI of Landau and Lifshitz), Pergamon, Oxford (1987).

- G. F. Lewis and L. A. Barnes, *A Fortunate Universe*, Cambridge University Press, Cambridge (2016).
- A. P. Lightman, W. H. Press, R. H. Price, and S. A. Teukolsky, *Problem Book in Relativity and Gravitation*, Princeton University Press, Princeton (1975).
- S. Lipschutz, *Schaum's Outline of Differential Geometry*, McGraw-Hill, New York (1969).
- M. Ludvigsen, *General Relativity*, CUP, Cambridge (1999).
- M. Maggiore, *Gravitational Waves*, OUP, Oxford (2007).
- C. W. Misner, K. S. Thorne, and J. A. Wheeler, *Gravitation*, W. H. Freeman and company, New York (1973).
- T. A. Moore, *A General Relativity Workbook*, University Science Books, Mill Valley, CA (2013).
- V. F. Mukhanov and S. Winitzki, *Introduction to Quantum Effects in Gravity*, CUP, Cambridge (2007).
- C. Nash and S. Sen *Topology and Geometry for Physicists*, Dover, New York (1983).
- M. Nakahara, *Geometry, Topology and Physics*, Adam Hilger, Bristol (1990).
- H. Năstase, *String theory methods for condensed matter physics*, CUP, Cambridge (2017).
- T. Needham, *Visual Complex Analysis*, OUP, Oxford (1997).
- T. Needham, *Visual Differential Geometry and Forms*, Princeton University Press, Princeton (2021).
- H. C. Ohanian and R. Ruffini, *Gravitation and Spacetime, 3rd edition*, CUP, Cambridge (2013).
- T. Padmanabhan, *Gravitation: Foundations and Frontiers*, CUP, Cambridge (2010).
- D. N. Page, New. J. Phys. **7**, 203 (2005).
- A. Pais, *Subtle Is the Lord: The Science and the Life of Albert Einstein*, OUP, Oxford (2005).
- J. A. Peacock, *Cosmological Physics*, CUP, Cambridge (1999).
- R. Penrose, *Techniques of Differential Topology in Relativity*, SIAM, Philadelphia (1973).
- R. Penrose, *The Road to Reality*, Vintage, London (2004).
- J. Plebański and A. Krasiński, *An Introduction to General Relativity and Cosmology*, CUP, Cambridge (2006).
- E. Poisson, *A Relativist's Toolkit*, CUP, Cambridge (2004).
- E. Poisson and C. M. Will, *Gravity*, CUP, Cambridge (2014).
- W. Rindler, *Relativity*, OUP, Oxford (2006).
- C. Rovelli, *General Relativity: The Essentials*, CUP, Cambridge (2021).
- C. Rovelli and F. Vidotto, *Covariant Loop Quantum Gravity*, CUP, Cambridge (2015).
- L. H. Ryder, *Quantum Field Theory*, CUP, Cambridge (1985).
- L. H. Ryder, *Introduction to General Relativity*, CUP, Cambridge (2009).
- D. W. Sciama, *The Physical Foundations of General Relativity*, Doubleday & Co., New York (1969).
- B. F. Schutz, *A First Course in General Relativity*, CUP, Cambridge (1985).
- B. F. Schutz, *Geometrical Methods of Mathematical Physics*, CUP, Cambridge (1980).

- M. Spivak, *Calculus on Manifolds*, Westview Press, Boulder (1971).
- M. Spivak, *A Comprehensive Introduction to Differential Geometry: Vol 1, 3rd Edition*, Publish or Perish, Houston (2005).
- M. Spivak, *A Comprehensive Introduction to Differential Geometry: Vol 2, 3rd Edition*, Publish or Perish, Houston (1999).
- J. L. Synge, *Relativity: The General Theory*, North-Holland, New York (1960).
- E. F. Taylor, J. A. Wheeler, and E. Bertshinger, *Exploring Black Holes, 2nd Edition*, available for free download from eftaylor.com/exploringblackholes (2017).
- K. S. Thorne and R. D. Blandford, *Modern Classical Physics*, Princeton University Press, Princeton (2017).
- R. M. Wald, *General Relativity*, University of Chicago Press, Chicago (1984).
- S. Weinberg, *Gravitation and Cosmology*, Wiley, New York (1972).
- S. Weinberg, *Cosmology*, CUP, Cambridge (2008).
- A. Zee, *Einstein Gravity in a Nutshell*, Princeton University Press, Princeton (2013).
- B. Zwiebach, *A First Course in String Theory, 2nd edition.*, CUP, Cambridge (2009).

B

Conventions and notation

This appendix contains a summary of some of the choices of conventions and notation we have made in the book.

B.1 Electromagnetic units

In SI units, Maxwell's equations in free space can be written as

$$\vec{\nabla} \cdot \vec{E} = \tfrac{\rho}{\epsilon_0}, \qquad \vec{\nabla} \times \vec{E} = -\tfrac{\partial \vec{B}}{\partial t},$$
$$\vec{\nabla} \cdot \vec{B} = 0, \qquad \vec{\nabla} \times \vec{B} = \mu_0 \vec{J} + \tfrac{1}{c^2}\tfrac{\partial \vec{E}}{\partial t}. \tag{B.1}$$

Although SI units are preferable for many applications in physics, the desire to make our (admittedly often complicated) equations as simple as possible motivates a different choice of units for the discussion of electromagnetism in field theory.[1] We therefore choose the **Heaviside–Lorentz**[2] system of units (also known as the 'rationalized Gaussian CGS' system) which can be obtained from SI by setting $\epsilon_0 = \mu_0 = 1$. Thus, the electrostatic potential $V(\vec{x}) = q/4\pi\epsilon_0|\vec{x}|$ of SI becomes $V(\vec{x}) = q/4\pi|\vec{x}|$ in Heaviside–Lorentz units, and Maxwell's equations can be written as

$$\vec{\nabla} \cdot \vec{E} = \rho, \qquad \vec{\nabla} \times \vec{E} = -\tfrac{1}{c}\tfrac{\partial \vec{B}}{\partial t},$$
$$\vec{\nabla} \cdot \vec{B} = 0, \qquad \vec{\nabla} \times \vec{B} = \tfrac{1}{c}(\vec{J} + \tfrac{\partial \vec{E}}{\partial t}). \tag{B.2}$$

Using our other choice of $c = 1$ obviously removes the factors of c too.

B.2 Vectors, 1-forms and tensors

In a particular basis, a vector is described by a set of components. If the basis is rotated, then the components will change, but the length of the vector will be unchanged.[3] Three-vectors (or 3-vectors) have three spatial components [such as (A^x, A^y, A^z) in a Cartesian coordinate system] and denoted by a letter with an arrow on top, such as $\vec{A}$ or $\vec{p}$. The components of 3-vectors are listed with a Roman index taken from the middle of the alphabet: e.g. A^i, with $i = 1, 2, 3$ so that we can write components $A^i = (A^1, A^2, A^3)$. We sometimes use the names of coordinates for the components: e.g. $A^i = (A^x, A^y, A^z)$. Component labels for vectors are always written in the upstairs position[4] (e.g. A^i) and never downstairs (A_i).

---

[1] Almost all books on classical and quantum field theories use Heaviside–Lorentz units, though the famous textbooks on electrodynamics by Landau and Lifshitz and by Jackson do not.

[2] These units are named after the English electrical engineer O. Heaviside (1850–1925) and the Dutch physicist H. A. Lorentz (1853–1928).

[3] We use passive transformations in this subject. That is to say, our transformations change the coordinates describing the position of an event, rather than changing the position of the event itself: the latter being an active transformation. Sidney Coleman notes that, in criminal circles, a passive transformation is analogous to an alias (the criminal is an event, after the transformation they remain at the position of the crime in spacetime, but they look different owing to the transformation), while the active transformation is like an alibi (the criminal/event is transformed to a different position in spacetime to the position of the crime).

[4] Upstairs components are sometimes called covariant components. We mostly avoid this terminology.

In most applications, we deal with $(3+1)$-dimensional spacetime. A four-vector (or 4-vector) that lives in this spacetime is a vector-valued object with a single timelike component and three spacelike components, which themselves form a three-vector. Four-vectors are displayed in bold script (e.g. $\boldsymbol{v}$). All bold-script quantities are coordinate free, existing independently of a specific basis. When referred to a basis, four-vector components are given a Greek index: for example, v^μ where $\mu = 0, 1, 2, 3$. We write $v^\mu = (v^0, v^1, v^2, v^3)$ or (v^0, v^i) or $(v^0, \vec{v})$. The zeroth component, v^0, is the timelike part. Basis vectors are written as $\boldsymbol{e}_\mu$ (and sometimes $\partial/\partial x^\mu$), so we can write a vector in terms of its components as $\boldsymbol{v} = v^\mu \boldsymbol{e}_\mu$, with a bold part on both sides of the equality.[5]

The vector's natural partner is the 1-form. These are written in frame-independent form using bold type with a tilde, e.g. $\tilde{\boldsymbol{A}}$. Like vectors they can be split into components and basis 1-forms, the latter written as $\boldsymbol{\omega}^\mu$ (and sometimes $\boldsymbol{d}x^\mu$). In terms of components and basis 1-forms, we write $\tilde{\boldsymbol{A}} = A_\mu \boldsymbol{\omega}^\mu$. Components of 1-forms always have the index written in the down position.[6] An example of a familiar 1-form is the gradient of a function $f(x^\mu)$, whose components are the derivatives $\frac{\partial f}{\partial x^\mu}$, which is sometimes written as $\partial_\mu f$ and sometimes written using the comma notation such that $\frac{\partial f}{\partial x^\mu} = f_{,\mu}$.

We use the Einstein convention that all indices repeated in both an up and down position are summed over. Inner products between 1-forms and tensors are written with angle brackets: $\langle \tilde{\boldsymbol{A}}, \boldsymbol{v} \rangle = A_\mu v^\mu$. Dot products (or, equivalently, scalar products) between two vectors are written as $\boldsymbol{v} \cdot \boldsymbol{u} = g_{\mu\nu} v^\mu u^\nu$, where $g_{\mu\nu}$ are the components of the metric.

Tensors are treated as slot machines and given bold symbols like $\boldsymbol{T}(\ ,\ )$. Their valence is specified separately in the form (n, m), meaning n slots for 1-forms and m slots for vectors.[7] When the slots are filled, the tensor outputs a number. Components can be extracted using the basis vectors $\boldsymbol{e}_\mu$ and basis 1-forms $\boldsymbol{\omega}^\nu$ via equations such as the following for a $(2,2)$ tensor: $\boldsymbol{S}(\boldsymbol{\omega}^\mu, \boldsymbol{\omega}^\nu, \boldsymbol{e}_\alpha, \boldsymbol{e}_\beta) = S^{\mu\nu}{}_{\alpha\beta}$. Tensors can be combined using outer products denoted $\otimes$, or wedge products denoted $\wedge$, with the relationship $\boldsymbol{v} \wedge \boldsymbol{u} = \boldsymbol{v} \otimes \boldsymbol{u} - \boldsymbol{u} \otimes \boldsymbol{v}$. Tensors can also be written in terms of their components using this notation

$$\boldsymbol{S} = S^{\mu\nu}{}_{\alpha\beta} \left(\boldsymbol{e}_\mu \otimes \boldsymbol{e}_\mu \otimes \boldsymbol{\omega}^\alpha \otimes \boldsymbol{\omega}^\beta \right). \tag{B.3}$$

Symmetrization of components is denoted with a round bracket, such as

$$T^{(\alpha\beta)} = \frac{1}{2}(T^{\alpha\beta} + T^{\beta\alpha}). \tag{B.4}$$

Antisymmetrization of components is denoted with a square bracket, such as

$$T^{[\alpha\beta]} = \frac{1}{2}(T^{\alpha\beta} - T^{\beta\alpha}). \tag{B.5}$$

The trace of a tensor is denoted by an italic letter,[8] e.g. $T = T^\mu{}_\mu$. Tensor components are sometimes denoted with the coordinates (e.g. $\mu = t, r, \theta, \phi$) and sometimes, equivalently, numbers (e.g. $\mu = 1...4$). Using the latter, ordered indices $|\mu\nu|$ are arranged such that $\mu < \nu$.

---

[5] Also in $(3+1)$ dimensions we generally use $\mathcal{V}$ to denote a 4-volume and V for a 3-volume. The invariant 4-volume is usually $\mathrm{d}\mathcal{V}$ and the invariant 3-volume is $\mathrm{d}\Sigma$. Some other texts use $\mathrm{d}\Omega$ for the invariant 4-volume, but we reserve $\mathrm{d}\Omega^2 = \mathrm{d}\theta^2 + \sin^2\theta \mathrm{d}\phi^2$ for the angular part of the spherical line element.

[6] These are sometimes called contravariant components.

[7] On the few occasions we want to make an argument about a general matrix, rather than about a tensor, we denote the matrix $\boldsymbol{X}$.

[8] The most important tensor in this subject is the metric. This is a $(0,2)$ tensor $\boldsymbol{g}(\ ,\ )$ with components $g_{\mu\nu} = g(\boldsymbol{e}_\mu, \boldsymbol{e}_\nu) = \boldsymbol{e}_\mu \cdot \boldsymbol{e}_\nu$. In an exception to our rules, the determinant of the metric (not the trace) is denoted g. We use the signature $(-+++)$ and specify the components of diagonal matrices by saying, for example, that the components of the Minkowski tensor are $\eta_{\mu\nu} = \mathrm{diag}(-1, 1, 1, 1)$. Indices are raised and lowered with the components of the metric.

Tensor fields are functions of position. We write a vector field $\boldsymbol{v}(x)$, meaning that at a point x we output a vector $\boldsymbol{v}$. The point here could be an abstract point in a manifold $\mathcal{P}$ or the coordinates of this point $x^\mu(\mathcal{P})$. This is intended to prevent any confusion with the slots carried by the tensor (e.g. the single slot of a vector field that takes a 1-form).

Position vectors, interpreted as pointing between points in spacetime, are not very useful in curved spacetime. Instead, we usually specify a general point in spacetime $\mathcal{P}$ or its coordinate $x^\mu(\mathcal{P})$, which are not treated as the components of a vector.[9] The most important vector field in relativity is the velocity, which provides the tangents to a world line $x^\mu(\tau)$, which is a curve parametrized by an affine parameter such as the proper time τ. The velocity field is given by $\boldsymbol{u}(x) = \left(\frac{\mathrm{d}x^\mu(\tau)}{\mathrm{d}\tau}\right)\boldsymbol{e}_\mu$, with the property $\boldsymbol{u} \cdot \boldsymbol{u} = -1$.

In the orthonormal frame, we write components with a hat. So a vector is written as $\boldsymbol{A} = A^{\hat\alpha}\boldsymbol{e}_{\hat\alpha}$. Indices in an orthonormal frame are raised and lowered with the Minkowski metric with components $\eta_{\mu\nu} = \mathrm{diag}(-1, 1, 1, 1)$. To translate between the orthonormal frame and a coordinate frame, we use the components of a vielbein, written using brackets in expressions such as

$$
\begin{aligned}
A^{\hat\alpha} = (\boldsymbol{e}_\mu)^{\hat\alpha} A^\mu, \quad & A^\mu = (\boldsymbol{e}_{\hat\alpha})^\mu A^{\hat\alpha}, \\
Z_\mu = (\boldsymbol{e}_\mu)^{\hat\alpha} Z_{\hat\alpha}, \quad & Z_{\hat\alpha} = (\boldsymbol{e}_{\hat\alpha})^\mu Z_\mu.
\end{aligned}
\tag{B.8}
$$

For a diagonal metric, with non-zero components $g_{\mu\mu}$, we have the useful square-root rule $(\boldsymbol{e}_\mu)^{\hat\mu} = \sqrt{|g_{\mu\mu}|}$, where no summation is implied.

B.3 Covariant derivatives

The most useful derivative in relativity is the covariant derivative, which is written in frame-independent form as $\boldsymbol{\nabla}_{\boldsymbol{u}}$, which is equivalent to $\boldsymbol{\nabla}_{\boldsymbol{u}} = \boldsymbol{u} \cdot \boldsymbol{\nabla}_{\boldsymbol{e}_\mu}$, where $\boldsymbol{u}$ is a vector. This is a directional derivative, taken along the direction of the vector $\boldsymbol{u}$. When the direction is given in terms of a basis vector we write $\boldsymbol{\nabla}_{\boldsymbol{e}_\mu} = \boldsymbol{\nabla}_\mu$. Confusingly, this is not a component expression: the μ subscript is short for $\boldsymbol{e}_\mu$, where μ labels the direction along which the derivative is taken. In terms of components and a basis, the covariant derivative can be written as

$$
\boldsymbol{\nabla}_\mu \boldsymbol{v} = (\boldsymbol{\nabla}_\mu \boldsymbol{v})^\alpha \boldsymbol{e}_\alpha = v^\alpha{}_{;\mu} \boldsymbol{e}_\alpha,
\tag{B.9}
$$

where the final expression uses semicolon notation. The latter is a component notation defined as

$$
v^\alpha{}_{;\mu} = \frac{\partial v^\alpha}{\partial x^\mu} + \Gamma^\alpha{}_{\mu\nu} v^\nu.
\tag{B.10}
$$

We also use a further notation for the covariant derivative made along a curve $x^\mu(\tau)$ with tangent $\boldsymbol{u}(x)$, which we write as $\mathrm{D}\boldsymbol{v}/\mathrm{d}\tau = \boldsymbol{\nabla}_{\boldsymbol{u}} \boldsymbol{v}$ and

$$
\left(\frac{\mathrm{D}\boldsymbol{v}}{\mathrm{d}\tau}\right)^\alpha = u^\mu\left(\frac{\partial v^\alpha}{\partial x^\mu} + \Gamma^\alpha{}_{\mu\nu} v^\nu\right) = u^\mu v^\alpha{}_{;\mu},
\tag{B.11}
$$

where the velocity $\boldsymbol{u}$ is tangent to the curve parametrized by τ.[10]

---

[9]If, as in Chapter 30, we do write points on the world line in terms of a set of displacement vectors $\boldsymbol{X} = X^\mu \boldsymbol{e}_\mu$ then we could write the tangent as

$$
\begin{aligned}
\boldsymbol{u} &= \frac{\mathrm{d}\boldsymbol{X}(\tau)}{\mathrm{d}\tau} \\
&= \frac{\mathrm{d}x^\mu}{\mathrm{d}\tau} \frac{\partial \boldsymbol{X}(\tau)}{\partial x^\mu},
\end{aligned}
\tag{B.6}
$$

so that the components are $u^\mu = \frac{\partial x^\mu}{\partial\tau}$ and the basis vectors on the curve are $\boldsymbol{e}_\mu = \frac{\partial \boldsymbol{X}(\tau)}{\partial x^\mu}$. However, the displacement vector does not transform according to the tensor transformation law, so is not useful for general relativity. The modern way of looking at vectors (Chapter 31) is not to invoke the displacement vectors and instead specify the tangent field as

$$
\boldsymbol{u}(x) = \frac{\mathrm{d}x^\mu}{\mathrm{d}\tau} \frac{\partial}{\partial x^\mu},
\tag{B.7}
$$

so that the basis vectors are written as $\boldsymbol{e}_\mu = \frac{\partial}{\partial x^\mu}$.

[10]The logic behind these definitions is that $\boldsymbol{\nabla}_{\boldsymbol{u}} \boldsymbol{v}$ is a vector written fully coordinate-free notation; a coordinate of this vector is written in mixed notation as $(\boldsymbol{\nabla}_{\boldsymbol{u}} \boldsymbol{v})^\mu$. On the other hand, $v^\alpha{}_{;\mu}$ is coordinate notation, equivalent to the mixed notation $(\boldsymbol{\nabla}_\mu \boldsymbol{v})^\alpha$. The notation $\mathrm{D}/\mathrm{d}\tau$ is used in different ways in the literature, where sometimes we see $\mathrm{D}\boldsymbol{v}/\mathrm{d}\tau = \boldsymbol{\nabla}_{\boldsymbol{u}} \boldsymbol{v}$ (which is our choice) and sometimes $\mathrm{D}v^\mu/\mathrm{d}\tau = u^\alpha v^\mu{}_{;\alpha}$. This latter expression is ambiguous, so we write it as $(\mathrm{D}\boldsymbol{v}/\mathrm{d}\tau)^\mu$.

Manifolds and bundles

Life is a short affair; We should try to make it smooth, and free from strife.
Euripides (c. 480–c. 406)

Given a shape or a space to understand, such as a two-dimensional spherical surface, we are usually tempted to embed it in a higher dimensional Euclidean space (e.g. three-dimensional space in this case) in order to examine its structure. However, in studying the spacetimes of general relativity, it is certainly not given that the spacetime of our Universe actually lives in some higher dimensional Euclidean space. We must therefore come to terms with a more intrinsic, geometrical description of the fabric of spacetime in terms of a **manifold**, without relying on the artifice of embedding it in a higher dimensional space. A manifold has only a little mathematical structure of its own. We only insist that a manifold be a **smooth** space without awkward discontinuities or unusual joints. Manifolds fit naturally with physical models described in terms of classical field theory which rely on this notion of smoothness. Thus, when we examine singularities, we can characterize them as places where the smooth manifold description breaks down.[1]

A good working definition of a manifold is a space that looks *locally* flat and Euclidean. This constrains the space to change smoothly, since a space full of discontinuities cannot look Euclidean at each of its points. Coordinates, functions, and curves can be defined on manifolds. We can perform calculus on manifolds, use them to define vectors and, if we choose, define metrics on them in order to work out lengths and angles. Nature, as explained by general relativity, seems to be based on the metric and so these two separate ingredients, manifold and metric, form the basis of the geometrical description of Nature.

The manifold point of view is a natural one for describing the geometry of physics. We call ordinary three-dimensional space, the manifold $\mathbb{R}^3$: it is simply the space in which 3-vectors live. *Space* can then be thought of as the manifold $\mathbb{R}^3$ with a flat metric $d(\ ,)$ defined on it that can be used to measure lengths. Special relativity asserts that *spacetime* is the manifold $\mathbb{R}^4$ (i.e. the space of 4-vectors) with a flat metric $\eta(\ ,)$ defined on it. In general relativity, spacetime is a manifold, usually called $\mathcal{M}$, on which a Lorentz metric $g(\ ,)$ is defined. The curvature of the spacetime is related to the matter distribution in spacetime via Einstein's equation.

In this book, we have been doing our physics on a (pseudo) Riemann manifold, which possesses a connection and a metric. In fact, the Rie-

[1]The material in this appendix lies behind the mathematics presented in this book, particularly the material on differential geometry. Although we have tried to minimize its use in the main body of the book, this mathematics provides the reason why modern general relativity looks the way it does, and is therefore used in many modern books on general relativity. For example, the study of singularities used to understand the structure of black holes and cosmology is especially reliant on the use of many of the ideas introduced here.

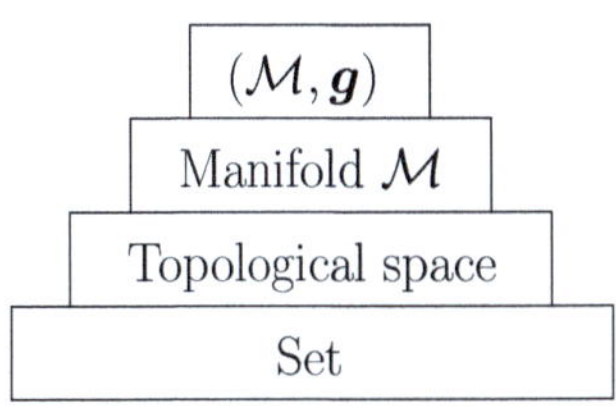

Fig. C.1 The Riemann manifold $(\mathcal{M}, g)$, with its metric structure, exists at the top of a pyramid of concepts in mathematics.

mann manifold is built upon a series of concepts, as shown in Fig. C.1. In this appendix, we take a step back, forgetting many of the mathematical notions we take for granted, such as the distances and times encoded in the metric. We shall deal with manifolds from the primitive point of view that everything needs to be built from scratch. Taking as little baggage as possible with us, we shall attempt to build a set of concepts suitable to describe the geometrical fabric of the Universe. This topic starts with simple notions of sets and intervals and then introduces the study of manifolds and their structure. We finish by describing some simple concepts of fibres and bundles that lie behind the tangent spaces of differential geometry and the physics of gauges.[2]

C.1   Preliminaries

A space with a metric defined on it is called a **metric space**. The metric allows us to work out how far points are from other points. We call a space without a metric a **topological space**. This has less structure, but we tend to describe points in the neighbourhood of other points in terms of parts of the space known as **open subsets**, which encode its topology. We introduce some of the relevant ideas here and in Fig. C.2. We start with some primitive notions of **sets** (or collections of objects or points) and intervals (i.e. a **subset** of points between two end points, or the **neighbourhood** of points between two end points). A manifold then turns out to be a set with some special properties. Let's start with some definitions:

- The **open interval** $a < x < b$, not including the endpoints a and b is written (a, b) [Fig. C.2(a)]. The **closed interval** $a \leq x \leq b$, which includes the endpoints a and b is written $[a, b]$ [Fig. C.2(b)].
- The expression $p \in A$ denotes that p is an element of the set A.
- The expression $A \cup B$ denotes the union of sets A and B: the set of objects belonging to A, B or both [Fig. C.2(c)].
- The expression $A \cap B$ denotes the intersection of sets A and B: the set of objects belonging to both A and B [Fig. C.2(d)].
- The expression $A \subset B$ denotes that A is a subset of B [Fig. C.2(e)].
- The expression $\varnothing$ denotes the empty set, containing no elements.
- $\mathbb{R}$ is the set of real numbers.
- We define $\mathbb{R}^n$ to be the n-dimensional Euclidean space we usually use for vector algebra. A point in $\mathbb{R}^n$ is a sequence of real numbers $(x^1, x^2, x^3, ...x^n)$, sometimes called an n-tuple. The space $\mathbb{R}^n$ is a metric space: with the distance between points is given by

$$|x - y| = \left[\sum_{\mu=1}^{n} (x^\mu - y^\mu)^2 \right]^{\frac{1}{2}}. \tag{C.1}$$

An **open ball** in $\mathbb{R}^n$ of radius r centred around a point y consists of the points x such that $|x - y| < r$. This is an example of an open subset of the set $\mathbb{R}^n$ [Fig. C.2(f)]. It is some region, usually assumed close to y, that doesn't include its boundary.

[2]For those not inclined to venture any further at this stage, here's an executive summary of the content of this appendix: general relativity takes place in smooth spacetime. The most basic mathematical structure that has this smoothness is a manifold. On a manifold, instead of a coordinate transformation we have the diffeomorphism and instead of the idea of a boundary we have the notion of compactness. Tangent vectors live in a manifold called a tangent space. The combination of a manifold and a tangent space is known as a fibre bundle.

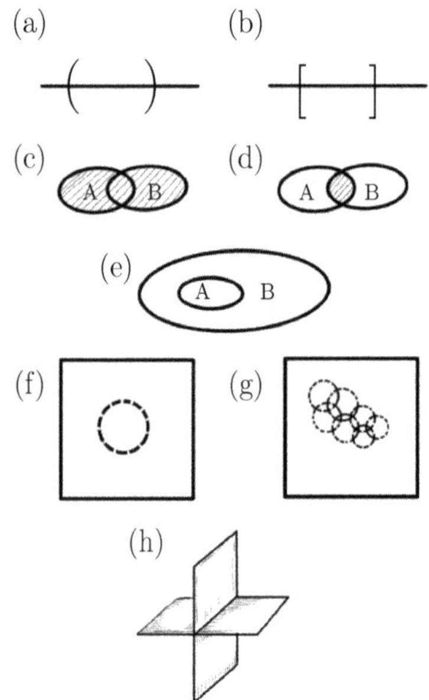

Fig. C.2 (a) Open interval; (b) closed interval; (c) the union of A and B; (d) the intersection of A and B; (e) A is a subset of B; (f) an open ball in $\mathbb{R}^2$; (g) an open cover of a set A, which is a subset of $\mathbb{R}^2$; (h) a non-Hausdorff space.

• A set of points S of $\mathbb{R}^n$ is **open** if every point in S has an open neighbourhood entirely within S. Such a set can be expressed as a union of open balls. Any reasonable chunk of $\mathbb{R}^n$ is open if we don't include its boundary in the set. A collection O of open sets is an **open cover** of a set A if every point in A is in the collection O [Fig. C.2(g)].

• The **Hausdorff property** of a set in $\mathbb{R}^n$ is the feature that any two distinct points have neighbourhoods that don't intersect (i.e. any line can be infinitely subdivided). A non-Hausdorff space is typified by branching [Fig. C.2(h)]. We shall only ever deal with Hausdorff spaces.

With the simple notions defined, we move on to discussing how to relate one element of a space to another element in another space.

C.2 Maps and functions

The basic tool for examining the properties of the various mathematical structures of use in physics is **mapping**. A **map** f from space $\mathcal{M}$ to space $\mathcal{N}$ is a rule that associates with an element x of $\mathcal{M}$ a unique element y of $\mathcal{N}$. The idea is shown in Fig. C.3. The simplest map is a[3] real **function**. For such a function, both $\mathcal{M}$ and $\mathcal{N}$ are elements of the set $\mathbb{R}$ (i.e. the set of real numbers). The function f takes an element x and spits out an element y. The notation saying that f maps elements in $\mathcal{M}$ to elements in $\mathcal{N}$ is written as

$$f : \mathcal{M} \to \mathcal{N}, \tag{C.2}$$

or in terms of the elements themselves

$$f : x \mapsto y = f(x). \tag{C.3}$$

When a map is a real-valued function of n variables we write $f : \mathbb{R}^n \to \mathbb{R}$. This simply amounts to saying that we input a n-tuple $(x^1, ..., x^n)$ to the function and output a single number.[4]

We can combine different mappings. If we have two maps f and g, $f : \mathcal{M} \to \mathcal{N}$ and $g : \mathcal{N} \to \mathcal{L}$, then there is a map called the **composition** of f and g denoted $g \circ f$, which maps $\mathcal{M}$ to $\mathcal{L}$. In ordinary algebra, $g \circ f$ would be written as $g(f(x))$.

C.3 One-to-one, into, and onto

The points, mapped from the subset of points S in $\mathcal{M}$ to points in $\mathcal{N}$, form a new set T called the **image** of S under f, or $f(S)$. The set S is called the inverse image, i.e. $S = f^{-1}(T)$. Using the notion of images, we can identify several sorts of mapping.

• If the map is **many-to-one**, then the inverse image of some point of $\mathcal{N}$ is not a single point in $\mathcal{M}$ [Fig. C.4(a)].

• If every point in $f(S)$ has a unique inverse image point in S, then f is said to be **one-to-one** or 1-1 [Fig. C.4(b) and (c)].

• If a map $\mathcal{M} \to \mathcal{N}$ is defined for all points in $\mathcal{M}$ (i.e. $S = \mathcal{M}$), then the mapping is from $\mathcal{M}$ **into** $\mathcal{N}$ [Fig. C.4(b and c)].

[3]In many texts, such as this one, the terms map and function are used interchangeable.

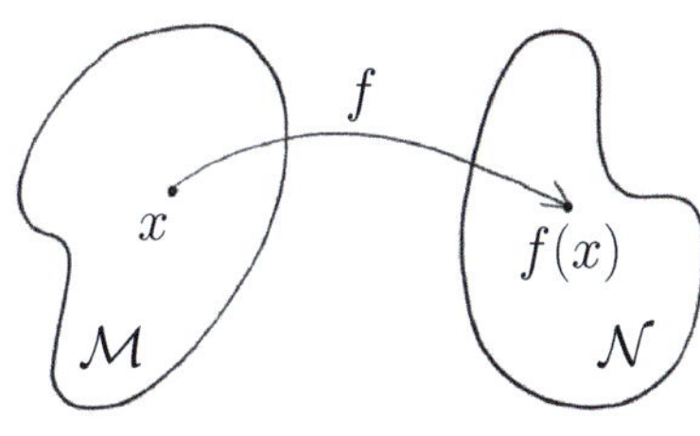

Fig. C.3 A function as a mapping: input an element x, output an element $y = f(x)$.

[4]Of course, we would usually write this as $f(x^1, ..., x^n) = y$.

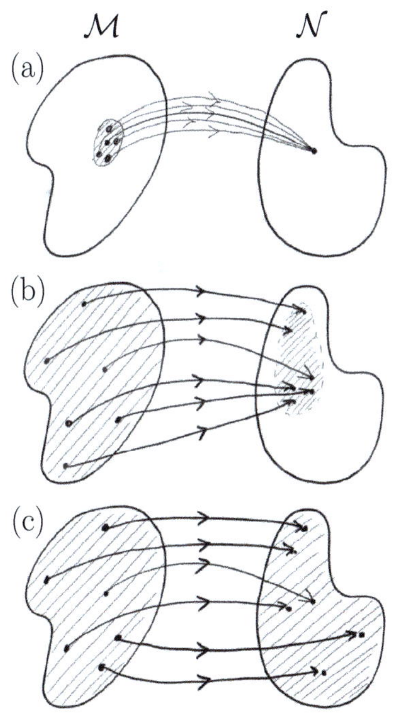

Fig. C.4 (a) *many-to-one* and (b) *into* mappings. (c) A bijection, which is both 1-1 and *onto*.

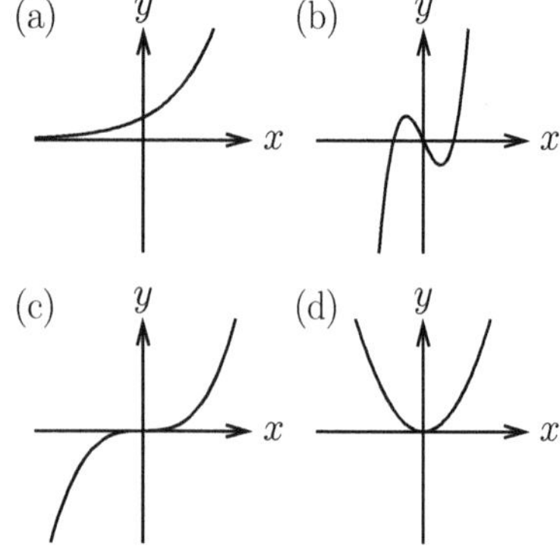

Fig. C.5 Functions discussed in Example C.5.

[6]This continuous deformation is only one example of a homeomorphism and the definition in terms of maps makes it rather more general.

- If every point in $\mathcal{N}$ has an inverse image (not necessarily a unique one), we say it is a mapping from $\mathcal{M}$ **onto** $\mathcal{N}$.
- A map that is both 1-1 and *onto* is called a **bijection**[5] [Fig. C.4(c)]. Only bijective maps have a unique inverse that is a map, which we denote $f^{-1}: \mathcal{N} \to S$. This map is then also bijective.

Example C.1

Figure C.5 shows examples of functions, $y = f(x)$, which can be thought of as maps from $\mathbb{R}$ to $\mathbb{R}$, i.e. $f : \mathbb{R} \to \mathbb{R}$.
(i) Figure C.5(a) is 1-1, but not onto.
(ii) Figure C.5(b) is onto but not 1-1.
(iii) Figure C.5(c) is a bijection (i.e. both 1-1 and onto).
(iv) Figure C.5(d) is neither 1-1 nor onto.

C.4 Continuous maps

The mathematician is often looking for ways to say that two spaces or systems look the same, or at least similar, since this constitutes a useful method of classifying the structure of a space. The way this is done in mathematical physics is via the use of **morphisms**. Roughly, a morphism is a type of map that preserves structure, allowing us to move between two spaces in order to compare them. Two sorts of morphisms are relevant in geometry: the homeomorphism and the diffeomorphism. The latter is the important morphism for general relativity. We shall first meet the homeomorphism which can be used, in this context, to say that two spaces share the same sort of *continuous* structure. The diffeomorphism is similar, but also includes the idea that the spaces are *differentiable*.

A map $\phi : \mathcal{M} \to \mathcal{N}$ is **continuous** at point x in $\mathcal{M}$ if any open set of $\mathcal{N}$ containing $\phi(x)$ contains the image of an open set of $\mathcal{M}$ containing x.

A **homeomorphism** is a 1-1, onto map from one space to another which is continuous and whose inverse is continuous.

Two spaces with a homeomorphism between them are said to be **homeomorphic**. Roughly speaking, a homeomorphism preserves the topological properties of a space, so that its 'overall shape' or 'overall structure' is preserved. In this sense, a homeomorphism allows us to say that a space 'looks like' another space. If this seems abstract, then one example of a homeomorphism to keep in mind is a continuous deformation. If you imagine that objects are made from a mouldable clay then if you can deform one object into another without breaking it, punching holes in the clay, gluing disparate parts or healing up holes already there, the objects are homeomorphic.[6]

Example C.2

Some examples of things that are homeomorphisms and things that aren't are the following. [Some terminology (in italics) is explained later in the chapter.]

I The unit disc is the interior of a unit circle. It is homeomorphic to the interior of a unit square, as the disc can be continuously deformed into the square.

II The graph of a differentiable function is homeomorphic to the domain of the function.

III A differentiable *parametrization of a curve* is a homeomorphism between the domain of parametrization and the curve.

IV A coffee mug and doughnut can be continuously deformed into one another (Fig. C.6). This continuous deformation is one example of a homeomorphism, so the coffee mug and doughnut are homeomorphic.

V The set $\mathbb{R}^m$ is not homeomorphic to $\mathbb{R}^n$ if $m \neq n$.

VI The Euclidean real line is not homeomorphic to the circle. (This is because the unit circle is *compact*, but the real line is not.)

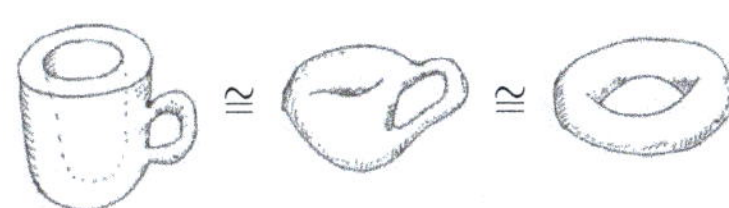

Fig. C.6 A coffee mug can be continuously deformed into a doughnut. Note that this only works because the coffee mug has a handle. A handleless coffee mug is not homeomorphic to the doughnut because, at some stage of the deformation, you would need to rip a hole in the 'deformable clay'. Hole-making is not a *continuous* deformation.

C.5   Manifolds, coordinates, and charts

As we have said, the symbol $\mathbb{R}^n$ represents the set of all n-tuples of real numbers $(x^1, x^2, x^3, ..., x^n)$. This is another way of saying it is the ordinary space in which vectors live. It is also known as flat, Euclidean space. We started with the idea that a manifold is a set of points that, locally, looks like $\mathbb{R}^n$. A more precise definition is as follows:

The set $\mathcal{M}$ is a manifold if each point of $\mathcal{M}$ has an open neighbourhood that is homeomorphic to an open set of $\mathbb{R}^n$ for some n.

If an object has some point that at no level of magnification can be made to look like the flat space of $\mathbb{R}^n$, then it is not a manifold. Note that a manifold, on its own, does not preserve lengths, angles, or other geometric quantities.

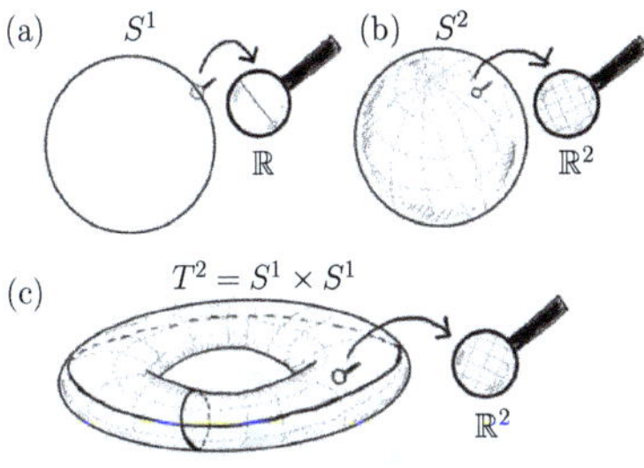

Fig. C.7 (a) A circle S^1 looks locally like $\mathbb{R}$. Note that by 'circle' we mean the one-dimensional space that is the boundary of a disc. (b) A sphere S^2 looks locally like $\mathbb{R}^2$. Note that by 'sphere' we mean the two-dimensional space that is the boundary of a ball (i.e. what is often called a 'spherical surface'). (c) A torus T^2 is the product space $S^1 \times S^1$ obtained from two circles (shown here as the two circles in bold). It looks locally like $\mathbb{R}^2$. A torus is the space describing the surface of an (edible) doughnut.

Example C.3

Some examples of manifolds are the following:
- The m-dimensional space $\mathbb{R}^m$ itself is a manifold. It looks locally like $\mathbb{R}^m$, after all!
- The circle S^1 is a manifold. It looks locally like $\mathbb{R}$ [see Fig. C.7(a)].
- The sphere S^2 is a manifold, looking locally like $\mathbb{R}^2$ [see Fig. C.7(b)].
- The torus T^2 is a manifold, looking locally like $\mathbb{R}^2$ [see Fig. C.7(c)].
- A plane with a line jutting out of it (Fig. 31.1 in Chapter 31) is not a manifold. The point of intersection never looks smooth at any level of magnification.
- The double cone (Fig. 31.1) is not a manifold. The position where the apex of one cone touches the other never looks smooth.

A point $\mathcal{P}$ in an m-dimensional manifold $\mathcal{M}$ exists independently of any coordinates. However, we want to be able to identify points like $\mathcal{P}$ on the manifold using our familiar coordinates $(x^1, ..., x^m)$ which, to remind you, live in $\mathbb{R}^m$. To do this, we need to map between the manifold $\mathcal{M}$ and $\mathbb{R}^m$. However, that map won't necessarily be a homeomorphism

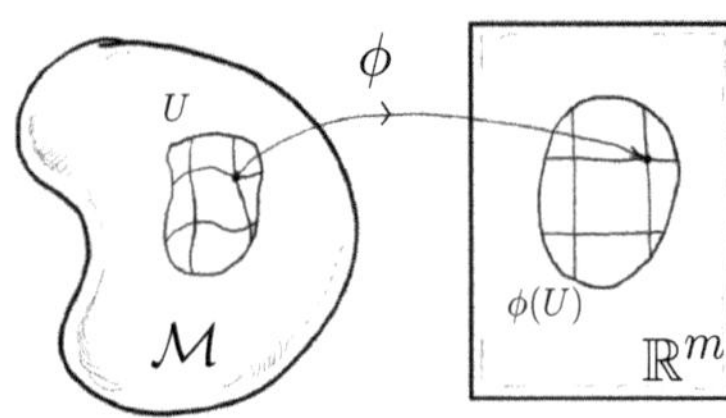

Fig. C.8 An open set U of the manifold $\mathcal{M}$ is mapped to the set $\phi(U)$ in $\mathbb{R}^m$.

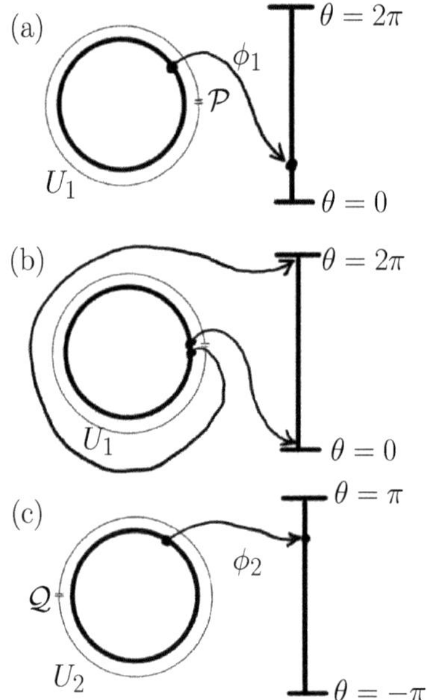

Fig. C.9 Coordinate neighbourhoods, chosen to cover the unit circle manifold. (a) The homeomorphism ϕ_1 maps a point from the subset U_1 on the manifold to a value θ. (b) We must exclude the point shown from U_1, since this could be mapped to both 0 and θ. (c) A different subset U_2 excludes a different point to that excluded from U_1.

since $\mathcal{M}$ and $\mathbb{R}^m$ might have different topologies. However, since $\mathcal{M}$ looks like $\mathbb{R}^m$ locally, a homeomorphism ϕ (which we call a **coordinate function**) can be constructed which maps between U and $\mathbb{R}^m$, where U is called a **coordinate neighbourhood**; this is an open set of the manifold, $U \subset \mathcal{M}$, which contains $\mathcal{P}$. In mathematical language, $\phi : U \to \mathbb{R}^m$, and this setup is shown pictorially in Fig. C.8. The coordinate function ϕ is represented by m real functions of the point $\mathcal{P}$, written as $\{x^1(\mathcal{P}), ..., x^m(\mathcal{P})\}$. This set is also often called a coordinate, for the sake of brevity. There is lots of scope for confusion here because we generally use x to represent the functions of $\mathcal{P}$, and the coordinates themselves. A simple shorthand equation to keep in mind is that the coordinates are given by

$$x^\mu = \phi(\mathcal{P}). \tag{C.4}$$

In words: Input a point $\mathcal{P}$ from $U \in \mathcal{M}$ and output a point x^μ in $\mathbb{R}^m$. Since ϕ is a homeomorphism, and therefore has a unique inverse, we can write things the other way round

$$\mathcal{P} = \phi^{-1}(x^\mu), \tag{C.5}$$

which, in words, says that we input a coordinate x^μ to ϕ^{-1} which outputs a point $\mathcal{P}$ on the open set U on the manifold.

We have to focus on the coordinate neighbourhood $U \subset \mathcal{M}$, rather than on the whole manifold, because the coordinate function ϕ often cannot be 1-1 over the entire manifold (because $\mathcal{M}$ only looks like $\mathbb{R}^m$ *locally*). We only need to be able map the region of $\mathcal{M}$ close to the point $\mathcal{P}$ to $\mathbb{R}^m$ using our particular homeomorphism ϕ. We can then map the manifold near other points to $\mathbb{R}^m$ using a different homeomorphism.

Example C.4

The unit circle is a manifold. We set up a map ϕ_1 which takes points on the manifold to the coordinate θ in $\mathbb{R}$, assumed to vary between 0 and 2π [see Fig. C.9(a)]. However, this won't work for all points in the manifold since the point $\mathcal{P}$ which we describe in $\mathbb{R}$ as $\theta = 0$ is also ascribed the point in $\mathbb{R}$ called $\theta = 2\pi$. The map ϕ that takes points on the manifold to $\mathbb{R}$ is not 1-1 if we include this point, so we are forced to drop it completely. The subset U_1 of $\mathcal{M}$, for which ϕ_1 is well defined, then encompasses all of the unit circle except this troublesome point $\mathcal{P}$ [see Fig. C.9(b)].

In general, we may need a collection of open sets, U_i, and a collection of maps ϕ_i, to complete our description. We want to be able to patch these U_is together to completely cover the manifold.

Example C.5

Returning to the unit circle, we can come up with an alternative subset of the manifold $\mathcal{M}$, called U_2. This one includes the whole of the circle except a point $\mathcal{Q} \neq \mathcal{P}$. As drawn in Fig. C.9(c), the missing point is the one that a map ϕ_2 would take to π and/or $-\pi$. The map ϕ_2 does however take all other points on U_2 to $\mathbb{R}$ in the open interval $-\pi$ to π. We see that by missing $\mathcal{Q}$ we do capture the point $\mathcal{P}$ that was not covered by U_1. Taken together, U_1 and U_2 are seen to cover $\mathcal{M}$.

Generally, then, the subsets U_i are a family of open subsets that, taken together, cover $\mathcal{M}$. The map ϕ_i maps from the subset U_i onto an open subset of $\mathbb{R}^m$. The subset U_i is called a coordinate neighbourhood. The pair (U_i, ϕ_i) is called a **chart**.[7] The whole family of charts $\{(U_i, \phi_i)\}$ is called an **atlas**.

[7]A chart is often called a coordinate system by physicists.

Example C.6

Consider two examples of spaces:
(i) m-dimensional Euclidean space. A single chart covers all of this space.
(ii) One-dimensional space. There are two possible manifolds: the line $\mathbb{R}^1$ and the circle S^1. A single chart covers the line. As we saw before, (at least) two charts are needed to cover the circle.

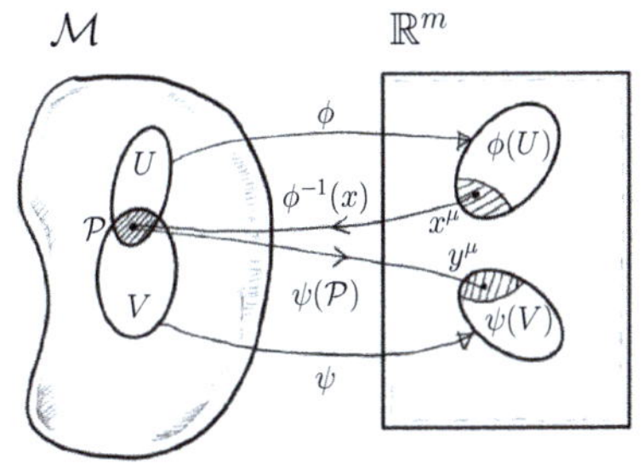

Fig. C.10 A coordinate transformation.

In having more than one set of coordinates (i.e. more than one chart), we do ask that they are **compatible**. Consider a manifold $\mathcal{M}$ with overlapping subsets U and V (Fig. C.10). The point $\mathcal{P}$ lies in the overlapping region. Homeomorphisms are defined such that $\phi : U \to \mathbb{R}^m$ and $\psi : V \to \mathbb{R}^m$. We have charts (U, ϕ) and (V, ψ) and write $\phi(\mathcal{P}) = x^\mu$ and $\psi(\mathcal{P}) = y^\mu$. By defining a **composite map** that combines the two homeomorphisms, we are able to recover the idea of a **coordinate transformation**. In order to get from x^μ to y^μ (and motivated by the diagram in Fig. C.10), we write

$$y^\mu = \psi \circ \phi^{-1}(x^\mu). \tag{C.6}$$

In words, input a coordinate x^μ that is taken to a point $\mathcal{P}$ on $\mathcal{M}$ and output a coordinate y^μ that corresponds to the same point.

We are now in the position to put everything together and write down a more technical description of a manifold. Of course, there's very little here we haven't seen earlier in words.

A manifold $\mathcal{M}$ has the following properties
- Each $\mathcal{P} \in \mathcal{M}$ lies in at least one open set U_i (i.e. the $\{U_i\}$ cover $\mathcal{M}$).
- For each i there is a homeomorphism $\phi : U_i \to \phi(U_i)$, where $\phi(U_i)$ is an open subset of $\mathbb{R}^m$.
- Where any two sets U_i and U_j overlap, we have a composite map $\phi_i \circ \phi_j^{-1}$ which takes points in $\phi_i(U_i \cap U_j) \subset \mathbb{R}^m$ to points in $\phi_j(U_i \cap U_j) \subset \mathbb{R}^m$.

C.6   Functions on the manifold

As introduced above, a **function** can be thought of as a map. In addition, a function can be thought of as living on a manifold. However, we really only have access to coordinates in $\mathbb{R}^m$ and so we often want to input and output these coordinates when interacting with objects defined on the manifold. Let's consider the function $f : \mathcal{M} \to \mathbb{R}$, that inputs some point $\mathcal{P}$ in the m-dimensional manifold $\mathcal{M}$, and outputs a number

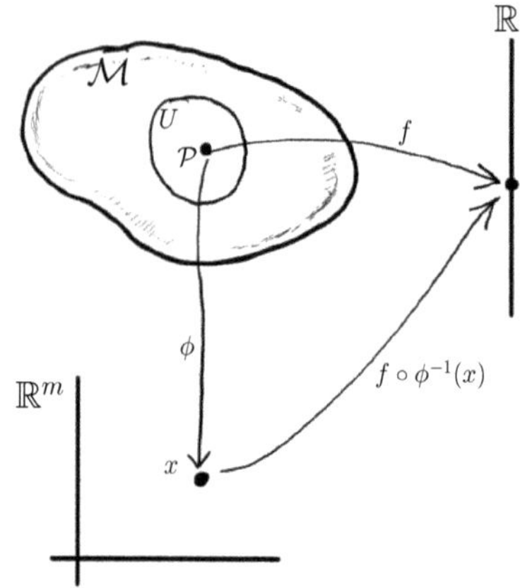

Fig. C.11 A function as a composite map $f \circ \phi^{-1} : \phi(U) \to \mathbb{R}$.

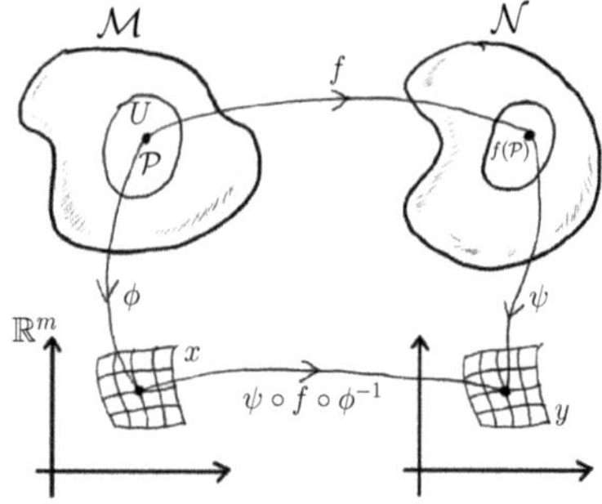

Fig. C.12 A function that maps between manifolds.

in $\mathbb{R}$. The point $\mathcal{P}$ lies in the subset U of $\mathcal{M}$. The question then is how can we tell how f assigns a real value to each point on $\mathcal{M}$, while we only have access to coordinates in $\mathbb{R}^m$.

The answer is shown in Fig. C.11. The homeomorphism ϕ takes $\mathcal{P}$ to coordinate $x^\mu = \phi(\mathcal{P})$ and U to coordinate neighbourhood $\phi(U) \subset \mathbb{R}^m$. This means we can write a composition $f \circ \phi^{-1} : \phi(U) \to \mathbb{R}$. In words, we input a coordinate x^μ in the region of $\mathbb{R}^m$ called $\phi(U)$ and output a point y^1 on real line $\mathbb{R}$. This is all we ask of this function, which is a machine that takes multidimensional points and outputs a number. The message then is that what we usually call $y = f(x^1, x^2, ..., x^m)$ should, when dealing with a function on a manifold that takes $\mathcal{M}$ to $\mathbb{R}$, be regarded as

$$y = f \circ \phi^{-1}(x^1, ..., x^m). \tag{C.7}$$

Another way of saying this is that $f \circ \phi^{-1}(x^\mu)$ is the **coordinate representation of the function**.

Example C.7

Consider a function that maps between two different m-dimensional manifolds $f : \mathcal{M} \to \mathcal{N}$ as shown in Fig. C.12. That is, it takes a point $\mathcal{P}$ in the m-dimensional manifold $\mathcal{M}$ to a point $f(\mathcal{P})$ on the m-dimensional manifold $\mathcal{N}$. Take a chart (U, ϕ) on $\mathcal{M}$ and a chart (V, ψ) on $\mathcal{N}$. Take $\mathcal{P}$ to be in U and $f(\mathcal{P})$ to be in V. The function has a coordinate representation

$$(y^1, ..., y^m) = \psi \circ f \circ \phi^{-1}(x^1, ..., x^m), \tag{C.8}$$

that is, it's an m-tuple-valued function $y^\mu = f(x^\mu)$ as shown in Fig. C.12.

C.7 Differentiation on the manifold

Let's consider a m-dimensional manifold $\mathcal{M}$ and a function $f : \mathcal{M} \to R$. We differentiate functions by varying them with respect to coordinates x^μ. It might seem like defining differentiation on manifolds should just rely on identifying a function f and a chart (U, ϕ) to map onto $\mathbb{R}^m$. This would allow us to vary $f \circ \phi^{-1}$ with respect to coordinates x^μ giving partial derivatives like $\frac{\partial}{\partial x^\nu}\left(f \circ \phi^{-1}\right)$. It is, unfortunately, not quite that simple. However, the fix we need provides the manifold with a rich structure that makes the extra effort involved in defining differentiation more than worth it.

First, the problem: it would seem reasonable that f should be differentiable if $f \circ \phi^{-1}$ is differentiable, but this is not the case. If ψ is another homeomorphism such that $\psi : V \to \mathbb{R}^m$ and $U \cap V \neq 0$ then it's not necessarily the case that $f \circ \psi^{-1}$ is also differentiable. Since we can write that

$$f \circ \psi^{-1} = f \circ \phi^{-1} \circ \left(\phi \circ \psi^{-1}\right), \tag{C.9}$$

we need the coordinate transformation $\phi \circ \psi^{-1}$ to be differentiable in order that $f \circ \psi^{-1}$ is also differentiable.

Therefore, in order to be able to differentiate on a manifold with charts (U_i, ϕ_i), we need that if a pair of regions U_i and U_j overlaps such that $U_i \cap U_j \neq \varnothing$, then the map

$$\phi_i \circ \phi_j^{-1} \tag{C.10}$$

should be infinitely differentiable (a property denoted C^∞).[8] We call such homeomorphisms C^∞-related. The idea of compatible charts allows us to construct a **maximal atlas**, which is the atlas that contains every compatible C^∞-related chart. (This allows us to ensure that two equivalent spaces with different atlases aren't actually two different manifolds.) A **differentiable manifold** is then specified by the set $\mathcal{M}$ and its (unique, maximal) atlas of C^∞-related charts $\{U\}$. Defined in this way, the differentiable manifold carries a significant amount of structure.

Example C.8

Let's consider the simplest possible differentiable manifold. Take a manifold $\mathcal{N}$ to be $\mathbb{R}$ and a homeomorphism $\eta : \mathbb{R} \to \mathbb{R}$ to be the identity $x \mapsto x$ (or, more simply, $\eta(x) = x$). The manifold $\mathcal{N}$ taken with the maximal atlas that contains the identity is a differentiable manifold. This is because[9] the identity $\eta(x) = x$ is indeed C^∞.

Having fixed up differentiation on a manifold in terms of its homeomorphisms, we can characterize a differentiable function between manifolds[10] (referring to Fig. C.12 again).

A function that maps between manifolds $f : \mathcal{M} \to \mathcal{N}$ is **differentiable** if for every coordinate system (ϕ, U) in $\mathcal{M}$ and (ψ, V) in $\mathcal{N}$, the map $\psi \circ f \circ \phi^{-1} : \mathbb{R}^m \to \mathbb{R}^n$ is differentiable.

We now have a concept of a differentiable manifold and a differentiable map between manifolds. If we further insist that the map $\psi \circ f \circ \phi^{-1}$ is *invertible* (i.e. that there exists a map $\phi \circ f^{-1} \circ \psi^{-1}$), and that both $y = \psi \circ f \circ \phi^{-1}(x)$ and $x = \phi \circ f^{-1} \circ \psi^{-1}(y)$ are C^∞, then $\mathcal{M}$ is said to be **diffeomorphic** to $\mathcal{N}$, and the map f is called a **diffeomorphism**. Because the map is invertible, the dimension of $\mathcal{M}$ has to equal that of $\mathcal{N}$, i.e. $m = n$. Two diffeomorphic manifolds can be regarded as essentially the same manifold.[11]

Example C.9

The map $f : \mathbb{R} \to \mathbb{R}$ given by $f(x) = x$ is a diffeomorphism. However, the map $g : \mathbb{R} \to \mathbb{R}$ given by $g(x) = x^2$ is not a diffeomorphism because it is not 1-1 (e.g. $g(2) = 4$, but also $g(-2) = 4$). The map $h : \mathbb{R} \to \mathbb{R}$ given by $h(x) = x^3$ is not a diffeomorphism either. Although h is a 1-1 map, its inverse $h^{-1}(x) = x^{1/3}$ is not sufficiently smooth (i.e. C^∞) at $x = 0$, since its first derivative is not defined. The map $\phi : \mathbb{R}^2 \to \mathbb{R}^2$ given by $\phi(x,y) = (x + \frac{y}{2}, y - \frac{x}{2})$ is a diffeomorphism; the determinant of the Jacobian of the map is non-zero everywhere[12] and so the map is invertible.

---

[8]If $f(x_1, ..., x_n)$ is a function defined on an open region S of $\mathbb{R}^n$, then it is **differentiable of class** C^k if all of the partial derivatives or order less than or equal to k exist and are continuous functions on S. A special case is the C^∞ (or smooth) function: a map is C^∞ if the coordinates of a point in $\mathcal{N}$ are infinitely differentiable functions of the coordinates of the inverse image of the point $\mathcal{M}$. All polynomial functions are C^∞. By contrast, a function like $x^{\frac{1}{3}}$ has a first derivative that is not continuous at the origin (where it blows up), so it is not C^∞.

[9]The identity map $x \mapsto x$, i.e. $\eta(x) = x$, is continuous, its first derivative is unity, so is continuous. Subsequent derivatives are zero, which is continuous too.

[10]We assume manifold $\mathcal{M}$ is m-dimensional and manifold $\mathcal{N}$ is n-dimensional.

[11]A diffeomorphism can only apply to manifolds. This is because of the local smoothness of a manifold that follows from it resembling $\mathbb{R}^n$ locally. In contrast, a homeomorphism can apply between things that aren't manifolds.

[12]The Jacobian matrix of the map is the matrix of partial derivatives $(\partial \phi^i / \partial x^j)$, where in this case $\phi^1 = x^1 + \frac{x^2}{2}$ and $\phi^2 = x^2 - \frac{x^1}{2}$. The determinant of this matrix is called the Jacobian, and in this case it is equal to $\frac{5}{4}$ which, crucially, is non-zero.

Note that a homeomorphism is basically a diffeomorphism without the differentiability requirement. One way to think about it is that calling a map between two spaces a homeomorphism means that you can deform one space to the other *continuously*. Calling a map between two spaces a diffeomorphism tells you something extra; it means that it is possible to deform one space to the other *smoothly*, and that smoothness of the coordinate transformations is independent of the coordinates chosen.

Why are diffeomorphisms so essential for general relativity? A diffeomorphism $\phi : \mathcal{M} \to \mathcal{M}$ maps a point to another point in the same manifold. Such diffeomorphisms are analogous to active coordinate transformations that transform from one point to another point. If the physics is unaffected by this transformation then this tells us about points that are the same (or indistinguishable). Diffeomorphisms therefore reveal the gauge symmetries of general relativity.[13] As a result, general relativity is sometimes called a diffeomorphism-invariant theory. Moreover, diffeomorphisms allow us to compare tensors defined at different points on a manifold and so the definition of the Lie derivative $\mathcal{L}_{\boldsymbol{u}}$ (Chapter 33) is most generally given in terms of diffeomorphisms.[14] In this case, the flow along the integral curves is represented by a diffeomorphism, where the vector field $\boldsymbol{u}$ encoding this flow is referred to as the *generator* of the diffeomorphism. We use this in the next example.

[13] Beyond general relativity, diffeomorphisms are useful in mechanics where they allow an insightful geometrical description of Hamiltonian mechanics. See Geroch's book *Geometrical Quantum Mechanics* for an introduction.

[14] The very important Killing vector fields that tell us about conserved quantities are therefore also most generally given in terms of diffeomorphisms.

Example C.10

We can use invariance with respect to diffeomorphisms to justify one of the most fundamental equations in relativity: $\boldsymbol{\nabla}\cdot\boldsymbol{T} = 0$. Let's use the machinery of Chapter 40 and examine the variation of the matter Lagrangian with the components of the metric

$$\delta S = \int \mathrm{d}^4 x\, \frac{\delta\left(\sqrt{-g}\mathcal{L}_{\mathrm{m}}\right)}{\delta g_{\mu\nu}}\, \delta g_{\mu\nu}. \tag{C.11}$$

If the variations in the metric components are generated by diffeomorphisms, we have[15] $\delta g_{\mu\nu} = (\mathcal{L}_{\boldsymbol{u}}\boldsymbol{g})_{\mu\nu} = 2u_{(\mu;\nu)}$. For a diffeomorphism-invariant theory we have $\delta S = 0$ and so, as a result, we write

$$0 = \int \mathrm{d}^4 x\, \frac{\delta\left(\sqrt{-g}\mathcal{L}_{\mathrm{m}}\right)}{\delta g_{\mu\nu}}\, u_{\mu;\nu}, \tag{C.13}$$

[15] Recall that we had

$$(\mathcal{L}_{\boldsymbol{u}}\boldsymbol{g})_{\mu\nu} = u^\sigma g_{\alpha\beta;\sigma} + g_{\alpha\sigma}u^\sigma{}_{;\beta} + g_{\sigma\beta}u^\sigma{}_{;\alpha}, \tag{C.12}$$

or $(\mathcal{L}_{\boldsymbol{u}}\boldsymbol{g})_{\mu\nu} = 2u_{(\alpha;\beta)}$.

where we drop the symmetrization of the covariant derivative of $\boldsymbol{u}$, as $\delta\left(\sqrt{-g}\mathcal{L}_{\mathrm{m}}\right)/\delta g_{\mu\nu}$ is symmetric, and so the integral is unaffected by the presence of the symmetrization. Integrating by parts, we find

$$0 = -\int \mathrm{d}^4 x\, \sqrt{-g}u_\nu \boldsymbol{\nabla}_\mu \left[\frac{1}{\sqrt{-g}}\frac{\delta\left(\sqrt{-g}\mathcal{L}_{\mathrm{m}}\right)}{\delta g_{\mu\nu}}\right]. \tag{C.14}$$

This must be true for diffeomorphisms generated by an arbitrary field $\boldsymbol{u}$ and so, using the definition of $\boldsymbol{T}$ in terms of the action from Chapter 40, i.e.

$$T^{\mu\nu} = \frac{2}{\sqrt{-g}}\frac{\delta\left(\sqrt{-g}\mathcal{L}_{\mathrm{m}}\right)}{\delta g_{\mu\nu}}, \tag{C.15}$$

we can see that the integrand is equivalent to $(\boldsymbol{\nabla}_\mu \boldsymbol{T})^{\mu\nu} = T^{\mu\nu}{}_{;\mu} = 0$ or $\boldsymbol{\nabla}\cdot\boldsymbol{T} = 0$.

C.8   Compact regions

We have stated that a closed interval between points a and b on the real line $\mathbb{R}$ is written $[a, b]$. It has the nice property that it encompasses a finite interval, doesn't have any holes in it, and includes its boundary. This notion can be generalized for a region of a manifold and the general term we will use is **compact**. If we say a region is compact, we mean that it doesn't do things like (i) go off to infinity; (ii) have bits removed; nor (iii) have bits of its boundary removed. One of the ways of achieving this is to insist that any sequence of points in our region must have a limit (or accumulation point) that also lies in the region.[16]

Example C.11

A closed interval $[0, 1]$ on $\mathbb{R}$ is compact, as can be seen by considering the sequence of points defined by $1/n$ where $n > 0$ is a positive integer, each of which lies within that interval, but crucially so does its limit (the point reached when $n \to \infty$, which is 0 and is a member of the set of points defined by $[0, 1]$). This would not work if we removed the point 0 on the boundary by considering $(0, 1]$ instead of $[0, 1]$. We conclude that $[0, 1]$ is compact, but $(0, 1]$ is not compact.

We have defined compactness as an 'upgrade' of the notion of a closed interval $[a, b]$ on $\mathbb{R}$, so it is not surprising that such a closed interval is compact, but this statement is termed the **Heine–Borel theorem**.[17] In fact, it is also possible to show that a subset of real numbers is compact if and only if it is closed and bounded.[18] A compact space can therefore be thought of as a space which, if it has a boundary, includes the boundary as part of the space, and it has no missing parts. Some examples of spaces that are and aren't compact are given below.

Example C.12

- The closed unit disc is compact, as is the sphere S^2 and the torus T^2.

- The Euclidean plane is not compact (it contains points that run off to infinity). Neither is the open unit disc (points on its boundary are not included in the space). Nor is the closed disc with a hole in it (it has a missing region).

C.9   Curves

A *curve* on a manifold can be described using a **parametrization**, with a real number λ telling us how far we are along the curve. Thus, if we take the map $c : [a, b] \to \mathcal{M}$ as shown, the real number λ on the real line $\mathbb{R}$ is mapped on to the point $c(\lambda) \in \mathcal{M}$, producing a curve on $\mathcal{M}$ as λ runs from a to b. As usual, the homeomorphism ϕ maps from

---

[16] An alternative, and rather grand and formal, way of defining a compact region is to say that a region $\mathcal{A}$ is compact if every open cover O contains a finite sub-collection of open sets which also cover $\mathcal{A}$.

[17] We state this theorem without proof here, but see e.g. Spivak's *Calculus on Manifolds* for the full story on this and other theorems about topological spaces. The theorem is named in honour of the German mathematician Eduard Heine (1821–1881) and the French mathematician Émile Borel (1871–1956).

[18] The term **bounded** means that the set doesn't run off to infinity but is enclosed within some finite region; to define a region of a manifold as bounded requires a notion of distance, i.e. it applies only to spaces endowed with a metric.

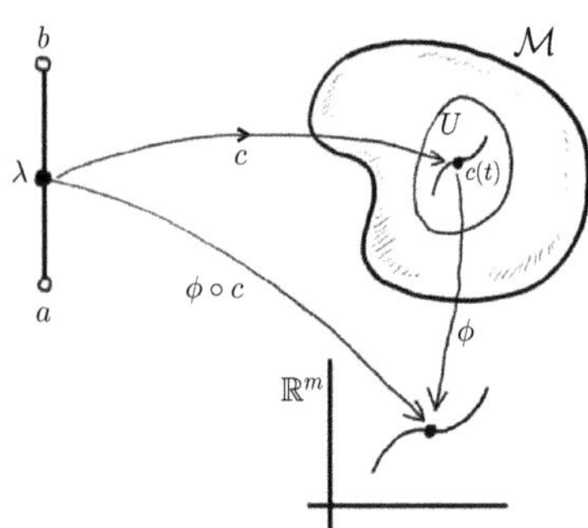

Fig. C.13 A curve $c(\lambda)$, expressed in $\mathbb{R}^m$ by a homeomorphism ϕ.

[19]This is the idea of a **fibre bundle**, examined in more detail in the next section.

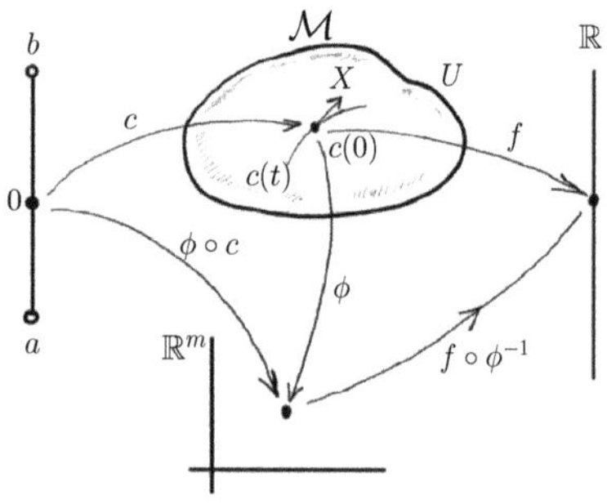

Fig. C.14 A vector field in terms of mappings.

$\mathcal{M}$ to a chart in $\mathbb{R}^m$. Therefore, the coordinate on the curve on $\mathcal{M}$ corresponding to some value of λ is given by the composite map

$$x^\mu = \phi \circ c(\lambda). \tag{C.16}$$

In short, input at parameter λ and output a point on the curve $(x^1, ... x^m) = \phi \circ c(\lambda)$ in $\mathbb{R}^m$ (see Fig. C.13).

C.10 Tangent spaces

Next, we build up to the notion of a **vector**. An arrow does not properly represent a vector on a manifold. There's no origin or concept of straightness, after all. The vectors we shall discuss are **tangent vectors**, that is, directional derivatives to curves that live in a manifold. A tangent vector does not live in the same manifold in which curves live, but instead lives in a manifold called a **tangent space**. There is not one tangent space but many: one, in fact, for each point on the manifold. The tangent spaces can be thought of as floating above the manifold.[19]

We shall generalize the procedure of finding a tangent vector as the directional derivative along a curve by declaring:

The tangent vector at $c(\lambda = 0)$ is defined as the directional derivative of a function $f(c(\lambda))$ along the curve $c(\lambda)$, evaluated at $\lambda = 0$.

A curve $c(\lambda)$ takes the parameter λ from $\mathbb{R}$ and puts it on the manifold. To take it from the manifold back to $\mathbb{R}$ we need a function $f : \mathcal{M} \to \mathbb{R}$. Once we have this, we can evaluate the rate of change of the curve with the parameter λ as follows:

$$\left(\begin{array}{c} \text{Rate of change of } f(c(\lambda)) \text{ along} \\ \text{the curve, evaluated at } \lambda = 0 \end{array} \right) = \left. \frac{\mathrm{d}f(c(\lambda))}{\mathrm{d}\lambda} \right|_{\lambda=0} = \left. \frac{\mathrm{d}(f \circ c)}{\mathrm{d}\lambda} \right|_{\lambda=0}, \tag{C.17}$$

where, in the last part, we've written out the composition.

Using the homeomorphism $\phi : \mathcal{M} \to \mathbb{R}^m$, we can map the point on the manifold into $\mathbb{R}^m$, as shown in Fig. C.14, and then the combination $f \circ c$ can be written as

$$f \circ c = (f \circ \phi^{-1}) \circ (\phi \circ c), \tag{C.18}$$

where the first bracket $(f \circ \phi^{-1})$ is a real-valued function of a point in $\mathbb{R}^m$ [that is $f \circ \phi^{-1} = f(x^\mu)$] and the second bracket $(\phi \circ c)$ takes a point λ from $\mathbb{R}$ and returns a point in $\mathbb{R}^m$ [that is, it maps out the curve in coordinate space and could therefore be written as $x^\mu(c(\lambda))$]. Since $f \circ \phi^{-1}$ and $\phi \circ c$ are coordinate representations of the function and curve respectively, we can write the derivative in terms of a Leibniz (or chain) rule

$$\begin{aligned} \frac{\mathrm{d}(f \circ c)}{\mathrm{d}\lambda} &= \frac{\partial}{\partial x^\mu}(f \circ \phi^{-1})\frac{\mathrm{d}}{\mathrm{d}\lambda}(\phi \circ c(\lambda)) \\ &= \left. \frac{\partial f(x^\mu)}{\partial x^\mu}\frac{\mathrm{d}x^\mu(c(\lambda))}{\mathrm{d}\lambda} \right|_{\lambda=0}. \end{aligned} \tag{C.19}$$

It is this expression that we use to define a vector $\boldsymbol{v}[f]$. It features the differential operator $\frac{\partial}{\partial x^\mu}$, which supplies the basis vectors, acting on a function f. It comes with a factor $\left.\frac{\mathrm{d}x^\mu(c(\lambda))}{\mathrm{d}\lambda}\right|_{\lambda=0}$, that we call the μth component and which tells us about the rate of change of the curve c with λ, when it is projected into $\mathbb{R}^m$. As a result, the tangent vector $\boldsymbol{v}$ is defined as

$$\boldsymbol{v}[f] = \left.\frac{\mathrm{d}x^\mu(c(\lambda))}{\mathrm{d}\lambda}\right|_{\lambda=0} \frac{\partial f}{\partial x^\mu}, \tag{C.20}$$

where the order of the terms on the right-hand side has been swapped to conform to our usual convention of writing $\boldsymbol{v} = v^\mu \boldsymbol{e}_\mu$.

Example C.13

The simplest example results if we let the function f be the coordinate function $x^\nu = \phi^\nu \circ c(\lambda)$ or

$$\phi^\nu \circ c(\lambda) = \phi^\nu(c(\lambda)) = x^\nu(\lambda). \tag{C.21}$$

The action of the vector on $x^\nu(\lambda)$ is

$$\boldsymbol{v}[x^\nu] = \frac{\mathrm{d}x^\mu}{\mathrm{d}\lambda}\frac{\partial x^\nu(\lambda)}{\partial x^\mu} = \left.\frac{\mathrm{d}x^\nu(\lambda)}{\mathrm{d}\lambda}\right|_{\lambda=0}, \tag{C.22}$$

thereby measuring the rate of change of the coordinate component x^ν with λ.

We can generalize and say that a tangent vector works on a number of functions (e.g. f and g) taken from the set $\mathcal{F} = C^\infty(\mathcal{M})$ of all smooth functions from $\mathcal{M}$ to $\mathbb{R}$ (or $f, g \in \mathcal{F}$). We may then firm up the definition a little at this point and say that a tangent vector $\boldsymbol{v}$ at a point $\mathcal{P} \in \mathcal{M}$ is a map $\boldsymbol{v} : \mathcal{F} \to R$, which is (i) linear and (ii) obeys the Leibniz rule, or

$$
\begin{array}{rll}
\text{(i)} & \boldsymbol{v}(\alpha f + \beta g) = & \alpha \boldsymbol{v}(f) + \beta \boldsymbol{v}(g), \\
\text{(ii)} & \boldsymbol{v}(fg) = & f \boldsymbol{v}(g) + g \boldsymbol{v}(f),
\end{array}
$$

where α and β are arbitrary numbers and the functions are evaluated at the point $\mathcal{P}$. In this way of viewing tangent vectors, there is a one-to-one correspondence between vectors and derivatives. It is consistent, therefore, to adopt the view that instead of vectors, we can work with derivatives.

There is a class of curves that all pass through $\mathcal{P}$. If they all have the same tangent vector, then we can identify them. If we have

$$c_1(\lambda = 0) = c_2(\lambda = 0) = \mathcal{P}, \tag{C.23}$$

and

$$\left.\frac{\mathrm{d}x^\mu(c_1(\lambda))}{\mathrm{d}\lambda}\right|_{\lambda=0} = \left.\frac{\mathrm{d}x^\mu(c_2(\lambda))}{\mathrm{d}\lambda}\right|_{\lambda=0}, \tag{C.24}$$

then these give the same vector component at $\mathcal{P}$. We identify the tangent vector $\boldsymbol{v}$ with the equivalence class of curves, rather than a single curve.

Lots of curves passing through $\mathcal{P}$ will have different tangent vectors. All of the tangent vectors at $\mathcal{P}$, one for each class of curves, form a vector space called a **tangent space** of the manifold $\mathcal{M}$ at point $\mathcal{P}$, denoted $\mathcal{T}_\mathcal{P}\mathcal{M}$. This is shown in Fig. C.15. We examine the generalization of the tangent space in the next section.

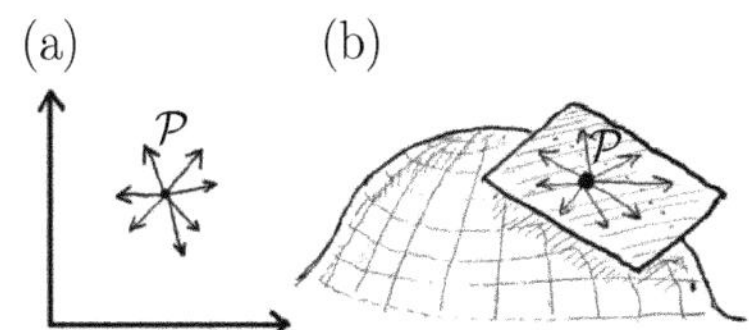

Fig. C.15 The tangent space contains all of the tangent vectors. (a) A set of vectors in the tangent space at a point $\mathcal{P}$. (b) The point of view used earlier in the book of a tangent plane to a surface relies on embedding the manifold in Euclidean space.

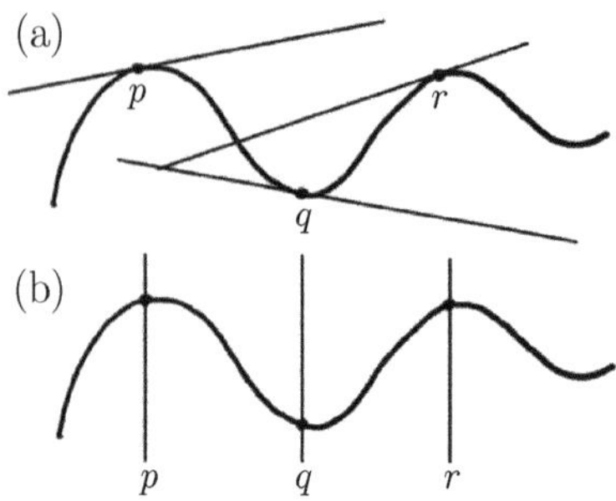

Fig. C.16 (a) The one-dimensional manifold with some tangent vectors identified at each point. (b) The tangent vectors represented as vertical lines. These are the fibres we use to form a bundle.

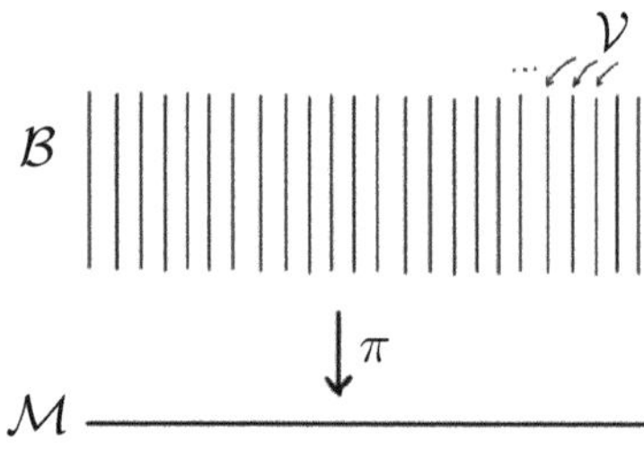

Fig. C.17 A fibre bundle $\mathcal{B}$ consisting of base space $\mathcal{M}$ and fibres $\mathcal{V}$. The projection π collapses a fibre down to a point.

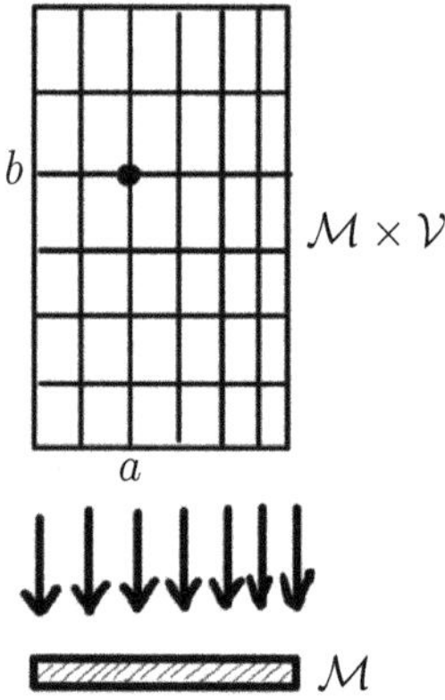

Fig. C.18 A fibre bundle and its projection π.

C.11   Fibre bundles

Consider a curve, which is itself a one-dimensional manifold $\mathcal{M}$. Take tangents at each point along the curve [some examples are shown in Fig. C.16(a)]. Drawn in this way, neighbouring tangent vectors intersect each other creating a certain amount of confusion. To avoid this, we could instead draw them as in Fig. C.16(b), rotating them around so they no longer keep bumping into each other. Of course, in this new drawing they no longer so obviously resemble tangent vectors, but let's imagine that we could somehow still encode that information in them. Now lift the tangent vectors for each point off the line so that we have the state of affairs shown in Fig. C.17. This shows the one-dimensional manifold $\mathcal{M}$. Above each and every point in $\mathcal{M}$ there is one (one-dimensional) manifold $\mathcal{V}$ floating above it. The particular tangent vector at a point in $\mathcal{M}$ is represented by a point on the particular manifold $\mathcal{V}$ (like beads on an abacus). We call the manifold $\mathcal{V}$ a **fibre**. By combining the manifold $\mathcal{M}$ (the points) and all of the $\mathcal{V}$s (the fibres or tangent spaces) we obtain a new, two-dimensional manifold $\mathcal{VM}$ known as a fibre bundle $\mathcal{B}$ or, in this special case of tangent spaces, a **tangent bundle** $\mathcal{TM}$.

More generally, then, a **fibre bundle** $\mathcal{B}$ is a manifold defined in terms of two other manifolds $\mathcal{M}$ and $\mathcal{V}$. Manifold $\mathcal{M}$ is called the **base space** and $\mathcal{V}$ is called the **fibre**. The dimension of the fibre bundle $\mathcal{B}$ is always the sum of the dimensions of $\mathcal{M}$ and $\mathcal{V}$. There are many copies of $\mathcal{V}$ in $\mathcal{B}$. One complete copy of $\mathcal{V}$ stands above each point in $\mathcal{M}$.

We can undo the previous construction if we define a continuous map from $\mathcal{B}$ back down on to $\mathcal{M}$. This is called the **canonical projection** π from $\mathcal{B}$ to $\mathcal{M}$. This map collapses each fibre down to the corresponding point in $\mathcal{M}$. If you like, it forgets about the information stored in the fibre, remembering only the point on $\mathcal{M}$ which the fibre was floating above.

The simplest example of a bundle is a product space of $\mathcal{M}$ with $\mathcal{V}$, which is written as $\mathcal{M} \times \mathcal{V}$. The points in $\mathcal{M} \times \mathcal{V}$ are pairs of elements (a, b) where a belongs to $\mathcal{M}$ and b belongs to $\mathcal{V}$ (see Fig. C.18). This is often called a **trivial bundle**.

Example C.14

Perhaps the simplest of all tangent bundles is formed by making the base manifold the unit circle. We shall consider $\mathcal{T}S^1$, the tangent bundle of the circle S^1 and its tangent vectors. The tangent bundle $\mathcal{T}S^1$ is identical to the product space $S^1 \times \mathbb{R}$, shown by the cylinder in Fig. C.19 (and so is a trivial bundle). Moreover, $\mathcal{T}S^1$ is a two-dimensional manifold which we can cover with coordinates. In S^1 a point is described by a coordinate θ. A tangent vector $\boldsymbol{v}$ at any point $\mathcal{P}$ can be written as $\boldsymbol{v} = y\frac{\partial}{\partial\theta} \equiv y\boldsymbol{e}_\theta$, where y is a coordinate in $\mathcal{T}_\theta$ (i.e. an amplitude taken from the tangent space floating above the particular angle θ). The coordinates (θ, y) then tell us about the position on the base space and the coordinate along the fibre.

Bundles are said to be **locally trivial** if they are formed from a product space. We can ask whether they're also **globally trivial**: whether the whole bundle can be represented by a product $\mathcal{M} \times \mathcal{V}$. The example in Fig. C.19, where the bundle resembles a cylinder, is indeed globally trivial. An interesting counterexample of a bundle that is not globally trivial is a **twisted bundle**. This resembles $\mathcal{M} \times \mathcal{V}$ locally, but as we move around $\mathcal{M}$, the fibres twist, so that globally $\mathcal{B}$ is different from $\mathcal{M} \times \mathcal{V}$.

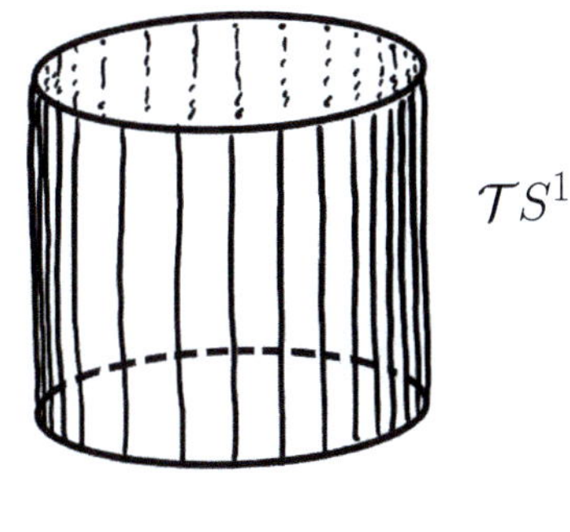

Fig. C.19 The bundle $\mathcal{T}S^1$ floating above its base space.

Example C.15

Once again, take $\mathcal{M}$ to be the circle S^1 and $\mathcal{V}$ to be the real line $\mathbb{R}$. The trivial bundle simply resembles a two-dimensional cylinder. Now construct a twisted bundle, by forming the fibres into a Möbius strip[20] (see Fig. C.20). Locally, this is the same as the cylinder; globally it is not. To see the local similarity, remove a point $\mathcal{P}$ from the base space. We then have a segment $S^1 - \mathcal{P}$ and the bundle above this segment can be deformed to look the same as the cylinder. It is only when we look at the whole of the base space that we notice the difference. To see *this*, consider two segments: $S^1 - \mathcal{P}$ and $S^1 - \mathcal{Q}$, where $\mathcal{P}$ and $\mathcal{Q}$ are different points. Each is locally trivial (i.e. each can be deformed to look like the cylinder). However, on gluing them together (with a twist) to make a whole we form the Möbius strip.

[20] August Ferdinand Möbius (1790–1868). The Möbius strip (a two-dimensional surface that, when embedded in three dimensions, has only one side) was discovered by Möbius and, independently, by Johann Benedict Listing.

Formally, we characterize a bundle by looking at its **cross section**. The cross section of the bundle $\mathcal{B}$ is a continuous image of the base space $\mathcal{M}$ in $\mathcal{B}$, which meets each fibre at a single point (Fig. C.21). This is called the **lift** of the base space into the bundle. So if we apply the lift via a continuous function $s : \mathcal{M} \to \mathcal{B}$, followed by the projection $\pi : \mathcal{B} \to \mathcal{M}$, we get the identity map from $\mathcal{M}$ into itself

$$\pi \circ s = I. \tag{C.25}$$

For the trivial bundle $\mathcal{M} \times \mathcal{V}$ the cross sections look like continuous functions on $\mathcal{M}$ which take values in the space $\mathcal{V}$. So a cross section of $\mathcal{M} \times \mathcal{V}$ assigns in a continuous way, a point of $\mathcal{V}$ to each point on $\mathcal{M}$. This is like an extension of the ordinary idea of a graph of a function.

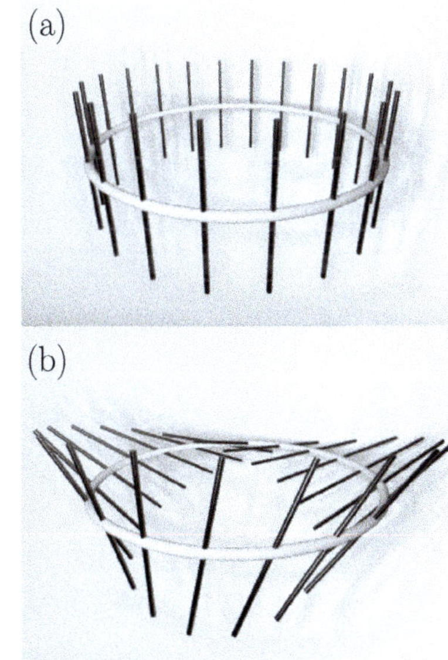

Fig. C.20 (a) The globally trivial bundle. (b) The bundle with a twist, forming a Möbius strip.

Example C.16

Consider the cylindrical product bundle $\mathcal{M} \times \mathcal{V}$. The cross section looks like a curve intersecting each fibre once as it goes around the cylinder. We can also mark out the curve featuring at the zeros of the vectors, called the zero section. There's no guarantee that any old curve intersects this line of zeros. The Möbius bundle is more complicated, but one thing that can be said is that a curve has to cross the zero section. This gives us a way of characterizing the difference between the Möbius strip and trivial bundle.

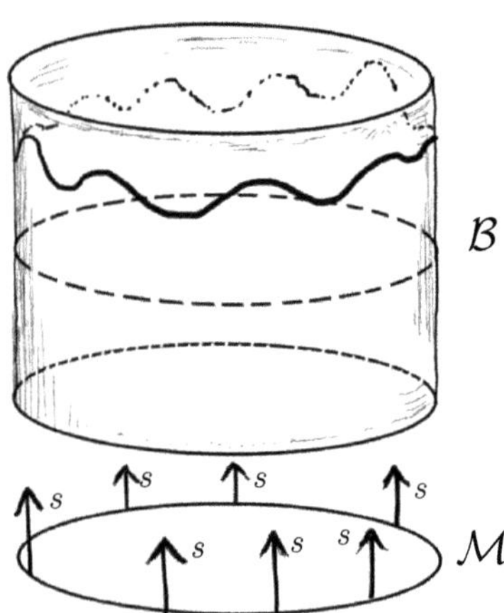

Fig. C.21 The cross section of the bundle $\mathcal{B}$, formed from the lift s of the base space $\mathcal{M}$. This can be thought of as a way of graphing a function on $\mathcal{M}$ in $\mathcal{B}$.

Chapter summary

- The set $\mathcal{M}$ is a manifold if each point in $\mathcal{M}$ has an open neighbourhood which has a continuous 1-1 map onto an open set of $\mathbb{R}^n$ for some n.

- Morphisms allow us to map between manifolds. A diffeomorphism relates two manifolds which are endowed with a differentiable structure (meaning that they are smooth) and is the most useful morphism in general relativity. A diffeomorphism that maps a manifold onto itself is equivalent to an active coordinate transformation.

- Compact regions don't go off to infinity, or have parts removed or on a boundary.

- Vectors can be defined in terms of derivatives using curves on the manifold and mappings using the expression

$$v[f] = \frac{\mathrm{d}(f \circ c)}{\mathrm{d}\lambda}. \tag{C.26}$$

- A fibre bundle $\mathcal{B}$ is the combination of a base space $\mathcal{M}$ and a fibre space $\mathcal{V}$, with a fibre defined at each point in the base space.

Embedding

When a blind beetle crawls over the surface of a curved branch, it doesn't notice that the track it has covered is indeed curved. I was lucky enough to notice what the beetle didn't notice.

Albert Einstein

In everyday life, if an object shows some degree of curvature (like the surface of a bowl, for example) we know about the curvature because the object is **embedded** in the three-dimensional Euclidean space of our everyday experience.[1] Mathematically, we can describe the geometry of the two-dimensional surface of the bowl using the three coordinates of the Euclidean space. It might therefore seem that a natural way to describe curved spacetime would be to embed it in a higher dimensional Euclidean space. However, there is absolutely no evidence that this is how Nature works and so, to avoid introducing unnecessary structure to the geometry of spacetime, we seek a way to describe curved space using only the coordinates available to a geometer **confined** to that space. Our fate, as geometers of a spacetime that we cannot step outside of is, therefore, a little like that of a beetle walking on the curved surface of the bowl. If the beetle is unable to escape the surface, then the coordinates it uses to measure lengths between points on the bowl will be two-dimensional. Despite this, we can use the notion of embedding to (i) work out the metric of a curved space in terms of those coordinates that a trapped geometer would use, and (ii) given a metric in terms of the coordinates within the surface, to visualize the space by embedding it in a Euclidean space. In order to embed a space, you need more than one more dimension in the Euclidean space than are used in the object's metric.[2]

In order to tackle embedding, we start by addressing a related problem. Say we already have an object in Euclidean space. We can then work out the metric on its surface in terms of the coordinates confined within that surface. We do this by relating the Euclidean coordinates to the internal, or *object* coordinates, and the expression for the metric we then obtain, in terms of the object's coordinates, is known as an **induced metric**.

Recall that the distance between points in Euclidean 3-space $\mathbb{R}^3$ with coordinates $X^\alpha = (X, Y, Z)$ is given by

$$ds^2 = dX^2 + dY^2 + dZ^2. \tag{D.1}$$

[1] Our discussion in this Appendix follows Zee, which can be consulted for further details.

[2] John Nash (1928–2015) is perhaps most famous for his Nobel Memorial Prize-winning work in Economics. His work in geometry includes the Nash embedding theorems, which demonstrate that any Riemannian manifold can be embedded in a Euclidean space. Silvia Nasar's biography of Nash, *A Beautiful Mind*, comes highly recommended, but should not be confused with the film of the same name, with which it bears little resemblance.

This expression generalizes to

$$\mathrm{d}s^2 = \sum_{\alpha=1}^{N} (\mathrm{d}X^\alpha)^2 \tag{D.2}$$

for N-dimensional Euclidean space $\mathbb{R}^N$. In what follows, we shall embed objects from D-dimensional spaces into $\mathbb{R}^N$. There are two methods to consider to find the induced metric.

Method I is suitable for when we have an equation for each of the Euclidean coordinates X^α in terms of the coordinates of the D-dimensional object x^μ. We write the coordinates $X^\alpha(x^1...x^D)$, where $\alpha = 1,...,N$. We start by saying

$$X^\alpha + \mathrm{d}X^\alpha = X^\alpha + \frac{\partial X^\alpha}{\partial x^\mu}\mathrm{d}x^\mu. \tag{D.3}$$

Then we use eqn D.1 to say

$$\begin{aligned}
\mathrm{d}s^2 &= \sum_\alpha (\mathrm{d}X^\alpha)^2 \\
&= \sum_\alpha \frac{\partial X^\alpha}{\partial x^\mu}\mathrm{d}x^\mu \frac{\partial X^\alpha}{\partial x^\nu}\mathrm{d}x^\nu \\
&= g_{\mu\nu}\mathrm{d}x^\mu\mathrm{d}x^\nu,
\end{aligned} \tag{D.4}$$

where, in the final line, we've noted the general rule that $\mathrm{d}s^2 = g_{\mu\nu}\mathrm{d}x^\mu\mathrm{d}x^\nu$. We deduce that the relationship between the metric in the object space $g_{\mu\nu}$ is related to the embedding coordinates via

$$g_{\mu\nu} = \sum_\alpha \frac{\partial X^\alpha}{\partial x^\mu}\frac{\partial X^\alpha}{\partial x^\nu}. \tag{D.5}$$

Method II is suitable for use when you can express the shape of the object you're trying to represent in Euclidean coordinates. Again we start with the metric line element in Euclidean space $\mathrm{d}s^2 = \mathrm{d}X^2 + \mathrm{d}Y^2 + \mathrm{d}Z^2$. We then write an equation describing the object $Z = f(X,Y)$ in terms of the embedding coordinates, which allows us to eliminate a coordinate Z. We then require an educated guess at some good object-space coordinates x^μ to reduce the metric to something convenient.

Example D.1

(a) Consider a circle. If we confine ourselves to the surface of the circle it is a one-dimensional space, so $D = 1$. The coordinate of particular points on the circle can be specified by giving an object coordinate θ.

Start with method I. We can mount a circle in Euclidean ($N = 2$)-space (see Fig. D.1) by writing

$$X = a\cos\theta \quad \text{and} \quad Y = a\sin\theta, \tag{D.6}$$

where a is the radius of the circle in the Euclidean space. We have therefore written a relationship between the circle's coordinate θ and the embedding coordinates X^α. Differentiating, we have

$$\frac{\partial X}{\partial \theta} = -a\sin\theta \quad \text{and} \quad \frac{\partial Y}{\partial \theta} = a\cos\theta. \tag{D.7}$$

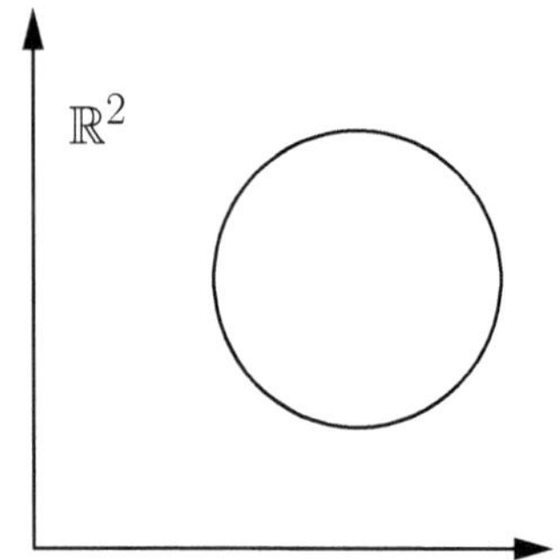

Fig. D.1 Circle embedded in $\mathbb{R}^2$.

This enables us to read out the metric component using eqn D.5 as

$$g_{\theta\theta} = a^2 \sin^2 \theta + a^2 \cos^2 \theta = a^2, \tag{D.8}$$

and we have the result

$$ds^2 = g_{\theta\theta}\, d\theta\, d\theta = a^2 d\theta^2. \tag{D.9}$$

Now try method II. We write

$$ds^2 = dX^2 + dY^2 = dX^2 \left[1 + \left(\frac{\partial Y}{\partial X} \right)^2 \right]. \tag{D.10}$$

The equation for the circle in embedding coordinates is $Y = (a^2 - X^2)^{\frac{1}{2}}$, so we have

$$\frac{\partial Y}{\partial X} = \frac{-X}{(a^2 - X^2)^{\frac{1}{2}}}, \tag{D.11}$$

leading to

$$ds^2 = dX^2 \left(1 + \frac{X^2}{(a^2 - X^2)} \right). \tag{D.12}$$

Now, if we let $X = a \cos\theta$ and $dX = -a \sin\theta\, d\theta$, we find (as before)

$$ds^2 = a^2 d\theta^2. \tag{D.13}$$

(b) Let's now go up a dimension and embed the unit $D = 2$ sphere in $\mathbb{R}^3$ (see Fig. D.2). We need two object-space coordinates $(x^1, x^2) = (\theta, \phi)$. We know a parametrization in $\mathbb{R}^3$ using spherical coordinates with $r = 1$ we write $(X, Y, Z) = (\sin\theta \cos\phi, \sin\theta \sin\phi, \cos\theta)$. We plug in to find the metric induced by the embedding

$$
\begin{aligned}
g_{\theta\theta} &= \left(\frac{\partial X}{\partial \theta} \right)^2 + \left(\frac{\partial Y}{\partial \theta} \right)^2 + \left(\frac{\partial Z}{\partial \theta} \right)^2 \\
&= \cos^2 \theta \cos^2 \phi + \cos^2 \theta \sin^2 \phi + \sin^2 \theta = 1, \tag{D.14} \\
g_{\phi\phi} &= \left(\frac{\partial X}{\partial \phi} \right)^2 + \left(\frac{\partial Y}{\partial \phi} \right)^2 + \left(\frac{\partial Z}{\partial \phi} \right)^2 = \sin^2 \theta \\
&= \sin^2 \theta \sin^2 \phi + \sin^2 \theta \cos^2 \phi = \sin^2 \theta, \tag{D.15}
\end{aligned}
$$

and find $g_{\phi\theta} = 0$. We conclude that the $D = 2$ sphere is described by the induced line element

$$ds^2 = d\theta^2 + \sin^2 \theta\, d\phi^2. \tag{D.16}$$

Let's now try method II. The equation of the spherical surface in $\mathbb{R}^3$ is $X^2 + Y^2 + Z^2 = 1$. Eliminating Z we have

$$ds^2 = dX^2 + dY^2 + \frac{(X dX + Y dY)^2}{1 - X^2 - Y^2}. \tag{D.17}$$

Now to check for consistency with some well-chosen coordinates. The metric looks cylindrically symmetric, so we replace (X, Y) with cylindrical coordinates. Use $X = r \cos\theta$, $Y = r \sin\theta$, $X^2 + Y^2 = r^2$ and $X dX + Y dY = r dr$. We find

$$ds^2 = \frac{dr^2}{1 - r^2} + r^2 d\phi^2. \tag{D.18}$$

This looks different to the method-I answer, until we realize we can write $r = \sin\theta$, then $dr^2 = \cos^2 \theta d\theta^2$ and so $ds^2 = d\theta^2 + \sin^2 \theta d\phi^2$ as we had before.

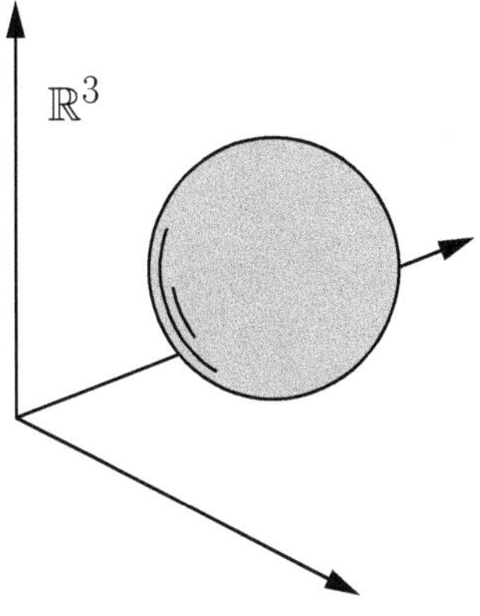

Fig. D.2 The 2-sphere, embedded in $\mathbb{R}^3$.

Let's now turn to the problem of visualizing metrics. Often we are faced with the situation where we are presented with a line element and we want to know what the surface looks like when it's embedded in a higher dimensional Euclidean space. The next example shows an example of how to do this for the Einstein–Rosen bridge.

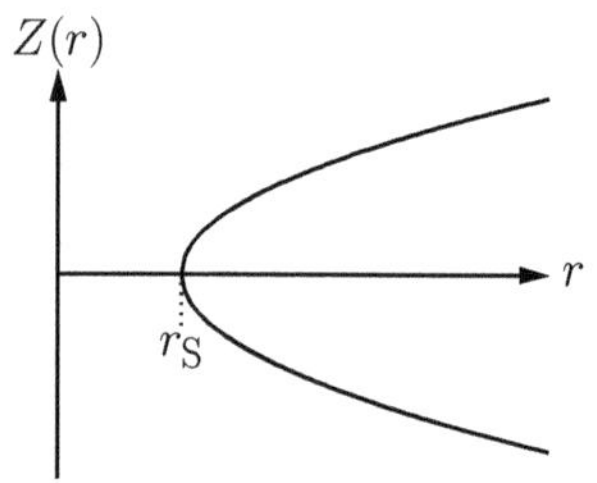

Fig. D.3 The function $Z(r) = \pm [4r_{\mathrm{S}}(r - r_{\mathrm{S}})]^{\frac{1}{2}}$.

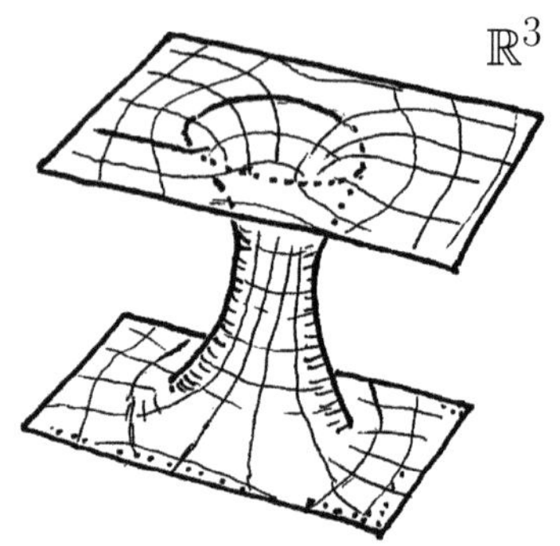

Fig. D.4 The wormhole embedded in $\mathbb{R}^3$.

[3]These regions are called **asymptotically flat**: they are flat at large distances away from the throat.

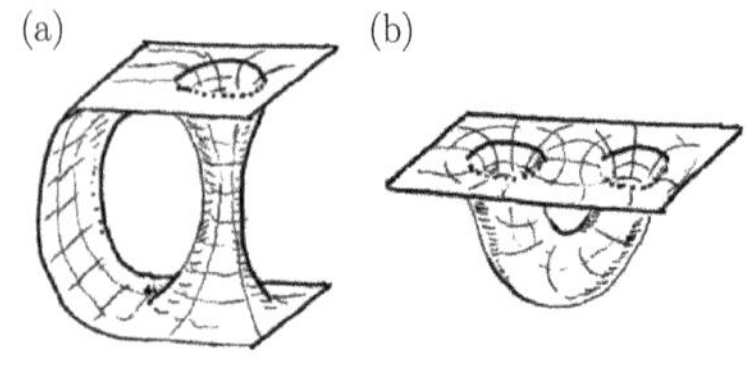

Fig. D.5 (a) The wormhole embedded in $\mathbb{R}^3$ showing a possible spacetime topology. (b) A topologically identical spacetime, showing the wormhole linking two distant points in spacetime.

[4]Like the Alcubierre warp drive from Chapter 5, the creation of a wormhole requires negative energies.

Example D.2

The Einstein–Rosen bridge (also known as a wormhole) is described by a $D = 2$ metric

$$ds^2 = \frac{r\,dr^2}{r - r_{\mathrm{S}}} + r^2 d\phi^2, \tag{D.19}$$

where r_{S} is a constant length. Let's embed this into $\mathbb{R}^3$ using method I. We require

$$\left(\frac{\partial X^1}{\partial r}\right)^2 + \left(\frac{\partial X^2}{\partial r}\right)^2 + \left(\frac{\partial X^3}{\partial r}\right)^2 = g_{rr} = \frac{r}{r - r_{\mathrm{S}}},$$

$$\left(\frac{\partial X^1}{\partial \theta}\right)^2 + \left(\frac{\partial X^2}{\partial \theta}\right)^2 + \left(\frac{\partial X^3}{\partial \theta}\right)^2 = g_{\phi\phi} = r^2. \tag{D.20}$$

We try a set of coordinates $(X^1, X^2, X^3) = (r\cos\theta, r\sin\theta, Z(r))$, which solve the second equation, as can be easily seen. The first equation then becomes

$$1 + \left(\frac{\partial Z(r)}{\partial r}\right)^2 = \frac{r}{r - r_{\mathrm{S}}}, \tag{D.21}$$

which is solved if we set $Z(r)^2 = 4r_{\mathrm{S}}(r - r_{\mathrm{S}})$. The function $Z(r)$ is graphed in Fig. D.3. If it is rotated about the z-axis it generates the spacetime shown in Fig. D.4. The wormhole metric, embedded in $\mathbb{R}^3$ appears to show two regions of flat[3] spacetime joined by a throat. (The narrow hole down the middle explains the name wormhole, of course.) The metric tells us about the detailed geometry of spacetime, but tells us nothing about the **topology** of the spacetime, that is, the large scale shape and connectivity of the space. As a result, the wormhole might link two completely different parts of a spacetime. One possibility is shown in Fig. D.5(a), where the wormhole links two distant points in a spacetime. A topologically identical diagram is shown in Fig. D.5(b), revealing that two distant patches of flat spacetime are linked by the wormhole. If the distance between the two regions via the flat space is greater than the distance through the throat, then we potentially have a shortcut between very distant points in spacetime. It has also been suggested that the wormhole might joint two disconnected parts of spacetime, allowing an explorer to visit a different Universe entirely.[4]

There are some surfaces that cannot be embedded in Euclidean space. The most important example for us is hyperbolic space.

Example D.3

Consider a space described by a metric

$$ds^2 = d\chi^2 + \sinh^2\chi\,d\phi^2. \tag{D.22}$$

Using the method-I rule, we need

$$1 = \left(\frac{\partial X}{\partial \chi}\right)^2 + \left(\frac{\partial Y}{\partial \chi}\right)^2 + \left(\frac{\partial Z}{\partial \chi}\right)^2,$$

$$\sinh^2\chi = \left(\frac{\partial X}{\partial \phi}\right)^2 + \left(\frac{\partial Y}{\partial \phi}\right)^2 + \left(\frac{\partial Z}{\partial \phi}\right)^2. \tag{D.23}$$

We can try some coordinates for X and Y that retain the cylindrical symmetry of this space, and determine the Z coordinate. We try

$$X = \sinh\chi\cos\phi, \quad Y = \sinh\chi\sin\phi. \tag{D.24}$$

Plugging in to the second equation, we see that $\partial Z/\partial\phi = 0$, so Z is indeed cylindrically symmetric. The first equation tells us that

$$\frac{\partial Z}{\partial \chi} = \left(1 - \cosh^2\chi\right)^{\frac{1}{2}} = \left(-\sinh^2\chi\right)^{\frac{1}{2}}. \tag{D.25}$$

This equation has no real solutions. The embedding has failed.

This embedding is indeed impossible. There are two options, the first is simply to embed part of the surface.[5] Perhaps a more satisfactory solution is not to mount the space in Euclidean space, but rather in some other space. For hyperbolic spaces, the natural higher dimensional space is known as pseudo-Euclidean space which has a metric

$$ds^2 = dX^2 + dY^2 - dW^2. \tag{D.26}$$

[5]See Exercise D.4.

[6]See exercise D.5 for a generalization of method I to this case.

Example D.4

(a) Consider a surface $X^2 + Y^2 - W^2 = -1$, known as a 2-sheet hyperboloid. This can be embedded in pseudo-Euclidean space.[6] Use the coordinates

$$X = \sinh\chi\cos\theta, \quad Y = \sinh\chi\sin\theta, \quad W = \cosh\chi, \tag{D.27}$$

which are consistent with the equation for the surface. Differentiating, we find that

$$\begin{aligned} ds^2 &= dX^2 + dY^2 - dW^2 \\ &= d\chi^2 + \sinh^2\chi\, d\theta^2. \end{aligned} \tag{D.28}$$

This is the metric we considered before. We conclude that the embedding is possible in the pseudo-Euclidean metric. The surface defined by these coordinates is shown in the Fig. D.6.

(b) Now try the surface $X^2 + Y^2 - W^2 = 1$, known as the one-sheet hyperboloid. This one is solved for the coordinate choice

$$X = \cosh\chi\cos\theta, \quad Y = \cosh\chi\sin\theta, \quad W = \sinh\chi, \tag{D.29}$$

resulting in the metric

$$ds^2 = -d\chi^2 + \cosh^2\chi d\theta^2. \tag{D.30}$$

Interestingly, we can turn this into something more familiar by setting $\cosh\chi = r$, from which we find

$$ds^2 = \frac{dr^2}{1 - r^2} + r^2 d\theta^2, \tag{D.31}$$

which is simply the metric for the 2-sphere we had before. In fact, we could have generated this metric directly with the choice of coordinates

$$X = r\cos\theta, \quad Y = r\sin\theta, \quad W = \left(r^2 - 1\right)^{\frac{1}{2}}. \tag{D.32}$$

The spherical line element that represents the one-sheet hyperboloid should not be misinterpreted to mean that the shape of the space is a sphere. We have eliminated W leaving a set of concentric circles behind, with distances encoded in the difference between the circles. (Put another way, we have sliced the spacetime leaving circular cross sections.) If we set $r = \sin\theta$ as we did for the 2-sphere then we would not be able to solve the equation that defines the surface: $r^2 - W^2 = 1$. Setting $r = \cosh\chi$ does not cause this problem. See Chapter 18 for more details on this space which is very useful in cosmology.

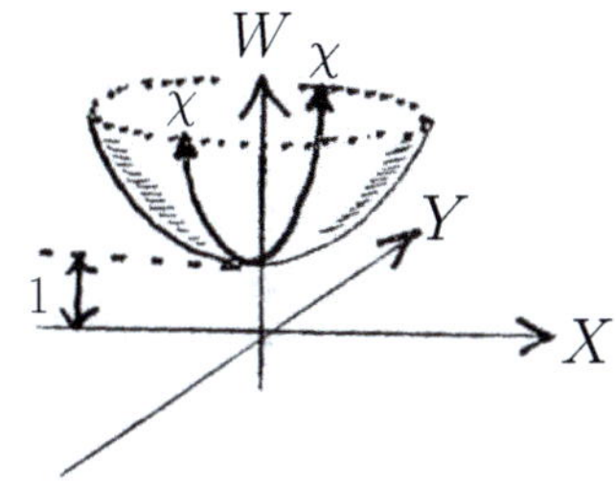

Fig. D.6 The 2-sheet hyperboloid embedded in pseudo-Euclidean space.

Chapter summary

- Embedding allows us to visualize a metric space by mounting it in a higher dimensional Euclidean space.

Exercises

(D.1) A parabolic 2-surface in Euclidean 3-space (see Fig. D.7) is described by

$$Z = \frac{a}{2}(X^2 + Y^2). \qquad \text{(D.33)}$$

Using method II, show that this leads to the metric

$$ds^2 = (1 + a^2 r^2)dr^2 + r^2 d\theta^2. \qquad \text{(D.34)}$$

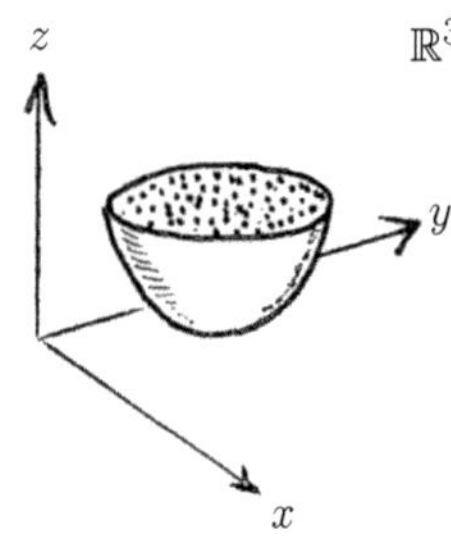

Fig. D.7 A parabolic 2-surface embedded in $\mathbb{R}^3$.

(D.2) Embed the $D = 3$ sphere in $\mathbb{R}^4$ using method II.

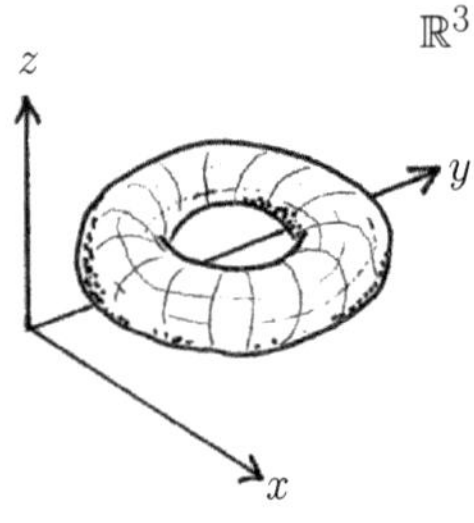

Fig. D.8 The torus embedded in $\mathbb{R}^3$.

(D.3) The $D = 2$ torus may be embedded in $\mathbb{R}^3$ (Fig. D.8) using the equation for its surface

$$\left[c - \left(X^2 + Y^2\right)^{\frac{1}{2}} \right]^2 + Z^2 = a^2. \qquad \text{(D.35)}$$

Use method I to determine its induced metric.

(D.4) *We saw in Example D.3 how embedding fails for a hyperbolic surface. However, we can embed part of this surface. Consider an alternative metric for a hyperbolic surface*

$$ds^2 = d\chi^2 + \cosh^2 \chi \, d\phi^2. \qquad \text{(D.36)}$$

Using method I show that we have

$$\frac{\partial Z}{\partial \chi} = \left(1 - \sinh^2 \chi\right)^{\frac{1}{2}}. \qquad \text{(D.37)}$$

This equation does enjoy real solutions for $\sinh \chi < 1$, corresponding to the range $0 \leq \chi \leq \sinh^{-1} 1$. So a partial embedding of this surface is possible. This wasn't the hyperbolic line element we started with, but like that line element, it does have a constant negative curvature. **Minding's theorem** *[named after Ferdinand Minding (1806–1885)] says that all constant-curvature surfaces have the same local geometry, and so the geometry of this line element is representative of the hyperbolic spacetime.*

(D.5) Consider the Rindler coordinate system $(x^1, x^2) = (t, x)$, embedded in a two-dimensional *Minkowski* space (T, X) via

$$T = x \sinh t, \quad X = x \cosh t. \qquad \text{(D.38)}$$

Compute the components of the induced metric. *Hint: Since we want to work in Minkowski space, rather than Euclidean space, we must use*

$$g_{\mu\nu} = \eta_{\alpha\beta} \frac{\partial X^\alpha}{\partial x^\mu} \frac{\partial X^\beta}{\partial x^\nu}. \qquad \text{(D.39)}$$

Answers to selected problems

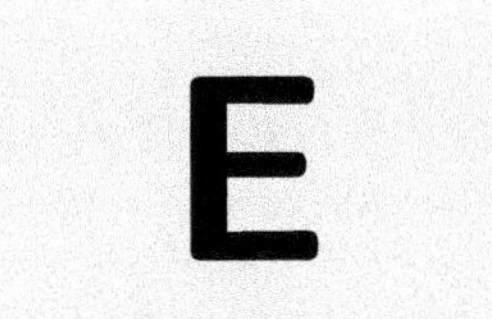

(0.1) Equating kinetic energy $\frac{1}{2}mv_{\text{esc}}^2$ to the gravitational potential energy GMm/r yields $v_{\text{esc}} = \sqrt{2GM/r}$ and since $\Phi = -GMm/r$ then we also have that $v_{\text{esc}} = \sqrt{2|\Phi|}$. Rearranging for $v_{\text{esc}} = c$ gives the Schwarzschild radius $r = 2GM/c^2$.

(0.2) (i) Earth: $9.8\,\text{m s}^{-2}$; $11.2\,\text{km s}^{-1}$ $(\approx 0.00004c)$; (ii) Sun: $274\,\text{m s}^{-2}$; $6.2{\times}10^5\,\text{m s}^{-1}$ $(\approx 0.002c)$; (iii) neutron star: $1.9 \times 10^{12}\,\text{m s}^{-2}$; $1.9 \times 10^8\,\text{m s}^{-1}$ $(\approx 0.64c)$.

(0.3) The inverse square law of force in Newtonian gravity implies that $|\Delta F|/|F| = 2|\Delta r|/|r|$ and so with $\Delta r = 1.8$ m we have that for the Earth the effect of the tidal force is $|\Delta F|/|F| = 2 \times 1.8\,\text{m}/R_\oplus$ which is around 0.6 ppm, i.e. the tidal force is less than a millionth of gravitational force, which is why you don't notice it. The Schwarzschild radius for a $3M_\odot$ black hole is about 9 km, so $|\Delta F|/|F| \approx 0.0004$ (though the tidal forces are enormous since the gravitational field is more than eleven orders of magnitude bigger than it is on Earth). The Schwarzschild radius for a $10^6 M_\odot$ black hole is about three million km, and hence $|\Delta F|/|F| \approx 4 \times 10^{-10}$. The tidal forces are smaller in this case since the gravitational field due to the black hole is only about six orders of magnitude greater than on Earth.

(At the Schwarzschild radius, since $r \propto M$ then $|\Delta F| \propto F/r$ and since $F \propto M/r^2$ then $|\Delta F| \propto M^{-2}$, so that tidal forces *at the Schwarzschild radius* are larger for small black holes than for supermassive black holes.)

(1.2) $\gamma = (1-0.995^2)^{-1/2} \approx 10$. (a) $0.995c{\times}2.2\,\mu\text{s}= 0.66$ km. (b) Multiply 0.66 km by γ: 6.6 km.

(1.3) The interval is given by $\Delta s^2 = -c^2\Delta t^2 + \Delta x^2$. The time interval Δt is given by the difference between the time of events q and p as measured by us using coordinates (t,x). Take the time coordinate of event r to be the origin. This means the event p takes place at a time $t_p = \tau_2$. Since the light takes a time $\tau_2 + \tau_1$ to reach q and return, we reason that the event q takes place at a time $t_q = (\tau_2 + \tau_1)/2$. As a result we have that

$$\Delta t = t_q - t_p = \frac{(\tau_2 + \tau_1)}{2} - \tau_2$$
$$= \frac{(\tau_1 - \tau_2)}{2}. \tag{E.1}$$

In order to work out the interval in position Δx, we note that the distance travelled from our world line to the event q is measured by the light pulse as $\Delta x = c(\tau_1 + \tau_2)/2$. Plugging into the expression for the interval Δs^2, we obtain

$$\Delta s^2 = - c^2\Delta t^2 + \Delta x^2$$
$$= -\frac{c^2\,(\tau_1 - \tau_2)^2}{4} + \frac{c^2\,(\tau_1 + \tau_2)^2}{4}$$
$$= c^2\tau_1\tau_2. \tag{E.2}$$

(1.4) (a) 0.140, (b) 0.417, (c) 0.866, (d) 0.996, (e) 0.99995.

(2.1) This is a straightforward application of eqn 2.38.

(2.2) Write $\boldsymbol{u} = \gamma(\vec{u})(1,\vec{u})$ and $\boldsymbol{v} = \gamma(\vec{v})(1,\vec{v})$, so that $\boldsymbol{u} \cdot \boldsymbol{v} = \gamma(\vec{u})\gamma(\vec{v})(-1 + \vec{u} \cdot \vec{v})$ and this is equal to $-\gamma$, yielding the result. In the non-relativistic limit, $\gamma = (1 - v_{\text{rel}}^2)^{-1/2} = 1 + v_{\text{rel}}^2/2 + \cdots$ and hence

$$1 + \frac{v_{\text{rel}}^2}{2} + \cdots = (1 + \frac{u^2}{2} + \cdots)(1 + \frac{v^2}{2} + \cdots)(1 - \vec{u} \cdot \vec{v}) \tag{E.3}$$

$$= 1 + \frac{(\vec{u} - \vec{v})^2}{2} + \cdots, \tag{E.4}$$

and the two results follow after taking the square root. This is just the Galilean relative velocity, and the two results are found because we have not specified whether the relative velocity is u relative to v or the other way around.

(2.3) The first answers follow by simple matrix multiplication using $\eta_{\mu\nu}$ given in eqn 2.15. Some of the later answers are given in the main text of the chapter.

(2.4) The proton has $\gamma = 10^{19}/10^9 = 10^{10}$ and so is travelling very close to the speed of light. It travels 10^5 light-years in 10^5 years. In its rest frame, the galaxy is contracted by a factor γ and so the journey is only 10^{-5} light-years long, taking 10^{-5} years, or about 5 minutes.

(2.5) The Lorentz transformation gives $p^\mu = (\gamma m, \gamma m v, 0, 0)$ and so one can just read off $E = \gamma m$ (i.e. $E = \gamma mc^2$ if you put the factors of c back in) and $p^x = \gamma m v$. Both sets of components give $\boldsymbol{p} \cdot \boldsymbol{p} = -(p^0)^2 + (p^1)^2 + (p^2)^2 + (p^3)^2 = -m^2$.

(2.6) Assume that we can have $\text{e} + \gamma \to \text{e}$, so that

$$\boldsymbol{p}_{\text{e}} + \boldsymbol{p}_\gamma = \boldsymbol{p}_{\text{e}}'. \tag{E.5}$$

If we square both sides, we get

$$\boldsymbol{p}_{\text{e}} \cdot \boldsymbol{p}_{\text{e}} + 2\boldsymbol{p}_{\text{e}} \cdot \boldsymbol{p}_\gamma + \boldsymbol{p}_\gamma \cdot \boldsymbol{p}_\gamma = \boldsymbol{p}_{\text{e}}' \cdot \boldsymbol{p}_{\text{e}}', \tag{E.6}$$

but $\boldsymbol{p}_e \cdot \boldsymbol{p}_e = \boldsymbol{p}'_e \cdot \boldsymbol{p}'_e = -m_e^2$ and $\boldsymbol{p}_\gamma \cdot \boldsymbol{p}_\gamma = 0$, so we are left with $2\boldsymbol{p}_e \cdot \boldsymbol{p}_\gamma = 0$. We can evaluate this in the rest frame of the electron where $\boldsymbol{p}_e = (m_e, \vec{0})$ and in general the photon has $\boldsymbol{p}_\gamma = (E, \vec{p})$ with $E = |\vec{p}|$. So we are left with $Em_e = 0$, which means that $E = 0$, i.e. only holds for (non-existent) zero-energy photons.

(2.7) Here the coordinate q of interest is $q = \phi$, the amplitude of the string's displacement. However, we're not interested in ϕ as a function of time, but rather of length x along the string. This is no problem, we simply proceed with $\dot{\phi} = \partial\phi/\partial x$. Instead of the action, the functional we want to minimize is the energy $E[\phi]$. There are two contributions to the energy, the strain $\propto d\phi/dx$ tells us it costs energy for the string to bend, the gravitational potential energy $V = -\rho g\phi(x)$ shows that the energy is reduced the large x. The energy functional is

$$E(\phi) = \int dx \left[\frac{T}{2} \left(\frac{d\phi(x)}{dx} \right)^2 - \rho g\phi(x) \right]. \quad \text{(E.7)}$$

Notice that this is in the same form as the mechanical action, with the first term playing the role of the kinetic energy. Feeding this through the Euler Lagrange equations, we find

$$\frac{\partial E}{\partial \left(\frac{d\phi}{dx} \right)} = T\frac{d\phi}{dx},$$

$$\frac{d}{dx}\frac{\partial E}{\partial \left(\frac{d\phi}{dx} \right)} = T\frac{d^2\phi}{dx^2}, \quad \text{(E.8)}$$

and

$$\frac{\partial E}{\partial \phi} = -\rho g. \quad \text{(E.9)}$$

This gives the equation of motion

$$\frac{d^2\phi}{dx^2} = -\frac{\rho g}{T}. \quad \text{(E.10)}$$

The solution tells us the shape of the hanging string, which is a parabola. This can be written as

$$\phi(x) = \frac{\rho g}{2T}\left[\left(\frac{L}{2}\right)^2 - x^2 \right]. \quad \text{(E.11)}$$

(2.8) We have that $p^i = p_i = \frac{\partial L}{\partial v^i} = \frac{\partial}{\partial v^i}(-m\sqrt{1 - v^j v^j}) = -\frac{m}{2}\frac{-2v^i}{\sqrt{1-v^j v^j}}$, and hence $p^i = m\gamma v^i$.

(2.9) $H = \vec{p}\cdot\vec{v} - L = m\gamma v^2 + \frac{m}{\gamma} = m\gamma(v^2 + 1 - v^2) = \gamma m$.

(3.2) We write

$$\frac{\partial \boldsymbol{e}_r}{\partial \theta} = \frac{\partial}{\partial\theta}(\cos\theta\boldsymbol{e}_x + \sin\theta\boldsymbol{e}_y)$$

$$= -\sin\theta\boldsymbol{e}_x + \cos\theta\boldsymbol{e}_y = \frac{\boldsymbol{e}_\theta}{r}. \quad \text{(E.12)}$$

We also work out that

$$\frac{\partial \boldsymbol{e}_\theta}{\partial \theta} = \frac{\partial}{\partial\theta}(-r\sin\theta\boldsymbol{e}_x + r\cos\theta\boldsymbol{e}_y)$$

$$= -r\cos\theta\boldsymbol{e}_x - r\sin\theta\boldsymbol{e}_y = -r\boldsymbol{e}_r. \quad \text{(E.13)}$$

The final two results follow from taking derivatives with respect to r

$$\frac{\partial \boldsymbol{e}_r}{\partial r} = 0, \quad \frac{\partial \boldsymbol{e}_\theta}{\partial r} = \frac{\boldsymbol{e}_\theta}{r}. \quad \text{(E.14)}$$

(3.3) Expanding $\sin(r/\sqrt{K})$ for $r \ll \sqrt{K}$ gives $C = 2\pi r = (1 - \frac{Kr^2}{6} + \dots)$ and rearranging gives the desired result.

(3.5) Substituting we obtain

$$ds^2 = -dt'^2\left[1 - \Omega^2(x'^2 + y'^2)\right] + dx'^2 + dy'^2 + dz'^2 - 2\Omega y'dx'dt' + 2\Omega x'dy'dt'. \quad \text{(E.15)}$$

(4.1) Since $\boldsymbol{u}$ and $\boldsymbol{v}$ transform as 4-vectors, the tensor transformation law gives $\Lambda^{\alpha'}{}_\alpha\Lambda^{\beta'}{}_\beta W^{\alpha\beta} = \Lambda^{\alpha'}{}_\alpha\Lambda^{\beta'}{}_\beta u^\alpha v^\beta = u^{\alpha'}v^{\beta'} = W^{\alpha'\beta'}$ as required.

(4.2) $\Lambda^{\alpha'}{}_\alpha\Lambda^\beta{}_{\beta'}\delta^\alpha{}_\beta = \Lambda^{\alpha'}{}_\beta\Lambda^\beta{}_{\beta'} = \delta^{\alpha'}{}_{\beta'}$ as required for it to transform properly. We can use the metric tensor to raise and lower the indices in $\delta^{\alpha\beta}$ and $\delta_{\alpha\beta}$ as appropriate to turn them into $\delta^\alpha{}_\beta$ (which we've just shown works as a tensor)

$$\delta^{\alpha\beta} = \eta^{\alpha\gamma}\delta^\beta{}_\gamma = \eta^{\alpha\beta},$$

$$\delta_{\alpha\beta} = \eta_{\alpha\gamma}\delta^\gamma{}_\beta = \eta_{\alpha\beta}. \quad \text{(E.16)}$$

Thus, if you try and write $\delta_{\alpha\beta}$ and $\delta^{\alpha\beta}$ as tensors, they end up being equal to the metric tensor and lose their Kronecker-delta property. Thus, it is best to keep $\delta_{\alpha\beta}$ and $\delta^{\alpha\beta}$ as non-tensor objects which are shorthands for quantities equalling one when $\alpha = \beta$ and zero otherwise. If you want the Kronecker delta to behave as a tensor, you have to use the mixed form $\delta^\alpha{}_\beta$.

(4.3) $\boldsymbol{S}$ is a tensor and so its components transforms properly

$$S^{\mu'}{}_{\nu'} = \Lambda^{\mu'}{}_\mu\Lambda^\nu{}_{\nu'}S^\mu{}_\nu. \quad \text{(E.17)}$$

Thus,

$$S^{\mu'}{}_{\mu'} = \Lambda^{\mu'}{}_\mu\Lambda^\nu{}_{\mu'}S^\mu{}_\nu = \delta^\nu{}_\mu S^\mu{}_\nu = S^\mu{}_\mu. \quad \text{(E.18)}$$

(4.4) (a) Start with the relationships

$$r = x^2 + y^2 + z^2,$$

$$\cos\theta = \frac{z}{r},$$

$$\tan\phi = \frac{y}{x}. \quad \text{(E.19)}$$

Using the transformation law

$$\boldsymbol{\omega}^\mu = \left(\frac{\partial x^\mu}{\partial x^\nu} \right)\boldsymbol{\omega}^\nu, \quad \text{(E.20)}$$

we find basis 1-forms

$$\boldsymbol{\omega}^r = \sin\theta\cos\phi\,\boldsymbol{\omega}^x + \sin\theta\sin\phi\,\boldsymbol{\omega}^y + \cos\theta\boldsymbol{\omega}^z,$$

$$\boldsymbol{\omega}^\theta = \frac{\cos\theta\cos\phi}{r}\boldsymbol{\omega}^x + \frac{\cos\theta\sin\phi}{r}\boldsymbol{\omega}^y - \frac{\sin\theta}{r}\boldsymbol{\omega}^z,$$

$$\boldsymbol{\omega}^\phi = -\frac{\sin\phi}{r\sin\theta}\boldsymbol{\omega}^x + \frac{\cos\phi}{r\sin\theta}\boldsymbol{\omega}^y. \quad \text{(E.21)}$$

(b) This time we start with

$$x = r\sin\theta\cos\phi,$$

$$y = r\sin\theta\sin\phi,$$

$$z = r\cos\theta. \quad \text{(E.22)}$$

Using the transformation law for basis vectors $e_\mu = (\partial x^\nu/\partial x^\mu)e_\nu$, we find that the basis vectors are

$$e_r = \sin\theta\cos\phi\, e_x + \sin\theta\sin\phi\, e_y + \cos\theta e_z,$$
$$e_\theta = \cos\theta\cos\phi\, e_x + \cos\theta\sin\phi\, e_y - \sin\theta\, e_z,$$
$$e_\phi = -r\sin\theta\sin\phi\, e_x + r\sin\theta\cos\phi\, e_y. \qquad \text{(E.23)}$$

(c) Notice that the components for the vector e_r and 1-form ω^r are the same; those for θ related by a factor r and those for ϕ by a factor $r^2\sin^2\theta$. The orthogonality condition $\langle\omega^\mu, e_\nu\rangle = \delta^\mu{}_\nu$ is then guaranteed by the orthogonality of the basis vectors e_ν (which can be checked).

(4.5) (a) Using $e_u = (\partial x^\mu/\partial u)e_\mu$, and so on, we find

$$e_u = e_x + e_y - 2ve_z,$$
$$e_v = -e_x + e_y - 2ue_z,$$
$$e_w = e_z. \qquad \text{(E.24)}$$

(b) Using $\omega^u = (\partial u/\partial x^\mu)\omega^\mu$ and so on, we have

$$\omega^u = \frac{1}{2}\omega^x + \frac{1}{2}\omega^y,$$
$$\omega^v = -\frac{1}{2}\omega^x + \frac{1}{2}\omega^y,$$
$$\omega^w = -x\omega^x + y\omega^y + \omega^z. \qquad \text{(E.25)}$$

Note that the system is not orthogonal.

(4.8) The transformation law is

$$\begin{pmatrix} \omega' \\ k^{x'} \\ k^{y'} \\ k^{z'} \end{pmatrix} = \begin{pmatrix} \gamma & -\beta\gamma & 0 & 0 \\ -\beta\gamma & \gamma & 0 & 0 \\ 0 & 0 & 1 & 0 \\ 0 & 0 & 0 & 1 \end{pmatrix} \begin{pmatrix} \omega \\ k^x \\ k^y \\ k^z \end{pmatrix}. \qquad \text{(E.26)}$$

We obtain

$$\omega' = \gamma(\omega - \beta k^x). \qquad \text{(E.27)}$$

Setting $k^x = |\vec{k}|\cos\alpha = \omega\cos\alpha$ for a light source whose direction of emission makes an angle α with the x direction, we have

$$\omega' = \gamma\omega(1 - \beta\cos\alpha). \qquad \text{(E.28)}$$

If the source is in the primed frame where it is measured to have a frequency ω_0, we have

$$\frac{\omega}{\omega_0} = \frac{(1-\beta^2)^{\frac{1}{2}}}{1-\beta\cos\alpha}. \qquad \text{(E.29)}$$

(5.3) (b) Dropping primes, setting $ds^2 = 0$ and dividing through by dt^2 we find, in the x-t plane, that

$$1 - v^2 = \left(\frac{dx}{dt}\right)^2 - 2v\frac{dx}{dt}. \qquad \text{(E.30)}$$

Solving this quadratic equation, we obtain an equation for the light cones of

$$\left(\frac{dx}{dt}\right) = v \pm 1, \qquad \text{(E.31)}$$

which tells us that the light cones make an angle $\tan^{-1}(v \pm 1)$ to the t axis.

(c) We obtain the usual expression for time dilation in flat spacetime $d\tau = dt\,(1 - v^2)^{\frac{1}{2}}$.

(d) We find

$$g^{\mu\nu} = \begin{pmatrix} -1 & -v & 0 & 0 \\ -v & 1-v^2 & 0 & 0 \\ 0 & 0 & 1 & 0 \\ 0 & 0 & 0 & 1 \end{pmatrix}. \qquad \text{(E.32)}$$

(5.4) (a) Setting $dt = 0$, we have a proper length interval

$$dl = \frac{a(t)dr}{(1-kr^2)^{\frac{1}{2}}}. \qquad \text{(E.33)}$$

(5.5) (a) $dl = dx$.
(b) $d\tau = x dt$, where x is the position of the event.

(6.1) The time dilation factor is $\sqrt{1 - 2GM/rc^2}$ and the numerical values come out to be (a) $1 - 1.1 \times 10^{-8}$ (b) $1 - 2.1 \times 10^{-6}$ (c) 0.84. All three values are a bit less than 1, but for (c) it's quite a bit less. Note that for the time dilation factor on the surface of the Earth (comparing it to zero gravitational field) you have to include the effect of the Sun's gravitational field at the Earth's surface, so that the answer is obtained by computing $\sqrt{1 - 2G/c^2(M_\oplus/R_\oplus + M_\odot/R)} \approx G/c^2(M_\oplus/R_\oplus + M_\odot/R)$, where R is an astronomical unit.

(6.2) Differentiating the $\sqrt{1 - 2GM_\oplus/rc^2}$ factor gives $-GM_\oplus/r^2c^2 = -g/c^2 = -1.1 \times 10^{-16}$ m^{-1} = -1.1×10^{-19} mm^{-1}, which is consistent with the value given. (Note that, in comparison with the previous exercise, the effect of the Sun is not needed. Although the Sun's gravitational field leads to a substantial time dilation on the Earth's surface, its gradient is much smaller and can be neglected.)

(6.4) (a) $\Delta\tau = \left(1 - \frac{2M}{r}\right)^{\frac{1}{2}} dt$.

(b) $\Delta\tau = \left(1 - \frac{2M}{r_2}\right)^{\frac{1}{2}} dt$, where dt is the coordinate time interval between wavefront events.
(c) Since the observers are parallel, the interval between the wavefronts being received is the same as the interval between which they are emitted. As a result, the observer at r_2 measures an interval $\Delta\tau = \left(1 - \frac{2M}{r_1}\right)^{\frac{1}{2}} dt$.

(6.6) (a) Figure 6.6 shows the set up. The time on A's world line will be

$$x^0 + \frac{dx_1^0 + dx_2^0}{2}. \qquad \text{(E.34)}$$

The two intervals are the two solutions in eqn 6.15, whose sum gives the quoted integrand.
(c) The optical path length is $2\pi r + 2\pi\Omega r^2$. For a counter-propagating beam, the optical path is $2\pi r - 2\pi\Omega r^2$. The fringe shift is the difference in optical path, divided by the wavelength of the light.

(7.1) (i) $\boldsymbol{\nabla}_u\boldsymbol{v} - \boldsymbol{\nabla}_v\boldsymbol{u}$.
(ii) $\boldsymbol{\nabla}_v\tilde{\boldsymbol{w}}(\boldsymbol{u})$.
(iii) In slot-machine notation, we could rewrite the expression $(\boldsymbol{\nabla}_u\boldsymbol{T})(\boldsymbol{v}, \boldsymbol{w})$, becoming $u^\mu T_{\alpha\beta;\mu}v^\alpha w^\beta$.

(7.2) (a) The components of the tangent are given by $u^\mu = (dx/ds, dy/ds)$ and so $u^\mu = (-\sin s, \cos s)$. This vector has unit length.
(b) The curve is not a geodesic, so we do not expect the components of the gradient to vanish. Differentiating, we find $(Du/ds)^\mu = (-\cos s, -\sin s)$, which gives zero

when dotted with $\boldsymbol{u}$.

(c) The tangent vector to the reparametrized curve has components $u^\mu = \left(-t/\sqrt{1-t^2}, 1\right)$, from which we find $\boldsymbol{u} \cdot \boldsymbol{u} = (1-t^2)^{-1}$. The gradient of this vector has components $(\mathrm{D}\boldsymbol{u}/\mathrm{d}t)^\mu = (-(1-t^2)^{-\frac{3}{2}}, 0)$ and so $\boldsymbol{u} \cdot \mathrm{D}\boldsymbol{u}/\mathrm{d}t = t/(1-t^2)^2$.

The fact that the tangent vector does not have a constant magnitude makes this parametrization less convenient than the (affine) length parametrization that uses s.

(7.3) (a) We find

$$\boldsymbol{\nabla}_x \boldsymbol{v} = \begin{pmatrix} 0 \\ C \end{pmatrix}, \quad \boldsymbol{\nabla}_y \boldsymbol{v} = \begin{pmatrix} 0 \\ 0 \end{pmatrix}. \tag{E.35}$$

(b) In cylindrical polars, we have

$$(v^r, v^\theta) = (Cr\sin\theta\cos\theta, C\cos^2\theta). \tag{E.36}$$

(c) The derivatives are

$$\boldsymbol{\nabla}_r \boldsymbol{v} = \begin{pmatrix} C\sin\theta\cos\theta \\ \frac{C}{r}\cos^2\theta \end{pmatrix}, \tag{E.37}$$

and

$$\boldsymbol{\nabla}_\theta \boldsymbol{v} = \begin{pmatrix} -Cr\sin^2\theta \\ -C\cos\theta\sin\theta \end{pmatrix}. \tag{E.38}$$

(d) One option is to compute

$$(\boldsymbol{\nabla}_{\mu'}\boldsymbol{v})^{\nu'} = \Lambda^\alpha{}_{\mu'}\Lambda^{\nu'}{}_\beta(\boldsymbol{\nabla}_\alpha\boldsymbol{v})^\beta, \tag{E.39}$$

for μ' and ν' each taking x and y. The only non-zero component is

$$\begin{aligned}
(\boldsymbol{\nabla}_x\boldsymbol{v})^y &= \Lambda^\alpha{}_x\Lambda^y{}_\beta(\boldsymbol{\nabla}_\alpha\boldsymbol{v})^\beta \\
&= \Lambda^r{}_x\Lambda^y{}_r(\boldsymbol{\nabla}_r\boldsymbol{v})^r + \Lambda^\theta{}_x\Lambda^y{}_r(\boldsymbol{\nabla}_\theta\boldsymbol{v})^r \\
&\quad + \Lambda^r{}_x\Lambda^y{}_\theta(\boldsymbol{\nabla}_r\boldsymbol{v})^\theta + \Lambda^\theta{}_x\Lambda^y{}_\theta(\boldsymbol{\nabla}_\theta\boldsymbol{v})^\theta \\
&= C. \tag{E.40}
\end{aligned}$$

However, it's simplest to work in the other direction and compute $(\boldsymbol{\nabla}_{\nu'}\boldsymbol{v})^{\mu'} = \Lambda^x{}_{\nu'}\Lambda^{\mu'}{}_y(\boldsymbol{\nabla}_x\boldsymbol{v})^y$ for $\mu', \nu' = r, \theta$ and check against the results in (c).

(8.1) (a) The E-L procedure gives

$$\frac{\partial L}{\partial r} = \frac{r}{L}\left(\frac{\mathrm{d}\theta}{\mathrm{d}\lambda}\right)^2, \quad \frac{\partial L}{\partial \theta} = 0.$$

$$\frac{\partial L}{\partial\left(\frac{\mathrm{d}r}{\mathrm{d}\lambda}\right)} = \frac{1}{L}\frac{\mathrm{d}r}{\mathrm{d}\lambda}, \quad \frac{\partial L}{\partial\left(\frac{\mathrm{d}\theta}{\mathrm{d}\lambda}\right)} = \frac{r^2}{L}\frac{\mathrm{d}\theta}{\mathrm{d}\lambda}. \tag{E.41}$$

At this stage we choose length parametrization

$$L = \left[\left(\frac{\mathrm{d}r}{\mathrm{d}\lambda}\right)^2 + r^2\left(\frac{\mathrm{d}\theta}{\mathrm{d}\lambda}\right)^2\right]^{\frac{1}{2}} = 1. \tag{E.42}$$

We obtain two equations,

$$\frac{\mathrm{d}}{\mathrm{d}\lambda}\frac{\partial L}{\partial\left(\frac{\mathrm{d}r}{\mathrm{d}\lambda}\right)} = \frac{\mathrm{d}^2 r}{\mathrm{d}\lambda^2}, \tag{E.43}$$

and

$$\frac{\mathrm{d}}{\mathrm{d}\lambda}\frac{\partial L}{\partial\left(\frac{\mathrm{d}\theta}{\mathrm{d}\lambda}\right)} = \frac{\mathrm{d}}{\mathrm{d}\lambda}\left(r^2\frac{\mathrm{d}\theta}{\mathrm{d}\lambda}\right). \tag{E.44}$$

The two equations of motion follow.

(b) We can integrate the second of these equations to find

$$r^2\frac{\mathrm{d}\theta}{\mathrm{d}\lambda} = a, \tag{E.45}$$

where a is a constant. Inserting this into the length parametrization eqn (E.42) we have

$$\left(\frac{\mathrm{d}r}{\mathrm{d}\lambda}\right)^2 + \frac{a^2}{r^2} = 1, \tag{E.46}$$

whose solution is

$$r^2 = \lambda^2 + a^2. \tag{E.47}$$

Returning to the first equation of motion we find

$$\frac{\mathrm{d}\theta}{\mathrm{d}\lambda} = \frac{a}{\lambda^2 + a^2}, \tag{E.48}$$

whose solution is

$$a\tan(\theta - \theta_0) = \lambda. \tag{E.49}$$

(8.2) (b) Using $x = r\cos\theta$ and $y = r\sin\theta$ we have

$$r = \frac{\gamma}{\alpha\cos\theta + \beta\sin\theta}. \tag{E.50}$$

If we make the substitutions

$$a = \frac{\gamma}{(\alpha^2+\beta^2)^{\frac{1}{2}}}, \quad \cos\theta_0 = \frac{\alpha}{(\alpha^2+\beta^2)^{\frac{1}{2}}},$$
$$\sin\theta_0 = \frac{\beta}{(\alpha^2+\beta^2)^{\frac{1}{2}}},$$

then eqn 8.48 becomes $r = a/\cos(\theta - \theta_0)$, which shows that this is, indeed, a description of a straight line.

(8.3) First evaluate velocity

$$\frac{\mathrm{d}\xi^\mu}{\mathrm{d}\tau} = \frac{\partial\xi^\mu}{\partial x^\nu}\frac{\mathrm{d}x^\nu}{\mathrm{d}\tau}. \tag{E.51}$$

This allows us to evaluate the acceleration

$$\begin{aligned}
\frac{\mathrm{d}^2\xi^\mu}{\mathrm{d}\tau^2} &= \frac{\mathrm{d}x^\sigma}{\mathrm{d}\tau}\frac{\partial}{\partial x^\sigma}\frac{\partial\xi^\mu}{\partial x^\nu}\frac{\mathrm{d}x^\nu}{\mathrm{d}\tau} \\
&= \frac{\mathrm{d}x^\sigma}{\mathrm{d}\tau}\frac{\partial^2\xi^\mu}{\partial x^\nu\partial x^\sigma}\frac{\mathrm{d}x^\nu}{\mathrm{d}\tau} + \frac{\mathrm{d}x^\sigma}{\mathrm{d}\tau}\frac{\partial\xi^\mu}{\partial x^\nu}\frac{\partial}{\partial x^\sigma}\frac{\mathrm{d}x^\nu}{\mathrm{d}\tau} \\
&= \frac{\mathrm{d}x^\sigma}{\mathrm{d}\tau}\frac{\partial^2\xi^\mu}{\partial x^\lambda\partial x^\sigma}\frac{\mathrm{d}x^\lambda}{\mathrm{d}\tau} + \frac{\partial\xi^\mu}{\partial x^\nu}\frac{\mathrm{d}^2 x^\nu}{\mathrm{d}\tau^2} = 0, \tag{E.52}
\end{aligned}$$

where, in the first term of the last line, we sum over λ instead of ν to avoid confusion in the next stage. Tidying, we have

$$\frac{\mathrm{d}^2 x^\nu}{\mathrm{d}\tau^2} + \frac{\partial x^\nu}{\partial\xi^\mu}\frac{\partial^2\xi^\mu}{\partial x^\lambda\partial x^\sigma}\frac{\mathrm{d}x^\lambda}{\mathrm{d}\tau}\frac{\mathrm{d}x^\sigma}{\mathrm{d}\tau} = 0. \tag{E.53}$$

This is

$$\frac{\mathrm{d}^2 x^\nu}{\mathrm{d}\tau^2} + \Gamma^\nu{}_{\lambda\sigma}\frac{\mathrm{d}x^\lambda}{\mathrm{d}\tau}\frac{\mathrm{d}x^\sigma}{\mathrm{d}\tau} = 0, \tag{E.54}$$

with

$$\Gamma^\nu{}_{\lambda\sigma} = \frac{\partial x^\nu}{\partial\xi^\mu}\frac{\partial^2\xi^\mu}{\partial x^\lambda\partial x^\sigma}. \tag{E.55}$$

(8.4) (a) Following the same steps yields

$$(\boldsymbol{\nabla}_{\boldsymbol{u}}\tilde{\boldsymbol{u}})_\mu = u^\alpha\left(\frac{\partial u_\mu}{\partial x^\alpha} - u_\lambda\Gamma^\lambda{}_{\alpha\mu}\right) = 0, \tag{E.56}$$

and so

$$\ddot{x}_\mu - \Gamma^\lambda{}_{\mu\alpha}\dot{x}^\alpha\dot{x}_\lambda = 0. \tag{E.57}$$

(b) Start with $\boldsymbol{u}\cdot\boldsymbol{u} = -1$, which implies $\mathrm{d}/\mathrm{d}\tau(\boldsymbol{u}\cdot\boldsymbol{u}) = 0$, from which we say

$$\dot{u}^\mu u_\mu + u^\mu \dot{u}_\mu = 0. \tag{E.58}$$

Now consider the equations of motion dotted with the velocity, in the forms

$$\begin{aligned}
u_\mu \dot{u}^\mu + u_\mu \Gamma^\mu{}_{\alpha\beta} u^\alpha u^\beta &= f^\mu u_\mu/m, \\
u^\mu \dot{u}_\mu - u^\mu \Gamma^\lambda{}_{\alpha\mu} u^\alpha u_\lambda &= f_\mu u^\mu/m.
\end{aligned} \tag{E.59}$$

Add these together to see that the right-hand side must vanish.

(9.4) (a) The Lagrangian, parametrized by the proper time, is

$$L = \left[x^2 \left(\frac{\mathrm{d}t}{\mathrm{d}\tau} \right)^2 - \left(\frac{\mathrm{d}x}{\mathrm{d}\tau} \right)^2 \right]^{\frac{1}{2}}. \tag{E.60}$$

Feeding this into the Euler–Lagrange equations we find

$$\frac{\partial L}{\partial \dot{t}} = \frac{x^2}{L} \dot{t}, \quad \frac{\partial L}{\partial \dot{x}} = -\frac{1}{L}\dot{x}, \quad \frac{\partial L}{\partial x} = \frac{1}{L} x \dot{t}^2. \tag{E.61}$$

We now choose length parametrization, which sets $L = 1$, and so

$$L^2 = x^2 \left(\frac{\mathrm{d}t}{\mathrm{d}\tau} \right)^2 - \left(\frac{\mathrm{d}x}{\mathrm{d}\tau} \right)^2 = 1. \tag{E.62}$$

The resulting equations of motion are

$$2\dot{x}\dot{t} + x\ddot{t} = 0, \quad \ddot{x} + x\dot{t}^2 = 0. \tag{E.63}$$

The connection coefficients are $\Gamma^t{}_{xt} = 1/x$ and $\Gamma^x{}_{tt} = x$.

(b) Using the chain rule we write

$$\frac{\mathrm{d}x}{\mathrm{d}\tau} = \frac{\mathrm{d}x}{\mathrm{d}t}\frac{\mathrm{d}t}{\mathrm{d}\tau}. \tag{E.64}$$

The condition $L = 1$ becomes

$$(x^2 - v^2)\dot{t}^2 = 1, \tag{E.65}$$

or

$$\dot{t} = (x^2 - v^2)^{-\frac{1}{2}}. \tag{E.66}$$

Substituting this into the second of the equations of motion, we find

$$\ddot{x} = -\frac{x}{x^2 - v^2}. \tag{E.67}$$

(9.5) (a) We have the inverse matrix

$$g^{\mu\nu} = \begin{pmatrix} -1 & -\Omega y & \Omega x & 0 \\ -\Omega y & 1 - \Omega^2 y^2 & \Omega^2 xy & 0 \\ \Omega x & \Omega^2 xy & 1 - \Omega^2 x^2 & 0 \\ 0 & 0 & 0 & 1 \end{pmatrix}. \tag{E.68}$$

(b) The non-zero connection coefficients are

$$\begin{aligned}
\Gamma^x{}_{tt} &= -\Omega^2 x, \quad \Gamma^y{}_{tt} = -\Omega^2 y, \\
\Gamma^x{}_{yt} &- \Omega, \quad \Gamma^y{}_{xt} = \Omega.
\end{aligned} \tag{E.69}$$

(9.6) (a) The rotating-frame line element is

$$\begin{aligned}
\mathrm{d}s^2 &= -\mathrm{d}t^2(1 - \Omega^2 r^2) + \mathrm{d}r^2 + r^2 \mathrm{d}\theta^2 \\
&\quad + \mathrm{d}z^2 + 2\Omega r^2 \mathrm{d}\theta \mathrm{d}t.
\end{aligned} \tag{E.70}$$

(b) The non-zero connection coefficients are

$$\begin{aligned}
\Gamma^r{}_{\theta\theta} &= -r, \quad \Gamma^r{}_{\theta t} = -\Omega r, \\
\Gamma^\theta{}_{r\theta} &= \frac{1}{r}, \quad \Gamma^\theta{}_{rt} = \frac{\Omega}{r}.
\end{aligned} \tag{E.71}$$

(9.7) We find

$$\begin{aligned}
\frac{\partial L}{\partial \dot{t}} &= \mathrm{e}^{2\Phi} \frac{1}{L}\frac{\mathrm{d}t}{\mathrm{d}\lambda}, \quad && \frac{\partial L}{\partial \dot{r}} = -\mathrm{e}^{2\Lambda} \frac{1}{L}\frac{\mathrm{d}r}{\mathrm{d}\lambda}, \\
\frac{\partial L}{\partial \dot{\theta}} &= -r^2 \frac{1}{L}\frac{\mathrm{d}\theta}{\mathrm{d}\lambda}, \quad && \frac{\partial L}{\partial \dot{\phi}} = -r^2 \sin^2\theta \frac{1}{L}\frac{\mathrm{d}\phi}{\mathrm{d}\lambda}.
\end{aligned} \tag{E.72}$$

This leads to E-L equation 1

$$\frac{\mathrm{d}^2 t}{\mathrm{d}\lambda^2} + 2\frac{\partial\Phi}{\partial r}\frac{\mathrm{d}r}{\mathrm{d}\lambda}\frac{\mathrm{d}t}{\mathrm{d}\lambda} = 0, \tag{E.73}$$

giving $\Gamma^t{}_{rt} = \frac{\partial\Phi}{\partial r}$. E-L equation 2 is

$$\begin{aligned}
\frac{\mathrm{d}^2 r}{\mathrm{d}\lambda^2} &+ \frac{\partial\Phi}{\partial r}\mathrm{e}^{2\Phi}\mathrm{e}^{-2\Lambda}\left(\frac{\mathrm{d}t}{\mathrm{d}\lambda}\right)^2 + \frac{\partial\Lambda}{\partial r}\left(\frac{\mathrm{d}r}{\mathrm{d}\lambda}\right)^2 \\
&- r\mathrm{e}^{-2\Lambda}\left(\frac{\mathrm{d}\theta}{\mathrm{d}\lambda}\right)^2 - r\mathrm{e}^{-2\Lambda}\sin^2\theta\left(\frac{\mathrm{d}\phi}{\mathrm{d}\lambda}\right)^2 = 0,
\end{aligned} \tag{E.74}$$

giving

$$\begin{aligned}
\Gamma^r{}_{tt} &= \Phi' \mathrm{e}^{2\Phi}\mathrm{e}^{-2\Lambda}, \quad && \Gamma^r{}_{rr} = \Lambda', \\
\Gamma^r{}_{\theta\theta} &= -r\mathrm{e}^{-2\Lambda}, \quad && \Gamma^r{}_{\phi\phi} = -r\mathrm{e}^{-2\Lambda}\sin^2\theta.
\end{aligned} \tag{E.75}$$

E-L equation 3 is

$$\frac{\mathrm{d}^2\theta}{\mathrm{d}\lambda^2} + \frac{2}{r}\frac{\mathrm{d}r}{\mathrm{d}\lambda}\frac{\mathrm{d}\theta}{\mathrm{d}\lambda} - \sin\theta\cos\theta\left(\frac{\mathrm{d}\phi}{\mathrm{d}\lambda}\right)^2 = 0, \tag{E.76}$$

giving

$$\Gamma^\theta{}_{r\theta} = \frac{1}{r}, \quad \Gamma^\theta{}_{\phi\phi} = -\sin\theta\cos\theta. \tag{E.77}$$

E-L equation 4 is

$$\frac{\mathrm{d}^2\phi}{\mathrm{d}\lambda^2} + \frac{2}{r}\frac{\mathrm{d}r}{\mathrm{d}\lambda}\frac{\mathrm{d}\phi}{\mathrm{d}\lambda} + 2\frac{\cos\theta}{\sin\theta}\frac{2}{r}\frac{\mathrm{d}\theta}{\mathrm{d}\lambda}\frac{\mathrm{d}\phi}{\mathrm{d}\lambda} = 0, \tag{E.78}$$

giving

$$\Gamma^\phi{}_{r\phi} = \frac{1}{r}, \quad \Gamma^\phi{}_{\theta\phi} = \frac{\cos\theta}{\sin\theta}. \tag{E.79}$$

(9.8) (b) Using the result from part (a) and rearranging we find

$$\mathrm{d}\lambda = \frac{\mathrm{d}r}{r\left(1 - \frac{r^2}{a^2}\right)^{\frac{1}{2}}}. \tag{E.80}$$

The substitution $r = a\sin t$ then gives the stated result.

(c) Write

$$\mathrm{d}x = \frac{r^2}{a}\mathrm{d}\lambda, \tag{E.81}$$

and substitute for $\mathrm{d}\lambda$ from part (b).

(e) The length is

$$\Delta\lambda = \int_a^b \frac{\mathrm{d}\lambda}{\mathrm{d}t}\mathrm{d}t = \int_a^b \frac{\mathrm{d}t}{\sin t} = -\ln\left| \frac{1 + \cos t}{\sin t} \right|_a^b. \tag{E.82}$$

The length of the so-called *maximal geodesic*, that starts at $t = 0$ and ends at $t = \pi$, is infinite.

(9.9) Call $y = (-g_{\mu\nu}\dot{x}^\mu\dot{x}^\nu)^{\frac{1}{2}}$. The Euler–Lagrange equation gives

$$\frac{\mathrm{d}^2 F}{\mathrm{d}y^2}\frac{\mathrm{d}y}{\mathrm{d}\lambda}\frac{\partial y}{\partial \dot{x}} + \frac{\mathrm{d}F}{\mathrm{d}y}\left[\frac{\mathrm{d}}{\mathrm{d}\lambda}\left(\frac{\partial y}{\partial \dot{x}} \right) - \frac{\partial y}{\partial x} \right] = 0. \tag{E.83}$$

For length parametrization we have $\mathrm{d}y/\mathrm{d}\lambda = 0$, wiping out the first term. We also have that $\mathrm{d}F/\mathrm{d}y \neq 0$, as the function is monotonic. This then results in the same Euler–Lagrange equation as obtained from varying y.

(9.11) We find connection coefficients

$$\Gamma^r{}_{rr} = \tfrac{1}{2(1+r)}, \quad \Gamma^r{}_{\phi\phi} = -\tfrac{r}{1+r}, \quad \Gamma^\phi{}_{r\phi} = \tfrac{1}{r}. \tag{E.84}$$

Differentiating with respect to λ, the velocity has components $u^\mu = (dr/d\lambda, d\phi/d\lambda)$ given by

$$u^\mu = \left((1+r)^{-\frac{1}{2}}, 0 \right). \tag{E.85}$$

We find

$$u^r{}_{;r} = \frac{\partial u^r}{\partial r} + \Gamma^r{}_{rr}(u^r)^2 = 0, \tag{E.86}$$

$$u^\phi{}_{;\phi} = r^{-1}(r+1)^{\frac{1}{2}}, \tag{E.87}$$

and the other components vanish. Contracting against the components of the velocity then gives a vanishing acceleration.

(10.1) (a) We obtain

$$A^{\hat\mu} = (A^t, aA^\chi, a\sinh\chi A^\theta, a\sinh\chi\sin\theta A^\phi). \tag{E.88}$$

(b) We find

$$G^{\hat\theta}{}_{\hat\chi\hat\phi} = G^\theta{}_{\chi\phi}(e_\theta)^{\hat\theta}(e_{\hat\chi})^\chi(e_{\hat\phi})^\phi = \frac{G^\theta{}_{\chi\phi}}{a\sin\theta}. \tag{E.89}$$

(10.2) (a) We find $\Gamma^{\hat r}{}_{\hat\theta\hat\theta} = -1/r$, $\Gamma^{\hat\theta}{}_{\hat\theta\hat r} = 1/r$ and $\Gamma^{\hat\theta}{}_{\hat r\hat\theta} = 0$.
(b) From Example 3.3, we have that $[e_{\hat r}, e_{\hat\theta}] = -e_{\hat\theta}/r$, and so $\left\langle \omega^{\hat\theta}, [e_{\hat r}, e_{\hat\theta}] \right\rangle = -1/r$, showing that the rule is obeyed for $\Gamma^{\hat\theta}{}_{\hat r\hat\theta} - \Gamma^{\hat\theta}{}_{\hat\theta\hat r}$.

(10.3) (a) Use $\boldsymbol{u} \cdot \boldsymbol{u} = -1$ to say

$$g_{\mu\mu}(u^\mu)^2 = -1. \tag{E.90}$$

The particle doesn't move in the θ and ϕ directions, so we have $u^t = g^{tt}a = a/g_{tt}$

$$a^2/g_{tt} + g_{rr}(u^r)^2 = -1, \tag{E.91}$$

and so

$$u^r = \left[\frac{-a^2 - g_{tt}}{g_{tt}g_{rr}} \right]^{\frac{1}{2}}. \tag{E.92}$$

(b) Since $\boldsymbol{u} = d\boldsymbol{x}/d\tau$, write

$$\frac{dr}{dt} = \frac{dr}{d\tau}\frac{d\tau}{dt} = \frac{u^r}{u^t} = \frac{g_{tt}u^r}{a}. \tag{E.93}$$

(c) A local observer measures time intervals $d\hat t = (e_t)^{\hat t}dt = \sqrt{-g_{tt}}\,dt$ and radial intervals $d\hat r = (e_r)^{\hat r}dr$, so that we have

$$\frac{d\hat r}{d\hat t} = \frac{(e_r)^{\hat r}}{(e_t)^{\hat t}}\frac{dr}{dt} = \frac{\sqrt{g_{rr}}}{\sqrt{-g_{tt}}}\frac{g_{tt}u^r}{a}$$

$$= \sqrt{-g_{rr}g_{tt}}\frac{u^r}{a} = \sqrt{1 + g_{tt}/a^2}. \tag{E.94}$$

(10.5) Observers at both the emission and observation points are at rest in the rotating frame, so will have velocity with non-zero component u^0 given by

$$g_{00}(u^0)^2 = -1, \tag{E.95}$$

or

$$u^0 = (1 - \Omega^2 r^2)^{-\frac{1}{2}}. \tag{E.96}$$

Since both emitter and observer are at the same value of r, then they both find the same value of $E = -\boldsymbol{p}\cdot\boldsymbol{u}$ for a photon of momentum $\boldsymbol{p}$.

(10.6) (a) Using $\eta_{\hat\mu\hat\nu} = \boldsymbol{e}_{\hat\mu} \cdot \boldsymbol{e}_{\hat\nu}$ we find

$$\eta_{\hat\mu\hat\nu} = \begin{pmatrix} 0 & 1 & 0 & 0 \\ 1 & 0 & 0 & 0 \\ 0 & 0 & 0 & -1 \\ 0 & 0 & -1 & 0 \end{pmatrix}. \tag{E.97}$$

(b) The 1-forms are

$$\omega^{\hat 1} = \boldsymbol{n}, \qquad \omega^{\hat 2} = \boldsymbol{l}, \tag{E.98}$$
$$\omega^{\hat 3} = -\bar{\boldsymbol{m}}, \qquad \omega^{\hat 4} = -\boldsymbol{m}.$$

(11.2) (a)(i) $R^t{}_{rrt} = \frac{2M}{r^2(2M-r)}$.
(ii) We have that $R_{rtrt} = R_{trtr} = g_{tt}R^t{}_{rtr}$, and so

$$R^r{}_{trt} = g^{rr}g_{tt}R^t{}_{rtr} = \frac{2M(2M-r)}{r^4}. \tag{E.99}$$

(12.1) The transformation is carried out using the transformation components

$$\begin{array}{ll} \Lambda^x{}_r = \sin\theta\cos\phi, & \Lambda^x{}_\theta = r\cos\theta\cos\phi, \\ \Lambda^x{}_\phi = -r\sin\theta\sin\phi, & \Lambda^y{}_r = \sin\theta\sin\phi, \\ \Lambda^y{}_\theta = r\cos\theta\sin\phi, & \Lambda^y{}_\phi = r\sin\theta\cos\phi, \\ \Lambda^z{}_r = \cos\theta, & \Lambda^z{}_\theta = -r\sin\theta, \\ \Lambda^z{}_\phi = 0. & \end{array}$$

Using these we find

$$T_{tt} = \rho, \quad T_{rr} = p, \quad T_{\theta\theta} = pr^2, \quad T_{\phi\phi} = pr^2\sin^2\theta. \tag{E.100}$$

(12.3) To apply the rule given in the question, write

$$T^{\mu\nu}(x) = \sum_n \int d\tau\, m\dot z_n^\mu \dot z_n^\nu \delta^{(3)}[x - z_n(\tau)]\delta[x^0 - z_n^0(\tau)], \tag{E.101}$$

then take $f(\tau_n) = x^0 - z_n^0(\tau_n)$, with zeros at τ_n when $x^0 = z_n^0(\tau_n)$ and derivative $|f'(\tau_n)| = |dz_n^0(\tau_n)/d\tau_n|$. Doing the integral, we obtain an energy-momentum tensor

$$T^{\mu\nu}(x) = \sum_n m\frac{\dot z_n^\mu(\tau_n)\dot z_n^\nu(\tau_n)}{\left|\frac{dz_n^0}{d\tau_n}\right|}\delta^{(3)}[x - z_n(\tau_n)]. \tag{E.102}$$

(a) Set $\mu = \alpha$ and $\nu = 0$ and we have

$$T^{\alpha 0}(x) = \sum_n m\frac{\dot z_n^\alpha(\tau_n)\dot z_n^0(\tau_n)}{|\dot z_n^0(\tau_n)|}\delta^{(3)}[x - z_n(\tau_n)]$$

$$= \sum_n p_n^\alpha(\tau_n)\delta^{(3)}[x - z_n(\tau_n)], \tag{E.103}$$

where we've written $p_n^\mu(\tau_n) = m\dot z_n^\mu(\tau_n)$.
(b) Set $\mu = \alpha$ and $\nu = i$ and we have

$$T^{\alpha i} = \sum_n m\frac{\dot z_n^\alpha(\tau_n)\dot z_n^i(\tau_n)}{|\dot z_n^0(\tau_n)|}\delta^{(3)}[x - z_n(\tau_n)]$$

$$= \sum_n p_n^\alpha(\tau_n)\delta^{(3)}[x - z_n(\tau_n)]\frac{dz^i(\tau_n)}{d\tau_n}\left|\frac{\partial\tau_n}{\partial z^0(\tau_n)}\right|, \tag{E.104}$$

and the answer follows from taking $z^0(\tau_n) = t$.
(c) Use the fact that $mdz_n^0/d\tau_n$ is equal to the energy $E_n(\tau_n)$ to write

$$T^{\alpha\beta} = \sum_n m^2\frac{\dot z_n^\alpha(\tau_n)\dot z_n^\beta(\tau_n)}{E_n(\tau_n)}\delta^{(3)}[x - z_n(\tau_n)], \tag{E.105}$$

then rewrite in terms of momentum.

(12.4) The rule we have to apply is

$$0 = T^{\mu\nu}{}_{,\mu} + \Gamma^{\mu}{}_{\mu\alpha}T^{\alpha\nu} + \Gamma^{\nu}{}_{\mu\alpha}T^{\alpha\mu}. \qquad \text{(E.106)}$$

The comma demands a derivative with respect to x^μ and not z^μ, so on contracting on the μ index, we obtain

$$\begin{aligned}
T^{\mu\nu}{}_{,\mu} =& m\frac{\partial(-g)^{-\frac{1}{2}}}{\partial x^\mu}\int \dot{z}^\mu \dot{z}^\nu \delta^{(4)}[x-z(\tau)]\mathrm{d}\tau \\
&+ \frac{m}{\sqrt{g}}\int \dot{z}^\mu \dot{z}^\nu \frac{\partial}{\partial x^\mu}\delta^{(4)}[x-z(\tau)]\mathrm{d}\tau.
\end{aligned} \qquad \text{(E.107)}$$

Use the identity $\frac{1}{\sqrt{-g}}\frac{\partial\sqrt{-g}}{\partial x^\mu} = \Gamma^\alpha{}_{\mu\alpha}$, to find that the second term becomes $-\Gamma^\alpha{}_{\mu\alpha}T^{\mu\nu}$ which, on relabelling indices, will cancel against the second term in eqn E.106. Since the delta function depends only on the separation of x and z, we are allowed to replace $\partial/\partial x^\mu$ with $-\partial/\partial z^\mu$ in the first term. We also note that the chain rule allows us to replace

$$\dot{z}^\mu \frac{\partial}{\partial z^\mu} = \frac{\mathrm{d}}{\mathrm{d}\tau}, \qquad \text{(E.108)}$$

leading to

$$\begin{aligned}
T^{\mu\nu}{}_{;\mu} =& -\frac{m}{\sqrt{-g}}\int \dot{z}^\nu \frac{\mathrm{d}}{\mathrm{d}\tau}\delta^{(4)}[x-z(\tau)]\mathrm{d}\tau \\
&+ \Gamma^\nu{}_{\mu\alpha}\frac{m}{\sqrt{-g}}\int \dot{z}^\mu \dot{z}^\alpha \delta^{(4)}[x-z(\tau)]\mathrm{d}\tau.
\end{aligned} \qquad \text{(E.109)}$$

Integrating the first term by parts and assuming that the velocity field dies at infinity, we find

$$\int \left(\ddot{z}^\mu + \Gamma^\mu{}_{\alpha\nu}\dot{z}^\alpha \dot{z}^\nu\right)\delta^{(4)}[x-z(\tau)]\mathrm{d}\tau = 0. \qquad \text{(E.110)}$$

This gives us the geodesic equation.

(12.5) (a) Insert a vector $\boldsymbol{v}$ to obtain

$$\boldsymbol{T}(\ ,\boldsymbol{v}) = -p n, \qquad \text{(E.111)}$$

because $\tilde{\boldsymbol{J}}(\boldsymbol{v}) = -n$. One of the eigenvectors is, therefore

$$\boldsymbol{T}(\ ,\boldsymbol{v}) = -\rho\boldsymbol{v}. \qquad \text{(E.112)}$$

(b) In an orthonormal basis with $\boldsymbol{v} = \boldsymbol{e}_{\hat{0}}$, we have $\boldsymbol{T}(\ ,\boldsymbol{e}_{\hat{0}}) = -\rho\boldsymbol{e}_{\hat{0}}$ and $\boldsymbol{T}(\ ,\boldsymbol{e}_{\hat{i}}) = p\boldsymbol{e}_{\hat{i}}$. As a consequence

$$\boldsymbol{T}(\ ,\boldsymbol{X}) = -X^{\hat{0}}\rho\boldsymbol{e}_{\hat{0}} + X^{\hat{i}}p\boldsymbol{e}_{\hat{i}}. \qquad \text{(E.113)}$$

As a result of part (a), when we input at 1-form $\tilde{\boldsymbol{Y}}$ we have

$$\begin{aligned}
\boldsymbol{T}(\tilde{\boldsymbol{Y}},\boldsymbol{X}) =& -X^{\hat{0}}Y_{\hat{0}}\rho + X^{\hat{i}}Y_{\hat{i}}p \\
=& -X^{\hat{0}}Y_{\hat{0}}\rho + X^{\hat{i}}Y_{\hat{i}}p - X^{\hat{0}}Y_{\hat{0}}p + X^{\hat{0}}Y_{\hat{0}}p \\
=& -X^{\hat{0}}Y_{\hat{0}}(\rho+p) + X^{\hat{\mu}}Y_{\hat{\mu}}p.
\end{aligned} \qquad \text{(E.114)}$$

(c) The right-hand side of the last equation is a number that can be rewritten as $X^{\hat{0}}Y^{\hat{0}}(\rho+p) + \eta_{\hat{\mu}\hat{\nu}}X^{\hat{\mu}}Y^{\hat{\nu}}p$, since $Y_{\hat{0}} = -Y^{\hat{0}}$ in flat space. Hence we can extract a (0,2) version of the tensor components. Using the fact that $\boldsymbol{v} = \boldsymbol{e}_{\hat{0}}$, we have that $\tilde{\boldsymbol{v}}$ can be used to extract the timelike components of vectors. The corresponding flat-space expression for $\boldsymbol{T}(\boldsymbol{Y},\boldsymbol{X})$ can then be reexpressed in vector notation using $\tilde{\boldsymbol{v}}$ and the metric $\boldsymbol{\eta}$ as

$$\boldsymbol{T}(\boldsymbol{Y},\boldsymbol{X}) = (\rho+p)\left[\tilde{\boldsymbol{v}}(\boldsymbol{Y})\tilde{\boldsymbol{v}}(\boldsymbol{X})\right] + p\boldsymbol{\eta}(\boldsymbol{Y},\boldsymbol{X}). \qquad \text{(E.115)}$$

In curved space, we swap $\boldsymbol{\eta} \to \boldsymbol{g}$. We therefore suggest that the form of the (0,2) tensor is

$$\boldsymbol{T}(\ ,\ ) = (\rho+p)\tilde{\boldsymbol{v}} \otimes \tilde{\boldsymbol{v}} + p\boldsymbol{g}. \qquad \text{(E.116)}$$

This tensor has components

$$T_{\mu\nu} = (\rho+p)v_\mu v_\nu + p g_{\mu\nu}. \qquad \text{(E.117)}$$

(15.1) Volume of a cylinder is $\pi r^2 h \approx 10^{11}$ in cubic parsecs, which is then the number of stars.

(15.2) (a) The invariant interval can be written in terms of the metric line element as

$$s = \int \mathrm{d}\tau \left\{\left(\frac{\mathrm{d}t}{\mathrm{d}\tau}\right)^2 - a(t)^2\left[\sum_{i=1}^{3}\left(\frac{\mathrm{d}x^i}{\mathrm{d}\tau}\right)^2\right]\right\}^{\frac{1}{2}}, \qquad \text{(E.118)}$$

where the interval has been parametrized using the proper time. We find

$$\begin{aligned}
&\frac{\partial L}{\partial \dot{t}} = \frac{\dot{t}}{L}, && \frac{\partial L}{\partial \dot{x}^i} = -\frac{a^2\dot{x}^i}{L}, \\
&\frac{\partial L}{\partial t} = -\frac{a\frac{\mathrm{d}a}{\mathrm{d}t}(\dot{x}^2+\dot{y}^2+\dot{z}^2)}{L}, && \frac{\partial L}{\partial x^i} = 0,
\end{aligned} \qquad \text{(E.119)}$$

where dots here indicate derivatives with respect to the proper time. This gives us equations of motion

$$\ddot{x}^i + \frac{2}{a}\frac{\mathrm{d}a}{\mathrm{d}t}\dot{x}^i\dot{t} = 0,$$

$$\ddot{t} + a\frac{\mathrm{d}a}{\mathrm{d}t}(\dot{x}^2 + \dot{y}^2 + \dot{z}^2) = 0. \qquad \text{(E.120)}$$

The connection coefficients can then be read off.
(b) The only derivatives that survive are

$$\Gamma^0{}_{ii,0} = (\dot{a}^2 + a\ddot{a}),$$

$$\Gamma^i{}_{0i,0} = \frac{\ddot{a}}{a} - \left(\frac{\dot{a}}{a}\right)^2, \qquad \text{(E.121)}$$

using dots here for derivatives with respect to t. Plugging into the equation for the components of $\boldsymbol{R}$ we find the non-zero parts are

$$\begin{aligned}
R^0{}_{i0i} &= \Gamma^0{}_{ii,0} - \Gamma^0{}_{ii}\Gamma^i{}_{i0} = a\ddot{a}, \\
R^i{}_{jij} &= \Gamma^i{}_{0i}\Gamma^0{}_{jj} = \dot{a}^2.
\end{aligned} \qquad \text{(E.122)}$$

These can be shifted into the orthonormal frame using the vielbein given in Chapter 15.

(15.3) (a) The Einstein tensor for the state of affairs described is given by

$$G_{\mu\nu} = R_{\mu\nu} - \frac{1}{2}R g_{\mu\nu} = -\frac{1}{4}R g_{\mu\nu}. \qquad \text{(E.123)}$$

(b) The components $G_{\mu\nu}$ must be equal to the energy-momentum contribution ($= -\Lambda g_{\mu\nu}$), so that the Einstein equation reads

$$-\frac{R}{4}g_{\mu\nu} = -\Lambda g_{\mu\nu}, \qquad \text{(E.124)}$$

and we have $\Lambda = R/4$.
(c) The Ricci scalar is given by

$$R = 6\left(\frac{\ddot{a}}{a} + \frac{\dot{a}^2}{a^2}\right), \qquad \text{(E.125)}$$

with $a(t) = e^{Ht}$. Substituting, we find that

$$R = 6(H^2 + H^2) = 12\frac{\Lambda}{3} = 4\Lambda. \qquad \text{(E.126)}$$

So fixing a Ricci scalar $R = 4\Lambda$ gives us the same flat, expanding Universe that we had before.

(15.5) (a) The energy density is of order 1 GeV$/(10^{-45}$ m$^3) \approx 1.602 \times 10^{35}$ Jm^{-3}.

(b) In SI units, the Einstein equation is

$$\boldsymbol{G} = \frac{8\pi G}{c^4}\boldsymbol{T}, \qquad (E.127)$$

where the factor $\frac{8\pi G}{c^4} = 2.07 \times 10^{-43}$ m^{-1}kg^{-1}s^2. The right-hand-side is therefore $\approx 3 \times 10^{-8}$ m^{-2}. Take the left-hand side to be equal to the curvature $1/a^2$, where a is the characteristic size of the Universe, we have $a \approx 5$ km. We conclude that the vacuum energy of the strong nuclear force would wind the Universe into a ball with a radius of a few km. We could barely go for a walk before finding we are back where we started!

(15.6) (a) The equation of motion for t is

$$\frac{\mathrm{d}^2 t}{\mathrm{d}\tau^2} + a\dot{a}\left(\frac{\mathrm{d}r}{\mathrm{d}\tau}\right)^2 = 0. \qquad (E.128)$$

The r-dependence can be eliminated by substituting using the length parametrization condition $L = 1$.

(c) Substituting into the velocity identity $g_{\mu\nu}u^\mu u^\nu = -1$, we find

$$-\left(\frac{\mathrm{d}t}{\mathrm{d}\tau}\right)^2 + g_{ii}(u^i)^2 = -1. \qquad (E.129)$$

If we take $g_{ii}(u^i)^2 = |\vec{u}|^2$ then using the solution of the equation of motion to substitute for $\mathrm{d}t/\mathrm{d}\tau$ yields the answer.

(15.7) (a) Take $\mathrm{d}s = 0$ and write $\mathrm{d}t/a(t) = \mathrm{d}r$, or

$$\int_0^\ell \mathrm{d}r = \int_{t_s}^{t_r} \mathrm{d}t\, e^{-Ht}, \qquad (E.130)$$

from which the answer follows.

(b) As ℓ increases, t_r also increases. The distance $\ell' = e^{-Ht_s}/H$ corresponds to an infinite receipt time $t_r \to \infty$. Therefore, an observer beyond a radius ℓ' will never receive the light signal sent from the origin, since the expansion of the Universe is fast enough to prevent this. Notice this scale is set by $1/H$.

(15.8) From Exercise 15.2, we start from one of the equations of motion, in the form,

$$\frac{\mathrm{d}}{\mathrm{d}\tau}(a^2 \dot{x}^i) = 0. \qquad (E.131)$$

This tells us that

$$e^{2Ht}\dot{x}^i = \text{const.}, \qquad (E.132)$$

or

$$\dot{x}^i = C^i e^{-2Ht}, \qquad (E.133)$$

with C^i a set of constants. To see that the solutions are straight lines, we write

$$\frac{\mathrm{d}x^i}{\mathrm{d}x^j} = \frac{C^i}{C^j}. \qquad (E.134)$$

This latter equation is solved with straight lines by saying, for example,

$$\begin{aligned} x^1 &= x_0^1 + C^1 x^3/C^3, \\ x^2 &= x_0^2 + C^2 x^3/C^3, \end{aligned} \qquad (E.135)$$

with x_0^i another set of constants set by the initial conditions.

(16.3) If computing directly from the metric elements, we obtain

$$\Gamma_{trr} = -\Gamma_{rrt} = -\frac{a\dot{a}}{1-kr^2},$$

$$\Gamma_{tii} = -\Gamma_{iit} = -\frac{\dot{a}}{a}g_{ii},$$

$$\Gamma_{rrr} = \frac{a^2 kr}{(1-kr^2)^2},$$

$$\Gamma_{r\theta\theta} = -\Gamma_{\theta\theta r} = -a^2 r,$$

$$\Gamma_{r\phi\phi} = -\Gamma_{\phi\phi r} = -a^2 r\sin^2\theta,$$

$$\Gamma_{\theta\phi\phi} = -\Gamma_{\phi\phi\theta} = -a^2 r^2 \sin\theta\cos\theta. \qquad (E.136)$$

The answers then follow by raising components with $g^{\mu\nu}$.

(16.5) Substitute the metric into the Bianchi identity to find

$$\begin{aligned} &C_{,\sigma}(g_{\mu\alpha}g_{\nu\beta} - g_{\mu\beta}g_{\nu\alpha}) \\ &+ C_{,\beta}(g_{\mu\sigma}g_{\nu\alpha} - g_{\mu\alpha}g_{\nu\sigma}) \\ &+ C_{,\alpha}(g_{\mu\beta}g_{\nu\sigma} - g_{\mu\sigma}g_{\nu\beta}) = 0, \qquad (E.137) \end{aligned}$$

where we've used the compatibility condition for the derivatives of the metric components (i.e. $g_{\mu\nu;\sigma} = 0$). Next, contract indices by (i) multiplying by $g^{\mu\alpha}$ and summing, then (ii) multiplying by $g^{\nu\beta}$ and summing. We find that $C_{,\sigma} = 0$, so C must be constant.

(16.6) We write

$$R_{\nu\beta} = Cg^{\mu\alpha}\left(g_{\mu\alpha}g_{\nu\beta} - g_{\mu\beta}g_{\nu\alpha}\right), \qquad (E.138)$$

from which we obtain, in $(3+1)$ dimensions,

$$R_{\nu\beta} = C(4g_{\nu\beta} - g_{\nu\beta}) = 3Cg_{\nu\beta}, \qquad (E.139)$$

and so $R = 12C$. This means that $R_{\nu\beta} = \frac{1}{4}Rg_{\nu\beta}$ and also that the components of the Einstein tensor are $G_{\nu\beta} = -\frac{1}{4}Rg_{\nu\beta}$. In general, we have for d-dimensional spacetime that $^{(d)}R_{\mu\nu} = (d-1)Cg_{\mu\nu}$ and $^{(d)}R = Cd(d-1)$. This allows us to write $^{(d)}R_{\mu\nu} = \frac{^{(d)}R}{d}g_{\mu\nu}$.

(16.7) We have

$$^{(3)}R = 3R_{\hat{i}\hat{i}} = 3\frac{\ddot{a}}{a} + 6\left(\frac{\dot{a}}{a}\right)^2 + 6\frac{k}{a}. \qquad (E.140)$$

This can be used to write $^{(3)}R_{ij} = \frac{1}{3}{}^{(3)}R\gamma_{ij}$, in a similar fashion to the previous question.

(16.8) We have that $\ell_{ob}/\ell_{em} = a(t_{ob})/a(t_{em})$. Since $a(t_{ob})/a(t_{em}) = (1+z)$, we have

$$\ell_{em} = \frac{\ell_{ob}}{1+z}. \qquad (E.141)$$

(17.1) (a) We have that

$$\Lambda^\chi{}_r = \frac{\partial\chi}{\partial r} = \frac{1}{(1-kr^2)^{\frac{1}{2}}}. \qquad (E.142)$$

(c) The transformation we want to check is

$$\begin{aligned} T_{\hat{r}\hat{r}} &= \left(\Lambda^{\hat{\chi}}{}_{\hat{r}}\right)^2 T_{\hat{\chi}\hat{\chi}} \\ &= \left[\Lambda^\chi{}_r(e_\chi)^{\hat{\chi}}(e_{\hat{r}})^r\right]^2 T_{\hat{\chi}\hat{\chi}}. \qquad (E.143) \end{aligned}$$

Using the results of parts (a) and (b) we find

$$\Lambda^\chi{}_r(e_\chi)^{\hat{\chi}}(e_{\hat{r}})^r = 1.$$

(17.2) (a) Using $U = -\partial(\ln Z)/\partial\beta$, with $\beta = 1/k_\mathrm{B}T$, we find that the internal energy per particle is $3k_\mathrm{B}T$.
(b) The pressure is found from $p = -(\partial F/\partial V)_T$, where $F = -\frac{1}{\beta}\ln Z$ is the free energy. We obtain a pressure $k_\mathrm{B}T/V$ per particle. The result in (c) follows from taking the ratio of these two results.

(17.3) Setting $k = 0$ we find
$$\dot{a}^2 = \text{const}, \qquad (\text{E.144})$$
so $a(t) \propto t$.

(17.4) Since $k = 0$, we have $\Omega_\mathrm{d}^{(0)} = 1$. Plugging in to eqn 17.33 we obtain
$$t(z) = \frac{1}{H_0}\int_0^{(z+1)^{-1}} \mathrm{d}x\, x^{\frac{1}{2}} = \frac{2}{3H_0}(z+1)^{-\frac{3}{2}}. \qquad (\text{E.145})$$

(18.1) (a) The first Friedmann equation is
$$\dot{a}^2 + k = 0, \qquad (\text{E.146})$$
which becomes
$$\dot{a} = 1, \qquad (\text{E.147})$$
so $a(t) = t$ and the expression for the line element follows on setting $k = -1$.
(b) We find
$$-\mathrm{d}T^2 + \mathrm{d}r^2 = -\mathrm{d}t^2 + t^2\mathrm{d}\chi^2, \qquad (\text{E.148})$$
and so
$$\mathrm{d}s^2 = -\mathrm{d}T^2 + \mathrm{d}r^2 + r^2(\mathrm{d}\theta^2 + \sin^2\theta\mathrm{d}\phi^2), \qquad (\text{E.149})$$
which is the line element for empty Minkowski space (Universe 0).

(18.5) We have $\mathrm{d}X = T\mathrm{d}T/X$, so
$$\mathrm{d}s^2 = -\mathrm{d}T^2 + \frac{T^2\mathrm{d}T^2}{1+T^2}$$
$$= -\frac{\mathrm{d}T^2}{1+T^2}. \qquad (\text{E.150})$$
Writing $T = \sinh\psi$, which implies $X = \cosh\psi$, we find $\mathrm{d}s^2 = -\mathrm{d}\psi^2$.
Usually we would call ψ the proper time τ. The embedding of a hyperbolic world line in Minkowski space therefore tells us that the intervals along the world line can be measured with the proper time, which is something we already knew.

(18.6) (a) Eliminate Y to find
$$\mathrm{d}s^2 = -\mathrm{d}T^2 + \mathrm{d}X^2 + \frac{(T\mathrm{d}T - X\mathrm{d}X)^2}{(1+T^2 - X^2)}. \qquad (\text{E.151})$$
Now define variables $T = v\cosh\psi$ and $X = v\sinh\psi$ to find
$$\mathrm{d}s^2 = -\mathrm{d}v^2 + v^2\mathrm{d}\psi^2 + \frac{v^2\mathrm{d}v^2}{1+v^2}, \qquad (\text{E.152})$$
which simplifies to the expression in the question.
(b) Set $v = \sinh\chi$ to find that the relationship is
$$T = \sinh\chi\cosh\psi,$$
$$X = \sinh\chi\sinh\psi,$$
$$Y = \cosh\chi. \qquad (\text{E.153})$$

(c) The line element $\mathrm{d}s^2 = -\mathrm{d}\chi^2 + \cosh^2\chi\mathrm{d}\theta^2$ is based on coordinates
$$T = \sinh\chi,$$
$$X = \cosh\chi\cos\theta,$$
$$Y = \cosh\chi\sin\theta. \qquad (\text{E.154})$$
This uses a set of circles in the X-Y plane as the basis for the coordinate system used. So if we fix the timelike variable χ, the coordinates describe a circle in the X-Y plane. In contrast, a fixed time χ in the other form of the line element gives us coordinates describing the hyperbolic curve $-T^2 + X^2 = 1$ in the $X - T$ plane.
(d) In the absence of matter, we have $\dot{a}^2 + k = \Lambda a^2/3$. In units where $\Lambda/3 = 1$, we have that $a(t) = \cosh t$ corresponds to a closed ($k = 1$) universe, and $a(t) = \sinh t$ corresponds to an open universe.

(18.7) (b) Adding, we find $T + X = \mathrm{e}^t$ or $t = \ln(T + X)$. So the coordinates only cover the region $T + X > 0$.
(c) Lines of constant t obey the equation $T = A - X$, where A is a constant. They also obey $T = B + CY^2$, with B and C constants.
(d) The transformation needed here is $u = \mathrm{e}^{-t}$.

(18.8) The embedding only works for $x < 1$. We have $T+X = (1-x)^{\frac{1}{2}}\mathrm{e}^t > 0$ and $-T+X = (1-x)^{\frac{1}{2}}\mathrm{e}^{-t} > 0$, so only the quarter of the hyperboloid with $X \geq |T|$ is covered by these coordinates.

(19.2) (b) For $\beta = 1$ we have $\mathrm{d}u/\mathrm{d}v = 0$, that we might call 'at rest' in terms of the coordinate velocity. For $\beta = 0$, the coordinate velocity $\mathrm{d}u/\mathrm{d}v = 1$ ('light coordinate speed') and for $\beta = -1$ we have infinite coordinate speed. In short, we should beware of interpretations of coordinate velocities.

(19.3) (a) In component form, we have
$$\cos\theta = \frac{g_{\mu\nu}u^\mu u^\nu}{(g_{\alpha\beta}u^\alpha u^\beta g_{\sigma\rho}v^\sigma v^\rho)^{\frac{1}{2}}}. \qquad (\text{E.155})$$
Substituting the transformation we find the factors of $f(x)$ cancel, so the angle is unchanged.
(b) A null curve with tangent $\boldsymbol{a}$ has $\boldsymbol{g}(\boldsymbol{a}, \boldsymbol{a}) = 0$. In component form, we have
$$0 = g_{\mu\nu}a^\mu a^\nu \to f(x)g_{\mu\nu}a^\mu a^\nu, \qquad (\text{E.156})$$
which will still vanish, so the curve remains null.

(19.5) (a) The form of the solution will be the same in the two spacetimes. The time dependence in terms of t is therefore written as
$$A_\mu \propto \mathrm{e}^{iB\int\frac{\mathrm{d}t}{a(t)}}. \qquad (\text{E.157})$$
(b) The instantaneous frequency $\omega(t)$ can be extracted by taking its derivative of the phase of the wave with respect to time t, which tells us that $\omega(t)a(t) = \text{const}$.

(19.6) Putting the steps together we find
$$t' = \tan^{-1}(t+r) + \tan^{-1}(t-r),$$
$$r' = \tan^{-1}(t+r) - \tan^{-1}(t-r). \qquad (\text{E.158})$$

(19.7) The geodesics are shown in Fig. E.1.

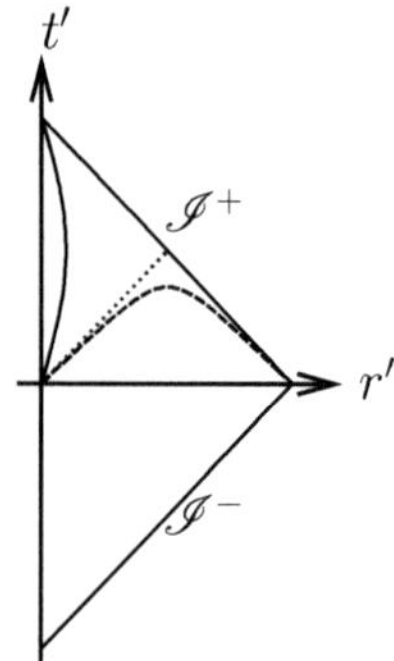

Fig. E.1 Penrose diagram for Exercise 19.7 showing (a) dotted, (b) solid and (c) dashed.

(19.8) (a) Use $\dot{a}^2 + k = C/a + \Lambda a^3/3$.
(b) Recall that a light ray travels a constant comoving distance
$$\int_{t_{\rm em}}^{t_{\rm ob}} \frac{{\rm d}t}{a(t)}, \tag{E.159}$$
and it is this distance that is measured by η.
(c) Differentiating, we find
$$\frac{{\rm d}\eta}{{\rm d}a} = -\frac{2}{C\sin\eta} = \frac{1}{a^{\frac{1}{2}}(C-a)^{\frac{1}{2}}}, \tag{E.160}$$
which is the equivalent to the equation of motion.
(d) To obtain the second equation, write
$${\rm d}t = a(t){\rm d}\eta = \frac{C}{2}(1-\cos\eta){\rm d}\eta, \tag{E.161}$$
and integrate.
(e) The equations for $a(\eta)$ and $t(\eta)$ describe a cycloid for $a(t)$. Starting at $\eta = 0$, the Universe reaches its maximum radius $a = C$ when $\eta = \pi$, and then reaches the Big Crunch at $\eta = 2\pi$. The variable η has a convenient interpretation in terms of the angle over which light traverses the unit circle, so at $\eta = \pi$ we have light travelling halfway around the circle from each direction. At the moment of the big crunch, when $\eta = 2\pi$, light can therefore, only just, traverse the whole of the circle.

(20.3) (a) We have potential energy
$$U = -\int {\rm d}M(r)\frac{Gm}{r}, \tag{E.162}$$
where ${\rm d}M(r)$ is an element of mass separated from the particle by a distance r. For the ring we have all of the elements of mass at a distance $r = (a^2+x^2)^{\frac{1}{2}}$ from m. We can at this point simply write the answer, or, if the mass per unit length is $\lambda = M/2\pi a$ we can integrate elements of length $a{\rm d}\theta$ to find
$$U = -\frac{GMm}{2\pi ar}\int_0^{2\pi} a{\rm d}\theta = -\frac{GMm}{r} = -\frac{GMm}{(a^2+x^2)^{\frac{1}{2}}}. \tag{E.163}$$

(b) By symmetry, the force can only act along the x-direction (meaning $F_y = F_z = 0$), so we differentiate with respect to x to find
$$F_x = -GMm\frac{x}{(a^2+x^2)^{\frac{3}{2}}}. \tag{E.164}$$
In the limit that $x \gg a$, this becomes $F_x = -GMm/x^2$, the attractive force between two point masses.
(c) The disc may be built from concentric annular rings of radius a, width ${\rm d}a$ and area $2\pi a{\rm d}a$. If we take the total mass per unit area of the disc to be $\sigma = \Omega/(\pi L^2)$, then each annular element has mass ${\rm d}\Omega = 2\pi a{\rm d}a\,\sigma$, and the contribution to the potential energy from each element is
$${\rm d}U = -\frac{Gm{\rm d}\Omega}{(a^2+x^2)^{\frac{1}{2}}} = -Gm2\pi\sigma\frac{a{\rm d}a}{(a^2+x^2)^{\frac{1}{2}}}. \tag{E.165}$$
Integrating and substituting for σ yields
$$U = -\frac{2Gm\Omega}{L^2}\left[(L^2+x^2)^{\frac{1}{2}} - x\right]. \tag{E.166}$$

(20.4) (a) Multiply through by a velocity component u^α and sum to find
$$mu^\alpha a_\alpha = -m\frac{\partial\Phi}{\partial x^\alpha}u^\alpha. \tag{E.167}$$
The right-hand side will not vanish in general.
(b) Multiplying through by u_α, the bracketed part becomes
$$u_\alpha(\eta^\alpha + u_\alpha u^\alpha u^\beta), \tag{E.168}$$
which, noting that $u_\alpha u^\alpha = -1$, gives $u^\beta - u^\beta = 0$.

(20.5) The equation of motion can be rewritten as
$$\frac{{\rm d}p^\alpha}{{\rm d}x^\beta}\frac{{\rm d}x^\beta}{{\rm d}\lambda}\frac{1}{m} = -m\left(\eta^{\alpha\beta} + u^\alpha u^\beta\right)\frac{\partial\Phi}{\partial x^\beta}$$
$$\left(\frac{{\rm d}p^\alpha}{{\rm d}x^\beta} + p^\alpha\frac{\partial\Phi}{\partial x^\beta}\right)\frac{{\rm d}x^\beta}{{\rm d}\lambda}\frac{1}{m} = -m\eta^{\alpha\beta}\frac{\partial\Phi}{\partial x^\beta}. \tag{E.169}$$
The right-hand side vanishes as $m \to 0$ leaving
$$\frac{{\rm d}p^\alpha}{{\rm d}x^\beta} + p^\alpha\frac{\partial\Phi}{\partial x^\beta} = 0, \tag{E.170}$$
whose solution is $p^\alpha e^\Phi = {\rm (const)}$ for all of the coordinates. As a result, the ratios of the components of the momentum remain constant along the world line. There can therefore be no deflection of the light rays by the field.

(21.1) We find
$$T^{tt} = \rho e^{-2\Phi}, \quad T^{rr} = p e^{-2\Lambda},$$
$$T^{\theta\theta} = \frac{p}{r^2}, \quad T^{\phi\phi} = \frac{p}{r^2\sin\theta}. \tag{E.171}$$

(21.5) (a) We find $g = r^2\sin\theta$ and so the area is $4\pi r^2$.
(b) The circumference is $2\pi r\sin\theta$.

(21.6) (a) Consider a spherical mass of fluid and a spherical-shell mass element of the fluid at a radius r, with surface area A, mass ${\rm d}m$ and width ${\rm d}r$. There is an outward-directed force on the inner surface of the element of $p(r)A$, resulting from the fluid closer to the centre. There are inward forces from (i) fluid outside the element, pushing back on the outer surface with force $p(r+{\rm d}r)A$; and (ii) the gravitational pull of the fluid closer to the centre of $Gm(r){\rm d}m/r^2$ (i.e. the fluid at a smaller radius acts as if it is all concentrated at the origin in Newtonian gravitation). Equating forces we have

$$Ap(r) = Ap(r + \mathrm{d}r) + \frac{Gm(r)\mathrm{d}m}{r^2}. \qquad \text{(E.172)}$$

Rearranging gives

$$A\mathrm{d}p = -\frac{Gm(r)\mathrm{d}m}{r^2}, \qquad \text{(E.173)}$$

where $\mathrm{d}p = p(r + \mathrm{d}r) - p(r)$. Finally we note that $A = 4\pi r^2$ and $\mathrm{d}m = \rho A \mathrm{d}r$, where ρ is the density and we write $m(r) = M$.

(21.7) (b) Set $G_{\hat{r}\hat{t}} = 0$ to find $\Lambda_{,t} = 0$.
(c) Set $G_{\hat{t}\hat{t}} = 0$ to find

$$\frac{\mathrm{d}r}{r} = \frac{2\mathrm{d}\Lambda}{1 - \mathrm{e}^{2\Lambda}}. \qquad \text{(E.174)}$$

Integrate this to find

$$r = \frac{C}{\mathrm{e}^{-2\Lambda} - 1}, \qquad \text{(E.175)}$$

and choose $C = -2M$.
(d) Consider $G_{\hat{r}\hat{r}} - G_{\hat{t}\hat{t}} = 0$.
(e) Rescale $\mathrm{d}t' = \mathrm{e}^{f(t)}\mathrm{d}t$ and the Schwarzschild metric is recovered.

(21.8) (a) Compare eqns 21.12 and 21.17, writing $\rho = -p = \zeta$.
(b) Substitution reveals we have a solution if $H^2 = 8\pi\zeta/3$, which is the result for the Hubble constant in the de Sitter Universe.

(22.1) (a) The velocity has components

$$\left[(-g_{00})^{-\frac{1}{2}}, 0, 0, 0\right], \qquad \text{(E.176)}$$

with g_{00} static, so the acceleration becomes

$$a^\mu = (u^0)^2 \Gamma^\mu{}_{00}, \qquad \text{(E.177)}$$

where the connection is given by $\Gamma^\mu{}_{00} = -\frac{1}{2}g_{00,\sigma}g^{\mu\sigma}$. We then have $a^0 = 0$ and

$$a^i = \frac{1}{2}\left(\frac{g^{ij}}{g_{00}}\right)g_{00,j}. \qquad \text{(E.178)}$$

(b) Plugging in the components of the Schwarzschild metric we obtain $g_{00,1} = -2M/r^2$ and so $a^\mu = (0, M/r^2, 0, 0)$. That is, despite not moving, the particle is accelerating outwards as it is not following a geodesic.
(c) Recall from Chapter 2 that the square of the proper acceleration α^2 is equal to $\boldsymbol{a}^2 = (1-2M/r)^{-1}(M^2/r^4)$, so α is given by

$$\alpha = \frac{M}{r^2}\left(\frac{r}{r - 2M}\right)^{\frac{1}{2}}. \qquad \text{(E.179)}$$

(22.2) (a) The equations of motion are as follows:
(i) For the t coordinate

$$\ddot{t} + \frac{2M}{r(r - 2M)}\dot{r}\dot{t} = 0. \qquad \text{(E.180)}$$

(ii) For the r coordinate

$$\ddot{r} - \frac{M}{r(r - 2M)}\dot{r}^2 + \frac{M(r - 2M)}{r^3}\dot{t}^2 - (r - 2M)\dot{\theta}^2$$
$$- (r - 2M)\sin^2\theta\,\dot{\phi}^2 = 0. \qquad \text{(E.181)}$$

(iii) For the θ coordinate

$$\ddot{\theta} + \frac{2}{r}\dot{r}\dot{\theta} - \sin\theta\cos\theta\,\dot{\phi}^2 = 0. \qquad \text{(E.182)}$$

(iv) For the ϕ coordinate

$$\ddot{\phi} + \frac{2}{r}\dot{r}\dot{\phi} + \frac{2\cos\theta}{\sin\theta}\dot{\theta}\dot{\phi} = 0. \qquad \text{(E.183)}$$

(b) Set $\theta = \pi/2$ and $\dot{\theta} = 0$. We obtain the following, in terms of the conserved quantities:
(i) For the t coordinate

$$\frac{\mathrm{d}}{\mathrm{d}\tau}\tilde{E} = 0. \qquad \text{(E.184)}$$

(ii) For the r coordinate

$$\ddot{r} - \frac{M}{r(r - 2M)}\dot{r}^2 + \frac{M\tilde{E}^2}{r(r - 2M)} - \frac{(r - 2M)\tilde{L}^2}{r^4} = 0. \qquad \text{(E.185)}$$

(iii) For the θ coordinate $\ddot{\theta} = 0$.
(iv) For the ϕ coordinate

$$\frac{\mathrm{d}}{\mathrm{d}\tau}\tilde{L} = 0. \qquad \text{(E.186)}$$

(c) Substituting $\dot{r}$ from the effective-energy equation into the equation of motion for $\ddot{r}$ gives a simplified equation of motion

$$\ddot{r} + \frac{M}{r^2} + \frac{3M\tilde{L}^2}{r^4} - \frac{\tilde{L}^2}{r^3} = 0. \qquad \text{(E.187)}$$

Differentiating the effective energy equation with respect to proper time gives an identical equation, so these are consistent.

(22.3) Using the conservation of p_t along the geodesic, we can relate the values of p^t at radius R and at ∞ by raising the index and writing g_{tt} on both sides

$$\left(1 - \frac{2M}{R}\right)p^t(R) = p^t(\infty). \qquad \text{(E.188)}$$

We are looking to write this in terms of measured values $p^{\hat{t}}$ [where $p^{\hat{t}} = (e_{\hat{t}})^t p^{\hat{t}}$], so we use the vielbein component $(e_{\hat{t}})^t = (1 - 2M/r)^{-\frac{1}{2}}$ so we have

$$\left(1 - \frac{2M}{R}\right)^{\frac{1}{2}}p^{\hat{t}}(R) = p^{\hat{t}}(\infty). \qquad \text{(E.189)}$$

Setting $p^{\hat{t}}(R) = \hbar\omega_R$ and $p^{\hat{t}}(\infty) = \hbar\omega_\infty$, the answer follows.

(22.5) (a) The escape velocity is defined such that the velocity u^r at $r = \infty$ is zero. At $r = \infty$ we have $\tilde{E} = 1$, which is then true along the geodesic. This gives $u^t = (1 - 2M/r)^{-1}$. The effective energy equation, or $u^\mu u_\mu = -1$, then gives $u^r = (2M/r)^{\frac{1}{2}}$. The coordinate velocity at $r = R$ is then

$$\frac{\mathrm{d}r}{\mathrm{d}t} = \frac{u^r}{u^t} = \left(\frac{2M}{R}\right)^{\frac{1}{2}}\left(1 - \frac{2M}{R}\right). \qquad \text{(E.190)}$$

The observer makes observations in their rest frame and so, using the vielbein components in the chapter, we have

$$\frac{\mathrm{d}\hat{r}}{\mathrm{d}\hat{t}} = \frac{(e_r)^{\hat{r}}}{(e_t)^{\hat{t}}}\frac{\mathrm{d}r}{\mathrm{d}t} = \left(1 - \frac{2M}{R}\right)^{-1}\frac{\mathrm{d}r}{\mathrm{d}t} = \left(\frac{2M}{R}\right)^{\frac{1}{2}}. \qquad \text{(E.191)}$$

This is the same as the non-relativistic result.
(b) The energy measured by an observer is $E = -m\boldsymbol{u} \cdot \boldsymbol{u}_{\mathrm{obs}}$ or (more simply) $p^{\hat{t}} = m u^{\hat{t}} = m u^t (\boldsymbol{e}_t)^{\hat{t}}$. These give

$$p^{\hat{t}} = m \left(1 - \frac{2M}{R}\right)^{-\frac{1}{2}}. \tag{E.192}$$

(22.6) (a) Write the angular speed of the earth as $\omega = \mathrm{d}\phi/\mathrm{d}t$ and we obtain, restoring factors,

$$\mathrm{d}\tau_1 = \left(1 - \frac{2GM}{c^2 r} - \frac{r^2 \omega^2}{c^2}\right)^{\frac{1}{2}} \mathrm{d}t. \tag{E.193}$$

The second and third terms are small and so the answer follows on expanding the bracket to first order.
(b) Make the substitution $r \to r + h$ and $\omega r \to \omega(r + h) + v$ and expand, noting that $h \ll r$.
(c) Restoring factors we have that

$$\Delta \approx -\frac{GMh}{c^2 r^2} + \frac{v}{2c^2}(v + 2\omega r). \tag{E.194}$$

Take $v = 250 \ \mathrm{ms}^{-1}$, $h = 10^4 \ \mathrm{m}$ and $\omega = 10^{-4} \ \mathrm{rads}^{-1}$, and note that $GM/r^2 \approx 10 \ \mathrm{ms}^{-2}$, to find $\Delta \approx 1 \times 10^{-12}$ for eastward flight.
(d) For westward flight substitute $v \to -v$ to find $\Delta \approx -2 \times 10^{-12}$.

(22.7) The observer at rest has velocity $\boldsymbol{u}$ with components $u^\mu = (u^t, 0, 0, 0)$, from which we can use the constraint on velocity to evaluate

$$g_{tt}(r_1)(u^t)^2 = -1, \tag{E.195}$$

giving

$$u^t = \left(1 - \frac{2M}{r_1}\right)^{-\frac{1}{2}}. \tag{E.196}$$

Since we are planning to take a dot product, we only need to evaluate the timelike component of the other observer's velocity $\boldsymbol{v}$. This observer is free falling, and so we have that $-\tilde{E} = v_t(r_2) = v_t(r_1)$. Since the observer is at rest at r_2 we have that $v^t(r_2) = \left(1 - \frac{2M}{r_2}\right)^{-\frac{1}{2}}$ and so

$$v_t(r_2) = -\left(1 - \frac{2M}{r_2}\right)^{\frac{1}{2}} = v_t(r_1), \tag{E.197}$$

giving

$$v^t(r_1) = g^{tt}(r_1) v_t(r_1) = \left(1 - \frac{2M}{r_1}\right)^{-1} \left(1 - \frac{2M}{r_2}\right)^{\frac{1}{2}}. \tag{E.198}$$

The dot product then yields

$$\begin{aligned} -\gamma(v_{\mathrm{rel}}) &= \boldsymbol{u} \cdot \boldsymbol{v} = g_{tt}(r_1) u^t(r_1) v^t(r_1) \\ &= -\left(1 - \frac{2M}{r_1}\right)\left(1 - \frac{2M}{r_1}\right)^{-\frac{1}{2}} \\ &\quad \times \left(1 - \frac{2M}{r_1}\right)^{-1}\left(1 - \frac{2M}{r_2}\right)^{\frac{1}{2}} \\ &= -\left(\frac{1 - 2M/r_2}{1 - 2M/r_1}\right)^{\frac{1}{2}}. \end{aligned} \tag{E.199}$$

Since $\gamma(v_{\mathrm{rel}}) = (1 - v_{\mathrm{rel}}^2)^{-\frac{1}{2}}$, the latter can be rearranged to access v_{rel}.

(23.1) (a) We have $\mathrm{d}r = \mathrm{d}\theta = 0$, and so the increment of proper time taken along the orbit is

$$\mathrm{d}\tau^2 = \left(1 - \frac{2M}{r}\right)\mathrm{d}t^2 - r^2 \mathrm{d}\phi^2. \tag{E.200}$$

Substituting $\mathrm{d}\phi^2 = \frac{v^2}{r^2}\mathrm{d}t^2$ we find

$$\mathrm{d}\tau = \pm \left[\left(1 - \frac{2M}{r}\right) - v^2\right]^{\frac{1}{2}} \mathrm{d}t. \tag{E.201}$$

Everything in the square bracket is constant, and the integral $\int \mathrm{d}t$ over a period T yields $\int \mathrm{d}t = 2\pi r/v$, giving

$$\Delta\tau = \frac{2\pi r}{v}\left[\left(1 - \frac{2M}{r}\right) - v^2\right]^{\frac{1}{2}}. \tag{E.202}$$

(b) Using the definition of $\tilde{L}$ and eqn 23.11 we find

$$(u^t)^2 = \tilde{L}^2/Mr. \tag{E.203}$$

Now use this with the definition of $\tilde{E}$, combined with eqn 23.5 to obtain

$$\mathrm{d}\tau = \left(1 - \frac{3M}{r}\right)^{\frac{1}{2}} \mathrm{d}t. \tag{E.204}$$

Kepler's law tells us Δt for an orbit is equal to $2\pi(r^3/M)^{\frac{1}{2}}$, from which we obtain the answer. Kepler's law also implies that $v = (M/r)^{\frac{1}{2}}$, which shows the answers from (a) and (b) are compatible.

(23.2) The method for solving problems of this type is explained in the answer to Exercise 23.1. Computing the proper time from the line element, we find $\Delta\tau = T(1 - v^2)^{\frac{1}{2}}$, where T is the period in coordinate time. We then integrate the equation for $v = e^t R \mathrm{d}\phi/\mathrm{d}t$, to find

$$T = \ln\left(1 - \frac{2\pi R}{v}\right)^{-1}. \tag{E.205}$$

(23.3) A free particle follows a geodesic with constant $\tilde{E} = (1 - 2M/r)u^t$. We therefore have

$$\frac{\mathrm{d}t}{\mathrm{d}\tau} = \left(1 - \frac{2M}{r}\right)^{-1}\tilde{E} = (\text{const.}). \tag{E.206}$$

We conclude that $t = c\tau + d$, with c and d constant. Note that a similar argument applies to the ϕ variable for motion at a constant value of θ.

(23.4) (a) We have $u^\mu = (C, 0, 0, \omega)$ and, using $\boldsymbol{u}^2 = -1$, we find

$$C^2 = \left(\frac{1 + \omega^2 r^2}{1 - 2M/r_0}\right). \tag{E.207}$$

(b) Taking a covariant derivative yields

$$\boldsymbol{\nabla}_{\boldsymbol{u}}\boldsymbol{u} = \left(\Gamma^\mu{}_{tt}C^2 + \Gamma^\mu{}_{t\phi}C\omega + \Gamma^\mu{}_{\phi\phi}\omega^2\right)\boldsymbol{e}_\mu. \tag{E.208}$$

The only connection coefficients we need are $\Gamma^r{}_{tt} = M(r_0 - 2M)/r_0^3$, and $\Gamma^r{}_{\phi\phi} = -(r_0 - 2M)$, giving

$$\boldsymbol{a} = \left(\frac{M}{r_0^2} + 3M\omega^2 - r_0\omega^2\right)\boldsymbol{e}_r. \tag{E.209}$$

(c) The acceleration vanishes when

$$\frac{M}{r_0^2} = (r_0 - 3M)\omega^2. \tag{E.210}$$

For a particle far from the source of the field we need $M/r_0^2 = r_0\omega^2$, i.e. the gravitational acceleration supplies the centripetal acceleration required to keep the particle in orbit. When this is the case the particle follows a geodesic and we have no acceleration.

(23.5) (b) Referring to the figure in the chapter, we cut out a wedge δ such that

$$(2\pi - \delta)R = 2\pi r, \qquad (\text{E.211})$$

where $r = R\cos\alpha$ and $\tan\alpha = dz/dr$. Treating α as a small angle, we have

$$(2\pi - \delta) \approx 2\pi \left[1 - \frac{1}{2} \left(\frac{dz}{dr} \right)^2 \right]. \qquad (\text{E.212})$$

Differentiating $z(r)$ and inserting into this equation, we find

$$\delta \approx \frac{2\pi M}{r - 2M} \approx \frac{2\pi M}{r}, \qquad (\text{E.213})$$

where in the final term we have approximated $r \gg 2M$. (d) One third of the effect is accounted from this contribution. As pointed out in Zee (2013), rubber-sheet models showing a constant-time slice of the Schwarzschild geometry can be misleading in trying to understand the origin of gravitational effects in relativity.

(23.6) (a) We find

$$\frac{\partial u}{\partial \phi} = \mp \left[\tilde{E}^2 (1 - 2Mu)^{-1} - 1 - \tilde{L}^2 u^2 \right]^{\frac{1}{2}}. \qquad (\text{E.214})$$

(c) In the limit of small u_c, we have $u_c \approx M\tilde{E}^2/\tilde{L}^2$.
(d) Substituting, we obtain

$$w'' + w + 1 \approx \frac{M\tilde{E}^2}{\tilde{L}^2} \left(\frac{1}{u_c} + 4M + 4Mw \right), \qquad (\text{E.215})$$

or, in the limit of small u_c

$$w'' + (1 - 4Mu_c)w \approx 0. \qquad (\text{E.216})$$

This is a simple harmonic oscillator equation with characteristic frequency $\omega_0^2 = 1 - 4Mu_c$.
(e) Using the hint we have

$$\omega_0 \Delta\phi = 2\pi, \qquad (\text{E.217})$$

which can be rearranged and expanded to find

$$\Delta\phi \approx 2\pi + \frac{4\pi M}{r_c}. \qquad (\text{E.218})$$

Note that the latter is 2/3 of the total perihelion shift.

(24.3) (a) From Fig. 24.8, we have

$$\theta_i D_s = \theta_s D_s + \alpha D_{\ell s}. \qquad (\text{E.219})$$

However, we also have that $\alpha = 4M/b$ is the deflection angle, where $b = \theta_i D_\ell$, so that

$$\theta_i = \theta_s + \frac{4M D_{\ell s}}{\theta_i D_\ell D_s}. \qquad (\text{E.220})$$

The answer then follows using the definition of θ_E.
(b) We have that $D_{\ell s}$ and D_s are both far larger than D_ℓ, which gives $\theta_E^2 \approx 4M/D_\ell$. For the case where the source, lens, and observer are collinear, we have $\theta_s = 0$ and $\theta_i = \theta_E$. The image is then a ring with this angular radius.

(24.4) (a) Write $d\chi^2 = \frac{dr^2}{1 - kr^2}$. For a radial ray $d\phi = d\theta = 0$ and the result follows.
(b) Metric coefficients are independent of χ along the ray and so we have a Killing vector $\boldsymbol{e}_\chi$ and so $u_\chi = C$ is conserved along the ray. Raising the index using $g^{\chi\chi}$ gives

$$u^\chi = C/a(t)^2. \qquad (\text{E.221})$$

Since $u^\chi = \frac{d\chi}{d\lambda}$, where λ is the affine parameter we have

$$d\lambda = C^{-1}a(t)^2 d\chi = \frac{1}{C} \frac{a(t)^2 dr}{(1 - kr^2)^{\frac{1}{2}}}. \qquad (\text{E.222})$$

(25.1) (a) Substituting yields

$$ds^2 = -(1 - v^2)dT^2 + \left[(1 - v^2)^{-1} - R^2(1 - v^2) \right] dr^2 + 2(1 - v^2)Rdr dT + r^2 d\Omega^2. \qquad (\text{E.223})$$

(b) Setting the square bracket to equal unity yields $R = v(1 - v^2)^{-1}$.
(d) A slice of space at a constant value of T is flat.
(e) The velocity $dr/d\tau$ is unchanged from the Schwarzschild value $u^r = -(2M/r)^{\frac{1}{2}}$. Making the substitution and rearranging we find that we also have

$$\frac{dr}{dT} = - \left(\frac{2M}{r} \right)^{\frac{1}{2}} = -|v|. \qquad (\text{E.224})$$

This is unity at the event horizon, diverging as $r \to 0$.
(f) Setting $ds^2 = 0$ we find that

$$\frac{dr}{dT} = \pm 1 - |v|. \qquad (\text{E.225})$$

(g) We have a ratio Q of the coordinate speed of the in-falling observer $(-|v|)$ to in-falling light $(-1 - |v|)$ of

$$Q = \frac{|v|}{1 + |v|} = \left(\frac{2M}{r} \right)^{\frac{1}{2}} \left[1 + \left(\frac{2M}{r} \right)^{\frac{1}{2}} \right]^{-1} < 1, \qquad (\text{E.226})$$

and so the speed appears subluminal.

(25.2) (a) We first want to evaluate the interval between two events at fixed r, measured by falling clocks. The interval is measured by one clock that falls through r and a second clock that falls through r an interval dt later. Each clock requires the same interval in t to fall to r, so the second clock must have been dropped dt seconds after the first. The difference in $d\check{t}$ registered between the arrival of the clocks is simply the difference dt between the clocks being dropped and so $d\check{t} = dt$.
(b) We are in the same situation as examined in Chapter 22 for radially falling observers. We saw there that $dr/d\tau = -(2M/r)^{\frac{1}{2}}$ and

$$\frac{dr}{dt} = - \left(1 - \frac{2M}{r} \right) \left(\frac{2M}{r} \right)^{\frac{1}{2}}. \qquad (\text{E.227})$$

(c) Write $dt_d = -dr/(dr/dt)$ and substitute from previous equation.
(d) Use eqn 22.31 to compute $d\tau$ via the integral

$$\Delta\tau = - \int_{r_A}^{r_B} dr \left(\frac{r}{2M} \right)^{\frac{1}{2}}. \qquad (\text{E.228})$$

(e) The total time delay is the sum of (i) the delay in drop time and (ii) the delay in proper time to reach their different radial positions at a time t.

(f) We evaluate

$$\frac{\partial \check{t}}{\partial r} = \frac{(\mathrm{d}\tau + \mathrm{d}t_{\mathrm{d}})}{\mathrm{d}r}$$

$$= \left(\frac{r}{2M}\right)^{\frac{1}{2}}\left[-1 + \left(1 - \frac{2M}{r}\right)^{-1}\right]$$

$$= \left(\frac{2M}{r}\right)^{\frac{1}{2}}\left(1 - \frac{2M}{r}\right)^{-1}. \tag{E.229}$$

(25.3) From the previous problem we have that

$$\mathrm{d}\check{t} = \mathrm{d}t + \frac{(2M/r)^{\frac{1}{2}}}{1 - 2M/r}\,\mathrm{d}r. \tag{E.230}$$

Substituting for $\mathrm{d}t$ in the Schwarzschild metric gives the desired answer.

(25.4) (a) The stationary observer has only one non-zero velocity component u^t, determined by the equation $g_{tt}(u^t)^2 = -1$, giving $u^t = (1 - 2M/r)^{-\frac{1}{2}}$. We can therefore write

$$-(1 - v_0^2)^{-\frac{1}{2}} = g_{tt}u^t v^t, \tag{E.231}$$

or

$$v^t = (1 - v_0^2)^{-\frac{1}{2}}(1 - 2M/r)^{-\frac{1}{2}}. \tag{E.232}$$

The only other non-zero component of v^μ is v^ϕ and so, using $\boldsymbol{g}(\boldsymbol{v}, \boldsymbol{v}) = -1$, we have

$$-(1 - v_0^2)^{-1} + r^2(v^\phi)^2 = -1 \tag{E.233}$$

or

$$v^\phi = \left(\frac{v_0^2/r_0^2}{1 - v_0^2}\right)^{\frac{1}{2}}. \tag{E.234}$$

(b) We compute

$$\tilde{E} = (1 - 2M/r)^{\frac{1}{2}}(1 - v_0^2)^{-\frac{1}{2}} \tag{E.235}$$

and

$$\tilde{L} = \frac{v_0 r}{(1 - v_0^2)^{\frac{1}{2}}}. \tag{E.236}$$

(c) Plugging in, we find that for $r = 4M$ we have $\tilde{E} = [2(1 - v_0^2)]^{-\frac{1}{2}}$ and therefore

$$\mathcal{E} = \frac{v_0^2 - \frac{1}{2}}{2 - 2v_0^2}. \tag{E.237}$$

We also have $\tilde{L} = \frac{4Mv_0}{(1 - v_0^2)^{\frac{1}{2}}}$ and

$$V_{\mathrm{eff}}(r) = \frac{v_0^2 - \frac{1}{2}}{2 - 2v_0^2}. \tag{E.238}$$

This makes sense if the particle is deflected at r, since we must have $\mathcal{E} = V_{\mathrm{eff}}(r)$ in order for the motion to be tangential at the deflection point. A value of $v_0 = 1/\sqrt{2}$ gives $\mathcal{E} = V_{\mathrm{eff}} = 0$ and also $\tilde{L}/M = 4M$. From Fig. 23.1 we see that the maximum value of the $\tilde{L}/M = 4M$ curve is $V_{\mathrm{eff}} = 0$, occurring exactly at $r = 4M$. We conclude that the particle is, only just, deflected from this point. In fact, $v_0 = 1/\sqrt{2}$ represents the minimum value of the relative velocity that allows the particle to escape the hole from this point.

(26.4) (a) Note that $\boldsymbol{e}_t = (\partial/\partial t)^\mu$ is a Killing vector, and so u_t is a constant of the motion, just as we've had previously. The constant can be rewritten using the metric components as

$$-\tilde{E} = g_{tt}u^t = -x^2\frac{\mathrm{d}t}{\mathrm{d}\lambda}, \tag{E.239}$$

which is also a constant of the motion.

(b) Substitute for x and t, set $u = \text{const.}$ and integrate to get the expression for the affine parameter for outgoing geodesics.

(27.4) Along with the world-line component $x^1(\tau) = x_0$, we use $\boldsymbol{u}^2 = -1$ to compute $g_{tt}(u^t)^2 = -1$ and find that the velocity has components $u^\mu = (1/x_0, 0)$. Taking a covariant derivative yields an acceleration with non-zero components

$$a^\mu = u^t\frac{\partial u^\mu}{\partial t} + \Gamma^\mu_{tt}(u^t)^2. \tag{E.240}$$

Using the connection coefficients from Exercise 9.4, we conclude $a^\mu = (0, 1/x_0)$. The proper acceleration α has the property $\alpha^2 = \boldsymbol{a}^2$, so we have $\alpha = 1/x_0$ too.

(27.5) (a) Identify τ and the observer's proper time and $\gamma = (1 - v^2)^{-\frac{1}{2}}$ and the result follows.

(b) Substituting into the exponential we obtain

$$A = \exp\left[i\Omega\gamma(1 - v)\tau\right], \tag{E.241}$$

which can be rewritten in terms of an effective frequency as

$$\Omega' = \Omega\left(\frac{1 - v}{1 + v}\right)^{\frac{1}{2}}. \tag{E.242}$$

It's worth noting that the instantaneous frequency can be extracted by operating on the phase with $-i\partial/\partial\tau$.

(c) Repeating we obtain

$$A = \exp\left[-i\frac{\Omega}{g}\left(e^{-g\tau}\right)\right]. \tag{E.243}$$

Acting on the phase with $-i\partial/\partial\tau$ yields up the frequency

$$\omega(\tau) = \Omega e^{-g\tau}. \tag{E.244}$$

(28.1) The area A is given by

$$A = \int \mathrm{d}A = \int \sqrt{g_{\theta\theta}g_{\phi\phi}}\,\mathrm{d}\theta\mathrm{d}\phi$$

$$= \int r_{\mathrm{S}}^2 \sin\theta\,\mathrm{d}\theta\mathrm{d}\phi$$

$$= 4\pi r_{\mathrm{S}}^2. \tag{E.245}$$

Since $r_{\mathrm{S}} = 2M$, we have $A = 16\pi M^2$.

(28.2) Use $M = \hbar/(8\pi k_{\mathrm{B}}T)$, $A = 16\pi M^2$, $S = Ak_{\mathrm{B}}/(4\hbar)$, and the result follows. The negative sign results from the fact that when a Schwarzschild black hole accretes matter, its energy and mass increases, but its temperature *decreases*. Similarly, when the black hole evaporates, it radiates energy, its mass decreases, and its temperature increases. Hence the negative heat capacity.

(28.3) (a) The required metric components, in matrix form for the coordinates (V, r), are

$$g_{\mu\nu} = \begin{pmatrix} -1 + \frac{2M}{r} & 1 \\ 1 & 0 \end{pmatrix}. \qquad \text{(E.246)}$$

Acting on ξ^μ with this matrix gives ξ_ν.

(c) The motivation for the acrobatics in this problem is to use $\xi^\mu = (1, 0, 0, 0)$ to simplify the expressions. Start by choosing $\sigma = V$ in $-\xi^\mu \xi_\mu{}^{;\sigma} = \kappa \xi^\sigma$ and then note that only the $\mu = V$ component survives in the contraction to obtain $\xi_V{}^{;V} = -\kappa$. There is a rule for dealing more directly with this equation for the divergence that we shall meet later. However, for now we can write

$$\xi_{V;\sigma} g^{\sigma V} = -\kappa. \qquad \text{(E.247)}$$

Then using

$$g^{\mu\nu} = \begin{pmatrix} 0 & 1 \\ 1 & 1 - \frac{2M}{r} \end{pmatrix}, \qquad \text{(E.248)}$$

we obtain the answer.

(28.4) (d) The angle ψ repeats after 2π radians so that $\psi = \psi + 2\pi n$, with n an integer. Since $\psi = t_{\rm E}/4M$, this gives the repeat period in imaginary time as $\Delta t_{\rm E} = 8\pi M$, corresponding to a temperature

$$k_{\rm B} T = \frac{\hbar c^3}{8\pi G M}. \qquad \text{(E.249)}$$

(29.6) We want to compute

$$\begin{pmatrix} g_{tt} & g_{t\phi} \\ g_{\phi t} & g_{\phi\phi} \end{pmatrix}^{-1} = \frac{1}{D} \begin{pmatrix} g_{\phi\phi} & -g_{t\phi} \\ -g_{\phi t} & g_{tt} \end{pmatrix}, \qquad \text{(E.250)}$$

where D is the determinant of the matrix in the question. Some algebra reveals $D = -\Delta \sin^2 \theta$, from which the desired equations follow straightforwardly.

(29.8) Using $d\tau^2 = -ds^2$, we find

$$d\tau^2 = \left[\left(1 - \frac{2M}{R}\right) + \frac{4Ma}{R^2} v \right.$$
$$\left. - \left(1 + \frac{a^2}{R^2} + \frac{2Ma^2}{R^3}\right) v^2 \right] dt^2. \qquad \text{(E.251)}$$

Since everything in the square braces is time independent, we have

$$d\tau = T \left[\left(1 - \frac{2M}{R}\right) + \frac{4Ma}{R^2} v \right.$$
$$\left. - \left(1 + \frac{a^2}{R^2} + \frac{2Ma^2}{R^3}\right) v^2 \right]^{\frac{1}{2}}, \qquad \text{(E.252)}$$

with $T = 2\pi R/v$.

(30.1) Use the equation $d\boldsymbol{t}/ds = \kappa \boldsymbol{p}$ and, for a time t, note that $d\boldsymbol{t}/ds = (d\boldsymbol{t}/dt)(dt/ds)$, with $ds/dt = 1$ for motion at unit speed. We therefore have

$$\frac{d\boldsymbol{t}}{dt} = \kappa \boldsymbol{p}, \qquad \text{(E.253)}$$

which is Newton's second law for a unit mass with velocity $\boldsymbol{t}$, subject to a force $\boldsymbol{F} = \kappa \boldsymbol{p}$, where $\boldsymbol{p}$ is a unit vector perpendicular to the velocity $\boldsymbol{t}$ and hence also to the curve.

(30.2) Consider the path from pole to equator and back shown in Fig. 11.2, where the vector ends up $90°$ out from where it started. We have

$$(\text{curvature}) = \frac{\pi/2}{(1/8)4\pi a^2} = \frac{1}{a^2}. \qquad \text{(E.254)}$$

(30.3) Start at the origin $(r, \theta) = (0, 0)$. The distance to point $(\varepsilon, 0)$ is

$$\int_{r=0}^{\varepsilon} ds = \int_0^\varepsilon dr = \varepsilon. \qquad \text{(E.255)}$$

The circumference of a circle with this radius is

$$\int_{\theta=0}^{2\pi} d\theta \, \sin \varepsilon = 2\pi \sin \varepsilon. \qquad \text{(E.256)}$$

We conclude

$$K = \lim_{\varepsilon \to 0} \frac{6}{\varepsilon^2} \left(1 - \frac{\sin \varepsilon}{\varepsilon}\right) = 1. \qquad \text{(E.257)}$$

As this does not vanish, the space is curved.

(30.5) (a) In the final step, we've used the result that

$$\hat{\boldsymbol{n}} \cdot \frac{d\boldsymbol{t}}{ds} = t^\mu t^\nu \frac{\partial \boldsymbol{e}_\mu}{\partial x^\nu} \cdot \hat{\boldsymbol{n}}$$
$$= t^\mu t^\nu K_{\mu\nu}.$$

(b) Write

$$g_{\mu\sigma} t^\mu \frac{\partial n^\sigma}{\partial x^\nu} \frac{dx^\nu}{ds} = -K_{\mu\nu} t^\mu t^\nu$$
$$\left(g_{\mu\sigma} \frac{\partial n^\sigma}{\partial x^\nu} + K_{\mu\nu}\right) t^\mu t^\nu = 0. \qquad \text{(E.258)}$$

(c) Weingarten's equation allows us to say

$$\frac{\partial \hat{\boldsymbol{n}}}{\partial x^1} \times \frac{\partial \hat{\boldsymbol{n}}}{\partial x^2} = K^\mu{}_1 \boldsymbol{e}_\mu \times K^\nu{}_2 \boldsymbol{e}_\nu$$
$$= \det(K^i{}_j)(\boldsymbol{e}_1 \times \boldsymbol{e}_2). \qquad \text{(E.259)}$$

Note then that $K^i{}_j = g^{ia} K_{aj}$ and so we obtain

$$\hat{\boldsymbol{n}} \cdot \left(\frac{\partial \hat{\boldsymbol{n}}}{\partial x^1} \times \frac{\partial \hat{\boldsymbol{n}}}{\partial x^2}\right) = \frac{K}{g} |\boldsymbol{e}_1 \times \boldsymbol{e}_2|. \qquad \text{(E.260)}$$

(30.7) (a) A computation yields $P^\mu{}_\nu$.
(b) Compute $\bar{X}^\mu n_\mu$, to show it vanishes. This actually follows from $P^\mu n_\mu = 0$.

(31.2) A suitable 1-form is $\tilde{\boldsymbol{W}}(\ ) = f_i dx^i(\ )$. Inserting $\vec{X}$ we find

$$\tilde{\boldsymbol{W}}(\vec{X}) = f_i X^i, \qquad \text{(E.261)}$$

which is the work done.

(31.3) (a) Expanding, we have

$$T^{(\alpha\beta\gamma)}{}_{\delta\epsilon\zeta} = \frac{1}{6}\left(T^{\alpha\beta\gamma}{}_{\delta\epsilon\zeta} + T^{\alpha\gamma\beta}{}_{\delta\epsilon\zeta} + T^{\beta\gamma\alpha}{}_{\delta\epsilon\zeta} \right.$$
$$\left. + T^{\beta\alpha\gamma}{}_{\delta\epsilon\zeta} + T^{\gamma\alpha\beta}{}_{\delta\epsilon\zeta} + T^{\gamma\beta\alpha}{}_{\delta\epsilon\zeta}\right). \qquad \text{(E.262)}$$

(b) We find

$$T^{\alpha\beta\gamma}{}_{[\delta\epsilon\zeta]} = \frac{1}{6}\left(T^{\alpha\beta\gamma}{}_{\delta\epsilon\zeta} - T^{\alpha\beta\gamma}{}_{\delta\zeta\epsilon} + T^{\alpha\beta\gamma}{}_{\epsilon\zeta\delta} \right.$$
$$\left. - T^{\alpha\beta\gamma}{}_{\epsilon\delta\zeta} + T^{\alpha\beta\gamma}{}_{\zeta\delta\epsilon} - T^{\alpha\beta\gamma}{}_{\zeta\epsilon\delta}\right). \qquad \text{(E.263)}$$

(31.4) Write

$$A^{\mu\nu}T_{\mu\nu} = \frac{1}{2}\left(A^{\mu\nu}T_{\mu\nu} + A^{\nu\mu}T_{\nu\mu}\right). \qquad \text{(E.264)}$$

For (a), this becomes $\frac{1}{2}A^{\mu\nu}\left(T_{\mu\nu}+T_{\nu\mu}\right)$ or $A^{\mu\nu}T_{(\mu\nu)}$. In the same way, for (b) we get $A^{\mu\nu}T_{[\mu\nu]}$.

(31.5) (b) For an antisymmetric tensor, the answer simplifies to

$$F_{[\mu\nu;\lambda]} = \frac{1}{3}\left(F_{\mu\nu;\lambda} + F_{\nu\lambda;\mu} + F_{\lambda\mu;\nu}\right). \qquad \text{(E.265)}$$

(31.6) Remove mention of the point ξ^μ on either side. Then interpret the combination $\mathrm{d}/\mathrm{d}\tau$ as a vector. This can be written as $\mathrm{d}/\mathrm{d}\tau = (\mathrm{d}x^\beta/\mathrm{d}\tau)(\partial/\partial x^\beta)$, interpreted as a set of components multiplied by a set of basis vectors, $\boldsymbol{e}_\beta = \partial/\partial x^\beta$:

$$\frac{\mathrm{d}}{\mathrm{d}\tau} = \frac{\mathrm{d}x^\mu}{\mathrm{d}\tau}\frac{\partial}{\partial x^\mu} = \frac{\mathrm{d}x^\mu}{\mathrm{d}\tau}\boldsymbol{e}_\mu. \qquad \text{(E.266)}$$

This allows us to write

$$\begin{aligned}
\frac{\partial^2 \xi^\mu}{\partial x^\lambda \partial x^\sigma} &= \Gamma^\nu{}_{\sigma\lambda}\frac{\partial \xi^\mu}{\partial x^\nu}\\
\frac{\partial^2}{\partial x^\lambda \partial x^\sigma} &= \Gamma^\nu{}_{\sigma\lambda}\frac{\partial}{\partial x^\nu}\\
\frac{\partial}{\partial x^\lambda}\frac{\partial}{\partial x^\sigma} &= \Gamma^\nu{}_{\sigma\lambda}\frac{\partial}{\partial x^\nu}\\
\frac{\partial \boldsymbol{e}_\sigma}{\partial x^\lambda} &= \Gamma^\nu{}_{\sigma\lambda}\boldsymbol{e}_\nu. \qquad \text{(E.267)}
\end{aligned}$$

(31.7) (b) Diagram (ii) represents $A^\mu{}_{\alpha\beta}B^\alpha$. Diagram (iii) represents $A^\mu{}_{\alpha\beta[\gamma}C^{\rho\beta}{}_{\sigma\lambda]}$.
(c) Briefly, (iv) represents $\boldsymbol{\nabla}_\mu F^{\mu\nu} = J^\nu$ and (v) is $F_{\mu\nu;\lambda} + F_{\nu\lambda;\mu} + F_{\lambda\mu;\nu} = 0$. See Chapter 42 for more detail.
(d) The diagram can be written as $\left(\boldsymbol{\nabla}_\alpha\boldsymbol{\nabla}_\beta - \boldsymbol{\nabla}_\beta\boldsymbol{\nabla}_\alpha\right)Z^\mu = R^\mu{}_{\nu\alpha\beta}Z^\nu$. This expression is examined in Chapter 35.

(32.2) (a) The first equation can be written as

$$u^\alpha a^\mu{}_{;\alpha} = -A^\mu(u^\beta a_\beta) + u^\mu(A^\sigma a_\sigma). \qquad \text{(E.268)}$$

(b) The acceleration $\boldsymbol{A} = 0$ along a geodesic, so Fermi transport becomes equivalent to parallel transport.
(c) Consider the dot product $\boldsymbol{a}\cdot\boldsymbol{b}$,

$$\begin{aligned}
\boldsymbol{\nabla}_u(\boldsymbol{a}\cdot\boldsymbol{b}) &= (\boldsymbol{\nabla}_u\boldsymbol{a})\cdot\boldsymbol{b} + \boldsymbol{a}\cdot(\boldsymbol{\nabla}_u\boldsymbol{b})\\
&= -(\boldsymbol{A}\wedge\boldsymbol{u})\left(\tilde{b},\tilde{a}\right) - (\boldsymbol{A}\wedge\boldsymbol{u})\left(\tilde{a},\tilde{b}\right)\\
&= 0. \qquad \text{(E.269)}
\end{aligned}$$

This implies that all angular relationships between the vectors that are Fermi transported are maintained.
(d) Consider the dot product $\boldsymbol{a}\cdot\boldsymbol{u}$,

$$\begin{aligned}
\boldsymbol{\nabla}_u(\boldsymbol{a}\cdot\boldsymbol{u}) &= (\boldsymbol{\nabla}_u\boldsymbol{a})\cdot\boldsymbol{u} + \boldsymbol{a}\cdot(\boldsymbol{\nabla}_u\boldsymbol{u})\\
&= -(\boldsymbol{A}\wedge\boldsymbol{u})\left(\tilde{u},\tilde{a}\right) + \boldsymbol{a}\cdot\boldsymbol{A}\\
&= -[A(\tilde{u})u(\tilde{a}) - u(\tilde{u})A(\tilde{a})] + \boldsymbol{a}\cdot\boldsymbol{A}\\
&= A(\tilde{a}) + \boldsymbol{a}\cdot\boldsymbol{A} = 0, \qquad \text{(E.270)}
\end{aligned}$$

where we have used the facts that: $A(\tilde{u}) = 0$ and $u(\tilde{u}) = -1$.
(e) Ordinary parallel transport does not maintain the angle between vectors, but Fermi transport does by virtue of the term $-(\boldsymbol{A}\wedge\boldsymbol{u})$, which projects the vector after parallel transport so that it is perpendicular to tangent and relative angles between $\boldsymbol{a}$, $\boldsymbol{b}$ and $\boldsymbol{c}$ are maintained.

(33.1) Write

$$\begin{aligned}
\mathcal{P}_4 - \mathcal{P}_3 &= [\boldsymbol{u}(\mathcal{P}_0) + \boldsymbol{v}(\mathcal{P}_1)] - [\boldsymbol{u}(\mathcal{P}_2) + \boldsymbol{v}(\mathcal{P}_0)]\\
&= [\boldsymbol{v}(\mathcal{P}_1) - \boldsymbol{v}(\mathcal{P}_0)] - [\boldsymbol{u}(\mathcal{P}_2) - \boldsymbol{u}(\mathcal{P}_0)]\\
&\approx \left(\frac{\partial v^\beta}{\partial x^\alpha}u^\alpha\boldsymbol{e}_\beta\right)_{\mathcal{P}_0} - \left(\frac{\partial u^\beta}{\partial x^\alpha}v^\alpha\boldsymbol{e}_\beta\right)_{\mathcal{P}_0}\\
&\approx [\boldsymbol{u},\boldsymbol{v}]_{\mathcal{P}_0}. \qquad \text{(E.271)}
\end{aligned}$$

(33.2) (a) We find

$$(\pounds_u\boldsymbol{A})^\mu_\nu = u^\alpha\frac{\partial A^\mu_\nu}{\partial x^\alpha} - A^\alpha_\nu\frac{\partial u^\mu}{\partial x^\alpha} + A^\mu_\alpha\frac{\partial u^\alpha}{\partial x^\nu}. \qquad \text{(E.272)}$$

(b) Using the previous result

$$\begin{aligned}
(\pounds_u\delta^\mu_\nu)^\mu_\nu &= u^\alpha\frac{\partial\left(\delta^\mu_\nu\right)}{\partial x^\alpha} - \delta^\alpha_\nu\frac{\partial u^\mu}{\partial x^\alpha} + \delta^\mu_\alpha\frac{\partial u^\alpha}{\partial x^\nu}\\
&= -\frac{\partial u^\mu}{\partial x^\nu} + \frac{\partial u^\mu}{\partial x^\nu} = 0. \qquad \text{(E.273)}
\end{aligned}$$

As a result the contraction (which is equivalent to the delta function) can go either side of the Lie derivative.

(33.4) If $\boldsymbol{u}$ and $\boldsymbol{v}$ are Killing fields vectors then

$$\pounds_u\boldsymbol{g} = \pounds_v\boldsymbol{g} = 0. \qquad \text{(E.274)}$$

Since $\pounds_u\pounds_v - \pounds_v\pounds_u = \pounds_{[u,v]}$, then we must have

$$\pounds_{[u,v]}\boldsymbol{g} = 0, \qquad \text{(E.275)}$$

so the commutator is also a Killing vector.

(34.1) We won't assume the commutators vanish. We have that

$$\begin{aligned}
\boldsymbol{\nabla}_{u+n}\boldsymbol{v} - \boldsymbol{\nabla}_v(\boldsymbol{u}+\boldsymbol{n}) &= [\boldsymbol{u}+\boldsymbol{n},\boldsymbol{v}] = [\boldsymbol{u},\boldsymbol{v}] + [\boldsymbol{n},\boldsymbol{v}]\\
&= \boldsymbol{\nabla}_u\boldsymbol{v} - \boldsymbol{\nabla}_v\boldsymbol{u} + \boldsymbol{\nabla}_n\boldsymbol{v} - \boldsymbol{\nabla}_v\boldsymbol{n}. \qquad \text{(E.276)}
\end{aligned}$$

Expanding the second term on the left and rearranging, we have

$$\begin{aligned}
\boldsymbol{\nabla}_{u+n}\boldsymbol{v} - \boldsymbol{\nabla}_v\boldsymbol{u} - \boldsymbol{\nabla}_v\boldsymbol{n} &= \boldsymbol{\nabla}_u\boldsymbol{v} - \boldsymbol{\nabla}_v\boldsymbol{u} + \boldsymbol{\nabla}_n\boldsymbol{v} - \boldsymbol{\nabla}_v\boldsymbol{n}\\
\boldsymbol{\nabla}_{u+n}\boldsymbol{v} &= \boldsymbol{\nabla}_u\boldsymbol{v} + \boldsymbol{\nabla}_n\boldsymbol{v}. \qquad \text{(E.277)}
\end{aligned}$$

This is a result that follows from the first rule for the covariant derivative given in the chapter.

(34.3) (a) We have

$$\begin{aligned}
\delta\ln\det\underline{M} &= \ln\det(\underline{M}+\delta\underline{M}) - \ln\det\underline{M}\\
&= \ln\frac{\det(\underline{M}+\delta\underline{M})}{\det\underline{M}}\\
&= \ln\det\underline{M}^{-1}(\underline{M}+\delta\underline{M})\\
&= \ln\det(I + \underline{M}^{-1}\delta\underline{M})\\
&\approx \ln(I + \mathrm{Tr}\underline{M}^{-1}\delta\underline{M})\\
&\approx \mathrm{Tr}\underline{M}^{-1}\delta\underline{M}. \qquad \text{(E.278)}
\end{aligned}$$

(b) Using the identity

$$\Gamma^\mu{}_{\lambda\mu} = \frac{1}{2}\frac{\partial}{\partial x^\lambda}\ln|g| = \frac{1}{\sqrt{|g|}}\frac{\partial}{\partial x^\lambda}\sqrt{|g|}. \qquad \text{(E.279)}$$

(c) Use

$$v^\mu{}_{;\mu} = \frac{\partial v^\mu}{\partial x^\mu} + \Gamma^\mu{}_{\mu\lambda}v^\lambda, \qquad \text{(E.280)}$$

and the answer follows.

(34.4) We have

$$u^\alpha \frac{\partial f}{\partial x^\alpha} = \boldsymbol{\nabla}_{\boldsymbol{u}} \langle \tilde{\boldsymbol{\sigma}}, \boldsymbol{v} \rangle$$

$$u^\alpha \frac{\partial (\sigma_\mu v^\mu)}{\partial x^\alpha} = \langle \boldsymbol{\nabla}_{\boldsymbol{u}} \tilde{\boldsymbol{\sigma}}, \boldsymbol{v} \rangle + \langle \tilde{\boldsymbol{\sigma}}, \boldsymbol{\nabla}_{\boldsymbol{u}} \boldsymbol{v} \rangle$$

$$u^\alpha v^\mu \frac{\partial \sigma_\mu}{\partial x^\alpha} + u^\alpha \sigma_\mu \frac{\partial v^\mu}{\partial x^\alpha} = v^\mu \left(\boldsymbol{\nabla}_{\boldsymbol{u}} \tilde{\boldsymbol{\sigma}} \right)_\mu + \sigma_\mu \left(\boldsymbol{\nabla}_{\boldsymbol{u}} \boldsymbol{v} \right)^\mu$$

$$= v^\mu \left(\boldsymbol{\nabla}_{\boldsymbol{u}} \tilde{\boldsymbol{\sigma}}_\mu \right)$$

$$+ \sigma_\mu \left(u^\alpha \frac{\partial v^\mu}{\partial x^\alpha} + \Gamma^\mu{}_{\alpha\gamma} v^\gamma u^\alpha \right)$$

$$u^\alpha v^\mu \frac{\partial \sigma_\mu}{\partial x^\alpha} = v^\mu \left(\boldsymbol{\nabla}_{\boldsymbol{u}} \tilde{\boldsymbol{\sigma}} \right)_\mu + \Gamma^\mu{}_{\alpha\gamma} v^\gamma u^\alpha \sigma_\mu$$

$$v^\mu \left(\boldsymbol{\nabla}_{\boldsymbol{u}} \tilde{\boldsymbol{\sigma}} \right)_\mu = u^\alpha v^\mu \frac{\partial \sigma_\mu}{\partial x^\alpha} - \Gamma^\lambda{}_{\alpha\mu} v^\mu u^\alpha \sigma_\lambda, \tag{E.281}$$

where, in the last equation, we've relabelled the indices in the final term.

(34.5) (a) Writing components, we find

$$(\boldsymbol{\nabla}_{\boldsymbol{u}} \boldsymbol{u}) \cdot \boldsymbol{e}_1 = u^\alpha u^\mu{}_{;\alpha} (\boldsymbol{e}_\mu \cdot \boldsymbol{e}_1)$$

$$= u^\alpha u^\mu{}_{;\alpha} g_{\mu 1}$$

$$= u^\alpha u_{1;\alpha}$$

$$= u^\alpha \left(\frac{\partial u_1}{\partial x^\alpha} - u_\sigma \Gamma^\sigma{}_{\alpha 1} \right)$$

$$= \frac{dx^\alpha}{d\lambda} \frac{\partial u_1}{\partial x^\alpha} - u^\alpha u_\sigma \Gamma^\sigma{}_{\alpha 1}$$

$$= \frac{du_1}{d\lambda} - u^\alpha u_\sigma \Gamma^\sigma{}_{\alpha 1}. \tag{E.282}$$

This quantity vanishes if $\boldsymbol{u}$ is the tangent field to a geodesic.

(b) We write $u^\alpha u_\sigma \Gamma^\sigma{}_{\alpha 1} = u^\alpha u^\sigma \Gamma_{\sigma\alpha 1}$, then substitute for the connection coefficient in terms of the metric components and make use of the symmetry of the products. A slick method is to directly symmetrize to say

$$\frac{du_1}{d\lambda} = u^\sigma u^\alpha \Gamma_{(\sigma\alpha)1}$$

$$= u^\sigma u^\alpha \frac{1}{2} \frac{\partial g_{0\alpha}}{\partial x^1} = 0, \tag{E.283}$$

where, in going between lines, we've used $g_{\alpha\beta,\gamma} = \Gamma_{\alpha\beta\gamma} + \Gamma_{\beta\alpha\gamma}$ and the fact that $\boldsymbol{g}$ is independent of x^1.

(34.6) (a) Combine the result from Example 34.8 with the general rule for forming inner products from eqn 32.21.

(34.7) To relate the two parametrizations we write

$$\frac{d}{d\lambda} = f' \frac{d}{ds},$$

$$\frac{d^2}{d\lambda^2} = f'' \frac{d}{ds} + (f')^2 \frac{d^2}{ds^2}, \tag{E.284}$$

where $f' = \frac{df}{d\lambda}$. The geodesic equation becomes

$$\frac{d^2 x^\mu}{ds^2} + \frac{f''}{(f')^2} \frac{dx^\mu}{ds} + \Gamma^\mu{}_{\alpha\beta} \frac{dx^\alpha}{ds} \frac{dx^\beta}{ds} = 0. \tag{E.285}$$

We see that we recover the geodesic equation if the second term vanishes, which is the case if s and λ are linearly related.

(34.8) If the magnitude of $\boldsymbol{u}$ is constant, then $\boldsymbol{g}(\boldsymbol{u}, \boldsymbol{u})$ is constant along the curve, and we must have

$$0 = u^\alpha \left(g_{\mu\nu} u^\mu u^\nu \right)_{;\alpha}$$

$$= u^\alpha u^\mu{}_{;\alpha} u_\mu + u^\alpha u_\nu u^\nu{}_{;\alpha}$$

$$= 2 u^\alpha u^\mu{}_{;\alpha} u_\mu, \tag{E.286}$$

where we have reindexed the second term in the sum and used $g_{\mu\nu;\alpha} = 0$. This implies $\boldsymbol{u} \cdot \boldsymbol{\nabla}_{\boldsymbol{u}} \boldsymbol{u} = u^\alpha u^\mu{}_{;\alpha} u_\mu = 0$, even though we don't have the geodesic condition $u^\alpha u^\mu{}_{;\alpha} = 0$.

(35.3) (a) In comma notation, we have

$$-A^\mu{}_{,\nu}{}^\nu + A^\nu{}_{,\nu}{}^\mu = J^\mu. \tag{E.287}$$

(b) Since we can write the second term $\partial^\mu \partial_\nu A^\nu$ or $\partial_\nu \partial^\mu A^\nu$ in flat space we can write

$$-A^\mu{}_{;\nu}{}^\nu + A^\nu{}_{;\nu}{}^\mu = J^\mu, \tag{E.288}$$

or

$$-A^\mu{}_{;\nu}{}^\nu + A^{\nu;\mu}{}_\nu = J^\mu. \tag{E.289}$$

These will not yield the same answer in curved space, where a non-zero $\boldsymbol{R}$ tells us that the covariant derivatives don't commute.

(c) The second equation becomes

$$-A^\mu{}_{;\nu}{}^\nu + A^\nu{}_{;\nu}{}^\mu + R^\mu{}_\nu A^\nu = J^\mu. \tag{E.290}$$

This gives us two versions, with coupling to curvature appearing in one of them through the Ricci tensor.

(35.4) (a) The geodesic equation says

$$\frac{d^2 \zeta^\mu}{d\lambda^2} + \Gamma^\mu{}_{\alpha\beta} \frac{d\zeta^\alpha}{d\lambda} \frac{d\zeta^\beta}{d\lambda} = 0. \tag{E.291}$$

Inserting $\zeta^\mu = \lambda u^\mu$ gives the quoted equation. Since $\Gamma^\mu{}_{\alpha\beta}$ at the origin can't be a function of the arbitrary directions u^μ, the connection coefficients must vanish at this point.

(c) Reexpress the covariant derivative as

$$\left(\frac{D\boldsymbol{n}_\nu}{d\lambda} \right)^\mu = u^\alpha \left(\frac{\partial n_\nu^\mu}{\partial x^\alpha} + \Gamma^\mu{}_{\alpha\beta} n_\nu^\beta \right)$$

$$= \frac{\partial}{\partial\lambda} n_\nu^\mu + u^\alpha \Gamma^\mu{}_{\alpha\beta} n_\nu^\beta, \tag{E.292}$$

and then substitute $n_\nu^\mu = \lambda \delta_\nu^\mu$ and the expression for the derivative of the connection coefficients.

(g) Permuting indices gives three expressions. Take one, subtract the second and add the third. Then use the symmetries of the Riemann tensor.

(35.6) (a) We can read off, using $\boldsymbol{\nabla}_\nu \boldsymbol{e}_\mu = \Gamma^\alpha{}_{\nu\mu} \boldsymbol{e}_\alpha$, that

$$\begin{aligned}
\Gamma^X{}_{XX} &= 0, & \Gamma^Y{}_{XX} &= 0, \\
\Gamma^X{}_{XY} &= 1, & \Gamma^Y{}_{XY} &= 1, \\
\Gamma^X{}_{YX} &= 1, & \Gamma^Y{}_{YX} &= -1, \\
\Gamma^X{}_{YY} &= 0, & \Gamma^Y{}_{YY} &= 0.
\end{aligned}$$

(b) We find non-zero components

$$\begin{aligned}
R^X{}_{XXY} &= -1, & R^Y{}_{XXY} &= -1, \\
R^X{}_{YXY} &= -1, & R^Y{}_{YXY} &= 1,
\end{aligned}$$

with other components vanishing.

(36.1) (a) We have

$$\omega^{\hat{t}}{}_{\hat{i}} = -\omega_{\hat{t}\hat{i}} = \omega_{\hat{i}\hat{t}} = \omega^{\hat{i}}{}_{\hat{t}}, \qquad (E.293)$$

and (b)

$$\omega^{\hat{i}}{}_{\hat{j}} = \omega_{\hat{i}\hat{j}} = -\omega_{\hat{j}\hat{i}} = -\omega^{\hat{j}}{}_{\hat{i}}. \qquad (E.294)$$

(36.2) See the answer to Exercise 10.2.

(36.3) Identify orthonormal components

$$\omega^{\hat{\sigma}} = d\sigma, \quad \omega^{\hat{\phi}} = r(\sigma)d\phi, \qquad (E.295)$$

and take derivatives

$$d\omega^{\hat{\sigma}} = 0,$$
$$d\omega^{\hat{\phi}} = \frac{dr}{d\sigma}d\sigma \wedge d\phi = -\frac{1}{r}\frac{dr}{d\sigma}\omega^{\hat{\phi}} \wedge \omega^{\hat{\sigma}}. \qquad (E.296)$$

From these, we extract

$$\omega^{\hat{\phi}}{}_{\hat{\sigma}} = \frac{1}{r}\frac{dr}{d\sigma}\omega^{\hat{\phi}} = -\omega^{\hat{\sigma}}{}_{\hat{\phi}}. \qquad (E.297)$$

The only thing to do is to take the exterior derivative of this

$$\begin{aligned} d\omega^{\hat{\sigma}}{}_{\hat{\phi}} &= d\left(-\frac{dr}{d\sigma}d\phi\right) \\ &= -\frac{d^2r}{d\sigma^2}d\sigma \wedge d\phi \\ &= -\frac{1}{r}\frac{d^2r}{d\sigma^2}\omega^{\hat{\sigma}} \wedge \omega^{\hat{\phi}}. \end{aligned} \qquad (E.298)$$

So we have

$$\mathcal{R}^{\hat{\sigma}}{}_{\hat{\phi}} = -\frac{1}{r}\frac{d^2r}{d\sigma^2}\omega^{\hat{\sigma}} \wedge \omega^{\hat{\phi}}. \qquad (E.299)$$

(36.4) We identify orthonormal components

$$\omega^{\hat{r}} = (1 + a^2 r^2)^{\frac{1}{2}}dr, \quad \omega^{\hat{\theta}} = rd\theta. \qquad (E.300)$$

We find connection 1-forms

$$d\omega^{\hat{r}} = 0, \quad d\omega^{\hat{\theta}} = dr \wedge d\theta = -\frac{\omega^{\hat{\theta}} \wedge \omega^{\hat{r}}}{r(1+a^2r^2)^{\frac{1}{2}}}. \qquad (E.301)$$

We read off that

$$\omega^{\hat{\theta}}{}_{\hat{r}} = \frac{\omega^{\hat{\theta}}}{r(1 + a^2r^2)^{\frac{1}{2}}} = \frac{d\theta}{(1 + a^2r^2)^{\frac{1}{2}}}. \qquad (E.302)$$

so we can calculate

$$\begin{aligned} d\omega^{\hat{\theta}}{}_{\hat{r}} &= -\frac{a^2 r}{(1 + a^2r^2)^{\frac{3}{2}}}dr \wedge d\theta \\ &= \frac{a^2}{(1 + a^2r^2)^2}\omega^{\hat{\theta}} \wedge \omega^{\hat{r}}. \end{aligned} \qquad (E.303)$$

We have that

$$\mathcal{R}^{\hat{\theta}}{}_{\hat{r}} = \frac{a^2}{(1 + a^2r^2)^2}\omega^{\hat{\theta}} \wedge \omega^{\hat{r}}, \qquad (E.304)$$

so we read off the non-zero component

$$R^{\hat{\theta}}{}_{\hat{r}|\hat{\theta}\hat{r}|} = R^{\hat{\theta}}{}_{\hat{r}\hat{\theta}\hat{r}} = \frac{a^2}{(1 + a^2r^2)^2}, \qquad (E.305)$$

where the $|\hat{\theta}\hat{r}|$ notation fixes the order of these two variables based on the order in the wedge product in eqn E.304. Note also that since $\omega^{\hat{r}}{}_{\hat{\theta}} = -\omega^{\hat{\theta}}{}_{\hat{r}}$, so we also have

$$\mathcal{R}^{\hat{r}}{}_{\hat{\theta}} = \frac{a^2}{(1 + a^2r^2)^2}\omega^{\hat{r}} \wedge \omega^{\hat{\theta}}, \qquad (E.306)$$

and

$$R^{\hat{r}}{}_{\hat{\theta}|\hat{r}\hat{\theta}|} = R^{\hat{r}}{}_{\hat{\theta}\hat{r}\hat{\theta}} = \frac{a^2}{(1 + a^2r^2)^2}. \qquad (E.307)$$

This gives a Ricci tensor with components

$$R_{\hat{r}\hat{r}} = \frac{a^2}{(1+a^2r^2)^2}, \quad R_{\hat{\theta}\hat{\theta}} = \frac{a^2}{(1+a^2r^2)^2}. \qquad (E.308)$$

and a Ricci scalar

$$R = \eta^{rr}R_{rr} + \eta^{\theta\theta}R_{\theta\theta} = 2a^2/(1 + a^2r^2)^2. \qquad (E.309)$$

Note that when $r = 0$, $R = 2a^2$.

(36.5) We identify obvious 1-forms

$$\omega^{\hat{u}} = (c + a\cos v)du, \quad \omega^{\hat{v}} = adv. \qquad (E.310)$$

Taking exterior derivatives we find

$$\begin{aligned} d\omega^{\hat{u}} &= -a\sin v dv \wedge du, \\ d\omega^{\hat{u}} &= 0. \end{aligned} \qquad (E.311)$$

We conclude that the non-zero curvature 1-form is

$$\omega^{\hat{u}}{}_{\hat{v}} = -\frac{\sin v}{c + a\cos v}\omega^{\hat{u}}. \qquad (E.312)$$

Now take a second exterior derivative to find

$$d\omega^{\hat{u}}{}_{\hat{v}} = -\frac{\cos v}{a(c + a\cos v)}\omega^{\hat{v}} \wedge \omega^{\hat{u}}. \qquad (E.313)$$

As a result we can construct the curvature 2-form, which is

$$\mathcal{R}^{\hat{u}}{}_{\hat{v}} = \frac{\cos v}{a(c + a\cos v)}\omega^{\hat{u}} \wedge \omega^{\hat{v}}. \qquad (E.314)$$

Assuming an ordering of components $(x^1, x^2) = (u, v)$ we then extract the components of the Riemann tensor

$$R^{\hat{u}}{}_{\hat{v}\hat{u}\hat{v}} = R^{\hat{v}}{}_{\hat{u}\hat{v}\hat{u}} = \frac{\cos v}{a(c + a\cos v)}. \qquad (E.315)$$

This gives us a Ricci tensor with components

$$R_{\hat{u}\hat{u}} = R_{\hat{v}\hat{v}} = \frac{\cos v}{a(c + a\cos v)}, \qquad (E.316)$$

and consequently a Ricci scalar $R = \frac{2\cos v}{a(c+a\cos v)}$.

(36.6) We make the obvious identification

$$\omega^{\hat{x}} = \frac{dx}{r}, \quad \omega^{\hat{r}} = \frac{dr}{r}. \qquad (E.317)$$

The second exterior derivative is

$$d\omega^{\hat{r}} = 0, \qquad (E.318)$$

so we guess that $\omega^{\hat{r}}{}_{\hat{x}} \propto \omega^{\hat{x}}$. The exterior derivative of the first basis 1-form is

$$d\omega^{\hat{x}} = -\frac{1}{r^2}dr \wedge dx = -\omega^{\hat{x}} \wedge \omega^{\hat{r}}, \qquad (E.319)$$

and we identify $\omega^{\hat{x}}{}_{\hat{r}} = -\omega^{\hat{x}}$ and $\omega^{\hat{r}}{}_{\hat{x}} = \omega^{\hat{x}}$. Taking another exterior derivative we have

$$d\omega^{\hat{r}}{}_{\hat{x}} = -\frac{1}{r^2}dr \wedge dx = -\omega^{\hat{r}} \wedge \omega^{\hat{x}}, \qquad (E.320)$$

which allows us to write

$$\mathcal{R}^{\hat{r}}{}_{\hat{x}} = -\boldsymbol{\omega}^{\hat{r}} \wedge \boldsymbol{\omega}^{\hat{x}}. \tag{E.321}$$

We find that $R^{\hat{r}}{}_{\hat{x}\hat{r}\hat{x}} = -1$ and then that $R^{\hat{x}}{}_{\hat{r}\hat{x}\hat{r}} = -1$. We conclude that the components of the Ricci tensor are

$$R_{\hat{x}\hat{x}} = -1, \quad R_{\hat{r}\hat{r}} = -1, \tag{E.322}$$

meaning that $R = -2$. Converting back to coordinate space

$$R_{xx} = -\frac{1}{r^2} \quad R_{rr} = -\frac{1}{r^2}. \tag{E.323}$$

(36.7) (a) We identify

$$\begin{aligned} \boldsymbol{\omega}^{\hat{t}} &= \boldsymbol{dt}, & \boldsymbol{\omega}^{\hat{x}} &= a\boldsymbol{d\chi}, \\ \boldsymbol{\omega}^{\hat{\theta}} &= a\sin\chi\,\boldsymbol{d\theta}, & \boldsymbol{\omega}^{\hat{\phi}} &= a\sin\chi\sin\theta\,\boldsymbol{d\phi}. \end{aligned} \tag{E.324}$$

Since $\boldsymbol{\omega}^{\hat{t}} = \boldsymbol{dt}$ we find $\boldsymbol{d\omega}^{\hat{t}} = 0$. We guess that $\boldsymbol{\omega}^{\hat{t}}{}_{\hat{k}} \propto \boldsymbol{\omega}^{\hat{k}}$, where $k = \theta, \phi$ or χ. (We also have $\boldsymbol{\omega}^{\hat{t}}{}_{\hat{k}} = \boldsymbol{\omega}^{\hat{k}}{}_{\hat{t}}$.) Another exterior derivative

$$\boldsymbol{d\omega}^{\hat{x}} = \dot{a}\,\boldsymbol{dt} \wedge \boldsymbol{d\chi} = -\frac{\dot{a}}{a}\boldsymbol{\omega}^{\hat{x}} \wedge \boldsymbol{\omega}^{\hat{t}}, \tag{E.325}$$

gives

$$\boldsymbol{\omega}^{\hat{x}}{}_{\hat{t}} = \frac{\dot{a}}{a}\boldsymbol{\omega}^{\hat{x}} = \dot{a}\,\boldsymbol{d\chi}. \tag{E.326}$$

This fixes the constant of proportionality, so that we assume $\boldsymbol{\omega}^{\hat{t}}{}_{\hat{k}} = (\dot{a}/a)\boldsymbol{\omega}^{\hat{k}}$ The next exterior derivative

$$\begin{aligned} \boldsymbol{d\omega}^{\hat{\theta}} &= \dot{a}\sin\chi\,\boldsymbol{dt} \wedge \boldsymbol{d\theta} + a\cos\chi\,\boldsymbol{d\chi} \wedge \boldsymbol{d\theta} \\ &= -\frac{\dot{a}}{a}\boldsymbol{\omega}^{\hat{\theta}} \wedge \boldsymbol{\omega}^{\hat{t}} - \frac{1}{a}\frac{\cos\chi}{\sin\chi}\boldsymbol{\omega}^{\hat{\theta}} \wedge \boldsymbol{\omega}^{\hat{x}}, \end{aligned} \tag{E.327}$$

gives

$$\boldsymbol{\omega}^{\hat{\theta}}{}_{\hat{\chi}} = \frac{1}{a}\frac{\cos\chi}{\sin\chi}\boldsymbol{\omega}^{\hat{\theta}} = \cos\chi\,\boldsymbol{d\theta}. \tag{E.328}$$

A final exterior derivative

$$\begin{aligned} \boldsymbol{d\omega}^{\hat{\phi}} &= \dot{a}\sin\chi\sin\theta\,\boldsymbol{dt} \wedge \boldsymbol{d\phi} + a\cos\chi\sin\theta\,\boldsymbol{d\chi} \wedge \boldsymbol{d\phi} \\ &\quad + a\sin\chi\cos\theta\,\boldsymbol{d\theta} \wedge \boldsymbol{d\phi}, \end{aligned} \tag{E.329}$$

gives

$$\boldsymbol{\omega}^{\hat{\phi}}{}_{\hat{\chi}} = \frac{1}{a}\frac{\cos\chi}{\sin\chi}\boldsymbol{\omega}^{\hat{\phi}} = \cos\chi\sin\theta\,\boldsymbol{d\phi}, \tag{E.330}$$

and

$$\boldsymbol{\omega}^{\hat{\phi}}{}_{\hat{\theta}} = \frac{1}{a}\frac{\cos\theta}{\sin\chi\sin\theta}\boldsymbol{\omega}^{\hat{\phi}} = \cos\theta\,\boldsymbol{d\phi}. \tag{E.331}$$

We're now in a position to compute the components of the curvature 2-form. These are

$$\begin{aligned} \mathcal{R}^{\hat{t}}{}_{\hat{\chi}} &= \boldsymbol{d\omega}^{\hat{t}}{}_{\hat{\chi}} + \boldsymbol{\omega}^{\hat{t}}{}_{\hat{\theta}} \wedge \boldsymbol{\omega}^{\hat{\theta}}{}_{\hat{\chi}} + \boldsymbol{\omega}^{\hat{t}}{}_{\hat{\phi}} \wedge \boldsymbol{\omega}^{\hat{\phi}}{}_{\hat{\chi}} \\ &= \frac{\ddot{a}}{a}\boldsymbol{\omega}^{\hat{t}} \wedge \boldsymbol{\omega}^{\hat{x}}, \end{aligned} \tag{E.332}$$

with $\mathcal{R}^{\hat{t}}{}_{\hat{k}} = \frac{\ddot{a}}{a}\boldsymbol{\omega}^{\hat{t}} \wedge \boldsymbol{\omega}^{\hat{k}}$ following. Continuing,

$$\begin{aligned} \mathcal{R}^{\hat{x}}{}_{\hat{\theta}} &= \boldsymbol{d\omega}^{\hat{x}}{}_{\hat{\theta}} + \boldsymbol{\omega}^{\hat{x}}{}_{\hat{\phi}} \wedge \boldsymbol{\omega}^{\hat{\phi}}{}_{\hat{\theta}} + \boldsymbol{\omega}^{\hat{x}}{}_{\hat{t}} \wedge \boldsymbol{\omega}^{\hat{t}}{}_{\hat{\theta}} \\ &= \frac{1}{a^2}\boldsymbol{\omega}^{\hat{x}} \wedge \boldsymbol{\omega}^{\hat{\theta}} + 0 + \left(\frac{\dot{a}}{a}\right)^2 \boldsymbol{\omega}^{\hat{x}} \wedge \boldsymbol{\omega}^{\hat{\theta}}, \end{aligned} \tag{E.333}$$

and also

$$\mathcal{R}^{\hat{k}}{}_{\hat{j}} = \frac{1 + \dot{a}^2}{a^2}\boldsymbol{\omega}^{\hat{k}} \wedge \boldsymbol{\omega}^{\hat{j}}. \tag{E.334}$$

The components of the Riemann tensor are then

$$R^{\hat{t}}{}_{\hat{i}\hat{t}\hat{i}} = \frac{\ddot{a}}{a}, \tag{E.335}$$

and

$$R^{\hat{i}}{}_{\hat{j}\hat{i}\hat{j}} = \frac{1 + \dot{a}^2}{a^2}, \tag{E.336}$$

where we've suspended the summation convention. Now we work out the Ricci tensor. Using $R^{\hat{t}}{}_{\hat{i}\hat{t}\hat{i}} = -R^{\hat{i}}{}_{\hat{t}\hat{i}\hat{t}}$ (no summation implied), we find

$$R_{\hat{t}\hat{t}} = \sum_{j=1}^{3} R^{\hat{j}}{}_{\hat{t}\hat{j}\hat{t}} = -3\frac{\ddot{a}}{a}, \tag{E.337}$$

and

$$R_{\hat{i}\hat{i}} = R^{\hat{t}}{}_{\hat{i}\hat{t}\hat{i}} + \sum_{\hat{j}=1}^{3} R^{\hat{j}}{}_{\hat{i}\hat{j}\hat{i}} = \frac{\ddot{a}}{a} + 2\left(\frac{1 + \dot{a}^2}{a^2}\right). \tag{E.338}$$

(b) Finally, reinstating the summation convention, we work out the R-scalar

$$\begin{aligned} R &= \eta^{\hat{\mu}\hat{\nu}} R_{\hat{\mu}\hat{\nu}} = 3\frac{\ddot{a}}{a} + 3\frac{\ddot{a}}{a} + 6\left(\frac{1 + \dot{a}^2}{a^2}\right) \\ &= 6\left(\frac{\ddot{a}}{a} + \frac{1}{a^2} + \frac{\dot{a}^2}{a^2}\right). \end{aligned} \tag{E.339}$$

(c) We can work out the Einstein tensor, whose components are

$$G_{\hat{t}\hat{t}} = 3\left(\frac{1}{a^2} + \frac{\dot{a}^2}{a^2}\right), \tag{E.340}$$

and

$$G_{\hat{i}\hat{i}} = -\left(\frac{2\ddot{a}}{a} + \frac{1}{a^2} + \frac{\dot{a}^2}{a^2}\right). \tag{E.341}$$

(36.8) The usual identification gives us 1-forms:

$$\begin{aligned} \boldsymbol{\omega}^{\hat{T}} &= e^{\Phi}\boldsymbol{dT}, & \boldsymbol{\omega}^{\hat{R}} &= e^{\Lambda}\boldsymbol{dR}, \\ \boldsymbol{\omega}^{\hat{\theta}} &= r\boldsymbol{d\theta}, & \boldsymbol{\omega}^{\hat{\phi}} &= r\sin\theta\,\boldsymbol{d\phi}. \end{aligned} \tag{E.342}$$

Taking derivatives, we obtain

$$\begin{aligned} \boldsymbol{d\omega}^{\hat{T}} &= -\Phi' e^{\Phi}e^{-\Lambda}\boldsymbol{dT} \wedge \boldsymbol{\omega}^{\hat{R}}, \\ \boldsymbol{d\omega}^{\hat{R}} &= -\dot{\Lambda}e^{-\Phi}e^{\Lambda}\boldsymbol{dR} \wedge \boldsymbol{\omega}^{\hat{T}}, \\ \boldsymbol{d\omega}^{\hat{\theta}} &= -\dot{r}e^{-\Phi}\boldsymbol{d\theta} \wedge \boldsymbol{\omega}^{\hat{T}} - r'e^{-\Lambda}\boldsymbol{d\theta} \wedge \boldsymbol{\omega}^{\hat{R}}, \\ \boldsymbol{d\omega}^{\hat{\phi}} &= -\dot{r}e^{-\Phi}\sin\theta\,\boldsymbol{d\phi} \wedge \boldsymbol{\omega}^{\hat{T}} - r'\sin\theta\,e^{-\Lambda}\boldsymbol{d\phi} \wedge \boldsymbol{\omega}^{\hat{R}} \\ &\quad - \cos\theta\,\boldsymbol{d\phi} \wedge \boldsymbol{\omega}^{\hat{\theta}}. \end{aligned} \tag{E.343}$$

In a slight departure from the other examples, we combine the first two of these to suggest

$$\boldsymbol{\omega}^{\hat{R}}{}_{\hat{T}} = \boldsymbol{\omega}^{\hat{T}}{}_{\hat{R}} = \Phi' e^{\Phi}e^{-\Lambda}\boldsymbol{dT} + \dot{\Lambda}e^{\Lambda}e^{-\Phi}\boldsymbol{dR}. \tag{E.344}$$

This will turn out to be the right choice. The others are then

$$\begin{aligned} \boldsymbol{\omega}^{\hat{\theta}}{}_{\hat{T}} &= \frac{\dot{r}}{r}e^{-\Phi}\boldsymbol{\omega}^{\hat{\theta}}, & \boldsymbol{\omega}^{\hat{\theta}}{}_{\hat{R}} &= \frac{r'}{r}e^{-\Lambda}\boldsymbol{\omega}^{\hat{\theta}}, \\ \boldsymbol{\omega}^{\hat{\phi}}{}_{\hat{T}} &= \frac{\dot{r}}{r}e^{-\Phi}\boldsymbol{\omega}^{\hat{\phi}}, & \boldsymbol{\omega}^{\hat{\phi}}{}_{\hat{R}} &= \frac{r'}{r}e^{-\Lambda}\boldsymbol{\omega}^{\hat{\phi}}, \\ \boldsymbol{\omega}^{\hat{\phi}}{}_{\hat{\theta}} &= \frac{\cos\theta}{r\sin\theta}\boldsymbol{\omega}^{\hat{\phi}}. \end{aligned}$$

The derivatives are

$$d\omega^{\hat{T}}{}_{\hat{R}} = \left\{ -\left[\Phi'' + (\Phi')^2 - \Phi'\Lambda'\right]e^{-2\Lambda} \right.$$
$$\left. + \left[\ddot{\Lambda} + (\dot{\Lambda}) - \dot{\Lambda}\dot{\Phi}e^{-2\Phi}\right] \right\}\omega^{\hat{\theta}} \wedge \omega^{\hat{R}}, \tag{E.345}$$

$$d\omega^{\hat{\theta}}{}_{\hat{T}} = \frac{1}{r}\left(-\ddot{r} + \dot{r}\dot{\Phi}\right)e^{-2\Phi}\omega^{\hat{\theta}} \wedge \omega^{\hat{T}}$$
$$+ \frac{1}{r}\left(-\dot{r}' + \dot{r}\Phi'\right)e^{-\Phi}e^{-\Lambda}\omega^{\hat{\theta}} \wedge \omega^{\hat{R}}, \tag{E.346}$$

$$d\omega^{\hat{\theta}}{}_{\hat{R}} = -\frac{1}{r}\left(\dot{r}' - r'\dot{\Lambda}\right)e^{-\Phi}e^{-\Lambda}\omega^{\hat{\theta}} \wedge \omega^{\hat{T}}$$
$$- \frac{1}{r}\left(r'' - r'\Lambda'\right)e^{-2\Lambda}\omega^{\hat{\theta}} \wedge \omega^{\hat{R}}, \tag{E.347}$$

$$d\omega^{\hat{\phi}}{}_{\hat{T}} = -\frac{1}{r}\left(\ddot{r} - \dot{r}\dot{\Phi}\right)e^{-2\Phi}\omega^{\hat{\phi}} \wedge \omega^{\hat{T}}$$
$$- \frac{1}{r}\left(\dot{r}' - \dot{r}\Phi'\right)e^{-\Lambda}e^{-\Phi}\omega^{\hat{\phi}} \wedge \omega^{\hat{R}}$$
$$- \frac{\dot{r}\cos\theta}{r^2\sin\theta}e^{-\Phi}\omega^{\hat{\phi}} \wedge \omega^{\hat{\theta}}, \tag{E.348}$$

$$d\omega^{\hat{\phi}}{}_{\hat{R}} = -\frac{1}{r}\left(\dot{r}' - r'\dot{\Lambda}\right)e^{-\Lambda}e^{-\Phi}\omega^{\hat{\phi}} \wedge \omega^{\hat{T}}$$
$$- \frac{1}{r}\left(r'' - r'\dot{\Lambda}\right)e^{-2\Lambda}\omega^{\hat{\phi}} \wedge \omega^{\hat{R}}$$
$$- \frac{r'\cos\theta}{r^2\sin\theta}e^{-\Lambda}\omega^{\hat{\phi}} \wedge \omega^{\hat{\theta}}, \tag{E.349}$$

and

$$d\omega^{\hat{\phi}}{}_{\hat{\theta}} = \frac{1}{r^2}\omega^{\hat{\phi}} \wedge \omega^{\hat{\theta}}. \tag{E.350}$$

We then have non-zero contributions to $\mathcal{R}^{\hat{\mu}}{}_{\hat{\nu}}$ of

$$\mathcal{R}^{\hat{T}}{}_{\hat{R}} = d\omega^{\hat{T}}{}_{\hat{R}},$$
$$\mathcal{R}^{\hat{T}}{}_{\hat{\theta}} = d\omega^{\hat{T}}{}_{\hat{\theta}} + \omega^{\hat{T}}{}_{\hat{R}} \wedge \omega^{\hat{R}}{}_{\hat{\theta}},$$
$$\mathcal{R}^{\hat{T}}{}_{\hat{\phi}} = d\omega^{\hat{T}}{}_{\hat{\phi}} + \omega^{\hat{T}}{}_{\hat{R}} \wedge \omega^{\hat{R}}{}_{\hat{\phi}} + \omega^{\hat{T}}{}_{\hat{\theta}} \wedge \omega^{\hat{\theta}}{}_{\hat{\phi}},$$
$$\mathcal{R}^{\hat{\theta}}{}_{\hat{\phi}} = d\omega^{\hat{\theta}}{}_{\hat{\phi}} + \omega^{\hat{\theta}}{}_{\hat{T}} \wedge \omega^{\hat{T}}{}_{\hat{\phi}} + \omega^{\hat{\theta}}{}_{\hat{R}} \wedge \omega^{\hat{R}}{}_{\hat{\phi}},$$
$$\mathcal{R}^{\hat{R}}{}_{\hat{\theta}} = d\omega^{\hat{R}}{}_{\hat{\theta}} + \omega^{\hat{R}}{}_{\hat{T}} \wedge \omega^{\hat{T}}{}_{\hat{\theta}},$$
$$\mathcal{R}^{\hat{R}}{}_{\hat{\phi}} = d\omega^{\hat{R}}{}_{\hat{\phi}} + \omega^{\hat{R}}{}_{\hat{T}} \wedge \omega^{\hat{T}}{}_{\hat{\phi}} + \omega^{\hat{R}}{}_{\hat{\theta}} \wedge \omega^{\hat{\theta}}{}_{\hat{\phi}}, \tag{E.351}$$

and substitution of these gives the required results.

(37.1) (a) Insert the velocity and we have $\tilde{\sigma}(u) = -Vu^\alpha u_\alpha = V$, as required for $\tilde{\sigma}$.
(b) The total momentum in the box is

$$\boldsymbol{p} = -V\boldsymbol{T}(\ ,\tilde{u}) = -VT^{\mu\nu}u_\nu \boldsymbol{e}_\mu. \tag{E.352}$$

(c) The total energy is

$$E = -\boldsymbol{p}\cdot\boldsymbol{u} = VT^{\mu\nu}u_\mu(g_{\sigma\nu}u^\sigma)$$
$$= VT^{\mu\nu}u_\mu u_\nu = V\boldsymbol{T}(\tilde{u},\tilde{u}). \tag{E.353}$$

(37.2) (a) Since the box lives in the observer's three space, the tetrad $(\boldsymbol{u}, \boldsymbol{A}, \boldsymbol{B}, \boldsymbol{C})$ spans spacetime.
(b) Inserting the velocity 1-form into the current gives $n = -\boldsymbol{J}(\tilde{u})$. We can then write the current vector as

$$\boldsymbol{J} = n\boldsymbol{u} + a\boldsymbol{A} + b\boldsymbol{B} + c\boldsymbol{C}. \tag{E.354}$$

The volume of the box is

$$V = \tilde{\omega}(\boldsymbol{u}, \boldsymbol{A}, \boldsymbol{B}, \boldsymbol{C}). \tag{E.355}$$

We then have

$$\tilde{\omega}(\boldsymbol{J}, \boldsymbol{A}, \boldsymbol{B}, \boldsymbol{C}) = n\tilde{\omega}(\boldsymbol{u}, \boldsymbol{A}, \boldsymbol{B}, \boldsymbol{C})$$
$$= nV = N. \tag{E.356}$$

(c) If we choose

$$\boldsymbol{A}' = \boldsymbol{A} + \alpha\boldsymbol{J},$$
$$\boldsymbol{B}' = \boldsymbol{B} + \beta\boldsymbol{J},$$
$$\boldsymbol{C}' = \boldsymbol{C} + \gamma\boldsymbol{J}, \tag{E.357}$$

where α, β and γ are constants, then each box captures the same number of particles.

(37.5) (a) We have, by definition of the covariant derivative,

$$\boldsymbol{\nabla}_\mu\tilde{\omega} = \langle d\tilde{\omega}, \boldsymbol{e}_\mu\rangle = 0. \tag{E.358}$$

(c) From part (b) we have that

$$-\theta_v = -\theta_v\varepsilon_{0123} = \varepsilon_{\mu123}v^\mu{}_{;0} + \varepsilon_{0\mu23}v^\mu{}_{;1}$$
$$+ \varepsilon_{01\mu3}v^\mu{}_{;2} + \varepsilon_{012\mu}v^\mu{}_{;3}$$
$$= v^\mu{}_{;\mu}. \tag{E.359}$$

(37.6) (a) Trivially, $\mathcal{L}_{\boldsymbol{J}}\boldsymbol{J} = 0$. The other equalities follow from the fact that the box is rigid, so its corners must continue to be connected after they are transported along their geodesics.
(b) This follows from the result of the previous problem.
(c) The quantity χ is simply the number of particles in the box. Since this is a constant, by construction, we have $\mathcal{L}_{\boldsymbol{J}}\chi = 0$.
(d) Since $\theta_{\boldsymbol{J}} = 0$, we also have from the previous problem that $\boldsymbol{\nabla}\cdot\boldsymbol{J} = 0$.

(38.2) (a) If $\tilde{\omega}$ is exact then $\tilde{\omega} = d\tilde{A}$, where $\tilde{A}$ is a 1-form. By Stokes theorem

$$\int_{S^2}\tilde{\omega} = \int_{S^2}d\tilde{A} = \int_{\partial S^2}\tilde{A} = 0, \tag{E.360}$$

since $\partial S^2 = 0$.
(b) We have $d\tilde{G} = dx^1 \wedge dx^2 \wedge dx^3$, which is the volume 3-form. Taken over the volume V of the ball from the last question, the integral

$$\int_V d\tilde{G} \neq 0, \tag{E.361}$$

since it must give the volume inside the ball. By Stokes' theorem this tell us that

$$\int_{S^2}\tilde{G} \neq 0. \tag{E.362}$$

Given the result from (a), this implies that $\boldsymbol{G}$ is not exact on S^2. It is, however, closed on S^2, as is the case for all 2-forms, since all 3-forms vanish in a two-dimensional space.

(38.4) (a) Using the results in Chapter 36 we find $\Gamma^{\hat{\phi}}{}_{\hat{\phi}\hat{\theta}} = \cot\theta$ and $\Gamma^{\hat{\theta}}{}_{\hat{\phi}\hat{\phi}} = -\cot\theta$. We then use the geodesic equation, noting that $u^{\hat{\phi}} = (e_{\phi})^{\hat{\phi}} u^{\phi} = \sin\theta\dot\phi$.

(c) A trick is needed here: we write the integral $\mathcal{I}$ as

$$\mathcal{I} = \int_0^T d\tau \left[\cos\theta(\tau)\dot\phi(\tau) + 0 \times \dot\theta(\tau)\right],$$

$$= \int_0^T \left[\cos\theta d\phi + 0 \times d\theta\right], \qquad (E.363)$$

which we interpret as taken around a path, parametrized by τ, that forms a closed loop that bounds a surface S. This allows us to use Green's theorem in the plane to write

$$\mathcal{I} = \int_S d\phi d\theta \left(\partial_\theta \cos\theta - \partial_\phi 0\right)$$

$$= -\int_S d\phi d\theta \sin\theta = -\mathcal{A}. \qquad (E.364)$$

This means we can write

$$\underline{R} = \exp\begin{pmatrix} 0 & -\mathcal{A} \\ \mathcal{A} & 0 \end{pmatrix}. \qquad (E.365)$$

Expanding the exponential $\exp\underline{M} = 1 + \underline{M} + \underline{M}^2/2! + ...$, we find the rotation matrix

$$\underline{R} = \begin{pmatrix} \cos\mathcal{A} & -\sin\mathcal{A} \\ \sin\mathcal{A} & \cos\mathcal{A} \end{pmatrix}. \qquad (E.366)$$

(39.1) Since $\rho gy + \frac{1}{2}\rho v^2 + p$ (where y is the height) is a constant, and the pressure is the same at the top of the reservoir as at the outlet. We have

$$\rho gy_2 = \rho gy_1 + \frac{1}{2}\rho v^2, \qquad (E.367)$$

since at the top of the reservoir (at height y_2) we assume the fluid has no velocity. We obtain $v = [2g(y_2 - y_1)]^{\frac{1}{2}}$, which is the same result for a particle having fallen through a distance $y_2 - y_1$ in a constant gravitational field g.

(39.2) (a) We find

$$\left[(\rho + p)u^0 u^\mu + p\eta^{0\mu}\right]_{,\mu}$$

$$= (\rho + p)_{,\mu}\gamma u^\mu + (\rho + p)\gamma_{,\mu} u^\mu$$

$$\quad + (\rho + p)\gamma u^\mu{}_{,\mu} + p_{,\mu}\eta^{0\mu}$$

$$= \frac{\partial(\rho + p)}{\partial t}\gamma u^0 + \frac{\partial(\rho + p)}{\partial x^i}\gamma u^i + (\rho + p)\frac{\partial\gamma}{\partial t}u^0$$

$$\quad + (\rho + p)\frac{\partial\gamma}{\partial x^i}u^i + (\rho + p)\gamma\frac{\partial u^0}{\partial t}$$

$$\quad + (\rho + p)\gamma\frac{\partial u^i}{\partial x^i} - \frac{\partial p}{\partial t}$$

$$= \gamma\frac{d}{d\tau}(\rho + p) + (\rho + p)\frac{d\gamma}{d\tau} + (p + \rho)\gamma\vec\nabla \cdot \vec u$$

$$= \frac{d}{d\tau}\left[\gamma(\rho + p)\right] + (p + \rho)\gamma\vec\nabla \cdot \vec u = 0, \qquad (E.368)$$

where we've assumed steady flow.

(b) Start with the previous result and substitute using conservation of rest mass

$$0 = \frac{d}{d\tau}\left[\gamma(\rho + p)\right] + (p + \rho)\gamma\vec\nabla \cdot \vec u$$

$$= \frac{d}{d\tau}\left[\gamma(\rho + p)\right] - \frac{(p + \rho)\gamma}{\rho_0}\frac{d\rho_0}{d\tau}$$

$$= \frac{d}{d\tau}\left[\frac{\gamma(\rho + p)}{\rho_0}\right] = \gamma\vec v \cdot \vec\nabla\left[\frac{\gamma(\rho + p)}{\rho_0}\right], \qquad (E.369)$$

where the final equality follows for the steady-flow condition $\frac{d}{d\tau} = \boldsymbol{u} \cdot \boldsymbol{\nabla} = u^0\frac{\partial}{\partial t} + \vec u \cdot \vec\nabla = \gamma\vec v \cdot \vec\nabla$.

(39.3) (a) Start with the conservation equation $(nu^\alpha)_{,\alpha} = 0$, where we use commas rather than semicolons since we're working in flat spacetime. We obtain

$$\frac{\partial}{\partial x^\alpha}\left[(n_* + \delta n)(u_*^\alpha + \delta u^\alpha)\right] = 0. \qquad (E.370)$$

For the first-order contribution, we only need consider

$$\frac{\partial}{\partial x^\alpha}\left[(u_*^\alpha \delta n + n_*\delta u^\alpha)\right] = 0. \qquad (E.371)$$

Plugging in components yields the desired expression.

(b) Start with the Euler equation in the form $(\rho + p)u_{\alpha,\nu}u^\nu + p_{,\alpha} + p_{,\nu}u^\nu u_\alpha = 0$, and repeat the same procedure as in part (a) to obtain the equation.

(c) If we have for a perfect fluid that

$$0 = \frac{ds}{d\tau} = u^\alpha s_{,\alpha}$$

$$= \left(\frac{\partial}{\partial t} + u^i\nabla_i\right)s = \frac{\partial}{\partial t}\delta s, \qquad (E.372)$$

then we have an unchanging entropy to first order. Plugging into the first law of thermodynamics then yields the stated equation.

(d) Using the definition, we can write $\delta p = c_s^2\delta\rho$, so that the first law gives $c_s^2(p_* + \rho_*)\delta n/n_* = \delta p$. Substituting into the result from (b) we find

$$\frac{\partial}{\partial t}\delta\vec v + \frac{c_s^2}{n_*}\vec\nabla\delta n = 0. \qquad (E.373)$$

Take the gradient of this latter expression and substitute into the result from (a) to obtain the wave equation.

(39.4) Using the identities given in the question we have

$$\boldsymbol{\nabla} \cdot \boldsymbol{J}_{\rm m} = \frac{m}{\sqrt{-g}}\int d\tau \frac{dz^\alpha}{d\tau}\frac{\partial}{\partial x^\alpha}\delta^{(4)}[x - z(\tau)]$$

$$= -\frac{m}{\sqrt{-g}}\int d\tau \frac{d}{d\tau}\delta^{(4)}[x - z(\tau)] = 0. \quad (E.374)$$

(39.5) The only components of $\boldsymbol{T}$ we need are $T^{tt} = \rho e^{-2\Phi}$ and $T^{rr} = pe^{-2\Lambda}$. We consider the component equation

$$T^{\mu\nu}{}_{;\mu} = T^{\mu\nu}{}_{,\mu} + \Gamma^\mu{}_{\mu\sigma}T^{\sigma\nu} + \Gamma^\nu{}_{\mu\sigma}T^{\mu\sigma}. \qquad (E.375)$$

First set $\mu = t$ and $\nu = r$ to find

$$T^{tr}{}_{;t} = 0 + \Gamma^t{}_{tr}T^{rr} + \Gamma^r{}_{tt}T^{tt}$$

$$= \Phi'(p + \rho)e^{-2\Lambda}. \qquad (E.376)$$

Then set $\mu = r$ and $\nu = r$ to find

$$T^{rr}{}_{;r} = \frac{\partial T^{rr}}{\partial r} + \Gamma^r{}_{rr}T^{rr} + \Gamma^r{}_{rr}T^{rr}$$

$$= \frac{\partial(pe^{-2\Lambda})}{\partial r} + 2\Lambda'pe^{-2\Lambda}$$

$$= \frac{\partial p}{\partial r}e^{-2\Lambda}. \qquad (E.377)$$

Adding the two contributions and equating them to zero then gives the answer.

(39.6) (b) We first note that $\langle \boldsymbol{d\tilde{\omega}}, \vec{v} \rangle = 0$. Then, plugging in, we have

$$\begin{aligned}
\pounds_{\vec{v}}(\rho_0 \tilde{\omega}) \\
=&\, \boldsymbol{d} \langle \tilde{\omega}, \rho_0 \vec{v} \rangle \\
=&\, \boldsymbol{d} \left(\rho_0 v^x \, \boldsymbol{dy} \wedge \boldsymbol{dz} - \rho_0 v^y \, \boldsymbol{dx} \wedge \boldsymbol{dz} + \rho_0 v^z \, \boldsymbol{dx} \wedge \boldsymbol{dy} \right) \\
=&\, \left(\frac{\partial(\rho_0 v^x)}{\partial x} + \frac{\partial(\rho_0 v^y)}{\partial y} + \frac{\partial(\rho_0 v^z)}{\partial z} \right) \boldsymbol{dx} \wedge \boldsymbol{dy} \wedge \boldsymbol{dz} \\
=&\, \left[\vec{\nabla} \cdot (\rho_0 \vec{v}) \right] \tilde{\omega}.
\end{aligned} \tag{E.378}$$

(39.7) Since $v_j = v^j$ in flat space, we can write a 1-form version of the given equation as

$$\begin{aligned}
(\pounds_{\vec{v}} \tilde{v})_i =&\, v^j \frac{\partial}{\partial x^j} v_i + v_j \frac{\partial}{\partial x^i} v^j \\
=&\, v^j \frac{\partial}{\partial x^j} v_i + \frac{1}{2} \frac{\partial}{\partial x^i} (v_j v^j),
\end{aligned} \tag{E.379}$$

and we have

$$v^j \frac{\partial}{\partial x^j} v_i = (\pounds_{\vec{v}} \tilde{v})_i - \frac{\partial}{\partial x^i} (v^2/2). \tag{E.380}$$

Substituting the other derivatives for (0,1) exterior derivatives for consistency, we get the answer.

(39.8) A suitable 2-form is $\tilde{f} = \varepsilon_{ijk} v^i \left[\boldsymbol{dx^j}(\ ) \wedge \boldsymbol{dx^k}(\ ) \right]$, where ε_{ijk} is the three-dimensional antisymmetric symbol. When the two displacement vectors are inserted into the slots, we obtain the triple scalar product between $\vec{v}$, $\vec{a}$ and $\vec{b}$.

(40.1) (a) Using the Euler–Lagrange equation, the equation of motion is found to be

$$\frac{\mathrm{d}}{\mathrm{d}\tau} (g_{\mu\nu} \dot{x}^\nu) = \frac{1}{2} \frac{\partial g_{\alpha\beta}}{\partial x^\mu} \dot{x}^\alpha \dot{x}^\beta. \tag{E.381}$$

The geodesic equation, in its usual form, follows from the steps used in the derivation in Chapter 9.

(c) The equations of motion are

$$\dot{x}^\mu = g^{\mu\nu} p_\nu, \quad \dot{p}_\mu = -\frac{1}{2} \frac{\partial g^{\alpha\beta}}{\partial x^\mu} p_\alpha p_\beta. \tag{E.382}$$

These can be used to derive the geodesic equation in the form given in (a), noting the minus sign going between $\delta g^{\mu\nu}$ and $\delta g_{\mu\nu}$.

(d) A metric component with no dependence on x^σ causes the right-hand side of the equation of motion to vanish, showing that momentum component p_σ is conserved.

(40.2) (a) We find

$$\begin{aligned}
p_t = -\left(1 - \frac{2M}{r}\right) \dot{t}, &\quad p_r = \left(1 - \frac{2M}{r}\right)^{-1} \dot{r}, \\
p_\theta = r^2 \dot{\theta}, &\quad p_\phi = r^2 \sin^2 \theta \dot{\phi}.
\end{aligned}$$

(b) We have that

$$\frac{\partial H}{\partial t} = \frac{\partial H}{\partial \phi} = 0, \tag{E.383}$$

implying that both p_t and p_ϕ are constants of the motion.

(40.4) Computing we find

$$\frac{\partial M^{\mu\nu\sigma}}{\partial x^\sigma} = \delta^\mu{}_\sigma T^{\nu\sigma} + x^\mu T^{\nu\sigma}{}_{,\sigma} - \delta^\nu{}_\sigma T^{\mu\sigma} - x^\nu T^{\mu\sigma}{}_{,\sigma}. \tag{E.384}$$

Since $T^{\mu\nu}{}_{,\mu} = 0$ we obtain

$$\frac{\partial M^{\mu\nu\sigma}}{\partial x^\sigma} = T^{\nu\mu} - T^{\mu\nu} = 0, \tag{E.385}$$

which requires $\boldsymbol{T}$ to be symmetrical.

(41.1) (a) We can compute

$$\Phi_{;\mu\nu} = (\partial_\mu \Phi)_{;\nu} = \partial_\nu \partial_\mu \Phi - \Gamma^\sigma{}_{\nu\mu} \partial_\sigma \Phi, \tag{E.386}$$

using the connection coefficients. We obtain diagonal components

$$\begin{aligned}
\Phi_{;tt} =&\, \partial_t^2 \Phi, \\
\Phi_{;rr} =&\, \partial_r^2 \Phi - \frac{a\dot{a}}{1 - kr^2} \partial_t \Phi - \frac{kr}{1 - kr^2} \partial_r \Phi, \\
\Phi_{;\theta\theta} =&\, \partial_\theta^2 \Phi - a\dot{a} r^2 \partial_t \Phi + r(1 - kr^2) \partial_r \Phi, \\
\Phi_{;\phi\phi} =&\, \partial_\phi^2 \Phi - a\dot{a} r^2 \sin^2 \theta \partial_t \Phi + r(1 - kr^2) \sin^2 \theta \partial_r \Phi \\
&\, + \sin \theta \cos \theta \partial_\theta \Phi.
\end{aligned} \tag{E.387}$$

These give us

$$\begin{aligned}
g^{\mu\nu} \Phi_{;\mu\nu} = &\, -\partial_t^2 \Phi + \frac{1 - k^2 r^2}{a^2} \left(\partial_r^2 \Phi - \frac{(a\dot{a} \partial_t \Phi + kr \partial_r \Phi)}{1 - kr^2} \right) \\
&\, + \frac{1}{a^2 r^2} \left[\partial_\theta^2 \Phi - a\dot{a} r^2 \partial_t \Phi + r(1 - kr^2) \partial_r \Phi \right] \\
&\, + \frac{1}{a^2 r^2 \sin^2 \theta} \left[\partial_\phi^2 \Phi - a\dot{a} r^2 \sin^2 \theta \partial_t \Phi + \right. \\
&\, \left. r(1 - kr^2) \sin^2 \theta \partial_r \Phi + \sin \theta \cos \theta \partial_\theta \Phi \right].
\end{aligned} \tag{E.388}$$

Collecting terms we obtain the equation of motion

$$\begin{aligned}
&\, -\partial_t^2 \Phi + \frac{1 - k^2 r^2}{a^2} \partial_r^2 \Phi + \frac{1}{a^2 r^2} \partial_\theta^2 \Phi + \frac{1}{a^2 r^2 \sin^2 \theta} \partial_\phi^2 \Phi \\
&\, - 3 \frac{\dot{a}}{a} \partial_t \Phi + \left[\frac{(2 - 3kr^2)}{a^2 r} \right] \partial_r \Phi + \frac{\cos \theta}{a^2 r^2 \sin \theta} \partial_\theta \Phi \\
&\, + \frac{\partial U}{\partial \phi} = 0.
\end{aligned} \tag{E.389}$$

(b) We can compare the previous expression against the rule for computing $\vec{\nabla}^2$ in spherical polars, which is

$$\frac{\partial^2}{\partial r^2} + \frac{1}{r^2} \frac{\partial^2}{\partial \theta^2} + \frac{1}{r^2 \sin^2 \theta} \frac{\partial^2}{\partial \phi^2} + \frac{2}{r} \frac{\partial}{\partial r} + \frac{\cos \theta}{r^2 \sin \theta} \frac{\partial}{\partial \theta}. \tag{E.390}$$

This allows us to check that the spatial parts behave as expected when $k = \dot{a} = 0$. If we are interested in slowly varying solutions in space, then the value of k does not enter at all.

(42.2) For the usual passive transformation using eqn 2.10 the components are

$$\begin{aligned}
(E')^x =&\, E^x, \\
(E')^y =&\, \gamma (E^y - \beta B^z), \\
(E')^z =&\, \gamma (E^z + \beta B^y), \\
(B')^x =&\, B^x, \\
(B')^y =&\, \gamma (B^y + \beta E^z), \\
(B')^z =&\, \gamma (B^z - \beta E^y).
\end{aligned} \tag{E.391}$$

For an active transformation, reverse the sign of β.

(42.4) For both (a) and (b) we have $\vec{B} = C\vec{e}_z$.

(42.13) (a) We find

$$L \approx -mc^2 + \frac{1}{2}mv^2 - \lambda\phi + \dots \tag{E.392}$$

(c) Invoking the equivalence principle for the equation of motion must remove the dependence on m. Scaling λ and rescaling the field, we find that the equation of motion must take the form

$$\frac{\mathrm{d}u_\mu}{\mathrm{d}\tau} = -\frac{1}{(1+\Phi)}\left(\Phi_{,\mu} + u_\mu u^\sigma \Phi_\sigma\right). \tag{E.393}$$

The action then takes the form suggested in the question for consistency.

(42.14) (b) A scalar that describes both matter and energy is the trace T of the energy-momentum tensor $\mathbf{T}$. Using this we can upgrade the Lagrangian to read

$$\mathcal{L} = -\frac{1}{8\pi G}\left(\partial_\mu\Phi\right)^2 + T\Phi. \tag{E.394}$$

(43.2) Using eqn 43.51 the equation in question can be rearranged and reduced to read

$$R^\beta{}_{\alpha\nu\mu;\beta} - R_{\alpha\mu;\nu} + R_{\alpha\nu;\mu} = 0. \tag{E.395}$$

This can be recreated from the contraction of the Bianchi identity $R^\beta{}_{\alpha[\beta\nu;\mu]} = 0$.

(43.3) Taking the dual with respect to the vector part of the (1,1) tensor (and raising an index) will result in a (1,3) tensor $\star\mathbf{T} = (\star T)^\nu{}_{\alpha\beta\gamma}\left(\mathbf{e}_\nu \otimes \boldsymbol{\omega}^\alpha \wedge \boldsymbol{\omega}^\beta \wedge \boldsymbol{\omega}^\gamma\right)$ with components

$$(\star T)^\nu{}_{\alpha\beta\gamma} = \frac{1}{3!}\sqrt{-g}\,\varepsilon_{\mu\alpha\beta\gamma}T^{\mu\nu}. \tag{E.396}$$

The expressions for the components of $\star\mathbf{G}$ are similar.

(43.4) Apply the operator $\mathbf{d}$ to $T^{\mu\nu}\left(\mathbf{e}_\mu \otimes \mathrm{d}\boldsymbol{\Sigma}_\nu\right)$. We obtain

$$\left[\mathbf{d}T^{\mu\nu}\mathbf{e}_\mu\sqrt{-g} + T^{\mu\nu}\mathbf{d}\mathbf{e}_\mu\sqrt{-g} + \mathbf{e}_\mu T^{\mu\nu}\mathbf{d}(\sqrt{-g})\right]$$
$$\otimes\,\varepsilon_{\nu|\alpha\beta\gamma|}\mathbf{d}x^\alpha \wedge \mathbf{d}x^\beta \wedge \mathbf{d}x^\gamma. \tag{E.397}$$

The first of the exterior derivatives yields (with some changes in contracted index names in the coefficients)

$$\sqrt{-g}\,\mathbf{e}_\mu\mathbf{d}T^{\mu\nu} = \sqrt{-g}\,\frac{\partial T^{\mu\lambda}}{\partial x^\lambda}\mathbf{e}_\mu \otimes \mathbf{d}x^\nu, \tag{E.398}$$

the second is

$$\sqrt{-g}\,T^{\mu\nu}\mathbf{d}\mathbf{e}_\mu = \sqrt{-g}\,\Gamma^\mu{}_{\lambda\sigma}T^{\sigma\lambda}\mathbf{e}_\mu \otimes \mathbf{d}x^\nu, \tag{E.399}$$

and the final one is

$$\mathbf{e}_\mu T^{\mu\nu}\mathbf{d}\sqrt{-g} = \sqrt{-g}\,\Gamma^\lambda{}_{\lambda\sigma}T^{\sigma\mu}\mathbf{e}_\mu \otimes \mathbf{d}x^\nu. \tag{E.400}$$

So we obtain $T^{\mu\lambda}{}_{;\lambda}\mathbf{e}_\mu \otimes \mathbf{d}x^\nu$. Consider the wedge product: the only parts that survive will have ν distinct from α, β and γ. We therefore obtain

$$\mathbf{d}\star\mathbf{T} = \mathbf{e}_\mu T^{\mu\nu}{}_{;\nu}\sqrt{-g}\,\mathbf{d}x^0 \wedge \mathbf{d}x^1 \wedge \mathbf{d}x^2 \wedge \mathbf{d}x^3. \tag{E.401}$$

The components are all zero by the usual conservation law $T^{\mu\nu}{}_{;\nu} = 0$.

(43.5) We run the argument from the chapter. The conservation law, written as a

$$\int_\mathcal{V} \mathbf{d}\star\mathbf{T} = 0. \tag{E.402}$$

Using the Einstein equation

$$\int_\mathcal{V} \mathbf{d}\star\mathbf{G} = 0, \tag{E.403}$$

and, using Stokes' theorem,

$$\int_{\partial\mathcal{V}} \star\mathbf{G} = 0. \tag{E.404}$$

(44.1) (a) Using the technique in the chapter, the effect of going round the loop is

$$\psi_{\mathcal{A}1} = (1 - ig\Delta S^{\mu\nu}G_{\mu\nu})\psi_{\mathcal{A}0}, \tag{E.405}$$

where $\Delta S^{\mu\nu} = \delta x^\mu \Delta x^\nu$ is the area $\mathcal{ABCD}$. (Here the subscripts give the point where the field is evaluated along with a number denoting which step we're doing in the computation.)
(b) The result here is simply

$$\psi_{\mathcal{P}2} = (1 + \mathrm{d}x^\rho D_\rho)\psi_{\mathcal{A}1}. \tag{E.406}$$

(c) Similarly to part (a) we have

$$\psi_{\mathcal{P}3} = (1 + ig\Delta S^{\mu\nu}G_{\mu\nu})\psi_{\mathcal{P}2}. \tag{E.407}$$

(d) Similarly to part (b) we have

$$\psi_{\mathcal{A}4} = (1 - \mathrm{d}x^\sigma D_\sigma)\psi_{\mathcal{P}3}. \tag{E.408}$$

(e) We find

$$\begin{aligned}\psi_{\mathcal{A}4} =&(1 - \mathrm{d}x^\sigma D_\sigma)(1 + ig\Delta S^{\mu\nu}G_{\mu\nu})\\ &\times (1 + \mathrm{d}x^\rho D_\rho)(1 - ig\Delta S^{\mu\nu}G_{\mu\nu})\psi_{\mathcal{A}0}\\ =&(1 - ig\Delta V^{\rho\mu\nu}[D_\rho, G_{\mu\nu}])\,\psi_{\mathcal{A}0}, \end{aligned} \tag{E.409}$$

where $V^{\rho\mu\nu} = \mathrm{d}x^\rho\delta x^\mu\Delta x^\nu$ is the volume of the cube. Since the differential operator acts on $G_{\mu\nu}$ and also $\psi_{\mathcal{A}0}$, we may write, approximately, that

$$\psi_{\mathcal{A}4} = (1 - ig\Delta V^{\rho\mu\nu}D_\rho G_{\mu\nu})\psi_{\mathcal{A}0}. \tag{E.410}$$

(g) The argument in Chapter 43 tells us that traversing the path makes zero contribution to the field, so the bracket in eqn 44.26 can be set to zero. Substitute for $G_{\mu\nu}$ and confirm that the expressions are identical.

(44.2) The symmetrized energy-momentum tensor has components

$$T_{\mu\nu} = \left(\psi^\dagger_{,\mu}\psi_{,\nu} + \psi_{,\mu}\psi^\dagger_{,\nu}\right) + g_{\mu\nu}\mathcal{L}. \tag{E.411}$$

(45.3) Use the fact that $T_{\mu\nu} = (\rho + p)u_\mu u_\nu + g_{\mu\nu}p$ to find the following:
(a) $\rho_\mathrm{g} = 2\rho + (-\rho + 3p)$.
(b) $\Pi_i = -(\rho + p)u_i u_0$ where, at low velocity, we have $u^\mu = (1, \vec{u})$ and $u_\mu = (-1, \vec{u})$.
(c) $\rho_\mathrm{c} = 2p - (-\rho + 3p)$.

(45.4) The key here is that we only retain terms up to order $\bar{u}^i$, ignoring those at order u^2. In the weak-field, low-velocity limit we have a geodesic equation

$$\frac{\partial^2 x^\mu}{\partial t^2} + \Gamma^\mu{}_{\alpha\beta} u^\alpha u^\beta = 0, \qquad (\text{E.412})$$

with $u^\alpha = (1, \vec{u}) = (1, u^i)$ and

$$\Gamma^\mu{}_{\alpha\beta} = \frac{1}{2}\eta^{\mu\nu}\left(h_{\alpha\nu,\beta} + h_{\beta\nu,\alpha} - h_{\alpha\beta,\nu}\right). \qquad (\text{E.413})$$

We find

$$\Gamma^i{}_{tt} = -\frac{1}{2}\eta^{ij} h_{tt,j}, \qquad (\text{E.414})$$

and

$$\Gamma^i{}_{tk} = \frac{1}{2}\eta^{ij}(h_{tj,k} - h_{tk,j}). \qquad (\text{E.415})$$

Substituting the connection coefficients into the geodesic equation and retaining terms up to order u^i we obtain the answer.

(b) From the tt-component equation

$$\nabla^2 h_{tt} = -16\pi\left(T_{tt} - \frac{1}{2}\eta_{tt}T\right), \qquad (\text{E.416})$$

we obtain Poisson's equation, whose solution is the expression given.

(45.5) (a) Start with

$$-\partial^2 h_{\mu\nu} = 8\pi\left(T_{\mu\nu} - \frac{1}{2}\eta_{\mu\nu}T\right) - \Lambda\eta_{\mu\nu}. \qquad (\text{E.417})$$

If, in the absence of matter, we were to take the effective energy density to be $T'_{\mu\nu} = -\frac{\Lambda}{8\pi}\eta_{\mu\nu}$, then we have a trace $T' = -\Lambda/2\pi$ and find an energy density $\rho_g = 2T'_{tt} - \eta_{tt}T' = -\Lambda/4\pi$. This is negative and so the force is repulsive.

(b) We find a field equation $\vec{\nabla}^2 h_{tt} = 2\Lambda$ and $h_{tt} = -2\Phi$, giving $\vec{\nabla}^2\Phi = -\Lambda$. The function $\Phi = \alpha r^2$ solves the resulting equation

$$\frac{1}{r^2}\frac{\partial}{\partial r}\left(r^2\frac{\partial\Phi}{\partial r}\right) = -\Lambda, \qquad (\text{E.418})$$

for $\alpha = -\Lambda/6$.

(45.6) (b) Using the usual formula for $\Gamma_{\mu\nu\sigma}$ in terms of the components $g_{\mu\nu}$, we find

$$\Gamma_{0jk} = \frac{1}{2}(\zeta_{j,k} + \zeta_{k,j}),$$

$$\Gamma_{i0k} = \frac{1}{2}(\zeta_{j,k} - \zeta_{k,j}),$$

$$\Gamma_{i00} = \phi_{,i},$$

$$\Gamma_{0i0} = -\phi_{,i},$$

$$\Gamma_{ijk} = = -\phi_{,k}\delta_{ij} - \phi_{,j}\delta_{ik} - \phi_{,i}\delta_{jk}. \qquad (\text{E.419})$$

To the order at which we're working, raising the index can be done with the Minkowski tensor. (Alternatively, use $g^{\mu\nu}$ and drop higher order terms.)

(45.7) (a) Use $S_\mu u^\mu = 0$ to say

$$S_0\frac{dt}{d\tau} = -S_i\frac{dx^i}{d\tau}$$

$$S_0 = -S_i\frac{dx^i}{d\tau}\frac{d\tau}{dt} = -S_i\frac{dx^i}{dt}. \qquad (\text{E.420})$$

(b) For parallel transport of a 1-form, we have

$$\left(\frac{\partial S_\mu}{\partial x^\nu} - \Gamma^\sigma{}_{\nu\mu}S_\sigma\right)u^\nu = 0. \qquad (\text{E.421})$$

Rearranging

$$\frac{\partial S_\mu}{\partial x^\nu}\frac{dx^\nu}{d\tau} = \Gamma^\sigma{}_{\nu\mu}S_\sigma\frac{dx^\nu}{d\tau}$$

$$\frac{dS_\mu}{d\tau} = \Gamma^\sigma{}_{\nu\mu}S_\sigma\frac{dx^\nu}{d\tau}$$

$$\frac{dS_\mu}{dt} = \Gamma^\sigma{}_{\nu\mu}S_\sigma\frac{dx^\nu}{d\tau}\frac{d\tau}{dt}$$

$$= \Gamma^\sigma{}_{\nu\mu}S_\sigma\frac{dx^\nu}{dt}. \qquad (\text{E.422})$$

(45.8) Parallel transport preserves the magnitude of a vector, so we can say that $g^{\mu\nu}S_\mu S_\nu$ is constant, which means

$$g^{00}(S_0)^2 + g^{ij}S_iS_j = \text{const}. \qquad (\text{E.423})$$

Then use $S_0 = -(dx^i/dt)S_i$ and $v^i = dx^i/dt$ and ignore the non-diagonal term with g^{ij}, since this is higher order. Note also that in writing the 3-vector, indices can be raised and lowered with the Minkowski metric tensor, so $v_i = v^i$.

(45.9) (a) Dot the equation for $\vec{S}$ with itself.
(b) Compute $\vec{v}\cdot\vec{S}$.
(c) Combine (a) and (b) with eqn 45.60.
(e) Taking a derivative, we have

$$\frac{d\vec{\Sigma}}{dt} = \frac{d\vec{S}}{dt} + (\vec{\nabla}\phi\cdot\vec{v})\vec{S} + \frac{1}{2}(\vec{v}\cdot\vec{S})\vec{\nabla}\phi$$
$$+ \frac{1}{2}(\vec{\nabla}\phi\cdot\vec{S})\vec{v}. \qquad (\text{E.424})$$

In vector notation,

$$\frac{d\vec{S}}{dt} = -2(\vec{v}\cdot\vec{S})\vec{\nabla}\phi + (\vec{\nabla}\phi\cdot\vec{S})\vec{v} - (\vec{\nabla}\phi\cdot\vec{v})\vec{S}$$
$$+ \frac{1}{2}\left[\vec{S}\times(\vec{\nabla}\times\vec{\zeta})\right]. \qquad (\text{E.425})$$

Combine these to find

$$\frac{d\vec{\Sigma}}{dt} = \frac{1}{2}\left[\vec{S}\times(\vec{\nabla}\times\vec{\zeta})\right]$$
$$- \frac{3}{2}\left[(\vec{v}\cdot\vec{S})\vec{\nabla}\phi - (\vec{\nabla}\phi\cdot\vec{S})\vec{v}\right]. \qquad (\text{E.426})$$

Using the rule for a double vector product, we can spot that the final two terms are equal to $\frac{3}{2}\left[\vec{S}\times(\vec{v}\times\vec{\nabla}\phi)\right]$ and the answer follows.

(f) First, $\vec{\nabla}\phi = M\vec{r}/r^3$. Then use components to say

$$(\vec{\nabla}\times\vec{\zeta})_a = \sum_{bc}\varepsilon_{abc}\nabla_b\zeta_c$$

$$= \sum_{bckm}\varepsilon_{abc}\frac{\partial}{\partial x^b}\frac{1}{r^3}\varepsilon_{ckm}x^k J^m$$

$$= -\sum_{bckm}\varepsilon_{cba}\varepsilon_{ckm}\frac{\partial}{\partial x^b}\left(\frac{x^k}{r^3}\right)J^m$$

$$= -\sum_{bckm}\varepsilon_{cba}\varepsilon_{ckm}\left(r^3\frac{\partial x^k}{\partial x^b} - 3x^k x^b r\right)\frac{J^m}{r^6}. \qquad (\text{E.427})$$

Then use the rule that $\sum_c \varepsilon_{cba}\varepsilon_{ckm} = \delta_k^b \delta_m^a - \delta_m^b \delta_k^a$ to obtain

$$(\vec{\nabla} \times \zeta)_a = -\sum_b \left[r^3 \frac{\partial x^b}{\partial x^b} - 3(x^b)^2 r \right] \frac{J^a}{r^6}$$
$$+ \sum_b \left[r^3 \frac{\partial x^a}{\partial x^b} - 3x^a x^b r \right] \frac{J^b}{r^6}, \qquad \text{(E.428)}$$

from which the dipolar expression follows on doing the sum over b and translating back into vector notation.

(46.2) (c) For example, term 5 can be written as

$$h^{\alpha\beta}\partial_\alpha\partial_\beta h_{\mu\nu} = \partial_\alpha(h^{\alpha\beta}\partial_\beta h_{\mu\nu}) - (\partial_\alpha h^{\alpha\beta})(\partial_\beta h_{\mu\nu}).$$
$$\text{(E.429)}$$

The first term on the right is a divergence which vanishes on the boundary. The second term is zero because $\partial_\alpha h^{\alpha\beta} = 0$. The other terms vanish through similar arguments.

(46.3) (c) The first term on the right-hand side of eqn 46.89 is a total derivative and we must have that $T^{\mu\nu}$ and its time derivative vanish at the boundary of the mass.

(46.4) (b) Taking the derivative of part (a), we have

$$\dot{h} \approx \frac{1}{R}\left(\frac{Gm}{r}\right)^{\frac{5}{2}}. \qquad \text{(E.430)}$$

(d) For (i) take the solar-system mass to be the solar mass, and the size to be 60 astronomical units (A.U.) where one A.U. $\approx 1.5 \times 10^{11}$ m. (This is about twice the radius of Neptune's orbit.). Plugging in yields ≈ 5 kW. For (ii), if we take the size of the system to be $\approx 4r_S$, then we obtain $\approx 10^{49}$ W. For (iii), taking $M = 0.5$ kg and $R = 0.2$ m, yields $\approx 10^{-81}$ W.

(46.5) (c) $\text{Tr}M_{k\ell}^{\text{TT}} = M_{kk}^{\text{TT}} = (P_{ik}P_{jk} - \frac{1}{2}P_{kk}P_{ij})M_{ij} = (1 - \frac{1}{2}P_{kk})M_{ij} = 0$ since $P_{kk} = \delta_{kk} - n_k n_k = 3 - 1 = 2$.
(e) Take $n_1 = \sin\theta\cos\phi$, $n_2 = \sin\theta\sin\phi$ and $n_3 = \cos\theta$ and $d\Omega = \sin\theta\,d\theta d\phi$. Then you can show that $\int n_1 n_2\,d\Omega = 0$, $\int n_3^2\,d\Omega = 4\pi/3$, $\int n_1^2 n_2^2\,d\Omega = 4\pi/15$, $\int n_1 n_2 n_3^2\,d\Omega = 0$, and $\int n_3^4\,d\Omega = 4\pi/5$. These are all the possibilities you need since other permutations of indices are related by symmetry.

(46.6) (a) Using Newtonian mechanics, the force between the stars is $GM^2/4r^2$ and equating this to the centripetal force we obtain $v^2 = GM/4r$. Identifying $\omega = v/r$, the answer follows.

(47.2) The combinations are

$$\epsilon_1 - i\epsilon_2 = e^{i\theta}(\epsilon_1' - i\epsilon_2'),$$
$$\epsilon_3 = \epsilon_3',$$
$$\epsilon_1 + i\epsilon_2 = e^{-i\theta}(\epsilon_1' + i\epsilon_2'), \qquad \text{(E.431)}$$

corresponding to $h = 1, 0$ and -1, respectively.

(47.3) (b) The combinations are

$$\epsilon_{11} - i\epsilon_{12} = e^{2i\theta}(\epsilon_{11}' - i\epsilon_{12}'),$$
$$\epsilon_{11} + i\epsilon_{12} = e^{-2i\theta}(\epsilon_{11}' + i\epsilon_{12}'), \qquad \text{(E.432)}$$

corresponding to $h = 2$ and -2, respectively.

(47.4) (a) The centre of mass is defined by $m_1 a_1 = m_2 a_2$ where the separation is $a = a_1 + a_2 = a_1(1 + m_1/m_2)$. The force on m_1 obeys $m_1\omega^2 a_1 = Gm_1 m_2/a^2$, so $\omega^2 = G(m_1 + m_2)/a^3$. The moment of inertia I is $I = m_1 a_1^2 + m_2 a_2^2 = m_1 m_2 a^2/(m_1 + m_2)$ after some algebra. The gravitational wave luminosity takes the form $(32G/5c^5)I^2\omega^6$ and hence the required answer is obtained. Plugging in numbers gives about 200 W of power. The energy of one graviton is $\hbar\omega$ so this means about 10^{43} gravitons per second are emitted.
(b) Handwaving answer: Very roughly, using a mass ~ 1 kg, a length of ~ 1 m, and a frantic waving frequency of 2 Hz, so that $\omega \sim 10\,\text{s}^{-1}$, yields a luminosity of only $\sim 10^{-45}$ W. The graviton energy is $\hbar\omega \sim 10^{-33}$ J, so you will need to wave your hands for tens of thousands of years before you emit a single graviton.

(48.1) (a) The exterior derivative yields

$$d\boldsymbol{\omega}^{\hat{5}} = 0 + \frac{\partial A_{\hat{\alpha}}}{\partial x^{\hat{\beta}}} dx^{\hat{\beta}} \wedge dx^{\hat{\alpha}}. \qquad \text{(E.433)}$$

Note that the derivative with respect to the fifth component vanishes, so we need only sum indices over the usual four dimensions. The expression can be rewritten as

$$d\boldsymbol{\omega}^{\hat{5}} = \frac{1}{2}F_{\hat{\alpha}\hat{\beta}}dx^{\hat{\alpha}} \wedge dx^{\hat{\beta}}, \qquad \text{(E.434)}$$

from which the answer follows on identifying $dx^{\hat{\beta}} = \boldsymbol{\omega}^{\hat{\beta}}$.
(b) Using idea 1, we have for the 1-forms in five-dimensional space

$$-d\boldsymbol{\omega}^{\hat{\alpha}} = \boldsymbol{\omega}^{\hat{\alpha}}{}_{\hat{a}} \wedge \boldsymbol{\omega}^{\hat{a}} = \boldsymbol{\omega}^{\hat{\alpha}}{}_{\hat{\beta}} \wedge \boldsymbol{\omega}^{\hat{\beta}} + \boldsymbol{\omega}^{\hat{\alpha}}{}_{\hat{5}} \wedge \boldsymbol{\omega}^{\hat{5}}, \quad \text{(E.435)}$$

where the sum over the Roman index $\hat{a}$ is assumed to range over all five dimensions. From part (a) we have $\boldsymbol{\omega}^{\hat{5}}{}_{\hat{\alpha}} = \frac{1}{2}F_{\hat{\alpha}\hat{\beta}}\boldsymbol{\omega}^{\hat{\beta}}$, from which we can deduce $\boldsymbol{\omega}^{\hat{\alpha}}{}_{\hat{5}} = -\frac{1}{2}F^{\hat{\alpha}}{}_{\hat{\beta}}\boldsymbol{\omega}^{\hat{\beta}}$. We therefore have

$$\boldsymbol{\omega}^{\hat{\alpha}}{}_{\hat{a}} \wedge \boldsymbol{\omega}^{\hat{a}} = \boldsymbol{\omega}^{\hat{\alpha}}{}_{\hat{\beta}} \wedge \boldsymbol{\omega}^{\hat{\beta}} + \boldsymbol{\omega}^{\hat{\alpha}}{}_{\hat{5}} \wedge \boldsymbol{\omega}^{\hat{5}}$$
$$= \boldsymbol{\omega}^{\hat{\alpha}}{}_{\hat{\beta}} \wedge \boldsymbol{\omega}^{\hat{\beta}} - \frac{1}{2}F^{\hat{\alpha}}{}_{\hat{\beta}}\boldsymbol{\omega}^{\hat{\beta}} \wedge \boldsymbol{\omega}^{\hat{5}}$$
$$= \left(\boldsymbol{\omega}^{\hat{\alpha}}{}_{\hat{\beta}} + \frac{1}{2}F^{\hat{\alpha}}{}_{\hat{\beta}}\boldsymbol{\omega}^{\hat{5}}\right) \wedge \boldsymbol{\omega}^{\hat{\beta}}. \qquad \text{(E.436)}$$

In four-dimensional space, we have basis 1-forms $\boldsymbol{\Omega}^{\hat{\alpha}}$ and so we write $-d\boldsymbol{\Omega}^{\hat{\alpha}} = \boldsymbol{\Omega}^{\hat{\alpha}}{}_{\hat{\beta}} \wedge \boldsymbol{\Omega}^{\hat{\beta}}$. However, it is also true that $\boldsymbol{\Omega}^{\hat{\alpha}} = \boldsymbol{\omega}^{\hat{\alpha}}$, so $d\boldsymbol{\omega}^{\hat{\alpha}} = d\boldsymbol{\Omega}^{\hat{\alpha}}$, and we conclude

$$\boldsymbol{\Omega}^{\hat{\alpha}}{}_{\hat{\beta}} = \boldsymbol{\omega}^{\hat{\alpha}}{}_{\hat{\beta}} + \frac{1}{2}F^{\hat{\alpha}}{}_{\hat{\beta}}\boldsymbol{\omega}^{\hat{5}}, \qquad \text{(E.437)}$$

which can be rearranged to give the answer.
(c) The ingredients are

$$d\boldsymbol{\omega}^{\hat{\alpha}}{}_{\hat{\beta}} = d\boldsymbol{\Omega}^{\hat{\alpha}}{}_{\hat{\beta}} - \frac{1}{2}F^{\hat{\alpha}}{}_{\hat{\beta},\hat{\gamma}}(\boldsymbol{\omega}^{\hat{\gamma}} \wedge \boldsymbol{\omega}^{\hat{5}}) - \frac{1}{2}F^{\hat{\alpha}}{}_{\hat{\beta}}d\boldsymbol{\omega}^{\hat{5}},$$
$$\text{(E.438)}$$
$$\boldsymbol{\omega}^{\hat{\alpha}}{}_{\hat{\mu}} \wedge \boldsymbol{\omega}^{\hat{\mu}}{}_{\hat{\beta}} = \left(\boldsymbol{\Omega}^{\hat{\alpha}}{}_{\hat{\mu}} - \frac{1}{2}F^{\hat{\alpha}}{}_{\hat{\mu}}\boldsymbol{\omega}^{\hat{5}}\right) \wedge \left(\boldsymbol{\Omega}^{\hat{\mu}}{}_{\hat{\beta}} - \frac{1}{2}F^{\hat{\mu}}{}_{\hat{\beta}}\boldsymbol{\omega}^{\hat{5}}\right),$$
$$\text{(E.439)}$$

and

$$\boldsymbol{\omega}^{\hat{\alpha}}{}_{\hat{5}} \wedge \boldsymbol{\omega}^{\hat{5}}{}_{\hat{\beta}} = \left(-\frac{1}{2}F^{\hat{\alpha}}{}_{\hat{\gamma}}\boldsymbol{\omega}^{\hat{\gamma}}\right) \wedge \left(\frac{1}{2}F_{\hat{\beta}\hat{\delta}}\boldsymbol{\omega}^{\hat{\delta}}\right). \qquad \text{(E.440)}$$

(f) The ingredients are

$$d\omega^{\hat{5}}{}_{\hat{\alpha}} = d\left(\frac{1}{2}F_{\hat{\alpha}\hat{\beta}}\omega^{\hat{\beta}}\right), \qquad \text{(E.441)}$$

and

$$\omega^{\hat{5}}{}_{\hat{\beta}} \wedge \omega^{\hat{\beta}}{}_{\hat{\alpha}} = \frac{1}{2}F_{\hat{\beta}\hat{\gamma}}\omega^{\hat{\gamma}} \wedge \left(\Omega^{\hat{\beta}}{}_{\hat{\alpha}} - \frac{1}{2}F^{\hat{\beta}}{}_{\hat{\alpha}}\omega^{\hat{5}}\right). \qquad \text{(E.442)}$$

(49.1) The wavelength λ can be estimated using $E = hc/\lambda$ and this yields $\lambda \sim \ell_\mathrm{P}$, ignoring small numerical constants. The gravitational self-energy for a particle of mass m_P and 'size'$\sim \lambda$ is $\approx Gm_\mathrm{P}^2/\lambda$, which also comes out to be $\sim E_\mathrm{P}$. By the same argument, this particle would have a Compton wavelength $h/m_\mathrm{P}c$ and Schwarzschild radius $2Gm_\mathrm{P}/c^2$ would also be of the order of ℓ_P.

(49.2) From eqn 49.17 the action is proportional to the area. Recall that we showed that elements formed by combinations like $d\tau d\sigma\sqrt{-\gamma}$ are invariants, meaning that we'll obtain the same answer if we use different coordinates. This is just what we mean by reparametrizing the string.

(49.5) (a) The Lagrangian is

$$\mathcal{L} = -T_0 \left[\left(\frac{\partial \vec{X}}{\partial \sigma} \cdot \frac{\partial \vec{X}}{\partial t}\right)^2 + \left(\frac{\partial \vec{X}}{\partial \sigma}\right)^2 \right.$$

$$\left. - \left(\frac{\partial \vec{X}}{\partial \sigma}\right)^2 \left(\frac{\partial \vec{X}}{\partial t}\right)^2 \right]^{\frac{1}{2}}. \qquad \text{(E.443)}$$

The components of the momentum are

$$P^i = \frac{\partial \mathcal{L}}{\partial(\partial_t X^i)}$$

$$= \frac{T_0^2}{\mathcal{L}} \left[\frac{\partial X^i}{\partial t} \frac{\partial \vec{X}}{\partial \sigma} - \frac{\partial X^i}{\partial \sigma} \left(\frac{\partial \vec{X}}{\partial \sigma} \cdot \frac{\partial \vec{X}}{\partial t}\right) \right]. \qquad \text{(E.444)}$$

If we then use $\partial X^i/\partial\sigma = (ds/d\sigma)(\partial X^i/\partial s)$ and $\left(\partial\vec{X}/\partial s\right)^2 = 1$, we have

$$\mathcal{L} = -T_0 \left(1 - v_\perp^2\right)^{\frac{1}{2}} \frac{ds}{d\sigma}, \qquad \text{(E.445)}$$

and the momentum can then be expressed as

$$\vec{P} = T_0 \frac{\vec{v}_\perp}{\sqrt{1 - v_\perp^2}} \frac{ds}{d\sigma}. \qquad \text{(E.446)}$$

(b) The Hamiltonian density is given by

$$\mathcal{H} = \vec{P} \cdot \frac{\partial \vec{X}}{\partial t} - \mathcal{L}. \qquad \text{(E.447)}$$

Using the property that

$$\vec{v}_\perp \cdot \frac{\partial \vec{X}}{\partial t} = v_\perp^2, \qquad \text{(E.448)}$$

we find

$$\mathcal{H} = T_0 \frac{ds}{d\sigma} \frac{1}{\sqrt{1 - v_\perp^2}}. \qquad \text{(E.449)}$$

The Hamiltonian can then be written as

$$H = \int d\sigma \mathcal{H} = \int \frac{T_0 ds}{\sqrt{1 - v_\perp^2}}. \qquad \text{(E.450)}$$

This has the form of $E = m/\sqrt{1 - v^2}$ for transverse motion, with a rest mass given by the string tension, so makes sense if interpreted as an energy.

(49.7) (a) The computation follows, e.g. Example 36.4. We obtain $R_{\hat{\theta}\hat{\theta}} = R_{\hat{\phi}\hat{\phi}} = 1/r_0^2$ and $R = 2/r_0^2$. The Einstein equation is solved with $8\pi\rho = 1/r_0^2$.

(b) The cross-sectional area of the string is

$$\int_0^{\theta_\mathrm{m}} r_0 d\theta \int_0^{2\pi} r_0 \sin\theta d\phi = 2\pi r_0^2 (1 - \cos\theta_\mathrm{m}). \qquad \text{(E.451)}$$

The mass per unit length is then $2\pi r_0^2(1 - \cos\theta_\mathrm{m})\rho$ or, using the previous part, $(1 - \cos\theta_\mathrm{m})/4$.

(49.8) (a) The flat cylindrical metric follows from the substitution

$$r' = r_0 \frac{\sin\theta}{\cos\theta_\mathrm{m}}, \qquad \phi' = \phi \cos\theta_\mathrm{m}. \qquad \text{(E.452)}$$

(b) The new variable ϕ' has a range $0 \leq \phi' \leq 2\pi\cos\theta_\mathrm{m}$. As a result, we have $\int d\phi' \sqrt{g_{\phi'\phi'}} = 2\pi a \cos\theta_\mathrm{m}$. Despite the spacetime looking flat, the fact that this circumference is less than $2\pi a$ gives a sense of the curvature caused by the string.

(49.9) $\vec{L}_1 = \frac{1}{2}\vec{a} \times \vec{b}, \ \vec{L}_2 = \frac{1}{2}\vec{b} \times \vec{c}, \ \vec{L}_3 = \frac{1}{2}\vec{c} \times \vec{a}$, and $\vec{L}_4 = \frac{1}{2}(\vec{b} - \vec{c}) \times (\vec{a} - \vec{c}) = \frac{1}{2}(\vec{b} \times \vec{a} - \vec{c} \times \vec{a} - \vec{b} \times \vec{c})$, and so the closure property is trivially satisfied.

(49.10) A tetrahedron is one sixth of the volume of the parallelepiped generated by $\vec{a} \times \vec{b} \cdot \vec{c}$. The modulus sign is because the scalar triple product can be negative depending on how the vectors are oriented, but volume can only be positive. Substitution of the results from the previous problem give

$$\vec{L}_1 \times \vec{L}_2 \cdot \vec{L}_3 = \frac{1}{8}[(\vec{a} \times \vec{b}) \times (\vec{b} \times \vec{c}) \cdot (\vec{c} \times \vec{a})], \qquad \text{(E.453)}$$

and using the vector identity

$$\vec{A} \times (\vec{B} \times \vec{C}) = (\vec{A} \cdot \vec{C})\vec{B} - (\vec{A} \cdot \vec{B})\vec{C}, \qquad \text{(E.454)}$$

the result follows.

(49.12) Substitute the given coordinates into the line element $ds^2 = -dT^2 + dX^2 + dY^2 - dW^2$. The answer follows after several lines of algebra.

(50.1) (a) $P_{\mu\nu} = g_{\mu\nu} + u_\mu u_\nu$.

(b) In components, we have $P_{\mu\nu}v^\nu = v_\mu + u_\mu u_\nu v^\nu$. Then

$$P_{\mu\nu}v^\nu u^\mu = v_\mu u^\mu + u_\mu u_\nu v^\nu u^\mu$$

$$= v_\mu u^\mu - u_\nu v^\nu = 0, \qquad \text{(E.455)}$$

since $u_\mu u^\mu = -1$.

(c) $P^{\mu\nu}P_{\mu\nu} = g^{\mu\nu}(g_{\mu\nu}u_\mu u_\nu) + u^\mu u^\nu(g_{\mu\nu}u_\mu u_\nu) = 4 - 1 - 1 + 1 = 3$.

(d) $P^{\alpha\beta}u_{\alpha;\beta} = g^{\alpha\beta}u_{\alpha;\beta} + u^\alpha u^\beta u_{\alpha;\beta}$ for a geodesic. The second term is $(\boldsymbol{\nabla_u u}) \cdot \boldsymbol{u} = 0$. The first term is $u^\beta{}_{;\beta} = \boldsymbol{\nabla} \cdot \boldsymbol{u}$.

(e) We find

$$P(\boldsymbol{n}, \boldsymbol{v}) = \boldsymbol{n} \cdot \boldsymbol{v} - |\boldsymbol{n}|^2 \boldsymbol{n} \cdot \boldsymbol{v} = 0, \qquad \text{(E.456)}$$

if $\boldsymbol{n} \cdot \boldsymbol{n} = 1$.

(50.2) (a) The vector $\boldsymbol{W}$ which links geodesics is Lie dragged, so we have $\pounds_{\boldsymbol{u}}\boldsymbol{W} = [\boldsymbol{u},\boldsymbol{W}] = 0$, and so

$$\boldsymbol{\nabla_u W} = \boldsymbol{\nabla_W u}. \qquad (\text{E.457})$$

We can therefore write

$$(\boldsymbol{\nabla_u W})^\mu = (\boldsymbol{\nabla_W u})^\mu = W^\nu(\boldsymbol{\nabla_\nu u})^\mu$$
$$= u^\mu{}_{;\nu}W^\nu. \qquad (\text{E.458})$$

(b) We write $u^\mu u_{\mu;\nu} = \frac{1}{2}(u^\mu u_\mu)_{;\nu} = 0$, since $u^\mu u_\mu = -1$. The other expression $u_{\mu;\nu}u^\nu = 0$ follows from the fact that $\boldsymbol{u}$ is tangent to a geodesic.

(c) We have

$$(\boldsymbol{\nabla_u B})_{\mu\nu} = u^\alpha u_{\mu;\nu\alpha}. \qquad (\text{E.459})$$

Note first that

$$u_{\mu;\nu\alpha} = u_{\mu;\alpha\nu} + R^\beta{}_{\mu\nu\alpha}u_\beta, \qquad (\text{E.460})$$

so we can write

$$(\boldsymbol{\nabla_u B})_{\mu\nu} = u^\alpha\left(u_{\mu;\alpha\nu} + R^\beta{}_{\mu\nu\alpha}u_\beta\right). \qquad (\text{E.461})$$

Consider the first term on the right, which we can cast as

$$u^\alpha u_{\mu;\alpha\nu} = (u^\alpha u_{\mu;\alpha})_{;\nu} - u^\alpha{}_{;\nu}u_{\mu;\alpha} = u^\alpha{}_{;\nu}u_{\mu;\alpha}. \qquad (\text{E.462})$$

The second term can be rearranged

$$R^\beta{}_{\mu\nu\alpha}u_\beta u^\alpha = -R^\beta{}_{\mu\alpha\nu}u_\beta u^\alpha$$
$$= -R_{\beta\mu\alpha\nu}u^\beta u^\alpha. \qquad (\text{E.463})$$

(d) This follows from the symmetry properties of the terms and the results of the previous problem.

(e) Use the result from (d) for the first three terms. For the final term, note $R_{\beta\mu\alpha\nu} = R_{\mu\beta\nu\alpha}$. The contraction with $g^{\mu\nu}$ then turns the Riemann tensor into the Ricci tensor.

(50.3) As in the text for the chapter, we substitute $R_{\mu\nu}u^\mu u^\nu = 8\pi\left(T_{\mu\nu} - \frac{1}{2}g_{\mu\nu}T\right)u^\mu u^\nu$. Considering the strong energy condition then guarantees that the right-hand side of the equation is negative, and so the congruence is forced to contract under the influence of gravity.

(D.1) We can embed a parabolic 2-surface in $\mathbb{R}^3$ using method II. The parabolic surface in Euclidean 3-space is described by

$$Z = \frac{a}{2}(X^2 + Y^2), \qquad (\text{E.464})$$

so we have

$$dZ = aX dX + aY dY. \qquad (\text{E.465})$$

We write

$$ds^2 = dX^2 + dY^2 + a^2(X dX + Y dY)^2, \qquad (\text{E.466})$$

and using the substitutions we have before: $(X, Y) = (r\cos\theta, r\sin\theta)$, we find

$$ds^2 = (1 + a^2 r^2)dr^2 + r^2 d\theta^2. \qquad (\text{E.467})$$

(D.2) The equation of the surface in $\mathbb{R}^4$ is given by $X^2 + Y^2 + Z^2 + W^2 = 1$. If we use spherical coordinates for X, Y and Z we have $X = r\sin\theta\cos\phi$, $Y = r\sin\theta\sin\phi$, $Z = r\cos\theta$ and we find that $W^2 = 1 - r^2$ and $W dW = -r dr$. We can use this to eliminate W from the line element

$$ds^2 = dX^2 + dY^2 + dZ^2 + dW^2$$
$$= dr^2 + r^2(d\theta^2 + \sin^2\theta d\phi^2) + \frac{(-r\,dr)^2}{1 - r^2}$$
$$= \frac{dr^2}{1 - r^2} + r^2(d\theta^2 + \sin^2\theta d\phi^2), \qquad (\text{E.468})$$

or

$$ds^2 = \frac{dr^2}{1 - r^2} + r^2 d\Omega_2^2, \qquad (\text{E.469})$$

with $d\Omega_2^2 = d\theta^2 + \sin^2\theta d\phi^2$. This looks very much like the $D = 2$ version, with the replacement of $d\phi$ with $d\Omega_2$. If we now write $r = \sin\psi$ we obtain

$$ds^2 = d\psi^2 + \sin^2\psi\left(d\theta^2 + \sin^2\theta d\phi^2\right), \qquad (\text{E.470})$$

and we guess that

$$d\Omega_3^2 = d\psi^2 + \sin^2\psi d\Omega_2^2. \qquad (\text{E.471})$$

(D.3) The equation for the surface may be expressed through the parametric equations

$$X = (c + a\cos v)\cos u,$$
$$Y = (c + a\cos v)\sin u,$$
$$Z = a\sin v. \qquad (\text{E.472})$$

We can use the method-I rules to determine the induced metric

$$\frac{\partial X}{\partial u} = -(c + a\cos v)\sin u, \qquad \frac{\partial X}{\partial v} = -a\sin v\cos u,$$
$$\frac{\partial Y}{\partial u} = (c + a\cos v)\cos u, \qquad \frac{\partial Y}{\partial v} = -a\sin v\sin u,$$
$$\frac{\partial Z}{\partial u} = 0, \qquad \frac{\partial Z}{\partial v} = a\cos v, \qquad (\text{E.473})$$

from which we obtain the induced metric

$$ds^2 = (c + a\cos v)^2 du^2 + a\,dv^2. \qquad (\text{E.474})$$

(D.5) Taking derivatives, we find non-zero components $g_{xx} = 1$ and $g_{tt} = -x^2$, giving the metric $ds^2 = -x^2 dt^2 + dx^2$.

Index